AF291259

# Quantum Resource Theories

In this original and modern book, the complexities of quantum phenomena and quantum resource theories are meticulously unraveled, from foundational entanglement and thermodynamics to the nuanced realms of asymmetry and beyond. Ideal for those aspiring to grasp the full scope of quantum resources, the text integrates advanced mathematical methods and physical principles within a comprehensive, accessible framework. Including over 760 exercises throughout, to develop and expand key concepts, readers will gain an unrivaled understanding of the topic. With its unique blend of pedagogical depth and cutting-edge research, it not only paves the way for a deep understanding of quantum resource theories but also illuminates the path toward innovative research directions.

Providing the latest developments in the field as well as established knowledge within a unified framework, this book will be indispensable to students, educators, and researchers interested in quantum science's profound mysteries and applications.

**Gilad Gour** is Professor at the Technion - Israel Institute of Technology, Haifa. Awarded the Katzir Prize for his PhD on black holes, he now pioneers in quantum information, significantly contributing to entanglement theory, quantum thermodynamics, and foundational science. Professor Gour has been recognized with the Early Career and Teaching Excellence Award.

# Quantum Resource Theories

**GILAD GOUR**

Technion - Israel Institute of Technology, Haifa

Shaftesbury Road, Cambridge CB2 8EA, United Kingdom

One Liberty Plaza, 20th Floor, New York, NY 10006, USA

477 Williamstown Road, Port Melbourne, VIC 3207, Australia

314–321, 3rd Floor, Plot 3, Splendor Forum, Jasola District Centre,
New Delhi – 110025, India

103 Penang Road, #05–06/07, Visioncrest Commercial, Singapore 238467

Cambridge University Press is part of Cambridge University Press & Assessment,
a department of the University of Cambridge.

We share the University's mission to contribute to society through the pursuit of
education, learning and research at the highest international levels of excellence.

www.cambridge.org
Information on this title: www.cambridge.org/9781009560917

DOI: 10.1017/9781009560870

© Gilad Gour 2025

This publication is in copyright. Subject to statutory exception and to the provisions
of relevant collective licensing agreements, no reproduction of any part may take
place without the written permission of Cambridge University Press & Assessment.

When citing this work, please include a reference to the DOI 10.1017/9781009560870

First published 2025

*A catalogue record for this publication is available from the British Library*

*A Cataloging-in-Publication data record for this book is available from the Library of Congress*

ISBN 978-1-009-56091-7 Hardback

Additional resources for this publication at www.cambridge.org/9781009560917.

Cambridge University Press & Assessment has no responsibility for the persistence
or accuracy of URLs for external or third-party internet websites referred to in this
publication and does not guarantee that any content on such websites is, or will
remain, accurate or appropriate.

Dedicated to the memory of my beloved father,
Gideon Gour
1949–2021

# Contents

# Contents

# Acknowledgments

I extend my deepest gratitude to the numerous colleagues who have profoundly influenced my understanding of quantum information and quantum resource theories. Through countless discussions, their insights have enriched my perspective and deepened my knowledge. I am particularly indebted to Fernando G. S. L. Brandão, S. Brandsen, Francesco Buscemi, Giulio Chiribella, Eric Chitambar, Nilanjana Datta, Julio De Vicente, Runyao Duan, Kun Fang, Shmuel Friedland, Yu Guo, Aram Harrow, Michał Horodecki, David Jennings, Amir Kalev, Barbara Kraus, Ludovico Lami, Iman Marvian, David A. Meyer, Markus P. Müller, Varun Narasimhachar, Jonathan Oppenheim, Carlo Maria Scandolo, Bartosz Regula, Robert W. Spekkens, Ryuji Takagi, Marco Tomamichel, Nolan Wallach, Xin Wang, Mark M. Wilde, Andreas Winter, and Nicole Yunger Halpern for their invaluable contributions.

My sincere appreciation goes to Julio Inigo De Vicente Majua for his meticulous review of the chapter on multipartite entanglement, offering numerous improvements and corrections. Special thanks to Thomas Theurer and Gershon Wolansky, whose relentless feedback on various drafts has been instrumental in refining this work. Additionally, I am grateful to Mark M. Wilde for his advice on enhancing the notations and clarity of presentation. I am also grateful to Ryuji Takagi for identifying errors and typos in the initial version of the book, as well as for offering a streamlined proof for Theorem 10.9.

I owe a debt of gratitude to the many students I've had the privilege of interacting with over the years. Their keen observations and identification of errors and typos in various drafts have been crucial in shaping the final content of this book. I thank John Burniston, Nuiok Decaire, Raz Firanko, Kimberly Golubeva, Michael Grabowecky, Alexander Hickey, Takla Nateeboon, Gaurav Saxena, Kuntal Sengupta, Guy Shemesh, Samuel Steakley, Goni Yoeli, and Elia Zanoni for their contributions.

Lastly, but most importantly, I wish to express my heartfelt appreciation to my family. To my parents, Iris and Gideon Gour, whose unwavering support and belief in my pursuits have been the bedrock of my resilience and determination. Their love and encouragement have been a constant source of strength throughout this journey. Thanks to my children, Sophia and Elijah Gour, who have been my greatest source of inspiration. Their curiosity and enthusiasm for life remind me daily of the joy of discovery and the importance of sharing knowledge. And thanks to my life partner, Eve Zhang, whose endless support and understanding have been nothing short of miraculous. Her presence and encouragement have been my guiding light, helping me navigate the challenges of this endeavor and crossing the finish line of completing this book. My journey is as much theirs as it is mine, and I am eternally grateful for their love, patience, and sacrifice.

First letters of the English alphabet, such as $A$, $B$, and $C$, are used to denote *both* quantum physical systems and their corresponding Hilbert spaces. The letter $R$ is used to denote a quantum *reference* physical system (and its corresponding Hilbert space), and sometimes the letter $E$ is used to denote the *environment* system. The last letters of the English alphabet, such as $X$, $Y$, and $Z$, are used to denote classical systems or classical registers. The dimension of a Hilbert space is denoted with vertical lines; for example, the dimensions of systems $A$, $B$, and $X$ are denoted respectively as $|A|$, $|B|$, and $|X|$. The tilde symbol above a system always represents a replica of the system. For example, $\tilde{A}$ represents another copy of system $A$ and in particular $|A| = |\tilde{A}|$.

We use $|\Omega^{A\tilde{A}}\rangle$ to denote the unnormalized maximally entangled state $\sum_{x\in[m]} |xx\rangle$, and $|\Phi^{A\tilde{A}}\rangle$ the *normalized* maximally entangled state $\frac{1}{\sqrt{|A|}} \sum_{x\in[m]} |xx\rangle$. We also use notations $\psi, \phi, \Omega, \Phi$ to denote respectively the rank-one pure states $|\psi\rangle\langle\psi|$, $|\phi\rangle\langle\phi|$, $|\Omega\rangle\langle\Omega|$, $|\Phi\rangle\langle\Phi|$.

# Symbols

| | |
|---|---|
| $[n]$ | The set $\{1, \ldots, n\}$ |
| $\lvert A \rvert$ | Dimension of the Hilbert space $A$ |
| $\mathbb{C}^{m \times n}$, $\mathbb{R}^{m \times n}$ | The set of all $m \times n$ complex and real matrices |
| $\mathbf{0}_{m,n}$ | The $m \times n$ zero matrix |
| $\mathbb{R}_+^{m \times n}$ | The set of all matrices in $\mathbb{R}^{m \times n}$ with nonnegative components |
| $[a, b]^n$ | The set of all vectors in $\mathbb{R}^n$ whose components are between $a$ and $b$ |
| $\mathrm{Prob}(n)$ | The set of all probability vectors in $\mathbb{R}_+^n$ |
| $\mathrm{Prob}_{>0}(n)$ | The vectors in $\mathrm{Prob}(n)$ whose components are all positive |
| $\mathrm{Prob}(m, n)$ | The set of matrices in $\mathbb{R}_+^{m \times n}$ whose components sum to one |
| $\mathrm{Prob}^\downarrow(n)$ | The set of all vectors in $\mathrm{Prob}(n)$ whose components are arranged in a nonincreasing order |
| $\mathrm{Pr}(X = x)$ | The probability that a random variable $X$ equals $x$ |
| $\mathrm{STOC}(m, n)$ | The set of all $m \times n$ column stochastic matrices |
| $\mathrm{Diag}(c_1, \ldots, c_n)$ | An $n \times n$ diagonal matrix with $c_1, \ldots, c_n$ on its diagonal |
| $\mathfrak{L}(A, B)$ | The set of all linear operators from the Hilbert space $A$ to $B$ |
| $\mathfrak{L}(A)$ | The set of all linear operators from the Hilbert space $A$ to itself |
| $\mathrm{Pos}(A)$ | The set of all positive semidefinite operators in $\mathfrak{L}(A)$ |
| $\mathrm{Pos}_{>0}(A)$ | The set of all positive definite operators in $\mathfrak{L}(A)$ |
| $\mathfrak{D}(A)$ | The set of all density matrices (mixed quantum states) in $\mathrm{Pos}(A)$ |
| $\mathfrak{D}_{\leqslant}(A)$ | The set of all the elements in $\mathrm{Pos}(A)$ with trace no greater than one |
| $\mathrm{Pure}(A)$ | The set of all pure (i.e. rank one) density matrices in $\mathfrak{D}(A)$ |
| $\mathrm{Herm}(A)$ | The set of all Hermitian operators in $\mathfrak{L}(A)$ |
| $\mathfrak{L}(A \to B)$ | The set of all linear operators from $\mathfrak{L}(A)$ to $\mathfrak{L}(B)$ |
| $\mathrm{Herm}(A \to B)$ | The subset $\{\mathcal{E} \in \mathfrak{L}(A \to B) : \mathcal{E}(\rho) \in \mathrm{Herm}(B) \quad \forall\, \rho \in \mathrm{Herm}(A)\}$ |
| $\mathrm{CP}(A \to B)$ | The set of all completely positive maps in $\mathfrak{L}(A \to B)$ |
| $\mathrm{CPTP}(A \to B)$ | The set of all quantum channels in $\mathfrak{L}(A \to B)$ |
| $\mathrm{Pos}(A \to B)$ | The set of all positive maps in $\mathfrak{L}(A \to B)$ |
| $\mathrm{id}^A$ | The identity element (channel) of $\mathfrak{L}(A \to A)$ |
| $\#_f$ | The Kubo–Ando operator mean (Definition B.2 of [online version]) |
| $\mathrm{Irr}(\pi)$ | The set of all irreps (up to equivalency) appearing in the decomposition of $\pi$ |
| $I_n$ | $n \times n$ identity matrix |
| $I^A$ | The identity operator in $\mathfrak{L}(A)$ |

| | |
|---|---|
| $\mathbf{u}^A$ | The maximally mixed state in $\mathfrak{L}(A)$ |
| $\mathbf{u}^{(n)}$ | The uniform probability vector in $\mathrm{Prob}(n)$ |
| $\mathbf{1}_n$ | The column vector $(1,\ldots,1)^T$ in $\mathbb{R}^n$ |
| $\lvert\Omega^{A\tilde{A}}\rangle$ | The *unnormalized* maximally entangled state $\sum_{x\in[m]}\lvert xx\rangle$ |
| $\lvert\Phi^{A\tilde{A}}\rangle$ | The *normalized* maximally entangled state $\frac{1}{\sqrt{\lvert A\rvert}}\sum_{x\in[m]}\lvert xx\rangle$ |
| $\mathrm{Eff}(A)$ | The set of all effects in $\mathrm{Pos}(A)$; that is, $\Lambda\in\mathrm{Eff}(A)$ if and only if $\mathbf{0}\leqslant\Lambda\leqslant I^A$ |
| $\mathrm{Im}(T)$ | The image of $T\in\mathfrak{L}(A,B)$ |
| $\mathrm{Ker}(T)$ | The kernel of $T\in\mathfrak{L}(A,B)$ |
| $\mathrm{supp}\,(T)$ | The support subspace of $T\in\mathfrak{L}(A,B)$ |
| $\mathrm{supp}\,(\mathbf{p})$ | The set $\{x\in[n]:p_x>0\}$, where $\mathbf{p}=(p_1,\ldots,p_n)^T\in\mathrm{Prob}(n)$ |
| $\rho\ll\sigma$ | Inclusion of supports; $\mathrm{supp}\,(\rho)\subseteq\mathrm{supp}\,(\sigma)$ for $\rho,\sigma\in\mathfrak{D}(A)$ |
| $\mathrm{spec}(H)$ | The set of all distinct eigenvalues of a Hermitian operator $H\in\mathrm{Herm}(A)$ |
| $\lVert\cdot\rVert_p$ | The Schatten $p$-norm with $p\in[1,\infty]$ (see (2.68)). |
| $\lVert\cdot\rVert_{(k)}$ | The Ky Fan norm with $k\in[n]$ (see Definition 2.4 and (2.78)). |
| $\mathfrak{B}_\varepsilon(\mathbf{p})$ | The set $\{\mathbf{p}'\in\mathrm{Prob}(n):\frac{1}{2}\lVert\mathbf{p}-\mathbf{p}'\rVert_1\leqslant\varepsilon\}$, where $\mathbf{p}\in\mathrm{Prob}(n)$ and $\varepsilon\in[0,1]$. |
| $\mathfrak{B}_\varepsilon(\rho)$ | The set $\{\rho'\in\mathfrak{D}(A):\frac{1}{2}\lVert\rho-\rho'\rVert_1\leqslant\varepsilon\}$, where $\rho\in\mathfrak{D}(A)$ and $\varepsilon\in[0,1]$. |
| $\sigma\approx_\varepsilon\rho$ | Short notation for $\sigma\in\mathfrak{B}_\varepsilon(\rho)$. |
| $\Pi_\rho$ | Projection to the support of $\rho$. |
| $\mathfrak{T}_\varepsilon(X^n)$ | The set of $\varepsilon$-typical sequences. |
| $\mathfrak{T}_\varepsilon(A^n)$ | The $\varepsilon$-typical subspace of $A^n$. |
| $\mathfrak{T}_\varepsilon^{\mathrm{st}}(X^n)$ | The set of strongly $\varepsilon$-typical sequences. |
| $\mathfrak{T}_\varepsilon^{\mathrm{st}}(A^n)$ | The strongly $\varepsilon$-typical subspace of $A^n$. |
| $\mathrm{PPT}(AB)$ | The set of density matrices in $\mathfrak{D}(AB)$ with positive partial transpose. |
| $\rho^\Gamma$ | The partial transpose of a bipartite state $\rho\in\mathfrak{D}(AB)$ w.r.t. system $B$. |
| $a\gtrapprox b$ | $a\geqslant\log\lfloor 2^b\rfloor,\,a,b\in\mathbb{R}_+$ |

# Introductory Material

## 1.1 Introduction

A recurring theme in the field of physics is the endeavor to unify a variety of distinct physical phenomena into a comprehensive framework that can offer both descriptions and explanations for each of them. One of the most astounding achievements in this endeavor is the unification of fundamental forces. When physicists realized that the forces of electricity and magnetism could be elegantly described using a single framework, it not only substantially enhanced our comprehension of these forces but also gave birth to the expansive domain of electromagnetism.

The remarkable success in unifying forces serves as a testament to the fact that seemingly unrelated phenomena can often be traced back to a common origin. This approach extends beyond the realm of forces and finds resonance in the burgeoning field of quantum information science. Within this field, a novel discipline has emerged, which seeks to identify shared characteristics among seemingly disparate quantum phenomena. The overarching theme of this approach lies in the recognition that various attributes of physical systems can be defined as "resources." This recognition not only alters our perspective on these phenomena but also seamlessly integrates them within a comprehensive framework known as "quantum resource theories."

For example, take the case of quantum entanglement. In the 1990s, it was transformed from a topic of philosophical debates and discussions into a valuable resource. This transformative shift revolutionized our perception of entanglement; it evolved from being an intriguing and nonintuitive phenomenon into the essential driving force behind numerous quantum information tasks. This new perspective on entanglement opened up a vast array of possibilities and applications, starting with its utilization in quantum teleportation and superdense coding. Today, entanglement stands as a fundamental resource in fields such as quantum communication, quantum cryptography, and quantum computing.

Given the success of entanglement theory, it is only natural to explore other physical phenomena that can also be recognized as valuable resources. Currently, there are several quantum phenomena that have been identified as such. These encompass areas such as quantum and classical communication, athermality (within the realm of quantum thermodynamics), asymmetry, magic (in the context of quantum computation), quantum coherence, Bell nonlocality, quantum contextuality, quantum steering, incompatibility of quantum measurements, and many more. The recognition of all these phenomena as resources enables us to unify them under the umbrella of quantum resource theories.

Resource theories serve as a crucial framework for addressing complex questions. They aim to unravel puzzles such as determining which sets of resources can be transformed into one another and the methods by which such conversions can occur. Additionally, they explore how to measure and detect different resources. If a direct transformation between particular resources is not feasible, resource theories examine the possibility of nondeterministic conversions and the computation of their associated probabilities. The introduction of catalysts into the equation further deepens the inquiry.

This investigative approach often yields profound insights into the underlying nature of the physical or information-theoretic phenomena under scrutiny – such as entanglement, asymmetry, athermality, and more. Furthermore, this perspective provides a structured framework for organizing theoretical findings pertaining to these phenomena. As demonstrated by the evolution of entanglement theory, the resource-theoretic perspective possesses the potential to revolutionize our understanding of familiar subjects.

In this context, chemistry exemplifies this framework, elucidating how abundant collections of chemicals can be converted into more valuable products. Similarly, thermodynamics fits this mold by addressing inquiries about the conversion of various types of nonequilibrium states – thermal, mechanical, chemical, and more – into one another, including the extraction of useful work from heat baths at differing temperatures.

Within the realm of quantum resource theories, a fundamental challenge arises in identifying equivalence classes of quantum systems that can be reversibly interconvert (or simulate each other) when considering an abundance of resource copies, and determining the rates at which these interconversions occur. The relative entropy of a resource plays a pivotal role in such reversible transformations, gauging the resourcefulness of a system by quantifying its deviation from the set of free (nonresourceful) systems. Remarkably, this function unifies essential (pseudo) metrics across seemingly disparate scientific domains. For instance, the relative entropy of a resource manifests as free energy in thermodynamics, entanglement entropy in pure state entanglement theory, and the entanglement-assisted capacity of a quantum channel in quantum communication; see Figure 1.1.

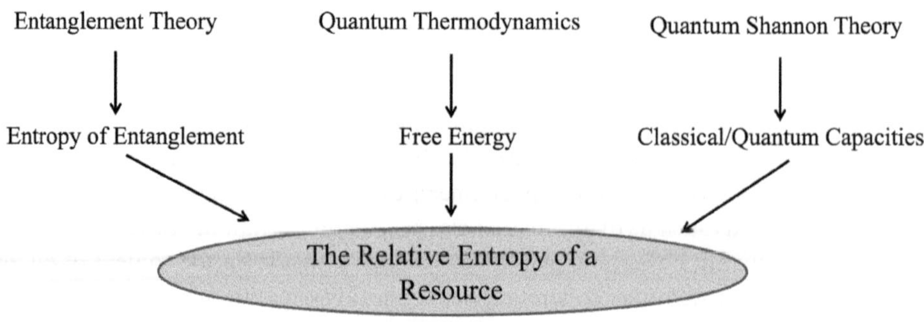


**Figure 1.1**     Unification of resources.

## 1.2 About this Book

As mentioned in the introduction, quantum resource theories have recently emerged as a vibrant research area within the quantum information science community. Initially, the emphasis was on understanding the resources used in quantum information processing tasks. However, it has become increasingly evident that quantum resources have broad relevance, extending from quantum computing to quantum thermodynamics and the fundamental principles of quantum physics. This realization has spurred rapid developments in the field, resulting in a proliferation of publications and the development of new tools and mathematical methods that firmly underpin this area of study.

In light of the extensive literature in the field of quantum information science, one might understandably question the need for yet another book on quantum resources. Isn't this territory already covered in existing quantum information textbooks? For instance, quantum Shannon theory can be seen as a theory of interconversions among different types of resources, and Wilde [235] and Watrous [233] have produced outstanding books delving into these topics. Additionally, detailed treatments of subjects covered in this book, such as quantum divergences and Rényi entropies, can be found in Tommamichel's noteworthy work [211].

While it is accurate to say that many of the topics covered in this book are available elsewhere, what distinguishes this book is its unique approach. It explores well-trodden subjects like entropy, uncertainty, divergences, nonlocality, entanglement, and energy from a fresh perspective rooted in resource theories. Specifically, the book adopts an axiomatic approach to rigorously introduce these concepts, providing illustrative examples. Only then does it transition to operational aspects that involve the examples discussed.

Take, for instance, the topic of conditional entropy, a subject widely covered in numerous textbooks in both classical and quantum information theory. This book, however, offers a distinctive approach by presenting this concept from three distinct perspectives: axiomatic, constructive, and operational. Notably, all three perspectives converge to the same notion of conditional entropy. This approach not only provides the reader with a deeper understanding of the concept but also underscores its robust foundation.

The primary goal of this book is pedagogical in nature, with the hope of providing readers with a contemporary perspective on quantum resource theories. It aspires to equip readers with the necessary physical principles and advanced mathematical techniques required to comprehend recent advancements in this field. Upon completing this book, readers should have the ability to explore open problems and research directions within the field, some of which will be highlighted in the text.

In anticipation of a diverse readership, this book is designed to be inclusive, targeting both graduate students and senior undergraduate students who possess a foundational understanding of linear algebra. It aims to provide them with a comprehensive resource for delving into this fascinating field. Simultaneously, the book serves as a reference, offering fresh insights and innovative approaches that researchers in the early stages of their careers may find valuable. With numerous examples and exercises, it aims to serve as a textbook for courses on the subject, enhancing the learning experience for students.

While the primary audience for this textbook consists of entry-level graduate students interested in pursuing research at the master's or PhD level in quantum resource theories, encompassing quantum information science, it may also prove valuable to researchers in fields influenced by quantum information and resource theories, such as quantum thermodynamics and condensed matter physics. They may find this book to be a useful and accessible reference source.

Although we have endeavored to make the book self-contained, a basic understanding of linear algebra is essential. The goal was to create a resource accessible to graduate students from diverse backgrounds in mathematics, physics, and computer science. As a result, the book includes preliminary chapters and several appendices (online version) that fill potential knowledge gaps, given the interdisciplinary nature of the subject matter.

Quantum resource theories constitute a vast research area, with new properties of physical systems continually being recognized as resources. Consequently, the aim of this book is not to exhaustively cover all resource theories but rather to select those that illustrate the techniques used in quantum resource theories effectively. On the technical front, we have chosen to begin with the modern single-shot approach and employ it to derive asymptotic rates. Historically, asymptotic rates were studied first, but from a pedagogical standpoint, it is more intuitive to start with the single-shot regime.

To the best of our knowledge, there are currently no dedicated books specifically focused on quantum resource theories. With this book, we hope to contribute to the field by providing a comprehensive overview and integrating both new and existing results within a unified framework. While we do not claim this book to be the ultimate authority, we believe it can serve as a valuable reference that consolidates ideas scattered across various journal articles, addressing the need for a centralized resource in the field of quantum resource theories.

## 1.3 The Structure of the Book

In this book, we delve deep into the comprehensive framework of quantum resource theories, offering a detailed study of their general principles and equipping readers with the necessary tools and methodologies. We extensively cover three illustrative examples of resource theories – Entanglement, Asymmetry, and Thermodynamics – chosen for their pedagogical value in showcasing the diverse facets of quantum resource theories. While we do not have a dedicated chapter solely focused on quantum coherence, this concept is seamlessly woven into our broader discussions. It serves as a recurring illustrative example that enriches our understanding of various aspects of quantum resource theories throughout the book.

The initial volume of this book is structured into five main parts, with an additional sixth part containing supplementary materials.

**Section 1** The opening section of this book is thoughtfully designed to cater to readers who may not possess prior knowledge of quantum mechanics or quantum information. Within this segment, we embark on a rigorous mathematical journey through quantum theory, emphasizing precise definitions and mathematical proofs of fundamental physical theorems. Key subjects covered in this section encompass quantum states, generalized quantum measurements, quantum channels, POVMs, and more. Moreover, this section extends its reach beyond the boundaries of quantum theory, delving into topics such as Ky Fan norms, the Strømer–Woronowicz theorem, the Pinching Inequality, the Reverse Hölder Inequality, certain hidden variable models, and other subjects that may not commonly cross the paths of graduate students in physics, mathematics, or computer science. Therefore, even those well-versed in these topics may find it beneficial to skim through this chapter briefly, as it has the potential to reveal previously undiscovered insights.

**Section 2** The second section delves deep into the methodologies and tools employed within the realm of quantum resource theories and quantum information. While it explores numerous quantum information concepts, it distinguishes itself from conventional quantum information theory textbooks. The introductory chapter of this section provides an all-encompassing mathematical review of majorization theory, encapsulating recent groundbreaking discoveries, such as relative majorization, conditional majorization, and the intersection of probability theory with this field.

Subsequent chapters in this section adopt a distinctive approach to elucidate concepts associated with metrics, divergences, and entropies. These notions are introduced and dissected using techniques and insights drawn from the framework of quantum resource theories. For instance, entropy, conditional entropy, relative entropies, and other divergences are introduced as additive functions that adhere to monotonicity under the set of free operations, a foundational concept in quantum resource theories.

The final chapter in this section is dedicated to the asymptotic regime, focusing on the consequences of the "law of large numbers" in quantum information and quantum resource theories. This chapter introduces concepts such as weak and strong typicality, the method of types, classical and quantum hypothesis testing, and the symmetric subspace. These tools prove particularly valuable in the asymptotic domain of quantum resource theories when exploring interconversion rates among infinitely many resources.

In summary, although the contents of this second section share some commonalities with conventional quantum information theory textbooks, they diverge significantly by presenting concepts and tools in a unique manner. Rather than employing Venn diagrams to define key concepts like entropy, this part of the book aims to provide a comprehensive and rigorous approach to precisely define these concepts by employing axiomatic, constructive, and operational approaches. Leveraging the framework of quantum resource theories, this section offers a fresh and innovative perspective on these familiar topics.

**Section 3** In the third section, we delve into the fundamental framework of quantum resource theories. Our journey begins with a meticulous mathematical elucidation of a quantum resource theory. We proceed to examine its foundational principles, including but not limited to the golden rule of free operations, resource nongenerating operations, physically implementable operations, convex and affine resource theories, state-based resource theories, as well as resource witnesses and their associated properties.

Next, we delve into the quantification of quantum resources. In this context, we introduce a plethora of resource measures and resource monotones, delving deep into their properties, which include additivity, sub-additivity, convexity, strong monotonicity, and asymptotic continuity. These concepts form the bedrock of quantum resource theories, and understanding them is pivotal.

Resource monotones and resource measures offer a valuable means of quantifying resources. Our emphasis is on divergence-based resource measures, such as the relative entropy of a resource, given their operational interpretations across various resource theories. We also explore techniques for computing these measures, including semidefinite programming, and delve into a practical approach for "smoothing" these measures, a technique commonly employed in single-shot quantum information science.

Concluding this section of the book, we introduce a rich array of resource interconversion scenarios. These encompass exact interconversions, stochastic (probabilistic) interconversions, approximate interconversions, and asymptotic interconversions. We delve into essential tools intricately linked to resource interconversions, such as the conversion distance within the single-shot regime, the asymptotic equipartition property, and the quantum Stein's lemma within the asymptotic domain. Additionally, we explore the uniqueness of the Umegaki relative entropy within the context of quantum resource theories. Our investigation extends to the evaluation of both the cost and distillation of resources, examining these processes within both the single-shot and asymptotic regimes. We have encapsulated the essence of this section of the book in Figure 1.2.

**Section 4** The fourth section is dedicated to the quintessential exemplar of quantum resource theories, often referred to as the "poster child" – entanglement theory. This section comprises three chapters, each focusing on distinct facets of entanglement. The first chapter delves into the realm of pure bipartite entanglement, followed by the second chapter, which explores mixed bipartite entanglement. The third chapter, in turn, delves into the intricacies of multipartite entanglement.

Within these chapters, we leverage the techniques and concepts developed in Sections 2 and 3 to delve into the theory of entanglement. This enables us to furnish a precise definition of quantum entanglement and undertake a comprehensive examination of its detection, manipulation, and quantification. Notably, the first of these three chapters serves as the cornerstone, offering an in-depth exploration of pure

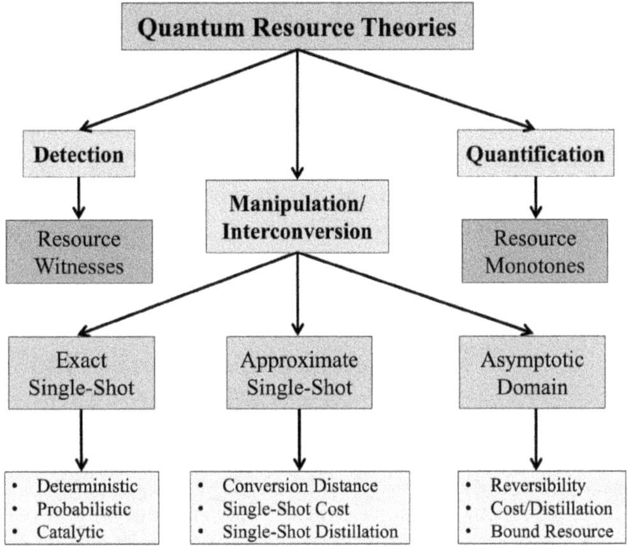


**Figure 1.2**    The structure of quantum resource theories

bipartite entanglement, which forms the foundational knowledge upon which the subsequent chapters on mixed and multipartite entanglement build.

**Section 5** The fifth section comprises three chapters, with the first two chapters focusing on asymmetry and nonuniformity, laying the groundwork for the third chapter on quantum thermodynamics. In this section, we reveal that athermality, the resource essential for thermodynamic tasks, consists of two components: time-translation asymmetry and nonuniformity.

The first chapter explores the resource theory of asymmetry, introducing an operational framework that arises from practical constraints when multiple parties lack a common shared reference frame. This theory has found numerous applications in quantum information and beyond.

The second chapter delves into the resource theory of nonuniformity. In this theory, maximally mixed states are considered free, while all other states are regarded as valuable resources. This theory can be seen as a unique variant of thermodynamics, involving completely degenerate Hamiltonians. Indeed, we introduce this chapter to serve as a gentle introduction to the world of quantum thermodynamics.

Finally, in the third chapter of this section, we dive into quantum thermodynamics. Throughout the book, whenever we introduce a new quantum resource theory, we adhere to the structured framework outlined in Figure 1.2.

**Section 6** The final section serves as a comprehensive resource aimed at ensuring the self-containment of the entire text. It exclusively includes material that directly complements the core content of the book.

In the initial three chapters, we delve into key subjects: convex analysis, operator monotonicity, and representation theory. It's important to note that each of these topics is vast in its own right, with numerous dedicated books solely focused on representation theory or convex analysis, for example. In this section, we have thoughtfully curated and presented the aspects of these topics that are pertinent to our book's core themes. Our approach emphasizes utilizing quantum notations and placing a strong emphasis on furnishing all the essential elements needed to ensure the book's self-contained nature.

**Appendices** The appendices can be downloaded from www.cambridge.org/9781009560917.

## 1.4 Resurrection of Quantum Entanglement: The Birth of a Fundamental Resource

In this section, we delve into the transformative protocols of quantum teleportation and superdense coding. These groundbreaking techniques marked a pivotal moment in the history of quantum physics, elevating entanglement from a purely theoretical curiosity to a precious resource with tangible applications. This paradigm shift, akin to the "resurrection" of quantum entanglement, carried profound implications for the burgeoning field of quantum information. In essence, it played a significant role in catalyzing the emergence of quantum information science. If you're new to the formalism of quantum mechanics, we recommend starting with Chapter 2 before delving into the following two sections.

Even in the early days of quantum mechanics, entanglement stood out as a distinctive and defining feature of the theory. As articulated by Schrödinger, he remarked, "I would not call [entanglement] one but rather the characteristic trait of quantum mechanics, the one that enforces its entire departure from classical lines of thought." This statement underscores the profound departure from classical physics that entanglement embodies. Significantly, the intriguing properties of entanglement were recognized well before Bell's seminal paper on the exclusion of local hidden variable models (as discussed in Section 2.6).

To illustrate this point, consider a composite system consisting of two 1/2-spin particles, such as electrons, in the singlet state:

$$|\Psi_-\rangle = \frac{1}{\sqrt{2}} \left( |\uparrow_\mathbf{n}\rangle|\downarrow_\mathbf{n}\rangle - |\downarrow_\mathbf{n}\rangle \uparrow_\mathbf{n}\rangle \right). \tag{1.1}$$

Here, $\{|\uparrow_\mathbf{n}\rangle, |\downarrow_\mathbf{n}\rangle\}$ forms an orthonormal basis in the complex vector space $\mathbb{C}^2$, representing the two eigenvectors of the spin observable corresponding to the "up" and "down" orientations along a direction $\mathbf{n} \in \mathbb{R}^3$. Notably, a remarkable property of this state is its independence from the specific spin direction $\mathbf{n}$ (see Chapter 2).

Now, if Alice performs a measurement in the $\mathbf{n}$ direction, it will instantaneously dictate Bob's post-measurement state to align with the opposite $\mathbf{n}$ direction. This peculiar phenomenon allows Alice to exert immediate influence on Bob's state by simply choosing whether to conduct a Stern–Gerlach measurement (as discussed in

Section 2.1) along the **n** or **m** direction. This nonintuitive behavior of entangled composite quantum systems prompted Einstein to describe it as a "spooky action at a distance."

Beyond its profound implications from a fundamental standpoint, entanglement has gained recognition as a valuable and indispensable resource for the realization of specific quantum information processing tasks. This shift in perspective has given rise to a substantial body of research, as entanglement is no longer solely a philosophical curiosity but a powerful tool with remarkable practical applications. These applications encompass protocols like quantum teleportation, superdense coding, and numerous innovations in quantum cryptography and quantum computing.

In this section, we embark on a journey through some of these protocols, known as unit protocols, as they exclusively rely on unit noiseless resources. These protocols serve as a testament to the versatility of entanglement and employ three distinct resource types: a noiseless quantum communication channel, a noiseless classical communication channel, and the entangled bit, abbreviated as ebit.

## 1.4.1 Quantum Teleportation

Quantum teleportation is a groundbreaking protocol enabling Alice to transmit an unknown quantum state $|\psi\rangle$ to Bob, all without the need for a dedicated quantum communication channel. Instead, it relies on the clever utilization of entanglement and a classical communication channel to achieve this remarkable feat, as illustrated in Figure 1.3. To elucidate, consider the scenario where Alice and Bob share a composite system comprising two electrons initially prepared in the singlet state:

$$|\Psi_-^{AB}\rangle = \frac{1}{\sqrt{2}}(|01\rangle - |10\rangle). \tag{1.2}$$

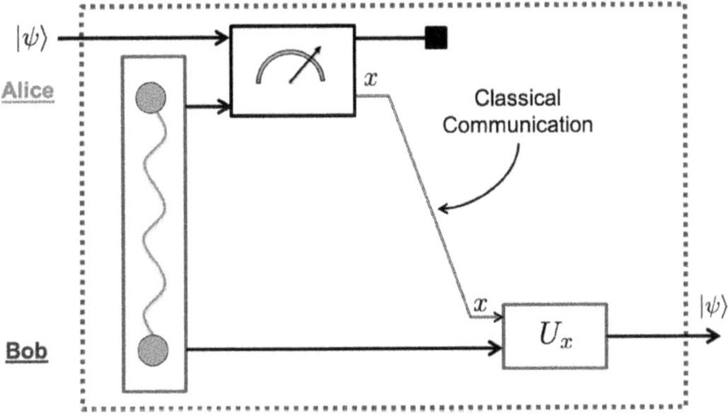


**Figure 1.3** Quantum teleportation. Single-line arrows correspond to quantum systems. Double line arrows correspond to classical systems.

Furthermore, let's consider the scenario where Alice possesses an additional electron in her system, characterized by a quantum state $|\psi^A\rangle = a|0\rangle + b|1\rangle$. Importantly, both Alice and Bob lack knowledge regarding the spin state of this electron, which means they are unaware of the specific values of $a$ and $b$. According to the principles of quantum mechanics, the collective quantum state of these three electrons – two under Alice's control and one under Bob's – is described by the tensor product:

$$|\psi^{\tilde{A}}\rangle \otimes |\Psi_-^{AB}\rangle = \frac{1}{\sqrt{2}}(a|0\rangle + b|1\rangle) \otimes (|01\rangle - |10\rangle)$$

$$\underset{\textbf{Parentheses}}{\overset{\textbf{Openning}}{\rightarrow}} = \frac{1}{\sqrt{2}}(a|001\rangle + b|101\rangle - a|010\rangle - b|110\rangle). \tag{1.3}$$

It's noteworthy that in our description in (1.3), we represented the state $|\psi^{\tilde{A}}\rangle \otimes |\Psi_-^{AB}\rangle$ using the computational basis of the vector space $\tilde{A}AB$. However, we can achieve a more insightful representation by substituting the computational basis $|00\rangle, |01\rangle, |10\rangle, |11\rangle$ of system $\tilde{A}A$ with the Bell basis consisting of $|\Phi_\pm^{\tilde{A}A}\rangle = \frac{1}{\sqrt{2}}(|00\rangle \pm |11\rangle)$ and $|\Psi_\pm^{\tilde{A}A}\rangle = \frac{1}{\sqrt{2}}(|01\rangle \pm |10\rangle)$. This substitution allows us to express the state as follows:

$$|\psi\rangle^{\tilde{A}} \otimes |\Psi_-^{AB}\rangle = \frac{1}{2}\Big[a\big(|\Phi_+^{\tilde{A}A}\rangle + |\Phi_-^{\tilde{A}A}\rangle\big)|1\rangle + b\big(|\Psi_+^{\tilde{A}A}\rangle - |\Psi_-^{\tilde{A}A}\rangle\big)|1\rangle$$

$$- a\big(|\Psi_+^{\tilde{A}A}\rangle + |\Psi_-^{\tilde{A}A}\rangle\big)|0\rangle - b\big(|\Phi_+^{\tilde{A}A}\rangle - |\Phi_-^{\tilde{A}A}\rangle\big)|0\rangle\Big]$$

$$\textbf{Collecting terms} \rightarrow = \frac{1}{2}\Big[|\Phi_+^{\tilde{A}A}\rangle(a|1\rangle - b|0\rangle) + |\Phi_-^{\tilde{A}A}\rangle(a|1\rangle + b|0\rangle)$$

$$+ |\Psi_+^{\tilde{A}A}\rangle(b|1\rangle - a|0\rangle) - |\Psi_-^{\tilde{A}A}\rangle(a|0\rangle + b|1\rangle)\Big]. \tag{1.4}$$

Therefore, if Alice performs the Bell measurement on her two qubits $\tilde{A}A$, that is the basis (projective) measurement

$$\left\{P_0 = |\Psi_-^{\tilde{A}A}\rangle\langle\Psi_-^{\tilde{A}A}|, \ P_1 = |\Phi_-^{\tilde{A}A}\rangle\langle\Phi_-^{\tilde{A}A}|, \ P_2 = |\Phi_+^{\tilde{A}A}\rangle\langle\Phi_+^{\tilde{A}A}|, \ P_3 = |\Psi_+^{\tilde{A}A}\rangle\langle\Psi_+^{\tilde{A}A}|\right\}, \tag{1.5}$$

she will get with equal probability four possible outcomes (denoted $x = 0, 1, 2, 3$, and global phase is ignored):

| Outcome | Post-Measurement State | Simplification (Up to a global phase) |
|---|---|---|
| $x = 0$ | $|\Psi_-^{\tilde{A}A}\rangle \otimes (a|0\rangle + b|1\rangle)$ | $|\Psi_-^{\tilde{A}A}\rangle \otimes |\psi\rangle$ |
| $x = 1$ | $|\Phi_-^{\tilde{A}A}\rangle \otimes (a|1\rangle + b|0\rangle)$ | $|\Phi_-^{\tilde{A}A}\rangle \otimes \sigma_1|\psi\rangle$ |
| $x = 2$ | $|\Phi_+^{\tilde{A}A}\rangle \otimes (a|1\rangle - b|0\rangle)$ | $|\Phi_+^{\tilde{A}A}\rangle \otimes \sigma_2|\psi\rangle$ |
| $x = 3$ | $|\Psi_+^{\tilde{A}A}\rangle \otimes (b|1\rangle - a|0\rangle)$ | $|\Psi_+^{\tilde{A}A}\rangle \otimes \sigma_3|\psi\rangle$ |

where we denote by $\{\sigma_x\}_{x=0,1,2,3}$ the identity matrix $\sigma_0 = I_2$, and the three Pauli matrices $\sigma_1, \sigma_2,$ and $\sigma_3$. Hence, up to a global phase, Bob's state after outcome $x$ occurred is $\sigma_x|\psi\rangle$. After Alice sends (via a classical communication channel) the measurement outcome $x$ to Bob, Bob can then perform the unitary operation $U_x = \sigma_x$ to obtain the state

$$\sigma_x\left(\sigma_x|\psi\rangle\right) = \sigma_x^2|\psi\rangle = |\psi\rangle. \tag{1.6}$$

Therefore, by using shared entanglement, and after transmitting two classical bits (cbits), Alice was able to transfer her unknown qubit state $|\psi\rangle$ to Bob's side.

If Bob did not receive the classical message from Alice, then his state is one of the four states $\{\sigma_x|\psi\rangle\}_{x=0,1,2,3}$. Since he does not know $x$, from his perspective his state is (see Exercise 1.1)

$$\rho = \frac{1}{4}\sum_{x=0}^{3}\sigma_x|\psi\rangle\langle\psi|\sigma_x = \frac{1}{2}I. \tag{1.7}$$

That is, without the knowledge of $x$, Bob's resulting state is the maximally mixed state, and contains no information about $|\psi\rangle$.

**Exercise 1.1.** *Show that for any density matrix $\rho \in \mathfrak{D}(\mathbb{C}^2)$,*

$$\frac{1}{4}\sum_{x=0}^{3}\sigma_x\rho\sigma_x = \frac{1}{2}I. \tag{1.8}$$

*Hint: Prove first that the left-hand side of (1.8) is invariant under a conjugation by $\sigma_x$.*

**Exercise 1.2.** *Show that if instead of the singlet state $|\Psi_-^{AB}\rangle$, Alice and Bob share another maximally entangled state $|\Phi^{AB}\rangle$ (i.e. the reduced density matrix of $|\Phi^{AB}\rangle$ is the maximally mixed state), then, by modifying slightly the protocol, they can still teleport an unknown quantum state from Alice to Bob.*

This protocol can be generalized in several different ways. First, in Exercise 1.3 you will generalize it to $d$-dimensions. Moreover, in general, if Alice and Bob do not share the singlet state, but instead their particles are prepared in some other nonseperable state (i.e. entangled state, but not maximally entangled state) $\rho^{AB} \in \mathfrak{D}(AB)$, then typically perfect/faithful teleportation will not be possible. Still in this case one can design a protocol achieving quantum teleportation with probability that is less than 1 (see Exercise 1.4), and/or in the end of the protocol the state in Bob's lab is not exactly equal to Alice's original state $|\psi\rangle^{\tilde{A}}$ but only close to it up to some treshold. Thus, the protocol described here is called *faithful* teleportation, since the protocol teleport perfectly $|\psi\rangle$ form Alice to Bob with 100% success rate.

**Exercise 1.3.** *Let $|\Phi^{AB}\rangle := \frac{1}{\sqrt{d}}\sum_{z\in[d]}|zz\rangle$ be a 2-qudit (normalized) maximally entangled state in $AB \cong \mathbb{C}^d \otimes \mathbb{C}^d$. Consider a family of $d^2$ states in $AB$ defined by*

$$|\psi_{xy}^{AB}\rangle = T^x \otimes S^y|\Phi^{AB}\rangle, \qquad x, y \in [d], \tag{1.9}$$

*where $T$ and $S$ are the phase and shift operators defined by $T|z\rangle = e^{i\frac{2\pi z}{d}}|z\rangle$ and $S|z\rangle = |z \oplus 1\rangle$, where $\oplus$ is the plus modulo $d$, and $z \in [d]$.*

1. *Show that $\{|\psi_{xy}^{AB}\rangle\}_{x,y\in[d]}$ is an orthonormal basis of $AB$.*
2. *Show that the reduced density matrix of $|\psi_{xy}^{AB}\rangle$ is the maximally mixed state for all $x, y \in [d]$.*
3. *Find a protocol for faithful teleportation of a qudit from Alice's lab to Bob's lab. Assume that the joint measurement that Alice performs on her two qudits is a basis measurement in the basis $\{|\psi_{xy}^{AB}\rangle\}_{x,y\in[d]}$. What are the unitary operators performed by Bob? How many classical bits (cbits) Alice transmits to Bob?*

**Exercise 1.4.** *Suppose Alice and Bob share the state $|\psi^{AB}\rangle = \frac{1}{2}|00\rangle + \frac{\sqrt{3}}{2}|11\rangle$. Show that there exists a 2-outcome (basis) measurement that Alice can perform, such that with some probability greater than zero, the state of Alice and Bob after the measurement becomes the maximally entangled state $|\Phi_+^{AB}\rangle = \frac{1}{\sqrt{2}}(|00\rangle + |11\rangle)$.*

So far we assumed that the teleported state is a pure state. However, the exact same protocol works even if the unknown state $|\psi\rangle$ is replaced with a mixed state $\rho$. This is because we can view any mixed state as some ensemble of pure states $\{p_x, |\psi_x\rangle\}$ in which the parameter $x$ is unknown. Irrespective of the value of $x$, the protocol in this section will teleport $|\psi_x\rangle$ from Alice to Bob, and thereby, given that the value of $x$ is unknown, Alice effectively teleported to Bob the mixed state $\rho := \sum_x p_x |\psi_x\rangle\langle\psi_x|$. Alternatively, note that the quantum teleportation protocol in Figure 1.3 can be described as a realization of the identity quantum channel $\mathrm{id} \in \mathrm{CPTP}(A \to B)$ (with $|A| = |B| := d$) given by

$$\mathrm{id}^{A \to B}(\rho^A) = \sum_{x\in[d^2]} \mathrm{Tr}_{A\tilde{A}}\left[\left(P_x^{\tilde{A}A} \otimes U_x^B\right)\left(\rho^A \otimes \Phi^{\tilde{A}B}\right)\left(P_x^{\tilde{A}A} \otimes U_x^B\right)^*\right], \quad (1.10)$$

where $\{P_x^{\tilde{A}A}\}_{x\in[d^2]}$ corresponds to the measurement on systems $\tilde{A}$ and $A$ in the maximally entangled basis, $U_x$ is the unitary performed by Bob after he received the value $x$ from Alice, and $\Phi^{AB}$ is the maximally entangled state on system $AB$. The quantum teleportation protocol states that there exists $\{P_x^{\tilde{A}A}\}$ and $\{U_x\}$ such that the quantum channel $\mathrm{id}^{A \to B}$ above is indeed the identity channel. Although, in the protocol above we proved it only for pure input states $|\psi\rangle\langle\psi|$, from the linearity of the quantum channel $\mathrm{id}^{A \to B}$, it follows that $\mathrm{id}^{A \to B}$ is the identity quantum channel on all mixed states.

**Exercise 1.5 (Entanglement Swapping).** *Consider four qubit systems A, B, C, and D, in the double-singlet state $|\Psi_-^{AB}\rangle \otimes |\Psi_-^{CD}\rangle$.*

1. *Show that a joint Bell measurement on system $BC$ generates a maximally entangled state on system $AD$ (along with another maximally entangled state on system $BC$) for all four possible outcomes of the measurement (see Figure 1.4).*
2. *Show that the singlet state in $AD$ can be generated by quantum teleportation between system $BC$ and system $D$.*
3. *Generalize the entanglement swapping protocol to four qudit systems each of dimension $d$.*

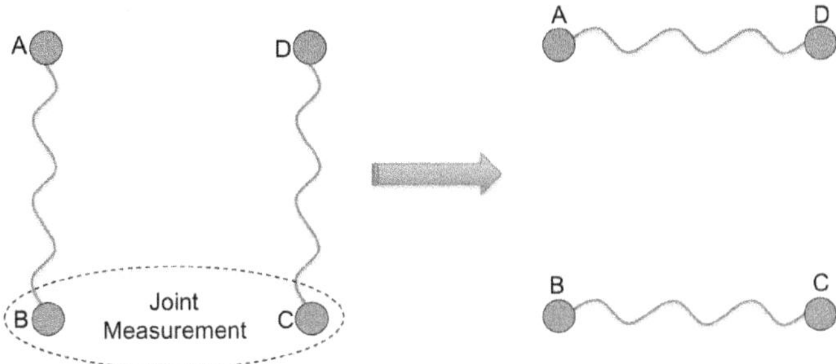


**Figure 1.4**    Entanglement swapping.

## 1.4.2 Superdense Coding

How much classical information can be transmitted with a single qubit? Suppose Alice wants to transmit a classical message to Bob, but all she has at her disposal is a single qubit (e.g. spin of an electron) and a perfect noiseless quantum communication channel that she can use to transmit the electron to Bob. She can prepare the single electron in the spin that she wants and send the electron to Bob over the noiseless quantum channel. Alice and Bob agree at the beginning that a message 0 corresponds to spin up in the $z$ direction, and a message 1 corresponds to spin down in the $z$ direction. In this way, Alice can transmit one cbit with one use of a perfectly noiseless quantum channel. Can they do better? We will see later on that it is not possible to encode more than one classical bit into a single electron, as long as Alice's electron is not entangled with another electron in Bob's system.

Suppose now that Alice's electron is maximally entangled with another electron in Bob's lab, so that Alice and Bob share the singlet state $|\Psi_-^{AB}\rangle$. In the first step of the protocol, Alice encodes a message $x$ into a her qubit. She is doing it by performing one out of several (unitary) rotations $\{U_x^A\}_{x=0}^{m-1}$ on her qubit (electron). If Alice chose to do the $x$ rotation, the state of the system after the rotation is

$$|\psi_x^{AB}\rangle := \left(U_x^A \otimes I^B\right)|\Psi_-\rangle^{AB}. \tag{1.11}$$

Taking $U_x = \sigma_x$ to be the four Pauli matrices (with $\sigma_0 = I_2$) we get that the four states $\{|\psi_x^{AB}\rangle\}_{x=0}^3$ are orthonormal and form a basis of $\mathbb{C}^2 \otimes \mathbb{C}^2$. In fact, this is the Bell basis we encountered in the previous subsection. In the next step of the protocol, Alice sends her electron (over a noiseless quantum communication channel) to Bob. Upon receiving Alice's electron, Bob has in his lab two electrons in the state $|\psi_x^{AB}\rangle$. Given that the set of states $\{|\psi_x^{AB}\rangle\}_{x=0}^3$ form an orthonormal basis, in the last step of the protocol, Bob performs a joint basis measurement on his two electrons, in the basis $\{|\psi_x^{AB}\rangle\}_{x=0}^3$, and thereby learns the outcome $x$. The outcome $x$ is the message that Alice intended to send Bob.

**Exercise 1.6.** *Show that the set of states $\{|\psi_x^{AB}\rangle\}_{x=0}^3$ is an orthonormal basis of $\mathbb{C}^2 \otimes \mathbb{C}^2$.*

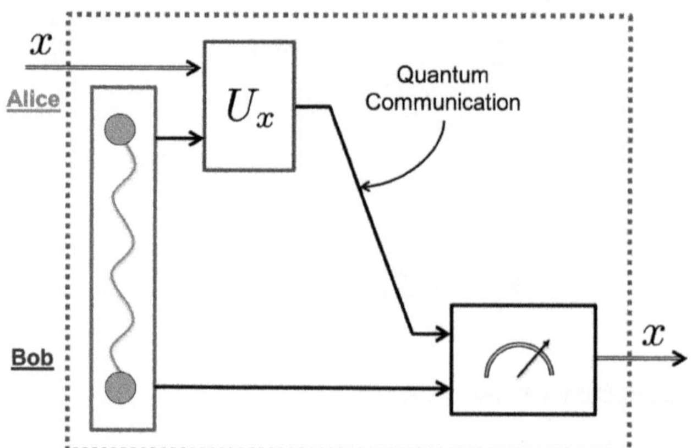


**Figure 1.5**   Superdense coding. Double lines correspond to classical systems, and single lines to quantum systems.

**Exercise 1.7.** *Let* $|\Phi^{AB}\rangle := \frac{1}{\sqrt{d}} \sum_{z \in [d]} |zz\rangle$ *be a maximally entangled state in* $\mathbb{C}^d \otimes \mathbb{C}^d$. *Show that Alice can use it to transmit to Bob* $2 \log_2(d)$ *cbits.*

## 1.5 Resource Analysis and Reversibility

The previous two protocols demonstrate that entanglement is a valuable resource with which certain tasks will not be possible without it. We have seen in the protocols above that entanglement can be converted to other types of resources such as quantum or classical communication channels. We will use the following notations to denote these resources:

1. $[qq]$ denotes one ebit; that is, a unit of a static noiseless resource comprising of two qubits in a maximally entangled state.
2. $[q \to q]$ denotes one use of an ideal (noiseless) qubit channel.
3. $[c \to c]$ denotes one use of a classical bit channel capable of transmitting perfectly one classical bit.
4. $[cc]$ denotes one bit of shared randomness.

Observe that these resource units can be classified into being classical or quantum, and static (such as $[qq]$ and $[cc]$) or dynamic (such as $[q \to q]$ or $[c \to c]$).

With these notations, the teleportation can be viewed as a process in which one ebit plus two uses of a classical bit channel are consumed to simulate a qubit channel. In resource symbols this can be characterized as the following resource inequality:

$$[qq] + 2[c \to c] \geqslant [q \to q]. \tag{1.12}$$

Note that the use of the inequality here is justified by the fact that a single use of a quantum channel cannot generate both an entangled state *and* a double use of a classical channel.

For superdense coding, an ebit plus one use of a quantum channel is used to simulate two uses of a classical channel. This can be expressed as the resource inequality

$$[qq] + [q \to q] \geqslant 2[c \to c]. \tag{1.13}$$

Note that if entanglement is not considered as a resource, that is, the parties are supplied with unlimited singlet states, then we can remove the ebit cost $[qq]$ in (1.12) and (1.13) and get that for teleportation $2[c \to c] \geqslant [q \to q]$ and for superdense coding $[q \to q] \geqslant 2[c \to c]$. This makes teleportation and superdense coding the dual protocols of each other, and in this case we can say that $[q \to q] = 2[c \to c]$.

However, in almost all practical scenarios, entanglement is an expensive resource that can be difficult to generate over long distances and that is also highly sensitive to decoherence and noise. Therefore, specifically pure maximally entanglement is scarce, and must be treated as a resource. The question then becomes if it is possible to change slightly the protocols of teleportation and superdense coding, making them more symmetric, in the sense that the two resource inequalities in (1.12) and (1.13) merge into a single resource equality. This is indeed possible if we replace $2[c \to c]$ in the right-hand side of (1.13) with two uses of an isometry channel known as the coherent bit channel.

## 1.5.1 The Coherent Bit Channel

We introduce here another unit resource that is called the *coherent bit channel*, or in short the *cobit* channel, and is denoted by $[q \to qq]$. As the symbol suggests, this unit resource represents one use of a channel. The channel is defined by the isometry, $V : A \to A \otimes B$, with $|A| = |B| = 2$, according to the following action on the basis of $A$:

$$V|x\rangle^A = |x\rangle^A |x\rangle^B \quad \forall\, x \in \{0, 1\} \quad \text{or equivalently} \quad V = \sum_{x=0}^{1} |x\rangle\langle x|^A \otimes |x\rangle^B. \tag{1.14}$$

We will denote by

$$\mathcal{V}_Z(\rho) := V\rho V^*, \qquad \forall\, \rho \in \mathfrak{L}(A), \tag{1.15}$$

where the subscript $Z$ indicates that the basis $\{|0\rangle, |1\rangle\}$ is an eigenbasis of the third Pauli operator (i.e. eigenvectors of the spin observable in the $z$-direction). One can define $V$ with respect to other bases. For example, we will denote by $\mathcal{V}_X(\cdot) = U(\cdot)U^*$ the coherent bit channel with respect to the basis $\{|+\rangle, |-\rangle\}$, where $U$ is the isometry defined by $U|\pm\rangle^A = |\pm\rangle^A |\pm\rangle^B$.

How is this resource related to other resources? First note that with such a resource Alice can transmit a classical bit to Bob. Indeed, Alice can encode a cbit $x \in \{0, 1\}$ in the state $|x\rangle^A$ and send it over the channel $\mathcal{V}_Z$. Then, Bob receives $|x\rangle^B$ on his system and performs a basis measurement to learn $x$. We therefore have

$$[q \to qq] \geqslant [c \to c]. \tag{1.16}$$

Exercise 1.8 shows that we also have $[q \to qq] \geq [qq]$. Among other things, this also implies that $[c \to c] \not\geq [q \to qq]$ or in other words, $[c \to c]$ is strictly less resourceful than $[q \to qq]$.

**Exercise 1.8.** *Show that* $\mathcal{V}_Z\left(|+\rangle\langle+|^A\right) = |\Phi_+^{AB}\rangle\langle\Phi_+^{AB}|.$

## 1.5.2 Coherent Superdense Coding

For superdense coding, we saw that an ebit, $[qq]$, plus one use of a quantum channel, $[q \to q]$, can be used to simulate two uses of a classical channel, $2[c \to c]$. We now show that the same resources can also be used to simulate two uses of the coherent map $\mathcal{V}$. That is, we will show that

$$[qq] + [q \to q] \geq 2[q \to qq]. \tag{1.17}$$

Note that due to (1.16), the above equation also implies the resource inequality (1.13). The quantum protocol that achieves this resource conversion is called *coherent* superdense coding.

Coherent superdense coding protocol (see Figure 1.6) consists of several steps. Initially, Alice and Bob share the maximally entangled state $|\Phi_+^{AB}\rangle$. Alice then prepares an input state $|x\rangle^{A_1}|y\rangle^{A_2}$ so that Alice and Bob's initial state (time $t_0$ is the figure) is

$$|x\rangle^{A_1}|y\rangle^{A_2}|\Phi_+^{AB}\rangle. \tag{1.18}$$

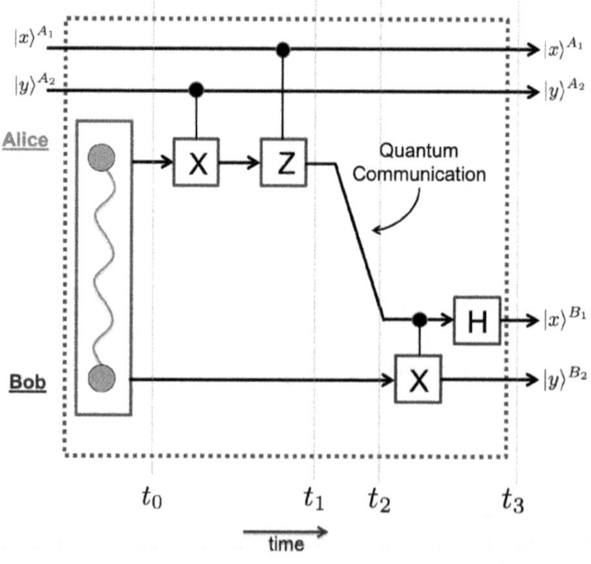


**Figure 1.6** Coherent superdense coding. One ebit plus one use of a noiseless qubit channel are implemented to realize two uses of the cobit channel.

Alice then performs a sequence of two controlled unitary gates, controlled $X$ on system $A_2$ and $A$, followed by controlled $Y$ gate on system $A_1$ and $A$. The resulting state at time $t_1$ is

$$|x\rangle^{A_1}|y\rangle^{A_2}|\phi_{xy}^{AB}\rangle, \quad \text{where} \quad |\phi_{xy}^{AB}\rangle := \left(Z^x X^y \otimes I^B\right)|\Phi_+^{AB}\rangle \quad \text{and} \quad x, y \in \{0, 1\}, \tag{1.19}$$

where $Z^x$ equals the identity matrix for $x = 0$, and the third Pauli matrix for $x = 1$ ($X^y$ is defined similarly). A key observation is that $\{|\phi_{xy}^{AB}\rangle\}_{x,y\in\{0,1\}}$ is precisely the Bell basis, and therefore forms an orthonormal basis for $\mathbb{C}^2 \otimes \mathbb{C}^2$. Note also that this encoding $(x, y) \to |\phi_{xy}^{AB}\rangle$ is done by Alice alone, and therefore essentially identical to the superdense coding protocol we encountered earlier.

In the next step Alice uses a noiseless qubit channel to transmit system $A$ to Bob. Therefore, at time $t_2$ the state of the system is $|x\rangle^{A_1}|y\rangle^{A_2}|\phi_{xy}^{B_1 B_2}\rangle$, where as before,

$$|\phi_{xy}^{B_1 B_2}\rangle := \left(Z^x X^y \otimes I^{B_2}\right)|\Phi_+^{B_1 B_2}\rangle. \tag{1.20}$$

Since both $\{|\phi_{xy}^{B_1 B_2}\rangle\}_{x,y\in\{0,1\}}$ and $\{|xy\rangle^{B_1 B_2}\}_{x,y\in\{0,1\}}$ are orthonormal bases of Bob's two qubit space $B_1 B_2$, we conclude that there exists a unitary matrix $U^{B_1 B_2}$ such that

$$|xy\rangle^{B_1 B_2} = U^{B_1 B_2}|\phi_{xy}^{B_1 B_2}\rangle \quad \forall \, x, y \in \{0, 1\}. \tag{1.21}$$

It turns out that the unitary $U^{AB}$ as already defined can be expressed as a CNOT gate followed by a Hadamard gate on system $B_1$ (see Bob's side in Figure 1.6 between time steps $t_2$ and $t_3$). Explicitly,

$$\begin{aligned} U^{B_1 B_2} &= \left(H|0\rangle\langle 0|^{B_1}\right) \otimes I^{B_2} + \left(H|1\rangle\langle 1|^{B_1}\right) \otimes X^{B_2} \\ &= |+\rangle\langle 0|^{B_1} \otimes I^{B_2} + |-\rangle\langle 1|^{B_1} \otimes X^{B_2}. \end{aligned} \tag{1.22}$$

Hence, after the application of the unitary $U^{B_1 B_2}$ on Bob's system, Alice and Bob state is $|x\rangle^{A_1}|y\rangle^{A_2}|x\rangle^{B_1}|y\rangle^{B_2}$. That is, the quantum circuit in Figure 1.6 simulates the linear transformation

$$|x\rangle^{A_1}|y\rangle^{A_2} \to |x\rangle^{A_1}|x\rangle^{B_1} \otimes |y\rangle^{A_2}|y\rangle^{B_2}, \tag{1.23}$$

which is equivalent to two coherent channels. The resources we used to simulate these two coherent channels are precisely the same ones used in superdense coding to simulate two noiseless classical channels.

**Exercise 1.9.** *Show that the unitary matrix $U^{B_1 B_2}$ above satisfies (1.21).*

**Exercise 1.10.** *Suppose the initial ebit shared between Alice and Bob was given in the singlet state $|\Psi_-^{AB}\rangle$ instead of $|\Phi_+^{AB}\rangle$, and consider the exact same protocol as in Figure 1.6, until time step $t_2$. Revise the unitary matrix $U^{B_1 B_2}$ after time step $t_2$ so that the protocol still simulates two coherent channels.*

## 1.5.3 Coherent Teleportation

The coherent teleportation protocol is the resource reversal of the coherent superdense coding protocol. Particularly, it reveals that two uses of a cobit channel are sufficient to generate one ebit and at the same time simulate one use of a qubit channel. That is, the coherent teleportation protocol demonstrates that

$$2[q \to qq] \geqslant [qq] + [q \to q]. \tag{1.24}$$

The protocol achieving this resource inequality is depicted in Figure 1.7.

In the first step of the protocol Alice sends a qubit $|\psi^A\rangle = a|0\rangle^A + b|1\rangle^A$ into the first cobit channel. The cobit channel $\mathcal{V}_Z$ transforms this state into the state

$$a|0\rangle^A|0\rangle^{B_1} + b|1\rangle^A|1\rangle^{B_1} = \frac{1}{\sqrt{2}}\left[|+\rangle^A\left(a|0\rangle^{B_1} + b|1\rangle^{B_1}\right) + |-\rangle^A\left(a|0\rangle^{B_1} - b|1\rangle^{B_1}\right)\right]. \tag{1.25}$$

In the next step, system $A$ goes through the second cobit channel, $\mathcal{V}_X$, yielding the state

$$\frac{1}{\sqrt{2}}\left[|+\rangle^A|+\rangle^{B_2}\left(a|0\rangle^{B_1} + b|1\rangle^{B_1}\right) + |-\rangle^A|-\rangle^{B_2}\left(a|0\rangle^{B_1} - b|1\rangle^{B_1}\right)\right]. \tag{1.26}$$

Finally, in the last step, Bob sends his systems through a CNOT gate. Note that $X|+\rangle = |+\rangle$ so that the CNOT gate only changes $|-\rangle^{B_2}|1\rangle^{B_1}$ to $-|-\rangle^{B_2}|1\rangle^{B_1}$ while keeping all the other terms intact. Hence, after Bob's CNOT gate, Alice and Bob share the state

$$\frac{1}{\sqrt{2}}\left[|+\rangle^A|+\rangle^{B_2}\left(a|0\rangle^{B_1} + b|1\rangle^{B_1}\right) + |-\rangle^A|-\rangle^{B_2}\left(a|0\rangle^{B_1} + b|1\rangle^{B_1}\right)\right] = |\Phi_+^{AB_2}\rangle|\psi^{B_1}\rangle. \tag{1.27}$$

That is, at the end of the protocol Alice teleported her quantum state $|\psi\rangle$ to Bob's system $B_1$, and also share with Bob's system $B_2$ the maximally entangled state $\Phi_+^{AB_2}$.

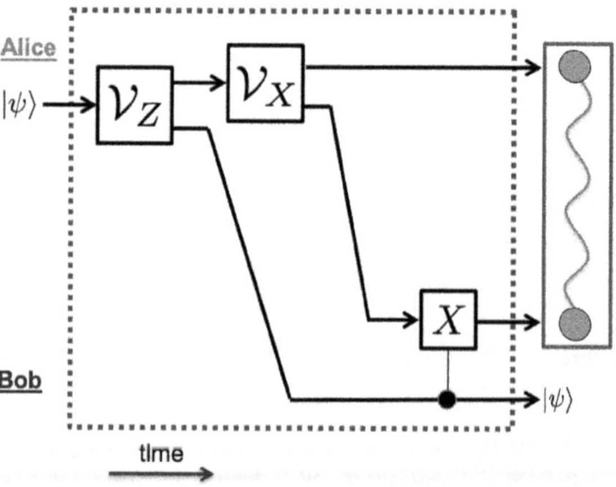


Figure 1.7    Coherent quantum teleportation. Two cobit channels produce one ebit plus one use of a noiseless qubit channel.

Coherent quantum teleportation and coherent superdense coding demonstrate that two cobit channels have the same resource value as one ebit and one use of a qubit channel:

$$[qq] + [q \to q] = 2[q \to qq]. \tag{1.28}$$

This means that coherent teleportation is the reversal process of coherent superdense coding and vice versa.

## 1.6 Notes and References

Quantum teleportation was discovered in Ref. [20], and quantum superdense coding in Ref. [23]. These seminal papers paved the way for the development of quantum Shannon theory and entanglement theory, as they demonstrated that entanglement, besides being interesting from a fundamental point of view, is a resource that can be consumed to achieve certain exotic tasks such as quantum teleportation. Moreover, these protocols are considered as the *unit* protocols (see, for example, Ref. [235]), since they form the building blocks with which one studies the capabilities of noisy quantum channels to transmit information in asymptotic settings involving many uses of the channels.

The resource analysis using notations such as $[q \to q]$ was first introduced in Ref. [62] where the rules of this "resource calculus" developed. The coherent bit, coherence teleportation, and coherent superdense coding are due to Ref. [112]. More details on coherent communication can be found in the book of Wilde [235].

PART I

# PRELIMINARIES

Quantum mechanics, which was discovered during the first quarter of the twentieth century, has profoundly transformed our understanding of the world around us. The theory predicts a plethora of nonintuitive phenomena, including entanglement, quantum nonlocality, wave-particle duality, coherence, the uncertainty principle, quantum contextuality, quantum steering, and no-cloning, to name a few. This remarkable departure from classical physics has led to numerous thought-provoking papers and interpretations of quantum mechanics. To date, there is no consensus on which view of quantum mechanics is the most natural one to adopt. Additionally, subfields of science, such as quantum logic, have arisen from these phenomena, particularly the inconsistency of classical logic and the uncertainty principle.

The development of quantum mechanics was a gradual process that involved many trials and errors. It began in 1900 with Max Planck's discretization of energy values used to solve the black body radiation. The process continued with Einstein's 1905 paper on the correspondence between energy and frequency, which provided a quantum explanation for the photoelectric effect. The process ultimately ended with the formalism that was developed in the mid 1920s by Erwin Schrödinger, Werner Heisenberg, Max Born, and others.

In this book, we will not review quantum mechanics from a historical or traditional perspective. Instead, we will study it in the context of the modern field of quantum information science, which emerged and developed in the early 1990s. We will discuss the theory's basic postulates, its corresponding mathematical structure, and its many consequences and applications, particularly to information theory, resource theories, and more broadly to physics and science.

The interplay between the concept of "information" and the field of quantum science is a complex and multifaceted one. The fundamental principles of quantum mechanics, such as the superposition of states and entanglement, have led to the development of quantum information theory, which studies the processing and transmission of information using quantum systems. Moreover, the very act of observing a quantum system can alter its state, and this observation is itself a form of information. This has profound implications for our understanding of the nature of reality and the limits of our ability to measure it. Thus, the relationship between "information" and quantum science is a rich and nuanced one, encompassing a broad range of topics from the foundations of quantum mechanics to the practical applications of quantum technologies.

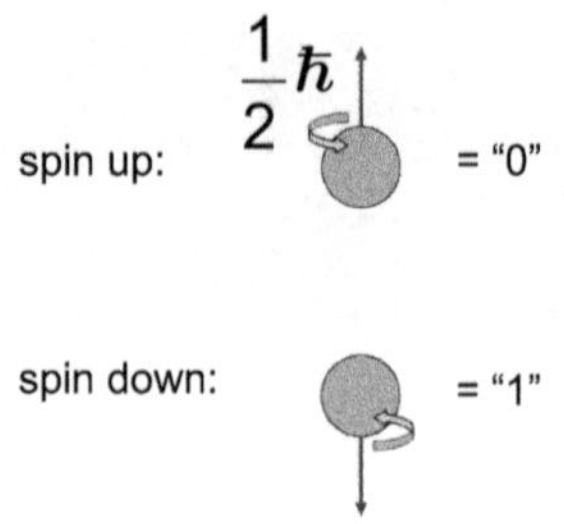

**Figure 2.1** Encoding information with the spin of an electron.

In Shannon's terms, information is defined as "that which can distinguish one thing from another." For instance, a coin has two sides, "head" and "tail," and the ability to differentiate between the two implies that the coin can store information. Typically, when referring to the distinguishability of two elements, we denote the options as 0 and 1, and information is then measured in bits, where the number of bits represents the number of distinguishable elements. For example, two bits correspond to four possible elements.

The definition of information is abstract and detached from any specific implementation or labeling. For instance, one bit may correspond to the head or tail of a coin, or the 5-volt versus 0-volt of an electrical circuit. All information processing, such as communication, computation, and manipulation of information, can be performed with either coins or electrical circuits. While storing information in coins is impractical (especially for large numbers of bits), from an information-theoretic perspective, any object can be used to implement classical bits, and we say that information is *fungible*.

However, what happens when we attempt to encode information in the spin of an electron? The electron, being an elementary particle, is uniquely determined by its mass and spin. The magnitude of its spin can only take one value, $\frac{1}{2}\hbar$ (where $\hbar$ is a unit of angular momentum), and its spin can point in any direction. The Stern–Gerlach (SG) experiment, discussed Section 2.1, demonstrates that it is possible to differentiate between "up" and "down" along the $z$-direction of the spin of an electron (see Figure 2.1). This implies that information can be encoded in the spin of an electron as well.

Are there any advantages to encoding information in the spins of quantum particles, such as electrons, as opposed to larger classical systems like coins or electrical circuits? If information is fungible, why should encoding information in quantum particles make any difference? Before addressing these questions, we will introduce the SG experiment and the necessary elements from linear algebra for the study of quantum physics.

## 2.1 The Stern–Gerlach Experiment

The SG experiment, conducted in Frankfurt, Germany, in 1922 by Otto Stern and Walther Gerlach, is a groundbreaking demonstration of the quantization of

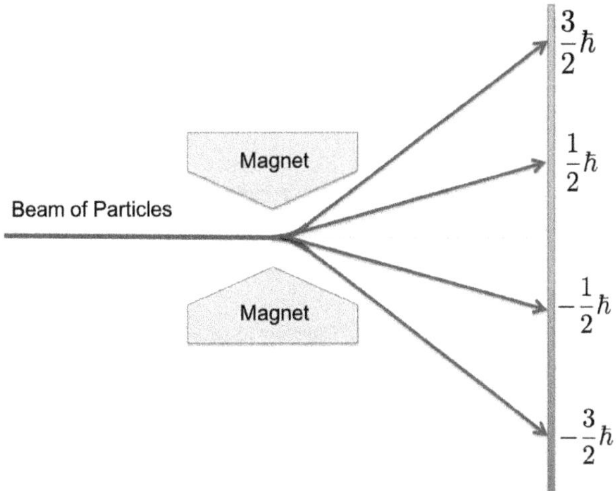


**Figure 2.2**   The Stern–Gerlach experiment.

the spatial orientation of angular momentum in physical systems. The experiment has become a paradigm of quantum measurement, and is considered one of the most significant experiments of the twentieth century. It paved the way for numerous subsequent experiments, complementing the double-slit experiment and contributing to our understanding of the behavior of quantum systems.

The SG experiment's relative simplicity and its ability to measure discrete properties of physical systems in finite dimensions make it an excellent example of how information can be extracted from a physical system. In the experiment, a beam of particles is passed through a spatially varying magnetic field, which causes the beam to split into discrete components that are detected on a screen. This behavior indicates that the angular momentum of the particles is quantized; that is, taking only discrete values. A heuristic illustration of the SG experiment is provided in Figure 2.2.

The SG experiment involves emitting electrons or atoms from an oven and passing them through a nonhomogeneous magnetic field. Their spin orientation determines whether they hit the screen above or below the horizontal line, with the distance from the line reflecting the magnitude of their spin. The experiment can be performed on any physical system, and always yields a discrete spectrum for the angular momentum, given by $\frac{1}{2}n\hbar$ where $n$ is an integer. Regions on the screen in Figure 2.2 where no particles hit indicate this quantization, which is true regardless of the type of particle used in the experiment.

The SG experiment has two noteworthy implications. Firstly, electrons always hit the screen at the same distance from the horizontal line, indicating a consistent magnitude of $\frac{1}{2}\hbar$ for their spin. Secondly, all electrons hit the screen in the same two areas, irrespective of their initial spin direction. Even if an electron's initial spin is pointing in the $x$-direction, it should have zero spin in the $z$-direction and therefore should not be deflected by the nonzero $z$-gradient of the magnetic field. However, the fact that electrons still hit the screen in the same two areas shows that measuring the spin in one direction affects its value in another direction.

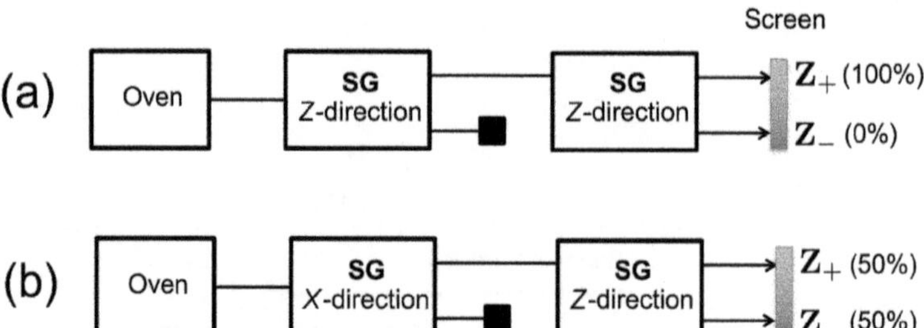


**Figure 2.3**  Combination of Stern–Gerlach (SG) experiments (a) An electron's spin remains unchanged by further measurements in the same direction. (b) The first SG box measures spin in the $x$-direction, resulting in the complete loss of information about the electron's spin in the $z$-direction.

To verify that an electron's spin remains intact after it is deflected by the SG experiment, we can concatenate two SG experiments in a variety of ways. For example, suppose we want to confirm that an electron deflected upwards in the first SG box has a spin pointing in the upward $z$-direction. We can set up the experiment as shown in Figure 2.3(a). After passing through the first SG box, only electrons deflected upwards will continue on to the second SG box, while those deflected downward are blocked. After passing through the second SG box, the electrons hit the screen. We observe that no electrons hit the screen below the horizontal line, indicating that all electrons deflected upwards in the first SG box also have a spin pointing in the upward $z$-direction. This implies that after an electron's spin has been measured, it remains intact by further measurements in the same direction.

Another interesting feature of the SG experiment is that it allows us to measure the spin of particles in different directions. For instance, we can measure the spin in the $x$-direction by using a modified version of the experiment. In this case, the electrons are deflected to the left or to the right, depending on their spin in the $x$-direction. Figure 2.3(b) shows a schematic of this experiment, where the first SG box measures the spin in the $x$-direction, and the second SG box measures the spin in the $z$-direction.

The results of this experiment are surprising from a classical point of view. According to classical physics, any physical system with its angular momentum pointing in the $x$-direction should have zero angular momentum pointing in the $z$-direction. However, in the SG experiment, we find that 50% of the electrons are deflected upward and 50% are deflected downward after passing through the second SG box. This shows that the spin in different directions of a quantum particle is not fully determined by its spin in one particular direction, and that measurements of spin along different axes can yield nontrivial and unexpected results.

What happens to a particle with spin in the $x$-direction when it passes through an SG experiment in the $z$-direction, followed by an SG experiment in the $x$-direction? Will its resulting spin in the $x$-direction remain the same? Figure 2.4 illustrates such a scenario with three SG boxes. The first box filters only particles with spins in the

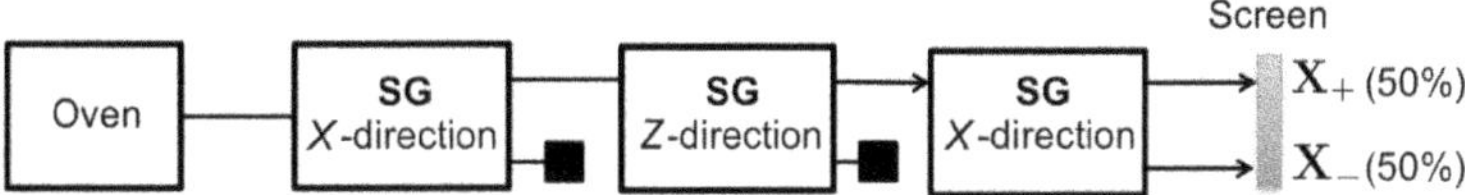


**Figure 2.4**     Erasing information with measurement.

positive (right) $x$-direction. The second box measures their spin in the $z$-direction, and the third box measures their spin in the $x$-direction. Remarkably, there is a 50% chance that the resulting particle will have spin in the negative $x$-direction. This implies that the measurement in the $z$-direction erases the information about the spin in the $x$-direction completely.

This phenomenon, known as quantum mechanical complementarity, is a fundamental feature of quantum mechanics and is one of the most profound concepts in physics. It implies that it is impossible to simultaneously measure certain pairs of physical quantities with arbitrary precision, such as position and momentum, or spin in different directions. Any measurement of one quantity necessarily disturbs the other, and the uncertainty principle sets a fundamental limit on the precision with which they can be measured simultaneously.

To summarize, the SG experiments discussed in this section demonstrate that measurements can disturb the state of a system. For instance, an electron with spin initially pointing upward may end up pointing downward after a sequence of SG experiments. However, this disturbance is not unique to a specific measurement apparatus; the same behavior and statistical outcomes will occur regardless of the device used. In other words, if a particle's spin is known to be pointing in the positive $z$-direction, then a measurement in the $x$- or $y$-directions will erase all knowledge about its spin in the $z$-direction, irrespective of which measurement apparatus is employed. Therefore, it is the *knowledge* of the spin in one direction that prevents its knowledge in another orthogonal direction. This phenomenon is a special instance of the uncertainty principle.

## 2.2 Inner Product Spaces and Hilbert Spaces

In this section and the following one, we discuss essential topics from linear algebra that serve as foundations for quantum theory. While most of the material is covered in standard textbooks on linear algebra, we introduce the Dirac notation for vectors and operators, as well as concepts such as quantum states and maximally entangled states that are used throughout this book. Therefore, even readers familiar with the topics covered here may benefit from reviewing and acquainting themselves with the notations we utilize.

We consider vector spaces over either the real or complex numbers and use the notation $\mathbb{F}$ when referring to either field.

**Inner Product Space**

**Definition 2.1.** An inner product space is a vector space $A$ over a field $\mathbb{F}$ equipped with a map

$$\langle\,|\,\rangle : A \times A \to \mathbb{F} \tag{2.1}$$

that satisfies for all vectors $\psi, \phi, \chi \in A$ and every scalar $c \in \mathbb{F}$ the following three axioms:

1. Conjugate symmetry: $\langle\psi|\phi\rangle = \overline{\langle\phi|\psi\rangle}$.
2. Linearity in the second argument: $\langle\psi|c\phi\rangle = c\langle\psi|\phi\rangle$ and $\langle\psi|\phi + \chi\rangle = \langle\psi|\phi\rangle + \langle\psi|\chi\rangle$.
3. Positive definiteness: $\langle\psi|\psi\rangle \geqslant 0$ with equality if and only if $|\psi\rangle = \mathbf{0}$.

*Remark.* We adopt the notation $\langle\,|\,\rangle$ instead of $\langle\,,\,\rangle$, as it conforms with the Dirac notation that we will define shortly. Additionally, while most mathematics textbooks define the linearity property with respect to the first argument, we will use the aforementioned notation to suitably align with the Dirac notation. We use the letters $A$, $B$, and $C$ to denote Hilbert spaces since, in quantum physics, Hilbert spaces correspond to physical systems that are operated on by parties such as Alice, Bob, Charlie, and so on.

The inner product induces a norm defined by

$$\|\psi\|_2 := \langle\psi|\psi\rangle^{1/2} \qquad \forall\, \psi \in A, \tag{2.2}$$

and a metric

$$d(\psi, \phi) := \|\psi - \phi\|_2 \qquad \forall\, \psi, \phi \in A. \tag{2.3}$$

**A Norm**

**Definition 2.2.** A norm is a real-valued function defined on the vector space, $A$, that has the following properties:

1. For all $\psi \in A$, $\|\psi\| \geqslant 0$ with equality if and only if $\psi = 0$.
2. For all $c \in \mathbb{F}$ and $\psi \in A$, $\|c\psi\| = |c|\|\psi\|$.
3. The triangle inequality: for all $\psi, \phi \in A$, $\|\psi + \phi\| \leqslant \|\psi\| + \|\phi\|$.

A vector space equipped with a norm is called a normed space.

**Exercise 2.1.** *Show that the norm defined by the inner product in (2.2) indeed satisfies the three fundamental properties of a norm.*

**Exercise 2.2.** *Let $A$ be a normed space, and let $\psi, \phi \in A$ with $\|\phi\| = 1$. Show that*

$$\left\|\frac{1}{\|\psi\|}\psi - \phi\right\| \leqslant 2\|\psi - \phi\|. \tag{2.4}$$

*Hint: Write $\frac{1}{\|\psi\|}\psi - \phi = \frac{1}{\|\psi\|}\psi - \psi + \psi - \phi$ and use the norm properties.*

It is important to note that not all norms are derived from an inner product, and as a result, not every metric necessarily originates from either an inner product or a norm.

**Exercise 2.3 (The $p$-Norms and the Hölder Inequality).** *The normed space $\ell^p(\mathbb{C}^n)$ (with $p \in [1,\infty]$) is the vector space $\mathbb{C}^n$ equipped with the $p$-norm, $\|\cdot\|_p$, defined on all $\psi = (a_1,\ldots,a_n)^T \in \mathbb{C}^n$ as*

$$\|\psi\|_p := \left(|a_1|^p + \cdots + |a_n|^p\right)^{1/p}. \tag{2.5}$$

*The Hölder inequality states that for any $p,q \in [1,\infty]$ with $\frac{1}{p} + \frac{1}{q} = 1$ and any $\psi = (a_1,\ldots,a_n)^T \in \mathbb{C}^n$ and $\phi = (b_1,\ldots,b_n)^T \in \mathbb{C}^n$, we have*

$$\sum_{x\in[n]} |a_x b_x| \leqslant \|\psi\|_p \|\phi\|_q. \tag{2.6}$$

*Use this to show that $\|\cdot\|_p$ is a norm, and that for $p \neq 2$ this norm is not induced from an inner product.*

The $p$-norms in Exercise 2.3 have numerous applications in many fields of science. Of particular importance are the extreme cases of $p = 1$ and $p = \infty$:

$$\|\phi\|_1 := |c_1| + \cdots + |c_n| \quad \text{and} \quad \|\phi\|_\infty := \max_{x\in[n]} |c_x|, \tag{2.7}$$

where throughout the book we use the notation $[n] := \{1,\ldots,n\}$ for every integer $n \in \mathbb{N}$. For these cases, the norms in Exercise 2.3 behave monotonically under stochastic matrices as we show now.

Let $S = (s_{xy}) \in \text{STOC}(m,n)$ be an $m \times n$ column stochastic matrix; that is, a matrix whose components are nonnegative real numbers, and all the columns sum to one. Then, the one-norm behaves monotonically under such matrices; that is,

$$\|S\mathbf{v}\|_1 \leqslant \|\mathbf{v}\|_1 \quad \forall \mathbf{v} \in \mathbb{C}^n. \tag{2.8}$$

Indeed, by definition,

$$\|S\mathbf{v}\|_1 = \sum_{x\in[m]} \left|\sum_{y\in[n]} s_{xy} v_y\right| \leqslant \sum_{x\in[m]}\sum_{y\in[n]} s_{xy} |v_y| = \sum_{y\in[n]} |v_y| = \|\mathbf{v}\|_1, \tag{2.9}$$

where we used the triangle inequality.

**Exercise 2.4.** *Let $R \in \mathbb{R}_+^{m\times n}$ be a row stochastic matrix (i.e. a matrix with nonnegative components whose rows sum to one). Show that*

$$\|R\mathbf{v}\|_\infty \leqslant \|\mathbf{v}\|_\infty \quad \forall\, \mathbf{v} \in \mathbb{C}^n. \tag{2.10}$$

## 2.2.1 Hilbert Spaces

A *Cauchy sequence* in an inner product space, $A$, is a sequence of vectors $\{\psi_n\}_{n\in\mathbb{N}}$ in $A$ with the property that for any $\varepsilon > 0$, there exists $N \in \mathbb{N}$ such that for any $n, m > N$

$$\|\psi_n - \psi_m\| < \varepsilon. \tag{2.11}$$

An inner product space $A$ is said to be *complete* if any Cauchy sequence in $A$ converges in $A$, with respect to the metric induced by the inner product. That is, $A$ is complete if for any Cauchy sequence, $\{\psi_n\}_{n\in\mathbb{N}} \subseteq A$, there exists a state $\psi \in A$ such that $\lim_{n\to\infty} \|\psi_n - \psi\| = 0$. Complete inner product spaces are called Hilbert spaces and complete normed spaces are called Banach spaces.

Common examples of Hilbert spaces are $\mathbb{R}^n$ and $\mathbb{C}^n$ with the standard inner products. In finite dimensions, all Hilbert spaces are isomorphic to these spaces. A common example in quantum mechanics is the Hilbert space of square integrable functions

$$L^2(\mathbb{R}) := \left\{ f : \mathbb{R} \to \mathbb{C} : \quad \int_{-\infty}^{\infty} |f(x)|^2 dx < \infty \right\}, \tag{2.12}$$

where the inner product between two functions $f, g \in L^2(\mathbb{R})$ is defined as

$$\langle g, f \rangle := \int_{-\infty}^{\infty} f(x)\overline{g(x)}dx. \tag{2.13}$$

**Exercise 2.5.** *Show that $L^2(\mathbb{R})$ satisfies all the axioms of a Hilbert space.*

To distinguish more clearly between Hilbert spaces and inner product spaces, consider the space $\mathfrak{C}([a,b])$ of continuous complex-valued functions. This infinite-dimensional vector space is equipped with an inner product given by

$$\langle g, f \rangle := \int_{a}^{b} f(t)\overline{g(t)}dt \qquad \forall\, f, g \in \mathfrak{C}([a,b]). \tag{2.14}$$

However, this space is not a Hilbert space since it is not complete with respect to the metric induced by this inner product. For example, consider the following sequence of continuous functions in $\mathfrak{C}([-1,1])$:

$$f_k(t) := \begin{cases} 0 & \text{if } t \in [-1,0) \\ kt & \text{if } t \in [0,\frac{1}{k}) \\ 1 & \text{if } t \in [\frac{1}{k},1] \end{cases}. \tag{2.15}$$

Note that while the sequence is Cauchy, its limit does not exist in $\mathfrak{C}([-1,1])$ as $f_k$ cannot converge to a continuous function. However, this book mostly considers finite-dimensional Hilbert spaces, in which all inner product spaces are complete and isomorphic to $\mathbb{C}^n$ (or $\mathbb{R}^n$). Thus, such examples are not relevant for our purposes, and we will use the terms "Hilbert space" and "inner product space" interchangeably. Similarly, in finite dimensions, all normed spaces are Banach spaces (i.e. complete normed spaces).

Another important example of a Hilbert space is the space of $n \times m$ complex matrices, denoted by $\mathbb{C}^{n \times m}$. In this space, the inner product is defined as follows: for all $M, N \in \mathbb{C}^{n \times m}$

$$\langle M | N \rangle := \mathrm{Tr}\left[ M^* N \right], \tag{2.16}$$

where $M^*$ denotes the adjoint of $M$. This specific inner product is referred to as the "Hilbert–Schmidt" inner product, and it is occasionally denoted with a subscript as $\langle \, | \, \rangle_{\mathrm{HS}}$.

**Exercise 2.6.** *Consider the space* $\mathbb{C}^{m \times n}$.

1. *Show that the definition in (2.16) satisfies the three axioms of an inner product.*
2. *Find an isometrical isomorphism* **vec**: $\mathbb{C}^{m \times n} \to \mathbb{C}^{mn}$ *such that for all* $M, N \in \mathbb{C}^{m \times n}$

$$\langle M | N \rangle_{HS} = \langle \mathbf{vec}(M) | \mathbf{vec}(N) \rangle_2, \tag{2.17}$$

   *where* $\langle \, | \, \rangle_{HS}$ *is the Hilbert–Schmidt inner product of* $\mathbb{C}^{m \times n}$, *and* $\langle \, | \, \rangle_2$ *is the standard inner product of* $\mathbb{C}^{mn}$.
3. *Express the norm induced by the inner product (2.16).*

**Exercise 2.7.** *Prove the following two properties of a Hilbert space A:*

1. *The Cauchy–Schwarz inequality: for all* $\psi, \phi \in A$

$$|\langle \phi, \psi \rangle| \leqslant \|\phi\| \|\psi\|. \tag{2.18}$$

2. *Pythagorean theorem: If* $\psi_1, \ldots, \psi_n$ *are orthogonal vectors, that is,* $\langle \psi_x | \psi_y \rangle = 0$ *for distinct indices* $x, y \in [n]$, *then*

$$\left\| \sum_{x \in [n]} \psi_x \right\|^2 = \sum_{x \in [n]} \|\psi_x\|^2. \tag{2.19}$$

## 2.2.2 The Dirac Notations

We will employ Dirac notation along with the ket symbol to represent vectors in a Hilbert space $A$. Consequently, going forward, we will represent the vectors $\psi$, $\phi$, and $\chi$ in $A$ as $|\psi\rangle$, $|\phi\rangle$, and $|\chi\rangle$. The standard basis, also known as the computational basis of $\mathbb{C}^n$, will be denoted as $\{|1\rangle, |2\rangle, \ldots, |n\rangle\}$ or sometimes as $\{|0\rangle, |1\rangle, \ldots, |n-1\rangle\}$. The latter notation is typically used for $n = 2$, in which case the standard basis is given by

$$|0\rangle := \begin{pmatrix} 1 \\ 0 \end{pmatrix} \quad \text{and} \quad |1\rangle := \begin{pmatrix} 0 \\ 1 \end{pmatrix}. \tag{2.20}$$

The *dual space*, $A^*$, of a Hilbert space, $A$, is defined as the set of all linear functionals on $A$. A linear functional is a function $f : A \to \mathbb{F}$ with the property that for all $|\psi\rangle, |\phi\rangle \in A$ and $a, b \in \mathbb{F}$, we have $f(a|\psi\rangle + b|\phi\rangle) = af(|\psi\rangle) + bf(|\phi\rangle)$. For a fixed vector $|\chi\rangle \in A$, the function $f_\chi : A \to \mathbb{F}$ defined by $f_\chi(|\psi\rangle) := \langle \chi | \psi \rangle$ is a linear functional, and every linear functional has this form. It is therefore convenient to denote the linear functionals with a "bra" notation. That is, instead of $f_\chi$, we denote this functional simply by $\langle \chi |$, so that its action on an element $|\psi\rangle$ is given by the inner product $\langle \chi | \psi \rangle$. Hence, $A^*$ consists of bra vectors.

As an example, the dual space of $\mathbb{C}^2$ is spanned by the standard bra basis

$$\langle 0 | := (1, 0) \quad \text{and} \quad \langle 1 | := (0, 1). \tag{2.21}$$

Note that there is a one-to-one correspondence between $A = \mathbb{C}^n$ and its dual $A^*$, via the bijective mapping

$$|\psi\rangle = \sum_{x \in [n]} c_x |x\rangle \quad \mapsto \quad \langle \psi | = \sum_{x \in [n]} \bar{c}_x \langle x |, \tag{2.22}$$

where $c_x \in \mathbb{C}$ and $\bar{c}_x$ denote the complex conjugate of $c_x$. Moreover, we will denote $\langle \psi | \rho | \phi \rangle := \langle \psi | \rho \phi \rangle$, where $\rho$ is a linear transformation (see Section 2.3).

## 2.2.3 Direct Sum of Hilbert Spaces

Given two subspaces $A$ and $B$ of a Hilbert space $C$, we say that $C$ is the *direct sum* of $A$ and $B$, and write $C = A \oplus B$ if $C = A + B$ and $A \cap B = \{0\}$. Note that in this case $|C| = |A| + |B|$, where we use the notation $|A|$ to denote the dimension of $A$. If $A$ and $B$ are two arbitrary Hilbert spaces, we can always construct a third Hilbert space, $C$, such that $C = A \oplus B$. For example, given two Hilbert spaces $A$ and $B$, define (abstractly)

$$C := \{(\psi, \phi): \ |\psi\rangle \in A, \quad |\phi\rangle \in B\} \tag{2.23}$$

with addition rule $(\psi_1, \phi_1) + (\psi_2, \phi_2) := (\psi_1 + \psi_2, \phi_1 + \phi_2)$, scalar multiplication $c(\psi, \phi) = (c\psi, c\phi)$, and inner product

$$\langle (\psi_1, \phi_1) | (\psi_2, \phi_2) \rangle := \langle \psi_1 | \psi_2 \rangle + \langle \phi_1 | \phi_2 \rangle. \tag{2.24}$$

Identifying an element $|\psi\rangle \in A$ with an element $(\psi, 0) \in C$ and an element $|\phi\rangle \in B$ with an element $(0, \phi) \in C$, we conclude that $C = A \oplus B$. For example, $\mathbb{R}^3$ can be decomposed as $\mathbb{R}^2 \oplus \mathbb{R}$ or as $\mathbb{R} \oplus \mathbb{R} \oplus \mathbb{R}$.

## 2.2.4 Tensor Product of Hilbert Spaces

Another way to combine two Hilbert spaces is through their *tensor product*. In quantum information, a composite physical system is often described by an element of a tensor product of Hilbert spaces. We will only consider tensor products of finite-dimensional Hilbert spaces, although it's worth noting that the tensor product can be defined in a basis-independent manner as a quotient space of a free vector space.

Let $A$ and $B$ be two finite-dimensional Hilbert spaces with dimensions $|A|$ and $|B|$, respectively. We define a *bilinear* function $\otimes$ that takes two vectors $|\psi\rangle \in A$ and $|\phi\rangle \in B$ and returns an element of the form $|\psi\rangle \otimes |\phi\rangle$. The bilinearity of $\otimes$ means that for all $c \in \mathbb{F}$, and all vectors $|\psi_1\rangle, |\psi_2\rangle \in A$, and $|\phi_1\rangle, |\phi_2\rangle \in B$,

1. $(|\psi_1\rangle + |\psi_2\rangle) \otimes |\phi\rangle = |\psi_1\rangle \otimes |\phi\rangle + |\psi_2\rangle \otimes |\phi\rangle$
2. $|\psi\rangle \otimes (|\phi_1\rangle + |\phi_2\rangle) = |\psi\rangle \otimes |\phi_1\rangle + |\psi\rangle \otimes |\phi_2\rangle$
3. $c\,(|\psi\rangle \otimes |\phi\rangle) = (c|\psi\rangle) \otimes |\phi\rangle = |\psi\rangle \otimes (c|\phi\rangle)$.

The span of all such objects, $|\psi\rangle \otimes |\phi\rangle$, with $|\psi\rangle \in A$ and $|\phi\rangle \in B$ is denoted by $A \otimes B$. Alternatively, let $\{|x\rangle^A\}_{x \in [m]}$ and $\{|y\rangle^B\}_{y \in [n]}$ be two corresponding orthonormal bases of $A$ and $B$, where $m := |A|$ and $n := |B|$ denote the dimensions of $A$ and $B$, respectively. Then, $A \otimes B$ can be defined as the collection

$$A \otimes B := \left\{ \sum_{x \in [m]} \sum_{y \in [n]} \mu_{xy} |x\rangle^A \otimes |y\rangle^B : \ \mu_{xy} \in \mathbb{F} \right\}. \tag{2.25}$$

From its definition, it follows that $A \otimes B$ is a vector space with an orthonormal basis $\{|x\rangle^A \otimes |y\rangle^B\}$. In particular, note that $|A \otimes B| = |AB| := |A||B|$ (we will therefore sometimes use the notation $AB$ to mean $A \otimes B$). The inner product between two elements $|\psi_1\rangle \otimes |\phi_1\rangle$ and $|\psi_2\rangle \otimes |\phi_2\rangle$ is simply given by the product of the inner products; that is, $\langle\psi_2|\psi_1\rangle\langle\phi_2|\phi_1\rangle$. More generally, given two states

$$|\psi^{AB}\rangle = \sum_{x \in [m]} \sum_{y \in [n]} \mu_{xy} |x\rangle^A \otimes |y\rangle^B \quad \text{and} \quad |\phi^{AB}\rangle = \sum_{x \in [m]} \sum_{y \in [n]} \nu_{xy} |x\rangle^A \otimes |y\rangle^B, \tag{2.26}$$

their inner product is given by

$$\langle\psi^{AB}|\phi^{AB}\rangle = \mathrm{Tr}\left[M^*N\right], \tag{2.27}$$

where the matrices $M := (\mu_{xy})$ and $N := (\nu_{xy})$. Using these definitions, the set $A \otimes B$ forms a Hilbert space. Notably, the inner product defined in (2.27) is the same as the one defined in (2.16). This is because each element $|\psi\rangle \in A \otimes B$ can be represented as a matrix $M = (\mu_{xy})$. It can be easily demonstrated that this mapping between bipartite vectors and matrices is an isometric isomorphism.

**Exercise 2.8.** *Show that* $\mathbb{C}^{m \times n} \cong \mathbb{C}^m \otimes \mathbb{C}^n$.

## 2.2.5 The Kronecker Tensor Product

We will use the notation $|\psi\rangle|\phi\rangle$ to mean $|\psi\rangle \otimes |\phi\rangle$. For basis elements, we will use interchangeably the notations $|xy\rangle^{AB} := |x\rangle^A|y\rangle^B := |x\rangle^A \otimes |y\rangle^B$. Since all Hilbert spaces in finite dimensions are isomorphic to $\mathbb{C}^n$ (for some $n \in \mathbb{N}$), any vector $|\psi^A\rangle = \sum_{x \in [m]} a_x |x\rangle^A \in A$ (where $m := |A|$) corresponds to a vector $\mathbf{a} := (a_1, \ldots, a_m)^T \in \mathbb{C}^m$, and any vector $|\phi^B\rangle = \sum_{y \in [n]} b_y |y\rangle^B \in B$ (where $n := |B|$) corresponds to a vector $\mathbf{b} := (b_1, \ldots, b_n)^T \in \mathbb{C}^n$. Hence, the vector $|\psi^A\rangle|\phi^B\rangle = \sum_{x \in [m]} \sum_{y \in [n]} a_x b_y |xy\rangle^{AB}$ corresponds to the vector $\mathbf{a} \otimes \mathbf{b}$ given by

$$\mathbf{a} \otimes \mathbf{b} := \begin{bmatrix} a_1 \mathbf{b} \\ \vdots \\ a_m \mathbf{b} \end{bmatrix} \in \mathbb{C}^{mn}. \tag{2.28}$$

The definition of a tensor product between vectors in $\mathbb{C}^m$ and $\mathbb{C}^n$ is the Kronecker product, which is also denoted with the symbol $\otimes$. The Kronecker product is defined on arbitrary matrices as follows. Let $M = (\mu_{xy}) \in \mathbb{C}^{k \times \ell}$ and $N \in \mathbb{C}^{p \times q}$. The Kronecker product $M \otimes N$ is a matrix in $\mathbb{C}^{kp \times \ell q}$ defined by

$$
M \otimes N = \begin{bmatrix} \mu_{11} N & \cdots & \mu_{1\ell} N \\ \vdots & \ddots & \vdots \\ \mu_{k1} N & \cdots & \mu_{k\ell} N \end{bmatrix}. \tag{2.29}
$$

It is simple to check that the tensor product is bilinear and associative; however, it is not commutative. Also, note that for $A = \mathbb{C}^{n \times m}$ and $B = \mathbb{C}^{p \times q}$, the tensor product given in Eq. (2.25) is equivalent to the Kronecker product. We will therefore use the terms tensor product and Kronecker product interchangeably.

**Exercise 2.9.** *Let M be an $m \times n$ matrix and N be an $n \times k$ matrix. Find all the values of $m, n, k$ for which $M \otimes N = MN$.*

**Exercise 2.10.** *Show that the Kronecker product is not commutative, and that there always exist permutation matrices P and Q in appropriate dimensions such that $M \otimes N = P(N \otimes M)Q$.*

**Exercise 2.11.** *Prove the following properties. For any matrices $K, L, M, N$ in appropriate dimensions (in some cases square matrices):*

1. $(K \otimes L)(M \otimes N) = KM \otimes LN$.
2. $K \otimes L$ is invertible if and only if $K$ and $L$ are invertible and in this case $(K \otimes L)^{-1} = K^{-1} \otimes L^{-1}$.
3. $(K \otimes L)^T = K^T \otimes L^T$ and $(K \otimes L)^* = K^* \otimes L^*$.
4. For $K \in \mathbb{C}^{m \times m}$ and $L \in \mathbb{C}^{n \times n}$, $\mathrm{Tr}[K \otimes L] = \mathrm{Tr}[K]\mathrm{Tr}[L]$ and $\det(K \otimes L) = (\det(K))^n (\det(L))^m$.
5. Ket $K \in \mathbb{C}^{m \times m}$ and $L \in \mathbb{C}^{n \times n}$, and let $\lambda_1, \ldots, \lambda_m$ be the eigenvalues of $K$ and $\mu_1, \ldots, \mu_n$ be those of $L$ (listed according to multiplicity). Then the eigenvalues of $K \otimes L$ are $\lambda_x \mu_y$ with $x \in [m]$ and $y \in [n]$.
6. $\mathrm{Rank}(K \otimes L) = \mathrm{Rank}(K)\mathrm{Rank}(L)$.

Kronecker also defined a direct sum that is closely related to the definition in this section. Given two matrices $M \in \mathbb{C}^{m \times m}$ and $N \in \mathbb{C}^{n \times n}$, their Kronecker sum is defined by

$$
M \oplus N := M \otimes I_n + I_m \otimes N. \tag{2.30}
$$

The sum appears naturally in physics, typically when describing the Hamiltonian of a composite system consisting of noninteracting subsystems. A celebrated result connecting the Kronecker product and Kronecker sum is given in the following exponential relation:

$$
e^{M \oplus N} = e^M \otimes e^N. \tag{2.31}
$$

**Exercise 2.12.** *Prove the equality in (2.31). Hint: Use the formula $e^M = \sum_{n=0}^{\infty} \frac{M^n}{n!}$ and the commutativity of $M \otimes I_n$ and $I_m \otimes N$.*

## 2.3 Linear Operators on Hilbert Spaces

An operator (i.e. a map) $M : A \to B$ is said to be *linear* if and only if for all $|\psi\rangle, |\phi\rangle \in A$ and $a, b \in \mathbb{F}$,

$$M(a|\psi\rangle + b|\phi\rangle) = aM|\psi\rangle + bM|\phi\rangle. \tag{2.32}$$

When $|B| = 1$, the linear operator $M$ is called a functional. Given a basis $\{|x\rangle^A\}_{x\in[m]}$ of $A$ and $\{|y\rangle^B\}_{y\in[n]}$ of $B$, $M$ can be represented in terms of a matrix $(\mu_{xy})$ with components $\mu_{yx} := {}^B\langle y|M|x\rangle^A$, which is a notation for the inner product between $M|x\rangle^A$ and $|y\rangle^B$. Specifically, $M$ can be expressed as

$$M = \sum_{x\in[m]} \sum_{y\in[n]} \mu_{yx} |y\rangle\langle x|. \tag{2.33}$$

For improved clarity in our exposition, we have omitted the superscripts $A$ and $B$ from $|x\rangle^A$ and $|y\rangle^B$. This use of the Dirac notations has the advantage that the action of $M$ on a vector $|\psi\rangle \in A$ becomes

$$M|\psi\rangle = \sum_{y\in[n]} \sum_{x\in[m]} \mu_{yx} |y\rangle\langle x|\psi\rangle = \sum_{y\in[n]} \left( \sum_{x\in[m]} \mu_{yx} \langle x|\psi\rangle \right) |y\rangle. \tag{2.34}$$

Note that the numbers $\mu_{yx}$ form a matrix, which is known as the matrix representation of $M$. Sometimes we will identify the matrix $(\mu_{yx})$ with the operator $M$ and write $M - (\mu_{yx})$. However, note that different choices of orthonormal bases $\{|x\rangle^A\}_{x\in[m]}$ and $\{|y\rangle^B\}_{x\in[n]}$ correspond to different matrix representations $(\mu_{yx})$ of the *same* linear operator $M$ (see Exercise 2.15).

**Exercise 2.13.** *Show that any linear operator* $M : A \to B$ *can be expressed as*

$$M = \sum_{z\in[k]} \lambda_z |v_z^B\rangle\langle u_z^A|, \tag{2.35}$$

*where* $\{\lambda_z\}_{z\in[k]}$ *are the singular values of* $M$, *and* $\{|u_z^A\rangle\}_{z\in[k]}$ *and* $\{|v_z^B\rangle\}_{z\in[k]}$ *are orthonormal sets of vectors in* $A$ *and* $B$, *respectively. Hint: Use the singular value decomposition of a complex matrix.*

The adjoint of a linear operator $M : A \to B$ is itself a linear operator $M^* : B \to A$ defined by the relation

$$\langle\phi|M\psi\rangle = \langle M^*\phi|\psi\rangle \qquad \forall\ |\psi\rangle \in A \ \text{ and } \ |\phi\rangle \in B. \tag{2.36}$$

If $M$ has the form (2.33), then

$$M^* = \sum_{x\in[m]} \sum_{y\in[n]} \overline{\mu}_{yx} |x\rangle\langle y|, \tag{2.37}$$

where the bar indicates complex conjugation.

**Exercise 2.14.** *Let $U : A \to A$ be a linear operator on a vector space A. Show that U is unitary, that is, $U^* = U^{-1}$ if and only if there exist two orthonormal bases of A, $\{|v_x\rangle\}_{x\in[m]}$ and $\{|u_x\rangle\}_{x\in[m]}$, such that*

$$U = \sum_{x\in[m]} |v_x\rangle\langle u_x|. \tag{2.38}$$

**Exercise 2.15.** *Let M be a linear operator as in (2.33), and denote by $\tilde{M}$ the matrix whose components are $\mu_{yx} := \langle y|M|x\rangle$ (i.e. M is an operator, whereas $\tilde{M}$ is a matrix). Let $\{|a_x\rangle\}$ and $\{|b_y\rangle\}$ be two orthonormal bases of A and B, respectively. Show that there exist two unitary matrices U and V (not necessarily of the same size) such that*

$$M = \sum_{x\in[m]} \sum_{y\in[n]} v_{yx}|b_y\rangle\langle a_x|, \tag{2.39}$$

*where $\{v_{yx}\}$ are the components of the matrix $N := V\tilde{M}U$.*

For any linear operator $T : A \to B$, its *kernel*, denoted by Ker(T), is the subspace of A consisting of all vectors $|\psi\rangle \in A$ such that $T|\psi\rangle = 0$. The *image* of T, denoted by Im(T), is the set of vectors $\{T|\psi\rangle\}$ over all vectors $|\psi\rangle \in A$. Finally, the *support* of T, denoted by supp(T), is also a subspace of A consisting of all the vectors that are orthogonal to all the elements in Ker(T). In particular, for any nonzero vector $|\psi\rangle \in \text{supp}(T)$, we have $T|\psi\rangle \neq 0$.

**Exercise 2.16.** *Let A and B be two Hilbert spaces, and let $T : A \to B$ be a linear transformation.*

1. *Show that if $\text{Ker}(T) = \{0\}$ and $\text{Im}(T) = B$, then $A = B$ and T is invertible.*
2. *Show that if $\text{Ker}(T) = A$, then $T = 0$.*

### 2.3.1 Isometries and Partial Isometries

A linear operator $V:A \to B$ is called an isometry if

$$\langle V\psi|V\psi\rangle = \langle\psi|\psi\rangle \qquad \forall\, |\psi\rangle \in A. \tag{2.40}$$

Using the dual $V^*:B \to A$, we can express (2.40) as

$$\langle\psi|V^*V|\psi\rangle = \langle\psi|\psi\rangle \qquad \forall\, |\psi\rangle \in A. \tag{2.41}$$

We therefore conclude (see Exercise 2.17) that V is an isometry if and only if $V^*V = I^A$, where $I^A$ is the identity operator on A. The condition that $V^*V = I^A$ implies that $|A| \leqslant |B|$ and if $|A| = |B|$ then V is necessarily a unitary operator. One can view isometries as "embeddings" of the Hilbert space A into B. In particular, they preserve the inner product; indeed, for any $|\phi\rangle, |\psi\rangle \in A$,

$$\langle V\phi|V\psi\rangle = \langle\phi|V^*V|\psi\rangle = \langle\phi|\psi\rangle. \tag{2.42}$$

We say that $V : A \to B$ is a partial isometry if V is an isometry when restricted to its support. Therefore, an isometry is a partial isometry but the converse is not necessarily true. Let $V: A \to B$ be a partial isometry, and let $\{|a_z\rangle\}_{z \in [k]}$ be an orthonormal basis of $\mathrm{supp}(V)$. Then, since $V$ is an isometry on $\mathrm{supp}(V) \subseteq A$, it follows that the vectors $|b_z\rangle := V|a_z\rangle$, with $z \in [k]$, form an orthonormal set of vectors in $B$. Therefore, $k$ cannot exceed $|B|$, and $V$ can be expressed as

$$V = \sum_{z \in [k]} |b_z\rangle\langle a_z|, \tag{2.43}$$

where $k := |\mathrm{supp}(V)| \leqslant \min\{|A|, |B|\}$, $\{|a_z\rangle\}_{z \in [k]}$ is an orthonormal set of vectors in $A$, and $\{|b_z\rangle\}_{[k]}$ is an orthonormal set of vectors in $B$. In other words, a linear operator $V: A \to B$ is a partial isometry if and only if there exist two sets of orthonormal vectors, $\{|a_z\rangle\}_{z \in [k]} \subset A$ and $\{|b_z\rangle\}_{z \in [k]} \subset B$ such that (2.43) holds.

**Exercise 2.17.** *Show that if Eq. (2.41) holds, then $V^*V = I^A$.*

**Exercise 2.18.** *Use (2.35) to show (2.43).*

**Exercise 2.19.** *A linear operator $\Pi: A \to A$ is called an orthogonal projection if and only if $\Pi^2 = \Pi = \Pi^*$.*

1. *Show that $\Pi: A \to A$ is an orthogonal projection if and only if $\Pi^*\Pi = \Pi$.*
2. *Show that $V: A \to B$ is a partial isometry if and only if $V^*V$ is an orthogonal projection in $A$, and $VV^*$ is an orthogonal projection in $B$.*

**Exercise 2.20.** *Let $A \subseteq B$ be a subspace of $B$, and let $V: A \to B$ be an isometry satisfying $V^*V = \Pi$, where $\Pi: A \to A$ is the projection onto the subspace $A$. Show that there exists a unitary matrix $U: B \to B$ such that*

$$U\Pi = V. \tag{2.44}$$

## 2.3.2 Hermitian and Positive Operators

A linear operator $H: A \to A$ ($A$ is a Hilbert space) is called Hermitian if $H = H^*$. Any Hermitian operator $H$ has a spectral decomposition

$$H = \sum_{x \in [m]} \lambda_x |v_x\rangle\langle v_x|, \tag{2.45}$$

where $\{|v_x\rangle\}_{x \in [m]}$ is an orthonormal basis of $A$. The coefficients $\{\lambda_x\}_{x \in [m]}$ are the eigenvalues of $H$. We denote by $\mathrm{Tr}[H]$ the trace of Hermitian operator $H: A \to A$. That is,

$$\mathrm{Tr}[H] := \sum_{x \in [m]} \langle x|H|x\rangle. \tag{2.46}$$

Note that (2.46) is independent on the choice of the orthonormal basis $\{|x\rangle\}$ of $A$.

We say that a linear operator $\rho: A \to A$ is positive semidefinite, and write $\rho \geqslant 0$, if and only if

$$\langle \psi|\rho|\psi\rangle \geqslant 0 \qquad \forall\, |\psi\rangle \in A. \tag{2.47}$$

If the inequality in (2.47) is strict for all nonzero $|\psi\rangle \in A$, then we say that $\rho$ is positive definite and write $\rho > 0$. We will also write $\rho \geqslant \sigma$ to mean $\rho - \sigma \geqslant 0$, and will use the Greek letters such as $\rho$ and $\sigma$ to denote linear operators that are positive semidefinite. The set of all positive semidefinite operators acting on Hilbert space $A$ will be denoted by $\mathrm{Pos}(A)$.

Every positive linear operator $\rho : A \to A$ is necessarily Hermitian. To see why, observe that the positivity property above implies

$$\langle\psi|\rho|\psi\rangle = \overline{\langle\psi|\rho|\psi\rangle} = \langle\psi|\rho^*|\psi\rangle. \tag{2.48}$$

Therefore, for all $|\psi\rangle \in A$ we have

$$\langle\psi|\rho - \rho^*|\psi\rangle = 0. \tag{2.49}$$

Now, observe that the operator $N := \rho - \rho^*$ satisfies $N^*N = NN^*$. Such operators are called *normal* operators and are known to be diagonalizable. Therefore, taking $|\psi\rangle$ above to be an eigenvector of $N$ we conclude that all the eigenvalues of $N$ are zero. Hence, $N = 0$ or equivalently $\rho = \rho^*$.

**Exercise 2.21.** *Let $\rho : A \to A$ be a linear operator. Show that the following are equivalent:*

*1. $\rho \geqslant 0$.*

*2. $\rho$ is Hermitian and all its eigenvalues are nonnegative.*

*3. There exists a linear map $M : A \to A$ such that $\rho = M^*M$.*

**Exercise 2.22.** *Let $\rho \in \mathrm{Pos}(A)$. Show that for any complex matrix $M : A \to B$ we have $M\rho M^* \in \mathrm{Pos}(B)$.*

**Exercise 2.23.** *Show that for any two vectors $|\psi\rangle, |\phi\rangle \in A$*

$$\mathrm{Tr}\left[|\psi\rangle\langle\phi|\right] = \langle\phi|\psi\rangle. \tag{2.50}$$

Note that the identity operator $I^A : A \to A$ is a positive operator (i.e. an operator with all eigenvalues strictly greater than zero) given by

$$I^A = \sum_{x \in [m]} |x\rangle\langle x|. \tag{2.51}$$

Note that for any orthonormal basis $\{|v_x\rangle\}_{x \in [m]}$ of $A$ we have

$$\left(\sum_{x \in [m]} |v_x\rangle\langle v_x|\right)|\psi\rangle = \sum_{x \in [m]} \langle v_x|\psi\rangle|v_x\rangle = |\psi\rangle. \tag{2.52}$$

Therefore, $\sum_{x \in [m]} |v_x\rangle\langle v_x| = I^A$ for any orthonormal basis $\{|v_x\rangle\}_{x \in [m]}$ of $A$.

## Decomposition of Hermitian Operators

Let $H : A \to A$ be a Hermitian operator. Since every Hermitian operator is diagonalizable, we can express $H$ as

$$H = \sum_{x \in [n]} \lambda_x |\phi_x\rangle\langle\phi_x|, \tag{2.53}$$

where $\{|\phi_x\rangle\}_{x \in [n]}$ is an orthonormal basis, and the eigenvalues $\{\lambda_x\}_{x \in [n]}$ are all real. Therefore, it is possible to decompose $H$ as

$$H = H_+ - H_-, \tag{2.54}$$

where

$$H_+ := \sum_{x:\,\lambda_x \geqslant 0} \lambda_x |\phi_x\rangle\langle\phi_x| \geqslant 0 \quad \text{and} \quad H_- := \sum_{x:\,\lambda_x < 0} |\lambda_x||\phi_x\rangle\langle\phi_x| \geqslant 0. \tag{2.55}$$

By definition $H_+, H_- \geqslant 0$ and $H_+ H_- = H_- H_+ = 0$. Furthermore, denote by $\Pi_- := \sum_{x:\,\lambda_x < 0} |\phi_x\rangle\langle\phi_x|$ the projection to the negative eigenspace of $H$, and by $\Pi_+ = I - \Pi_-$ the projection to the nonnegative eigenspace of $H$. Then, $H_\pm = H\Pi_\pm = \Pi_\pm H$.

**Exercise 2.24.** *Let $H : A \to A$ be a Hermitian operator. Show that*

$$H_\pm = \frac{|H| \pm H}{2}, \tag{2.56}$$

*where $H_\pm$ are the positive and negative parts of $H$ as defined in (2.55).*

### 2.3.3 Quantum States

A positive semidefinite operator $\rho : A \to A$ (which may be denoted as $\rho^A$ to indicate its underlying Hilbert space $A$) is known as a *quantum state* or a *density matrix* if its trace is equal to 1. In this case, we denote its eigenvalues as $\{p_x\}_{x \in [m]}$, as they represent the components of a probability distribution.

A quantum state $\rho$ is called *pure* if its rank is equal to 1. Consequently, pure states are projections onto one-dimensional subspaces of $A$, and the normalized vector they project onto is also referred to as a quantum state. It should be noted that a positive semidefinite matrix $\rho$ is a pure state if and only if $\text{Tr}[\rho^2] = \text{Tr}[\rho] = 1$. We often denote pure states by $\phi := |\phi\rangle\langle\phi|$ or $\psi := |\psi\rangle\langle\psi|$, where $|\psi\rangle$ and $|\phi\rangle$ are normalized vectors in $A$.

**Exercise 2.25.** *Show that any Hermitian linear operator $\rho : A \to A$ (with $A$ being a finite dimensional Hilbert space) is a pure quantum state if and only if*

$$\text{Tr}[\rho^3] = \text{Tr}[\rho^2] = \text{Tr}[\rho] = 1. \tag{2.57}$$

*Give an example of Hermitian operator $H : A \to A$ such that $\text{Tr}[H^2] = \text{Tr}[H] = 1$ but $H$ is not a pure quantum state.*

**Exercise 2.26.** *A linear operator $G \in \mathfrak{L}(A)$ is called a contraction if $G^* G \leqslant I^A$.*

1. *Show that $G \in \mathcal{L}(A)$ is a contraction if and only if $G^*$ is a contraction.*
2. *Show that $G$ is a contraction if and only ig $\big\| G|\psi\rangle \big\|_2 \leqslant \big\| |\psi\rangle \big\|_2$ for all $|\psi\rangle \in A$.*
3. *Show that if $G$ is a contraction, then the real part of $G$, $P := \frac{1}{2}(G + G^*)$, satisfies $P \leqslant I^A$. Hint: Use the triangle inequality to show that $\big\| P|\psi\rangle \big\|_2 \leqslant \big\| |\psi\rangle \big\|_2$.*

Every collection of $m$ normalized vectors $\{|\psi_x\rangle\}_{x\in[m]}$ in $A$, along with a probability distribution $\{p_x\}_{x\in[m]}$, is called an *ensemble* of states. To every ensemble of states, $\{|\psi_x\rangle,\ p_x\}_{x\in[m]}$, there is a corresponding quantum state defined by

$$\rho = \sum_{x\in[m]} p_x |\psi_x\rangle\langle\psi_x|. \tag{2.58}$$

Note that the above *pure state decomposition* of $\rho$ is not necessarily the spectral decomposition since the pure states $|\psi_x\rangle$ are not necessarily orthogonal. In fact, any quantum state corresponds to infinitely many ensembles of quantum states. For example, consider a quantum state $\rho: \mathbb{C}^2 \to \mathbb{C}^2$ defined by

$$\rho = \frac{1}{4}|0\rangle\langle 0| + \frac{3}{4}|1\rangle\langle 1|. \tag{2.59}$$

Clearly, this is the spectral decomposition of $\rho$. Now, it is simple to check that $\rho$ can also be expressed as

$$\rho = \frac{1}{2}|u\rangle\langle u| + \frac{1}{2}|v\rangle\langle v|, \tag{2.60}$$

where

$$|u\rangle := \sqrt{\frac{1}{4}}|0\rangle + \sqrt{\frac{3}{4}}|1\rangle \quad , \quad |v\rangle := \sqrt{\frac{1}{4}}|0\rangle - \sqrt{\frac{3}{4}}|1\rangle. \tag{2.61}$$

Note that $|u\rangle$ and $|v\rangle$ are not orthogonal, and both ensembles in (2.59) and (2.60) correspond to the *same* quantum state $\rho$.

**Exercise 2.27.** *Let $\{|\psi_x\rangle,\ p_x\}_{x\in[m]}$ and $\{|\phi_y\rangle,\ q_y\}_{y\in[n]}$ be two ensembles of quantum states in $A$ with $m \geqslant n$. Show that they correspond to the same density matrix*

$$\rho := \sum_{x\in[m]} p_x |\psi_x\rangle\langle\psi_x| = \sum_{y\in[n]} q_y |\phi_y\rangle\langle\phi_y| \tag{2.62}$$

*if and only if there exists an $m \times n$ isometry matrix $V = (v_{xy})$ (i.e. $V^*V = I_n$) such that*

$$\sqrt{p_x}|\psi_x\rangle = \sum_{y\in[n]} v_{xy}\sqrt{q_y}|\phi_y\rangle \qquad \forall\, x \in [m]. \tag{2.63}$$

## 2.3.4 The Space of Linear Operators

We will denote by $\mathcal{L}(A, B)$ the set of all linear operators from $A$ to $B$, and set $\mathcal{L}(A) := \mathcal{L}(A, A)$. The space $\mathcal{L}(A, B)$ is itself a Hilbert space. Let $m := |A|$ and $n := |B|$, and observe that $\mathcal{L}(A, B)$ is isomorphic to $\mathbb{C}^{n \times m}$ with inner product defined for all $M, N \in \mathcal{L}(A, B)$ by

$$\langle M|N\rangle := \mathrm{Tr}\big[M^*N\big]. \tag{2.64}$$

Note that $M^*N \in \mathfrak{L}(A)$ for which the trace is well defined ($\mathrm{Tr}[M]$ is not well defined if $m \neq n$). The standard basis of $\mathfrak{L}(A, B)$ is given by $\{|y\rangle\langle x|\}$, with indices running over $x \in [m]$ and $y \in [n]$, where $|x\rangle \in A$ and $|y\rangle \in B$. Hence, the dimension of $\mathfrak{L}(A, B)$ is $mn$.

Most of the objects in quantum mechanics, like quantum states, observables, positive operator valued measures (POVM), and so on, are described with Hermitian operators. It is therefore useful to characterize the space of Hermitian operators in $\mathfrak{L}(A)$. We denote this space by $\mathrm{Herm}(A)$. That is, $H \in \mathrm{Herm}(A)$ if and only if $H \in \mathfrak{L}(A)$ with $H^* = H$. Note that $\mathrm{Herm}(A)$ cannot be a vector space over the complex numbers. This is because if $c \in \mathbb{C}$ is a nonreal complex number, and $0 \neq H \in \mathrm{Herm}(A)$, then

$$(cH)^* = \bar{c}H^* = \bar{c}H \neq cH. \tag{2.65}$$

Therefore, $cH \notin \mathrm{Herm}(A)$ so that $\mathrm{Herm}(A)$ cannot be a vector space over the complex numbers.

However, it is simple to verify that $\mathrm{Herm}(A)$ is a vector space over the real numbers. Particularly, if $r_1, r_2$ are real numbers and $H_1, H_2 \in \mathrm{Herm}(A)$ are two Hermitian operators, then also $r_1 H_1 + r_2 H_2$ is a Hermitian operator. The inner product between two elements of $\mathrm{Herm}(A)$ is induced from $\mathfrak{L}(A)$, and is given by

$$\langle H_1 | H_2 \rangle := \mathrm{Tr}\,[H_1 H_2]. \tag{2.66}$$

To summarize, $\mathrm{Herm}(A)$ is a real Hilbert space.

**Exercise 2.28.** *Show that the dimension of* $\mathrm{Herm}(A)$ *is* $|A|^2$.

**Exercise 2.29.** *Show that* $\mathfrak{L}(A)$ *has an orthonormal basis consisting of Hermitian matrices.*

**Exercise 2.30.** *Let* $\mathfrak{L}_0(A)$ *be the space of all matrices in* $\mathfrak{L}(A)$ *with zero trace.*

1. *Show that* $\mathfrak{L}_0(A)$ *is an inner product space (under the Hilbert–Schmidt inner product) of dimension* $d := |A|^2 - 1$.
2. *Show that if* $\{\eta_1, \ldots, \eta_d\}$ *is an orthonormal basis of* $\mathfrak{L}_0(A)$, *then* $\left\{\frac{1}{\sqrt{|A|}} I^A, \eta_1, \ldots, \eta_d\right\}$ *is an orthonormal basis of* $\mathfrak{L}(A)$.
3. *Let* $\omega \in \mathfrak{L}(A)$. *Show that* $\omega$ *is proportional to the identity matrix if and only if* $\mathrm{Tr}\,[\omega\eta] = 0$ *for all traceless matrices* $\eta \in \mathfrak{L}_0(A)$.

**Exercise 2.31.** *Let* $A$ *be a two-dimensional Hilbert space (e.g.* $A = \mathbb{C}^2$) *with an orthonormal basis* $\{|0\rangle, |1\rangle\}$ *and let*

$$\sigma_0 := I^A = |0\rangle\langle 0| + |1\rangle\langle 1|, \quad \sigma_1 - |0\rangle\langle 1| + |1\rangle\langle 0|,$$
$$\sigma_2 = -i|0\rangle\langle 1| + i|1\rangle\langle 0|, \quad \sigma_3 = |0\rangle\langle 0| - |1\rangle\langle 1|. \tag{2.67}$$

*The three operators* $\sigma_1, \sigma_2$, *and* $\sigma_3$ *are known as the Pauli operators.*

1. *Show that all the operators above are Hermitian, unitary, and have a norm equal to $\sqrt{2}$.*
2. *Show that the set $\{\sigma_0, \sigma_1, \sigma_2, \sigma_3\}$ forms an orthogonal basis of* Herm$(A)$.
3. *Show that the commutator $[\sigma_i, \sigma_j] = 2i \sum_{k \in [3]} \varepsilon_{ijk} \sigma_k$, where the structure constant $\varepsilon_{ijk}$ is the Levi-Civita symbol, and $i, j, k \in \{1, 2, 3\}$.*
4. *Show that the anti-commutator $\{\sigma_i, \sigma_j\} = 2\delta_{ij} I$.*

Like the inner product, also norms in $\mathbb{C}^n$ have a natural extension to the space $\mathfrak{L}(A, B)$. In particular, the $\ell^p$ norms as defined in Exercise (2.3) on vectors in $\mathbb{C}^n$ can be extended to elements of $\mathfrak{L}(A, B)$.

> **The Schatten Norms**
>
> **Definition 2.3.** Let $A$ and $B$ be two Hilbert spaces, $M \in \mathfrak{L}(A, B)$, and $p \in [1, \infty]$. The Schatten $p$-norm of $M$ is defined as
>
> $$\|M\|_p := \left(\mathrm{Tr}|M|^p\right)^{\frac{1}{p}}, \quad \text{where} \quad |M| := \sqrt{M^*M}. \tag{2.68}$$

The case $p = 1$ is often called the *trace norm* and we will discuss it in detail in Section 5.4.1. The case $p = \infty$ is understood in terms of the limit $p \to \infty$. It is often called the *operator norm* and is given by

$$\|M\|_\infty = \lambda_{\max}(|M|), \tag{2.69}$$

where $\lambda_{\max}(|M|)$ is the largest eigenvalue of $|M|$, or equivalently, the largest singular value of $M$. The Schatten norms appear quite often in quantum Shannon theory due to their relation to the Rényi entropies that we will study later on. We leave it as an exercise for the reader to prove some of their key properties.

**Exercise 2.32.** *Let $A$ and $B$ be two Hilbert spaces, $M, N \in \mathfrak{L}(A, B)$, and $p, q \in [1, \infty]$ such that $\frac{1}{p} + \frac{1}{q} = 1$. Show that the p-Schatten norm is indeed a norm satisfying the following properties:*

1. **Invariance.** *For any two Hilbert spaces $A', B'$ with $|A'| \geqslant |A|$ and $|B'| \geqslant |B|$, and any isometries $V \in \mathfrak{L}(B, B')$ and $U \in \mathfrak{L}(A, A')$*

$$\|VMU^*\|_p = \|M\|_p. \tag{2.70}$$

1. **Hölder Inequality.**

$$\|MN\|_1 \leqslant \|M\|_p \|N\|_q. \tag{2.71}$$

2. **Sub-multiplicativity.**

$$\|MN\|_p \leqslant \|M\|_p \|N\|_p. \tag{2.72}$$

3. **Monotonicity.** *If $p \leqslant q$*

$$\|M\|_1 \geqslant \|M\|_p \geqslant \|M\|_q \geqslant \|M\|_\infty. \tag{2.73}$$

**4. Duality.**

$$\|M\|_p = \sup\left\{ \left|\mathrm{Tr}\left[M^*L\right]\right| : \|L\|_q = 1, \ \ L \in \mathfrak{L}(A,B) \right\}. \tag{2.74}$$

**Exercise 2.33 (Young's Inequality).** *Let $A$ and $B$ be two Hilbert spaces, $M, N \in \mathrm{Pos}(A)$, and $p, q \in [1, \infty)$ such that $\frac{1}{p} + \frac{1}{q} = 1$. Use the Hölder inequality of the Schatten norm to show that*

$$\mathrm{Tr}[MN] \leqslant \frac{1}{p}\mathrm{Tr}[M^p] + \frac{1}{q}\mathrm{Tr}[N^q]. \tag{2.75}$$

*with equality if and only if $M^p = N^q$. Hint: Take the logarithm on both sides of Hölder inequality and use the concavity property of the logarithm.*

**Exercise 2.34.** *Show that for any $M \in \mathrm{Herm}(A)$, the operator norm and the trace norm can be expressed as*

$$\|M\|_1 = \max_{\substack{\eta \in \mathrm{Herm}(A) \\ \|\eta\|_\infty \leqslant 1}} \mathrm{Tr}[\eta M] \quad and \quad \|M\|_\infty = \max_{\substack{\eta \in \mathrm{Herm}(A) \\ \|\eta\|_1 \leqslant 1}} \mathrm{Tr}[\eta M]. \tag{2.76}$$

The Ky Fan Norms

**Definition 2.4.** Let $A$ and $B$ be two Hilbert spaces, $M \in \mathfrak{L}(A,B)$, $n := \min\{|A|, |B|\}$, and $k \in [n]$. The Ky Fan $k$-norm of $M$ is defined as

$$\|M\|_{(k)} := s_1 + s_2 + \cdots + s_k, \tag{2.77}$$

where $s_1 \geqslant s_2 \geqslant \cdots \geqslant s_k$ are the $k$ largest singular values of $M$.

*Remark.* When restricting $M$ to be diagonal real matrix we get the following definition of the Ky Fan norm on $\mathbb{R}^n$: For any $\mathbf{r} \in \mathbb{R}^n$ and $k \in [n]$ the $k$th-Ky Fan norm of $\mathbf{r}$ is defined as

$$\|\mathbf{r}\|_{(k)} := \sum_{x \in [k]} |r_x^\downarrow|, \tag{2.78}$$

where $\{r_x^\downarrow\}_{x \in [n]}$ are the components of $\mathbf{r}$ arranged such that $|r_1^\downarrow| \geqslant |r_2^\downarrow| \geqslant \cdots \geqslant |r_n^\downarrow|$.

The Ky Fan norms play an important role in the resource theory of entanglement. We leave it as an exercise to prove that the Ky Fan norms are indeed norms.

**Exercise 2.35.** *Show that the Ky Fan norms are indeed norms that have the following invariance property. Using the same notations as in the definition 2.4, show that for any two Hilbert spaces $A', B'$ with $|A'| \geqslant |A|$ and $|B'| \geqslant |B|$, and any isometries $V \in \mathfrak{L}(B, B')$ and $U \in \mathfrak{L}(A, A')$*

$$\|VMU^*\|_{(k)} = \|M\|_{(k)}. \tag{2.79}$$

**Exercise 2.36.** *Show that the Ky Fan $k$-norm can be expressed as*

$$\|M\|_{(k)} = \sup \mathrm{Tr}\left[\Pi|M|\right], \tag{2.80}$$

*where the supremum is over all orthogonal projections $\Pi$ with rank no greater than $k$.*

**Exercise 2.37 (The Ky Fan norms on $\mathbb{R}^n$).** *Consider the variant of the Ky Fan norm as defined in (2.78).*

1. *Show that the Ky Fan norms are indeed norms in $\mathbb{R}^n$.*
2. *Let D be an $n \times n$ matrix that can be written as a convex combination of permutation matrices (such matrices are known as doubly stochastic matrices; see Appendix A.5 of online version for more details). Show that for all $k \in [n]$ and all $\mathbf{p} \in \mathbb{R}^n_+$*

$$\|\mathbf{p}\|_{(k)} \geqslant \|D\mathbf{p}\|_{(k)}. \tag{2.81}$$

3. *Show that for any two probability vectors $\mathbf{p}, \mathbf{q} \in Prob(n)$ and any $k \in [n]$ we have*

$$\|\mathbf{p} - \mathbf{q}\|_{(k)} \leqslant \frac{1}{2}\|\mathbf{p} - \mathbf{q}\|_1 \tag{2.82}$$

*and conclude that*

$$\|\mathbf{p}\|_{(k)} - \|\mathbf{q}\|_{(k)} \leqslant \frac{1}{2}\|\mathbf{p} - \mathbf{q}\|_1. \tag{2.83}$$

*Hint: For the first inequality use the fact that $\frac{1}{2}\|\mathbf{p} - \mathbf{q}\|_1 = \sum_{x \in [n]} (p_x - q_x)_+$, and for the second inequality we use the properties of a norm. For any $r \in \mathbb{R}$ the symbol $(r)_+ := r$ if $r \geqslant 0$ and otherwise $(r)_+ := 0$.*

## 2.3.5 Linear Operators as Bipartite Vectors

As seen in Exercise 2.8, given two finite dimensional Hilbert spaces $A$ and $B$ of dimensions $m := |A|$ and $n := |B|$, the space $A \otimes B$ is (isometrically) isomorphic to the Hilbert space of $m \times n$ matrices $\mathbb{C}^{m \times n}$. For convenience, we will denote $AB := A \otimes B$ and call it a bipartite Hilbert space and its elements bipartite vectors. Moreover, we will use the tilde notation to indicate spaces with the same dimension as without the tilde. For example, $A$ and $\tilde{A}$ both will have the same dimension $m := |A|$, and consequently $A\tilde{A} := A \otimes \tilde{A}$ will have dimension $m^2$. The isomorphism $A \otimes B \cong \mathbb{C}^{m \times n} \cong \mathfrak{L}(B, A)$ indicates that we can think of bipartite vectors in $AB$ as linear operators from $B$ to $A$. This correspondence will be very useful later on, so we discuss it in more detail here.

Any bipartite vector $|\psi\rangle \in AB$ can be expressed in terms of the orthonormal basis $\{|x\rangle^A \otimes |y\rangle^B\}$ as

$$|\psi^{AB}\rangle = \sum_{x \in [m]} \sum_{y \in [n]} \mu_{xy}|x\rangle^A \otimes |y\rangle^B = \sum_{y \in [n]} \left( \sum_{x \in [m]} \mu_{xy}|x\rangle^A \right) \otimes |y\rangle^B, \tag{2.84}$$

where $\mu_{xy} \in \mathbb{C}$. Let $M_\psi$ be a linear map from $\tilde{B}$ to $A$ defined here by its action on the basis elements $\{|y\rangle^{\tilde{B}}\}$ of $\tilde{B}$:

$$M_\psi|y\rangle^{\tilde{B}} := \sum_{x \in [m]} \mu_{xy}|x\rangle^A. \tag{2.85}$$

With this definition we get that

$$|\psi^{AB}\rangle = \sum_{y \in [n]} \left( M_\psi|y\rangle^{\tilde{B}} \right) \otimes |y\rangle^B = \left( M_\psi \otimes I^B \right) \sum_{y \in [n]} |yy\rangle^{\tilde{B}B}. \tag{2.86}$$

Denoting by

$$|\Omega^{\tilde{B}B}\rangle := \sum_{y\in[n]} |yy\rangle^{\tilde{B}B}, \tag{2.87}$$

we conclude that

$$|\psi^{AB}\rangle = M_\psi \otimes I^B |\Omega^{\tilde{B}B}\rangle. \tag{2.88}$$

Therefore, for any bipartite vector $|\psi\rangle^{AB}$ there is a corresponding linear map $M_\psi : \tilde{B} \to A$ and vice versa. In other words, the mapping

$$|\psi^{AB}\rangle \mapsto M_\psi \tag{2.89}$$

is an (isometrically) isomorphism map from the space $AB$ and the space $\mathbb{C}^{m\times n}$. The vector $|\Omega^{\tilde{B}B}\rangle$ has many interesting properties, and later on we will see that, physically, its normalized version corresponds to a composite system of two maximally entangled subsystems.

**Exercise 2.38.** *Prove the following properties of $|\Omega^{A\tilde{A}}\rangle$:*

1. *For any matrix $N \in \mathfrak{L}(\tilde{A})$,*

$$\langle\Omega^{A\tilde{A}}|I \otimes N|\Omega^{A\tilde{A}}\rangle = \mathrm{Tr}\,[N]. \tag{2.90}$$

2. *Let $M: B \to A$ be a linear map and denote its transpose map by $M^T: A \to B$. Show that*

$$I^A \otimes M^T|\Omega^{A\tilde{A}}\rangle = M \otimes I^B|\Omega^{\tilde{B}B}\rangle. \tag{2.91}$$

3. *Show that for any invertible matrices $M, N \in \mathfrak{L}(A)$,*

$$M \otimes N|\Omega^{A\tilde{A}}\rangle = |\Omega^{A\tilde{A}}\rangle \quad \Longleftrightarrow \quad M = \left(N^{-1}\right)^T. \tag{2.92}$$

4. *Show that if $|\psi\rangle = L \otimes R|\varphi\rangle$ for some matrices $L$ and $R$, then*

$$M_\psi = LM_\varphi R^T. \tag{2.93}$$

5. ***Schmidt Decomposition:*** *Show that for any normalized vector $|\psi^{AB}\rangle \in AB$ there exists orthonormal sets of $k \leqslant \min\{|A|,|B|\}$ vectors, $\{|u_z^A\rangle\}_{z\in[k]}$, and $\{|v_z^B\rangle\}_{z\in[k]}$, in $A$ and $B$, respectively, such that*

$$|\psi^{AB}\rangle = \sum_{z\in[k]} \sqrt{p_z}|u_z^A\rangle \otimes |v_z^B\rangle, \tag{2.94}$$

*where $p_z > 0$ for all $z \in [k]$, and $\sum_{z\in[k]} p_z = 1$. Hint: Use the singular value decomposition of the matrix $M_\psi = UDV$, where $U, V$ are unitary matrices and $D$ is a $|B| \times |A|$ diagonal matrix with diagonal consisting of the singular values of $M_\psi$.*

## The Reduced Density Matrix

For any bipartite state $|\psi^{AB}\rangle$ as in (2.88), the density matrix $\rho_\psi^A := M_\psi M_\psi^*$ is called the *reduced density matrix* on system $A$ of the bipartite state $|\psi^{AB}\rangle$, and the density matrix $\rho_\psi^B := \left( M_\psi^* M_\psi \right)^T$ is called the *reduced density matrix* on system $B$ of the bipartite state $|\psi^{AB}\rangle$. Note that $\rho_\psi^A$ is an $|A| \times |A|$ matrix while $\rho_\psi^B$ is an $|B| \times |B|$ matrix.

**Exercise 2.39.** *Show that the two reduced density matrices, $\rho_\psi^A$ and $\rho_\psi^B$, have the same nonzero eigenvalues.*

---

**The Partial Trace**

**Exercise 2.40.** *Let* $\mathrm{Tr}_B : \mathcal{L}(AB) \to \mathcal{L}(A)$ *be a linear map defined by its action on the basis elements of* $\mathcal{L}(AB)$ *as:*

$$\mathrm{Tr}_B \left[ |x\rangle\langle y|^A \otimes |z\rangle\langle w|^B \right] := |x\rangle\langle y|^A \, \mathrm{Tr}\left[ |z\rangle\langle w|^B \right] = \delta_{zw} |x\rangle\langle y|^A. \tag{2.95}$$

*Show that the reduced density matrix $\rho_\psi^A$ of a bipartite pure state $|\psi\rangle \in AB$ can be expressed as*

$$\rho_\psi^A = \mathrm{Tr}_B \left[ |\psi^{AB}\rangle\langle\psi^{AB}| \right]. \tag{2.96}$$

---

It is well known that the trace remains invariant under cyclic permutation of product of matrices. The following exercise states that this remains true also for the partial trace.

**Exercise 2.41.** *Show that for any $\rho \in \mathcal{L}(AB)$ and any two matrices $\eta, \zeta \in \mathcal{L}(B)$,*

$$\mathrm{Tr}_B \left[ (I^A \otimes \eta^B)\rho^{AB}(I^A \otimes \zeta^B) \right] = \mathrm{Tr}_B \left[ (I^A \otimes \zeta^B \eta^B)\rho^{AB} \right]. \tag{2.97}$$

For any pure bipartite state as in (2.88) there is a unique reduced density matrix $\rho_\psi^A$. On the other hand, for any density matrix $\rho \in \mathfrak{D}(A)$ there are many bipartite pure states $|\psi^{AB}\rangle$ with the same reduced density matrix $\rho$ (see Exercise 2.44).

**Exercise 2.42.** *Let $\rho \in \mathcal{L}(AB)$. Show that if for all $\eta \in \mathcal{L}(B)$ the matrix*

$$\mathrm{Tr}_B \left[ \left( I^A \otimes \eta^B \right) \rho^{AB} \right] \tag{2.98}$$

*is proportional to the identity matrix, then $\rho^{AB} = \mathbf{u}^A \otimes \rho^B$, where $\mathbf{u}^A := \frac{1}{|A|} I^A$ is the uniform density matrix also known as the maximally mixed state. Hint: Use Part 3 of Exercise 2.30.*

**Exercise 2.43.** *Let $A, B, A'$, and $B'$ be four Hilbert spaces and let $\Lambda \in \mathcal{L}(AB, A'B')$. Show that if*

$$\mathrm{Tr}_B \left[ \left( I^{A'} \otimes T \right) \Lambda \right] = 0 \qquad \forall \, T \in \mathcal{L}(B', B), \tag{2.99}$$

*then $\Lambda = 0$. Observe that the operator $\left( I^{A'} \otimes T \right) \Lambda$ belongs to $\mathfrak{L}(AB, A'B)$, so the partial trace above over $B$ is well defined. Hint: Let $\{N_x\}$ be an orthonormal basis (w.r.t. the Hilbert–Schmidt inner product) of $\mathfrak{L}(B, B')$ and write $\Lambda = \sum_x M_x \otimes N_x$, where $\{M_x\}$ are some matrices in $\mathfrak{L}(A, A')$. Then show that by taking $T$ above to be $N_y$ you get $M_y = 0$.*

> **Definition 2.5.** Let $\rho \in \mathfrak{D}(A)$ be a density operator. A normalized bipartite pure quantum state $|\psi^{AB}\rangle \in A \otimes B$ is called a *purification* of $\rho^A$, if $\rho^A$ is the reduced density matrix of $|\psi^{AB}\rangle$.

**Exercise 2.44.** *Let $\rho \in \mathfrak{D}(A)$ be a density matrix.*

1. *Show that $\sqrt{\rho} \otimes I^{\tilde{A}} |\Omega^{A\tilde{A}}\rangle$ is a purification of $\rho$.*
2. *Show that $|\psi^{AB}\rangle \in AB$ and $|\phi^{AC}\rangle \in AC$ (assuming $|B| \leqslant |C|$) are two purifications of $\rho \in \mathfrak{D}(A)$ if and only if there exists an isometry matrix $V : B \to C$ such that*

$$|\phi^{AC}\rangle = I^A \otimes V^{B \to C} |\psi^{AB}\rangle. \tag{2.100}$$

3. *Use Part 2 to provide alternative (simpler!) proof of the claim in Exercise 2.27.*

**Exercise 2.45.** *Operator Schmidt Decomposition: Let $A$ and $B$ be two Hilbert spaces of dimensions $m := |A|$ and $n := |B|$ and denote by $k := \min\{m^2, n^2\}$. Show that for every $\rho \in$ Herm$(AB)$ there exists $k$ nonnegative real numbers $\{\lambda_z\}_{z \in [k]}$, and two orthonormal sets of Hermitian matrices (w.r.t. the Hilbert Schmidt inner product) $\{\eta_z\}_{z \subset [k]} \subset$ Herm$(A)$ and $\{\zeta_z\}_{z \in [k]} \subset$ Herm$(B)$ such that*

$$\rho^{AB} = \sum_{z \in [k]} \lambda_z \eta_z^A \otimes \zeta_z^B. \tag{2.101}$$

*Hint: Use similar lines as in part five of Exercise 2.38.*

## 2.4 Encoding Information in Quantum States

The first postulate of quantum mechanics states that to any physical system there is a (separable) Hilbert space $A$ that is associated with it, and that the information about the system is completely described by quantum states; that is, unit-trace positive semidefinite operators in $\mathfrak{L}(A)$. Furthermore, for isolated physical systems, the information is described by a pure state $|\psi\rangle\langle\psi| \in \mathfrak{L}(A)$. Such pure states can be described by *rays* of the form $\{e^{i\theta}|\psi\rangle : \theta \in [0, 2\pi]\}$, where $|\psi\rangle$ is a unit vector in $A$. Therefore, aside from an irrelevant phase, isolated systems are described with unit vectors in a Hilbert space. As an example, we start with the building block of quantum information: the quantum bit.

## 2.4.1 The Quantum Bit

The quantum bit, or in short the *qubit*, is the quantum generalization of the classical bit. We will use here the example of a spin of an electron to describe the qubit. The spin of an electron in some fixed direction can take two possible values. We therefore associate with it a two-dimensional Hilbert space $A \cong \mathbb{C}^2$. If the spin of the electron is pointing in the positive $z$-direction, it is described by a pure state, say $|0\rangle\langle 0|$. The question now is how to represent the spin of the electron in all other directions if we choose $|0\rangle\langle 0|$ to correspond to the positive $z$-direction. To answer this question we will make use of a representation of the rotation group SO(3) on the space $\mathbb{C}^2$.

Consider a counterclockwise rotation along an axis of rotation that is described by the unit vector $\mathbf{n} \in \mathbb{R}^3$. In $\mathbb{R}^3$, such a rotation by an angle $\theta$ is described by an orthogonal $3 \times 3$ matrix, $R_\theta^{(\mathbf{n})}$, rotating a vector $\mathbf{v} \in \mathbb{R}^3$ to the vector $R_\theta^{(\mathbf{n})}\mathbf{v}$ (see (C.7) for the explicit form of $R_\theta^{(\mathbf{n})}$ in terms of $\mathbf{n}$ and $\theta$). For example, $\mathbf{x} = R_{\frac{\pi}{2}}^{(\mathbf{y})}\mathbf{z}$, where $\mathbf{x}, \mathbf{y}, \mathbf{z}$ are the unit vectors in the $x$-, $y$-, and $z$-directions. Our goal now is to find a $2 \times 2$ complex matrix $T_\theta^{(\mathbf{n})}$ such that if $|0\rangle \in \mathbb{C}^2$ corresponds to the spin in the $z$-direction, then $T_\theta^{(\mathbf{n})}|0\rangle$ corresponds to the spin in the direction $R_\theta^{(\mathbf{n})}\mathbf{z}$. More generally, we look for a matrix $T_\theta^{(\mathbf{n})}$ that has the following property: If the qubit state $|\psi\rangle \in \mathbb{C}^2$ corresponds to a spin in a direction $\mathbf{m}$, the state $T_\theta^{(\mathbf{n})}|\psi\rangle$ corresponds to the spin in the direction $R_\theta^{(\mathbf{n})}\mathbf{m}$. Note that in this definition we made an assumption that it is the same matrix $T_\theta^{(\mathbf{n})}$ that is applied to describe a rotation by an angle $\theta$ along the $\mathbf{n}$-axis irrespective if the initial state of the system was in the $z$-direction or any other $\mathbf{m}$-direction. This is justified physically since the physical process that causes the spin of the particle to rotate is *external* to the particle and therefore is independent on the direction that the initial spin of the particle is pointing at. This assumption is also related to the fourth (evolution) axiom of quantum systems that we will discuss later on.

The matrices $T_\theta^{(\mathbf{n})}$ are not unique since for any phase $\alpha \in [0, 2\pi]$ the matrices $e^{i\alpha}T_\theta^{(\mathbf{n})}$ also transform $\psi := |\psi\rangle\langle\psi|$ to $T_\theta^{(\mathbf{n})}\psi(T_\theta^{(\mathbf{n})})^*$. Hence, $T_\theta^{(\mathbf{n})}$ is determined uniquely up to a phase. Moreover, since $\|T_\theta^{(\mathbf{n})}|\psi\rangle\| = 1$ for all normalized vectors $|\psi\rangle \in \mathbb{C}^2$, $T_\theta^{(\mathbf{n})}$ must be a unitary matrix. This is a special case of Wigner theorem that states that physical symmetries act on the Hilbert space of quantum states unitarily (or antiunitarily). Moreover, repeating the previous arguments, we conclude that the state $T_{\theta_2}^{(\mathbf{n}_2)}T_{\theta_1}^{(\mathbf{n}_1)}|\psi\rangle$ is the quantum state that corresponds to the spin in the direction $R_{\theta_2}^{(\mathbf{n}_2)}R_{\theta_1}^{(\mathbf{n}_1)}\mathbf{m}$. Combining everything we conclude that the mapping

$$R_\theta^{(\mathbf{n})} \mapsto \mathcal{T}_\theta^{(\mathbf{n})}, \quad \text{where} \quad \mathcal{T}_\theta^{(\mathbf{n})}(\rho) := T_\theta^{(\mathbf{n})}\rho(T_\theta^{(\mathbf{n})})^* \quad \forall \rho \in \mathfrak{L}(A), \tag{2.102}$$

is a group representation of $SO(3)$ on the Hilbert space $\mathfrak{L}(A \to A)$ (i.e. the space of linear operators from $\mathfrak{L}(A)$ to $\mathfrak{L}(A)$).

It will be more convenient to work with a unitary representation on the space of $\mathfrak{L}(A)$ itself rather than the Hilbert space $\mathfrak{L}(A \to A)$. For this purpose we need to eliminate the freedom in the choice of the phase so that $R_\theta^{(\mathbf{n})}$ is mapped to a unique $T_\theta^{(\mathbf{n})}$. We therefore assume without loss of generality that $\det\left(T_\theta^{(\mathbf{n})}\right) = 1$ so that $T_\theta^{(\mathbf{n})} \in$ SU(2). This almost eliminates completely the ambiguity in the phase, although note that if $T_\theta^{(\mathbf{n})} \in$ SU(2), then also $-T_\theta^{(\mathbf{n})} \in$ SU(2). This would mean that both $\pm T_\theta^{(\mathbf{n})}$ correspond to the same $R_\theta^{(\mathbf{n})}$. To summarize, up to a sign factor, the collection of matrices $\{T_\theta^{(\mathbf{n})}\}_{\mathbf{n},\theta}$ form a group representation of SO(3). Such a 2:1 and onto homomorphism $h : $ SU(2) $\to$ SO(3) with the property that $h(T) = h(-T)$ for any $T \in$ SU(2) was found by Cornwell in 1984 (see the Exercise C.3). We now discuss the explicit form of $T_\theta^{(\mathbf{n})}$.

In Appendix C of the online version we show that the most general unitary matrix in SU(2) has the form $e^{-i\frac{\theta}{2}(\mathbf{n}\cdot\boldsymbol{\sigma})}$ (see (C.15)), where the factor $1/2$ implies that under a $2\pi$ addition to $\theta$ we get $e^{-i\frac{\theta+2\pi}{2}(\mathbf{n}\cdot\boldsymbol{\sigma})} = -e^{-i\frac{\theta}{2}(\mathbf{n}\cdot\boldsymbol{\sigma})}$. This property will be consistent with the identification $T_\theta^{(\mathbf{n})} = e^{-i\frac{1}{2}\theta(\mathbf{n}\cdot\boldsymbol{\sigma})}$ which we motivate below (recall that $T_\theta^{(\mathbf{n})} \in$ SU(2)) since a rotation by $\theta$ or by $\theta + 2\pi$ along any axis $\mathbf{n}$ should have the same effect on any qubit state; that is,

$$e^{-i\frac{\theta+2\pi}{2}(\mathbf{n}\cdot\boldsymbol{\sigma})}|\psi\rangle\langle\psi|e^{i\frac{\theta+2\pi}{2}(\mathbf{n}\cdot\boldsymbol{\sigma})} = e^{-i\frac{\theta}{2}(\mathbf{n}\cdot\boldsymbol{\sigma})}|\psi\rangle\langle\psi|e^{i\frac{\theta}{2}(\mathbf{n}\cdot\boldsymbol{\sigma})}. \tag{2.103}$$

On the other hand, if we didn't include the factor $1/2$, then the identification $T_\theta^{(\mathbf{n})} = e^{-i\theta(\mathbf{n}\cdot\boldsymbol{\sigma})}$ would imply an undesired property that a rotation by $\theta$ or by $\theta + \pi$ (along any fixed axis $\mathbf{n}$) would have the same effect on a qubit $|\psi\rangle\langle\psi|$.

To justify the identification $T_\theta^{(\mathbf{n})} = e^{-i\frac{1}{2}\theta(\mathbf{n}\cdot\boldsymbol{\sigma})}$, recall that any rotation around the $z$-axis should not change the state $|0\rangle\langle 0|$, as it represents spin in the $z$-direction. Taking $\mathbf{n} = \mathbf{z}$ we get

$$e^{-i\frac{\theta}{2}(\mathbf{z}\cdot\boldsymbol{\sigma})}|0\rangle = \big(\cos(\theta/2)\,I - i\sin(\theta/2)\,\sigma_3\big)|0\rangle = e^{-i\frac{\theta}{2}}|0\rangle, \tag{2.104}$$

where we used (C.14). Recall that the vector $e^{i\frac{\theta}{2}}|0\rangle$ corresponds to the same quantum state $|0\rangle\langle 0|$. Therefore, although there are many possible representations for $SO(3)$ in $\mathbb{C}^2$ (such as $e^{i0.7\theta(\mathbf{n}\cdot\boldsymbol{\sigma})}$, for example), the representation $T_\theta^{(\mathbf{n})} = e^{-i\frac{1}{2}\theta(\mathbf{n}\cdot\boldsymbol{\sigma})}$ is the only one that has the following essential properties:

1. The mapping $T_\theta^{(\mathbf{n})} \mapsto R_\theta^{(\mathbf{n})}$ is an onto homomorphism between $SU(2)$ and $SO(3)$.
2. For any $|\psi\rangle \in \mathbb{C}^2$ we have $T_{\theta+2\pi}^{(\mathbf{n})}|\psi\rangle\langle\psi|\left(T_{\theta+2\pi}^{(\mathbf{n})}\right)^* = T_\theta^{(\mathbf{n})}|\psi\rangle\langle\psi|\left(T_\theta^{(\mathbf{n})}\right)^*$.
3. $T_\theta^{(\mathbf{z})}|0\rangle\langle 0|\left(T_\theta^{(\mathbf{z})}\right)^* = |0\rangle\langle 0|$ for all $\theta \in \mathbb{R}$.

With this representation at hand, we are ready to identify spins in different directions. We start with a few examples. The spin in the negative $z$-direction can be obtained by rotating $|0\rangle$ by $180°$ along the $x$ (or $y$) axis. It is therefore given by

$$T_\pi^{(\mathbf{x})}|0\rangle = \left(\cos\left(\frac{\pi}{2}\right)I - i\sin\left(\frac{\pi}{2}\right)\mathbf{x}\cdot\boldsymbol{\sigma}\right)|0\rangle = -i|1\rangle. \tag{2.105}$$

Therefore, the quantum state $|1\rangle\langle 1|$ corresponds to the negative $z$-direction. Recall from the previous section that using the SG experiment one can determine with certainty if an electron was prepared in the positive $z$-direction or negative $z$-direction. This ability to distinguish between the two possible spins of the electron is reflected mathematically by the orthogonality of the vectors $|0\rangle$ and $|1\rangle$. This is a general property of quantum mechanics that any two distinguishable states of a physical system are described mathematically by orthogonal vectors. Other examples include the following:

- Spin in the positive $x$-direction

$$T^{(y)}_{\frac{\pi}{2}}|0\rangle = \left(\cos\left(\frac{\pi}{4}\right) I - i \sin\left(\frac{\pi}{4}\right)\sigma_2\right)|0\rangle = \frac{1}{2}\left(|0\rangle + |1\rangle\right) := |+\rangle. \qquad (2.106)$$

- Spin in the negative $x$-direction

$$T^{(y)}_{-\frac{\pi}{2}}|0\rangle = \left(\cos\left(\frac{\pi}{4}\right) I + i \sin\left(\frac{\pi}{4}\right)\sigma_2\right)|0\rangle = \frac{1}{2}\left(|0\rangle - |1\rangle\right) := |-\rangle. \qquad (2.107)$$

- Spin in the positive $y$-direction

$$T^{(x)}_{-\frac{\pi}{2}}|0\rangle = \left(\cos\left(\frac{\pi}{4}\right) I + i \sin\left(\frac{\pi}{4}\right)\sigma_1\right)|0\rangle = \frac{1}{2}\left(|0\rangle + i|1\rangle\right) := |+i\rangle. \qquad (2.108)$$

- Spin in the negative $y$-direction

$$T^{(x)}_{\frac{\pi}{2}}|0\rangle = \left(\cos\left(\frac{\pi}{4}\right) I - i \sin\left(\frac{\pi}{4}\right)\sigma_1\right)|0\rangle = \frac{1}{2}\left(|0\rangle - i|1\rangle\right) := |-i\rangle. \qquad (2.109)$$

In general, rotations along the n-axis do not change a spin that points in the positive or negative n-direction. We can use this physical property to compute the qubit representing a spin in the n-direction. Specifically, a quantum state $|\psi\rangle$ represents an electron with spin in the positive or negative n-direction if and only if $T^{(\mathbf{n})}_\theta |\psi\rangle = e^{i\alpha}|\psi\rangle$ for some phase $e^{i\alpha}$. Now, since $T^{(\mathbf{n})}_\theta = \cos(\theta/2)I - i\sin(\theta/2)\mathbf{n}\cdot\boldsymbol{\sigma}$ we get that $|\psi\rangle$ is an eigenvector of the matrix $T^{(\mathbf{n})}_\theta$ if and only if it is an eigenvector of the *spin matrix* $S_\mathbf{n} := \frac{1}{2}\mathbf{n}\cdot\boldsymbol{\sigma}$.

**Exercise 2.46.** *Let $S_\mathbf{n}$ be the spin matrix in direction $\mathbf{n} = (\sin(\alpha)\cos(\beta), \sin(\alpha)\sin(\beta), \cos(\alpha))^T$, with $\alpha$ and $\beta$ being its spherical coordinates.*

*1. Show that $S^2_\mathbf{n} = \frac{1}{4}I$.*
*2. Show that the eigenvalues of $S_\mathbf{n}$ are $\pm\frac{1}{2}$.*
*3. Show that if $S_\mathbf{n}|\psi\rangle = \frac{1}{2}|\psi\rangle$, then up to global phase*

$$|\psi\rangle = \cos\left(\frac{\alpha}{2}\right)|0\rangle + e^{i\beta}\sin\left(\frac{\alpha}{2}\right)|1\rangle. \qquad (2.110)$$

From Exercise (2.46) it follows that any qubit is characterized as in (2.110) and corresponds to spin in the positive direction of $\mathbf{n} = (\sin(\alpha)\cos(\beta), \sin(\alpha)\sin(\beta), \cos(\alpha))^T$. This correspondence between the point on the sphere and a qubit is known in the community as the Bloch representation of a qubit. In Figure 2.5 we show some of the popular qubit states and their location on the Bloch sphere.

Note that although we focused here on the spin of an electron, the qubit corresponds to any two-level quantum system. For example, one can implement a qubit with a photon, using, say, $|0\rangle$ to correspond to positive (or left) circular polarization, and the $|1\rangle$ to correspond to the negative (or right) circular polarization. Any linear combination of $|0\rangle$ and $|1\rangle$ will then correspond to different types of polarizations. Other

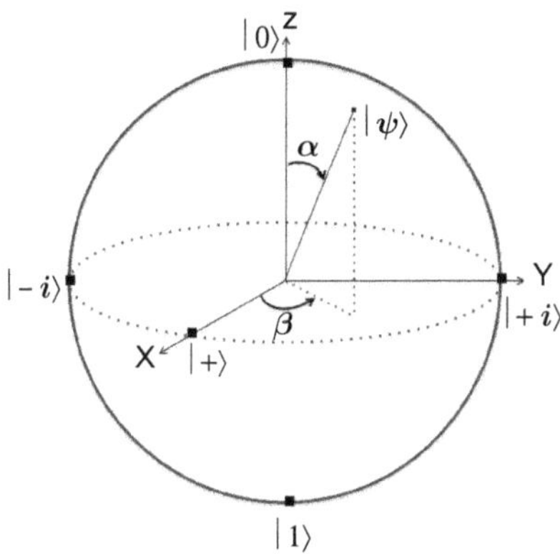



Figure 2.5    The Bloch sphere.

examples are atoms, molecules, and nucleuses, with two energy levels (excited state vs ground state). All these examples demonstrate that the qubit can be implemented in many different ways, and in this sense, we can claim that quantum information is fungible!

## 2.4.2 The Quantum Dit and Observables

The quantum dit (qudit) represents any $d$-level quantum system with $d > 2$. To any such physical system we associate a $d$-dimensional Hilbert space $A \cong \mathbb{C}^d$. How should we interpret the quantum states in $A$? Recall that in the qubit case we associated to any orthonormal basis of $\mathbb{C}^2$ a spin in some direction $\mathbf{n}$, with the two possibilities of "spin up" and "spin down" corresponding to the two elements of the basis. Moreover, we were able to construct the "spin matrix" $S_\mathbf{n}$ whose eigenvectors are the basis elements, and its eigenvalues give the spins (i.e. $\pm 1/2$) associated with the two basis elements.

In a similar way, any orthonormal basis of $A \cong \mathbb{C}^d$ corresponds to $d$ possible outcomes that can be, at least in principle, observed in some experiment. Moreover, the second postulate of quantum mechanics states that any observable (a dynamic variable that can be measured, like position, momentum, spin, energy, etc.) is represented with a Hermitian operator whose eigenvalues correspond to the values of the observable. Recall that for any Hermitian operator $H$, there exists an orthonormal basis of $A$ consisting only of eigenvectors of $H$. This basis corresponds to the possible outcomes in the measurement of the observable $H$.

For example, in the qubit case, the spin matrix $S_\mathbf{n}$ is an observable corresponding to the measurement of spin with the SG experiment. In physical systems of $d$ energy levels, the Hamiltonian $H = \sum_{x \in [d]} E_x |x\rangle\langle x|$ is an observable corresponding to the measurement of energy. This particular observable states that the values for the energy of the system (i.e. $E_x$) are discrete, and that the eigenvectors $\{|x\rangle\}_{x \in [d]}$ correspond to these energy levels.

## 2.4.3 Composite Systems

The third postulate of quantum mechanics states that the Hilbert space associated with a composite system is the Hilbert space formed by the tensor product of the state spaces associated with the component subsystems. For example, the Hilbert space associated with three electrons is given by $ABC = \mathbb{C}^2 \otimes \mathbb{C}^2 \otimes \mathbb{C}^2$. Note that although $\mathbb{C}^2 \otimes \mathbb{C}^2 \otimes \mathbb{C}^2 \cong \mathbb{C}^8$, the tensor product structure has a physical significance as the component subsystems correspond to individual particles. For example, the state

$$|0\rangle^A |-\rangle^B |+i\rangle^C \tag{2.111}$$

corresponds to three spin particles (e.g. electrons) with spin A pointing in the positive $z$-direction, spin B pointing in the negative $x$-direction, and spin C pointing in the positive $y$-direction. Of course, not all states have the same tensor product form as the state above. For example, the Greenberger–Horne–Zeilinger (GHZ) state of three qubits

$$|GHZ\rangle := \frac{1}{\sqrt{2}} \left( |0\rangle^A |0\rangle^B |0\rangle^C + |1\rangle^A |1\rangle^B |1\rangle^C \right) \tag{2.112}$$

cannot be written as a tensor product of three vectors. States like these will be called entangled.

**Exercise 2.47.** *Show that the GHZ state above cannot be written as a tensor product of three vectors; that is,*

$$|GHZ\rangle \neq |\psi^A\rangle |\phi^B\rangle |\chi^C\rangle \tag{2.113}$$

*for any three qubit states $|\psi^A\rangle$, $|\phi^B\rangle$, and $|\chi^C\rangle$.*

**Exercise 2.48.** *Show that for any unit vector $\mathbf{n} \in \mathbb{R}^3$ the singlet state $|\Psi^{AB}_-\rangle := (|01\rangle - |10\rangle)/\sqrt{2}$ can be expressed as*

$$|\Psi^{AB}_-\rangle = \frac{1}{\sqrt{2}} \left( |\uparrow_\mathbf{n}\rangle |\downarrow_\mathbf{n}\rangle - |\downarrow_\mathbf{n}\rangle |\uparrow_\mathbf{n}\rangle \right), \tag{2.114}$$

*where $|\uparrow_\mathbf{n}\rangle$ and $|\downarrow_\mathbf{n}\rangle$ are the eigenvalues of the spin matrix $S_\mathbf{n}$. In other words, for any $2 \times 2$ unitary matrix $U$ we have $U \otimes U |\Psi_-\rangle \langle \Psi_-| U^* \otimes U^* = |\Psi_-\rangle \langle \Psi_-|$.*

**Exercise 2.49.** *Consider the Hilbert space of two electrons $AB \cong \mathbb{C}^2 \otimes \mathbb{C}^2$. For any unit vector $\mathbf{n} \in \mathbb{R}^3$ denoted by $J_\mathbf{n} := S_\mathbf{n} \otimes I + I \otimes S_\mathbf{n}$, and by $J^2 := J_\mathbf{x}^2 + J_\mathbf{y}^2 + J_\mathbf{z}^2$:*

*1. Show that the eigenvalues of $J_\mathbf{n}$ are 1, 0, and $-1$.*
*2. Find the eigenvalues of $J^2$.*
*3. Show that $[J^2, J_z] = 0$.*
*4. Show that each of the following four 2-qubit states are eigenvectors of both $J^2$ and $J_z$:*

$$|00\rangle, \quad |11\rangle, \quad and \quad |\Psi_\pm\rangle := \frac{1}{\sqrt{2}} \left( |01\rangle \pm |10\rangle \right). \tag{2.115}$$

**Exercise 2.50.** *Let $S_{\mathbf{n}} = \frac{1}{2}\mathbf{n} \cdot \boldsymbol{\sigma}$ and $S_{\mathbf{m}} = \frac{1}{2}\mathbf{m} \cdot \boldsymbol{\sigma}$ be the spin matrices (observables) in the directions of the unit vectors $\mathbf{n}$ and $\mathbf{m}$, respectively.*

1. *Show that the commutator $[S_{\mathbf{n}}, S_{\mathbf{m}}] = i\, S_{\mathbf{r}}$, where $\mathbf{r} \in \mathbb{R}^3$ is a unit vector. What is the direction of $\mathbf{r}$?*
2. *Calculate*

$$\langle \Psi_-|S_{\mathbf{n}} \otimes S_{\mathbf{m}}|\Psi_- \rangle, \quad where \quad |\Psi_-\rangle = \frac{1}{\sqrt{2}}\left(|0\rangle \otimes |1\rangle - |1\rangle \otimes |0\rangle\right). \tag{2.116}$$

3. *Let $\mathbf{n}'$ and $\mathbf{m}'$ be two more unit vectors, and let*

$$B := S_{\mathbf{n}} \otimes S_{\mathbf{m}} + S_{\mathbf{n}} \otimes S_{\mathbf{m}'} + S_{\mathbf{n}'} \otimes S_{\mathbf{m}} - S_{\mathbf{n}'} \otimes S_{\mathbf{m}'}. \tag{2.117}$$

*Show that*

$$B^2 = \frac{1}{4}I - [S_{\mathbf{n}}, S_{\mathbf{n}'}] \otimes [S_{\mathbf{m}}, S_{\mathbf{m}'}], \tag{2.118}$$

*and use it to prove that*

$$|\langle \psi|B|\psi \rangle| \leqslant \frac{1}{\sqrt{2}}, \tag{2.119}$$

*for any state in $|\psi\rangle \in \mathbb{C}^2 \otimes \mathbb{C}^2$.*

## 2.5 Quantum Measurements

Since quantum mechanics aim to study the behavior of subatomic particles, the process of measurement is essential to the theory and requires a rigorous treatment. The SG experiment demonstrates that physical systems are not separated from the apparatuses that are measuring them. From a philosophical standpoint, "the observer is not separated from that which is being observed." The effect of observation on physical systems is not unique to quantum mechanics. The following story from Jostein Gaarder's novel, "Sophie's World," shows, among many other things, that even the behavior of a centipede is effected by measurements and observations.

> Once upon a time there was a centipede that was amazingly good at dancing with all hundred legs. All the creatures of the forest gathered to watch every time the centipede danced, and they were all duly impressed by the exquisite dance. But there was one creature that didn't like watching the centipede dance - that was a tortoise.
>
> How can I get the centipede to stop dancing? thought the tortoise. He couldn't just say he didn't like the dance. Neither could he say he danced better himself, that would obviously be untrue. So he devised a fiendish plan.
>
> He sat down and wrote a letter to the centipede. "O incomparable centipede," he wrote, "I am a devoted admirer of your exquisite dancing. I must know how you go about it when you dance. Is it that you lift your left leg number 28 and then your right leg number 39? Or do you begin by lifting your right leg number 17 before you lift your left leg number 44? I await your answer in breathless anticipation. Yours truly, Tortoise."

You can easily guess now what happened next!

## 2.5.1 Born's Rule and von-Neumann Projective Measurements

Consider the SG experiment, which involves measuring the spin of an electron along an arbitrary direction denoted as $\mathbf{n}$. It is well-established that the act of making this measurement can impact the state of the electron. This change occurs when the electron's initial spin is not aligned with either the positive or negative directions of $\mathbf{n}$.

Consider the initial quantum state $|\psi\rangle = a|0\rangle + b|1\rangle$, where $a$ and $b$ are complex numbers, subject to the normalization condition $|a|^2 + |b|^2 = 1$. Additionally, let $|\uparrow_{\mathbf{n}}\rangle$ and $|\downarrow_{\mathbf{n}}\rangle$ denote the eigenvectors of the spin operator $S_{\mathbf{n}}$, corresponding to the electron's spin aligned with the upward and downward directions along $\mathbf{n}$, respectively. Under an SG experiment in the direction $\mathbf{n}$, the transformation of the state $|\psi\rangle$ follows this simple rule:

1. If the SG experiment yields an outcome in the upward direction along $\mathbf{n}$, the state evolves to $|\uparrow_{\mathbf{n}}\rangle$.
2. If the SG experiment yields an outcome in the downward direction along $\mathbf{n}$, the state evolves to $|\downarrow_{\mathbf{n}}\rangle$.

It's crucial to emphasize that this transformation is independent of the specific form of $|\psi\rangle$. What varies with $|\psi\rangle$ is the probability associated with each possible outcome. For instance, consider the case where $\mathbf{n} = \mathbf{z}$ and $|\psi\rangle$ is initially prepared as $|\uparrow_{x}\rangle$. In this case, both the upward and downward outcomes are equally probable, each with a 50% chance.

The general rule governing the probability of obtaining a particular outcome in the measurement is known as Born's rule. According to Born's rule, the probability, denoted as $\Pr(\psi, \mathbf{n})$, of observing the outcome $\uparrow_{\mathbf{n}}$ (i.e. the electron's spin aligned with the positive direction of $\mathbf{n}$) in an SG experiment along the $\mathbf{n}$ direction, when the electron is initially prepared in the state $|\psi\rangle$, is given by

$$\Pr(\psi, \mathbf{n}) = |\langle\psi|\uparrow_{\mathbf{n}}\rangle|^2. \tag{2.120}$$

This fundamental principle provides a mathematical framework for determining the likelihood of various outcomes in quantum measurements, and it plays a central role in quantum mechanics. For example, suppose an electron in the state $|\psi\rangle = a|0\rangle + b|1\rangle$ is sent through an SG experiment in the $z$-direction. Then, using the Born's rule (2.120) we get that $|\langle\psi|\uparrow_{z}\rangle|^2 = |a|^2$ is the probability to obtain spin up (in the $z$-direction), and $|\langle\psi|\downarrow_{\mathbf{n}}\rangle|^2 = |b|^2$ is the probability to obtain spin down.

Similarly, we can extend the Born's rule for any qudit $|\psi\rangle \in A \cong \mathbb{C}^m$, and any quantum measurement that corresponds to an orthonormal basis $\{|\phi_x\rangle\}_{x\in[m]}$ of $A$. The probability to obtain an outcome $x$ is given by

$$\Pr(\psi, \phi_x) = |\langle\psi|\phi_x\rangle|^2. \tag{2.121}$$

Note that the above assignment of probability to each $\phi_x$ is indeed a probability; that is,

$$\sum_{x\in[m]} \Pr(\psi,\phi_x) = \sum_{x\in[m]} |\langle\psi|\phi_x\rangle|^2 = \sum_{x\in[m]} \langle\psi|\phi_x\rangle\langle\phi_x|\psi\rangle$$
$$= \langle\psi|\Big(\sum_{x\in[m]} |\phi_x\rangle\langle\phi_x|\Big)|\psi\rangle = \langle\psi|\psi\rangle = 1. \tag{2.122}$$

We call every such a measurement that corresponds to an orthonormal basis a *basis measurement*.

To establish a connection between a basis measurement and a physical observable, let's consider the energy of a physical system. Energy is a fundamental observable in quantum mechanics and therefore can be measured. As previously discussed, any observable in quantum mechanics is represented by a Hermitian operator acting on the Hilbert space $A$. We denote the energy operator, often referred to as the Hamiltonian, as

$$H = \sum_{x\in[m]} a_x|\phi_x\rangle\langle\phi_x|, \tag{2.123}$$

where $\{|\phi_x\rangle\}_{x\in[m]}$ is an orthonormal basis of $A$. Therefore, in order to measure the energy, one has to perform a basis measurement corresponding to the orthonormal basis $\{|\phi_x\rangle\}_{x\in[m]}$, since the energy $a_x$ is determined by the value of $x$. However, this system of $m$ energy levels can be degenerated, as it happens quite often in many physical systems. In this case, not all of the energy values $\{a_x\}_{x\in[m]}$ are distinct. Suppose for example that $a_1 = a_2 < a_3 < \cdots < a_m$; that is, the state with minimum energy (the ground state) is degenerate. In this case, both outcomes 1 and 2 correspond to the same ground state, so that the probability that the energy is equal $a_1 = a_2 := b_1$ is given by

$$\Pr(\psi,\phi_1) + \Pr(\psi,\phi_2) = \langle\psi|\big(|\phi_1\rangle\langle\phi_1| + |\phi_2\rangle\langle\phi_2|\big)|\psi\rangle := \langle\psi|\Pi|\psi\rangle, \tag{2.124}$$

where $\Pi := |\phi_1\rangle\langle\phi_1| + |\phi_2\rangle\langle\phi_2|$. More generally, if we have degeneracy in other energy levels, we can always express the observable $H$ as

$$H = \sum_{y\in[r]} b_y\Pi_y, \tag{2.125}$$

where $b_1 < b_2 < \cdots < b_r$, and each $\Pi_y$ is a sum of rank 1 projections from $\{|\phi_x\rangle\langle\phi_x|\}$ that corresponds to the same energy level $b_y$. With this at hand, the probability to measure an energy of value $b_y$ is given by

$$\Pr\big(\psi,\Pi_y\big) = \langle\psi|\Pi_y|\psi\rangle. \tag{2.126}$$

Therefore, the basis measurement that we considered so far can be extended to *projective von-Neumann measurement*, which is defined as follows.

> **Definition 2.6.** A von-Neumann projective measurement (or, in short, projective measurement) on a Hilbert space $A$, is a collection of mutually orthogonal projections $\{\Pi_x\}_{x\in[r]}$ satisfying $\sum_{x\in[r]}\Pi_x = I^A$ and for all $x,y \in [r]$
>
> $$\Pi_x\Pi_y = \delta_{xy}\Pi_x. \tag{2.127}$$

Historically, the Born's rule (see (2.126)) was determined essentially from consistency with experiments. That is, one can perform many experiments, like the SG experiment for example, collect the data, and find a rule that is consistent with the data. Later on, however, Gleason came up with a theorem showing how to calculate probabilities in quantum mechanics, and loosely speaking *derived* the Born's rule from a few fundamental principles involving measures of a Hilbert space. Gleason's theorem is applicable for general (separable) Hilbert spaces in any dimension, but for us only the finite dimensional case, that is, the qudit, will be relevant. We postpone the discussion on Gleason's theorem for the next chapter, after we discuss other types of quantum measurements, in order to prove a slightly more generalized version of Gleason's theorem, which will be applicable to all types of measurements (not only to projective von-Neumann measurements).

**Exercise 2.51.** *Let $\Pi$ be a projection on a Hilbert space A. Show that $\{\Pi, I - \Pi\}$ is a two-outcome von-Neumann projective measurement.*

**Exercise 2.52.** *Let $\{\Pi_x\}_{x\in[r]}$ be a projective von-Neumann measurement on a finite dimensional Hilbert space A. Show that the collection of all the linearly independent normalized eigenvectors, of all the projections $\{\Pi_x\}_{x\in[r]}$, form an orthonormal basis of A.*

## 2.5.2 The Post-measurement State

Recall that the basis measurements we discussed earlier lead to a change in the system's state according to the rule $|\psi\rangle \rightarrow |\phi_x\rangle$ if outcome $x$ occurs. However, when it comes to projective measurements, it's possible to encounter a scenario where a degenerate energy value, such as $b_1 = a_1 = a_2$, occurs, as illustrated in (2.124). In this case, after the measurement yields outcome $b_1$, all we can ascertain is that the system's state belongs to the subspace $B$, defined as $B := \text{span}\{|\phi_1\rangle, |\phi_2\rangle\}$, since any state within this subspace is an eigenvector of the Hamiltonian $H$ associated with the same energy eigenvalue $b_1$. However, it raises the question of which specific state within this subspace will become the post-measurement state.

Notably, $B$ is a subspace of $A$, and $\Pi_1 = |\phi_1\rangle\langle\phi_1| + |\phi_2\rangle\langle\phi_2|$ projects states from $A$ to $B$. Furthermore, if the pre-measurement state $|\psi\rangle \in B$, then $|\psi\rangle$ is already an eigenvector of $H$ and should remain unaffected by the measurement of $H$. Consequently, if we denote by $\Lambda$ the transformation that converts the pre-measurement state $|\psi\rangle$ into the post-measurement state (which may not be normalized), it follows that $\Lambda|\psi\rangle = |\psi\rangle$ for $|\psi\rangle \in B$.

On the other hand, if $|\psi\rangle \in B^\perp$ (the orthogonal complement of $B$ in $A$), the energy $b_1$ has zero probability of occurring, and therefore, we assume that its corresponding post-measurement state is $\Lambda|\psi\rangle = 0$. In this context, $\Lambda$ must be equivalent to $\Pi_1$. However, unless $|\psi\rangle$ is within $B$, the state $\Pi_1|\psi\rangle$ is not normalized. Thus, the rule stipulates that the pre-measurement state $|\psi\rangle$ undergoes transformation to the normalized post-measurement state $\Pi_1\|\psi\rangle/|\Pi_1|\psi\rangle\|$.

To summarize, for any physical system that is prepared in a state $|\psi\rangle \in A$, and for any von-Neumann projective measurement, $\{\Pi_x\}_{x\in[r]}$, of the observable $H = \sum_{x\in[r]} a_x \Pi_x$, the rules of quantum mechanics state that the probability to obtain a value $a_x$ is given by the Born's rule

$$\Pr\left(\psi, \Pi_x\right) = \langle\psi|\Pi_x|\psi\rangle, \tag{2.128}$$

and the quantum state of the system after the outcome $x$ occurred is

$$\frac{1}{\||P_x|\psi\rangle\|_2} P_x |\psi\rangle. \tag{2.129}$$

**Exercise 2.53.** *Consider the space $A \cong \mathbb{C}^3$.*

1. *Find the projection $\Pi_0$ to the two-dimensional subspace $B := \mathrm{span}\{|0\rangle + |1\rangle, |0\rangle + |2\rangle\}$.*
2. *Use the projection $\Pi_0$ that you found in Part 1 to construct a two-outcome projective measurement $\{\Pi_0, \Pi_1\}$, with $\Pi_1 = I - \Pi_0$. If a physical system was prepared initially in the state $|\psi\rangle = \frac{1}{\sqrt{3}}\left(|0\rangle + |1\rangle + |2\rangle\right)$, what is the probability that this projective measurement yields an outcome 0 (i.e. corresponds to $\Pi_0$)? What will be the post-measurement state in this case?*

**Exercise 2.54.** *Let $A$ be a $d$-dimensional Hilbert space, and let $|\psi\rangle, |\phi\rangle \in A$ be two quantum states.*

1. *Show that if $|\psi\rangle$ and $|\phi\rangle$ are orthogonal, then there exists a projective measurement that distinguishes them. That is, there exists a two-outcome projective measurement $\{\Pi_0, \Pi_1\}$ such that*

$$\Pr(\psi, \Pi_0) = 1 \quad and \quad \Pr(\phi, \Pi_1) = 1. \tag{2.130}$$

2. *Show that if $|\psi\rangle$ and $|\phi\rangle$ are not orthogonal, then there is no projective measurement that distinguishes them.*

## 2.6 Hidden Variable Models

The axioms of quantum mechanics that we considered so far have profound consequences. For example, suppose an electron has been prepared with a spin in the positive $z$-direction. The rules of quantum mechanics tells us that if we where to measure its spin in other directions (including $x$- or $y$-directions), there is a nonzero probability to find out the spin pointing in those directions. This raises the question of whether it is possible to model the quantum system (i.e. the electron in this case) with a (possibly uncountable) collection of classical random variables, sometimes called *hidden* variables, each contains information about the spin of the electron in some direction. Remarkably, such attempts to model physical systems with local random variables instead of quantum states lead to inconsistencies with experiments. That is, the axioms of quantum mechanics are inconsistent with local hidden variable models of reality.

## 2.6.1 The CHSH Inequality and Local Realism

Let 0 and 1 represent positive and negative directions of a spin of an electron, respectively, and let $\mathbf{n}$ be a unit vector in $\mathbb{R}^3$. Denote by $A_\mathbf{n} \in \{0, 1\}$ the value of the spin in the $\mathbf{n}$ direction. That is, we replace the Hilbert space $A$ of the electron, with a collection of classical variables $\{A_\mathbf{n}\}_\mathbf{n}$, where $\mathbf{n}$ is running over all unit vectors in $\mathbb{R}^3$. Suppose the electron is prepared at the positive $\mathbf{z}$-direction. This means that $A_\mathbf{z} = 0$, while from the SG experiment we saw that $A_\mathbf{x} = 0$ with probability 50% and $A_\mathbf{x} = 1$ also with 50% probability. That is, $A_\mathbf{n}$ is a random bit, and the probability that $A_\mathbf{n} = a$ (with $a \in \{0, 1\}$) is denoted by $p(a|\mathbf{n})$. The conditional probabilities $p(a|\mathbf{n})$ provide all the information about the spin of an electron that was prepared initially in the $\mathbf{z}$-direction.

Aside from being somewhat artificial (i.e. the model does not provide a mechanism to derive $p(a|\mathbf{n})$ from a set of axioms), a priori, it seems to provide a valid description of the information about the spin. One can view $A_\mathbf{n}$ as a *hidden variable* that the observer does not know (unless $\mathbf{n} = \pm\mathbf{z}$) until she/he performs an SG experiment (or any other experiment for that matter). Note that after every measurement the observer will need to update the conditional probability $p(a|\mathbf{n})$.

For any such a *hidden variable model*, there is an inherent assumption that the values of the hidden variables are fixed, predetermined, and corresponds to an *element of reality*. It is just the observer's lack of knowledge about this element of reality that leads to statistical behaviors. Historically, hidden variable theories were promoted by some physicists who argued that the formulation of quantum mechanics (as we will discuss in the rest of this book) does not provide a complete description for the system. Along with Albert Einstein, they argued that quantum mechanics is ultimately incomplete, and that a complete theory would avoid any indeterminism. Indeed, hidden variable models as the one described in this section for the spin of one electron cannot be ruled out, although, as we discuss now, *local* hidden variable models can!

Consider two friends, Alice and Bob, who are located far from each other, and each one possess an electron in their lab. How can we describe the spins of the two electrons? Following the same line of thoughts as above, we denote by $A_\mathbf{n}$ the random variable associated with the spin of Alice's electron in the $\mathbf{n}$-direction, and by $B_\mathbf{m}$ the random variable associated with the spin of Bob's electron in the $\mathbf{m}$-direction. We denote by $p(ab|\mathbf{nm})$, with $a, b \in \{0, 1\}$, the joint probability that the two SG experiments in Alice's lab and Bob's lab will yield, respectively, $A_\mathbf{n} = a$ and $B_\mathbf{m} = b$.

Since it is possible that the spins are correlated in some way, we are *not* assuming that $p(ab|\mathbf{nm})$ has the form $p^A(a|\mathbf{n})p^B(b|\mathbf{m})$, where $p^A(a|\mathbf{n})$ is the probability that Alice will get the value $a$ in an SG experiment in the $\mathbf{n}$-direction (and $p^B(b|\mathbf{m})$ is defined similarly). Instead, since in general $A_\mathbf{n}$ and $B_\mathbf{m}$ can be correlated, there exists a parameter $\lambda$ ($\lambda$ can describe a collection of variables) and a probability distribution $q_\lambda$ over it, such that

$$p(ab|\mathbf{nm}) = \int d\lambda \; q_\lambda \; p_\lambda^A(a|\mathbf{n}) \; p_\lambda^B(b|\mathbf{m}), \tag{2.131}$$

where $p_\lambda^A(a|\mathbf{n})$ and $p_\lambda^B(b|\mathbf{m})$ are probability distributions that depend on the correlating parameter $\lambda$. The parameter $\lambda$ can be either continuous or discrete, and for the latter the integral in (2.131) is replaced with a sum. Note that the distribution is more general than the form $p^A(a|\mathbf{n})p^B(b|\mathbf{m})$ as it allows for correlations between Alice's and Bob's spins. Yet, it is a *local* probability distribution depending only on the local variables $A_\mathbf{n}$ and $B_\mathbf{m}$. We now discuss a crucial consequence of this local hidden variable model for the spin of two electrons.

**Exercise 2.55.** *Let* $\mathbf{n}$, $\mathbf{n}'$, $\mathbf{m}$, *and* $\mathbf{m}'$, *be four unit vectors in* $\mathbb{R}^3$, *and use the tilde symbol over a random variable* $X$ *to mean* $\tilde{X} = 2X - 1$ *(i.e.* $\tilde{X}$ *takes values* $\pm 1$ *while* $X$ *takes values* $0, 1$*). Show that*

$$\left| \tilde{A}_\mathbf{n} \tilde{B}_\mathbf{m} + \tilde{A}_{\mathbf{n}'} \tilde{B}_\mathbf{m} + \tilde{A}_\mathbf{n} \tilde{B}_{\mathbf{m}'} - \tilde{A}_{\mathbf{n}'} \tilde{B}_{\mathbf{m}'} \right| \leqslant 2. \tag{2.132}$$

**Exercise 2.56.** *Denote by*

$$\langle AB \rangle_\mathbf{nm} := \langle A_\mathbf{n} B_\mathbf{m} \rangle := \sum_{a,b \in \{0,1\}} ab\, p(ab|\mathbf{nm}). \tag{2.133}$$

*Show that any local probability distribution* $p(ab|\mathbf{nm})$ *as in* (2.131) *satisfies*

$$\left| \langle \tilde{A}\tilde{B} \rangle_\mathbf{nm} + \langle \tilde{A}\tilde{B} \rangle_{\mathbf{nm}'} + \langle \tilde{A}\tilde{B} \rangle_{\mathbf{n}'\mathbf{m}} - \langle \tilde{A}\tilde{B} \rangle_{\mathbf{n}'\mathbf{m}'} \right| \leqslant 2. \tag{2.134}$$

The inequality in the Exercise 2.56 is called the CHSH inequality after Clauser, Horne, Shimony, and Holt, and it generalizes a similar inequality that was proved in a seminal paper by John Bell from 1964. As we will see in the following exercise, not all probability distributions $p(ab|\mathbf{nm})$ satisfy this inequality. One obvious property of the local distribution (2.131) is that if we sum over $a$, the dependance on $\mathbf{n}$ disappears, and similarly if we sum over $b$, the dependance in $\mathbf{m}$ disappears. This property is called "no-signaling," because by choosing different directions of $\mathbf{n}$, Alice cannot signal Bob, since the marginal distribution on his side remains intact. The no-signaling property can be stated as follows:

$$\sum_a p(ab|\mathbf{nm}) = \sum_a p(ab|\mathbf{n}'\mathbf{m}) := p^B(b|\mathbf{m}) \quad \forall\, b, \mathbf{n}, \mathbf{n}', \mathbf{m},$$

$$\sum_b p(ab|\mathbf{nm}) = \sum_b p(ab|\mathbf{nm}') := p^A(a|\mathbf{n}) \quad \forall\, a, \mathbf{n}, \mathbf{m}, \mathbf{m}'. \tag{2.135}$$

The following exercise shows that there exists a probability distribution that on the one hand is nonsignaling, and on the other hand is violating the CHSH inequality (2.134).

**Exercise 2.57.** *Denote the two directions in Alice's side by* $\mathbf{n}_0 := \mathbf{n}$ *and* $\mathbf{n}_1 := \mathbf{n}'$, *and the two direction vectors in Bob's side by* $\mathbf{m}_0 := \mathbf{m}$ *and* $\mathbf{m}_1 := \mathbf{m}'$. *Denote also by* $p(ab|xy) = p(ab|\mathbf{n}_x\mathbf{m}_y)$ *with* $x, y \in \{0, 1\}$. *Consider the probability distribution given by*

$$p(ab|xy) = \begin{cases} \frac{1}{2} & if\ \ a \oplus b = xy \\ 0 & otherwise \end{cases}, \tag{2.136}$$

*where the* $\oplus$ *denotes addition modulo 2.*

1. *Show that $p(ab|xy)$ is nonsignaling; that is, satisfies (2.135).*
2. *Show that $p(ab|xy)$ is nonlocal by showing that it violates the CHSH inequality (2.134).*
3. *Show that no other probability distribution (even a signaling distribution) can provide a higher violation than the one achieved by the distribution (2.136).*

To summarize, any local hidden variable model has two main assumptions. The first one is called the *realism* assumption, corresponding to our assumption that the spins of the electrons in all directions have definite values which exist independently of observation. The second assumption is called the *locality* assumption corresponding to our implicit assumption that if, say, Alice is performing a measurement on her electron, it does not influence the result of Bob's measurement (on the spin of the electron in his lab). The following violation of the CHSH inequality demonstrates that *local realism* does not hold!

## 2.6.2 Quantum Violation of the CHSH Inequality

The violation of the Bell and CHSH inequalities is one of the most profound results of the twentieth century. It states that the formalism of quantum mechanics allows for a violation of the inequality in (2.134). This means that a local hidden variable model cannot account for quantum correlations. To see the violation, consider two electrons, one located in Alice's lab and the other in Bob's lab, that are prepared in some state $|\psi^{AB}\rangle \in AB \cong \mathbb{C}^2 \otimes \mathbb{C}^2$. Both Alice and Bob perform an SG experiment with Alice in the direction **n** and Bob in the direction **m** (see Figure 2.6). As in the previous section, denote by $p(ab|\mathbf{nm})$ the probability that Alice obtains an outcome $a$ (with $a = 0$ for positive **n**-direction and $a = 1$ for negative **n**-direction) and Bob obtains an outcome $b$ (again with the same correspondence of positive and negative directions for $b = 0$ and $b = 1$, respectively). According to Born's rule, this probability is given by

$$p_\psi(ab|\mathbf{nm}) = \left|\langle\psi^{AB}|\phi_a^A \otimes \varphi_b^B\rangle\right|^2, \tag{2.137}$$

where $|\phi_a^A\rangle \in \mathbb{C}^2$ (with $a = 0, 1$) are the eigenvectors of the spin matrix $S_\mathbf{n}$, and $|\varphi_b^B\rangle \in \mathbb{C}^2$ (with $b = 0, 1$) are the eigenvectors of the spin matrix $S_\mathbf{m}$. The corresponding eigenvalues are given by $\frac{1}{2} - a$, for $|\phi_a^A\rangle$, and by $\frac{1}{2} - b$, for $|\varphi_b^B\rangle$. Note the relation with the previous notations; for example, $|\phi_0^A\rangle = |\uparrow_\mathbf{n}\rangle$ and $|\phi_1^A\rangle = |\downarrow_\mathbf{n}\rangle$.

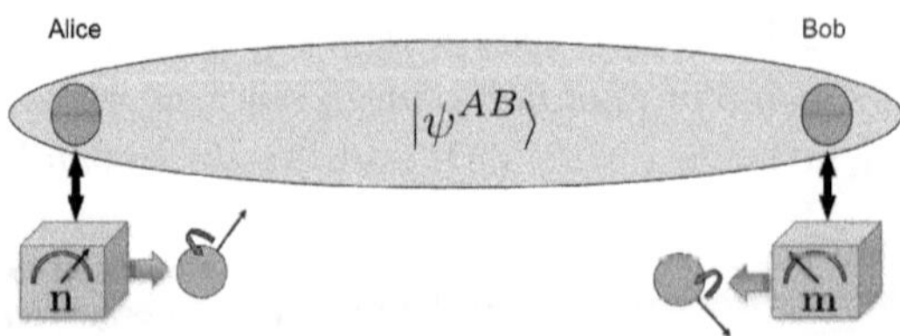


    Measurements of the spins of two electrons.

**Exercise 2.58.** *Show that the product spin average is given by*

$$\sum_{a,b\in\{0,1\}} \left(\frac{1}{2} - a\right)\left(\frac{1}{2} - b\right) p_\psi(ab|\mathbf{nm}) = \langle\psi^{AB}|S_\mathbf{n} \otimes S_\mathbf{m}|\psi^{AB}\rangle. \tag{2.138}$$

Recall that $\tilde{A}_\mathbf{n}$ and $\tilde{B}_\mathbf{m}$ in (2.134) are random variables taking the values $\pm 1$, whereas the eigenvalues of $S_\mathbf{n}$ and $S_\mathbf{m}$ are $\pm\frac{1}{2}$. Keeping this in mind, from the exercise above we conclude that the probability distribution $p_\psi(ab|\mathbf{nm})$ as given in (2.137) violates the CHSH inequality (2.134) if

$$|\langle\psi|B|\psi\rangle| > \frac{1}{2}, \tag{2.139}$$

where $B$ is the Bell/CHSH operator from Exercise 2.50. From Exercise 2.50 it follows that for any state $|\psi\rangle \in \mathbb{C}^2 \otimes \mathbb{C}^2$, we have $|\langle\psi|B|\psi\rangle| \leq 1/\sqrt{2}$. From the next exercise it follows that there exists directions $\mathbf{n}, \mathbf{m}, \mathbf{n}'$ and $\mathbf{m}'$ such that this bound is saturated, thereby violating the CHSH inequality since $\frac{1}{\sqrt{2}} > \frac{1}{2}$. This bound is called the Tsirelson bound.

**Exercise 2.59.** *Prove that the Tsirelson bound can be achieved by taking $|\psi^{AB}\rangle$ to be the singlet state $|\Psi_-^{AB}\rangle = (|01\rangle - |10\rangle)/\sqrt{2}$. That is, find four directions $\mathbf{n}, \mathbf{n}', \mathbf{m}$, and $\mathbf{m}'$ such that*

$$|\langle\Psi_-|B|\Psi_-\rangle| = \frac{1}{\sqrt{2}}. \tag{2.140}$$

Note that the violation of the CHSH inequality implies that the quantum probability distribution $p_\psi(ab|\mathbf{nm})$ is in general not local; that is, not of the form (2.131). Such nonlocal probability distributions have other nonintuitive consequences, as we discuss in the following subsections.

## 2.6.3 John Preskill's Example: Quantum Coins

Consider two electrons, one in Alice's lab and the other in Bob's lab, prepared in the singlet state $|\Psi_-^{AB}\rangle = (|01\rangle - |10\rangle)/\sqrt{2}$. Suppose Alice wants to measure the spin of her electron in two directions $\mathbf{n}_1$ and $\mathbf{n}_2$. She knows that if she performs a measurement in the $\mathbf{n}_1$-direction, that will affect the state of her electron, and she will not be able to determine what would have happened if she did the $\mathbf{n}_2$-measurement instead. Therefore, she asks Bob to perform the $\mathbf{n}_2$-measurement on his electron, while she performs the $\mathbf{n}_1$-measurement on her electron. From Exercise 2.48, the singlet state can be written as $(|\uparrow_{\mathbf{n}_2}\rangle|\downarrow_{\mathbf{n}_2}\rangle - |\downarrow_{\mathbf{n}_2}\rangle|\uparrow_{\mathbf{n}_2}\rangle)/\sqrt{2}$. This means that if Bob's measurement output is $\uparrow_{\mathbf{n}_2}$, then Alice would have measured $\downarrow_{\mathbf{n}_2}$ had she chose to do the $\mathbf{n}_2$ measurement instead. Similarly, if Bob's measurement output is $\downarrow_{\mathbf{n}_2}$, then Alice would have measured $\uparrow_{\mathbf{n}_2}$. This gives a way for Alice to determine what would be the output of the measurement if she chose to perform the $\mathbf{n}_2$-measurement, and thereby know simultaneously the values of the spins in directions $\mathbf{n}_1$ and $\mathbf{n}_2$ of her electron. This idea of determining the outputs of several possibly *counterfactual* measurements by the use of entangled

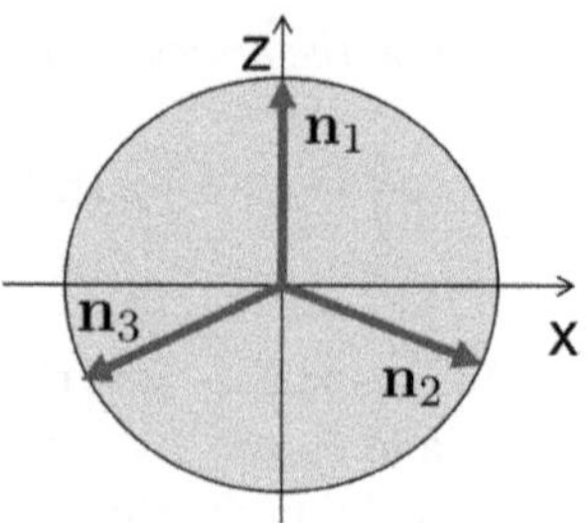


**Figure 2.7**  Three directions with an angle of 120° between any two.

states (like the singlet) is the key technique used in many experiments, including the famous *delayed choice quantum eraser experiment*.

**Exercise 2.60.** *Using the method described above, show that the probability $p_{same}(\mathbf{n}_1, \mathbf{n}_2)$, that Alice will obtain both spins of her (single) electron pointing in the same direction (i.e. both positive $\uparrow_{\mathbf{n}_1}^A \uparrow_{\mathbf{n}_2}^A$ or both negative $\downarrow_{\mathbf{n}_1}^A \downarrow_{\mathbf{n}_2}^A$), is*

$$p_{same}(\mathbf{n}_1, \mathbf{n}_2) := \left| \langle \Psi_-^{AB} | \uparrow_{\mathbf{n}_1}^A \downarrow_{\mathbf{n}_2}^B \rangle \right|^2 + \left| \langle \Psi_-^{AB} | \downarrow_{\mathbf{n}_1}^A \uparrow_{\mathbf{n}_2}^B \rangle \right|^2 = \frac{1}{2}(1 + \cos(\theta)), \qquad (2.141)$$

*where $\theta$ is the angle between the unit vectors $\mathbf{n}_1$ and $\mathbf{n}_2$.*

Consider now three unit vectors $\mathbf{n}_1, \mathbf{n}_2, \mathbf{n}_3 \in \mathbb{R}^3$ with an angle of 120° between any two; see Figure 2.7. From the exercise above we get that $p_{same}(\mathbf{n}_1, \mathbf{n}_2) = p_{same}(\mathbf{n}_1, \mathbf{n}_3) = p_{same}(\mathbf{n}_2, \mathbf{n}_3) = \frac{1}{4}$ since $\cos(120) = -1/2$. Therefore,

$$p_{same}(\mathbf{n}_1, \mathbf{n}_2) + p_{same}(\mathbf{n}_1, \mathbf{n}_3) + p_{same}(\mathbf{n}_2, \mathbf{n}_3) = \frac{3}{4} < 1. \qquad (2.142)$$

On the other hand, suppose it was possible to describe Alice's electron $\mathbf{n}_1$-spin, $\mathbf{n}_2$-spin, and $\mathbf{n}_3$-spin, with three random variables $X_1$, $X_2$, and $X_3$ (with some underlying probability distribution over the three variables). Each of the three random variables can take the values $\pm\frac{1}{2}$ determining if the spin is pointing in the positive or negative direction. Then, irrespective of the underlying probability distribution, the probabilities $\Pr(X_j = X_k)$ (with $j \neq k$ and $j, k \in \{1, 2, 3\}$) must satisfy

$$\Pr(X_1 = X_2) + \Pr(X_1 = X_3) + \Pr(X_2 = X_3) \geqslant 1. \qquad (2.143)$$

This problem is analogous to the problem of flipping three coins and asking what is the probability that at least two of them are the same (either two heads or two tails). Clearly, flipping three coins will always yield two that show the same symbol (either head or tail). Equation (2.142) shows that this is not the case for quantum coins (i.e. spins of an electron).

So far we have seen a contradiction between quantum mechanics and local realism through the violation of the CHSH inequality (2.134), and the inequality in (2.142). The next two paradoxes show that this inconsistency between quantum mechanics and local realism can be expressed without inequalities. In the literature they are referred to as "Bell nonlocality without inequalities."

## 2.6.4 Hardy's Paradox

Using the same notations as in Exercise (2.57), we consider as before two electrons, one on Alice's side and the other on Bob's side. We denote two directions in Alice's side by $\mathbf{n}_0$ and $\mathbf{n}_1$, and the two directions in Bob's side by $\mathbf{m}_0$ and $\mathbf{m}_1$. Furthermore, we denote by $p(ab|xy) = p(ab|\mathbf{n}_x\mathbf{m}_y)$ (with $x, y \in \{0, 1\}$) the probability that two SG experiments, one on Alice's side and the other on Bob's side, yield an outcome $a$ for Alice's spin measurement in the direction $\mathbf{n}_x$, and $b$ for Bob's spin measurement in the direction $\mathbf{m}_y$.

Recall that the probability distribution $p(ab|xy)$ is said to be local if there exists conditional probabilities $p_\lambda^A(a|x)$ and $p_\lambda^B(b|y)$ such that

$$p(ab|xy) = \int d\lambda \; q_\lambda \; p_\lambda^A(a|x) \; p_\lambda^B(b|y). \tag{2.144}$$

Suppose now that the probability distribution $p(ab|xy)$ satisfies

$$p(00|00) = p(01|10) = p(10|01) = 0. \tag{2.145}$$

**Exercise 2.61.** *Show that if $p(ab|xy)$ is local and satisfies (2.145), then $p(00|11) = 0$.*

We now show that the logical implication of Exercise (2.61) does not hold for quantum mechanics. Unlike the use of the singlet in the previous examples, here we consider a bipartite state $|\psi_\theta^{AB}\rangle$ that has the form

$$|\psi_\theta^{AB}\rangle = \frac{\tan(\theta)}{\sqrt{1 + 2\tan^2(\theta)}}(|01\rangle + |10\rangle) - \frac{1}{\sqrt{1 + 2\tan^2(\theta)}}|11\rangle, \tag{2.146}$$

with $\theta \in [0, 2\pi]$ being some angle. Note that this state is normalized for all $\theta$. Suppose that Alice and Bob perform the same measurements, and in particular $\mathbf{n}_0 = \mathbf{m}_0 = \mathbf{z}$ corresponds to a measurement in the computational basis, while $\mathbf{n}_1 = \mathbf{m}_1$ corresponds to a measurement in the orthonormal basis $|u_0\rangle := \cos(\theta)|0\rangle + \sin(\theta)|1\rangle$ and $|u_1\rangle := \sin(\theta)|0\rangle - \cos(\theta)|1\rangle$.

**Exercise 2.62.** *Verify that the above choices satisfy:*

$$p_\psi(00|00) = |\langle\psi^{AB}|0\rangle|0\rangle|^2 = 0$$

$$p_\psi(01|10) = |\langle\psi^{AB}|u_0\rangle|1\rangle|^2 = 0 \tag{2.147}$$

$$p_\psi(10|01) = |\langle\psi^{AB}|1\rangle|u_0\rangle|^2 = 0,$$

*while*

$$P_{Hardy}(\theta) := p_\psi(00|11) - |\langle\psi^{AB}|u_0\rangle|u_0\rangle|^2 = \frac{\sin^4(\theta)}{1 + 2\tan^2(\theta)}. \tag{2.148}$$

We therefore see that for this example $p_{\text{Hardy}}(\theta) > 0$ for all $0 < \theta < \frac{\pi}{2}$. Interestingly, $p_{\text{Hard}} > 0$ for all nonproduct states in $\{|\psi_\theta^{AB}\rangle\}_\theta$ except for the maximally entangled state $|\psi_{\theta=\pi/2}^{AB}\rangle = (|01\rangle + |10\rangle)/\sqrt{2}$ for which $p_{\text{Hard}}(\pi/2) = 0$. The maximum value of the function $p_{\text{Hard}}(\theta)$ can easily be computed to give

$$\max_{\theta \in [0,2\pi]} p_{\text{Hardy}}(\theta) = \frac{1}{2}(5\sqrt{5} - 11) \approx 0.09. \tag{2.149}$$

## 2.6.5 The GHZ Paradox

The Hardy paradox shows that the inconsistency of local realism with quantum mechanics can be demonstrated without inequalities as the CHSH inequality, but it is still probabilistic; that is, provide constraints on $p_\psi(ab|xy)$. Our final example of this inconsistency is due to Greenberger, Horne, and Zeilinger. Perhaps this is the example for which the contradiction between local realism and quantum mechanics is the sharpest.

Consider three electrons shared between Alice's, Bob's, and Charlie's labs, and prepared in the state

$$|\text{GHZ}\rangle := \frac{1}{\sqrt{2}} (|000\rangle + |111\rangle). \tag{2.150}$$

The state $|\text{GHZ}\rangle$ in written above in the $zzz$-basis; that is, $|0\rangle := |\uparrow_z\rangle$ and $|1\rangle := |\downarrow_z\rangle$ are the eigenvectors of $S_z$. We can also rewrite this vector in many other bases such as the $yyx$-basis or the $xxx$-basis.

**Exercise 2.63.** *Show that the GHZ state as defined in* (2.150) *can be expressed in the yyx-basis as*

$$|GHZ\rangle = \frac{1}{2}\left[\left(|\uparrow_y^A \uparrow_y^B\rangle + |\downarrow_y^A \downarrow_y^B\rangle\right) \otimes |\downarrow_x^C\rangle + \left(|\uparrow_y^A \downarrow_y^B\rangle + |\downarrow_y^A \uparrow_y^B\rangle\right) \otimes |\uparrow_x^C\rangle\right] \tag{2.151}$$

*and in the xxx-basis as*

$$|GHZ\rangle = \frac{1}{2}\left[\left(|\uparrow_x^A \uparrow_x^B\rangle + |\downarrow_x^A \downarrow_x^B\rangle\right) \otimes |\uparrow_x^C\rangle + \left(|\uparrow_x^A \downarrow_x^B\rangle + |\downarrow_x^A \uparrow_x^B\rangle\right) \otimes |\downarrow_x^C\rangle\right]. \tag{2.152}$$

Denote by $A_x$ (and similarly $A_y$) the random variables that take the value $+1$ if the spin of the first electron in the x-direction is positive, and take the value $-1$ if it is in the negative x-direction. The random variables $B_x$, $B_y$, $C_x$, and $C_y$ are defined similarly.

Now, according to (2.152), if Alice, Bob, and Charlie perform the $xxx$-measurement, the results of their measurements, given by $A_x$, $B_x$, and $C_x$, must satisfy

$$A_x B_x C_x = 1. \tag{2.153}$$

On the other hand, if they chose to do the $yyx$-measurement, according to (2.151) they would get

$$A_y B_y C_x = -1. \tag{2.154}$$

Moreover, since the GHZ state (2.150) is invariant under any permutation of the three subsystems, we conclude also that

$$A_y B_x C_y = -1 \quad \text{and} \quad A_x B_y C_y = -1. \tag{2.155}$$

With all this at hand, we get the following contradiction:

$$\begin{aligned}
-1 = (-1)(-1)(-1) &= (A_y B_y C_x)(A_y B_x C_y)(A_x B_y C_y) \\
&= (A_x B_x C_x) A_y^2 B_y^2 C_y^2 \\
&= A_x B_x C_x = 1,
\end{aligned} \tag{2.156}$$

where we used $A_x^2 = B_y^2 = C_y^2 = 1$ since these variable can only take the two values $\pm 1$. To summarize, according to quantum mechanics, an $xxx$-measurement can only yield one of the four possible outcomes:

$$|{\uparrow_x^A}{\uparrow_x^B}{\uparrow_x^C}\rangle, \quad |{\downarrow_x^A}{\downarrow_x^B}{\uparrow_x^C}\rangle, \quad |{\uparrow_x^A}{\downarrow_x^B}{\downarrow_x^C}\rangle, \quad |{\downarrow_x^A}{\uparrow_x^B}{\downarrow_x^C}\rangle. \tag{2.157}$$

On the other hand, local realism predicts that an $xxx$-measurement yields the four possible outcomes:

$$|{\uparrow_x^A}{\uparrow_x^B}{\downarrow_x^C}\rangle, \quad |{\downarrow_x^A}{\downarrow_x^B}{\downarrow_x^C}\rangle, \quad |{\uparrow_x^A}{\downarrow_x^B}{\uparrow_x^C}\rangle, \quad |{\downarrow_x^A}{\uparrow_x^B}{\uparrow_x^C}\rangle, \tag{2.158}$$

in maximal contrast with quantum mechanics. One may argue that we used quantum mechanics to express the GHZ state in the form (2.151), but this does not affect the conclusion that cannot coexist with the quantum mechanical formalism.

### 2.6.6 The CHSH Game

Consider the following game known as the CHSH game played by two players, Alice and Bob, along with a referee. The referee chooses at random (sampled from a uniform distribution) two bits $x$ and $y$, and sends $x$ to Alice and $y$ to Bob. After receiving the bits from the referee, Alice sends back to the referee the number $a$, and Bob sends back the number $b$ (see Figure 2.8).

The rule of the game is that Alice and Bob win the game if $a \oplus b = xy$, where $\oplus$ is addition modulus 2. The following table summarizes the desired value for $a \oplus b$ for each of the values of $x$ and $y$:

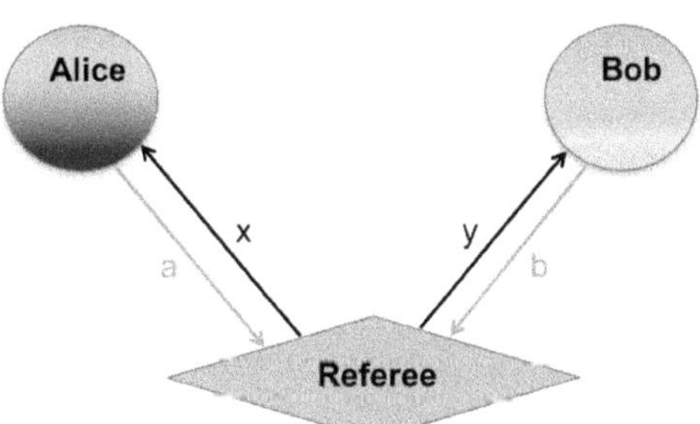


Figure 2.8    The CHSH game.

| $x$ | $y$ | $a \oplus b$ |
|---|---|---|
| 0 | 0 | 0 |
| 0 | 1 | 0 |
| 1 | 0 | 0 |
| 1 | 1 | 1 |

Clearly, from the table above it is obvious that if Alice and Bob always choose $a = b = 0$ (no matter what the values of $x$ and $y$), then they will win the game 3/4 of the times. Can they do better?

**Exercise 2.64.** *Show that Alice and Bob cannot win more than 3/4 of the times even if they use some randomness (i.e. they share some correlated random variable).*

Suppose now that Alice and Bob share quantum correlations; in particular, suppose they each posses an electron in their lab, and that the two electrons are prepared in the some bipartite state $|\psi^{AB}\rangle$. With this state at hand, they use the following strategy. Based on the bits $x$ and $y$ that they receive from the referee, they choose to perform spin-measurements in the direction $\mathbf{n}_x$ for Alice, and in the direction $\mathbf{m}_y$ for Bob. They then send to the referee the outcomes of their corresponding measurements. The probability that Alice and Bob win this CHSH game is given by

$$p_{\text{win}} := \frac{1}{4} \sum_{x,y,a,b} p_\psi(ab|xy)\, \delta_{xy,a\oplus b}, \tag{2.159}$$

where the factor $1/4$ represents the (uniform) probability that the referee sends $x$ to Alice and $y$ to Bob. From the following exercise it follows that for appropriate choices of $\mathbf{n}_x$, $\mathbf{m}_y$, and $\psi^{AB}$, Alice and Bob can win the game with a probability greater than $3/4$.

**Exercise 2.65.** *Consider the quantum strategy described above, and denote by $p_{lose} = 1 - p_{win}$ the probability that Alice and Bob lose the game. Recall the Bell operator $B$ as defined in Exercise 2.50 with $\mathbf{n} := \mathbf{n}_0$, $\mathbf{n}' := \mathbf{n}_1$, $\mathbf{m} := \mathbf{m}_0$, and $\mathbf{m}' := \mathbf{m}_1$.*

*1. Show that*

$$\langle \psi | B | \psi \rangle = p_{win} - p_{lose}. \tag{2.160}$$

*2. Use Part 1 together with the Tsirelson bound to show that there exists a quantum strategy (i.e. directions $\mathbf{n}_x, \mathbf{m}_y$ and a quantum state $|\psi^{AB}\rangle$) such that*

$$p_{win} = \frac{1}{2} + \frac{1}{2\sqrt{2}} > \frac{3}{4}. \tag{2.161}$$

## 2.6.7 All Bell Inequalities

The CHSH inequality is one out of many similar inequalities known as Bell inequalities. Any Bell inequality can be expressed in the following form:

$$\mathbf{s} \cdot \mathbf{p} := \sum_{a,b,x,y} s_{abxy}\, p(ab|xy) \leqslant c, \tag{2.162}$$

where $0 < c \in \mathbb{R}$ and $\mathbf{p}$ is the vector whose components are the conditional probabilities $p(ab|xy)$ and $\mathbf{s}$ is any real vector with the same dimension as $\mathbf{p}$. Note that for any real vector $\mathbf{s}$ one can take $c$ in (2.162) to be

$$c = \max_{\mathbf{p} \in \mathcal{L}(n)} \mathbf{s} \cdot \mathbf{p}. \tag{2.163}$$

where $\mathcal{L}(n) \subset R^n$ is the set of all vectors $\mathbf{p} = \{p(ab|xy)\}$ whose components have the form (cf. (2.131))

$$p(ab|xy) = \int d\lambda\; q_\lambda\; p_\lambda^A(a|x)\; p_\lambda^B(b|y). \tag{2.164}$$

We can therefore identify any Bell inequality with a single real vector $\mathbf{s}$ (since the constant $c$ is determined from above). In the general case, $x = 1, \ldots, |X|$, $y = 1, \ldots, |Y|$, $a = 1, \ldots, |A|$, and $b = 1, \ldots, |B|$, can take more than two values. We also denote by $n := |X| \cdot |Y| \cdot |A| \cdot |B|$ the dimension of the vectors $\mathbf{p}$ and $\mathbf{s}$. This corresponds to higher dimensional systems, and in the quantum case $a$ corresponds to the outcome of a projective von-Neumann measurement that is labeled by $x$ on Alice's subsystem, and similarly $b$ corresponds to the outcome of a projective measurement that is labeled by $y$ on Bob's subsystem. Note that the definition of local distribution as in (2.164) remains unchanged in higher dimensions. Therefore, there are many Bell inequalities, and in recent years much effort has been made to characterize and understand the structure of all them.

The Bell inequalities that we consider here are those that can be used to test if a given distribution vector $\mathbf{p}$ is local (i.e. has the form (2.164)). If a given distribution vector $\mathbf{p}$ violates a Bell inequality $\mathbf{s}$ (i.e. a Bell inequality of the form (2.162)), then we learn from it that $\mathbf{p}$ is nonlocal. However, if a probability distribution does not violate a particular Bell inequality, $\mathbf{s}$, this alone does not mean that the distribution is local.

Given a probability vector $\mathbf{p}$, how can we decide if it is local (i.e. can be written in the form (2.164))? To answer this question, we first discuss the convexity property of local distributions.

**Exercise 2.66.** *Denote by $\mathcal{P}(n) \subset \mathbb{R}^n$ the space of all real vectors in dimension $n = |ABXY|$ whose components are given in terms of conditional probabilities $\{p(ab|xy)\}$, and let $\mathcal{L}(n) \subset \mathcal{P}(n)$ be the set of all local vectors as in (2.164). Show that $\mathcal{L}(n)$ is a convex set.*

**Exercise 2.67.** *Show that $\mathcal{P}(n)$ is a polytope in $\mathbb{R}^n$. Hint: Recall the definition of a polytope in Sec. A.2 online version.*

Consider now a vector $\mathbf{p} \in \mathcal{P}(n)$, and define the set $\{\mathbf{p}\}$ consisting of exactly one vector. As such, it is (trivially) a convex set in $\mathbb{R}^n$. Suppose now that $\mathbf{p} \notin \mathcal{L}(n)$. This means that $\{\mathbf{p}\} \cap \mathcal{L}(n) = \emptyset$, or in other words, $\{\mathbf{p}\}$ and $\mathcal{L}(n)$ are two disjoint convex sets. Therefore, from the hyperplane separation theorem (see Theorem A.2 of the online version) it follows that there exists a vector $\mathbf{s} \in \mathbb{R}^n$ and a real number $r$ such that

$$\mathbf{s} \cdot \mathbf{q} \leqslant r < \mathbf{s} \cdot \mathbf{p} \qquad \forall\, \mathbf{q} \in \mathcal{L}(n). \tag{2.165}$$

The above equation can be interpreted as follows. If $\mathbf{p} \notin \mathcal{L}(n)$, then there exists a Bell inequality s that it violates. We summarize it in the following theorem.

> **Theorem 2.1.** Let $\mathcal{L}(n) \subset \mathcal{P}(n)$ be the set of all local probability vectors as in (2.164) with fixed cardinalities $|X|$, $|Y|$, $|A|$, $|B|$, and $n = |ABXY|$. Then, $\mathbf{p} \notin \mathcal{L}(n)$ if and only if it violates at least one Bell inequality.

This theorem implies that in order to determine if a probability distribution is local, one has to check that it doesn't violate all Bell inequalities as in (2.162). This, at first, may seem as an impossible task as one will have to check infinite number of inequalities corresponding to each vector $\mathbf{s} \in \mathbb{R}^n$. However, as we know from Sec. A online version (convex analysis) this is not necessary. First, note that there are many cases that are redundant (e.g., if we checked the Bell inequality for s there is no need to check it for 2s). More importantly, it follows that $\mathcal{L}(n)$ in addition to being convex, is in fact a *polytope*.

**Exercise 2.68.** *Show that $\mathcal{L}(n)$ is a polytope in $\mathbb{R}^n$ (i.e. a convex hull of a finite number of points). Hint: Show first that the set of vectors $\mathbf{p}^A$, whose components are any conditional probabilities $\{p(a|x)\}$, is itself a polytope (i.e. find its extreme points and show that there are a finite number of them).*

From Theorem A.8 online version, the polytope $\mathcal{L}(n)$ can be represented as an intersection of finitely many half-spaces. Denoting by $\mathbf{s}^{(j)}$ (with $j = 1, \ldots, m$) the normal vectors to these half spaces, we therefore conclude that $\mathbf{p} \in \mathcal{L}(n)$ if and only if

$$\mathbf{s}^{(j)} \cdot \mathbf{p} \leqslant c_j \quad \forall j = 1, \ldots, m, \tag{2.166}$$

where $c_j := \max_{\mathbf{q} \in \mathcal{L}(n)} \mathbf{s}^{(j)} \cdot \mathbf{q}$. In other words, there exists finitely many Bell inequalities that can determine if a vector $\mathbf{p} \in \mathcal{L}(n)$.

This analysis may give the impression that deciding if $\mathbf{p}$ is local is easy. Therefore, it is important to note first that the computation of $\mathbf{s}^{(j)}$ may be hard, and that the number $m$ may grow exponentially with the cardinalities $|X|, |Y|, |A|$, and $|B|$. In particular, already for the case that $|A| = |B| = 2$ with arbitrary large $|X| = |Y|$, it was shown that the decision problem of whether $\mathbf{p}$ is local is NP-complete [11]. On the other hand, the simplest case in which $|A| = |B| = |X| = |Y| = 2$ was fully character-ized in [81], and independently by [78], and, in particular, it was shown that the only nontrivial Bell inequality is the CHSH inequality. That is, for bits $x, y, a, b \in \{0, 1\}$, the 16-dimensional vector $\mathbf{p} = (p(ab|xy))$ is local if and only if it does not violate any of the CHSH inequalities.

## 2.7 Unitary Evolution and the Schrödinger Equation

The last postulate of quantum mechanics is about time evolution of physical systems. It states that the state of a closed physical system evolves unitarily in time. That is, if $|\psi(t)\rangle$ is the state describing the system at time $t$, then there exists a parameterized family of unitary matrices $U(t)$ with parameter $t \in \mathbb{R}$ such that

$$|\psi(t)\rangle = U(t)|\psi(0)\rangle. \tag{2.167}$$

We emphasize here that $U(t)$ (with $t > 0$) does not depend on the initial state (i.e. on the preparation of the system at time $t = 0$). It is important also to note that the formalism of quantum mechanics does not propose which unitary family $U(t)$ one should choose to describe a particular evolution of a quantum system. It just states the evolution (whatever the specific causes for it) is described with a unitary matrix. We saw earlier something similar about quantum states of the spin of an electron. The first postulate of quantum mechanics did not tell us which states to assign to a spe-cific system. It only stated that all the information about the system is encoded in a quantum state. We then used, as in the example of the spin of an electron, further symmetry properties to assign the physical interpretation of any qubit state like $|0\rangle$, $|1\rangle$, $|+\rangle$ $|-i\rangle$.

One can view the unitary evolution postulate as a principle of *distinguishability pre-serving*. Recall from Exercise 2.54 that if two quantum states are orthogonal then they can be perfectly distinguished by a suitable projective measurement. The principle of distinguishability preserving asserts that if a closed system is prepared in one out of two or more distinguishable states, then the ability to distinguish between them remains intact throughout the evolution, unless some type of external noise is pumped into the system. Therefore, one can view a unitary evolution as a distinguishability preserving map. Alternatively, since information quantifies the ability to distinguish between one thing from another, the unitary evolution postulate of quantum mechanics, loosely speaking, is the statement that closed systems don't lose information (i.e. the ability to distinguish) if they don't interact with the external world.

We now discuss the form of the parametrized family of the unitaries $U(t)$ given in (2.167). We will assume here that the function $t \mapsto U(t)$ is continuous, and even differentiable. Moreover, $U(0) = I$ is the identity matrix so that we can express for a small $t = \varepsilon > 0$:

$$U(\varepsilon) = I - iH\varepsilon + O(\varepsilon^2), \tag{2.168}$$

where $H$ is some Hermitian matrix. Note that $H$ must be Hermitian since otherwise $U(\varepsilon)$ will not be a unitary matrix; that is,

$$U^*(\varepsilon)U(\varepsilon) = (I + iH\varepsilon)(I - iH\varepsilon) + O(\varepsilon^2) = I + O(\varepsilon^2), \qquad (2.169)$$

where we assumed that $H$ is Hermitian. Therefore, taking the derivative on both sides of (2.167) and setting $t = 0$ gives

$$\frac{d}{dt}|\psi(t)\rangle\Big|_{t=0} = -iH|\psi(0)\rangle. \qquad (2.170)$$

Now, since the system is isolated, the state $|\psi(t)\rangle$ must evolve according to the same rule as the state $|\psi(0)\rangle$. Hence, this homogeneity assumption implies that for all $t > 0$,

$$\frac{d}{dt}|\psi(t)\rangle = -iH|\psi(t)\rangle. \qquad (2.171)$$

**Exercise 2.69.** *Show that from (2.171) it follows that*

$$U(t) = e^{-iHt}. \qquad (2.172)$$

As we discussed in this chapter, in quantum mechanics, any Hermitian operator corresponds to an observable. The observable $H$ above is known as the Hamiltonian of the system and it corresponds to the energy of the system. There are many books in physics from which you can learn how to construct the Hamiltonian $H$ for specific physical systems, but generally speaking, quantum mechanics itself does not provide the prescription on how to construct the Hamiltonian of a specific physical system. Hamiltonians are also constructed in classical physics.

The Hamiltonian has the units of energy. Therefore, when incorporating the physical dimensions, (2.171) takes the form of the celebrated Schrödinger equation,

**The Schrödinger Equation**

$$i\hbar\frac{d}{dt}|\psi(t)\rangle = H|\psi(t)\rangle, \qquad (2.173)$$

where the Plank's constant $\hbar = h/2\pi$ has the units of energy$\times$time so that both sides of the equation have the same dimensions. Since the Hamiltonian $H$ is a Hermitian matrix it can be diagonalized as $H = \sum_x E_x|\varphi_x\rangle\langle\varphi_x|$, where $\{E_x\}$ are the energy levels of the system, and $\{|\varphi_x\rangle\}$ are the corresponding eigenstates. The eigenstate $|\varphi_x\rangle$ that corresponds to the lowest energy level is called the *ground state* of the system.

Finally, we assumed that the system is closed, that is, does not interact with the environment in any way. This led us to assume a continuous uniform evolution. However, many physical systems are not closed, and even we, the experimenters, can change the Hamiltonian by changing parameters in the lab at different times. We leave this discussion to the next chapter that covers evolution of open systems.

## 2.7.1 The Measurement Problem

Quantum mechanics allows for two types of evolutions for isolated systems. One is a probabilistic evolution, in which quantum measurements such as the projective von-Neumann measurement, transform a state $|\psi\rangle \in A$ to another post-measurement state $|\psi_x\rangle$ with some probability $p_x$. The other is a deterministic evolution, in which a quantum state $|\psi\rangle$ evolves unitarily and deterministically to another state $U|\psi\rangle$, where the unitary matrix $U$ is determined from the Hamiltonian of the system. It is therefore natural to ask if both evolutions can coexist or if they lead to inconsistencies within quantum theory.

We have already learned that a physical system is not really isolated if it is being measured, since the measurement apparatus can be viewed as external system that interacts with the system. In fact, we saw that the measurement can change the state of the system. Therefore, at first glance it seems that there is no contradiction between quantum measurements and the assertion that closed systems evolve unitarily. However, we can consider both the system and its measuring device as a single composite system.

Any measuring device (including the device itself and we, the experimenters) consists of numerous atoms and molecules. This means that practically it is impossible to write down its Hamiltonian as one will need to include the contributions from all the $10^{23}$ (even more) particles constituting the device. Yet, according to the rules of quantum mechanics, there exists a Hamiltonian, $H^{AE}$, associated with the measuring device (environment system E) + the quantum system (system A) that is being measured. Then, according to Schrödinger equation, since the system + environment form a closed composite system, they must evolve unitarily according to the joint unitary matrix $U^{AE} = e^{-iH^{AE}t}$. On the other hand, according to Born's rule they must evolve probabilistically. Which evolution will occur, the deterministic one or the probabilistic one?

Let's consider the SG experiment for measuring the spin in the $z$-direction of a single electron. In this case, the quantum system $A$ is described with a unit vector $|\psi^A\rangle = a|0\rangle + b|1\rangle$ in $A \cong \mathbb{C}^2$, where $|a|^2 + |b|^2 = 1$. Denote by $E$ the Hilbert space associated with the measuring device (plus the experimenters and the rest of the universe for that matter). If the measurement apparatus is treated externally, according to Born's rule one obtains the outcome $|0\rangle$ with probability $|a|^2$ and the outcome $|1\rangle$ with probability $|b|^2$.

If the measurement apparatus is treated internally, then we can assume that prior to the experiment, the measuring device was given in some "ready" state. That is, according to the first postulate of quantum mechanics there exists a vector $|\text{ready}\rangle \in E$ containing all the information about the measuring device prior to the measurement. Now, suppose first that the state of the system was $|0\rangle^A$. Then, the initial state of the system+device is $|0\rangle^A|\text{ready}\rangle^E$. After the measurement, the joint system evolves to

$$|0\rangle^A|\text{ready}\rangle^E \, \rightarrow \, U^{AE}|0\rangle^A|\text{ready}\rangle^E = |0\rangle^A|\text{output "0"}\rangle^E, \tag{2.174}$$

where the equality follows from the fact that the measurement is performed in the $z$-direction, so the system state $|0\rangle^A$ must remain intact, while the vector $|\text{ready}\rangle^E$ of the measuring device is transformed to another vector $|\text{output "0"}\rangle^E$ in $E$, containing the information that the output was 0. Similarly, if the initial state of the system was $|1\rangle^A$, then the initial state of the system+device is $|1\rangle^A|\text{ready}\rangle^E$, and after the measurement, the joint system would evolve to

$$|1\rangle^A|\text{ready}\rangle^E \to U^{AE}|1\rangle^A|\text{ready}\rangle^E = |1\rangle^A|\text{output "1"}\rangle^E. \qquad (2.175)$$

Now, let's consider the case in which the initial state of the system is $|\psi\rangle^A = a|0\rangle + b|1\rangle$. In this case, as before, the initial state of the system+device is $|\psi\rangle^A|\text{ready}\rangle^E$. However, after the measurement, the system evolves unitarily to the state

$$|\psi\rangle^A|\text{ready}\rangle^E \to U^{AE}\left[|\psi\rangle^A|\text{ready}\rangle^E\right] = U^{AE}\left[(a|0\rangle^A + b|1\rangle^A)|\text{ready}\rangle^E\right]$$
$$= aU^{AE}\left[|0\rangle^A|\text{ready}\rangle^E\right] + bU^{AE}\left[|1\rangle^A|\text{ready}\rangle^E\right]$$
$$= a|0\rangle^A|\text{output "0"}\rangle^E + b|1\rangle^A|\text{output "1"}\rangle^E, \qquad (2.176)$$

where in the last equality we used (2.174) and (2.175). The above state is an entangled state between the system and the measuring device representing quantum correlations between the two.

We therefore see a sharp contrast between the two types of evolution of a quantum state. Although this problem haunt quantum mechanics right from its early formulation at the beginning of the twentieth century, there is a controversy on how to resolve this problem. This is also related to the different interpretations of quantum mechanics. For example, the Everett "many worlds" interpretation adopts the unitary evolution, whereas others adopt the probabilistic nature of it. This is a fascinating topic, but it goes far beyond the scope of this book.

## 2.7.2 The No-Cloning Theorem

One of the very useful properties of classical information is that it can be cloned; that is, it can be copied and for example broadcast to several other parties (see Figure 2.9a). We now explore if quantum information also has this property. Quantum information is encoded in quantum states, so cloning of quantum information corresponds to copying of an unknown quantum state $|\psi\rangle \in A$ (see Figure 2.9b).

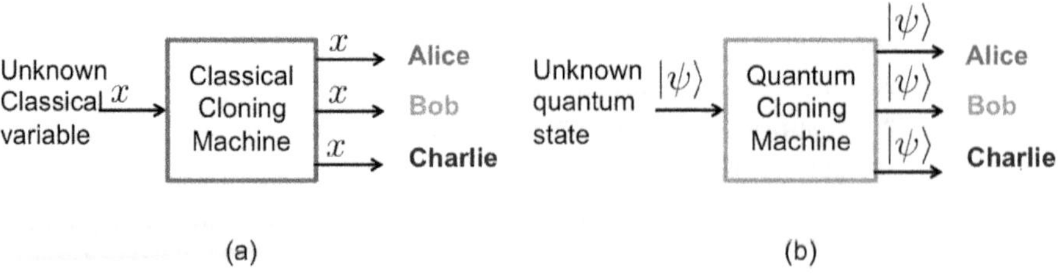


**Figure 2.9**  Classical and quantum cloning machines.

Suppose there exists a quantum machine that takes an unknown state $|\psi\rangle \in A$ and output two copies of it $|\psi\rangle|\psi\rangle$. Since it maps any normalized vector to a normalized vector, it can be modeled by an isometry $V: A \to A \otimes A$ with the property that $V^*V = I^A$. Consider now two arbitrary states $|\psi\rangle, |\phi\rangle \in A$. Then, from our assumption

$$V|\psi\rangle = |\psi\rangle|\psi\rangle \quad \text{and} \quad V|\phi\rangle = |\phi\rangle|\phi\rangle. \tag{2.177}$$

Therefore,

$$\langle\psi|\phi\rangle = \langle\psi|V^*V|\phi\rangle = \left(\langle\psi|\phi\rangle\right)^2 \tag{2.178}$$

But any complex number that satisfies $c = c^2$ must be equal to 0 or 1. Hence, $\langle\psi|\phi\rangle \in \{0, 1\}$ which means that either $|\psi\rangle = |\phi\rangle$ or that $|\psi\rangle$ is orthogonal to $|\phi\rangle$. Therefore, there is no quantum machine that is capable of generating two copies of an arbitrary unknown quantum state. This result is known by the term *the no-cloning theorem*.

### 2.7.3 The Controlled Unitary Operation

The control unitary operation is a particular quantum evolution that is a key building block in quantum circuits. It is used quite often in quantum computing, and in particular the controlled not (CNOT) gate, which is a key component in the construction of a quantum computer, is a special type of a controlled unitary map. In its most general form, a unitary map $U: AB \to AB$ (recall that $AB = A \otimes B$) is a *controlled* unitary map (or gate) if $U^{AB}$ can be written as

$$U^{AB} = \sum_{x \in [m]} |x\rangle\langle x|^A \otimes U_x^B, \tag{2.179}$$

where $\{|x\rangle^A\}_{x \in [m]}$ is some orthonormal basis of $A$, and $\{U_x^B\}$ is a collection of $|A|$ unitary matrices in $B$. Note that $U^{AB}\left(|x\rangle^A|\psi^B\rangle\right) = |x\rangle^A \otimes \left(U_x^B|\psi^B\rangle\right)$, so that by choosing the input $|x\rangle^A$, Alice controls the unitary that is acted on $|\psi\rangle^B$.

**Exercise 2.70.** *Verify that $U^{AB}$ in the equation above is indeed a unitary matrix.*

In quantum circuits, the controlled unitary is depicted as in Figure 2.10a. The CNOT gate is the controlled unitary map

$$U^{AB} = |0\rangle\langle 0| \otimes I + |1\rangle\langle 1| \otimes \sigma_1, \tag{2.180}$$

where $\sigma_1$ is the first Pauli (unitary) matrix. The CNOT gate is depicted in Figure 2.10b.

Figure 2.10 (a) Controlled unitary gate. (b) Controlled NOT (CNOT) gate.

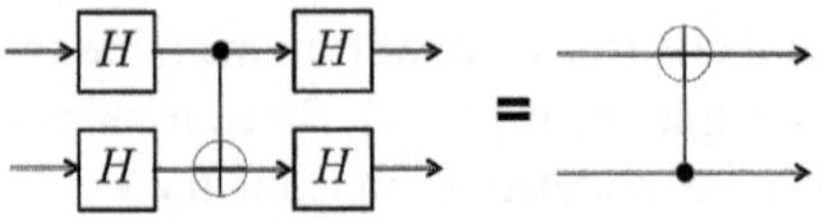

**Figure 2.11** CNOT gate in Hadamard basis.

**Exercise 2.71.** *Show that the CNOT gate can generate the maximally entangled state $(|00\rangle + |11\rangle)/\sqrt{2}$ from a tensor product of two vectors (i.e. from a product state of the form $|\psi\rangle^A|\phi\rangle^B$).*

**Exercise 2.72.** *Show the equivalence of the two circuits in Figure 2.11, where*

$$H := \frac{1}{\sqrt{2}}\begin{bmatrix} 1 & 1 \\ 1 & -1 \end{bmatrix} \tag{2.181}$$

*is the Hadamard unitary matrix.*

## 2.8 Notes and References

Many books on quantum mechanics contain much of the material presented here. More details on the SG experiment can be found for example in traditional books on quantum mechanics such as in Refs. [164] and [198].

For topics on inner product spaces in finite dimensions that include many of the concepts used in this book, we refer to Refs. [25, 125]. The treatment of linear algebra with Dirac notations can be found in many text books on quantum physics and quantum information, including, for example, Refs. [173, 235, 233]. Each of these books also provide a review on quantum mechanics.

The example of the three quantum coins was taken from [185], and the Hardy paradox can be found in Ref. [111]. More details on Bell nonlocality and many related references can be found in the review article [38]. Readers interested to learn more about the measurement problem and the different interpretations of quantum mechanics may find the review article [201] as useful starting point.

Much more detail on the no-cloning theorem can be found in Ref. [200].

Open physical systems are systems that have interactions with other external systems. These external systems, which we will refer to as "the environment," can either be correlated with the system, and/or exchange information, energy, or matter, with it. Consequently, the description and evolution of such systems can be very different than those we discussed for isolated systems. Yet, there is no need to introduce new postulates in order to develop the theory of open quantum systems. Instead, we will see that *all* the postulates of quantum mechanics on isolated systems are sufficient to determine the evolution, the measurements, and the description of open systems.

## 3.1 Generalized Measurements

In the previous chapter we saw that isolated physical systems can undergo two types of evolutions: the unitary (Schrödinger) evolution and the probabilistic measurement evolution. The combination of these two evolutions yield another type of evolution. Explicitly, let $|\psi\rangle \in A$ be a pure state of an isolated physical system, $U$ be a unitary operator, and $\{P_x\}_{x\in[m]}$ be a projective measurement. Applying the projective measurement after the state $|\psi\rangle$ evolved to the state $U|\psi\rangle$ yields the state $|\psi_x\rangle := \frac{1}{\sqrt{p_x}}P_x U|\psi\rangle$ with probability $p_x := \langle\psi|U^*P_x U|\psi\rangle$. Denoting by $M_x := P_x U$ we get

$$|\psi_x\rangle = \frac{1}{\sqrt{p_x}}M_x|\psi\rangle \quad \text{and} \quad p_x = \langle\psi|M_x^*M_x|\psi\rangle. \tag{3.1}$$

Therefore, the combination of a unitary evolution followed by a projective measurement can be modeled by a collection of complex matrices $\{M_x = P_x U\}_{x\in[m]}$. Note that $M_x$ are not projections, although they have a very special form given by $P_x U$. If additional ancillary systems are available (i.e. the system is not closed), then the combination of a unitary evolution on both the system and the ancilla, followed by a projective (or basis) measurement, yields an even more general type of evolution known by the name *generalized measurement* as it generalizes the von-Neumann projective measurement.

In Figure 3.1, we describe the following evolution of a quantum state $|\psi\rangle \in A$. In the first step of the evolution, an ancillary system is introduced which is prepared in some state $|1\rangle \in R$. Consequently, the state of the joint system is $|1\rangle^R|\psi^A\rangle$. Next, a joint unitary evolution, $U^{RA}$, is applied to the joint state $|1\rangle|\psi\rangle$, yielding the bipartite

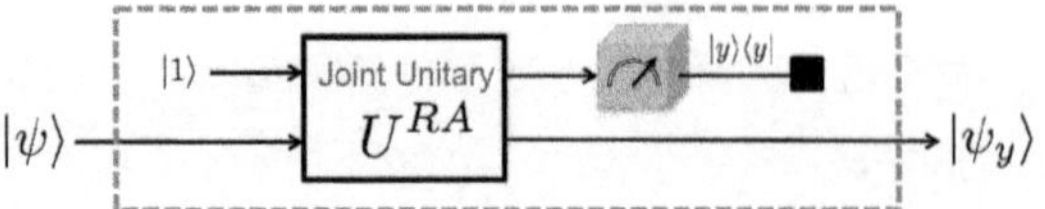

**Figure 3.1**    Realization of a generalized measurement.

state $U^{RA}|1\rangle|\psi\rangle$. Finally, a basis measurement $\{|y\rangle\langle y|^R\}_{y\in[n]}$ is applied on the reference system $R$.

We discuss now how the output state $|\psi_y\rangle$ is related to the input state $|\psi\rangle$, and what is the probability to obtain an outcome $y$. Denote by $m := |A|$ and $n := |R|$. As an operator in the vector space $\mathfrak{L}(R \otimes A)$, the unitary matrix $U^{RA}$ can be expressed as

$$U^{RA} = \sum_{y,y'\in[n]} |y\rangle\langle y'| \otimes \Lambda_{yy'} \quad \text{where} \quad \Lambda_{yy'} \in \mathfrak{L}(A). \tag{3.2}$$

Note that any operator in $\mathfrak{L}(R \otimes A)$ has the above form, but since $U^{RA}$ is unitary, we have the equivalence

$$U^*U = I^{RA} \quad \Longleftrightarrow \quad \sum_{y\in[n]} \Lambda_{yz}^* \Lambda_{yx} = \delta_{xz} I^A \quad \forall\, x, z \in [n]. \tag{3.3}$$

Now, observe that

$$\left(|y\rangle\langle y|^R \otimes I^A\right) U^{RA}|1\rangle^R|\psi^A\rangle = |y\rangle^R\left(\Lambda_{y1}|\psi^A\rangle\right). \tag{3.4}$$

Therefore, denoting by $M_y := \Lambda_{y1}$, we get

$$|\psi_y\rangle = \frac{1}{\sqrt{p_y}} M_y|\psi\rangle \quad \text{with} \quad p_y := \langle\psi|M_y^* M_y|\psi\rangle. \tag{3.5}$$

Moreover, from (3.3) it follows that $\sum_{y\in[n]} M_y^* M_y = I^A$, so that $\sum_{y\in[n]} p_y = 1$, where $p_y$ is the probability to obtain an outcome $y$. Note that the post-measurement state $|\psi_y\rangle$, with its associated probability $p_y$, has a very similar form to the form of $|\psi_x\rangle$ and $p_x$ in (3.9). However, unlike the form $P_x U$ of $M_x$ in (3.1), the only condition on $\{M_y := \Lambda_{y1}\}$ is that they can be extended to a family of matrices $\{\Lambda_{yy'}\}$ that satisfies (3.3). This will ensure that $U^{AB}$ is unitary. We now show that any set of complex matrices $\{M_y\}_{y\in[n]}$ with the property $\sum_{y\in[n]} M_y^* M_y = I^A$ can be completed to a full family of matrices $\{\Lambda_{yy'}\}$ that satisfies (3.3). To see this, observe that the matrix $U^{RA}$ can be expressed in the following block form

$$U^{RA} = \begin{bmatrix} \Lambda_{11} & \Lambda_{12} & \cdots & \Lambda_{1n} \\ \Lambda_{21} & \Lambda_{22} & \cdots & \Lambda_{2n} \\ \vdots & \vdots & \ddots & \vdots \\ \Lambda_{n1} & \Lambda_{n2} & \cdots & \Lambda_{nn} \end{bmatrix} \tag{3.6}$$

with the matrices $\{M_y := \Lambda_{y1}\}_{y\in[n]}$ appearing in the first block column. Moreover, this first column satisfies

$$\begin{bmatrix} M_1^* & M_2^* & \cdots & M_n^* \end{bmatrix} \begin{bmatrix} M_1 \\ M_2 \\ \vdots \\ M_n \end{bmatrix} = \sum_{y \in [n]} M_y^* M_y = I^A. \tag{3.7}$$

Therefore, the first column block in (3.6) consists of $m := |A|$ orthonormal vectors. Any such set of $m$ orthonormal vectors in $\mathbb{C}^{mn}$ can be completed to a full orthonormal basis of $\mathbb{C}^{mn}$ (e.g. by the Gram–Schmidt process). Therefore, it is always possible to construct a unitary matrix $U^{RA}$ as above, from a set of matrices $\{\Lambda_{y1}\}_{y \in [n]}$ that satisfies $\sum_{y \in [n]} \Lambda_{y1}^* \Lambda_{y1} = I^A$.

**Generalized Measurement**

**Definition 3.1.** A generalized measurement is a collection of $m \in \mathbb{N}$ complex matrices $\{M_x\}_{x \in [m]} \subset \mathfrak{L}(A)$ with the property that

$$\sum_{x \in [m]} M_x^* M_x = I^A. \tag{3.8}$$

When a generalized measurement is applied to a physical system, it transforms the state of the system, $|\psi\rangle$, to the post-measurement state

$$|\psi_x\rangle := \frac{1}{\sqrt{p_x}} M_x |\psi\rangle \quad \text{with probability} \quad p_x := \langle \psi | M_x^* M_x | \psi \rangle. \tag{3.9}$$

We have already shown that a generalized measurement can always be realized as in Figure 3.1. Furthermore, both projective measurements and the measurements described in (3.9) with $M_x = P_x U$ are special types of generalized measurements.

How do we know that the generalized measurement described above is indeed the most general one? Can we construct another circuit like the one in Figure 3.1 which would yield perhaps a more general measurement? Note that the generalized measurement described in Figure 3.1 makes use of the two types of evolutions in quantum mechanics: a unitary evolution followed by a projective measurement. Since these are the *only* two types of evolution in quantum mechanics, any evolution can be decomposed into a sequence of these two types of processes. Since both unitary evolution and projective measurements are themselves generalized measurement, we conclude that the most general measurement on a quantum system can be described as a sequence of generalized measurements. In the following exercise it is argued that any such sequence of generalized measurements can be simulated by a single generalized measurement. Hence, the generalized measurement described above is indeed general enough to describe the most general measurement in quantum mechanics.

**Exercise 3.1.** *Show that if $\{M_x\}_{x \in [m]}$ and $\{N_y\}_{y \in [n]}$ are two generalized measurements, then $\{M_x N_y\}$ is also a generalized measurement. Use this to show that a sequence of generalized measurements can be simulated by a single generalized measurement.*

**Exercise 3.2.** *Show that the matrices (operators) $M_x$ do not have to be square. That is, show that any collection of m operators $\{M_x\}_{x\in[m]} \subset \mathfrak{L}(A, B)$ that satisfy (3.8) can also be realized as a generalized measurement as depicted in Figure 3.1. Hint: Consider a unitary operator $U: RA \to R'B$ where the reference systems $R$ and $R'$ are such that $|RA| = |R'B|$.*

**Exercise 3.3.** *Let $A = \mathbb{C}^2$, and let $M_0 = a|+\rangle\langle 0|$ and $M_1 = b|0\rangle\langle +|$ be two operators in $\mathfrak{L}(A)$ with $a, b \in \mathbb{C}$. Find the precise conditions on $a$ and $b$ for the existence of a third operator $M_2 \in \mathfrak{L}(A)$ such that $\{M_0, M_1, M_2\}$ form a generalized measurement.*

**Exercise 3.4.** *Consider $d$ (rank 1) operators $\{M_x = |\psi_x\rangle\langle\phi_x|\}_{x\in[d]}$ in $\mathfrak{L}(\mathbb{C}^d)$, where $\{|\psi_x\rangle\}_{x\in[d]}$ and $\{|\phi_x\rangle\}_{x\in[d]}$ are some normalized states in $\mathbb{C}^d$. Show that $\{M_x\}_{x\in[d]}$ is a generalized measurement if and only if $\{|\phi_x\rangle\}_{x\in[d]}$ is an orthonormal basis of $\mathbb{C}^d$.*

## 3.2 The Mixed Quantum State

The first postulate of quantum mechanics states that the information about closed physical systems is encoded in pure quantum states. Here we show how this postulate implies that for open quantum systems, mixed quantum states encode all the information that can be extracted from the system. We derive this conclusion in two different ways, one by considering a system that is correlated *classically* to another ancillary system, and the other by considering *quantum* correlations with the ancillary system.

### 3.2.1 The Emergence of Density Operators from Classical Correlations

So far we considered isolated systems that we described with a pure state $|\psi\rangle \in A$, or more precisely, with the rank 1 matrix $|\psi\rangle\langle\psi| \in \mathfrak{D}(A)$. Suppose now that in addition to the quantum system, Alice also has access to classical systems, like coins or dice, that can generate random numbers. In this case, Alice can roll a dice with $m$ possible outcomes, and based on the outcome, prepare one of the $m$ states $\{|\psi_x\rangle\langle\psi_x|\}_{x\in[m]} \subset \mathfrak{D}(A)$. This way, Alice can prepare the state $|\psi_x\rangle$ with some probability $p_x$ (the classical systems, i.e. the coins or dice, do not have to be unbiased). Now, suppose that Alice forgot the value of $x$. Then, Alice knows that her state is one out of the $m$ states in the *ensemble* of states $\{|\psi_x\rangle\langle\psi_x|, p_x\}_{x\in[m]}$. How should we characterize the ensemble $\{|\psi_x\rangle\langle\psi_x|, p_x\}_{x\in[m]}$? We will see that there exist many other ensemble of states that contains the exact same information as the ensemble $\{|\psi_x\rangle\langle\psi_x|, p_x\}_{x\in[m]}$. Therefore, instead of characterizing the information with a particular ensemble (such as $\{|\psi_x\rangle\langle\psi_x|, p_x\}_{x\in[m]}$), we will characterize it with a mathematical object that remains invariant under exchanges of such equivalent ensembles.

To gather information about her system, Alice can execute a generalized measurement, denoted as $\{M_y\}_{y\in[n]}$, on her system, characterized by the ensemble

$\{|\psi_x\rangle\langle\psi_x|, p_x\}_{x\in[m]}$. This measurement results in an outcome $y$ with a corresponding probability denoted as $q_y$. Furthermore, following the occurrence of outcome $y$, there emerges a post-measurement ensemble that describes the state of Alice's system. We will now delve into these details to demonstrate that the dependencies of these quantities rely solely on a density matrix associated with the ensemble $\{|\psi_x\rangle\langle\psi_x|, p_x\}_{x\in[m]}$.

If the pre-measurement state was $|\psi_x\rangle$, then the post-measurement state after outcome $y$ occurred is

$$|\phi_{xy}\rangle := \frac{1}{\sqrt{p_{y|x}}} M_y |\psi_x\rangle \tag{3.10}$$

with probability

$$p_{y|x} := \langle\psi_x| M_y^* M_y |\psi_x\rangle = \mathrm{Tr}\left[ M_y^* M_y |\psi_x\rangle\langle\psi_x| \right]. \tag{3.11}$$

However, since Alice does not know the value of $x$, if she performs a measurement $\{M_y\}_{y\in[n]}$ on her system, she will get the outcome $y$ with probability

$$q_y := \sum_{x\in[m]} p_{y|x} p_x = \sum_{x\in[m]} p_x \mathrm{Tr}\left[ M_y^* M_y |\psi_x\rangle\langle\psi_x| \right] = \mathrm{Tr}\left[ M_y^* M_y \rho \right] \tag{3.12}$$

where

$$\rho := \sum_{x\in[m]} p_x |\psi_x\rangle\langle\psi_x| \tag{3.13}$$

is the density matrix associated with the ensemble $\{|\psi_x\rangle\langle\psi_x|, p_x\}_{x\in[m]}$.

Note that $p_{y|x} p_x$ is the probability that both the pre-measurement state is $|\psi_x\rangle$ *and* that the outcome of its measurement is $y$. Therefore, using the Bayesian rule of probabilities, the probability that the pre-measurement state is $|\psi_x\rangle$, given that the measurement outcome is $y$, can be expressed as $q_{x|y} := p_{y|x} p_x / q_y$. Consequently, after outcome $y$ occurred, the ensemble $\{p_x, |\psi_x\rangle\langle\psi_x|\}$ changes to

$$\{q_{x|y}, |\phi_{xy}\rangle\langle\phi_{xy}|\}_{x\in[m]}. \tag{3.14}$$

Note that the density operator, $\sigma_y$, that is associated with the above ensemble is given by

$$\sigma_y := \sum_{x\in[m]} q_{x|y} |\phi_{xy}\rangle\langle\phi_{xy}|$$

$$\boxed{q_{x|y} := p_{y|x} p_x / q_y} \longrightarrow \quad = \frac{1}{q_y} \sum_{x\in[m]} p_x p_{y|x} |\phi_{xy}\rangle\langle\phi_{xy}|$$

$$(3.10)\rightarrow \quad = \frac{1}{q_y} \sum_{x\in[m]} p_x M_y |\psi_x\rangle\langle\psi_x| M_y^* \tag{3.15}$$

$$= \frac{1}{q_y} M_y \rho M_y^*.$$

To summarize, the outcome $y$, of any generalized measurement $\{M_y\}_{y\in[n]}$, occurs with probability $q_y = \mathrm{Tr}\left[ M_y^* M_y \rho \right]$, when applied to an ensemble $\{p_x, |\psi_x\rangle\langle\psi_x|\}_{x\in[m]}$.

Recall from Exercise 2.27 that aside from the ensemble $\{p_x, |\psi_x\rangle\langle\psi_x|\}_{x\in[m]}$, there are infinitely many other ensembles that also correspond to the same density operator $\rho$. Therefore, the dependance of $q_y$ only on $\rho$ demonstrates that the statistics of any measurement outcome depends only on the density operator, and not on the particular ensemble that realizes it. To clarify, suppose $\{p_x, |\psi_x\rangle\langle\psi_x|\}_{x\in[m]}$ and $\{r_z, |\varphi_z\rangle\langle\varphi_z|\}_{z\in[k]}$ are two ensembles that correspond to the same density operator $\rho$. Then, the probability to obtain an outcome $y$ is the same for both ensembles, and therefore there is no way to distinguish between the two ensembles. One may argue that maybe there is a way to distinguish between the post-measurement ensembles; however, as can be seen in (3.15), any post-measurement ensemble is also associated with a unique density operator $\frac{1}{q_y}M_y\rho M_y^*$ that depends only on $\rho$ and not on the particular ensemble $\{p_x, |\psi_x\rangle\langle\psi_x|\}_{x\in[m]}$ or $\{r_z, |\varphi_z\rangle\langle\varphi_z|\}_{z\in[k]}$ that realizes $\rho$.

Given that the formalism of quantum mechanics lacks the means to differentiate between ensembles of states corresponding to the same density operator, all the information about the physical system accessible to us as observers is encapsulated within the density operator. Consequently, instead of characterizing physical systems using ensembles of states, we shall henceforth employ density operators for their descriptions.

> **The Rules of Quantum Measurements on Density Operators**
>
> To any physical system (open or closed) there is a Hilbert space $A$ that is associated with it. The information about a physical system is encoded in a density operator $\rho \in \mathfrak{D}(A)$. Information about the system can be extracted with an $m$-output generalized measurement, $\{M_x\}_{x\in[m]}$. The probability to obtain an outcome $x$ is given by
>
> $$p_x = \mathrm{Tr}\left[M_x^* M_x \rho\right], \tag{3.16}$$
>
> and the post-measurement state of the system, $\sigma_x$, after output $x$ occurred is
>
> $$\sigma_x = \frac{1}{p_x}M_x\rho M_x^*. \tag{3.17}$$

As an example, consider a qubit state $\rho \in \mathfrak{D}(\mathbb{C}^2)$. That is, $\rho \geqslant 0$ and $\mathrm{Tr}[\rho] = 1$. Any such qubit state can be expressed as a linear combination of the Pauli basis of $\mathrm{Herm}(\mathbb{C}^2)$

$$\rho = r_0\sigma_0 + r_1\sigma_1 + r_2\sigma_2 + r_3\sigma_3, \tag{3.18}$$

where $\sigma_0 := I_2$. Now, since the Pauli matrices $\sigma_1, \sigma_2$, and $\sigma_3$ are traceless, the condition $\mathrm{Tr}[\rho] = 1$ gives $r_0 = 1/2$. What are the conditions on $\mathbf{r} := (r_1, r_2, r_3)^T \in \mathbb{R}^3$ that ensure that $\rho \geqslant 0$? Since $\rho$ has two eigenvalues, say $\lambda$ and $1 - \lambda$, it follows that $\rho \geqslant 0$ if and only if $0 \leqslant \lambda \leqslant 1$. This condition is equivalent to $\mathrm{Tr}[\rho^2] = \lambda^2 + (1-\lambda)^2 \leqslant 1$. Therefore, $\rho \geqslant 0$ if and only if

$$1 \geqslant \mathrm{Tr}[\rho^2] = \frac{1}{2} + \sum_{j,k} r_j r_k \mathrm{Tr}[\sigma_j\sigma_k] = \frac{1}{2} + 2\|\mathbf{r}\|_2^2. \tag{3.19}$$

That is, $\|\mathbf{r}\|_2 \leqslant 1/2$. Therefore, after the renaming $\mathbf{r} \to \frac{1}{2}\mathbf{r}$ we conclude that all qubit quantum states have the form:

$$\rho = \frac{1}{2}(I_2 + \mathbf{r} \cdot \boldsymbol{\sigma}), \tag{3.20}$$

with $\|\mathbf{r}\|_2 \leqslant 1$. Moreover, since $\mathrm{Tr}[\rho^2] = 1$ if and only if $\|r\|_2 = 1$, we get that $\rho$ above is pure if and only if $\|r\|_2 = 1$. Hence, a qubit can be represented by the *Bloch Sphere* (see Figure 2.5) with the pure states represented on the boundary of the sphere and mixed states in the interior of the sphere. Note that the center of the sphere, that is, $\mathbf{r} = 0$, corresponds to the state $\rho = \frac{1}{2}I$, which is called the *maximally mixed state*.

**Exercise 3.5.** *Show that for* $\mathbf{r} = (\sin(\alpha)\cos(\beta), \sin(\alpha)\sin(\beta), \cos(\alpha))^T$, $\rho$ *in (3.20) is given by the state* $\rho = |\psi\rangle\langle\psi|$ *with* $|\psi\rangle$ *as in (2.110).*

**Exercise 3.6.** *Consider a density operator for a qutrit; that is,* $\rho \in \mathfrak{D}(\mathbb{C}^3)$, $\rho \geqslant 0$, *and* $\mathrm{Tr}[\rho] = 1$. *Let* $\boldsymbol{\lambda} = (\lambda_1, \lambda_2, \ldots, \lambda_8)$ *be a vector of matrices with* $\{\lambda_j\}_{j\in[8]}$ *being some Hermitian traceless* $3 \times 3$ *matrices satisfying the condition* $\mathrm{Tr}(\lambda_i\lambda_j) = 2\delta_{ij}$ *(note that also the Pauli matrices satisfy this orthogonality condition).*

*1. Show that* $\rho$ *can be written as*

$$\rho = \frac{1}{3}I_3 + \mathbf{t} \cdot \boldsymbol{\lambda}, \tag{3.21}$$

*where* $\mathbf{t} \in \mathbb{R}^8$ *and* $I_3$ *is the* $3 \times 3$ *identity matrix.*
*2. Show that* $\|\mathbf{t}\|_2 \leqslant \frac{1}{\sqrt{3}}$.
*3. Show that if* $\rho$ *is a pure state, then* $\|\mathbf{t}\|_2 = \frac{1}{\sqrt{3}}$.

*4. Is it true that for every* $\mathbf{t}$ *with* $\|\mathbf{t}\|_2 \leqslant \frac{1}{\sqrt{3}}$, $\rho$ *above corresponds to a density matrix? If yes prove it, otherwise give a counter example.*

## 3.2.2 The Emergence of Density Operators from Quantum Correlations

The density operator can be interpreted from a different perspective, distinct from the ensemble-based interpretation discussed earlier. Let's delve into this interpretation within the context of a composite system comprising two particles distributed between two separate entities, namely Alice and Bob. If the system is prepared in a pure state $|\psi^{AB}\rangle \in AB$, how should Alice represent the marginal state that corresponds to the electron in her lab? We assume here that the labs are very far from each other (perhaps on two different galaxies!) and Alice and Bob cannot even communicate.

Without loss of generality, we will assume that $|A| \leqslant |B|$ and write $|\psi^{AB}\rangle$ as (see Exercise 2.44)

$$|\psi^{AB}\rangle = \sqrt{\rho} \otimes V |\Omega^{A\tilde{A}}\rangle, \tag{3.22}$$

where $V : \tilde{A} \to B$ is an isometry and $\rho := \mathrm{Tr}_B\left[\psi^{AB}\right]$ is the reduced density matrix of $\psi^{AB} = |\psi^{AB}\rangle\langle\psi^{AB}|$. Suppose now that Alice performs a generalized measurement, $\{N_x\}_{x\in[m]} \subset \mathcal{L}(A)$, on her subsystem. Then, the probability that outcome $x$ occurs is given by

$$p_x := \langle\psi^{AB}|N_x^* N_x \otimes I^B|\psi^{AB}\rangle$$

$$(3.22)\!\rightarrow \; = \langle\Omega^{A\tilde{A}}|\sqrt{\rho}N_x^* N_x \sqrt{\rho} \otimes V^* V|\Omega^{A\tilde{A}}\rangle$$

$$\boxed{V^* V = I^{\tilde{A}}} \longrightarrow \; = \langle\Omega^{A\tilde{A}}|\sqrt{\rho}N_x^* N_x \sqrt{\rho} \otimes I^{\tilde{A}}|\Omega^{A\tilde{A}}\rangle \tag{3.23}$$

$$\textbf{Part 1 of Exercise 2.38}\!\rightarrow \; = \mathrm{Tr}\left[N_x^* N_x \rho^A\right].$$

Thus, the outcome probability $p_x$ depends only on the reduced density matrix $\rho^A$ and not (directly) on the bipartite state $|\psi^{AB}\rangle$. Moreover, the post-measurement state after outcome $x$ occurred is given by

$$|\psi_x^{AB}\rangle = \frac{1}{\sqrt{p_x}}N_x \otimes I^B|\psi^{AB}\rangle$$

$$(3.22)\!\rightarrow \; = \frac{1}{\sqrt{p_x}}N_x \sqrt{\rho} \otimes V|\Omega^{A\tilde{A}}\rangle. \tag{3.24}$$

Therefore, the reduced density matrix $\sigma_x^A$ of $\psi_x^{AB} = |\psi_x^{AB}\rangle\langle\psi_x^{AB}|$ is given by

$$\sigma_x^A := \mathrm{Tr}_B\left[\psi_x^{AB}\right] = \frac{1}{p_x}N_x \sqrt{\rho}\,\mathrm{Tr}_B\left[\left(I^A \otimes V\right)\Omega^{A\tilde{A}}\left(I^A \otimes V\right)^*\right]\sqrt{\rho}N_x^*, \tag{3.25}$$

where we substitute the expression in (3.24) for $\psi_x^{AB}$. Now, from the cyclic property of the partial trace (see Exercise 2.41) we have that

$$\mathrm{Tr}_B\left[\left(I^A \otimes V\right)\Omega^{A\tilde{A}}\left(I^A \otimes V\right)^*\right] = \mathrm{Tr}_{\tilde{A}}\left[\left(I^A \otimes V^* V\right)\Omega^{A\tilde{A}}\right] = \mathrm{Tr}_{\tilde{A}}\left[\Omega^{A\tilde{A}}\right] = I^A. \tag{3.26}$$

Combining this with the previous equation we conclude that

$$\sigma_x^A = \frac{1}{p_x}N_x \rho^A N_x^*. \tag{3.27}$$

This demonstrates that the reduced density matrix $\rho^A$ along with the measurement operators $\{N_x\}_{x\in[m]}$ determine the post-measurement reduced density matrices in the exact same way as we saw in the previous section. For the same reasons as before, we conclude that all the information that can be extracted from Alice's subsystem (via quantum generalized measurements) is encoded in the marginal state $\rho^A$. Therefore, if Alice has no accesses to Bob's subsystem, then from her perspective, the state of her subsystem can be characterized by the marginal density operator $\rho^A$, and the fact that her subsystem is entangled with Bob's can be ignored.

### 3.2.3 The Classical-Quantum State

Any ensemble of quantum states $\{p_x, |\psi_x\rangle\langle\psi_x|\}_{x\in[m]}$ in $\mathfrak{D}(A)$ can be viewed from two distinct perspectives, depending on the treatment of the variable $x$. This duality emerges when considering whether $x$ remains unknown and unrecorded, or if it is explicitly stored in a classical system.

When $x$ remains both unknown and unrecorded, as discussed previously, the complete characterization of the system is encapsulated by the density operator $\rho^A = \sum_{x \in [m]} p_x |\psi_x\rangle\langle\psi_x|$. Conversely, when $x$ is recorded within the classical system $X$ using the mapping $x \mapsto |x\rangle\langle x|^X$, the description of the system adopts a classical-quantum state, abbreviated as a "cq-state," represented by

$$\rho^{XA} := \sum_{x \in [m]} p_x |x\rangle\langle x|^X \otimes |\psi_x\rangle\langle\psi_x|^A. \tag{3.28}$$

To establish the equivalence between $\rho^{XA}$ and the ensemble $\{p_x, |\psi_x\rangle\langle\psi_x|\}\{x \in [m]\}$, we demonstrate that it is possible to transform $\rho^{XA}$ into $\{p_x, |\psi_x\rangle\langle\psi_x|\}_{x \in [m]}$ and vice versa. Firstly, consider performing a measurement in the $|x\rangle$ basis on system $X$ of a composite system $XA$ in the cq-state $\rho^{XA}$. This measurement yields the state $|\psi_x\rangle$ with a probability of $p_x$. Consequently, this process reconstructs the ensemble $\{p_x, |\psi_x\rangle\langle\psi_x|\}_{x \in [m]}$ from $\rho^{XA}$.

Conversely, imagine that we have a state $|\psi_x\rangle$ randomly selected from the ensemble $\{p_x, |\psi\rangle\langle\psi_x|\}_{x \in [m]}$. If Alice possesses knowledge of which state was selected (i.e. she knows the value of $x$), she can encode this information by introducing $|x\rangle\langle x|^X$, resulting in her state transitioning to $|x\rangle\langle x|^X \otimes |\psi_x\rangle\langle\psi_x|^A$. When Alice opts to forget the specific value of $x$, her quantum state becomes identical to $\rho^{XA}$. Furthermore, it is worth noting that when we only have access to the marginal state $\rho^A = \mathrm{Tr}_X\left[\rho^{XA}\right] = \sum_{x \in [m]} p_x |\psi_x\rangle\langle\psi_x|$, it is generally impossible to perfectly recover the value of $x$.

Cq-states play a pivotal role in quantum information science, particularly when describing the outcomes of quantum measurements. Let's consider a physical system characterized by the density operator $\rho \in \mathfrak{D}(A)$ and a generalized quantum measurement $\{M_x\}_{x \in [m]}$. As already discussed, the application of this generalized measurement to the state $\rho^A$ results in the state $\sigma_x^A$, as outlined in (3.17), with the associated probability $p_x$ as defined in (3.16). Because we know the outcome $x$, we have the option to record it within a classical system denoted as $X$. In this context, we can perceive the measurement's effect as a transformation, mapping the state $\rho^A$ to a cq-state, represented by $\sigma^{XA}$, defined as follows:

$$\sigma^{XA} := \sum_{x \in [m]} p_x |x\rangle\langle x|^X \otimes \sigma_x^A = \sum_{x \in [m]} |x\rangle\langle x|^X \otimes M_x \rho^A M_x^*, \tag{3.29}$$

where we substitute $p_x \sigma_x^A = M_x \rho^A M_x^*$. In essence, by employing a cq-state, we can describe the impact of a generalized measurement $\{M_x\}_{x \in [m]}$ on $\rho^A$ as a "deterministic" process, transforming $\rho^A$ into $\sigma^{XA}$. It's important to note that we use the term "deterministic" to describe the transformation $\rho^A \to \sigma^{XA}$, not to characterize the mapping $\rho^A \to \sigma_x^A$, which is inherently indeterministic in nature.

If the information about the measurement outcome $x$ is lost, or if Alice has no access to the value of $x$, then the post-measurement state is given by

$$\sigma^A = \mathrm{Tr}_X\left[\sigma^{XA}\right] = \sum_{x \in [m]} M_x \rho^A M_x^*. \tag{3.30}$$

We will see later on that in this case, the measurement acts as a quantum channel, converting one density operator, $\rho^A$, to another, $\sigma^A$ (see Figure 3.3c).

## 3.2.4 Separable Density Operators

Density operators that are acting on a bipartite Hilbert space $A \otimes B$ can be divided into two types: separable and entangled. A separable density matrix can be prepared in the following way (see Figure (3.2)). A referee samples a number $x$ with a probability distribution $p_x$ (e.g. roll a possiblly biased dice, or flip a coin) and sends the number $x$ to Alice and Bob who are spatially separated. Based on this value, Alice prepares the state $\rho_x^A$, and Bob prepares the state $\tau_x^B$. Then, if Alice and Bob forget the value of $x$, but still know the distribution $p_x$ from which $x$ was sampled, the state of their composite system becomes

$$\sigma^{AB} = \sum_{x \in [m]} p_x \rho_x^A \otimes \tau_x^B \in \mathfrak{D}(A \otimes B). \tag{3.31}$$

Note that the role of the referee above is to provide Alice and Bob with a shared randomness. Therefore, any separable state as in (3.31) can be prepared by local operations assisted with shared randomness. Bipartite density matrices that do not have this form are called entangled and we will discuss them in detail in the following chapters on entanglement theory.

**Exercise 3.7.** *Show that the maximally entangled state* $\Phi^{AB} := |\Phi^{AB}\rangle\langle\Phi^{AB}| \in \mathfrak{D}(A \otimes B)$ *is not separable.*

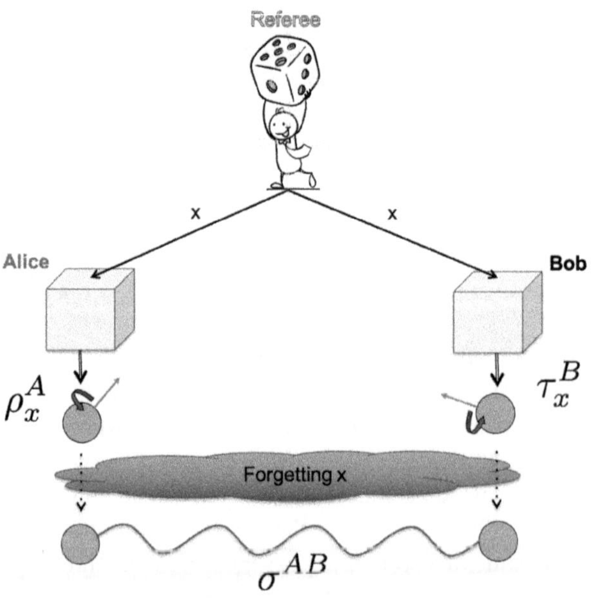


Figure 3.2　Preparation of a separable state with shared randomness.

**Exercise 3.8.** *Show that if $\sigma \in \mathfrak{D}(A \otimes B)$ is separable, then there exists an integer $k \in \mathbb{N}$, a probability distribution $\{q_z\}_{z\in[k]}$, a set of $k$ pure states $\{\psi_z\}_{z\in[k]}$ in Alice's Hilbert space (i.e. each $\psi_z \in \mathrm{Pure}(A)$), and a set of $k$ pure states $\{\phi_z\}_{z\in[k]}$ on Bob's Hilbert space $B$, such that*

$$\sigma^{AB} = \sum_{z\in[k]} q_z \psi_z^A \otimes \phi_z^B. \tag{3.32}$$

## 3.3 Positive Operator Valued Measure (POVM)

Every quantum measurement can be viewed as a box (see Figure 3.3a) that takes as its input a quantum state $\rho$, and outputs a classical variable $x$ and a post-measurement state $\sigma_x$. In the SG experiment that we discussed in Section 2.1, the electrons get absorbed by the screen, and all there is left after the measurements are the spots on the screen. Therefore, the SG experiment can be viewed as a special type of measurement in which the quantum output is "traced out" (see Figure 3.3b). Such quantum measurements with only classical output are called positive operator valued measures, or in short POVM.

Recall that the Born's rule (adapted to density matrices and generalized measurements) states that the probability to obtain an outcome $x$, when a measurement $\{M_x\}_{x\in[m]}$ is performed on a system described by a density operator $\rho$, is given by $p_x = \mathrm{Tr}\left[M_x^* M_x \rho\right]$. Therefore, to describe a POVM we only need to consider the operators, $\{\Lambda_x := M_x^* M_x\}_{x\in[m]}$, since we are only interested in the statistics of the measurement and not the post-measurement state. The POVM operators $\Lambda_x$ are called *effects*, and have the following two properties:

$$\Lambda_x \geqslant 0 \quad \text{and} \quad \sum_{x\in[m]} \Lambda_x = I^A. \tag{3.33}$$

To every generalized measurement there exists a *unique* POVM that corresponds to it via the relation $\Lambda_x = M_x^* M_x$. However, for every POVM there are many quantum measurements corresponding to it.

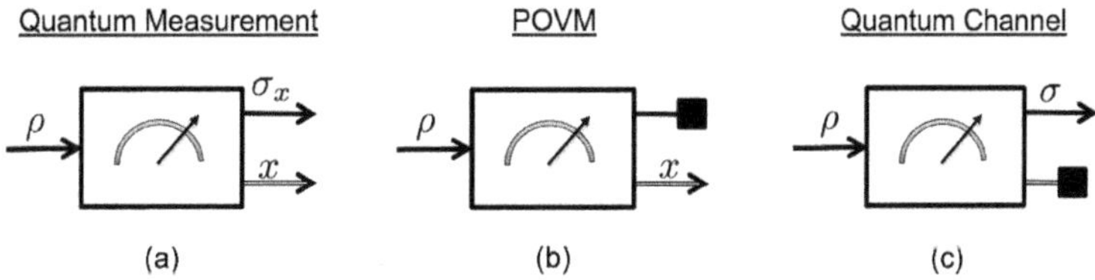


 Three types of generalized quantum measurements.

**Exercise 3.9.**  [Polar Decomposition]

1. *Show that for any $n \times n$ complex matrix $A$ there exists an $n \times n$ unitary matrix $U$ such that*

$$A = U|A| \quad where \quad |A| := \sqrt{A^* A}. \tag{3.34}$$

2. *Show that for $A$ and $U$ as above,*

$$\max_{U} \mathrm{Tr}\,[AU] = \mathrm{Tr}[|A|], \tag{3.35}$$

*where the maximum is over all unitary matrices $U$.*

**Exercise 3.10.**  *Let $\{\Lambda_x\}_{x\in[m]}$ be a POVM in $\mathrm{Pos}(A)$. Show that a generalized measurement $\{M_x\}_{x\in[m]} \subset \mathcal{L}(A)$ corresponds to the POVM $\{\Lambda_x\}_{x\in[m]}$ if and only if there exists $m$ unitary matrices, $\{U_x\}_{x\in[m]}$, in $\mathcal{L}(A)$ such that*

$$M_x = U_x\sqrt{\Lambda_x}. \tag{3.36}$$

*Hint: Use the polar decomposition of a complex matrix.*

**Exercise 3.11.**  *Consider the following POVM in $\mathrm{Pos}(\mathbb{C}^2)$*

$$\Lambda_1 = a|1\rangle\langle 1| \quad , \quad \Lambda_2 = b|-\rangle\langle -| \quad , \quad \Lambda_3 = I - \Lambda_1 - \Lambda_2 \quad a, b \in \mathbb{R}. \tag{3.37}$$

1. *Find all the possible values of $a$ and $b$ for which the set $\{\Lambda_1, \Lambda_2, \Lambda_3\}$ is a POVM.*
2. *Which values of $a$ and $b$ that you found in Part I correspond to a rank 1 POVM (i.e. all the POVM elements have rank 1)?*

**Exercise 3.12.**  *Suppose Alice and Bob share a composite quantum system in the state $\rho^{AB}$. Alice performs a measurement on her system described by a POVM $\{\Lambda_x\}_{x\in[m]}$, and record the outcome $x$ in a classical system $X$. Show that the post-measurement state can be expressed as a cq-state of the form*

$$\sigma^{XB} = \sum_{x\in[m]} p_x |x\rangle\langle x|^X \otimes \sigma_x^B. \tag{3.38}$$

*Express the probabilities $p_x$, and density matrices $\sigma_x^B$, in terms of $\Lambda_x$ and $\rho^{AB}$.*

### 3.3.1 Informationally Complete POVMs and Quantum Tomography

Consider a scenario where you have access to a machine that consistently produces an unknown quantum state $\rho \in \mathfrak{D}(A)$. Your objective is to learn the identity of this state $\rho$ by employing quantum measurements. Since we assume that this machine can be used repeatedly, generating an abundance of copies of $\rho$, it is sufficient to focus on POVMs. This is due to the fact that post-measurement states cannot reveal more information about $\rho$ than $\rho$ itself.

One effective strategy for learning $\rho$ is to carry out basis measurements $\{|x\rangle\langle x|\}_{x\in[d]}$ (where $d \equiv |A|$). By conducting these measurements multiple times, you can approximate the probabilities associated with each outcome $x$. Given that $p_x = \mathrm{Tr}[|x\rangle\langle x|\rho] = $

$\langle x|\rho|x\rangle$, this approach allows you to estimate the diagonal elements of $\rho$ within the basis $\{|x\rangle\}_{x\in[d]}$. Repeating this procedure using various bases enables you to ultimately learn the complete structure of $\rho$ by capturing its diagonal elements with respect to multiple bases. As we illustrate now, opting for POVMs over projective measurements in specific bases enables the construction of a single POVM that can be employed to fully identify the state $\rho$.

> **Definition 3.2.** A POVM $\{\Lambda_x\}_{x\in[m]}$ in Herm($A$) is said to be *informationally complete* if
>
> $$\text{span}_{\mathbb{R}}\{\Lambda_1, \Lambda_2, \ldots, \Lambda_m\} = \text{Herm}(A). \tag{3.39}$$

The span above is with respect to the real numbers since Herm($A$) is a real vector space. By definition, if a POVM $\{\Lambda_x\}_{x\in[m]}$ is informationally complete, then $m \geqslant d^2$, where $d := |A|$. Moreover, if $m = d^2$, then the informationally complete POVM form a basis of Herm($A$). Clearly, the basis is *not* orthonormal since $\Lambda_x \geqslant 0$ for all $x \in [d^2]$. A theorem from linear algebra states that for any basis of a vectors space there exists a *dual basis*. That is, if $\{\Lambda_1, \Lambda_2, \ldots, \Lambda_{d^2}\}$ is a basis of Herm($A$), then there exists another basis $\{\Gamma_1, \Gamma_2, \ldots, \Gamma_{d^2}\}$ of Herm($A$), such that the Hilbert Schmidt inner products

$$\text{Tr}\left[\Lambda_x \Gamma_y\right] = \delta_{xy} \quad \forall x, y \in [d^2]. \tag{3.40}$$

**Exercise 3.13.** *Consider the following four elements of* Pos($\mathbb{C}^2$):

$$\Lambda_0 := |0\rangle\langle 0| \quad , \quad \Lambda_1 := |1\rangle\langle 1| \quad , \quad \Lambda_2 := |+\rangle\langle +| \quad , \quad and \ \Lambda_3 := |+i\rangle\langle +i|. \tag{3.41}$$

1. *Show that* $\{\Lambda_0, \Lambda_1, \Lambda_2, \Lambda_3\}$ *is a basis of* Herm($\mathbb{C}^2$).
2. *Find its dual basis* $\{\Gamma_0, \Gamma_1, \Gamma_2, \Gamma_3\}$.
3. *Set* $\Lambda := \sum_{x=0}^{3} \Lambda_x$ *and show that it is invertible, and that the operators* $\{\tilde{\Lambda}_0, \tilde{\Lambda}_1, \tilde{\Lambda}_2, \tilde{\Lambda}_3\}$, *with* $\tilde{\Lambda}_x := \Lambda^{-1/2}\Lambda_x\Lambda^{-1/2}$, *form a rank 1 informationally complete POVM.*

**Exercise 3.14.**

1. *Construct a rank 1 informationally complete POVM in* Herm($A$). *Hint: Try to generalize the qubit example in the previous exercise to any (finite) dimension.*
2. *Let* $\Gamma \in$ Herm($AB$). *Show that if*

$$\text{Tr}\left[\Gamma^{AB}\left(\psi^A \otimes \phi^B\right)\right] = 0 \quad \forall \ \psi \in \text{Pure}(A) \quad \forall \ \phi \in \text{Pure}(B), \tag{3.42}$$

*then* $\Gamma^{AB} = 0$.

Informationally complete POVM can be used to learn an unknown quantum state. Let $\rho \in \mathfrak{D}(A)$ be an unknown quantum state, and let $\{\Lambda_x\}_{x\in[m]}$ be an informationally complete POVM in Herm($A$). If $m > d^2$ than $\{\Lambda_x\}_{x\in[m]}$ is an over-complete basis, also referred to, in linear algebra, as a *frame*. Every frame, like a basis, has a unique dual frame, $\{\Gamma_y\}_{y\in[m]}$, defined by the condition

$$\sum_{x\in[m]} \mathrm{Tr}[\Gamma_x M]\Lambda_x = \sum_{x\in[m]} \mathrm{Tr}[\Lambda_x M]\Gamma_x \qquad \forall\, M \in \mathrm{Herm}(A). \tag{3.43}$$

However, unlike a basis, in general a frame does not satisfy the relation (3.40) with its dual.

Consider now the case that $m = d^2$ so that $\{\Lambda_x\}_{x\in[m]}$ is a basis of $\mathrm{Herm}(A)$. Since its dual $\{\Gamma_y\}_{y\in[m]}$ also span $\mathrm{Herm}(A)$, it follows that the density matrix $\rho$ can be written in terms of the linear combination

$$\rho = \sum_{y\in[m]} p_y \Gamma_y \tag{3.44}$$

with some *real* coefficients $p_y$. A key observation is that

$$\mathrm{Tr}\,[\Lambda_x \rho] = \sum_{y\in[m]} p_y \mathrm{Tr}\,[\Lambda_x \Gamma_y]$$

$$(3.40)\rightarrow\ = \sum_{y\in[m]} p_y \delta_{xy} = p_x. \tag{3.45}$$

That is, the coefficients $p_y$ in (3.44) are given by $p_y = \mathrm{Tr}[\Lambda_y \rho] \geqslant 0$, and can be interpreted as the probability to obtain an outcome $y$. The significance of (3.44) with $p_y = \mathrm{Tr}[\Lambda_y \rho] \geqslant 0$, is that by repeating the POVM $\{\Lambda_x\}_{x\in[m]}$ on many copies of $\rho$, one can estimate from the measurement outcomes the values of the $p_y$s, and thereby learn $\rho$ due to the relation (3.44).

**Exercise 3.15.** *Let $\{\Lambda_x\}_{x\in[m]}$ be an informationally complete POVM in $\mathrm{Herm}(A)$, and let its dual frame be $\{\Gamma_y\}_{y\in[m]}$.*

1. *Show that at least one of the matrices in the dual frame is not positive semidefinite. That is, there exists $y \in [m]$ such that $\Gamma_y \not\geqslant 0$.*
2. *Show that $\mathrm{Tr}[\Gamma_y] = 1$ for all $y \in [m]$.*

**Exercise 3.16. [Symmetric Informationally Complete (SIC) POVM]**
*Let $d := |A|$, $m := d^2$, and $\{\Lambda_x\}_{x\in[m]}$ be an informationally complete POVM in $\mathrm{Herm}(A)$, with the following properties:*

1. *Each $\Lambda_x$ is rank 1.*
2. *$\mathrm{Tr}\,[\Lambda_x] = \mathrm{Tr}\,[\Lambda_y]$ for all $x, y \in [m]$.*
3. *$\mathrm{Tr}\,[\Lambda_x \Lambda_y] = \mathrm{Tr}\,[\Lambda_{x'} \Lambda_{y'}]$ for all $x, x', y, y' \in [m]$ with $x \neq y$ and $x' \neq y'$.*

*Such a POVM is called* symmetric *informationally complete POVM or in short SIC-POVM.*

1. *Show that*

$$\mathrm{Tr}\,[\Lambda_x] = \frac{1}{d} \quad and \quad \mathrm{Tr}\,[\Lambda_x \Lambda_y] = \frac{d\delta_{xy} + 1}{d^2(d+1)} \qquad \forall\, x, y \in [m]. \tag{3.46}$$

*Hint: Denote $a_x := \mathrm{Tr}[\Lambda_x]$ and $b := \mathrm{Tr}[\Lambda_x\Lambda_y]$ for any $x \neq y$. Then, $a_x^2 = \mathrm{Tr}[\Lambda_x^2]$ (why?) so that $a_x = \mathrm{Tr}[\Lambda_x I] = \sum_{y\in[m]} \mathrm{Tr}[\Lambda_x\Lambda_y] = a_x^2 + (d^2 - 1)b$ and conclude that $a_x = a_y$ for all $x, y \in [m]$.*

2. *Find the dual frame $\{\Gamma_x\}_{x\in[m]}$. Hint: Express each $\Gamma_x$ as a linear combination of $\Lambda_x$ and the identity.*

3. *Show that any density operator $\rho \in \mathrm{Herm}(A)$ can be expressed as*

$$\rho = \sum_{x\in[m]} \big(d(d+1)p_x - 1\big)\Lambda_x \quad \text{with} \quad p_x := \mathrm{Tr}[\Lambda_x\rho]. \tag{3.47}$$

4. *Show that*

$$\Lambda_1 = \frac{1}{12\sqrt{3}}\begin{pmatrix} 3\sqrt{3}+1 & -5+i \\ -5-i & 3\sqrt{3}-1 \end{pmatrix}, \Lambda_2 = \frac{1}{12\sqrt{3}}\begin{pmatrix} 3\sqrt{3}+1 & 1-5i \\ 1+5i & 3\sqrt{3}-1 \end{pmatrix}$$

$$\Lambda_3 = \frac{1}{12\sqrt{3}}\begin{pmatrix} 3\sqrt{3}-5 & 1+i \\ 1-i & 3\sqrt{3}+5 \end{pmatrix}, \Lambda_4 = \frac{1}{4\sqrt{3}}\begin{pmatrix} \sqrt{3}+1 & 1+i \\ 1-i & \sqrt{3}-1 \end{pmatrix} \tag{3.48}$$

*form a SIC POVM in $\mathrm{Herm}(\mathbb{C}^2)$. Moreover, show that the four pure state $\{2\Lambda_x\}_{x\in[4]}$ are the vertices of a tetrahedron in the Bloch sphere.*

## 3.3.2 Gleason's Theorem

One of the enigmas of quantum mechanics concerns with the emergence of probabilities at a very fundamental level. One may wonder if the Born's rule, which specifies how to assign probabilities to different measurement outcomes, could have been different. Since we are only interested in the statistics of the measurement we will consider here a POVM $\{\Lambda_x\}_{x\in[m]}$. According to Born's rule, the probability to obtain an outcome $x$ is given by the formula $p_x = \mathrm{Tr}[\Lambda_x\rho]$. Is this formula unique?

Specifically, consider the convex set of all effects

$$\mathrm{Eff}(A) := \left\{\Lambda \in \mathrm{Pos}(A): \; \Lambda \leqslant I^A\right\}. \tag{3.49}$$

A measure on the set $\mathrm{Eff}(A)$ is a function $\mu: \mathrm{Eff}(A) \to \mathbb{R}$ with the following three properties:

1. $0 \leqslant \mu(\Lambda) \leqslant 1$ for all $\Lambda \in \mathrm{Eff}(A)$.
2. $\mu(I^A) = 1$.
3. For any possibly incomplete POVM $\{\Lambda_x\}_{x\in[m]}$ with $\sum_{x\in[m]} \Lambda_x \leqslant I^A$

$$\mu\left(\sum_{x\in[m]} \Lambda_x\right) = \sum_{x\in[m]} \mu(\Lambda_x). \tag{3.50}$$

We now show that any function $\mu$ with these properties *must* have the form $\mu(\Lambda) = \mathrm{Tr}[\rho\Lambda]$ for some fixed density operator $\rho \in \mathfrak{D}(A)$. This means that there is no way to assign probabilities to effects other than the Born's rule. Originally, this remarkable result was proved by Gleason for the case of projective measurements, that is, the set $\mathrm{Eff}(A)$ was replaced with the set of all projections on $A$, and consequently the requirement on $\mu$ was much weaker, assuming only that (3.50) holds for orthogonal projections. Nonetheless, Gleason was able to derive the probability formula for systems of dimension $d \geqslant 3$, and in the qubit case he showed that there are counter examples. Gleason also considered in his proof the infinite dimensional case, and derived similar results. Gleason's proof for projective measurements goes beyond the scope of this book, and we will follow here a much simpler proof of the above generalization of Gleason theorem (see the section on Notes and References for more details).

> **Theorem 3.1.** Let $A$ be a finite dimensional Hilbert space, and $\mu$ a measure on the set $\mathrm{Eff}(A)$. Then, there exists $\rho \in \mathfrak{D}(A)$ such that
>
> $$\mu(\Lambda) = \mathrm{Tr}[\rho\Lambda] \qquad \forall \, \Lambda \in \mathrm{Eff}(A). \tag{3.51}$$

The idea of the proof is as follows. First we will show that for any $r \in [0, 1]$ $\mu(r\Lambda) = r\mu(\Lambda)$, and use it to show that $\mu$ can be extended to a linear functional on the space of Hermitian operators $\mathrm{Herm}(A)$. Then, as a linear functional, it can be expressed as $\mu(\Lambda) = \mathrm{Tr}[\Lambda\rho]$, and we end by showing that $\rho$ must be a density operator.

**Proof**   Note that for any effect $\Lambda \in \mathrm{Eff}(A)$ and any integer $n$, we have $\Lambda = \frac{1}{n}\Lambda + \cdots \frac{1}{n}\Lambda \leqslant I^A$, where the sum contains $n$ terms. Therefore, from (3.50), $\mu(\Lambda) = n\mu(\frac{1}{n}\Lambda)$. Multiplying this equation by an integer $m \leqslant n$ and dividing by $n$ gives $\frac{m}{n}\mu(\Lambda) = m\mu(\frac{1}{n}\Lambda) = \mu\left(\frac{m}{n}\Lambda\right)$, where we used the third property of a measure $\mu$ as defined above. So far we have shown that for any rational number $p \in [0, 1]$, we must have $\mu(p\Lambda) = p\mu(\Lambda)$. Let $r \in [0, 1]$ be a real number and let $\{p_j\}$ and $\{q_k\}$ be two sequences of *rational* numbers in $[0, 1]$ that converge to $r$ and have the property that $p_j \leqslant r \leqslant q_k$ for all $j$ and $k$. We therefore have

$$p_j\mu(\Lambda) = \mu(p_j\Lambda) \leqslant \mu(r\Lambda) \leqslant \mu(q_k\Lambda) = q_k\mu(\Lambda), \tag{3.52}$$

where the inequalities above follows from the fact that if two effects satisfy $\Lambda \geqslant \Gamma$ (i.e. $\Lambda - \Gamma \geqslant 0$), then

$$\mu(\Lambda) = \mu\big(\Gamma + (\Lambda - \Gamma)\big) = \mu(\Gamma) + \mu(\Lambda - \Gamma) \geqslant \mu(\Gamma). \tag{3.53}$$

Taking the limit $j, k \to \infty$ gives $\mu(r\Lambda) = r\mu(\Lambda)$ for all $r \in [0, 1]$.

We now extend the definition of $\mu$ to any element in $\mathrm{Herm}(A)$. First, for any positive semidefinite matrix $P \geqslant 0$ that is *not* in $\mathrm{Eff}(A)$, there always exists $r > 1$ such that $\frac{1}{r}P \in \mathrm{Eff}(A)$. Define $\mu(P) := r\mu(\frac{1}{r}P)$. To show that $\mu(P)$ is well defined, let $r' > 1$ be another number such that $\frac{1}{r'}P \in \mathrm{Eff}(A)$ and assume without loss of generality that $r' > r$ so that $\frac{r}{r'} < 1$. Then,

$$r\mu\left(\frac{1}{r}P\right) = r'\frac{r}{r'}\mu\left(\frac{1}{r}P\right) = r'\mu\left(\frac{1}{r'}P\right). \tag{3.54}$$

Note that this extension of the domain of $\mu$ to any element of Pos($A$) preserves the linearity of $\mu$; that is, for any two matrices $M, N \in$ Pos($A$) and large enough $r$ such that $\frac{1}{r}(M + N) \in$ Eff($A$),

$$\mu(M + N) = r\mu\left(\frac{1}{r}(M + N)\right) = r\mu\left(\frac{1}{r}M\right) + r\mu\left(\frac{1}{r}N\right) = \mu(M) + \mu(N). \quad (3.55)$$

Finally, we extend the definition of $\mu$ to include in its domain any matrix $L \in$ Herm($A$). Any such matrix can be expressed as $L = M - N$, with $M, N \in$ Pos($A$). We therefore define

$$\mu(L) := \mu(M) - \mu(N). \quad (3.56)$$

To show that $\mu(L)$ is well defined, we need to show that for any other decomposition of $L = M' - N'$ with $M', N' \in$ Pos($A$), we have

$$\mu(M) - \mu(N) = \mu(M') - \mu(N'). \quad (3.57)$$

This is indeed the case since the equality $L = M - N = M' - N'$ implies that $M + N' = M' + N$ and from the additivity property in (3.55) we conclude that

$$\mu(M) + \mu(N') = \mu(M') + \mu(N), \quad (3.58)$$

which is equivalent to (3.57).

To summarize, we where able to extend $\mu$ to a linear functional $\mu : $ Herm($A$) $\to \mathbb{R}$. Since the dual space of Herm($A$) is itself, any linear functional on Herm($A$) can be expressed as $\mu(\Lambda) = \text{Tr}[\rho\Lambda]$, where $\rho \in$ Herm($A$) is a fixed matrix. Now, from the fact that $\mu(\Lambda) \geqslant 0$ for all effects $\Lambda \in$ Eff($A$) we conclude that $\text{Tr}[\rho\Lambda] \geqslant 0$ for all $\Lambda \geqslant 0$, which implies that $\rho \geqslant 0$ (see the following exercise). Finally, to show that $\rho$ is a density operator, note that the condition $\mu(I) = 1$ gives $\text{Tr}[\rho] = 1$. This completes the proof. $\blacksquare$

**Exercise 3.17.** *Show that $\rho \geqslant 0$ if and only if $\text{Tr}[\rho\Lambda] \geqslant 0$ for all $\Lambda \geqslant 0$.*

### 3.3.3 Naimark's Dilation Theorem

Naimark's dilation theorem reveals that any POVM can be implemented with a von-Neumann projective measurement on a larger Hilbert space. This is not much of a surprise to us as we have seen that generalized measurements can be implemented by a bipartite unitary map followed by a projective measurement. However, Naimark's theorem deals directly with POVMs and makes the connection between POVMs and projective measurements more transparent. Moreover, Naimark's dilation theorem is applicable to infinite dimensional systems, although we will prove it here only in the finite dimensional case.

> **Naimark's Theorem**
>
> **Theorem 3.2.** Let $A$ be a Hilbert space and $\{\Lambda_x\}_{x\in[m]} \subset \mathrm{Eff}(A)$ be a POVM. Then, there exists an extended Hilbert space $B$ (i.e. $|B| \geqslant |A|$), an isometry $V:A \to B$, and a von-Neumann projective measurement $\{P_x\}_{x\in[m]} \subset \mathrm{Eff}(B)$, such that $\Lambda_x = V^*P_xV$ for all $x \in [m]$.

**Proof** Every POVM element $\Lambda_x$ can be expressed as $\Lambda_x = M_x^*M_x$, where $\{M_x\}$ is a generalized measurement. In Section 3.1 we saw that for every generalized measurement there exists an ancillary system $R$, and unitary matrix $U^{RA}$, such that (cf. (3.2))

$$M_x^A = {}^R\langle x|U^{RA}|1\rangle^R, \tag{3.59}$$

where $|1\rangle^R$ is some fixed state in $R$. Define $B := RA$ and define the operator $V : A \to B$ by $V := U^{RA}|1\rangle^R$. That is, for any $|\psi\rangle \in A$

$$V|\psi\rangle^A := U^{RA}|1\rangle^R|\psi\rangle^A \in B. \tag{3.60}$$

Clearly, $V^*V = I^A$ so that $V$ is an isometry. With this definition we get that

$$\begin{aligned}\Lambda_x &= M_x^*M_x \\ (3.59)\to &= {}^R\langle 1|U^{*RA}\left(|x\rangle\langle x|^R \otimes I^A\right)U^{RA}|1\rangle^R \\ &= V^*\left(|x\rangle\langle x|^R \otimes I^A\right)V.\end{aligned} \tag{3.61}$$

Denoting by $P_x^B := |x\rangle\langle x|^R \otimes I^A$ we conclude that $\Lambda_x = V^*P_x^BV$. This completes the proof. $\blacksquare$

We point out that any POVM $\{\Lambda_x\}_{x\in[m]} \subset \mathrm{Eff}(A)$ can be implemented with a rank 1 POVM in the following simple way. Since each $\Lambda_x$ is positive semidefinite, it can be expressed in terms of its (unnormalized) eigenvectors $\{|\phi_{xy}\rangle\}_{y\in[n]}$ as

$$\Lambda_x = \sum_{y\in[n]} \phi_{xy}. \tag{3.62}$$

Moreover, the set of rank 1 matrices $\{\phi_{xy}\}$ also form a POVM (i.e. $\sum_{x,y}\phi_{xy} = I^A$). Therefore, one can implement the POVM $\{\Lambda_x\}_{x\in[m]}$ by first implementing the rank 1 POVM $\{\phi_{xy}\}$, with corresponding $(x, y)$ outcomes, and then forgetting/ignoring the outcome $y$.

**Exercise 3.18.** *Show that if* $\{\Lambda_x\}_{x\in[m]}$ *is a rank 1 POVM in* $\mathrm{Eff}(A)$, *where* $m := |A|$, *then* $\{\Lambda_x\}_{x\in[m]}$ *is a basis measurement (i.e. rank 1 von-Neumann projective measurement).*

**Exercise 3.19.** *Consider the rank 1 matrices*

$$\Gamma_x = \frac{2}{3}|\uparrow_{\mathbf{n}_x}\rangle\langle\uparrow_{\mathbf{n}_x}|, \quad x = 1, 2, 3 \;;\; |\uparrow_{\mathbf{n}_x}\rangle \in \mathbb{C}^2 \tag{3.63}$$

*were the unit vectors* $\mathbf{n}_x \in \mathbb{R}^3$ *satisfies* $\mathbf{n}_1 + \mathbf{n}_2 + \mathbf{n}_3 = 0$.

1.  *Show that the set $\{\Gamma_x\}_{x\in[3]}$ is a POVM.*
2.  *Find an orthonormal basis $\{|\varphi_x\rangle\}_{x\in[3]}$ in $\mathbb{C}^3$ such that $\{\Gamma_x\}_{x\in[3]}$ is realized by $\{\varphi_x\}_{x\in[3]}$.*

## 3.4 Evolution of Open Systems

What is the most general evolution that a (possibly open) quantum system can undergo? We will first tackle this problem axiomatically, using minimal physical assumptions, and then provide several ways to demonstrate how to physically realize such an evolution. We will see that any evolution of a physical system can be described with a quantum channel. Quantum channels lie at the heart of quantum information theory, and we will devote much of this section to describe their different representations.

### 3.4.1 The Axiomatic Approach

For closed systems, a pure state $|\psi\rangle$ can evolve either deterministically (described by a unitary matrix $U$) to $U|\psi\rangle$ or probabilistically to $|\phi_x\rangle$ with some probability $p_x$. The latter can be described as an evolution from the density matrix $|\psi\rangle\langle\psi|$ to the classical quantum state $\sigma^{XA} = \sum_x p_x |x\rangle\langle x| \otimes |\phi_x\rangle\langle\phi_x|$. Since any open system is described with a density operator, any evolution of a quantum system can be described with a transformation $\mathcal{E}$ that takes density operators in $\mathfrak{D}(A)$ to density operators in $\mathfrak{D}(B)$. Note that since the systems are open, the transformation $\mathcal{E}$ can change the dimension (e.g. particles added or discarded) so that the input dimension $|A|$ can be different than the output dimension $|B|$.

Recall that we use the notation $\mathfrak{L}(A, B)$ to indicate linear transformations from $A$ to $B$. However, our current focus lies in the set $\mathfrak{L}(A \to B)$, which denotes transformations from $\mathfrak{L}(A)$ to $\mathfrak{L}(B)$. We will typically employ calligraphic letters like $\mathcal{E}, \mathcal{F}, \mathcal{N}$, and $\mathcal{M}$ to denote the elements of $\mathfrak{L}(A \to B)$ and sometimes include a superscript, such as $\mathcal{E}^{A\to B}$, to emphasize the underlying input and output Hilbert spaces. The identity element of $\mathfrak{L}(A \to A)$ will be denoted as $\mathrm{id}^A$ or $\mathrm{id}^{A\to A}$.

**Exercise 3.20.** *Let $A$ and $B$ be two Hilbert spaces, $m := |A|$, and $\{\eta_j\}_{j\in[m^2]}$ be an orthonormal basis of $\mathfrak{L}(A)$ (in the Hilbert–Schmidt inner product). For any two elements $\mathcal{E}, \mathcal{F} \in \mathfrak{L}(A \to B)$ define*

$$\langle\mathcal{E},\mathcal{F}\rangle := \sum_{j\in[m^2]} \big\langle\mathcal{E}(\eta_j),\mathcal{F}(\eta_j)\big\rangle_{HS}, \tag{3.64}$$

*where $\langle\,,\,\rangle_{HS}$ denotes the Hilbert–Schmidt inner product between matrices.*

1.  *Show that the function above is well defined in the sense that it is independent on the choice of the orthonormal basis $\{\eta_j\}_{j\in[m^2]}$ of $\mathfrak{L}(A)$.*
2.  *Show that the function in (3.64) is an inner product in the vector space $\mathfrak{L}(A \to B)$. Hence, $\mathfrak{L}(A \to B)$ is a Hilbert space.*

We are now ready to introduce the axiomatic approach describing a physical evolution from system $A$ into $B$. Below each axiom we provide the physical justification.

**Axiom 1:** *A physical evolution can be described with a linear transformation.*

Consider the scenario where Alice rolls a dice to obtain a classical variable $x \in [m]$, with associated probabilities $\{p_x\}_{x \in [m]}$. If she obtains the value $x$, she prepares her system in the state $\rho_x$. After a quantum evolution takes place, if the initial state was $\rho_x$, it will evolve to $\mathcal{E}(\rho_x)$. Now, if Alice forgets which state she prepared (i.e. she forgets the outcome of the dice roll), from her new perspective, the input state is given by $\sum_{x \in [m]} p_x \rho_x$. In the same vein, the post-evolution state of the system becomes $\sum_{x \in [m]} p_x \mathcal{E}(\rho_x)$. Consequently, we must have

$$\mathcal{E}\left( \sum_{x \in [m]} p_x \rho_x \right) = \sum_{x \in [m]} p_x \mathcal{E}(\rho_x). \tag{3.65}$$

Equation (3.65) signifies that $\mathcal{E}$ is convex-linear (i.e. linear under convex combinations) on the set of density matrices and can be extended linearly to act on the entire space $\mathfrak{L}(A)$ (not limited to $\mathfrak{D}(A)$ and not limited to convex combinations). Therefore, we infer that every map describing a physical evolution is an element of $\mathfrak{L}(A \to B)$.

**Axiom 2:** *A physical evolution is trace preserving.*

Since a physical evolution $\mathcal{E} \in \mathfrak{L}(A \to B)$ takes one density matrix to another density matrix, it must satisfy for all $\rho \in \mathfrak{D}(A)$, $\mathrm{Tr}[\mathcal{E}(\rho)] = 1$, since $\mathcal{E}(\rho)$ is a density matrix. Now, let $\eta \in \mathrm{Herm}(A)$ be an arbitrary Hermitian matrix. Then, $\eta$ can always be written as $\eta = t\rho - s\sigma$, where $t, s \geqslant 0$ and $\rho, \sigma \in \mathfrak{D}(A)$. Therefore, from the linearity of $\mathcal{E}$ we get that

$$\mathrm{Tr}\left[\mathcal{E}(\eta)\right] = t\,\mathrm{Tr}\left[\mathcal{E}(\rho)\right] - s\,\mathrm{Tr}\left[\mathcal{E}(\sigma)\right] = t - s = \mathrm{Tr}[\eta], \tag{3.66}$$

so that $\mathcal{E}$ preserves the trace of hermitian matrices. Moreover, if $M \in \mathfrak{L}(A)$ is not Hermitian, it can still be expressed as $M = \eta_0 + i\eta_1$, where both $\eta_0 := (M + M^*)/2$ and $\eta_1 := (M - M^*)/2i$ are Hermitian matrices, so that

$$\mathrm{Tr}\left[\mathcal{E}(M)\right] = \mathrm{Tr}\left[\mathcal{E}(\eta_0)\right] + i\,\mathrm{Tr}\left[\mathcal{E}(\eta_1)\right] = \mathrm{Tr}\left[\eta_0\right] + i\,\mathrm{Tr}\left[\eta_1\right] = \mathrm{Tr}[M]. \tag{3.67}$$

We therefore conclude that any physical evolution $\mathcal{E}$ is a *trace preserving* (TP) linear map.

**Axiom 3:** *A physical evolution is completely positive.*

Since a physical evolution $\mathcal{E}$ takes one density matrix to another density matrix, it also preserves positivity . That is, if $\rho \in \mathrm{Pos}(A)$ is positive semidefinite matrix, then also $\mathcal{E}(\rho)$ is a positive semidefinite matrix in $\mathrm{Pos}(B)$. We call such linear maps *positive* maps. There is yet one more property that $\mathcal{E}$ has to satisfy if it describes an evolution a physical system.

Consider a composite system consisting of two subsystems $A$ and $B$. Such a system is described by a bipartite density operator $\rho^{AB} \in \mathfrak{D}(A \otimes B)$. If the subsystem $B$ undergoes a physical evolution described by a linear map $\mathcal{E} \in \mathcal{L}(B \to B')$, while system $A$ does not evolve and remain intact, then the state $\rho^{AB}$ will evolve to the state

$$\sigma^{AB'} := \mathrm{id}^A \otimes \mathcal{E}^{B \to B'}\left(\rho^{AB}\right). \tag{3.68}$$

Therefore, if $\mathcal{E}$ represents a physical evolution, then both $\mathcal{E}$ and $\mathrm{id}^A \otimes \mathcal{E}$ must take one density matrix to another density matrix. In particular, the linear map $\mathrm{id} \otimes \mathcal{E} \in \mathcal{L}(AB \to AB')$ must also be a positive map for any system $A$. It turns out that there are linear maps $\mathcal{E}$ that are positive while $\mathrm{id}^A \otimes \mathcal{E}$ is not positive. One such example is the transposition map.

Consider the linear map $\mathcal{T} \in \mathcal{L}(A \to A)$ defined by

$$\mathcal{T}(\rho) := \rho^T \qquad \forall\, \rho \in \mathcal{L}(A). \tag{3.69}$$

The transpose map preserves the eigenvalues and therefore is a trace preserving positive map. Now, take $A = \mathbb{C}^2$ and consider the matrix $\Omega^{A\tilde{A}} := |\Omega^{A\tilde{A}}\rangle\langle\Omega^{A\tilde{A}}| \in \mathcal{L}(\mathbb{C}^2 \otimes \mathbb{C}^2)$. The matrix $\Omega^{A\tilde{A}}$ is a rank 1, positive semidefinite matrix that can be expressed as

$$\Omega^{A\tilde{A}} = |00\rangle\langle00| + |00\rangle\langle11| + |11\rangle\langle00| + |11\rangle\langle11| = \begin{bmatrix} 1 & 0 & 0 & 1 \\ 0 & 0 & 0 & 0 \\ 0 & 0 & 0 & 0 \\ 1 & 0 & 0 & 1 \end{bmatrix}. \tag{3.70}$$

On the other hand, its *partial transpose* on system $\tilde{A}$ is given by

$$\mathcal{T}^{\tilde{A} \to \tilde{A}}\left(\Omega^{A\tilde{A}}\right) = |00\rangle\langle00| + |01\rangle\langle10| + |10\rangle\langle01| + |11\rangle\langle11| = \begin{bmatrix} 1 & 0 & 0 & 0 \\ 0 & 0 & 1 & 0 \\ 0 & 1 & 0 & 0 \\ 0 & 0 & 0 & 1 \end{bmatrix}. \tag{3.71}$$

It is relatively easy to check (see Exercise (3.21)) that $\mathcal{T}^{\tilde{A} \to \tilde{A}}\left(\Omega^{A\tilde{A}}\right)$ has three eigenvalues equals to 1, and one eigenvalue equals $-1$. Hence, $\mathcal{T}^{\tilde{A} \to \tilde{A}}\left(\Omega^{A\tilde{A}}\right)$ is not positive semidefinite!

**Exercise 3.21.** *Let* $\psi^{A\tilde{A}} := |\psi\rangle\langle\psi|$ *with* $|\psi^{A\tilde{A}}\rangle = \sum_{x\in[d]} \sqrt{p_x}|x\rangle|x\rangle \in A \otimes \tilde{A}$. *Find the eigenvalues and eigenvectors of* $\mathcal{T}^{\tilde{A} \to \tilde{A}}(\psi^{A\tilde{A}})$. *For which values of* $p_x$*, the matrix* $\mathcal{T}^{\tilde{A} \to \tilde{A}}(\psi^{A\tilde{A}})$ *is positive semidefinite?*

**Complete Positivity**

> **Definition 3.3.** A linear map $\mathcal{E} \in \mathcal{L}(A \to B)$ is called $k$-positive if $\mathrm{id}_k \otimes \mathcal{E}$ is positive, where $\mathrm{id}_k \in \mathcal{L}(\mathbb{C}^k \to \mathbb{C}^k)$ is the identity map. Furthermore, $\mathcal{E}$ is called completely positive (CP) if it is $k$-positive for all $k \in \mathbb{N}$.

By definition, every map that is $k$-positive is also $k'$-positive if $k' \leqslant k$. On the other hand, there are maps that are $k$-positive but not $(k+1)$-positive. The canonical example of such a map is the map $\mathcal{E}^{(k)} \in \mathfrak{L}(A \to A)$ (with $d := |A|$) defined by

$$\mathcal{E}^{(k)}(\eta) = k\mathrm{Tr}[\eta]I_d - \eta \qquad \forall\, \eta \in \mathfrak{L}(A). \tag{3.72}$$

From Exercise 3.34 in the next few subsections it follows that $\mathcal{E}^{(k)}$ is $k$-positive, and yet, if $k < d$, then $\mathcal{E}^{(k)}$ is *not* $(k+1)$-positive. If $k \geqslant d$, then this map is completely positive. More generally, we will see later that a map is $d$-positive if and only if it is completely positive.

In conclusion, any evolution of a physical system has to be (1) linear, (2) trace preserving, and (3) completely positive (CP). Such a linear CPTP map is called a quantum channel. The set of all quantum channels in $\mathfrak{L}(A \to B)$ will be denoted by CPTP$(A \to B)$. In the next subsections we discuss several representations of quantum channels, and along the way show that *any* quantum channel has a physical realization; that is, it can be implemented by physical processes. Therefore, the axiomatic approach discussed in Section 3.4.1 led us to the precise conditions on the evolution of a physical system that are both necessary and sufficient for the existence of its physical realization.

**Exercise 3.22.** *Let $\Lambda \in \mathfrak{L}(A)$ and define a map $\mathcal{F}_\Lambda : \mathfrak{L}(A) \to \mathfrak{L}(A)$ via*

$$\mathcal{F}_\Lambda(\omega) := \Lambda \omega \Lambda^* \qquad \forall\, \omega \in \mathfrak{L}(A). \tag{3.73}$$

*Show that $\mathcal{F}_\Lambda$ is a completely positive* linear *map. Hint: Prove first that $\mathcal{F}_\Lambda^{A \to A}\left(\psi^{RA}\right) \geqslant 0$ for a pure state $|\psi^{RA}\rangle = M \otimes I^A |\Omega^{\tilde{A}A}\rangle$.*

## The Dual of a Linear Map in $\mathfrak{L}(A \to B)$

We have already seen that any vector in a Hilbert space has a dual vector. In particular, we saw that any matrix $M \in \mathfrak{L}(A, B)$ has a dual or adjoint matrix $M^* \in \mathfrak{L}(B, A)$. Since the space $\mathfrak{L}(A \to B)$ is also a Hilbert space with respect to the inner product given in (3.64), it follows that any map in $\mathfrak{L}(A \to B)$ has a dual map in $\mathfrak{L}(B \to A)$.

---

**Definition 3.4.** Let $\mathcal{E} \in \mathfrak{L}(A \to B)$ be a linear map. Its dual or adjoint map is the map $\mathcal{E}^* \in \mathfrak{L}(B \to A)$ that satisfies

$$\mathrm{Tr}\left[\sigma^* \mathcal{E}(\rho)\right] = \mathrm{Tr}\left[\left(\mathcal{E}^*(\sigma)\right)^* \rho\right] \qquad \forall\, \rho \in \mathfrak{L}(A) \quad \text{and} \quad \forall\, \sigma \in \mathfrak{L}(B). \tag{3.74}$$

Furthermore, we say that $\mathcal{E}$ is self-adjoint if $|A| = |B|$ and $\mathcal{E} = \mathcal{E}^*$.

---

Definition 3.4 of the dual map is analogous to the definition of a dual map in $\mathfrak{L}(A, B)$. Recall that the dual of a matrix $M \in \mathfrak{L}(A, B)$ is defined via the relation

$$\langle \phi | M \psi \rangle = \langle M^* \phi | \psi \rangle \qquad \forall\, |\psi\rangle \in A,\ |\phi\rangle \in B. \tag{3.75}$$

Similarly, the definition of $\mathcal{E}^*$ in (3.74) can be expressed as

$$\langle \sigma, \mathcal{E}(\rho)\rangle_{HS} = \langle \mathcal{E}^*(\sigma), \rho\rangle_{HS} \qquad \forall\, \rho \in \mathfrak{L}(A),\ \sigma \in \mathfrak{L}(B), \tag{3.76}$$

where $\langle \cdot, \cdot\rangle_{HS}$ is the Hilbert–Schmidt inner product.

**Exercise 3.23.** *Let $\mathcal{E} \in \mathfrak{L}(A \to B)$ be a linear map.*

1. *Show that $\mathcal{E}$ is trace preserving if and only if its dual $\mathcal{E}^*$ is unital; that is, $\mathcal{E}^*(I^B) = I^A$.*
2. *Show that $\mathcal{E}$ is trace nonincreasing (i.e. $\mathrm{Tr}[\mathcal{E}(\eta)] \leqslant \mathrm{Tr}[\eta]$ for all $\eta \in \mathrm{Pos}(A)$) if and only if its dual $\mathcal{E}^*$ is sub-unital; that is, $\mathcal{E}^*(I^B) \leqslant I^A$.*
3. *Show that $\mathcal{E}$ is positive if and only if $\mathcal{E}^*$ is positive.*
4. *Show that $\mathcal{E}$ is completely positive if and only if $\mathcal{E}^*$ is completely positive.*

**Exercise 3.24.** *Show that a unitary evolution is a CPTP map. Specifically, show that a unitary map $\mathcal{U} \in \mathfrak{L}(A \to A)$, defined by*

$$\mathcal{U}(\rho) := U\rho U^* \qquad \forall\, \rho \in \mathfrak{L}(A), \tag{3.77}$$

*where $U \in \mathfrak{U}(A)$, is a unitary operator and a quantum channel.*

**Exercise 3.25.** *The replacement map is a map $\mathcal{E} \in \mathfrak{L}(A \to B)$ defined by*

$$\mathcal{E}(\rho) := \mathrm{Tr}[\rho]\,\sigma \qquad \forall\, \rho \in \mathfrak{L}(A), \tag{3.78}$$

*where $\sigma \in \mathfrak{D}(B)$ is some fixed density matrix.*

1. *Show that $\mathcal{E}$ is a quantum channel.*
2. *Show that for any two quantum states $\rho \in \mathfrak{D}(A)$ and $\sigma \in \mathfrak{D}(B)$ there exists a quantum channel $\mathcal{E}$ such that $\mathcal{E}(\rho) = \sigma$.*

**Exercise 3.26.** *Let $A$ and $B$ be two finite dimensional Hilbert spaces, and denote the set of positive maps in $\mathfrak{L}(A \to B)$ by*

$$\mathrm{Pos}(A \to B) := \Big\{ \mathcal{E} \in \mathfrak{L}(A \to B) : \mathcal{E}(\rho) \in \mathrm{Pos}(B) \quad \forall\, \rho \in \mathrm{Pos}(A) \Big\}. \tag{3.79}$$

*Denote also by $\mathcal{R} \in \mathrm{CPTP}(A \to B)$ the replacement channel $\mathcal{R}(\rho^A) := \mathrm{Tr}[\rho^A]\frac{1}{|B|}I^B$ for all $\rho \in \mathfrak{L}(A)$.*

1. *Show that $\mathrm{Pos}(A \to B)$ is a convex cone in the Hilbert space $\mathfrak{L}(A \to B)$.*
2. *Prove the equivalence of the following properties of a map $\mathcal{E} \in \mathrm{Pos}(A \to B)$:*

   *(a) $\mathcal{E}$ belongs to the interior of the cone $\mathrm{Pos}(A \to B)$.*
   *(b) $\mathcal{E} = (1 - t)\mathcal{F} + t\mathcal{R}$ for some $t \in (0, 1]$ and some $\mathcal{F} \in \mathrm{Pos}(A \to A)$.*
   *(c) $\mathcal{E}^*$ belongs to the interior of the cone $\mathrm{Pos}(A \to B)$.*
   *(d) For any nonzero $\rho \in \mathrm{Pos}(A)$, we have $\mathcal{E}(\rho) > 0$.*

## 3.4.2 The Matrix Representation

We begin our exploration of linear maps with the most familiar representation: the matrix representation. In the context of linear transformations between two Hilbert spaces, every element in the set $\mathcal{L}(A \rightarrow B)$ possesses a corresponding matrix representation. While this matrix representation can prove valuable in certain applications, especially when dealing with low-dimensional cases, we will discover later on that it is, in fact, the least intuitive representation and finds limited use in the realm of quantum information. The reason for this lies in the fact that the concept of complete positivity, a crucial property of quantum channels, does not translate naturally and tends to become cumbersome within this matrix-based framework.

Let $\mathcal{E} \in \mathcal{L}(A \rightarrow B)$ be a linear map, $\rho \in \mathcal{L}(A)$, and $\sigma := \mathcal{E}(\rho) \in \mathcal{L}(B)$. The relationship $\sigma = \mathcal{E}(\rho)$ can be expressed in matrix form as $\mathbf{s}_\sigma = M_\mathcal{E} \mathbf{r}_\rho$, where $\mathbf{r}_\rho$ and $\mathbf{s}_\sigma$ denote column vectors representing $\rho$ and $\sigma$, respectively, and $M_\mathcal{E}$ is a complex matrix representing the linear map $\mathcal{E}$. To illustrate it explicitly, let $\{\Lambda_x\}_{x \in [m^2]}$ with $m := |A|$ be a fixed orthonormal basis of $\mathcal{L}(A)$, and $\{\Gamma_y\}_{y \in [n^2]}$ with $n := |B|$ be a fixed orthonormal basis of $\mathcal{L}(B)$. Any operator $\rho \in \mathcal{L}(A)$ can be expressed as a linear combination of the basis elements

$$\rho = \sum_{x \in [m^2]} r_x \Lambda_x, \tag{3.80}$$

with $\mathbf{r}_\rho := (r_1, \ldots, r_{m^2})^T \in \mathbb{C}^{m^2}$. Observe that the mapping $\rho \mapsto \mathbf{r}_\rho$ defines an isomorphism between $\mathcal{L}(A)$ and $\mathbb{C}^{m^2}$. Similarly, for any $\sigma = \sum_{x \in [n^2]} s_y \Gamma_y \in \mathcal{L}(B)$ we define $\mathbf{s}_\sigma := (s_1, \ldots, s_{n^2})^T$. Finally, for any linear map $\mathcal{E} \in \mathcal{L}(A \rightarrow B)$ we define the $n^2 \times m^2$ matrix $M_\mathcal{E}$ whose components are given by

$$(M_\mathcal{E})_{yx} := \langle \Gamma_y | \mathcal{E}(\Lambda_x) \rangle_{HS} := \mathrm{Tr}\left[ \Gamma_y^* \mathcal{E}(\Lambda_x) \right]. \tag{3.81}$$

With these notations we have

$$\sigma = \mathcal{E}(\rho) \iff \mathbf{s}_\sigma = M_\mathcal{E} \mathbf{r}_\rho. \tag{3.82}$$

Therefore, $M_\mathcal{E}$ is the matrix representation of the linear map $\mathcal{E}$.

Suppose now that the map $\mathcal{E}$ is a quantum channel, and that the orthonormal bases $\{\Lambda_x\}_{x \in [m^2]}$ and $\{\Gamma_y\}_{y \in [n^2]}$ consist of Hermitian matrices (see Exercise 2.29). Then, $\rho$ is Hermitian if and only if $\mathbf{r}_\rho$ is a real vector, and similarly $\sigma$ is Hermitian if and only if $\mathbf{s}_\sigma$ is a real vector. Therefore, $\mathcal{E}$ is Hermitian preserving if and only if $M_\mathcal{E}$ is a real matrix. Moreover, since $\mathcal{E}$ is trace preserving, it will be convenient to choose the orthonormal bases $\{\Lambda_x\}_{x \in [m^2]}$ and $\{\Gamma_y\}_{y \in [n^2]}$ with the first element proportional to the identity. That is, we choose

$$\Lambda_1 := \frac{1}{\sqrt{m}} I^A \quad \text{and} \quad \Gamma_1 := \frac{1}{\sqrt{n}} I^B, \tag{3.83}$$

so that the remaining elements of the orthonormal bases, $\{\Lambda_x\}_{x=2}^{m^2}$ and $\{\Gamma_y\}_{y=2}^{n^2}$, are all traceless (since they have to be orthogonal in the Hilbert–Schmidt inner product to the first element). With these choices, an operator $\rho \in \mathcal{L}(A)$ has trace 1 if and only if $\mathbf{r}_\rho$

has the form $(\frac{1}{\sqrt{m}}, r_2, \ldots, r_{m^2})^T$. We therefore conclude that the linear map $\mathcal{E}$ is both trace preserving and Hermitian preserving if and only if its matrix representation has the form

$$M_{\mathcal{E}} = \begin{bmatrix} \sqrt{\frac{m}{n}} & \mathbf{0} \\ \mathbf{t} & N_{\mathcal{E}} \end{bmatrix}, \quad \text{where} \quad \mathbf{t} \in \mathbb{R}^{n^2-1} \quad \text{and} \quad N_{\mathcal{E}} \in \mathbb{R}^{(n^2-1)\times(m^2-1)}. \tag{3.84}$$

What are the conditions on the matrix $N_{\mathcal{E}}$ and the vector $\mathbf{t}$ that correspond to the condition that $\mathcal{E}$ is completely positive? These conditions can be very complicated since, in general, even the set of all vectors $\mathbf{r}_\rho$ for which $\rho \geqslant 0$ doesn't have a simple characterization.

**Exercise 3.27.** *Let $\mathcal{E} \in \mathfrak{L}(A \to B)$ be a linear map.*

*1. Show that*

$$M_{\mathcal{E}^*} = M_{\mathcal{E}}^*. \tag{3.85}$$

*2. Show that $\mathcal{E}$ is a self-adjoint map (i.e. $\mathcal{E} = \mathcal{E}^*$) that is both trace preserving and Hermitian preserving if and only if the matrix $M_{\mathcal{E}}$ in (3.84) has $\mathbf{t} = 0$ and $N_{\mathcal{E}}$ is a real symmetric matrix (i.e. $N_{\mathcal{E}}^T = N_{\mathcal{E}}$).*

## The Qubit Case

In the qubit case, where $|A| = |B| = 2$, the Bloch representation simplifies the situation. In particular, $\rho \in \mathfrak{L}(\mathbb{C}^2)$ is a density matrix if and only if $\mathbf{r}_\rho = \frac{1}{\sqrt{2}}(1, \mathbf{r})^T$, where $\mathbf{r}$ corresponds to the Bloch vector belonging to $\mathbb{R}^3$ and its length $\|\mathbf{r}\| \leqslant 1$. Therefore, in this case, $\sigma = \mathcal{E}(\rho)$ if and only if the Bloch vector of $\sigma$, denoted by $\mathbf{s}$, is related to $\mathbf{r}$ via

$$\mathbf{s} = \mathbf{t} + (N_{\mathcal{E}})\mathbf{r}. \tag{3.86}$$

From the relation above, $\mathcal{E}$ is a positive map if the matrices $\mathbf{t}$ and $N_{\mathcal{E}}$ are such that whenever $|\mathbf{r}| \leqslant 1$, $|\mathbf{s}| \leqslant 1$ also holds. It's important to note, however, that this criterion pertains solely to the positivity of $\mathcal{E}$ and does not address its complete positivity. In the specific scenario of qubits, it is feasible to articulate the conditions governing $N_{\mathcal{E}}$ and $\mathbf{t}$ for $\mathcal{E}$ to be completely positive. Nevertheless, these conditions tend to be rather intricate, and we direct the interested reader to the pertinent literature found in the Notes and References section at the end of this chapter. In the following exercises, you will demonstrate that these conditions become more straightforward in the case of doubly stochastic maps.

Doubly stochastic maps encompass mappings that possess two key properties: trace preservation and unitality, meaning they preserve the identity operator. One of the simplest examples of such maps is the unitary map $\mathcal{U}(\rho) := U\rho U^*$, where $U$ is a unitary matrix. Another example includes convex combinations of unitary maps in the form $\sum_{x \in [m]} p_x \mathcal{U}_x$, where $\{p_x\}_{x \in [m]}$ forms a probability distribution, and each $\mathcal{U}_x$ represents a unitary quantum channel. These instances exemplify completely positive maps that also qualify as doubly stochastic.

Conversely, the transpose map $\mathcal{T}(\rho) = \rho^T$ and its combination with a unitary map, denoted as $\mathcal{U} \circ \mathcal{T}$, serve as examples of doubly stochastic maps that are positive but not completely positive. In the subsequent set of problems, we will see that for the qubit case, all positive doubly stochastic maps can be expressed as convex combinations of such maps.

**Exercise 3.28.** *Let $\mathcal{E} \in \mathrm{Pos}\left(\mathbb{C}^2 \to \mathbb{C}^2\right)$ be a positive linear map.*

1. *Show that $\mathcal{E}$ is both trace preserving and unital (i.e. doubly stochastic) if and only if*

$$M_{\mathcal{E}} = \begin{bmatrix} 1 & \mathbf{0} \\ \mathbf{0} & N_{\mathcal{E}} \end{bmatrix}, \tag{3.87}$$

*where $N_{\mathcal{E}} \in \mathbb{R}^{3 \times 3}$ has the property that $\|N_{\mathcal{E}}\mathbf{r}\|_2 \leqslant 1$ for all $\mathbf{r} \in \mathbb{R}^3$ with $\|\mathbf{r}\|_2 = 1$ (in particular, the absolute value of the eigenvalues of $N_{\mathcal{E}}$ cannot exceed 1).*

2. *Suppose $\mathcal{E} = \mathcal{U}$ is the doubly stochastic unitary map given by $\mathcal{U}(\rho) = U\rho U^*$, where $U \in SU(2)$ can be expressed as $U = wI_2 + i(x\sigma_1 + y\sigma_2 + z\sigma_3)$ with $w, x, y, z \in \mathbb{R}$ and $w^2 + x^2 + y^2 + z^2 = 1$ (cf. (C.10) of online version). Show that the matrix $N_{\mathcal{U}}$ of (3.87) is an orthogonal matrix in $SO(3)$ given by*

$$N_{\mathcal{U}} = \begin{pmatrix} 1 - 2y^2 - 2z^2 & 2xy + 2zw & 2xz - 2yw \\ 2xy - 2zw & 1 - 2x^2 - 2z^2 & 2yz + 2xw \\ 2xz + 2yw & 2yz - 2xw & 1 - 2x^2 - 2y^2 \end{pmatrix} \tag{3.88}$$

*(cf. (C.7) of online version). Hint: Calculate directly the components $\frac{1}{2}\mathrm{Tr}\left[\sigma_i U \sigma_j U^*\right]$ for $i, j \in \{1, 2, 3\}$.*

3. *Use the previous parts to show that for any map of the form $\mathcal{U}(\rho) = U\rho U^*$, we have that $U \in U(2)$ if and only if the matrix $N_{\mathcal{U}} \in SO(3)$. Hint: Every unitary $U \in U(2)$ can be written as $U = \exp(i\theta)\tilde{U}$, where $\tilde{U} \in SU(2)$ and $\theta \in [0, 2\pi)$.*

**Exercise 3.29.** *Let $\mathcal{T} \in \mathfrak{L}\left(\mathbb{C}^2 \to \mathbb{C}^2\right)$ be the transpose map defined by $\mathcal{T}(\rho) = \rho^T$ for all $\rho \in \mathfrak{L}(\mathbb{C}^2)$.*

1. *Show that the matrix representation of the transpose map with respect to the Pauli basis of $\mathfrak{L}(\mathbb{C}^2)$ is given by*

$$M_{\mathcal{T}} = \begin{bmatrix} 1 & 0 & 0 & 0 \\ 0 & 1 & 0 & 0 \\ 0 & 0 & -1 & 0 \\ 0 & 0 & 0 & 1 \end{bmatrix}. \tag{3.89}$$

2. *Show that if $\mathcal{E} \in \mathfrak{L}(\mathbb{C}^2 \to \mathbb{C}^2)$ is a positive doubly stochastic linear map with $N_{\mathcal{E}} \in O(3)$ and with $\det(N_{\mathcal{E}}) = -1$, then $\mathcal{E} = \mathcal{T} \circ \mathcal{U}$ for some unitary map $\mathcal{U}$. Hint: Use the fact that any $3 \times 3$ orthogonal matrix can be expressed as a matrix product of an element in $SO(3)$ with $N_{\mathcal{T}}$.*

**Exercise 3.30.** *Use the exercises above in conjunction with Exercise A.8 of online version to conclude that any doubly stochastic positive map $\mathcal{E} \in \mathrm{Pos}(\mathbb{C}^2 \to \mathbb{C}^2)$ can be expressed as*

$$\mathcal{E} = t\mathcal{N}_1 + (1-t)\mathcal{T} \circ \mathcal{N}_2, \tag{3.90}$$

*where $t \in [0,1]$ and both $\mathcal{N}_1$ and $\mathcal{N}_2$ are mixtures of unitary maps; that is, maps of the form $\sum_{j \in [m]} p_j \, \mathcal{U}_j$ with each $\mathcal{U}_j$ being a unitary map and $\{p_j\}_{j \in [m]}$ is a probability distribution. Hint: Use Exercise A.8 of online version to show that $N_\mathcal{E}$ can be expressed as a finite convex combination of orthogonal matrices.*

## Positivity versus Complete Positivity in Low Dimensions

We have seen before that the transpose map $\mathcal{T}$ is positive, but not 2-positive (and consequently, the transpose map is not completely positive). The following theorem that was proved originally by Strømer and Woronowicz shows essentially that in low dimensions combinations of the transpose map with completely positive maps are the only maps that can be positive but not completely positive.

> **Størmer–Woronowicz Theorem**
>
> **Theorem 3.3.** Let $\mathcal{E} \in \mathrm{Pos}(A \to B)$ be a positive linear map. If $|A| = 2$ and $|B| \leqslant 3$, then there exists two CP maps $\mathcal{N}_1, \mathcal{N}_2 \in \mathrm{CP}(A \to B)$ such that
>
> $$\mathcal{E} = \mathcal{N}_1 + \mathcal{T} \circ \mathcal{N}_2, \tag{3.91}$$
>
> where $\mathcal{T} \in \mathrm{Pos}(B \to B)$ is the transpose map.

*Remark.* The case $|B| = 2$ was proven by Størmer, and the case $|B| = 3$ was proven by Woronowicz. Here, we will only prove Størmer theorem (i.e. $|A| = |B| = 2$) and refer the reader to the section "Notes and References" (at the end of this chapter) for more details.

**Proof**  We prove the theorem for the case that $\mathcal{E}$ is in the interior of $\mathrm{Pos}(A \to A)$ (the more general case will then follow from a continuity argument; see Exercise 3.32). From Exercise 3.26 it follows that also $\mathcal{E}^*$ is in the interior of $\mathrm{Pos}(A \to A)$, and furthermore, $\mathcal{E}(\rho) > 0$ for any nonzero $\rho \in \mathrm{Pos}(A)$.

The key idea of the proof is to find two positive definite operators $\Lambda, \Gamma > 0$ with the property that the channel

$$\mathcal{D} := \mathcal{F}_\Lambda \circ \mathcal{E} \circ \mathcal{F}_\Gamma \tag{3.92}$$

is doubly stochastic, where

$$\mathcal{F}_\Lambda(\omega) := \Lambda \omega \Lambda \quad \text{and} \quad \mathcal{F}_\Gamma(\omega) := \Gamma \omega \Gamma \qquad \forall \, \omega \in \mathfrak{L}(A). \tag{3.93}$$

From Exercise 3.22 (see also the discussion on operator sum representation in Section 3.4.4) it follows that the above maps are completely positive. A priori it is not clear if such positive definite matrices $\Lambda$ and $\Gamma$ exists, but if they do, then from (3.92) we have

$\mathcal{E} = \mathcal{F}_{\Lambda^{-1}} \circ \mathcal{D} \circ \mathcal{F}_{\Gamma^{-1}}$, and since all doubly stochastic positive maps have the form (3.91) (see Exercise 3.30), it follows that also $\mathcal{E}$ has the form (3.91). It is therefore left to show that such $\Lambda$ and $\Gamma$ do exist.

By definition, the channel $\mathcal{D}$ is doubly stochastic if and only if both $\mathcal{D}$ and its dual $\mathcal{D}^*$ are unital channels. Since the dual of $\mathcal{D}$ is given by $\mathcal{D}^* = \mathcal{F}_\Gamma \circ \mathcal{E}^* \circ \mathcal{F}_\Lambda$ (see Exercise 3.31), we conclude that $\mathcal{D}$ is doubly stochastic if and only if the matrices $\Lambda$ and $\Gamma$ satisfy

$$I = \mathcal{D}(I) = \Lambda\mathcal{E}(\Gamma^2)\Lambda \quad \text{and} \quad I = \mathcal{D}^*(I) = \Gamma\mathcal{E}^*(\Lambda^2)\Gamma. \tag{3.94}$$

By conjugating with the inverses of $\Lambda$ and $\Gamma$, the two equations above can be expressed as

$$\Lambda^{-2} = \mathcal{E}(\Gamma^2) \quad \text{and} \quad \Gamma^{-2} = \mathcal{E}^*(\Lambda^2). \tag{3.95}$$

It is therefore left to show that there exists $\rho := \Lambda^{-2} > 0$ and $\sigma := \Gamma^2 > 0$ such that

$$\rho = \mathcal{E}(\sigma) \quad \text{and} \quad \sigma^{-1} = \mathcal{E}^*\left(\rho^{-1}\right). \tag{3.96}$$

Observe that if $\rho$ and $\sigma$ satisfy the two equations above, then for any $s > 0$ also $s\rho$ and $s\sigma$ satisfy the two equations. Hence, without loss of generality we can assume that if there exist $\rho$ and $\sigma$ that satisfy the equation above, then $\sigma$ is normalized, and since $\mathcal{E}$ is trace preserving, this implies that also $\rho := \mathcal{E}(\sigma)$ is normalized.

The equation $\rho = \mathcal{E}(\sigma)$ can be taken to be the definition of $\rho$. Substituting this $\rho$ into the second equality of (3.96) implies that

$$\sigma^{-1} = \mathcal{E}^*\left(\left(\mathcal{E}(\sigma)\right)^{-1}\right). \tag{3.97}$$

To show that such a $\sigma$ exists, define the function $f : \mathfrak{D}(A) \to \text{Pos}(A)$ via

$$f(\omega) := \left(\mathcal{E}^*\left(\left(\mathcal{E}(\omega)\right)^{-1}\right)\right)^{-1} \quad \forall\, \omega \in \mathfrak{D}(A), \tag{3.98}$$

and observe that (3.97) is equivalent to $f(\sigma) = \sigma$. We also define the normalized version of $f$, the function $g : \mathfrak{D}(A) \to \mathfrak{D}(A)$, as

$$g(\omega) := \frac{f(\omega)}{\text{Tr}\left[f(\omega)\right]} \quad \forall\, \omega \in \mathfrak{D}(A). \tag{3.99}$$

Then, from Brouwer's fixed-point theorem (see Theorem A.13) there exists a density matrix $\sigma \in \mathfrak{D}(A)$ such that $g(\sigma) = \sigma$. Denoting by $t := \text{Tr}[f(\sigma)] > 0$, this is equivalent to

$$f(\sigma) = t\sigma. \tag{3.100}$$

It is therefore left to show that $t = 1$. For this purpose, observe first that with the definition $\rho := \mathcal{E}(\sigma)$ we can express the above equation as

$$(t\sigma)^{-1} = \mathcal{E}^*\left(\rho^{-1}\right), \tag{3.101}$$

so that

$$t^{-1}I = \sigma^{\frac{1}{2}}\mathcal{E}^*\left(\rho^{-1}\right)\sigma^{\frac{1}{2}} = \mathcal{D}^*(I), \tag{3.102}$$

where $\mathcal{D}$ is defined in (3.92) with $\Lambda = \rho^{-1/2}$ and $\Gamma = \sigma^{1/2}$. On the other hand, the relation $\rho = \mathcal{E}(\sigma)$ can be written as

$$I = \rho^{-\frac{1}{2}} \mathcal{E}(\sigma) \rho^{-\frac{1}{2}} = \mathcal{D}(I), \tag{3.103}$$

which implies that the linear map $\mathcal{D}$ is unital. Since the dual of a unital map is trace preserving (see the first part of Exercise 3.23), we conclude that $\mathcal{D}^*$ is trace preserving. Hence, by taking the trace on both sides of (3.102), and using the fact that $\mathcal{D}^*$ is trace preserving, we conclude that $t = 1$. This completes the proof. ∎

**Exercise 3.31.** *Show that if $\mathcal{D} := \mathcal{F}_\Gamma \circ \mathcal{E} \circ \mathcal{F}_\Lambda$ as in the proof above, then*

$$\mathcal{D}^* := \mathcal{F}_\Lambda \circ \mathcal{E}^* \circ \mathcal{F}_\Gamma. \tag{3.104}$$

**Exercise 3.32.** *Use a continuity argument to prove that if all maps in the interior of* $\mathrm{Pos}(A \to A)$ *have the form (3.91), then all the maps in* $\mathrm{Pos}(A \to A)$ *have this form.*

## 3.4.3 The Choi Representation

The Choi representation, also known as the Choi-Jamiolkowski isomorphism, is another method to characterize linear maps using matrices, offering a simpler way to identify complete positivity. In this representation, quantum channels are associated with positive semidefinite matrices. This characteristic makes the Choi representation particularly useful in various applications within quantum information science, especially because it allows for the translation of certain optimization problems into semidefinite programs.

In subsequent discussions and throughout the rest of the book, we will adopt the shorthand notation $\mathcal{E}^{B \to B'}(\rho^{AB})$ to denote $\left(\mathrm{id}^A \otimes \mathcal{E}^{B \to B'}\right)(\rho^{AB})$. This notation simplifies expressions and discussions involving the application of a quantum channel $\mathcal{E}$ from system $B$ to system $B'$ on part of a bipartite state $\rho^{AB}$.

Given a linear map $\mathcal{E} \in \mathfrak{L}(A \to B)$, the Choi matrix is defined by the action of $\mathcal{E}$ on one subsystem of a maximally entangled state. Setting $m := |A|$ and denoting by

$$\Omega^{A\tilde{A}} = |\Omega^{A\tilde{A}}\rangle\langle\Omega^{A\tilde{A}}| = \sum_{x,y \in [m]} |x\rangle\langle y|^A \otimes |x\rangle\langle y|^{\tilde{A}}, \tag{3.105}$$

the Choi matrix of $\mathcal{E}$ is defined by

$$J_{\mathcal{E}}^{AB} := \mathcal{E}^{\tilde{A} \to B}\left(\Omega^{A\tilde{A}}\right) = \sum_{x,y \in [m]} |x\rangle\langle y| \otimes \mathcal{E}\left(|x\rangle\langle y|\right). \tag{3.106}$$

One of the key properties of the Choi matrix is that it satisfies the relation

$$\mathcal{E}(\rho) = \mathrm{Tr}_A\left[J_{\mathcal{E}}^{AB}\left(\rho^T \otimes I\right)\right] \qquad \forall \, \rho \in \mathfrak{L}(A). \tag{3.107}$$

To see this, let $\rho = \sum_{x,y \in [m]} r_{xy}|x\rangle\langle y|$ be a linear operator in $\mathfrak{L}(A)$ with components $r_{xy} \in \mathbb{C}$ and $m := |A|$. From the definition of the Choi matrix in (3.106) we get

$$\mathrm{Tr}_A\left[J_{\mathcal{E}}^{AB}\left(\rho^T\otimes I\right)\right] = \sum_{x,y\in[m]} \mathrm{Tr}\left[\rho^T|x\rangle\langle y|\right]\mathcal{E}\left(|x\rangle\langle y|\right)$$

$$= \sum_{x,y\in[m]} r_{xy}\mathcal{E}\left(|x\rangle\langle y|\right) \tag{3.108}$$

$$= \mathcal{E}\left(\sum_{x,y\in[m]} r_{xy}|x\rangle\langle y|\right) = \mathcal{E}(\rho).$$

The two relations (3.106,3.107) demonstrate that the mapping $\mathcal{E}\mapsto J_{\mathcal{E}}$ is a linear bijection (i.e. an isomorphism). This isomorphism is between the vector space $\mathfrak{L}(A\to B)$ of linear operators from $\mathfrak{L}(A)$ to $\mathfrak{L}(B)$, and the space of bipartite matrices/operators $\mathfrak{L}(AB)$. In the following exercise, we will show that the mapping $\mathcal{E}\mapsto J_{\mathcal{E}}$ is in fact isometrically isomorphism between these two spaces.

**Exercise 3.33.** *Let $A$ and $B$ be two finite dimensional Hilbert spaces and consider the Hilbert space $\mathfrak{L}(A\to B)$ equiped with the inner product defined in (3.64). Show that this inner product can be expressed as follows. For all $\mathcal{E},\mathcal{F}\in\mathfrak{L}(A\to B)$ we have*

$$\langle\mathcal{E},\mathcal{F}\rangle = \langle J_{\mathcal{E}}^{AB}, J_{\mathcal{F}}^{AB}\rangle_{HS}, \tag{3.109}$$

*where on the right-hand side we have the Hilbert–Schmidt inner product between the two Choi matrices of $\mathcal{E}$ and $\mathcal{F}$.*

---

**Theorem 3.4.** A linear map $\mathcal{E}\in\mathfrak{L}(A\to B)$ is completely positive if and only if $J_{\mathcal{E}}^{AB}\geqslant 0$.

---

**Proof** If $\mathcal{E}$ is completely positive, then by definition $J_{\mathcal{E}}^{AB} := \mathcal{E}^{\tilde{A}\to B}(\Omega^{A\tilde{A}})\geqslant 0$. Suppose now that $J_{\mathcal{E}}^{AB}\geqslant 0$. Let $k\in\mathbb{N}$, and $|\psi^{RA}\rangle\in\mathbb{C}^k\otimes\mathbb{C}^d$, where $R$ is a $k$-dimensional (reference) system. Recall that any bipartite vector $|\psi\rangle^{RA}$ can be expressed as

$$|\psi\rangle^{RA} = M\otimes I^A|\Omega^{\tilde{A}A}\rangle, \tag{3.110}$$

where $M:\tilde{A}\to R$ is some linear operator. We therefore have

$$(\mathrm{id}_k\otimes\mathcal{E})\left(|\psi^{RA}\rangle\langle\psi^{RA}|\right) = (\mathrm{id}_k\otimes\mathcal{E})\left((M\otimes I^A)\Omega^{A\tilde{A}}(M^*\otimes I^A)\right)$$

$$= M\otimes I^B\left(\mathcal{E}^{\tilde{A}\to B}(\Omega^{A\tilde{A}})\right)M^*\otimes I^B \tag{3.111}$$

$$= \left(M\otimes I^B\right)J_{\mathcal{E}}^{AB}\left(M^*\otimes I^B\right)$$

$$\text{Exercise 2.22}\to\ \geqslant 0.$$

Finally, any operator $\rho^{RA}\geqslant 0$ can be diagonalized as $\rho^{RA} = \sum_{x\in[m]}|\psi_x^{RA}\rangle\langle\psi_x^{RA}|$, where $|\psi_x^{RA}\rangle\in\mathbb{C}^k\otimes\mathbb{C}^d$ are some (possibly unnormalized) pure states. Since $|\psi^{RA}\rangle$ above was arbitrary, we conclude that

$$(\mathrm{id}_k \otimes \mathcal{E})\left(\rho^{RA}\right) = \sum_{x\in[m]} (\mathrm{id}_k \otimes \mathcal{E})\left(|\psi_x^{RA}\rangle\langle\psi_x^{RA}|\right) \geqslant 0, \qquad (3.112)$$

since each term in the sum is positive semidefinite. This completes the proof. ∎

> **Corollary 3.1.** A linear map $\mathcal{E} \in \mathfrak{L}(A \to B)$ is $|A|$-positive if and only if it is completely positive.

**Proof**   By definition if $\mathcal{E}$ is a CP map then it is $|A|$-positive. Conversely, if $\mathcal{E}$ is $|A|$-positive, then its Choi matrix $\mathcal{E}^{\tilde{A}\to B}(\Omega^{A\tilde{A}})$ is positive semidefinite. Hence, from the theorem above $\mathcal{E}$ is a CP map. ∎

> **Theorem 3.5.** A linear map $\mathcal{E} \in \mathfrak{L}(A \to B)$ is trace preserving if and only if the marginal state $J_{\mathcal{E}}^A := \mathrm{Tr}_B\left[J_{\mathcal{E}}^{AB}\right] = I^A$.

**Proof**   Suppose $\mathcal{E}$ is trace preserving and set $m := |A|$. Then, from (3.106)

$$\mathrm{Tr}_B\left[J_{\mathcal{E}}^{AB}\right] = \sum_{x,y\in[m]} |x\rangle\langle y|\mathrm{Tr}\left[\mathcal{E}\left(|x\rangle\langle y|\right)\right]$$

$$\mathcal{E} \text{ is trace preserving} \to \; = \sum_{x,y\in[m]} |x\rangle\langle y|\mathrm{Tr}\left[|x\rangle\langle y|\right] \qquad (3.113)$$

$$= \sum_{x,y\in[m]} |x\rangle\langle y|\delta_{xy} = I^A.$$

Conversely, suppose $J_{\mathcal{E}}^A = I^A$, then from (3.107) for every $\rho \in \mathfrak{L}(A)$

$$\mathrm{Tr}\left[\mathcal{E}(\rho)\right] = \mathrm{Tr}\left[J_{\mathcal{E}}^{AB}\left(\rho^T \otimes I\right)\right] = \mathrm{Tr}\left[J_{\mathcal{E}}^A \rho^T\right] = \mathrm{Tr}[\rho^T] = \mathrm{Tr}[\rho]. \qquad (3.114)$$

This completes the proof. ∎

We therefore conclude that a linear map $\mathcal{E}$ is a quantum channel if and only if its Choi matrix $J_{\mathcal{E}}^{AB} \geqslant 0$, and its marginal $J_{\mathcal{E}}^A = I^A$. In particular, the Choi matrix has trace $|A|$ so that $\frac{1}{|A|}J_{\mathcal{E}}^{AB} \in \mathfrak{D}(A \otimes B)$. Hence, the Choi representation reveals that quantum channels can be represented with bipartite quantum states. This equivalence between quantum channels and bipartite quantum states is used very often in quantum information science.

**Exercise 3.34.** *Show that the linear map, $\mathcal{E}^{(k)}$, defined in (3.72), with $k < d$, is $k$-positive but not $(k+1)$-positive.*

**Exercise 3.35.** *Let $\mathcal{E} \in \mathfrak{L}(A \to A)$ be a quantum channel with the property that $\mathcal{E}(U\rho U^*) = U\mathcal{E}(\rho)U^*$ for any unitary matrix $U \in \mathfrak{U}(A)$. Show that its Choi matrix $J_{\mathcal{E}}^{A\tilde{A}}$ must satisfy*

$$\left(\bar{U} \otimes U\right) J_{\mathcal{E}}^{A\tilde{A}} \left(\bar{U} \otimes U\right)^* = J_{\mathcal{E}}^{A\tilde{A}} \tag{3.115}$$

*for all unitary matrices $U$.*

**Exercise 3.36.** *Show that any density matrix $\rho \in \mathfrak{D}(AB)$ can be expressed as*

$$\rho^{AB} = \mathcal{E}^{\tilde{A} \to B}\left(\psi^{A\tilde{A}}\right), \tag{3.116}$$

*where $\psi \in \mathrm{Pure}(A\tilde{A})$ is pure bipartite state that has a marginal $\psi^A = \rho^A$, and $\mathcal{E} \in \mathrm{CPTP}(A \to B)$ is a quantum channel. Hint: Look at $(\rho^A)^{-\frac{1}{2}}\rho^{AB}(\rho^A)^{-\frac{1}{2}}$.*

**Exercise 3.37.** *Show that if instead of the standard Choi representation above, one defines*

$$J_{\mathcal{E}}^{AB} := \mathcal{E}^{\tilde{A} \to B}\left(\psi^{A\tilde{A}}\right), \quad \text{where} \quad |\psi^{A\tilde{A}}\rangle := \sum_{x \in [m]} \sqrt{p_x}|x\rangle^A |x\rangle^{\tilde{A}} \tag{3.117}$$

*with $\{p_x\}$ a probability distribution, then*

$$\mathcal{E}(\rho) = \mathrm{Tr}_A\left[J_{\mathcal{E}}^{AB}\left(\sigma^{-1/2}U\rho^T U^* \sigma^{-1/2} \otimes I^B\right)\right], \tag{3.118}$$

*where $U$ is some fixed diagonal unitary on system $A$ and $\sigma := \mathrm{Tr}_{\tilde{A}}[\psi^{A\tilde{A}}]$.*

### 3.4.4 The Operator Sum Representation

In Figure 3.3c we considered a generalized measurement $\{M_x\}_{x \in [m]}$ on a physical system described by the state $\rho \in \mathfrak{D}(A)$. If outcome $x$ occurred, then the state of the system $\rho$ changes to $\left(M_x \rho M_x^*\right)/p_x$, where $p_x$ is the probability that outcome $x$ occurred. However, if the value of $x$ is erased after the measurement, then the post-measurement state is given by the average over all the possible outcomes; that is,

$$\sum_{x \in [m]} p_x \left(\frac{M_x \rho M_x^*}{p_x}\right) = \sum_{x \in [m]} M_x \rho M_x^*. \tag{3.119}$$

We now show that the mapping $\rho \mapsto \sum_{x \in [m]} M_x \rho M_x^*$ is a quantum channel and that every quantum channel can be realized in this way. This representation of a quantum channel is called the *operator sum representation*, and the elements $\{M_x\}_{x \in [m]}$ (with $\sum_{x \in [m]} M_x^* M_x = I^A$) are called the *Kraus operators*.

---

**Theorem 3.6.** A linear map $\mathcal{E} \in \mathfrak{L}(A \to B)$ is a quantum channel if and only if it has an operator sum representation. That is, $\mathcal{E}$ is a quantum channel if and only if there exists a set of Kraus operators $\{M_x\}_{x \in [m]} \subset \mathfrak{L}(A, B)$, such that

$$\mathcal{E}(\rho) = \sum_{x \in [m]} M_x \rho M_x^*. \tag{3.120}$$

**Proof** Suppose $\mathcal{E}$ is a quantum channel. Since the Choi matrix of a quantum channel is positive semidefinite, we can always express it as

$$J_{\mathcal{E}}^{AB} = \sum_{x \in [m]} |\psi_x^{AB}\rangle\langle\psi_x^{AB}| \tag{3.121}$$

for some integer $m$ and some (possibly unnormalized) vectors $|\psi_x^{AB}\rangle \in A \otimes B$. Recall that any bipartite state $|\psi_x^{AB}\rangle$ can be expressed as

$$|\psi_x^{AB}\rangle = I^A \otimes M_x |\Omega^{A\tilde{A}}\rangle = M_x^T \otimes I^B |\Omega^{\tilde{B}B}\rangle, \tag{3.122}$$

where $M_x \in \mathfrak{L}(A, B)$ is a linear operator. Moreover, since the marginal Choi matrix $J_{\mathcal{E}}^A = I^A$, we get

$$
\begin{aligned}
I^A = \operatorname{Tr}_B\left[J_{\mathcal{E}}^{AB}\right] &= \operatorname{Tr}_B\left[\sum_{x \in [m]} \left(M_x^T \otimes I^B\right) \Omega^{\tilde{B}B} \left((M_x^*)^T \otimes I^B\right)\right] \\
&= \sum_{x \in [m]} M_x^T \operatorname{Tr}_B\left[\Omega^{\tilde{B}B}\right] (M_x^*)^T = \sum_{x \in [m]} M_x^T (M_x^*)^T.
\end{aligned}
\tag{3.123}
$$

By taking the transpose on both sides of the equation above we get

$$I^A = \sum_{x \in [m]} M_x^* M_x. \tag{3.124}$$

Moreover, substituting (3.121) into (3.107) gives

$$
\begin{aligned}
\mathcal{E}(\rho) &= \sum_{x \in [m]} \operatorname{Tr}_A\left[|\psi_x^{AB}\rangle\langle\psi_x^{AB}| \left(\rho^T \otimes I\right)\right] \\
&= \sum_{x \in [m]} \operatorname{Tr}_A\left[\left(I^A \otimes M_x\right) |\Omega^{A\tilde{A}}\rangle\langle\Omega^{A\tilde{A}}| \left(I^A \otimes M_x^*\right) \left(\rho^T \otimes I\right)\right] \\
&= \sum_{x \in [m]} M_x \left(\operatorname{Tr}_A\left[|\Omega^{A\tilde{A}}\rangle\langle\Omega^{A\tilde{A}}| \left(\rho^T \otimes I\right)\right]\right) M_x^*.
\end{aligned}
\tag{3.125}
$$

To simplify the last term, note that

$$
\begin{aligned}
\operatorname{Tr}_A\left[|\Omega^{A\tilde{A}}\rangle\langle\Omega^{A\tilde{A}}| \left(\rho^T \otimes I\right)\right] &= \operatorname{Tr}_A\left[|\Omega^{A\tilde{A}}\rangle\langle\Omega^{A\tilde{A}}| (I \otimes \rho)\right] \\
&= \operatorname{Tr}_A\left[|\Omega^{A\tilde{A}}\rangle\langle\Omega^{A\tilde{A}}|\right]\rho = I^{\tilde{A}}\rho = \rho.
\end{aligned}
\tag{3.126}
$$

We therefore conclude that $\mathcal{E}(\rho) = \sum_{x \in [m]} M_x \rho M_x^*$.

To prove the converse, suppose that $\mathcal{E}$ has the form (3.120). Then, clearly the Choi matrix $J_{\mathcal{E}}^{AB} := \mathcal{E}^{\tilde{A} \to B}(\Omega^{A\tilde{A}})$ has the form (3.121) with $|\psi_x^{AB}\rangle := I^A \otimes M_x |\Omega^{A\tilde{A}}\rangle$. Hence, $J_{\mathcal{E}}^{AB} \geqslant 0$ and $J_{\mathcal{E}}^A = I^A$ since $\sum_{x \in [m]} M_x^* M_x = I^A$. This completes the proof. $\blacksquare$

**Exercise 3.38.** *Let $\mathcal{E} \in \mathfrak{L}(A \to B)$ be a linear map.*

1. *Show that there exists two sets of matrices $\{M_x\}_{x \in [m]}$ and $\{N_x\}_{x \subset [m]}$ such that*

$$\mathcal{E}(\rho) = \sum_{x \in [m]} M_x \rho N_x^*. \tag{3.127}$$

*Hint: Start by showing that it is possible to express the complex Choi matrix as $J_{\mathcal{E}}^{AB} = \sum_{x\in[m]} |\psi_x\rangle\langle\phi_x|$, and then follow similar lines as in the proof above.*

2. *Show that the dual (adjoint) map $\mathcal{E}^*: \mathfrak{L}(B \to A)$ is given by*

$$\mathcal{E}^*(\rho) = \sum_{x\in[m]} M_x^* \rho N_x. \tag{3.128}$$

3. *Show that $\mathcal{E} \in \mathfrak{L}(A \to B)$ is completely positive if and only if its dual map $\mathcal{E}^* \in \mathfrak{L}(B \to A)$ is completely positive.*

We next prove a uniqueness theorem of operator sum representations.

---

**Theorem 3.7.** Let $m, n \in \mathbb{N}$ with $m \leqslant n$. Consider a generalized measurement $\{M_x\}_{x\in[m]} \subset \mathfrak{L}(A, B)$ to be a generalized measurement, and a set of matrices $\{N_y\}_{y\in[n]} \subset \mathfrak{L}(A, B)$. The following statements are equivalent:

1. The sets $\{M_x\}_{x\in[m]}$ and $\{N_y\}_{y\in[n]}$ constitute two operator sum representations of the same quantum channel.
2. There exists an $n \times m$ isometry $V = (v_{yx})$ such that for all $y \in [n]$:

$$N_y := \sum_{x\in[m]} v_{yx} M_x. \tag{3.129}$$

---

**Proof**   We first prove the implication $2 \Rightarrow 1$. From (3.129) it follows that

$$\sum_{y\in[n]} N_y^* N_y = \sum_{x,x'\in[m]} \sum_{y\in[n]} \bar{v}_{yx} v_{yx'} M_x^* M_{x'}$$

$$\boxed{V^*V = I_m} \longrightarrow = \sum_{x,x'\in[m]} \delta_{xx'} M_x^* M_{x'} = \sum_{x\in[m]} M_x^* M_x = I^A, \tag{3.130}$$

so that $\{N_y\}_{y\in[n]}$ is a generalized measurement. Similarly, for any $\rho \in \mathfrak{L}(A)$ we have

$$\sum_{y\in[n]} N_y \rho N_y^* = \sum_{x,x'\in[m]} \sum_{y\in[n]} v_{yx} \bar{v}_{yx'} M_x \rho M_{x'}^*$$

$$\boxed{V^*V = I_m} \longrightarrow = \sum_{x,x'\in[m]} \delta_{xx'} M_x \rho M_{x'}^* = \sum_{x\in[m]} M_x \rho M_x^*. \tag{3.131}$$

Hence, the sets $\{M_x\}_{x\in[m]}$ and $\{N_y\}_{y\in[n]}$ are two operator sum representations of the same quantum channel.

Next, we prove the implication $1 \Rightarrow 2$. From the assumption we have in particular that for every $\psi \in \text{Pure}(A)$ we have

$$\rho := \sum_{x\in[m]} M_x |\psi\rangle\langle\psi| M_x^* = \sum_{y\in[n]} N_y |\psi\rangle\langle\psi| N_y^*. \tag{3.132}$$

Since both $\{M_x|\psi\rangle\}_{x\in[m]}$ and $\{N_y|\psi\rangle\}_{y\in[n]}$ form an unnormalized pure-state decomposition of the same density matrix $\rho$, we get from Exercise 2.27 that there exists an $n \times m$ isometry matrix $V = (v_{yx})$ such that for all $y \in [n]$

$$N_y|\psi\rangle = \sum_{x\in[m]} v_{yx} M_x|\psi\rangle. \tag{3.133}$$

Since the above equality holds for all $\psi \in \mathrm{Pure}(A)$, the relation (3.129) must hold. This completes the proof. $\blacksquare$

**Exercise 3.39.** *Show that for every quantum channel $\mathcal{E} \in \mathrm{CPTP}(A \to B)$ there exists an operator sum representation with no more than $|AB|$ elements.*

## The Canonical Operator Sum Representation

Recall from Exercise 2.27 that $J_{\mathcal{E}}^{AB}$ has pure state decompositions that are all related via some isometry in the same manner as the two sets of the Kraus operators above are related. Therefore, each operator sum representations of $\mathcal{E}$ corresponds to a particular pure state decomposition of $J_{\mathcal{E}}^{AB}$ as in (3.121). The *canonical* operator sum representation of the quantum channel $\mathcal{E}$ is the one corresponding to the diagonalization of $J_{\mathcal{E}}^{AB}$. That is, in the canonical representation we take the vectors $\{|\psi_x^{AB}\rangle\}_{x\in[m]}$ in (3.121) to be orthogonal. This means that they are linearly independent and consequently we must have $m \leqslant |AB|$ since the rank of $J_{\mathcal{E}}^{AB}$ cannot exceed $|A \otimes B| = |AB|$. Moreover, the orthogonality of the vectors $\{|\psi_x^{AB}\rangle\}_{x\in[m]}$ implies that for $x \neq x'$

$$0 = \langle\psi_x^{AB}|\psi_{x'}^{AB}\rangle = \langle\Omega^{A\tilde{A}}|M_x^* M_{x'} \otimes I^{\tilde{A}}|\Omega^{A\tilde{A}}\rangle = \mathrm{Tr}\left[M_x^* M_{x'}\right]. \tag{3.134}$$

That is, the Kraus operators are also orthogonal in the Hilbert–Schmidt inner product. We therefore arrived at the following corollary.

> **Corollary 3.2.** Let $\mathcal{E} \in \mathfrak{L}(A \to B)$ be a quantum channel. Then $\mathcal{E}$ has a *canonical* operator sum representation $\{M_x\}_{x\in[m]}$, with $m \leqslant |AB|$, $\sum_{x\in[m]} M_x^* M_x = I^A$, and for each $x \neq x'$
>
> $$\mathrm{Tr}\left[M_x^* M_{x'}\right] = 0. \tag{3.135}$$

In particular, there are always operator sum representations with no more than $|AB|$ Kraus operators.

**Exercise 3.40.** *Let $\{M_x\}_{x\in[m]}$ be a canonical Kraus decomposition of $\mathcal{E} \in \mathrm{CPTP}(A \to B)$. Show that for any $m \times m$ unitary matrix $U = (u_{yx})$, also $\{N_y\}_{y\in[n]}$ with*

$$N_y := \sum_{x\in[m]} u_{yx} M_x \tag{3.136}$$

*is a canonical Kraus decomposition of $\mathcal{E}$.*

## 3.4.5 The Unitary Representation

In the previous section, we saw that quantum measurement can be used to realize any CPTP map. Here we show that a deterministic unitary evolution can be used to realize a quantum channel. Specifically, in Figure 3.4 a system $A$ is assumed to be initially uncorrelated with the environment. The initial state of the system is denoted by $\rho^A$ and the initial state of the environment by $|0\rangle\langle 0|^E$. Then, the system+environment undergoes a unitary evolution which converts the initial state $\rho^A \otimes |0\rangle\langle 0|^E$ to the state

$$U\left(\rho^A \otimes |0\rangle\langle 0|^E\right)U^* \quad ; \quad U^*U = I^{AE}. \tag{3.137}$$

Finally, the environment system is traced out yielding the final state

$$\mathcal{E}(\rho^A) := \mathrm{Tr}_E\left[U\left(\rho^A \otimes |0\rangle\langle 0|^E\right)U^*\right]. \tag{3.138}$$

We show now that every quantum channel can be realized in this way, giving a new interpretation for quantum channels as joint unitary evolutions on the system plus environment. Recall that typically the degrees of freedom of the environment are not accessible and therefore they are traced out at the end of the process.

> **Stinespring Dilation Theorem**
>
> **Theorem 3.8.** A linear map $\mathcal{E} \in \mathfrak{L}(A \to B)$ is a quantum channel if and only if there exist an environment (ancillary) system (and corresponding Hilbert space) $E$ of dimension $|E| \leqslant |AB|$, and an isometry $V: A \to B \otimes E$ with $V^*V = I^A$ such that
>
> $$\mathcal{E}(\rho^A) := \mathrm{Tr}_E\left[V\rho^A V^*\right] \qquad \forall\, \rho \in \mathfrak{L}(A). \tag{3.139}$$

*Remark.* Theorem 3.8 is an adaptation of Stinespring Dilation Theorem to the finite dimensional case.

**Proof** Suppose $\mathcal{E}$ has the form (3.139). To show that $\mathcal{E}$ is a quantum channel we denote by $\{|\phi_z^E\rangle\}_{z\in[k]}$ an orthonormal basis of $E$, where $k := |E|$, and by $M_z := \langle\phi_z^E|V$. By definition, for every $z \in [k]$, $M_z: A \to B$, and from (3.139) we get

$$\mathcal{E}(\rho^A) = \sum_{z\in[k]} \langle\phi_z^E|V\rho^A V^*|\phi_z^E\rangle = \sum_{z\in[k]} M_z\rho M_z^*. \tag{3.140}$$

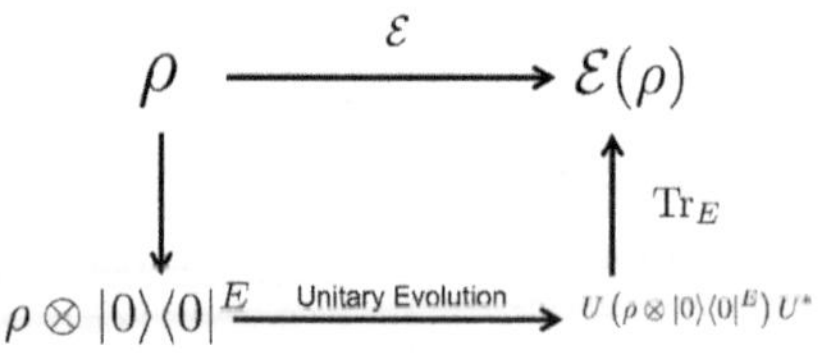

**Figure 3.4**   Stinespring dilation.

Moreover, since $V^*V = I^A$ we obtain

$$\sum_{z\in[k]} M_z^* M_z = \sum_{z\in[k]} V^*|\phi_z^E\rangle\langle\phi_z^E|V = V^*I^EV = V^*V = I^A. \tag{3.141}$$

Conversely, suppose $\mathcal{E}$ is a quantum channel. Then, from the previous section it has an operator sum representation $\mathcal{E}(\rho) = \sum_{z\in[k]} M_z\rho M_z^*$, with $k \leqslant |A||B|$ and $\sum_{z\in[k]} M_z^* M_z = I^A$. Set $E := \mathbb{C}^k$ and define the map $V : A \to BE$ via

$$V = \sum_{z\in[k]} M_z \otimes |\phi_z^E\rangle. \tag{3.142}$$

From its definition,

$$V^*V = \sum_{z',z\in[k]} M_{z'}^* M_z\langle\phi_{z'}^E|\phi_z^E\rangle = \sum_{z\in[k]} M_z^* M_z = I^A, \tag{3.143}$$

so that $V$ is an isometry. Moreover, from the definition of $V$ we get $M_z = \langle\phi_z^E|V$, so that

$$\begin{aligned}
\mathrm{Tr}_E\left[V\rho^A V^*\right] &= \sum_{z\in[k]} \langle\phi_z^E|V\rho^A V^*|\phi_z^E\rangle \\
&= \sum_{z\in[k]} M_z\rho M_z^* = \mathcal{E}(\rho).
\end{aligned} \tag{3.144}$$

Hence, there exists an isometry $V : A \to BE$ such that (3.139) holds. This completes the proof. ∎

**Exercise 3.41.** *Consider the isometry $V : A \to BE$ as expressed in (3.142), where each $M_z : A \to B$. Let $\{|x\rangle^A\}_{x\in[m]}$ be an orthonormal basis of $A$, and $\{|\psi_x^{BE}\rangle\}_{x\in[m]}$ be an orthonormal set of vectors in $BE$, such that*

$$V = \sum_{x\in[m]} |\psi_x^{BE}\rangle{}^A\langle x|. \tag{3.145}$$

*Finally, express each $|\psi_x^{BE}\rangle$ as*

$$|\psi_x^{BE}\rangle = \sum_{z\in[k]} |\phi_{z|x}^B\rangle|\phi_z^E\rangle \tag{3.146}$$

*with some vectors $\{|\phi_{z|x}^B\rangle\}_{z\in[k]} \subset B$. Show that with these notations*

$$M_z = \sum_{x\in[m]} |\phi_{z|x}^B\rangle{}^A\langle x|. \tag{3.147}$$

**Exercise 3.42.** *Show that a linear map $\mathcal{E} \in \mathfrak{L}(A \to A)$ is a quantum channel if and only if there exists an environment system $E$ and a unitary matrix $U : AE \to AE$ such that $\mathcal{E}$ has the form (3.138). Hint: Complete the isometry $V$ in the above theorem into a unitary operator.*

Note that in the proof above we defined the isometry $V$ in (3.142) using the Kraus operators. Therefore, (3.142) provides a direct relationship between the Stinespring representation and the operator sum representation. Moreover, the operator sum representation is directly related to the Choi representation via the relationship in (3.121,3.122). Therefore, together with (3.142) we can establish a direct relationship among all three representations. We will use these relationships quite often in the next sections.

## How Unique Is Stinespring Dilation?

In the following theorem we address the uniqueness of the isometry $V$ that appears in Stinespring dilation theorem.

---

**Theorem 3.9.** Let $\mathcal{E} \in \mathrm{CPTP}(A \to B)$, $V : A \to BE$ be an isometry satisfying (3.139), and let $W : A \to BE$ be a linear operator. The following statements are equivalent:

1. There exists a unitary matrix $U^E : E \to E$ such that

$$W = \left( I^B \otimes U^E \right) V. \tag{3.148}$$

2. For all $\rho \in \mathcal{L}(A)$

$$\mathcal{E}(\rho) = \mathrm{Tr}_E \left[ W \rho W^* \right]. \tag{3.149}$$

---

**Proof**   The proof of the implication $1 \Rightarrow 2$ is straightforward and is left as an exercise. We will now focus on proving the implication $2 \Rightarrow 1$. Following the methodology used in Stinesprings' dilation theorem, let us denote by $\{|\phi_z^E\rangle\}_{z \in [k]}$ an orthonormal basis of $E$, where $k := |E|$. For each $z \in [k]$, we define $M_z := \langle \phi_z^E | V$ and $N_z := \langle \phi_z^E | W$. Note that for all $z \in [k]$, $M_z$ and $N_z$ are linear operators from $A$ to $B$. In the proof of Stinesprings' theorem, particularly in (3.144), it was demonstrated that $\{M_z\}_{z \in [k]}$ constitutes an operator sum representation of $\mathcal{E}$. Similarly, from (3.149), we get

$$\mathcal{E}(\rho^A) = \sum_{z \in [k]} \langle \phi_z^E | W \rho^A W^* | \phi_z^E \rangle = \sum_{z \in [k]} N_z \rho N_z. \tag{3.150}$$

Therefore, $\{N_z\}_{z \in [k]}$ also forms an operator sum representation of $\mathcal{E}$. Observe in particular that the property that $\mathcal{E}$ is trace preserving implies that

$$I^A = \sum_{z \in [k]} N_z^* N_z$$

$$\boxed{N_z := \langle \phi_z^E | W} \longrightarrow \quad = \sum_{z \in [k]} W^* | \phi_z^E \rangle \langle \phi_z^E | W \tag{3.151}$$

$$\boxed{\sum_{z \in [k]} \phi_z^E = I^E} \longrightarrow \quad = W^* W.$$

Thus, $W$ is an isometry. Given that both $\{M_z\}_{z\in[k]}$ and $\{N_z\}_{z\in[k]}$ are operator sum representations of $\mathcal{E}$, Theorem 3.7 implies the existence of a $k \times k$ unitary matrix $V = (v_{zw})$, such that for every $z \in [k]$

$$N_z = \sum_{w\in[k]} u_{zw} M_w. \tag{3.152}$$

This leads us to (cf. (3.142))

$$\begin{aligned}
W &= \sum_{z\in[k]} N_z \otimes |\phi_z^E\rangle \\
&= \sum_{w,z\in[k]} v_{zw} M_w \otimes |\phi_z^E\rangle \\
&= \sum_{w\in[k]} M_w \otimes \sum_{z\in[k]} v_{zw} |\phi_z^E\rangle.
\end{aligned} \tag{3.153}$$

Defining the matrix $U^E := V^T$, we find for all $w \in [k]$

$$U^E |\phi_w^E\rangle = \sum_{z\in[k]} v_{zw} |\phi_z^E\rangle. \tag{3.154}$$

Thus, we conclude

$$\begin{aligned}
W &= \left(I^B \otimes U^E\right) \sum_{w\in[k]} M_w \otimes |\phi_w^E\rangle \\
(3.142)\rightarrow &= \left(I^B \otimes U^E\right) V.
\end{aligned} \tag{3.155}$$

This concludes the proof. $\blacksquare$

## 3.5 Examples of Quantum Channels

In this section we list and discuss briefly several examples of quantum channels. These channels have many interesting properties that make the interplay among the different representations of a quantum channel more apparent. These channels appear quite often in the field of quantum information, and we will also encounter some of them later on in the book.

### 3.5.1 Qubit Channels

We discuss a few common examples of qubit channels; that is, CPTP maps from $\mathfrak{L}(\mathbb{C}^2)$ to itself. Unlike quantum channels in higher dimensions, qubit channels can be characterized by their effect on the Bloch vector. This is a convenient property that is useful for some applications. We start with a communication channel that represents the most basic error in information theory; namely, the bit flip.

## The Classical Bit Flip

The classical bit-flip channel (see Figure 3.5) is a process that flips a classical bit according to some probability distribution. Specifically, the zero is flipped to 1 with some probability $p$ and remains unchanged with probability $1 - p$. The bit-flip channel is symmetric if the probability to flip 1 is given by the same probability $p$.

## The Quantum Bit Flip

The quantum bit-flip channel acts in a similar way. The bit-flip operator is the Pauli first matrix $X = \begin{pmatrix} 0 & 1 \\ 1 & 0 \end{pmatrix}$, since $X|0\rangle\langle 0|X = |1\rangle\langle 1|$ and $X|1\rangle\langle 1|X = |0\rangle\langle 0|$. Denote by

$$M_0 = \sqrt{1-p}\,I \quad \text{and} \quad M_1 = \sqrt{p}\,X, \tag{3.156}$$

and note that $M_0^* M_0 + M_1^* M_1 = I$. The qubit channel $\mathcal{E} \in \mathcal{L}(A \to B)$ (with $|A| = |B| = 2$) defined by $\mathcal{E}(\rho) = M_0 \rho M_0^* + M_1 \rho M_1^*$ is called the quantum bit-flip channel. Its Choi representation is given by

$$J_{\mathcal{E}}^{AB} = \mathcal{E}^{\tilde{A}\to B}(\Omega^{A\tilde{A}}) = 2(1-p)\Phi_+^{AB} + 2p\,\Psi_+^{AB}, \tag{3.157}$$

where $|\Phi_\pm^{AB}\rangle = \frac{1}{\sqrt{2}}(|00\rangle \pm |11\rangle)$ and $|\Psi_\pm^{AB}\rangle = \frac{1}{\sqrt{2}}(|01\rangle \pm |10\rangle)$ form the *Bell basis* of the two qubit system.

## The Depolarizing Channel

Another very common example of a qubit channel is the depolarizing channel. It is defined by

$$\mathcal{E}(\rho) = \frac{p}{2}\mathrm{Tr}[\rho]I + (1-p)\rho \qquad \forall\, \rho \in \mathcal{L}(\mathbb{C}^2). \tag{3.158}$$

Note that this channel has the unique property that for any unitary matrix $U \in \mathcal{L}(\mathbb{C}^2)$

$$\mathcal{E}(U\rho U^*) = U\mathcal{E}(\rho)U^*. \tag{3.159}$$

**Exercise 3.43.** *Show that the depolarizing channel is indeed a quantum channel, by showing that it has an operator sum representation that is given in terms of the following four Kraus operators:*

$$M_0 = \sqrt{1 - \frac{3p}{4}}\,I \quad \text{and for } j \in [3], \ M_j = \frac{\sqrt{p}}{2}\sigma_j, \tag{3.160}$$

*where $\sigma_1, \sigma_2,$ and $\sigma_3$ are the three Pauli matrices.*

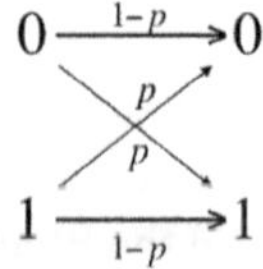


Figure 3.5 The symmetric classical bit-flip channel.

From the exercise above it also follows that the normalized version of the Choi matrix of the depolarizing channel has the form

$$\frac{1}{2} J_{\mathcal{E}}^{AB} = \mathcal{E}^{\tilde{A} \to B}\left(\Phi_+^{A\tilde{A}}\right) = \left(1 - \frac{3p}{4}\right) \Phi_+^{AB} + \frac{p}{4}\left(\Phi_-^{AB} + \Psi_+^{AB} + \Psi_-^{AB}\right). \tag{3.161}$$

The state above is known (up to local unitary) as the 2-qubit isotropic state and is used quite often in quantum information as it has several interesting properties. We will discuss it in more detail later on.

The Stinespring isometry of the depolarizing channel can also be computed from (3.142) and the exercise above; it is given by

$$V = \sum_{j=0}^{3} M_j \otimes |j\rangle^E = \sqrt{1 - \frac{3p}{4}} I \otimes |0\rangle + \frac{\sqrt{p}}{2} \sum_{j\in[3]} \sigma_j \otimes |j\rangle, \tag{3.162}$$

where $\sigma_1$, $\sigma_2$, and $\sigma_3$ are the three Pauli matrices. Interestingly, the following exercise shows that the Bloch representation is in some sense the simplest representation of the depolarizing channel.

**Exercise 3.44.** *Let* $\rho = \frac{1}{2}(I + \mathbf{r} \cdot \boldsymbol{\sigma})$ *and* $\rho' = \frac{1}{2}(I + \mathbf{r}' \cdot \boldsymbol{\sigma})$ *be two Bloch representations of two quantum states, and let* $\mathcal{E}$ *be the depolarizing channel* (3.158). *Show that if* $\rho' = \mathcal{E}(\rho)$, *then* $\mathbf{r}' = (1 - p)\mathbf{r}$.

## 3.5.2 The Completely Dephasing Channel

The completely dephasing map, sometimes also referred to as the completely decohering map, is the channel $\Delta \in \mathrm{CPTP}(A \to A)$ that removes the off-diagonal terms from any matrix $\rho \in \mathcal{L}(A)$, with respect to some fixed orthonormal basis $\{|x\rangle\}_{x\in[m]}$, where $m := |A|$. Specifically, its Kraus representation is given by $\{M_x := |x\rangle\langle x|\}_{x\in[m]}$. Denoting by $\{r_{xy}\}_{x,y\in[m]}$ the matrix elements of $\rho$ with respect to the basis $\{|x\rangle\}_{x\in[m]}$, we get from the operator sum representation of $\Delta$ that

$$\Delta(\rho) = \sum_{x\in[m]} |x\rangle\langle x|\rho|x\rangle\langle x| = \sum_{x\in[m]} r_{xx}|x\rangle\langle x|. \tag{3.163}$$

From the above representation of $\Delta$, it is clear that the completely dephasing map is *idempotent*, that is, it satisfies

$$\Delta^2 := \Delta \circ \Delta = \Delta. \tag{3.164}$$

Its Choi matrix is

$$J_\Delta = \Delta^{\tilde{A} \to \tilde{A}}(\Omega^{A\tilde{A}}) = \sum_{x\in[m]} |x\rangle\langle x| \otimes |x\rangle\langle x|. \tag{3.165}$$

Its Stinespring isometry $V_\Delta : A \to A \otimes E$ is given by:

$$V_\Delta |x\rangle^A = |x\rangle^A |x\rangle^E, \tag{3.166}$$

where $\{|x\rangle^E\}$ is an orthonormal basis of $E$.

**Exercise 3.45.** *A generalized dephasing channel $\mathcal{N} \in \mathrm{CPTP}(A \to A)$ is a channel that transmit some preferred basis $\{|x\rangle\}_{x\in[m]}$ of $A$ without error. That is,*

$$\mathcal{N}(|x\rangle\langle x|) = |x\rangle\langle x|. \tag{3.167}$$

*1. Show that the completely dephasing channel $\Delta$ is a generalized dephasing channel.*
*2. Show that $\mathcal{N}$ above has a Stinespring isometry*

$$V_{\mathcal{N}}|x\rangle^A = |x\rangle^A|\phi_x^E\rangle \qquad \forall\, x \in [m], \tag{3.168}$$

*where $\{|\phi_x^E\rangle\}_{x\in[m]}$ are some normalized vectors in $E$ (note that if $\{|\phi_x^E\rangle\}$ is an orthonormal set then $\mathcal{N} = \Delta$).*
*3. Show that*

$$\Delta \circ \mathcal{N} = \mathcal{N} \circ \Delta = \Delta. \tag{3.169}$$

### 3.5.3 Classical Channels

A classical channel is a stochastic process that converts an input (classical) variable $X = x$ into an output variable $Y = y$. The classical "noise" is modeled by some conditional probability distribution $\Pr(Y = y|X = x)$ that determines the probability that the output variable $Y$ is equal to $y$ given the input variable $X$ equals $x$. Denote by $p_x := \Pr(X = x)$ with $x \in [m]$, and $q_y = \Pr(Y = y)$ with $y \in [n]$, the (marginal) distributions associated with the random variables $X$ and $Y$. Then the probability vectors $\mathbf{p} = (p_1, \ldots, p_m)^T$ and $\mathbf{q} = (p_1, \ldots, p_n)^T$ are related by

$$\mathbf{q} = T\mathbf{p}, \tag{3.170}$$

where the $n \times m$ evolution matrix $T = (t_{y|x})$, also known as the *transition matrix*, has components $t_{y|x} = \Pr(Y = y|X = x)$. This is the Shannon's model for a classical channel.

A classical channel can be viewed as a very special type of a quantum channel. In particular, we say that a quantum channel $\mathcal{E} \in \mathrm{CPTP}(A \to B)$ is a classical channel if

$$\Delta^B \circ \mathcal{E} \circ \Delta^A = \mathcal{E}, \tag{3.171}$$

where $\Delta^A \in \mathrm{CPTP}(A \to A)$ and $\Delta^B \in \mathrm{CPTP}(B \to B)$ are the completely dephasing channels with respect to some fixed classical bases $\{|x\rangle\}_{x\in[m]}$ and $\{|y\rangle\}_{y\in[n]}$ of $A$ and $B$, respectively. From its definition it is clear that $\mathcal{E}(\rho) = \mathcal{E}(\Delta^A(\rho))$. Therefore, it is sufficient to consider only diagonal input states. Moreover, if $\sigma = \mathcal{E}(\rho)$, then $\Delta^B(\sigma) = \sigma$; that is, the output state is always diagonal in the basis $\{|y\rangle\langle y|\}_{y\in[n]}$. Hence, taking $\rho = \sum_{x\in[m]} p_x|x\rangle\langle x|^A$ we get

$$\sigma := \mathcal{E}(\rho) = \Delta^B\left((\mathcal{E}(\rho)\right) = \sum_{y\in[n]} \langle y|\mathcal{E}(\rho)|y\rangle|y\rangle\langle y|^B. \tag{3.172}$$

Denote by $t_{y|x} := \langle y|\mathcal{E}(|x\rangle\langle x|)|y\rangle \geqslant 0$, and note that $\sum_{y\in[n]} t_{y|x} = \mathrm{Tr}[\mathcal{E}(|x\rangle\langle x|)] = 1$ since $\mathcal{E}$ is trace preserving. We therefore have

$$q_y := \langle y|\sigma|y\rangle = \langle y|\mathcal{E}(\rho)|y\rangle = \sum_{x\in[m]} p_x\langle y|\mathcal{E}(|x\rangle\langle x|)|y\rangle = \sum_{x\in[m]} t_{y|x}\, p_x. \tag{3.173}$$

In other words, we get that $\mathbf{q} = T\mathbf{p}$ as in (3.170), with $\mathbf{p}$ and $\mathbf{q}$ being the vectors whose components are the diagonal elements of $\rho$ and $\sigma$, respectively, and $T = (t_{y|x})$ being the column stochastic matrix whose components are $\langle y|\mathcal{E}(|x\rangle\langle x|)|y\rangle$.

**Exercise 3.46.** *Let $\{\rho_z\}_{z\in[k]}$ and $\{\sigma_z\}_{z\in[k]}$ be two sets of $k$ **diagonal** density operators with respect to two fixed bases $\{|x\rangle\}_{x\in[m]}$ and $\{|y\rangle\}_{y\in[n]}$ of $A$ and $B$, respectively. Show that there exist a quantum channel $\mathcal{E} \in \mathrm{CPTP}(A \to B)$ such that $\sigma_z = \mathcal{E}(\rho_z)$ for all $z \in [k]$ if and only if there exists a classical channel with the same property.*

### 3.5.4 POVM Channels

A quantum channel $\mathcal{E} \in \mathrm{CPTP}(A \to B)$ is called a *quantum to classical* channel, or POVM channel, if

$$\Delta^B \circ \mathcal{E}^{A\to B} = \mathcal{E}^{A\to B}, \tag{3.174}$$

where $\Delta^B$ is the completely dephasing map on the output space with respect to some fixed basis $\{|y\rangle\}_{y\in[n]}$ of $B$, where $n := |B|$. Such channels have the property that for any $\rho \in \mathfrak{D}(A)$

$$\mathcal{E}(\rho) = \Delta\left(\mathcal{E}(\rho)\right) = \sum_{y\in[n]} \langle y|\mathcal{E}(\rho)|y\rangle |y\rangle\langle y|. \tag{3.175}$$

Note that

$$\langle y|\mathcal{E}(\rho)|y\rangle = \mathrm{Tr}\left[|y\rangle\langle y|\mathcal{E}(\rho)\right] = \mathrm{Tr}\left[\mathcal{E}^*(|y\rangle\langle y|)\rho\right] := \mathrm{Tr}\left[\Lambda_y\rho\right], \tag{3.176}$$

where $\{\Lambda_y := \mathcal{E}^*(|y\rangle\langle y|)\}_{y\in[n]}$ are positive semidefinite matrices in $\mathrm{Pos}(A)$ satisfying

$$\sum_{y\in[n]} \Lambda_y = \sum_{y\in[n]} \mathcal{E}^*(|y\rangle\langle y|) = \mathcal{E}^*(I^B) = I^A, \tag{3.177}$$

since the dual map of any CPTP map is unital (see Exercise 3.23). We therefore get that a POVM channel has the form

$$\mathcal{E}^{A\to B}\left(\rho^A\right) = \sum_{y\in[n]} \mathrm{Tr}\left[\Lambda_y^A\rho^A\right] |y\rangle\langle y|^B. \tag{3.178}$$

**Exercise 3.47.** *Let $\mathcal{E} \in \mathrm{CPTP}(A \to B)$ be a POVM channel as in (3.178).*

1. *Show that its Choi matrix is given by*

$$J_{\mathcal{E}}^{AB} = \sum_{y\in[n]} \Lambda_y^T \otimes |y\rangle\langle y|. \tag{3.179}$$

2. *Show that the operators $M_{xy}: A \to B$ with*

$$M_{xy} := |y^B\rangle\langle x^A|\sqrt{\Lambda_y} \tag{3.180}$$

*can be used to form an operator sum representation of $\mathcal{E}$.*

3. *Show that the map* $V: A \to B \otimes \tilde{A}\tilde{B}$ *given by*

$$V = \sum_{y \in [n]} |y\rangle^B \otimes \sqrt{\Lambda_y^{A \to \tilde{A}}} \otimes |y\rangle^{\tilde{B}} \tag{3.181}$$

*is a Stinespring isometry (the environment system is $\tilde{A}\tilde{B}$) satisfying*

$$\mathrm{Tr}_{\tilde{A}\tilde{B}}\left[V\rho V^*\right] = \sum_{y \in [n]} \mathrm{Tr}\left[\Lambda_y \rho\right] |y\rangle\langle y|^B = \mathcal{E}(\rho) \quad and \quad V^*V = I^A. \tag{3.182}$$

*Hint: Use (3.142) and show that* $\sum_x {}^A\langle x| \sqrt{\Lambda_y^{A \to A}} \otimes |x\rangle^{\tilde{A}} = \sqrt{\Lambda_y^{A \to \tilde{A}}}$.

## 3.5.5 Preparation Channels

A preparation channel, commonly referred to as a classical-quantum (cq) channel, is a linear map that transforms classical inputs into quantum states. To elaborate, a channel $\mathcal{E} \in \mathrm{CPTP}(A \to B)$ is classified as a cq-channel if it fulfills the following criterion:

$$\mathcal{E}^{A \to B} \circ \Delta^{A \to A} = \mathcal{E}^{A \to B}. \tag{3.183}$$

Hence, for every $\rho \in \mathfrak{D}(A)$ we get

$$\mathcal{E}(\rho) = \sum_{x \in [m]} p_x \mathcal{E}(|x\rangle\langle x|) = \sum_{x \in [m]} p_x \sigma_x, \tag{3.184}$$

where $m := |A|$, $p_x := \langle x|\rho|x\rangle$ for all $x \in [m]$, and each $\sigma_x := \mathcal{E}(|x\rangle\langle x|)$ is a fixed density matrix in $\mathfrak{D}(B)$. We can therefore view $\mathcal{E}$ as the mapping $x \mapsto \sigma_x$. The Choi matrix of a cq-channel has the form

$$\begin{aligned} J_{\mathcal{E}}^{AB} = \mathcal{E}^{\tilde{A} \to B}\left(\Omega^{A\tilde{A}}\right) &= \mathcal{E}^{\tilde{A} \to B} \circ \Delta^{\tilde{A} \to \tilde{A}}\left(\Omega^{A\tilde{A}}\right) \\ &= \mathcal{E}^{\tilde{A} \to B}\left(\sum_{x \in [m]} |xx\rangle\langle xx|^{A\tilde{A}}\right) \\ &= \sum_{x \in [m]} |x\rangle\langle x|^A \otimes \sigma_x^B. \end{aligned} \tag{3.185}$$

Therefore, $\mathcal{E} \in \mathrm{CPTP}(A \to B)$ is a cq-channel if and only if its Choi matrix $J_{\mathcal{E}}^{AB}$ is a cq-state.

**Exercise 3.48.** *Let $\mathcal{E} \in \mathrm{CPTP}(A \to B)$ be a cq-channel as above, and set $m := |A|$ and $n := |B|$.*

1. *Show that the set of Kraus operators $\{M_{xy}\}$ with $M_{xy}: A \to B$*

$$M_{xy} := \sqrt{\sigma_x}|y^B\rangle\langle x^A| \quad x \in [m], \ y \in [n], \tag{3.186}$$

*forms an operator sum representation of $\mathcal{E}$.*

2. *Show that the map $V: A \to B \otimes \tilde{A}\tilde{B}$ given by*

$$V = \sum_{x \in [m]} \left(\sqrt{\sigma_x^B} \otimes I^{\tilde{B}}\right) |\Omega^{B\tilde{B}}\rangle \otimes |x^{\tilde{A}}\rangle\langle x^A| \tag{3.187}$$

*is a Stinespring isometry (the environment system is $\tilde{A}\tilde{B}$) satisfying*

$$\text{Tr}_{\tilde{A}\tilde{B}}\left[V\rho V^*\right] = \sum_{x\in[m]} p_x\sigma_x = \mathcal{E}(\rho) \quad and \quad V^*V = I^A. \tag{3.188}$$

## 3.5.6 Measurement-Prepare Channels and Entanglement Breaking Channels

A measurement-prepare channel represents a process where a quantum system is first measured, and then, depending on the outcome of this measurement, a specific quantum state is prepared. More specifically, a measurement-prepare channel, $\mathcal{E} \in \text{CPTP}(A \to B)$, is a type of quantum channel characterized by the following form:

$$\mathcal{E}(\rho^A) = \sum_{z\in[k]} \text{Tr}\left[\Lambda_z^A \rho^A\right]\sigma_z^B, \tag{3.189}$$

where $\{\Lambda_z\}_{z\in[k]} \subset \text{Pos}(A)$ is a POVM with $m$ outcomes, and $\{\sigma_z^B\}_{z\in[k]}$ are $k$ quantum states in $\mathfrak{D}(B)$. Clearly, the map $\mathcal{E}$ above is linear and trace preserving. Moreover, setting $m := |A|$ we get that its Choi matrix is given by

$$\begin{aligned}
J_{\mathcal{E}}^{AB} = \mathcal{E}^{\tilde{A}\to B}(\Omega^{A\tilde{A}}) &= \sum_{x,x'\in[m]} |x\rangle\langle x'| \otimes \mathcal{E}(|x\rangle\langle x'|) \\
&= \sum_{z\in[m]}\sum_{x,x'\subset[m]} \text{Tr}\left[\Lambda_z|x\rangle\langle x'|\right] |x\rangle\langle x'| \otimes \sigma_z^B \\
&= \sum_{z\in[k]} \Lambda_z^T \otimes \sigma_z^B \geqslant 0.
\end{aligned} \tag{3.190}$$

Therefore, $\mathcal{E}$ in (3.189) is a quantum channel. Observe that the Choi matrix above is separable.

Measurement-prepare channels can be viewed as a combination of a POVM channel (i.e. qc-channel) followed by a preparation Channel (i.e. cq-channel). Specifically, let $\mathcal{M} \in \text{CPTP}(A \to Z)$ be a POVM Channel given as in (3.178) by

$$\mathcal{M}^{A\to Z}(\rho^A) := \sum_{z\in[k]} \text{Tr}\left[\Lambda_z^A \rho^A\right] |z\rangle\langle z|^Z, \tag{3.191}$$

for some POVM $\{\Lambda_z\}_{z\in[k]}$ in $\text{Pos}(A)$, and a classical system $Z$ of dimension $|Z| = k$. Let $\mathcal{P} \in \text{CPTP}(Z \to B)$ be a preparation Channel given as in (3.184) via

$$\mathcal{P}^{Z\to B}(|z\rangle\langle z|^Z) = \sigma_z^B \qquad \forall\, z \in [k], \tag{3.192}$$

where $\sigma_z \in \mathfrak{D}(B)$ are some density matrices. Then, the measurement-prepare channel $\mathcal{E}$ given in (3.189) can be expressed as

$$\mathcal{E}^{A\to B} = \mathcal{P}^{Z\to B} \circ \mathcal{M}^{A\to Z}. \tag{3.193}$$

**Exercise 3.49.** *Let A, B, and R be three Hilbert spaces and let $\Delta \in \mathrm{CPTP}(R \to R)$ be the completely dephasing map with respect to some fixed basis of R. Show that for any two quantum channels $\mathcal{E} \in \mathrm{CPTP}(A \to R)$ and $\mathcal{F} \in \mathrm{CPTP}(R \to B)$ the channel*

$$\mathcal{F}^{R \to B} \circ \Delta^R \circ \mathcal{E}^{A \to R} \tag{3.194}$$

*is a measurement-prepare channel in* $\mathrm{CPTP}(A \to B)$.

**Exercise 3.50.** *Let $\mathcal{E} \in \mathrm{CPTP}(B \to B')$ be a measurement-prepare channel as above, and let $\rho^{AB} \in \mathfrak{D}(A \otimes B)$ be a bipartite density operator. Show that the density operator*

$$\sigma^{AB'} := \mathcal{E}^{B \to B'}(\rho^{AB}) \tag{3.195}$$

*is separable. Hint: Use (3.193) and consider the cq-state $\tau^{ZB} := \mathcal{M}^{A \to Z}(\rho^{AB})$.*

The above exercise demonstrates that a measurement-prepare channel breaks the entanglement when applied to a subsystem of a composite bipartite system. Channels with this property are called *entanglement breaking channels*. Therefore, measurement-prepare channels are entanglement breaking. It turns out that the converse is also true! That is, any entanglement breaking channel can be represented as a measurement-prepare channel. For this reason, we will use, depending on the context, both terms interchangeably.

**Exercise 3.51.** *Show that any entanglement breaking channel is a measurement-prepare channel. Hint: Start by observing that the Choi matrix of any entanglement breaking channel must be separable.*

**Exercise 3.52.** *Let $\mathcal{E}^{A \to B}$ be the quantum channel defined in (3.189). Show that its dual is given by*

$$\mathcal{E}^{*B \to A}\left(\eta^B\right) = \sum_{z \in [m]} \mathrm{Tr}\left[\eta^B \sigma_z^B\right] \Lambda_z^A. \tag{3.196}$$

### 3.5.7 The Partial Trace

The partial trace as defined in Exercise 2.40 is a linear map $\mathrm{Tr}_B \in \mathfrak{L}(AB \to A)$. What is its Choi matrix? Let

$$\Omega^{(AB)(\tilde{A}\tilde{B})} = \Omega^{A\tilde{A}} \otimes \Omega^{B\tilde{B}} \tag{3.197}$$

be the unnormalized maximally entangled state between system $AB$ and system $\tilde{A}\tilde{B}$. Then, by definition, its Choi matrix is

$$J_{\mathcal{E}}^{(AB)\tilde{A}} = \mathrm{id}^{AB} \otimes \mathrm{Tr}_{\tilde{B}}\left(\Omega^{(AB)(\tilde{A}\tilde{B})}\right)$$

$$(\mathbf{3.197}) \to = \Omega^{A\tilde{A}} \otimes I^B \geqslant 0. \tag{3.198}$$

Moreover, note that the marginal $\mathrm{Tr}_{\tilde{A}}\left[J_{\mathcal{E}}^{(AB)\tilde{A}}\right] = I^{AB}$. We therefore conclude that the partial trace is a quantum channel. Physically, it represents the process of discarding a subsystem from a composite system.

**Exercise 3.53.** *Let $A$ and $B$ be two Hilbert spaces, and let $\{|y\rangle^B\}_{y\in[n]}$ be an orthonormal basis of $B$ (with $n := |B|$). Show that the set of operators $\{M_y\}_{y\in[n]} \subset \mathfrak{L}(AB, A)$ given by*

$$M_y = I^A \otimes \langle y|^B \tag{3.199}$$

*forms an operator sum representation of the partial trace.*

## 3.5.8 Isometry Channels

Let $V: A \to B$ be an isometry matrix; that is, $V^*V = I^A$ and $|A| \leqslant |B|$. We define an isometry channel $\mathcal{V} \in \mathrm{CPTP}(A \to B)$ as

$$\mathcal{V}\left(\rho^A\right) = V\rho^A V^* \qquad \forall\, \rho \in \mathfrak{L}(A). \tag{3.200}$$

Such an isometry channel can be viewed as an *embedding* of system $A$ into $B$. Note that like unitary channels, isometry channels have an operator sum representation with a single Kraus operator.

Inrestingly, isometry channels have inverses. Specifically, for any $\tau \in \mathfrak{D}(A)$ define

$$\mathcal{V}_\tau^{-1}\left(\sigma^B\right) := V^*\sigma^B V + \mathrm{Tr}\left[\left(I^B - VV^*\right)\sigma^B\right]\tau^A \qquad \forall\, \sigma \in \mathfrak{L}(B). \tag{3.201}$$

The linear map above is a quantum channel in $\mathrm{CPTP}(B \to A)$ and it is an inverse of the isometry channel $\mathcal{V}$ above (see the following exercise).

**Exercise 3.54.** *Show that for all $\tau \in \mathfrak{D}(A)$ the linear map $\mathcal{V}_\tau^{-1}$ as defined above is a channel in $\mathrm{CPTP}(B \to A)$ that satisfies*

$$\mathcal{V}_\tau^{-1} \circ \mathcal{V} = \mathrm{id}^A. \tag{3.202}$$

## 3.5.9 Unital Channels

A map $\mathcal{E} \in \mathfrak{L}(A \to A)$ is termed a unital quantum channel, or more specifically, a *doubly stochastic* channel, if it satisfies two conditions: It is a CPTP map, and it is a unital map (meaning it preserves the identity, as in $\mathcal{E}(I) = I$). In Exercise 3.23, we demonstrated that a linear map is unital if and only if its dual is trace preserving. Consequently, $\mathcal{E}$ is a doubly stochastic channel if and only if both $\mathcal{E}$ and its dual $\mathcal{E}^*$ are quantum channels.

If $\mathcal{E} \in \mathrm{CPTP}(X \to X)$ is a *classical* doubly stochastic channel, then $\mathcal{E}$ can be represented by the $m \times m$ evolution column-stochastic matrix $E$, where $m := |X|$, and the unital condition $\mathcal{E}(I^X) = I^X$ translates to $E\mathbf{1}_m = \mathbf{1}_m$, with $\mathbf{1}_m$ is the $m$-dimensional vector $\mathbf{1}_m := (1, \ldots, 1)^T$. This means that the sum of each row of $E$ is equal to 1. Given that $E$ is inherently a column-stochastic matrix, it follows that classical unital channels correspond to doubly stochastic matrices. A doubly stochastic matrix is characterized as a square, real matrix with nonnegative entries, where the sum of the elements in each row and each column is 1. Consequently, in a classical framework, the input distribution $\mathbf{p}$

transitions to the output distribution $\mathbf{q}$ through the doubly stochastic matrix relationship $\mathbf{q} = D\mathbf{p}$. Here, the matrix $D$ is used to denote the evolution matrix, emphasizing its nature as doubly stochastic.

In the quantum case there is also a similar relation between the input state $\rho$ and the output state $\sigma = \mathcal{E}(\rho)$ of a doubly stochastic quantum channel $\mathcal{E} \in \mathrm{CPTP}(A \to A)$. To see this relation, set $m := |A|$, and denote by $\{|\psi_x\rangle\}_{x \in [m]}$ and $\{|\phi_y\rangle\}_{y \in [m]}$ the eigenvectors of $\rho$ and $\sigma$, respectively, and by $\{p_x\}_{x \in [m]}$ and $\{q_y\}_{y \in [m]}$, their corresponding eigenvalues. With these notations we have

$$\rho = \sum_{x \in [m]} p_x |\psi_x\rangle\langle\psi_x| \quad \text{and} \quad \sigma = \sum_{y \in [m]} q_y |\phi_y\rangle\langle\phi_y|. \tag{3.203}$$

The relation $\sigma = \mathcal{E}(\rho)$ is equivalent to

$$q_y = \langle\phi_y|\sigma|\phi_y\rangle = \langle\phi_y|\mathcal{E}(\rho)|\phi_y\rangle = \sum_{x \in [m]} p_x \langle\phi_y|\mathcal{E}(|\psi_x\rangle\langle\psi_x|)|\phi_y\rangle. \tag{3.204}$$

Let $D = (d_{xy})$ be the $m \times m$ matrix whose components are $d_{yx} := \langle\phi_y|\mathcal{E}(|\psi_x\rangle\langle\psi_x|)|\phi_y\rangle$, and note that $d_{yx} \geqslant 0$ for all $x$ and $y$. Moreover,

$$\sum_{x \in [m]} d_{yx} = \langle\phi_y|\mathcal{E}(I)|\phi_y\rangle = \langle\phi_y|I|\phi_y\rangle = 1 \quad \text{and}$$

$$\sum_{y \in [m]} d_{yx} = \mathrm{Tr}\left[\mathcal{E}(|\psi_x\rangle\langle\psi_x|)\right] = \mathrm{Tr}\left[|\psi_x\rangle\langle\psi_x|\right] = 1. \tag{3.205}$$

Hence, $D$ is a doubly stochastic matrix and (3.204) becomes $\mathbf{q} = D\mathbf{p}$, where $\mathbf{p}$ and $\mathbf{q}$ are the probability vectors consisting of the eigenvalues of $\rho$ and $\sigma$, respectively.

**Exercise 3.55.** *Let $\mathcal{E} \in \mathrm{CPTP}(A \to B)$ be a quantum channel. Show that if $\mathcal{E}(I^A) = I^B$, then we must have $|A| = |B|$.*

## No Quantum Analogue to Birkhoff Theorem

Birkhoff theorem (see Theorem A.5 of online version) states that any doubly stochastic matrix can be expressed as a convex combination of permutation matrices. For example,

$$\begin{bmatrix} \frac{5}{6} & \frac{1}{6} & 0 \\ 0 & \frac{1}{2} & \frac{1}{2} \\ \frac{1}{6} & \frac{1}{3} & \frac{1}{2} \end{bmatrix} = \frac{1}{3}\begin{bmatrix} 1 & 0 & 0 \\ 0 & 1 & 0 \\ 0 & 0 & 1 \end{bmatrix} + \frac{1}{2}\begin{bmatrix} 1 & 0 & 0 \\ 0 & 0 & 1 \\ 0 & 1 & 0 \end{bmatrix} + \frac{1}{6}\begin{bmatrix} 0 & 1 & 0 \\ 0 & 0 & 1 \\ 1 & 0 & 0 \end{bmatrix}. \tag{3.206}$$

In general, from Birkhoff theorem (Theorem A.5 of online version), any doubly stochastic matrix can be expressed as $D = \sum_{w \in [k]} t_w \Pi_w$, where $\{t_w\}_{w \in [k]}$ is a probability distribution and $\{\Pi_w\}_{w \in [k]}$ are permutation matrices. Therefore, the relation $\mathbf{q} = D\mathbf{p} = \sum_{w \in [k]} t_w \Pi_w \mathbf{p}$ implies that $\mathbf{q}$ is a convex combination of permuted versions of $\mathbf{p}$. We now discuss whether there exists a quantum analogue to this property.

A fundamental example of a unital quantum channel is the unitary evolution. This is analogous to the permutation evolution matrix of a classical channel. Indeed, if $\sigma = U\rho U^*$, then the vector $\mathbf{q}$, consisting of eigenvalues of $\sigma$, is related by a permutation matrix to the vector $\mathbf{p}$, consisting of the eigenvalues of $\rho$. Clearly, the unitary channel $\mathcal{E}(\rho) = U\rho U^*$ satisfies $\mathcal{E}(I) = UIU^* = UU^* = I$, where $U$ is a unitary matrix. One can extend this definition to include mixture of unitaries. Such channels are called *mixed-unitary channels* (or random-unitary channels). That is, a mixed-unitary channel is a quantum channel $\mathcal{E} \in \mathrm{CPTP}(A \to A)$ that has the form

$$\mathcal{E}(\rho) = \sum_{w \in [k]} t_w U_w \rho U_w^*, \qquad (3.207)$$

where $\{U_w\}_{w \in [k]}$ is a set of $k$ unitary matrices, and $\{t_w\}_{w \in [k]}$ is a probability distribution. This is the quantum version of a convex combination of permutation matrices. One can implement such a quantum channel, for example, by rolling a dice and based on the outcome $w$ of the dice apply the evolution $\rho \mapsto U_w \rho U_w^*$. After forgetting the value of $w$, such a process can be described as in (3.207).

**Exercise 3.56.** *Find the operator sum representation of the mixed-unitary channel (3.207).*

One may wonder whether all unital channels can be expressed as mixed-unitary channels. To answer this question, consider the following example given by Peter Shor (2010). Let $A = \mathbb{C}^3$ and $\mathcal{E} \in \mathrm{CPTP}(A \to A)$ be the quantum channel

$$\mathcal{E}(\omega) = M_1 \omega M_1^* + M_2 \omega M_2^* + M_3 \omega M_3^* \qquad \forall\, \omega \in \mathfrak{D}(A), \qquad (3.208)$$

where the Kraus operators

$$M_z := \frac{|z\rangle\langle z+1| + |z+1\rangle\langle z|}{\sqrt{2}} \qquad \forall\, z \in [3], \qquad (3.209)$$

with $|4\rangle := |1\rangle$ (i.e. the summation in $|z+1\rangle$ is modulo 3).

**Exercise 3.57.** *Show that the quantum channel $\mathcal{E} \in \mathrm{CPTP}(A \to A)$ as defined in (3.208) and (3.209) is unital.*

The unital channel $\mathcal{E}$ as defined in (3.208) and (3.209) is *not* a mixed-unitary channel. To see this, suppose by contradiction that $\mathcal{E}$ can be expressed as in (3.207). Then, $\{M_z\}_{z \in [3]}$ and $\{\sqrt{t_w} U_w\}_{w \in [k]}$ are operator sum representations of the *same* channel $\mathcal{E}$. Therefore, there exists an $m \times 3$ isometry channel $V = (v_{wz})$ such that

$$\sqrt{t_w} U_w = \sum_{z \in [3]} v_{wz} M_z. \qquad (3.210)$$

Multiplying each side of the equation above by its conjugate we get

$$t_w I^A = \sum_{z,z' \in [3]} \bar{v}_{wz'} v_{wz} M_{z'}^* M_z. \qquad (3.211)$$

Now, taking all summations to be modulo 3, we have by definition that $M_z M_{z+1} = |z\rangle\langle z+2|$ and $M_z M_{z+2} = |z+1\rangle\langle z+2|$ (by definition $M_z = M_z^*$). Hence, the equation above implies that

$$\bar{v}_{wz} v_{w(z+1)} = \bar{v}_z v_{w(z+2)} = 0 \qquad \forall\, z \in [3]. \qquad (3.212)$$

In other words, we must have $v_{wz'}v_{wz} = 0$ for all $z \neq z' \in [3]$. This, in turn, implies that $v_{wz} = 0$ for at least two values of $z \in [3]$. Hence, the relation (3.210) implies that $U_z$ is proportional to one of the three matrices $\{M_z\}_{z\in[3]}$, in contradiction with the fact that all $\{M_z\}_{z\in[3]}$ have rank 2, whereas $U_w$ has a full rank. Therefore, the channel $\mathcal{E}$ is *not* a mixed-unitary channel.

From the exmple above it follows that there is no quantum analogue to Birkhoff theorem, as there are unital channels that are not mix-unitary channels. What is the distinction between unital channels and mix-unitary channels in the Choi representation? Consider a unital quantum channel $\mathcal{E} \in \mathrm{CPTP}(A \to A)$, and let

$$J_{\mathcal{E}}^{A\tilde{A}} = \mathcal{E}^{\tilde{A}\to\tilde{A}}(\Omega^{A\tilde{A}}) \tag{3.213}$$

be its Choi matrix. Then, since it is trace preserving, $J_{\mathcal{E}}^{A} = I^{A}$. On the other hand, since it is unital,

$$J_{\mathcal{E}}^{\tilde{A}} = \mathrm{Tr}_A\left[J_{\mathcal{E}}^{A\tilde{A}}\right] = \mathcal{E}(I^{\tilde{A}}) = I^{\tilde{A}}. \tag{3.214}$$

Therefore, $\mathcal{E}$ is unital if and only if both marginals of its Choi matrix are equal to the identity matrix.

If $\mathcal{E} \in \mathrm{CPTP}(A \to A)$ is a mixture of unitaries, then its Choi matrix is proportional to a convex combination of maximally entangled states. A maximally entangled state $\phi^{A\tilde{A}} \in \mathrm{Pure}(A\tilde{A})$ is a normalized vector with the property that its reduced density matrix is the maximally mixed state; that is, $\phi^{A} = \mathbf{u}^{A}$. From Exercise 2.44 it follows that all maximally entangled states in $\mathrm{Pure}(A\tilde{A})$ must have the form

$$\frac{1}{\sqrt{m}}\left(I^{A} \otimes U^{\tilde{A}}\right)|\Omega^{A\tilde{A}}\rangle, \tag{3.215}$$

where $m := |A|$, and $U$ is some unitary matrix. Now, observe that the Choi matrix of the mix-unitary channel (3.207) is given by

$$J_{\mathcal{E}}^{A\tilde{A}} = \mathcal{E}^{\tilde{A}\to\tilde{A}}(\Omega^{A\tilde{A}}) = \sum_{w\in[k]} t_w(I^{A} \otimes U_w)\Omega^{A\tilde{A}}(I^{A} \otimes U_w^*) = m \sum_{w\in[k]} t_w|\phi_w^{A\tilde{A}}\rangle\langle\phi_w^{A\tilde{A}}|, \tag{3.216}$$

where each

$$|\phi_w^{A\tilde{A}}\rangle := \frac{1}{\sqrt{m}}\left(I^{A} \otimes U_w^{\tilde{A}}\right)|\Omega^{A\tilde{A}}\rangle \tag{3.217}$$

is a maximally entangled state.

**Exercise 3.58.** *Let $\mathcal{E} \in \mathrm{CPTP}(A \to A)$ be a quantum channel of the form*

$$\mathcal{E}^{A\to A}(\rho^{A}) := \mathrm{Tr}_B\left[U^{AB}\left(\rho^{A} \otimes \mathbf{u}^{B}\right)U^{*AB}\right], \tag{3.218}$$

*where $U \in \mathfrak{L}(AB)$ is a unitary operator.*

1.  *Show that the above map is a unital quantum channel.*
2.  *Show that every mixed-unitary channel, such as in (3.207), with **rational** probabilities $\{t_w\}_{w\in[k]}$, can be expressed as in (3.218). That is, there exists a system $B$ and a joint unitary matrix $U^{AB}$ such that the expression for $\mathcal{E}(\rho)$ in (3.218) becomes (3.207).*
3.  *Determine if the Shor example above can be expressed in the form (3.218).*

## 3.5.10 Quantum Instruments

A quantum instrument is a quantum channel that takes a quantum state as its input and outputs a cq-state. Consequently, it can be viewed as a mathematical abstraction of a quantum measurement with the classical output as the recorded measurement outcome, and with the quantum outcome as the post-measurement state. Mathematically, let $A$ be the input system, $X$ the classical register of the output system, and $B$ be the quantum output. Then a quantum channel $\mathcal{N} \in \mathrm{CPTP}(A \to XB)$ is a quantum instrument if

$$\Delta^{X \to X} \circ \mathcal{N}^{A \to XB} = \mathcal{N}^{A \to XB}, \tag{3.219}$$

where $\Delta \in \mathrm{CPTP}(X \to X)$ is the completely dephasing channel with respect to the classical basis of $X$ (see Figure 3.6).

Set $m := |X|$ and observe that for all $\omega \in \mathfrak{D}(A)$

$$\begin{aligned}
\mathcal{N}^{A \to XB}(\omega^A) &= \Delta^{X \to X}\left(\mathcal{N}^{A \to XB}\left(\omega^A\right)\right) \\
&= \sum_{x \in [m]} |x\rangle\langle x|^X \otimes \mathcal{N}_x^{A \to B}\left(\omega^A\right),
\end{aligned} \tag{3.220}$$

where for each $x \in [m]$ we define the linear map $\mathcal{N}_x \in \mathfrak{L}(A \to B)$ via

$$\mathcal{N}_x^{A \to B}\left(\omega^A\right) := \mathrm{Tr}_X\left[\left(|x\rangle\langle x|^X \otimes I^B\right)\mathcal{N}^{A \to XB}\left(\omega^A\right)\right] \qquad \forall\, \omega \in \mathfrak{L}(A). \tag{3.221}$$

By definition, the marginal channel

$$\mathcal{N}^{A \to B} := \mathrm{Tr}_X \circ \mathcal{N}^{A \to XB} = \sum_{x \in [m]} \mathcal{N}_x^{A \to B}. \tag{3.222}$$

Observe further that $\mathcal{N}^{A \to B}$ is a quantum channel since it can be expressed as a combination of the two quantum channels $\mathrm{Tr}_X$ and $\mathcal{N}^{A \to BX}$. Moreover, each $\mathcal{N}_x^{A \to B}$ is a CP map. Indeed, let $J_{\mathcal{N}}^{AXB} = \mathcal{N}^{\tilde{A} \to BX}(\Omega^{A\tilde{A}})$ be the Choi matrix of the quantum instrument $\mathcal{N}$, then the Choi matrix of $\mathcal{N}_x$ is given by

$$J_{\mathcal{N}_x}^{AB} = \mathrm{Tr}_X\left[\left(I^A \otimes |x\rangle\langle x|^X \otimes I^B\right)J_{\mathcal{N}}^{AXB}\right], \tag{3.223}$$

which is positive semidefinite since $J_{\mathcal{N}}^{AXB} \geqslant 0$. We therefore conclude that $\{\mathcal{N}_x^{A \to B}\}_{x \in [m]}$ are trace nonincreasing CP maps that sum up to a CPTP map.

**Exercise 3.59.** *Let $\mathcal{E} \in \mathrm{CP}(A \to B)$ be a trace nonincreasing map.*

*1. Show that any operator sum representation $\{M_x\}_{x \in [m]}$ of $\mathcal{E}$ satisfies $\sum_{x \in [m]} M_x^* M_x \leqslant I$.*

*2. Show that the marginal of the Choi matrix $J_{\mathcal{E}}^{AB}$ satisfies $J_{\mathcal{E}}^A := \mathrm{Tr}_B\left[J_{\mathcal{E}}^{AB}\right] \leqslant I^A$.*

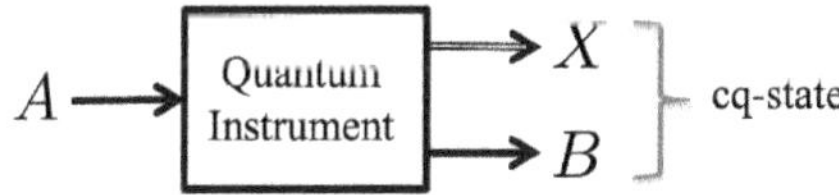

**Exercise 3.60.** *Find an operator sum representation of the quantum instrument $\mathcal{N}^{A\to XB}$ discussed above.*

## 3.5.11 Complementary Channel

Due to the Stinespring dilation theorem, we can associate to any quantum channel $\mathcal{N} \in \mathrm{CPTP}(A \to B)$ an isometry channel $\mathcal{V} \in \mathrm{CPTP}(A \to BE)$ such that

$$\mathcal{N}^{A\to B} = \mathrm{Tr}_E \circ \mathcal{V}^{A\to BE}. \tag{3.224}$$

The complementary channel of $\mathcal{N}$, denoted by $\mathcal{N}_c$, is a channel in $\mathrm{CPTP}(A \to E)$ obtained by tracing system $B$ from $\mathcal{V}^{A\to BE}$ (see Figure 3.7); that is,

$$\mathcal{N}_c^{A\to E} := \mathrm{Tr}_B \circ \mathcal{V}^{A\to BE}. \tag{3.225}$$

As an example, let $\mathcal{N}^{A\to B}$ be the qubit depolarizing channel with $|A| = |B| = 2$. From it's Stinespring isometry in (3.162) we can get its complementary channel. For this purpose, we denote by $t_0 := \sqrt{1 - \frac{3p}{4}}$ and $t_1 = t_2 = t_3 = \frac{\sqrt{p}}{2}$, and by $\{\sigma_j\}_{j=0}^3$ the four Pauli matrices (including $\sigma_0 := I$). We therefore get

$$\mathcal{N}_c^{A\to E}(\rho^A) = \mathrm{Tr}_B\left[\mathcal{V}\rho^A V^*\right] = \sum_{j,k=0}^{3} t_j t_k \, \mathrm{Tr}\left[\sigma_j \sigma_k \rho\right] |j\rangle\langle k|^E \qquad \forall\, \rho \in \mathfrak{L}(A). \tag{3.226}$$

**Exercise 3.61.** *Use the Bloch representation $\rho = \frac{1}{2}(I + \mathbf{r}\cdot\boldsymbol{\sigma})$ to simplify the expression in (3.226) for the complementary channel of the depolarizing channel.*

## 3.5.12 The Pinching Channel

The pinching channel is a generalization of the completely dephasing channel. For any observable $H \in \mathrm{Herm}(A)$ with $m$ distinct eigenvalues $\mathrm{spec}(H) = \{\lambda_1, \ldots, \lambda_m\}$ we define the pinching channel $\mathcal{P}_H \in \mathrm{CPTP}(A \to A)$ associated with $H$ as

$$\mathcal{P}_H(\rho) := \sum_{x\in[m]} P_x \rho P_x \quad \forall\, \rho \in \mathfrak{L}(A), \tag{3.227}$$

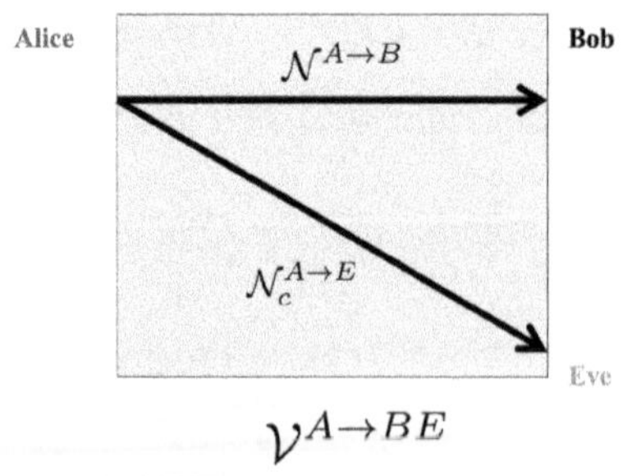


**Figure 3.7** The complementary channel of $\mathcal{N}^{A\to B}$.

where for each $x \in [m]$, $P_x$ is the projector the eigenspace of $\lambda_x$. Note that $\mathcal{P}_H$ is indeed a quantum channel since $\{P_x\}_{x \in [m]}$ forms an orthogonal set of projectors that sum to the identity. Moreover, if $m = |A|$ (i.e. $H$ has $|A|$ distinct eigenvalues), then $\mathcal{P}_H = \Delta$, where $\Delta$ is the completely dephasing quantum channel in the basis comprising of the eigenvectors of $H$.

**Exercise 3.62.** *Let $H \in \mathrm{Herm}(A)$ and $\rho \in \mathfrak{D}(A)$. Show that $\rho = \mathcal{P}_H(\rho)$ if and only if $[\rho, H] = 0$.*

**Exercise 3.63.** *Let $A$ be a quantum system and $\rho, \sigma \in \mathfrak{D}(A)$ be two density matrices, with $\sigma = \sum_{y \in [n]} \lambda_y \Pi_y$, where $\{\Pi_y\}_{y \in [n]}$ forms an orthogonal projective von-Neumann measurement on system A, and $\{\lambda_y\}_{y \in [n]}$ is the set of* distinct *eigenvalues of $\sigma$. For each $y \in [n]$, let $m_y := \mathrm{Tr}[\Pi_y]$ be the multiplicity of the eigenvalue $\lambda_y$.*

*1. Show that $\sigma$ commutes with $\mathcal{P}_\sigma(\rho)$; that is,*

$$[\mathcal{P}_\sigma(\rho), \sigma] = 0. \tag{3.228}$$

*2. Show that there exists an orthonormal basis of A, denoted by $\{|\phi_{xy}\rangle\}_{x,y}$, with the property that*

$$\Pi_y = \sum_{x \in [m_y]} \phi_{xy} \quad and \quad \mathcal{P}_\sigma(\rho) = \sum_{y \in [n]} \sum_{x \in [m_y]} r_{xy} \phi_{xy} \tag{3.229}$$

*for some $r_{xy} \geqslant 0$ with $\sum_{y \in [n]} \sum_{x \in [m_y]} r_{xy} = 1$.*
*3. Let $y \in [n]$ and $x, x' \in [m_y]$. Show that if $x \neq x'$, then*

$$\langle \phi_{x'y} | \rho | \phi_{xy} \rangle = 0. \tag{3.230}$$

*4. Let $\Delta \in \mathrm{CPTP}(A \to A)$ be the completely dephasing channel in the basis $\{|\phi_{xy}\rangle\}_{x,y}$. Show that*

$$\Delta(\rho) = \mathcal{P}_\sigma(\rho) \quad and \quad \Delta(\sigma) = \sigma. \tag{3.231}$$

**Exercise 3.64.** *Let $A$ be a quantum system and $\rho, \sigma \in \mathfrak{D}(A)$ be two density matrices satisfying $\rho \not\ll \sigma$ (i.e. $\mathrm{supp}\,(\rho) \not\subseteq \mathrm{supp}\,(\sigma)$). Show also that*

$$\mathcal{P}_\sigma(\rho) \not\ll \sigma. \tag{3.232}$$

The following exercise demonstrates that the pinching channel is a special type of mixture of unitaries.

**Exercise 3.65.** *Let $H \in \mathrm{Herm}(A)$ be an observable with $\mathrm{spec}(H) = \{\lambda_1, \ldots, \lambda_k\}$ as above, and let $\mathcal{P}_H$ be its associated pinching channel as given in (3.227). Show that for any $\rho \subset \mathfrak{L}(A)$*

$$\mathcal{P}_H(\rho) = \frac{1}{m} \sum_{y \in [m]} U_y \rho U_y^*, \quad where \quad U_y := \sum_{x \in [m]} e^{i \frac{2\pi xy}{m}} P_x. \tag{3.233}$$

The above exercise demonstrates that the pinching channel is a mixed unitary channel. Observe also that for $y = m$ we have $U_m = I_m$ so that

$$\mathcal{P}_H(\rho) = \frac{1}{m}\rho + \frac{1}{m}\sum_{y\in[m-1]} U_y\rho U_y^* \geqslant \frac{1}{m}\rho. \tag{3.234}$$

Since $m := |\mathrm{spec}(H)|$, we get the following inequality known as the *pinching inequality*.

---

**The Pinching Inequality**

For any $H \in \mathrm{Herm}(A)$ and $\rho \in \mathfrak{D}(A)$,

$$\mathcal{P}_H(\rho) \geqslant \frac{1}{|\mathrm{spec}(H)|}\rho. \tag{3.235}$$

---

Since we got the inequality above by removing $m-1$ terms, it may give the impression that this inequality is not very useful since it is never saturated and not tight. However, we will see that in applications, this inequality can be used to provide good enough approximation to $\mathcal{P}_H(\rho)$ when we consider the asymptotic case in which $H = \sigma^{\otimes n}$, where $\sigma$ is some quantum state and $n$ is a very large integer. In particular, we will see in Chapter 8 (particularly Sec. 8.4.1) that in this case $m = |\mathrm{spec}(\sigma^{\otimes n})|$ grows polynomially with $n$. The fact that it is not an exponential growth with $n$ is one of the key reasons why the pinching inequality is quite useful.

The pinching map can also be used to prove the reverse Hölder inequality with $p \in (0,1)$. In this case, we still use the notation $\|M\|_p := \left(\mathrm{Tr}\left[|M|^p\right]\right)^{\frac{1}{p}}$ for any $M \in \mathcal{L}(A)$. However, one has to be careful since for $p \in (0,1)$, $\|\cdot\|_p$ is not a norm.

---

**Reverse Hölder Inequality**

**Lemma 3.1.** Let $\tau, \omega \in \mathrm{Pos}(A)$ and $p \in (0,1)$. If $\tau \ll \omega$, then

$$\mathrm{Tr}[\omega\tau] \geqslant \frac{\|\tau\|_p}{\left\|\omega^{-1}\right\|_{\frac{p}{1-p}}}, \tag{3.236}$$

where the inverse is a generalized inverse.

---

**Proof** We first prove the theorem for the case that $\tau$ and $\omega$ commute. In this case,

$$\|\tau\|_p^p = \mathrm{Tr}[\tau^p] = \mathrm{Tr}\left[\tau^p\omega^p\omega^{-p}\right] = \left\|\tau^p\omega^p\omega^{-p}\right\|_1$$

$$\text{Hölder inequality (2.71)} \rightarrow \ \leqslant \left\|\tau^p\omega^p\right\|_{\frac{1}{p}}\left\|\omega^{-p}\right\|_{\frac{1}{1-p}} \tag{3.237}$$

$$= (\mathrm{Tr}[\tau\omega])^p \left\|\omega^{-p}\right\|_{\frac{1}{1-p}}.$$

This proves the theorem for the case that $\omega$ and $\tau$ commute. On the other hand, from Exericse 3.63 we get that $\mathcal{P}_\omega(\tau)$ and $\omega$ commute so that

$$\mathrm{Tr}[\tau\omega] = \mathrm{Tr}[\mathcal{P}_\omega(\tau)\omega] \geqslant \frac{\|\mathcal{P}_\omega(\tau)\|_p}{\left\|\omega^{-1}\right\|_{\frac{p}{1-p}}}. \tag{3.238}$$

Finally, since $t \mapsto t^p$ is operator concave for $p \in (0, 1)$ (see Table B.1 of online version), we conclude that

$$\|\mathcal{P}_\omega(\tau)\|_p^p = \text{Tr}\left[(\mathcal{P}_\omega(\tau))^p\right] \geqslant \text{Tr}\left[\mathcal{P}_\omega(\tau^p)\right] = \text{Tr}[\tau^p]. \tag{3.239}$$

That is, $\|\mathcal{P}_\omega(\tau)\|_p \geqslant \|\tau\|_p$. Substituting this into (3.238) completes the proof. ∎

**Exercise 3.66 (Reverse Young's Inequality).** *Let $A$ and $B$ be two Hilbert spaces, $M, N \in$ Pos$(A)$, $p \in (0, 1)$, and $q$ defined via $\frac{1}{p} + \frac{1}{q} = 1$ (hence $q < 0$). Use the reverse Hölder inequality of the Schatten norm to show that*

$$\text{Tr}[MN] \geqslant \frac{1}{p}\text{Tr}[M^p] + \frac{1}{q}\text{Tr}[N^q] \tag{3.240}$$

*with equality if and only if $M^p = N^q$. Hint: Take the logarithm on both sides of the reverse Hölder inequality and use the concavity property of the logarithm.*

### 3.5.13 Twirling Channels

In Chapter 15, particularly in Section 15.2.1, we will explore a channel associated with any compact Lie group through its Haar measure (refer to Appendix C of online version for more definitions and details on Haar measures and compact Lie groups). This channel is known as the **G**-twirling operation. A simple example of such a channel, defined for all $\rho \in \mathfrak{L}(A)$, is

$$\mathcal{G}(\rho) := \int_{\mathfrak{U}(A)} dU \; U\rho U^*, \tag{3.241}$$

where $dU$ is the Haar measure of the unitary group $\mathfrak{U}(A)$. In other words, $\rho$ is "twirled" over all unitary matrices $U \in \mathfrak{U}(A)$.

**Exercise 3.67.** *Consider $\rho \in \mathfrak{L}(A)$ and $\sigma := \mathcal{G}(\rho)$, where $\mathcal{G}$ is the channel defined in (3.241). Demonstrate that the commutator $[U, \sigma] = 0$. Hint: Utilize the properties of the Haar measure described in online C.4 of the appendix of online version.*

The exercise above illustrates that the channel defined in (3.241) is not unfamiliar to us. In fact, it can be represented as the replacement channel

$$\mathcal{G}(\rho) := \int_{\mathfrak{U}(A)} dU \; U\rho U^* = \text{Tr}[\rho]\mathbf{u}^A, \tag{3.242}$$

where $\mathbf{u}^A = I^A/|A|$ is the maximally mixed state. To understand why, remember from the previous exercise that $\sigma := \mathcal{G}(\rho)$ commutes with all unitary matrices in $\mathfrak{U}(A)$ and, thus, must be proportional to the identity matrix. We now consider a more complex example with numerous applications in quantum information science.

Let $B$ be a replica of $A$, and set $m := |A| = |B|$. Consider the twirling map $\mathcal{G} \in$ CPTP$(AB \to AB)$, defined for all $\rho \in \mathfrak{L}(AB)$ as

$$\mathcal{G}(\rho^{AB}) := \int_{\mathfrak{U}(m)} dU \; (U \otimes U)\rho^{AB}(U \otimes U)^*. \tag{3.243}$$

As in the previous example, for every $\rho \in \mathfrak{L}(AB)$, the matrix $\sigma^{AB} := \mathcal{G}(\rho^{AB})$ commutes with $U \otimes U$ for all $U \in \mathfrak{U}(A)$. The ensuing question is which matrices $\sigma^{AB}$ commute with all matrices of the form $U \otimes U$, where $U$ is a unitary matrix? Clearly, any matrix proportional to the identity matrix satisfies this criterion. However, there exists another type of operator that fulfills this property, known as the *swap* (or flip) operator:

$$F^{AB} := \sum_{x,y \in [m]} |x\rangle\langle y|^A \otimes |y\rangle\langle x|^B. \tag{3.244}$$

**Exercise 3.68.** *Prove that the swap operator $F^{AB}$ satisfies*

$$\left[U \otimes U, F^{AB}\right] = 0 \qquad \forall\, U \in \mathfrak{U}(m). \tag{3.245}$$

Our analysis in Sec. C.10.2 of online version indicates that any operator commuting with all matrices in the set $\{U \otimes U\}_{U \in \mathfrak{U}(m)}$ can be expressed as a linear combination of the identity and swap operators. Consequently, $\mathcal{G}(\rho^{AB}) = aI^{AB} + bF^{AB}$ for some $a, b \in \mathbb{R}$, determinable from the requirement that $\mathcal{G}$ is CPTP (see the following exercise). Furthermore, in Sec. C.10.2 of Appendix C of online version, we demonstrate that the representation $U \mapsto U \otimes U$ decomposes into two irreps, corresponding to the symmetric and antisymmetric subspaces, each with a multiplicity of one. Hence, from Theorem C.5 in Appendix C of online version, it follows that for all $\rho \in \mathfrak{L}(AB)$,

$$\mathcal{G}(\rho) = \text{Tr}\left[\rho \Pi_{\text{Sym}}\right] \frac{\Pi_{\text{Sym}}}{\text{Tr}\left[\Pi_{\text{Sym}}\right]} + \text{Tr}\left[\rho \Pi_{\text{Asy}}\right] \frac{\Pi_{\text{Asy}}}{\text{Tr}\left[\Pi_{\text{Asy}}\right]}, \tag{3.246}$$

where $\Pi_{\text{Sym}} := \frac{1}{2}(I^{AB} + F^{AB})$ and $\Pi_{\text{Asy}} := \frac{1}{2}(I^{AB} - F^{AB})$ are the projections onto the symmetric and antisymmetric subspaces of $AB$ (see (C.186) and (C.188) of online version). When $\rho \in \mathfrak{D}(AB)$ is a density matrix, the right-hand side above represents a density matrix known as the *Werner quantum state*.

**Exercise 3.69.** *Let $F^{AB}$ be the swap operator, with $m := |A| = |B|$.*

*1. Starting with $\mathcal{G}(\rho^{AB}) = aI^{AB} + bF^{AB}$ for some $a, b \in \mathbb{R}$, prove that $\mathcal{G}$ is CPTP if and only if for all $\rho \in \mathfrak{L}(AB)$,*

$$\mathcal{G}\left(\rho^{AB}\right) = \frac{m\text{Tr}[\rho^{AB}] - \text{Tr}\left[\rho^{AB} F^{AB}\right]}{m(m^2 - 1)} I^{AB} + \frac{m\text{Tr}\left[\rho^{AB} F^{AB}\right] - \text{Tr}[\rho^{AB}]}{m(m^2 - 1)} F^{AB}. \tag{3.247}$$

*2. Derive (3.247) from (3.246) by expressing the projections onto the symmetric and antisymmetric subspaces in terms of the swap operator.*

**Exercise 3.70.** *Show that for all $M \in \mathfrak{L}(A)$ and $B \cong A$ we have*

$$\text{Tr}\left[M^2\right] = \text{Tr}\left[(M \otimes M)\, F^{AB}\right]. \tag{3.248}$$

**Exercise 3.71.** *Let $B := \tilde{A}$, $\mathcal{G} \in \text{CPTP}(AB \to AB)$ be as in (3.243), and $\mathcal{E} \in \text{CPTP}$ $(AB \to AB)$.*

1. Show that $\mathcal{E} = \mathcal{E} \circ \mathcal{G}$ if and only if there exists $\omega_1, \omega_2 \in \mathrm{Eff}(AB)$ such that for all $\rho \in \mathfrak{L}(AB)$

$$\mathcal{E}\left(\rho^{AB}\right) = \omega_1^{AB} \mathrm{Tr}\left[\rho \Pi_{\mathrm{Sym}}^{AB}\right] + \omega_2^{AB} \mathrm{Tr}\left[\rho \Pi_{\mathrm{Asy}}^{AB}\right]. \tag{3.249}$$

2. Show that $\mathcal{E} = \mathcal{G} \circ \mathcal{E}$ if and only if there exists $\Lambda \in \mathrm{Eff}(AB)$ such that for all $\rho \in \mathfrak{L}(AB)$

$$\mathcal{E}\left(\rho^{AB}\right) = \frac{1}{\mathrm{Tr}\left[\Pi_{\mathrm{Sym}}^{AB}\right]} \Pi_{\mathrm{Sym}}^{AB} \mathrm{Tr}\left[\rho^{AB} \Lambda^{AB}\right] + \frac{1}{\mathrm{Tr}\left[\Pi_{\mathrm{Asy}}^{AB}\right]} \Pi_{\mathrm{Asy}}^{AB} \mathrm{Tr}\left[\rho^{AB} \Lambda^{AB}\right]. \tag{3.250}$$

**Exercise 3.72.** *Let* $\mathcal{G}\colon \mathrm{CPTP}(A^2 \to A^2)$ *be the twirling channel*

$$\mathcal{G}(\rho) := \int_{U(m)} dU \, (U \otimes \overline{U})\rho(U \otimes \overline{U})^* \qquad \forall \, \rho \in \mathfrak{L}(A^2), \tag{3.251}$$

*where* $\overline{U} := (U^*)^T$.

1. *Show that*

$$\mathcal{G}(\Phi^{A\tilde{A}}) = \Phi^{A\tilde{A}}, \tag{3.252}$$

*where* $\Phi \in \mathfrak{D}(A\tilde{A})$ *is the maximally entangled state.*
2. *Show that for* $m := |A| = |B|$ *and* $\rho \in \mathfrak{D}(AB)$

$$\mathcal{G}(\rho^{AB}) = m^2 \frac{1 - \mathrm{Tr}\left[\Phi^{AB}\rho^{AB}\right]}{m^2 - 1} \mathbf{u}^{AB} + \frac{m^2 \mathrm{Tr}\left[\Phi^{AB}\rho^{AB}\right] - 1}{m^2 - 1} \Phi^{AB}. \tag{3.253}$$

*Hint: Use (3.247) and observe that* $\Phi^{AB} = \mathcal{T}^{B \to B}(F^{AB})$, *where* $\mathcal{T}^{B \to B}$ *is the transpose map.*

**Exercise 3.73.** *Let* $m := |A| = |B|$, $\mathcal{G} \in \mathrm{CPTP}(AB \to AB)$ *be as in (3.251), and* $\mathcal{E} \in \mathrm{CPTP}(AB \to AB)$. *Define also the state*

$$\tau^{AB} := \frac{I^{AB} - \Phi^{AB}}{m^2 - 1}. \tag{3.254}$$

1. *Show that for all* $\rho \in \mathfrak{D}(AB)$

$$\mathcal{G}(\rho^{AB}) = \left(1 - \mathrm{Tr}\left[\Phi^{AB}\rho^{AB}\right]\right)\tau^{AB} + \mathrm{Tr}\left[\Phi^{AB}\rho^{AB}\right]\Phi^{AB}. \tag{3.255}$$

2. *Show that* $\mathcal{E} = \mathcal{E} \circ \mathcal{G}$ *if and only if there exists* $\omega_1, \omega_2 \in \mathrm{Eff}(AB)$ *such that for all* $\rho \in \mathfrak{D}(AB)$

$$\mathcal{E}\left(\rho^{AB}\right) = \left(1 - \mathrm{Tr}\left[\Phi^{AB}\rho^{AB}\right]\right)\omega_1^{AB} + \mathrm{Tr}\left[\Phi^{AB}\rho^{AB}\right]\omega_2^{AB}. \tag{3.256}$$

3. *Show that* $\mathcal{E} = \mathcal{G} \circ \mathcal{E}$ *if and only if there exists* $\Lambda \in \mathrm{Eff}(AB)$ *such that for all* $\rho \in \mathfrak{D}(AB)$

$$\mathcal{E}\left(\rho^{AB}\right) = \left(1 - \mathrm{Tr}\left[\Lambda^{AB}\rho^{AB}\right]\right)\tau^{AB} + \mathrm{Tr}\left[\Lambda^{AB}\rho^{AB}\right]\Phi^{AB}. \tag{3.257}$$

**Exercise 3.74.** *Let A be a Hilbert space of dimension* $m := |A|$, *and consider the channel given* (3.241). *Show that for all* $\rho \in \mathfrak{L}(A)$

$$\mathcal{E}(\rho) = \frac{1}{m^2} \sum_{p,q \in [m]} W_{p,q} \rho W_{p,q}^*, \tag{3.258}$$

*where* $W_{p,q}$ *are the are the Hiesenberg–Weyl operators defined in* (C.35 *of online version*).

## 3.6 Notes and References

Many books on quantum information contain much of the material presented here. Informationally complete POVMs were first introduced in Refs. [186, 42] and since then they became an integral part of quantum tomography. In particular, SIC POVMs have been studied intensively and more details and references about them can be found in the review article in Ref. [189].

Gleason's theory was proved originally for projective von-Neumann measurements in [85]. The proof in this case holds only in dimension greater than 2, and there are counterexamples in the qubit case. The version of Gleason's theorem that we considered here is due to Ref. [41]. In this case, the proof is much simpler (and holds in all dimensions) since *effects* replace orthogonal projections (and effects are simpler to work with). Gleason's theorem is of particular importance in the foundations of quantum physics as well as the field of quantum logic in its effort to minimize the number of axioms needed to formulate quantum mechanics. It also closes the bridge between some of the axioms of quantum mechanics and Born's rule. More details, references, and History of Gleason's theorem can be found in a Wikipedia article entitled "Gleason's theorem."

Naimark's and Stinspring's dilation theorems are results from operator theory and are valid also in infinite dimensional Hilbert spaces. Here we only studied their adaptation to the finite dimensional case. More details on their infinite dimensional version can be found in many books on operator theory; for example, the book in Ref. [179].

The Størmer–Woronowicz theorem was first proved for the qubit-to-qubit case in Ref. [208] and later on for the qubit-to-qutrit case in Ref. [239]. Counterexamples exists in higher dimensions so these dimensions are optimal. Both proofs involve somewhat complicated calculations, and the simplified proof of the Størmer case presented here is in Ref. [7]. It is an open problem to find a simpler proof of Woronowicz theorem.

More information on the pinching channel and its properties can be found in Ref. [211].

# PART II

# TOOLS AND METHODS

A preorder is a binary relation between objects that are *reflexive* and *transitive*. For example, consider the inclusion relation $\supseteq$ between subsets of $[n] := \{1, \ldots, n\}$ for some $n \in \mathbb{N}$. Then, $\supseteq$ is reflexive since for any subset $\mathcal{A}$ of $[n]$ we have $\mathcal{A} \supseteq \mathcal{A}$. The relation $\supseteq$ is transitive since for any three subsets $\mathcal{A}, \mathcal{B}, \mathcal{C} \subseteq [n]$ with $\mathcal{A} \supseteq \mathcal{B}$ and $\mathcal{B} \supseteq \mathcal{C}$ it follows that $\mathcal{A} \supseteq \mathcal{C}$. Furthermore, the relation $\supseteq$ has yet another property known as *symmetry*. That is, if $\mathcal{A}, \mathcal{B} \subseteq [n]$ satisfies both $\mathcal{A} \supseteq \mathcal{B}$ and $\mathcal{B} \supseteq \mathcal{A}$, then necessarily $\mathcal{A} = \mathcal{B}$. A preorder that satisfies this additional symmetry property is called a *partial order*.

Partial orders play a fundamental role in quantum resource theories. They typically stem from a set of restrictions imposed on quantum operations. For example, we saw in quantum teleportation that Alice and Bob are restricted to act locally and cannot communicate quantum particles. We will see in Chapter 12 that this restriction imposes a partial order between two entangled states, determining if one entangled state can be converted to another under operations that are restricted to be local. It turns out that there is one partial order with variants that appear in many resource theories. This partial order, known as majorization, has been studied extensively in quantum information and other fields, particularly in the field of matrix analysis, and there are several books on the topic (see section "Notes and References" at the end of this chapter).

## 4.1 Majorization between Probability Vectors

Consider a gambling game in which a host rolls a biased dice, and a player has to guess the outcome. Denote by $\mathbf{p} = (p_1, \ldots, p_n)^T \in \mathrm{Prob}(n)$ the probability vector corresponding to the $n$ possible outcomes. The player win the game if he or she guesses correctly the outcome. Clearly, the player will guess the value $x \in [n]$ that satisfies $p_x = \max\{p_1, \ldots, p_n\}$. In other games, the player is allowed to provide more than one guess, and wins the game if the outcome belongs to the set of numbers (guesses) he or she provided to the host. For example, if the player is allowed to provide a set with two numbers as his or her guesses, then in order to have the highest odds to win, he or she will choose the two numbers out of the $n$ numbers that have the highest probability to occur. Hence, the player will win the game with probability $p_1^{\downarrow} + p_2^{\downarrow}$, where we denote by $\mathbf{p}^{\downarrow} = \left(p_1^{\downarrow}, \ldots, p_n^{\downarrow}\right)^T$ the vector obtained from $\mathbf{p}$ by rearranging its components in nonincreasing order. We call a game in which the player is allowed to provide a set with $k$-numbers as guesses, a *k-gambling game*. Note that the highest probability to win a $k$-gambling game is given by $\sum_{x \in [k]} p_x^{\downarrow}$.

Suppose now that at the beginning of each game, the player is allowed to choose between two dice with corresponding probabilities $\mathbf{p}$ and $\mathbf{q}$. Clearly, the player will choose the dice that has better odds to win the game. For a $k$-game the player will choose the $\mathbf{p}$-dice if

$$\sum_{x \in [k]} p_x^{\downarrow} \geqslant \sum_{x \in [k]} q_x^{\downarrow}. \tag{4.1}$$

If the relation above holds for all $k \in [n]$, then the player will choose the $\mathbf{p}$-dice for any $k$-gambling game. In this case we say that $\mathbf{p}$ majorizes $\mathbf{q}$ and write $\mathbf{p} \succ \mathbf{q}$.

> **Majorization**
>
> **Definition 4.1.** Let $\mathbf{p}, \mathbf{q} \in \mathbb{R}^n$. We say that $\mathbf{p}$ majorizes $\mathbf{q}$ and write $\mathbf{p} \succ \mathbf{q}$ if (4.1) holds for all $k \in [n]$ with equality for $k = n$.

*Remark.* Note that in the definition above we did not assume that $\mathbf{p}$ and $\mathbf{q}$ are probability vectors, however, in the applications we consider in this book, $\mathbf{p}$ and $\mathbf{q}$ will always be probability vectors.

Majorization is a preorder. That is, given three real vectors $\mathbf{p}, \mathbf{q}, \mathbf{r} \in \mathbb{R}^n$ we have $\mathbf{p} \succ \mathbf{p}$ (reflexivity), and if $\mathbf{p} \succ \mathbf{q}$ and $\mathbf{q} \succ \mathbf{r}$, then $\mathbf{p} \succ \mathbf{r}$ (transitivity). Moreover, if both $\mathbf{p} \succ \mathbf{q}$ and $\mathbf{q} \succ \mathbf{p}$, then $\mathbf{p}$ and $\mathbf{q}$ are related by a permutation matrix. Therefore, using the notation $\mathrm{Prob}^{\downarrow}(n)$ to denote the subset of $\mathrm{Prob}(n)$ consisting of all vectors $\mathbf{p} \in \mathrm{Prob}(n)$ with the property that $\mathbf{p} = \mathbf{p}^{\downarrow}$, we get that the majorization relation $\succ$ is a partial order on the set $\mathrm{Prob}^{\downarrow}(n)$.

**Exercise 4.1.** *Show that for any n-dimensional probability vector* $\mathbf{p}$ *we have*

$$(1, 0, \ldots, 0)^T \succ \mathbf{p} \succ (1/n, \ldots, 1/n)^T. \tag{4.2}$$

**Exercise 4.2.** *Let* $\mathbf{p} \in \mathrm{Prob}(n)$, $\mathbf{u}^{(n)} := (1/n, \ldots, 1/n)^T \in \mathrm{Prob}(n)$ *be the uniform probability vector, and* $t \in [0, 1]$. *Show that*

$$\mathbf{p} \succ t\mathbf{p} + (1 - t)\mathbf{u}^{(n)}. \tag{4.3}$$

**Exercise 4.3.** *Find an example of two vectors* $\mathbf{p}, \mathbf{q} \in \mathrm{Prob}(3)$ *such that* $\mathbf{p}$ *does not majorize* $\mathbf{q}$, *and* $\mathbf{q}$ *does not majorize* $\mathbf{p}$. *Vectors with such a property that* $\mathbf{p} \nsucc \mathbf{q}$ *and* $\mathbf{q} \nsucc \mathbf{p}$ *are said to be incomparable.*

**Exercise 4.4.** *Let* $n \in \mathbb{N}$ *and* $\mathbf{p} \in \mathrm{Prob}(n)$. *Show that for sufficiently large* $m \in \mathbb{N}$ *we have*

$$\mathbf{p} \succ \left(\mathbf{u}^{(2)}\right)^{\otimes m}, \tag{4.4}$$

*where* $\mathbf{u}^{(2)} := \frac{1}{2}(1, 1)^T$ *is the two-dimensional uniform distribution.*

**Exercise 4.5.** *Let $m \in \mathbb{N}$, $\mathbf{p}, \mathbf{q} \in \mathrm{Prob}(m)$, and $L$ be the $m \times m$ lower triangular matrix*

$$L := \begin{pmatrix} 1 & 0 & 0 & \cdots & 0 \\ 1 & 1 & 0 & \cdots & 0 \\ 1 & 1 & 1 & \cdots & 0 \\ \vdots & \vdots & & \ddots & \vdots \\ 1 & 1 & 1 & \cdots & 1 \end{pmatrix}. \tag{4.5}$$

1. *Show that $\mathbf{p} \succ \mathbf{q}$ if and only if $L\mathbf{p}^{\downarrow} \geqslant L\mathbf{q}^{\downarrow}$, where the inequality is entrywise.*
2. *Show that $L$ is invertible and find $L^{-1}$.*
3. *Show that $L^{-1}\mathbf{p} \geqslant 0$ (entrywise) if and only if $\mathbf{p} = \mathbf{p}^{\uparrow}$.*
4. *Show that $(L^{T})^{-1}\mathbf{p} \geqslant 0$ (entrywise) if and only if $\mathbf{p} = \mathbf{p}^{\downarrow}$.*

**Exercise 4.6.** *Let $\mathbf{p}_1, \ldots, \mathbf{p}_k \in \mathrm{Prob}^{\downarrow}(m)$ and $\mathbf{q}_1, \ldots, \mathbf{q}_k \in \mathrm{Prob}(m)$ be such that $\mathbf{p}_x \succ \mathbf{q}_x$ for all $x \in [k]$. Show that for any $(r_1, \ldots, r_k)^{T} \in \mathrm{Prob}(k)$ we have*

$$\sum_{x \in [k]} r_x \mathbf{p}_x \succ \sum_{x \in [k]} r_x \mathbf{q}_x. \tag{4.6}$$

**Exercise 4.7.** *Let $\mathbf{p}, \mathbf{q} \in \mathrm{Prob}^{\downarrow}(n)$ and suppose $\mathbf{p} \neq \mathbf{q}$ and $\mathbf{p} \succ \mathbf{q}$. Show that the largest integer $z \in [n]$ for which $p_z \neq q_z$ must satisfy $q_z > p_z$.*

**Exercise 4.8.** *Let $\mathbf{p}, \mathbf{q} \in \mathrm{Prob}^{\downarrow}(n)$ and $k, m \in [n-1]$ be such that $k \leqslant m$. Suppose that $\mathbf{q}$ has the form*

$$\mathbf{q} = (\underbrace{a, a, \ldots, a}_{k\text{-times}}, \ q_{k+1}, \ldots, q_m, \underbrace{b, b, \ldots, b}_{(n-m)\text{-times}})^{T}. \tag{4.7}$$

*Show that $\mathbf{p} \succ \mathbf{q}$ if and only if*

$$\|\mathbf{p}\|_{(\ell)} \geqslant \|\mathbf{q}\|_{(\ell)} \qquad \forall\, \ell \in \{k, k+1, \ldots, m\} \tag{4.8}$$

*(i.e. there is no need to consider the cases $\ell < k$ or $\ell > m$). Hint: Use the fact that $\mathbf{p} = \mathbf{p}^{\downarrow}$ and $\mathbf{q} = \mathbf{q}^{\downarrow}$.*

## 4.1.1 Characterization of Majorization

Equation (4.2) corresponds to the fact that majorization can be used to determine if one probability distribution is more spread out than another one. This intuition can be deepened with further analysis of the dice games. Specifically, consider the $\mathbf{p}$-dice and $\mathbf{q}$-dice of the discussion above with $\mathbf{p}, \mathbf{q} \in \mathrm{Prob}(n)$. We saw earlier that if $\mathbf{p} \succ \mathbf{q}$, then a player has better odds to win any $k$-gambling game with the $\mathbf{p}$-dice than with the $\mathbf{q}$-dice. One can therefore conclude that the outcomes obtained by rolling the $\mathbf{q}$-dice are more uncertain than those obtained by rolling the $\mathbf{p}$-dice.

To make this notion of *uncertainty* precise, consider a game of chance in which the player is allowed to permute the symbols on the dice (e.g., permute the stickers on the dice). Clearly, such a permutation cannot change the odds in any $k$-gambling game. This *relabeling* of the outcomes is equivalently described by a permutation matrix $P$

that is acting on **p**; that is, after the relabeling (permutation) of the symbols on the **p**-dice, the new probability vector is given by $P$**p**.

Consider now a (somewhat unrealistic) scenario in which the player chooses to perform the relabeling at random. For example, the player can flip an unbiased coin and if the outcome is "head" the player does nothing to the **p**-dice, whereas if the outcome is a "tail" the player performs relabeling described by the permutation matrix $P$. Moreover, suppose also that the player forgets the outcome of the coin flipping. Hence, the player changes his odds in winning the game since with probability 1/2 he did nothing to the **p**-dice and with probability 1/2 he changed the order of the stickers on the dice. This means that now, effectively, the player holds a **q**-dice with

$$\mathbf{q} := \frac{1}{2}\mathbf{p} + \frac{1}{2}P\mathbf{p}. \tag{4.9}$$

Since by "forgetting" the outcome of the coin flip the player cannot decrease the uncertainty associated with the **p**-dice, we must conclude that the new **q**-dice is more uncertain than the initial **p**-dice.

The relation above between the **p** and **q** can be expressed as

$$\mathbf{q} = D\mathbf{p}, \tag{4.10}$$

where $D := \frac{1}{2}I_n + \frac{1}{2}P$. More generally, if instead of an unbiased coin the player uses a random device that produces the outcome $j \in [m]$ with probability $t_j$ and a relabeling corresponding to a permutation matrix $P_j$, then the matrix $D$ can be expressed as

$$D = \sum_{j \in [m]} t_j P_j. \tag{4.11}$$

We therefore conclude that for any such matrix and any probability vector **p**, the vector $\mathbf{q} := D\mathbf{p}$ corresponds to more uncertainty than **p**. The matrix $D$ above has the property that all its components are nonnegative and each row and column sums to 1. Such matrices are called *doubly stochastic* (see Appendix A.5 of online version).

**Exercise 4.9.** *Show that the matrix D in (4.11) is doubly stochastic.*

The converse of the statement of Exercise 4.9 is equally valid. Owing to Birkhoff's theorem (see Theorem A.5 of online version), we know that any $n \times n$ doubly stochastic matrix can be expressed as a convex combination of no more than $m \leqslant (n-1)^2 + 1$ permutation matrices (see Exercise A.13 of online version). By integrating this theorem with our prior analysis, we deduce that **q** is more uncertain than **p** if and only if $\mathbf{q} = D\mathbf{p}$. Shortly, we will demonstrate that the relationship $\mathbf{q} = D\mathbf{p}$ corresponds to majorization. However, before we present this, it's essential to introduce the concept of a $T$-transform.

## $T$-Transform

A $T$-transform is a special kind of a linear transformation from $\mathbb{R}^n$ to itself. The matrix representation of a $T$-transform is an $n \times n$ matrix

$$T = tI_n + (1-t)P, \tag{4.12}$$

where $t \in [0, 1]$ and $P$ is a permutation matrix that just exchanges two components of the vector it acts upon. Therefore, for $T$ as above there exist $x, y \in [n]$ with $x < y$ such that for every $\mathbf{p} = (p_1, \ldots, p_n)^T \in \mathrm{Prob}(n)$ and every $z \in [n]$ the $z$-component of the vector $\mathbf{r} := T\mathbf{p}$ is given by

$$r_z = \begin{cases} p_z & \text{for } z \notin \{x, y\} \\ tp_x + (1 - t)p_y & \text{for } z = x \\ tp_y + (1 - t)p_x & \text{for } z = y \end{cases} \qquad . \qquad (4.13)$$

**Exercise 4.10.** *Let $\mathbf{p}$ and $\mathbf{r} := T\mathbf{p}$ be as above. Show that $\mathbf{p} \succ \mathbf{r}$.*

**Lemma 4.1.** Let $\mathbf{p}, \mathbf{q} \in \mathbb{R}^n$ be such that $\mathbf{p} \succ \mathbf{q}$. Then, there exists a finite $m \in \mathbb{N}$ and $n \times n$ $T$-transforms $T_1, \ldots, T_m$ such that

$$\mathbf{q} = T_1 \cdots T_m \mathbf{p}. \qquad (4.14)$$

**Proof**   If $\mathbf{p}$ and $\mathbf{q}$ are related by a permutation matrix, then the lemma follows from the fact that any permutation matrix on $n$ elements is a product of transposition matrices (i.e. matrices that only exchange two elements and keep the rest unchanged). We therefore assume now that $\mathbf{p}$ is not a permutation of $\mathbf{q}$, and without loss of generality assume that $\mathbf{p} = \mathbf{p}^\downarrow$ and $\mathbf{q} = \mathbf{q}^\downarrow$.

The main idea of the proof is to construct a $T$-transform of the form given in (4.12) such that the vector $\mathbf{r} := T\mathbf{p}$ as given in (4.13) has the following three properties:

1. $\mathbf{p} \succ \mathbf{r} \succ \mathbf{q}$.
2. $r_x \neq p_x$ and $r_y \neq p_y$.
3. $r_x = q_x$ or $r_y = q_y$.

From the third property at least one of the components of $\mathbf{r}$ is equal to one of the components of $\mathbf{q}$. Therefore, if such a $T$-transform exists, by a repetition of the above process, $\mathbf{q}$ can be obtained from $\mathbf{p}$ by a finite number of such $T$-transforms. It is therefore left to show that a $T$-transform with the above three properties exists.

Given that both $\mathbf{p}$ and $\mathbf{q}$ are probability vectors satisfying $\mathbf{p} \neq \mathbf{q}$, it is impossible for their components to satisfy $p_x \geqslant q_x$ for all $x \in [n]$. If this were the case, it would lead to a contradiction, as the sum of the components of both $\mathbf{p}$ and $\mathbf{q}$ must equal 1, implying $p_x = q_x$ for all $x \in [n]$. Consequently, we define $x \in [n]$ as the largest integer for which $p_x > q_x$.

Similar arguments to those discussed above imply that the reverse scenario, where $p_x \leqslant q_x$ for all $x \in [n]$, is also not feasible. Moreover, since $\mathbf{p} \neq \mathbf{q}$ and $\mathbf{p} \succ \mathbf{q}$, it follows from Exercise 4.7 that the largest integer $z \in [n]$ for which $p_z \neq q_z$ must satisfy $q_z > p_z$. Therefore, there exists an integer $y \in [n]$ with the property that $y > x$ and $q_y > p_y$. We take $y$ to be the *smallest* integer that satisfies these two criteria.

Since $x \in [n]$ is the largest integer for which $p_x > q_x$ we get that $p_w \leqslant q_w$ for all $w > x$. Similarly, since $y$ is the *smallest* integer that satisfies $y > x$ and $q_y > p_y$, we get that for all $x < w < y$ we have $p_w \leqslant q_w$. Combining these two observations we conclude that

$$p_w = q_w \qquad \forall\, w \in [n] \text{ such that } x < w < y. \tag{4.15}$$

Moreover, from the definitions of $x$ and $y$, along with the fact that $x < y$ and $\mathbf{p} = \mathbf{p}^{\downarrow}$ and $\mathbf{q} = \mathbf{q}^{\downarrow}$, we deduce the following inequality:

$$p_x > q_x \geqslant q_y > p_y. \tag{4.16}$$

Utilizing (4.13), we find that $r_x = t p_y + (1-t) p_x$ and $r_y = t p_x + (1-t) p_y$. By choosing $t \in (0, 1)$, which means $t$ is strictly between 0 and 1, we ensure that the second condition, $r_x \neq p_x$ and $r_y \neq p_y$, is satisfied, given that $p_x \neq p_y$. It is important to note that $\mathbf{p} \succ \mathbf{r}$ for any $T$-transform (as per Exercise 4.10). Consequently, our remaining task is to demonstrate the existence of a $t \in (0, 1)$ such that $\mathbf{r} \succ \mathbf{q}$, and either $r_x = q_x$ or $r_y = q_y$ is true.

Set $\varepsilon := \min\{p_x - q_x, q_y - p_y\} > 0$, and define $t := 1 - \frac{\varepsilon}{p_x - p_y}$. By definition, $0 < t < 1$ (due to (4.16)), and if $\varepsilon = p_x - q_x$, then $t = \frac{q_x - p_y}{p_x - p_y}$ so that

$$\begin{aligned}
r_x = t p_x + (1-t) p_y &= p_y + t(p_x - p_y) \\
&= p_y + q_x - p_y = q_x.
\end{aligned} \tag{4.17}$$

Similarly, if $\varepsilon = q_y - p_y$, then $t = \frac{p_x - q_y}{p_x - p_y}$ so that

$$\begin{aligned}
r_y = t p_y + (1-t) p_x &= p_x - t(p_x - p_y) \\
&= p_x - p_x + q_y = q_y.
\end{aligned} \tag{4.18}$$

We therefore conclude that for this choice of $t \in (0, 1)$ we get that $r_x = q_x$ or $r_y = q_y$.

Before showing that $\mathbf{r} \succ \mathbf{q}$, we argue that $r_x$ can never be strictly smaller that $q_x$. Indeed, we saw above that if $\varepsilon = p_x - q_x$, then $r_x = q_x$. Moreover, for the second option that $\varepsilon = q_y - p_y$ we have $t = \frac{p_x - q_y}{p_x - p_y}$ so that

$$\begin{aligned}
r_x = t p_x + (1-t) p_y &= p_y + t(p_x - p_y) \\
&= p_y + p_x - q_y \\
\boxed{\varepsilon = q_y - p_y} \longrightarrow &= p_x - \varepsilon \\
\boxed{\varepsilon \leqslant p_x - q_x} \longrightarrow &\geqslant p_x - (p_x - q_x) = q_x.
\end{aligned} \tag{4.19}$$

We therefore conclude that for both options $r_x \geqslant q_x$.

Finally, to show that $\mathbf{r} \succ \mathbf{q}$ we show that $\|\mathbf{r}\|_{(k)} \geqslant \|\mathbf{q}\|_{(k)}$ for all $k \in [n]$. We show it in three cases:

1. For $1 \leqslant k < x$ we have $\|\mathbf{r}\|_{(k)} = \|\mathbf{p}\|_{(k)} \geqslant \|\mathbf{q}\|_{(k)}$ since $\mathbf{p} \succ \mathbf{q}$ and $r_w = p_w$ for $w \in [k]$.
2. For $x \leqslant k < y$, we have the following relation:

$$\|\mathbf{r}\|_{(k)} \geqslant \sum_{w \in [k]} r_w = \|\mathbf{p}\|_{(x-1)} + r_x + \sum_{w = x+1}^{k} p_w. \tag{4.20}$$

The first term on the right-hand side, $\|\mathbf{p}\|_{(x-1)}$, satisfies $\|\mathbf{p}\|_{(x-1)} \geqslant \|\mathbf{q}\|_{(x-1)}$ since $\mathbf{p} \succ \mathbf{q}$. For the second term, we have already established that $r_x \geqslant q_x$, and for the third term we get from (4.15) that $\sum_{w=x+1}^{k} p_w = \sum_{w=x+1}^{k} q_w$. Incorporating these three relations into (4.20) yields $\|\mathbf{r}\|_{(k)} \geqslant \|\mathbf{q}\|_{(k)}$.

3. For $y \leqslant k \leqslant n$ we use the fact that $r_x + r_y = p_x + p_y$ so that

$$\|\mathbf{r}\|_{(k)} \geqslant \sum_{w \in [k]} r_w = \sum_{w \in [k]} p_w = \|\mathbf{p}\|_{(k)} \geqslant \|\mathbf{q}\|_{(k)}, \tag{4.21}$$

where the last inequality follows from the fact that $\mathbf{p} \succ \mathbf{q}$.

Hence, $\mathbf{r} \succ \mathbf{q}$. This completes the proof. ∎

**Exercise 4.11.** *Prove the converse of the statement presented in Lemma 4.1. Specifically, demonstrate that for any vector $\mathbf{p} \in \mathbb{R}^n$ and a sequence of $m$ $n \times n$ $T$-transforms, denoted as $T_1, \ldots, T_m$, the resulting vector $\mathbf{q} := T_1 \cdots T_m \mathbf{p}$ fulfills the condition $\mathbf{p} \succ \mathbf{q}$. Hint: Refer to Exercise 4.10 for guidance.*

## The Fundamental Theorem of Majorization

We are now prepared to introduce the fundamental theorem of majorization, which delineates the relationship between doubly stochastic matrices and majorization. For clarity, we adopt the notation $\mathbf{1}_n := (1, \ldots, 1)^T \in \mathbb{R}^n$. This allows for a succinct expression of the sum of the components of any vector $\mathbf{p} \in \mathbb{R}^n$ as the dot product $\mathbf{1}_n \cdot \mathbf{p}$.

> **Characterization**
>
> **Theorem 4.1.** Let $\mathbf{p}, \mathbf{q} \in \mathbb{R}^n$. The following are equivalent:
>
> 1. $\mathbf{p} \succ \mathbf{q}$.
> 2. There exists an $n \times n$ doubly stochastic matrix $D$ such that $\mathbf{q} = D\mathbf{p}$.
> 3. $\mathbf{1}_n \cdot \mathbf{p} = \mathbf{1}_n \cdot \mathbf{q}$ and for every $t \in \mathbb{R}$
>
> $$\left\| \mathbf{p} - t\mathbf{u}^{(n)} \right\|_1 \geqslant \left\| \mathbf{q} - t\mathbf{u}^{(n)} \right\|_1. \tag{4.22}$$

*Remark.* In some books the last condition is expressed as

$$\sum_{x \in [n]} (p_x - t)_+ \geqslant \sum_{x \in [n]} (q_x - t)_+, \tag{4.23}$$

where for all $r \in \mathbb{R}$ the notation $(r)_+ = r$ if $r \geqslant 0$ and otherwise $(r)_+ = 0$. To see the equivalence note that $(r)_+ = \frac{1}{2}(|r| + r)$ and "absorb" the factor $1/n$ into $t$.

**Proof** We divide the proof into three parts:

*The implication $1 \Rightarrow 2$:* From Lemma 4.1 we get that if $\mathbf{p} \succ \mathbf{q}$, then $\mathbf{q} = T_1 \cdots T_m \mathbf{p}$ for some $T$-transforms $T_1, \ldots, T_m$. Since the product of the $T$-transforms $D := T_1 \cdots T_m$ is doubly stochastic (see Exercise 4.12), it follows that $\mathbf{q} = D\mathbf{p}$ and $D$ is doubly stochastic.

*The implication* $2 \Rightarrow 3$*:* Suppose $\mathbf{q} = D\mathbf{p}$ for some doubly stochastic matrix $D$. Then, since $D\mathbf{u}^{(n)} = \mathbf{u}^{(n)}$ we get for all $t \in \mathbb{R}$

$$\left\| \mathbf{q} - t\mathbf{u}^{(n)} \right\|_1 = \left\| D\mathbf{p} - t\mathbf{u}^{(n)} \right\|_1 = \left\| D\mathbf{p} - tD\mathbf{u}^{(n)} \right\|_1$$

$$\mathbf{(2.8)} \rightarrow \; \leqslant \left\| \mathbf{p} - t\mathbf{u}^{(n)} \right\|_1 , \tag{4.24}$$

where in the last inequality we used the property (2.8) of the norm $\| \cdot \|_1$, in conjunction with the fact that the doubly stochastic matrix $D$ is particularly column stochastic.

*The implication* $3 \Rightarrow 1$*:* Suppose (4.23) holds for all $t \in \mathbb{R}$. Without loss of generality suppose that $\mathbf{p} = \mathbf{p}^{\downarrow}$ and $\mathbf{q} = \mathbf{q}^{\downarrow}$. Fix $k \in [n-1]$ and observe that for $t = p_{k+1}$ the left-hand side of (4.23) can be expressed as:

$$\sum_{x \in [n]} (p_x - t)_+ = \sum_{x \in [n]} (p_x - p_{k+1})_+$$

$$\boxed{\mathbf{p} = \mathbf{p}^{\downarrow}} \longrightarrow \; = \|\mathbf{p}\|_{(k)} - k p_{k+1}. \tag{4.25}$$

Furthermore, the right-hand side of (4.23) satisfies

$$\sum_{x \in [n]} (q_x - p_{k+1})_+ \geqslant \sum_{x \in [k]} (q_x - p_{k+1})_+$$

$$\boxed{(q_x - p_{k+1})_+ \geqslant q_x - p_{k+1}} \longrightarrow \; \geqslant \sum_{x \in [k]} (q_x - p_{k+1}) \tag{4.26}$$

$$\boxed{\mathbf{q} = \mathbf{q}^{\downarrow}} \longrightarrow \; = \|\mathbf{q}\|_{(k)} - k p_{k+1}.$$

Hence, the combination of the two equations above with our assumption that (4.23) holds for all $t \in \mathbb{R}$, and in particular for $t = p_{k+1}$, gives $\|\mathbf{p}\|_{(k)} \geqslant \|\mathbf{q}\|_{(k)}$. Since $k \in [n-1]$ was arbitrary we conclude that $\mathbf{p} \succ \mathbf{q}$. This completes the proof. $\blacksquare$

**Exercise 4.12.** *Show that the product of two $n \times n$ doubly stochastic matrices is itself a doubly stochastic matrix.*

## 4.1.2 Three Equivalent Approaches

We have seen that majorization is a preorder relationship between vectors in Prob($m$). Given that a probability vector can represent a classical system $X$, we can understand majorization in terms of *mixing operations* applied to system $X$. To approach this, let's set aside our initial definition of majorization and attempt to redefine it as follows: We state that a vector $\mathbf{p}$ majorizes another vector $\mathbf{q}$ (denoted as $\mathbf{p} \succ \mathbf{q}$) if there exists a mixing operation $M \in \mathrm{STOC}(m, m)$ that fulfills the condition:

$$\mathbf{q} = M\mathbf{p}. \tag{4.27}$$

The pivotal question then becomes how to define these mixing operations. Conceptually, mixing operations are processes that increase the uncertainty of system $X$. In this context, we propose that the mixing operation $M$ can be conceptualized in three distinct ways.

1. *The Axiomatic Approach:* Since mixing operations can only increase the uncertainty of system $X$, the uniform distribution remains invariant under mixing operations, as its uncertainty cannot be increased further. Hence, $M \in \text{STOC}(n, n)$ can be defined as a mixing operation if

$$M\mathbf{u}^X = \mathbf{u}^X. \tag{4.28}$$

That is, $M = D$ is doubly stochastic.

2. *The Constructive Approach:* In this approach the mixing operations are defined intuitively as a convex combination of permutation matrices. Indeed, mixing a pack of cards literally corresponds to the action of a random permutation. Therefore, in this approach, $M \in \text{STOC}(n, n)$ corresponds to a mixing operation if there exists $k \in \mathbb{N}$ $m \times m$ permutation matrices $\{P_j\}_{j \in [k]}$ such that

$$M = \sum_{j \in [k]} s_j P_j, \tag{4.29}$$

for some $\mathbf{s} \in \text{Prob}(k)$.

3. *Operational Approach:* In this approach, a mixing operation is defined as a stochastic process $M \in \text{STOC}(m, m)$ that cannot increase the chances to win a game of chance. Specifically, $M \in \text{STOC}(m, m)$ is a mixing operation if and only if for all $\mathbf{p} \in \text{Prob}(m)$ and for all $k \in [m]$ the probability, $\text{Pr}_k$, to win a $k$-gambling game satisfies

$$\text{Pr}_k(\mathbf{p}) \geqslant \text{Pr}_k(M\mathbf{p}). \tag{4.30}$$

As we proved in the preceding subsections, all the three approaches are equivalent, leading to the same preorder given in (4.1). Furthermore, the established equivalence of these approaches solidifies the conceptual foundation of uncertainty. This, in turn, validates functions that exhibit monotonic behavior under majorization as reliable quantifiers of uncertainty. Such measures of uncertainty are known as Schur concave functions.

## 4.1.3 Schur Convexity

A function $f : \mathbb{R}^n \to \mathbb{R}$ that preserves the majorization order is said to be *Schur convex*. That is, $f$ is Schur convex if and only if for any two vector $\mathbf{p}, \mathbf{q} \in \mathbb{R}^n$ such that $\mathbf{p} \succ \mathbf{q}$ we have

$$f(\mathbf{p}) \geqslant f(\mathbf{q}). \tag{4.31}$$

We also say that $f$ is Schur concave if $-f$ is Schur convex. Note that a Schur convex function $f$ is symmetric with respect to permutations, since if $\mathbf{q} = P\mathbf{p}$ for some permutation matrix $P$ then we have $\mathbf{p} \succ \mathbf{q}$ and $\mathbf{q} \succ \mathbf{p}$ so that $f(\mathbf{p}) = f(\mathbf{q})$. Moreover, from the fundamental theorem of majorization if $\mathbf{p} \succ \mathbf{q}$ then there exists a set of permutation matrices $\{P_j\}_{j \in [m]}$ and a probability distribution $\{t_j\}_{j \in [m]}$ such that

$$\mathbf{q} = \sum_{j \in [m]} t_j P_j \mathbf{p}. \tag{4.32}$$

Therefore, if a function $f : \mathbb{R}^n \to \mathbb{R}$ is symmetric (under permutations) and convex, then it is necessarily Schur convex. To see this, observe that

$$f(\mathbf{q}) = f\left( \sum_{j \in [m]} t_j P_j \mathbf{p} \right)$$

$$f \text{ is convex} \to \; \leqslant \sum_{j \in [m]} t_j f \left( P_j \mathbf{p} \right) \tag{4.33}$$

$$f \text{ is symmetric} \to \; = \sum_{j \in [m]} t_j f \left( \mathbf{p} \right) = f(\mathbf{p}).$$

As an example, consider the Shannon entropy, defined for any probability vector $\mathbf{p} \in \text{Prob}(n)$ as

$$H(\mathbf{p}) := - \sum_{x \in [n]} p_x \log_2 (p_x). \tag{4.34}$$

This function is clearly symmetric under any permutation of the components of $\mathbf{p}$, and it is also concave. Therefore, the Shannon entropy is an example of a Schur concave function.

**Exercise 4.13.** *Show that the geometric mean function*

$$G(\mathbf{p}) := \left( p_1 p_2 \cdots p_n \right)^{\frac{1}{n}} \qquad \forall \, \mathbf{p} \in \text{Prob}(n) \tag{4.35}$$

*is Schur concave. Hint: Show first that $\log G(\mathbf{p})$ is Schur concave by showing that it is both symmetric and concave (what is its Hessian matrix?).*

From the following exercise it follows that not all Schur convex functions are symmetric and convex. Therefore, in this sense, the notion of Schur convexity is weaker than (standard) convexity.

**Exercise 4.14.** *Consider the function*

$$f(\mathbf{p}) = \log \max\{p_1, \ldots, p_n\} \qquad \forall \, \mathbf{p} \in \text{Prob}(n). \tag{4.36}$$

1. *Show that $f$ is symmetric.*
2. *Show that $f$ is not convex.*
3. *Show that $f$ is Schur convex.*

One way to test if a given multivariable function is convex is to check if its Hessian matrix is positive semidefinite. Since Schur convex functions are not necessarily convex, this test cannot always be used to determine if a function is Schur convex. Instead, we can use the theorem below to test if a given symmetric function is Schur convex.

> **Schur's Test**
>
> **Theorem 4.2.** Let $f : \text{Prob}(n) \to \mathbb{R}$ be a continuous function that is also continuously differentiable on the interior of $\text{Prob}(n)$. Then, $f$ is Schur convex if and only if the following two conditions hold:
>
> 1. $f$ is symmetric in $\text{Prob}(n)$; that is, for every $n \times n$ permutation matrix $P$
>
> $$f(\mathbf{p}) = f(P\mathbf{p}) \qquad \forall\, \mathbf{p} \in \text{Prob}(n). \tag{4.37}$$
>
> 2. For all $0 < \mathbf{p} \in \text{Prob}(n)$
>
> $$(p_1 - p_2)\left(\frac{\partial f(\mathbf{p})}{\partial p_1} - \frac{\partial f(\mathbf{p})}{\partial p_2}\right) \geq 0. \tag{4.38}$$

*Remark.* Since the function $f$ is symmetric, the condition in (4.38) is equivalent to the following condition. For all $0 < \mathbf{p} \in \text{Prob}(n)$ and all $x \neq y \in [n]$

$$(p_x - p_y)\left(\frac{\partial f(\mathbf{p})}{\partial p_x} - \frac{\partial f(\mathbf{p})}{\partial p_y}\right) \geq 0. \tag{4.39}$$

**Proof** Suppose $f$ is Schur convex. We need to show that (4.38) holds. Let $\mathbf{p} > 0$ and observe that if $p_1 = p_2$ the condition clearly holds. Therefore, without loss of generality we assume that $p_1 > p_2$ (and recall that $p_2 > 0$ since $\mathbf{p} > 0$). Let $0 < \varepsilon < p_1 - p_2$ and define

$$\tilde{\mathbf{p}}_\varepsilon := (p_1 - \varepsilon, p_2 + \varepsilon, p_3, \ldots, p_n)^T. \tag{4.40}$$

Let $a := \frac{\varepsilon}{p_1 - p_2} < 1$ and observe that $\tilde{\mathbf{p}}_\varepsilon = D\mathbf{p}$, where $D = D_2 \oplus I_{n-2}$ with

$$D_2 := \begin{pmatrix} 1 - a & a \\ a & 1 - a \end{pmatrix}. \tag{4.41}$$

Since $D$ is doubly stochastic, we get from Theorem 4.1 that $\mathbf{p} \succ \tilde{\mathbf{p}}_\varepsilon$. Therefore, from the assumption that $f$ is Schur convex we conclude that for all $0 < \varepsilon < p_1 - p_2$

$$
\begin{aligned}
0 &\leq \frac{f(\mathbf{p}) - f(\tilde{\mathbf{p}}_\varepsilon)}{\varepsilon} \\
&= \frac{f(\mathbf{p}) - f(p_1 - \varepsilon, p_2, \ldots, p_n)}{\varepsilon} + \frac{f(p_1 - \varepsilon, p_2, \ldots, p_n) - f(\tilde{\mathbf{p}}_\varepsilon)}{\varepsilon} \\
&\xrightarrow{\varepsilon \to 0^+} \frac{\partial f(\mathbf{p})}{\partial p_1} - \frac{\partial f(\mathbf{p})}{\partial p_2}.
\end{aligned}
\tag{4.42}
$$

Hence, we get that $\frac{\partial f(\mathbf{p})}{\partial p_1} \geq \frac{\partial f(\mathbf{p})}{\partial p_2}$, which is equivalent to (4.38) (since $p_1 > p_2$).

For the converse, let $0 < \mathbf{p}, \mathbf{q} \in \text{Prob}(n)$ and suppose $\mathbf{p} \succ \mathbf{q}$. We need to show that $f(\mathbf{p}) \geq f(\mathbf{q})$. From Lemma 4.1 we know that $\mathbf{q}$ can be obtained from $\mathbf{p}$ by a sequence of $T$-transforms. Therefore, it is sufficient to show that $f$ is nonincreasing under a single action of $T$-transform. Explicitly, it is sufficient to show that $f(T\mathbf{p}) \leq f(\mathbf{p})$ for any

$T = tI_n + (1-t)P$ where $t \in [0,1]$ and $P$ is a permutation matrix exchanging only two components. Moreover, we can assume that $t \in [\frac{1}{2},1]$ since $f$ is symmetric under permutations and therefore $f(T\mathbf{p}) = f(PT\mathbf{p})$, where $PT$ is the same $T$ transform but with $1-t$ replacing $t$ (since $P^2 = I_n$). For simplicity of the exposition, we also assume that $P$ is the matrix exchanging the first and second components, so that

$$T\mathbf{p} = (tp_1 + (1-t)p_2, tp_2 + (1-t)p_1, p_3, \ldots, p_n)^T. \tag{4.43}$$

Note that if $p_1 = p_2$, then the transformation does not effect $\mathbf{p}$. Therefore, without loss of generality suppose that $p_1 > p_2$ (recall that $f$ is symmetric, so we can exchange between $p_1$ and $p_2$ if necessary). Now, from (4.38) we get that

$$\frac{d}{d\varepsilon} f(p_1 - \varepsilon, p_2 + \varepsilon, p_3, \ldots, p_n) \leqslant 0, \tag{4.44}$$

for any $0 \leqslant \varepsilon \leqslant \frac{1}{2}(p_1 - p_2)$ (note that in this domain $p_1 - \varepsilon \geqslant p_2 + \varepsilon$). We therefore conclude that the function

$$g(\varepsilon) := f(p_1 - \varepsilon, p_2 + \varepsilon, p_3, \ldots, p_n) \tag{4.45}$$

is nonincreasing in the domain $0 \leqslant \varepsilon \leqslant \frac{1}{2}(p_1 - p_2)$. Taking $\varepsilon := (1-t)(p_1 - p_2)$ we conclude that

$$f(\mathbf{p}) = g(0) \geqslant g(\varepsilon) = f(p_1 - \varepsilon, p_2 + \varepsilon, p_3, \ldots, p_n).$$
$$\textbf{By definition of } \varepsilon \rightarrow = f(tp_1 + (1-t)p_2, tp_2 + (1-t)p_1, p_3, \ldots, p_n) \tag{4.46}$$
$$= f(T\mathbf{p}).$$

This completes the proof.  ∎

As an example, consider the family of Rényi entropies defined for any $\alpha \in [0, \infty]$ and all $\mathbf{p} \in \mathrm{Prob}(n)$ as

$$H_\alpha(\mathbf{p}) := \frac{1}{1-\alpha} \log \sum_{x \in [n]} p_x^\alpha, \tag{4.47}$$

where the cases $\alpha = 0, 1, \infty$ are defined in terms of their limits. In the next chapter we will study these functions in more detail. Here we show that for all $\alpha \in (0, \infty)$ the Rényi entropies are Schur concave. Due to the monotonicity of the log function, it is enough to show that $f_\alpha(\mathbf{p}) := \sum_{x \in [n]} p_x^\alpha$ is Schur convex for $\alpha \in (0,1)$ and Schur concave for $\alpha \in (1, \infty)$. Indeed,

$$(p_1 - p_2)\left(\frac{\partial f(\mathbf{p})}{\partial p_1} - \frac{\partial f(\mathbf{p})}{\partial p_2}\right) = \alpha(p_1 - p_2)\left(p_1^{\alpha-1} - p_2^{\alpha-1}\right), \tag{4.48}$$

which is always nonnegative for $\alpha > 1$ and nonpositive for $\alpha \in (0,1)$. Hence, from the Schur's test it follows that $H_\alpha(\mathbf{p})$ is Schur concave for all $\alpha \in [0, \infty]$ (the cases $\alpha = 0, 1, \infty$ follow from the continuity of $H_\alpha$ in $\alpha$).

As another example, consider the elementary symmetric function defined for each $k \in [n]$ by

$$f_k(\mathbf{p}) := \sum_{\substack{x_1 < \cdots < x_k \\ x_1, \ldots, x_k \in [n]}} p_{x_1} \cdots p_{x_k} \quad \forall\, \mathbf{p} \in \mathrm{Prob}(n). \tag{4.49}$$

For example, for $k = 2$ they take the form

$$f_2(\mathbf{p}) = \sum_{\substack{x < y \\ x,y \in [n]}} p_x p_y, \tag{4.50}$$

and for $k = n$ we have $f_n(\mathbf{p}) = p_1 \cdots p_n$. From Schur's test it follows that the elementary symmetric functions are Schur concave.

**Exercise 4.15.** *Use Schur's test to verify that the elementary symmetric functions are Schur concave.*

## 4.2 Approximate Majorization

We saw in the previous section that given two probability vectors $\mathbf{p}, \mathbf{q} \in \text{Prob}(n)$, it is possible to have both $\mathbf{p} \not\succ \mathbf{q}$ and $\mathbf{q} \not\succ \mathbf{p}$ (i.e. $\mathbf{p}$ and $\mathbf{q}$ are incomparable). For some applications in resource theories, it is often useful to know how much one has to perturb $\mathbf{p}$ so that $\mathbf{p} \succ \mathbf{q}$. To make this idea rigour, we first introduce several notions of "greatest" and "least" elements in a subset of probability vectors.

> **Maximal and Minimal Elements**
>
> **Definition 4.2.** Let $\mathfrak{C} \subseteq \text{Prob}(n)$ be a subset of $n$-dimensional probability vectors.
>
> - A vector $\mathbf{p} \in \mathfrak{C}$ is said to be a maximal element of $\mathfrak{C}$ if for all $\mathbf{q} \in \mathfrak{C}$ we have $\mathbf{p} \succ \mathbf{q}$.
> - A vector $\mathbf{p} \in \mathfrak{C}$ is said to be a minimal element of $\mathfrak{C}$ if for all $\mathbf{q} \in \mathfrak{C}$ we have $\mathbf{q} \succ \mathbf{p}$.

In general, partial orders don't always have maximal and minimal elements. For example, the set

$$\mathfrak{C} := \left\{ \begin{bmatrix} 1/2 \\ 1/4 \\ 1/4 \end{bmatrix}, \begin{bmatrix} 2/5 \\ 2/5 \\ 1/5 \end{bmatrix} \right\} \tag{4.51}$$

has no maximal nor minimal elements since none of the two vectors majorize the other. On the other hand, one can define upper and lower *bounds* on a set of probability vectors. Specifically, given a subset $\mathfrak{C} \subseteq \text{Prob}(n)$:

- A vector $\mathbf{p} \in \text{Prob}(n)$ is said to be an upper bound of $\mathfrak{C}$ if for all $\mathbf{q} \in \mathfrak{C}$ we have $\mathbf{p} \succ \mathbf{q}$.
- A vector $\mathbf{p} \in \text{Prob}(n)$ is said to be a lower bound of $\mathfrak{C}$ if for all $\mathbf{q} \in \mathfrak{C}$ we have $\mathbf{q} \succ \mathbf{p}$.

Note that lower and upper bounds always exist since the vector $\mathbf{e}_1 = (1, 0, \ldots, 0)^T$ is always an upper bound and the vector $\mathbf{u}^{(n)}$ is always a lower bound. Less trivial bounds are those that are optimal: Given a subset $\mathfrak{C} \subseteq \text{Prob}(n)$,

- An upper bound $\mathbf{p} \in \mathrm{Prob}(n)$ of $\mathfrak{C}$ is said to be *optimal* if for any other upper bound $\mathbf{p}' \in \mathrm{Prob}(n)$ of $\mathfrak{C}$ we have $\mathbf{p}' \succ \mathbf{p}$.
- A lower bound $\mathbf{p} \in \mathrm{Prob}(n)$ of $\mathfrak{C}$ is said to be *optimal* if for any other lower bound $\mathbf{p}' \in \mathrm{Prob}(n)$ of $\mathfrak{C}$ we have $\mathbf{p} \succ \mathbf{p}'$.

**Exercise 4.16.** *Let*

$$\mathfrak{C} := \{\mathbf{p}_1, \dots, \mathbf{p}_m\} \subset \mathrm{Prob}(n) \tag{4.52}$$

*be a set consisting of $m$ probability vectors, and for each $z \in [n]$ denote by*

$$s_z := \max_{y \in [m]} \|\mathbf{p}_y\|_{(z)}, \tag{4.53}$$

*where $\| \cdot \|_{(z)}$ is the Ky Fan norm (see (2.78)). Finally, denote by*

$$q_z := s_z - s_{z-1} \qquad \forall\, z \in [n], \tag{4.54}$$

*with $s_0 := 0$. Show that the vector $\mathbf{q} = (q_1, \dots, q_n)^T$ is a probability vector in $\mathrm{Prob}(n)$ that is also an upper bound of $\mathfrak{C}$. Is it optimal?*

The study of the above notions of extrema under majorization of a set $\mathfrak{C}$ goes beyond the scope of this book. Here, we are only interested in a particular set of vectors, namely, a ball of a small radius around some probability vector. Let $\mathbf{p} \in \mathrm{Prob}(n)$ be a probability vector and for any $\varepsilon \in [0, 1]$ define a "ball" of radius $\varepsilon$ around it as

$$\mathfrak{B}_\varepsilon(\mathbf{p}) := \left\{ \mathbf{p}' \in \mathrm{Prob}(n) : \frac{1}{2}\|\mathbf{p} - \mathbf{p}'\|_1 \leqslant \varepsilon \right\}. \tag{4.55}$$

Remarkably, the above subset of $\mathrm{Prob}(n)$ has both minimal and maximal elements, known as the *flattest* and *steepest* $\varepsilon$-approximations of $\mathbf{p}$.

## 4.2.1 The Steepest $\varepsilon$-Approximation

In this subsection we find the maximal element (under majorization) of $\mathfrak{B}_\varepsilon(\mathbf{p})$, where $\mathbf{p} \in \mathrm{Prob}(n)$. Recall that the vector $\mathbf{e}_1 := (1, 0, \dots, 0)^T \in \mathrm{Prob}(n)$ satifies $\mathbf{e}_1 \succ \mathbf{q}$ for all $\mathbf{q} \in \mathrm{Prob}(n)$. Therefore, if $\frac{1}{2}\|\mathbf{p} - \mathbf{e}_1\|_1 \leqslant \varepsilon$, that is, $\mathbf{e}_1 \in \mathfrak{B}_\varepsilon(\mathbf{p})$, then the maximal element of $\mathfrak{B}_\varepsilon(\mathbf{p})$ is unique up to permutation of the components, and is given by $\mathbf{e}_1$. We will therefore assume now that $\frac{1}{2}\|\mathbf{p} - \mathbf{e}_1\|_1 > \varepsilon$ and that the components of $\mathbf{p}$ are arranged in nonincreasing order; that is, $\mathbf{p} = \mathbf{p}^\downarrow$.

**Exercise 4.17.** *Let $\{\mathbf{e}_z\}_{z=1}^n$ be the elementary basis of $\mathbb{R}^n$. Show that if $\mathbf{p} = \mathbf{p}^\downarrow$ and $\mathbf{e}_1 \notin \mathfrak{B}_\varepsilon(\mathbf{p})$, then for all $z \in [n]$ we have $\mathbf{e}_z \notin \mathfrak{B}_\varepsilon(\mathbf{p})$. Hint: Show first that $\frac{1}{2}\|\mathbf{p} - \mathbf{e}_z\|_1 = 1 - p_z$.*

Fix $\varepsilon \in (0, 1)$ and let $k \in [n]$ be the integer satisfying

$$\|\mathbf{p}\|_{(k)} \leqslant 1 - \varepsilon < \|\mathbf{p}\|_{(k+1)}. \tag{4.56}$$

With this index $k$ we define the steepest $\varepsilon$-approximation of $\mathbf{p}$, denoted by $\overline{\mathbf{p}}^{(\varepsilon)}$, whose components are

$$
\overline{p}_x^{(\varepsilon)} := \begin{cases} p_1 + \varepsilon & \text{if } x = 1 \\ p_x & \text{if } x \in \{2, \ldots, k\} \\ 1 - \varepsilon - \|\mathbf{p}\|_{(k)} & \text{if } x = k+1 \\ 0 & \text{otherwise} \end{cases}. \tag{4.57}
$$

Note that from its definition above, $\overline{\mathbf{p}}^{(\varepsilon)}$ is indeed a probability vector whose components are arranged in nonincreasing order.

**Exercise 4.18.** *Utilize the definition of $k$ as provided in (4.56) and the definition of $\overline{\mathbf{p}}^{(\varepsilon)}$ as outlined in (4.57) to demonstrate the following two properties:*

1. *The components of $\overline{\mathbf{p}}^{(\varepsilon)}$ are arranged in nonincreasing order. In particular, $\overline{p}_k^{(\varepsilon)} > \overline{p}_{k+1}^{(\varepsilon)}$.*
2. *The components of $\overline{\mathbf{p}}^{(\varepsilon)}$ satisfy*

$$
\overline{p}_x^{(\varepsilon)} \leqslant p_x \qquad \forall\, x \in \{2, \ldots, n\}. \tag{4.58}
$$

*In particular, $\overline{p}_{k+1}^{(\varepsilon)} < p_{k+1}$.*

The intuition behind the definition above is that we want to alter $\mathbf{p}$ in a way that it becomes more similar to $\mathbf{e}_1$. However, since $\overline{\mathbf{p}}^{(\varepsilon)}$ must be close to $\mathbf{p}$ we cannot increase $p_1$ by too much. Indeed, the vector $\overline{\mathbf{p}}^{(\varepsilon)}$ as defined above is $\varepsilon$-close to $\mathbf{p}$. To see this, observe that from its definition

$$
\frac{1}{2} \left\| \overline{\mathbf{p}}^{(\varepsilon)} - \mathbf{p} \right\|_1 = \sum_{x \in [n]} \left( \overline{p}_x^{(\varepsilon)} - p_x \right)_+ \tag{4.59}
$$

$$
(\mathbf{4.58}) \to \,= p_1 + \varepsilon - p_1 = \varepsilon.
$$

Therefore, the vector $\overline{\mathbf{p}}^{(\varepsilon)}$ is indeed in $\mathfrak{B}_\varepsilon(\mathbf{p})$.

---

**Theorem 4.3.** Let $\mathbf{p} \in \mathrm{Prob}^{\downarrow}(n)$ be such that $\frac{1}{2}\|\mathbf{p} - \mathbf{e}_1\|_1 > \varepsilon$. Then, the vector $\overline{\mathbf{p}}^{(\varepsilon)}$ as defined in (4.57) is the maximal element (under majorization) of $\mathfrak{B}_\varepsilon(\mathbf{p})$.

---

**Proof**   Since we already showed that $\overline{\mathbf{p}}^{(\varepsilon)} \in \mathfrak{B}_\varepsilon(\mathbf{p})$ it is left to show that for any $\mathbf{q} \in \mathfrak{B}_\varepsilon(\mathbf{p})$ we have $\overline{\mathbf{p}}^{(\varepsilon)} \succ \mathbf{q}$. Indeed, since $\mathbf{q} \in \mathfrak{B}_\varepsilon(\mathbf{p})$ it follows from (2.83) that for every $\ell \in [n]$

$$
\|\mathbf{q}\|_{(\ell)} - \|\mathbf{p}\|_{(\ell)} \leqslant \frac{1}{2}\|\mathbf{q} - \mathbf{p}\| \leqslant \varepsilon. \tag{4.60}
$$

Therefore, for $\ell \in [k]$ we get

$$
\|\mathbf{q}\|_{(\ell)} \leqslant \|\mathbf{p}\|_{(\ell)} + \varepsilon
$$
$$
(\mathbf{4.57}) \to \,= \|\overline{\mathbf{p}}^{(\varepsilon)}\|_{(\ell)}. \tag{4.61}
$$

Combining this with the fact that for $k+1 \leqslant \ell \leqslant n$, $\|\overline{\mathbf{p}}^{(\varepsilon)}\|_{(\ell)} = 1$, we conclude that $\overline{\mathbf{p}}^{(\varepsilon)} \succ \mathbf{q}$. This completes the proof.     ■

**Exercise 4.19.** *Let $m, n \in \mathbb{N}$ be such that $m < n$, and let $\mathbf{p} \in \mathrm{Prob}^{\downarrow}(n)$ be such that $\mathbf{u}^{(m)} \not\succ \mathbf{p}$. Show that a minimal element of the set*

$$\mathfrak{C}_{\mathbf{p},m} := \left\{ \mathbf{q}' \in \mathrm{Prob}(m) : \mathbf{q}' \succ \mathbf{p} \right\} \tag{4.62}$$

*is given by a probability vector $\mathbf{q} \in \mathrm{Prob}^{\downarrow}(m)$ of the form*

$$\mathbf{q} = \left( p_1, \ldots, p_\ell, p_\ell, \ldots p_\ell, 1 - (m - \ell - 1)p_\ell \right)^T, \tag{4.63}$$

*where $\ell \in [m]$ is the largest integer satisfying*

$$(m - \ell)p_\ell \geqslant 1 - \|\mathbf{p}\|_{(\ell)}. \tag{4.64}$$

One can use the steepest $\varepsilon$-approximation to compute the distance of a vector $\mathbf{p} \in \mathrm{Prob}(n)$ to the set of all vectors $\mathbf{r} \in \mathrm{Prob}(n)$ that majorizes $\mathbf{q}$. Specifically, let

$$\mathrm{Majo}(\mathbf{q}) := \{\mathbf{r} \in \mathrm{Prob}(n) : \mathbf{r} \succ \mathbf{q}\} \tag{4.65}$$

denote the set of all vectors in $\mathrm{Prob}(n)$ that majorize $\mathbf{q}$, and define the distance between $\mathbf{p} \in \mathrm{Prob}(n)$ and the set $\mathrm{Majo}(\mathbf{q})$ as

$$T\big(\mathbf{p}, \mathrm{Majo}(\mathbf{q})\big) := \min_{\mathbf{r} \in \mathrm{Majo}(\mathbf{q})} \frac{1}{2} \|\mathbf{p} - \mathbf{r}\|_1. \tag{4.66}$$

> **Theorem 4.4.** Using the same notations as above, for all $\mathbf{p}, \mathbf{q} \in \mathrm{Prob}(n)$
> $$T\big(\mathbf{p}, \mathrm{Majo}(\mathbf{q})\big) = \max_{\ell \in [n]} \left\{ \|\mathbf{q}\|_{(\ell)} - \|\mathbf{p}\|_{(\ell)} \right\}. \tag{4.67}$$

**Proof**  Without loss of generality we will assume that $\mathbf{p}, \mathbf{q} \in \mathrm{Prob}^{\downarrow}(n)$. For any $\varepsilon \in (0, 1)$, let $\overline{\mathbf{p}}^{(\varepsilon)}$ be the steepest $\varepsilon$-approximation of $\mathbf{p}$; see (4.57). Observe that by definition, for any $m \in [d]$ we have $\left\| \overline{\mathbf{p}}^{(\varepsilon)} \right\|_{(m)} \leqslant \|\mathbf{p}\|_{(m)} + \varepsilon$ with equality if $m \in [k]$. In Theorem 4.3 we showed that $\overline{\mathbf{p}}^{(\varepsilon)}$ is the maximal element of $\mathfrak{B}_\varepsilon(\mathbf{p})$ as long as $\varepsilon < \frac{1}{2}\|\mathbf{p} - \mathbf{e}_1\|_1$ (otherwise, $\mathbf{e}_1$ is the maximal element). Hence,

$$T\big(\mathbf{p}, \mathrm{Majo}(\mathbf{q})\big) := \min \left\{ \frac{1}{2} \|\mathbf{p} - \mathbf{r}\|_1 : \mathbf{r} \succ \mathbf{q}, \quad \mathbf{r} \in \mathrm{Prob}(n) \right\}$$

$$= \min \left\{ \varepsilon \in [0, 1] : \mathbf{r} \succ \mathbf{q}, \quad \mathbf{r} \in \mathfrak{B}_\varepsilon(\mathbf{p}) \right\} \tag{4.68}$$

$$\boxed{\overline{\mathbf{p}}^{(\varepsilon)} \succ \mathbf{r} \quad \forall \, \mathbf{r} \in \mathfrak{B}_\varepsilon(\mathbf{p})} \longrightarrow \quad = \min \left\{ \varepsilon \in [0, 1] : \overline{\mathbf{p}}^{(\varepsilon)} \succ \mathbf{q} \right\}.$$

That is, it is left to compute the smallest $\varepsilon$ that satisfy $\overline{\mathbf{p}}^{(\varepsilon)} \succ \mathbf{q}$. We will show that this smallest $\varepsilon$ equals

$$\delta := \max_{\ell \in [n]} \left\{ \|\mathbf{q}\|_{(\ell)} - \|\mathbf{p}\|_{(\ell)} \right\}. \tag{4.69}$$

We first show that $\overline{\mathbf{p}}^{(\delta)} \succ \mathbf{q}$. Let $k$ be the integer satisfying (4.56) but with $\delta$ replacing $\varepsilon$. Then, from the definition of $\overline{\mathbf{p}}^{(\delta)}$ it follows that for $m > k$ we have

$$\left\| \overline{\mathbf{p}}^{(\delta)} \right\|_{(m)} = 1 \geqslant \|\mathbf{q}\|_{(m)}. \tag{4.70}$$

Moreover, for $m \in [k]$ the definition in (4.100) gives $\delta \geqslant \|\mathbf{q}\|_{(m)} - \|\mathbf{p}\|_{(m)}$ so that

$$\left\|\overline{\mathbf{p}}^{(\delta)}\right\|_{(m)} = \|\mathbf{p}\|_{(m)} + \delta \geqslant \|\mathbf{q}\|_{(m)}. \tag{4.71}$$

Hence, $\overline{\mathbf{p}}^{(\delta)} \succ \mathbf{q}$. To prove the optimality of $\delta$, we show that for any $0 < \delta' < \delta$ we must have $\overline{\mathbf{p}}^{(\delta')} \not\succ \mathbf{q}$. Indeed, since $\delta' < \delta$, there exists $m \in [d]$ such that

$$\delta' < \|\mathbf{q}\|_{(m)} - \|\mathbf{p}\|_{(m)}. \tag{4.72}$$

Combining this with the observation that $\left\|\overline{\mathbf{p}}^{(\delta')}\right\|_{(\ell)} \leqslant \|\mathbf{p}\|_{(\ell)} + \delta'$ for all $\ell \in [n]$, we get for this $m$

$$\begin{aligned}
\left\|\overline{\mathbf{p}}^{(\delta')}\right\|_{(m)} &\leqslant \|\mathbf{p}\|_{(m)} + \delta' \\
\mathbf{(4.72)} &\rightarrow \ < \|\mathbf{q}\|_{(m)}.
\end{aligned} \tag{4.73}$$

Hence, $\overline{\mathbf{p}}^{(\delta')} \not\succ \mathbf{q}$. This concludes the proof.   ∎

## 4.2.2 The Flattest $\varepsilon$-Approximation

In this subsection, we aim to identify the minimal element within the set $\mathfrak{B}_\varepsilon(\mathbf{p})$, given that $\mathbf{p} \in \mathrm{Prob}(n)$. Specifically, our objective is to locate the vector in $\mathfrak{B}_\varepsilon(\mathbf{p})$ that exhibits the most uniform (or "flattest") distribution. It's evident that if the uniform distribution vector $\mathbf{u}^{(n)}$ is a member of $\mathfrak{B}_\varepsilon(\mathbf{p})$, then it is the minimal element of $\mathfrak{B}_\varepsilon(\mathbf{p})$. Therefore, our analysis will focus on the scenario where $\mathbf{u}^{(n)} \notin \mathfrak{B}_\varepsilon(\mathbf{p})$. In this context, the parameter $\varepsilon$ satisfies the following condition:

$$0 < c < \frac{1}{2}\left\|\mathbf{p} - \mathbf{u}^{(n)}\right\|_1. \tag{4.74}$$

Additionally, we will assume that the components of the vector $\mathbf{p}$ are sorted in a nonincreasing order; that is, $\mathbf{p} = \mathbf{p}^\downarrow$.

**Exercise 4.20.** *Let $\varepsilon \in (0, 1)$, $\mathbf{p} \in \mathrm{Prob}^\downarrow(n)$, and $\ell \in [n]$ be the integer satisfying $p_\ell \geqslant \frac{1}{n} > p_{\ell+1}$. Show that the inequality in (4.74) holds if and only if*

$$\varepsilon < \|\mathbf{p}\|_{(\ell)} - \frac{\ell}{n}. \tag{4.75}$$

*Hint: Start by expressing $\frac{1}{2}\left\|\mathbf{p} - \mathbf{u}^{(n)}\right\|_1$ as $\sum_{x \in [n]} \left(p_x - 1/n\right)_+$.*

The minimal element of $\mathfrak{B}_\varepsilon(\mathbf{p})$ can be found by "flattening" the tip of $\mathbf{p}$ (i.e. first few components of $\mathbf{p}$) and tail of $\mathbf{p}$ (i.e. the last few components of $\mathbf{p}$). The intuition behind this idea is to alter the vector $\mathbf{p}$ so that it becomes more similar to the uniform distribution $\mathbf{u}^{(n)}$. This process involves replacing the first $k$ components of $\mathbf{p}$ with a constant $a$, and substituting the last $n - m$ components with another constant $b$. We denote by $\underline{\mathbf{p}}^{(\varepsilon)} \in \mathrm{Prob}(n)$ the resulting vector. Its components are given by

$$\underline{p}_x^{(\varepsilon)} := \begin{cases} a & \text{if } x \in [k] \\ p_x & \text{if } k < x \leqslant m \\ b & \text{if } x \in \{m+1, \ldots, n\} \end{cases}. \tag{4.76}$$

The objective is to select suitable values for $a$, $b$, $k$, and $m$, ensuring that $\underline{\mathbf{p}}^{(\varepsilon)}$ forms the flattest $\varepsilon$-approximation of $\mathbf{p}$.

To find the coefficients $a, b, k, m$ we outline the properties that $\underline{\mathbf{p}}^{(\varepsilon)}$ has to satisfy:

1. The vector $\underline{\mathbf{p}}^{(\varepsilon)}$ is a probability vector in $\mathrm{Prob}(n)$. Since all of its components are nonnegative, we just need to require that they sum to one. Using the relation $\sum_{x=k+1}^{m} p_x = \|\mathbf{p}\|_{(m)} - \|\mathbf{p}\|_{(k)}$ we get that the coefficients $a, b, k, m$ must satisfy

$$1 = \sum_{x \in [n]} \underline{p}_x^{(\varepsilon)} = ka + \|\mathbf{p}\|_{(m)} - \|\mathbf{p}\|_{(k)} + (n - m)b. \tag{4.77}$$

2. The vector $\underline{\mathbf{p}}^{(\varepsilon)} \in \mathrm{Prob}^{\downarrow}(n)$; that is, its components are arranged in nondecreasing order. Since $\mathbf{p} = \mathbf{p}^{\downarrow}$ it is sufficient to require that $a > p_{k+1}$ and $b < p_m$ (these inequalities are strict since we want $k$ and $m$ to mark the indices in (4.76) in which the "flattening" process ends and begins, respectively). We therefore conclude that

$$a \in (p_{k+1}, p_k] \quad \text{and} \quad b \in [p_{m+1}, p_m). \tag{4.78}$$

3. The vector $\underline{\mathbf{p}}^{(\varepsilon)} \in \mathfrak{B}^{(\varepsilon)}(\mathbf{p})$. Moreover, since $\underline{\mathbf{p}}^{(\varepsilon)}$ is an optimal vector, we would expect it to be $\varepsilon$-close (and not $\delta$-close with $\delta < \varepsilon$) to $\mathbf{p}$. Therefore, we require that

$$\varepsilon = \frac{1}{2} \left\| \mathbf{p} - \underline{\mathbf{p}}^{(\varepsilon)} \right\|_1 = \sum_{x \in [n]} \left( p_x - \underline{p}_x^{(\varepsilon)} \right)_+$$

$$\boxed{\begin{matrix} a \in (p_{k+1}, p_k] \\ b \in [p_{m+1}, p_m) \end{matrix}} \longrightarrow = \sum_{x \in [k]} (p_x - a) \tag{4.79}$$

$$= \|\mathbf{p}\|_{(k)} - ka.$$

In addition to these three conditions, we need to require that $\underline{\mathbf{p}}^{(\varepsilon)}$ is the minimal element of $\mathfrak{B}_\varepsilon(\mathbf{p})$. However, we first show that the three conditions above already determine uniquely the coefficients $a, b, k, m$. Indeed, from (4.77) it follows that

$$\|\mathbf{p}\|_{(k)} - ka = (n - m)b + \|\mathbf{p}\|_{(m)} - 1. \tag{4.80}$$

Comparing this equality with (4.79) implies that $\varepsilon = \|\mathbf{p}\|_{(k)} - ka$ and $\varepsilon = (n - m)b + \|\mathbf{p}\|_{(m)} - 1$. We therefore conclude that

$$a = \frac{\|\mathbf{p}\|_{(k)} - \varepsilon}{k} \quad \text{and} \quad b = \frac{1 + \varepsilon - \|\mathbf{p}\|_{(m)}}{n - m}. \tag{4.81}$$

That is, the equation above can be viewed as the definitions of $a$ and $b$, and it is left to determine $k$ and $m$.

Substituting the above definitions of $a$ and $b$ into (4.78) and isolating $\varepsilon$ gives

$$\varepsilon \in [r_k, r_{k+1}) \quad \text{and} \quad \varepsilon \in [s_{m+1}, s_m), \tag{4.82}$$

where for all $z \in [n]$

$$r_z := \|\mathbf{p}\|_{(z)} - z p_z \quad \text{and} \quad s_z := (n - z)p_z + \|\mathbf{p}\|_{(z)} - 1. \tag{4.83}$$

Moreover, in the exercise below you show that the components of the vectors $\mathbf{r} := (r_1, \ldots, r_n)^T$ and $\mathbf{s} \in (s_1, \ldots, s_n)^T$ are nonnegative and satisfy $\mathbf{r} = \mathbf{r}^{\uparrow}$ and $\mathbf{s} = \mathbf{s}^{\downarrow}$.

Thus, the relations in (4.82) uniquely specify $k$ and $m$. However, it is left to show that $k \leqslant m$ since otherwise $\mathbf{p}^{(\varepsilon)}$ would not be well defined.

For this purpose, let $\bar{\ell} \in [n]$ be the integer defined in Exercise 4.20. We will show that $k \leqslant \ell \leqslant m$. To prove $k \leqslant \ell$, suppose by contradiction that $k \geqslant \ell + 1$. Since $\varepsilon \in [r_k, r_{k+1})$ we have $\varepsilon \geqslant r_k \geqslant r_{\ell+1}$, where the second inequality follows from the fact that $\mathbf{r} = \mathbf{r}^{\uparrow}$ and our assumption that $k \geqslant \ell + 1$. Combining this with the definition of $r_{\ell+1}$ in (4.83), we get

$$
\begin{aligned}
\varepsilon &\geqslant \|\mathbf{p}\|_{(\ell+1)} - (\ell + 1)p_{\ell+1} \\
&= \|\mathbf{p}\|_{(\ell)} - \ell p_{\ell+1} \\
&\xrightarrow{\boxed{p_{\ell+1} < \frac{1}{n}}} > \|\mathbf{p}\|_{(\ell)} - \frac{\ell}{n},
\end{aligned}
\tag{4.84}
$$

which is in contradiction with (4.75). Therefore the assumption that $k \geqslant \ell + 1$ cannot hold and we conclude that $k \leqslant \ell$.

Similarly, to prove that $m \geqslant \ell$, suppose by contradiction that $m \leqslant \ell - 1$. Since $\varepsilon \in [s_{m+1}, s_m)$, we have $\varepsilon \geqslant s_{m+1} \geqslant s_\ell$, where the second inequality follows from the fact that $\mathbf{s} = \mathbf{s}^{\downarrow}$ and our assumption that $m + 1 \geqslant \ell$. Combining this with the definition of $s_\ell$ in (4.83), we get

$$
\begin{aligned}
\varepsilon &\geqslant (n - \ell)p_\ell + \|\mathbf{p}\|_{(\ell)} - 1 \\
&\xrightarrow{\boxed{p_\ell \geqslant \frac{1}{n}}} \geqslant \|\mathbf{p}\|_{(\ell)} - \frac{\ell}{n},
\end{aligned}
\tag{4.85}
$$

which is again in contradiction with (4.75). Therefore, the assumption that $m \leqslant \ell - 1$ cannot hold and we conclude that $m \geqslant \ell$. Combining this with our earlier result that $k \leqslant \ell$ we conclude that $k \leqslant m$.

**Exercise 4.21.** *Show that the vectors* $\mathbf{r}$ *and* $\mathbf{s}$, *whose components are given in* (4.83), *satisfy*

$$
0 = r_1 \leqslant r_2 \leqslant \cdots \leqslant r_n = 1 - np_n \quad \text{and} \quad np_1 - 1 = s_1 \geqslant s_2 \geqslant \cdots \geqslant s_n = 0.
\tag{4.86}
$$

It's important to note that the index $k$ is characterized by its role as the maximizer of the function $\ell \mapsto t_\ell := \frac{\|\mathbf{p}\|_{(\ell)} - \varepsilon}{\ell}$. To put it another way, $t_k = \max_{\ell \in [n]}\{t_\ell\}$. This implies that the coefficient $a$ can be straightforwardly defined as

$$
a := \max_{\ell \in [n]} \left\{ \frac{\|\mathbf{p}\|_{(\ell)} - \varepsilon}{\ell} \right\}.
\tag{4.87}
$$

To understand this, let $\ell$ be the largest integer that satisfies $t_\ell = \max_{\ell' \in [n]}\{t_{\ell'}\}$. The inequality $t_\ell > t_{\ell+1}$ leads to (see Exercise 4.22)

$$
0 < t_\ell - t_{\ell+1} = \frac{r_{\ell+1} - \varepsilon}{\ell(\ell + 1)},
\tag{4.88}
$$

where $r_\ell := \|\mathbf{p}\|_{(\ell)} - \ell p_\ell$, as previously defined. This implies that $r_{\ell+1} > \varepsilon$. Conversely, by following a similar reasoning, the condition $t_\ell \geqslant t_{\ell-1}$ yields $r_\ell \leqslant \varepsilon$. Therefore, we conclude that $\ell$ is the integer for which $\varepsilon$ falls in the interval $[r_\ell, r_{\ell+1})$, which leads us to deduce that $\ell = k$.

**Exercise 4.22.** *Verify the equality $t_\ell - t_{\ell+1} = \frac{r_{\ell+1}-\varepsilon}{\ell(\ell+1)}$.*

**Exercise 4.23.** *Using the same notations as above, show that for every $\varepsilon \in (0,1)$ and $\mathbf{p} \in \mathrm{Prob}(n)$ the coefficient $b$ can be expressed as*

$$b = \min_{\ell \in [n-1]} \frac{1 + \varepsilon - \|\mathbf{p}\|_{(\ell)}}{n - \ell}. \tag{4.89}$$

> **Theorem 4.5.** Let $\varepsilon \in (0,1)$ and $\mathbf{p} \in \mathrm{Prob}^{\downarrow}(n)$ be a probability vector such that (4.74) holds. Let $k, m \in [n-1]$ be the integers satisfying (4.82), and $a$ and $b$ be the numbers defined in (4.81). Then, for these choices of $k$, $m$, $a$, and $b$, the vector $\underline{\mathbf{p}}^{(\varepsilon)}$ as defined in (4.76) is the minimal element (under majorization) of $\mathfrak{B}_\varepsilon(\mathbf{p})$.

**Proof** We already showed that $\underline{\mathbf{p}}^{(\varepsilon)} \in \mathfrak{B}_\varepsilon(\mathbf{p})$. It is therefore left to show that if $\mathbf{q} \in \mathfrak{B}_\varepsilon(\mathbf{p})$, then $\mathbf{q} \succ \underline{\mathbf{p}}^{(\varepsilon)}$. To establish that $\|\mathbf{q}\|_{(\ell)} \geqslant \left\|\underline{\mathbf{p}}^{(\varepsilon)}\right\|_{(\ell)}$ for every $\ell \in [n]$, we partition the proof into three distinct cases.

1. The case $\ell \in [k]$. In this scenario, since $\left\|\underline{\mathbf{p}}^{(\varepsilon)}\right\|_{(\ell)} = \ell a$, the condition $\|\mathbf{q}\|_{(\ell)} \geqslant \left\|\underline{\mathbf{p}}^{(\varepsilon)}\right\|_{(\ell)}$ simplifies to $\frac{1}{\ell}\|\mathbf{q}\|_{(\ell)} \geqslant a$. Since the components of $\mathbf{q}^{\downarrow}$ are arranged in nonincreasing order, for every integer $\ell \leqslant k$ the average $\frac{1}{\ell}\|\mathbf{q}\|_{(\ell)}$ is no smaller than the average $\frac{1}{k}\|\mathbf{q}\|_{(k)}$. Therefore, it is sufficient to show that $\frac{1}{k}\|\mathbf{q}\|_{(k)} \geqslant a$. Indeed, since $\mathbf{q} \in \mathfrak{B}_\varepsilon(\mathbf{p})$, we get from (2.83) that

$$\|\mathbf{p}\|_{(k)} - \|\mathbf{q}\|_{(k)} \leqslant \frac{1}{2}\|\mathbf{p} - \mathbf{q}\| \leqslant \varepsilon. \tag{4.90}$$

Isolating $\|\mathbf{q}\|_{(k)}$ gives

$$\begin{aligned}
\|\mathbf{q}\|_{(k)} &\geqslant \|\mathbf{p}\|_{(k)} - \varepsilon \\
(4.81)\rightarrow &= ka,
\end{aligned} \tag{4.91}$$

so that $\frac{1}{k}\|\mathbf{q}\|_{(k)} \geqslant a$.

2. The case $k < \ell \leqslant m$. We use again (2.83) (with $\ell$ replacing $k$) to get

$$\begin{aligned}
\|\mathbf{q}\|_{(\ell)} &\geqslant \|\mathbf{p}\|_{(\ell)} - \varepsilon \\
&= \|\mathbf{p}\|_{(k)} - \varepsilon + \sum_{x=k+1}^{\ell} p_x \\
(4.81)\rightarrow &= ka + \sum_{x=k+1}^{\ell} p_x \\
(4.76)\rightarrow &= \left\|\underline{\mathbf{p}}^{(\varepsilon)}\right\|_{(\ell)}.
\end{aligned} \tag{4.92}$$

3. The case $m < \ell \leqslant n$. We use once more (2.83) (with $m$ replacing $k$) to get $\|\mathbf{q}\|_{(m)} \geqslant \|\mathbf{p}\|_{(m)} - \varepsilon$. Moreover, observe that in this case we have for all $m < \ell \leqslant n$

$$\|\mathbf{q}\|_{(\ell)} = 1 - \sum_{x=\ell+1}^{n} q_x^{\downarrow} \quad \text{and} \quad \left\|\underline{\mathbf{p}}^{(\varepsilon)}\right\|_{(\ell)} = 1 - (n-\ell)b. \tag{4.93}$$

Therefore, in order to prove that $\|\mathbf{q}\|_{(\ell)} \geqslant \left\|\underline{\mathbf{p}}^{(\varepsilon)}\right\|_{(\ell)}$ it is sufficient to show that

$$\frac{1}{n-\ell} \sum_{x=\ell+1}^{n} q_x^{\downarrow} \leqslant b. \tag{4.94}$$

The inequality in (4.94) is equivalent to the statement that the average of the last $n - \ell$ components $\mathbf{q}^{\downarrow}$ is no greater than $b$. Since the components of $\mathbf{q}^{\downarrow}$ are arranged in nonincreasing order, this average is no greater than the average of the last $n - m$ components of $\mathbf{q}^{\downarrow}$ (recall that $n - m > n - \ell$). Hence,

$$\frac{1}{n-\ell} \sum_{x=\ell+1}^{n} q_x^{\downarrow} \leqslant \frac{1}{n-m} \sum_{x=m+1}^{n} q_x^{\downarrow}$$

$$= \frac{1 - \|\mathbf{q}\|_{(m)}}{n-m} \tag{4.95}$$

$$\boxed{\|\mathbf{q}\|_{(m)} \geqslant \|\mathbf{p}\|_{(m)} - \varepsilon} \longrightarrow \quad = \frac{1 + \varepsilon - \|\mathbf{p}\|_{(m)}}{n-m}$$

$$(4.81)\rightarrow \; = b.$$

We therefore concludes that $\mathbf{q} \succ \underline{\mathbf{p}}^{(\varepsilon)}$.      ∎

One can use the flatest $\varepsilon$-approximation to compute the distance of a vector $\mathbf{p} \in \mathrm{Prob}(n)$ to the set of all vectors $\mathbf{r} \in \mathrm{Prob}(n)$ that are majorized by $\mathbf{q}$. Specifically, let

$$\mathrm{majo}(\mathbf{q}) := \{\mathbf{r} \in \mathrm{Prob}(n) : \mathbf{q} \succ \mathbf{r}\} \tag{4.96}$$

denote the set of all vectors in $\mathrm{Prob}(n)$ that are majorized by $\mathbf{q}$, and define the distance between $\mathbf{p} \in \mathrm{Prob}(n)$ and the set $\mathrm{majo}(\mathbf{q})$ as

$$T\big(\mathbf{p}, \mathrm{majo}(\mathbf{q})\big) := \min_{\mathbf{r} \in \mathrm{majo}(\mathbf{q})} \frac{1}{2} \|\mathbf{p} - \mathbf{r}\|_1. \tag{4.97}$$

---

**Theorem 4.6.** Using the same notations as above, for all $\mathbf{p}, \mathbf{q} \in \mathrm{Prob}(n)$,

$$T\big(\mathbf{p}, \mathrm{majo}(\mathbf{q})\big) = \max_{\ell \in [n]} \big\{ \|\mathbf{p}\|_{(\ell)} - \|\mathbf{q}\|_{(\ell)} \big\}. \tag{4.98}$$

---

**Proof**   Without loss of generality we will assume that $\mathbf{p}, \mathbf{q} \in \mathrm{Prob}^{\downarrow}(n)$ and $\mathbf{q} \not\succ \mathbf{p}$. For any $\varepsilon \in (0, 1)$, let $\underline{\mathbf{p}}^{(\varepsilon)}$ be the flattest $\varepsilon$-approximation of $\mathbf{p}$; see (4.76). By definition,

$$T\big(\mathbf{p}, \mathrm{majo}(\mathbf{q})\big) := \min \left\{ \frac{1}{2} \|\mathbf{p} - \mathbf{r}\|_1 : \mathbf{q} \succ \mathbf{r}, \quad \mathbf{r} \in \mathrm{Prob}(n) \right\}$$

$$= \min \left\{ \varepsilon \in [0, 1] : \mathbf{q} \succ \mathbf{r}, \quad \mathbf{r} \in \mathfrak{B}_{\varepsilon}(\mathbf{p}) \right\} \tag{4.99}$$

$$\boxed{\mathbf{r} \succ \underline{\mathbf{p}}^{(\varepsilon)} \quad \forall\, \mathbf{r} \in \mathfrak{B}_{\varepsilon}(\mathbf{p})} \longrightarrow \quad = \min \left\{ \varepsilon \in [0, 1] : \mathbf{q} \succ \underline{\mathbf{p}}^{(\varepsilon)} \right\}.$$

That is, it is left to compute the smallest $\varepsilon$ that satisfies $\mathbf{q} \succ \underline{\mathbf{p}}^{(\varepsilon)}$. We will show that this smallest $\varepsilon$ equals

$$\delta := \max_{\ell \in [n]} \left\{ \|\mathbf{p}\|_{(\ell)} - \|\mathbf{q}\|_{(\ell)} \right\}. \tag{4.100}$$

We first show that $\mathbf{q} \succ \underline{\mathbf{p}}^{(\delta)}$. Let $k, m \in [n-1]$ be the integers satisfying (4.82), and $a$ and $b$ be the numbers defined in (4.81), but with $\delta$ replacing $\varepsilon$. From Exercise 4.8 we have $\mathbf{q} \succ \underline{\mathbf{p}}^{(\delta)}$ if and only if

$$\|\mathbf{q}\|_{(\ell)} \geqslant \left\| \underline{\mathbf{p}}^{(\delta)} \right\|_{(\ell)} \qquad \forall\, \ell \in \{k, k+1, \ldots, m\}. \tag{4.101}$$

Now, for $k \leqslant \ell \leqslant m$,

$$\|\mathbf{q}\|_{(\ell)} - \left\| \underline{\mathbf{p}}^{(\delta)} \right\|_{(\ell)} = \|\mathbf{q}\|_{(k)} - ka + \sum_{x=k+1}^{\ell} (q_x - p_x)$$

$$\boxed{ka = \|\mathbf{p}\|_{(k)} - \delta} \longrightarrow \quad = \delta + \|\mathbf{q}\|_{(k)} - \|\mathbf{p}\|_{(k)} + \sum_{x=k+1}^{\ell} (q_x - p_x) \tag{4.102}$$

$$= \delta + \|\mathbf{q}\|_{(\ell)} - \|\mathbf{p}\|_{(\ell)}.$$

Hence, $\mathbf{q} \succ \underline{\mathbf{p}}^{(\delta)}$ if and only if for all $\ell \in \{k, \ldots, m\}$ we have $\delta \geqslant \|\mathbf{p}\|_{(\ell)} - \|\mathbf{q}\|_{(\ell)}$. From its definition, $\delta \geqslant \|\mathbf{p}\|_{(\ell)} - \|\mathbf{q}\|_{(\ell)}$ for all $\ell \in [n]$. Hence, $\mathbf{q} \succ \underline{\mathbf{p}}^{(\delta)}$.

To prove the optimality of $\delta$, we use the fact that $\underline{\mathbf{p}}^{(\delta)}$ is $\delta$-close to $\mathbf{p}$ so that from (2.83) we get for any $\ell \in [n]$

$$\delta \geqslant \|\mathbf{p}\|_{(\ell)} - \left\| \underline{\mathbf{p}}^{(\delta)} \right\|_{(\ell)}$$

$$\mathbf{q} \succ \underline{\mathbf{p}}^{(\delta)} \rightarrow \quad \geqslant \|\mathbf{p}\|_{(\ell)} - \|\mathbf{q}\|_{(\ell)} . \tag{4.103}$$

Since the above inequality holds for all $\ell \in [n]$ we conclude that $\delta$ is optimal. This concludes the proof. $\blacksquare$

## Two Key Functions

We end the section by introducing two functions frequently used in majorization theory, enabling the study of approximate majorization more effectively. For a given $\mathbf{p} \in \mathrm{Prob}(n)$ we define $f_{\mathbf{p}} : [0,1] \to [0,1]$ and $g_{\mathbf{p}} : [0,1] \to [0, n-1]$ via

$$f_{\mathbf{p}}(t) := \sum_{x \in [n]} (p_x - t)_+ \quad \text{and} \quad g_{\mathbf{p}}(t) := \sum_{x \in [n]} (t - p_x)_+. \tag{4.104}$$

Recall that $(s - t)_+ = \frac{1}{2}(|s - t| + s - t)$, and since the absolute value is a continuous function, these functions are continuous (although not differentiable). Observe that $f_{\mathbf{p}}(t) = 0$ for $t \geqslant p_1$, whereas $g_{\mathbf{p}}(t) = 0$ for $t \in [0, p_n]$ and $g_{\mathbf{p}}(t) = nt - 1$ for $t \geqslant p_1$. The function $f_{\mathbf{p}}(t)$ is nonincreasing in $t$, while $g_{\mathbf{p}}(t)$ is nondecreasing in $t$. See Figure 4.1 for examples of $f_{\mathbf{p}}(t)$ and $g_{\mathbf{p}}(t)$.

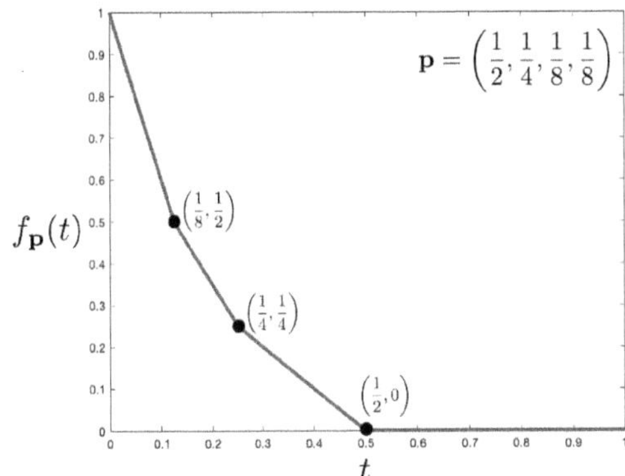


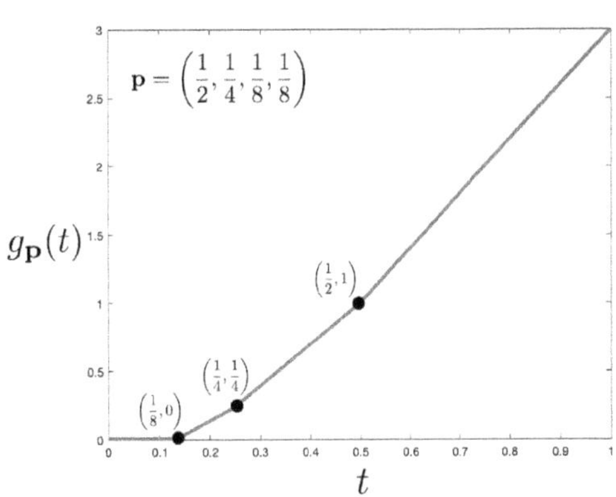


**Figure 4.1**    The functions $f_{\mathbf{p}}(t)$ and $g_{\mathbf{p}}(t)$. The dots indicate the points at which the slop of the functions changes.

**Exercise 4.24.** *In this exercise we use the same notations used in this subsection with a fix* $\mathbf{p} \in \mathrm{Prob}^{\downarrow}(n)$.

1. *Show that for any* $z \in [n]$

$$f_{\mathbf{p}}(p_z) = r_z \quad and \quad g_{\mathbf{p}}(p_z) = s_z. \tag{4.105}$$

2. *Use Part 1 to provide an alternative proof that* $\mathbf{r} = \mathbf{r}^{\uparrow}$ *and* $\mathbf{s} = \mathbf{s}^{\downarrow}$.
3. *Show that* $f_{\mathbf{p}}(a) = g_{\mathbf{p}}(b) = \varepsilon$.
4. *Show that* $f_{\mathbf{p}}(1/n) = g_{\mathbf{p}}(1/n) = \|\mathbf{p}\|_{(\ell)} - \frac{\ell}{n}$, *where* $\ell \in [n]$ *is the largest integer satisfying* $p_\ell \geqslant \frac{1}{n}$.
5. *Show that*

$$\|\mathbf{p}\|_{(\ell)} - \frac{\ell}{n} \leqslant 1 - np_n \quad and \quad \|\mathbf{p}\|_{(\ell)} - \frac{\ell}{n} \leqslant np_1 - 1. \tag{4.106}$$

The functions $f_{\mathbf{p}}(t)$ and $g_{\mathbf{p}}(t)$ are one-to-one when restricted to the domains $[0, p_1]$ and $[p_n, 1]$, respectively (and therefore in these domains they are monotonically decreasing and monotonically increasing, respectively). Therefore, the functions $f_{\mathbf{p}}: [0, p_1] \to [0, 1]$ and $g_{\mathbf{p}}: [p_n, 1] \to [0, n - 1]$ have inverse functions. The inverse function $f_{\mathbf{p}}^{-1}: [0, 1] \to [0, p_1]: r \mapsto f_{\mathbf{p}}^{-1}(r)$ is given by (see Exercise 4.25)

$$f_{\mathbf{p}}^{-1}(r) = \begin{cases} \frac{\|\mathbf{p}\|_{(k)} - r}{k} & \text{if } r > 0 \\ 1 & \text{if } r = 0 \end{cases}, \tag{4.107}$$

where $k \in [n]$ is the integer satisfying $r \in (r_k, r_{k+1}]$. The inverse function $g_{\mathbf{p}}^{-1}: [0, n - 1] \to [p_n, 1]: s \mapsto g_{\mathbf{p}}^{-1}(s)$ is given by (see Exercise 4.25)

$$g_{\mathbf{p}}^{-1}(s) = \begin{cases} \frac{1 + s - \|\mathbf{p}\|_{(m)}}{n - m} & \text{if } s > 0 \\ 0 & \text{if } s = 0 \end{cases}, \tag{4.108}$$

where $m \in \{0, 1, \ldots, n - 1\}$ is the integer satisfying $s \in (s_{m+1}, s_m]$.

**Exercise 4.25.** *Consider the functions* $f_{\mathbf{p}} \colon [0, p_1] \to [0, 1]$ *and* $g_{\mathbf{p}} \colon [p_n, 1] \to [0, n-1]$ *as defined above and let* $f_{\mathbf{p}}^{-1}$ *and* $g_{\mathbf{p}}^{-1}$ *be as defined in* (4.107) *and* (4.108).

*1. Show that for any* $t \in [0, p_1]$ *and* $r \in [0, 1]$

$$f_{\mathbf{p}}^{-1}\left(f_{\mathbf{p}}(t)\right) = t \quad \text{and} \quad f_{\mathbf{p}}\left(f_{\mathbf{p}}^{-1}(r)\right) = r. \tag{4.109}$$

*2. Show that for any* $t \in [p_n, 1]$ *and* $s \in [0, n-1]$

$$g_{\mathbf{p}}^{-1}\left(g_{\mathbf{p}}(t)\right) = t \quad \text{and} \quad g_{\mathbf{p}}\left(g_{\mathbf{p}}^{-1}(s)\right) = s. \tag{4.110}$$

## 4.3 Relative Majorization

In the first section of this chapter we compared between a **p**-dice and a **q**-dice via the degree of uncertainty they each posses. In this section, we study the degree of distinguishability between the two dices. Unlike arbitrary vectors, objects like probability vectors (as well as quantum states, quantum channels, etc.) contain information about physical systems and therefore their distinguishability is typically quantified with functions that are sensitive to this information. For example, suppose a player receives a biased dice whose probability distribution is either $\mathbf{p} := \{p_x\}_{x \in [n]}$ or $\mathbf{q} := \{q_x\}_{x \in [n]}$. The player can estimate which of the two distributions corresponds to the dice by rolling the dice many times. The intuition is that if **p** and **q** are very distinguishable, it would be easier (i.e. quicker) to determine which one of them corresponds to the dice.

One of the key observations in any distinguishability task is that by sending the information source (i.e. the outcomes of the dice) through a communication channel, the player cannot increase his or her ability to distinguish between the two distributions **p** and **q**. This means that if $E$ is the column stochastic matrix that corresponds to a classical communication channel, then the resulting distributions $E\mathbf{p}$ and $E\mathbf{q}$ are less distinguishable than **p** and **q**.

> **Relative Majorization**
>
> **Definition 4.3.** Let $\mathbf{p}, \mathbf{q} \in \mathrm{Prob}(n)$ and $\mathbf{p}', \mathbf{q}' \in \mathrm{Prob}(m)$ be two pairs of probability distributions. We say that $(\mathbf{p}, \mathbf{q})$ relatively majorize $(\mathbf{p}', \mathbf{q}')$ and write
>
> $$(\mathbf{p}, \mathbf{q}) \succ (\mathbf{p}', \mathbf{q}') \quad \Longleftrightarrow \quad \mathbf{p}' = E\mathbf{p} \text{ and } \mathbf{q}' = E\mathbf{q}, \tag{4.111}$$
>
> where $E$ is an $m \times n$ column stochastic matrix. Furthermore, if $(\mathbf{p}, \mathbf{q}) \succ (\mathbf{p}', \mathbf{q}') \succ (\mathbf{p}, \mathbf{q})$ we then write
>
> $$(\mathbf{p}, \mathbf{q}) \sim (\mathbf{p}', \mathbf{q}'). \tag{4.112}$$

Relative majorization is a preorder. The property $(\mathbf{p}, \mathbf{q}) \succ (\mathbf{p}, \mathbf{q})$ (i.e. reflexivity) follows by taking $E$ in Definition (4.3) to be the identity matrix. The transitivity of relative majorization follows from the fact that the product of two column stochastic matrices is also a column stochastic matrix.

**Exercise 4.26.** *Consider the equivalence relation $\sim$ as defined in (4.112).*

*1. Let $\mathbf{p}, \mathbf{q} \in \mathrm{Prob}(n)$ be two n-dimensional probability vectors and let $P$ be an $n \times n$ permutation matrix. Show that*

$$(P\mathbf{p}, P\mathbf{q}) \sim (\mathbf{p}, \mathbf{q}) \quad and \quad (\mathbf{p} \oplus 0, \mathbf{q} \oplus 0) \sim (\mathbf{p}, \mathbf{q}). \tag{4.113}$$

*2. Show that for any $m, n \in \mathbb{N}$ and any probability vectors, $\mathbf{p} \in \mathrm{Prob}(n)$ and $\mathbf{q} \in \mathrm{Prob}(m)$,*

$$(\mathbf{p}, \mathbf{p}) \sim (\mathbf{q}, \mathbf{q}). \tag{4.114}$$

*3. Let $\mathbf{p}_1, \mathbf{p}_2 \in \mathrm{Prob}(n)$ and $\mathbf{q}_1, \mathbf{q}_2 \in \mathrm{Prob}(m)$. Show that if $\mathbf{p}_1 \cdot \mathbf{p}_2 = \mathbf{q}_1 \cdot \mathbf{q}_2 = 0$, then*

$$(\mathbf{p}_1, \mathbf{p}_2) \sim (\mathbf{q}_1, \mathbf{q}_2). \tag{4.115}$$

From the second relation in (4.113) it follows that without loss of generality we can always assume that there is no $x \in [n]$ such that $p_x = q_x = 0$ (since any $x$-component with $p_x = q_x = 0$ can be removed from the vectors $\mathbf{p}$ and $\mathbf{q}$ without changing the equivalency).

> **Standard Form**
>
> **Definition 4.4.** A pair $(\mathbf{p}, \mathbf{q})$ of probability vectors in $\mathrm{Prob}(n)$ is said to be given in a *standard form* if there is no $x \in [n]$ such that $p_x = q_x = 0$, and the components of the vectors $\mathbf{p}$ and $\mathbf{q}$ are arranged such that
>
> $$\frac{p_1}{q_1} \geqslant \frac{p_2}{q_2} \geqslant \cdots \geqslant \frac{p_n}{q_n}, \tag{4.116}$$
>
> where we used the convention $p_x/q_x = \infty$ for $x \in [n]$ with $p_x > 0$ and $q_x = 0$.

Observe that since $(P\mathbf{p}, P\mathbf{q}) \sim (\mathbf{p}, \mathbf{q})$ for every permutation matrix $P$, any pair of vectors is equivalent (under relative majorization) to its standard form. The choice of the order given in (4.116) will be clear later on when we characterize relative majorization with testing regions.

**Exercise 4.27.** *Let $\{\mathbf{e}_1, \mathbf{e}_2\}$ be the standard basis of $\mathbb{R}^2$. Express the pair $(\mathbf{e}_1, \mathbf{e}_2)$ in the standard form.*

**Exercise 4.28.** *Show that if $\mathbf{p}, \mathbf{q} \in \mathrm{Prob}(n)$ and $\mathbf{q}$ has the form $\mathbf{q} = (q_1, \ldots, q_r, 0, \ldots, 0)$ for some $r < n$, then*

$$(\mathbf{p}, \mathbf{q}) \sim (\mathbf{p}', \mathbf{q}) \tag{4.117}$$

*for any $\mathbf{p}' \in \mathrm{Prob}(n)$ whose first r components equal the first r components of $\mathbf{p}$.*

## 4.3.1 Lower and Upper Bounds

We saw earlier that the maximal and minimal elements of $\mathrm{Prob}(n)$ under majorization are $\mathbf{e}_1$ and $\mathbf{u}^{(n)}$, respectively. In the following exercise you find the maximal and minimal elements of $\mathrm{Prob}(n) \times \mathrm{Prob}(n)$ under relative majorization.

**Exercise 4.29.** *Let* $\mathbf{p}, \mathbf{q} \in \mathrm{Prob}(n)$ *be two n-dimensional probability vectors, and let* $\mathbf{e}_1, \mathbf{e}_2 \in \mathrm{Prob}(m)$ *be two m-dimensional probability vectors with orthogonal support; that is,* $\mathbf{e}_1 \cdot \mathbf{e}_2 = 0$. *Show that for any* $k \in \mathbb{N}$ *and any* $\mathbf{r} \in \mathrm{Prob}(k)$,

$$(\mathbf{e}_1, \mathbf{e}_2) \succ (\mathbf{p}, \mathbf{q}) \succ (\mathbf{r}, \mathbf{r}). \tag{4.118}$$

In the following theorem we bound any pair of probability vectors by pairs of two dimensional vectors. For any $n \in \mathbb{N}$ and $\mathbf{p}, \mathbf{q} \in \mathrm{Prob}(n)$, we denote by

$$\lambda_{\min} := \sum_{x \in \mathrm{supp}\,(\mathbf{p})} q_x \quad \text{and} \quad \lambda_{\max} := \min_{x \in \mathrm{supp}\,(\mathbf{p})} \frac{q_x}{p_x}, \tag{4.119}$$

where $\mathrm{supp}\,(\mathbf{p}) := \{x \in [n] : p_x \neq 0\}$. Later in the book we will see that $\lambda_{\max}$ and $\lambda_{\min}$ are related to the min and max relative entropies. In the following theorem, we use the notations $\mathbf{e}_1 := (1, 0)^T$ and $\mathbf{e}_2 := (0, 1)^T$, and in addition denote by

$$\mathbf{v}_{\max} := \lambda_{\max}\mathbf{e}_1 + (1 - \lambda_{\max})\mathbf{e}_2 \quad \text{and} \quad \mathbf{v}_{\min} := \lambda_{\min}\mathbf{e}_1 + (1 - \lambda_{\min})\mathbf{e}_2. \tag{4.120}$$

**Theorem 4.7.** Using the same notations as above we have for any $\mathbf{p}, \mathbf{q} \in \mathrm{Prob}(n)$

$$(\mathbf{e}_1, \mathbf{v}_{\max}) \succ (\mathbf{p}, \mathbf{q}) \succ (\mathbf{e}_1, \mathbf{v}_{\min}). \tag{4.121}$$

*Remark.* Note that the bounds are not symmetric, meaning that if we swap $\mathbf{p}$ with $\mathbf{q}$, the vector $\mathbf{e}_1 := (1, 0)^T$ will appear second in the bounding pairs, and $\lambda_{\min}$ and $\lambda_{\max}$ will also change. Moreover, if $\mathbf{p} > 0$ (i.e. all the components of $\mathbf{p}$ are strictly positive), then $\lambda_{\min} = 1$, and consequently the lower bound becomes trivial.

**Proof** We first prove the upper bound. For this purpose, we need to find an $n \times 2$ column stochastic channel $E \in \mathrm{STOC}(n, 2)$ with the property that

$$E\mathbf{e}_1 = \mathbf{p} \quad \text{and} \quad E\mathbf{v}_{\max} = \mathbf{q}, \tag{4.122}$$

where $\mathbf{e}_1 := (1, 0)^T$ and $\mathbf{e}_2 := (0, 1)^T$. Observe that the first condition above implies that the first column of $E$ must be equal to $\mathbf{p}$. Combining this with the definition of $\mathbf{v}_{\max}$ and with the second condition in (4.122) we get

$$E\mathbf{v}_{\max} = \lambda_{\max}\mathbf{p} + (1 - \lambda_{\max})E\mathbf{e}_2 = \mathbf{q} \quad \Rightarrow \quad E\mathbf{e}_2 = \frac{\mathbf{q} - \lambda_{\max}\mathbf{p}}{1 - \lambda_{\max}}. \tag{4.123}$$

By definition, $\lambda_{\max} \in [0, 1]$ and it has the property that $\mathbf{q} \geqslant \lambda_{\max}\mathbf{p}$ (i.e. $q_x \geqslant \lambda_{\max}p_x$ for each $x \in [n]$). Therefore, the right-hand side of the equation above is a probability vector. To summarize, the $n \times 2$ column stochastic matrix $E$, whose first column is $\mathbf{p}$, and its second column is $\frac{\mathbf{q}-\lambda_{\max}\mathbf{p}}{1-\lambda_{\max}}$ satisfies (4.122) so that by definition the upper bound in (4.121) holds.

We now prove the lower bound. By definition, it is sufficient to show that there exists a channel $E \in \mathrm{STOC}(2,n)$ such that

$$E\mathbf{p}=\mathbf{e}_1 \quad\text{and}\quad E\mathbf{q}=\mathbf{v}_{\min}=\lambda_{\min}\mathbf{e}_1 + (1-\lambda_{\min})\mathbf{e}_2. \tag{4.124}$$

Since $E$ must be a column stochastic matrix with two rows, it follows that if its first row is $\mathbf{t}^T$, then its second row is $(\mathbf{1}_n - \mathbf{t})^T$, where $\mathbf{1}_n^T = (1,\ldots,1)$. Hence, $E$ satisfies the above conditions if and only if

$$\mathbf{t}\cdot\mathbf{p}=1 \quad\text{and}\quad \mathbf{t}\cdot\mathbf{q}=\lambda_{\min}. \tag{4.125}$$

Note also that we must have $0 \leqslant \mathbf{t} \leqslant \mathbf{1}_n$ (element-wise) since $E$ is column stochastic. We therefore choose $\mathbf{t}=(t_1,\ldots,t_n)^T$ with

$$t_x = \begin{cases} 1 & \text{if } p_x > 0 \\ 0 & \text{if } p_x = 0 \end{cases} \quad \forall\, x \in [n]. \tag{4.126}$$

It is simple to check that this $\mathbf{t}$ satisfies (4.125). This completes the proof.   ■

Note that when $\mathrm{supp}\,(\mathbf{p}) = \mathrm{supp}\,(\mathbf{q})$, the value of $\lambda_{\max}$, as defined in (4.119), is constrained to the range $0 < \lambda_{\max} < 1$, ensuring that $\mathbf{v}_{\max} > 0$. In this case we can improve the upper bound $(\mathbf{e}_1, \mathbf{v}_{\max})$. Indeed, let $0 < s < t \leqslant 1$ be such that

$$\frac{1-t}{1-s}\mathbf{p} \leqslant \mathbf{q} \leqslant \frac{t}{s}\mathbf{p}. \tag{4.127}$$

Note that such $s$ and $t$ exist since we assume that $\mathbf{p}$ and $\mathbf{q}$ have the same support, and we can take $s$ close enough to 0 and $t$ close enough to 1. Now, define a stochastic evolution matrix $E = [\mathbf{v}_1\ \mathbf{v}_2] \in \mathrm{STOC}(m,2)$, with the two columns $\mathbf{v}_1,\mathbf{v}_2 \in \mathrm{Prob}(m)$ given by

$$\mathbf{v}_1 := \frac{(1-s)\mathbf{q}-(1-t)\mathbf{p}}{t-s} \quad\text{and}\quad \mathbf{v}_2 := \frac{t\mathbf{p}-s\mathbf{q}}{t-s}. \tag{4.128}$$

Note that the conditions in (4.127) imply that $E$ is indeed a column stochastic matrix since $t\mathbf{p} - s\mathbf{q} \geqslant 0$ (entrywise) and $(1-s)\mathbf{q} - (1-t)\mathbf{p} \geqslant 0$. Moreover, denoting by $\mathbf{s} := (s, 1-s)^T$ and $\mathbf{t} := (t, 1-t)^T$ we have by direct calculation (Exercise 4.30)

$$E\mathbf{s}=\mathbf{p} \quad\text{and}\quad E\mathbf{t}=\mathbf{q}. \tag{4.129}$$

We therefore conclude that for every $\mathbf{p},\mathbf{q} \in \mathrm{Prob}(n)$ with equal support, that is, $\mathrm{supp}\,(\mathbf{p}) = \mathrm{supp}\,(\mathbf{q})$, there exist $\mathbf{s},\mathbf{t} \in \mathrm{Prob}_{>0}(2)$ satisfying the relation:

$$(\mathbf{s},\mathbf{t}) \succ (\mathbf{p},\mathbf{q}). \tag{4.130}$$

Observe that the relation above hold as long as $s,t \in [0,1]$ satisfies $s < t$ and

$$(1-s)\mathbf{q} \geqslant (1-t)\mathbf{p} \quad and \quad t\mathbf{p} \geqslant s\mathbf{q}. \tag{4.131}$$

**Exercise 4.30.** *Verify by direct calculation the relations in* (4.129).

**Exercise 4.31.** *Show that by taking $t = 1 - \lambda_{\max}$ and $s = 0$ the relation $(\mathbf{s},\mathbf{t}) \succ (\mathbf{p},\mathbf{q})$ is equivalent to the upper bound in* (4.121).

## 4.3.2 Majorization versus Relative Majorization

Majorization and relative majorization are interrelated concepts, rather than independent ones. Specifically, they become equivalent when one of the probability vectors in each pair has a uniform distribution. In this section, we will explore the deep connection between these two concepts. This exploration will allow us, in the following section, to understand and provide a geometrical characterization of relative majorization. Moreover, the insights gained will be used later in the book to prove and explore other related findings.

Relative majorization generalizes majorization between vectors. Specifically, for any two probability vectors $\mathbf{p}, \mathbf{q} \in \mathrm{Prob}(n)$

$$(\mathbf{p}, \mathbf{u}) \succ (\mathbf{q}, \mathbf{u}) \quad \Longleftrightarrow \quad \mathbf{p} \succ \mathbf{q}, \tag{4.132}$$

where $\mathbf{u} := \left(\frac{1}{n}, \ldots, \frac{1}{n}\right)^T$ is the uniform distribution. In the following exercise we will prove this assertion using Theorem 4.1.

**Exercise 4.32.** *Use the equivalence between the first two conditions in Theorem 4.1 to prove* (4.132).

## The Special Case of Vectors with Rational Components

If one of the vectors has positive rational components, the relationship between majorization and relative majorization becomes even closer than what we have seen so far. Consider a pair of vectors $(\mathbf{p}, \mathbf{q})$, where $\mathbf{p} \in \mathrm{Prob}(n)$ and

$$\mathbf{q} := \left(\frac{k_1}{k}, \ldots, \frac{k_n}{k}\right)^T \qquad k_1, \ldots, k_n \in \mathbb{N}, \tag{4.133}$$

and $k := k_1 + \cdots + k_n$. Define the vector $\mathbf{r} \in \mathrm{Prob}(k)$ via

$$\mathbf{r} := \bigoplus_{x \in [n]} p_x \mathbf{u}^{(k_x)} = \left( \underbrace{\frac{p_1}{k_1}, \ldots, \frac{p_1}{k_1}}_{k_1\text{-times}}, \underbrace{\frac{p_2}{k_2}, \ldots, \frac{p_2}{k_2}}_{k_2\text{-times}}, \ldots, \underbrace{\frac{p_n}{k_n}, \ldots, \frac{p_n}{k_n}}_{k_n\text{-times}} \right)^T, \tag{4.134}$$

where $\mathbf{u}^{(k_x)}$ is the uniform probability vector in $\mathrm{Prob}(k_x)$. We then have the following theorem.

**Theorem 4.8.** Let $\mathbf{p}, \mathbf{q} \in \mathrm{Prob}(n)$ and $\mathbf{r} \in \mathrm{Prob}(k)$ be as above with $\mathbf{q}$ having positive rational components. Then,

$$(\mathbf{p}, \mathbf{q}) \sim (\mathbf{r}, \mathbf{u}^{(k)}). \tag{4.135}$$

*Remark.* Observe that any vector $0 < \mathbf{q} \in \mathrm{Prob}(n) \cap \mathbb{Q}^n$ can be expressed as in (4.133) for sufficiently large $k$. This $k$ is a common denominator for all the components of $\mathbf{q}$.

**Proof** We first show that $(\mathbf{p}, \mathbf{q}) \succ (\mathbf{r}, \mathbf{u}^{(k)})$. For any $x \in [n]$, let $E^{(x)}$ be the $k_x \times n$ matrix whose $x$-column is $\mathbf{u}^{(k_x)}$ and all the remaining $n - 1$ columns are zero. Moreover, let $E$ be the $k \times n$ matrix given by

$$E := \begin{bmatrix} E^{(1)} \\ E^{(2)} \\ \vdots \\ E^{(n)} \end{bmatrix}. \tag{4.136}$$

By definition, $E^{(x)}\mathbf{p} = p_x \mathbf{u}^{(k_x)}$ so that $E\mathbf{p} = \mathbf{r}$. Similarly, $E^{(x)}\mathbf{q} = \frac{k_x}{k}\mathbf{u}^{(k_x)} = \frac{1}{k}\mathbf{1}_{k_x}$ so that $E\mathbf{q} = \mathbf{u}^{(k)}$. Therefore, since $E$ is column stochastic we get that $(\mathbf{p}, \mathbf{q}) \succ (\mathbf{r}, \mathbf{u}^{(k)})$.

For the converse, let $F^{(x)}$ be the $n \times k_x$ matrix whose $x$-row is $[1, \ldots, 1]$ and all the remaining $n - 1$ rows are zero. Moreover, denote by $F$ the $n \times k$ column stochastic matrix given by

$$F := \begin{bmatrix} F^{(1)} & F^{(2)} & \cdots & F^{(n)} \end{bmatrix}. \tag{4.137}$$

Observe that $FE = I_n$. Therefore, $F\mathbf{r} = FE\mathbf{p} = \mathbf{p}$ and similarly $F\mathbf{u}^{(k)} = FE\mathbf{q} = \mathbf{q}$. In other words, $(\mathbf{r}, \mathbf{u}^{(k)}) \succ (\mathbf{p}, \mathbf{q})$. But since we have already proved that $(\mathbf{p}, \mathbf{q}) \succ (\mathbf{r}, \mathbf{u}^{(k)})$, we conclude that $(\mathbf{p}, \mathbf{q}) \sim (\mathbf{r}, \mathbf{u}^{(k)})$. $\blacksquare$

**Exercise 4.33.** *Verify all steps in the proof above; that is, show that $E$ and $F$ are indeed column stochastic and $FE = I_n$.*

We have seen that relative majorization reduces to majorization when one of the vectors is the uniform vector (see (4.132)). In the following excerise, you show that the remarkable equivalence between $(\mathbf{p}, \mathbf{q})$ and $(\mathbf{r}, \mathbf{u}^{(k)})$ implies that relative majorization reduces to majorization if one of the vectors has positive rational components.

**Exercise 4.34.** *Let $\mathbf{p} \in \mathrm{Prob}(n)$, $0 < \mathbf{q} \in \mathrm{Prob}(n) \cap \mathbb{Q}^n$, $\mathbf{p}' \in \mathrm{Prob}(m)$, and $0 < \mathbf{q}' \in \mathrm{Prob}(m) \cap \mathbb{Q}^m$. Show that there exists $k \in \mathbb{N}$ and vectors $\mathbf{r}, \mathbf{r}' \in \mathrm{Prob}(k)$ such that*

$$(\mathbf{p}, \mathbf{q}) \succ (\mathbf{p}', \mathbf{q}') \iff \mathbf{r} \succ \mathbf{r}'. \tag{4.138}$$

**Exercise 4.35.** *Let $\mathbf{p} \in \mathrm{Prob}(n)$ and $0 < \mathbf{q} \in \mathrm{Prob}(n) \cap \mathbb{Q}^n$. Show that the pair $(\mathbf{p}, \mathbf{q})$ is given in the standard form (i.e. satisfies (4.116)) if and only if the vector $\mathbf{r}$ as defined in (4.134) satisfies $\mathbf{r} = \mathbf{r}^{\downarrow}$.*

### 4.3.3 Testing Regions

Relative majorization can be characterized geometrically in terms of testing regions (also known as zonotopes). Testing regions are regions in $\mathbb{R}^2$ that have several applications in statistics, particularly the area of hypothesis testing. The testing region associated with a pair of probability vectors $\mathbf{p}, \mathbf{q} \in \mathrm{Prob}(n)$ is a region in $\mathbb{R}^2$ defined by

$$\mathfrak{T}(\mathbf{p}, \mathbf{q}) := \left\{ (\mathbf{p} \cdot \mathbf{t}, \mathbf{q} \cdot \mathbf{t}) : \mathbf{t} \in [0, 1]^n \right\}, \tag{4.139}$$

where **t** (also known as a probabilistic hypothesis test) is an $n$-dimensional vector with entries between 0 and 1. Note that for any pair of probability vectors the points $(0,0)$ and $(1,1)$ belong to its testing region. Explicitly, $(0,0)$ is obtained by taking **t** to be the zero vector, and $(1,1)$ is obtained by taking $\mathbf{t}=(1,\ldots,1)^{T}$. An example of a testing region is plotted in Figure 4.2.

**Exercise 4.36.** *Show that the testing region is convex, and it has the symmetry that if* $(x,y)\in\mathfrak{T}(\mathbf{p},\mathbf{q})$, *then also* $(1-x,1-y)\in\mathfrak{T}(\mathbf{p},\mathbf{q})$. *Hint: For the latter property, consider the vector* $\mathbf{t}'=(1,\ldots,1)-\mathbf{t}$.

The testing region is bounded by two curves known as lower and upper Lorenz curves. Due to the symmetry that $(1-x,1-y)\in\mathfrak{T}(\mathbf{p},\mathbf{q})$ for any $(x,y)\in\mathfrak{T}(\mathbf{p},\mathbf{q})$, the upper Lorenz curve can be obtained from the lower Lorenz curve through a 180-degree rotation centered at the midpoint $(1/2,1/2)$. Consequently, either the lower or the upper Lorenz curve is sufficient to uniquely define the entire testing region.

Since the testing region is convex it can be characterized by its extreme points. It is tempting to draw a parallel with the convex set $[0,1]^n$, which possesses $2^n$ extreme points encapsulated within the set $\{0,1\}^n$. However, this analogy can be misleading in the context of our testing region. In reality, only $2n$ points are necessary to fully characterize the testing region. We will focus on the extreme points characterizing the lower Lorenz curve. Note that the lower Lorenz curve is a convex curve (while the upper Lorenz curve is concave).

**Exercise 4.37.** *Let* $\mathbf{p},\mathbf{q}\in\mathrm{Prob}(n)$ *and let* $P$ *be an* $n\times n$ *permutation matrix.*

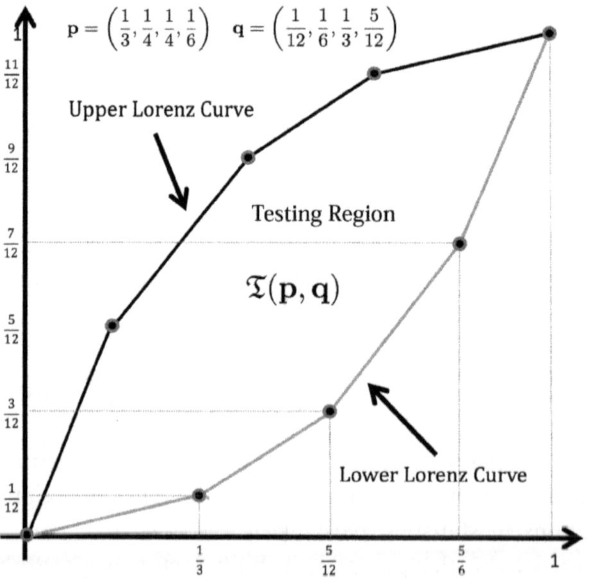

1. *Show that*

$$\mathfrak{T}(P\mathbf{p}, P\mathbf{q}) = \mathfrak{T}(\mathbf{p}, \mathbf{q}). \tag{4.140}$$

2. *Show that*

$$\mathfrak{T}(\mathbf{p} \oplus 0, \mathbf{q} \oplus 0) = \mathfrak{T}(\mathbf{p}, \mathbf{q}). \tag{4.141}$$

Since for any permutation matrix $P$, $(\mathbf{p}, \mathbf{q})$ and $(P\mathbf{p}, P\mathbf{q})$ have the same testing region, we can assume without loss of generality that $(\mathbf{p}, \mathbf{q})$ are always given in the standard form. This is also justified by the fact that under relative majorization we have $(\mathbf{p}, \mathbf{q}) \sim (P\mathbf{p}, P\mathbf{q})$ for any permutation matrix $P$.

---

**Theorem 4.9.** Given $\mathbf{p}, \mathbf{q} \in \mathrm{Prob}(n)$ in standard form, the extreme points on the lower boundary of the testing region $\mathfrak{T}(\mathbf{p}, \mathbf{q})$ (specifically, on its lower Lorenz curve) are the $n + 1$ vertices:

$$(a_k, b_k) := \left( \sum_{x \in [k]} p_x, \ \sum_{x \in [k]} q_x \right) \qquad k = 0, 1, \ldots, n, \tag{4.142}$$

where $a_0 := 0$ and $b_0 := 0$.

---

*Remark.* In general, the sum $\sum_{x \in [k]} p_x$ does not equal to $\|\mathbf{p}\|_{(k)}$ since the components of $\mathbf{p}$ are not necessarily arranged in a nonincreasing order. Instead, the components of $\mathbf{p}$ and $\mathbf{q}$ are arranged such that the order in (4.116) holds.

**Proof** Let $f : [0, 1] \to [0, 1]$ be the function whose graph is the lower Lorenz curve of $(\mathbf{p}, \mathbf{q})$. Then, by definition, for every $a \in [0, 1]$, $(a, f(a))$ is the lowest point in $\mathfrak{T}(\mathbf{p}, \mathbf{q})$ whose $x$-coordinate is $a$. Therefore, for any $r \in [0, 1]$ we can express $f(r)$ as

$$f(r) = \min \left\{ \mathbf{q} \cdot \mathbf{t} \ : \ \mathbf{t} \in [0, 1]^n, \ \mathbf{p} \cdot \mathbf{t} = r \right\} \tag{4.143}$$

Our objective is to demonstrate that the function $f(r)$ defines the segment connecting two adjacent vertices. Specifically, consider a fixed $k \in \{0, 1, \ldots, n\}$. Our aim is to establish that for any $r$ in the interval $[a_k, a_{k+1})$, the function $f(r)$ corresponds to the line segment joining the points $(a_k, b_k)$ and $(a_{k+1}, b_{k+1})$. Mathematically, this means that for all $r \in [a_k, a_{k+1})$, we have

$$f(r) = s_{k+1}(r - a_k) + b_k, \tag{4.144}$$

where $s_x := q_x / p_x$ for all $x \in [n]$, adhering to the convention that $s_x := \infty$ if $p_x = 0$ and $q_x > 0$ (recall that we assume that there is no $x \in [n]$ such that both $p_x$ and $q_x$ are equal to zero). Successfully proving this relationship implies that the set of points $\{(a_k, b_k)\}_{k=0}^{n}$ are indeed the extreme points on the lower Lorenz curve of the testing region $\mathfrak{T}(\mathbf{p}, \mathbf{q})$.

To prove (4.144), observe that the optimization problem in (4.143) is a linear program. In Exercise 4.38, you will apply methods discussed in Sec. A.9 of online version, specifically the dual problem framework, to express $f(r)$ as

$$f(r) = \max \left\{ rs - \mathbf{v} \cdot \mathbf{1}_n : \mathbf{v} \in \mathbb{R}_+^n, \ s\mathbf{p} - \mathbf{q} \leqslant \mathbf{v}, s \in \mathbb{R}_+ \right\}, \tag{4.145}$$

where $\mathbf{1}_n := (1, \ldots, 1)^T$ and the inequality is entry-wise. The maximization in (4.145) can be simplified since the vector $\mathbf{v}$ with the smallest nonnegative components that satisfies the constraint $\mathbf{v} \geqslant s\mathbf{p} - \mathbf{q}$ is given by $\mathbf{v} = (s\mathbf{p} - \mathbf{q})_+$ (the components of $(s\mathbf{p} - \mathbf{q})_+$ are $\{(sp_x - q_x)_+\}_{x \in [n]}$). Hence,

$$f(r) = \max_{s \geqslant 0} \left\{ sr - (s\mathbf{p} - \mathbf{q})_+ \cdot \mathbf{1}_n \right\}. \tag{4.146}$$

To simplify further, note that

$$(s\mathbf{p} - \mathbf{q})_+ \cdot \mathbf{1}_n = \sum_{x \in [n]} (sp_x - q_x)_+ = \sum_{x \in [n]} p_x (s - s_x)_+ . \tag{4.147}$$

Now, from (4.116) it follows that $s_1 \leqslant s_2 \leqslant \cdots \leqslant s_n$. Therefore, for any $s \geqslant 0$ there exists $\ell \in \{0, 1, \ldots, n\}$ with the property that $s_\ell \leqslant s < s_{\ell+1}$, where we added the definitions $s_0 := 0$ and $s_{n+1} := \infty$. With this definition of $\ell$ we get

$$(s\mathbf{p} - \mathbf{q})_+ \cdot \mathbf{1}_n = \sum_{x \in [\ell]} p_x (s - s_x) = sa_\ell - b_\ell . \tag{4.148}$$

Therefore, by splitting the maximization in (4.146) into maximization over all $\ell \in \{0, 1, \ldots, n\}$ and all $s \in [s_\ell, s_{\ell+1})$ we get

$$f(r) = \max_{\ell \in \{0,1\ldots,n\}} \sup_{s \in [s_\ell, s_{\ell+1})} \left\{ s(r - a_\ell) + b_\ell \right\}. \tag{4.149}$$

If the optimal $\ell$ above satisfies $\ell \leqslant k$, then $r - a_\ell \geqslant 0$ so that

$$\sup_{s \in [s_\ell, s_{\ell+1})} \left\{ s(r - a_\ell) + b_\ell \right\} = s_{\ell+1}(r - a_\ell) + b_\ell . \tag{4.150}$$

Moreover, among all $\ell \in \{0, 1, \ldots, k\}$ the choice $\ell = k$ yields the greatest value since the right-hand side above is increasing in $\ell$ as long as $\ell \leqslant k$ (see Exercise 4.39); hence,

$$\max_{\ell \in \{0,1,\ldots,k\}} \sup_{s \in [s_\ell, s_{\ell+1})} \left\{ s(r - a_\ell) + b_\ell \right\} = s_{k+1}(r - a_k) + b_k . \tag{4.151}$$

On the other hand, if $\ell > k$, then $r - a_\ell < 0$ so that

$$\sup_{s \in [s_\ell, s_{\ell+1})} \left\{ s(r - a_\ell) + b_\ell \right\} = s_\ell(r - a_\ell) + b_\ell . \tag{4.152}$$

Furthermore, among all $\ell \in \{k+1, \ldots, n\}$ the choice $\ell = k+1$ yields the greatest value since the right-hand side above is decreasing in $\ell$ as long as $\ell \geqslant k+1$ (see Exercise 4.39); that is,

$$\max_{\ell \in \{k+1,\ldots,n\}} \sup_{s \in [s_\ell, s_{\ell+1})} \left\{ s(r - a_\ell) + b_\ell \right\} = s_{k+1}(r - a_{k+1}) + b_{k+1} . \tag{4.153}$$

The right-hand side of (4.151) is in fact equal to the right-hand side of (4.153) (see Exercise 4.39). We therefore conclude that for any $k \in \{0, 1, \ldots, n\}$ and any $r \in [a_k, a_{k+1})$ we have $f(r) = s_{k+1}(r - a_k) + b_k$. This completes the proof. $\blacksquare$

**Exercise 4.38.** *Consider the function $f(r)$ as defined in (4.143).*

1. *Show that the condition* $\mathbf{p} \cdot \mathbf{t} = r$ *in (4.143) can be replaced with* $\mathbf{p} \cdot \mathbf{t} \geqslant r$. *Hint: Observe that any* $\mathbf{t}$ *satisfying* $\mathbf{p} \cdot \mathbf{t} > r$ *can be rescaled to give* $\mathbf{p} \cdot \mathbf{t} = r$ *(and this rescaling can only decrease* $\mathbf{q} \cdot \mathbf{t}$*).*
2. *Prove the equality in (4.145). Hint: First express the minimization in (4.143) (after replacing* $\mathbf{p} \cdot \mathbf{t} = r$ *with* $\mathbf{p} \cdot \mathbf{t} \geqslant r$*) as a conic linear programming of the form (A.52) of online version (with vectors in* $\mathbb{R}^n$ *replacing Hermitian matrices, and the dot product replacing the Hilbert–Schmidt inner product). Then use (A.57) of online version and the strong duality to get (4.145).*

**Exercise 4.39.** *Show that the right-hand side of (4.150) is increasing in* $\ell \in [k]$*, and the right-hand side of (4.152) is decreasing in* $\ell \in \{k + 1, \dots, n\}$*. Moreover, show that the two expressions are the same for* $\ell = k$ *and* $\ell = k + 1$ *(i.e. show that the right-hand side of (4.151) is equal to the right-hand side of (4.153)).*

**Exercise 4.40.** *Let* $\mathbf{p}, \mathbf{q} \in \mathrm{Prob}(n)$ *and* $t \in \mathbb{R}$*. Prove the following equalities:*

$$
\begin{aligned}
1. \quad & (\mathbf{p} - t\mathbf{q})_+ \cdot \mathbf{1}_n = \frac{1}{2}\big(\|\mathbf{p} - t\mathbf{q}\|_1 + 1 - t\big) \\
2. \quad & (t\mathbf{p} - \mathbf{q})_+ \cdot \mathbf{1}_n = \frac{1}{2}\big(\|t\mathbf{p} - \mathbf{q}\|_1 + t - 1\big).
\end{aligned}
\tag{4.154}
$$

*Hint: Use the relation* $(a - b)_+ = \frac{1}{2}|a - b| + \frac{1}{2}(a - b)$.

**Exercise 4.41.** *Compute the vertices of the lower Lorenz curve of the example given in Figure 4.2. If necessary, rearrange the components of* $\mathbf{p}$ *and* $\mathbf{q}$ *so that (4.116) holds.*

**Exercise 4.42.** *For a given* $\mathbf{p}, \mathbf{q}, \subset \mathrm{Prob}(n)$*, find the vertices of* $\mathfrak{T}(\mathbf{p}, \mathbf{q})$ *that are located on the upper Lorenz curve of* $(\mathbf{p}, \mathbf{q})$*.*

**Exercise 4.43.** *Let* $\mathbf{p} \in \mathrm{Prob}(n)$*. Show that the vertices of the lower Lorenz curve of the pair* $(\mathbf{p}, \mathbf{u}^{(n)})$ *are given by*

$$
\left( \|\mathbf{p}\|_{(k)}, \frac{k}{n} \right) \qquad k = 0, 1, \dots, n,
\tag{4.155}
$$

*with the convention that for* $k = 0$*,* $\|\mathbf{p}\|_{(0)} := 0$*.*

Theorem 4.9 has the following interesting corollary.

> **Corollary 4.1.** Let $\mathbf{p}, \mathbf{q} \in \mathrm{Prob}(n)$ and $\mathbf{p}', \mathbf{q}' \in \mathrm{Prob}(n')$. If for all $t \geqslant 1$ we have $\|\mathbf{p} - t\mathbf{q}\|_1 \geqslant \|\mathbf{p}' - t\mathbf{q}'\|_1$, then $\mathfrak{T}(\mathbf{p}, \mathbf{q}) \supseteq \mathfrak{T}(\mathbf{p}', \mathbf{q}')$.

*Remark.* We will see shortly that the converse to the statement in Corollary (4.1) is also true.

**Proof**  The proof follows immediately from the expression for the lower Lorenz curve in (4.146). Explicitly, by using the variable $t := \frac{1}{s}$ in (4.146), and using the notation $f_{\mathbf{p},\mathbf{q}}(r)$ for $f(r)$ we get that for all $r \in [0, 1]$

$$f_{\mathbf{p},\mathbf{q}}(r) = \max_{t \geqslant 1} \left\{ \frac{r - (\mathbf{p} - t\mathbf{q})_+ \cdot \mathbf{1}_n}{t} \right\}$$

$$\textbf{(4.154)} \rightarrow \; = \max_{t \geqslant 1} \left\{ \frac{2r - \|\mathbf{p} - t\mathbf{q}\|_1 + t - 1}{2t} \right\}. \qquad (4.156)$$

Therefore, if $\|\mathbf{p} - t\mathbf{q}\|_1 \geqslant \|\mathbf{p}' - t\mathbf{q}'\|_1$ for all $t \geqslant 1$, then $f_{\mathbf{p},\mathbf{q}}(r) \leqslant f_{\mathbf{p}',\mathbf{q}'}(r)$ for all $r \in [0, 1]$; that is, the lower Lorenz curve of the pair $(\mathbf{p}, \mathbf{q})$ is nowhere above the lower Lorenz curve of $(\mathbf{p}', \mathbf{q}')$ so that $\mathfrak{T}(\mathbf{p}, \mathbf{q}) \supseteq \mathfrak{T}(\mathbf{p}', \mathbf{q}')$. ∎

**Exercise 4.44.** *Let* $\mathbf{p}, \mathbf{q} \in \mathrm{Prob}(n)$ *and* $\mathbf{p}', \mathbf{q}' \in \mathrm{Prob}(n')$. *Show that if* $(\mathbf{p}, \mathbf{q}) \succ (\mathbf{p}', \mathbf{q}')$, *then for all* $t \in \mathbb{R}$ *we have* $\|\mathbf{p} - t\mathbf{q}\|_1 \geqslant \|\mathbf{p}' - t\mathbf{q}'\|_1$. *Hint: Use the property in Eq 2.8 of the 1-norm* $\| \cdot \|_1$.

**Exercise 4.45.** *Let* $\mathbf{p}, \mathbf{q} \in \mathrm{Prob}(n)$ *and* $\mathbf{p}', \mathbf{q}' \in \mathrm{Prob}(n')$.

1. *Show that if* $(\mathbf{p}, \mathbf{q}) \succ (\mathbf{p}', \mathbf{q}')$, *then* $\mathfrak{T}(\mathbf{p}', \mathbf{q}') \subseteq \mathfrak{T}(\mathbf{p}, \mathbf{q})$.
2. *Show that if* $(\mathbf{p}, \mathbf{q}) \sim (\mathbf{p}', \mathbf{q}')$, *then* $\mathfrak{T}(\mathbf{p}', \mathbf{q}') = \mathfrak{T}(\mathbf{p}, \mathbf{q})$.

*Hint: For the first part, let* $E \in \mathrm{STOC}(n', n)$ *be such that* $\mathbf{p}' = E\mathbf{p}$ *and* $\mathbf{q}' = E\mathbf{q}$, *and show first that for any* $\mathbf{t}' \in [0, 1]^n$ *the vector* $\mathbf{t} := E^T\mathbf{t}'$ *belongs to* $[0, 1]^n$ *and satisfies* $(\mathbf{t}' \cdot \mathbf{p}', \mathbf{t}' \cdot \mathbf{q}') = (\mathbf{t} \cdot \mathbf{p}, \mathbf{t} \cdot \mathbf{q})$.

### 4.3.4 Characterization of Relative Majorization

Relative majorization possesses several valuable characterizations, all of which are succinctly summarized in the theorem that follows. In Appendix D.9 of online version, we present a more extensive and traditional proof of this theorem, employing concepts from convex analysis such as support functions and sublinear functionals. In contrast to this comprehensive approach, we also provide a considerably shorter proof here. This brief proof avoids reliance on the aforementioned concepts from convex analysis, and instead leverages the intricate relationship that we previously examined between majorization and relative majorization.

**Characterization**

**Theorem 4.10.** Let $n, n' \in \mathbb{N}$, $\mathbf{p}, \mathbf{q} \in \mathrm{Prob}(n)$, and $\mathbf{p}', \mathbf{q}' \in \mathrm{Prob}(n')$. Then, the following are equivalent:

1. $(\mathbf{p}, \mathbf{q}) \succ (\mathbf{p}', \mathbf{q}')$.
2. For all $t \in \mathbb{R}$ we have $\|\mathbf{p} - t\mathbf{q}\|_1 \geqslant \|\mathbf{p}' - t\mathbf{q}'\|_1$.
3. $\mathfrak{T}(\mathbf{p}, \mathbf{q}) \supseteq \mathfrak{T}(\mathbf{p}', \mathbf{q}')$.

*Remark.* The equivalence between 1 and 3 in Theorem 4.10 provides a very simple geometrical characterization of relative majorization. Denoting by $\mathrm{LC}(\mathbf{p}, \mathbf{q})$ and $\mathrm{LC}(\mathbf{p}', \mathbf{q}')$ the two lower Lorenz curves associated with the two testing regions, we have that $(\mathbf{p}, \mathbf{q}) \succ (\mathbf{p}', \mathbf{q}')$ if and only if $\mathrm{LC}(\mathbf{p}, \mathbf{q})$ is nowhere above $\mathrm{LC}(\mathbf{p}', \mathbf{q}')$. An example illustrating this property is depicted in Figure 4.3.

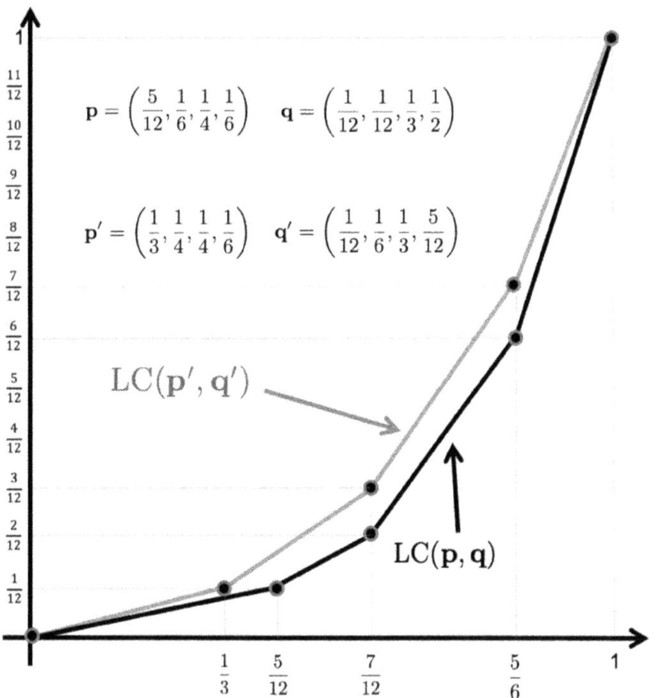


**Figure 4.3** Lower Lorenz curves. The lower Lorenz curve $\mathrm{LC}(\mathbf{p}, \mathbf{q})$ is nowhere above the lower Lorenz curve $\mathrm{LC}(\mathbf{p}', \mathbf{q}')$. This means that the pair $(\mathbf{p}, \mathbf{q})$ relatively majorizes the pair $(\mathbf{p}', \mathbf{q}')$. Note that aside from the vertices $(0, 0)$ and $(1, 1)$, the vertices of $\mathrm{LC}(\mathbf{p}, \mathbf{q})$ are $\left(\frac{5}{12}, \frac{1}{12}\right)$, $\left(\frac{7}{12}, \frac{1}{6}\right)$, $\left(\frac{5}{6}, \frac{1}{2}\right)$, and the vertices of $\mathrm{LC}(\mathbf{p}', \mathbf{q}')$ are $\left(\frac{1}{3}, \frac{1}{12}\right)$, $\left(\frac{7}{12}, \frac{1}{4}\right)$, $\left(\frac{5}{6}, \frac{7}{12}\right)$.

The implication $1 \Rightarrow 2$ can be easily deduced from the monotonicity property of the norm $\|\cdot\|_1$, as discussed in (2.8). This part of the proof is straightforward and hence is suggested as an exercise for the reader (refer to Exercise 4.44). Having previously established the implication $2 \Rightarrow 3$ in Corollary 4.1, our remaining task is to demonstrate that $3 \Rightarrow 1$. We begin this proof by focusing on the case where both $\mathbf{q}$ and $\mathbf{q}'$ consist of positive rational components.

> **Lemma 4.2.** Let $\mathbf{q}, \mathbf{p} \in \mathrm{Prob}(n)$ and $\mathbf{p}', \mathbf{q}' \in \mathrm{Prob}(n')$, and suppose that $\mathbf{q}$ and $\mathbf{q}'$ have positive rational components. If $\mathfrak{T}(\mathbf{p}, \mathbf{q}) \supseteq \mathfrak{T}(\mathbf{p}', \mathbf{q}')$, then $(\mathbf{p}, \mathbf{q}) \succ (\mathbf{p}', \mathbf{q}')$.

**Proof** From Theorem 4.8 we get that there exist vectors $\mathbf{r}, \mathbf{r}' \in \mathrm{Prob}(k)$ such that $(\mathbf{p}, \mathbf{q}) \sim (\mathbf{r}, \mathbf{u}^{(k)})$ and $(\mathbf{p}', \mathbf{q}') \sim (\mathbf{r}', \mathbf{u}^{(k)})$, where $k$ is a common denominator of all the components of $\mathbf{q}$ and $\mathbf{q}'$. Therefore, from the second part of Exercise 4.45 we get that $\mathfrak{T}(\mathbf{p}, \mathbf{q}) = \mathfrak{T}(\mathbf{r}, \mathbf{u}^{(k)})$ and $\mathfrak{T}(\mathbf{p}', \mathbf{q}') = \mathfrak{T}(\mathbf{r}', \mathbf{u}^{(k)})$. Moreover, since we assume that

$\mathfrak{T}(\mathbf{p},\mathbf{q}) \supseteq \mathfrak{T}(\mathbf{p}',\mathbf{q}')$ we get that $\mathfrak{T}(\mathbf{r},\mathbf{u}^{(k)}) \supseteq \mathfrak{T}(\mathbf{r}',\mathbf{u}^{(k)})$. Hence, the Lorenz curve $LC(\mathbf{r},\mathbf{u}^{(k)})$ is nowhere above the Lorenz curve $LC(\mathbf{r}',\mathbf{u}^{(k)})$. In addition, the nonzero vertices of $LC(\mathbf{r},\mathbf{u}^{(k)})$ and $LC(\mathbf{r}',\mathbf{u}^{(k)})$ are given, respectively by (cf. (4.155))

$$\left\{\left(\|\mathbf{r}\|_{(\ell)}, \frac{\ell}{k}\right)\right\}_{\ell \in [k]} \quad \text{and} \quad \left\{\left(\|\mathbf{r}'\|_{(\ell)}, \frac{\ell}{k}\right)\right\}_{\ell \in [k]}. \tag{4.157}$$

Therefore, since the vertex $(\|\mathbf{r}\|_{(\ell)}, \ell/k)$ has the same $y$-coordinate as the vertex $(\|\mathbf{r}\|_{(\ell)}, \ell/k)$, and since the convex curve $LC(\mathbf{r},\mathbf{u}^{(k)})$ is nowhere above the convex curve $LC(\mathbf{r}',\mathbf{u}^{(k)})$, we get that $\|\mathbf{r}\|_{(\ell)} \geqslant \|\mathbf{r}'\|_{(\ell)}$ for all $\ell \in [k]$. That is, $\mathbf{r} \succ \mathbf{r}'$ and from (4.138) this is equivalent to $(\mathbf{p},\mathbf{q}) \succ (\mathbf{p}',\mathbf{q}')$. This completes the proof. ∎

In order to complete the proof of Theorem 4.10 we will need a continuity argument that extends Lemma (4.2) to the general case of arbitrary $\mathbf{q}$ and $\mathbf{q}'$.

---

**Lemma 4.3.** Let $\mathbf{q},\mathbf{p} \in \text{Prob}(n)$ and $\mathbf{p}',\mathbf{q}' \in \text{Prob}(n')$ and suppose $\mathfrak{T}(\mathbf{p},\mathbf{q}) \supseteq \mathfrak{T}(\mathbf{p}',\mathbf{q}')$. Then, for every $\varepsilon \in (0,1)$ there exist two vectors $\mathbf{q}^{(\varepsilon)} \in \text{Prob}(n)$ and $\mathbf{q}'^{(\varepsilon)} \in \text{Prob}(n')$ with positive rational components such that $\mathbf{q}^{(\varepsilon)} \approx_\varepsilon \mathbf{q}$, $\mathbf{q}'^{(\varepsilon)} \approx_\varepsilon \mathbf{q}'$, and

$$\mathfrak{T}\left(\mathbf{p},\mathbf{q}^{(\varepsilon)}\right) \supseteq \mathfrak{T}\left(\mathbf{p}',\mathbf{q}'^{(\varepsilon)}\right). \tag{4.158}$$

---

**Proof**  We provide a geometrical proof using Figure 4.4. By keeping $\mathbf{p}'$ unchanged, we can raise slightly and vertically the vertices of $LC(\mathbf{p}',\mathbf{q}')$ to get the lower Lorenz curve of $LC(\mathbf{p}',\mathbf{q}'^{(\varepsilon)})$, where $\mathbf{q}'^{(\varepsilon)}$ has positive rational components and is $\varepsilon$-close to $\mathbf{q}'$; see Figure 4.4(a) for an illustration. Explicitly, let $\varepsilon_1, \ldots, \varepsilon_{n-1}$ be small enough positive numbers such that for all $x \in [n-1]$, $q'^{(\varepsilon)}_x := q'_x + \varepsilon_x$ is a rational number.

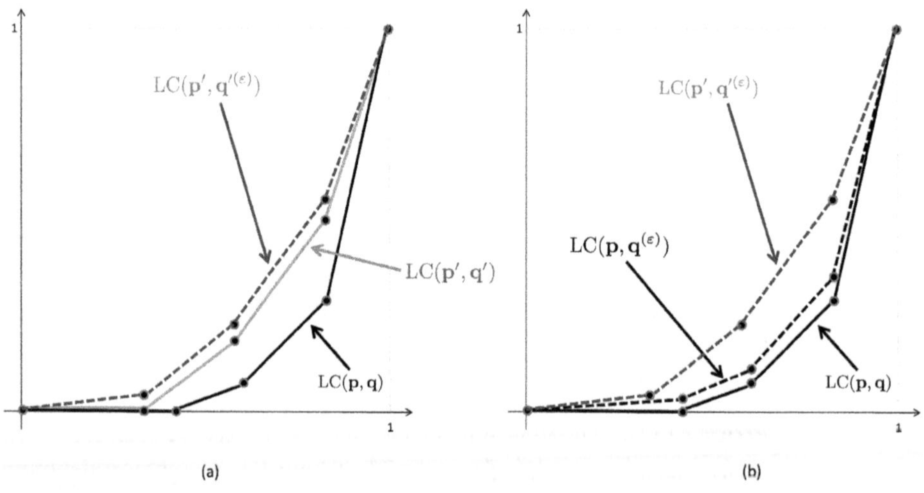



**Figure 4.4**  Geometrical proof of lemma 4.3.

Furthermore, we can always choose $\varepsilon_1, \ldots, \varepsilon_{n-1}$ to be small enough such that their sum $\delta := \sum_{x \in [n-1]} \varepsilon_x < \varepsilon$ and satisfies $q'^{(\varepsilon)}_n := q'_n - \delta > 0$ (due to the standard form of $(\mathbf{p}', \mathbf{q}')$ we have $q'_n > 0$). For these choices, the $\mathbf{q}'^{(\varepsilon)} := (q'^{(\varepsilon)}_1, \ldots, q'^{(\varepsilon)}_n)^T$ has positive rational numbers, and $\mathbf{q}'^{(\varepsilon)}$ is also $\varepsilon$-close to $\mathbf{q}$. Furthermore, by construction, $\mathrm{LC}(\mathbf{p}', \mathbf{q}'^{(\varepsilon)})$ is everywhere above $\mathrm{LC}(\mathbf{p}', \mathbf{q}')$.

Once we have established $\mathrm{LC}(\mathbf{p}', \mathbf{q}'^{(\varepsilon)})$ we construct $\mathbf{q}^{(\varepsilon)}$ in a similar way; see Figure 4.4(b). Explicitly, let $\nu_1, \ldots, \nu_{n-1}$ be small enough positive numbers such that for all $x \in [n-1]$, $q^{(\varepsilon)}_x := q_x + \nu_x$ is a rational number. Furthermore, we can always choose $\nu_1, \ldots, \nu_{n-1}$ to be small enough such that their sum $\nu := \sum_{x \in [n-1]} \nu_x < \varepsilon$ and satisfies $q^{(\varepsilon)}_n := q_n - \nu > 0$. For these choices, the vector $\mathbf{q}^{(\varepsilon)} := (q^{(\varepsilon)}_1, \ldots, q^{(\varepsilon)}_n)^T$ has positive rational numbers, is $\varepsilon$-close to $\mathbf{q}$, and as long as $\nu_1, \ldots, \nu_{n-1}$ are sufficiently small, $\mathrm{LC}(\mathbf{p}, \mathbf{q}^{(\varepsilon)})$ is everywhere below $\mathrm{LC}(\mathbf{p}', \mathbf{q}'^{(\varepsilon)})$. This completes the proof. $\blacksquare$

With these lemmas at hand, we are now ready to prove the theorem.

## Proof of Theorem 4.10

**Proof** Recall that it is left to prove that $3 \Rightarrow 1$. Since we assume that $\mathfrak{T}(\mathbf{p}, \mathbf{q}) \supseteq \mathfrak{T}(\mathbf{p}', \mathbf{q}')$, Lemma 4.3 implies that for any $\varepsilon \in (0, 1)$ there exist two vectors $\mathbf{q}^{(\varepsilon)} \in \mathrm{Prob}(n)$ and $\mathbf{q}'^{(\varepsilon)} \in \mathrm{Prob}(n')$ with positive rational components such that $\mathfrak{T}(\mathbf{p}, \mathbf{q}^{(\varepsilon)}) \supseteq \mathfrak{T}(\mathbf{p}', \mathbf{q}'^{(\varepsilon)})$. Thus, applying Lemma 4.2 with the two pairs $(\mathbf{p}, \mathbf{q}^{(\varepsilon)})$ and $(\mathbf{p}', \mathbf{q}'^{(\varepsilon)})$ replacing $(\mathbf{p}, \mathbf{q})$ and $(\mathbf{p}', \mathbf{q}')$ gives

$$(\mathbf{p}, \mathbf{q}^{(\varepsilon)}) \succ (\mathbf{p}', \mathbf{q}'^{(\varepsilon)}). \tag{4.159}$$

Now, let $\{\varepsilon_\ell\}_{\ell \in \mathbb{N}}$ be a sequence of numbers in $(0, 1)$ with zero limit, and denote by $\mathbf{q}_\ell := \mathbf{q}^{(\varepsilon_\ell)}$ and $\mathbf{q}'_\ell := \mathbf{q}'^{(\varepsilon_\ell)}$. With these notations for each $\ell \in \mathbb{N}$ we have $(\mathbf{p}, \mathbf{q}_\ell) \succ (\mathbf{p}', \mathbf{q}'_\ell)$ and $\lim_{\ell \to \infty} \mathbf{q}_\ell = \mathbf{q}$ and $\lim_{\ell \to \infty} \mathbf{q}'_\ell = \mathbf{q}'$. The condition $(\mathbf{p}, \mathbf{q}_\ell) \succ (\mathbf{p}', \mathbf{q}'_\ell)$ implies that for each $\ell \in \mathbb{N}$ there exists a matrix $E^{(\ell)} \in \mathrm{STOC}(n', n)$ such that

$$\mathbf{p}' = E^{(\ell)}\mathbf{p} \quad \text{and} \quad \mathbf{q}'_\ell = E^{(\ell)}\mathbf{q}_\ell. \tag{4.160}$$

Since $\mathrm{STOC}(n', n)$ is a compact set, the sequence $\{E^{(\ell)}\}_{\ell \in \mathbb{N}}$ has a converging subsequence. For simplicity of the exposition we assume that $\{E^{(\ell)}\}_{\ell \in \mathbb{N}}$ is itself a convergent sequence (otherwise, just replace everywhere $\{\ell\}_{\ell \in \mathbb{N}}$ with a subsequence of elements $\{\ell_j\}_{j \in \mathbb{N}}$). Therefore, there exists $E \in \mathrm{STOC}(n', n)$ such that $E = \lim_{\ell \to \infty} E^{(\ell)}$. By taking the limit $\ell \to \infty$ on both of the equations in (4.160) we get $\mathbf{p}' = E\mathbf{p}$ and $\mathbf{q}' = E\mathbf{q}$. That is, $(\mathbf{p}, \mathbf{q}) \succ (\mathbf{p}', \mathbf{q}')$. This completes the proof. $\blacksquare$

**Exercise 4.46.** *Consider the second statement of Theorem 4.10.*

1. *Show that $t \in \mathbb{R}$ can be restricted to $t \geqslant 1$. Hint: Recall Corollary 4.1.*

2. *Show that $t \in \mathbb{R}$ can be restricted to $t \in [0, 1]$. Hint: Recall that $(\mathbf{p}, \mathbf{q}) \succ (\mathbf{p}', \mathbf{q}')$ if and only if $(\mathbf{q}, \mathbf{p}) \succ (\mathbf{q}', \mathbf{p}')$.*

**Exercise 4.47.** *Let* $\mathbf{p} = (p, 1-p)^T$ *and* $\mathbf{q} = (q, 1-q)^T$ *be two probability vectors satisfying* $p > q$. *Let* $\mathbf{p}' = (p', 1-p')^T$ *and* $\mathbf{q}' = (q', 1-q')^T$ *be another pair of probability vectors with* $p' \geqslant q$ *and* $, q' \leqslant p$. *Show that*

$$
\begin{aligned}
(\mathbf{p}', \mathbf{q}) \prec (\mathbf{p}, \mathbf{q}) &\iff p' \leqslant p \\
(\mathbf{p}, \mathbf{q}') \prec (\mathbf{p}, \mathbf{q}) &\iff q \leqslant q'.
\end{aligned}
\tag{4.161}
$$

**Exercise 4.48.** *Let* $0 < \mathbf{q} \in \mathrm{Prob}(n)$ *and suppose* $\mathbf{q} = \mathbf{q}^{\downarrow}$. *Show that*

$$
(\mathbf{e}_n, \mathbf{q}) \succ (\mathbf{p}, \mathbf{q}) \qquad \forall\, \mathbf{p} \in \mathrm{Prob}(n),
\tag{4.162}
$$

*where* $\mathbf{e}_n = (0, \ldots, 0, 1)^T$.

## 4.3.5 Continuity of Relative Majorization

In this subsection we investigate several continuity properties of relative majorization. We start by noting that in the case that $\mathbf{q}$ doesn't have positive rational components, the equivalence of the form (4.135) does not hold in general. However, it is still possible to approximate this relation as the following exercise demonstrates.

**Exercise 4.49.** *Let* $\varepsilon > 0$, *and* $\mathbf{p}, \mathbf{q} \in \mathrm{Prob}(n)$. *Show that:*

1. *If* $\mathbf{p} \neq \mathbf{q}$, *then there exists* $0 < \mathbf{q}' \in \mathrm{Prob}(n) \cap \mathbb{Q}^n$ *such that* $\mathbf{q}'$ *is* $\varepsilon$-*close to* $\mathbf{q}$ *and* $(\mathbf{p}, \mathbf{q}) \succ (\mathbf{p}, \mathbf{q}')$.
2. *There exists a vector* $\mathbf{q}' \in \mathrm{Prob}(n) \cap \mathbb{Q}^n$ *such that* $\mathbf{q}'$ *is* $\varepsilon$-*close to* $\mathbf{q}$, $\mathrm{supp}(\mathbf{q}') = \mathrm{supp}(\mathbf{q})$, *and* $(\mathbf{p}, \mathbf{q}') \succ (\mathbf{p}, \mathbf{q})$.

*Hint: Take* $\mathbf{q}'$ *to be* $\mathbf{q}^{(\varepsilon)}$ *or* $\mathbf{q}'^{(\varepsilon)}$ *as defined in Lemma 4.3.*

In this exercise we approximated $(\mathbf{p}, \mathbf{q})$ with a pair of vectors $(\mathbf{p}, \mathbf{q}')$, where $\mathbf{q}'$ has some desired properties (particularly, rational components), and $\mathbf{p}$ is fixed. In the next two lemmas we remove some of the assumptions on $\mathbf{q}$ by allowing $\mathbf{p}$ to vary. We will only assume that $\mathbf{q}'$ is close to $\mathbf{q}$.

Specifically, consider two probability vectors $\mathbf{p}, \mathbf{q} \in \mathrm{Prob}(n)$ and let $\varepsilon \in (0, 1)$ be a sufficiently small number to be determined later. Our objective is to show that for every $\mathbf{p}' \in \mathfrak{B}_\varepsilon(\mathbf{p})$ there exists $\delta \in (0, 1)$ and $\mathbf{q}' \in \mathfrak{B}_\delta(\mathbf{q})$ such that $(\mathbf{p}', \mathbf{q}') \succ (\mathbf{p}, \mathbf{q})$. We will be able to show that such a $\mathbf{q}'$ exists if we assume that $\mathbf{q} > 0$ and that $\mathrm{supp}(\mathbf{p}') \subseteq \mathrm{supp}(\mathbf{p})$. In the following lemma we will use the notation $\varepsilon_0 := \frac{1}{2} p_{\min} q_{\min}$, where $p_{\min}$ and $q_{\min}$ are the smallest nonzero components of $\mathbf{p}$ and $\mathbf{q}$, respectively.

> **Lemma 4.4.** Let $\mathbf{p}, \mathbf{q} \in \mathrm{Prob}(n)$, $\mathbf{q} > 0$, $\varepsilon \in (0, \varepsilon_0)$, $\delta := \varepsilon/q_{\min}$, and $\mathbf{p}' \in \mathfrak{B}_\varepsilon(\mathbf{p})$ with $\mathrm{supp}(\mathbf{p}') \subseteq \mathrm{supp}(\mathbf{p})$. Then, there exists $\mathbf{q}' \in \mathfrak{B}_\delta(\mathbf{q})$ such that $(\mathbf{p}', \mathbf{q}') \succ (\mathbf{p}, \mathbf{q})$.

**Proof**   We need to define $\mathbf{q}' \in \mathfrak{B}_\delta(\mathbf{q})$ and a channel $E \in \mathrm{STOC}(n,n)$ such that $E\mathbf{p}' = \mathbf{p}$ and $E\mathbf{q}' = \mathbf{q}$. The key idea is to look for a matrix $E$ of the form

$$E\mathbf{t} := \mathbf{p} + s\left(\mathbf{t} - \mathbf{p}'\right) \qquad \forall\, \mathbf{t} \in \mathrm{Prob}(n), \tag{4.163}$$

where $s \in \mathbb{R}_+$ is some coefficient. Observe that we define the matrix $E$ by its action on probability vectors. Clearly, by construction $E\mathbf{p}' = \mathbf{p}$. However, $E$ above is not necessarily a stochastic matrix since for arbitrary $s \in \mathbb{R}$ the vector $\mathbf{p} + s\left(\mathbf{t} - \mathbf{p}'\right)$ could have negative components. We therefore choose $s \geqslant 0$ to be such that $\mathbf{p} \geqslant s\mathbf{p}'$ (entry-wise) so that $E$, as defined above, is indeed a column stochastic matrix. We therefore take $s$ to be the optimal $s$ that satisfies $\mathbf{p} \geqslant s\mathbf{p}'$; that is, we define

$$s := \min_{x \in \mathrm{supp}(\mathbf{p}')} \frac{p_x}{p'_x}. \tag{4.164}$$

Observe that $s \leqslant 1$ (since we cannot have $p_x > p'_x$ for all $x \in \mathrm{supp}(\mathbf{p}')$), and since $\mathrm{supp}(\mathbf{p}') \subseteq \mathrm{supp}(\mathbf{p})$, we have $s > 0$. In fact, since $\mathbf{p}'$ is $\varepsilon$-close to $\mathbf{p}$ we must have $p'_x \leqslant p_x + \varepsilon$ for all $x \in [n]$ so that

$$s \geqslant \min_{x \in \mathrm{supp}(\mathbf{p}')} \frac{p_x}{p_x + \varepsilon} \geqslant \frac{p_{\min}}{p_{\min} + \varepsilon}. \tag{4.165}$$

Our next goal is to define $\mathbf{q}'$ such that $E\mathbf{q}' = \mathbf{q}$. Observe that according to (4.163) we have $E\mathbf{q}' = \mathbf{p} + s\left(\mathbf{q}' - \mathbf{p}'\right)$ so that $E\mathbf{q}' = \mathbf{q}$ if $\mathbf{p} + s\left(\mathbf{q}' - \mathbf{p}'\right) = \mathbf{q}$. Isolating $\mathbf{q}'$ we get that $\mathbf{q}'$ should have the form

$$\mathbf{q}' := \mathbf{p}' + \frac{1}{s}(\mathbf{q} - \mathbf{p}) \tag{4.166}$$

(indeed, check that with this $\mathbf{q}'$ we have $E\mathbf{q}' = \mathbf{q}$). However, it is not obvious that $\mathbf{q}'$ has nonnegative components. Therefore, we next show that $\varepsilon$ is small enough so that $\mathbf{q}' \geqslant 0$ (i.e. $\mathbf{q}'$ is a probability vector).

Observe that $\mathbf{q}' \geqslant 0$ if and only if for all $x \in [n]$ we have $q_x \geqslant p_x - sp'_x$. Now, since $\mathbf{p}$ is $\varepsilon$-close to $\mathbf{p}'$ we get that

$$p_x - sp'_x \leqslant p_x - s(p_x - \varepsilon) = (1-s)p_x + s\varepsilon$$

$$\boxed{p_x, s \leqslant 1} \longrightarrow \leqslant 1 - s + \varepsilon \tag{4.167}$$

$$\mathbf{(4.165)} \rightarrow \leqslant \frac{\varepsilon}{p_{\min} + \varepsilon} + \varepsilon \leqslant \frac{2\varepsilon}{p_{\min}} \leqslant q_{\min}.$$

Hence, $\mathbf{p}'$ is a probability vector. By definition,

$$\frac{1}{2}\left\|\mathbf{q}' - \mathbf{q}\right\|_1 = \frac{1}{2s}\left\|(1-s)\mathbf{q} - (\mathbf{p} - s\mathbf{p}')\right\|_1$$

$$\textbf{Triangle inequality} \rightarrow \leqslant \frac{1}{2s}\left(1 - s + \left\|\mathbf{p} - s\mathbf{p}'\right\|_1\right)$$

$$\boxed{\mathbf{p} \geqslant s\mathbf{p}'} \longrightarrow = \frac{1-s}{s} = \frac{1}{s} - 1 \tag{4.168}$$

$$\mathbf{(4.165)} \rightarrow \leqslant \frac{\varepsilon}{p_{\min}} = \delta.$$

This completes the proof. ∎

**Exercise 4.50.**  *Using the same notations as in Lemma 4.4, let*

$$\varepsilon_1 := \min_{x \in \text{supp}(\mathbf{p})} \frac{p_{\min} q_x}{p_{\min} + (p_x - q_x)_+}.$$  (4.169)

*1. Show that $\varepsilon_0 < \varepsilon_1$.*
*2. Show that the Lemma 4.4 still holds if we replace $\varepsilon_0$ with $\varepsilon_1$.*

> **Lemma 4.5.**  Let $\mathbf{p}, \mathbf{q}, \mathbf{q}' \in \text{Prob}(n)$ and denote by $\delta := 1 - \min_{x \in [n]} \frac{q'_x}{q_x}$. Then, there exists $\mathbf{p}' \in \mathfrak{B}_\delta(\mathbf{p})$ such that $(\mathbf{p}, \mathbf{q}) \succ (\mathbf{p}', \mathbf{q}')$.

*Remark.*  Note that if $\frac{1}{2}\|\mathbf{q} - \mathbf{q}'\|_1 \leqslant \varepsilon$ for some $\varepsilon > 0$, then

$$\delta = \max_{x \in [n]} \frac{q_x - q'_x}{q_x} \leqslant \frac{\varepsilon}{q_\ell},$$  (4.170)

where $\ell$ is the integer satisfying $\delta = 1 - q'_\ell/q_\ell$. Therefore, if $\mathbf{q}$ and $\mathbf{q}'$ are very close to each other so are $\mathbf{p}$ and $\mathbf{p}'$.

**Proof**   Let $s := 1 - \delta = \min_{x \in [n]} \frac{q'_x}{q_x}$. From the definition of $s$ we have $\mathbf{q}' \geqslant s\mathbf{q}$ (entry-wise), so that the mapping

$$E\mathbf{t} := \mathbf{q}' + s(\mathbf{t} - \mathbf{q}) \qquad \forall\, \mathbf{t} \in \text{Prob}(n)$$  (4.171)

is a channel. By definition, $E\mathbf{q} = \mathbf{q}'$. Define

$$\mathbf{p}' := E\mathbf{p}$$
$$(4.171)\rightarrow\ = \mathbf{q}' + s(\mathbf{p} - \mathbf{q}),$$  (4.172)

so that $(\mathbf{p}, \mathbf{q}) \succ (\mathbf{p}', \mathbf{q}')$. Then,

$$\frac{1}{2}\|\mathbf{p}' - \mathbf{p}\|_1 = \frac{1}{2}\|\mathbf{q}' - s\mathbf{q} - (1-s)\mathbf{p}\|_1$$
$$\text{Triangle inequality}\rightarrow\ = \frac{1}{2}\|\mathbf{q}' - s\mathbf{q}\|_1 + \frac{1}{2}(1 - s)$$  (4.173)
$$\boxed{q' \geqslant sq}\ \longrightarrow\ = 1 - s.$$

This completes the proof.                                                      ∎

**Exercise 4.51.**  *Prove the following theorem: Let $\mathbf{p}, \mathbf{q}, \mathbf{p}' \in \text{Prob}(n)$ and denote by $\delta := 1 - \min_{x \in [n]} \frac{p'_x}{p_x}$. Then, there exists $\mathbf{q}' \in \mathfrak{B}_\delta(\mathbf{q})$ such that $(\mathbf{p}, \mathbf{q}) \succ (\mathbf{p}', \mathbf{q}')$.*

## 4.4 The Trumping Relation

Another preorder that plays an important role in several resource theories is the following variant of majorization.

---

> **The Trumping Relation**
>
> **Definition 4.5.** For any $p, q \in \text{Prob}(n)$ we say that $p$ *trumps* $q$ and write
>
> $$p \succ_* q \qquad (4.174)$$
>
> if there exists an integer $m \in \mathbb{N}$ and a vector $r \in \text{Prob}(m)$, known as the catalyst vector, such that
>
> $$p \otimes r \succ q \otimes r. \qquad (4.175)$$

*Remark.* Observe that while we require that the dimension $m < \infty$, it is still unbounded. We will see later that this implies that the trumping relation is very sensitive to perturbations, and also leads to a phenomenon know as "embezzlement" of entanglement.

By definition, if $p \succ q$, then necessarily $p \succ_* q$. The following example demonstrates that the opposite direction does not hold in general. In this sense, the trumping relation imposes a weaker constraint than majorization.

**Example.** *Consider the probability vectors*

$$p := \begin{bmatrix} 2/5 \\ 2/5 \\ 1/10 \\ 1/10 \end{bmatrix} \quad and \quad q = \begin{bmatrix} 1/2 \\ 1/4 \\ 1/4 \\ 0 \end{bmatrix}.$$

*It is simple to check (see the following exercise) that $p \not\succ q$ and $q \not\succ p$. Yet, $p \succ_* q$ since the vector $r = (3/5, 2/5)^T$ satisfies (4.175).*

**Exercise 4.52.** *Let $p, q, r$ be as in the example above.*

1. *Verify that $p \not\succ q$ and $q \not\succ p$.*
2. *Verify that (4.175) holds.*

**Exercise 4.53.** *Show that if $p, q \in \text{Prob}(3)$ and $p \succ_* q$, then $p \succ q$.*

**Exercise 4.54.** *Show that the uniform probability vector $u$ cannot act as a catalyst verctor; that is, show that if $p \in \text{Prob}(n)$ and $q \in \text{Prob}(m)$ are such that $p \not\succ q$, then for any $k \in \mathbb{N}$, $p \otimes u^{(k)} \not\succ q \otimes u^{(k)}$.*

**Exercise 4.55.** *Let $f : \text{Prob}(n) \to \mathbb{R}$ be a Schur convex function that is additive under tensor product; that is, for all $p \in \text{Prob}(n)$ and $q \in \text{Prob}(m)$*

$$f(p \otimes q) = f(p) + f(q). \qquad (4.175)$$

*Show that*

$$p \succ_* q \quad \Rightarrow \quad f(p) \geqslant f(q). \qquad (4.176)$$

The tramping relation can also be extended to pairs of probability vectors.

Relative Trumping

**Definition 4.6.** Let $\mathbf{p}, \mathbf{q} \in \mathrm{Prob}(n)$ and $\mathbf{p}', \mathbf{q}' \in \mathrm{Prob}(m)$. We say that the pair $(\mathbf{p}, \mathbf{q})$ relatively trumps the pair $(\mathbf{p}', \mathbf{q}')$, and write

$$(\mathbf{p}, \mathbf{q}) \succ_* (\mathbf{p}', \mathbf{q}') \tag{4.177}$$

if there exists $k \in \mathbb{N}$ and vectors $\mathbf{r}, \mathbf{s} \in \mathrm{Prob}(k)$ with $\mathbf{s} > 0$ such that

$$(\mathbf{p} \otimes \mathbf{r}, \mathbf{q} \otimes \mathbf{s}) \succ (\mathbf{p}' \otimes \mathbf{r}, \mathbf{q}' \otimes \mathbf{s}). \tag{4.178}$$

**Exercise 4.56.** *Show that if we did not impose $\mathbf{s} > 0$ (or alternatively that $\mathbf{r} > 0$) in the definition above, then there would always exists a catalyst. Hint: Take $\mathbf{r}$ and $\mathbf{s}$ to be orthogonal.*

*Remark.* In Definition (4.6), an alternative approach could have been to require that the vectors $\mathbf{r}$ and $\mathbf{s}$ are not orthogonal instead of enforcing $\mathbf{s} > 0$. Nevertheless, opting for the stricter criterion of $\mathbf{s} > 0$ brings two distinct benefits. Firstly, within the rational field, where vectors like $\mathbf{q}, \mathbf{q}', \mathbf{r}$ and others comprise rational components, Theorem 4.8 implies that $(\mathbf{r}, \mathbf{s}) \sim (\mathbf{t}, \mathbf{u})$, for some vector $\mathbf{t}$ of higher dimensionality. This equivalence effectively simplifies relative trumping to standard trumping in this scenario. Secondly, when applying the concept of relative trumping to thermodynamic contexts, the vector $\mathbf{s}$ typically represents a Gibbs state, which is inherently positive by nature. This correspondence ensures that the mathematical model is in harmony with the underlying physical principles of Gibbs states in thermodynamics.

In the next chapter we will study functions that behave monotonically under relative trumping. A well-known family of such functions are Rényi divergences. Rényi divergences are defined for any $\alpha \in [0, \infty]$ and any $\mathbf{p}, \mathbf{q} \in \mathrm{Prob}(n)$ as

$$D_\alpha(\mathbf{p}\|\mathbf{q}) := \begin{cases} \frac{1}{\alpha-1} \log \sum_{x \in [m]} p_x^\alpha q_x^{1-\alpha} & \text{if } \mathrm{supp}\,(\mathbf{p}) \subseteq \mathrm{supp}\,(\mathbf{q}). \\ \infty & \text{otherwise} \end{cases} \tag{4.179}$$

The cases $\alpha = 0, 1, \infty$ are defined by taking the appropriate limits (more details will be given in the next chapter). Both the trumping and relative trumping relations can be characterized with the above family of Rényi divergences.

Characterization of the Trumping Relation

**Theorem 4.11.** Let $\mathbf{p}, \mathbf{q} \in \mathrm{Prob}(n)$ with either $\mathbf{p} > 0$ or $\mathbf{q} > 0$ and $\mathbf{p} \neq \mathbf{q}$. Then,

$$\mathbf{p} \succ_* \mathbf{q} \tag{4.180}$$

if and only if for all $\alpha \geqslant \frac{1}{2}$

$$D_\alpha(\mathbf{p}\|\mathbf{u}) > D_\alpha(\mathbf{q}\|\mathbf{u}) \quad \text{and} \quad D_\alpha(\mathbf{u}\|\mathbf{p}) > D_\alpha(\mathbf{u}\|\mathbf{q}). \tag{4.181}$$

The proof of the theorem above is rather complicated and goes beyond the scope of this book. In the section "Notes and References" at the end of this chapter, we discuss its history and provide relevant references for further reading. In the following corollary we show that this theorem can be extended to relative tramping for the case that one of the vectors in each pair has positive rational components. We use the notation

$$(\mathbf{p}, \mathbf{q}) \otimes (\mathbf{p}' \otimes \mathbf{q}') := (\mathbf{p} \otimes \mathbf{p}', \mathbf{q} \otimes \mathbf{q}') \qquad \forall\, \mathbf{p}, \mathbf{q} \in \mathrm{Prob}(n), \;\; \forall\, \mathbf{p}', \mathbf{q}' \in \mathrm{Prob}(m). \quad (4.182)$$

---

**Corollary 4.2.** Let $m, n \in \mathbb{N}$, $\mathbf{p}, \mathbf{q} \in \mathrm{Prob}(n)$, $\mathbf{p}', \mathbf{q}' \in \mathrm{Prob}(m)$, and suppose that both $\mathbf{q}$ and $\mathbf{q}'$ have positive rational components. Then, the following are equivalent:

1. There exists $k \in \mathbb{N}$ and a vector $\mathbf{s} \in \mathrm{Prob}(k)$ such that

$$(\mathbf{p}, \mathbf{q}) \otimes \left( \mathbf{s}, \mathbf{u}^{(k)} \right) \succ (\mathbf{p}', \mathbf{q}') \otimes \left( \mathbf{s}, \mathbf{u}^{(k)} \right). \qquad (4.183)$$

2. For all $\alpha \geqslant \frac{1}{2}$

$$D_\alpha(\mathbf{p} \| \mathbf{q}) > D_\alpha(\mathbf{p}' \| \mathbf{q}') \quad \text{and} \quad D_\alpha(\mathbf{q} \| \mathbf{p}) > D_\alpha(\mathbf{q}' \| \mathbf{p}'). \qquad (4.184)$$

---

Note that the condition 1 above implies in particular that $(\mathbf{p}, \mathbf{q}) \succ_* (\mathbf{p}', \mathbf{q}')$. We leave the proof of the corollary as an exercise.

**Exercise 4.57.** *Prove Corollary 4.2 using the combination of Theorem 4.11 and Theorem 4.8.*

# 4.5 Catalytic Majorization

In this section, we study a variant of the trumping relation that we call catalytic majorization. We will see that catalytic majorization is robust under small perturbations and therefore is more useful in some applications, particularly in thermodynamics.

## 4.5.1 Robustness Under Small Perturbations

Both majorization and relative majorization are robust under small perturbations. To be precise, let $\{\mathbf{p}_k\}_{k \in \mathbb{N}}$ and $\{\mathbf{q}_k\}_{k \in \mathbb{N}}$ be sequences in $\mathrm{Prob}(n)$ with limits $\mathbf{p}$ and $\mathbf{q}$, respectively, and let $\{\mathbf{p}'_k\}_{k \in \mathbb{N}}$ and $\{\mathbf{q}'_k\}_{k \in \mathbb{N}}$ be sequences in $\mathrm{Prob}(m)$ with limits $\mathbf{p}'$ and $\mathbf{q}'$. Suppose now that $(\mathbf{p}_k, \mathbf{q}_k) \succ (\mathbf{p}'_k, \mathbf{q}'_k)$ for all $k \in \mathbb{N}$ and recall from Theorem 4.10 that this means that $\mathfrak{T}(\mathbf{p}_k, \mathbf{q}_k) \supseteq \mathfrak{T}(\mathbf{p}'_k, \mathbf{q}'_k)$. Since this inclusion of testing regions is robust under taking the limit, we conclude that $\mathfrak{T}(\mathbf{p}, \mathbf{q}) \supseteq \mathfrak{T}(\mathbf{p}', \mathbf{q}')$ so that necessarily $(\mathbf{p}, \mathbf{q}) \succ (\mathbf{p}', \mathbf{q}')$.

The above argument cannot be applied to the trumping and relative trumping relations. To see why, consider two sequences $\{\mathbf{p}_k\}_{k\in\mathbb{N}} \subseteq \mathrm{Prob}(n)$ and $\{\mathbf{q}_k\}_{k\in\mathbb{N}} \subseteq \mathrm{Prob}(m)$ with limits $\mathbf{p}$ and $\mathbf{q}$, and suppose that $\mathbf{p}_k \succ_* \mathbf{q}_k$ for all $k \in \mathbb{N}$. This means that for each $k \in \mathbb{N}$ there exists a catalyst vector $\mathbf{r}_k \in \mathrm{Prob}(\ell_k)$ such that $\mathbf{p}_k \otimes \mathbf{r}_k \succ \mathbf{q}_k \otimes \mathbf{r}_k$, where $\ell_k \in \mathbb{N}$ is the dimension of $\mathbf{r}_k$ that can depend on $k$. Without invoking additional arguments, one cannot rule out the possibility that the dimension $\ell_k$ goes to infinity as $k$ goes to infinity. This means that we cannot conclude that $\mathbf{p} \succ_* \mathbf{q}$. However, at the time of writing this book, it is left open to find an example with convergent sequences satisfying $\mathbf{p}_k \succ_* \mathbf{q}_k$ for all $k \in \mathbb{N}$, whereas their limits satisfy $\mathbf{p} \not\succ_* \mathbf{q}$.

**Exercise 4.58.** *Show that if there exists an example as above, then there is also a similar example for relative trumping. That is, there exists sequences $\{\mathbf{p}_k\}_{k\in\mathbb{N}}$, $\{\mathbf{q}_k\}_{k\in\mathbb{N}}$, $\{\mathbf{p}'_k\}_{k\in\mathbb{N}}$, and $\{\mathbf{q}'_k\}_{k\in\mathbb{N}}$, in $\mathrm{Prob}(n)$, with limits $\mathbf{p}$, $\mathbf{q}$, $\mathbf{p}'$, and $\mathbf{q}'$, respectively, such that $(\mathbf{p}_k, \mathbf{q}_k) \succ_*$ $(\mathbf{p}'_k, \mathbf{q}'_k)$ for all $k \in \mathbb{N}$ and $(\mathbf{p}, \mathbf{q}) \not\succ_* (\mathbf{p}', \mathbf{q}')$.*

## 4.5.2 Robust Version of Relative Trumping

In the following definition we define a robust version of relative trumping. Similarly, one can define a robust version for trumping itself; however, we do not do it here since the robust version of relative trumping is the concept that we will use later on in applications to resource theories.

> **Catalytic Majorization**
>
> **Definition 4.7.** Let $m, n \in \mathbb{N}$ and $\mathbf{p}, \mathbf{q} \in \mathrm{Prob}(n)$ and $\mathbf{p}', \mathbf{q}' \in \mathrm{Prob}(m)$. We say that $(\mathbf{p}, \mathbf{q})$ catalytically majorizes $(\mathbf{p}', \mathbf{q}')$ and write
>
> $$(\mathbf{p}, \mathbf{q}) \succ_c (\mathbf{p}', \mathbf{q}') \tag{4.185}$$
>
> if for any $\varepsilon > 0$ there exists four vector $\mathbf{p}_\varepsilon \in \mathcal{B}_\varepsilon(\mathbf{p})$, $\mathbf{q}_\varepsilon \in \mathcal{B}_\varepsilon(\mathbf{q})$, $\mathbf{p}'_\varepsilon \in \mathcal{B}_\varepsilon(\mathbf{p}')$, and $\mathbf{q}'_\varepsilon \in \mathcal{B}_\varepsilon(\mathbf{q}')$ such that $(\mathbf{p}_\varepsilon, \mathbf{q}_\varepsilon) \succ_* (\mathbf{p}'_\varepsilon, \mathbf{q}'_\varepsilon)$.

**Exercise 4.59.** *Show that $\succ_c$ is indeed a preorder, and if $(\mathbf{p}, \mathbf{q}) \succ_* (\mathbf{p}', \mathbf{q}')$, then necessarily $(\mathbf{p}, \mathbf{q}) \succ_c (\mathbf{p}', \mathbf{q}')$.*

The relation $\succ_c$ is robust under perturbation essentially by definition.

**Exercise 4.60 (Robustness of Catalytic Majorization).** *Show that if for any $\varepsilon > 0$ there exists four vector $\mathbf{p}_\varepsilon \in \mathcal{B}_\varepsilon(\mathbf{p})$, $\mathbf{q}_\varepsilon \in \mathcal{B}_\varepsilon(\mathbf{q})$, $\mathbf{p}'_\varepsilon \in \mathcal{B}_\varepsilon(\mathbf{p}')$, and $\mathbf{q}'_\varepsilon \in \mathcal{B}_\varepsilon(\mathbf{q}')$ such that $(\mathbf{p}_\varepsilon, \mathbf{q}_\varepsilon) \succ_c (\mathbf{p}'_\varepsilon, \mathbf{q}'_\varepsilon)$, then $(\mathbf{p}, \mathbf{q}) \succ_c (\mathbf{p}', \mathbf{q}')$.*

In the definition of catalytic majorization, we considered four balls of radius $\varepsilon$ around $\mathbf{p}$, $\mathbf{q}$, $\mathbf{p}'$, and $\mathbf{q}'$. We now show that under certain support conditions, it is sufficient to consider only two such balls.

> **Lemma 4.6.** Let $m, n \in \mathbb{N}, \mathbf{p}, \mathbf{q} \in \mathrm{Prob}(n)$ and $\mathbf{p}', \mathbf{q}' \in \mathrm{Prob}(m)$. Suppose further that both $\mathbf{p}' > 0$ and $\mathbf{q} > 0$. Then, the following are equivalent:
>
> 1. $(\mathbf{p}, \mathbf{q}) \succ_c (\mathbf{p}', \mathbf{q}')$.
> 2. For any $\varepsilon > 0$ there exist $\mathbf{q}_\varepsilon \in \mathfrak{B}_\varepsilon(\mathbf{q})$ and $\mathbf{q}'_\varepsilon \in \mathfrak{B}_\varepsilon(\mathbf{q}')$ such that $(\mathbf{p}, \mathbf{q}_\varepsilon) \succ_*$ $(\mathbf{p}', \mathbf{q}'_\varepsilon)$.

**Proof** Clearly, if for all $\varepsilon > 0$ the two vectors in the second statement exist, then $(\mathbf{p}, \mathbf{q}) \succ_c (\mathbf{p}', \mathbf{q}')$ since we can define $\mathbf{p}_\varepsilon := \mathbf{p}$ and $\mathbf{p}'_\varepsilon := \mathbf{p}'$, so that the four vectors $\mathbf{p}_\varepsilon$, $\mathbf{q}_\varepsilon$, $\mathbf{p}'_\varepsilon$, and $\mathbf{q}'_\varepsilon$ satisfy the conditions in Definition 4.7. It thus remains to show the converse implication.

Suppose $(\mathbf{p}, \mathbf{q}) \succ_c (\mathbf{p}', \mathbf{q}')$ and let $\mathbf{p}_\varepsilon$, $\mathbf{q}_\varepsilon$, $\mathbf{p}'_\varepsilon$, and $\mathbf{q}'_\varepsilon$ be as in Definition 4.7. In particular, $(\mathbf{p}_\varepsilon, \mathbf{q}_\varepsilon) \succ_* (\mathbf{p}'_\varepsilon, \mathbf{q}'_\varepsilon)$ and we choose $\varepsilon > 0$ to be sufficiently small so that $\mathbf{q}_\varepsilon > 0$ (recall that $\mathbf{q} > 0$ and $\mathbf{q}_\varepsilon$ is $\varepsilon$-close to $\mathbf{q}$) and $\mathbf{p}'_\varepsilon > 0$. From Lemma 4.4 it follows that there exists $\mathbf{r} \in \mathfrak{B}^\delta(\mathbf{q}_\varepsilon)$ with $\delta := \frac{\varepsilon}{q_{\varepsilon,\min}}$ ($q_{\varepsilon,\min}$ being the smallest component of $\mathbf{q}_\varepsilon$) such that $(\mathbf{p}, \mathbf{r}) \succ (\mathbf{p}_\varepsilon, \mathbf{q}_\varepsilon)$. Similarly, from the version of Lemma (4.5) that is given in Exercise 4.51 it follows that there exists a vector $\mathbf{r}' \in \mathfrak{B}^{\delta'}(\mathbf{q}'_\varepsilon)$ with $\delta' := 1 - \min_{x \in [n]} \frac{p'_x}{(\mathbf{p}'_\varepsilon)_x}$, such that $(\mathbf{p}'_\varepsilon, \mathbf{q}'_\varepsilon) \succ (\mathbf{p}', \mathbf{r}')$. Observe that $\mathbf{r}$ and $\mathbf{r}'$ satisfy

$$(\mathbf{p}, \mathbf{r}) \succ (\mathbf{p}_\varepsilon, \mathbf{q}_\varepsilon) \succ_* (\mathbf{p}'_\varepsilon, \mathbf{q}'_\varepsilon) \succ (\mathbf{p}', \mathbf{r}'). \tag{4.186}$$

Hence, in particular $(\mathbf{p}, \mathbf{r}) \succ_* (\mathbf{p}', \mathbf{r}')$. Now, recall that $\mathbf{r}$ is $\delta$-close to $\mathbf{q}_\varepsilon$ and therefore $(\delta + \varepsilon)$-close to $\mathbf{q}$. Similarly, $\mathbf{r}'$ is $(\delta' + \varepsilon)$-close to $\mathbf{q}'$. The proof is therefore concluded by the observation that both $\delta$ and $\delta'$ go to zero in the limit $\varepsilon \to 0$ so that $\mathbf{r}$ and $\mathbf{r}'$ can be made arbitrarily close to $\mathbf{q}$ and $\mathbf{q}'$, respectively. $\blacksquare$

**Exercise 4.61.** *Show that under the assumption that $\mathbf{q} > 0$ and $\mathbf{p}' > 0$ both $\delta$ and $\delta'$ in Lemma (4.6) go to zero as $\varepsilon$ goes to zero.*

### 4.5.3 Characterization of Catalytic Majorization

This subsection concludes with an insightful characterization of catalytic majorization. This characterization is articulated through Rényi divergences, a topic we will delve into in greater detail in subsequent chapters. We have previously encountered Rényi divergences in Theorem 4.11 and Corollary 4.2. Both Theorem 4.11 and Corollary 4.2 incorporate strict inequalities among Rényi divergences. These strict inequalities highlight the fact that relative trumping may not be resilient to noise, though a counterexample has not yet been identified. The upcoming theorem expands upon Theorem 4.11, replacing the concept of the trumping relation with the robust relative trumping relation, what we define as catalytic majorization.

**Characterization of Catalytic Majorization**

**Theorem 4.12.** Let $m, n \in \mathbb{N}$, $\mathbf{p}, \mathbf{q} \in \mathrm{Prob}(n)$, and $\mathbf{p}', \mathbf{q}' \in \mathrm{Prob}(m)$ such that either $\mathbf{p} > 0$ or $\mathbf{q} > 0$. Then the following statements are equivalent:

1. $(\mathbf{p}, \mathbf{q}) \succ_c (\mathbf{p}', \mathbf{q}')$
2. For all $\alpha \geqslant \frac{1}{2}$ we have $D_\alpha(\mathbf{p}\|\mathbf{q}) \geqslant D_\alpha(\mathbf{p}'\|\mathbf{q}')$ and $D_\alpha(\mathbf{q}\|\mathbf{p}) \geqslant D_\alpha(\mathbf{q}'\|\mathbf{p}')$.

**Proof**   The proof of the theorem for the special case that either $\mathbf{p} = \mathbf{q}$ or $\mathbf{p}' = \mathbf{q}'$ is very simple and is left as an exercise. Therefore, we assume now that $\mathbf{p} \neq \mathbf{q}$ and $\mathbf{p}' \neq \mathbf{q}'$. Due to the symmetry in the roles of $\mathbf{p}$ and $\mathbf{q}$, and since one of them has full support, we assume without loss of generality (without loss of generality) that it is $\mathbf{q}$; that is, we assume $\mathbf{q} > 0$. The proof of the monotonicity property of $D_\alpha$ under catalytic majorization will be detailed in Chapter 6, where we extensively study the properties of the Rényi divergences. Consequently, this section will only cover the proof of the implication $2 \Rightarrow 1$.

From Exercise 4.49 it follows that for any $\varepsilon > 0$ there exist $\mathbf{q}_\varepsilon \in \mathfrak{B}_\varepsilon(\mathbf{q})$ and $\mathbf{q}'_\varepsilon \in \mathfrak{B}_\varepsilon(\mathbf{q}')$ with $0 < \mathbf{q}_\varepsilon \in \mathrm{Prob}(n) \cap \mathbb{Q}^n$ and $0 < \mathbf{q}'_\varepsilon \in \mathrm{Prob}(m) \cap \mathbb{Q}^m$ such that

$$(\mathbf{p}, \mathbf{q}_\varepsilon) \succ (\mathbf{p}, \mathbf{q}) \quad \text{and} \quad (\mathbf{p}', \mathbf{q}') \succ (\mathbf{p}', \mathbf{q}'_\varepsilon). \tag{4.187}$$

In Chapter 6, we will see that $D_\alpha$ behaves monotonically under both relative majorization and catalytic majorization. Therefore, the relations above, combined with our assumption that $(\mathbf{p}, \mathbf{q}) \succ_c (\mathbf{p}', \mathbf{q}')$, lead to the conclusion that

$$D_\alpha(\mathbf{p}\|\mathbf{q}_\varepsilon) \geqslant D_\alpha(\mathbf{p}\|\mathbf{q}) \geqslant D_\alpha(\mathbf{p}'\|\mathbf{q}') \geqslant D_\alpha(\mathbf{p}'\|\mathbf{q}'_\varepsilon), \tag{4.188}$$

and similarly

$$D_\alpha(\mathbf{q}_\varepsilon\|\mathbf{p}) \geqslant D_\alpha(\mathbf{q}\|\mathbf{p}) \geqslant D_\alpha(\mathbf{q}'\|\mathbf{p}') \geqslant D_\alpha(\mathbf{q}'_\varepsilon\|\mathbf{p}'). \tag{4.189}$$

Now, since both $\mathbf{q}_\varepsilon$ and $\mathbf{q}'_\varepsilon$ have positive rational components, there exists two finite dimensional probability vectors $\mathbf{r}_\varepsilon, \mathbf{r}'_\varepsilon \in \mathrm{Prob}(m_\varepsilon)$, with $m_\varepsilon \in \mathbb{N}$, such that (see Theorem 4.8)

$$\left(\mathbf{r}_\varepsilon, \mathbf{u}^{(m_\varepsilon)}\right) \sim (\mathbf{p}, \mathbf{q}_\varepsilon) \quad \text{and} \quad \left(\mathbf{r}'_\varepsilon, \mathbf{u}^{(m_\varepsilon)}\right) \sim (\mathbf{p}', \mathbf{q}'_\varepsilon). \tag{4.190}$$

Hence, for all $\alpha \geqslant \frac{1}{2}$

$$D_\alpha\left(\mathbf{r}_\varepsilon\|\mathbf{u}^{(m_\varepsilon)}\right) \geqslant D_\alpha\left(\mathbf{r}'_\varepsilon\|\mathbf{u}^{(m_\varepsilon)}\right) \quad \text{and} \quad D_\alpha\left(\mathbf{u}^{(m_\varepsilon)}\|\mathbf{r}_\varepsilon\right) \geqslant D_\alpha\left(\mathbf{u}^{(m_\varepsilon)}\|\mathbf{r}'_\varepsilon\right). \tag{4.191}$$

Our strategy is to use Theorem 4.11 in conjunction with the inequalities above in order to obtain a majorization relation between $\mathbf{r}'_\varepsilon$ and $\mathbf{r}_\varepsilon$. However, since the inequalities above are not strict, we cannot use Theorem 4.11 and will need to tweak a bit the vector $\mathbf{r}'_\varepsilon$.

We first rule out the possibility that $\mathbf{r}'_\varepsilon = \mathbf{u}^{(m_\varepsilon)}$. Consulting the construction in Theorem 4.8, we see that $\mathbf{r}'_\varepsilon = \mathbf{u}^{(m_\varepsilon)}$ implies $\mathbf{p}' = \mathbf{q}'_\varepsilon$. However, this cannot occur for sufficiently small $\varepsilon > 0$ since $\mathbf{q}'_\varepsilon \xrightarrow{\varepsilon \to 0^+} \mathbf{q}' \neq \mathbf{p}'$ by our assumption. Hence, we can assume $\mathbf{r}'_\varepsilon \neq \mathbf{u}^{(m_\varepsilon)}$ for sufficiently small $\varepsilon > 0$. Moreover, observe that for any $\varepsilon \in (0, 1)$, we have (see Exercise 4.2)

$$\mathbf{r}'_\varepsilon \succ \mathbf{s}_\varepsilon := (1 - \varepsilon)\,\mathbf{r}'_\varepsilon + \varepsilon \mathbf{u}^{(m_\varepsilon)}. \tag{4.192}$$

A combination of the relation $\mathbf{r}'_\varepsilon \succ \mathbf{s}_\varepsilon$ (note that $\mathbf{s}_\varepsilon > 0$) with Theorem 4.11 and (4.191), gives the following strict inequalities for all $\alpha \geqslant \frac{1}{2}$:

$$D_\alpha\big(\mathbf{r}_\varepsilon \big\| \mathbf{u}^{(m_\varepsilon)}\big) > D_\alpha\big(\mathbf{s}_\varepsilon \big\| \mathbf{u}^{(m_\varepsilon)}\big) \quad \text{and} \quad D_\alpha\big(\mathbf{u}^{(m_\varepsilon)} \big\| \mathbf{r}_\varepsilon\big) > D_\alpha\big(\mathbf{u}^{(m_\varepsilon)} \big\| \mathbf{s}_\varepsilon\big). \tag{4.193}$$

Since the condition above is equivalent to the condition given in Theorem 4.11, it follows that $\mathbf{r}_\varepsilon \succ_* \mathbf{s}_\varepsilon$. Hence,

$$(\mathbf{p}, \mathbf{q}_\varepsilon) \sim (\mathbf{r}_\varepsilon, \mathbf{u}^{(m_\varepsilon)}) \succ_* (\mathbf{s}_\varepsilon, \mathbf{u}^{(m_\varepsilon)}) \sim (\mathbf{p}'_\varepsilon, \mathbf{q}'_\varepsilon), \tag{4.194}$$

where

$$\mathbf{p}'_\varepsilon := (1 - \varepsilon)\,\mathbf{p}' + \varepsilon \mathbf{q}'_\varepsilon. \tag{4.195}$$

The equivalence $(\mathbf{s}_\varepsilon, \mathbf{u}^{(m_\varepsilon)}) \sim (\mathbf{p}'_\varepsilon, \mathbf{q}'_\varepsilon)$ can be verified from the construction in Theorem 4.8. Since catalytic majorization is robust to small perturbations (see Exercise 4.60), the relation $(\mathbf{p}, \mathbf{q}_\varepsilon) \succ_c (\mathbf{p}'_\varepsilon, \mathbf{q}'_\varepsilon)$ that follows from (4.194) (and holds for any sufficiently small $\varepsilon > 0$) gives $(\mathbf{p}, \mathbf{q}) \succ_c (\mathbf{p}', \mathbf{q}')$. This concludes the proof. $\blacksquare$

**Exercise 4.62.** *Prove the theorem above for the special case that $\mathbf{p} = \mathbf{q}$. Similarly, prove it also for the case that $\mathbf{p}' = \mathbf{q}'$.*

**Exercise 4.63.** *Prove the equivalence $(\mathbf{s}_\varepsilon, \mathbf{u}^{(m_\varepsilon)}) \sim (\mathbf{p}'_\varepsilon, \mathbf{q}'_\varepsilon)$. Hint: Use the construction in Theorem 4.8.*

# 4.6 Conditional Majorization

Conditional majorization is a concept that helps us understand the uncertainty inherent in a physical system when we have access to another correlated system. This idea is formalized as a preorder relationship within the set of probability distributions denoted as $\text{Prob}(mn)$, where $m$ represents the dimension of a classical system denoted as $X$ (held by Alice) and $n$ is the dimension of the "conditioning" system referred to as $Y$ (held by Bob). In essence, this preorder quantifies the uncertainty associated with system $X$ when we possess information about system $Y$. As we delve further into this topic in the book, we'll find that conditional majorization serves as the foundation for defining conditional entropy.

We will use the notation $\mathbf{p}^{XY}$ to denote a probability vector in $\text{Prob}(mn)$, which is $mn$-dimensional. It's important to note that we view $\mathbf{p}^{XY}$ as a probability distribution associated with the *joint* system $XY$. To clarify this perspective, we denote the components of $\mathbf{p}^{XY}$ as $\{p_{xy}\}$ and express $\mathbf{p}^{XY}$ as follows:

$$\mathbf{p}^{XY} = \sum_{x \in [m]} \sum_{y \in [n]} p_{xy} \mathbf{e}_x^X \otimes \mathbf{e}_y^Y, \tag{4.196}$$

where $\{\mathbf{e}_x^X\}_{x \in [m]}$ is the standard basis of $\mathbb{R}^m$, and $\{\mathbf{e}_y^Y\}_{y \in [n]}$ is the standard basis of $\mathbb{R}^n$. While mathematically $\mathbf{p}^{XY}$ is a vector in $\mathrm{Prob}(mn)$, conceptually, we treat it as a joint probability distribution.

To introduce the concept of conditional majorization, we build upon the foundation of conditionally mixing operations. Our objective is to characterize a set of evolution matrices that possess a specific property: they increase the conditional uncertainty associated with system $X$ when provided access to system $Y$. To embark on this journey, consider three classical systems: $X$, $Y$, and $Y'$, each with dimensions $m$, $n$, and $n'$, respectively. Additionally, consider two probability vectors $\mathbf{p}^{XY} \in \mathrm{Prob}(mn)$ and $\mathbf{q}^{XY'} \in \mathrm{Prob}(mn')$. We say that $\mathbf{p}^{XY}$ conditionally majorizes $\mathbf{q}^{XY'}$ and denote it as $\mathbf{p}^{XY} \succ_X \mathbf{q}^{XY'}$ when there exists a conditionally mixing operation (to be define shortly), denoted as $M \in \mathrm{STOC}(mn', mn)$, such that

$$\mathbf{q}^{XY'} = M\mathbf{p}^{XY}. \tag{4.197}$$

The challenge lies in crafting a meaningful definition for $M$ that aligns with the concept of "conditionally mixing." In this context, we demonstrate that there exist three distinct approaches to defining $M$, mirroring the three methodologies introduced in Section 4.1.2. Remarkably, all of these approaches converge to the same definition of conditionally mixing and conditional majorization, thereby establishing a solid foundation for the notion of conditional majorization.

This section is structured as follows: Initially, we introduce both the axiomatic and constructive approaches, showing that they both lead to the identical definition of a conditionally mixing operation. We then use this definition to establish conditional majorization and to examine some of its key properties. Following that, we explore a useful characterization of conditional majorization, paying special attention to cases in smaller dimensions. Lastly, in the final subsection on this topic, we investigate the operational approach to conditional majorization and demonstrate its consistency with the axiomatic and constructive approaches.

## 4.6.1 The Axiomatic Approach

In this approach, we build our foundation upon two minimalistic axioms, which we regard as fundamental for any reasonable definition of a conditionally mixing operation. Much like in the previous context of mixing operations, our framework operates under the assumption that the conditionally mixing operation can be represented by a stochastic matrix denoted by $M$, which belongs to the set $\mathrm{STOC}(mn', mn)$. This matrix $M$ serves as a representation of a classical channel that transforms the joint system $XY$ into $XY'$. In a conceptual sense, we can envision Alice as the possessor of system $X$, while Bob holds systems $Y$ and $Y'$.

## Axiom 1: No-Signaling from Alice to Bob

A key condition that the conditionally mixing operation $M \in \text{STOC}(mn, mn')$ must adhere to in order to avoid decreasing conditional uncertainty relates to the prevention of information leakage from subsystem $X$ to subsystem $Y$. Such leakage could potentially reduce uncertainty about system $X$. Conditional uncertainty specifically pertains to the notion of uncertainty about system $X$ when one has access to system $Y$. To address this, we introduce a minimalistic causality assumption that accounts for the property that system $X$ has no causal effect on system $Y'$. In mathematical terms, this assumption implies that the components of the stochastic matrix $M = (\mu_{x'y'|xy})$ satisfy the following equation for all $x \in [m]$, $y \in [n]$, and $y' \in [n']$:

$$\sum_{x' \in [m]} \mu_{x'y'|xy} = r_{y'|y},  \tag{4.198}$$

where $\{r_{y'|y}\}$ (with $y \in [n]$, and $y' \in [n']$) is some conditional probability distribution independent on $x$. We refer to this condition as nonsignaling from $X$ to $Y'$ or in short $X \not\to Y'$-signalling (see (2.135) for a similar definition).

Matrices that satisfy the above nonsignaling condition have a relatively simple form. Specifically, for every $y \in [n]$ and $y' \in [n']$ let $T^{(y,y')} \in \mathbb{R}_+^{m \times m}$ be the matrix whose components are

$$t_{x'|x}^{(y,y')} := \frac{\mu_{x'y'|xy}}{r_{y'|y}}  \qquad \forall x, x' \in [m].  \tag{4.199}$$

From (4.198) it then follows that $T^{(y,y')}$ is column stochastic; that is, for every $y \in [n]$ and $y' \in [n']$ we have $T^{(y,y')} \in \text{STOC}(m, m)$. With these notations we can express $M$ as (summations run over all $x, x' \in [m]$ and all $y \in [n]$ and $y' \in [n']$)

$$\begin{aligned} M &= \sum_{x,x',y,y'} \mu_{x'y'|xy} |x'\rangle\langle x| \otimes |y'\rangle\langle y| \\ &= \sum_{y,y'} r_{y'|y} \sum_{x,x'} t_{x'|x}^{(y,y')} |x'\rangle\langle x| \otimes |y'\rangle\langle y|, \end{aligned}  \tag{4.200}$$

where we employed quantum notations by denoting $|x'\rangle\langle x|$ (and similarly $|y'\rangle\langle y|$) as the $m \times m$ rank one matrix $\mathbf{e}_{x'}^T \mathbf{e}_x$, which is a matrix with a 1 at the $(x', x)$-position and zeros elsewhere. We therefore conclude that $M \in \text{STOC}(mn, mn')$ is $X \not\to Y'$-signalling if and only if there exists $nn'$ stochastic matrices $T^{(y,y')} \in \text{STOC}(m, m)$ (with $y \in [n]$ and $y' \in [n']$), and another stochastic matrix $R = (r_{y'|y}) \in \text{STOC}(n', n)$ such that

$$M = \sum_{y,y'} r_{y'|y} T^{(y,y')} \otimes |y'\rangle\langle y|.  \tag{4.201}$$

In the following exercise you show that the above form of $M$ represents a bipartite channel that can be realized with one-way communication from Bob to Alice.

**Exercise 4.64.** *Let $M \in \text{STOC}(mn', mn)$. Show that the following two statements are equivalent:*

1. *M is $X \not\rightarrow Y'$-signaling.*

2. *M can be realized with one-way communication from Bob to Alice. That is, M can be expressed as (refer to Figure 4.5)*

$$M = \sum_{j\in[k]} T^{(j)} \otimes R_j, \tag{4.202}$$

*where $k \in \mathbb{N}$, and for each $j \in [k]$, $T^{(j)} \in \mathrm{STOC}(m,m)$, $R_j \in \mathbb{R}_+^{n'\times n}$, and $R := \sum_{j\in[k]} R_j \in \mathrm{STOC}(n',n)$.*

The term $X \not\rightarrow Y'$-signaling is sometimes referred to as $X \not\rightarrow Y'$ semi-causal. The following exercise introduces the notion of $X \not\rightarrow Y'$ semi-causal.

**Exercise 4.65.** *Let $M = (\mu_{x'y'|xy}) \in \mathrm{STOC}(mn',mn)$, and $N = (\nu_{y'|xy}) \in \mathrm{STOC}(n',mn)$, where $\nu_{y'|xy} := \sum_{x'\in[m]} \mu_{x'y'|xy}$.*

1. *Show that if M satisfies (4.198), then for every evolution matrix $E \in \mathrm{STOC}(m,m)$ the marginal channel N satisfies*

$$N(E \otimes I_n) = N. \tag{4.203}$$

*This condition ensure that any operation E that Alice (system X) may chose to apply to her system cannot be detected by Bob (system Y). Such a condition is also called $X \not\rightarrow Y'$ semi-causal. See Figure 4.6 for an illustration of a semi-causal channel.*

2. *Show that if N satisfies (7.14) for all $E \in \mathrm{STOC}(m,m)$, then M satisfies (4.198).*

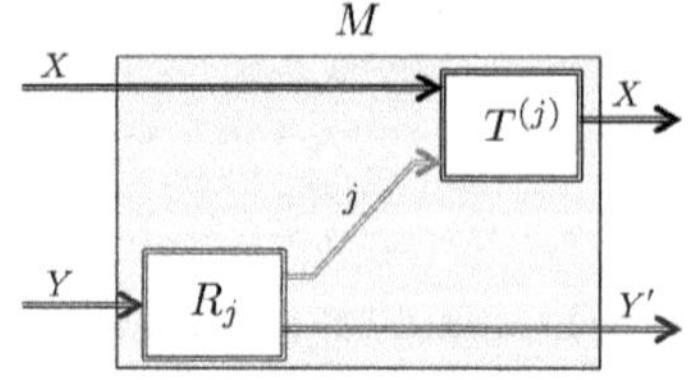



**Figure 4.5** An $X \not\rightarrow Y'$-signaling bipartite channel $M \in \mathrm{STOC}(mn',mn)$.

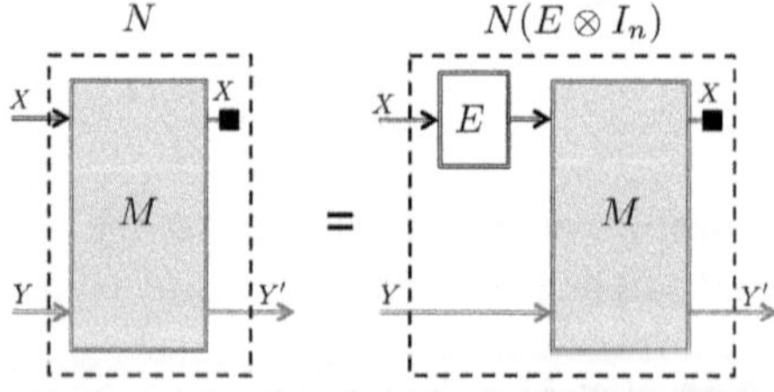



**Figure 4.6** An illustration of a semi-causal classical bipartite channel $M$. The marginal channel $N$ equals $N(E \otimes I_n)$ for any choice of $E \in \mathrm{STOC}(m,m)$.

## Axiom 2: Preservation of Maximal Uncertainty

Consider a joint probability vector in the form of $\mathbf{u}^X \otimes \mathbf{p}^Y$, where $\mathbf{u}^X \in \text{Prob}(m)$ represents the uniform probability vector, and $\mathbf{p}^Y$ is some probability vector in $\text{Prob}(n)$. Since this joint probability distribution is uncorrelated, having access to $Y$ cannot aid in reducing the uncertainty associated with $X$. Therefore, we can assume that such a joint probability vector exhibits the highest degree of conditional uncertainty concerning $X|Y$ (i.e. $X$ given access to $Y$). Therefore, any stochastic (evolution) matrix $M \in \text{STOC}(mn, mn')$ that does not decrease conditional uncertainty must map states with maximal conditional uncertainty to states that also possess maximal conditional uncertainty. In explicit terms, such a stochastic map $M \in \text{STOC}(mn, mn')$ must satisfy, for all $\mathbf{p}^Y \in \text{Prob}(n)$:

$$M\left(\mathbf{u}^X \otimes \mathbf{p}^Y\right) = \mathbf{u}^X \otimes \mathbf{q}^{Y'}, \tag{4.204}$$

where $\mathbf{q}^{Y'}$ is a probability vector in $\text{Prob}(n')$. It is important to note that this assumption is exceedingly minimalistic, as it merely posits that if $X$ is initially maximally uncertain, then applying a conditionally mixing operation should not diminish this maximal uncertainty.

The vector $\mathbf{q}^{Y'}$ in (4.204) is not independent of $M$ and $\mathbf{p}^Y$. In fact, by left-multiplying both sides of the equation with $\mathbf{1}_m^T \otimes I^{Y'}$, we can express the vector $\mathbf{q}^{Y'}$ as follows:

$$\mathbf{q}^{Y'} = \left(\mathbf{1}_m^T \otimes I^{Y'}\right) M \left(\mathbf{u}^X \otimes \mathbf{p}^Y\right). \tag{4.205}$$

Note that the multiplication by row vector $\mathbf{1}_m^T$ in (4.205) effectively functions as the "tracing out" of system $X$.

## Definition of Conditionally Mixing Operations (CMO)

By combining the two axioms presented above, we arrive at the following definition for conditionally mixing operations. In this definition, we consider three classical systems: $X$, $Y$, and $Y'$, with dimensions $m$, $n$, and $n'$, respectively.

> **Definition 4.8.** A stochastic matrix $M \in \text{STOC}(mn', mn)$ is called a conditionally mixing operation if it satisfies the two conditions given in (4.204) and (4.198). The set of all conditionally mixing operations in $\text{STOC}(mn', mn)$ is denoted by $\text{CMO}(mn', mn)$.

Before we delve into characterizing the maps within $\text{CMO}(mn', mn)$, let's explore an alternative approach, which we refer to as the constructive approach. We will demonstrate that this approach ultimately yields the same set of conditionally mixing operations.

## 4.6.2 The Constructive Approach

In the constructive approach we propose to construct conditionally mixing operation as it is intuitively suggests. More precisely, a conditionally mixing operation $M$ is a stochastic map obtained by Alice applying a mixing operation to her system (i.e. doubly stochastic map) conditioned on information received from Bob (see Figure 4.7). Mathematically,

$$M = \sum_{j \in [k]} D^{(j)} \otimes R_j, \tag{4.206}$$

where $j \in [k]$ is the information Bob sends to Alice after he processes his input $y$ via $R_j = (r_{y'j|y})$. Upon receiving $j$ Alice applies a mixing operation to her input $x$ described by the $m \times m$ doubly stochastic matrix $D^{(j)}$.

**Conditionally Doubly Stochastic (CDS)**

**Definition 4.9.** A stochastic matrix $M \in \mathrm{STOC}(mn', mn)$ is said to be conditionally doubly stochastic (CDS) if it has the form given in (4.206) with $k \in \mathbb{N}$, and for each $j \in [k]$, $D^{(j)}$ is an $m \times m$ doubly stochastic matrix, and $R_j \in \mathbb{R}_+^{n' \times n}$ is a sub-stochastic matrix such that $\sum_{j \in [k]} R_j \in \mathrm{STOC}(n', n)$. The set of all conditionally doubly stochastic matrices in $\mathrm{STOC}(mn', mn)$ is denoted by $\mathrm{CDS}(mn', mn)$.

It is important to note that expression in (4.206) is very similar to the one given in (4.202) except that for any fixed $j \in [k]$ the stochastic matrix $T^{(j)}$ is replaced with the doubly stochastic matrix $D^{(j)}$. Therefore, CDS channels are necessarily $X \not\to Y'$ semi-causal. Moreover, since each $D^{(j)}$ is doubly stochastic we get that for any $\mathbf{p}^Y \in \mathrm{Prob}(n)$,

$$M\left(\mathbf{u}^X \otimes \mathbf{p}^Y\right) = \sum_{j \in [k]} D^{(j)}\mathbf{u}^X \otimes R_j\mathbf{p}^Y$$

$$\boxed{D^{(j)}\mathbf{u}^X = \mathbf{u}^X} \longrightarrow = \mathbf{u}^X \otimes R\mathbf{p}^Y, \tag{4.207}$$

where $R = \sum_{j \in [k]} R_j$. That is, $M$ satisfies the condition given in (4.204). We therefore conclude that every CDS channel $M$ is necessarily a CMO. In the next theorem

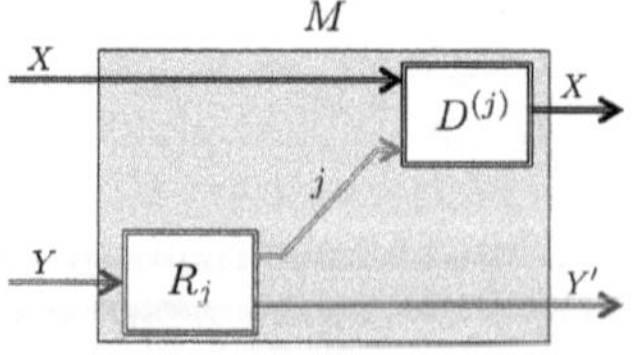

Figure 4.7    A conditionally doubly stochastic (CDS) channel $M \in \mathrm{STOC}(mn', mn)$.

we prove that the converse also holds. Therefore, we get both the axiomatic and the constructive approaches leading to the same set of conditionally mixing operations.

**Exercise 4.66.** *Let $M$ be as in* (4.206). *Show that without loss of generality we can assume that the matrices $D^{(j)}$ are permutation matrices. Hint: Recall that every doubly stochastic matrix is a convex combination of permutation matrices.*

**Exercise 4.67.** *Show that for any doubly stochastic matrix $D \in \mathrm{STOC}(m,m)$ and any stochastic matrix $R \in \mathrm{STOC}(n',n)$ we have that $R \otimes D$ is CDS.*

---

**Theorem 4.13.** Let $m, n, n' \in \mathbb{N}$. Then,

$$\mathrm{CDS}(mn', mn) = \mathrm{CMO}(mn', mn). \tag{4.208}$$

---

**Proof** We have already proved the inclusion $\mathrm{CDS}(mn', mn) \subseteq \mathrm{CMO}(mn', mn)$ (see the discussion in Definition 4.9). To prove the opposite inclusion, suppose $M \in \mathrm{CMO}(mn', mn)$. We want to show that $M \in \mathrm{CDS}(mn', mn)$. For this purpose we will use the form given in (4.201) for $X \not\to Y'$ semi-causal, and show that each $T^{(y,y')}$ is an $m \times m$ doubly stochastic matrix. Observe that if $r_{y'|y} = 0$ for some $y' \in [n']$ and $y \in [n]$, then replacing $T^{(y,y')}$ with the identity matrix will not affect $M$, since $r_{y'|y} = 0$. Consequently, it suffices to demonstrate that $T^{(y,y')}$ is doubly stochastic for those indices $y$ and $y'$ where $r_{y'|y} \neq 0$.

Let $\{\mathbf{e}_y^Y\}_{y \in [n]}$ be the standard basis of $\mathbb{R}^n$ and consider the condition given in (4.204). Fix $y \in [n]$ and observe that from (4.201) we get

$$M\left(\mathbf{u}^X \otimes \mathbf{e}_y^Y\right) = \sum_{y' \in [n']} r_{y'|y} T^{(y,y')} \mathbf{u}^X \otimes \mathbf{e}_{y'}^{Y'}. \tag{4.209}$$

On the other hand, by taking $\mathbf{p}^Y = \mathbf{e}_y^Y$ in (4.204) we get

$$\begin{aligned} M\left(\mathbf{u}^X \otimes \mathbf{e}_y^Y\right) &= \mathbf{u}^X \otimes \mathbf{q}_y^{Y'} \\ &= \sum_{y' \in [n']} q_{y'|y} \mathbf{u}^X \otimes \mathbf{e}_{y'}^{Y'}, \end{aligned} \tag{4.210}$$

for some vector $\mathbf{q}_y^{Y'} := \sum_{y' \in [n']} q_{y'|y} \mathbf{e}_{y'}^{Y'}$ in $\mathrm{Prob}(n')$. Since $\{\mathbf{e}_y^{Y'}\}_{y' \in [n']}$ is an orthonarmal basis of $\mathbb{R}^{n'}$, a comparison of (4.209) and (4.210) reveals that for all $y \in [n]$ and $y' \in [n']$

$$r_{y'|y} T^{(y,y')} \mathbf{u}^X = q_{y'|y} \mathbf{u}^X. \tag{4.211}$$

Observe that since $T^{(y,y')}$ is column stochastic, $T^{(y,y')} \mathbf{u}^X$ is a probability vector so its dot product with the vector $\mathbf{1}_m$ equals 1. Thus, by taking the dot product on both sides of (4.211) with the vector $\mathbf{1}_m$ we get $r_{y'|y} = q_{y'|y}$ for all $y \in [n]$ and $y' \in [n']$. We therefore conclude that $T^{(y,y')} \mathbf{u}^X = \mathbf{u}^X$ (recall we assume $r_{y'|y} \neq 0$). Hence, $T^{(y,y')}$ is doubly stochastic. This completes the proof. ∎

In this proof, we showed that the matrices $T^{(y,y')}$ as appearing in (4.201) are doubly stochastic. We therefore conclude that $M \in \text{STOC}(mn', mn)$ is a conditionally mixing operation if and only if there exists a column stochastic matrix $R \in \text{STOC}(n', n)$ and $nn'$ doubly stochastic matrices $D^{(y,y')} \in \text{STOC}(m, m)$, with $y \in [n]$ and $y' \in [n']$, such that

$$M = \sum_{y \in [n]} \sum_{y' \in [n']} r_{y'|y} D^{(y,y')} \otimes |y'\rangle\langle y|. \tag{4.212}$$

With the above conclusion we arrive at the definition of conditional majorization.

**Conditional Majorization**

**Definition 4.10.** Let $X$, $Y$, and $Y'$, be three classical systems of dimensions $m$, $n$, and $n'$, respectively. Furthermore, let $\mathbf{p}^{XY} \in \text{Prob}(mn)$ and $\mathbf{q}^{XY'} \in \text{Prob}(mn')$. We say that $\mathbf{p}^{XY}$ conditionally majorizes $\mathbf{q}^{XY'}$ with respect to $X$, and write

$$\mathbf{p}^{XY} \succ_X \mathbf{q}^{XY'}, \tag{4.213}$$

if there exists $M \in \text{CMO}(mn', mn)$ such that $\mathbf{q}^{XY'} = M\mathbf{p}^{XY}$. We further write $\mathbf{q}^{XY'} \sim_X \mathbf{p}^{XY}$ if both $\mathbf{p}^{XY} \succ_X \mathbf{q}^{XY'}$ and $\mathbf{q}^{XY'} \succ_X \mathbf{p}^{XY}$.

## 4.6.3 Basic Properties

In this subsection, we study a few basic properties of conditional majorization. For this purpose, we will express $\mathbf{p}^{XY}$ that appears in (4.196) in a more concise form as

$$\mathbf{p}^{XY} = \sum_{y \in [n]} \mathbf{p}_y^X \otimes \mathbf{e}_y^Y, \quad \text{where} \quad \mathbf{p}_y^X := \sum_{x \in [m]} p_{xy} \mathbf{e}_x^X. \tag{4.214}$$

Likewise, if we denote the components of $\mathbf{q}^{XY'}$ as $\{q_{xy'}\}$, we can represent $\mathbf{q}^{XY'}$ as

$$\mathbf{q}^{XY'} = \sum_{y' \in [n']} \mathbf{q}_{y'}^X \otimes \mathbf{e}_{y'}^{Y'}, \quad \text{where} \quad \mathbf{q}_{y'}^X := \sum_{x \in [m]} q_{xy'} \mathbf{e}_x^X. \tag{4.215}$$

Using these notations, the preorder $\mathbf{p}^{XY} \succ_X \mathbf{q}^{XY'}$ can be expressed as a relationship between the two sets of vectors $\{\mathbf{p}_y^X\}_{y \in [n]}$ and $\{\mathbf{q}_{y'}^X\}_{y' \in [n']}$. Observe further that all these vectors have nonnegative components and their sums are given by the marginal probability vectors:

$$\mathbf{p}^X := \sum_{y \in [n]} \mathbf{p}_y^X \in \text{Prob}(m) \quad \text{and} \quad \mathbf{q}^X := \sum_{y' \in [n']} \mathbf{q}_{y'}^X \in \text{Prob}(m). \tag{4.216}$$

By definition, if $\mathbf{p}^{XY} \succ_X \mathbf{q}^{XY'}$, then there exists a matrix $M$ of the form (4.212) such that $\mathbf{q}^{XY'} = M\mathbf{p}^{XY}$. Using the notations above this relation can be expressed as (see Exercise 4.68)

$$\mathbf{q}_{y'} = \sum_{y \in [n]} r_{y'|y} D^{(y,y')} \mathbf{p}_y \qquad \forall \, y' \in [n'], \tag{4.217}$$

where $R = (r_{y'|y}) \in \mathrm{STOC}(n', n)$ and each $D^{(y,y')}$ is an $m \times m$ doubly stochastic matrix.

**Exercise 4.68.** *Prove the relation* (4.217) *using the above form of* $\mathbf{p}^{XY}$ *and* $\mathbf{q}^{XY'}$, *and the form* (4.212) *of* $M$.

The relation (4.217) implies that if $\tilde{\mathbf{p}}^{XY} = \sum_{y \in [n]} \tilde{\mathbf{p}}_y^X \otimes \mathbf{e}_y^Y$ is a probability vector in $\mathrm{Prob}(mn)$ obtained from $\mathbf{p}^{XY}$ by permutation of the components of the vectors $\{\mathbf{p}_y^X\}_{y \in [n]}$, that is, for each $y \in [n]$, $\tilde{\mathbf{p}}_y^X = \Pi^{(y)} \mathbf{p}_y^X$ for some $m \times m$ permutation matrix $\Pi^{(y)}$, then

$$\tilde{\mathbf{p}}^{XY} \sim_X \mathbf{p}^{XY}. \tag{4.218}$$

**Exercise 4.69.** *Prove the relation* (4.218). *Hint: Take in* (4.217), $Y' = Y$, *and for each* $y', y \in [n]$ *take* $r_{y'y} = \delta_{y'y}$ *and* $D^{(y',y')} = \Pi^{(y')}$.

The relation (4.218) implies that without loss of generality we can assume that the components of the vectors $\{\mathbf{p}_y^X\}_{y \in [n]}$ and $\{\mathbf{q}_{y'}^X\}_{y' \in [n']}$ are arranged in nonincreasing order. We will therefore assume this order in the rest of this section.

---

**Theorem 4.14.** Let $\mathbf{p}^{XY} \in \mathrm{Prob}(mn)$ and $\mathbf{q}^{XY'} \in \mathrm{Prob}(mn')$ with $m := |X|$, $n := |Y|$, and $n := |Y'|$. Then, $\mathbf{p}^{XY} \succ_X \mathbf{q}^{XY'}$ if and only if there exists $R = (r_{y'|y}) \in \mathrm{STOC}(n', n)$ such that

$$\sum_{y \in [n]} r_{y'|y} \mathbf{p}_y^X \succ \mathbf{q}_{y'}^X \qquad \forall\, y' \in [n'], \tag{4.219}$$

where $\{\mathbf{p}_y^X\}_{y \in [n]}$ and $\{\mathbf{q}_{y'}^X\}_{y' \in [n']}$ are defined in (4.214) and (4.215), respectively.

---

*Remarks*

1. We assume in Theorem 4.14 that for each $y \in [n]$ we have $\left(\mathbf{p}_y^X\right)^{\downarrow} = \mathbf{p}_y^X$.

2. The vectors $\mathbf{p}_y^X$ and $\mathbf{q}_{y'}^X$ are not necessarily probability vectors since the sum of their components $p_y := \mathbf{1}_m \cdot \mathbf{p}_y^X$ and $q_{y'} := \mathbf{1}_m \cdot \mathbf{q}_{y'}^X$ are in general smaller than 1. Therefore, the majorization relation in (4.219) implies in particular that

$$\sum_{y \in [n]} r_{y'|y} p_y = q_{y'} \qquad \forall\, y' \in [n']. \tag{4.220}$$

We obtained this equality by summing the components on both sides of (4.219).

3. Observe that the relation (4.219) can be expressed also as

$$\sum_{y \in [n]} r_{y'|y} L \mathbf{p}_y^X \geqslant L \mathbf{q}_{y'}^X \qquad \forall\, y' \in [n'], \tag{4.221}$$

where the inequality is entry-wise, and $L$ is the $m \times m$ matrix defined in Exercise 4.5. We will later demonstrate that the inequality (4.221) is instrumental in characterizing conditional majorization as a semidefinite program.

**Proof**  Suppose $\mathbf{p}^{XY} \succ_X \mathbf{q}^{XY'}$ so that the relation (4.217) holds. Since each $D^{(y,y')}$ is doubly stochastic, we have $\mathbf{p}_y^X \succ D^{(y,y')}\mathbf{p}_y^X$. Multiplying both sides of this relation by $r_{y'|y}$ and summing over $y \in [n]$ gives (see Exercise 4.6)

$$\sum_{y\in[n]} r_{y'|y}\mathbf{p}_y \succ \sum_{y\in[n]} r_{y'|y} D^{(y,y')}\mathbf{p}_y \tag{4.222}$$

$$(\mathbf{4.217}){\rightarrow} = \mathbf{q}_{y'}^X.$$

Conversely, suppose (4.219) holds. Then, from Theorem 4.1 for every $y' \in [n']$ there exists a doubly stochastic matrix $D^{(y')} \in \text{STOC}(m,m)$ such that

$$\mathbf{q}_{y'}^X = D^{(y')} \sum_{y\in[n]} r_{y'|y}\mathbf{p}_y^X = \sum_{y\in[n]} r_{y'|y} D^{(y')}\mathbf{p}_y^X. \tag{4.223}$$

Defining $D^{(y,y')} := D^{(y')}$ we conclude that the above relation is a special case of the relation (4.217). Hence, $\mathbf{p}^{XY} \succ \mathbf{q}^{XY'}$. This completes the proof.  ∎

To get a better intuition about conditional majorization, we first consider the cases in which one of the systems $X$, $Y$, and $Y'$ is trivial:

1. The Case $|X| = 1$. This is a trivial case in which there is no uncertainty about system $X$. We therefore expect the preorder to be trivial as well. Indeed, in this case $\mathbf{p}^{XY} = \mathbf{p}^Y \in \text{Prob}(n)$ and $\mathbf{q}^{XY'} = \mathbf{q}^{Y'} \in \text{Prob}(n')$, so the relation (4.219) becomes $\mathbf{q}^{Y'} = R\mathbf{p}^Y$. Since for any $\mathbf{p}^Y \in \text{Prob}(n)$ and any $\mathbf{q}^{Y'} \in \text{Prob}(n')$ there exists a row stochastic matrix $R$ that satisfies $\mathbf{q}^{Y'} = R\mathbf{p}^Y$, we conclude that $\mathbf{p}^{XY} \sim_X \mathbf{q}^{XY'}$ for any probability vectors $\mathbf{p}^{XY}$ and $\mathbf{q}^{XY'}$ with $|X| = 1$.
2. The Case $|Y| = 1$. In this case, $\mathbf{p}^{XY} = \mathbf{p}^X \in \text{Prob}(m)$, and the stochastic matrix $R$ that appears in Theorem 4.14 is a vector $R = \mathbf{r} := (r_1, \ldots, r_{n'})^T \in \text{Prob}(n')$. Therefore, the relation (4.219) becomes $r_{y'}\mathbf{p}^X \succ \mathbf{q}_{y'}^X$ for all $y' \in [n']$. Moreover, denoting by $q_{y'} := \mathbf{1}_{n'} \cdot \mathbf{q}_{y'}$ the sum of the components of $\mathbf{q}_{y'}$, we get from (4.220) that $r_{y'} = q_{y'}$ for all $y' \in [n']$. We therefore conclude that for $|Y| = 1$,

$$\mathbf{p}^X \succ_X \mathbf{q}^{XY'} \qquad \Longleftrightarrow \qquad \mathbf{p}^X \succ \mathbf{q}_{|y'}^X \qquad \forall\, y' \in [n'], \tag{4.224}$$

where $\mathbf{q}_{|y'}^X := \frac{1}{q_{y'}}\mathbf{q}_{y'}^X$ is the vector whose components are $\{q_{x|y'}\}_{x\in[m]}$.
3. The Case $|Y'| = 1$. In this case, $\mathbf{q}^{XY'} = \mathbf{q}^X \in \text{Prob}(m)$, and since $R$ in Theorem 4.14 has to be an $1 \times n$ column stochastic matrix, it must be equal to the row vector $(1, \ldots, 1)$. Therefore, denoting by $\mathbf{p}^X := \sum_{y\in[n]} \mathbf{p}_y^X$ we get from Theorem 4.14 that

$$\mathbf{p}^{XY} \succ_X \mathbf{q}^X \qquad \Longleftrightarrow \qquad \mathbf{p}^X \succ \mathbf{q}^X. \tag{4.225}$$

**Exercise 4.70.** *Let* $\mathbf{p}^{XY} \in \text{Prob}(mn)$ *and* $\mathbf{q}^{XY'} \in \text{Prob}(mn')$. *Show that if* $\mathbf{p}^{XY} \succ_X \mathbf{q}^{XY'}$, *then* $\mathbf{p}^X \succ \mathbf{q}^X$.

## 4.6.4 Standard Form

Consider the majorization relation between probability vectors in $\text{Prob}(n)$. This relation is not a partial order since if two vectors $\mathbf{p}, \mathbf{q} \in \text{Prob}(n)$ satisfy both $\mathbf{p} \succ \mathbf{q}$ and

$\mathbf{q} \succ \mathbf{p}$ it is still possible that $\mathbf{p} \neq \mathbf{q}$. However, the majorization relation becomes a partial order if we assume that the components of the vectors are given in some *standard form*. For example, we can define the standard form of a vector $\mathbf{p} \in \mathrm{Prob}(n)$ to be $\mathbf{p}^{\downarrow}$. Then, the majorization is a partial order when restricted to this standard form; that is, to vectors in $\mathrm{Prob}^{\downarrow}(n)$.

Similarly, we would like to define a standard form for conditional majorization. For this purpose, we will consider two probability vectors $\mathbf{p}^{XY} \in \mathrm{Prob}(mn)$ and $\mathbf{q}^{XY'} \in \mathrm{Prob}(mn')$ and characterize the relation $\mathbf{p}^{XY} \sim_X \mathbf{q}^{XY'}$, meaning $\mathbf{p}^{XY} \succ_X \mathbf{q}^{XY'}$ and $\mathbf{q}^{XY'} \succ_X \mathbf{p}^{XY}$. We start by discussing three examples of vectors in $\mathrm{Prob}(mn)$ that are equivalent under conditional majorization to the vector $\mathbf{p}^{XY}$ as defined in (4.214):

1. In (4.218), we saw that if $\tilde{\mathbf{p}}^{XY}$ is obtained from $\mathbf{p}^{XY}$ by applying an arbitrary permutation matrix to each of the vectors $\{\mathbf{p}_y^X\}_{y \in [n]}$, we get that $\tilde{\mathbf{p}}^{XY} \sim_X \mathbf{p}^{XY}$.
2. Similarly, for any permutation/bijection $\pi : [n] \to [n]$ we get that (see Exercise 4.71)

$$\mathbf{p}^{XY} \sim_X \sum_{y \in [n]} \mathbf{p}_{\pi(y)}^X \otimes \mathbf{e}_y^Y. \tag{4.226}$$

3. Suppose that there exists $\lambda \in \mathbb{R}_+$ such that $\mathbf{p}_1^X = \lambda \mathbf{p}_2^X$. Then (see Exercise 4.72),

$$\mathbf{p}^{XY} \sim_X (\mathbf{p}_1^X + \mathbf{p}_2^X) \otimes \mathbf{e}_1^{Y'} + \sum_{y=3}^{n} \mathbf{p}_y^X \otimes \mathbf{e}_{y-1}^{Y'}, \tag{4.227}$$

where $Y'$ is a system of dimension $|Y'| = n - 1$.

**Exercise 4.71.** *Prove (4.226). Hint: Take $r_{y'|y} = \delta_{y'\pi(y)}$ and $D^{(y,y')} = I_m$.*

**Exercise 4.72.** *Suppose that there exists $\lambda \in \mathbb{R}_+$ such that $\mathbf{p}_1^X = \lambda \mathbf{p}_2^X$. Prove the equivalence relation given in (4.227). Hint: To show $LHS \succ_X RHS$, find $R \in \mathrm{STOC}(n-1, n)$ that satisfies for $j \in \{1, 2\}$, $R\mathbf{e}_j^Y = \mathbf{e}_1^{Y'}$, and for $j \in \{3, \ldots, n\}$, $R\mathbf{e}_j^Y = \mathbf{e}_{j-1}^{Y'}$. To show $RHS \succ_X LHS$, find $R \in \mathrm{STOC}(n, n-1)$ that satisfies $R\mathbf{e}_1^{Y'} = \frac{1}{1+\lambda}\mathbf{e}_1^Y + \frac{\lambda}{1+\lambda}\mathbf{e}_2^Y$ and for $j \in \{2, \ldots, n-1\}$, $R\mathbf{e}_j^{Y'} = \mathbf{e}_{j+1}^Y$.*

Recall that the vectors $\{\mathbf{p}_y^X\}_{y \in [n]}$ associated with $\mathbf{p}^{XY}$ are not probability vectors. Hence, it will be convenient to denote for every $y \in [n]$ $p_y := \mathbf{1}_m \cdot \mathbf{p}_y$ and $p_{x|y} := p_{xy}/p_y$. Since the re-ordering of the vectors $\{\mathbf{p}_y^X\}_{y \in [n]}$ is a reversible CMO operation, we can choose a particular order to be the order of the "standard form." The choice of this order is somewhat arbitrary, but it will be useful (in particular for the spacial case in which $|X| = 2$ as we will see later) to order the vectors $\{\mathbf{p}_y^X\}_{y \in [n]}$ such that

$$p_{1|1} \geqslant p_{1|2} \geqslant \cdots \geqslant p_{1|n}. \tag{4.228}$$

Moreover, if there exists $y \in [n-1]$ such that $p_{1|y} = p_{1|y+1}$ then, if necessary, we will exchange the vectors $\mathbf{p}_y^X$ and $\mathbf{p}_{y+1}^X$ so that $p_{2|y} \geqslant p_{2|y+1}$. If the latter inequality is also an equality, we continue by induction until we get a $k \in [m-1]$ such that $p_{x|y} = p_{x|y+1}$ for all $x \in [k]$ but $p_{k+1|y} > p_{k+1|y+1}$. Combining these observations with the exercise above we are ready to define the standard form.

> **Standard Form**
>
> Let $\mathbf{p}^{XY} \in \mathrm{Prob}(mn)$ be as defined in (4.214). We say that $\mathbf{p}^{XY}$ is given in the standard form if the vectors $\{\mathbf{p}_y^X\}_{y\in[n]}$ satisfy the following three conditions:
>
> 1. For all $y \in [n]$, $\mathbf{p}_y^X = \left(\mathbf{p}_y^X\right)^{\downarrow}$.
> 2. The vectors $\{\mathbf{p}_y^X\}_{y\in[n]}$ are arranged as discussed around (4.228).
> 3. There is no $\lambda \in \mathbb{R}$ such that $\mathbf{p}_y^X = \lambda \mathbf{p}_w^X$ for some $y, w \in [n]$ with $y \neq w$.

**Exercise 4.73.** *Let $\mathbf{p}^{XY} \in \mathrm{Prob}(mn)$ and $\mathbf{q}^{XY'} \in \mathrm{Prob}(mn')$ be two probability vectors given in their standard form, and $L$ be the $m \times m$ matrix defined in Exercise 4.5. Use Theorem 4.14 to show that $\mathbf{p}^{XY} \succ_X \mathbf{q}^{XY'}$ if and only if there exists $R \in \mathrm{STOC}(n',n)$ such that*

$$(L \otimes R)\mathbf{p}^{XY} \geqslant (L \otimes I_{n'})\,\mathbf{q}^{XY'}, \tag{4.229}$$

*where the inequality is entrywise.*

Based on the preceding discussion, particularly the three examples provided, we can deduce that any $\mathbf{p}^{XY}$ is under conditional majorization, equivalent to its standard form. Consequently, without loss of generality, we may always assume that $\mathbf{p}^{XY} \in \mathrm{Prob}(mn)$ is presented in its standard form. We will now demonstrate that conditional majorization between vectors in standard form indeed constitutes a partial order.

> **Theorem 4.15.** Let $\mathbf{p}^{XY} \in \mathrm{Prob}(mn)$ and $\mathbf{q}^{XY'} \in \mathrm{Prob}(mn')$ be two probability vectors given in their standard forms. Suppose further that $\mathbf{p}^{XY} \sim_X \mathbf{q}^{XY'}$. Then, $\mathbf{p}^{XY} = \mathbf{q}^{XY'}$ (in particular, $Y = Y'$ and $n = n'$).

**Proof** From Exercise 4.73, the relation $\mathbf{p}^{XY} \sim_X \mathbf{q}^{XY'}$ implies that there exists $R \in \mathrm{STOC}(n',n)$ and $R' \in \mathrm{STOC}(n,n')$ such that

$$(L \otimes R)\mathbf{p}^{XY} \geqslant (L \otimes I_{n'})\,\mathbf{q}^{XY'} \quad \text{and} \quad (L \otimes R')\mathbf{q}^{XY'} \geqslant (L \otimes I_n)\,\mathbf{p}^{XY}. \tag{4.230}$$

Denote by $S = R'R$ and $S' := RR'$, and observe that the two equations above imply that (see Exercise 4.74)

$$(L \otimes S)\mathbf{p}^{XY} \geqslant (L \otimes I_n)\,\mathbf{p}^{XY} \quad \text{and} \quad (L \otimes S')\mathbf{q}^{XY'} \geqslant (L \otimes I_n)\,\mathbf{q}^{XY'}. \tag{4.231}$$

Denoting by $s_{y|w}$ the $(w, y)$-component of the matrix $S$ we get that the equation above is equivalent to

$$\sum_{w\in[n]} s_{y|w} L\mathbf{p}_w \geqslant L\mathbf{p}_y \qquad \forall\, y \in [n]. \tag{4.232}$$

On the other hand, observe that by taking the sum over $y \in [n]$ on both sides of the equation above we get an equality between the two sides. Therefore, all the $n$ inequalities above must be equalities! Multiplying both sides by the inverse of $L$ gives

$$\mathbf{p}_y = \sum_{w \in [n]} s_{y|w} \mathbf{p}_w \qquad \forall\, y \in [n]. \tag{4.233}$$

Observe that these equalities can be expressed as

$$(1 - s_{y|y})\mathbf{p}_y = \sum_{\substack{w \in [n] \\ w \neq y}} s_{y|w} \mathbf{p}_w \qquad \forall\, y \in [n]. \tag{4.234}$$

We now argue that $s_{y|y} = 1$ for all $y \in [n]$ (which is equivalent to $s_{y|w} = \delta_{yw}$ and $S = I_n$). Otherwise, suppose by contradiction that there exists $y \in [n]$ such that $s_{y|y} < 1$. Without loss of generality suppose that this $y$ is $n$. We then get that

$$\mathbf{p}_n = \sum_{w \in [n-1]} \frac{s_{n|w}}{1 - s_{n|n}} \mathbf{p}_w. \tag{4.235}$$

Substituting this into (4.233) gives for all $y \in [n-1]$

$$\mathbf{p}_y = \sum_{w \in [n-1]} s_{y|w} \mathbf{p}_w + s_{y|n}\mathbf{p}_n = \sum_{w \in [n-1]} s_{y|w} \mathbf{p}_w + s_{y|n} \sum_{w \in [n-1]} \frac{s_{n|w}}{1 - s_{n|n}} \mathbf{p}_w. \tag{4.236}$$

Denoting by

$$t_{y|w} := s_{y|w} + \frac{s_{y|n} s_{n|w}}{1 - s_{n|n}} \tag{4.237}$$

we conclude that

$$\mathbf{p}_y = \sum_{w \in [n-1]} t_{y|w} \mathbf{p}_w \qquad \forall\, y \in [n-1]. \tag{4.238}$$

Observe that $\sum_{y \in [n-1]} t_{y|w} = 1$. Next, we rule out the case $t_{y|w} = \delta_{yw}$ for all $y, w \in [n-1]$. Otherwise, this relation implies in particular that for all $y, w \in [n-1]$ with $y \neq w$ we have $s_{y|w} = 0$ and $s_{y|n}s_{n|w} = 0$. Now, recall that we assumed that $s_{n|n} < 1$ so there exists $y \in [n-1]$ such that $s_{y|n} > 0$. For this choice of $y \in [n-1]$ the relation $s_{y|n}s_{n|w} = 0$ gives $s_{n|w} = 0$ for all $w \neq y$. Substituting this into (4.235) gives

$$\mathbf{p}_n = \frac{s_{n|y}}{1 - s_{n|n}} \mathbf{p}_y \tag{4.239}$$

in contradiction with the third property of the standard form of the vector $\mathbf{p}^{XY}$. We therefore conclude that there must exist $y \in [n-1]$ such that $t_{y|y} < 1$. Observe that we started with the relation (4.233) with the condition that there exists $s_{y|y} < 1$ for some $y \in [n]$, and we reduced it to the relation (4.238) with the condition that there exists $t_{y|y} < 1$ for some $y \in [n-1]$. Continuing by induction until we have only one term in the sum on the right-hand side of (4.233) (or of (4.238)), we conclude that one of the vectors of $\{\mathbf{p}_y^X\}_{y \in [n]}$ is proportional to another vector in the same set, in contradiction with the standard form of $\mathbf{p}^{XY}$. Therefore, the assumption that there exists $y \in [n]$ such that $s_{y|y} < 1$ is incorrect, and we conclude that $S = I_n$ or equivalently $R'R = I_n$.

Moreover, following the same arguments as above we conclude that also $S' := RR' = I_{n'}$. Combining this with $R'R = I_n$ we must have $n' = n$ and $R' = R^{-1}$. However, the only stochastic matrix whose inverse is also stochastic is a permutation matrix (i.e. doubly stochastic and orthogonal). We therefore conclude that the sets

$\{\mathbf{p}_y^X\}_{y\in[n]}$ and $\{\mathbf{q}_y^X\}_{y\in[n]}$ can only differ up to a permutation; that is, for all $y \in [n]$, $\mathbf{p}_y^X = \mathbf{q}_{\pi(y)}^Y$ for some permutation $\pi: [n] \to [n]$. However, since the $\{\mathbf{p}_y^X\}_{y\in[n]}$ and $\{\mathbf{q}_y^X\}_{y\in[n]}$ are ordered in a specific way given in the second property of the standard form, we conclude that $\pi(y) = y$ for all $y \in [n]$. This completes the proof. $\blacksquare$

**Exercise 4.74.** *Prove the relations in (4.231). Hint: Multiply both sides of the first inequality in (4.230) by $R'$ and the second inequality by $R$.*

## 4.6.5 Conditional Schur Convex Functions

Conditionally Schur convex functions are functions from Prob($mn$) to the real line that behave monotonically under conditional majorization. Such functions generalize Schur convex functions, and can be used to quantify the amount of conditional uncertainty contained in a correlated source; that is, they are measures of conditional uncertainty. In Chapter 7, we will study a class of conditionally Schur concave functions known as conditional entropies.

---

**Definition 4.11.** A function $f: \bigcup_{n,m\in\mathbb{N}} \text{Prob}(mn) \to \mathbb{R}$ is said to be conditionally Schur-convex if for every $\mathbf{p}^{XY} \in \text{Prob}(mn)$ and $\mathbf{q}^{XY'} \in \text{Prob}(mn')$

$$\mathbf{p}^{XY} \succ_X \mathbf{q}^{XY'} \quad \Rightarrow \quad f\left(\mathbf{p}^{XY}\right) \geqslant f\left(\mathbf{q}^{XY'}\right). \tag{4.240}$$

---

*Remark.* In addition to Definition 4.11, $f$ is said to be conditionally Schur-concave if $-f$ is conditionally Schur-convex.

Observe that the conditionally Schur convex functions reduce to Schur convex functions when restricted to Prob($m$) (i.e. $n = 1$). Conversely, in the following theorem we show that every convex symmetric function on the set of probability vectors can be extended to a conditionally Schur convex function (see Theorem 4.16). Remember that in Subsection 4.1.3, we established that such symmetric convex functions are in particular Schur convex.

In Theorem 4.16, for every convex symmetric function $f: \text{Prob}(m) \to \mathbb{R}$ we define its extension, $H_f$, to any vector $\mathbf{p}^{XY} \in \text{Prob}(mn)$ via

$$H_f(\mathbf{p}^{XY}) := \sum_{y\in[n]} p_y f\left(\mathbf{p}_{|y}^X\right), \tag{4.241}$$

where $\mathbf{p}_{|y} := \frac{1}{p_y}\mathbf{p}_y$ is the probability vector whose components are $\{p_{x|y}\}_{x\in[m]}$.

---

**Theorem 4.16.** Let $f: \bigcup_{m\in\mathbb{N}} \text{Prob}(m) \to \mathbb{R}$ be a symmetric convex function. Then, the function $H_f$, as defined in (4.241), is conditionally Schur concave.

---

**Proof**   Recall that $\mathbf{p}^{XY} \succ_X \mathbf{q}^{XY}$ if and only if there exists a stochastic matrix $R \in$ STOC$(n', n)$ such that (4.219) holds. By rewriting (4.219) with $\mathbf{p}_y^X = p_y \mathbf{p}_{|y}^X$ and $\mathbf{q}_{y'}^X = q_{y'} \mathbf{q}_{|y'}^X$ we get that

$$\sum_{y \in [n]} \frac{r_{y'|y} p_y}{q_{y'}} \mathbf{p}_{|y}^X \succ \mathbf{q}_{|y'}^X \qquad \forall \, y' \in [n']. \tag{4.242}$$

Therefore, if $\mathbf{p}^{XY} \succ_X \mathbf{q}^{XY}$, then

$$
\begin{aligned}
H_f\left(\mathbf{q}^{XY'}\right) &= \sum_{y' \in [n']} q_{y'} f\left(\mathbf{q}_{|y'}^X\right) \\
\xrightarrow{\;\boxed{f \text{ is Schur convex}}\;} &\leqslant \sum_{y'=1}^{n'} q_{y'} f\left(\sum_{y \in [n']} \frac{r_{y'|y} p_y}{q_{y'}} \mathbf{p}_{|y}^X\right),
\end{aligned}
\tag{4.243}
$$

where we used (4.242). Moreover, from (4.220) we have

$$\sum_{y \in [n]} \frac{r_{y'|y} p_y}{q_{y'}} = 1. \tag{4.244}$$

Thus, since $f$ is convex we get from (4.243)

$$
\begin{aligned}
H_f\left(\mathbf{q}^{XY'}\right) &\leqslant \sum_{y' \in [n']} \sum_{y \in [n]} r_{y'|y} p_y f\left(\mathbf{p}_{|y}^X\right) \\
\xrightarrow{\;\boxed{R \text{ is stochastic}}\;} &= \sum_{y \in [n]} p_y f\left(\mathbf{p}_{|y}^X\right) = H_f(\mathbf{p}^{XY}).
\end{aligned}
\tag{4.245}
$$

This completes the proof.     ∎

## 4.6.6 Two-Dimensional Cases

We study here the relatively simpler cases in which the dimension of the systems $X$ or $Y$ is 2. In these cases, it is possible to get the exact analytical expressions that determine if one given state is conditionally majorized by another. We will see in the next section that in the more general case, conditional majorization can be characterized with a linear program.

### The Case $|X| = 2$

In this case, the vectors $\{\mathbf{p}_y^X\}_{y \in [n]}$ and $\{\mathbf{q}_w^X\}_{w \in [n']}$ are all two dimensional. It will be convenient to denote their components as follows:

$$\mathbf{p}_y^X := \begin{pmatrix} a_y \\ p_y - a_y \end{pmatrix} \quad \text{and} \quad \mathbf{q}_{y'}^X := \begin{pmatrix} b_{y'} \\ q_{y'} - b_{y'} \end{pmatrix}, \tag{4.246}$$

where $y \in [n]$, $y' \in [n']$, $a_y := p_{1y}$, $b_{y'} := q_{1y'}$, and $p_y := p_{1y} + p_{2y}$ and $q_{y'} := q_{1y'} + q_{2y'}$ are the sums of the components of $\mathbf{p}_y^X$ and $\mathbf{q}_{y'}^X$, respectively. With these notations we get for all $y \in [n]$ and $y' \in [n']$

$$Lp_y^X = \begin{pmatrix} a_y \\ p_y \end{pmatrix} \quad \text{and} \quad Lq_{y'}^X = \begin{pmatrix} b_{y'} \\ q_{y'} \end{pmatrix}, \tag{4.247}$$

where $L := \begin{pmatrix} 1 & 0 \\ 1 & 1 \end{pmatrix}$. Moreover, since we assume that $\mathbf{p}^{XY}$ and $\mathbf{q}^{XY'}$ are given in their standard form, we have (see (4.228))

$$\frac{a_1}{p_1} \geqslant \cdots \geqslant \frac{a_n}{p_n} \quad \text{and} \quad \frac{b_1}{q_1} \geqslant \cdots \geqslant \frac{b_{n'}}{q_{n'}}. \tag{4.248}$$

From Theorem 4.14 we have that $\mathbf{p}^{XY} \succ_X \mathbf{q}^{XY'}$ if and only if there exists a row stochastic matrix $R \in \mathrm{STOC}(n', n)$ such that for all $y' \in [n']$ we have

$$b_{y'} \leqslant \sum_{y \in [n]} r_{y'|y} a_y \quad \text{and} \quad q_{y'} \leqslant \sum_{y \in [n]} r_{y'|y} p_y. \tag{4.249}$$

Observe that since $\sum_{y' \in [n']} q_{y'} = \sum_{y \in [n]} p_y = 1$, the second inequality above must hold with equality (in fact, we know it already from (4.220)).

Let $\mathbf{a}, \mathbf{b}, \mathbf{p},$ and $\mathbf{q}$ be the vectors whose components are, respectively, $\{a_y\}_{y \in [n]}$, $\{b_{y'}\}_{y' \in [n']}$, $\{p_y\}_{y \in [n]}$, and $\{q_{y'}\}_{y' \in [n']}$. To streamline our analysis, we omitted the superscripts $Y$ and $Y'$ when referring to the vectors $\mathbf{a}, \mathbf{b}, \mathbf{p},$ and $\mathbf{q}$. It is important for the reader to bear in mind that the vectors $\mathbf{a} := \mathbf{a}^Y$ and $\mathbf{p} := \mathbf{p}^Y$ correspond to a system of dimension $n$ (referred to as system $Y$), while $\mathbf{b} := \mathbf{b}^{Y'}$ and $\mathbf{q} := \mathbf{q}^{Y'}$ pertain to a system of dimension $n'$ (referred to as system $Y'$). With these notations we get that $\mathbf{p}^{XY} \succ_X \mathbf{q}^{XY'}$ if and only if there exists $R \in \mathrm{STOC}(n', n)$ such that

$$R\mathbf{a} \geqslant \mathbf{b} \quad \text{and} \quad R\mathbf{p} = \mathbf{q}. \tag{4.250}$$

This relation is closely related to relative majorization; however, note that the vectors $\mathbf{a}$ and $\mathbf{b}$ are not probability vectors since their components in general don't sum to 1. We therefore say in this case that the pair $(\mathbf{a}, \mathbf{p})$ *relatively submajorizes* the pair $(\mathbf{b}, \mathbf{q})$. Observe also that if $a := \|\mathbf{a}\|_1$ equals $b := \|\mathbf{b}\|_1$, then the inequality $R\mathbf{a} \geqslant \mathbf{b}$ can be replaced with $R\mathbf{a} = \mathbf{b}$ so that (4.250) becomes equivalent to relative majorization; that is,

$$\left( \frac{1}{a}\mathbf{a}, \mathbf{p} \right) \succ \left( \frac{1}{b}\mathbf{b}, \mathbf{q} \right). \tag{4.251}$$

We therefore assume now that $a > b$ (the case $a < b$ is not possible since $R\mathbf{a} \geqslant \mathbf{b}$, and $R$ is column stochastic).

Even though the components of $\mathbf{a}$ do not sum to 1 (in general), we can still define its testing region as (see (4.139))

$$\mathfrak{T}(\mathbf{a}, \mathbf{p}) := \left\{ (\mathbf{a} \cdot \mathbf{t}, \mathbf{p} \cdot \mathbf{t}) : \mathbf{t} \in [0, 1]^n \right\}. \tag{4.252}$$

By taking $\mathbf{t} = \mathbf{1}_n$ we get the point $(a, 1) \in \mathfrak{T}(\mathbf{a}, \mathbf{p})$ as opposed to the point $(1, 1)$ that one would get if $\mathbf{a}$ was a probability vector. In fact, the testing region of the pair of probability vectors $(\frac{1}{a}\mathbf{a}, \mathbf{p})$ is almost identical to that of $(\mathbf{a}, \mathbf{p})$ except for a rescaling of the $x$-axis by a factor of $a$; that is, $(r, s) \in \mathfrak{T}(\frac{1}{a}\mathbf{a}, \mathbf{p})$ if and only if $(ar, s) \in \mathfrak{T}(\mathbf{a}, \mathbf{p})$. Therefore, if $(r, s)$ is an extreme point of $\mathfrak{T}(\frac{1}{a}\mathbf{a}, \mathbf{p})$ then $(ar, s)$ is an extreme point of

$\mathfrak{T}(\mathbf{a},\mathbf{p})$. That is, there are $n+1$ extreme points on the lower Lorenz curve of $\mathfrak{T}(\mathbf{a},\mathbf{p})$ given by $(0,0)$ and the $n$ points $\{(\mu_\ell,\nu_\ell)\}_{\ell\in[n]}$, where

$$\mu_\ell := \sum_{x\in[\ell]} a_x \quad \text{and} \quad \nu_\ell := \sum_{x\in[\ell]} p_x. \tag{4.253}$$

Recall that since we assume that $\mathbf{p}^{XY}$ is given in its standard form, the components of $\mathbf{a}$ and $\mathbf{p}$ satisfy (4.248). See the red line in Figure 4.8 for an example of the lower Lorenz curve of the pair $(\mathbf{a},\mathbf{p})$.

In Figure 4.8, we also depicted another Lorenz curve, by taking it to be identical to the Lorenz curve of $(\mathbf{a},\mathbf{p})$ if the $x$-coordinate is no greater than $b$, and a vertical line if the $x$-coordinate equals $b$. This curve is a Lorenz curve of some pair of vectors $(\tilde{\mathbf{a}},\tilde{\mathbf{p}})$ for which $\|\tilde{\mathbf{a}}\|_1 = b$ and $\tilde{\mathbf{p}}$ is a probability vector. Moreover, the Lorenz curve of $(\tilde{\mathbf{a}},\tilde{\mathbf{p}})$ has the property that *any* other Lorenz curve $\mathrm{LC}(\mathbf{b},\mathbf{q})$ (see the top curve in Figure 4.8) that is no where below the Lorenz curve of $(\mathbf{a},\mathbf{p})$ is also nowhere below the Lorenz curve of $(\tilde{\mathbf{a}},\tilde{\mathbf{p}})$. We will see shortly that this implies that the relation $\mathbf{p}^{XY} \succ_X \mathbf{q}^{XY'}$ is equivalent to $(\tilde{\mathbf{a}},\tilde{\mathbf{p}}) \succ (\mathbf{b},\mathbf{p})$.

The vectors $\tilde{\mathbf{a}}$ and $\tilde{\mathbf{p}}$ that correspond to the purple Lorenz curve of Figure 4.8 can be expressed as follows. Let $k \in [n-1]$ be the integer satisfying $\mu_k \leqslant b < \mu_{k+1}$, or equivalently

$$b - a_{k+1} < \mu_k \leqslant b. \tag{4.254}$$

Such an index $k$ exists since we assume that $a > b$. The line connecting the vertices $\mathbf{v}_k := (\mu_k,\nu_k)$ with $\mathbf{v}_{k+1} := (\mu_{k+1},\nu_{k+1})$ contains the point $(b,\lambda)$ (see Figure 4.8), where

$$\lambda := \frac{p_{k+1}}{a_{k+1}}(b - \mu_k) + \nu_k. \tag{4.255}$$

The point $(1,\lambda)$ is therefore a vertex of the curve $\mathrm{LC}(\tilde{\mathbf{a}},\tilde{\mathbf{p}})$ in Figure 4.8. With this notation, $\tilde{\mathbf{a}}$ and $\tilde{\mathbf{p}}$ are given by

$$\begin{aligned}
\tilde{\mathbf{a}} &:= (a_1,\dots,a_k,b-\mu_k,0)^T \in \mathbb{R}_+^{k+2} \\
\tilde{\mathbf{p}} &:= (p_1,\dots,p_k,\lambda-\nu_k,1-\lambda)^T \in \mathrm{Prob}(k+2).
\end{aligned} \tag{4.256}$$

With these notations we have the following characterization of conditional majorization.

> **Theorem 4.17.** Using the same notations as above, for $|X|=2$ the following statements are equivalent:
>
> 1. $\mathbf{p}^{XY} \succ_X \mathbf{q}^{XY'}$
> 2. The curve $\mathrm{LC}(\mathbf{a},\mathbf{p})$ is nowhere above the curve $\mathrm{LC}(\mathbf{b},\mathbf{p})$.
> 3. $(\tilde{\mathbf{a}},\tilde{\mathbf{p}}) \succ (\mathbf{b},\mathbf{q})$.

**Proof** The case $a=b$ is relatively simple and is left as an exercise. We therefore prove the theorem for the case $a > b$. Suppose $\mathbf{p}^{XY} \succ_X \mathbf{q}^{XY'}$. Then, there exists

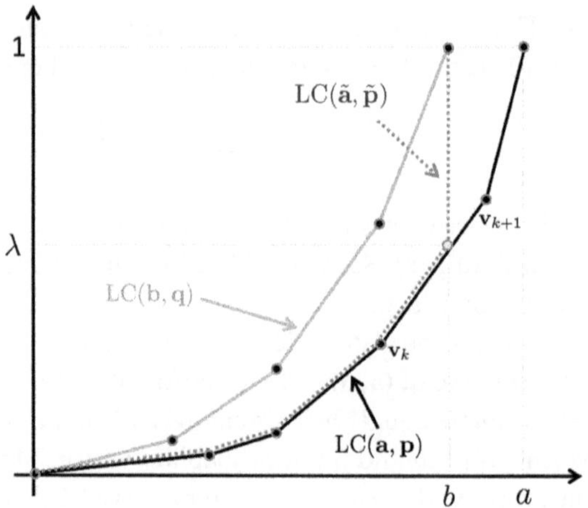


**Figure 4.8** Submajorization. Given the Lorenz curve LC(a, p), we can always construct another Lorenz curve, LC(ã, p̃), with $\|\tilde{a}\|_1 = b$ such that *any* Lorenz curve LC(b, q) that is no where below LC(a, p) is also nowhere below LC(ã, p̃). In this example $n = 5$ and $k = 3$.

$R \in \mathrm{STOC}(n', n)$ such that (4.250) holds. Let $(\mathbf{b} \cdot \mathbf{t}', \mathbf{q} \cdot \mathbf{t}') \in \mathrm{LC}(\mathbf{b}, \mathbf{q})$ be a point on the lower Lorenz curve of the testing region of $(\mathbf{b}, \mathbf{q})$, where $\mathbf{t}'$ is some vector in $[0, 1]^{n'}$. Then, the vector $\mathbf{t} := R^T \mathbf{t}' \in [0, 1]^n$ has the property that

$$\mathbf{a} \cdot \mathbf{t} = \mathbf{a}^T R^T \mathbf{t}' \geqslant \mathbf{b} \cdot \mathbf{t}' \quad \text{and} \quad \mathbf{p} \cdot \mathbf{t} = \mathbf{p}^T R^T \mathbf{t}' = \mathbf{q} \cdot \mathbf{t}', \tag{4.257}$$

where we used the relations in (4.250). The above relations imply that for any point $(\mathbf{b} \cdot \mathbf{t}', \mathbf{q} \cdot \mathbf{t}')$ in $\mathrm{LC}(\mathbf{b}, \mathbf{q})$, there exists a point $(\mathbf{a} \cdot \mathbf{t}, \mathbf{p} \cdot \mathbf{t})$ in the testing region of $(\mathbf{a}, \mathbf{p})$ that is located to its right (i.e. a point with the same $y$-coordinate and no smaller $x$-coordinate). Since the lower Lorenz curve is convex, this means that $\mathrm{LC}(\mathbf{b}, \mathbf{q})$ is a nowhere below $\mathrm{LC}(\mathbf{a}, \mathbf{p})$. That is, the second statement of the theorem holds.

To prove that the second statement implies the third statement of the theorem, observe that by construction of $\mathrm{LC}(\tilde{\mathbf{a}}, \tilde{\mathbf{p}})$ (see Figure 4.8), the curve $\mathrm{LC}(\mathbf{b}, \mathbf{q})$ is nowhere below the curve $\mathrm{LC}(\tilde{\mathbf{a}}, \tilde{\mathbf{p}})$. Since $\|\tilde{a}\|_1 = b$ we get from Theorem 4.10 that $(\tilde{\mathbf{a}}, \tilde{\mathbf{p}}) \succ (\mathbf{b}, \mathbf{q})$.

It is therefore left to prove that the third statement in the theorem implies the first one. Since we assume that $(\tilde{\mathbf{a}}, \tilde{\mathbf{p}}) \succ (\mathbf{b}, \mathbf{q})$, there exists a matrix $S \in \mathrm{STOC}(n', k + 2)$ such that $S\tilde{\mathbf{a}} = \mathbf{b}$ and $S\tilde{\mathbf{p}} = \mathbf{q}$. On the other hand, the matrix

$$T := \begin{pmatrix} I_k & \mathbf{0}_{k,n-k} \\ \mathbf{0}_{2,k} & M \end{pmatrix} \in \mathrm{STOC}(k+2, n), \quad \text{where} \quad M := \begin{pmatrix} \frac{\lambda - v_k}{p_{k+1}} & 0 & \cdots & 0 \\ \frac{v_{k+1} - \lambda}{p_{k+1}} & 1 & \cdots & 1 \end{pmatrix} \tag{4.258}$$

satisfies $T\mathbf{a} \geqslant \tilde{\mathbf{a}}$ and $T\mathbf{p} = \tilde{\mathbf{p}}$ (see Exercise 4.75). Therefore,

$$ST\mathbf{a} \geqslant S\tilde{\mathbf{a}} = \mathbf{b} \quad \text{and} \quad ST\mathbf{p} = S\tilde{\mathbf{p}} = \mathbf{q}, \tag{4.259}$$

so that the matrix $R := ST \in \mathrm{STOC}(n', n)$ satisfies $R\mathbf{a} \geqslant \mathbf{b}$ and $R\mathbf{p} = \mathbf{q}$. That is, the condition in (4.250) holds and therefore $\mathbf{p}^{XY} \succ_X \mathbf{q}^{XY'}$. This completes the proof. $\blacksquare$

**Exercise 4.75.** *Using the same notations as the proof above:*

1. *Prove the theorem above for the case $a = b$.*
2. *Show that (4.254) implies that $\lambda < v_{k+1}$.*
3. *Verify that the matrix $T$ in (4.258) is column stochastic and satisfies $T\mathbf{a} \geqslant \tilde{\mathbf{a}}$ and $T\mathbf{p} = \tilde{\mathbf{p}}$.*

## The Case $|Y| = 2$

In this case, $\mathbf{p}^{XY}$ has the form

$$\mathbf{p}^{XY} = \mathbf{p}_1 \otimes \mathbf{e}_1 + \mathbf{p}_2 \otimes \mathbf{e}_2, \tag{4.260}$$

where $\mathbf{p}_1, \mathbf{p}_2 \in \mathrm{Prob}(m)$ and $\mathbf{e}_1, \mathbf{e}_2 \in \mathbb{R}^2$ form the standard basis of $\mathbb{R}^2$ (for simplicity, we omitted the superscript $X$ from $\mathbf{p}_1$ and $\mathbf{p}_2$). The matrix $R$ that appears in Theorem 4.14 has the form $R = [\mathbf{a} \ \mathbf{b}]$, where $\mathbf{a} := (a_1, \dots, a_{n'})^T$ and $\mathbf{b} := (b_1, \dots, b_{n'})^T$ are probability vectors. Therefore, the relation in (4.219) can be expressed as

$$a_w \mathbf{p}_1 + b_w \mathbf{p}_2 \succ \mathbf{q}_w \qquad \forall \, w \in [n']. \tag{4.261}$$

Moreover, from (4.220) we get that

$$q_w = a_w p_1 + b_w p_2 \qquad \forall \, w \in [n'], \tag{4.262}$$

where $p_y := \mathbf{1}_m \cdot \mathbf{p}_y$ for all $y \in [n]$, and $q_w := \mathbf{1}_m \cdot \mathbf{q}_w$ for all $w \in [n']$.

Let $\mathbf{p}_{|y} := \frac{1}{p_y} \mathbf{p}_y \in \mathrm{Prob}(m)$ for $y = 1, 2$, and $\mathbf{q}_{|w} := \frac{1}{q_w} \mathbf{q}_w \in \mathrm{Prob}(m)$ for $w \in [n']$. With these notations, for all $w \in [n']$, the two equations above give

$$t_w \mathbf{p}_{|1} + (1 - t_w)\mathbf{p}_{|2} \succ \mathbf{q}_{|w}, \quad \text{where} \quad t_w := \frac{a_w p_1}{a_w p_1 + b_w p_2} = \frac{a_w p_1}{q_w}. \tag{4.263}$$

Observe that if we find a vector $\mathbf{t} \in [0, 1]^{n'}$, whose components satisfy $t_w \mathbf{p}_{|1} + (1 - t_w)\mathbf{p}_{|2} \succ \mathbf{q}_{|w}$ for all $w \in [n']$, it could still be the case that there are no $\mathbf{a}, \mathbf{b} \in \mathrm{Prob}(n')$ that relates to $\mathbf{t}$ as above. To find the condition on $\mathbf{t}$ that ensures the existence of such $\mathbf{a}, \mathbf{b} \in \mathrm{Prob}(n')$, we rearrange the expression above for $t_w$ to get for all $w \in [n']$

$$q_w t_w = a_w p_1. \tag{4.264}$$

Using the fact that $\mathbf{a} \in \mathrm{Prob}(n')$, by summing over $w \in [n']$ both sides of the equation above we get that $\mathbf{t}$ must satisfy

$$\sum_{w \in [n']} q_w t_w = p_1. \tag{4.265}$$

We therefore conclude that $\mathbf{p}^{XY} \succ_X \mathbf{q}^{XY'}$ if and only if there exists $\mathbf{t} \in [0, 1]^{n'}$ that satisfies (4.265) and for all $w \in [w']$, $t_w \mathbf{p}_{|1} + (1 - t_w)\mathbf{p}_{|2} \succ \mathbf{q}_{|w}$.

In order to determine when such $\mathbf{t} \in [0, 1]^{n'}$ exists, we assume that $\mathbf{p}^{XY}$ is given in its standard form so that both $\mathbf{p}_{|1} = \mathbf{p}_{|1}^{\downarrow}$ and $\mathbf{p}_{|2} = \mathbf{p}_{|2}^{\downarrow}$. With this property, the majorization relation given in (4.263) is equivalent to

$$t_w \|\mathbf{p}_{|1}\|_{(k)} + (1 - t_w)\|\mathbf{p}_{|2}\|_{(k)} \geqslant \|\mathbf{q}_{|w}\|_{(k)} \qquad \forall \, w \in [n'], \ \forall \, k \in [m]. \tag{4.266}$$

Our next goal is to characterize the constraints that the equation above impose on $t_w$. For this purpose, we denote by $\mathfrak{I}_+$, $\mathfrak{I}_0$, and $\mathfrak{I}_-$ the set of all indices $k \in [m]$ for which $\|\mathbf{p}_{|1}\|_{(k)} - \|\mathbf{p}_{|2}\|_{(k)}$ is positive, zero, and negative, respectively. With these notations, if $k \in \mathfrak{I}_0$, then (4.266) takes the form $\|\mathbf{p}_{|2}\|_{(k)} \geqslant \|\mathbf{q}_{|w}\|_{(k)}$. On the other hand, if $k \in \mathfrak{I}_+$ we can isolate $t_w$ to get

$$t_w \geqslant \frac{\|\mathbf{q}_{|w}\|_{(k)} - \|\mathbf{p}_{|2}\|_{(k)}}{\|\mathbf{p}_{|1}\|_{(k)} - \|\mathbf{p}_{|2}\|_{(k)}}. \tag{4.267}$$

Therefore, since this inequality holds for all $k \in \mathfrak{I}_+$ and since $t_w \geqslant 0$ we conclude that $t_w \geqslant \mu_w$ for all $w \in [n']$, where

$$\mu_w := \max \left\{ 0, \max_{k \in \mathfrak{I}_+} \frac{\|\mathbf{q}_{|w}\|_{(k)} - \|\mathbf{p}_{|2}\|_{(k)}}{\|\mathbf{p}_{|1}\|_{(k)} - \|\mathbf{p}_{|2}\|_{(k)}} \right\}. \tag{4.268}$$

Similarly, by isolating $t_w$ in (4.266) for the cases that $k \in \mathfrak{I}_-$ we get $t_w \leqslant v_w$ for all $w \in [n']$, where

$$v_w := \min \left\{ 1, \min_{k \in \mathfrak{I}_-} \frac{\|\mathbf{p}_{|2}\|_{(k)} - \|\mathbf{q}_{|w}\|_{(k)}}{\|\mathbf{p}_{|2}\|_{(k)} - \|\mathbf{p}_{|1}\|_{(k)}} \right\}. \tag{4.269}$$

We therefore arrive at the following theorem.

---

**Theorem 4.18.** Using the same notations as above, for the case $|Y| = 2$ we have $\mathbf{p}^{XY} \succ_X \mathbf{q}^{XY'}$ if and only if the following conditions hold:

1. For all $w \in [n']$ we have $v_w \geqslant \mu_w$.
2. For all $k \in \mathfrak{I}_0$ we have $\|\mathbf{p}_{|2}\|_{(k)} \geqslant \max_{w \in [n']} \|\mathbf{q}_{|w}\|_{(k)}$.
3. $\sum_{w \in [n']} q_w \mu_w \leqslant p_1 \leqslant \sum_{w \in [n']} q_w v_w$.

---

**Exercise 4.76.** *Use the arguments above to prove Theorem 4.18.*

**Exercise 4.77.** *Simplify the conditions in Theorem 4.18 for the case that $\mathfrak{I}_+ = [m]$.*

**Exercise 4.78.** *Consider the case $|X| = |Y| = |Y'| = 2$ and let $\mathbf{p}^{XY}, \mathbf{q}^{XY} \in \mathrm{Prob}(4)$ be such that $\mathbf{p}^Y = \mathbf{q}^Y = \mathbf{u}^{(2)}$. Simplify the necessary and sufficient conditions given in the theorem above for this case.*

## 4.6.7 Conditional Majorization with Linear Programming

In Theorem 4.14 we presented a useful characterization of conditional majorization. This characterization posits that $\mathbf{p}^{XY} \succ \mathbf{q}^{XY'}$ if and only if a stochastic matrix $R \in$ STOC$(n', n)$ exists, satisfying the condition specified in equation (4.219). Furthermore, Theorem 4.14 suggests that determining whether $\mathbf{p}^{XY} \succ_X \mathbf{q}^{XY}$ can be accomplished through linear programming.

To see it explicitly, suppose that $\mathbf{p}^{XY}$ and $\mathbf{q}^{XY'}$ are given in their standard form, and recall that $\mathbf{p}^{XY} \succ_X \mathbf{q}^{XY'}$ if and only if there exists $R \in \mathrm{STOC}(n', n)$ such that

$$\sum_{y \in [n]} r_{y'|y} L \mathbf{p}_y^X \geqslant L \mathbf{q}_{y'}^X \qquad \forall\, y' \in [n'], \tag{4.270}$$

where $L$ is the $m \times m$ matrix defined in Exercise 4.5. Denote the rows of $R$ by $\mathbf{r}_1, \ldots, \mathbf{r}_{n'} \in \mathbb{R}_+^n$; that is, $\mathbf{r}_{y'} := (r_{y'|1}, \ldots, r_{y'|n})^T$, and denote by $\mathbf{r} \in \mathrm{Prob}(nn')$ the probability vector

$$\mathbf{r} := \begin{bmatrix} \mathbf{r}_1 \\ \vdots \\ \mathbf{r}_{n'} \end{bmatrix}. \tag{4.271}$$

Note that in this vector form of $R$, the condition that $R$ is stochastic is equivalent to $\mathbf{r} \geqslant 0$ and $\sum_{y' \in [n']} \mathbf{r}_{y'} = \mathbf{1}_n$. Let $P := \begin{bmatrix} \mathbf{p}_1^X & \cdots & \mathbf{p}_n^X \end{bmatrix} \in \mathbb{R}^{m \times n}$, and denote by

$$M := \begin{bmatrix} -LP & 0 & \cdots & 0 \\ 0 & -LP & \cdots & 0 \\ \vdots & \vdots & \ddots & \vdots \\ 0 & 0 & \cdots & -LP \\ I_n & I_n & \cdots & I_n \end{bmatrix} \in \mathbb{R}^{(mn'+n) \times nn'} \quad \text{and} \quad \mathbf{b} := \begin{bmatrix} -L\mathbf{q}_1^X \\ \vdots \\ -L\mathbf{q}_{n'}^X \\ \mathbf{1}_n \end{bmatrix} \in \mathbb{R}^{mn'+n}. \tag{4.272}$$

It is then straight forward to check that the inequalities given in (4.270) can be expressed compactly as $M\mathbf{r} \leqslant \mathbf{b}$. The only other constraint is that $\mathbf{r} \in \mathbb{R}_+^{nn'}$. The problem of determining if such a vector $\mathbf{r}$ exists is known as a linear programming feasibility problem, and there are several algorithms that can be used to solve it.

**Exercise 4.79 (Farkas Lemma).** *Show that there exists $\mathbf{r} \in \mathbb{R}_+^{nn'}$ satisfying $M\mathbf{r} \leqslant \mathbf{b}$ if and only if for every $\mathbf{v} \in \mathbb{R}_+^{mn'+n}$ that satisfies $\mathbf{v}^T M \geqslant 0$ (entrywise) we have $\mathbf{v} \cdot \mathbf{b} \geqslant 0$. Hint: For the harder direction, use the hyperplane separation theorem (Theorem A.2 of online version).*

## Dual Characterization

So far, we saw that the condition $\mathbf{p}^{XY} \succ_X \mathbf{q}^{XY'}$ is equivalent to the existence of a vector $\mathbf{r} \in \mathbb{R}_+^{nn'}$ such that $M\mathbf{r} \leqslant \mathbf{b}$. Moreover, from Exercise (4.79) it follows that such an $\mathbf{r}$ exists if and only if for every $\mathbf{v} \in \mathbb{R}_+^{mn'+n}$ that satisfies $\mathbf{v}^T M \geqslant 0$ we have $\mathbf{v} \cdot \mathbf{b} \geqslant 0$. We now express this later condition in terms of sub-linear functionals.

For this purpose, we express $\mathbf{v}$ as

$$\mathbf{v} := \begin{bmatrix} \mathbf{v}_1 \\ \vdots \\ \mathbf{v}_{n'} \\ \mathbf{t} \end{bmatrix}, \tag{4.273}$$

where $\mathbf{v}_1, \ldots, \mathbf{v}_{n'} \in \mathbb{R}_+^m$ and $\mathbf{t} \in \mathbb{R}_+^n$. From the definition of $M$ in (4.272) we get that the condition $\mathbf{v}^T M \geqslant 0$ can be expressed as

$$\mathbf{t} - \mathbf{v}_w^T L P \geqslant 0 \qquad \forall\, w \in [n'], \tag{4.274}$$

which in terms of the components $\{t_y\}_{y \in [n]}$ of $\mathbf{t}$ can be expressed as

$$t_y \geqslant \max_{w \in [n']} \mathbf{v}_w^T L \mathbf{p}_y^X \qquad \forall\, y \in [n]. \tag{4.275}$$

Similarly, from the definition of $\mathbf{b}$ in (4.272) we get that the condition $\mathbf{v} \cdot \mathbf{b} \geqslant 0$ is equivalent to

$$\sum_{y \in [n]} t_y \geqslant \sum_{y' \in [n']} \mathbf{v}_{y'}^T L \mathbf{q}_{y'}^X. \tag{4.276}$$

Hence, the condition that $\mathbf{v}^T M \geqslant 0$ implies $\mathbf{v} \cdot \mathbf{b} \geqslant 0$ is equivalent to

$$\sum_{y \in [n]} \max_{w \in [n']} \mathbf{v}_w^T L \mathbf{p}_y^X \geqslant \sum_{w \in [n']} \mathbf{v}_w^T L \mathbf{q}_w^X, \tag{4.277}$$

where we took $t_y$ in (4.276) to be equal to its smallest possible value as given in (4.275). Finally, for each $w \in [n']$ let $\mathbf{s}_w := L^T \mathbf{v}_w$ and observe that the inequality above can be written as

$$\sum_{y \in [n]} \max_{w \in [n']} \mathbf{s}_w \cdot \mathbf{p}_y^X \geqslant \sum_{w \in [n']} \mathbf{s}_w \cdot \mathbf{q}_{y'}^X. \tag{4.278}$$

Note that since each $\mathbf{v}_w \in \mathbb{R}_+^m$ we get that $\mathbf{s}_w \in \mathbb{R}_+^m$ and $\mathbf{s}_w = \mathbf{s}_w^\downarrow$ (see Exercise 4.5). Finally, by dividing both sides of the inequality above by a sufficiently large number and absorbing it into each $\mathbf{s}_w$ we can assume without loss of generality that the matrix $S := [\mathbf{s}_1 \cdots \mathbf{s}_{n'}] \in \mathrm{STOC}_{\leqslant}(m, n')$ is sub-stochastic (i.e. the components of each column sums to a number smaller or equal to 1). We therefore arrived at the following theorem.

---

**Theorem 4.19.** Let $\mathbf{p}^{XY} \in \mathrm{Prob}(mn)$ and $\mathbf{q}^{XY'} \in \mathrm{Prob}(mn')$ be given in their standard form. Then, $\mathbf{p}^{XY} \succ_X \mathbf{q}^{XY'}$ if and only if for every sub-stochastic matrix $S := [\mathbf{s}_1 \cdots \mathbf{s}_{n'}] \in \mathrm{STOC}_{\leqslant}(m, n')$, whose columns satisfy $\mathbf{s}_w = \mathbf{s}_w^\downarrow$ for all $w \in [n']$, we have

$$\sum_{y \in [n]} \max_{w \in [n']} \mathbf{s}_w \cdot \mathbf{p}_y^X \geqslant \sum_{w \in [n']} \mathbf{s}_w \cdot \mathbf{q}_{y'}^X. \tag{4.279}$$

---

**Exercise 4.80.** *Consider the theorem above without the assumption that $\mathbf{s}_w = \mathbf{s}_w^\downarrow$ for all $w \in [n']$ and without the assumption that $\mathbf{p}^{XY}$ and $\mathbf{q}^{XY'}$ are given in their standard form. Show that $\mathbf{p}^{XY} \succ_X \mathbf{q}^{XY'}$ if and only if for every sub-stochastic matrix $S := [\mathbf{s}_1 \cdots \mathbf{s}_{n'}] \in \mathrm{STOC}_{\leqslant}(m, n')$*

$$\sum_{y \in [n]} \max_{w \in [n']} \mathbf{s}_w^\downarrow \cdot \mathbf{p}_y^\downarrow \geqslant \sum_{w \in [n']} \mathbf{s}_w^\downarrow \cdot \mathbf{q}_{y'}^\downarrow, \tag{4.280}$$

*where for simplicity we removed the superscript $X$ from $\mathbf{p}_y^X$ and $\mathbf{q}_w^X$.*

## 4.6.8 The Operational Approach: Games of Chance with a Correlated Source

In this subsection we delve into an alternative approach that leads to an equivalent definition of conditional majorization. This particular approach is grounded in operational principles, drawing parallels with games of chance. As a result, it presents a particularly persuasive rationale for the definition of conditional majorization as introduced in the previous subsections.

In the beginning of this chapter we introduced the concept of majorization using games of chance. We saw that two probability vectors, $\mathbf{p}, \mathbf{q} \in \text{Prob}(n)$ satisfy $\mathbf{p} \succ \mathbf{q}$ if and only if in all games of chance a player has better odds to win the game with the $\mathbf{p}$-dice rather than with the $\mathbf{q}$-dice. Similarly, we will see that our definition of conditional majorization as given in Definition 4.10 can be characterized with games of chance that involve a correlated source.

We can think about a correlated source $XY$ as two dice that are connected with a gum. Rolling the two dice results in an outcome $x$ for system $X$ and a correlated outcome $y$ for system $Y$. As before, we denote by $\mathbf{p}^{XY} \in \text{Prob}(mn)$ the probability matrix whose $(x, y)$-entry, $p_{xy}$, represents the probability that $X = x$ and $Y = y$. It will be convenient to denote by $p_{x|y} = \frac{p_{xy}}{p_y}$, the conditional probability that $X = x$ given that $Y = y$, where $p_y := \sum_{x \in [m]} p_{xy}$ for any $y \in [n]$.

Consider now a gambling game with such two correlated dice, in which a player, say Alice, has to provide $k \leqslant m$ numbers as her guesses for the value of $X$. If Alice has access to the value $y$ of $Y$, then she will choose the $k$ numbers that has the largest probability to occur relative to the conditional probability $\{p_{x|y}\}_{x \in [m]}$. Therefore, the maximum probability to win such a $k$-gambling game is given by

$$\sum_{y \in [n]} p_y \sum_{x \in [k]} p_{x|y}^{\downarrow}. \tag{4.281}$$

That is, Alice chooses the $k$ numbers that has the largest probability to occur *after* she learns the value of $Y = y$, which occur with probability $p_y$.

The example provided earlier is not the only kind of gambling game that Alice can engage in with a correlated source, like the two-dice system. More expansively, we can envisage a game where the host randomly determines the value of $k$ according to a certain distribution. In line with our aim to explore the widest range of scenarios in a gambling game with a correlated source, we allow the player a degree of control in choosing which $k$-gambling game will be played. This control is exercised through the player selecting a number $w \in [\ell]$ and communicating it to the game host. Subsequently, the host decides the value of $k$ based on a distribution $T := (t_{k|w}) \in \mathbb{R}_+^{m \times \ell}$, a detail known to the player. This distribution adheres to the conditions that $t_{k|w} \geqslant 0$ and $\sum_{k \in [m]} t_{k|w} \leqslant 1$ for all $w \in [\ell]$. Notably, we also accommodate the scenario where the set $\{t_{k|w}\}_{k \in [m]}$ does not sum to one, reflecting the possibility of no $k$ value occurring, resulting in the player losing the game from the onset. The procedural steps of such a $T$-gambling game are illustrated in Figure 4.9.

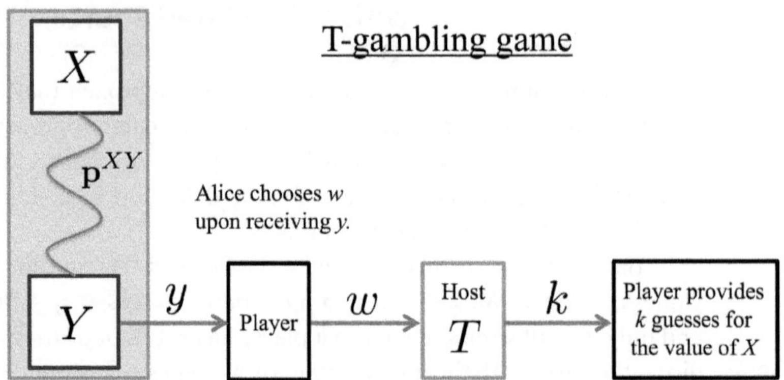


**Figure 4.9**  A $T$-gambling game with correlated source $\mathbf{p}^{XY}$. Upon learning the value of $Y$, the player provides the host a number $w$. Then, the host chooses $k$ (at random) according to the distribution $\{t_{k|w}\}_{k\in[m]}$. After that, the player provides her $k$ guesses with the highest probability to occur.

Note that the set encompassing all $T$-gambling games includes all $k$-gambling games as well. This is evident when we consider $T = (t_{k|w})$ with $t_{k|w} = \delta_{kk_0}$, where $k_0$ is a specific integer within $[m]$. In this scenario, the game essentially becomes a $k_0$-gambling game, meaning the host selects $k = k_0$ regardless of $w$. In another example, where $t_{k|w} = \frac{1}{m}$, the host picks $k$ from a uniform distribution, also independent of $w$.

Generally, for a given $Y = y$ and a chosen $w \in [\ell]$, the optimal chance Alice has to win a $T$-gambling game can be calculated by the probability:

$$\sum_{k\in[m]} t_{k|w} \sum_{x\in[k]} p^{\downarrow}_{x|y}. \tag{4.282}$$

Consequently, for each $Y = y$, Alice will select the number $w$ that maximizes this probability. Therefore, the maximum likelihood of winning a $T$-gambling game, as outlined above, is given by

$$\Pr_T\left(\mathbf{p}^{XY}\right) = \sum_{y\in[n]} p_y \max_{w\in[\ell]} \sum_{k\in[m]} t_{k|w} \sum_{x\in[k]} p^{\downarrow}_{x|y}. \tag{4.283}$$

The formula for calculating the winning probability, sometimes referred to as the reward function in game theory, can be simplified. Consider the following transformation:

$$\sum_{k\in[m]} \sum_{x\in[k]} t_{k|w} p^{\downarrow}_{x|y} = \sum_{x\in[m]} \sum_{k=x}^{m} t_{k|w} p^{\downarrow}_{x|y}. \tag{4.284}$$

Let's introduce the matrix $S = (s_{xw}) \in \mathbb{R}^{m\times\ell}$, whose coefficients are defined by

$$s_{xw} := \sum_{k=x}^{m} t_{k|w}. \tag{4.285}$$

It's important to note that the columns of $S$ are in nondecreasing order; that is, for every $w \in [\ell]$,

$$1 \geqslant s_{1w} \geqslant s_{2w} \geqslant \cdots \geqslant s_{mw}. \tag{4.286}$$

With this notation, the probability of winning can be expressed as

$$\mathrm{Pr}_T\left(\mathbf{p}^{XY}\right) = \sum_{y \in [n]} p_y \max_{w \in [\ell]} \sum_{x \in [m]} s_{xw} p^{\downarrow}_{x|y}. \tag{4.287}$$

Finally, denoting the columns of $S$ by $\mathbf{s}_1, \ldots, \mathbf{s}_\ell$ we conclude that

$$\mathrm{Pr}_T\left(\mathbf{p}^{XY}\right) = \sum_{y \in [n]} \max_{w \in [\ell]} \mathbf{s}_w \cdot \mathbf{p}^{\downarrow}_y, \tag{4.288}$$

where $\mathbf{p}_y \in \mathbb{R}^m_+$ is the vector with components $\{p_{xy}\}_{x \in [m]}$. Observe that the formula (4.288) for calculating the winning probability coinsides with the left-hand side of (4.279). This observation provides our initial clue about the connection between games of chance and conditional majorization. Another key insight is that conditionally mixing operations cannot increase the maximum probability of winning the game, as the following lemma demonstrates.

> **Lemma 4.7.** Let $T \in \mathrm{STOC}_\leqslant(m, \ell)$, $\mathbf{p}^{XY} \in \mathrm{Prob}(mn)$, and $\mathbf{q}^{XY'} \in \mathrm{Prob}(mn')$. If $\mathbf{p}^{XY} \succ \mathbf{q}^{XY'}$ then the maximal probability to win a $T$-gambling game satisfies
>
> $$\mathrm{Pr}_T\left(\mathbf{p}^{XY}\right) \geqslant \mathrm{Pr}_T\left(\mathbf{q}^{XY'}\right). \tag{4.289}$$

**Proof**    Let $S$ be the $m \times \ell$ matrix whose components are as defined in (4.285). Due to (4.286) the columns of $S = [\mathbf{s}_1 \cdots \mathbf{s}_\ell]$ satisfy $\mathbf{s}^{\downarrow}_w = \mathbf{s}_w$ for all $w \in [\ell]$. According to (4.288), the function $\mathrm{Pr}_T(\mathbf{p}^{XY})$ has the form (4.241) with $f : \mathrm{Prob}(m) \to \mathbb{R}$ being the sublinear functional

$$f(\mathbf{p}) := \max_{w \in [n']} \mathbf{s}_w \cdot \mathbf{p}^{\downarrow} \qquad \forall\, \mathbf{p} \in \mathrm{Prob}(m). \tag{4.290}$$

Since the function above is convex and symmetric (under permutations) we get from Theorem 4.16 that $\mathrm{Pr}_T(\mathbf{p}^{XY})$ is conditionally Schur concave and in particular $\mathrm{Pr}_T(\mathbf{p}^{XY}) \geqslant \mathrm{Pr}_T(\mathbf{q}^{XY'})$.        ■

The subsequent exercise establishes a one-to-one correspondence (bijection) between the set of all $m \times \ell$ $T$-matrices and all $m \times \ell$ $S$-matrices.

**Exercise 4.81.** *Use (4.285) to find an $m \times m$ matrix $U$ such that $S = UT$. Show that $U$ is invertible by computing its inverse, and use that to show that for any matrix $S \in \mathbb{R}^{m \times \ell}_+$ whose components satisfy (4.286), the matrix $U^{-1}S$ has nonnegative entries.*

## Operational Characterization of Conditional Majorization

The reward function above can serve as the basis for defining conditional majorization. Recall that majorization can be formally established through the concept of $k$-gambling games: $\mathbf{p} \succ \mathbf{q}$ if, for every $k \in [m]$, a player has a higher probability of winning a $k$-gambling game using the $\mathbf{p}$-dice rather than the $\mathbf{q}$-dice. In a similar vein, we can operationally define conditional majorization as follows: $\mathbf{p}^{XY} \succ_X \mathbf{q}^{XY}$ if, for all sub-stochastic matrices $T$ (over all dimensions $\ell \in \mathbb{N}$), a player enjoys superior odds of winning a $T$-gambling game when using the $\mathbf{p}^{XY}$ dice-pair as opposed to the $\mathbf{q}^{XY'}$ dice-pair. We show now that this operational definition of conditional majorization coincides with Definition 4.10 that follows from the axiomatic and constructive approaches.

**Conditional Majorization and Games of Chance**

**Theorem 4.20.** Let $\mathbf{p} \in \mathrm{Prob}(mn)$ and $\mathbf{q} \in \mathrm{Prob}(mn')$. We have $\mathbf{p}^{XY} \succ_X \mathbf{q}^{XY'}$ if and only if

$$\mathrm{Pr}_T\left(\mathbf{p}^{XY}\right) \geqslant \mathrm{Pr}_T\left(\mathbf{q}^{XY'}\right) \qquad \forall\, T \in \mathrm{STOC}_{\leqslant}(m, n'), \tag{4.291}$$

where $\mathrm{Pr}_T$ is the maximal probability to win a $T$-gambling game.

*Remark.* We emphasize that Theorem (4.20) states that $\mathbf{p}^{XY} \succ_X \mathbf{q}^{XY'}$ if and only if with the $\mathbf{p}^{XY}$-dice pair Alice has better odds to win all $T$-gambling games than with the $\mathbf{q}^{XY'}$-dice pair. Moreover, observe that instead of considering $T$-gambling games with $T \in \mathrm{STOC}_{\leqslant}(m, \ell)$ over all $\ell \in \mathbb{N}$, it is sufficient to consider $\ell = n'$. That is, the dimensions of $T$ are completely determined by $X$ and $Y'$.

**Proof**   Due to Lemma 4.7, it is sufficient to prove that (4.291) implies $\mathbf{p}^{XY} \succ_X \mathbf{q}^{XY'}$. Let $S := [\mathbf{s}_1 \cdots \mathbf{s}_{n'}] \in \mathrm{STOC}_{\leqslant}(m, n')$, whose columns satisfy $\mathbf{s}_w = \mathbf{s}_w^{\downarrow}$ for all $w \in [n']$. From Exercise 4.81 it follows that there exists a sub-stochastic matrix $T \in \mathrm{STOC}_{\leqslant}(m, n')$ that satisfies the relation (4.285). Therefore,

$$\sum_{y \in [n]} \max_{w \in [n']} \mathbf{s}_w \cdot \mathbf{p}_y = \mathrm{Pr}_T\left(\mathbf{p}^{XY}\right)$$

$$(\mathbf{4.291}) \rightarrow\ \geqslant \mathrm{Pr}_T\left(\mathbf{q}^{XY'}\right)$$

$$= \sum_{y' \in [n']} \max_{w \in [n']} \mathbf{s}_w \cdot \mathbf{q}_{y'} \tag{4.292}$$

$$\geqslant \sum_{y' \in [n']} \mathbf{s}_{y'} \cdot \mathbf{q}_{y'}.$$

Since the above inequality holds for all $S := [\mathbf{s}_1 \cdots \mathbf{s}_{n'}] \in \mathrm{STOC}_{\leqslant}(m, n')$, whose columns satisfy $\mathbf{s}_w = \mathbf{s}_w^{\downarrow}$ for all $w \in [n']$, we conclude from Theorem 4.19 that $\mathbf{p}^{XY} \succ_X \mathbf{q}^{XY'}$. This completes the proof.                    ∎

## 4.7 Notes and References

The book in Ref. [155] is dedicated solely to the theory of majorization. Also the book in Ref. [25] is a good source, particularly, the second chapter covers majorization. Lemma 4.1 goes back to Refs. [169] and [110], and the Schur's test (Theorem 4.2) is due to Refs. [202] and [178].

Approximate majorization was introduced in Ref. [134], and we will see later on that the concept of the flattest $\varepsilon$-approximation plays a useful role in several resource theories.

Relative majorization is the backbone of the resource theoretic approach to quantum thermodynamics. It was studied under different names such as $d$-majorization in Ref. [6], matrix majorization in Ref. [57], and thermo-majorization in Ref. [133]. The ideas of the proof of main characterization theorem of relative majorization (Theorem 4.10) goes back to Ref. [28]. More details were given in Refs. [196] and [142]. Independent proof was also given more recently in Ref. [57] by employing techniques from convex analysis. To the author's knowledge, the proof we presented here did not appear elsewhere.

In the proof of Theorem 4.9 we followed Ref. [192]. Theorems 4.7 and 4.8 as well as Exercise 4.49 are due to Ref. [99]. The vector $\mathbf{r}$ that appears in Theorem 4.8 was first introduced in Ref. [29] in the context of thermodynamics.

The characterization of the trumping relation (Theorem 4.11) was proved in Ref. [215] and independently in Ref. [145]. Both proofs are very complicated and it is an open problem to find a simpler/shorter proof of the theorem. The symmetric rewrite of this theorem in terms of Rényi divergences, and the characterization of catalytic majorization (Theorem 4.12), are due to Ref. [99].

Conditional majorization was first introduced in Ref. [90] in the context of the quantum uncertainty principle. More recently, its relation to games of chance was introduced in Ref. [36], and its quantum version was given in Ref. [103]. In Section 4.6.6 we saw that for the case $|X| = 2$ conditional majorization becomes equivalent to relative submajorization. Relative submajorization has some applications in thermodynamics, and it was first introduced and studied in Ref. [192].

# 5

# Divergences and Distance Measures

This chapter explores methods to quantify the distinguishability between entities such as probability distributions and quantum states. Unlike generic vectors, mathematical objects like probability vectors and quantum states embody information about physical systems. Consequently, their distinguishability is typically measured using functions attuned to this inherent information. Consider this example: Alice possesses a system in her laboratory that is either in state $\rho$ (for instance, an electron with its spin oriented in the $z$-direction) or in state $\sigma$ (such as the same electron with spin in the $x$-direction). Alice can attempt to discern the state of her system (whether it is $\rho$ or $\sigma$) by performing a quantum measurement on it. The underlying principle is that the greater the distinguishability between $\rho$ and $\sigma$, the easier (or more likely) it is for Alice to accurately identify which of the two states her system is in.

In any task involving distinguishability, such as the one mentioned in Section 4.3, a key observation is that sending a system (like the electron in Alice's lab) through a quantum communication channel does not enhance Alice's ability to differentiate between the two states, $\rho$ and $\sigma$. This implies that if $\mathcal{E} \in \mathrm{CPTP}(A \to B)$ represents a quantum channel, the states $\mathcal{E}(\rho)$ and $\mathcal{E}(\sigma)$ that result from this channel are less distinguishable than the original states $\rho$ and $\sigma$ (this concept is visually illustrated in Figure 5.1). In essence, any measure that quantifies the distinguishability between two quantum states $\rho$ and $\sigma$ must decrease (or at most stay the same) under any quantum process that transforms the pair $(\rho, \sigma)$ into $(\mathcal{E}(\rho), \mathcal{E}(\sigma))$. Functions that adhere to this principle are known as quantum divergences. Their characteristic of reducing in value under such transformations is often referred to as the *data processing inequality* (DPI).

Quantum divergence extends the concept of divergences from classical to quantum realms. In a classical context, divergences are functions that behave monotonically under transformations that map a pair of probability vectors $(\mathbf{p}, \mathbf{q})$ to $(E\mathbf{p}, E\mathbf{q})$, with $E$ being a column stochastic matrix. As a result, many metrics in $\mathbb{R}^n$, like the Euclidean distance, do not serve well for quantifying distinguishability between two probability vectors. It's also noteworthy that divergences are functions that behave monotonically under relative majorization. Therefore, the tools developed in Chapter 4 will be very useful in this context as well.

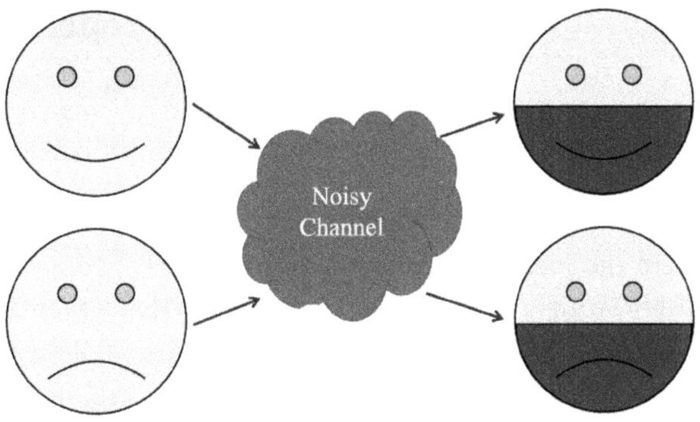


**Figure 5.1**    A noisy process makes objects look more similar.

## 5.1 Classical Divergences

We begin by presenting the formal definition of a classical divergence. Let $\mathbb{D}$ represent a function defined as

$$\mathbb{D}: \bigcup_{n \in \mathbb{N}} \left\{ \mathrm{Prob}(n) \times \mathrm{Prob}(n) \right\} \to \mathbb{R} \cup \{\infty\}, \tag{5.1}$$

which operates on pairs of probability vectors across all finite dimensions.

**Classical Divergence**

**Definition 5.1.** The function $\mathbb{D}$, as defined in (5.1), is termed a *divergence* provided it fulfills these two conditions:

1. Data Processing Inequality (DPI)

$$\mathbb{D}\big(E\mathbf{p} \big\| E\mathbf{q}\big) \leqslant \mathbb{D}(\mathbf{p}\|\mathbf{q}) \tag{5.2}$$

     for all $n, m \in \mathbb{N}$, all $\mathbf{p}, \mathbf{q} \in \mathrm{Prob}(n)$, and all $E \in \mathrm{STOC}(m, n)$.
2. Normalization, $\mathbb{D}(1\|1) = 0$.

Note that for the trivial dimension $n = 1$, $\mathrm{Prob}(n)$ contains only the number 1. In this dimension, we require the divergence to be zero. Functions as above that satisfy the DPI but with $\mathbb{D}(1\|1) \neq 0$ will be called *unnormalized divergences*. Moreover, observe that the DPI property of a divergence $\mathbb{D}$ can also be viewed as monotonicity under relative majorization. That is, we can state the first property above as follows: For all $\mathbf{p}, \mathbf{q} \subset \mathrm{Prob}(n)$ and $\mathbf{p}', \mathbf{q}' \in \mathrm{Prob}(m)$ such that $(\mathbf{p}, \mathbf{q}) \succ (\mathbf{p}', \mathbf{q}')$ we have

$$\mathbb{D}(\mathbf{p}\|\mathbf{q}) \geqslant \mathbb{D}(\mathbf{p}'\|\mathbf{q}'). \tag{5.3}$$

We now discuss a few basic properties of classical divergences.

## 5.1.1 Basic Properties

Divergences are nonnegative since the one-row matrix $[1, \ldots, 1]$ is column stochastic matrix in $\mathrm{STOC}(1, n)$, so that for any pair of probability vectors $\mathbf{p}, \mathbf{q} \in \mathrm{Prob}(n)$

$$\mathbb{D}(\mathbf{p}\|\mathbf{q}) \geqslant \mathbb{D}\big([1, \ldots, 1]\mathbf{p}\,\big\|\,[1, \ldots, 1]\mathbf{q}\big) = \mathbb{D}(1\|1) = 0, \tag{5.4}$$

where the inequality follows from the DPI. Moreover, any probability vector $\mathbf{p} \in \mathrm{Prob}(n)$ with dimension $n > 1$ can be viewed as a preparation channel (i.e. one-column stochastic matrix) $\mathbf{p} \in \mathrm{STOC}(n, 1)$, so that

$$\mathbb{D}(\mathbf{p}\|\mathbf{p}) = \mathbb{D}\big(\mathbf{p} \cdot 1 \,\big\|\, \mathbf{p} \cdot 1\big) \leqslant \mathbb{D}(1\|1) = 0, \tag{5.5}$$

where again we used the DPI for divergences. Combining this with the nonnegativity property of divergences, we conclude that for any state $\mathbf{p} \in \mathrm{Prob}(n)$ in any dimension $n \in \mathbb{N}$

$$\mathbb{D}(\mathbf{p}\|\mathbf{p}) = 0. \tag{5.6}$$

This is consistent with the intuition that divergences quantify the distinguishability between two states. An interesting question remaining is whether the converse of the above property also holds. That is, the question is whether $\mathbb{D}(\mathbf{p}\|\mathbf{q}) = 0$ necessarily implies that $\mathbf{p} = \mathbf{q}$. This property is called *faithfulness* and we will see later on that there exists divergences that are not faithful.

Another interesting property that divergences satisfy is the following lower and upper bounds for any $\mathbf{p}, \mathbf{q} \in \mathrm{Prob}(n)$:

$$\mathbb{D}\big(\mathbf{e}_1 \,\big\|\, \lambda_{min}\mathbf{e}_1 + (1 - \lambda_{min})\mathbf{e}_2\big) \leqslant \mathbb{D}(\mathbf{p}\|\mathbf{q}) \leqslant \mathbb{D}\big(\mathbf{e}_1 \,\big\|\, \lambda_{max}\mathbf{e}_1 + (1 - \lambda_{max})\mathbf{e}_2\big), \tag{5.7}$$

where $\lambda_{\max}$ and $\lambda_{\min}$ are defined in (4.119), and $\{\mathbf{e}_1, \mathbf{e}_2\}$ is the computational basis of $\mathbb{R}^2$. This property follows from a combination of the DPI with Theorem 4.7.

We now move to discuss some examples of divergences. In the previous chapter, particularly Theorem 4.10, we found several characterizations for relative majorization. Some of the characterizations are given in terms of divergences (although not explicitly). For example, fix $t \geqslant 1$ and define for all $n \in \mathbb{N}$ and all $\mathbf{p}, \mathbf{q} \in \mathrm{Prob}(n)$ the function

$$D_t(\mathbf{p}\|\mathbf{q}) := \sum_{x \in [n]} (p_x - tq_x)_+. \tag{5.8}$$

Since we assume $t \geqslant 1$ we have $D_t(\mathbf{p}\|\mathbf{p}) = 0$ for all $\mathbf{p} \in \mathrm{Prob}(n)$. To show that $D_t$ satisfies the DPI we can use the relation $(r)_+ = (|r| + r)/2$ for all $r \in \mathbb{R}$ to express $D_t$ as

$$D_t(\mathbf{p}\|\mathbf{q}) = \frac{1}{2}\left(\|\mathbf{p} - t\mathbf{q}\|_1 + 1 - t\right). \tag{5.9}$$

Consequently, the property outlined in (2.8) implies that $\{D_t\}_{t \geqslant 1}$ is a family of divergences. Moreover, from Theorem 4.10 we learn that this family of classical divergences can be used to characterize relative majorization; that is, for all $\mathbf{p}, \mathbf{q} \in \mathrm{Prob}(n)$ and $\mathbf{p}', \mathbf{q}' \in \mathrm{Prob}(n')$ we have (see Exercise 5.1)

$$(\mathbf{p}, \mathbf{q}) \succ (\mathbf{p}', \mathbf{q}') \qquad \Longleftrightarrow \qquad D_t(\mathbf{p}\|\mathbf{q}) \geqslant D_t(\mathbf{p}'\|\mathbf{q}') \qquad \forall\, t \geqslant 1. \tag{5.10}$$

**Exercise 5.1.** *Prove (5.10). Hint: Use Theorem 4.10 in conjunction with Corollary 4.1.*

**Exercise 5.2.** *Let $\mathbb{D}$ be a divergence, and $P$ be an $n \times n$ permutation matrix. Show that*

$$\mathbb{D}(\mathbf{p}\|\mathbf{q}) = \mathbb{D}(P\mathbf{p}\|P\mathbf{q}) \qquad \forall\, \mathbf{p}, \mathbf{q} \in \mathrm{Prob}(n). \tag{5.11}$$

There is another family of divergences, consisting of numerous divergences (that appear in applications), and also includes the above example as a special case. This family is called the $f$-Divergence.

## 5.1.2 The $f$-Divergence

**Definition 5.2.** Let $f : (0, \infty) \to \mathbb{R}$ be a convex function that satisfy $f(1) = 0$. Then, the $f$-*Divergence* is defined for any $\mathbf{p}, \mathbf{q} \in \mathrm{Prob}(n)$ by

$$D_f(\mathbf{p}\|\mathbf{q}) := \sum_{x \in [n]} q_x f\left(\frac{p_x}{q_x}\right) \tag{5.12}$$

with the following conventions:

$$f(0) := \lim_{r \to 0^+} f(r), \quad 0f\left(\frac{0}{0}\right) := 0, \quad 0f\left(\frac{a}{0}\right) := \lim_{r \to 0^+} rf\left(\frac{a}{r}\right) = a \lim_{s \to 0^+} sf\left(\frac{1}{s}\right).$$

*Remark.* We do not assume that $f$ is positive nor that

$$\tilde{f}(0) := \lim_{s \to 0^+} sf\left(\frac{1}{s}\right) \tag{5.13}$$

is finite. Therefore, for some convex functions as in Definition 5.2, we can have $D_f(\mathbf{p}\|\mathbf{q}) = \infty$ for some choices of $\mathbf{p}, \mathbf{q} \in \mathrm{Prob}(n)$. From Theorem 5.1 it will follow that $D_f$ is a divergence and therefore is always nonnegative (even if $f(x)$ is negative for some $x \in (0, \infty)$). Furthermore, observe that the $f$-Divergence can be expressed for any $\mathbf{p}, \mathbf{q} \in \mathrm{Prob}(n)$ as

$$D_f(\mathbf{p}\|\mathbf{q}) = \tilde{f}(0) \sum_{x \notin \mathrm{supp}(\mathbf{q})} p_x + \sum_{x \in \mathrm{supp}(\mathbf{q})} q_x f\left(\frac{p_x}{q_x}\right), \tag{5.14}$$

where we split the sum in (5.12) into a sum over all $x \in [n]$ with $q_x = 0$ and over all $x \in [n]$ with $q_x \neq 0$.

**Exercise 5.3.** *Show that for every $t \geqslant 1$ the function $D_t$ as defined in (5.8) is an $f$-Divergence.*

**Exercise 5.4.** *Show that the definition above for the $f$-Divergence is equivalent to the following definition. Let $f : (0, \infty) \to \mathbb{R}$ be a convex function and define $f(0) := \lim_{\varepsilon \to 0^+} f(\varepsilon)$. Then, the $f$-Divergence is defined as in (5.12) for $\mathbf{p}, \mathbf{q} \in \mathrm{Prob}(n)$ with $\mathbf{q} > 0$, and for $\mathbf{q} \not> 0$,*

$$D_f(\mathbf{p}\|\mathbf{q}) := \lim_{\varepsilon \to 0^+} D_f\left(\mathbf{p}\|(1 - \varepsilon)\mathbf{q} + \varepsilon\mathbf{u}\right), \tag{5.15}$$

*where $\mathbf{u}$ is the uniform distribution in $\mathrm{Prob}(n)$.*

**Exercise 5.5.** *Let $f: (0, \infty) \to \mathbb{R}$ be a convex function that satisfies $f(1) = 0$, and let $\tilde{f}(r) := rf\left(\frac{1}{r}\right)$. Show that $\tilde{f}$ is also convex with $\tilde{f}(1) = 0$ and prove that*

$$D_f(\mathbf{p}\|\mathbf{q}) = D_{\tilde{f}}(\mathbf{q}\|\mathbf{p}). \tag{5.16}$$

> **Theorem 5.1.** Let $f: (0, \infty) \to \mathbb{R}$ be a convex function that satisfies $f(1) = 0$. Then, $D_f$ as defined in Definition 5.2 is a classical divergence.

**Proof**   The normalization condition $D_f(1|1) = 0$ is directly derived from the requirement that $f(1) = 0$. To illustrate the data processing inequality, consider $m, n \in \mathbb{N}$, a stochastic matrix $E \in \text{STOC}(m, n)$, and probability vectors $\mathbf{p}, \mathbf{q} \in \text{Prob}(n)$. Define $\mathbf{r} := E\mathbf{p}$ and $\mathbf{s} := E\mathbf{q}$. For each $x \in [m]$ and $y \in [n]$, let $e_{x|y}$ represent the $(x, y)$-component of $E$.

With these definitions, the $x$-components of $\mathbf{r}$ and $\mathbf{s}$ are, respectively, $r_x = \sum_{y \in [n]} e_{x|y} p_y$ and $s_x = \sum_{y \in [n]} e_{x|y} q_y$. Assuming initially that $\mathbf{q} > 0$, we find that if $s_x = 0$, then $e_{x|y} = 0$ for all $y \in [n]$. Consequently, if $s_x = 0$, it follows that $r_x$ must also be 0. This leads to the conclusion that

$$D_f(E\mathbf{p}\|E\mathbf{q}) = D_f(\mathbf{r}\|\mathbf{s}) = \sum_{x \in \text{supp}(\mathbf{s})} s_x f\left(\frac{r_x}{s_x}\right), \tag{5.17}$$

where the summation is limited to all $x \in [m]$, for which $s_x \neq 0$. If $s_x = 0$, then $r_x$ is also 0, which contributes $0 f(\frac{0}{0}) = 0$ to the sum in (5.12), with $(\mathbf{r}, \mathbf{s})$ replacing $(\mathbf{p}, \mathbf{q})$.

The strategy of the proof involves representing $r_x / s_x$, as seen on the right-hand side of (5.17), as a convex combination of the ratios $p_y / q_y \, y \in [n]$. This is achieved by defining, for each $x \in \text{supp}(\mathbf{s})$,

$$t_{y|x} := \frac{e_{x|y} q_y}{\sum_{y'} e_{x|y'} q_{y'}} = \frac{e_{x|y} q_y}{s_x}. \tag{5.18}$$

It is important to note that for every $x \in \text{supp}(\mathbf{s})$, the set $\{t_{y|x}\}_{y \in [n]}$ forms a probability vector, and for all $x \in \text{supp}(\mathbf{s})$, it holds that

$$\frac{r_x}{s_x} = \sum_{y \in [n]} t_{y|x} \frac{p_y}{q_y}. \tag{5.19}$$

Inserting this expression into (5.17), we obtain

$$D_f(E\mathbf{p}\|E\mathbf{q}) = \sum_{x \in \text{supp}(\mathbf{r})} s_x f\left(\sum_{y \in [n]} t_{y|x} \frac{p_y}{q_y}\right)$$

$$\text{by convexity} \to \; \leqslant \sum_{x \in \text{supp}(\mathbf{s})} \sum_{y \in [n]} t_{y|x} s_x f\left(\frac{p_y}{q_y}\right)$$

$$\textbf{(5.18)} \to \; = \sum_{y \in [n]} \sum_{x \in \text{supp}(\mathbf{s})} e_{x|y} q_y f\left(\frac{p_y}{q_y}\right) \tag{5.20}$$

$$\boxed{\sum_{x \in \text{supp}(\mathbf{s})} e_{x|y} = 1} \longrightarrow \; = \sum_{y \in [n]} q_y f\left(\frac{p_y}{q_y}\right) = D_f(\mathbf{p}\|\mathbf{q}),$$

where the equality $\sum_{x \in \mathrm{supp}(s)} e_{x|y} = 1$ is valid because $e_{x|y} = 0$ if $s_x = 0$.

For the case $\mathbf{q} \geqslant 0$ define $\mathbf{q}_\varepsilon := (1 - \varepsilon)\mathbf{q} + \varepsilon\mathbf{u}$. We then get that $\mathbf{q}_\varepsilon > 0$ so that for any $\varepsilon > 0$

$$D_f(E\mathbf{p}\|E\mathbf{q}_\varepsilon) \leqslant D_f(\mathbf{p}\|\mathbf{q}_\varepsilon). \tag{5.21}$$

Taking the limit $\varepsilon \to 0^+$ on both sides of the equation above and using continuity of $D_f(\mathbf{p}\|\mathbf{q})$ in $\mathbf{q}$ (see Exercise 5.6) completes the proof. ∎

**Exercise 5.6 (Continuity of $D_f(\mathbf{p}\|\mathbf{q})$ in q).** *In the last part of the proof above we used the fact that the $f$-Divergence is continuous in* $\mathbf{q}$. *Show that in general, if* $\{\mathbf{q}_k\}_{k \in \mathbb{N}}$ *is a sequence of probability vectors in* $\mathrm{Prob}(n)$ *that satisfies* $\mathbf{q}_k \to \mathbf{q}$ *as* $k \to \infty$, *then*

$$\lim_{k \to \infty} D_f(\mathbf{p}\|\mathbf{q}_k) = D_f(\mathbf{p}\|\mathbf{q}). \tag{5.22}$$

*Hint: Observe that $f$ is continuous in $(0, \infty)$ since it is convex.*

---

**Corollary 5.1.** Let $n, m \in \mathbb{N}$, $\mathbf{p}, \mathbf{q} \in \mathrm{Prob}(n)$, and $\mathbf{p}', \mathbf{q}' \in \mathrm{Prob}(m)$. The condition $(\mathbf{p}, \mathbf{q}) \succ (\mathbf{p}', \mathbf{q}')$ holds if and only if for every convex function $f : (0, \infty) \to \mathbb{R}$ with $f(1) = 0$

$$D_f(\mathbf{p}\|\mathbf{q}) \geqslant D_f(\mathbf{p}'\|\mathbf{q}'). \tag{5.23}$$

---

**Exercise 5.7.** *Prove Corollary (5.1). Hint: Use (5.10), Exercise 5.3, and Theorem 5.1 to prove the corollary.*

### 5.1.3 Examples

In this subsection, we give several examples of $f$-Divergences that play important role in applications.

## Kullback–Leibler Divergence

The Kullback–Leibler divergence (also known as the KL-divergence or the relative entropy) is perhaps the most well known divergence which appears in numerous applications in statistics, information theory, and as we will see in resource theories. For this reason, it is the only divergence that we will denote simply by $D$ without any subscript. It is the $f$-Divergence that corresponds to the function $f(r) = r \log r$. For this choice, we get,

$$D(\mathbf{p}\|\mathbf{q}) = \begin{cases} \sum_{x \in [n]} p_x\big(\log p_x - \log q_x\big) & \text{if } \mathbf{p} \ll \mathbf{q} \\ \infty & \text{otherwise,} \end{cases} \tag{5.24}$$

where $\mathbf{p} \ll \mathbf{q}$ denotes $\mathrm{supp}(\mathbf{p}) \subseteq \mathrm{supp}(\mathbf{q})$, and we use the convention $0 \log 0 = 0$. In the next chapter we will study the many properties of this divergence.

## The Trace Distance

The trace distance, also known as the total variation distance (sometimes also called statistical distance), is an $f$-Divergence with $f(r) = \frac{1}{2}|r - 1|$. For this convex function we get

$$D_f(\mathbf{p}\|\mathbf{q}) = \sum_{x\in[n]} q_x \frac{1}{2}\left|\frac{p_x}{q_x} - 1\right| = \frac{1}{2}\sum_{x\in[n]} |p_x - q_x| = \frac{1}{2}\|\mathbf{p} - \mathbf{q}\|_1. \tag{5.25}$$

This $f$-Divergence, which also functions as a metric, will be examined in detail in the subsequent sections.

## The Hellinger Distance

This is another distance measure that is closely related to an $f$-Divergence with $f(r) = \frac{1}{2}(\sqrt{r} - 1)^2$. For this $f$ we get

$$D_f(\mathbf{p}\|\mathbf{q}) = \sum_{x\in[n]} q_x \frac{1}{2}\left(\sqrt{\frac{p_x}{q_x}} - 1\right)^2 = \frac{1}{2}\sum_{x\in[n]} \left(\sqrt{p_x} - \sqrt{q_x}\right)^2. \tag{5.26}$$

The Hellinger distance is defined as the square root of the above expression:

$$H(\mathbf{p}, \mathbf{q}) := \sqrt{\frac{1}{2}\sum_{x\in[n]} \left(\sqrt{p_x} - \sqrt{q_x}\right)^2} \tag{5.27}$$

We will see later on that the above divergence is also a metric that is closely related to a quantity known as the fidelity.

## The $\alpha$-Divergence

The $\alpha$-Divergence is an $f$-Divergence corresponding to $f_\alpha(r) = \frac{r^\alpha - r}{\alpha(\alpha - 1)}$, where $\alpha \in [0, \infty)$, where the case $\alpha = 1$ is defined by the limit $\lim_{\alpha\to 1} \frac{r^\alpha - r}{\alpha(\alpha - 1)} = r \ln(r)$ (which yields the KL-divergence), and similarly the case $\alpha = 0$ is given by $-\ln(r)$. For this choice of $f$ we get

$$D_{f_\alpha}(\mathbf{p}\|\mathbf{q}) = \sum_{x\in[n]} q_x \frac{1}{\alpha(\alpha - 1)}\left(\left(\frac{p_x}{q_x}\right)^\alpha - \frac{p_x}{q_x}\right) = \frac{1}{\alpha(\alpha - 1)}\left(\sum_{x\in[n]} p_x^\alpha q_x^{1-\alpha} - 1\right). \tag{5.28}$$

The $\alpha$-Divergence can be expressed as a function of the Rényi divergences that we will study in the Chapter 6.

**Exercise 5.8.** *Show that all the functions $f$ above are convex and satisfy $f(1) = 0$.*

**Exercise 5.9 (The Jensen–Shannon Divergence).** *Let $f : (0, \infty) \to \mathbb{R}$ given by*

$$f(r) = (r + 1)\log\left(\frac{2}{r + 1}\right) + r \log r \qquad \forall\, r \in \mathbb{R}. \tag{5.29}$$

*Show that $f$ is convex with $f(1) = 0$ and compute its $f$-Divergence.*

## 5.1.4 Continuity of Divergences

Divergences are generally not continuous over $\mathrm{Prob}(n) \times \mathrm{Prob}(n)$, even when they exhibit continuity in their first and/or second arguments. For example, consider the two sequences of probability vectors

$$\mathbf{p}_k := \left(\frac{1}{k}, 1 - \frac{1}{k}\right)^T \quad \text{and} \quad \mathbf{q}_k := \left(\frac{1}{k^2}, 1 - \frac{1}{k^2}\right)^T \quad \forall\, k \in \mathbb{N}. \tag{5.30}$$

Clearly, we have

$$\mathbf{p} := \lim_{k \to \infty} \mathbf{p}_k = \begin{bmatrix} 0 \\ 1 \end{bmatrix} \quad \text{and} \quad \mathbf{q} := \lim_{k \to \infty} \mathbf{q}_k = \begin{bmatrix} 0 \\ 1 \end{bmatrix}. \tag{5.31}$$

Consider now the $\alpha$-Divergence with $\alpha = 2$ as defined in (5.28). Denote this divergence by

$$D'(\mathbf{p}\|\mathbf{q}) = \frac{1}{2}\left(\sum_{x \in [n]} p_x^2 q_x^{-1} - 1\right). \tag{5.32}$$

It is simple to check (see the following exercise) that

$$\lim_{k \to \infty} D'(\mathbf{p}_k\|\mathbf{q}_k) = \frac{1}{2} \neq 0 = D'(\mathbf{p}\|\mathbf{q}). \tag{5.33}$$

For the $\alpha$ divergences, this discontinuity only happens when the limits $\mathbf{p}, \mathbf{q}$ are on the boundary. In fact, we will see in Chapter 6 that many divergences are continuous in the interior of $\mathrm{Prob}(n) \times \mathrm{Prob}(n)$.

**Exercise 5.10.** *Let $D'$ be the $\alpha$-Divergence for $\alpha = 2$, and let $\mathbf{p}_k$ and $\mathbf{q}_k$ be as in (5.31). Prove (5.33).*

We show now that by utilizing the data processing inequality, if a divergence is continuous in one of its arguments it is necessarily continuous in the second argument as well.

> **Theorem 5.2.** Let $\mathbb{D}$ be a divergence. The following statements are equivalent:
>
> 1. For every fixed $\mathbf{q} \in \mathrm{Prob}(n)$ the function $\mathbf{p} \mapsto \mathbb{D}(\mathbf{p}\|\mathbf{q})$ is continuous in $\mathrm{Prob}(n)$.
> 2. For every fixed $\mathbf{p} \in \mathrm{Prob}(n)$ the function $\mathbf{q} \mapsto \mathbb{D}(\mathbf{p}\|\mathbf{q})$ is continuous in $\mathrm{Prob}(n)$.

**Proof** We will demonstrate the implication of $1 \Rightarrow 2$. The converse, $2 \Rightarrow 1$, will be established using a similar approach.

Let $\mathbf{q}, \mathbf{q}' \in \mathrm{Prob}(n)$, and define the channel $E \in \mathrm{STOC}(n, n)$ by its action on every $\mathbf{v} \in \mathrm{Prob}(n)$ as

$$E\mathbf{v} := (1 - \varepsilon)(\mathbf{v} - \mathbf{q}) + \mathbf{q}', \quad \text{where} \quad \varepsilon := 1 - \min_{x \in [n]} \frac{q_x'}{q_x}. \tag{5.34}$$

By definition, $E\mathbf{q} = \mathbf{q}'$, and $\varepsilon \in (0, 1)$ can be reduced to an arbitrarily small value by taking $\mathbf{q}$ and $\mathbf{q}'$ to have sufficiently small trace distance $\frac{1}{2}\|\mathbf{q} - \mathbf{q}'\|_1$. Note also that from the definition of $\varepsilon$ it follows that $\mathbf{q}' - (1 - \varepsilon)\mathbf{q} \geqslant 0$. Therefore, for every $\mathbf{v} \in \mathrm{Prob}(n)$,

$$E\mathbf{v} = (1 - \varepsilon)\mathbf{v} + \left(\mathbf{q}' - (1 - \varepsilon)\mathbf{q}\right) \geqslant 0. \tag{5.35}$$

The inequality (5.35) implies that $E$ is indeed column stochastic. Moreover, by definition,

$$\|\mathbf{p} - E\mathbf{p}\|_1 = \left\|\varepsilon(\mathbf{p} - \mathbf{q}) + \mathbf{q}' - \mathbf{q}\right\|_1 \leqslant \varepsilon + \|\mathbf{q} - \mathbf{q}'\|_1. \tag{5.36}$$

This inequality demonstrates that as $\mathbf{q}$ converges toward $\mathbf{q}'$, the vector $E\mathbf{p}$ approaches $\mathbf{p}$ (recall that $\varepsilon$ goes to zero as $\mathbf{q}$ approaches $\mathbf{q}'$). Applying the DPI with $E$ we can thus bound

$$\mathbb{D}(\mathbf{p}\|\mathbf{q}) - \mathbb{D}(\mathbf{p}\|\mathbf{q}') \geqslant \mathbb{D}(E\mathbf{p}\|E\mathbf{q}) - \mathbb{D}(\mathbf{p}\|\mathbf{q}')$$
$$= \mathbb{D}(E\mathbf{p}\|\mathbf{q}') - \mathbb{D}(\mathbf{p}\|\mathbf{q}'). \tag{5.37}$$

Since $\mathbb{D}(\mathbf{p}\|\mathbf{q})$ is continuous in $\mathbf{p}$, the expression on the right-hand side of (5.37) vanishes when $\mathbf{q} \to \mathbf{q}'$.

Subsequently, we establish a comparable upper bound for $\mathbb{D}(\mathbf{p}|\mathbf{q}) - \mathbb{D}(\mathbf{p}|\mathbf{q}')$ by introducing $\tilde{E} \in \mathrm{STOC}(n, n)$ and $\tilde{\varepsilon} \in (0, 1)$. These are defined identically to how $E$ and $\varepsilon$ were defined, but with the roles of $\mathbf{q}$ and $\mathbf{q}'$ reversed. Specifically, $\tilde{E}$ is defined by its action on every $\mathbf{v} \in \mathrm{Prob}(n)$ as

$$\tilde{E}\mathbf{v} := (1 - \tilde{\varepsilon})(\mathbf{v} - \mathbf{q}') + \mathbf{q}, \quad \text{where} \quad \tilde{\varepsilon} := 1 - \min_{x \in [n]} \frac{q_x}{q'_x}. \tag{5.38}$$

By definition, $\tilde{E}\mathbf{q}' = \mathbf{q}$ and $\tilde{\varepsilon} \in (0, 1)$. Furthermore, continuing with similar steps, it can be verified that $\tilde{E}$ is indeed column stochastic, and $\tilde{E}\mathbf{p}$ approaches $\mathbf{p}$ as $\mathbf{q}$ approaches $\mathbf{q}'$. Utilizing the DPI we also have

$$\mathbb{D}(\mathbf{p}\|\mathbf{q}) - \mathbb{D}(\mathbf{p}\|\mathbf{q}') \leqslant \mathbb{D}(\mathbf{p}\|\mathbf{q}) - \mathbb{D}(\tilde{E}\mathbf{p}\|\tilde{E}\mathbf{q}')$$
$$= \mathbb{D}(\mathbf{p}\|\mathbf{q}) - \mathbb{D}(\tilde{E}\mathbf{p}\|\mathbf{q}). \tag{5.39}$$

Therefore, as before, due to the continuity of $\mathbb{D}(\mathbf{p}\|\mathbf{q})$ in $\mathbf{p}$, the expression on the right-hand side of (5.39) vanishes when $\mathbf{q}' \to \mathbf{q}$. Combining this with the lower bound in (5.37), we conclude that $\mathbb{D}(\mathbf{p}\|\mathbf{q})$ is continuous in $\mathbf{q}$. $\blacksquare$

## 5.1.5 Divergences from Measures of Nonuniformity

In this subsection, we demonstrate a one-to-one correspondence between divergences and Schur convex functions that fulfill specific criteria. This relationship allows for the construction of divergences from Schur convex functions, thereby expanding the class of $f$-Divergences. This correspondence proves particularly valuable in Chapter 6, where we utilize it to establish a bijection between entropies and relative entropies.

> **A Measure of Nonuniformity**
>
> **Definition 5.3.** A function
>
> $$g : \bigcup_{n \in \mathbb{N}} \mathrm{Prob}(n) \to \mathbb{R} \cup \{\infty\} \tag{5.40}$$
>
> is called a measure of nonuniformity if it satisfies the following three properties:
>
> 1. For every $n \in \mathbb{N}$, the restriction of $g$ to $\mathrm{Prob}(n)$ is Schur convex and continuous on $\mathrm{Prob}(n)$.
> 2. For $n = 1$, it is normalized to $g(1) = 0$.
> 3. For all $k, n \in \mathbb{N}$ and $\mathbf{r} \in \mathrm{Prob}(n)$, $g\big(\mathbf{r} \otimes \mathbf{u}^{(k)}\big) = g(\mathbf{r})$.

In Chapter 16, we will study the resource theory of nonuniformity in which the functions above quantify the resource of this theory. Specifically, these functions quantify how different a probability vector $\mathbf{p}$ is from the uniform distribution $\mathbf{u}$. As an indication of this, note that if $\mathbb{D}$ is a classical divergence that is continuous in its first argument, then the function

$$g_{\mathbb{D}}(\mathbf{p}) := \mathbb{D}\left(\mathbf{p} \| \mathbf{u}^{(n)}\right) \qquad \forall\, \mathbf{p} \in \mathrm{Prob}(n) \tag{5.41}$$

is a measure of nonuniformity.

**Exercise 5.11.** *Verify that $g_{\mathbb{D}}$ indeed satisfies all the three properties above. Hint: For the third property show that*

$$\mathbb{D}(\mathbf{p} \| \mathbf{u}^{(n)}) = \mathbb{D}\big(\mathbf{p} \otimes \mathbf{u}^{(k)} \| \mathbf{u}^{(nk)}\big) \tag{5.42}$$

*as a consequence of the DPI applied twice for channels introducing and removing an independent distribution $\mathbf{u}^{(k)}$.*

**Exercise 5.12.** *Let $f : (0, \infty) \to \mathbb{R}$ be convex with $f(1) = 0$. Show that for the $f$-Divergence, $D_f$, we have*

$$g_{D_f}(\mathbf{p}) = \frac{1}{n} \sum_{x \in [n]} f(n\mathbf{p}_x) \qquad \forall\, \mathbf{p} \in \mathrm{Prob}(n). \tag{5.43}$$

*Verify by direct calculation that this expression satisfies the three properties of $g$.*

In Theorem 4.8, we established that for every $n \in \mathbb{N}$, $\mathbf{p} \in \mathrm{Prob}(n)$, and $\mathbf{q} \in \mathrm{Prob}_{>0}(n) \cap \mathbb{Q}^n$, there is a vector $\mathbf{r} \in \mathrm{Prob}(k)$ with the property that $(\mathbf{p}, \mathbf{q}) \sim (\mathbf{r}, \mathbf{u}^{(k)})$. To elaborate, let $\mathbf{q} = (\frac{k_1}{k}, \ldots, \frac{k_n}{k})^T$, where each $k_x \in \mathbb{N}$ and $k := k_1 + \cdots + k_n$. The vector $\mathbf{r}$ is then expressed as

$$\mathbf{r} := \bigoplus_{x \in [n]} p_x \mathbf{u}^{(k_x)}. \tag{5.44}$$

Building upon this equivalency, the following theorem demonstrates a bijective relationship between divergences and measures of nonuniformity. However, prior to this, it's essential to explore the uniqueness of the vector $\mathbf{r}$ in (5.44).

The vector $\mathbf{r}$ is not unique. To see why, let $\{m_x\}_{x\in[n]}$ be a set of $n$ integers satisfying

$$q_x = \frac{k_x}{k} = \frac{m_x}{m}, \quad \text{where} \quad m := \sum_{x\in[n]} m_x. \tag{5.45}$$

Given any such set, we can define the probability vector $\mathbf{s} := \bigoplus_{x\in[n]} p_x \mathbf{u}^{(m_x)}$ so that $(\mathbf{p},\mathbf{q}) \sim (\mathbf{s},\mathbf{u}^{(m)})$. This demonstrates that $\mathbf{r}$ is not unique. However, since $km_x = mk_x$, we observe that

$$\mathbf{u}^{(m)} \otimes \mathbf{r} = \bigoplus_{x\in[n]} p_x \mathbf{u}^{(mk_x)} = \bigoplus_{x\in[n]} p_x \mathbf{u}^{(km_x)} = \mathbf{u}^{(k)} \otimes \mathbf{s}. \tag{5.46}$$

Equation (5.46) highlights that for any measure of nonuniformity $g$, following the third property of Definition 5.3, we obtain

$$\begin{aligned} g(\mathbf{r}) &= g\left(\mathbf{u}^{(m)} \otimes \mathbf{r}\right) \\ \textbf{(5.46)} \rightarrow &= g\left(\mathbf{u}^{(k)} \otimes \mathbf{s}\right) \\ &= g(\mathbf{s}), \end{aligned} \tag{5.47}$$

where the last equality again utilizes the third property of Definition 5.3. Thus, in this context, $\mathbf{r}$ and $\mathbf{s}$ have the same nonuniformity.

---

**Bijection between Divergences and Measures of Nonuniformity**

**Theorem 5.3.** Let $g$ be a measure of nonuniformity. For any $n \in \mathbb{N}, \mathbf{p} \in \mathrm{Prob}(n)$ and $\mathbf{q} \in \mathrm{Prob}_{>0}(n) \cap \mathbb{Q}^n$, define

$$D_g(\mathbf{p}\|\mathbf{q}) := g(\mathbf{r}), \tag{5.48}$$

where $\mathbf{r} \in \mathrm{Prob}(k)$ is the vector defined in (5.44). For general $\mathbf{q} \in \mathrm{Prob}(n)$, $D_g$ is defined via a continuous extension. Then, $D_g$ is a divergence and is continuous in $\mathbf{q}$ for any fixed $\mathbf{p} \in \mathrm{Prob}(n)$.

---

*Remark.* Observe that $D_g$ in (5.48) is well defined since $g(\mathbf{r}) = g(\mathbf{s})$ for any other vector $\mathbf{s}$ as already defined. We will also show in the proof next that the continuous extension for general $\mathbf{q} \in \mathrm{Prob}(n)$ is well defined.

**Proof** We first show that $D_g$ is a divergence on the restricted space in which $\mathbf{q} \in \mathrm{Prob}_{>0}(n) \cap \mathbb{Q}^n$. The normalization of $D_g$ holds since $D_g(1\|1) = g(1) = 0$. To show the DPI, let $\mathbf{p} \in \mathrm{Prob}(n)$, $\mathbf{q} \in \mathrm{Prob}_{>0}(n) \cap \mathbb{Q}^n$, and $E \in \mathrm{STOC}(m,n) \cap \mathbb{Q}^{m\times n}$ be a stochastic matrix (channel) with rational components. Let further $k \in \mathbb{N}$ be large enough such that we can express

$$\mathbf{q} = \left(\frac{k_1}{k},\dots,\frac{k_n}{k}\right)^T, \quad \mathbf{q}' := E\mathbf{q} = \left(\frac{k_1'}{k},\dots,\frac{k_m'}{k}\right)^T, \tag{5.49}$$

where $k_x, k_x' \in \mathbb{N}$ with $\sum_{x \in [n]} k_x = \sum_{x \in [m]} k_x' = k$. Due to Theorem 4.8 there exists $\mathbf{r}, \mathbf{s} \in \mathrm{Prob}(k)$ such that $(\mathbf{p}, \mathbf{q}) \sim (\mathbf{r}, \mathbf{u}^{(k)})$ and $(E\mathbf{p}, E\mathbf{q}) \sim (\mathbf{s}, \mathbf{u}^{(k)})$. By definition, $(\mathbf{p}, \mathbf{q}) \succ (E\mathbf{p}, E\mathbf{q})$ so that $(\mathbf{r}, \mathbf{u}^{(k)}) \succ (\mathbf{s}, \mathbf{u}^{(k)})$, or equivalently, $\mathbf{r} \succ \mathbf{s}$. Hence,

$$D_g(E\mathbf{p} \| E\mathbf{q}) = g(\mathbf{s}) \leqslant g(\mathbf{r}) = D_g(\mathbf{p} \| \mathbf{q}), \tag{5.50}$$

where the inequality follows from the Schur convexity of $g$.

Next, we must demonstrate continuity in the second argument within $\mathrm{Prob}_{>0}(n) \cap \mathbb{Q}^n$ to ensure the continuous extension is well defined. As inferred from (5.48), for any fixed $\mathbf{q} \in \mathrm{Prob}_{>0}(n) \cap \mathbb{Q}^n$, the function $D_g(\mathbf{p} \| \mathbf{q})$ is continuous in $\mathbf{p} \in \mathrm{Prob}(n)$ due to the continuity of $g$. Combining this with Theorem 5.2 we conclude that indeed the function $D_g(\mathbf{p} \| \mathbf{q})$ is also continuous in the second argument $\mathbf{q} \in \mathrm{Prob}_{>0}(n) \cap \mathbb{Q}^n$. Finally, observe that the DPI remains valid when we define the quantity for irrational $\mathbf{q}$ via continuous extension. This completes the proof. $\blacksquare$

**Exercise 5.13.** *Describe explicitly the bijection between divergences that are continuous in their second argument and measures of nonuniformity. That is, for any $g$ express the corresponding $\mathbb{D}$ and vice versa.*

## 5.2 Quantum Divergences

In this section, we study the quantum version of a classical divergence. We start with their formal definition and some of their basics properties, and then move to study a systematic approach to extend classical divergences to the quantum domain. We then give examples focusing on the quantum extension of the $f$-Divergence. Similarly to the definition of a classical divergence, a quantum divergence is also defined in terms of the DPI.

In the following definition we consider a function that is acting on pairs of quantum states in all finite dimensions:

$$\mathbb{D} : \bigcup_A \left\{ \mathfrak{D}(A) \times \mathfrak{D}(A) \right\} \to \mathbb{R} \cup \{\infty\}. \tag{5.51}$$

---

**Definition 5.4.** The function $\mathbb{D}$, as presented in (5.51), is termed a *quantum divergence* if it fulfills the following two criteria:

1. Data Processing Inequality (DPI): $\mathbb{D}\big(\mathcal{E}(\rho) \big\| \mathcal{E}(\sigma)\big) \leqslant \mathbb{D}(\rho \| \sigma)$ for all $\mathcal{E} \in \mathrm{CPTP}(A \to B)$ and all $\rho, \sigma \in \mathfrak{D}(A)$.
2. Normalization: $\mathbb{D}(1 \| 1) = 0$.

---

*Remark.* Note that a classical divergence can be viewed as a quantum divergence whose domain is restricted to classical systems. The union in (5.51) is over all systems $A$ and particularly over all finite dimensions $|A|$. Therefore, the domain of $\mathbb{D}$ consists of pairs of density matrices $(\rho, \sigma)$ in any dimension $|A| \in \mathbb{N}$. For the case of a trivial system $A$ with $|A| = 1$ the only density matrix in $\mathfrak{D}(A)$ is the number 1. In this case, divergences satisfy $\mathbb{D}(1\|1) = 0$.

Like classical divergences, quantum divergences are nonnegative since for any pair of states $\rho, \sigma \in \mathfrak{D}(A)$ we have the trace $\mathrm{Tr} \in \mathrm{CPTP}(A \to 1)$, so that

$$\mathbb{D}(\rho\|\sigma) \geqslant \mathbb{D}\big(\mathrm{Tr}[\rho]\big\|\mathrm{Tr}[\sigma]\big) = \mathbb{D}(1\|1) = 0, \tag{5.52}$$

where the inequality follows from the DPI. Moreover, recall that any state $\rho \in \mathfrak{D}(A)$ with dimension $|A| > 1$ can be viewed as a preparation channel $\rho^{1 \to A} \in \mathrm{CPTP}(1 \to A)$. Hence,

$$\mathbb{D}(\rho^A\|\rho^A) = \mathbb{D}\big(\rho^{1 \to A}(1)\big\|\rho^{1 \to A}(1)\big) \leqslant \mathbb{D}(1\|1) = 0, \tag{5.53}$$

where again we used the DPI for divergences. Combining this with the nonnegativity of divergences we conclude that for any state $\rho \in \mathfrak{D}(A)$ in any dimension $|A| \in \mathbb{N}$

$$\mathbb{D}(\rho\|\rho) = 0. \tag{5.54}$$

This is consistent with the intuition that divergences quantify the distinguishability between two states.

An interesting question remaining is whether the converse of the property in (5.54) also holds. A quantum divergence, $\mathbb{D}$, is said to be *faithful* if for any $\rho, \sigma \in \mathfrak{D}(A)$, the condition $\mathbb{D}(\rho\|\sigma) = 0$ implies $\rho = \sigma$. We will see later on that not all quantum divergences are faithful. However, in the following lemma we show that a quantum divergence is faithful if and only if its reduction to classical systems is faithful.

> **Lemma 5.1.** Let $\mathbb{D}$ be a quantum divergence. Then, $\mathbb{D}$ is faithful if and only if its reduction to classical (diagonal) states is faithful.

**Proof**   Clearly, if $\mathbb{D}$ is faithful on quantum states it is also faithful on classical states as the latter is a subset of the former. Suppose now that $\mathbb{D}$ is faithful on classical states, and suppose by contradiction that there exists $\rho \neq \sigma \in \mathfrak{D}(A)$ such that $\mathbb{D}(\rho\|\sigma) = 0$. Then, there exists a basis of $A$ such that the diagonal of $\rho$ in this basis does not equal to the diagonal of $\sigma$ (see the following exercise). Let $\Delta \in \mathrm{CPTP}(A \to A)$ be the completely dephasing channel in this basis. Then, $\Delta(\rho) \neq \Delta(\sigma)$ and we get

$$\mathbb{D}\big(\Delta(\rho)\big\|\Delta(\sigma)\big) \leqslant \mathbb{D}(\rho\|\sigma) = 0. \tag{5.55}$$

But since $\mathbb{D}$ is faithful on diagonal states we get the contradiction that $\Delta(\rho) = \Delta(\sigma)$. Hence, $\mathbb{D}$ is faithful also on quantum states.   ∎

**Exercise 5.14.** *Let $\rho, \sigma \in \mathfrak{D}(A)$. Show that if $\rho \neq \sigma$, then there exists a basis of $A$ such that the diagonal of $\rho$ in this basis does not equal to the diagonal of $\sigma$ in the same basis.*

The data processing inequality implies that quantum divergences are invariant under isometries. That is, for any isometry channel $\mathcal{V} \in \text{CPTP}(A \to B)$ we have

$$\mathbb{D}\big(\mathcal{V}(\rho)\big\|\mathcal{V}(\sigma)\big) = \mathbb{D}(\rho\|\sigma) \qquad \forall\, \rho, \sigma \in \mathfrak{D}(A). \tag{5.56}$$

To see why, recall that every isometry channel has a left inverse channel $\mathcal{R} \in \text{CPTP}(B \to A)$ that satisfies $\mathcal{R}^{B \to A} \circ \mathcal{V}^{A \to B} = \text{id}^A$ (see Section 3.5.8). Hence, by definition of $\mathcal{R}$

$$
\begin{aligned}
D(\rho\|\sigma) &= \mathbb{D}\big(\mathcal{R} \circ \mathcal{V}(\rho) \big\| \mathcal{R} \circ \mathcal{V}(\sigma)\big) \\
\textbf{DPI} &\to \;\leqslant \mathbb{D}\big(\mathcal{V}(\rho)\big\|\mathcal{V}(\sigma)\big) \\
\textbf{DPI} &\to \;\leqslant \mathbb{D}(\rho\|\sigma).
\end{aligned}
\tag{5.57}
$$

That is, both inequalities must be equalities so that (5.56) holds.

**Exercise 5.15.** *Use the invariance property of quantum divergences under isometries to show that any classical divergence, $\mathbb{D}$, satisfies*

$$\mathbb{D}(\mathbf{p} \oplus 0\|\mathbf{q} \oplus 0) = \mathbb{D}(\mathbf{p}\|\mathbf{q}). \tag{5.58}$$

**Exercise 5.16.** *Let $\mathbb{D}$ be a quantum divergence, $\rho, \sigma \in \mathfrak{D}(A)$, and $\mathcal{E} \in \text{CPTP}(A \to B)$.*

1. *Show that if there exists a channel $\mathcal{F} \in \text{CPTP}(B \to A)$ such that $\mathcal{F} \circ \mathcal{E}(\rho) = \rho$ and $\mathcal{F} \circ \mathcal{E}(\sigma) = \sigma$, then*

$$\mathbb{D}(\rho\|\sigma) = \mathbb{D}\big(\mathcal{E}(\rho)\big\|\mathcal{E}(\sigma)\big). \tag{5.59}$$

2. *Show that for any $\omega \in \mathfrak{D}(B)$*

$$\mathbb{D}(\rho \otimes \omega\|\sigma \otimes \omega) = \mathbb{D}(\rho\|\sigma). \tag{5.60}$$

**Exercise 5.17.** *Let $\rho, \sigma \in \mathfrak{D}(A)$ and denote by*

$$\sup(\rho/\sigma) := \sup_{0 \leqslant \Lambda \leqslant I^A} \frac{\text{Tr}[\rho\Lambda]}{\text{Tr}[\sigma\Lambda]}. \tag{5.61}$$

1. *Show that the function above satisfies the DPI; that is,*

$$\sup\big(\mathcal{E}(\rho)/\mathcal{E}(\sigma)\big) \leqslant \sup(\rho/\sigma) \qquad \forall\, \mathcal{E} \in \text{CPTP}(A \to B). \tag{5.62}$$

2. *Show that*

$$\sup(\rho/\sigma) = \inf\big\{\lambda \in \mathbb{R}: \lambda\sigma - \rho \geqslant 0\big\}. \tag{5.63}$$

3. *Show that the function $f(\rho\|\sigma) := \sup(\rho/\sigma) - 1$ is a quantum divergence.*

## Joint Convexity

We say that a quantum divergence $\mathbb{D}$ is jointly convex if for any quantum system $A$, $m \in \mathbb{N}$, $\mathbf{p} \in \text{Prob}(m)$, and two sets, $\{\rho_x\}_{x \in [m]}$ and $\{\sigma_x\}_{x \in [m]}$ of $m$ density matrices in $\mathfrak{D}(A)$ we have

$$\mathbb{D}\Big(\sum_{x \in [n]} p_x \rho_x \Big\| \sum_{x \in [n]} p_x \sigma_x\Big) \leqslant \sum_{x \in [n]} p_x \mathbb{D}(\rho_x\|\sigma_x). \tag{5.64}$$

Although not every quantum divergence exhibits joint convexity, the combination of joint convexity with both the property described in (5.60), and the invariance under isometries, results in a condition that is more stringent than DPI.

> **Lemma 5.2.** Let $\mathbb{D}$ be a function with the same domain and range as a quantum divergence that is invariant under isometries. Suppose further that $\mathbb{D}$ is jointly convex and satisfies (5.60) for any quantum systems $A$ and $B$, and quantum states $\rho, \sigma \in \mathfrak{D}(A)$ and $\omega \in \mathfrak{D}(B)$. Then $\mathbb{D}$ satisfies the DPI.

**Proof** Due to Stinespring dilation theorem, the invariance under isometries implies that it is sufficient to prove that for any two bipartite states $\rho, \sigma \in \mathfrak{D}(AB)$

$$\mathbb{D}\left(\rho^{AB} \middle\| \sigma^{AB}\right) \geqslant \mathbb{D}\left(\rho^{A} \middle\| \sigma^{A}\right). \tag{5.65}$$

Let $k := |B|^2$ and $\mathcal{R} \in \mathrm{CPTP}(B \to B)$ be the completely randomizing (or equivalently depolarizing) channel. Such a channel takes all states to the maximally mixed state, and in Chapter 8 (particularly, Exercise 3.74) we will see that it can be expressed as the following uniform mixture of unitary channels, $\mathcal{R} = \frac{1}{n} \sum_{x \in [n]} \mathcal{U}_x$, where $n := |B|^2$ and each $\mathcal{U}_x \in \mathrm{CPTP}(B \to B)$ is a unitary channel. We then get from (5.60) that

$$
\begin{aligned}
\mathbb{D}\left(\rho^{A} \middle\| \sigma^{A}\right) &= \mathbb{D}\left(\rho^{A} \otimes \mathbf{u}^{B} \middle\| \sigma^{A} \otimes \mathbf{u}^{B}\right) \\
&= \mathbb{D}\left(\mathcal{R}^{B \to B}\left(\rho^{AB}\right) \middle\| \mathcal{R}^{B \to B}\left(\sigma^{AB}\right)\right) \\
\text{Joint convexity} \rightarrow &\leqslant \frac{1}{n} \sum_{x \in [n]} \mathbb{D}\left(\mathcal{U}_x^{B \to B}\left(\rho^{AB}\right) \middle\| \mathcal{U}_x^{B \to B}\left(\sigma^{AB}\right)\right) \\
\text{Invariance under unitaries} \rightarrow &= \mathbb{D}\left(\rho^{AB} \middle\| \sigma^{AB}\right).
\end{aligned}
\tag{5.66}
$$

This completes the proof. ∎

## 5.2.1 The Quantum $f$-Divergence

In previous subsections, we observed that a wide range of classical divergences can be represented as an $f$-Divergence. This subsection delves into their extension within the quantum realm. We will explore how this quantum extension gives rise to a variety of divergences that are significant in practical applications. However, it is important to note that extending classical divergences to the quantum domain generally do not yield unique results. In particular, the quantum $f$-Divergence is not the only possible quantum extension. We will explore other quantum extensions of this concept in the following subsection.

To motivate the formal definition of the quantum $f$-Divergence, we first discuss a useful correspondence between pairs of density matrices and pairs of probability distributions. Consider two quantum states $\rho, \sigma \in \mathfrak{D}(A)$ with spectral decomposition (here $m := |A|$)

$$\rho = \sum_{x\in[m]} p_x |a_x\rangle\langle a_x| \quad \text{and} \quad \sigma = \sum_{y\in[m]} q_y |b_y\rangle\langle b_y|, \tag{5.67}$$

where $\{|a_x\rangle\}_{x\in[m]}$ and $\{|b_y\rangle\}_{y\in[m]}$ are orthonormal bases consisting of the eigenvectors of $\rho$ and $\sigma$, respectively. Define the probability vectors $\tilde{\mathbf{p}}, \tilde{\mathbf{q}} \in \mathrm{Prob}(m^2)$ whose components are given by

$$\tilde{p}_{xy} := p_x |\langle a_x|b_y\rangle|^2 \quad \text{and} \quad \tilde{q}_{xy} = q_y |\langle a_x|b_y\rangle|^2 \quad \forall\, x, y \in [m]. \tag{5.68}$$

Now, if $\mathbb{D}$ is a classical divergence, then we can extend it to $\rho, \sigma \in \mathfrak{D}(A)$ by

$$\mathbb{D}^q(\rho\|\sigma) := \mathbb{D}\left(\tilde{\mathbf{p}}\|\tilde{\mathbf{q}}\right). \tag{5.69}$$

Clearly, $\mathbb{D}^q(\rho\|\sigma)$ is zero if $\rho = \sigma$, and in Exercise (5.18) we will show that if $\rho$ and $\sigma$ are diagonal, then $\mathbb{D}^q(\rho\|\sigma) := \mathbb{D}\left(\mathbf{p}\|\mathbf{q}\right)$, where $\mathbf{p}$ and $\mathbf{q}$ are the diagonals of $\rho$ and $\sigma$. In the following lemma we show that $\mathbb{D}^q$ is invariant under isometries.

> **Lemma 5.3.** Let $\mathbb{D}$ be a classical divergence and define $\mathbb{D}^q$ as in (5.69). Then, for any isometry channel $\mathcal{V} \in \mathrm{CPTP}(A \to B)$ and any $\rho, \sigma \in \mathfrak{D}(A)$
>
> $$\mathbb{D}^q\left(\mathcal{V}(\rho)\|\mathcal{V}(\sigma)\right) = \mathbb{D}^q(\rho\|\sigma). \tag{5.70}$$

**Proof** The nonzero components of $\tilde{\mathbf{p}}$ and $\tilde{\mathbf{q}}$ remain unchanged if $\rho$ and $\sigma$ are replaced with $\mathcal{V}(\rho)$ and $\mathcal{V}(\sigma)$, for any isometry $\mathcal{V} \in \mathrm{CPTP}(A \to B)$. Moreover, note that with $n := |A|$

$$\mathcal{K} := \mathrm{span}\{V|a_x\rangle : x \in [m]\} = \mathrm{span}\{V|b_y\rangle : y \in [m]\}, \tag{5.71}$$

since both $\{|a_x\rangle\}_{x\in[m]}$ and $\{|b_y\rangle\}_{y\in[m]}$ are bases of $A$. Hence, denoting by $n := |B|$, the additional $n - m$ zero eigenvalues of $\mathcal{V}(\rho)$ (and similarly of $\mathcal{V}(\sigma)$), corresponds to eigenvectors that are in the orthogonal complement of $\mathcal{K}$. Hence, if $\tilde{\mathbf{p}}, \tilde{\mathbf{q}}$ corresponds to $\rho$ and $\sigma$ as in (5.68), then $\tilde{\mathbf{p}} \oplus \mathbf{0}_k$ and $\tilde{\mathbf{q}} \oplus \mathbf{0}_k$ correspond to $\mathcal{V}(\rho)$ and $\mathcal{V}(\sigma)$, respectively, where $\mathbf{0}_k$ is the zero vector in dimension $k := n^2 - m^2$. Hence,

$$\mathbb{D}^q\left(\mathcal{V}(\rho)\|\mathcal{V}(\sigma)\right) = \mathbb{D}\left(\tilde{\mathbf{p}} \oplus \mathbf{0}_k \| \tilde{\mathbf{q}} \oplus \mathbf{0}_k\right) = \mathbb{D}\left(\tilde{\mathbf{p}}\|\tilde{\mathbf{q}}\right) = \mathbb{D}^q(\rho\|\sigma), \tag{5.72}$$

where the second equality follows from the fact that classical divergences are invariant under embedding (see Exercise 5.15). $\blacksquare$

**Exercise 5.18.** *Show that if $\rho$ and $\sigma$ are diagonal, then $\mathbb{D}^q(\rho\|\sigma) := \mathbb{D}\left(\mathbf{p}\|\mathbf{q}\right)$, where $\mathbf{p}$ and $\mathbf{q}$ are the diagonals of $\rho$ and $\sigma$.*

Due to Lemma 5.3, the Stinespring delation implies that $\mathbb{D}^q(\rho\|\sigma)$ as defined in (5.69) is a quantum divergence if and only if it is nonincreasing under the partial trace. This later property does not hold in general; however, it does hold when $\mathbb{D}$ belongs to a large class of $f$-Divergences.

**Quantum $f$-Divergence**

**Definition 5.5.** Let $f : (0, \infty) \to \mathbb{R}$ be an operator convex function satisfying $f(1) = 0$. Let $D_f$ be its corresponding classical $f$-Divergence as defined in Definition 5.2. The quantum $f$-Divergence, $D_f^q$, is defined on any $\rho, \sigma \in \mathfrak{D}(A)$ as

$$D_f^q(\rho \| \sigma) := D_f(\tilde{\mathbf{p}} \| \tilde{\mathbf{q}}), \tag{5.73}$$

where $\tilde{\mathbf{p}}$ and $\tilde{\mathbf{q}}$ are the probability vectors in $\mathrm{Prob}(m^2)$ as defined in (5.68).

*Remark.* We will see next that the requirement that $f$ is operator convex (vs just convex) ensures that $D_f^q$ is indeed a quantum divergence. Moreover, from (5.15), for any $\rho, \sigma \in \mathfrak{D}(A)$ with spectral decomposition as in (5.67) we have (see the following exercise)

$$D_f(\rho \| \sigma) = \lim_{\varepsilon \to 0^+} \sum_{x,y} (q_y + \varepsilon) f\left(\frac{p_x}{q_y + \varepsilon}\right) |\langle a_x | b_y \rangle|^2. \tag{5.74}$$

**Exercise 5.19.** *Prove (5.74) and use it to show that for any $\rho, \sigma \in \mathfrak{D}(A)$*

$$D_f(\rho \| \sigma) = \sum_{\substack{x \in \mathrm{supp}\,(\mathbf{p}) \\ y \in \mathrm{supp}\,(\mathbf{q})}} q_y f\left(\frac{p_x}{q_y}\right) |\langle a_x | b_y \rangle|^2 + f(0)\mathrm{Tr}\left[(I - \rho^0)\sigma\right] + \tilde{f}(0)\mathrm{Tr}\left[(I - \sigma^0)\rho\right],$$
$$\tag{5.75}$$

*where $\tilde{f}(0) := \lim_{r \to \infty} \frac{f(r)}{r}$, and $\rho^0$ and $\sigma^0$ are the projections to the supports of $\rho$ and $\sigma$. Hint: For the first part, note that the components of $(1 - \varepsilon)\mathbf{q} + \varepsilon \mathbf{u}$ (see (5.15)) can be written as $q_y + \varepsilon \left(\frac{1}{n} - q_y\right)$. Now $\varepsilon\left(\frac{1}{n} - q_y\right)$ is positive if $q_y = 0$, and if $q_y > \frac{1}{n}$ then since $f$ is continuous we can still replace $\varepsilon\left(\frac{1}{n} - q_y\right)$ with $\varepsilon$ in the limit $\varepsilon \to 0^+$.*

The quantum $f$-Divergence has the following quantum formula.

**Quantum Formula**

**Theorem 5.4.** Let $f : (0, \infty) \to \mathbb{R}$ be an operator convex function satisfying $f(1) = 0$. For any $\rho, \sigma \in \mathfrak{D}(A)$ with $\sigma > 0$ the quantum $f$-Divergence can be expressed as

$$D_f(\rho \| \sigma) = \mathrm{Tr}\left[\phi_\sigma^{A\tilde{A}} f\left(\sigma^{-1} \otimes \rho^T\right)\right], \tag{5.76}$$

where $|\phi_\sigma^{A\tilde{A}}\rangle := \sigma^{1/2} \otimes I^{\tilde{A}} |\Omega^{A\tilde{A}}\rangle$ is a purification of $\sigma$. For $\sigma \geq 0$ the $f$-Divergence satisfies (with $\mathbf{u} \in \mathfrak{D}(A)$ denotes the maximally mixed state)

$$D_f(\rho \| \sigma) = \lim_{\varepsilon \to 0^+} D_f\left(\rho \| (1 - \varepsilon)\sigma + \varepsilon \mathbf{u}\right). \tag{5.77}$$

**Proof**  Suppose first that $\sigma > 0$. Then, by definition

$$
\begin{aligned}
D_f(\rho\|\sigma) := D_f\left(\tilde{\mathbf{p}}\|\tilde{\mathbf{q}}\right) &= \sum_{x,y\in[m]} \tilde{q}_{xy}\, f\left(\frac{\tilde{p}_{xy}}{\tilde{q}_{xy}}\right) \\
&= \sum_{x,y\in[m]} q_y|\langle a_x|b_y\rangle|^2 f\left(\frac{p_x}{q_y}\right).
\end{aligned}
\tag{5.78}
$$

Now, for every $x, y \in [m]$ we can express $q_y|\langle a_x|b_y\rangle|^2$ as follows:

$$
\begin{aligned}
q_y|\langle a_x|b_y\rangle|^2 &= \left\langle \Omega^{A\tilde{A}}\middle| q_y|b_y\rangle\langle b_y| \otimes |a_x\rangle\langle a_x|^T \middle|\Omega^{A\tilde{A}}\right\rangle \\
&= \left\langle \phi_\sigma^{A\tilde{A}}\middle| |b_y\rangle\langle b_y| \otimes |a_x\rangle\langle a_x|^T \middle|\phi_\sigma^{A\tilde{A}}\right\rangle.
\end{aligned}
\tag{5.79}
$$

$$\boxed{q_y|b_y\rangle\langle b_y| = \sigma^{\frac{1}{2}}|b_y\rangle\langle b_y|\sigma^{\frac{1}{2}}}$$

Substituting this expression into (5.78) we obtain

$$
\begin{aligned}
D_f(\rho\|\sigma) &= \left\langle \phi_\sigma^{A\tilde{A}}\middle| \sum_{x,y\in[m]} f\left(\frac{p_x}{q_y}\right) |b_y\rangle\langle b_y| \otimes |a_x\rangle\langle a_x|^T \middle|\phi_\sigma^{A\tilde{A}}\right\rangle \\
&= \left\langle \phi_\sigma^{A\tilde{A}}\middle| f\left(\sum_{x,y\in[m]} \frac{p_x}{q_y} |b_y\rangle\langle b_y| \otimes |a_x\rangle\langle a_x|^T\right) \middle|\phi_\sigma^{A\tilde{A}}\right\rangle \\
&= \left\langle \phi_\sigma^{A\tilde{A}}\middle| f\left(\sigma^{-1} \otimes \rho^T\right) \middle|\phi_\sigma^{A\tilde{A}}\right\rangle,
\end{aligned}
\tag{5.80}
$$

$$\boxed{\left\{|b_y\rangle\langle b_y|\otimes |a_x\rangle\langle a_x|^T\right\}_{x,y\in[m]} \text{ is orthonormal}}$$

The case $\sigma \geqslant 0$ follows directly from Exercise 5.4 and is left as an exercise. ∎

We now demonstrate that the expression for the $f$-Divergence, as outlined in the preceding theorem, satisfies the data processing inequality when $f$ is operator convex.

> **Theorem 5.5.** Let $f : (0,\infty) \to \mathbb{R}$ be an operator convex function satisfying $f(1) = 0$. Then, $D_f^q$ as defined in Definition 5.5 is a quantum divergence.

**Proof**  Due to Stinespring dilation theorem, any $\mathcal{E} \in \mathrm{CPTP}(A \to B)$ can be expressed as an isometry followed by a partial trace. From its definition and Lemma 5.3 it follows that the quantum $f$-Divergence is invariant under isometries. It is therefore left to show that it is monotonic under partial trace. For this purpose, let $\rho, \sigma \in \mathfrak{D}(AB)$ and without loss of generality we assume that $\sigma^{AB} > 0$ since the case $\sigma^{AB} \geqslant 0$ follows from the continuity of $D_f$ in the limit $\varepsilon \to 0^+$.

We need to show that

$$
D_f(\rho^A\|\sigma^A) \leqslant D_f(\rho^{AB}\|\sigma^{AB}).
\tag{5.81}
$$

From the quantum formula in (5.76) we see that the left-hand side of (5.81) depends on $(\sigma^A)^{-1} \otimes (\rho^A)^T$, whereas the right-hand side depends on $(\sigma^{AB})^{-1} \otimes (\rho^{AB})^T$. In the following exercise we will show that there exists an isometry that relates between these two expressions. Explicitly, we will show that there exists an isometry $V : A\tilde{A} \to A\tilde{A}B\tilde{B}$ (with $V^*V = I^{A\tilde{A}}$) such that

$$V^*\left(\left(\sigma^{AB}\right)^{-1} \otimes \left(\rho^{\tilde{A}\tilde{B}}\right)^T\right)V = \left(\sigma^A\right)^{-1} \otimes \left(\rho^{\tilde{A}}\right)^T. \tag{5.82}$$

Combining this with the operator Jensen's inequality (B.30 – please see online appendix) for operator convex functions we get

$$f\left(\left(\sigma^A\right)^{-1} \otimes \left(\rho^{\tilde{A}}\right)^T\right) = f\left(V^*\left(\sigma^{AB}\right)^{-1} \otimes \left(\rho^{\tilde{A}\tilde{B}}\right)^T V\right)$$

$$(\mathbf{B.30}) \to \;\; \leqslant V^* f\left(\left(\sigma^{AB}\right)^{-1} \otimes \left(\rho^{\tilde{A}\tilde{B}}\right)^T\right)V. \tag{5.83}$$

Finally, multiplying both sides by $\phi_\sigma^{A\tilde{A}}$ and taking the trace gives

$$D_f\left(\rho^A \| \sigma^A\right) \leqslant \mathrm{Tr}\left[V\phi_\sigma^{A\tilde{A}}V^* f\left(\left(\sigma^{AB}\right)^{-1} \otimes \left(\rho^{\tilde{A}\tilde{B}}\right)^T\right)\right]$$

$$(\mathbf{5.86}) \to \;\; = \mathrm{Tr}\left[\phi_\sigma^{AB\tilde{A}\tilde{B}} f\left(\left(\sigma^{AB}\right)^{-1} \otimes \left(\rho^{\tilde{A}\tilde{B}}\right)^T\right)\right] \tag{5.84}$$

$$= D_f\left(\rho^{AB} \| \sigma^{AB}\right).$$

This completes the proof. ∎

**Exercise 5.20.** *Let* $V : A\tilde{A} \to AB\tilde{A}\tilde{B}$ *be the matrix*

$$V = \sum_{y\in[m]} \left(\sigma^{AB}\right)^{\frac{1}{2}}\left(\left(\sigma^A\right)^{-\frac{1}{2}} \otimes |y\rangle^B\right) \otimes I^{\tilde{A}} \otimes |y\rangle^{\tilde{B}}. \tag{5.85}$$

1. *Show that* $V$ *is an isometry, that is,* $V^*V = I^{A\tilde{A}}$.
2. *Show that* $V$ *satisfies (5.82).*
3. *Show that*

$$V\left(\left(\sigma^A\right)^{\frac{1}{2}} \otimes I^{\tilde{A}}\right)|\Omega^{A\tilde{A}}\rangle = \left(\left(\sigma^{AB}\right)^{\frac{1}{2}} \otimes I^{\tilde{A}\tilde{B}}\right)|\Omega^{AB\tilde{A}\tilde{B}}\rangle, \tag{5.86}$$

## Examples

1. The Umegaki Divergence. For the function $f(r) = r\log r$

$$D_f(\rho\|\sigma) = \lim_{\varepsilon\to 0^+} \sum_{x,y}(q_y + \varepsilon)f\left(\frac{p_x}{q_y + \varepsilon}\right)|\langle a_x|b_y\rangle|^2$$

$$= \lim_{\varepsilon\to 0^+} \sum_{x,y} p_x \log\left(\frac{p_x}{q_y + \varepsilon}\right)|\langle a_x|b_y\rangle|^2 \tag{5.87}$$

$$= \sum_x p_x \log p_x - \sum_{x,y} p_x|\langle a_x|b_y\rangle|^2 \log q_y$$

$$= \mathrm{Tr}[\rho\log\rho] - \mathrm{Tr}[\rho\log\sigma],$$

where in the last line we used the relation $\langle b_y|\rho|b_y\rangle = \sum_x p_x|\langle a_x|b_y\rangle|^2$. Since $f(r) = r\log r$ is operator convex, the above expression is a quantum divergence. It is known as the Umegaki divergence or sometimes referred to as *the* relative entropy. We will discuss many of its properties in the following chapters.

2. The Trace Distance? The function $f(r) = \frac{1}{2}|r-1|$ is convex but it is not operator convex (on any domain that includes 1). Therefore, we cannot conclude that for this choice $D_f^q$ is a quantum divergence. Moreover, note that for this case

$$
\begin{aligned}
D_f(\rho\|\sigma) &= \lim_{\varepsilon\to 0^+} \sum_{x,y}(q_y+\varepsilon)\frac{1}{2}\left|\frac{p_x}{q_y+\varepsilon}-1\right| |\langle a_x|b_y\rangle|^2 \\
&= \frac{1}{2}\sum_{x,y}\left| p_x|\langle a_x|b_y\rangle|^2 - q_x|\langle a_x|b_y\rangle|^2 \right|,
\end{aligned}
\tag{5.88}
$$

which cannot be expressed as a simple function of $\rho$ and $\sigma$.

3. The Quantum $\alpha$-Divergence. The function $f(r) = \frac{r^\alpha - r}{\alpha(\alpha-1)}$ is known to be operator convex for $\alpha \in [0,2]$ (cf. Table B.1 of online version). For any $\rho,\sigma \in \mathfrak{D}(A)$ let $\tilde{\mathbf{p}},\tilde{\mathbf{q}} \in \mathrm{Prob}(m^2)$ (with $m := |A|$) be the probability vectors defined in (5.68). Then, for this $f$ we have

$$
\begin{aligned}
D_f^q(\rho\|\sigma) = D_f(\tilde{\mathbf{p}}\|\tilde{\mathbf{q}}) &= \frac{1}{\alpha(\alpha-1)}\left(\sum_{x,y}\tilde{p}_{xy}^\alpha\tilde{q}_{xy}^{1-\alpha} - 1\right) \\
&= \frac{1}{\alpha(\alpha-1)}\left(\sum_{x,y}\left(p_x|\langle a_x|b_y\rangle|^2\right)^\alpha\left(q_y|\langle a_x|b_y\rangle|^2\right)^{1-\alpha} - 1\right) \\
&= \frac{1}{\alpha(\alpha-1)}\left(\sum_{x,y}|\langle a_x|b_y\rangle|^2 p_x^\alpha q_y^{1-\alpha} - 1\right) \\
&= \frac{1}{\alpha(\alpha-1)}\left(\mathrm{Tr}\left[\rho^\alpha\sigma^{1-\alpha}\right] - 1\right).
\end{aligned}
\tag{5.89}
$$

**Exercise 5.21.** *Use the quantum formula given in Theorem 5.4 to compute the Umegaki divergence and the quantum $\alpha$-Divergence.*

## 5.3 Optimal Quantum Extensions of Divergences

In addition to the method for extension introduced Section 5.2, sometimes it is possible to guess a quantum extension of a classical divergence simply by replacing $\mathbf{p}$ with $\rho$ and $\mathbf{q}$ with $\sigma$. For example, the trace distance divergence $\frac{1}{2}\|\mathbf{p}-\mathbf{q}\|_1$ can be replaced with its quantum version $\frac{1}{2}\|\rho-\sigma\|_1$. However, this method is not always simple and, in general, it is not even unique. Moreover, one still needs to verify that the proposed quantum divergence indeed satisfies the DPI. We therefore investigate in this subsection a more systematic approach for quantum extensions.

Given a *classical* divergence $\mathbb{D}$ we denote by $\underline{\mathbb{D}}$ its minimal quantum extension, and by $\overline{\mathbb{D}}$ its maximal quantum extension. That is, $\underline{\mathbb{D}}$ and $\overline{\mathbb{D}}$ are quantum divergences that reduce to $\mathbb{D}$ on classical systems, and any other quantum divergence, $D'$, that reduces to $\mathbb{D}$ on classical states, necessarily satisfies

$$\underline{\mathbb{D}}(\rho\|\sigma) \leqslant D'(\rho\|\sigma) \leqslant \overline{\mathbb{D}}(\rho\|\sigma) \qquad \forall\, \rho,\sigma \in \mathfrak{D}(A). \tag{5.90}$$

One may wonder whether such optimal quantum extensions exist. In the next theorem we will prove that they do exist, using the following construction.

Let $\mathbb{D}$ be a classical divergence, and for any $\rho,\sigma \in \mathfrak{D}(A)$ define

$$\underline{\mathbb{D}}(\rho^A\|\sigma^A) := \sup \mathbb{D}\left(\mathcal{E}^{A\to X}\left(\rho^A\right) \middle\| \mathcal{E}^{A\to X}\left(\sigma^A\right)\right), \tag{5.91}$$

$$\overline{\mathbb{D}}(\rho\|\sigma) := \inf\left\{\mathbb{D}\left(\mathbf{p}^X\|\mathbf{q}^X\right) : \rho^A = \mathcal{F}^{X\to A}(\mathbf{p}^X),\ \sigma^A = \mathcal{F}^{X\to A}(\mathbf{q}^X)\right\}, \tag{5.92}$$

where the optimizations are over the classical system $X$, the channels $\mathcal{E} \in \mathrm{CPTP}(A\to X)$, and $\mathcal{F} \in \mathrm{CPTP}(X \to A)$, as well as the probability distributions (diagonal density matrices) $\mathbf{p},\mathbf{q} \in \mathfrak{D}(X)$. Note that $\mathcal{E}$ is a POVM Channel and therefore $\mathbb{D}\big(\mathcal{E}(\rho)\big\|\mathcal{E}(\sigma)\big)$ is well defined since $\mathcal{E}(\rho)$ and $\mathcal{E}(\sigma)$ are classical states; that is, they can be viewed as probability vectors or diagonal density matrices. Similarly, $\mathbf{p}^X$ and $\mathbf{q}^X$ can be viewed either as diagonal density matrices or as probability vectors. Moreover, the supremum and infimum are taken over all dimensions $|X| \in \mathbb{N}$.

Optimal Extensions

> **Theorem 5.6.** Let $\mathbb{D}$ be a classical divergence, and let $\underline{\mathbb{D}}$ and $\overline{\mathbb{D}}$ be as in (5.91) and (5.92), respectively. Then, both $\underline{\mathbb{D}}$ and $\overline{\mathbb{D}}$ are quantum divergences that reduce to $\mathbb{D}$ on classical states. In addition, any other quantum divergence $D'$ that reduces to $\mathbb{D}$ on classical states satisfies (5.90).

**Proof**  We first prove the reduction property. Let $\rho,\sigma \in \mathfrak{D}(A)$ be classical states. Then, for $\underline{\mathbb{D}}$ we can take $X$ in (5.91) to be a classical system with $|X| = |A|$ and $\mathcal{E}$ to be the identity channel. Since this identity channel is not necessarily the optimal channel, we get that

$$\underline{\mathbb{D}}(\rho\|\sigma) \geqslant \mathbb{D}(\rho\|\sigma). \tag{5.93}$$

Conversely, since $\rho$ and $\sigma$ are classical, any $\mathcal{E}$ in (5.91) can be assumed to be classical since

$$\mathcal{E}(\rho) = \mathcal{E} \circ \Delta(\rho) \quad \text{and} \quad \mathcal{E}(\sigma) = \mathcal{E} \circ \Delta(\sigma), \tag{5.94}$$

where $\Delta$ is the completely dephasing channel. Therefore, if $\mathcal{E}$ is not classical we can replace it with $\mathcal{E} \circ \Delta$, which is classical (recall that the output of $\mathcal{E}$ is classical). Now, by the DPI property of the classical divergence $\mathbb{D}$ we have for all such classical $\mathcal{E}$, $\mathbb{D}\big(\mathcal{E}(\rho)\big\|\mathcal{E}(\sigma)\big) \leqslant \mathbb{D}(\rho\|\sigma)$. Hence, we must have

$$\underline{\mathbb{D}}(\rho\|\sigma) \leqslant \mathbb{D}(\rho\|\sigma). \tag{5.95}$$

Combining (5.93) with (5.95) we conclude that $\underline{\mathbb{D}}(\rho\|\sigma) = \mathbb{D}(\rho\|\sigma)$. Similarly, for $\overline{\mathbb{D}}$ we can assume that $\mathcal{F}$ in (5.92) is a classical channel since $\rho$ and $\sigma$ are classical. Hence, by the DPI of $\mathbb{D}$ we get the lower bound $\overline{\mathbb{D}}(\rho\|\sigma) \geqslant \mathbb{D}(\rho\|\sigma)$, and this bound can be saturated since we can take $\mathcal{F}$ in (5.92) to be the identity channel.

We next prove that $\underline{\mathbb{D}}$ and $\overline{\mathbb{D}}$ both satisfy the DPI. Let $\mathcal{N} \in \mathrm{CPTP}(A \to B)$. Then,

$$\underline{\mathbb{D}}\big(\mathcal{N}(\rho)\big\|\mathcal{N}(\sigma)\big) = \sup_X \big\{\mathbb{D}\big(\mathcal{E} \circ \mathcal{N}(\rho)\big\|\mathcal{E} \circ \mathcal{N}(\sigma)\big) : \mathcal{E} \in \mathrm{CPTP}(B \to X)\big\}$$

$$\boxed{\mathcal{E}' \text{ replaces } \mathcal{E} \circ \mathcal{N} \to} \longrightarrow \leqslant \sup_X \big\{\mathbb{D}\big(\mathcal{E}'(\rho)\big\|\mathcal{E}'(\sigma)\big) : \mathcal{E}' \in \mathrm{CPTP}(A \to X)\big\}$$

$$= \underline{\mathbb{D}}(\rho\|\sigma). \tag{5.96}$$

For $\overline{\mathbb{D}}$ we have

$$\overline{\mathbb{D}}(\rho\|\sigma) := \inf_X \big\{\mathbb{D}(\mathbf{p}\|\mathbf{q}): \rho = \mathcal{F}(\mathbf{p}),\ \sigma = \mathcal{F}(\mathbf{q}),\ \mathcal{F} \in \mathrm{CPTP}(X \to A)\big\}$$

$$\geqslant \inf_X \big\{\mathbb{D}(\mathbf{p}\|\mathbf{q}): \mathcal{N}(\rho) = \mathcal{N} \circ \mathcal{F}(\mathbf{p}),\ \mathcal{N}(\sigma) = \mathcal{N} \circ \mathcal{F}(\mathbf{q}),\ \mathcal{F} \in \mathrm{CPTP}(X \to A)\big\}$$

$$\geqslant \inf_X \big\{\mathbb{D}(\mathbf{p}\|\mathbf{q}): \mathcal{N}(\rho) = \mathcal{F}'(\mathbf{p}),\ \mathcal{N}(\sigma) = \mathcal{F}'(\mathbf{q}),\ \mathcal{F}' \in \mathrm{CPTP}(X \to B)\big\}$$

$$= \overline{\mathbb{D}}\big(\mathcal{N}(\rho)\big\|\mathcal{N}(\sigma)\big), \tag{5.97}$$

where the first inequality follows from the fact that if $\rho = \mathcal{F}(\mathbf{p})$, then necessarily $\mathcal{N}(\rho) = \mathcal{N} \circ \mathcal{F}(\mathbf{p})$ (but the converse is not necessarily true), and in the second inequality we replaced $\mathcal{N} \circ \mathcal{F}$ with $\mathcal{F}'$.

Finally, we prove the optimality of $\underline{\mathbb{D}}$ and $\overline{\mathbb{D}}$. First observe that from the DPI of $D'$ we have for any $\rho, \sigma \in \mathfrak{D}(A)$ and any $\mathcal{E} \in \mathrm{CPTP}(A \to X)$

$$D'(\rho\|\sigma) \geqslant D'\big(\mathcal{E}(\rho)\big\|\mathcal{E}(\sigma)\big) = \mathbb{D}\big(\mathcal{E}(\rho)\big\|\mathcal{E}(\sigma)\big), \tag{5.98}$$

where the last equality follows from the fact that $D'$ reduces to $\mathbb{D}$ on classical states. Since the above inequality holds for all $\mathcal{E} \in \mathrm{CPTP}(A \to X)$ it also holds for the supremum over such $\mathcal{E}$. We therefore conclude that $D'(\rho\|\sigma) \geqslant \underline{\mathbb{D}}(\rho\|\sigma)$. For the second inequality, let $\rho, \sigma \in \mathfrak{D}(A)$ and $\mathbf{p}, \mathbf{q} \in \mathfrak{D}(X)$, and suppose there exists $\mathcal{F} \in \mathrm{CPTP}(X \to A)$ such that $\rho = \mathcal{F}(\mathbf{p})$ and $\sigma = \mathcal{F}(\mathbf{q})$. Then, from the DPI of $D'$ we get

$$D'(\rho\|\sigma) = D'\big(\mathcal{F}(\mathbf{p})\big\|\mathcal{F}(\mathbf{q})\big) \leqslant D'(\mathbf{p}\|\mathbf{q}) = \mathbb{D}(\mathbf{p}\|\mathbf{q}), \tag{5.99}$$

where the last equality follows from the fact that $D'$ reduces to $\mathbb{D}$ on classical states. Since the above inequality holds for all such $\mathbf{p}, \mathbf{q}$ for which there exists an $\mathcal{F}$ that takes them to $\rho$ and $\sigma$, it must also hold for the infimum over all such $\mathbf{p}, \mathbf{q}$. Hence, $D'(\rho\|\sigma) \leqslant \overline{\mathbb{D}}(\rho\|\sigma)$. $\blacksquare$

Since the maximal and minimal extensions provide upper and lower bounds on all extensions, it can be useful to have a closed formula for them. Remarkably, a closed formula for the maximal extension exists if one of the input states is pure, or for the $f$-Divergences if $f$ is operator convex. On the other hand, at the time of writing this book, a closed formula for the minimal extension of the $f$-Divergence is not known. However, for specific examples such as the trace distance and fidelity, the minimal extension can be computed (see the next section), and as we will see in Chapter 6 the regularized minimal extension can also be computed for all relative entropies.

**Exercise 5.22.** *Let $f : [0, \infty) \to [0, \infty)$ be an operator convex function, and for all $\rho, \sigma \in \mathfrak{D}(A)$ with $\sigma > 0$ define*

$$D'_f(\rho \| \sigma) := \mathrm{Tr}\left[\rho \#_f \sigma\right] = \mathrm{Tr}\left[\sigma f \left(\sigma^{-\frac{1}{2}} \rho \sigma^{-\frac{1}{2}}\right)\right], \tag{5.100}$$

*where $\#_f$ is the Kubo-Ando operator mean (see Definition B.2 of online version). Finally, let $\overline{D}_f$ be the maximal $f$-Divergence.*

1. *Show that $D'_f$ reduces to the classical $f$-Divergence when $\rho$ and $\sigma$ are classical (i.e. diagonal in the same basis).*
2. *Show that $D'_f$ satisfies the DPI in the domain $\mathfrak{D}(A) \times \mathfrak{D}_{>0}(A)$. Hint: Show that $D'_f$ satisfies all the conditions of Lemma 5.2.*
3. *Show that for any $\rho, \sigma \in \mathfrak{D}(A)$ with $\sigma > 0$*

$$\overline{D}_f(\rho \| \sigma) \geqslant D'_f(\rho \| \sigma). \tag{5.101}$$

*Hint: Use Theorem 6.4*

## 5.3.1 The Maximal Quantum Extension

The maximal extension can be expressed as

$$\overline{\mathbb{D}}(\rho \| \sigma) = \inf \mathbb{D}(\mathbf{p} \| \mathbf{q}) \tag{5.102}$$

subject to the conditions

$$\rho = \sum_{x \in [n]} p_x \omega_x \quad \text{and} \quad \sigma = \sum_{x \in [n]} q_x \omega_x, \tag{5.103}$$

where $n \in \mathbb{N}$, $\mathbf{p}, \mathbf{q} \in \mathrm{Prob}(n)$, and for each $x \in [n]$, $\omega_x \in \mathfrak{D}(A)$. Note that we replaced $\mathcal{F}(|x\rangle\langle x|)$ with $\omega_x$. The infimum above can include vectors $\mathbf{p}$ and $\mathbf{q}$ with zero components. We now show that the number of zeros in each of these vectors can be restricted to be at most one.

> **Lemma 5.4.** The infimum in (5.102) can be restricted to vectors $\mathbf{p}, \mathbf{q} \in \mathrm{Prob}(n)$ that has at most one zero component.

**Proof**   We first show that $\mathbf{q}$ can have this property. Since divergences are invariant under (joint) permutation of the components of $\mathbf{p}$ and $\mathbf{q}$, without loss of generality we can assume that

$$q_1 \geqslant \cdots \geqslant q_r > q_{r+1} = \cdots = q_n = 0, \tag{5.104}$$

where $r$ is the number of nonzero components of $\mathbf{q}$. With this order of $\mathbf{q}$ we have (see Exercise 4.28) $(\mathbf{p}, \mathbf{q}) \sim (\mathbf{p}', \mathbf{q})$, where

$$\mathbf{p}' = \left(p_1, \ldots, p_r, p'_{r+1}, 0, \ldots, 0\right)^T \quad \text{where} \quad p'_{r+1} := \sum_{x=r+1}^{n} p_x. \tag{5.105}$$

Note that the relations in (5.103) can be expressed as

$$\rho = \sum_{x \in [r]} p_x \omega_x + p'_{r+1}\tau \quad \text{and} \quad \sigma = \sum_{x \in [r]} q_x \omega_x, \tag{5.106}$$

where $\tau := \frac{1}{p'_{r+1}} \sum_{x=r+1}^{n} p_x \omega_x$. Therefore, the vectors

$$\tilde{\mathbf{p}} = \left(p_1, \ldots, p_r, p'_{r+1}\right)^T \quad \text{and} \quad \tilde{\mathbf{q}} = (q_1, \ldots, q_r, 0) \tag{5.107}$$

satisfy both (5.103) with $n$ replaced by $r + 1$, and $\mathbb{D}(\mathbf{p}\|\mathbf{q}) = \mathbb{D}(\tilde{\mathbf{p}}\|\tilde{\mathbf{q}})$. Repeating the same arguments for $\tilde{\mathbf{p}}$ completes the proof. ∎

It's important to note that Lemma 5.4 aids in simplifying the optimization problem described in (5.102). This simplification is achieved by assuming, without any loss of generality, that $\mathbf{p}$ and $\mathbf{q}$ have forms similar to $\tilde{\mathbf{p}}$ and $\tilde{\mathbf{q}}$ as specified in (5.107). Consequently, we can redefine the infimum in (5.102) as an infimum over all $1 < n \in \mathbb{N}$, $\mathbf{p} \in \mathrm{Prob}(n)$, and $0 < \mathbf{q} \in \mathrm{Prob}(n - 1)$, provided there are $n - 1$ density matrices $\{\omega_x\}_{x \in [n-1]} \subset \mathfrak{D}(A)$ meeting the following criteria:

$$\rho \geqslant \sum_{x \in [n-1]} p_x \omega_x \quad \text{and} \quad \sigma = \sum_{x \in [n-1]} q_x \omega_x, \tag{5.108}$$

where it is understood that the inequality in the first relation is satisfied if and only if there exists a density matrix $\omega_n \in \mathfrak{D}(A)$ such that

$$\rho = \sum_{x \in [n-1]} p_x \omega_x + p_n \omega_n. \tag{5.109}$$

As we will see, in certain cases, working with the expression in (5.108) becomes more manageable because $\mathbf{q} > 0$. In other situations, it might be preferable to work with $\mathbf{p} > 0$. It is worth noting that by applying the same reasoning but substituting $\mathbf{p}$ for $\mathbf{q}$ and vice versa, we can also express the infimum in (5.102) as an infimum over all $1 < n \in \mathbb{N}, 0 < \mathbf{p} \in \mathrm{Prob}(n - 1)$ and $\mathbf{q} \in \mathrm{Prob}(n)$, with the requirement of having $n - 1$ density matrices $\{\omega_x\}_{x \in [n-1]} \subset \mathfrak{D}(A)$ that satisfy

$$\rho = \sum_{x \in [n-1]} p_x \omega_x \quad \text{and} \quad \sigma \geqslant \sum_{x \in [n-1]} q_x \omega_x. \tag{5.110}$$

In the next theorem we employ this property to calculate the maximal divergence when one of the input states is pure.

---

**Theorem 5.7.** Let $\mathbb{D}$ be a classical divergence, $\psi \in \mathrm{Pure}(A)$, and $\sigma \in \mathfrak{D}(A)$. Then,

$$\overline{\mathbb{D}}(\psi\|\sigma) = \mathbb{D}\left((1,0)^T \,\big\|\, (\lambda_{\max}, 1 - \lambda_{\max})^T\right), \tag{5.111}$$

where

$$\lambda_{\max} := \max\left\{\lambda \in \mathbb{R} : \lambda\psi \leqslant \sigma\right\}. \tag{5.112}$$

**Proof**   Consider the relation (5.110) with the pure state $\psi$ replacing $\rho$. Since $\rho := \psi$ is a pure state, the first relation in (5.110) can hold if and only if for any $x \in [n-1]$ we have $\omega_x = \psi$. Substituting this into the second relation in (5.110) we obtain

$$\sigma \geqslant \sum_{x \in [n-1]} q_x \psi = (1 - q_n)\psi. \tag{5.113}$$

Observe that this condition is equivalent to $(1 - q_n) \leqslant \lambda_{\max}$. Consequently, we have reached the following expression:

$$\overline{\mathbb{D}}(\psi \| \sigma) = \inf_{1 < n \in \mathbb{N}} \left\{ \mathbb{D}(\mathbf{p} \oplus 0 \| \mathbf{q}) : 0 < \mathbf{p} \in \text{Prob}(n-1), \ \mathbf{q} \in \text{Prob}(n), \ q_n \geqslant 1 - \lambda_{\max} \right\}. \tag{5.114}$$

Finally, we simplify the expression (5.114) by demonstrating that we can confine the value of $n$ in the optimization above to be equal to two. To achieve this, let $E$ be the $2 \times n$ column stochastic matrix

$$E := \begin{bmatrix} 1 & \cdots & 1 & 0 \\ 0 & \cdots & 0 & 1 \end{bmatrix}. \tag{5.115}$$

Observe that $E\mathbf{p} = (1, 0)^T$ and $E\mathbf{q} = (1 - q_n, q_n)^T$ so that

$$(\mathbf{p} \oplus 0, \mathbf{q}) \succ (E(\mathbf{p} \oplus 0), E\mathbf{q}) = \left( (1,0)^T, (1 - q_n, q_n)^T \right)$$

$$\boxed{1 - q_n \leqslant \lambda_{\max}} \longrightarrow \ \succ \left( (1,0)^T, (\lambda_{\max}, 1 - \lambda_{\max})^T \right). \tag{5.116}$$

Therefore, the minimum is obtained with $n = 2$ and with the pair $(\mathbf{p}, \mathbf{q})$ being equal to the pair on the right-hand side of (5.116).   ∎

## 5.3.2 The Maximal $f$-Divergence

In section, we calculate the maximal $f$-Divergence $\overline{D}_f(\rho|\sigma)$ for $\rho, \sigma \in \mathfrak{D}(A)$ with the condition that $\sigma > 0$. The special case where $\sigma$ possesses zero eigenvalues will be deferred to the appendix online version.

When $\mathbb{D}$ equals the $f$-Divergence, the expression given in (5.14), along with (5.108), indicates that the infimum in (5.102) can be represented as follows:

$$\overline{D}_f(\rho \| \sigma) := \inf_{1 < n \in \mathbb{N}} \left\{ \sum_{x \in [n-1]} q_x f\left(\frac{p_x}{q_x}\right) + \tilde{f}(0) p_n \right\}, \tag{5.117}$$

where $\tilde{f}(0)$ is defined in (5.13), and the infimum above is over all $1 < n \in \mathbb{N}, \mathbf{p} \in \text{Prob}(n)$ and $0 < \mathbf{q} \in \text{Prob}(n-1)$, such that there exists $n-1$ density matrices $\{\omega_x\}_{x \in [n-1]} \subset \mathfrak{D}(A)$ satisfying (5.108). Denoting by $\Lambda_x := q_x \sigma^{-\frac{1}{2}} \omega_x \sigma^{-\frac{1}{2}}$, and applying the conjugation $\sigma^{-\frac{1}{2}}(\cdot)\sigma^{-\frac{1}{2}}$ to both sides of (5.108) gives the relations:

$$\sigma^{-\frac{1}{2}} \rho \sigma^{-\frac{1}{2}} \geqslant \sum_{x \in [n-1]} \frac{p_x}{q_x} \Lambda_x \quad \text{and} \quad \sum_{x \in [n-1]} \Lambda_x = I^A. \tag{5.118}$$

With these new notations, the infimum in (5.117) is taken over all $1 < n \in \mathrm{Prob}(n)$, all $\mathbf{p} \in \mathrm{Prob}(n)$, and all POVMs $\{\Lambda_x\}_{x \in [n-1]}$ for which the inequality (5.118) holds with $q_x := \mathrm{Tr}[\Lambda_x \sigma] > 0$.

One natural choice/guess for the optimal $n$, $\mathbf{p}$ and $\{\Lambda_x\}_{x \in [n-1]}$, is to choose them such that the inequality in (5.118) becomes an equality. This is possible for example by taking $n = |A| + 1$, and for any $x \in [n-1]$ to take $\Lambda_x = \psi_x \in \mathrm{Pure}(A)$ with $|\psi_x\rangle$ being the $x$-eigenvector of $\sigma^{-\frac{1}{2}} \rho \sigma^{-\frac{1}{2}}$ corresponding to the eigenvalue $p_x/q_x$ (i.e. $\mathbf{p}$ is chosen such that $p_x/\mathrm{Tr}[\sigma \Lambda_x]$ is the $x$-eigenvalue of $\sigma^{-\frac{1}{2}} \rho \sigma^{-\frac{1}{2}}$). For this choice we have

$$\sigma^{-\frac{1}{2}} \rho \sigma^{-\frac{1}{2}} = \sum_{x \in [n-1]} \frac{p_x}{q_x} |\psi_x\rangle\langle\psi_x|, \tag{5.119}$$

which forces $p_n$ to be

$$p_n = 1 - \sum_{x \in [n-1]} p_x = 1 - \sum_{x \in [n-1]} \frac{p_x}{q_x} \langle\psi_x|\sigma|\psi_x\rangle = 1 - \mathrm{Tr}[\rho] = 0, \tag{5.120}$$

where the last equality follows by multiplying both sides of (5.119) by $\sigma$ and taking the trace. Moreover, for these choices of $n$, $\mathbf{p}$, and $\{\Lambda_x\}$, we have

$$\sum_{x \in [n-1]} q_x f\left(\frac{p_x}{q_x}\right) = \sum_{x \in [n-1]} \mathrm{Tr}[\sigma |\psi_x\rangle\langle\psi_x|] f\left(\frac{p_x}{q_x}\right)$$

$$\boxed{\forall t \geqslant 0 \;\; f(t|\psi_x\rangle\langle\psi_x|) = f(t)|\psi_x\rangle\langle\psi_x|} \longrightarrow = \sum_{x \in [n-1]} \mathrm{Tr}\left[\sigma f\left(\frac{p_x}{q_x}|\psi_x\rangle\langle\psi_x|\right)\right] \tag{5.121}$$

$$\boxed{\{|\psi_x\rangle\}_{x \in [n-1]} \text{ is orthonormal}} \longrightarrow = \mathrm{Tr}\left[\sigma f\left(\sum_{x \in [n-1]} \frac{p_x}{q_x}|\psi_x\rangle\langle\psi_x|\right)\right]$$

$$\mathbf{(5.119)} \rightarrow \; = \mathrm{Tr}\left[\sigma f\left(\sigma^{-\frac{1}{2}} \rho \sigma^{-\frac{1}{2}}\right)\right].$$

Note that we obtained the formula above for a particular choice of $n$, $\mathbf{p}$, and $\{\Lambda_x\}_{x \in [n-1]}$. Therefore, since this is not necessarily the optimal choice (recall $\overline{D}_f$ is defined in terms of an infimum), we must have

$$\overline{D}_f(\rho\|\sigma) \leqslant \mathrm{Tr}\left[\sigma f\left(\sigma^{-\frac{1}{2}} \rho \sigma^{-\frac{1}{2}}\right)\right] = \mathrm{Tr}[\rho \#_f \sigma], \tag{5.122}$$

where $\#_f$ is the Kubo-Ando operator mean (see Definition B.2 of online version). Interestingly, to get this upper bound we did not even assume that $f$ is convex, but if $f$ is operator convex we get an equality above.

Closed Formula of the Maximal $f$-Divergence

**Theorem 5.8.** Let $\rho, \sigma \in \mathfrak{D}(A)$ with $\sigma > 0$, and let $f := (0, \infty) \to \mathbb{R}$ be operator convex, with $f(0) := \lim_{\varepsilon \to 0^+} f(\varepsilon)$ and $f(1) = 0$. Then,

$$\overline{D}_f(\rho\|\sigma) = \mathrm{Tr}[\rho \#_f \sigma] = \mathrm{Tr}\left[\sigma f\left(\sigma^{-\frac{1}{2}} \rho \sigma^{-\frac{1}{2}}\right)\right]. \tag{5.123}$$

*Remark.*

1. In the theorem above we restricted the domain of $\overline{D}_f$ to $\mathfrak{D}(A) \times \mathfrak{D}_{>0}(A)$. To incorporate the singular case it is necessary to employ additional more delicate arguments (see, for example, D.41 in the appendix of online version). Explicitly, for the case that $\sigma$ is singular, we can write $\sigma = \begin{pmatrix} \tilde{\sigma} & 0 \\ 0 & 0 \end{pmatrix}$ in a block matrix form with $\tilde{\sigma} > 0$, and the formula for $\overline{D}_f$ becomes (see Appendix D.2 of online version)

$$\overline{D}_f(\rho\|\sigma) = \mathrm{Tr}\left[\tilde{\rho}\#_f\tilde{\sigma}\right] + (1 - \mathrm{Tr}[\tilde{\rho}])\tilde{f}(0), \tag{5.124}$$

where $\tilde{f}(0) := \lim_{\varepsilon \to 0^+} \varepsilon f(\frac{1}{\varepsilon})$, and $\tilde{\rho} := \rho/\rho_{22} := \rho_{11} - \zeta\rho_{22}^{-1}\zeta^*$ is the Schur complement (see (B.75) of online version) of the block $\rho_{22}$ of $\rho = \begin{pmatrix} \rho_{11} & \zeta \\ \zeta^* & \rho_{22} \end{pmatrix}$.

2. In Theorem B.6 of online version we proved that for a continuous function $f : [0, \infty) \to [0, \infty)$ the Kubo-Ando operator mean $\#_f$ is operator convex if and only if it is jointly convex. Therefore, at least in the domain $\mathfrak{D}(A) \times \mathfrak{D}_{>0}(A)$ the maximal $f$-divergence is jointly convex for any operator convex $f$.

**Proof**   The proof of the theorem follows immediately from the inequality (5.122) combined with the opposite inequality (5.101).     ■

## Examples:

1. The Belavkin–Staszewski divergence. Consider the function $f(r) = r\log r$. In this case, we have $\tilde{f}(0) = \lim_{\varepsilon \to 0^+} \varepsilon f(1/\varepsilon) = \lim_{\varepsilon \to 0^+} \log(1/\varepsilon) = \infty$. According to the closed form in (D.42 of online version), this means that unless $\mathrm{supp}\,(\rho) \subseteq \mathrm{supp}\,(\sigma)$ we have $\overline{D}_f(\rho\|\sigma) = \infty$. For the case $\mathrm{supp}\,(\rho) \subseteq \mathrm{supp}\,(\sigma)$ we have

$$\overline{D}_f(\rho\|\sigma) = \mathrm{Tr}\left[\sigma\left(\sigma^{-\frac{1}{2}}\rho\sigma^{-\frac{1}{2}}\right)\log\left(\sigma^{-\frac{1}{2}}\rho\sigma^{-\frac{1}{2}}\right)\right]. \tag{5.125}$$

This expression can be simplified by using the relation $Mf(M^*M) = f(MM^*)M$ from Exercise B.1 of online version. Denoting $M := \rho^{\frac{1}{2}}\sigma^{-\frac{1}{2}}$ we get

$$\begin{aligned}
\overline{D}_f(\rho\|\sigma) &= \mathrm{Tr}\left[\sigma M^*M \log\left(M^*M\right)\right] \\
\textbf{(B.1)}\to &= \mathrm{Tr}\left[\sigma M^* \log\left(MM^*\right)M\right] \\
\boxed{M := \rho^{\frac{1}{2}}\sigma^{-\frac{1}{2}}} \longrightarrow &= \mathrm{Tr}\left[\rho\log\left(\rho^{\frac{1}{2}}\sigma^{-1}\rho^{\frac{1}{2}}\right)\right].
\end{aligned} \tag{5.126}$$

This divergence is known as the Belavkin–Staszewski divergence.

2. The maximal $\alpha$-Divergences. Consider the function $f_\alpha(r) = \frac{r^\alpha - r}{\alpha(\alpha-1)}$ which is known to be operator convex for $\alpha \in (0, 2]$. Therefore, for $\alpha \in (0, 2]$ we have (recall we assume $\sigma > 0$)

$$\begin{aligned}
\overline{D}_{f_\alpha}(\rho\|\sigma) &= \frac{1}{\alpha(\alpha-1)}\mathrm{Tr}\left[\sigma\left(\left(\sigma^{-\frac{1}{2}}\rho\sigma^{-\frac{1}{2}}\right)^\alpha - \sigma^{-\frac{1}{2}}\rho\sigma^{-\frac{1}{2}}\right)\right] \\
&= \frac{1}{\alpha(\alpha-1)}\left(\mathrm{Tr}\left[\sigma\left(\sigma^{-\frac{1}{2}}\rho\sigma^{-\frac{1}{2}}\right)^\alpha\right] - 1\right).
\end{aligned} \tag{5.127}$$

# 5.4 Divergences that Are Metrics

We saw that every norm induces a corresponding metric. Metrics are used to measure how close two vectors are, but since there are many norms (e.g. the family of $p$-norms) there are also numerous metrics that can be used to measure distance. Mathematically, in finite dimensions, all metrics are topologically equivalent. Yet, physically, only very few have a known operational meaning. In this subsection we study metrics that are also divergences.

We first consider norms that induce metrics that are also divergences. We saw in Chapter 1 that the 1-norm satisfies the monotonicity condition

$$\|E\mathbf{v}\|_1 \leqslant \|\mathbf{v}\|_1 \quad \forall\, \mathbf{v} \in \mathbb{C}^m \quad \forall\, E \in \mathrm{STOC}(n,m). \tag{5.128}$$

This monotonicity property ensures that the metric $D(\mathbf{p}, \mathbf{q}) := \frac{1}{2}\|\mathbf{p} - \mathbf{q}\|_1$ is a divergence. Remarkably, for vectors in $\mathbb{R}_+^n$, up to a multiplication by a constant, the 1-norm is the only norm with this monotonicity property!

To see this, let $\|\cdot\|$ be a norm satisfying the same monotonicity property 5.128, and suppose $\|(1,0)^T\| = 1$. From (5.128) with $\|\cdot\|$ replacing $\|\cdot\|_1$, it follows that $\|\mathbf{v}\|$ is invariant under permutation of its components, and further more it also implies that $\|\mathbf{v}\| = \|\mathbf{v} \oplus 0\|$. Hence, if $\{\mathbf{e}_1, \ldots, \mathbf{e}_m\}$ is the standard basis of $\mathbb{C}^m$, then $\|\mathbf{e}_j\| = 1$ for all $j = 1, \ldots, m$. Now, let $E$ be the column stochastic matrix whose first row is $[1, 1, \ldots, 1]$ and all other rows are zero. Then,

$$|v_1 + \cdots + v_m| = \|E\mathbf{v}\| \leqslant \|\mathbf{v}\| = \left\| \sum_{j \in [m]} v_j \mathbf{e}_j \right\| \leqslant \sum_{j \in [m]} |v_j| = \|\mathbf{v}\|_1. \tag{5.129}$$

Note that if all the components of $\mathbf{v}$ are nonnegative real numbers, then all the inequalities above must be equalities and we get in particular that $\|\mathbf{v}\| = \|\mathbf{v}\|_1$.

## 5.4.1 The Trace Norm

The trace norm is the Schatten 1-norm as introduced in Definition 2.3. Specifically, for any operator $M : A \to B$ it is defined by

$$\|M\|_1 := \mathrm{Tr}\sqrt{M^*M}. \tag{5.130}$$

Therefore, the trace norm is the sum of the singular values of $M$. Recall that all Schatten norms satisfy the invariance property under isometries (see (2.70)). Particularly, that for any two isometries $U : B \to B'$ and $V : A \to A'$ we have

$$\left\| UMV^* \right\|_1 = \|M\|_1. \tag{5.131}$$

**Exercise 5.23.**

1. *Show that the trace norm is indeed a norm.*
2. *Show that the trace norm is always bigger than the norm induced by the inner product (2.16).*

**Exercise 5.24.** *Show that for any three Hermitian operators* $M, N, \sigma \in \mathrm{Herm}(A)$, *with* $\sigma > 0$, *the following holds:*

1. $\mathrm{Tr}(MN) \leqslant \|M\|_2 \|N\|_2$
2. $\|\sqrt{\sigma} M \sqrt{\sigma}\|_1 \leqslant \|M\|_2 \|\sigma\|_2$
3. $\|M\|_1 \leqslant \sqrt{\mathrm{Tr}[\sigma]} \, \|\sigma^{-1/4} M \sigma^{-1/4}\|_2$.
   *Hint: Use part (2) with $M$ replaced by $\sigma^{-1/4} M \sigma^{-1/4}$ and $\sigma$ replaced by $\sqrt{\sigma}$.*

*where $\| \cdot \|_2$ is the norm induced by the Hilbert–Schmidt inner product.*

The subsequent two lemmas establish that the trace norm can be formulated as optimization problems. These formulations are instrumental in proving various properties of the trace norm. We begin with an expression that is particularly useful for Hermitian matrices.

> **Lemma 5.5.** Let $M : A \to A$ be a Hermitian operator. The trace norm of $M$ can be expressed as
>
> $$\|M\|_1 = \max\left\{ \mathrm{Tr}\,[M\Pi] : \, -I^A \leqslant \Pi \leqslant I^A, \; \Pi \in \mathrm{Herm}(A) \right\}. \qquad (5.132)$$

**Proof** Let $M_+$ and $M_-$ be the positive and negative parts of $M$ (see (2.54)), and let $\Pi_-$ and $\Pi_+ = I - \Pi_-$ the projections to the negative, and nonnegative eigenspaces of $M$. With these notations we have $|M| = M_+ + M_-$, so that the trace norm of $M$ can be expressed as

$$\|M\|_1 = \mathrm{Tr}[M_+] + \mathrm{Tr}[M_-]$$
$$= \mathrm{Tr}\left[ M\,(\Pi_+ - \Pi_-) \right] \qquad (5.133)$$
$$\textbf{Exercise 5.25} \rightarrow \;\; = \max_{-I \leqslant \Pi \leqslant I} \mathrm{Tr}\,[M\Pi],$$

where the maximum is over all matrices $\Pi \in \mathrm{Herm}(A)$ with eigenvalues between $-1$ and $1$. $\blacksquare$

**Exercise 5.25.** *Prove the last equality in (5.133).*

**Exercise 5.26.** *Show that for any two (normalized) pure states $|\psi\rangle, |\phi\rangle \in A$ we have*

$$T(\psi, \phi) := \frac{1}{2}\big\||\psi\rangle\langle\psi| - |\phi\rangle\langle\phi|\big\|_1 = \sqrt{1 - |\langle\psi|\phi\rangle|^2}. \qquad (5.134)$$

*Hint: Denote $|0\rangle := |\psi\rangle$ and express $|\phi\rangle := a|0\rangle + b|1\rangle$, where $|1\rangle$ is some (normalized) orthogonal vector to $|0\rangle$.*

**Exercise 5.27.** *Let $A$ be a Hilbert space and let $|\psi\rangle, |\phi\rangle \in A$ be two (normalized) states in A. Denote $\psi := |\psi\rangle\langle\psi|$ and $\phi := |\phi\rangle\langle\phi|$. Show that*

$$\frac{1}{2}\|\psi - \phi\|_1 \leqslant \big\||\psi\rangle - |\phi\rangle\big\|, \qquad (5.135)$$

*where the norm on the right-hand side is the induced inner-product norm* $\||\chi\rangle\| := \langle\chi|\chi\rangle^{1/2}$. *Hint: Use the previous exercise.*

The trace norm can also be expressed as an optimization over partial isometries.

> **Lemma 5.6.** Let $A$ and $B$ be two finite dimensional Hilbert spaces, and let $M\colon A \to B$ be a linear operator. Then, the trace norm of $M$ can be expressed as
>
> $$\|M\|_1 = \max_{V:B\to A} |\operatorname{Tr}[VM]|, \tag{5.136}$$
>
> where the maximum is over all partial isometries $V\colon B \to A$.

**Proof**  Express $M = \sum_{x\in[n]} \lambda_x |\phi_x^B\rangle\langle\psi_x^A|$, where $\{\lambda_x\}_{x\in[n]}$ are the singular values of $M$, and $\{|\psi_x^A\rangle\}_{x\in[n]}$ and $\{|\phi_x^B\rangle\}_{x\in[n]}$ are orthonormal sets of vectors in $A$ and $B$, respectively. Let $U\colon B \to A$ be the partial isometry $U = \sum_{x\in[n]} |\psi_x^A\rangle\langle\phi_x^B|$. We then have

$$\|M\|_1 = \left|\operatorname{Tr}[UM]\right| \leqslant \max_{V:B\to A} \left|\operatorname{Tr}[VM]\right|, \tag{5.137}$$

where the maximum is over all partial isometries $V\colon B \to A$. Substituting into the right-hand side $M = \sum_{x\in[n]} \lambda_x |\phi_x^B\rangle\langle\psi_x^A|$ gives

$$\|M\|_1 \leqslant \max_{V:B\to A} \left| \sum_{x\in[n]} \lambda_x \operatorname{Tr}\left[V|\phi_x^B\rangle\langle\psi_x^A|\right] \right|$$

$$\leqslant \max_{V:B\to A} \sum_{x\in[n]} \lambda_x \left|\langle\psi_x^A|V|\phi_x^B\rangle\right| \tag{5.138}$$

$$\textbf{see (5.139) below} \to \ \leqslant \sum_{x\in[n]} \lambda_x = \|M\|_1,$$

where we used the Cauchy–Schwarz inequality to get

$$\left|\langle\psi_x^A|V|\phi_x^B\rangle\right| \leqslant \sqrt{\langle\psi_x|\psi_x\rangle\langle\phi_x|V^*V|\phi_x\rangle}$$

$$= \sqrt{\langle\phi_x|V^*V|\phi_x\rangle} \tag{5.139}$$

$$\boxed{V^*V \text{ is a projection}} \ \longrightarrow \ \leqslant 1.$$

Hence, all the inequalities in (5.138) must be equalities. This completes the proof. ∎

**Exercise 5.28.** *Show that if* $|A| \geqslant |B|$ *in Lemma (5.6), then the maximization over partial isometries in (5.136) can be replaced with maximization over isometries* $V\colon B \to A$. *Similarly, show that if* $|A| \leqslant |B|$, *then*

$$\|M\|_1 = \max_{U:A\to B} \operatorname{Tr}\left[U^*M\right], \tag{5.140}$$

*where the maximum is over all isometries* $U\colon A \to B$.

## Strong Monotonicity Property

We show here that the trace norm behaves monotonically under a positive linear map (not necessarily CPTP). This monotonicity property is stronger than what we discussed so far.

> **Monotonicity of the Trace Norm**
>
> **Theorem 5.9.** Let $\mathcal{E} \in \mathfrak{L}(A \to B)$ be a trace nonincreasing positive linear map, and let $M \in \mathfrak{L}(A)$. Then,
>
> $$\|\mathcal{E}(M)\|_1 \leq \|M\|_1. \tag{5.141}$$

**Proof**   Express $M = \sum_{x \in [n]} \lambda_x |\phi_x^A\rangle\langle\psi_x^A|$, where $\{\lambda_x\}_{x \in [n]}$ are the singular values of $M$, and $\{|\psi_x^A\rangle\}_{x \in [n]}$ and $\{|\phi_x^A\rangle\}_{x \in [n]}$ are orthonormal sets of vectors in $A$. Then, from the triangle inequality of the trace norm we have

$$\|\mathcal{E}(M)\|_1 = \left\| \sum_{x \in [n]} \lambda_x \mathcal{E}\left(|\phi_x^A\rangle\langle\psi_x^A|\right) \right\|_1 \leq \sum_{x \in [n]} \lambda_x \left\| \mathcal{E}\left(|\phi_x^A\rangle\langle\psi_x^A|\right) \right\|_1. \tag{5.142}$$

Hence, it will be sufficient to prove that $\left\| \mathcal{E}\left(|\phi_x^A\rangle\langle\psi_x^A|\right) \right\|_1 \leq 1$ for all $x$ since this would imply that $\|\mathcal{E}(M)\|_1 \leq \sum_{x \in [n]} \lambda_x = \|M\|_1$. For simplicity of the exposition we remove the sub-index $x$ from the rest of the proof, since nothing will depend on it.

Now, the square matrix $\mathcal{E}(|\phi\rangle\langle\psi|)$ has a polar decomposition

$$\mathcal{E}(|\phi\rangle\langle\psi|) = U \, |\mathcal{E}(|\phi\rangle\langle\psi|)|. \tag{5.143}$$

Since $U$ is a unitary matrix in $\mathfrak{L}(B)$ it is diagonalizable and can be expressed as

$$U = \sum_{y \in [n]} e^{i\theta_y} |\varphi_y\rangle\langle\varphi_y|, \tag{5.144}$$

where $\{|\varphi_y\rangle\}$ is an orthonormal basis of $B$, and $\{\theta_y\}$ are some phases. Hence,

$$
\begin{aligned}
\|\mathcal{E}(|\phi\rangle\langle\psi|)\|_1 = \operatorname{Tr}|\mathcal{E}(M)| &= \operatorname{Tr}\left[U^*\mathcal{E}(|\phi\rangle\langle\psi|)\right] \\
\mathbf{(5.144)} \to \; &= \sum_{y \in [n]} e^{-i\theta_y} \operatorname{Tr}\left[|\varphi_y\rangle\langle\varphi_y|\mathcal{E}\left(|\phi\rangle\langle\psi|\right)\right] \\
&= \sum_{y \in [n]} e^{-i\theta_y} \operatorname{Tr}\left[\mathcal{E}^*\left(|\varphi_y\rangle\langle\varphi_y|\right)|\phi\rangle\langle\psi|\right].
\end{aligned}
\tag{5.145}
$$

From the Exercise 3.23, it follows that $\mathcal{E}^*$ is positive and sub-unital (i.e. $\mathcal{E}^*(I) \leq I$). Therefore, the matrices $\Lambda_y := \mathcal{E}^*\left(|\varphi_y\rangle\langle\varphi_y|\right) \geq 0$ form an incomplete POVM since

$$\sum_{y \in [n]} \Lambda_y = \mathcal{E}^*(I^B) \leq I^A. \tag{5.146}$$

With this notation,

$$
\|\mathcal{E}(|\phi\rangle\langle\psi|)\|_1 = \sum_{y\in[n]} e^{-i\theta_y}\operatorname{Tr}\left[\Lambda_y|\phi\rangle\langle\psi|\right] = \sum_{y\in[n]} e^{-i\theta_y}\langle\psi|\Lambda_y|\phi\rangle
$$

$$
\leqslant \sum_{y\in[n]} |\langle\psi|\Lambda_y|\phi\rangle| = \sum_{y\in[n]} |\langle\psi|\sqrt{\Lambda_y}\sqrt{\Lambda_y}|\phi\rangle|
$$

$$
\textbf{Cauchy–Schwarz inequality}\rightarrow \;\leqslant\; \sum_{y\in[n]} \sqrt{\langle\psi|\Lambda_y|\psi\rangle\langle\phi|\Lambda_y|\phi\rangle}
$$

$$
\textbf{geometric-arithmetic inequality}\rightarrow \;\leqslant\; \frac{1}{2}\sum_{y\in[n]} \left(\langle\psi|\Lambda_y|\psi\rangle + \langle\phi|\Lambda_y|\phi\rangle\right) \leqslant 1.
$$

$$(5.147)$$

This completes the proof. ∎

**Exercise 5.29.** *Provide an alternative (simpler) proof of the theorem above for the case that M is Hermitian. Hint: Use the previous lemma and prove first that if $-I^B \leqslant \Pi \leqslant I^B$, then $-I^A \leqslant \mathcal{E}^*(\Pi) \leqslant I^A$.*

## 5.4.2 The Trace Distance

The trace distance between two states $\rho,\sigma \in \mathfrak{D}(A)$ is defined by

$$
T(\rho,\sigma) := \frac{1}{2}\|\rho - \sigma\|_1. \qquad (5.148)
$$

The inclusion of the one-half factor is for normalization purposes, specifically to ensure that the distance reaches its maximum value of 1 when the two states, $\rho$ and $\sigma$, are orthogonal (refer to Exercise 5.30 for more details).

Consider the case in which $\rho$ and $\sigma$ are classical, or equivalently commute, and therefore diagonal in the same basis. In this case, denoting $\rho = \sum_{x\in[n]} p_x|x\rangle\langle x|$ and $\sigma = \sum_{x\in[n]} q_x|x\rangle\langle x|$ we get

$$
T(\rho,\sigma) := \frac{1}{2}\left\|\sum_{x\in[n]}(p_x - q_x)|x\rangle\langle x|\right\|_1 = \frac{1}{2}\sum_{x\in[n]}|p_x - q_x| =: T(\mathbf{p},\mathbf{q}), \qquad (5.149)
$$

where $\mathbf{p} := (p_1,\dots,p_n)^T$, $\mathbf{q} = (q_1,\dots,q_n)^T$, and $T(\mathbf{p},\mathbf{q})$ as defined above denotes the trace distance between the classical probability vectors $\mathbf{p}$ and $\mathbf{q}$.

In the general case, since both $\rho$ and $\sigma$ have the same trace

$$
0 = \operatorname{Tr}[\rho - \sigma] = \operatorname{Tr}[(\rho - \sigma)_+] - \operatorname{Tr}[(\rho - \sigma)_-], \qquad (5.150)
$$

where $(\rho - \sigma)_\pm$ are the positive and negative parts of $\rho - \sigma$. Therefore, denoting by $\Pi_+$ the projection to the positive eigenspace of $\rho - \sigma$, we conclude that

$$
T(\rho,\sigma) := \frac{1}{2}\left(\operatorname{Tr}[(\rho - \sigma)_+] + \operatorname{Tr}[(\rho - \sigma)_-]\right) = \operatorname{Tr}[(\rho - \sigma)_+] = \operatorname{Tr}[(\rho - \sigma)\Pi_+]. 
$$

$$(5.151)$$

That is, the trace distance can be written as

$$T(\rho, \sigma) = \max_{0 \leqslant \Pi \leqslant I} \text{Tr}[(\rho - \sigma)\Pi], \tag{5.152}$$

where the maximization is over any matrix $\Pi \in \text{Pos}(A)$ (not necessarily a projection) with eigenvalues between 0 and 1. The expression above for the trace distance will be useful in some of applications we discuss later on.

The monotonicity of the trace norm under quantum channels (in fact positive maps) implies the monotonicity of the trace distance as well. We summarize it in the following theorem.

> **Monotonicity of the Trace Distance**
>
> **Theorem 5.10.** The trace distance is a quantum divergence. In particular, for any $\rho, \sigma \in \mathfrak{D}(A)$ and $\mathcal{E} \in \text{CPTP}(A \to B)$
>
> $$T(\mathcal{E}(\rho), \mathcal{E}(\sigma)) \leqslant T(\rho, \sigma). \tag{5.153}$$

**Exercise 5.30.** *Let* $\rho, \sigma \in \mathfrak{D}(A)$. *Show that*

$$T(\rho, \sigma) = 1 \quad \Longleftrightarrow \quad \rho\sigma = \sigma\rho = 0. \tag{5.154}$$

**Exercise 5.31.** *Let* $\mathbf{u} \in \mathfrak{D}(A)$ *be the maximally mixed state and* $\psi \in \text{Pure}(A)$ *be a pure state. Show that the trace distance between these two states is given by*

$$T\left(\mathbf{u}^A, \psi^A\right) = \frac{m-1}{m}, \tag{5.155}$$

*where* $m := |A|$.

## A Relationship between the Trace Distance and the Ky Fan Norm

There is an interesting relationship between the trace distance and the Ky Fan norm (see Definition 2.4) that we will use in our study of entanglement theory. Consider a quantum state $\rho \in \mathfrak{D}(A)$ and let $m < |A|$ be an integer. Furthermore, we denote by $\rho^{(m)}$ the $m$-pruned version of $\rho$:

$$\rho^{(m)} := \frac{\Pi_m \rho \Pi_m}{\text{Tr}\left[\rho \Pi_m\right]}, \tag{5.156}$$

where $\Pi_m \in \text{Pos}(A)$ is the projection to the subspace spanned by the $m$ eigenvectors of the $m$ largest eigenvalues of $\rho$. By definition, $\text{Tr}\left[\rho\Pi_m\right] = \|\rho\|_{(m)}$ and $\rho$ commutes with $\rho^{(m)}$. In the following exercise we will use these properties to show that the trace distance between $\rho$ and $\rho^{(m)}$ is related to the Ky Fan norm.

**Exercise 5.32.** *Using the same notations as above, show that*

$$T\left(\rho, \rho^{(m)}\right) = 1 - \|\rho\|_{(m)}, \tag{5.157}$$

*where* $\|\cdot\|_{(m)}$ *is the Ky Fan norm. Hint: Use the relations* $T\left(\rho, \rho^{(m)}\right) = \text{Tr}\left(\rho^{(m)} - \rho\right)_+$ *and* $\text{Tr}\left[\rho\Pi_m\right] = \|\rho\|_{(m)}$, *and the fact that* $\rho$ *commutes with* $\rho^{(m)}$.

In the following theorem we use the notation $\mathfrak{D}_m(A)$ to denote the set of all density matrices in $\mathfrak{D}(A)$, whose rank is not greater than $m$.

**Theorem 5.11.** Using the same notations as above, the trace distance of $\rho$ to the set $\mathfrak{D}_m(A)$ is given by

$$T\left(\rho, \mathfrak{D}_m(A)\right) := \min_{\sigma \in \mathfrak{D}_m(A)} T(\rho, \sigma) = T\left(\rho, \rho^{(m)}\right) = 1 - \|\rho\|_{(m)}. \qquad (5.158)$$

**Proof**  Let $\sigma \in \mathfrak{D}_m(A)$ and observe that since $\mathrm{Rank}(\sigma) \leqslant m$ we have

$$\|\rho\|_{(m)} \geqslant \mathrm{Tr}\left[\Pi_\sigma \rho\right]$$
$$= \mathrm{Tr}\left[\Pi_\sigma \sigma\right] + \mathrm{Tr}\left[\Pi_\sigma(\rho - \sigma)\right]$$
$$= 1 + \mathrm{Tr}\left[\Pi_\sigma\left((\rho - \sigma)_+ - (\rho - \sigma)_-\right)\right]$$
$$\boxed{\mathrm{Tr}\left[\Pi_\sigma(\rho - \sigma)_+\right] \geqslant 0} \longrightarrow \geqslant 1 - \mathrm{Tr}\left[\Pi_\sigma(\rho - \sigma)_-\right]$$
$$\boxed{\Pi_\sigma \leqslant I^A} \longrightarrow \geqslant 1 - \mathrm{Tr}(\rho - \sigma)_-$$
$$= 1 - T(\rho, \sigma).$$

$(5.159)$

Therefore, for every $\sigma \in \mathfrak{D}_m(A)$ we get that

$$T(\rho, \sigma) \geqslant 1 - \|\rho\|_{(m)}$$
$$\textbf{Exercise 5.32} \rightarrow\ = T\left(\rho, \rho^{(m)}\right).$$

$(5.160)$

On the other hand, since $\rho^{(m)} \in \mathfrak{D}_m(A)$ by taking above $\sigma = \rho^{(m)}$ we can achieve an equality. Hence,

$$T\left(\rho, \mathfrak{D}_m(A)\right) = T\left(\rho, \rho^{(m)}\right) = 1 - \|\rho\|_{(m)}. \qquad (5.161)$$

This completes the proof. ∎

## Optimality of the Trace Distance

In (5.25), we saw that for $f(r) = \frac{1}{2}|r - 1|$ we get that the classical $f$-Divergence is equal to the classical trace distance. We show now that the minimal quantum extension of the classical trace distance equals to the quantum trace distance. Explicitly, let $T_c$ be the classical trace distance, and let $\underline{T}_c$ be its minimal quantum extension. That is, for any $\rho, \sigma \in \mathfrak{D}(A)$ (cf. (5.91))

$$\underline{T}_c(\rho, \sigma) := \sup_{\mathcal{E} \in \mathrm{CPTP}(A \to X)} T_c\left(\mathcal{E}(\rho), \mathcal{E}(\sigma)\right), \qquad (5.162)$$

where the supremum is over all classical systems X, POVM channels $\mathcal{E} \in \mathrm{CPTP}(A \to X)$, and the diagonal matrices $\mathcal{E}(\rho)$ and $\mathcal{E}(\sigma)$ are viewed as probability vectors.

> **Theorem 5.12.** Using the same notations as in (5.162), for all $\rho, \sigma \in \mathfrak{D}(A)$
>
> $$\underline{T}_c(\rho, \sigma) = T(\rho, \sigma) := \frac{1}{2}\|\rho - \sigma\|_1 \qquad (5.163)$$

*Remark.* Theorem 5.12 demonstrates that the quantum trace distance is the smallest divergence that reduces to the classical trace distance on classical states.

**Proof**  From Section (5.3), particularly Theorem 6.4, it follows that for any $\rho, \sigma \in \mathfrak{D}(A)$ we have $T(\rho, \sigma) \geqslant \underline{T}_c(\rho, \sigma)$ since $\underline{T}_c$ is the minimal quantum divergence that reduces to the classical trace distance when the input restricted to classical states. To prove the converse inequality $T(\rho, \sigma) \leqslant \underline{T}_c(\rho, \sigma)$, let $\Pi_\pm$ be the two projections to the positive and negative eigenspaces of $\rho - \sigma$, and let $\mathcal{E} \in \mathrm{CPTP}(A \to X)$ with $|X| = 2$ be its corresponding POVM Channel; that is, $\mathcal{E}(\omega) = \mathrm{Tr}[\omega \Pi_+]|0\rangle\langle 0| + \mathrm{Tr}[\omega \Pi_-]|1\rangle\langle 1|$ for all $\omega \in \mathfrak{D}(A)$. Then, by definition,

$$\underline{T}_c(\rho, \sigma) \geqslant T_c\big(\mathcal{E}(\rho), \mathcal{E}(\sigma)\big) = \frac{1}{2}\left(\mathrm{Tr}[(\rho - \sigma)_+] + \mathrm{Tr}[(\rho - \sigma)_-]\right) = T(\rho, \sigma). \quad (5.164)$$

This completes the proof.  ∎

## Joint Convexity of the Trace Distance

The monotonicity of the trace distance under quantum channels implies that the trace distance is jointly convex.

> **Theorem 5.13.** Let $\mathbf{p} = (p_1, \ldots, p_m)^T$ and $\mathbf{q} = (q_1, \ldots, q_m)^T$ be two probability vectors, and let $\{\rho_x\}_{x\in[m]}$ and $\{\sigma_x\}_{x\in[m]}$ be two sets of $m$ density matrices in $\mathfrak{D}(A)$. Then,
>
> $$T\left(\sum_{x\in[m]} p_x\rho_x, \sum_{x\in[m]} q_x\sigma_x\right) \leqslant T(\mathbf{p}, \mathbf{q}) + \sum_{x\in[m]} p_x T(\rho_x, \sigma_x), \qquad (5.165)$$
>
> where $T(\mathbf{p}, \mathbf{q}) := \frac{1}{2}\sum_{x\in[m]}|p_x - q_x|$ is the classical trace distance between two probability vectors.

We provide two proofs for this theorem to illustrate different techniques.

**Proof**  Let $X$ be a classical system (register) and define the classical-quantum states

$$\rho^{XA} := \sum_{x\in[m]} p_x |x\rangle\langle x|^X \otimes \rho_x^A \quad \text{and} \quad \sigma^{XA} := \sum_{x\in[m]} q_x |x\rangle\langle x|^X \otimes \sigma_x^A. \qquad (5.166)$$

Then,

$$T\left(\rho^{XA},\sigma^{XA}\right) = \frac{1}{2}\left\|\sum_{x\in[m]}|x\rangle\langle x|\otimes(p_x\rho_x - q_x\sigma_x)\right\|_1$$

$$= \frac{1}{2}\sum_{x\in[m]}\|p_x\rho_x - q_x\sigma_x\|_1$$

$$= \frac{1}{2}\sum_{x\in[m]}\|p_x\rho_x - p_x\sigma_x + p_x\sigma_x - q_x\sigma_x\|_1 \tag{5.167}$$

$$\textbf{Triangle inequality}\rightarrow\; \leqslant \frac{1}{2}\sum_{x\in[m]}\left(\|p_x\rho_x - p_x\sigma_x\|_1 + \|p_x\sigma_x - q_x\sigma_x\|_1\right)$$

$$= \sum_{x\in[m]}p_x T(\rho_x,\sigma_x) + \frac{1}{2}\sum_{x\in[m]}|p_x - q_x|.$$

Hence,

$$T\left(\sum_{x\in[m]}p_x\rho_x^A,\; \sum_{x\in[m]}q_x\sigma_x^A\right) = T\left(\mathrm{Tr}_X[\rho^{XA}],\mathrm{Tr}_X[\sigma^{XA}]\right)$$

$$\textbf{monotonicity (Theorem 5.10)}\rightarrow\; \leqslant T\left(\rho^{XA},\sigma^{XA}\right)$$

$$\leqslant \sum_{x\in[m]}p_x T(\rho_x^A,\sigma_x^A) + T(\mathbf{p},\mathbf{q}).$$

This completes the proof.    ■

**Alternative Proof**   Let $\Pi$ be the optimal projection such that

$$T\left(\sum_{x\in[m]}p_x\rho_x,\; \sum_{x\in[m]}q_x\sigma_x\right) = \mathrm{Tr}\left[\Pi\sum_{x\in[m]}(p_x\rho_x - q_x\sigma_x)\right]. \tag{5.168}$$

Therefore,

$$T\left(\sum_{x\in[m]}p_x\rho_x,\; \sum_{x\in[m]}q_x\sigma_x\right) = \sum_{x\in[m]}\left(p_x\mathrm{Tr}\left[\Pi(\rho_x - \sigma_x)\right] + (p_x - q_x)\mathrm{Tr}[\Pi\sigma_x]\right)$$

$$\leqslant \sum_{x\in[m]}p_x T(\rho_x,\sigma_x) + \sum_{x\in[m]}(p_x - q_x)_+ \tag{5.169}$$

$$= \sum_{x\in[m]}p_x T(\rho_x,\sigma_x) + T(\mathbf{p},\mathbf{q}),$$

where we used the fact that $\mathrm{Tr}[\Pi\sigma_x] \leqslant 1$ and the fact that $p_x - q_x \leqslant (p_x - q_x)_+$. This completes the proof.    ■

We conclude this subsection by discussing a nuanced yet crucial property of the trace distance. This property is highly relevant to certain applications in quantum information, though it is often overlooked. Consider $\rho,\sigma \in \mathfrak{D}(A)$ and let us define $\varepsilon := T(\rho,\sigma)$. If $\varepsilon$ is very small, it implies that $\rho$ and $\sigma$ are nearly identical states. This concept can be articulated as follows: Decompose $\rho - \sigma$ into positive and negative parts, written as $\rho - \sigma = (\rho - \sigma)_+ - (\rho - \sigma)_-$. Then, define two states $\omega_\pm := \frac{1}{\varepsilon}(\rho - \sigma)_\pm$. Given that $\varepsilon = T(\rho, \sigma) = \mathrm{Tr}(\rho - \sigma)_+ = \mathrm{Tr}(\rho - \sigma)_-$, it follows that $\omega_\pm$ are valid density matrices in $\mathfrak{D}(A)$. Furthermore, we can express

$$\rho - \sigma = \varepsilon(\omega_+ - \omega_-). \tag{5.170}$$

The importance of this equation lies in the fact that the matrix $H := \omega_+ - \omega_-$ is bounded, satisfying $-I \leqslant H \leqslant I$. Additionally, the equation $\rho = \sigma + \varepsilon H$ does not depend explicitly on the underlying dimension $|A|$.

To further elucidate this point, consider the following straightforward example involving the Schatten 2-norm (the norm induced by the Hilbert–Schmidt inner product). Let $\rho_n = \frac{1}{n} I_n$ denote the $n \times n$ maximally mixed state. Observe that its 2-norm is calculated as follows:

$$\|\rho_n\|_2 := \sqrt{\mathrm{Tr}[\rho_n^2]} = \frac{1}{\sqrt{n}}. \tag{5.171}$$

Consequently, as $n$ approaches infinity, $\|\rho_n\|_2$ tends toward zero, while the trace norm $\|\rho_n\|_1 = 1$ for all $n \in \mathbb{N}$.

**Exercise 5.33.** *Using the same notations as above, show that if a set of Hermitian matrices $\{H_n\}$, with each $H_n \in \mathrm{Herm}(\mathbb{C}^n)$, satisfies $\lim_{n \to \infty} \|H_n\|_1 = 0$, then there exists a sequence of positive numbers $\{\varepsilon_n\}$ with a limit $\lim_{n \to \infty} \varepsilon_n = 0$ and a set of bounded matrices $\{M_n\}$ with $-I_n \leqslant M_n \leqslant I_n$ such that $H_n = \varepsilon_n M_n$.*

### 5.4.3 The Fidelity

The fidelity is another distance like measure between two quantum states; however, unlike the trace distance or any other divergence, it reaches its maximal value when the two states are the same, and its minimal value when they are orthogonal. For any $\rho, \sigma \in \mathfrak{D}(A)$ it is defined by

$$F(\rho, \sigma) := \|\sqrt{\rho}\sqrt{\sigma}\|_1 = \mathrm{Tr}\left[|\sqrt{\rho}\sqrt{\sigma}|\right] = \mathrm{Tr}\sqrt{\sqrt{\sigma}\rho\sqrt{\sigma}}. \tag{5.172}$$

Consider first the simpler case that $\rho$ and $\sigma$ commutes. In this case, there exists a basis $\{|x\rangle\}_{x \in [n]}$ of $A$ such that $\rho = \sum_{x \in [n]} p_x |x\rangle\langle x|$ and $\sigma = \sum_{x \in [n]} q_x |x\rangle\langle x|$. We then have

$$F(\rho, \sigma) = \|\sqrt{\rho}\sqrt{\sigma}\|_1 = \left\| \sum_{x \in [n]} \sqrt{p_x q_x} |x\rangle\langle x| \right\|_1 = \sum_{x \in [n]} \sqrt{p_x q_x} := F(\mathbf{p}, \mathbf{q}), \tag{5.173}$$

where $\mathbf{p} := (p_1, \ldots, p_n)^T$, $\mathbf{q} := (q_1, \ldots, q_n)^T$, and $F(\mathbf{p}, \mathbf{q})$ as defined above denotes the fidelity between the classical probability vectors $\mathbf{p}$ and $\mathbf{q}$. If $\rho = \sigma$ we get that $F(\rho, \rho) = \|\sqrt{\rho}\sqrt{\rho}\|_1 = \|\rho\|_1 = \mathrm{Tr}[\rho] = 1$. Moreover, for any $\rho, \sigma \in \mathfrak{D}(A)$, the fidelity $F(\rho, \sigma)$ cannot be greater than 1. This will follow trivially from the following Uhlmann's theorem, and can also be seen from the following argument:

$$|\sqrt{\rho}\sqrt{\sigma}|^2 = \sqrt{\sigma}\rho\sqrt{\sigma}$$

$$\boxed{\rho = I - (I - \rho)} \longrightarrow = \sigma - \sqrt{\sigma}(I - \rho)\sqrt{\sigma} \tag{5.174}$$

$$\boxed{\sqrt{\sigma}(I - \rho)\sqrt{\sigma} \geqslant 0} \longrightarrow \leqslant \sigma \leqslant I^A.$$

Therefore, $|\sqrt{\rho}\sqrt{\sigma}| \leqslant I^A$.

**Exercise 5.34.** *Let $\rho, \sigma \in \mathfrak{D}(A)$.*

1. *Show that the fidelity is symmetric: $F(\rho, \sigma) = F(\sigma, \rho)$. Hint: Use the fact that for any complex matrix $M$, the matrix $M^*M$ has the same nonzero eigenvalues as $MM^*$.*
2. *Show that if $\sigma = |\psi\rangle\langle\psi|$ is pure, then $F(\rho, \sigma) = \sqrt{\langle\psi|\rho|\psi\rangle}$.*

**Exercise 5.35.** *Let $\rho, \sigma \in \mathfrak{D}(A)$. Show that if $\lambda$ is an eigenvalue of the matrix $|\sqrt{\rho}\sqrt{\sigma}|$, then $\lambda^2$ is an eigenvalue of the non-Hermitian matrix $\rho\sigma$. Hint: Let $M = \sqrt{\sigma}\rho\sqrt{\sigma}$ and $N = \rho\sigma$ and find a matrix $\eta \geqslant 0$ such that $M = \eta^{-1}N\eta$, where $\eta^{-1}$ is the generalized inverse of $\eta$.*

## Uhlmann's Theorem

The last part in the exercise above also implies that if both $\rho = |\psi\rangle\langle\psi|$ and $\sigma = |\phi\rangle\langle\phi|$ are pure, then the fidelity becomes the absolute value of the inner product between the two states; that is, $F(\rho, \sigma) = |\langle\psi|\phi\rangle|$. The following theorem by Uhlmann's shows that this can be extended to mixed states by considering all the possible purifications of $\rho$ and $\sigma$.

> **Uhlmann's Theorem**
>
> **Theorem 5.14.** Let $\rho, \sigma \in \mathfrak{D}(A)$ be two density matrices, and let $|\psi^{AB}\rangle$ and $|\phi^{AC}\rangle$ be two purifications of $\rho^A$ and $\sigma^A$, respectively. Then,
>
> $$F(\rho^A, \sigma^A) = \max_{V^{B \to C}} |\langle\psi^{AB}|V^*|\phi^{AC}\rangle|, \tag{5.175}$$
>
> where the maximum is over all partial isometries $V : B \to C$.

*Remark.* We emphasize that the purifying systems $B$ and $C$ are not necessarily isomorphic. That is, we can have $|B| \neq |C|$.

**Proof** From Exercise (2.44) it follows that the purifications $|\psi^{AB}\rangle$ and $|\phi^{AC}\rangle$ must have the form

$$|\psi^{AB}\rangle = \sqrt{\rho} \otimes U^{\tilde{A} \to B}|\Omega^{A\tilde{A}}\rangle \quad \text{and} \quad |\phi^{AB}\rangle = \sqrt{\sigma} \otimes W^{\tilde{A} \to C}|\Omega^{A\tilde{A}}\rangle, \tag{5.176}$$

where $U, W$ are two isometries. Thus,

$$\max_{V^{B \to C}} \left| \langle \psi^{AB} | V^* | \phi^{AC} \rangle \right| = \max_{V^{B \to C}} \left| \langle \Omega^{A\tilde{A}} | \sqrt{\rho}\sqrt{\sigma} \otimes U^* V^* W | \Omega^{A\tilde{A}} \rangle \right|$$

$$\textbf{part 2 of Exercise 2.38} \to \quad = \max_{V^{B \to C}} \left| \mathrm{Tr}\left[ \sqrt{\rho}\sqrt{\sigma} \left( U^* V^* W \right)^T \right] \right|$$

$$= \max_{V^{B \to C}} \left| \mathrm{Tr}\left[ \bar{U} \sqrt{\rho}\sqrt{\sigma} W^T \bar{V} \right] \right| \tag{5.177}$$

$$\textbf{Lemma 5.6} \to \quad = \left\| \bar{U} \sqrt{\rho}\sqrt{\sigma} W^T \right\|_1$$

$$\textbf{(5.131)} \to \quad = \left\| \sqrt{\rho}\sqrt{\sigma} \right\|_1$$

$$= F(\rho^A, \sigma^A).$$

This completes the proof. ∎

Uhlmann's theorem has numerous applications in quantum information, and we will use it quite often later on in the book. The following corollary is an immediate consequence of Uhlmann's theorem. We leave its proof as an exercise.

**Corollary 5.2.** Let $\rho, \sigma \in \mathfrak{D}(A)$. Then, $F(\rho, \sigma) \leqslant 1$ with equality if and only if $\rho = \sigma$.

The next consequence of Uhlmann's theorem is the monotonicity of the fidelity under quantum channels.

---

**Monotonicity of the Fidelity**

**Corollary 5.3.** Let $\rho, \sigma \in \mathfrak{D}(A)$, and let $\mathcal{E} \in \mathrm{CPTP}(A \to B)$ be a quantum channel. Then,

$$F(\rho, \sigma) \leqslant F(\mathcal{E}(\rho), \mathcal{E}(\sigma)). \tag{5.178}$$

---

**Proof**  Let $|\psi^{AC}\rangle$ and $|\phi^{AC}\rangle$ be optimal purifications of $\rho$ and $\sigma$ such that the fidelity $F(\rho, \sigma) = |\langle \psi^{AC} | \phi^{AC} \rangle|$ (i.e. we are using Uhlmann's theorem). Now, from Stinespring dilation theorem there exists an isometry $V : A \to BE$ such that

$$\mathcal{E}(\rho) = \mathrm{Tr}_E \left[ V\rho V^* \right] = \mathrm{Tr}_{EC} \left[ \left( V \otimes I^C \right) |\psi^{AC}\rangle\langle\psi^{AC}| \left( V^* \otimes I^C \right) \right]$$
$$\mathcal{E}(\sigma) = \mathrm{Tr}_E \left[ V\sigma V^* \right] = \mathrm{Tr}_{EC} \left[ \left( V \otimes I^C \right) |\phi^{AC}\rangle\langle\phi^{AC}| \left( V^* \otimes I^C \right) \right]. \tag{5.179}$$

Denote by $|\tilde{\psi}^{BEC}\rangle := \left( V^{A \to BE} \otimes I^C \right) |\psi^{AC}\rangle$ and by $|\tilde{\phi}^{BEC}\rangle := \left( V^{A \to BE} \otimes I^C \right) |\phi^{AC}\rangle$. Therefore, the above equation implies that $|\tilde{\psi}^{BEC}\rangle$ and $|\tilde{\phi}^{BEC}\rangle$ are purifications of $\mathcal{E}(\rho)$ and $\mathcal{E}(\sigma)$. We therefore get from Uhlmann's theorem that

$$F\left(\mathcal{E}(\rho), \mathcal{E}(\sigma)\right) \geqslant \left| \langle \tilde{\psi}^{BEC} | \tilde{\phi}^{BEC} \rangle \right| = \left| \langle \psi^{AC} | \phi^{AC} \rangle \right| = F(\rho, \sigma). \tag{5.180}$$

This completes the proof. ∎

Note that since the partial trace is a quantum channel it follows that for any two bipartite states $\rho, \sigma \in \mathfrak{D}(AB)$ we have

$$F(\rho^A, \sigma^A) \geqslant F(\rho^{AB}, \sigma^{AB}). \tag{5.181}$$

Another corollary of Uhlmann's theorem is the joint concavity of the fidelity.

> **Joint Concavity of the Fidelity**
>
> **Corollary 5.4.** Let $\{p_x\}_{x\in[m]}$ and $\{q_x\}_{x\in[m]}$ be two probability distributions, and let $\{\rho_x\}_{x\in[m]}$ and $\{\sigma_x\}_{x\in[m]}$ be two sets of $m$ density matrices in $\mathfrak{D}(A)$. Then,
>
> $$F\left(\sum_{x\in[m]} p_x\rho_x, \sum_{x\in[m]} q_x\sigma_x\right) \geqslant \sum_{x\in[m]} \sqrt{p_x q_x}\, F(\rho_x,\sigma_x). \qquad (5.182)$$

*Remark.* Note that from the corollary above it follows in particular that

$$F\left(\sum_{x\in[m]} p_x\rho_x, \sum_{x\in[m]} p_x\sigma_x\right) \geqslant \sum_{x\in[m]} p_x F(\rho_x,\sigma_x). \qquad (5.183)$$

Hence, the fidelity is jointly concave.

**Proof**  Let $\{|\psi_x^{AB}\rangle\}_{x\in[m]}$ and $\{|\phi_x^{AB}\rangle\}_{x\in[m]}$ be optimal purifications, respectively, of $\{\rho_x^A\}_{x\in[m]}$ and $\{\sigma_x^A\}_{x\in[m]}$ such that

$$F(\rho_x,\sigma_x) = \langle\psi_x^{AB}|\phi_x^{AB}\rangle \qquad \forall\, x \in [m]. \qquad (5.184)$$

Define also the pure states

$$|\tilde{\psi}^{ABC}\rangle := \sum_{x\in[m]} \sqrt{p_x}|\psi_x^{AB}\rangle|x\rangle^C \quad \text{and} \quad |\tilde{\phi}^{ABC}\rangle := \sum_{x\in[m]} \sqrt{q_x}|\phi_x^{AB}\rangle|x\rangle^C, \qquad (5.185)$$

where $C$ is some $m$ dimensional system. Note that the two states above are purifications of $\sum_{x\in[m]} p_x\rho_x^A$ and $\sum_{x\in[m]} q_x\sigma_x^A$, respectively. Therefore, we must have

$$F\left(\sum_{x\in[m]} p_x\rho_x^A, \sum_{x\in[m]} q_x\sigma_x^A\right) \geqslant \left|\langle\tilde{\psi}^{ABC}|\tilde{\phi}^{ABC}\rangle\right|$$

$$(5.185)\rightarrow = \sum_{x\in[m]} \sqrt{p_x q_x}\langle\psi_x^{AB}|\phi_x^{AB}\rangle \qquad (5.186)$$

$$= \sum_{x\in[m]} \sqrt{p_x q_x}\, F(\rho_x,\sigma_x).$$

This completes the proof.  ∎

**Exercise 5.36.** *Prove the joint concavity of the fidelity by defining classical quantum states as in (5.166) and using the monotonicity of the Fidelity under the partial trace.*

**Exercise 5.37.** *The square fidelity on* $\mathrm{Prob}(n) \times \mathrm{Prob}(n)$ *is defined for all* $\mathbf{p}, \mathbf{q} \in \mathrm{Prob}(n)$ *as*

$$F(\mathbf{p},\mathbf{q})^2 = \left(\sum_{x\in[n]} \sqrt{p_x q_x}\right)^2 = \mathbf{p}\cdot\mathbf{q} + \sum_{\substack{x,y\in[n]\\x\neq y}} \sqrt{p_x q_x}\sqrt{p_y q_y}. \qquad (5.187)$$

*Show that the square of the fidelity is concave on each of its arguments; that is, show that for any $k \in \mathbb{N}$, $\{\mathbf{q}_z\}_{z \in [k]} \subset \mathrm{Prob}(n)$, and $\mathbf{t} \in \mathrm{Prob}(k)$ we have*

$$\sum_{z \in [k]} t_z F(\mathbf{p}, \mathbf{q}_z)^2 \leqslant F\left(\mathbf{p}, \sum_{z \in [k]} t_z \mathbf{q}_z\right)^2. \tag{5.188}$$

*Similarly, show that the square fidelity is concave with respect to the first argument.*

**Exercise 5.38.** *Let $\rho, \sigma \in \mathfrak{D}(A)$ and let $\tau, \omega \in \mathfrak{D}(B)$. Show that*

$$F\left(\rho^A \otimes \tau^B, \sigma^A \otimes \omega^B\right) = F\left(\rho^A, \sigma^A\right) F\left(\tau^B, \omega^B\right). \tag{5.189}$$

## Optimal Extensions of the Classical Fidelity

The Corollary 5.2 also implies that the function $1 - F(\rho, \sigma)$ is a faithful quantum divergence. We can therefore study the optimality of the fidelity by exploring the minimal and maximal quantum extensions of the classical divergence $1 - F(\mathbf{p}, \mathbf{q})$. In particular, the maximal quantum extension of the classical fidelity $F(\mathbf{p}, \mathbf{q})$, which corresponds to the minimal quantum extension of $1 - F(\mathbf{p}, \mathbf{q})$, is given by (cf. (5.91))

$$\overline{F}(\rho^A, \sigma^A) := \inf_{\mathcal{E} \in \mathrm{CPTP}(A \to X)} F\left(\mathcal{E}^{A \to X}(\rho^A), \mathcal{E}^{A \to X}(\sigma^A)\right) \quad \forall\, \rho, \sigma \in \mathfrak{D}(A), \tag{5.190}$$

where the infimum is over all classical systems $X$ and all POVM channels $\mathcal{E} \in$ CPTP$(A \to X)$. By applying Theorem 6.4 to the classical divergence $1 - F(\mathbf{p}, \mathbf{q})$ we get that any function $f$ that satisfies the same monotonicity property (5.178) as the fidelity and that reduces to the fidelity on classical states must satisfy $f(\rho, \sigma) \leqslant \overline{F}(\rho, \sigma)$ for all $\rho, \sigma \in \mathfrak{D}(A)$. Remarkably, the Uhlmann's theorem implies that the fidelity in fact equals to this maximal quantum extension.

> **Optimality**
>
> **Corollary 5.5.** For any $\rho, \sigma \in \mathfrak{D}(A)$,
>
> $$\overline{F}(\rho, \sigma) = F(\rho, \sigma). \tag{5.191}$$

**Proof** As already discussed, the inequality $\overline{F}(\rho, \sigma) \geqslant F(\rho, \sigma)$ follows by applying Theorem 6.4 to the classical divergence $1 - F(\mathbf{p}, \mathbf{q})$. To prove the converse, let $\{|x\rangle\langle x|\}_{x \in [m]}$ be the orthonormal eigenbasis of the Hermitian matrix

$$\Lambda = \sigma^{-1/2} \left(\sigma^{1/2} \rho \sigma^{1/2}\right)^{1/2} \sigma^{-1/2} \tag{5.192}$$

such that $\Lambda |x\rangle = \lambda_x |x\rangle$, with $\{\lambda_x\}_{x \in [m]}$ being the eigenvalues of $\Lambda$. The key reason for this choice of basis is that the matrix $\Lambda$ satisfies

$$\Lambda \sigma \Lambda = \rho. \tag{5.193}$$

Taking $\mathcal{E} \in \mathrm{CPTP}(A \to X)$ to be the completely dephasing channel in the basis $\{|x\rangle\}_{x\in[m]}$, we obtain

$$
\begin{aligned}
F\big(\mathcal{E}(\rho),\mathcal{E}(\sigma)\big) &= \sum_x \sqrt{\langle x|\rho|x\rangle \langle x|\sigma|x\rangle} = \sum_x \sqrt{\langle x|\Lambda\sigma\Lambda|x\rangle \langle x|\sigma|x\rangle} \\
&= \sum_x \sqrt{\lambda_x^2 \langle x|\sigma|x\rangle \langle x|\sigma|x\rangle} = \sum_x \lambda_x \langle x|\sigma|x\rangle \\
&= \sum_x \lambda_x \operatorname{Tr}\left[\sigma|x\rangle\langle x|\right] = \operatorname{Tr}[\Lambda\sigma] \\
&= \operatorname{Tr}\left[\left(\sigma^{1/2}\rho\sigma^{1/2}\right)^{1/2}\right] = F(\rho,\sigma).
\end{aligned}
$$

$$(5.194)$$

This completes the proof. $\blacksquare$

**Exercise 5.39.** *Prove* (5.193).

Since the function $1 - F(\mathbf{p},\mathbf{q})$ is a classical divergence, we can also define its maximal quantum extension. This maximal extension corresponds to the minimal quantum extension of the fidelity. The minimal quantum extension of the classical fidelity is given by

$$
\underline{F}(\rho,\sigma) = \sup F(\mathbf{p},\mathbf{q}),
\tag{5.195}
$$

where the supremum is over all classical systems $X$, and over all $\mathbf{p},\mathbf{q} \in \mathfrak{D}(X)$ for which there exists a channel $\mathcal{E} \in \mathrm{CPTP}(X \to A)$ such that $\rho = \mathcal{E}(\mathbf{p})$ and $\sigma = \mathcal{E}(\mathbf{q})$ (depending on the context, we are using the notation $\mathbf{p},\mathbf{q}$ to indicate either diagonal density matrices in $\mathfrak{D}(A)$ or probability vectors in $\mathrm{Prob}(n)$ ).

**The Minimal Fidelity**

**Theorem 5.15.** The minimal quantum extension of the classical fidelity is given by

$$
\underline{F}(\rho,\sigma) = F_M(\rho,\sigma) := \operatorname{Tr}\left[\tilde{\sigma}\left(\tilde{\sigma}^{-\frac{1}{2}}\tilde{\rho}\tilde{\sigma}^{-\frac{1}{2}}\right)^{\frac{1}{2}}\right],
\tag{5.196}
$$

where $\tilde{\sigma} = \sigma_{11}$ and $\tilde{\rho} := \rho_{11} - \zeta\rho_{22}^{-1}\zeta^*$ are as given in (5.124).

**Proof** Define

$$
D_\star(\mathbf{p}\|\mathbf{q}) := 1 - F(\mathbf{p},\mathbf{q}) = \sum_{x\in[n]} \left(q_x - \sqrt{p_x q_x}\right) = \sum_{x\in\mathrm{supp}(\mathbf{q})} q_x\left(1 - \sqrt{\frac{p_x}{q_x}}\right).
\tag{5.197}
$$

Therefore, $D_\star$ equals the $f$-Divergence, $D_f$, with $f(r) := 1 - \sqrt{r}$. Note that the function $f : [0,\infty) \to \mathbb{R}$ is continuous and operator convex, so we can apply the formula (5.124) to get a closed form for the maximal quantum extension of $D_f$. Explicitly, from (5.124) and the fact that $\ddot{f}(0) := \lim_{\varepsilon\to 0^+} \varepsilon f(1/\varepsilon) = \lim_{\varepsilon\to 0^+} (\varepsilon + \sqrt{\varepsilon}) = 0$ we get

$$\overline{D}_f(\rho\|\sigma) = \operatorname{Tr}\left[\tilde{\sigma}\, f\left(\tilde{\sigma}^{-\frac{1}{2}}\tilde{\rho}\tilde{\sigma}^{-\frac{1}{2}}\right)\right]$$

$$= \operatorname{Tr}\left[\tilde{\sigma}\left(1 - \left(\tilde{\sigma}^{-\frac{1}{2}}\tilde{\rho}\tilde{\sigma}^{-\frac{1}{2}}\right)^{\frac{1}{2}}\right)\right] \tag{5.198}$$

$$= 1 - \operatorname{Tr}\left[\tilde{\sigma}\left(\tilde{\sigma}^{-\frac{1}{2}}\tilde{\rho}\tilde{\sigma}^{-\frac{1}{2}}\right)^{\frac{1}{2}}\right].$$

Hence,

$$\underline{F}(\rho,\sigma) = 1 - \overline{D}_f(\rho\|\sigma) = \operatorname{Tr}\left[\tilde{\sigma}\left(\tilde{\sigma}^{-\frac{1}{2}}\tilde{\rho}\tilde{\sigma}^{-\frac{1}{2}}\right)^{\frac{1}{2}}\right]. \tag{5.199}$$

This completes the proof.  ∎

Note that if $\sigma > 0$, then $\tilde{\sigma} = \sigma$ and $\tilde{\rho} = \rho$, and if $\sigma \not> 0$, then one can express the minimal fidelity in terms of the limit (cf. Exercise D.2 of online version)

$$F_M(\rho,\sigma) = \lim_{\varepsilon\to 0^+} F_M(\rho,\sigma + \varepsilon I). \tag{5.200}$$

**Exercise 5.40.** *Prove the following properties of the minimal fidelity:*

1. *Ranges from zero to 1; that is, $0 \leqslant F_M(\rho,\sigma) \leqslant 1$ for all $\rho,\sigma \in \mathfrak{D}(A)$.*
2. *Symmetry; $F_M(\rho,\sigma) = F_M(\sigma,\rho)$ for all $\rho,\sigma \in \mathfrak{D}(A)$ (Hint: Use the symmetry of the classical divergence).*
3. *Attains zero for orthogonal states.*
4. *Joint concavity; satisfies (5.183) with $F$ replaced by $F_M$.*
5. *Multiplicativity over tensor products:*

$$F_M(\rho_1 \otimes \rho_2, \sigma_1 \otimes \sigma_2) = F_M(\rho_1,\sigma_1)F_M(\rho_2,\sigma_2) \tag{5.201}$$

*for all $\rho_1,\sigma_1 \in \mathfrak{D}(A)$ and all $\rho_2,\sigma_2 \in \mathfrak{D}(B)$.*

## 5.4.4 The Relation between the Trace Distance and the Fidelity

The trace distance and the fidelity satisfy the following inequalities.

> **Theorem 5.16.** Let $\rho,\sigma \in \mathfrak{D}(A)$, $F$ be the fidelity, and $T$ the trace distance. Then,
>
> $$1 - F(\rho,\sigma) \leqslant T(\rho,\sigma) \leqslant \sqrt{1 - F(\rho,\sigma)^2}. \tag{5.202}$$

This relation reveals that if the fidelity is close to 1 then the trace distance is close to zero, and if the fidelity is close to zero then the trace distance is close to 1.

**Proof**  We first prove the upper bound. Let $\psi^{AB}$ and $\phi^{AB}$ be purifications or $\rho^A$ and $\sigma^A$, such that $F(\rho^A,\sigma^A) = |\langle\psi^{AB}|\phi^{AB}\rangle|$. Such purifications exists due to Uhlmann's theorem. We then have from monotonicity of the trace distance under partial trace

$$T\left(\rho^A,\sigma^A\right) \leqslant T\left(\psi^{AB},\phi^{AB}\right)$$

$$\textbf{Exercise 5.26}\rightarrow \; = \sqrt{1 - |\langle\psi^{AB}|\phi^{AB}\rangle|^2} \tag{5.203}$$

$$= \sqrt{1 - F\left(\rho^A,\sigma^A\right)^2},$$

where the last equality follows from the definition of $\psi^{AB}$ and $\phi^{AB}$.

To get the lower bound of (5.202), we start by observing that from Corollary 5.5 there exists a POVM $\{\Lambda_x\}_{x\in[n]}$ such that

$$F(\rho,\sigma) = \sum_{x\in[n]} \sqrt{p_x q_x} \quad \text{where} \quad p_x := \mathrm{Tr}[\Lambda_x\rho], \quad q_x := \mathrm{Tr}[\Lambda_x\sigma]. \tag{5.204}$$

Hence, from the equality $2\sqrt{p_x q_x} = p_x + q_x - (\sqrt{p_x} - \sqrt{q_x})^2$ it follows

$$F(\rho,\sigma) = \frac{1}{2}\sum_{x\in[n]}\left(p_x + q_x - (\sqrt{p_x} - \sqrt{q_x})^2\right) = 1 - \frac{1}{2}\sum_x (\sqrt{p_x} - \sqrt{q_x})^2. \tag{5.205}$$

To bound the last term, observe that

$$\frac{1}{2}\sum_{x\in[n]} (\sqrt{p_x} - \sqrt{q_x})^2 = \frac{1}{2}\sum_x |\sqrt{p_x} - \sqrt{q_x}||\sqrt{p_x} - \sqrt{q_x}|$$

$$\boxed{|\sqrt{p_x} - \sqrt{q_x}| \leqslant \sqrt{p_x} + \sqrt{q_x}} \longrightarrow \; \leqslant \frac{1}{2}\sum_{x\in[n]} |\sqrt{p_x} - \sqrt{q_x}|(\sqrt{p_x} + \sqrt{q_x}) \tag{5.206}$$

$$= \frac{1}{2}\sum_{x\in[n]} |p_x - q_x| = T(\mathbf{p},\mathbf{q})$$

$$\textbf{Theorem 5.12}\rightarrow \; \leqslant T(\rho,\sigma).$$

Combining (5.205,5.206) gives

$$F(\rho,\sigma) \geqslant 1 - T(\rho,\sigma), \tag{5.207}$$

which is equivalent to the lower bound of (5.202).      ∎

We say that two states $\rho,\sigma \in \mathfrak{D}(A)$ are $\varepsilon$-close in trace distance if $T(\rho,\sigma) \leqslant \varepsilon$ for some $\varepsilon > 0$. Similarly, $\rho$ and $\sigma$ are $\varepsilon$-close in fidelity if $F(\rho,\sigma) \geqslant 1 - \varepsilon$. In the exercise below you will show that the relation between the trace distance and the fidelity can be used to show that the two notions of "$\varepsilon$-close" are essentially the same.

**Exercise 5.41.**

1. *Show that if two states $\rho,\sigma \in \mathfrak{D}(A)$ are $\varepsilon$-close in trace distance, then they are $\varepsilon$-close in fidelity.*
2. *Show that if two states $\rho,\sigma \in \mathfrak{D}(A)$ are $\varepsilon$-close in fidelity, then they are $\sqrt{2\varepsilon}$-close in fidelity.*

The relation between the trace distance and the fidelity can also be used to derive some additional bounds on the trace distance. For example, consider a *pure* state $\rho^{AB}$ whose marginal (mixed) state $\rho^A$ is $\varepsilon$-close to a pure state $\psi^A$. Since $\psi^A$ is pure, this means that the marginal state $\rho^A$ is itself close to being pure, and this in turn means that the pure state $\rho^{AB}$ should be close to a produce state $\psi^A \otimes \rho^B$. We make this intuition rigorous in the following lemma.

> **Lemma 5.7.** Let $\rho \in \mathrm{Pure}(AB)$ be a *pure* state and let $\psi \in \mathrm{Pure}(A)$ be another pure state. If the marginal of $\rho^{AB}$ satisfies $T\left(\rho^A, \psi^A\right) \leqslant \varepsilon$, then
>
> $$T\left(\rho^{AB}, \psi^A \otimes \rho^B\right) \leqslant 2\sqrt{2\varepsilon}. \tag{5.208}$$

**Proof**   From the relation (5.202)

$$F(\rho^A, \psi^A) \geqslant 1 - T\left(\rho^A, \psi^A\right) \geqslant 1 - \varepsilon. \tag{5.209}$$

From Uhlmann's theorem there exists a pure state $|\phi\rangle \in B$ such that

$$F(\rho^A, \psi^A) = F(\rho^{AB}, \psi^A \otimes \phi^B). \tag{5.210}$$

Hence, applying again (5.202) gives

$$
\begin{aligned}
T\left(\rho^{AB}, \psi^A \otimes \phi^B\right) &\leqslant \sqrt{1 - F\left(\rho^{AB}, \psi^A \otimes \phi^B\right)^2} \\
&= \sqrt{1 - F(\rho^A, \psi^A)^2} \\
\textbf{(5.209)} \rightarrow \;\; &\leqslant \sqrt{1 - (1 - \varepsilon)^2} \leqslant \sqrt{2\varepsilon}.
\end{aligned}
\tag{5.211}
$$

From the monotonicity of the trace distance we also have

$$T\left(\rho^B, \phi^B\right) \leqslant T\left(\rho^{AB}, \psi^A \otimes \phi^B\right) \leqslant \sqrt{2\varepsilon}. \tag{5.212}$$

Hence, from the triangle inequality of the trace distance we get

$$
\begin{aligned}
T\left(\rho^{AB}, \psi^A \otimes \rho^B\right) &\leqslant T\left(\rho^{AB}, \psi^A \otimes \phi^B\right) + T\left(\psi^A \otimes \phi^B, \psi^A \otimes \rho^B\right) \\
&= T\left(\rho^{AB}, \psi^A \otimes \phi^B\right) + T\left(\phi^B, \rho^B\right) \\
&\leqslant \sqrt{2\varepsilon} + \sqrt{2\varepsilon} = 2\sqrt{2\varepsilon}.
\end{aligned}
\tag{5.213}
$$

This completes the proof.      ∎

**Exercise 5.42.** *Using the same notations as in the theorem above, suppose $F(\rho^A, \psi^A) \geqslant 1 - \varepsilon$. What is the best lower bound that you can find for $F(\rho^{AB}, \psi^A \otimes \rho^B)$?*

## The Gentle Measurement Lemma

We now use the relationship between the trace distance and fidelity to prove an intuitive phenomenon related to the disturbance of states under quantum measurements. Specifically, let $\rho \in \mathfrak{D}(A)$, and let $\Lambda \in \mathrm{Eff}(A)$ be one element of a POVM (i.e. $\Lambda$ is an effect). From Born's rule, the quantity $\mathrm{Tr}[\rho\Lambda]$ can be interpreted as the probability that the outcome associated with $\Lambda$ will occur. The gentle measurement lemma asserts that if this probability is high then the post-measurement state will not change much and remain very close to $\rho$. This fundamental property lies at the heart of several applications of quantum information and quantum resource theories.

---

**Lemma 5.8.** Let $\varepsilon \in (0, 1)$, $\rho \in \mathfrak{D}(A)$, and $\Lambda \in \text{Eff}(A)$. Suppose $\text{Tr}[\rho\Lambda] \geqslant 1-\varepsilon$. Then, the post-measurement state is $\sqrt{\varepsilon}$-close to $\rho$; that is,

$$\frac{1}{2}\|\rho - \tilde{\rho}\|_1 \leqslant \sqrt{\varepsilon} \quad \text{where} \quad \tilde{\rho} := \frac{\sqrt{\Lambda}\rho\sqrt{\Lambda}}{\text{Tr}[\Lambda\rho]}. \tag{5.214}$$

---

**Proof**    Let $|\psi^{AR}\rangle$ be a purification of $\rho^A$, and observe that (see Exercise 5.43)

$$|\tilde{\psi}^{AR}\rangle := \frac{\sqrt{\Lambda} \otimes I^R |\psi^{AR}\rangle}{\sqrt{\text{Tr}[\rho\Lambda]}} \tag{5.215}$$

is a purification of $\tilde{\rho}^A$. From Uhlmann's theorem, and the fact that $\Lambda \leqslant \sqrt{\Lambda}$ for any effect $0 \leqslant \Lambda \leqslant I^A$ we have

$$F(\rho, \tilde{\rho}) \geqslant \left|\langle\psi^{AR}|\tilde{\psi}^{AR}\rangle\right| = \frac{\langle\psi^{AR}|\sqrt{\Lambda} \otimes I^R|\psi^{AR}\rangle}{\sqrt{\text{Tr}[\rho\Lambda]}} \geqslant \frac{\langle\psi^{AR}|\Lambda \otimes I^R|\psi^{AR}\rangle}{\sqrt{\text{Tr}[\rho\Lambda]}}$$

$$= \frac{\text{Tr}[\rho\Lambda]}{\sqrt{\text{Tr}[\rho\Lambda]}} = \sqrt{\text{Tr}[\rho\Lambda]} \geqslant \sqrt{1-\varepsilon}. \tag{5.216}$$

Therefore, from the relation (5.202) between the trace distance and the fidelity we conclude that

$$\frac{1}{2}\|\rho - \tilde{\rho}\|_1 \leqslant \sqrt{1 - F(\rho, \tilde{\rho})^2} \leqslant \sqrt{1 - (\sqrt{1-\varepsilon})^2} = \sqrt{\varepsilon}. \tag{5.217}$$

$\blacksquare$

**Exercise 5.43.** *Prove that the state defined in (5.215) is indeed a purification of $\tilde{\rho}^A$.*

## 5.5 Distance between Subnormalized States

Subnormalized states are positive semidefinite matrices with a trace less than or equal to 1. These states, acting on a Hilbert space $A$, are denoted as

$$\mathfrak{D}_{\leqslant}(A) := \left\{\rho \in \text{Pos}(A): \text{Tr}[\rho] \leqslant 1\right\}. \tag{5.218}$$

Note that $\mathfrak{D}(A) \subset \mathfrak{D}_{\leqslant}(A)$.

Although subnormalized states do not represent physical systems, they arise in quantum measurements. For instance, consider a quantum state $\rho \in \mathfrak{D}(A)$ and a quantum instrument $\mathcal{E} = \sum_{x \in [m]} \mathcal{E}_x^{A \to B} \otimes |x\rangle\langle x|^X \in \text{CPTP}(A \to BX)$, as discussed in Section 3.5.10. The quantum state $\mathcal{E}(\rho)$ is given by

$$\mathcal{E}^{A \to BX}(\rho^A) = \sum_{x \in [m]} \mathcal{E}_x^{A \to B}(\rho^A) \otimes |x\rangle\langle x|^X, \tag{5.219}$$

where $\{\mathcal{E}_x\}_{x \in [m]}$ are trace nonincreasing CP maps, and $\{\mathcal{E}_x(\rho)\}_{x \in [m]}$ are subnormalized states. These states provide both the information about the probability $p_x := \mathrm{Tr}[\mathcal{E}_x(\rho)]$ that an outcome $x \in [m]$ occurs during the quantum measurement, and the post-measurement state $\frac{1}{p_x}\mathcal{E}_x(\rho)$.

We previously saw that distance measures for normalized states are monotonic under quantum channels and satisfy the DPI, a crucial aspect in applications. Quantum channels map normalized states with normalized states, while trace nonincreasing (TNI) CP maps, including CPTP maps, take subnormalized states to subnormalized states. Therefore, it's beneficial to define a distance measure for subnormalized states that is monotonic under TNI-CP maps. We denote by $\mathrm{CP}_{\leqslant}(A \to B)$ the set of all TNI maps in $\mathrm{CP}(A \to B)$.

In Section 5.3, we explored extending divergences from the classical to the quantum domain. We now apply a similar approach to extend divergences from normalized to sub-normalized states. However, unlike classical-to-quantum extensions, we will see that there is no analogous "minimal" extension from normalized to subnormalized states. Thus, we begin by introducing the maximal extension of a quantum divergence to the subnormalized domain.

---

**The Maximal Extension**

**Definition 5.6.** Let $\mathbb{D}$ be a quantum divergence. The maximal extension of $\mathbb{D}$ to subnormalized states, $\overline{\mathbb{D}}$, is defined for any $\rho, \sigma \in \mathfrak{D}_{\leqslant}(A)$ by

$$\overline{\mathbb{D}}(\rho\|\sigma) := \inf \mathbb{D}(\tilde{\rho}\|\tilde{\sigma}), \tag{5.220}$$

where the infimum is over all systems $R$, and all density matrices $\tilde{\rho}, \tilde{\sigma} \in \mathfrak{D}(R)$ for which there exists $\mathcal{E} \in \mathrm{CP}_{\leqslant}(R \to A)$ such that

$$\rho = \mathcal{E}(\tilde{\rho}) \quad \text{and} \quad \sigma = \mathcal{E}(\tilde{\sigma}). \tag{5.221}$$

---

*Remark.* Note that earlier we used the same notation $\overline{\mathbb{D}}$ to denote the maximal extension of a classical divergence to a quantum one. The bar symbol over $\mathbb{D}$ in our notations will always indicate maximal extensions from one domain to a larger one, whereas the domain of a given extension should be clear from the context.

The maximal extension $\overline{\mathbb{D}}$ have the following three properties:

1. *Reduction.* For any $\rho, \sigma \in \mathfrak{D}(A)$ we have

$$\overline{\mathbb{D}}(\rho\|\sigma) = \mathbb{D}(\rho\|\sigma). \tag{5.222}$$

2. *Monotonicity.* For any $\mathcal{E} \in \mathrm{CP}_{\leqslant}(A \to B)$ and any subnormalized states $\rho, \sigma \in \mathfrak{D}_{\leqslant}(A)$,

$$\overline{\mathbb{D}}\big(\mathcal{E}(\rho)\|\mathcal{E}(\sigma)\big) \leqslant \overline{\mathbb{D}}(\rho\|\sigma). \tag{5.223}$$

3. *Optimality.* If $f : \mathfrak{D}_{\leqslant}(A) \times \mathfrak{D}_{\leqslant}(A) \to \mathbb{R}_+$ is a function that reduces to $\mathbb{D}$ when restricted to normalized states, and behaves monotonically under TNI-CP maps (as in (5.223)), then

$$f(\rho\|\sigma) \leqslant \overline{\mathbb{D}}(\rho\|\sigma) \qquad \forall\, \rho, \sigma \in \mathfrak{D}_{\leqslant}(A). \tag{5.224}$$

The last property justifies the name for $\overline{\mathbb{D}}$ as the maximal extension of $\mathbb{D}$ to subnormalized states.

**Exercise 5.44.** *Prove the three properties above using the same techniques that were used to prove Theorem 6.4. Hint: As you follow the same lines used in the proof of Theorem 6.4, replace "classical states" with "normalized quantum states" and "quantum states" with "subnormalized quantum states."*

Remarkably, the maximal extension has the following closed formula.

> **Closed Formula**
>
> **Theorem 5.17.** Let $\mathbb{D}$ be a quantum divergence and $\overline{\mathbb{D}}$ be its maximal extension to subnormalized states as defined in (5.220). For any pair of subnormalized states $\rho, \sigma \in \mathfrak{D}_{\leqslant}(A)$,
>
> $$\overline{\mathbb{D}}(\rho\|\sigma) = \mathbb{D}\big(\rho \oplus (1 - \mathrm{Tr}[\rho]) \big\| \sigma \oplus (1 - \mathrm{Tr}[\sigma])\big). \tag{5.225}$$

**Proof** Let $\tilde{\rho}, \tilde{\sigma} \in \mathfrak{D}(R)$ and $\mathcal{E} \in \mathrm{CP}_{\leqslant}(R \to A)$ be a TNI-CP map such that $\rho = \mathcal{E}(\tilde{\rho})$ and $\sigma = \mathcal{E}(\tilde{\sigma})$. Moreover, define $\mathcal{N} \in \mathrm{CPTP}(R \to A \oplus \mathbb{C})$ as

$$\mathcal{N}(\omega) := \mathcal{E}(\omega) \oplus \big(\mathrm{Tr}[\omega] - \mathrm{Tr}[\mathcal{E}(\omega)]\big) \qquad \forall\, \omega \in \mathfrak{L}(A). \tag{5.226}$$

Then, since $\mathcal{N}$ is a CPTP map,

$$\begin{aligned}
\mathbb{D}(\tilde{\rho}\|\tilde{\sigma}) &\geqslant \mathbb{D}\big(\mathcal{N}(\tilde{\rho})\|\mathcal{N}(\tilde{\sigma})\big) \\
&= \mathbb{D}\big(\mathcal{E}(\tilde{\rho}) \oplus (1 - \mathrm{Tr}[\mathcal{E}(\tilde{\rho})]) \big\| \mathcal{E}(\tilde{\sigma}) \oplus (1 - \mathrm{Tr}[\mathcal{E}(\tilde{\sigma})])\big) \\
&= \mathbb{D}\big(\rho \oplus (1 - \mathrm{Tr}[\rho]) \big\| \sigma \oplus (1 - \mathrm{Tr}[\sigma])\big).
\end{aligned} \tag{5.227}$$

Since the above inequality holds for all such $\tilde{\rho}, \tilde{\sigma}, \mathcal{E}$ we must have that $\overline{\mathbb{D}}(\tilde{\rho}\|\tilde{\sigma})$ is no smaller than the right-hand side on (5.225). To prove the converse inequality, take $R = A \oplus \mathbb{C}$, $\tilde{\rho} = \rho \oplus (1 - \mathrm{Tr}[\rho])$, $\tilde{\sigma} = \sigma \oplus (1 - \mathrm{Tr}[\sigma])$, and $\mathcal{E}(\cdot) := P(\cdot)P^{\dagger}$, where $P$ is the projection to the subspace $A$ in $R$. Then, $\rho = \mathcal{E}(\tilde{\rho})$ and $\sigma = \mathcal{E}(\tilde{\sigma})$ so that by definition (see (5.220)) we must have $\overline{\mathbb{D}}(\rho\|\sigma) \leqslant \mathbb{D}(\tilde{\rho}\|\tilde{\sigma})$. Together with the previous inequality, this completes the proof of the equality in (5.225). $\blacksquare$

If $\mathbb{D}$ is a quantum divergence, its minimal extension $\underline{\mathbb{D}}$ can be defined in analogy with (5.91) as

$$\underline{\mathbb{D}}(\rho\|\sigma) := \sup \mathbb{D}(\mathcal{E}(\rho)\|\mathcal{E}(\sigma)) \qquad \forall\, \rho, \sigma \in \mathfrak{D}_{\leqslant}(A), \tag{5.228}$$

where the supremum is over all systems $R$ and all $\mathcal{E} \in \mathrm{CP}_{\leqslant}(A \to R)$ such that $\mathcal{E}(\rho)$ and $\mathcal{E}(\sigma)$ are normalized states. However, such $\mathcal{E}$ does not exist if either $\rho$ or $\sigma$ has trace strictly smaller than 1. Hence, the minimal extension of $\mathbb{D}$ must satisfy

$$\underline{\mathbb{D}}(\rho\|\sigma) = 0 \tag{5.229}$$

for all subnormalized states $\rho, \sigma \in \mathfrak{D}_{\leqslant}(A)$ with either $\mathrm{Tr}[\rho] < 1$ or $\mathrm{Tr}[\sigma] < 1$. Therefore, this extension is rather pathological and not useful in applications. The following corollary applies specifically to the case where $\mathbb{D}$ functions as both a divergence and a metric.

> **Corollary 5.6.** Let $\mathbb{D}$ be a quantum divergence that is also a metric. Then, its maximal extension to subnormalized states, $\overline{\mathbb{D}}$, is also a metric.

**Proof** We need to show that $\overline{\mathbb{D}}$ is symmetric and satisfies the trumping. To see it, let $\rho, \sigma, \omega \in \mathfrak{D}_{\leqslant}(A)$. The symmetry of $\overline{\mathbb{D}}$ follows from the symmetry of $\mathbb{D}$:

$$\begin{aligned}
\overline{\mathbb{D}}(\rho\|\sigma) &= \mathbb{D}\big(\rho \oplus (1 - \mathrm{Tr}[\rho]) \,\big\|\, \sigma \oplus (1 - \mathrm{Tr}[\sigma])\big) \\
\mathbb{D} \text{ is symmetric} \to &= \mathbb{D}\big(\sigma \oplus (1 - \mathrm{Tr}[\sigma])\big\|\rho \oplus (1 - \mathrm{Tr}[\rho])\big) \\
&= \overline{\mathbb{D}}(\sigma\|\rho).
\end{aligned} \tag{5.230}$$

Similarly, the triangle inequality of $\overline{\mathbb{D}}$ follows from the triangle inequality of $\mathbb{D}$:

$$\begin{aligned}
\overline{\mathbb{D}}(\rho\|\sigma) &= \mathbb{D}\big(\rho \oplus (1 - \mathrm{Tr}[\rho])\big\|\sigma \oplus (1 - \mathrm{Tr}[\sigma])\big) \\
&\leqslant \mathbb{D}\big(\rho \oplus (1 - \mathrm{Tr}[\rho])\big\|\omega \oplus (1 - \mathrm{Tr}[\omega])\big) + \mathbb{D}\big(\omega \oplus (1 - \mathrm{Tr}[\omega])\big\|\sigma \oplus (1 - \mathrm{Tr}[\sigma])\big) \\
&= \overline{\mathbb{D}}(\rho\|\omega) + \overline{\mathbb{D}}(\omega\|\sigma).
\end{aligned}$$

$$\tag{5.231}$$

∎

## 5.5.1 Examples

### The Generalized Trace Distance

The maximal extension of the trace distance to subnormalized states is known as the *genaralized* trace distance. From Theorem 5.17 we get that the generalized trace distance has the following simple form.

> **Corollary 5.7.** The generalized trace distance can be expressed for any $\rho, \sigma \in \mathfrak{D}_{\leqslant}(A)$ as
>
> $$\overline{T}(\rho, \sigma) = \frac{1}{2}\|\rho - \sigma\|_1 + \frac{1}{2}\big|\mathrm{Tr}[\rho - \sigma]\big|. \tag{5.232}$$

**Exercise 5.45.** *Prove the corollary above using the formula given in Theorem 5.17 when $\mathbb{D}$ is replaced by the trace distance.*

*Remark.* The generalized trace distance can also be expressed as

$$\overline{T}(\rho,\sigma) = \max\left\{\mathrm{Tr}(\rho-\sigma)_+, \mathrm{Tr}(\rho-\sigma)_-\right\}. \tag{5.233}$$

To see why, set $a := \mathrm{Tr}(\rho-\sigma)_+$ and $b := \mathrm{Tr}(\rho-\sigma)_-$, and use the relation $\max\{a,b\} = \frac{1}{2}(a+b+|a-b|)$. The formula (5.233) is consistent with the fact that the trace distance is the largest extension of the trace distance to subnormalized states that satisfies the monotonicity property under TNI-CP maps.

**Exercise 5.46.** *Show that for any $\rho,\sigma \in \mathfrak{D}_{\leqslant}(A)$ the function $f(\rho,\sigma) = \frac{1}{2}\|\rho-\sigma\|_1$ is also an extension of D to subnormalized states that satisfies the exact same properties satisfied by $\overline{T}$ except for the optimality. Give an example showing that $f(\rho,\sigma)$ can be strictly smaller than $\overline{T}(\rho,\sigma)$.*

**Exercise 5.47.** *Show that for two subnormalized pure states $\psi,\phi \in \mathfrak{D}_{\leqslant}(A)$ the generalized trace distance can be expressed as*

$$\overline{T}(\psi,\phi) = \sqrt{\frac{1}{4}(\mathrm{Tr}[\psi+\phi])^2 - |\langle\psi|\phi\rangle|^2} + \frac{1}{2}|\mathrm{Tr}[\psi-\phi]|. \tag{5.234}$$

*Hint: Use similar techniques as in Exercise 5.26.*

## The Gentle Operator Lemma

The gentle operator lemma is a variant of the gentle measurement lemma (Lemma 5.8) in which the post-measurement state is taken to be unnormalized. Specifically, in the following Lemma 5.9 we let $\varepsilon \in (0,1)$, $\rho \in \mathfrak{D}(A)$, $\Lambda \in \mathrm{Eff}(A)$, and consider the subnormalized state $\tilde{\rho} := \sqrt{\Lambda}\rho\sqrt{\Lambda}$.

**Lemma 5.9.** Using the notations above, if $\mathrm{Tr}[\rho\Lambda] \geqslant 1 - \varepsilon$, then $\frac{1}{2}\|\rho-\tilde{\rho}\|_1 \leqslant \sqrt{\varepsilon}$.

*Remark.* The gentle operator lemma's extension to cases where $\tilde{\rho} = G\rho G^*$, with $G \in \mathfrak{L}(A,B)$ being an arbitrary element of a generalized measurement and $\Lambda \equiv G^*G \in \mathrm{Eff}(A)$, may seem promising. However, without imposing further constraints on $G$, such an extension could result in noninformative bounds. Consider, for instance, the scenario where $G$ is a unitary matrix, making $\Lambda = G^*G = I^A$. Here, $\mathrm{Tr}[\Lambda\rho] = 1 \geqslant 1-\varepsilon$ for any $\varepsilon \geqslant 0$. But, if we choose $\rho = |0\rangle\langle0|$ and a unitary $G$ such that $G|0\rangle = |1\rangle$, it follows that $\frac{1}{2}\|\rho-\tilde{\rho}\|_1 = 1$. This example illustrates that extending the gentle operator lemma to encompass arbitrary elements of generalized measurements is impractical without specific additional constraints on $G$.

**Proof**   Let $|\psi^{A\tilde{A}}\rangle = \sqrt{\rho} \otimes I^{\tilde{A}}|\Omega^{A\tilde{A}}\rangle$ and $|\tilde{\psi}^{A\tilde{A}}\rangle := \sqrt{\Lambda} \otimes I^{\tilde{A}}|\psi^{A\tilde{A}}\rangle$ be purifications $\rho^A$ and $\tilde{\rho}^A$, respectively. Denote by $t := \mathrm{Tr}[\rho\Lambda] \geqslant 1 - \varepsilon$ and observe that

$$\langle\psi^{A\tilde{A}}|\tilde{\psi}^{A\tilde{A}}\rangle = \langle\psi^{A\tilde{A}}|\sqrt{\Lambda} \otimes I^{\tilde{A}}|\psi^{A\tilde{A}}\rangle$$

$$\boxed{\sqrt{\Lambda} \geqslant \Lambda} \longrightarrow \; \geqslant \langle\psi^{A\tilde{A}}|\Lambda \otimes I^{\tilde{A}}|\psi^{A\tilde{A}}\rangle \tag{5.235}$$

$$\boxed{|\psi^{A\tilde{A}}\rangle = \sqrt{\rho} \otimes I^{\tilde{A}}|\Omega^{A\tilde{A}}\rangle} \longrightarrow \; = \mathrm{Tr}[\rho\Lambda] = t.$$

Due to the DPI of the generalized trace distance it follows that

$$\overline{T}(\rho^A, \tilde{\rho}^A) \leqslant \overline{T}(\psi^{A\tilde{A}}, \tilde{\psi}^{A\tilde{A}}). \tag{5.236}$$

From (5.232) we get that both sides on the equation above contain the same term $\frac{1}{2}|\mathrm{Tr}[\rho - \tilde{\rho}]| = \frac{1}{2}|\mathrm{Tr}[\psi - \tilde{\psi}]|$, so we can cancel it. Combining this with Exercise 5.47 we conclude that

$$\frac{1}{2}\|\rho - \tilde{\rho}\|_1 \leqslant \sqrt{\frac{1}{4}\left(\mathrm{Tr}[\psi + \tilde{\psi}]\right)^2 - |\langle\psi|\tilde{\psi}\rangle|^2}$$

$$\textbf{(5.235)}\rightarrow \; \leqslant \sqrt{\frac{1}{4}(1 + t)^2 - t^2} \tag{5.237}$$

$$\textbf{Exercise 5.48}\rightarrow \; \leqslant \sqrt{1 - t}$$

$$\boxed{t \geqslant 1 - \varepsilon} \longrightarrow \; \leqslant \sqrt{\varepsilon}.$$

This completes the proof.      ∎

**Exercise 5.48.** *Show that the function* $f(t) := \sqrt{\frac{1}{4}(1 + t)^2 - t^2}$ *is smaller than* $\sqrt{1 - t}$ *for all* $t \in [0, 1]$.

**Exercise 5.49.** *Show that one can use the gentle measurement lemma (Lemma 5.8) to prove a slightly weaker version of the gentle operator lemma (Lemma 5.9). Use only Lemma 5.8 and the triangle inequality of the trace norm to show that*

$$\frac{1}{2}\|\rho - \tilde{\rho}\|_1 \leqslant \sqrt{\varepsilon} + \frac{1}{2}\varepsilon. \tag{5.238}$$

*Hint: Set* $\rho' := \frac{\sqrt{\Lambda}\rho\sqrt{\Lambda}}{\mathrm{Tr}[\Lambda\rho]}$ *and write* $\rho - \tilde{\rho} = \rho - \rho' + \rho' - \tilde{\rho}$.

## The Generalized Fidelity

We can use the techniques already developed to extend the fidelity to subnormalized states. However, since the fidelity achieves its maximum for identical states, the infimum of (5.220) will be replaced with a supremum.

> **Definition 5.7.** Let $\rho, \sigma \in \mathfrak{D}_{\leqslant}(A)$ be two subnormalized states. We define the *generalized fidelity* $\underline{F}: \mathfrak{D}_{\leqslant}(A) \times \mathfrak{D}_{\leqslant}(A) \to \mathbb{R}_+$ to be
>
> $$\underline{F}(\rho, \sigma) := \sup F(\tilde{\rho}, \tilde{\sigma}), \tag{5.239}$$
>
> where the supremum is over all systems $R$, and all density matrices $\tilde{\rho}, \tilde{\sigma} \in \mathfrak{D}(R)$ for which there exists $\mathcal{E} \in \mathrm{CP}_{\leqslant}(R \to A)$ with the property that $\rho = \mathcal{E}(\tilde{\rho})$ and $\sigma = \mathcal{E}(\tilde{\sigma})$.

Since $1 - F(\rho, \sigma)$ is a quantum divergence we can use Theorem 5.17 to get a closed formula for the generalized fidelity.

> **Corollary 5.8.** The generalized fidelity can be expressed as
>
> $$\underline{F}(\rho, \sigma) = \|\sqrt{\rho}\sqrt{\sigma}\|_1 + \sqrt{(1 - \mathrm{Tr}[\rho])(1 - \mathrm{Tr}[\sigma])} \qquad \forall\, \rho, \sigma \in \mathfrak{D}_{\leqslant}(A). \tag{5.240}$$

*Remark.* The formula (5.240) for the generalized fidelity reveals that we have $\underline{F}(\rho, \sigma) = \|\sqrt{\rho}\sqrt{\sigma}\|_1$ even if only one of the states is normalized. Note that for the trace distance $\overline{T}(\rho, \sigma) = T(\rho, \sigma)$ only if both states are normalized.

The generalized fidelity has the following properties:

1. *Reduction.* For any $\rho, \sigma \in \mathfrak{D}(A)$ we have $\underline{F}(\rho, \sigma) = F(\rho, \sigma)$.
2. *Monotonicity.* For every $\mathcal{E} \in \mathrm{CP}_{\leqslant}(A \to B)$ and every subnormalized states $\rho, \sigma \in \mathfrak{D}_{\leqslant}(A)$ we have $\underline{F}\big(\mathcal{E}(\rho), \mathcal{E}(\sigma)\big) \leqslant \underline{F}(\rho, \sigma)$.
3. *Faithfulness.* For every $\rho, \sigma \in \mathfrak{D}_{\leqslant}(A)$, $\underline{F}(\rho, \sigma) = 1 \iff \rho = \sigma$.
4. *Symmetry.* For any $\rho, \sigma \in \mathfrak{D}_{\leqslant}(A)$, $\underline{F}(\rho, \sigma) = \underline{F}(\sigma, \rho)$.
5. *Optimality.* If $f: \mathfrak{D}_{\leqslant}(A) \times \mathfrak{D}_{\leqslant}(A) \to \mathbb{R}_+$ is a function that reduces to $F$ when restricted to normalized states, and behaves monotonically under TNI-CP maps, then $f(\rho, \sigma) \geqslant \underline{F}(\rho, \sigma)$ for all $\rho, \sigma \in \mathfrak{D}_{\leqslant}(A)$.

The last property indicates that the generalized fidelity is the minimal extension of the fidelity to subnormalized states.

**Exercise 5.50.** *Prove the above five properties of the generalized fidelity.*

**Exercise 5.51.** *Show that for two subnormalized pure states $\psi \in \mathfrak{D}_{\leqslant}(A)$ and $\phi \in \mathfrak{D}_{\leqslant}(A)$ the generalized fidelity is given by*

$$\underline{F}(\psi, \phi) = |\langle \psi | \phi \rangle| + \sqrt{(1 - \langle \psi | \psi \rangle)(1 - \langle \phi | \phi \rangle)}. \tag{5.241}$$

Generalization of Uhlmann's theorem

**Theorem 5.18.** Let $\rho, \sigma \in \mathfrak{D}_{\leqslant}(A)$ be two subnormalized states, and let $|\psi^{AB}\rangle$ and $|\phi^{AC}\rangle$ be two purifications of $\rho^A$ and $\sigma^A$, respectively. Suppose also that $|B| \leqslant |C|$. Then,

$$\underline{F}\left(\rho^A, \sigma^A\right) = \max_{\mathcal{V}^{B \to C}} \underline{F}\left(\mathcal{V}^{B \to C}\left(\psi^{AB}\right), \phi^{AC}\right), \tag{5.242}$$

where the maximum is over all CPTP maps $\mathcal{V}(\cdot) := V(\cdot)V^*$, where $V : B \to C$ is an isometry.

**Proof**   Note first that

$$\mathrm{Tr}\left[\mathcal{V}^{B \to C}\left(\psi^{AB}\right)\right] = \mathrm{Tr}\left[\psi^{AB}\right] = \mathrm{Tr}[\rho^A], \tag{5.243}$$

and similarly $\mathrm{Tr}\left[\mathcal{V}^{B \to C}\left(\phi^{AB}\right)\right] = \mathrm{Tr}[\sigma^A]$. Therefore, it is sufficient to show that

$$\left\|\sqrt{\rho}\sqrt{\sigma}\right\|_1 = \max_{\mathcal{V}^{B \to C}} \left\|\sqrt{\mathcal{V}^{B \to C}\left(\psi^{AB}\right)}\sqrt{\phi^{AC}}\right\|_1. \tag{5.244}$$

Since $\mathcal{V}^{B \to C}\left(\psi^{AB}\right)$ and $\phi^{AC}$ are rank 1 subnormalized states we have (Exercise (5.52))

$$\left\|\sqrt{\mathcal{V}^{B \to C}\left(\psi^{AB}\right)}\sqrt{\phi^{AC}}\right\|_1 = \left|\langle\psi^{AB}|I^A \otimes V^*|\phi^{AC}\rangle\right|. \tag{5.245}$$

Hence, the rest of the proof follows the exact same lines as in the proof of Uhlmann's theorem (Theorem 5.14). In particular, note that all the steps in (5.177) hold even if $\rho$ and $\sigma$ are subnormalized. $\blacksquare$

**Exercise 5.52.** *Prove the equality in* (5.245).

## 5.6 The Purified Distance

The trace distance and fidelity are both highly valuable in numerous applications. The trace distance benefits from being a metric, especially due to its compliance with the triangle inequality. Conversely, the fidelity's advantage lies in its compatibility with Uhlmann's theorem, allowing the simplification of expressions involving fidelity through the use of quantum state purifications. Given the desirability of both these properties in quantum information applications (which will be further explored in this book), it naturally leads to the question: Is there a distance measure that embodies both qualities? As we will discuss now, such a measure does indeed exist.

For any two pure state $\psi, \phi \in \mathrm{Pure}(A)$ the trace distance is given by (see Exercise 5.26)

$$T(\psi, \phi) = \sqrt{1 - |\langle\psi|\phi\rangle|^2} = \sqrt{1 - F(\psi, \phi)^2}. \tag{5.246}$$

Considering this close relationship between trace distance and fidelity when applied to pure states, we will explore all possible extensions of the trace distance from pure states to mixed states. In this context, we define the purified distance as the maximal extension among all such extensions.

> **Definition 5.8.** Let $T$ be the trace distance. The purified distance is defined for all $\rho, \sigma \in \mathfrak{D}(A)$ as
>
> $$P(\rho,\sigma) := \inf_{\psi,\phi\in\text{Pure}(R)} \left\{ T(\psi,\phi) : \rho = \mathcal{E}(\psi),\ \sigma = \mathcal{E}(\phi),\ \mathcal{E} \in \text{CPTP}(R \to A) \right\},$$
>
> $$(5.247)$$
>
> where the infimum is also over all systems $R$.

*Remarks*

1. Observe that the extension of the trace distance in the definition above is reminiscent to the maximal quantum extension of the trace distance that was discussed in the previous sections. Later on we will develop a framework to extend certain functions (specifically resource monotones) from one domain to a larger one. This framework is very general and all extensions discussed in this chapter (including the above extension of the trace distance, that is, the purified distance) are just specific applications of the framework.

2. We will see that the purified distance has a closed formula. Historically, this closed formula has been used as its definition. However, the definition above emphasizes its operational meaning as the largest mixed-state extension of the trace distance (see Theorem 5.19).

3. The justification for the name "purified distance" will become clear from the properties discussed in this section.

We start by showing that the purified distance is an optimal divergence.

> **Theorem 5.19.** The purified distance is a quantum divergence that reduces to the trace distance on pure states. Moreover, if $\mathbb{D}$ is another quantum divergence that reduces to the trace distance on pure states, then for any $\rho, \sigma \in \mathfrak{D}(A)$ we have
>
> $$\mathbb{D}(\rho\|\sigma) \leqslant P(\rho,\sigma). \qquad (5.248)$$

The proof follows very similar lines as in the proof of Theorem 6.4 and is left as an exercise.

**Exercise 5.53.** *Prove Theorem (5.19). Hint: Adopt the methodology used in Theorem 6.4 related to $\overline{\mathbb{D}}$. In this process, substitute each occurrence of a classical state on system X with a pure state on system R.*

The upcoming lemma demonstrates that the purified distance is derived from a purification process, which justifies its name. We will utilize this lemma to derive a closed formula for the purified distance.

---

**Lemma 5.10.** Let $P$ be the purified distance and $T$ the trace distance. Then, for all $\rho, \sigma \in \mathfrak{D}(A)$

$$P(\rho^A, \sigma^A) = \inf_{\psi, \phi} T(\psi^{AB}, \phi^{AB}), \qquad (5.249)$$

where the infimum is over all purifications of $\rho^A$ and $\sigma^A$.

---

**Proof**  Let $\psi^{AB}$ and $\phi^{AB}$ be purifications of $\rho^A$ and $\sigma^A$, respectively, and denote by $\mathcal{E}^{AB \to A} := \mathrm{Tr}_B$. By definition, $\mathcal{E}^{AB \to A}(\psi^{AB}) = \rho^A$ and $\mathcal{E}^{AB \to A}(\phi^{AB}) = \sigma^A$ so that $\psi^{AB}$ and $\phi^{AB}$ satisfies the conditions in (5.247) with $R := AB$. Therefore, $P(\rho^A, \sigma^A)$ cannot be greater than the right-hand side of (5.249). To get the other direction, recall the definition (5.247) and let $\rho^A = \mathcal{E}(\psi^R)$ and $\sigma^A = \mathcal{E}(\phi^R)$ for some $\psi, \phi \in \mathrm{Pure}(R)$ and $\mathcal{E} \in \mathrm{CPTP}(R \to A)$. Let $\mathcal{V} \in \mathrm{CPTP}(R \to AB)$ be the isometry purifying $\mathcal{E}^{R \to A}$. Therefore, $\rho^A = \mathrm{Tr}_B\left[\mathcal{V}^{R \to AB}\left(\psi^R\right)\right]$ and $\sigma^A = \mathrm{Tr}_B\left[\mathcal{V}^{R \to AB}\left(\phi^R\right)\right]$. Finally, since the trace distance is invariant under isometries, denoting by $\chi^{AB} := \mathcal{V}^{R \to AB}\left(\psi^R\right)$ and $\varphi^{AB} := \mathcal{V}^{R \to AB}\left(\phi^R\right)$ we get

$$T(\psi^R, \phi^R) = T(\chi^{AB}, \varphi^{AB}) \geqslant \inf_{\psi', \phi'} T(\psi'^{AB}, \phi'^{AB}), \qquad (5.250)$$

where the infimum is over all purifications $\psi'^{AB}$ and $\phi'^{AB}$ of $\rho^A$ and $\sigma^A$. Hence, since $\psi^R$ and $\phi^R$ were arbitrary pure states that satisfy the conditions in (5.247), we conclude that $P(\rho^A, \sigma^A)$ is no smaller than the right-hand side of (5.249). This completes the proof. ∎

---

Closed Formula

**Theorem 5.20.** Let $P$ be the purified distance. Then, for all $\rho, \sigma \in \mathfrak{D}(A)$,

$$P(\rho, \sigma) = \sqrt{1 - F(\rho, \sigma)^2}. \qquad (5.251)$$

---

**Proof**  From Lemma 5.10 we get by direct computation

$$P(\rho, \sigma) = \min_{\psi, \phi} T(\psi, \phi)$$

$$\text{Exercise 5.26} \to \quad = \min_{\psi, \phi} \sqrt{1 - |\langle \psi | \phi \rangle|^2}$$

$$= \sqrt{1 - \max_{\psi, \phi} |\langle \psi | \phi \rangle|^2} \qquad (5.252)$$

$$\text{Uhlamnn's Theorem} \to \quad = \sqrt{1 - F(\rho, \sigma)^2}.$$

This completes the proof. ∎

The maximal extension of purified distance from density matrices to subnormalized states follows trivially from Theorem 5.17. We therefore extend the definition of the purified distance to subnormalized states in the following way.

> **The Purified Distance**
>
> **Definition 5.9.** Let $\rho, \sigma \in \mathfrak{D}_{\leqslant}(A)$ be two subnormalized states. The purified distance is defined as
>
> $$P(\rho, \sigma) := \sqrt{1 - \underline{F}(\rho, \sigma)^2}, \tag{5.253}$$
>
> where $\underline{F}$ is the generalized fidelity as given in (5.240).

*Remark.* The purified distance on normalized states has been defined earlier as the maximal extension of the trace distance from pure states to mixed states. Therefore, the purified distance on subnormalized states can be viewed as the maximal extension of the trace distance from pure states to mixed subnormalized states. Moreover, observe that the purified distance can also be expressed as

$$P(\rho, \sigma) := \sqrt{1 - F(\tilde{\rho}, \tilde{\sigma})^2}, \tag{5.254}$$

where $\tilde{\rho} := \rho \oplus (1 - \mathrm{Tr}[\rho])$ and $\tilde{\sigma} := \sigma \oplus (1 - \mathrm{Tr}[\sigma])$.

Finally, we show that the purified distance is a metric.

> **Theorem 5.21.** The purified distance is a metric on the set of subnormalized states.

**Proof** Since $\underline{F}(\rho, \sigma) \leqslant 1$ the purified distance is nonnegative. Since $\underline{F}(\rho, \sigma) = 1$ if and only if $\rho = \sigma$ the purified distance $P(\rho, \sigma) = 0$ if and only if $\rho = \sigma$. Since $\underline{F}$ is symmetric also $P$ is symmetric. It is therefore left to show that the purified distance satisfies the triangle inequality.

Let $\rho, \sigma, \omega \in \mathfrak{D}_{\leqslant}(A)$ and set $\tilde{\rho} := \rho \oplus (1 - \mathrm{Tr}[\rho])$, $\tilde{\sigma} := \sigma \oplus (1 - \mathrm{Tr}[\sigma])$, and $\tilde{\omega} := \omega \oplus (1 - \mathrm{Tr}[\omega])$. Moreover, let $\psi, \phi, \varphi \in \mathfrak{D}(B\tilde{B})$ be the purifications of $\tilde{\rho}$, $\tilde{\sigma}$, and $\tilde{\omega}$ such that $F(\tilde{\rho}, \tilde{\omega}) = F(\psi, \varphi)$ and $F(\tilde{\omega}, \tilde{\sigma}) = F(\varphi, \phi)$. Such purifications exist due to Uhlmann's theorem. Moreover, note that from the Uhlmann's theorem we also have $F(\tilde{\rho}, \tilde{\sigma}) \geqslant F(\psi, \phi)$. Hence,

$$P(\rho, \omega) + P(\omega, \sigma) = \sqrt{1 - F(\tilde{\rho}, \tilde{\omega})^2} + \sqrt{1 - F(\tilde{\omega}, \tilde{\sigma})^2}$$

$$= \sqrt{1 - F(\psi, \varphi)^2} + \sqrt{1 - F(\varphi, \phi)^2}$$

$$\textbf{(5.134)} \rightarrow = T(\psi, \varphi) + T(\varphi, \phi)$$

$$\textbf{Triangle inequality of } T \rightarrow \geqslant T(\psi, \phi) \tag{5.255}$$

$$\textbf{(5.134)} \rightarrow = \sqrt{1 - F(\psi, \phi)^2}$$

$$\boxed{F(\tilde{\rho}, \tilde{\sigma}) \geqslant F(\psi, \phi)} \longrightarrow \geqslant \sqrt{1 - F(\tilde{\rho}, \tilde{\sigma})^2}$$

$$= P(\rho, \sigma).$$

This completes the proof. ∎

Note that the purified distance is monotonic under TNI-CP maps. That is, for every map $\mathcal{E} \in \mathrm{CP}_{\leqslant}(A \to B)$, and any two subnormalized states $\rho, \sigma \in \mathfrak{D}_{\leqslant}(A)$ we have

$$P\big(\mathcal{E}(\rho), \mathcal{E}(\sigma)\big) \leqslant P(\rho, \sigma). \tag{5.256}$$

This follows trivially from Theorem 5.17 and the monotonicity property in (5.223) (or equivalently from the monotonicity of the generalized fidelity). Moreover, note that from Theorem 5.18 it follows that for any $\rho, \sigma \in \mathfrak{D}_{\leqslant}(A)$ and any purification $\psi^{AB}$ of $\rho^A$, there exists a purification $\phi^{AB}$ of $\sigma^A$ such that

$$P(\rho^A, \sigma^A) = P(\psi^{AB}, \phi^{AB}). \tag{5.257}$$

We end this subsection by showing that the purified distance is bounded by the generalized trace distance, $\overline{T}$.

---

**Theorem 5.22.** Let $\rho, \sigma \in \mathfrak{D}_{\leqslant}(A)$ be two subnormalized states. The generalized trace distance and the purified distance satisfy the following inequalities:

$$\overline{T}(\rho, \sigma) \leqslant P(\rho, \sigma) \leqslant \sqrt{2\overline{T}(\rho, \sigma)}. \tag{5.258}$$

---

**Proof**  Set $\tilde{\rho} = \rho \oplus (1 - \mathrm{Tr}\rho)$ and $\tilde{\sigma} = \sigma \oplus (1 - \mathrm{Tr}\sigma)$ and observe

$$\begin{aligned}
P(\rho, \sigma) &= \sqrt{1 - F(\tilde{\rho}, \tilde{\sigma})^2} \\
\mathbf{(5.202)} \to\ &\geqslant T(\tilde{\rho}, \tilde{\sigma}) = \overline{T}(\rho, \sigma).
\end{aligned} \tag{5.259}$$

To get the other inequality observe that

$$\begin{aligned}
P(\rho, \sigma)^2 &= 1 - F(\tilde{\rho}, \tilde{\sigma})^2 \\
\mathbf{(5.202)} \to\ &\leqslant 1 - \big(1 - T(\tilde{\rho}, \tilde{\sigma})\big)^2 \\
&\leqslant 2T(\tilde{\rho}, \tilde{\sigma}) \\
&= 2\overline{T}(\rho, \sigma).
\end{aligned} \tag{5.260}$$

This completes the proof.  ∎

**Exercise 5.54.** *Let $\rho, \sigma \in \mathfrak{D}_{\leqslant}(A)$ be two subnormalized states. Let $\tau \in \mathfrak{D}_{\leqslant}(AB)$ be an extension of $\rho^A$. That is, $\mathrm{Tr}_B[\tau^{AB}] = \rho^A$. Show that there exists an extension $\omega \in \mathfrak{D}_{\leqslant}(AB)$ of $\sigma^A$ such that*

$$P\left(\rho^A, \sigma^A\right) = P\left(\tau^{AB}, \omega^{AB}\right). \tag{5.261}$$

## 5.7 Notes and References

We followed the definition and basic properties of classical and quantum divergences as given in Refs. [99] and [98]. The classical $f$-Divergences goes back to the work in Ref. [193] followed by the independent works in Refs. [55], [167], and [5]. The quantum version of the $f$-Divergences are a special case of Petz' quasi-entropies defined in Ref. [181]. Extensive details on their properties, applications in quantum information, and additional references can be found in the papers in Refs. [121] and [120] (see also the appendix in Ref. [213] of online version for a similar derivation of the closed formula in Theorem 5.4). The maximal extension of the classical $f$-Divergence (Theorem D.2 of online version) is due to Ref. [163]. The trace distance and the fidelity of subnormalized states were introduced in [211] and later on developed further in Ref. [98].

# 6

# Entropies and Relative Entropies

In Chapter 5, we introduced the concept of a divergence as a measure quantifying the distinguishability between probability vectors or quantum states, emphasizing that any measure of distinguishability should satisfy the data processing inequality. This chapter builds upon that foundation by integrating the principle of additivity. The additivity of certain functions under the tensor product of states is a recurring theme in physics. For instance, entropy's additivity under tensor products is intimately linked to several characteristics of thermal systems, including the second law of thermodynamics.

This chapter approaches the definition of entropies and relative entropies axiomatically. This method is particularly beneficial, as it reveals a multitude of properties common to all entropies and relative entropies. Furthermore, it distinguishes unique properties of certain relative entropies from those that are universally applicable. For instance, we will discover that the KL-divergence, introduced in Chapter 5, is the sole relative entropy characterized by asymptotic continuity. This distinction underpins the significant role of KL-divergence in information theory.

## 6.1 Entropy

Entropy is pivotal in numerous fields, including statistical mechanics, thermodynamics, information theory, black hole physics, cosmology, chemistry, and even economics. This wide range of applications has led to diverse interpretations of entropy. In thermodynamics, it's seen as a measure of energy dispersal at a specific temperature. In contrast, information theory views it as a rate of compression. Other perspectives, explored extensively in literature, link entropy to disorder, chaos, system randomness, and the concept of time's arrow. These varying attributes and contexts give rise to different measures of entropy, such as Gibbs and Boltzmann entropy, Tsallis entropies, Rényi entropies, and von-Neumann and Shannon entropies, along with other entropy functions like molar entropy, entropy of mixing, and loop entropy.

The multifaceted nature of entropy calls for a systematic and unifying approach, where entropy is defined rigorously and context-independently. This requires identifying common characteristics across all forms of entropy. One such universal trait is *uncertainty*, whether it's about the state of a physical system or the output of a compression scheme. In various contexts, this uncertainty also encompasses concepts like disorder and randomness. For instance, uncertainty about a system's state correlates with its disorder level.

In Chapter 4, especially in Section 4.1, we delved into the role of majorization in defining uncertainty. We employed three different methodologies – axiomatic, constructive, and operational – to determine that every measure of uncertainty should inherently be a Schur concave function. Consequently, it is reasonable to anticipate that entropy functions will exhibit monotonic behavior under majorization.

Besides uncertainty, entropy embodies other attributes. A second key feature, related to the second law of thermodynamics – especially the Clausius and Kelvin–Planck statements – involves cyclic processes where a system undergoes a thermodynamic transition while all other systems, including the environment and heat baths, return to their original state. Recent developments in quantum information's approach to small-scale thermodynamics, as referenced in [29], categorize these as catalytic processes. Consider a thermodynamical evolution where a physical system $A$ in state $\rho^A$ transitions into system $B$ in state $\sigma^B$. The encompassing thermal machine, including heat baths, environment, and so on, can be represented as an additional system $C$ in state $\tau^C$. Thus, for cyclic processes, the thermodynamic transition can be described as

$$\rho^A \otimes \tau^C \to \sigma^B \otimes \tau^C. \tag{6.1}$$

In this framework, the second law asserts not just that system $A$'s entropy is no greater than that of system $B$, but also that this holds true only if the entropy of the combined state $\rho^A \otimes \tau^C$ increases or remains unchanged in such a thermodynamic cyclic process where $\tau^C$ is preserved. If entropy is measured with an additive function (under tensor products), then the entropy of $\rho^A$ being no greater than that of $\sigma^B$ implies the same relationship between $\rho^A \otimes \tau^C$ and $\sigma^B \otimes \tau^C$. Thus, we define entropies as *additive* measures of uncertainty.

## 6.1.1 Classical Entropy

In this section, we study entropy in the classical domain. We first introduce the formal definition of an entropy in terms of two axioms, and then provide further justification for these axioms. We will consider a function

$$\mathbb{H} \colon \bigcup_{n \in \mathbb{N}} \mathrm{Prob}(n) \to \mathbb{R} \tag{6.2}$$

that maps probability vectors in all finite dimensions to the real numbers.

**Entropy**

**Definition 6.1.** The function $\mathbb{H}$ as given in (6.2) is called an entropy if it is not equal to the constant zero function and it satisfies the following two axioms:

1. **Monotonicity under mixing.** For every $n, m \in \mathbb{N}$, and every $\mathbf{p} \in \mathrm{Prob}(n)$ and $\mathbf{q} \in \mathrm{Prob}(m)$ that satisfy $\mathbf{p} \succ \mathbf{q}$, we have $\mathbb{H}(\mathbf{p}) \leqslant \mathbb{H}(\mathbf{q})$.
2. **Additivity.** For any $n, m \in \mathbb{N}$, and any $\mathbf{p} \in \mathrm{Prob}(n)$ and $\mathbf{q} \in \mathrm{Prob}(m)$,

$$\mathbb{H}(\mathbf{p} \otimes \mathbf{q}) = \mathbb{H}(\mathbf{p}) + \mathbb{H}(\mathbf{q}). \tag{6.3}$$

The first axiom ensures that an entropy quantifies uncertainty. In Section 4.1, we arrived at the definition of majorization from a game of chance, indicating that $\mathbf{p} \succ \mathbf{q}$ if $\mathbf{q}$ is more uncertain than $\mathbf{p}$. Note however that we extend here the definition of majorization to vectors that are not necessarily of the same dimension. This can be done by adding zeros to the vector with the smaller dimension to make the vectors in the same dimension. This means in particular that for any $\mathbf{p} \in \mathrm{Prob}(n)$, any entropy $\mathbb{H}$ satisfies $\mathbb{H}(\mathbf{p} \oplus 0) = \mathbb{H}(\mathbf{p})$. Note also that this axiom also implies that $\mathbb{H}$ is Schur concave.

The additivity axiom distinguishes entropy functions from arbitrary measures of uncertainty. For example, in Section 4.1.3 we encounter several Schur concave functions, such as the symmetric elementary functions (see (4.49)) that are in general not additive. Therefore, such functions cannot be entropies. The additivity property is consistent with the extensivity property of entropy in thermodynamics, and particularly, the monotonicity of entropy under cyclic thermodynamical processes. As already mentioned, in such cycles, all degrees of freedom other than the degrees of freedom of the system remain intact at the end of the cycle. Therefore, suppose the system at the beginning and end of the cycle is characterized with some probability vectors $\mathbf{p}$ and $\mathbf{q}$, respectively. If the initial state of the system was described by $\mathbf{p} \otimes \mathbf{r}$, where $\mathbf{r}$ corresponds to the remaining degrees of freedom, then at the end of the cycle the system+environment are described by $\mathbf{q} \otimes \mathbf{r}$ (i.e. with the same $\mathbf{r}$). Since entropy should be monotonic under such cycle in which $\mathbf{p} \otimes \mathbf{r} \succ \mathbf{q} \otimes \mathbf{r}$, it motivates the additivity property of an entropy function so that it is monotonic under the trumping relation. That is, the monotonicity under mixing can be strengthened using the additivity property such that

$$\mathbf{p} \succ_* \mathbf{q} \quad \Rightarrow \quad \mathbb{H}(\mathbf{p}) \leqslant \mathbb{H}(\mathbf{q}). \tag{6.4}$$

There are other arguments to motivate the additivity axiom that comes from information theory and we will discuss them as we go along.

In Definition 6.1, we allow for the case that $n = 1$. In this trivial case, $\mathrm{Prob}(n) = \mathrm{Prob}(1)$ contains only the one-dimensional vector (i.e. number) 1. Observe that for any $\mathbf{p} \in \mathrm{Prob}(n)$ we get from the additivity axiom that $\mathbb{H}(\mathbf{p}) = \mathbb{H}(\mathbf{p} \otimes 1) = \mathbb{H}(\mathbf{p}) + \mathbb{H}(1)$, so that $\mathbb{H}(1) = 0$. From the fact that for all $x \in [n]$ $1 \succ \mathbf{e}_x \succ 1$ (i.e. $1 \sim \mathbf{e}_x$), where $\{\mathbf{e}_x\}_{x \in [n]}$ is the standard (elementary) basis of $\mathbb{R}^n$, we conclude that also $\mathbb{H}(\mathbf{e}_x) = 0$ for all $x \in [n]$. Moreover, since for every $n \in \mathbb{N}$ and every $\mathbf{p} \in \mathrm{Prob}(n)$ we have $\mathbf{e}_x \succ \mathbf{p}$ we get from the monotonicity axiom that $\mathbb{H}(\mathbf{p}) \geqslant \mathbb{H}(\mathbf{e}_x) = 0$. That is, entropy functions cannot be negative.

In the definition of entropy here we assumed that the entropy $\mathbb{H}$ is not the zero function. This means that there exists $n \in \mathbb{N}$ and $\mathbf{p} \in \mathrm{Prob}(n)$ such that $\mathbb{H}(\mathbf{p}) \neq 0$. Since entropy cannot be negative, this means that $\mathbb{H}(\mathbf{p}) > 0$. On the other hand, for sufficiently large $m \in \mathbb{N}$ we have $\mathbf{p} \succ \left(\mathbf{u}^{(2)}\right)^{\otimes m}$ (see Exercise 4.4) so that

$$\mathbb{H}(\mathbf{u}^{(2)}) = \frac{1}{m} \mathbb{H}\left(\left(\mathbf{u}^{(2)}\right)^{\otimes m}\right)$$

$$\boxed{\mathbf{p} \succ \left(\mathbf{u}^{(2)}\right)^{\otimes m}} \longrightarrow \geqslant \frac{1}{m} \mathbb{H}(\mathbf{p}) > 0. \tag{6.5}$$

Therefore, all entropy functions take strictly positive values on $\mathbf{u}^{(2)} := \frac{1}{2}(1,1)^T \in$ Prob(2). It will be convenient to normalize all entropy functions such that

$$\mathbb{H}\left(\mathbf{u}^{(2)}\right) = 1. \tag{6.6}$$

Throughout the remainder of the book, we will focus exclusively on entropy functions that are normalized as in (6.6).

---

**Lemma 6.1.** Let $\mathbb{H}$ be an entropy normalized as in (6.6), $n \in \mathbb{N}$, and $\mathbf{u}^{(n)} := \frac{1}{n}(1,\ldots,1)^T \in \text{Prob}(n)$. Then, for all $\mathbf{p} \in \text{Prob}(n)$ we have

$$\mathbb{H}(\mathbf{p}) \leqslant \mathbb{H}(\mathbf{u}^{(n)}) = \log n. \tag{6.7}$$

---

**Proof**　The inequality follow from the Schur concavity of $\mathbb{H}$ and the fact that $\mathbf{p} \succ \mathbf{u}^{(n)}$. To prove the equality, define $f : \mathbb{N} \to \mathbb{R}$, via $f(n) := \mathbb{H}(\mathbf{u}^{(n)})$. From the normalization (6.6) we have $f(2) = 1$ and from the additivity $f(2^k) = k$ for all $k \in \mathbb{N}$. More generally, for any $m, n \in \mathbb{N}$ the additivity gives

$$f(n^m) = \mathbb{H}\left(\mathbf{u}^{(n^m)}\right) = \mathbb{H}\left(\left(\mathbf{u}^{(n)}\right)^{\otimes m}\right) = m\mathbb{H}\left(\mathbf{u}^{(n)}\right) = mf(n). \tag{6.8}$$

Moreover, from the monotonicity property of $\mathbb{H}$ and the fact that $\mathbf{u}^{(n)} \succ \mathbf{u}^{(n+1)}$ we get that $f$ is monotonically nondecreasing. Using these properties of $f$ we get for all $k, m \in \mathbb{N}$

$$
\begin{aligned}
f(n) = \frac{1}{m}f(n^m) &= \frac{1}{m}f\left(2^{m\log(n)}\right) \\
\boxed{f \text{ is nondecreasing}} \longrightarrow &\leqslant \frac{1}{m}f\left(2^{\lceil m\log(n)\rceil}\right) \\
&= \frac{1}{m}\lceil m\log(n)\rceil.
\end{aligned}
\tag{6.9}
$$

Similarly, taking the floor instead of the ceiling above gives $f(n) \geqslant \frac{1}{m}\lfloor m\log(n)\rfloor$. In the limit $m \to \infty$ both of these bounds converge to $\log n$. This concludes the proof.　∎

**Exercise 6.1.**　*Show that any convex combination of entropies is itself an entropy. That is, if $\{\mathbb{H}_x\}_{x=1}^{k}$ is a set of entropies and $\mathbf{s} \in \text{Prob}(k)$, then $\sum_{x \in [k]} s_x \mathbb{H}_x$ is itself an entropy.*

## The Rényi Entropies

An important class of entropies is the class of Rényi entropies. The Rényi entropies are defined for any $n \in \mathbb{N}$, $\mathbf{p} \in \text{Prob}(n)$, and $\alpha \in [0, \infty]$, as

$$H_\alpha(\mathbf{p}) := \frac{1}{1-\alpha}\log \sum_{x \in [n]} p_x^\alpha, \tag{6.10}$$

where the cases $\alpha = 0, 1, \infty$ are defined by the appropriate limits. That is, for $\alpha = 0$ the Rényi entropy is also known by the name *the max-entropy* and is given by

$$H_{\max}(\mathbf{p}) := H_0(\mathbf{p}) := \lim_{\alpha \to 0} H_\alpha(\mathbf{p}) = \log |\operatorname{supp}(\mathbf{p})|, \tag{6.11}$$

where $|\operatorname{supp}(\mathbf{p})|$ is the number of nonzero components in $\mathbf{p}$. For $\alpha = 1$ the Rényi entropy reduces to the Shannon entropy

$$H(\mathbf{p}) = \lim_{\alpha \to 1} H_\alpha(\mathbf{p}) = -\sum_{x \in [n]} p_x \log p_x. \tag{6.12}$$

Finally, for the case $\alpha = \infty$ the Rényi entropy is also known by the name the min-entropy and is given by

$$H_{\min}(\mathbf{p}) := \lim_{\alpha \to \infty} H_\alpha(\mathbf{p}) = -\log \max_{x \in [n]}\{p_x\}. \tag{6.13}$$

**Exercise 6.2.** *Prove the three limits in* (6.11)*,* (6.12)*, and* (6.13)*.*

The Rényi entropies are indeed entropies since they satisfy the three axioms of Definition 6.1. To show the monotonicity under mixing (i.e. Schur concavity) observe that for $x \neq y \in [n]$ and $\alpha \in (0, \infty)$

$$(p_x - p_y)\left(\frac{\partial H_\alpha(\mathbf{p})}{\partial p_x} - \frac{\partial H_\alpha(\mathbf{p})}{\partial p_y}\right) = \frac{\alpha}{1-\alpha}\frac{1}{\|\mathbf{p}\|_\alpha^\alpha}(p_x - p_y)\left(p_x^{\alpha-1} - p_y^{\alpha-1}\right) < 0. \tag{6.14}$$

Hence, from the Schur's test in (4.39) this implies that $H_\alpha$ are strictly Schur concave for $\alpha \in (0, \infty)$. The Schur concavity of $H_\alpha$ for the cases $\alpha = 0$ and $\alpha = \infty$ follows by taking the limits $\alpha \to 0^+$ and $\alpha \to \infty$, respectively.

**Exercise 6.3.** *Prove directly, without taking the limits on $\alpha$, that $H_{\min}$ and $H_{\max}$ are Schur concave.*

Note that the Rényi entropy can be expressed as

$$H_\alpha(\mathbf{p}) := \frac{\alpha}{1-\alpha} \log \|\mathbf{p}\|_\alpha, \tag{6.15}$$

where $\|\mathbf{p}\|_\alpha$ is the $p$-norm with $p = \alpha$. Hence, for $\alpha > 1$ the function $\|\mathbf{p}\|_\alpha$ is convex and also symmetric, so that $\|\mathbf{p}\|_\alpha$ is in particular Schur convex. Combining this with the fact that the log is a monotonically increasing function and recalling our assumption that $1 - \alpha < 0$, this provides an alternative proof that $H_\alpha$ is Schur concave for $\alpha > 1$.

**Exercise 6.4.**

1. *Show that the Rényi entropy satisfies the additivity axiom of an entropy.*
2. *Show by direct calculation that $H_\alpha(\mathbf{u}^{(n)}) = \log n$ for all $n \in \mathbb{N}$.*
3. *Show that for any fixed $\mathbf{p} \in \mathrm{Prob}(n)$, $H_\alpha(\mathbf{p})$ is monotonically decreasing in $\alpha$.*

The following is a very interesting result proved in [168]. It essentially states that all the entropy functions are Rényi entropies. We refer the reader to [168] for the proof as it goes beyond the scope of this book.

> **Theorem 6.1.** Let $\mathbb{H}$ be an entropy as defined in Definition 6.1. Then, $\mathbb{H}$ can be expressed as a convex combination of the Rényi entropies.

## 6.1.2 Quantum Entropies

In the following definition, we make use of the notation $\rho \succ \sigma$ to indicate that the eigenvalues of $\rho$ form a vector that majorizes the vector consisting of the eigenvalues of $\sigma$. With this notation the extension of Definition 6.1 to the quantum domain is straightforward. Specifically, we will consider a function

$$\mathbb{H} : \bigcup_A \mathfrak{D}(A) \to \mathbb{R} \tag{6.16}$$

that maps density matrices in all finite dimensions to the real numbers.

**Quantum Entropy**

**Definition 6.2.** Let $\mathbb{H}$ be as in (6.16) and suppose it is not equal to the constant zero function. Then, $\mathbb{H}$ is called an entropy if it satisfies the following two axioms:

1. **Monotonicity.** For every $\rho \in \mathfrak{D}(A)$ and $\sigma \in \mathfrak{D}(B)$ that satisfy $\rho^A \succ \sigma^B$ we have $\mathbb{H}(\rho^A) \leqslant \mathbb{H}(\sigma^B)$.
2. **Additivity.** For every $\rho \in \mathfrak{D}(A)$ and $\sigma \in \mathfrak{D}(B)$, $\mathbb{H}(\rho^A \otimes \sigma^B) = \mathbb{H}(\rho^A) + \mathbb{H}(\sigma^B)$.

There is a one-to-one correspondence between quantum entropies and classical entropies. Explicitly, note that the monotonicity property of entropies implies that they are invariant under unitaries; that is, let $\mathbb{H}$ be an entropy function and let $U \in \mathfrak{U}(A)$ be a unitary matrix. Then,

$$\mathbb{H}(\rho) = \mathbb{H}\left(U\rho U^*\right) \qquad \forall \, \rho \in \mathfrak{D}(A). \tag{6.17}$$

The above invariance property implies that $\mathbb{H}(\rho)$ depends only on the eigenvalues of $\rho$ and therefore the quantum entropy of $\rho$ can be viewed as the classical entropy of the probability vector consisting of the eigenvalues of $\rho$. Therefore, any classical entropy $\mathbb{H}_{\text{classical}}$ can be extended to the quantum domain via ($m := |A|$)

$$\mathbb{H}_{\text{quantum}}(\rho) := \mathbb{H}_{\text{classical}}(\lambda_1, \ldots, \lambda_m) \qquad \forall \rho \in \mathfrak{D}(A), \tag{6.18}$$

where $\{\lambda_x\}_{x \in [m]}$ are the eigenvalues of $\rho$. It is left as a simple exercise to show that $\mathbb{H}_{\text{quantum}}$ is indeed a quantum entropy that satisfies the two axioms of Definition 6.2.

As an example, consider the classical Rényi entropies as defined in (6.10). By replacing the components $\{p_x\}_{x \in [n]}$ with the eigenvalues $\{\lambda_x\}_{x \in [n]}$ of $\rho$, we get the quantum version of the Rényi entropies. For any $\alpha \in [0, \infty]$ they are given by

$$H_\alpha(\rho) := \frac{1}{1-\alpha} \log \sum_{x \in [n]} \lambda_x^\alpha = \frac{1}{1-\alpha} \log \mathrm{Tr}[\rho^\alpha]. \tag{6.19}$$

Similarly, from the classical case we get that the limits $\alpha = 0, 1, \infty$ are given for all $\rho \in \mathfrak{D}(A)$ by:

1. The max-entropy ($\alpha = 0$),

$$H_{\max}(\rho) = \log \mathrm{Tr}\left[\Pi_\rho\right], \tag{6.20}$$

   where $\Pi_\rho$ is the projector to the support of $\rho$.

2. The von-Neumann entropy ($\alpha = 1$),

$$H(\rho) = -\mathrm{Tr}[\rho \log \rho]. \tag{6.21}$$

3. The min-entropy ($\alpha = \infty$),

$$H_{\min}(\rho) = -\log \|\rho\|_\infty. \tag{6.22}$$

**Exercise 6.5.** *Let* $\mathbb{H}$ *be an entropy function, and let* $\mathcal{E} \in \mathrm{CPTP}(A \to B)$ *be a random isometry channel; that is,*

$$\mathcal{E}^{A \to B} = \sum_{x \in [n]} p_x \mathcal{V}_x^{A \to B}, \tag{6.23}$$

*where* $\{p_x\}_{x \in [n]}$ *is a probability distribution and each* $\mathcal{V}_x \in \mathrm{CPTP}(A \to B)$ *is an isometry. Show that*

$$\mathbb{H}(\rho) \leqslant \mathbb{H}\big(\mathcal{E}(\rho)\big) \qquad \forall \, \rho \in \mathfrak{D}(A). \tag{6.24}$$

**Exercise 6.6.** *Let* $\mathbb{H}$ *be a quantum entropy, and let* $\mathcal{U} \in \mathrm{CPTP}(A \to A)$ *be a unital channel. Show that*

$$\mathbb{H}(\rho) \leqslant \mathbb{H}\big(\mathcal{U}(\rho)\big) \qquad \forall \, \rho \in \mathfrak{D}(A). \tag{6.25}$$

## 6.2 Classical Relative Entropies

Let

$$\mathbb{D} \colon \bigcup_{n \in \mathbb{N}} \Big\{ \mathrm{Prob}(n) \times \mathrm{Prob}(n) \Big\} \to \mathbb{R} \cup \{\infty\} \tag{6.26}$$

be a function acting on pairs of probability vectors in all finite dimensions.

---

**Relative Entropy**

**Definition 6.3.** The function $\mathbb{D}$ in (6.26) is called a *relative entropy* if it satisfies the following three conditions:

1. *Data Processing Inequality.* See (5.2).
2. *Additivity.* For any $n, n' \in \mathbb{N}$, $\mathbf{p}, \mathbf{q} \in \mathrm{Prob}(n)$, and $\mathbf{p}', \mathbf{q}' \in \mathrm{Prob}(n')$,

$$\mathbb{D}(\mathbf{p} \otimes \mathbf{p}' \| \mathbf{q} \otimes \mathbf{q}') = \mathbb{D}(\mathbf{p}\|\mathbf{q}) + \mathbb{D}(\mathbf{p}'\|\mathbf{q}'). \tag{6.27}$$

3. *Normalization.* $\mathbb{D}(\mathbf{e}_1 \| \mathbf{u}^{(2)}) = 1$, where $\mathbf{e}_1 = (1,0)^T$ and $\mathbf{u}^{(2)} := (\tfrac{1}{2}, \tfrac{1}{2})^T$.

In Definition 6.3 we did not include the normalization condition $\mathbb{D}(1\|1) = 0$ (as satisfied by all normalized divergences) since it follows from the additivity property. Indeed, let $\mathbf{p}, \mathbf{q} \in \mathrm{Prob}(n)$ and observe that

$$\mathbb{D}(\mathbf{p}\|\mathbf{q}) = \mathbb{D}(\mathbf{p} \otimes 1 \| \mathbf{q} \otimes 1) = \mathbb{D}(\mathbf{p}\|\mathbf{q}) + \mathbb{D}(1\|1). \tag{6.28}$$

Therefore, we must have $\mathbb{D}(1\|1) = 0$. Hence, relative entropies are divergences.

The normalization condition $\mathbb{D}(\mathbf{e}_1 | \mathbf{u}^{(2)}) = 1$ eliminates the possibility of scaling a relative entropy by a fixed constant. Notably, this normalization condition is asymmetric with respect to its two inputs. Indeed, any alternative normalization would disrupt this symmetry, as we cannot select a symmetric input (bearing in mind that all divergences satisfy $\mathbb{D}(\mathbf{p}|\mathbf{p}) = 0$ for every $\mathbf{p} \in \mathrm{Prob}(n)$).

**Exercise 6.7.** *Show that any relative entropy $\mathbb{D}$ must satisfy*

$$\mathbb{D}(\mathbf{e}_1 \| \mathbf{e}_2) = \infty, \tag{6.29}$$

*where $\mathbf{e}_1 = (1,0)^T$ and $\mathbf{e}_2 = (0,1)^T$. Hint: Show first that $\mathbb{D}(\mathbf{e}_1 \| \mathbf{e}_2) \geqslant 1$ and then use the additivity property together with the DPI to show that for any $n \in \mathbb{N}$, $\mathbb{D}(\mathbf{e}_1 \| \mathbf{e}_2) \geqslant n$.*

## 6.2.1 The Rényi Relative Entropies

In his seminal paper, Rényi introduced a one parameter family of relative entropies. We already encountered them in Section 4.4, and here we will study some of their properties.

**The Rényi Relative Entropies**

**Definition 6.4.** The Rényi relative entropy of order $\alpha \in [0, \infty]$ is defined for all $\mathbf{p}, \mathbf{q} \in \mathrm{Prob}(n)$ as

$$D_\alpha(\mathbf{p}\|\mathbf{q}) := \begin{cases} \frac{1}{\alpha-1} \log \sum_{x \in [n]} p_x^\alpha q_x^{1-\alpha} & \text{if } \mathbf{p} \ll \mathbf{q} \text{ or } \alpha \in [0,1) \text{ and } \mathbf{p} \cdot \mathbf{q} \neq 0. \\ \infty & \text{otherwise.} \end{cases}$$

The cases $\alpha = 0, 1, \infty$ are defined in terms of the appropriate limits.

*Remark.* We use the convention that if $q_x = p_x = 0$, then $p_x^\alpha q_x^{1-\alpha} = 0$ even for $\alpha > 1$. With this convention, the conditions that $\text{supp}(\mathbf{p}) \subseteq \text{supp}(\mathbf{q})$ or $\alpha \in [0,1)$ and $\mathbf{p} \cdot \mathbf{q} \neq 0$ are precisely the conditions that the expression $\frac{1}{\alpha-1} \log \sum_{x \in [n]} p_x^\alpha q_x^{1-\alpha}$ is well defined. Otherwise, if it is not well defined, the Rényi relative entropy is set to be infinity.

For $\alpha = 0$ the relative Rényi entropy is called the min-relative entropy. It is given by

$$D_{\min}(\mathbf{p}\|\mathbf{q}) := \lim_{\alpha \to 0^+} D_\alpha(\mathbf{p}\|\mathbf{q}) = -\log \sum_{x \in \text{supp}(\mathbf{p})} q_x. \tag{6.30}$$

Observe that if $D_{\min}(\mathbf{p}\|\mathbf{q}) \neq 0$, then $\mathbf{p}$ must have zero components. For $\alpha = \infty$ the relative Rényi entropy is called the max-relative entropy. It is given by

$$D_{\max}(\mathbf{p}\|\mathbf{q}) := \lim_{\alpha \to \infty} D_\alpha(\mathbf{p}\|\mathbf{q}) = \log \max_{x \in [n]} \left\{ \frac{p_x}{q_x} \right\}. \tag{6.31}$$

Finally, for $\alpha = 1$ the Rényi relative entropy is called the Kullback–Leibler divergence, or in short the KL-divergence. It is given by

$$D(\mathbf{p}\|\mathbf{q}) := \lim_{\alpha \to 1} D_\alpha(\mathbf{p}\|\mathbf{q}) = \sum_{x \in [n]} p_x (\log p_x - \log q_x), \tag{6.32}$$

with the convention $0 \log 0 = 0$.

**Exercise 6.8.** *Prove the three limits in (6.30), (6.31), and (6.32)*

**Exercise 6.9.** *Show that the Rényi entropies are related to the Rényi relative entropies via*

$$H_\alpha(\mathbf{p}) = \log n - D_\alpha(\mathbf{p}\|\mathbf{u}^{(n)}) \quad \forall \, \mathbf{p} \in \text{Prob}(n). \tag{6.33}$$

**Exercise 6.10.** *Let* $\mathbf{p}, \mathbf{q} \in \text{Prob}(n)$ *and* $r \in \mathbb{R}_+$.

1. *Show that* $r\mathbf{q} - \mathbf{p} \geq 0$ *(entry-wise) if and only if* $r \geq 2^{D_{\max}(\mathbf{p}\|\mathbf{q})}$.
2. *Show that* $\mathbf{p} - r\mathbf{q} \geq 0$ *if and only if* $r \leq 2^{-D_{\max}(\mathbf{q}\|\mathbf{p})}$.
3. *Why in the first inequality above $r$ must be greater than 1, whereas in the second it must be smaller than 1?*

**Exercise 6.11.** *Show that all the Rényi entropies satisfy the additivity and normalization properties of a relative entropy as given in Definition 6.3.*

**Exercise 6.12.** *A relative entropy $\mathbb{D}$ is said to be pathological if $\mathbb{D}(\mathbf{u}^{(2)}\|\mathbf{e}_1^{(2)}) = 0$, where $\mathbf{u}^{(2)} := (\frac{1}{2}, \frac{1}{2})^T$ is the uniform distribution in $\text{Prob}(2)$, and $\mathbf{e}_1^{(2)} := (1,0)^T$. Show that $D_{\min}$ is pathological and use it to show that $D_{path}$, which is defined for any $n \in \mathbb{N}$ and $\mathbf{p}, \mathbf{q} \in \text{Prob}(n)$ as*

$$D_{path}(\mathbf{p}\|\mathbf{q}) := D_{\min}(\mathbf{p}\|\mathbf{q}) + D_{\min}(\mathbf{q}\|\mathbf{p}), \tag{6.34}$$

*is a relative entropy.*

We now show that in addition to the additivity and normalization, the Rényi relative entropies also satisfy the DPI.

> **Theorem 6.2.** The Rényi relative entropy of any order $\alpha \in [0, \infty]$ is a relative entropy; that is, it satisfies the axioms of DPI, additivity, and normalization, as given in Definition 6.3.

**Proof** The additivity and normalization you proved in Exercise 6.11. To show DPI recall the $\alpha$ divergences given in (5.28) by

$$D_{f_\alpha}(\mathbf{p}\|\mathbf{q}) = \frac{1}{\alpha(\alpha - 1)} \left( \sum_{x \in [n]} p_x^\alpha q_x^{1-\alpha} - 1 \right). \tag{6.35}$$

Since the above expression has been derived from the convex function $f_\alpha(r) = \frac{r^\alpha - r}{\alpha(\alpha - 1)}$ it is an $f$-Divergence and in particular satisfies the DPI. For $\alpha = 1$ the above expression coincides with the Rényi relative entropy of that order (i.e. the KL-divergence), so in this case the DPI property follows. For $\alpha \neq 1$ we denote by

$$Q_\alpha(\mathbf{p}\|\mathbf{q}) := \sum_{x \in [n]} p_x^\alpha q_x^{1-\alpha}. \tag{6.36}$$

Observe that from the DPI of $D_{f_\alpha}$ we get that for $\alpha > 1$ the function $Q_\alpha(\mathbf{p}\|\mathbf{q})$ is monotonically nondecreasing under maps $(\mathbf{p}, \mathbf{q}) \mapsto (E\mathbf{p}, E\mathbf{q})$ with $E \in \text{STOC}(m, n)$, and for $\alpha < 1$ it is monotonically nonincreasing under such maps. Since the Rényi relative entropy can be expressed as $D_\alpha(\mathbf{p}\|\mathbf{q}) := \frac{1}{\alpha-1} \log Q_\alpha(\mathbf{p}\|\mathbf{q})$ and the log is monotonically increasing function, we conclude that $D_\alpha(\mathbf{p}\|\mathbf{q})$ satisfies the DPI. ∎

**Exercise 6.13.** *Show that for any $\alpha \in (0, 1)$ and $\mathbf{p}, \mathbf{q} \in \text{Prob}(n)$*

$$D_\alpha(\mathbf{p}\|\mathbf{q}) = \frac{\alpha}{1 - \alpha} D_{1-\alpha}(\mathbf{q}\|\mathbf{p}). \tag{6.37}$$

**Exercise 6.14.** *Show that if $\mathbf{p}, \mathbf{q} \in \text{Prob}(n)$ and $\rho, \sigma \in \mathfrak{D}(X)$ are two diagonal density matrices with diagonals $\mathbf{p}$ and $\mathbf{q}$, respectively, then*

$$D_\alpha(\mathbf{p}\|\mathbf{q}) = \frac{1}{\alpha - 1} \log \text{Tr}[\rho^\alpha \sigma^{1-\alpha}]. \tag{6.38}$$

## 6.2.2 Properties of Relative Entropies

The three axioms of a relative entropy provides it with enough structure that yields many interesting properties. In this subsection we explore some of these key properties that hold for *all* relative entropies.

> **Theorem 6.3.** Let $\mathbb{D}$ be a relative entropy, and let $\{e_x\}_{x \in [n]}$ be the standard (elementary) basis of $\mathbb{R}^n$. Then, for any $\mathbf{p} \in \mathrm{Prob}(n)$ and $x \in [n]$ we have
>
> $$\mathbb{D}(e_x \| \mathbf{p}) = -\log p_x, \qquad (6.39)$$
>
> where $p_x$ is the $x$-component of $\mathbf{p}$.

The proof of the theorem above is based on the following lemma by Erdös.

**Erdös Theorem**

> **Lemma 6.2.** Let $g \colon \mathbb{N} \to \mathbb{R}$ be a function from the set of natural numbers to the real line. Suppose $g$ is nondecreasing and is additive; that is, $g(mn) = g(n) + g(m)$ for all $n, m \in \mathbb{N}$. Then, there exists a constant $c \in \mathbb{R}$ such that $g(n) = c \log(n)$ for all $n \in \mathbb{N}$.

**Proof**  Suppose by contradiction that $\frac{g(n)}{\log n}$ is not a constant. Therefore, there exists $m, n \in \mathbb{N}$ such that

$$\frac{g(m)}{\log m} > \frac{g(n)}{\log n}. \qquad (6.40)$$

Denote by $a := \frac{g(m)}{\log m}$ and $b := \frac{g(n)}{\log n}$ and observe that $a > b$ or equivalently $\frac{b}{a} < 1$. Multiplying both sides of the inequality $\frac{b}{a} < 1$ by the positive number $\frac{\log n}{\log m} k$, where $k$ is any integer, gives

$$\frac{b}{a} \frac{\log n}{\log m} k < \frac{\log n}{\log m} k. \qquad (6.41)$$

Therefore, for sufficiently large $k \in \mathbb{N}$ there must exist an integer between the above two numbers; that is, there exists $\ell \in \mathbb{N}$ such that

$$\frac{b}{a} \frac{\log n}{\log m} k < \ell < \frac{\log n}{\log m} k. \qquad (6.42)$$

The above two inequalities can be expressed as

$$k \log n > \ell \log m \quad \text{and} \quad kb \log n < \ell a \log m. \qquad (6.43)$$

The first expression in (6.43) implies that the integers $n^k$ and $m^\ell$ satisfy $n^k > m^\ell$, and the second equation implies that $kg(n) < \ell g(m)$. From the additivity of $g$ we therefore conclude that $g(n^k) < g(m^\ell)$. To summarize, we have

$$n^k > m^\ell \quad \text{and} \quad g(n^k) < g(m^\ell). \qquad (6.44)$$

These two inequalities are in contradiction with the assumption that $g$ is nondecreasing. This completes the proof. ∎

**Proof of Theorem 6.3** Since divergences (and therefore relative entropies) are invariant under permutations (see (5.11)), it is sufficient to show that $\mathbb{D}(\mathbf{e}_1\|\mathbf{p}) = -\log p_1$. We first show that for any vector $\mathbf{r} =, (r_1,\ldots,r_n)^T \in \mathrm{Prob}(n)$ with $r_1 = 0$ we have

$$(\mathbf{e}_1,\mathbf{p}) \sim \big(\mathbf{e}_1, p_1\mathbf{e}_1 + (1 - p_1)\mathbf{r}\big), \tag{6.45}$$

where the symbol $\sim$ corresponds to the equivalence relation under relative majorization. Define $E := [\mathbf{e}_1,\mathbf{r},\ldots,\mathbf{r}] \in \mathrm{STOC}(n,n)$ to be the column stochastic matrix whose first column is $\mathbf{e}_1$ and the remaining $n - 1$ columns equals $\mathbf{r}$. We then have

$$(\mathbf{e}_1,\mathbf{p}) \succ (E\mathbf{e}_1, E\mathbf{p}) = \big(\mathbf{e}_1, p_1\mathbf{e}_1 + (1 - p_1)\mathbf{r}\big). \tag{6.46}$$

Conversely, define $\tilde{\mathbf{p}} := \frac{1}{1-p_1}(0, p_2,\ldots, p_n)^T \in \mathrm{Prob}(n)$ and $\tilde{E} := [\mathbf{e}_1,\tilde{\mathbf{p}},\ldots,\tilde{\mathbf{p}}] \in \mathrm{STOC}(n,n)$. Then,

$$\big(\mathbf{e}_1, p_1\mathbf{e}_1 + (1 - p_1)\mathbf{r}\big) \succ \big(\tilde{E}\mathbf{e}_1, p_1\tilde{E}\mathbf{e}_1 + (1 - p_1)\tilde{E}\mathbf{r}\big) = (\mathbf{e}_1,\mathbf{p}). \tag{6.47}$$

Combining (6.46) and (6.47) gives (6.45).

The relation in (6.45) implies that

$$\mathbb{D}(\mathbf{e}_1\|\mathbf{p}) = \mathbb{D}\big(\mathbf{e}_1\big\| p_1\mathbf{e}_1 + (1 - p_1)\mathbf{r}\big) \tag{6.48}$$

so that the function $f(p_1) := \mathbb{D}(\mathbf{e}_1\|\mathbf{p})$ is independent on $p_2,\ldots, p_n$. Moreover, the function $f : [0, 1] \to \mathbb{R}_+ \cup \{\infty\}$ has the following two properties:

1. $f$ is monotonically nonincreasing.
2. $f$ is additive; that is, $f(st) = f(s) + f(t)$ for all $s, t \in [0, 1]$.

The first property of $f$ follows from the fact that for any $E \in \mathrm{STOC}(n,n)$ with $E\mathbf{e}_1 = \mathbf{e}_1$ we get $E\mathbf{p} = p_1\mathbf{e}_1 + (1-p_1)\mathbf{s}$ for some $\mathbf{s} \in \mathrm{Prob}(n)$. This means that the first component of $\mathbf{q} := E\mathbf{p}$ satisfies $q_1 \geqslant p_1$. From the DPI we have

$$f(p_1) := \mathbb{D}(\mathbf{e}_1\|\mathbf{p}) \geqslant \mathbb{D}(E\mathbf{e}_1\|E\mathbf{p}) = \mathbb{D}(\mathbf{e}_1\|\mathbf{q}) = f(q_1), \tag{6.49}$$

so that $f$ is monotonically nonincreasing. The additivity of $f$ follows trivially from the additivity of $\mathbb{D}$ under tensor products (see the following exercise).

Define now the function $g : \mathbb{N} \to \mathbb{R}_+ \cup \{\infty\}$ via the relation $g(m) := f\left(\frac{1}{m}\right)$. This function is nondecreasing and additive. Therefore, from Erdös theorem there exists a constant $c \in \mathbb{R}$ such that $g(m) = c\log m$ for all $m \in \mathbb{N}$. The condition $g(2) = f(1/2) = \mathbb{D}(\mathbf{e}_1\|\mathbf{u}^{(2)}) = 1$ gives $c = 1$. Therefore, for any $m \in \mathbb{N}$ we have $f(1/m) = \log m$. Furthermore, observe that for any $k \leqslant m$ the additivity of $f$ gives

$$\log k + f\left(\frac{k}{m}\right) = f\left(\frac{1}{k}\right) + f\left(\frac{k}{m}\right) = f\left(\frac{1}{m}\right) = \log m. \tag{6.50}$$

Hence, $f\left(\frac{k}{m}\right) = \log m - \log k = -\log(k/m)$. Hence, $f(r) = -\log r$ for all rationals in $[0, 1]$. To prove that this relation holds for any $r \in [0, 1]$ (possibly irrational), let $\{s_k\}$ and $\{t_k\}$ be two sequences of rational numbers in $[0, 1]$ both with limit $r$ and with $s_k \leqslant r \leqslant t_k$ for all $k \in \mathbb{N}$. Then, the monotonicity property of $f$ gives for any $k \in \mathbb{N}$

$$-\log s_k = f(s_k) \geqslant f(r) \geqslant f(t_k) = -\log t_k. \tag{6.51}$$

Taking the limit $k \to \infty$ on both sides and using the continuity of the log function gives $f(r) = -\log r$. This completes the proof. ∎

The theorem above has the following interesting corollary that justifies the terminology of the max and min relative entropies.

> **Corollary 6.1.** Let $\mathbb{D}$ be a relative entropy. Then for any $n \in \mathbb{N}$ and any $\mathbf{p}, \mathbf{q} \in$ Prob$(n)$,
>
> $$D_{\min}(\mathbf{p}\|\mathbf{q}) \leqslant \mathbb{D}(\mathbf{p}\|\mathbf{q}) \leqslant D_{\max}(\mathbf{p}\|\mathbf{q}). \tag{6.52}$$

**Proof**　In Theorem 4.7 we proved that for any $\mathbf{p}, \mathbf{q} \in$ Prob$(n)$

$$\left((1,0)^T, (\lambda_{\max}, 1 - \lambda_{\max})^T\right) \succ (\mathbf{p}, \mathbf{q}) \succ \left((1,0)^T, (\lambda_{\min}, 1 - \lambda_{\min})^T\right), \tag{6.53}$$

where $\lambda_{\max} := \min_{x \in [n]} q_x/p_x = 2^{-D_{\max}(\mathbf{p}\|\mathbf{q})}$ and $\lambda_{\min} := \sum_{x \in \text{supp}(\mathbf{p})} q_x = 2^{-D_{\min}(\mathbf{p}\|\mathbf{q})}$. Hence, the monotonicity of $\mathbb{D}$ under relative majorization gives

$$\mathbb{D}(\mathbf{p}\|\mathbf{q}) \leqslant \mathbb{D}\left(\mathbf{e}_1 \,\Big\|\, \left(2^{-D_{\max}(\mathbf{p}\|\mathbf{q})}, 1 - 2^{-D_{\max}(\mathbf{p}\|\mathbf{q})}\right)^T\right)$$
$$\text{Theorem 6.3} \to \; = D_{\max}(\mathbf{p}\|\mathbf{q}), \tag{6.54}$$

and

$$\mathbb{D}(\mathbf{p}\|\mathbf{q}) \geqslant \mathbb{D}\left(\mathbf{e}_1 \,\Big\|\, \left(2^{-D_{\min}(\mathbf{p}\|\mathbf{q})}, 1 - 2^{-D_{\min}(\mathbf{p}\|\mathbf{q})}\right)^T\right)$$
$$\text{Theorem 6.3} \to \; = D_{\min}(\mathbf{p}\|\mathbf{q}). \tag{6.55}$$

This completes the proof. ∎

**Exercise 6.15.** *Use the corollary above to show that any relative entropy $\mathbb{D}$ satisfies for all $\mathbf{p}, \mathbf{q} \in$ Prob$(n)$:*

- $\mathbb{D}(\mathbf{p}\|\mathbf{q}) < \infty$ *if* supp$(\mathbf{p}) \subseteq$ supp$(\mathbf{q})$, *and*
- $\mathbb{D}(\mathbf{p}\|\mathbf{q}) = \infty$ *if* $\mathbf{p} \cdot \mathbf{q} = 0$.

Relative entropies are not metrics but we show now that they satisfy the following variant of the triangle inequality.

> **Triangle Inequality**
>
> **Theorem 6.4.** Let $\mathbb{D}$ be a relative entropy. Then, for any $\mathbf{p}, \mathbf{q}, \mathbf{r} \in$ Prob$(n)$,
>
> $$\mathbb{D}(\mathbf{p}\|\mathbf{q}) \leqslant \mathbb{D}(\mathbf{p}\|\mathbf{r}) + D_{\max}(\mathbf{r}\|\mathbf{q}). \tag{6.56}$$

**Proof** The key idea of the proof is to denote by $\varepsilon := 2^{-D_{\max}(\mathbf{r}\|\mathbf{q})}$ and observe that the right-hand side of (6.56) can be expressed as

$$\mathbb{D}(\mathbf{p}\|\mathbf{r}) + D_{\max}(\mathbf{r}\|\mathbf{q}) = \mathbb{D}(\mathbf{p}\|\mathbf{r}) - \log\varepsilon$$

$$\textbf{Theorem 6.3} \rightarrow = \mathbb{D}(\mathbf{p}\|\mathbf{r}) + \mathbb{D}\left(\mathbf{e}_1 \,\big\|\, (\varepsilon, 1-\varepsilon)^T\right) \tag{6.57}$$

$$\textbf{Additivity} \rightarrow = \mathbb{D}\left(\mathbf{p}\otimes\mathbf{e}_1 \,\big\|\, \mathbf{r}\otimes(\varepsilon, 1-\varepsilon)^T\right).$$

Therefore, to prove the inequality (6.56) it is sufficient to show that

$$\left(\mathbf{p}\otimes\mathbf{e}_1 \,\big\|\, \mathbf{r}\otimes(\varepsilon, 1-\varepsilon)^T\right) \succ (\mathbf{p}\|\mathbf{q}). \tag{6.58}$$

To prove the above relation, we define a channel $E \in \text{STOC}(n, 2n)$ that acts on $\text{Prob}(n) \otimes \text{Prob}(2)$ as an identity upon detecting $\mathbf{e}_1 := (1, 0)^T$ in the second register, and produces a constant output $\mathbf{t} \in \text{Prob}(n)$ (to be determined shortly) upon detecting $\mathbf{e}_2 := (0, 1)^T$ in the second register; explicitly, for any $\mathbf{s} \in \text{Prob}(n)$

$$E\,(\mathbf{s}\otimes\mathbf{e}_1) = \mathbf{s} \quad , \quad E\,(\mathbf{s}\otimes\mathbf{e}_2) = \mathbf{t}. \tag{6.59}$$

By definition, $E\,(\mathbf{p}\otimes\mathbf{e}_1) = \mathbf{p}$. Our objective is to choose $\mathbf{t}$ such that $E(\mathbf{r}\otimes(\varepsilon, 1-\varepsilon)^T) = \mathbf{q}$. From the definition of $E$ we have

$$E\left(\mathbf{r}\otimes(\varepsilon, 1-\varepsilon)^T\right) = \varepsilon E\,(\mathbf{r}\otimes\mathbf{e}_1) + (1-\varepsilon)E\,(\mathbf{r}\otimes\mathbf{e}_2)$$
$$= \varepsilon\mathbf{r} + (1-\varepsilon)\mathbf{t}. \tag{6.60}$$

Therefore, it is left to show that there exists $\mathbf{t} \in \text{Prob}(n)$ such that $\varepsilon\mathbf{r} + (1-\varepsilon)\mathbf{t} = \mathbf{q}$. This means that we need to show that the vector

$$\mathbf{t} := \frac{\mathbf{q} - \varepsilon\mathbf{r}}{1 - \varepsilon} \tag{6.61}$$

has nonnegative components. Indeed, from the definition of $\varepsilon$ we have $\mathbf{q} - \varepsilon\mathbf{r} \geqslant 0$. This completes the proof. ∎

**Exercise 6.16.** *Use the theorem above to show that the function*

$$D_T(\mathbf{p}\|\mathbf{q}) := \max\left\{D_{\max}(\mathbf{p}\|\mathbf{q}), D_{\max}(\mathbf{q}\|\mathbf{p})\right\} \tag{6.62}$$

*is a divergence that is also a metric. Show that it satisfies $D_T(\mathbf{p}\|\mathbf{q}) < \infty$ if and only if $\text{supp}\,(\mathbf{p}) = \text{supp}\,(\mathbf{q})$. This metric is known as the Thompson's metric.*

The theorem above can be expressed in terms of the Thompson's metric.

**Corollary 6.2.** Any relative entropy $\mathbb{D}$ satisfies for all $\mathbf{p}, \mathbf{q}, \mathbf{q}' \in \text{Prob}(n)$

$$\left|\mathbb{D}(\mathbf{p}\|\mathbf{q}) - \mathbb{D}(\mathbf{p}\|\mathbf{q}')\right| \leqslant D_T(\mathbf{q}\|\mathbf{q}'). \tag{6.63}$$

**Exercise 6.17.** *Let* $\mathbf{p}, \mathbf{q}, \mathbf{r} \in \text{Prob}(n)$ *be 3 probability vectors. Show that*

$$D(\mathbf{r}\|\mathbf{p}) + D(\mathbf{r}\|\mathbf{q}) \geqslant D_{1/2}(\mathbf{p}\|\mathbf{q}) \tag{6.64}$$

*where $D$ is the KL-divergence and $D_{1/2}$ is the Rényi relative entropy of order $\alpha = 1/2$. Moreover, show that for any choice of $\mathbf{p}$ and $\mathbf{q}$ there exists $\mathbf{r}$ that achieves the equality.*

## Continuity of Relative Entropies

The corollary above demonstrates that any relative entropy is continuous in the second argument when $\mathbf{q}, \mathbf{q}' > 0$. We will now use it to explore the continuity of a relative entropy in $\text{Prob}(n) \times \text{Prob}(n)$. We say that $\mathbb{D}$ is upper semi-continuous at $(\mathbf{p}, \mathbf{q}) \in \text{Prob}(n) \times \text{Prob}(n)$ if for any sequence $\{(\mathbf{p}_k, \mathbf{q}_k)\}_{k \in \mathbb{N}} \subset \text{Prob}(n) \times \text{Prob}(n)$ that converges to $(\mathbf{p}, \mathbf{q})$ we have

$$\limsup_{k \to \infty} \mathbb{D}(\mathbf{p}_k \| \mathbf{q}_k) \leqslant \mathbb{D}(\mathbf{p}\|\mathbf{q}). \tag{6.65}$$

We say that $\mathbb{D}$ is lower semi-continuous at $(\mathbf{p}, \mathbf{q}) \in \text{Prob}(n) \times \text{Prob}(n)$ if

$$\liminf_{k \to \infty} \mathbb{D}(\mathbf{p}_k \| \mathbf{q}_k) \geqslant \mathbb{D}(\mathbf{p}\|\mathbf{q}). \tag{6.66}$$

Note that $\mathbb{D}$ is both lower and upper semi-continuous at $(\mathbf{p}, \mathbf{q})$ if and only if it is continuous at $(\mathbf{p}, \mathbf{q})$.

**Exercise 6.18.**

1. *Show that the max relative entropy, $D_{\max}(\mathbf{p}\|\mathbf{q})$, is not upper semi-continuous when $\mathbf{q}$ does not have full support. Hint: Consider the sequences $\{\mathbf{p}_k\}_{k \in \mathbb{N}}$ and $\{\mathbf{q}_k\}_{k \in \mathbb{N}}$ with*
   $\mathbf{p}_k := \left( \frac{1}{k}, 1 - \frac{1}{k} \right)^T$ *and* $\mathbf{q}_k := \left( \frac{1}{k^2}, 1 - \frac{1}{k^2} \right)^T$.
2. *Show that $D_{path}(\mathbf{p}\|\mathbf{q}) := D_{\min}(\mathbf{p}\|\mathbf{q}) + D_{\min}(\mathbf{q}\|\mathbf{p})$ is not lower semi-continuous at the boundary of* $\text{Prob}(n) \times \text{Prob}(n)$.

From the exercise above it is clear that we cannot expect relative entropies to be continuous everywhere in $\text{Prob}(n) \times \text{Prob}(n)$. However, if we remove some of the points in the boundary, we get the following continuity property.

> **Continuity of Relative Entropies**
>
> **Theorem 6.5.** Let $\mathbb{D}$ be a relative entropy. Then, $\mathbb{D}$ is upper semi-continuous at any point in $\text{Prob}(n) \times \text{Prob}_{>0}(n)$, and is continuous at any point in $\text{Prob}_{>0}(n) \times \text{Prob}_{>0}(n)$.

**Proof**  Let $(\mathbf{p}_k, \mathbf{q}_k)_{k \in \mathbb{N}}$ be a sequence in $\text{Prob}(n) \times \text{Prob}(n)$ that converges to $(\mathbf{p}, \mathbf{q})$. For any $k \in \mathbb{N}$, define a column stochastic matrix $E_k \in \text{STOC}(n, n)$ by its action on every $\mathbf{s} \in \text{Prob}(n)$ as

$$E_k \mathbf{s} := \mathbf{p}_k + 2^{-D_{\max}(\mathbf{p}\|\mathbf{p}_k)}(\mathbf{s} - \mathbf{p}). \tag{6.67}$$

Since $\lim_{k\to\infty} \mathbf{p}_k = \mathbf{p}$, for sufficiently large $k$ we have $2^{-D_{\max}(\mathbf{p}\|\mathbf{p}_k)} > 0$ (see the following exercise). Moreover, from the definition of $D_{\max}$ we get that $\mathbf{p}_k - 2^{-D_{\max}(\mathbf{p}\|\mathbf{p}_k)}\mathbf{p} \geqslant 0$ so that $E_k$ is indeed a column stochastic matrix. Using these notations, we derive the following from the DPI:

$$
\begin{aligned}
\mathbb{D}(\mathbf{p}\|\mathbf{q}) &\geqslant \mathbb{D}(E_k\mathbf{p}\|E_k\mathbf{q}) \\
(\mathbf{6.67}) \to \; &= \mathbb{D}(\mathbf{p}_k\|E_k\mathbf{q}) \\
\textbf{Theorem 6.4} \to \; &\geqslant \mathbb{D}(\mathbf{p}_k\|\mathbf{q}_k) - D_{\max}(E_k\mathbf{q}\|\mathbf{q}_k).
\end{aligned}
\tag{6.68}
$$

Moving the term involving $D_{\max}$ to the other side and taking the supremum limit on both sides gives

$$
\limsup_{k\to\infty} \mathbb{D}(\mathbf{p}_k\|\mathbf{q}_k) \leqslant \mathbb{D}(\mathbf{p}\|\mathbf{q}) + \limsup_{k\to\infty} D_{\max}(E_k\mathbf{q}\|\mathbf{q}_k). \tag{6.69}
$$

The second term on the right-hand side above vanishes since the vector

$$
\tilde{\mathbf{q}}_k := E_k\mathbf{q} = 2^{-D_{\max}(\mathbf{p}\|\mathbf{p}_k)}\mathbf{q} + \left(\mathbf{p}_k - 2^{-D_{\max}(\mathbf{p}\|\mathbf{p}_k)}\mathbf{p}\right) \tag{6.70}
$$

has a limit $\lim_{k\to\infty} \tilde{\mathbf{q}}_k = \mathbf{q}$ so that

$$
\limsup_{k\to\infty} D_{\max}(\tilde{\mathbf{q}}_k\|\mathbf{q}_k) = 0. \tag{6.71}
$$

Note that we used indirectly the fact that $\mathbf{q} > 0$, since for sufficiently large $k$ we must have $\mathbf{q}_k > 0$ so the limit above is indeed zero. This completes the proof that $\mathbb{D}$ is upper semi-continuous on $\mathrm{Prob}(n) \times \mathrm{Prob}_{>0}(n)$.

We now prove the lower semi-continuity on $\mathrm{Prob}_{>0}(n) \times \mathrm{Prob}_{>0}(n)$. Note that since we already proved upper semi-continuity in this domain, this will imply that $\mathbb{D}$ is continuous on $\mathrm{Prob}_{>0}(n) \times \mathrm{Prob}_{>0}(n)$. For any $k \in \mathbb{N}$, we define $E_k$ as before but with the role of $\mathbf{p}_k$ and $\mathbf{p}$ interchanged; that is, $E_k \in \mathrm{STOC}(n,n)$ is defined by its action on any $\mathbf{s} \in \mathrm{Prob}(n)$ as

$$
E_k\mathbf{s} := \mathbf{p} + 2^{-D_{\max}(\mathbf{p}_k\|\mathbf{p})}(\mathbf{s} - \mathbf{p}_k). \tag{6.72}
$$

Note that for all $k$, $2^{-D_{\max}(\mathbf{p}_k\|\mathbf{p})} > 0$ since we assume $\mathbf{p} > 0$. Moreover, from the definition of $D_{\max}$ we have $\mathbf{p} - 2^{-D_{\max}(\mathbf{p}_k\|\mathbf{p})}\mathbf{p}_k \geqslant 0$, so that $E_k$ is indeed a column stochastic matrix. With the above notations we get from the DPI

$$
\begin{aligned}
\mathbb{D}(\mathbf{p}_k\|\mathbf{q}_k) &\geqslant \mathbb{D}(E_k\mathbf{p}_k\|E_k\mathbf{q}_k) \\
(\mathbf{6.72}) \to \; &= \mathbb{D}(\mathbf{p}\|E_k\mathbf{q}_k) \\
\textbf{Theorem 6.4} \to \; &\geqslant \mathbb{D}(\mathbf{p}\|\mathbf{q}) - D_{\max}(E_k\mathbf{q}_k\|\mathbf{q}).
\end{aligned}
\tag{6.73}
$$

Taking the infimum limit on both sides gives

$$
\liminf_{k\to\infty} \mathbb{D}(\mathbf{p}_k\|\mathbf{q}_k) \geqslant \mathbb{D}(\mathbf{p}\|\mathbf{q}), \tag{6.74}
$$

where we used the fact that

$$
E_k\mathbf{q}_k = 2^{-D_{\max}(\mathbf{p}_k\|\mathbf{p})}\mathbf{q}_k + \left(\mathbf{p} - 2^{-D_{\max}(\mathbf{p}_k\|\mathbf{p})}\mathbf{p}_k\right) \tag{6.75}
$$

has a limit $\lim_{k \to \infty} E_k \mathbf{q}_k = \mathbf{q}$ so that

$$\limsup_{k \to \infty} D_{\max}(E_k \mathbf{q}_k \| \mathbf{q}) = 0. \tag{6.76}$$

This completes the proof. ∎

**Exercise 6.19.**

1. *Show that if $\{\mathbf{p}_k\}_{k \in \mathbb{N}}$ is a sequences in $\mathrm{Prob}(n)$ that converges to $\mathbf{p} \in \mathrm{Prob}(n)$, then for sufficiently large $k$ we have $D_{\max}(\mathbf{p} \| \mathbf{p}_k) < \infty$. Hint: Show that for sufficiently large $k$, $\mathrm{supp}(\mathbf{p}) \subseteq \mathrm{supp}(\mathbf{p}_k)$.*
2. *Prove the limits in (6.71) and (6.76).*

## Faithfulness of Relative Entropies

We have seen before that any divergence $\mathbb{D}$ has the property that $\mathbb{D}(\mathbf{p} \| \mathbf{q}) = 0$ if $\mathbf{p} = \mathbf{q}$. Faithfulness of a divergence refers to the property that this equality holds if and only if $\mathbf{p} = \mathbf{q}$. The minimal relative entropy, $D_{\min}$, provides an example of a relative entropy that is not faithful. Particularly, $D_{\min}(\mathbf{p} \| \mathbf{q}) = 0$ for any $\mathbf{p}$ with $\mathrm{supp}(\mathbf{q}) \subseteq \mathrm{supp}(\mathbf{p})$. However, from the following theorem it follows that $D_{\min}$ is a very unique relative entropy and almost all relative entropies are faithful.

> **Faithfulness**
>
> **Theorem 6.6.** Let $\mathbb{D}$ be a relative entropy. The following statements are equivalent:
>
> 1. $\mathbb{D}$ is not faithful.
> 2. $\mathbb{D}(\mathbf{p} \| \mathbf{q}) = 0$ for all $m \in \mathbb{N}$ and all $\mathbf{p}, \mathbf{q} \in \mathrm{Prob}(m)$ with $\mathrm{supp}(\mathbf{p}) = \mathrm{supp}(\mathbf{q})$.

**Proof**　The direction $2 \Rightarrow 1$ is trivial. We therefore prove that $1 \Rightarrow 2$. Since $\mathbb{D}$ is not faithful there exists $\mathbf{p}, \mathbf{q} \in \mathrm{Prob}(m)$ such that $\mathbf{p} \neq \mathbf{q}$ and $\mathbb{D}(\mathbf{p} \| \mathbf{q}) = 0$. For any $n \in \mathbb{N}$ it follows from the additivity property of $\mathbb{D}$ that also $\mathbb{D}\left(\mathbf{p}^{\otimes n} \| \mathbf{q}^{\otimes n}\right) = 0$. Now, in Corollary 8.1 we will see that for any $\mathbf{s}, \mathbf{t} \in \mathrm{Prob}_{>0}(2)$ and large enough $n$ we have

$$(\mathbf{p}^{\otimes n}, \mathbf{q}^{\otimes n}) \succ (\mathbf{s}, \mathbf{t}). \tag{6.77}$$

Therefore,

$$0 = \mathbb{D}\left(\mathbf{p}^{\otimes n} \| \mathbf{q}^{\otimes n}\right) \geqslant \mathbb{D}(\mathbf{s} \| \mathbf{t}). \tag{6.78}$$

We therefore conclude that $\mathbb{D}(\mathbf{s} \| \mathbf{t}) = 0$ for all $\mathbf{s}, \mathbf{t} \in \mathrm{Prob}_{>0}(2)$. It is left to show that this also holds in dimensions higher than 2.

Indeed, let $\mathbf{p}, \mathbf{q} \in \mathrm{Prob}(m)$ with $\mathrm{supp}(\mathbf{p}) = \mathrm{supp}(\mathbf{q})$, and recall from (4.130) that there exists $\mathbf{s}, \mathbf{t} \in \mathrm{Prob}_{>0}(2)$ such that $(\mathbf{s}, \mathbf{t}) \succ (\mathbf{p}, \mathbf{q})$. We therefore get that

$$\mathbb{D}(\mathbf{p} \| \mathbf{q}) \leqslant \mathbb{D}(\mathbf{s} \| \mathbf{t}). \tag{6.79}$$

Since we already proved that $\mathbb{D}(\mathbf{s}\|\mathbf{t}) = 0$ for all $\mathbf{s}, \mathbf{t} \in \mathrm{Prob}_{>0}(2)$ we conclude that $\mathbb{D}(\mathbf{p}\|\mathbf{q}) = 0$ for all $\mathbf{p}$ and $\mathbf{q}$ with the same support. This completes the proof. ∎

> **Continuity Implies Faithfulness**
>
> **Corollary 6.3.** Let $\mathbb{D}$ be a relative entropy, $2 \leqslant m \in \mathbb{N}$, and $\mathbf{q} \in \mathrm{Prob}(m)$ be a probability vector whose first component satisfies $q_1 \in (0, 1)$. If $\mathbb{D}$ is not faithful, then the function $f_{\mathbf{q}}(\mathbf{p}) := \mathbb{D}(\mathbf{p}\|\mathbf{q})$ is not lower semi-continuous at $\mathbf{p} = \mathbf{e}_1^{(m)}$.

*Remark.* Note that the corollary above in particular implies that relative entropies that are continuous in the first argument must be faithful.

**Proof** Let $\{\mathbf{p}_k\}_{k \in \mathbb{N}}$ be a sequence in $\mathrm{Prob}(m)$ such that $\mathrm{supp}(\mathbf{q}) = \mathrm{supp}(\mathbf{p}_k)$ and $\mathbf{p}_k \to \mathbf{e}_1$ and $k \to \infty$. Such a sequence exists since $q_1 \in (0, 1)$. From the theorem above it follows that $\mathbb{D}(\mathbf{p}_k \| \mathbf{q}) = 0$ so that

$$\lim_{k \to \infty} \mathbb{D}(\mathbf{p}_k \| \mathbf{q}) = 0 < -\log(q_1) = \mathbb{D}(\mathbf{e}_1 \| \mathbf{q}). \tag{6.80}$$

Therefore, $f_{\mathbf{q}}(\mathbf{p})$ cannot be lower-semi-continuous at $\mathbf{p} = \mathbf{e}_1$. ∎

## 6.2.3 Bijection between Entropies and Relative Entropies

The Rényi entropies are related to the Rényi relative entropies via (see Exercise (6.9))

$$H_\alpha(\mathbf{p}) = \log n - D_\alpha(\mathbf{p}\|\mathbf{u}^{(n)}) \quad \forall \, \mathbf{p} \in \mathrm{Prob}(n). \tag{6.81}$$

More generally, every relative entropy $\mathbb{D}$ can be used to define an entropy $\mathbb{H}$ via

$$\mathbb{H}(\mathbf{p}) := \log n - \mathbb{D}(\mathbf{p}\|\mathbf{u}^{(n)}) \quad \forall \, \mathbf{p} \in \mathrm{Prob}(n). \tag{6.82}$$

**Exercise 6.20.** *Show that if $\mathbb{D}$ is a relative entropy, then $\mathbb{H}$ as defined in (6.82) satisfies the normalization and additivity axioms of an entropy.*

To show that $\mathbb{H}$ as defined in (6.82) is indeed an entropy, we need to prove the monotonicity property (in addition to the properties you proved in this exercise). Recall that if $\mathbf{p}, \mathbf{q} \in \mathrm{Prob}(n)$ and $\mathbf{p} \succ \mathbf{q}$, then there exists a doubly stochastic matrix $D \in \mathrm{STOC}(n, n)$ such that $\mathbf{q} = D\mathbf{p}$. Therefore, in this case we get that

$$\begin{aligned}
\mathbb{H}(\mathbf{q}) = \mathbb{H}(D\mathbf{p}) &= \log n - \mathbb{D}(D\mathbf{p}\|\mathbf{u}^{(n)}) \\
\mathbf{D \text{ is doubly-stochastic}} \to &= \log n - \mathbb{D}(D\mathbf{p}\|D\mathbf{u}^{(n)}) \\
\mathbf{DPI} \to &\geqslant \log n - \mathbb{D}(\mathbf{p}\|\mathbf{u}^{(n)}) \\
&= \mathbb{H}(\mathbf{p}).
\end{aligned} \tag{6.83}$$

That is, $\mathbb{H}$ satisfies the monotonicity property of an entropy if the two vectors have the same dimension. If $\mathbf{p} \in \mathrm{Prob}(n)$ and $\mathbf{q} \in \mathrm{Prob}(m)$ have different dimensions (i.e. $n \neq m$), then the relation $\mathbf{p} \succ \mathbf{q}$ is equivalent to a majorization relation between two vectors with the same dimension $\max\{n, m\}$ in which one of the vectors is padded with zeros to make the dimensions equal. Therefore, to show that $\mathbb{H}$ (6.82) satisfies the monotonicity property of an entropy, it is left to show that it is invariant under embedding; that is, $\mathbb{H}(\mathbf{p} \oplus 0) = \mathbb{H}(\mathbf{p})$ for all $\mathbf{p} \in \mathrm{Prob}(n)$. For this purpose, note that

$$\mathbb{H}(\mathbf{e}_1^{(n)}) = \log n - \mathbb{D}(\mathbf{e}_1^{(n)} \| \mathbf{u}^{(n)}) = \log n - \log n = 0, \tag{6.84}$$

where we used Theorem 6.3. Therefore, from the additivity property that you proved in Exercise 6.20 it follows that for any $\mathbf{p} \in \mathrm{Prob}(n)$,

$$\mathbb{H}(\mathbf{p}) = \mathbb{H}\left(\mathbf{p} \otimes \mathbf{e}_1^{(n+1)}\right) = \mathbb{H}\left((\mathbf{p} \oplus 0) \otimes \mathbf{e}_1^{(n)}\right) = \mathbb{H}(\mathbf{p} \oplus 0). \tag{6.85}$$

Hence, $\mathbb{H}$ satisfies the monotonicity property of an entropy, and when combined with the Exercise 6.20 we conclude that (6.82) demonstrates that for any relative entropy there is a corresponding entropy. Remarkably, the next theorem shows that the converse is also true.

> **One-To-One Correspondence**
>
> **Theorem 6.7.** There exists a bijection $\mathfrak{f}$ with inverse $\mathfrak{f}^{-1}$ mapping between relative entropies that are continuous in the second argument and entropies.

**Proof**   For any relative entropy $\mathbb{D}$ we define

$$\mathfrak{f}(\mathbb{D}) := \mathbb{H}, \tag{6.86}$$

where $\mathbb{H}$ is defined as in (6.82). Since we have already established that $\mathbb{H}$ is an entropy it follows that $\mathfrak{f}$ is indeed a mapping between relative entropies and entropies. We therefore need to show that when the domain of $\mathfrak{f}$ is restricted to relative entropies that are continuous in the second arguments, then it has an inverse.

Let $\mathbb{H}$ be an entropy and define the function

$$D_{\mathbb{H}} : \bigcup_{n \in \mathbb{N}} \mathrm{Prob}(n) \times (\mathrm{Prob}_{>0}(n) \cap \mathbb{Q}^n) \to \mathbb{R} \tag{6.87}$$

via

$$D_{\mathbb{H}}(\mathbf{p}\|\mathbf{q}) := \log n - \mathbb{H}\left(\bigoplus_{x=1}^{n} p_x \mathbf{u}^{(k_x)}\right), \tag{6.88}$$

for all $n \in \mathbb{N}$, $\mathbf{p} \in \mathrm{Prob}(n)$, and $\mathbf{q} \in \mathrm{Prob}_{>0}(n) \cap \mathbb{Q}^n$ with $\mathbf{q} = (\frac{k_1}{k}, \ldots, \frac{k_n}{k})^T$ for $k_x \in \mathbb{N}$ and $k = k_1 + \cdots + k_n$. Note that this construction is equivalent to the one given in Theorem 5.3 with $g(\mathbf{p}) := \log n - \mathbb{H}(\mathbf{p})$, although we do not assume here that $\mathbb{H}$ is continuous, and therefore also $g$ is not assumed to be continuous. Still, the same arguments given in the proof of Theorem 5.3 imply that $D_{\mathbb{H}}$ is a divergence in the restricted domain in which the second argument has positive rational components. Moreover, since $\mathbb{H}$ is additive also $D_{\mathbb{H}}$ as defined (6.20) is additive under tensor products; that is, $D_{\mathbb{H}}$ is a relative entropy with a restricted domain (see Exercise 6.21). This restricted domain will not change the arguments leading to (6.123) and we therefore conclude that for any fixed $n \in \mathbb{N}$ and $\mathbf{p} \in \mathrm{Prob}(n)$, $D_{\mathbb{H}}(\mathbf{p}\|\mathbf{q})$ is continuous in $\mathbf{q} \in \mathrm{Prob}_{>0}(n) \cap \mathbb{Q}^n$. Therefore, the continuous extension of $D_{\mathbb{H}}$ to $\mathrm{Prob}(n) \times \mathrm{Prob}(n)$ is well defined. We therefore define $\mathfrak{f}^{-1}(\mathbb{H}) := D_{\mathbb{H}}$, where $D_{\mathbb{H}}$ is the continuous extension of the expression in (B.3) of [online version] to the full domain $\mathrm{Prob}(n) \times \mathrm{Prob}(n)$. Note that data processing inequality and additivity are preserved under continuous extensions and thus the resulting quantity $D_{\mathbb{H}}$ is indeed a relative entropy, concluding the proof. ∎

**Exercise 6.21.** *Show that $D_{\mathbb{H}}$ as defined in* (B.3 of online version) *is a relative entropy on the restricted domain*

$$\bigcup_{n \in \mathbb{N}} \mathrm{Prob}(n) \times (\mathrm{Prob}_{>0}(n) \cap \mathbb{Q}^n). \tag{6.89}$$

*Explicitly, show that*

1. *Normalization:* $D_{\mathbb{H}}(\mathbf{e}_1^{(2)}\|\mathbf{u}^{(2)}) = 1$.
2. *DPI: for all* $\mathbf{p} \in \mathrm{Prob}(n)$, $\mathbf{q} \in \mathrm{Prob}_{>0}(n) \cap \mathbb{Q}^n$, *and* $E \in \mathrm{STOC}(m,n) \cap \mathbb{Q}_{>0}^{m \times n}$,

$$D_{\mathbb{H}}(E\mathbf{p}\|E\mathbf{q}) \leqslant D_{\mathbb{H}}(\mathbf{p}\|\mathbf{q}). \tag{6.90}$$

   *Hint: Look at the proof of Theorem 5.3.*
3. *Additivity: For all* $\mathbf{p}_1, \mathbf{p}_2 \in \mathrm{Prob}(n)$ *and* $\mathbf{q}_1, \mathbf{q}_2 \in \mathrm{Prob}_{>0}(n) \cap \mathbb{Q}^n$,

$$D_{\mathbb{H}}(\mathbf{p}_1 \otimes \mathbf{p}_2\|\mathbf{q}_1 \otimes \mathbf{q}_2) = D_{\mathbb{H}}(\mathbf{p}_1\|\mathbf{q}_1) + D_{\mathbb{H}}(\mathbf{p}_2\|\mathbf{q}_2). \tag{6.91}$$

In Exercise 6.18, we showed that $D_{\mathrm{path}}(\mathbf{p}\|\mathbf{q}) := D_{\min}(\mathbf{p}\|\mathbf{q}) + D_{\min}(\mathbf{q}\|\mathbf{p})$ provides a counterexample to lower semi-continuity. Note that $\mathfrak{f}(D_{\mathrm{path}}) = H_{\max}$, and $H_{\max}$ is in turn mapped to $D_{\min}(\mathbf{p}\|\mathbf{q})$ by its inverse $\mathfrak{f}^{-1}$; that is, $\mathfrak{f}^{-1}(H_{\max}) = D_{\min}$ so that the contribution $D_{\min}(\mathbf{q}\|\mathbf{p})$ that is discontinuous in $\mathbf{q}$ is lost in the process. This underscores why the continuity of relative entropies in the second argument is essential for the existence of the bijection $\mathfrak{f}$. Finally, observe that the correspondence between relative entropies and entropies allows to import certain results from relative entropies to entropies.

---

**Corollary 6.4.** Let $\mathbb{H}$ be an entropy and $n \in \mathbb{N}$. Then, $\mathbb{H}$ is continuous on $\mathrm{Prob}_{>0}(n)$ and lower semi-continuous everywhere. Moreover, for all $\mathbf{p} \in \mathrm{Prob}(n)$, we have

$$H_{\min}(\mathbf{p}) \leqslant H(\mathbf{p}) \leqslant H_{\max}(\mathbf{p}), \tag{6.92}$$

where the min and max relative entropies have been defined in (6.13) and (6.11), respectively.

---

**Exercise 6.22.** *Prove Corollary 6.4.*

We end this section by recalling Theorem 6.1 proved by [168]. This theorem states that any entropy function can be expressed as a convex combination of Rényi entropies.

Combining this with the one-to-one correspondence between entropies and relative entropies we get the following uniqueness result.

**Uniqueness of Rényi Divergences**

**Corollary 6.5.** Let $\mathbb{D}$ be a relative entropy as defined in Definition 6.3 and that is continuous in its second argument. Then, $\mathbb{D}$ can be expressed as a convex combination of the Rényi divergences.

Observe the crucial need for continuity in the second argument. This is highlighted by the fact that $D_{\text{path}}(\mathbf{p}|\mathbf{q}) := D_{\min}(\mathbf{p}|\mathbf{q}) + D_{\min}(\mathbf{q}|\mathbf{p})$ lacks continuity in its second argument and is not a convex combination of Rényi divergences.

## 6.3 Quantum Relative Entropies

Quantum relative entropies are defined analogously to their classical counterpart. Replacing probability vectors with quantum states we will define a quantum divergence as a function:

$$\mathbb{D}: \bigcup_A \left\{ \mathfrak{D}(A) \times \mathfrak{D}(A) \right\} \to \mathbb{R} \cup \{\infty\} \tag{6.93}$$

that is acting on pairs of quantum states in all finite dimensions $|A| < \infty$.

**Quantum Relative Entropy**

**Definition 6.5.** The function $\mathbb{D}$ as given in (6.93) is called a *quantum relative entropy* if it satisfies the following three conditions:

1. DPI: For all $\rho, \sigma \in \mathfrak{D}(A)$ and all $\mathcal{E} \in \text{CPTP}(A \to B)$ $\mathbb{D}\big(\mathcal{E}(\rho)\|\mathcal{E}(\sigma)\big) \leqslant \mathbb{D}(\rho\|\sigma)$.
2. Additivity: For any $\rho, \sigma \in \mathfrak{D}(A)$ and $\rho', \sigma' \in \mathfrak{D}(B)$,

$$\mathbb{D}\big(\rho \otimes \rho' \| \sigma \otimes \sigma'\big) = \mathbb{D}(\rho\|\sigma) + \mathbb{D}(\rho'\|\sigma'). \tag{6.94}$$

3. Normalization: For $\mathbb{D}\big(|0\rangle\langle 0| \big\| \mathbf{u}^{(2)}\big) = 1$, where $|0\rangle\langle 0|, \mathbf{u}^{(2)} \in \mathfrak{D}(\mathbb{C}^2)$.

*Remark.* Quantum relative entropies can be viewed as generalizations of classical relative entropies. Particularly, a quantum relative entropy reduces to a classical relative entropy when the domain is restricted to diagonal states in a fixed basis, and the diagonals identified with probability vectors.

From the remark above it follows that some of the properties of a quantum relative entropy follow trivially from their classical counterpart.

**Exercise 6.23.** *Let $\mathbb{D}$ be a quantum relative entropy.*

1. *Show that $\mathbb{D}(\rho\|\rho) = 0$.*
2. *Show that if $|0\rangle \in A$ is an eigenvector of $\rho$ then*

$$\mathbb{D}\left(|0\rangle\langle 0|\,\|\,\rho\right) = -\log\langle 0|\rho|0\rangle. \tag{6.95}$$

*Hint: Use Theorem 6.3.*

Theorem 6.3 has an interesting consequence in the quantum domain.

---

**Theorem 6.8.** Let $\mathbb{D}$ be a quantum relative entropy, $n \in \mathbb{N}$, $\mathbf{p} \in \mathrm{Prob}(n)$, $\rho \in \mathfrak{D}(A)$, and $\sigma^{AX} = \sum_{x\in[n]} p_x \sigma_x^A \otimes |x\rangle\langle x|^X$ a cq-state. Then, for any $x \in [n]$,

$$\mathbb{D}\left(\rho^A \otimes |x\rangle\langle x|^X \,\|\, \sigma^{AX}\right) = \mathbb{D}\left(\rho^A \,\|\, \sigma_x^A\right) - \log p_x. \tag{6.96}$$

---

**Proof**  Fix $x \in [n]$ and let $\mathcal{E} \in \mathrm{CPTP}(AX \to AX)$ be a quantum channel that acts as the identity channel if the input on the classical system $X$ is $|x\rangle\langle x|^X$, and otherwise acting as a replacement channel on system $A$ with output $\sigma_x^A$. Explicitly, for all $\tau \in \mathfrak{D}(A)$ and $w \in [n]$

$$\mathcal{E}^{AX\to AX}\left(\tau^A \otimes |w\rangle\langle w|^X\right) := \begin{cases} \tau^A \otimes |x\rangle\langle x|^X & \text{if } w = x \\ \sigma_x^A \otimes |w\rangle\langle w|^X & \text{otherwise.} \end{cases} \tag{6.97}$$

Then, denoting by $\mathbf{p}^X := \sum_{w\in[n]} p_w |w\rangle\langle w|^X$, we get from the DPI of $\mathbb{D}$:

$$\mathbb{D}\left(\rho^A \otimes |x\rangle\langle x|^X \,\|\, \sigma^{AX}\right) \geqslant \mathbb{D}\left(\mathcal{E}^{AX\to AX}\left(\rho^A \otimes |x\rangle\langle x|^X\right) \,\|\, \mathcal{E}^{AX\to AX}\left(\sigma^{AX}\right)\right)$$

$$\mathbf{(6.97)} \to\; = \mathbb{D}\left(\rho^A \otimes |x\rangle\langle x|^X \,\|\, \sigma_x^A \otimes \mathbf{p}^X\right)$$

$$\mathbf{Additivity} \to\; = \mathbb{D}\left(\rho^A \,\|\, \sigma_x^A\right) + \mathbb{D}\left(|x\rangle\langle x|^X \,\|\, \mathbf{p}^X\right) \tag{6.98}$$

$$\mathbf{Theorem\ 6.3} \to\; = \mathbb{D}\left(\rho^A \,\|\, \sigma_x^A\right) - \log p_x.$$

Conversely, for all $w \in [n]$, define $\mathcal{F} \in \mathrm{CPTP}(AX \to AX)$ as

$$\mathcal{F}^{AX\to AX}\left(\tau^A \otimes |w\rangle\langle w|^X\right) := \begin{cases} \tau^A \otimes |x\rangle\langle x|^X & \text{if } w = x \\ \sigma_w^A \otimes |w\rangle\langle w|^X & \text{otherwise.} \end{cases} \tag{6.99}$$

We then get

$$\mathbb{D}\left(\rho^A \otimes |x\rangle\langle x|^X \,\|\, \sigma_x^A \otimes \mathbf{p}^X\right) \geqslant \mathbb{D}\left(\mathcal{F}^{AX\to AX}\left(\rho^A \otimes |x\rangle\langle x|^X\right) \,\|\, \mathcal{F}^{AX\to AX}\left(\sigma_x^A \otimes \mathbf{p}^X\right)\right)$$

$$\mathbf{(6.99)} \to\; = \mathbb{D}\left(\rho^A \otimes |x\rangle\langle x|^X \,\|\, \sigma^{AX}\right). \tag{6.100}$$

The combination of (6.100) with (6.98) concludes the proof.  ∎

**Exercise 6.24.** *Let $\mathbb{D}$ be a quantum relative entropy, $\rho, \sigma \in \mathfrak{D}(A)$, $\omega \in \mathfrak{D}(B)$, and $t \in [0,1]$. In addition, let $Z$ be a $|A| \times |B|$ complex matrix such that $\begin{bmatrix} t\sigma & Z \\ Z^* & (1-t)\omega \end{bmatrix}$ is a density matrix in $\mathfrak{D}(A \oplus B)$. Show that*

$$\mathbb{D}\left( \begin{bmatrix} \rho & 0 \\ 0 & 0 \end{bmatrix} \middle\| \begin{bmatrix} t\sigma & Z \\ Z^* & (1-t)\omega \end{bmatrix} \right) \geq \mathbb{D}(\rho \| \sigma) - \log t, \tag{6.101}$$

*with equality if $Z = 0$.*

**Exercise 6.25.** *Let $\mathbb{D}$ be a quantum relative entropy and $\mathbf{u} \in \mathfrak{D}(A)$ be the maximally mixed state.*

1. *Show that*

$$\mathbb{H}(\rho^A) := \log |A| - \mathbb{D}(\rho^A \| \mathbf{u}^A) \qquad \forall \, \rho \in \mathfrak{D}(A) \tag{6.102}$$

   *is a quantum entropy.*
2. *Show that if $\mathbb{D}$ is jointly convex, then $\mathbb{H}$, as defined above, is concave.*

Before we discuss additional properties of quantum relative entropies, we first consider an example of a family of relative entropies that generalizes the Rényi relative entropies.

## 6.3.1 The Petz Quantum Rényi Divergence

The first generalization of the Rényi divergences that we consider here is perhaps the most straightforward one, in the sense that the expression $\sum_x p_x^\alpha q_x^{1-\alpha}$ is simply replaced by $\mathrm{Tr}\left[\rho^\alpha \sigma^{1-\alpha}\right]$, which looks most reminiscent to its classical counterpart.

> **The Petz Quantum Rényi Divergence**
>
> **Definition 6.6.** For any $\alpha \in [0,2]$ and $\rho, \sigma \in \mathfrak{D}(A)$ the Petz quantum Renyi divergence is defined as
>
> $$D_\alpha(\rho \| \sigma) := \begin{cases} \frac{1}{\alpha-1} \log \mathrm{Tr}\left[\rho^\alpha \sigma^{1-\alpha}\right] & \text{if } \mathrm{supp}\,(\rho) \subseteq \mathrm{supp}\,(\sigma) \text{ or } \alpha < 1 \text{ and } \rho\sigma \neq 0 \\ \infty & \text{otherwise.} \end{cases}$$
>
> The cases $\alpha = 0, 1$ are defined by appropriate limits.

*Remark.* If $\mathrm{supp}\,(\rho) \subseteq \mathrm{supp}\,(\sigma)$, then the trace in the definition above is strictly positive for all $\alpha \in [0, \infty]$. Also, if $\alpha < 1$ and $\rho\sigma \neq 0$ (i.e. $\rho$ and $\sigma$ are not orthogonal), then also $\rho^\alpha \sigma^{1-\alpha} \neq 0$ and we have in this case $\mathrm{Tr}\left[\rho^\alpha \sigma^{1-\alpha}\right] > 0$. In all other cases, the trace in the definition above is either zero or not well defined. One can also extend the definition to $\alpha > 2$; however, we will see Theorem 6.9 that the DPI only holds for $\alpha \in [0, 2]$.

**Exercise 6.26.** *Prove all the statements in the remark above (except for the very last one about $\alpha > 2$).*

**Exercise 6.27.** *Let $\rho, \sigma \in \mathfrak{D}(A)$ and consider their spectral decomposition as given in (5.67). Set $m := |A|$, and let $\mathbf{p}^{XY}, \mathbf{q}^{XY} \in \mathrm{Prob}(m^2)$ be the probability vectors whose components are $\{p_x |\langle a_x | b_y \rangle|^2\}_{x,y \in [m]}$ and $\{q_y |\langle a_x | b_y \rangle|^2\}_{x,y \in [m]}$, respectively (cf. (5.68)). Show that*

$$D_\alpha(\rho \| \sigma) = D_\alpha\left(\mathbf{p}^{XY} \| \mathbf{q}^{XY}\right), \tag{6.103}$$

*where the right-hand side is the classical Rényi divergence between $\mathbf{p}^{XY}$ and $\mathbf{q}^{XY}$.*

The Petz-Rényi divergence satisfies all the properties of a relative entropy. The normalization and additivity properties you will prove in the following exercise, and we now prove the data processing inequality.

> **Theorem 6.9.** The Petz quantum $\alpha$-Rényi divergence is a relative entropy for any $\alpha \in [0, 2]$.

**Proof** In (5.89), we proved that the quantum $\alpha$-Divergence

$$D^q_{f_\alpha}(\rho \| \sigma) = \frac{1}{\alpha(\alpha - 1)} \left( \mathrm{Tr}\left[ \rho^\alpha \sigma^{1-\alpha} \right] - 1 \right) \tag{6.104}$$

is a divergence for $\alpha \in [0, 2]$. Therefore, for $\alpha \in [0, 1)$ the expression $\mathrm{Tr}\left[ \rho^\alpha \sigma^{1-\alpha} \right]$ is monotonically increasing under mappings $(\rho, \sigma) \mapsto (\mathcal{E}(\rho), \mathcal{E}(\sigma))$ with $\mathcal{E} \in \mathrm{CPTP}$ $(A \to B)$. Similarly, for $\alpha \in (1, 2]$ the expression $\mathrm{Tr}\left[ \rho^\alpha \sigma^{1-\alpha} \right]$ is monotonically decreasing under such mappings. Therefore, for any $1 \neq \alpha \in [0, 2]$ the Petz quantum Rényi $\alpha$-Divergence satisfies the DPI. The DPI for the case $\alpha = 1$ has been proven in (5.87). $\blacksquare$

**Exercise 6.28.** *Show that for any $\alpha \geqslant 0$, the Petz quantum Rényi entropy $D_\alpha$ satisfies the normalization and additivity properties of a relative entropy.*

The calculation of the limits $\alpha \to 0, 1$ of the Petz quantum Rényi divergence is a bit more subtle than the classical case. For this purpose, we will use the expression (6.103) in Exercise 6.27. For the limit $\alpha \to 0$ observe that

$$\lim_{\alpha \to 0^+} D_\alpha(\rho \| \sigma) = \lim_{\alpha \to 0^+} D_\alpha\left(\mathbf{p}^{XY} \| \mathbf{q}^{XY}\right) \tag{6.105}$$

$$= -\log \sum_{\substack{x,y \in [m] \\ p_x |\langle u_x | v_y \rangle|^2 > 0}} q_y |\langle u_x | v_y \rangle|^2 \tag{6.106}$$

$$= -\log \sum_{\substack{x,y \in [m] \\ p_x > 0}} q_y |\langle u_x | v_y \rangle|^2 \tag{6.107}$$

$$= -\log \mathrm{Tr}[\Pi_\rho \sigma], \tag{6.108}$$

where $\Pi_\rho$ is the projection to the support of $\rho$. The quantity above is also known as the min quantum relative entropy and is denoted by

> **The Min Quantum Relative Entropy**
>
> $$D_{\min}(\rho\|\sigma) := \begin{cases} -\log \mathrm{Tr}[\Pi_\rho \sigma] & \text{if } \rho\sigma \neq 0 \\ \infty & \text{otherwise.} \end{cases} \tag{6.109}$$

Note that the quantum min relative entropy reduces to the classical min relative entropy when the states are classical (i.e. diagonal).

For the limit $\alpha \to 1$ we use again (6.103) to get

$$\lim_{\alpha\to 1} D_\alpha(\rho\|\sigma) = \lim_{\alpha\to 1} D_\alpha\left(\mathbf{p}^{XY}\|\mathbf{q}^{XY}\right) = D\left(\mathbf{p}^{XY}\|\mathbf{q}^{XY}\right)$$

$$= \sum_{x,y} p_x |\langle u_x|v_y\rangle|^2 \log\left(\frac{p_x}{q_y}\right) \tag{6.110}$$

$$= \mathrm{Tr}[\rho \log \rho] - \mathrm{Tr}[\rho \log \sigma] = D(\rho\|\sigma),$$

where $D(\rho\|\sigma)$ is the Umegaki relative entropy.

**Exercise 6.29.** *Prove the last two lines in the equation above; particularly, show that* $\sum_{x,y} p_x |\langle u_x|v_y\rangle|^2 \log\left(\frac{p_x}{q_y}\right) = \mathrm{Tr}[\rho \log \rho] - \mathrm{Tr}[\rho \log \sigma]$.

**Exercise 6.30 (Quasi-Convexity).** *Show that for any* $\alpha \in [0,2]$, $\rho, \omega_0, \omega_1 \in \mathfrak{D}(A)$, *and* $t \in [0,1]$ *we have*

$$D_\alpha\left(\rho\|t\omega_0 + (1-t)\omega_1\right) \leqslant \max\left\{D_\alpha(\rho\|\omega_0), D_\alpha(\rho\|\omega_1)\right\}. \tag{6.111}$$

Similar to the definition of the min quantum relative entropy, we can extend the max relative entropy to the quantum domain.

> **The Max Quantum Relative Entropy**
>
> The max quantum relative entropy is defined for all $\rho, \sigma \in \mathfrak{D}(A)$ as
>
> $$D_{\max}(\rho\|\sigma) := \log\min\left\{t \in \mathbb{R} : t\sigma \geqslant \rho\right\} \tag{6.112}$$
>
> for the case that $\mathrm{supp}(\rho) \subseteq \mathrm{supp}(\sigma)$ and otherwise it is set to $\infty$.

**Exercise 6.31.** *Show that:*

1. *The max quantum relative entropy is indeed a relative entropy.*
2. $D_{\max}(\rho\|\sigma)$ *reduces to the classical max relative entropy when* $\rho$ *and* $\sigma$ *commutes.*
3. *For the case that* $\mathrm{supp}(\rho) \subseteq \mathrm{supp}(\sigma)$ $D_{\max}(\rho\|\sigma) = \log\|\sigma^{-\frac{1}{2}}\rho\sigma^{-\frac{1}{2}}\|_\infty$. *Hint: Conjugate both sides of* $t\sigma \geqslant \rho$ *by* $\sigma^{-\frac{1}{2}}(\,\cdot\,)\sigma^{-\frac{1}{2}}$.

4. $D_{\max}(\rho\|\sigma) = \lim_{\alpha\to\infty} D_\alpha(\rho\|\sigma)$ *if $\rho$ and $\sigma$ commute, and give an example for which* $D_{\max}(\rho\|\sigma) \neq \lim_{\alpha\to\infty} D_\alpha(\rho\|\sigma)$. *Here $D_\alpha$ refers to the same formula as the Petz quantum Rényi divergence but with $\alpha > 2$.*

## 6.3.2 Basic Properties

In this subsection, we will see that several of the properties of classical relative entropies carry over to the quantum domain. However, the proofs of these properties have to be adjusted to incorporate the larger domain.

> **Theorem 6.10.** Let $\mathbb{D}$ be a relative entropy. Then for any quantum system $A$ and any $\rho, \sigma, \omega \in \mathfrak{D}(A)$:
>
> 1. Bounds:
> $$D_{\min}(\rho\|\sigma) \leqslant \mathbb{D}(\rho\|\sigma) \leqslant D_{\max}(\rho\|\sigma). \tag{6.113}$$
>
> 2. Triangle inequality:
> $$\mathbb{D}(\rho\|\sigma) \leqslant \mathbb{D}(\rho\|\omega) + D_{\max}(\omega\|\sigma). \tag{6.114}$$

**Proof**    Let $\Pi_\rho$ denote the projector to the support of $\rho$. Define the POVM Channel $\mathcal{E} \in \mathrm{CPTP}(A \to X)$ with $|X| = 2$ as

$$\mathcal{E}(\sigma) := \mathrm{Tr}\big[\sigma\,\Pi_\rho\big]|0\rangle\langle 0|^X + \mathrm{Tr}\big[\sigma\,(I - \Pi_\rho)\big]|1\rangle\langle 1|^X. \tag{6.115}$$

Then,

$$
\begin{aligned}
\mathbb{D}(\rho\|\sigma) &\geqslant \mathbb{D}\big(\mathcal{E}(\rho)\|\mathcal{E}(\sigma)\big) \\
\textbf{(6.115)} \to &= \mathbb{D}\left(|0\rangle\langle 0|\,\Big\|\,\mathrm{Tr}\big[\sigma\,\Pi_\rho\big]|0\rangle\langle 0| + \mathrm{Tr}\big[\sigma\,(I - \Pi_\rho)\big]|1\rangle\langle 1|\right) \\
\textbf{Theorem 6.3} \to &= -\log \mathrm{Tr}\big[\sigma\,\Pi_\rho\big] \\
&= D_{\min}(\rho\|\sigma).
\end{aligned}
\tag{6.116}
$$

For the second inequality, denote by $t = 2^{D_{\max}(\rho\|\sigma)}$, and note that in particular, $t\sigma \geqslant \rho$ (i.e. $t\sigma - \rho \geqslant 0$). Define a channel $\mathcal{E} \in \mathrm{CPTP}(X \to A)$ with $|X| = 2$ by

$$\mathcal{E}(|0\rangle\langle 0|) := \rho \quad \text{and} \quad \mathcal{E}(|1\rangle\langle 1|) := \frac{t\sigma - \rho}{t - 1}. \tag{6.117}$$

Furthermore, denote

$$\mathbf{q}^X := \frac{1}{t}|0\rangle\langle 0|^X + \frac{t-1}{t}|1\rangle\langle 1|^X, \tag{6.118}$$

and observe that $\mathcal{E}(\mathbf{q}^X) = \sigma$. Hence,

$$\mathbb{D}(\rho\|\sigma) = \mathbb{D}\Big(\mathcal{E}(|0\rangle\langle 0|^X)\big\|\mathcal{E}(\mathbf{q}^X)\Big)$$

$$\mathbf{DPI}\to \ \leqslant \mathbb{D}\Big(|0\rangle\langle 0|^X\big\|\mathbf{q}^X\Big) \tag{6.119}$$

$$\mathbf{Theorem\ 6.3}\to \ = -\log\frac{1}{t} = D_{\max}(\rho\|\sigma).$$

This completes the proof of (6.113).

To prove the triangle inequality (6.114), note first that for $|A| = 1$ the statement is trivial so we can assume $|A| \geqslant 2$. Let $\varepsilon := 2^{-D_{\max}(\omega\|\sigma)} \in (0,1)$, and observe that $\sigma \geqslant \varepsilon\omega$ so that the matrix $\tau := (\sigma - \varepsilon\omega)/(1-\varepsilon)$ is a density matrix satisfying

$$\sigma = \varepsilon\omega + (1-\varepsilon)\tau. \tag{6.120}$$

From the definition of $\varepsilon$ we have

$$\begin{aligned}
\mathbb{D}(\rho\|\omega) + D_{\max}(\omega\|\sigma) &= \mathbb{D}(\rho\|\omega) - \log\varepsilon\\
\mathbf{Theorem\ 6.3}\to &= \mathbb{D}(\rho\|\omega) + \mathbb{D}\big(|0\rangle\langle 0|\,\big\|\,\varepsilon|0\rangle\langle 0| + (1-\varepsilon)|1\rangle\langle 1|\big)\\
\mathbf{Additivity}\to &= \mathbb{D}\Big(\rho\otimes|0\rangle\langle 0|\,\Big\|\,\omega\otimes\big(\varepsilon|0\rangle\langle 0| + (1-\varepsilon)|1\rangle\langle 1|\big)\Big)\\
\mathbf{DPI}\to &\geqslant \mathbb{D}(\rho\|\sigma),
\end{aligned} \tag{6.121}$$

where in the last inequality we used the DPI property of $\mathbb{D}$ with a quantum channel that acts as an identity upon measuring $|0\rangle\langle 0|$ in the second register, and produces a constant output $\tau$ upon measuring $|1\rangle\langle 1|$ in the second register. ∎

**Exercise 6.32.** *The quantum Thompson's metric is defined for any $\rho,\sigma \in \mathrm{Prob}(n)$ by*

$$D_T(\rho\|\sigma) := \max\Big\{D_{\max}(\rho\|\sigma), D_{\max}(\sigma\|\rho)\Big\}. \tag{6.122}$$

1. *Prove that the quantum Thompson's metric is both a quantum divergence and a metric in $\mathfrak{D}(A) \times \mathfrak{D}(A)$.*
2. *Prove that any quantum relative entropy $\mathbb{D}$ satisfies for all $\rho,\sigma,\sigma' \in \mathfrak{D}(A)$*

$$\big|\mathbb{D}(\rho\|\sigma) - \mathbb{D}(\rho\|\sigma')\big| \leqslant D_T(\sigma\|\sigma'). \tag{6.123}$$

This exercise demonstrates that quantum relative entropies are continuous in their second argument. One can also get a continuity property in the first argument.

---

**Lemma 6.3.** Let $\rho,\rho',\sigma \in \mathfrak{D}(A)$ be quantum states. Then, we have

$$\mathbb{D}(\rho\|\sigma) - \mathbb{D}(\rho'\|\sigma) \leqslant \min_{0\leqslant s\leqslant 2^{-D_{\max}(\rho'\|\rho)}} D_{\max}\big(\rho + s(\sigma-\rho')\big\|\sigma\big) \tag{6.124}$$

$$\leqslant \log\left(1 + \frac{\|\rho-\rho'\|_\infty}{\lambda_{\min}(\rho')\lambda_{\min}(\sigma)}\right), \tag{6.125}$$

where the second inequality holds if $\sigma > 0$ and $\lambda_{\min}(\rho') > \|\rho-\rho'\|_\infty$.

**Proof** In somewhat of a variation of the previous theorem, fix $0 \leqslant s \leqslant 2^{-D_{\max}(\rho'\|\rho)}$ and denote by $\varepsilon := 2^{-D_{\max}(\rho + s(\sigma - \rho')\|\sigma)}$. Then,

$$\mathbb{D}(\rho'\|\sigma) + D_{\max}\left(\rho + s(\sigma - \rho')\|\sigma\right) = \mathbb{D}(\rho'\|\sigma) - \log \varepsilon$$

$$\textbf{Theorem 6.3} \rightarrow = \mathbb{D}(\rho'\|\sigma) + \mathbb{D}\left(|0\rangle\langle 0|\,\Big\|\,\varepsilon|0\rangle\langle 0| + (1 - \varepsilon)|1\rangle\langle 1|\right)$$

$$\textbf{Additivity} \rightarrow = \mathbb{D}\left(\rho' \otimes |0\rangle\langle 0|\,\Big\|\,\sigma \otimes \left(\varepsilon|0\rangle\langle 0| + (1 - \varepsilon)|1\rangle\langle 1|\right)\right)$$

$$\textbf{DPI} \rightarrow \geqslant \mathbb{D}\left(\mathcal{N}(\rho')\,\big\|\,\varepsilon\mathcal{N}(\sigma) + (1 - \varepsilon)\omega\right),$$

$$\tag{6.126}$$

where in the last inequality we used the DPI with a channel that acts as some channel $\mathcal{N} \in \mathrm{CPTP}(A \to A)$ when measuring $|0\rangle\langle 0|$ in the second register and outputs some state $\omega \in \mathfrak{D}(A)$ when measuring $|1\rangle\langle 1|$. In other words, the inequality above holds for all $\mathcal{N} \in \mathrm{CPTP}(A \to A)$ and all $\omega \in \mathfrak{D}(A)$. It is therefore left to show that there exists such $\mathcal{N}$ and $\omega$ that satisfy $\mathcal{N}(\rho') = \rho$ and $\varepsilon\mathcal{N}(\sigma) + (1 - \varepsilon)\omega = \sigma$. The latter implies that we can define $\omega$ to be

$$\omega := \frac{\sigma - \varepsilon\mathcal{N}(\sigma)}{1 - \varepsilon}. \tag{6.127}$$

Note that we need to choose $\mathcal{N}$ such that $\sigma - \varepsilon\mathcal{N}(\sigma) \geqslant 0$ so that $\omega \in \mathfrak{D}(A)$. We take $\mathcal{N} \in \mathrm{CPTP}(A \to A)$ to be a measurement-prepare channel of the form

$$\mathcal{N}(\eta) := s\eta + (1 - s)\tau \qquad \forall \eta \in \mathfrak{L}(A), \tag{6.128}$$

where we want to choose $\tau$ such that both $\mathcal{N}(\rho') = \rho$ and $\sigma - \varepsilon\mathcal{N}(\sigma) \geqslant 0$. The condition $\mathcal{N}(\rho') = \rho$ can be expressed as $\rho = s\rho' + (1 - s)\tau$. Isolating $\tau$ we get that

$$\tau = \frac{\rho - s\rho'}{1 - s}. \tag{6.129}$$

The above matrix is positive semidefinite if and only if $\rho \geqslant s\rho'$, which holds since $s \leqslant 2^{-D_{\max}(\rho'\|\rho)}$. We therefore choose $\tau$ as above so that $\mathcal{N}(\rho') = \rho$. It is left to check that $\sigma - \varepsilon\mathcal{N}(\sigma) \geqslant 0$. Indeed, since $\mathcal{N}(\sigma) := s\sigma + (1 - s)\tau$ we have

$$\sigma - \varepsilon\mathcal{N}(\sigma) = (1 - \varepsilon s)\sigma - \varepsilon(1 - s)\tau$$

$$\textbf{(6.129)} \rightarrow = (1 - \varepsilon s)\sigma - \varepsilon(\rho - s\rho')$$

$$= \sigma - \varepsilon\left(\rho + s(\sigma - \rho')\right)$$

$$\tag{6.130}$$

$$\textbf{By definition of } \varepsilon \rightarrow \geqslant 0.$$

To summarize, we showed that for any $0 \leqslant s \leqslant 2^{-D_{\max}(\rho'\|\rho)}$ we have

$$\mathbb{D}(\rho'\|\sigma) + D_{\max}\left(\rho + s(\sigma - \rho')\|\sigma\right) \geqslant D(\rho\|\sigma). \tag{6.131}$$

Since the above equation holds for all $s \leqslant 2^{-D_{\max}(\rho'\|\rho)}$ we conclude that the inequality (6.124) holds.

To prove the second inequality, observe first that the inequality (6.124) can be expressed as

$$\mathbb{D}(\rho\|\sigma) - \mathbb{D}(\rho'\|\sigma) \leqslant \log \min\left\{r \geqslant 0 : (r - s)\sigma \geqslant \rho - s\rho' \geqslant 0, \, s \geqslant 0\right\}. \tag{6.132}$$

Since we assume now that $\mu := \lambda_{\min}(\sigma) > 0$, we can take $r = 1 + \frac{1-s}{\mu}$. Note that for this choice of $r$ we have

$$(r - s)\sigma = (1 - s)(1 + \mu)\frac{\sigma}{\mu} \geqslant (1 - s)(1 + \mu)I^A \geqslant \rho - s\rho', \qquad (6.133)$$

since $\rho - s\rho'$ is a subnormalized state with trace $1 - s$. Moreover, if $\lambda_{\min}(\rho') \geqslant \|\rho - \rho'\|_\infty$, then we can take $s = 1 - \frac{\|\rho - \rho'\|_\infty}{\lambda_{\min}(\rho')}$ since in this case $s \leqslant 2^{-D_{\max}(\rho'\|\rho)}$ (or equivalently $\rho \geqslant s\rho'$, see Exercise 6.33). We therefore get for these choices of $r$ and $s$

$$\mathbb{D}(\rho\|\sigma) - \mathbb{D}(\rho'\|\sigma) \leqslant \log r = \log\left(1 + \frac{\|\rho - \rho'\|_\infty}{\lambda_{\min}(\rho')\lambda_{\min}(\sigma)}\right). \qquad (6.134)$$

This completes the proof. ∎

**Exercise 6.33.** *Show that if $\lambda_{\min}(\rho') \geqslant \|\rho - \rho'\|_\infty > 0$, then $\rho \geqslant s\rho'$, where $s = 1 - \frac{\|\rho - \rho'\|_\infty}{\lambda_{\min}(\rho')}$.*

**Exercise 6.34.** *Show that if $\rho, \sigma \in \mathfrak{D}(A)$ and $\lambda_{\min}(\rho) > \|\sigma - \rho\|_\infty$, then*

$$D_{\max}(\rho\|\sigma) \leqslant -\log\left(1 - \frac{\|\sigma - \rho\|_\infty}{\lambda_{\min}(\rho)}\right). \qquad (6.135)$$

*Use this to get a bound on $D_T(\sigma\|\sigma')$ in (6.123).*

> ### Continuity of Quantum Relative Entropies
>
> **Theorem 6.11.** Let $\mathbb{D}$ be a quantum relative entropy. Then, $\mathbb{D}$ is upper semi-continuous at any point in $\mathfrak{D}(A) \times \mathfrak{D}_{>0}(A)$, and is continuous at any point in $\mathfrak{D}_{>0}(A) \times \mathfrak{D}_{>0}(A)$.

**Proof**　Let $(\rho_k, \sigma_k)_{k \in \mathbb{N}}$ be a sequence in $\mathfrak{D}(A) \times \mathfrak{D}(A)$ that converges to $(\rho, \sigma)$. For any $k \in \mathbb{N}$, define a quantum channel $\mathcal{E}_k \in \mathrm{CPTP}(A \to A)$ by its action on any $\omega \in \mathfrak{D}(A)$ as

$$\mathcal{E}_k(\omega) := \rho_k + 2^{-D_{\max}(\rho\|\rho_k)}(\omega - \rho). \qquad (6.136)$$

Note that for sufficiently large $k$, $2^{-D_{\max}(\rho\|\rho_k)} > 0$ (see the following exercise). Moreover, observe that $\rho_k - 2^{-D_{\max}(\rho\|\rho_k)}\rho \geqslant 0$ so that $\mathcal{E}_k$ is indeed a quantum channel. With the above notations we get from the DPI

$$\begin{aligned}
\mathbb{D}(\rho\|\sigma) &\geqslant \mathbb{D}\big(\mathcal{E}_k(\rho)\big\|\mathcal{E}_k(\sigma)\big) \\
(6.136) \to {}&= \mathbb{D}\big(\rho_k\big\|\mathcal{E}_k(\sigma)\big) \\
(6.114) \to {}&\geqslant \mathbb{D}(\rho_k\|\sigma_k) - D_{\max}\big(\mathcal{E}_k(\sigma)\big\|\sigma_k\big).
\end{aligned} \qquad (6.137)$$

Moving the term involving $D_{\max}$ to the other side and taking the supremum limit on both sides gives

$$\limsup_{k\to\infty} \mathbb{D}(\rho_k\|\sigma_k) \leqslant \mathbb{D}(\rho\|\sigma) + \limsup_{k\to\infty} D_{\max}\big(\mathcal{E}_k(\sigma)\big\|\sigma_k\big). \qquad (6.138)$$

The second term on the right-hand side above vanishes since the density matrix

$$\tilde{\sigma}_k := \mathcal{E}_k(\sigma) = 2^{-D_{\max}(\rho\|\rho_k)}\sigma + \left(\rho_k - 2^{-D_{\max}(\rho\|\rho_k)}\rho\right) \tag{6.139}$$

has a limit $\lim_{k\to\infty}\tilde{\sigma}_k = \sigma$ so that

$$\limsup_{k\to\infty} D_{\max}(\tilde{\sigma}_k\|\sigma_k) = 0. \tag{6.140}$$

Note that we used indirectly the fact that $\sigma > 0$, since for sufficiently large $k$ we must have $\sigma_k > 0$ so the limit above is indeed zero. This completes the proof that $\mathbb{D}$ is upper semi-continuous on $\mathfrak{D}(A) \times \mathfrak{D}_{>0}(A)$.

We now prove the lower semi-continuity on $\mathfrak{D}_{>0}(A) \times \mathfrak{D}_{>0}(A)$. Note that since we already proved upper semi-continuity in this domain, this will imply that $\mathbb{D}$ is continuous on $\mathfrak{D}_{>0}(A) \times \mathfrak{D}_{>0}(A)$. For any $k \in \mathbb{N}$, we define $\mathcal{E}_k$ as before but with the role of $\rho_k$ and $\rho$ interchanged; that is, $\mathcal{E}_k \in \mathrm{CPTP}(A \to A)$ is defined by its action on any $\omega \in \mathfrak{D}(A)$ as

$$\mathcal{E}_k(\omega) := \rho + 2^{-D_{\max}(\rho_k\|\rho)}(\omega - \rho_k). \tag{6.141}$$

Since we assume that $\rho > 0$ we get that $2^{-D_{\max}(\rho_k\|\rho)} > 0$ for all $k$. Moreover, observe that $\rho - 2^{-D_{\max}(\rho_k\|\rho)}\rho_k \geqslant 0$ so that $\mathcal{E}_k$ is indeed a quantum channel. With the above notations we get from the DPI

$$\begin{aligned}
\mathbb{D}(\rho_k\|\sigma_k) &\geqslant \mathbb{D}\big(\mathcal{E}_k(\rho_k)\big\|\mathcal{E}_k(\sigma_k)\big) \\
\textbf{(6.141)} \rightarrow \; &= \mathbb{D}\big(\rho\big\|\mathcal{E}_k(\sigma_k)\big) \\
\textbf{(6.114)} \rightarrow \; &\geqslant \mathbb{D}(\rho\|\sigma) - D_{\max}\big(\mathcal{E}_k(\sigma_k)\big\|\sigma\big).
\end{aligned} \tag{6.142}$$

Taking the infimum limit on both sides gives

$$\liminf_{k\to\infty} \mathbb{D}(\rho_k\|\sigma_k) \geqslant \mathbb{D}(\rho\|\sigma), \tag{6.143}$$

where we used the fact that

$$\mathcal{E}_k(\sigma_k) = 2^{-D_{\max}(\rho_k\|\rho)}\sigma_k + \left(\rho - 2^{-D_{\max}(\rho_k\|\rho)}\rho_k\right) \tag{6.144}$$

has a limit $\lim_{k\to\infty}\mathcal{E}_k(\sigma_k) = \sigma$ so that

$$\limsup_{k\to\infty} D_{\max}(\mathcal{E}_k(\sigma_k)\|\sigma) = 0. \tag{6.145}$$

This completes the proof. ∎

**Exercise 6.35.**

1. *Show that if $\{\rho_k\}_{k\in\mathbb{N}}$ is a sequences in $\mathfrak{D}(A)$ that converges to $\rho \in \mathfrak{D}(A)$, then for sufficiently large $k$ we have $D_{\max}(\rho\|\rho_k) < \infty$. Hint: Show that for sufficiently large $k$, $\mathrm{supp}(\rho) \subseteq \mathrm{supp}(\rho_k)$.*
2. *Prove the limits in (6.140) and (6.145).*

# 6.4 Optimal Quantum Extensions of Relative Entropies

The minimal and maximal quantum extensions, $\underline{\mathbb{D}}$ and $\overline{\mathbb{D}}$, of a classical divergence $\mathbb{D}$ are the smallest and largest quantum divergences that reduce to $\mathbb{D}$ on classical states. We encountered them in Section 5.3 particularly through (5.91,5.92). However, the expressions given in (5.91,5.92) for $\underline{\mathbb{D}}$ and $\overline{\mathbb{D}}$ are, in general, not additive under tensor products even if the classical divergence $\mathbb{D}$ is additive. Therefore, in order to get the optimal extensions of relative entropies, we will use regularization to make the quantum extensions at least partially additive.

Suppose $\mathbb{D}$ is a classical relative entropy and define $\underline{\mathbb{D}}$ and $\overline{\mathbb{D}}$ as in (5.91); that is,

$$\underline{\mathbb{D}}(\rho\|\sigma) := \sup \mathbb{D}\big(\mathcal{E}(\rho)\|\mathcal{E}(\sigma)\big), \tag{6.146}$$

$$\overline{\mathbb{D}}(\rho\|\sigma) := \inf\big\{\mathbb{D}(\mathbf{p}\|\mathbf{q}) : \rho = \mathcal{F}(\mathbf{p}),\ \sigma = \mathcal{F}(\mathbf{q})\big\}, \tag{6.147}$$

where the optimizations are over the classical system $X$, the channels $\mathcal{E} \in \mathrm{CPTP}(A \to X)$ and $\mathcal{F} \in \mathrm{CPTP}(X \to A)$ as well as the diagonal density matrices $\mathbf{p}, \mathbf{q} \in \mathfrak{D}(X)$. The functions $\underline{\mathbb{D}}$ and $\overline{\mathbb{D}}$ are in general not additive even if the $\mathbb{D}$ is a classical relative entropy (and therefore additive). However, in the following lemma we show that in this case $\underline{\mathbb{D}}$ is super-additive, while $\overline{\mathbb{D}}$ is sub-additive.

> **Lemma 6.4.** Let $\mathbb{D}$ be a classical relative entropy, and let $\overline{\mathbb{D}}$ and $\underline{\mathbb{D}}$ be its maximal and minimal quantum extensions as defined in (5.91). Then, for all $\rho_1, \sigma_1 \in \mathfrak{D}(A_1)$ and $\rho_2, \sigma_2 \in \mathfrak{D}(A_2)$ we have:
>
> 1. Super-additivity: $\underline{\mathbb{D}}\left(\rho_1 \otimes \rho_2 \| \sigma_1 \otimes \sigma_2\right) \geq \underline{\mathbb{D}}(\rho_1\|\sigma_1) + \underline{\mathbb{D}}(\rho_2\|\sigma_2)$.
> 2. Sub-additivity: $\overline{\mathbb{D}}\left(\rho_1 \otimes \rho_2 \| \sigma_1 \otimes \sigma_2\right) \leq \overline{\mathbb{D}}(\rho_1\|\sigma_1) + \overline{\mathbb{D}}(\rho_2\|\sigma_2)$.

**Proof**   We will prove the sub-additivity property and leave it as an exercise to prove the super-additivity using similar lines. By definition we have

$$\underline{\mathbb{D}}\left(\rho_1 \otimes \rho_2 \| \sigma_1 \otimes \sigma_2\right) = \sup_{\mathcal{E}\in\mathrm{CPTP}(A_1 A_2 \to X)} \mathbb{D}\big(\mathcal{E}(\rho_1 \otimes \rho_2)\|\mathcal{E}(\sigma_1 \otimes \sigma_2)\big)$$

$$\boxed{\text{Restricting } \mathcal{E} = \mathcal{E}_1 \otimes \mathcal{E}_2 \to} \quad \geq \sup_{\substack{\mathcal{E}_1\in\mathrm{CPTP}(A_1 \to X_1) \\ \mathcal{E}_2\in\mathrm{CPTP}(A_2 \to X_2)}} \mathbb{D}\big(\mathcal{E}_1(\rho_1) \otimes \mathcal{E}_2(\rho_2)\|\mathcal{E}_1(\sigma_1) \otimes \mathcal{E}_2(\sigma_2)\big)$$

$$\mathbf{Additivity\ of\ } \mathbb{D} \to\ = \sup_{\mathcal{E}_1} \mathbb{D}\big(\mathcal{E}_1(\rho_1)\|\mathcal{E}_1(\sigma_1)\big) + \sup_{\mathcal{E}_2} \mathbb{D}\big(\mathcal{E}_2(\rho_2)\|\mathcal{E}_2(\sigma_2)\big)$$

$$= \underline{\mathbb{D}}(\rho_1\|\sigma_1) + \underline{\mathbb{D}}(\rho_2\|\sigma_2). \tag{6.148}$$

This completes the proof.   ∎

**Exercise 6.36.** *Prove the sub-additivity of* $\overline{\mathbb{D}}$.

Since $\underline{\mathbb{D}}$ and $\overline{\mathbb{D}}$ are not necessarily additive, we define their regularization as

$$\underline{\mathbb{D}}^{\text{reg}}(\rho\|\sigma) := \lim_{n\to\infty} \frac{1}{n}\underline{\mathbb{D}}\left(\rho^{\otimes n}\|\sigma^{\otimes n}\right) \quad \text{and} \quad \overline{\mathbb{D}}^{\text{reg}}(\rho\|\sigma) := \lim_{n\to\infty} \frac{1}{n}\overline{\mathbb{D}}\left(\rho^{\otimes n}\|\sigma^{\otimes n}\right).$$

$$(6.149)$$

In Exercise 6.37 we will show that the limits above exist and that in general $\underline{\mathbb{D}}^{\text{reg}}(\rho\|\sigma) \geqslant \underline{\mathbb{D}}(\rho\|\sigma)$ and $\overline{\mathbb{D}}^{\text{reg}}(\rho\|\sigma) \leqslant \overline{\mathbb{D}}(\rho\|\sigma)$. Moreover, note that by definition, $\underline{\mathbb{D}}^{\text{reg}}$ and $\overline{\mathbb{D}}^{\text{reg}}$ are at least partially additive in the sense that for any $n \in \mathbb{N}$ and any $\rho,\sigma \in \mathfrak{D}(A)$,

$$\underline{\mathbb{D}}^{\text{reg}}\left(\rho^{\otimes n}\|\sigma^{\otimes n}\right) = n\underline{\mathbb{D}}^{\text{reg}}\left(\rho\|\sigma\right) \quad \text{and} \quad \overline{\mathbb{D}}^{\text{reg}}\left(\rho^{\otimes n}\|\sigma^{\otimes n}\right) = n\overline{\mathbb{D}}^{\text{reg}}\left(\rho\|\sigma\right). \quad (6.150)$$

It is an open problem to determine if $\underline{\mathbb{D}}^{\text{reg}}$ and $\overline{\mathbb{D}}^{\text{reg}}$ are fully additive. We will see that in many examples, $\underline{\mathbb{D}}^{\text{reg}}$ and $\overline{\mathbb{D}}^{\text{reg}}$ turn out to be fully additive so that they are in fact relative entropies. The following theorem shows that these functions remains optimal.

> **Theorem 6.12.** Let $\mathbb{D}$ be a classical relative entropy, and let $\underline{\mathbb{D}}^{\text{reg}}$ and $\overline{\mathbb{D}}^{\text{reg}}$ be as in (6.149). Then, both $\underline{\mathbb{D}}^{\text{reg}}$ and $\overline{\mathbb{D}}^{\text{reg}}$ are partially additive quantum divergences that reduce to $\mathbb{D}$ on classical states. In addition, any other quantum relative entropy $D'$ that reduces to $\mathbb{D}$ on classical states satisfies for all $\rho,\sigma \in \mathfrak{D}(A)$
>
> $$\underline{\mathbb{D}}^{\text{reg}}(\rho\|\sigma) \leqslant D'(\rho\|\sigma) \leqslant \overline{\mathbb{D}}^{\text{reg}}(\rho\|\sigma). \quad (6.151)$$

*Remark.* Observe that since in general $\underline{\mathbb{D}}^{\text{reg}}(\rho\|\sigma) \geqslant \underline{\mathbb{D}}(\rho\|\sigma)$ and $\overline{\mathbb{D}}^{\text{reg}}(\rho\|\sigma) \leqslant \overline{\mathbb{D}}(\rho\|\sigma)$, the bounds on $D'$ above are tighter than the bounds given in (5.90). We are able to get tighter bounds since $\mathbb{D}$ is additive.

**Proof**    We already saw that $\underline{\mathbb{D}}^{\text{reg}}$ and $\overline{\mathbb{D}}^{\text{reg}}$ are partially additive quantum divergences that reduce to $\mathbb{D}$ on classical states. It is therefore left to prove the inequality (6.151). From (5.90) we have for all $n \in \mathbb{N}$

$$\underline{\mathbb{D}}\left(\rho^{\otimes n}\|\sigma^{\otimes n}\right) \leqslant D'\left(\rho^{\otimes n}\|\sigma^{\otimes n}\right) \leqslant \overline{\mathbb{D}}\left(\rho^{\otimes n}\|\sigma^{\otimes n}\right). \quad (6.152)$$

Since $D'$ is additive under tensor product we get after dividing the equation above by $n$

$$\frac{1}{n}\underline{\mathbb{D}}\left(\rho^{\otimes n}\|\sigma^{\otimes n}\right) \leqslant D'\left(\rho\|\sigma\right) \leqslant \frac{1}{n}\overline{\mathbb{D}}\left(\rho^{\otimes n}\|\sigma^{\otimes n}\right). \quad (6.153)$$

The proof is concluded by taking the limit $n \to \infty$ in the equation above.      ∎

**Exercise 6.37.** *Let $\rho,\sigma \in \mathfrak{D}(A)$ and let $\mathbb{D}$ be a classical relative entropy with maximal and minimal quantum extensions $\overline{\mathbb{D}}$ and $\underline{\mathbb{D}}$. Denote by*

$$a_n := \overline{\mathbb{D}}\left(\rho^{\otimes n}\|\sigma^{\otimes n}\right) \quad \text{and} \quad b_n := \underline{\mathbb{D}}\left(\rho^{\otimes n}\|\sigma^{\otimes n}\right). \quad (6.154)$$

*1. Show that the sequences $\{a_n\}$ and $\{b_n\}$ satisfy for all $n,m \in \mathbb{N}$*

$$a_{n+m} \leqslant a_n + a_m \quad \text{and} \quad b_{n+m} \geqslant b_n + b_m. \quad (6.155)$$

2. *Use the inequalities above to show that*

$$\lim_{n\to\infty}\frac{a_n}{n}=\alpha:=\inf\left\{\frac{a_n}{n}:n\in\mathbb{N}\right\}$$

$$\lim_{n\to\infty}\frac{b_n}{n}=\beta:=\sup\left\{\frac{b_n}{n}:n\in\mathbb{N}\right\}.\qquad(6.156)$$

*Hint: Let $\varepsilon>0$, choose $k$ such that $\alpha+\varepsilon>\frac{a_k}{k}$, and observe that for any integers $n,m\in\mathbb{N}$ that satisfies $nk\leqslant m<(n+1)k$ we have $a_m\leqslant a_{nk}+a_{m-nk}\leqslant na_k+c$, where $c:=\max\{a_j\}_{j\in[k]}$. Use this to bound $\limsup_{m\to\infty}\frac{a_m}{m}$.*

## 6.4.1 The Minimal Quantum Extension

This section illuminates the remarkable aspect of Rényi relative entropies, specifically the existence of a closed formula for the minimal quantum extension of the Rényi divergence. It is noteworthy that if $D_\alpha$ represents the *classical* Rényi relative entropy of order $\alpha$, then the quantum extension, denoted as $D_\alpha^{\mathrm{reg}}(\rho\|\sigma)$, can be expressed as follows:

$$D_\alpha^{\mathrm{reg}}(\rho\|\sigma):=\lim_{n\to\infty}\frac{1}{n}\sup_{\mathcal{E}_n\in\mathrm{CPTP}(A^n\to X)}D_\alpha\left(\mathcal{E}_n(\rho^{\otimes n})\big\|\mathcal{E}_n(\sigma^{\otimes n})\right),\qquad(6.157)$$

where the supremum encompasses all dimensions of the classical system $X$.

To derive a single-letter closed formula for the expression above, a two-step approach is required:

1. First, identify a function $\mathcal{E}_n$ that approaches optimality as $n\to\infty$.
2. Then, use this selected $\mathcal{E}_n$ to compute the limit as $n\to\infty$, which will lead to the desired closed formula.

This approach enables the development of a precise and concise formula representing the minimal quantum extension for the Rényi divergence.

A natural guess for optimal POVM channels $\mathcal{E}_n$ are the pinching channels discussed in Section 3.5.12. Recall that for any $\rho,\sigma\in\mathfrak{D}(A)$, and a pinching channel $\mathcal{P}_\sigma\in\mathrm{CPTP}(A\to A)$, we have that $\mathcal{P}_\sigma(\rho)$ and $\sigma$ commutes. Therefore, $\mathcal{P}_\sigma(\rho)$ and $\sigma$ have a common eigenbasis $\{|x\rangle\}_{x\in[m]}$ (with $m:=|X|=|A|$) that spans $A$. Let $\Delta\in\mathrm{CPTP}(A\to X)$ be the completely dephasing channel in this basis. Then, the channel $\Delta\in\mathrm{CPTP}(A\to X)$ is a POVM Channel that we can take to be $\mathcal{E}_1$. From Exercise 3.63 it follows that $\Delta(\sigma)=\sigma$ and $\Delta(\rho)=\mathcal{P}_\sigma(\rho)$ (see (3.231)).

In general, for any $n\in\mathbb{N}$, we can choose $\mathcal{E}_n=\Delta_n$, where $\Delta_n\in\mathrm{CPTP}(A^n\to X^n)$ is the completely dephasing channel in the common eigenbasis of $\mathcal{P}_{\sigma^{\otimes n}}\left(\rho^{\otimes n}\right)$ and $\sigma^{\otimes n}$. We will see shortly that this choice is indeed optimal in the limit $n\to\infty$.

Before we continue with the derivation of the closed formula, we first give a snapshot of what one can expect the formula to be. With $\{|x\rangle\}_{x\in[m]}$ being the common eigenbasis of $\mathcal{P}_\sigma(\rho)$ and $\sigma$ we get (cf. (3.231)) that

$$D_\alpha\left(\mathcal{P}_\sigma(\rho)\,\big\|\,\sigma\right)=D_\alpha\left(\Delta(\rho)\,\big\|\,\sigma\right)=\frac{1}{\alpha-1}\log\sum_{x\in[m]}\langle x|\rho|x\rangle^\alpha\langle x|\sigma|x\rangle^{1-\alpha},\qquad(6.158)$$

where we used the fact that each $|x\rangle$ is a common eigenvector of both $\sigma$ and $\mathcal{P}_\sigma(\rho)$. In particular, for any $\lambda \in \mathbb{R}$ we have $\langle x|\sigma^\lambda|x\rangle = \langle x|\sigma|x\rangle^\lambda$. Therefore, the term inside the sum above can be expressed as

$$\langle x|\rho|x\rangle^\alpha \langle x|\sigma|x\rangle^{1-\alpha} = \left(\langle x|\sigma^{\frac{1-\alpha}{2\alpha}}|x\rangle\langle x|\rho|x\rangle\langle x|\sigma^{\frac{1-\alpha}{2\alpha}}|x\rangle\right)^\alpha$$

$$\boxed{|x\rangle\langle x|\sigma^{\frac{1-\alpha}{2\alpha}}|x\rangle = \sigma^{\frac{1-\alpha}{2\alpha}}|x\rangle} \longrightarrow = \left(\langle x|\sigma^{\frac{1-\alpha}{2\alpha}}\rho\,\sigma^{\frac{1-\alpha}{2\alpha}}|x\rangle\right)^\alpha . \tag{6.159}$$

We therefore conclude that

$$D_\alpha\left(\mathcal{P}_\sigma(\rho)\,\middle\|\,\sigma\right) = \frac{1}{\alpha-1}\log\sum_{x\in[m]}\left(\langle x|\sigma^{\frac{1-\alpha}{2\alpha}}\rho\,\sigma^{\frac{1-\alpha}{2\alpha}}|x\rangle\right)^\alpha . \tag{6.160}$$

Since the function $x \mapsto x^\alpha$ is concave for $\alpha \in (0,1)$ and convex for $\alpha \geqslant 1$ it follows from the Jensen's inequality (B.31 of online version) that

$$D_\alpha\left(\mathcal{P}_\sigma(\rho)\,\middle\|\,\sigma\right) \leqslant \frac{1}{\alpha-1}\log\sum_{x\in[m]}\langle x|\left(\sigma^{\frac{1-\alpha}{2\alpha}}\rho\,\sigma^{\frac{1-\alpha}{2\alpha}}\right)^\alpha|x\rangle$$

$$= \frac{1}{\alpha-1}\log\mathrm{Tr}\left(\sigma^{\frac{1-\alpha}{2\alpha}}\rho\,\sigma^{\frac{1-\alpha}{2\alpha}}\right)^\alpha . \tag{6.161}$$

The expression on the right-hand side is known as the sandwiched Rényi relative entropy. Remarkably, we will see next that the regularization of the left-hand side equals the right-hand side in the equation above. For this purpose, it will be convenient to denote the trace in the equation above as

$$\tilde{Q}_\alpha(\rho\|\sigma) := \mathrm{Tr}\left(\sigma^{\frac{1-\alpha}{2\alpha}}\rho\,\sigma^{\frac{1-\alpha}{2\alpha}}\right)^\alpha . \tag{6.162}$$

**Exercise 6.38.** *Show that for any isometry channel $\mathcal{V} \in \mathrm{CPTP}(A \to B)$, any $\rho,\sigma \in \mathfrak{D}(A)$, and any $\omega \in \mathfrak{D}(C)$,*

$$Q_\alpha\left(\mathcal{V}(\rho)\,\middle\|\,\mathcal{V}(\sigma)\right) = Q_\alpha(\rho\|\sigma) \quad and \quad Q_\alpha(\rho\otimes\omega\|\sigma\otimes\omega) = Q_\alpha(\rho\|\sigma). \tag{6.163}$$

**The Sandwiched Rényi Relative Entropy**

**Definition 6.7.** The *sandwiched Rényi relative entropy* of order $\alpha \in [0,\infty]$ is defined on any quantum system $A$ and $\rho,\sigma \in \mathfrak{D}(A)$ as

$$\tilde{D}_\alpha(\rho\|\sigma) = \begin{cases} \frac{1}{\alpha-1}\log\tilde{Q}_\alpha(\rho\|\sigma) & \text{if } \left(\frac{1}{2}\leqslant\alpha<1 \text{ and } \rho\not\perp\sigma\right) \text{ or } \rho\ll\sigma \\ \frac{1}{\alpha-1}\log\tilde{Q}_{1-\alpha}(\sigma\|\rho) & \text{if } 0\leqslant\alpha<\frac{1}{2} \text{ and } \rho\not\perp\sigma \\ \infty & \text{otherwise.} \end{cases}$$

The cases $\alpha = 0,1,\infty$ are understood in terms of limits.

We first show that $\tilde{D}_\alpha$ is indeed a relative entropy. It's additivity and normalization properties are relatively easy to show and are left as an exercise.

**Exercise 6.39.** *Show that the sandwiched Rényi relative entropy of order $\alpha \in [0, \infty]$ satisfies the additivity and normalization properties of a quantum relative entropy.*

**Exercise 6.40.** *Show that for any $\rho, \sigma \in \mathfrak{D}(A)$,*

$$\mathrm{Tr}\left(\sigma^{\frac{1-\alpha}{2\alpha}} \rho \sigma^{\frac{1-\alpha}{2\alpha}}\right)^{\alpha} = \mathrm{Tr}\left(\rho^{\frac{1}{2}} \sigma^{\frac{1-\alpha}{\alpha}} \rho^{\frac{1}{2}}\right)^{\alpha}. \tag{6.164}$$

*Hint: Recall that for any complex matrix $M$, the matrices $MM^*$ and $M^*M$ have the same nonzero eigenvalues.*

---

> **Theorem 6.13.** The sandwiched Rényi relative entropy of any order $\alpha \in [0, \infty]$ is a quantum relative entropy; that is, it satisfies the three relative entropy axioms of DPI, additivity, and normalization.

---

**Proof** Since $\tilde{D}_\alpha(\rho \| \sigma)$ fulfills both additivity and normalization properties (as shown in Exercise 6.39), our task is to demonstrate its compliance with the DPI. For $\alpha > 1$, the DPI of $\tilde{D}_\alpha$ is derived from that of $\tilde{Q}_\alpha$. For $\alpha \in \left[\frac{1}{2}, 1\right)$, it follows from the DPI of $-\tilde{Q}_\alpha$. Based on Exercise 6.38 and Lemma 5.2, we know that if $\tilde{Q}_\alpha$ is jointly convex for $\alpha > 1$, then it satisfies the DPI. Similarly, for $\alpha \in [\frac{1}{2}, 1)$, if $\tilde{Q}_\alpha$ is jointly concave, then $-\tilde{Q}_\alpha$ satisfies the DPI. Our objective is therefore to show that for $\alpha > 1$, $\tilde{Q}_\alpha$ is jointly convex, and for $\alpha \in [\frac{1}{2}, 1)$, $\tilde{Q}_\alpha$ is jointly concave. The case $\alpha \in (0, \frac{1}{2}]$ is effectively covered by the case $\alpha \in [\frac{1}{2}, 1)$ when we swap $\rho$ with $\sigma$, thus it need not be considered separately.

Firstly, consider $\alpha > 1$ and define $\beta := \frac{\alpha - 1}{2\alpha}$. The proof's central strategy is to decompose the trace $\mathrm{Tr}\left[(\sigma^{-\beta} \rho \sigma^{-\beta})^{\alpha}\right]$ into two terms, one dependent only on $\rho$ and the other solely on $\sigma$. This decomposition allows us to separately assess the convexity in $\rho$ and $\sigma$. To obtain this decomposition, we utilize Young's inequality (2.75), choosing $M = \sigma^{-\beta} \rho \sigma^{-\beta}$, $N = \sigma^\beta \eta \sigma^\beta$, $p = \alpha$, and $q = \frac{\alpha}{\alpha - 1} = \frac{1}{2\beta}$, where $\eta$ is an arbitrary positive semidefinite matrix in $\mathrm{Pos}(A)$. With these choices, $\mathrm{Tr}[MN] = \mathrm{Tr}[\rho\eta]$, leading to the inequality (cf. (2.75))

$$\mathrm{Tr}\left[\rho\eta\right] \leqslant \frac{1}{\alpha}\mathrm{Tr}\left(\sigma^{-\beta}\rho\sigma^{-\beta}\right)^{\alpha} + \frac{\alpha - 1}{\alpha}\mathrm{Tr}\left(\sigma^\beta \eta \sigma^\beta\right)^{\frac{1}{2\beta}}. \tag{6.165}$$

Rearranging terms and recalling $\tilde{Q}_\alpha(\rho \| \sigma) = \mathrm{Tr}\left(\sigma^{-\beta} \rho \sigma^{-\beta}\right)^{\alpha}$, we obtain

$$\tilde{Q}_\alpha(\rho \| \sigma) \geqslant \alpha\mathrm{Tr}[\rho\eta] - (\alpha - 1)\mathrm{Tr}\left(\sigma^\beta \eta \sigma^\beta\right)^{\frac{1}{2\beta}}. \tag{6.166}$$

This inequality holds for all $\eta \in \mathrm{Pos}(A)$, with equality if $M^p = N^q$, which translates to (Exercise (6.41))

$$\eta = \sigma^{-\beta}\left(\sigma^{-\beta}\rho\sigma^{-\beta}\right)^{\alpha - 1}\sigma^{-\beta}. \tag{6.167}$$

Therefore, $\tilde{Q}_\alpha(\rho \| \sigma)$ can be expressed as

$$\tilde{Q}_\alpha(\rho \| \sigma) = \sup_{\eta \geqslant 0}\left\{\alpha\mathrm{Tr}[\rho\eta] - (\alpha - 1)\mathrm{Tr}\left(\sigma^\beta \eta \sigma^\beta\right)^{\frac{1}{2\beta}}\right\}. \tag{6.168}$$

With this expression, we can now analyze the convexity of each term independently.

A consequence of Lieb's concavity theorem, given in Corollary B.6 of online version, establishes the concavity of the function

$$\sigma \mapsto \mathrm{Tr}\left(\sigma^\beta \eta \sigma^\beta\right)^{\frac{1}{2\beta}} = \mathrm{Tr}\left(\eta^{\frac{1}{2}}\sigma^{2\beta}\eta^{\frac{1}{2}}\right)^{\frac{1}{2\beta}}, \tag{6.169}$$

where we used the fact that $LL^*$ and $L^*L$ have the same nonzero eigenvalues, where $L := \sigma^\beta \eta^{\frac{1}{2}}$. Therefore, the term $-(\alpha - 1)\mathrm{Tr}\left(\sigma^\beta \eta \sigma^\beta\right)^{\frac{1}{2\beta}}$ is convex in $\sigma$. Furthermore, the linearity of $\alpha\mathrm{Tr}[\rho\eta]$ in $\rho$ ensures its convexity in $\rho$. As a result, for any $\mathbf{p} \in \mathrm{Prob}(n)$ and two sets of $n$ density matrices in $\mathfrak{D}(A)$, $\{\rho_x\}_{x\in[n]}$ and $\{\sigma_x\}_{x\in[n]}$, it follows that

$$\tilde{Q}_\alpha\left(\sum_{x\in[n]} p_x \rho_x \,\Big\|\, \sum_{x\in[n]} p_x \sigma_x\right) \leq \sup_{\eta \geq 0}\left\{\alpha \sum_{x\in[n]} p_x \mathrm{Tr}[\rho_x \eta] - (\alpha - 1)\sum_{x\in[n]} p_x \mathrm{Tr}\left(\eta^{\frac{1}{2}}\sigma_x^{2\beta}\eta^{\frac{1}{2}}\right)^{\frac{1}{2\beta}}\right\}$$

$$\leq \sum_{x\in[n]} p_x \sup_{\eta \geq 0}\left\{\alpha \mathrm{Tr}[\rho_x \eta] - (\alpha - 1)\mathrm{Tr}\left(\eta^{\frac{1}{2}}\sigma_x^{2\beta}\eta^{\frac{1}{2}}\right)^{\frac{1}{2\beta}}\right\}$$

$$(6.168)\rightarrow = \sum_{x\in[n]} p_x \tilde{Q}_\alpha(\rho_x \| \sigma_x). \tag{6.170}$$

This proves the case for $\alpha > 1$. For $\alpha \in \left[\frac{1}{2}, 1\right)$, we apply similar reasoning using the reverse Young's inequality (3.240). Using the same substitutions for $M$ and $N$, we obtain (6.166) but with the inequality reversed. Consequently, we have

$$\tilde{Q}_\alpha(\rho \| \sigma) = \inf_{\eta \geq 0}\left\{\alpha \mathrm{Tr}[\rho\eta] + (1 - \alpha)\mathrm{Tr}\left(\sigma^\beta \eta \sigma^\beta\right)^{\frac{1}{2\beta}}\right\}. \tag{6.171}$$

Observe that $\beta < 0$ since $\alpha < 1$. As with the previous case, the joint concavity of $\tilde{Q}_\alpha$ follows from the concavity of the function in (6.169), completing the proof. ∎

**Exercise 6.41.** *Using the same notations as in the proof above, show that $M^p = N^q$ if and only if $\eta$ has the form given in (6.167).*

**Exercise 6.42.** *Show that for any $\rho, \sigma \in \mathfrak{D}(A)$, the function $\alpha \mapsto \tilde{D}_\alpha(\rho \| \sigma)$ is continuous for all $\alpha \in [0, \infty]$.*

We are now ready to prove the closed formula for the minimal quantum Rényi relative entropy.

**Single Letter Formula**

**Theorem 6.14.** For any $\alpha \in [0, \infty]$, the regularized minimal quantum extension of the Rényi relative entropy, $D_\alpha$, is given by the sandwiched Rényi relative entropy of order $\alpha$. That is, for all $\alpha \in [0, \infty]$, quantum system $A$, and $\rho, \sigma \in \mathfrak{D}(A)$, we have

$$\underline{D}_\alpha^{\mathrm{reg}}(\rho \| \sigma) = \tilde{D}_\alpha(\rho \| \sigma). \tag{6.172}$$

*Remark.* Recall that a priori, $\underline{D}_\alpha^{\mathrm{reg}}$ is only known to be partially additive; however, the theorem above implies that it is fully additive.

**Proof** It is sufficient to prove the theorem for all $\alpha \geqslant \frac{1}{2}$, since if the theorem holds for this case, then the case $\alpha \in (0, \frac{1}{2})$ simply follows from Exercise 6.13 via the relation

$$\underline{D}_\alpha^{\mathrm{reg}}(\rho\|\sigma) = \frac{\alpha}{1-\alpha}\underline{D}_{1-\alpha}^{\mathrm{reg}}(\rho\|\sigma)$$

$$(6.172) \rightarrow \ = \frac{\alpha}{1-\alpha}\tilde{D}_{1-\alpha}(\rho\|\sigma) \tag{6.173}$$

$$= \tilde{D}_\alpha(\rho\|\sigma).$$

We will therefore assume in the rest of the proof that $\alpha \geqslant \frac{1}{2}$. Since $\tilde{D}_\alpha$ is a relative entropy that reduces to the Rényi relative entropy in the classical domain, it follows from Theorem 6.12 that

$$\underline{D}_\alpha^{\mathrm{reg}}(\rho\|\sigma) \leqslant \tilde{D}_\alpha(\rho\|\sigma). \tag{6.174}$$

For the reversed inequality, we first show that

$$\underline{D}_\alpha^{\mathrm{reg}}(\rho\|\sigma) \geqslant \lim_{n\to\infty} \frac{1}{n} D_\alpha\left(\mathcal{P}_{\sigma^{\otimes n}}\left(\rho^{\otimes n}\right)\big\|\sigma^{\otimes n}\right). \tag{6.175}$$

Indeed, since $\mathcal{P}_{\sigma^{\otimes n}}\left(\rho^{\otimes n}\right)$ commutes with $\sigma^{\otimes n}$, they have a common eigenbasis that spans $A^n$. Let $\Delta_n \in \mathrm{CPTP}(A^n \to A^n)$ be the completely dephasing channel in this basis. From Exercise 3.63 we have $\mathcal{P}_{\sigma^{\otimes n}}\left(\rho^{\otimes n}\right) = \Delta_n\left(\rho^{\otimes n}\right)$. Therefore,

$$\sup_{\mathcal{E}\in\mathrm{CPTP}(A^n\to X)} D_\alpha\left(\mathcal{E}\left(\rho^{\otimes n}\right)\big\|\mathcal{E}\left(\sigma^{\otimes n}\right)\right) \geqslant D_\alpha\left(\Delta_n\left(\rho^{\otimes n}\right)\big\|\Delta_n\left(\sigma^{\otimes n}\right)\right)$$

$$= D_\alpha\left(\mathcal{P}_{\sigma^{\otimes n}}\left(\rho^{\otimes n}\right)\big\|\sigma^{\otimes n}\right). \tag{6.176}$$

Dividing both sides of the equation above by $n$ and taking the limit $n \to \infty$ proves (6.175).

It is left to show that the right-hand side of (6.175) is no smaller than $\tilde{D}_\alpha(\rho\|\sigma)$. We will divide this part of the proof into several cases:

1. The case $\alpha > 1$ and $\rho \not\ll \sigma$. Recall that for every $n \in \mathbb{N}$ the states $\mathcal{P}_{\sigma^{\otimes n}}\left(\rho^{\otimes n}\right)$ and $\sigma^{\otimes n}$ have a common eigenbasis. Therefore, both of these states are diagonal in this eigenbasis and the condition that $\rho \not\ll \sigma$ implies that these diagonal states also satisfy $\mathcal{P}_{\sigma^{\otimes n}}\left(\rho^{\otimes n}\right) \not\ll \sigma^{\otimes n}$ (see Exercise 3.64). Hence, we must have that $D_\alpha\left(\mathcal{P}_{\sigma^{\otimes n}}\left(\rho^{\otimes n}\right)\big\|\sigma^{\otimes n}\right) = \infty$ for all $n \in \mathbb{N}$.

2. The case $\alpha > 1$ and $\rho \ll \sigma$. Observe first that since $\mathcal{P}_\sigma(\rho)$ and $\sigma$ commute, we have (cf. Exercise 6.14)

$$D_\alpha\left(\mathcal{P}_\sigma\left(\rho\right)\big\|\sigma\right) = \frac{1}{\alpha - 1}\log\mathrm{Tr}\left[\left(\sigma^{\frac{1-\alpha}{2\alpha}}\mathcal{P}_\sigma(\rho)\sigma^{\frac{1-\alpha}{2\alpha}}\right)^\alpha\right]. \tag{6.177}$$

Now, from the pinching inequality (3.235) we have $\rho \leqslant |\mathrm{spec}(\sigma)|\mathcal{P}_\sigma(\rho)$, so that (cf. Exercise B.7 of online version)

$$D_\alpha \left( \mathcal{P}_\sigma \left( \rho \right) \| \sigma \right) \geqslant \frac{1}{\alpha - 1} \log \mathrm{Tr} \left[ \left( \sigma^{\frac{1-\alpha}{2\alpha}} \frac{\rho}{|\mathrm{spec}(\rho)|} \sigma^{\frac{1-\alpha}{2\alpha}} \right)^\alpha \right]$$

$$= \frac{1}{\alpha - 1} \log \mathrm{Tr} \left[ \left( \sigma^{\frac{1-\alpha}{2\alpha}} \rho \sigma^{\frac{1-\alpha}{2\alpha}} \right)^\alpha \right] - \frac{\alpha}{\alpha - 1} \log |\mathrm{spec}(\sigma)|. \tag{6.178}$$

Hence, replacing $\rho$ and $\sigma$ above with $\rho^{\otimes n}$ and $\sigma^{\otimes n}$, and recalling from (8.102) that $|\mathrm{spec}(\sigma^{\otimes n})| \leqslant (n+1)^{|A|}$ we get in the limit $n \to \infty$

$$\lim_{n \to \infty} \frac{1}{n} D_\alpha \left( \mathcal{P}_{\sigma^{\otimes n}} \left( \rho^{\otimes n} \right) \| \sigma^{\otimes n} \right) \geqslant \frac{1}{\alpha - 1} \log \mathrm{Tr} \left[ \left( \sigma^{\frac{1-\alpha}{2\alpha}} \rho \sigma^{\frac{1-\alpha}{2\alpha}} \right)^\alpha \right]. \tag{6.179}$$

3. The case $\alpha \in [\frac{1}{2}, 1)$ and $\rho \perp \sigma$. In this case $D_\alpha \left( \mathcal{P}_{\sigma^{\otimes n}} \left( \rho^{\otimes n} \right) \| \sigma^{\otimes n} \right) = \infty$ since $\mathcal{P}_{\sigma^{\otimes n}} \left( \rho^{\otimes n} \right) = \rho^{\otimes n}$ (and note also that $\rho$ and $\sigma$ commute; i.e. classical).

4. The case $\alpha \in [\frac{1}{2}, 1)$ and $\rho \not\perp \sigma$. In this case, the first inequality in (6.178) holds in the opposite direction since the factor $\frac{1}{\alpha-1}$ is negative. We therefore need another argument or trick. First, observe that

$$\left( \sigma^{\frac{1-\alpha}{2\alpha}} \rho \sigma^{\frac{1-\alpha}{2\alpha}} \right)^\alpha = \left( \sigma^{\frac{1-\alpha}{2\alpha}} \rho \sigma^{\frac{1-\alpha}{2\alpha}} \right)^{\alpha-1} \sigma^{\frac{1-\alpha}{2\alpha}} \rho \sigma^{\frac{1-\alpha}{2\alpha}}. \tag{6.180}$$

Then, using the pinching inequality $\rho \leqslant |\mathrm{spec}(\sigma)| \mathcal{P}_\sigma (\rho)$ we get

$$\sigma^{\frac{1-\alpha}{2\alpha}} \rho \sigma^{\frac{1-\alpha}{2\alpha}} \leqslant |\mathrm{spec}(\sigma)| \sigma^{\frac{1-\alpha}{2\alpha}} \mathcal{P}_\sigma (\rho) \sigma^{\frac{1-\alpha}{2\alpha}}. \tag{6.181}$$

Combining this with the fact that the function $t \mapsto t^{\alpha-1}$ is anti-operator monotone for $\alpha \in [\frac{1}{2}, 1)$ (see Table B.1 of online version) we get that

$$\left( \sigma^{\frac{1-\alpha}{2\alpha}} \rho \sigma^{\frac{1-\alpha}{2\alpha}} \right)^{\alpha-1} \geqslant |\mathrm{spec}(\sigma)|^{\alpha-1} \left( \sigma^{\frac{1-\alpha}{2\alpha}} \mathcal{P}_\sigma (\rho) \sigma^{\frac{1-\alpha}{2\alpha}} \right)^{\alpha-1}. \tag{6.182}$$

Combining the above inequality with (6.180) gives

$$\mathrm{Tr} \left( \sigma^{\frac{1-\alpha}{2\alpha}} \rho \sigma^{\frac{1-\alpha}{2\alpha}} \right)^\alpha \geqslant |\mathrm{spec}(\sigma)|^{\alpha-1} \mathrm{Tr} \left[ \left( \sigma^{\frac{1-\alpha}{2\alpha}} \mathcal{P}_\sigma (\rho) \sigma^{\frac{1-\alpha}{2\alpha}} \right)^{\alpha-1} \sigma^{\frac{1-\alpha}{2\alpha}} \rho \sigma^{\frac{1-\alpha}{2\alpha}} \right]$$

$$\textbf{Exercise 6.43} \to = |\mathrm{spec}(\sigma)|^{\alpha-1} \mathrm{Tr} \left[ \left( \sigma^{\frac{1-\alpha}{2\alpha}} \mathcal{P}_\sigma (\rho) \sigma^{\frac{1-\alpha}{2\alpha}} \right)^\alpha \right]. \tag{6.183}$$

Using the above inequality in (6.177) gives

$$D_\alpha \left( \mathcal{P}_\sigma \left( \rho \right) \| \sigma \right) \geqslant \frac{1}{\alpha - 1} \log \mathrm{Tr} \left[ \left( \sigma^{\frac{1-\alpha}{2\alpha}} \rho \sigma^{\frac{1-\alpha}{2\alpha}} \right)^\alpha \right] - \log |\mathrm{spec}(\sigma)|. \tag{6.184}$$

Finally, replacing $\rho$ and $\sigma$ above with $\rho^{\otimes n}$ and $\sigma^{\otimes n}$, and recalling that $|\mathrm{spec}(\sigma^{\otimes n})| \leqslant (n+1)^{|A|}$, we get (6.179) in the limit $n \to \infty$.

This completes the proof. ∎

**Exercise 6.43.** *Show that for any $a, b \in \mathbb{R}$ and any $\rho, \sigma \in \mathfrak{D}(A)$,*

$$\mathrm{Tr} \left[ \left( \sigma^a \mathcal{P}_\sigma (\rho) \sigma^a \right)^b \sigma^a \rho \sigma^a \right] = \mathrm{Tr} \left[ \left( \sigma^a \mathcal{P}_\sigma (\rho) \sigma^a \right)^{b+1} \right]. \tag{6.185}$$

*Hint: Recall that $\mathcal{P}_\sigma (\rho)$ commutes with $\sigma$ and that all the pinching projectors commute with all the operators above except for the single $\rho$.*

> **Corollary 6.6.** Let $A$ be a quantum system and $\rho, \sigma \in \mathfrak{D}(A)$. For $\alpha \geqslant 1/2$,
> $$\underline{D}_\alpha^{\mathrm{reg}}(\rho\|\sigma) = \lim_{n\to\infty} \frac{1}{n} D_\alpha\left(\mathcal{P}_{\sigma^{\otimes n}}\left(\rho^{\otimes n}\right)\big\|\sigma^{\otimes n}\right). \tag{6.186}$$

**Proof**   Follows trivially from a combination of the theorem above with (6.175,6.179).

∎

Note that the corollary above demonstrates that an optimizer for (6.157) is $\mathcal{E}_n = \Delta_n$.

**Exercise 6.44.** *Show that for all $\alpha \in [0, \infty]$,*

$$\underline{D}_\alpha^{\mathrm{reg}}(\rho\|\sigma) \geqslant \limsup_{n\to\infty} \frac{1}{n} D_\alpha\left(\rho^{\otimes n}\big\|\mathcal{P}_{\rho^{\otimes n}}\left(\sigma^{\otimes n}\right)\right). \tag{6.187}$$

*Furthermore, show that the equality above holds for all $\alpha \in (0, \frac{1}{2})$.*

## 6.4.2 The Maximal Quantum Extension

In this subsection we apply the results of Section 5.3.1 to relative entropies. In particular, we will see that the maximal quantum extension of the Rényi relative entropy has a closed formula for Rényi order parameter $\alpha \in [0, 2]$. We start with the following corollary of Theorem 5.7.

> **Corollary 6.7.** Let $\mathbb{D}$ be a classical relative entropy, $\rho, \sigma \in \mathfrak{D}(A)$, and suppose $\rho = \psi := |\psi\rangle\langle\psi|$. Then,
> $$\overline{\mathbb{D}}(\psi\|\sigma) = D_{\max}(\psi\|\sigma), \tag{6.188}$$
> where $D_{\max}$ is the quantum max relative entropy.

**Proof**   Let $\{e_1, e_2\}$ be the standard basis of $\mathbb{R}^2$, and observe that $\lambda_{\max}$ in (5.112) is precisely $2^{-D_{\max}(\psi\|\sigma)}$. Therefore, Theorem 5.7 gives

$$\overline{\mathbb{D}}(\psi\|\sigma) = \mathbb{D}\left(e_1 \big\| \big\| 2^{-D_{\max}(\psi\|\sigma)} e_1 + \left(1 - 2^{-D_{\max}(\psi\|\sigma)}\right) e_2\right)$$
$$\mathbf{Theorem\ 6.3} \rightarrow\ = D_{\max}(\psi\|\sigma), \tag{6.189}$$

where the last equality holds since $\mathbb{D}$ is a relative entropy.

∎

The corollary above demonstrates that the maximal quantum extension is closely related to $D_{\max}$. The corollary is universal in the sense that it holds for any classical relative entropy $\mathbb{D}$; however, it is quite limited as it holds only for pure $\rho$. In Corollary 6.8, we will see that for some of the Rényi relative entropies there exists a closed formula for the maximal quantum extension without any restriction on $\rho$ and $\sigma$. This closed formula is given in terms of the family of *geometric relative entropies*.

> ### The Geometric Relative Entropy
>
> **Definition 6.8.** The geometric relative entropy of order $\alpha \in [0, 2]$ is defined for any $\rho \in \mathfrak{D}(A)$ and $0 < \sigma \in \mathrm{Pos}(A)$ as
>
> $$\widehat{D}_\alpha(\rho\|\sigma) := \frac{1}{\alpha - 1} \log \mathrm{Tr}\left[\sigma \left(\sigma^{-\frac{1}{2}}\rho\sigma^{-\frac{1}{2}}\right)^\alpha\right] \tag{6.190}$$
>
> and for singular $\sigma \in \mathrm{Pos}(A)$ is defined by
>
> $$\widehat{D}_\alpha(\rho\|\sigma) := \lim_{\varepsilon \to 0^+} \widehat{D}_\alpha(\rho\|\sigma + \varepsilon I). \tag{6.191}$$

*Remarks*

1. Alternatively, one can define the geometric relative entropy for any $\rho, \sigma \in \mathfrak{D}(A)$ using the decomposition (D.27 of online version) with $\tilde{\rho} := \rho_{11} - \zeta\rho_{22}^{-1}\zeta^*$ and $\tilde{\sigma} := \sigma_{11}$. Then, the geometric relative entropy of order $\alpha \in [0, 2]$ is given by

$$\widehat{D}_\alpha(\rho\|\sigma) = \begin{cases} \frac{1}{\alpha-1} \log \mathrm{Tr}\left[\tilde{\sigma} \left(\tilde{\sigma}^{-\frac{1}{2}}\tilde{\rho}\tilde{\sigma}^{-\frac{1}{2}}\right)^\alpha\right] & \text{if } \alpha \in [0, 1) \text{ or } \rho \ll \sigma \\ \infty & \text{otherwise.} \end{cases} \tag{6.192}$$

2. The geometric relative entropy can be written differently using the relation $Mf(M^*M) = f(MM^*)M$ given in Exercise B.1 of online version. Denoting $M := \rho^{\frac{1}{2}}\sigma^{-\frac{1}{2}}$ we get

$$\begin{aligned} \widehat{D}_\alpha(\rho\|\sigma) &= \frac{1}{\alpha - 1} \log \mathrm{Tr}\left[\sigma M^*M \left(M^*M\right)^{\alpha-1}\right] \\ &= \frac{1}{\alpha - 1} \log \mathrm{Tr}\left[\sigma M^* \left(M^*M\right)^{\alpha-1} M\right] \\ &= \frac{1}{\alpha - 1} \log \mathrm{Tr}\left[\rho \left(\rho^{\frac{1}{2}}\sigma^{-1}\rho^{\frac{1}{2}}\right)^{\alpha-1}\right]. \end{aligned} \tag{6.193}$$

3. In the limit $\alpha \to 1$, we get for $\rho \ll \sigma$

$$\widehat{D}(\rho\|\sigma) := \lim_{\alpha \to 1} \widehat{D}_\alpha(\rho\|\sigma) = \mathrm{Tr}\left[\rho \log\left(\rho^{\frac{1}{2}}\sigma^{-1}\rho^{\frac{1}{2}}\right)\right], \tag{6.194}$$

   which is known as the Belavkin–Staszewski relative entropy.
4. Observe that for $\alpha = 2$ the definition of the geometric relative entropy coincides with the Petz quantum Rényi divergence of the same order.

**Exercise 6.45.** *Show that the two definitions (Definition 6.8 and the definition in equation (6.192)) for the geometric relative entropy are equivalent (for any two density matrices $\rho, \sigma \in \mathfrak{D}(A)$); that is, prove (6.192).*

**Exercise 6.46.** *Show that the geometric relative entropy satisfies the properties (axioms) of additivity and normalization of a quantum relative entropy.*

**Exercise 6.47.** *Show that the geometric relative entropy reduces to the Rényi relative entropy in the classical domain.*

Instead of proving directly that the geometric relative entropy satisfies the DPI, we will show that it is equal to the maximal quantum extension of the Rényi relative entropy. Since the latter satisfies the DPI, this will imply that geometric relative entropy also satisfies the DPI.

> **Corollary 6.8.** The regularized maximal quantum extension of the Rényi divergence, $D_\alpha$, with $\alpha \in [0, 2]$, is given by the geometric relative entropy; specifically, for any $\rho, \sigma \in \mathfrak{D}(A)$ and $\alpha \in [0, 2]$
> $$\overline{D}_\alpha^{\mathrm{reg}}(\rho\|\sigma) = \widehat{D}_\alpha(\rho\|\sigma). \tag{6.195}$$

**Proof** The proof follows directly from Theorem 5.8 for the case $\sigma > 0$ and from Theorem D.2 of online version for the general case. We leave the details as an exercise.

∎

**Exercise 6.48.** *Provide the full details of the proof of Corollary 6.8.*

## 6.5 Notes and References

Axiomatic derivations of entropies and relative entropies have a plentiful literature, starting with the seminal work given in Ref. [203] followed and refined in Refs. [73], [64], and [3], among others. These early papers focused on the derivation of the Shannon entropy until the scope was extended in Ref. [193]. Detailed reviews on the various axiomatic derivations can be found in Refs. [2] and [71], and for a more recent guide on the topic, see Ref. [56]. The axiomatic approach presented here was formally introduced in Refs. [98] and [99]. We point out that other functions that were studied in literature, like the Tsallis entropies, are not entropies according to the definition adopted here, as they are not additive in general. Moreover, historically, the terminology of "relative entropy" has been reserved only to the KL-divergence or to the Umegaki relative entropy in the quantum case, whereas we used this terminology to include all additive, monotonic functions, as given in Definitions 6.3 and 6.5.

There are several proofs that can be found in the literature on Erdös theorem (Lemma 6.2). The elementary proof we adopted here is from Ref. [136].

There is a rich literature on both classical and quantum Rényi divergences. A recent guide on classical Rényi divergences with thorough details of their properties can be found in the review article in Ref. [218]. In the quantum domain, Ref. [211], and more recently Ref. [209], devotes significant portion to the study of the quantum Rényi divergences, and we refer the reader to these books for more details on the history and developments of quantum Rényi divergences.

In this chapter, we focused on three types of extensions of the Rényi relative entropy to the quantum domain: the Petz quantum Rényi divergence (introduced in Ref. [181]), the minimal (sandwiched) quantum Rényi divergence (introduced in Ref. [170] and independently in Ref. [236]), and the maximal quantum Rényi divergence (introduced in Ref. [163]). These three extensions are by no means the only quantum extensions of the classical Rényi relative entropy. Examples include the two parameter family studied in Refs. [138] and [10], and the more recent divergence that was introduced in Ref. [77].

The min and max quantum relative entropies were first introduced in Ref. [58]. Given that all quantum relative entropies are bounded by these two divergences, it is not a surprise that the min and max relative entropies have be used extensively in quantum information.

# 7

# Conditional Entropy

In this chapter, we delve into a variant of the entropy function, widely prevalent in information theory and quantum resource theories, especially in the realm of dynamical resources, which is the focus of the second volume of this book. This variant is known as conditional entropy, which pertains to the entropy associated with a physical system $A$ that shares a correlation with another system $B$. When an observer, say Bob, has access to system $B$, he can reduce his uncertainty about system $A$ by performing a quantum measurement on his subsystem. In essence, conditional entropy quantifies the residual uncertainty of system $A$ when such access to system $B$ is available.

Traditionally, the conditional entropy of a bipartite state $\rho^{AB}$ is defined in terms of the von-Neumann entropy associated with system $AB$ minus the von-Neumann entropy associated with system $B$. This is given by

$$H(A|B)_\rho := H\left(\rho^{AB}\right) - H\left(\rho^B\right). \tag{7.1}$$

See Figure 7.1 for a heuristic description of this definition in terms of a Venn diagram. However, in this chapter, we take a different approach. Here, conditional entropy is defined axiomatically, similar to how we defined entropy and relative entropy. This approach provides a more rigorous definition of conditional entropy, placing the intuitive Venn diagram interpretation on a more solid theoretical foundation.

## 7.1 Quantum Conditional Majorization

We start this chapter with a section in which we extend the definition of conditional majorization to the quantum domain. Recall that conditional majorization, as defined in Section 4.6, characterizes the uncertainty associated with a physical system given access to another system that is correlated with it. Therefore, as we will see shortly, conditional majorization provides the foundation for the definition of conditional entropy.

The majorization relation between two probability vectors can be generalized to the quantum domain in a straightforward manner. That is, for any two density matrices $\rho \in \mathfrak{D}(A)$ and $\sigma \in \mathfrak{D}(A')$ we say that $\rho$ majorizes $\sigma$ and write $\rho^A \succ \sigma^{A'}$ if the probability vector consisting of the eigenvalues of $\rho$ majorizes the probability vector consisting of the eigenvalues of $\sigma$. For conditional majorization, such a straightforward extension from the classical to the quantum domain is more complex as it involves two systems

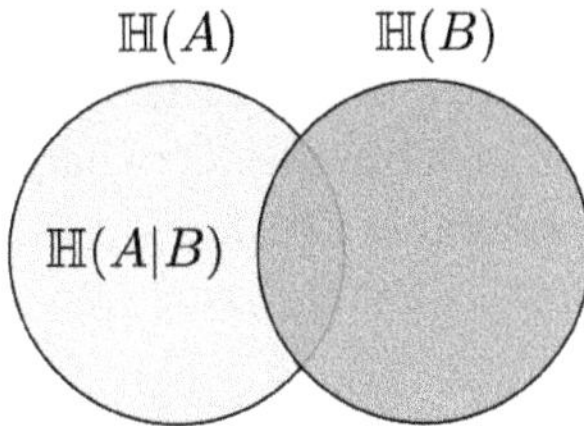


**Figure 7.1**   A Venn diagram for the conditional entropy.

that can be correlated quantumly (i.e. entangled). For this reason we employ the axiomatic approach to introduce quantum conditional majorization, and then discuss some of its key properties.

As already discussed, intuitively, conditional majorization is a preorder on the set $\mathfrak{D}(AB)$ that characterizes the uncertainty of system $A$ given access to system $B$. To make this intuition more precise, we employ the axiomatic approach determining which set of operations in $\mathrm{CPTP}(AB \to AB')$ can only increase the uncertainty of system $A$ (even if one has access to system $B$). We will now examine two highly intuitive axioms that these channels must adhere to. These two axioms extend the principles explored in Section 4.6.1 to the quantum domain.

## 7.1.1 Conditionally Unital Channels

What types of channels are expected to increase the conditional uncertainty associated with a system $A$ given access to $B$? To address this question, let's consider a bipartite quantum state in the form $\mathbf{u}^A \otimes \rho^B$, where $\mathbf{u}^A$ represents the maximally mixed state and $\rho^B$ is a certain density matrix. Given that such a product state is uncorrelated, access to $B$ does not aid in reducing the uncertainty about $A$. Consequently, we can infer that this state exhibits the highest degree of conditional uncertainty on $A|B$ (i.e. $A$ given access to $B$). Now, any channel $\mathcal{N} \in \mathrm{CPTP}(AB \to AB')$ that preserves or increases this conditional uncertainty should transform states with maximal conditional uncertainty into states retaining this property. Specifically, for all $\rho \in \mathfrak{D}(B)$, such a channel $\mathcal{N} \in \mathrm{CPTP}(AB \to AB')$ must satisfy

$$\mathcal{N}^{AB \to AB'}\left(\mathbf{u}^A \otimes \rho^B\right) = \mathbf{u}^A \otimes \sigma^{B'}, \tag{7.2}$$

where $\sigma^{B'}$ is some density matrix in $\mathfrak{D}(B')$. By tracing out system $A$, we can express $\sigma^{B'}$ as

$$\sigma^{B'} = \mathcal{N}^{AB \to B'}\left(\mathbf{u}^A \otimes \rho^B\right), \tag{7.3}$$

where $\mathcal{N}^{AB \to B'} := \mathrm{Tr}_A \circ \mathcal{N}^{AB \to AB'}$. Such a channel $\mathcal{N} \in \mathrm{CPTP}(AB \to AB')$ is referred to as *conditionally unital*. It is important to note that both the input and output systems on Alice's side remain the same, whereas on Bob's side, the systems $B$ and $B'$ can be different.

> **Lemma 7.1.** Let $\mathcal{N} \in \mathrm{CPTP}(AB \to \tilde{A}B')$ be a bipartite quantum channel and let $J_{\mathcal{N}}^{AB\tilde{A}B'}$ be its Choi matrix. Then, $\mathcal{N}^{AB \to \tilde{A}B}$ is conditionally unital if and only if its Choi matrix satisfies
>
> $$J_{\mathcal{N}}^{B\tilde{A}B'} = J_{\mathcal{N}}^{BB'} \otimes \mathbf{u}^{\tilde{A}}. \tag{7.4}$$

**Proof**   We begin by proving that the channel $\mathcal{N}^{AB \to \tilde{A}B'}$ is conditionally unital if its Choi matrix has the form in (7.4). To this end, let $\rho \in \mathfrak{D}(B)$ and consider that

$$
\begin{aligned}
\mathcal{N}\left(\mathbf{u}^A \otimes \rho^B\right) &= \mathrm{Tr}_{AB}\left[ J_{\mathcal{N}}^{AB\tilde{A}B'} \left(\mathbf{u}^A \otimes \left(\rho^B\right)^T \otimes I^{\tilde{A}B'}\right)\right] \\
&= \frac{1}{|A|} \mathrm{Tr}_B\left[ J_{\mathcal{N}}^{B\tilde{A}B'} \left(\left(\rho^B\right)^T \otimes I^{\tilde{A}B'}\right)\right] \\
(7.4)\to &= \frac{1}{|A|} \mathrm{Tr}_B\left[ \mathbf{u}^{\tilde{A}} \otimes J_{\mathcal{N}}^{BB'} \left(\left(\rho^B\right)^T \otimes I^{\tilde{A}B'}\right)\right] \\
&= \mathbf{u}^{\tilde{A}} \otimes \sigma^B,
\end{aligned}
\tag{7.5}
$$

where $\sigma^B := \frac{1}{|A|} \mathrm{Tr}_B\left[ J_{\mathcal{N}}^{BB'} \left(\left(\rho^B\right)^T \otimes I^{B'}\right)\right]$.

We next prove that the Choi matrix of $\mathcal{N}^{AB \to \tilde{A}B'}$ has the form in (7.4) if $\mathcal{N}^{AB \to \tilde{A}B'}$ is conditionally unital. Recall that when the defining property of a conditionally unital channel is that for every state $\rho \in \mathfrak{D}(B)$, there exists a state $\sigma \in \mathfrak{D}(B')$ such that

$$\mathcal{N}^{AB \to \tilde{A}B'}(\mathbf{u}^A \otimes \rho^B) = \mathbf{u}^{\tilde{A}} \otimes \sigma^{B'}. \tag{7.6}$$

By taking the trace over $\tilde{A}$ on both sides of the equation above, we get that

$$
\begin{aligned}
\sigma^{B'} = \mathcal{N}^{AB \to B'}(\mathbf{u}^A \otimes \rho^B) &= \mathrm{Tr}_{AB}\left[ J_{\mathcal{N}}^{ABB'} \left(\mathbf{u}^A \otimes \left(\rho^B\right)^T \otimes I^{B'}\right)\right] \\
&= \frac{1}{|A|} \mathrm{Tr}_B\left[ J_{\mathcal{N}}^{BB'} \left(\left(\rho^B\right)^T \otimes I^{B'}\right)\right].
\end{aligned}
\tag{7.7}
$$

On the other hand, observe that

$$
\begin{aligned}
\mathcal{N}^{AB \to \tilde{A}B'}(\mathbf{u}^A \otimes \rho^B) &= \mathrm{Tr}_{AB}\left[ J_{\mathcal{N}}^{AB\tilde{A}B'} \left(\mathbf{u}^A \otimes \left(\rho^B\right)^T \otimes I^{\tilde{A}B'}\right)\right] \\
&= \frac{1}{|A|} \mathrm{Tr}_B\left[ J_{\mathcal{N}}^{B\tilde{A}B'} \left(\left(\rho^B\right)^T \otimes I^{\tilde{A}B'}\right)\right].
\end{aligned}
\tag{7.8}
$$

Therefore, from (7.7) and (7.8) for $\sigma^B$ and $\mathcal{N}(\mathbf{u}^A \otimes \rho^B)$ we conclude that (7.6) can be expressed as

$$\mathrm{Tr}_B\left[ J_{\mathcal{N}}^{B\tilde{A}B'} \left(\left(\rho^B\right)^T \otimes I^{\tilde{A}B'}\right)\right] = \mathrm{Tr}_B\left[ (\mathbf{u}^{\tilde{A}} \otimes J_{\mathcal{N}}^{BB'}) \left(\left(\rho^B\right)^T \otimes I^{\tilde{A}B'}\right)\right]. \tag{7.9}$$

Denote by $\eta^{B\tilde{A}B'} := J_{\mathcal{N}}^{B\tilde{A}B'} - \mathbf{u}^{\tilde{A}} \otimes J_{\mathcal{N}}^{BB'}$ and observe that (7.9) can be written as

$$\mathrm{Tr}\left[ \eta^{B\tilde{A}B'} \left(\left(\rho^B\right)^T \otimes I^{\tilde{A}B'}\right)\right] = 0 \quad \forall \, \rho \in \mathfrak{D}(B). \tag{7.10}$$

Due to the existence of bases of density operators that span the space of linear operators acting on $B$, we conclude from (7.10) that for every operator $\zeta \in \mathfrak{L}(B)$ we have

$$\mathrm{Tr}_B\left[ \eta^{B\tilde{A}B'} \left( \zeta^B \otimes I^{\tilde{A}B'} \right) \right] = 0. \tag{7.11}$$

Note that by multiplying both sides of the equation above by any element $\xi \in \mathfrak{L}(\tilde{A}B')$ and taking the trace we get that

$$\mathrm{Tr}\left[ \eta^{B\tilde{A}B'} \left( \zeta^B \otimes \xi^{\tilde{A}B'} \right) \right] = 0. \tag{7.12}$$

Since (7.12) holds for all $\zeta \in \mathfrak{L}(B)$ and all $\xi \in \mathfrak{L}(\tilde{A}B')$ it also holds for any linear combinations of matrices of the form $\zeta^B \otimes \xi^{\tilde{A}B'}$. Since matrices of the form $\zeta^B \otimes \xi^{\tilde{A}B'}$ span the whole space $\mathfrak{L}(B\tilde{A}B')$ we conclude that $\eta^{B\tilde{A}B'}$ is orthogonal (in the Hilbert-Schmidt inner product) to all the elements of $\mathfrak{L}(B\tilde{A}B')$. Therefore, we must have $\eta^{B\tilde{A}B'} = 0$, which is equivalent to (7.4). This completes the proof. ∎

## 7.1.2 Semi-causal Channels

An additional requirement for the channel $\mathcal{N} \in \mathrm{CPTP}(AB \to AB')$ to not diminish conditional uncertainty is related to preventing information leakage from Alice's subsystem to Bob's. This is crucial since any leaked information could reduce the uncertainty about system $A$. Conditional uncertainty specifically pertains to the uncertainty about system $A$ when only Bob's system is accessible. To support this, we introduce a causality assumption ensuring that system $A$ does not causally influence system $B'$. This is mathematically represented as follows: for all $\mathcal{M} \in \mathrm{CPTP}(A \to A)$,

$$\mathcal{N}^{AB \to B'} \circ \mathcal{M}^{A \to A} = \mathcal{N}^{AB \to B'}, \tag{7.13}$$

where $\mathcal{N}^{AB \to B'} := \mathrm{Tr}_A \circ \mathcal{N}^{AB \to AB'}$. This condition guarantees that any operation $\mathcal{M}^{A \to A}$ applied by Alice to her system remains undetected by Bob. We refer to such a condition as $A \not\to B'$ semi-causal. For a visual representation, see Figure 7.2 depicting a semi-causal channel.

**Exercise 7.1.** *Let $\mathcal{N} \in \mathrm{CPTP}(AB \to \tilde{A}B')$. Show that $\mathcal{N}^{AB \to \tilde{A}B'}$ is $A \not\to B'$ semi-causal if and only if the marginals of its Choi matrix satisfy*

$$J_{\mathcal{N}}^{ABB'} = \mathbf{u}^A \otimes J_{\mathcal{N}}^{BB'}. \tag{7.14}$$

*Hint: In one direction, take $\mathcal{M} \in \mathrm{CPTP}(A)$ in (7.13) to be the replacer channel that always outputs the maximally mixed state irrespective of the input state (i.e. take $\mathcal{M}$ to be the completely depolarizing channel, also know as the completely randomizing channel), and compute the Choi matrix for the channels on both sides of (7.13).*

The following theorem establishes the equivalence between $A \not\to B'$ semi-causal channels and $A \not\to B'$ signaling channels. Specifically, a bipartite quantum channel $\mathcal{N} \in \mathrm{CPTP}(AB \to AB')$ is classified as $A \not\to B'$ signaling if it meets

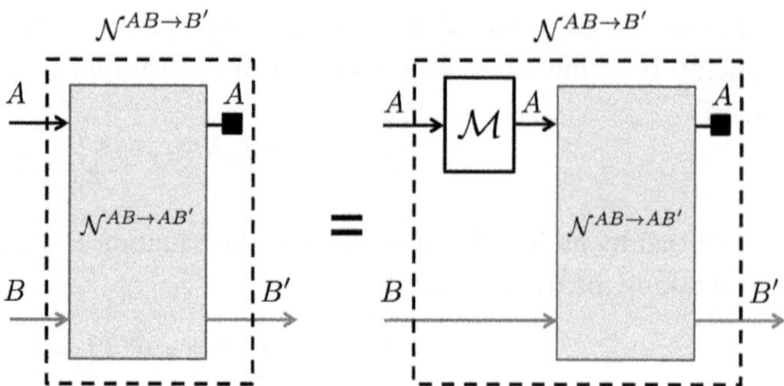



**Figure 7.2**  An illustration of an $A \not\to B'$ semi-causal bipartite channel $\mathcal{N}^{AB\to AB'}$. The marginal channel $\mathcal{N}^{AB\to B'}$ equals $\mathcal{N}^{AB\to B'} \circ \mathcal{M}^{A\to A}$ for any choice of $\mathcal{M} \in \mathrm{CPTP}(A\to A)$.

the following criteria: there exists a reference system $R$, alongside a quantum channel $\mathcal{E} \in \mathrm{CPTP}(AR \to A)$ and an isometry $\mathcal{F} \in \mathrm{CPTP}(B \to RB')$ such that

$$\mathcal{N}^{AB\to AB'} = \mathcal{E}^{RA\to A} \circ \mathcal{F}^{B\to RB'}. \tag{7.15}$$

In essence, channels that are $A \not\to B'$ signaling are those that can be implemented via one-way communication from Bob to Alice. An illustration of this concept can be found in Figure 7.3.

> **Theorem 7.1.** Let $\mathcal{N} \in \mathrm{CPTP}(AB \to AB')$ be a bipartite quantum channel. The following two statements are equivalent:
>
> 1. The channel $\mathcal{N}^{AB\to AB'}$ is an $A \not\to B'$ signaling, as defined in (7.15).
> 2. The channel $\mathcal{N}^{AB\to AB'}$ is an $A \not\to B$ semi-causal, as defined in (7.13).

*Remark.*  The Theorem 7.1 demonstrates the intuitive assertion that semi-causal bipartite channels are channels that can be realized with one-way communication from Bob to Alice. With such channels, Alice cannot influence Bob's system. We also point out that the relation in (7.15) has been written in a compact form; that is, we removed identity channels so that

$$\mathcal{E}^{RA\to A} \circ \mathcal{F}^{B\to RB'} := \left(\mathcal{E}^{RA\to A} \otimes \mathrm{id}^{B'\to B'}\right) \circ \left(\mathrm{id}^{A\to A} \otimes \mathcal{F}^{B\to RB'}\right). \tag{7.16}$$

**Proof**  We begin by proving the implication $1 \Rightarrow 2$. Consider the following marginal of the channel $\mathcal{N}^{AB\to AB'}$:

$$
\begin{aligned}
\mathcal{N}^{AB\to B'} &:= \mathrm{Tr}_A \circ \mathcal{N}^{AB\to AB'} \\
\textbf{(7.15)}\to\ &= \mathrm{Tr}_A \circ \mathcal{E}^{RA\to A} \circ \mathcal{F}^{B\to RB'} \\
&= \mathrm{Tr}_{RA} \circ \mathcal{F}^{B\to RB'} \\
&= \mathrm{Tr}_A \circ \mathcal{F}^{B\to B'},
\end{aligned}
\tag{7.17}
$$

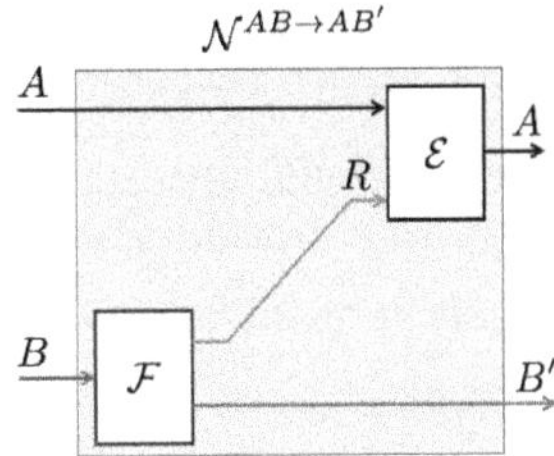



**Figure 7.3** A bipartite channel that is $A \not\to B'$ signaling.

where we have utilized the trace-preserving property of quantum channels and denoted $\mathcal{F}^{B \to B'} := \mathrm{Tr}_R \circ \mathcal{F}^{B \to RB'}$. It is evident that this channel satisfies (7.13) with any trace preserving map $\mathcal{M}^{A \to A}$, establishing that $\mathcal{N}^{AB \to AB'}$ is $A \not\to B'$ semi-causal.

Moving on to the implication $2 \Rightarrow 1$, we examine two distinct purifications of the marginal Choi matrix $J_{\mathcal{N}}^{ABB'}$:

1. Consider a physical system $C$ and a pure (unnormalized) state $\psi^{ABAB'C}$, which acts as a purification of both $J_{\mathcal{N}}^{ABAB'}$ and its marginal state $J_{\mathcal{N}}^{ABB'}$.

2. Denoting by $\varphi^{BB'R}$ an (unnormalized) purification of the operator $\frac{1}{|A|}J_{\mathcal{N}}^{BB'}$, we get from (7.14) that $\Omega^{A\tilde{A}} \otimes \varphi^{BB'R}$ is another purification $J_{\mathcal{N}}^{ABB'}$.

Since the marginal of $J_{\mathcal{N}}^{AB}$ equals $I^{AB}$, it implies that $\varphi^B = \frac{1}{|A|}J_{\mathcal{N}}^B = I^B$. This property implies the existence of an isometry $\mathcal{F} \in \mathrm{CPTP}(B \to B'R)$ that satisfies

$$\varphi^{BB'R} = \mathcal{F}^{\tilde{B} \to B'R}(\Omega^{B\tilde{B}}). \tag{7.18}$$

Moreover, given that two purifications of the same positive semi-definite matrix are connected by an isometry (as per Exercise 2.44), there must be an isometry $\mathcal{V} \in \mathrm{CPTP}(R\tilde{A} \to AC)$ satisfying

$$\psi^{ABAB'C} = \mathcal{V}^{R\tilde{A} \to AC}\left(\Omega^{A\tilde{A}} \otimes \phi^{BB'R}\right). \tag{7.19}$$

Finally, tracing out system $C$ and denoting by $\mathcal{E}^{R\tilde{A} \to A} := \mathrm{Tr}_C \circ \mathcal{V}^{R\tilde{A} \to AC}$ gives

$$\begin{aligned}
J_{\mathcal{N}}^{ABAB'} &= \mathcal{E}^{R\tilde{A} \to A}\left(\Omega^{A\tilde{A}} \otimes \phi^{BB'R}\right) \\
\mathbf{(7.18)} \to &= \mathcal{E}^{R\tilde{A} \to A}\left(\Omega^{A\tilde{A}} \otimes \mathcal{F}^{\tilde{B} \to B'R}(\Omega^{B\tilde{B}})\right) \\
&= \mathcal{E}^{R\tilde{A} \to A} \circ \mathcal{F}^{\tilde{B} \to B'R}\left(\Omega^{(AB)(\tilde{A}\tilde{B})}\right).
\end{aligned} \tag{7.20}$$

Equation (7.20) implies that (7.15) holds. This completes the proof. ∎

**Exercise 7.2.** *Show that if $|B'| = 1$, then any channel in $\mathrm{CPTP}(AB \to AB')$ is $A \not\to B'$ semi-causal.*

## 7.1.3 Quantum Conditionally Mixing Operations

We do not expect conditional uncertainty to decrease under operations that are both conditionally unital and $A \not\to B'$ semi-causal. Such operations provide a quantum extension to the operations introduced in Definition 4.8.

> **Definition 7.1.** A quantum channel $\mathcal{N} \in \mathrm{CPTP}(AB \to AB')$ is called a *conditionally mixing operation* (CMO) if it is both conditionally unital and $A \not\to B'$ semi-causal. The set of all such conditionally mixing operations in $\mathrm{CPTP}(AB \to AB')$ is denoted by $\mathrm{CMO}(AB \to AB')$.

The Choi matrix, $J_{\mathcal{N}}^{AB\tilde{A}B'}$, of a channel $\mathcal{N} \in \mathrm{CMO}(AB \to \tilde{A}B')$ that is both conditionally unital and $A \not\to B'$ semi-causal must satisfy the following:

1. $J_{\mathcal{N}}^{B\tilde{A}B'} = J_{\mathcal{N}}^{BB'} \otimes \mathbf{u}^{\tilde{A}}$ (i.e. $\mathcal{N}$ is conditionally unital; see (7.4)).
2. $J_{\mathcal{N}}^{ABB'} = \mathbf{u}^{A} \otimes J_{\mathcal{N}}^{BB'}$ (i.e. $\mathcal{N}$ is semi-causal; see (7.14)).
3. $J_{\mathcal{N}}^{AB} = I^{AB}$ (i.e. $\mathcal{N}$ is trace preserving).
4. $J_{\mathcal{N}}^{AB\tilde{A}B'} \geqslant 0$ (i.e. $\mathcal{N}$ is completely positive).

Observe the symmetry of the first two conditions above under exchange of the local input system $A$ and the local output system $\tilde{A}$.

As a straightforward example of a CMO, consider that the reference system $R$ in Theorem 7.1 is classical. In such a scenario, we define $X := R$ and the channel $\mathcal{N}^{AB \to AB'}$ can be expressed as

$$\mathcal{N}^{AB \to AB'} = \mathcal{E}^{XA \to A} \circ \mathcal{F}^{B \to XB'} = \sum_{x \in [m]} \mathcal{E}_{(x)}^{A \to A} \otimes \mathcal{F}_x^{B \to B'}. \tag{7.21}$$

Here, $\{\mathcal{F}_x^{B \to B'}\}_{x \in [m]}$ constitutes a quantum instrument, and for each $x \in [m]$, the map $\mathcal{E}_{(x)}^{A \to A}$ is a quantum channel in $\mathrm{CPTP}(A \to A)$. We will explore later that this channel typifies one-way LOCC (Local Operations and Classical Communication). Notably, if each $\mathcal{E}_{(x)}$ is a unital channel, this one-way LOCC is also conditionally unital. This channel essentially represents Bob performing a quantum measurement on his system and conveying the result to Alice, who then applies a unital channel to her system.

**Exercise 7.3.** *Let* $\omega \in \mathfrak{D}(B)$ *and* $\mathcal{N} \in \mathrm{CMO}(AB \to \tilde{A}B')$. *Show that the channel* $\mathcal{E} \in \mathrm{CPTP}(A \to \tilde{A}B')$ *defined for any* $\rho \in \mathfrak{L}(A)$ *by*

$$\mathcal{E}^{A \to \tilde{A}B'}(\rho^A) := \mathcal{N}^{AB \to \tilde{A}B'}(\rho^A \otimes \omega^B) \tag{7.22}$$

*is also* CMO; *that is, show that* $\mathcal{E} \in \mathrm{CMO}(A \to \tilde{A}B')$.

**Exercise 7.4.** *Let* $\Upsilon \in \mathfrak{L}(AB\tilde{A}B' \to AB\tilde{A}B')$ *be a linear map defined for all* $\omega \in \mathfrak{L}(AB\tilde{A}B')$ *as*

$$\Upsilon\left(\omega^{AB\tilde{A}B'}\right) := \mathbf{u}^A \otimes \left(\omega^{B\tilde{A}B'} - \omega^{BB'} \otimes \mathbf{u}^{\tilde{A}}\right) + \mathbf{u}^{\tilde{A}} \otimes \left(\omega^{ABB'} - \mathbf{u}^A \otimes \omega^{BB'}\right). \tag{7.23}$$

1. *Show that a channel $\mathcal{N} \in \mathrm{CPTP}(AB \to \tilde{A}B')$ is CMO if and only if $\Upsilon(J_{\mathcal{N}}^{AB\tilde{A}B'}) = 0$.*

2. *Show that $\Upsilon$ is self-adjoint; that is, show that $\Upsilon = \Upsilon^*$.*

3. *Show that $\Upsilon$ is idempotent; that is, $\Upsilon \circ \Upsilon = \Upsilon$.*

## 7.1.4 Definition of Conditional Majorization

We are now prepared to define quantum conditional majorization, grounded on the two aforementioned axioms, for channels that preserve or increase conditional uncertainty.

> **Quantum Conditional Majorization**
>
> **Definition 7.2.** Let $A$, $B$, and $B'$ be three quantum systems, and let $\rho \in \mathfrak{D}(AB)$ and $\sigma \in \mathfrak{D}(AB')$. We say that $\rho^{AB}$ conditionally majorizes $\sigma^{AB'}$, and write $\rho^{AB} \succ_A \sigma^{AB'}$, if there exists a channel $\mathcal{N} \in \mathrm{CMO}(AB \to AB')$ such that
>
> $$\sigma^{AB'} = \mathcal{N}^{AB \to AB'}\left(\rho^{AB}\right). \tag{7.24}$$

**Exercise 7.5.** *Show that quantum conditional majorization as defined above is a preorder.*

This definition effectively extends the concept of majorization. Specifically, when $|B| = |B'| = 1$, the set $\mathrm{CMO}(A \to A)$ coincides with the set of unital channels. Consequently, the relation $\rho^A \succ_A \sigma^A$ as already defined (under the condition $|B| = |B'| = 1$) transforms into the well-known majorization relation $\rho^A \succ \sigma^A$. Expanding on this concept, quantum conditional majorization exhibits the following notable property.

> **Lemma 7.2.** Let $\rho \in \mathfrak{D}(AB)$ and $\sigma \in \mathfrak{D}(AB')$ be two product states; that is, $\rho^{AB} = \rho^A \otimes \rho^B$ and $\sigma^{AB'} = \sigma^A \otimes \sigma^{B'}$. Then,
>
> $$\rho^{AB} \succ_A \sigma^{AB'} \iff \rho^A \succ \sigma^A. \tag{7.25}$$

**Proof** If $\rho^A \succ \sigma^A$, then there exists a unital channel such that $\sigma^A = \mathcal{U}^{A \to A}(\rho^A)$. Let $\mathcal{E} \in \mathrm{CPTP}(B \to B')$ be a replacement channel that always outputs $\sigma^{B'}$. Then, the channel

$$\mathcal{N}^{AB \to AB'} := \mathcal{U}^{A \to A} \otimes \mathcal{E}^{B \to B'} \tag{7.26}$$

is CMO and satisfies $\sigma^A \otimes \sigma^{B'} = \mathcal{N}^{AB \to AB'}\left(\rho^A \otimes \rho^B\right)$. Hence, $\rho^{AB} \succ_A \sigma^{AB'}$. Conversely, if $\rho^{AB} \succ_A \sigma^{AB'}$ then there exists a semi-causal quantum channel $\mathcal{N}^{AB \to AB'} = \mathcal{E}^{RA \to A} \circ \mathcal{F}^{B \to RB'}$ (that is also conditionally unital) that satisfies

$$\sigma^A \otimes \sigma^{B'} = \mathcal{E}^{RA \to A} \circ \mathcal{F}^{B \to RB'}\left(\rho^A \otimes \rho^B\right). \tag{7.27}$$

Tracing system $B'$ on both sides, and denoting $\tau^R := \mathrm{Tr}_{B'} \circ \mathcal{F}^{B \to RB'}\left(\rho^B\right)$ gives

$$\sigma^A = \mathcal{E}^{RA \to A}\left(\rho^A \otimes \tau^R\right). \tag{7.28}$$

Finally, note that the channel

$$\mathcal{U}^{A \to A}(\omega^A) := \mathcal{E}^{RA \to A}\left(\omega^A \otimes \tau^R\right) \qquad \forall\, \omega \in \mathfrak{D}(A) \tag{7.29}$$

is a unital channel since $\mathcal{N}^{AB \to AB'}$ is conditionally unital. Since by definition, $\sigma^A = \mathcal{U}^{A \to A}\left(\rho^A\right)$, we conclude that $\rho^A \succ \sigma^A$. This completes the proof. $\blacksquare$

Determining whether $\rho^{AB} \succ_A \sigma^{AB'}$ for two given bipartite quantum states $\rho^{AB}$ and $\sigma^{AB'}$ can be challenging. This task is essentially about verifying the existence of a Choi matrix $J^{AB\tilde{A}B'}_{\mathcal{N}}$ that satisfies both the four initial conditions outlined at the start of this subsection (characteristic of a CMO) and the additional criterion:

$$\mathrm{Tr}_{AB}\left[J^{AB\tilde{A}B'}_{\mathcal{N}}\left(\left(\rho^{AB}\right)^T \otimes I^{\tilde{A}B'}\right)\right] = \sigma^{\tilde{A}B'}. \tag{7.30}$$

These five conditions imposed on $J^{AB\tilde{A}B'}_{\mathcal{N}}$ constitute an SDP (semi-definite programming) feasibility problem that can be solved efficiently and algorithmically on a computer.

## 7.1.5 Specific Conditional Majorization Relations

Generally, as already mentioned, determining if one state conditionally majorizes another can be resolved through an SDP. However, there are key examples that are useful for later discussions and can be resolved without an SDP. For instance, for any state $\rho \in \mathfrak{D}(ABB')$, it holds that

$$\rho^{ABB'} \succ_A \rho^{AB} \tag{7.31}$$

since partial tracing over system $B'$ qualifies as a CMO. Another straightforward case is for any $\rho \in \mathfrak{D}(AB)$ and $\sigma \in \mathfrak{D}(B)$, where

$$\rho^{AB} \succ_A \rho^A \otimes \sigma^B. \tag{7.32}$$

A more nuanced example is the fact that the maximally entangled state conditionally majorizes all states of the same dimensions.

---

**Theorem 7.2.** Let $\rho \in \mathfrak{D}(AB)$. Then,

$$\Phi^{AB} \succ_A \rho^{AB}. \tag{7.33}$$

---

**Proof**  In Section 1.4.1, we discussed how the maximally entangled state $\Phi^{AB}$ can be utilized for teleporting an unknown quantum state. Specifically, a teleportation protocol from Bob to Alice (note that in Section 1.4.1, we examined teleportation from Alice to Bob) involves Bob performing a joint quantum measurement on his part of $\Phi^{AB}$ and the state to be teleported. This is followed by classical communication to Alice, who then applies a unitary operation. Such a protocol conforms to the structure of CMO channels described in (7.21), where each $\mathcal{E}_{(x)}^{A \to A}$ is a unitary channel. This implies that for any bipartite state $\rho^{AB}$, there exists a channel $\mathcal{N} \in \mathrm{CMO}(AB \to AB)$ such that

$$\rho^{AB} = \mathcal{N}^{AB \to AB}\left(\Phi^{AB}\right). \tag{7.34}$$

We emphasize that in the realization of the channel $\mathcal{N}^{AB \to AB}$, Bob locally prepares the state $\rho^{B'B}$ with $B' \cong A$ and then employs the maximally entangled state $\Phi^{AB}$ to teleport the $B'$ subsystem to Alice, resulting in the state $\rho^{AB}$. The equation above implies that $\Phi^{AB} \succ_A \rho^{AB}$.   ■

The next theorem characterizes the states in $\mathfrak{D}(AB)$ that are conditionally majorized by a state in Pure($A$). Note that all the states in Pure($A$) are equivalent under majorization.

---

**Theorem 7.3.** Let $\rho \in \mathfrak{D}(AB)$. Then, the following statements are equivalent:

1. The state $\rho^{AB}$ satisfies

$$I^A \otimes \rho^B \geq \rho^{AB}. \tag{7.35}$$

2. For every pure state $\psi \in \mathrm{Pure}(A)$

$$\psi^A \succ_A \rho^{AB}. \tag{7.36}$$

---

**Proof** To demonstrate that the first statement implies the second, we search for a channel $\mathcal{N} \in \mathrm{CMO}(A \to AB)$ satisfying $\rho^{AB} = \mathcal{N}^{A \to AB}(\psi^A)$. Our strategy involves considering a binary measurement-and-prepare channel defined as

$$\mathcal{N}^{A \to AB}(\omega^A) := \mathrm{Tr}[\psi^A \omega^A]\rho^{AB} + \mathrm{Tr}[(I^A - \psi^A)\omega^A]\tau^{AB} \quad \forall; \omega \in \mathfrak{L}(A), \tag{7.37}$$

where $\tau^{AB}$ is a density matrix to be determined. By definition, we have $\rho^{AB} = \mathcal{N}^{A \to AB}(\psi^A)$. The remaining task is to identify a $\tau \in \mathfrak{D}(AB)$ that ensures $\mathcal{N}^{A \to AB}$ is a CMO (both $A \not\to B$ semi-causal and conditionally unital).

For $\mathcal{N}^{A \to AB}$ to be $A \not\to B$ semi-causal, it must satisfy the condition $\mathcal{N}^{A \to B} \circ \mathcal{M}^{A \to A} = \mathcal{N}^{A \to B}$ for any $\mathcal{M} \in \mathrm{CPTP}(A \to A)$ (cf. (7.13)). That is, $\mathcal{N}^{A \to B}$ must be a replacement channel, independent of its input. Observing that for every input state $\omega \in \mathfrak{D}(A)$,

$$\mathcal{N}^{A \to B}(\omega^A) = \mathrm{Tr}[\psi^A \omega^A]\rho^B + \mathrm{Tr}\left[\left(I^A - \psi^A\right)\omega^A\right]\tau^B, \tag{7.38}$$

and given that the left-hand side is independent of $\omega^A$, we conclude $\mathcal{N}^{A \to AB}$ is $A \not\to B$ semi-causal if and only if $\tau^B = \rho^B$. This is the first condition for $\tau^{AB}$.

The second condition arises from $\mathcal{N}^{A \to AB}$ being conditionally unital. This is satisfied if and only if (cf. (7.2, 7.3))

$$\mathcal{N}^{A\to AB}(I^A) = \rho^{AB} + (|A| - 1)\,\tau^{AB} \tag{7.39}$$

equals

$$I^A \otimes \mathcal{N}^{A\to B}(\mathbf{u}^A) = I^A \otimes \rho^B, \tag{7.40}$$

using (7.38) with $\tau^B = \rho^B$. Equating these two operators dictates that $\tau^{AB}$ must be

$$\tau^{AB} := \frac{I^A \otimes \rho^B - \rho^{AB}}{|A| - 1}. \tag{7.41}$$

$\tau^{AB}$ is positive semi-definite, as we assume $I^A \otimes \rho^B \geqslant \rho^{AB}$, and has unit trace, qualifying as a density matrix. Also, it satisfies $\tau^B = \rho^B$. This concludes the proof that the first statement implies the second.

Conversely, if $\mathcal{N} \in \mathrm{CMO}(A \to AB)$ is such that $\rho^{AB} = \mathcal{N}^{A\to AB}(\psi^A)$, then

$$\rho^B = \mathcal{N}^{A\to B}(\psi^A), \tag{7.42}$$

where $\mathcal{N}^{A\to B} := \mathrm{Tr}_A \circ \mathcal{N}^{A\to AB}$ is the marginal channel. Since $\mathcal{N}^{A\to AB}$ is $A \nrightarrow B$ semi-causal, the marginal channel $\mathcal{N}^{A\to B}$ must satisfy

$$\mathcal{N}^{A\to B}\left(\mathbf{u}^A\right) = \mathcal{N}^{A\to B}\left(\psi^A\right)$$
$$(7.42)\to\ = \rho^B. \tag{7.43}$$

Moreover, since $\mathcal{N}^{A\to AB}$ is conditionally unital we have

$$\mathcal{N}^{A\to AB}\left(I^A\right) = I^A \otimes \mathcal{N}^{A\to B}(\mathbf{u}^A)$$
$$(7.43)\to\ = I^A \otimes \rho^B. \tag{7.44}$$

Combining everything we get

$$I^A \otimes \rho^B - \rho^{AB} = \mathcal{N}^{A\to AB}\left(I^A\right) - \mathcal{N}^{A\to AB}\left(\psi^A\right)$$
$$= \mathcal{N}^{A\to AB}\left(I^A - \psi^A\right) \geqslant 0, \tag{7.45}$$

where the last inequality follows from the fact that $I^A - \psi^A \geqslant 0$ and $\mathcal{N}^{A\to AB}$ is a completely positive map. This concludes the proof. ∎

In Theorem 7.2, we saw that under conditional majorization $\Phi^{AB}$ is the maximal element of $\mathfrak{D}(AB)$. On the other hand, the maximally mixed state $\mathbf{u}^A$ satisfies the opposite inequality that $\rho^{AB} \succ_A \mathbf{u}^A$ for all $\rho \in \mathfrak{D}(AB)$. Combining this maximal and minimal elements gives the state

$$\Phi^{AB} \otimes \mathbf{u}^{\tilde{A}}. \tag{7.46}$$

Remarkably, the following theorem shows that under conditional majorization, this state is equivalent to any pure state in $\mathrm{Pure}(A\tilde{A})$.

> **Theorem 7.4.** For all $\psi \in \mathrm{Pure}(AB)$ with $|B| = |A|$ we have,
>
> $$\psi^{A\tilde{A}} \succ_{A\tilde{A}} \Phi^{AB} \otimes \mathbf{u}^{\tilde{A}} \succ_{A\tilde{A}} \psi^{A\tilde{A}}. \tag{7.47}$$

**Proof** We first prove that $\Phi^{AB} \otimes \mathbf{u}^{\tilde{A}} \succ_{A\tilde{A}} \psi^{A\tilde{A}}$. We will denote by $\tau^{A\tilde{A}}$ the density matrix

$$\tau^{A\tilde{A}} := \frac{I^{A\tilde{A}} - \psi^{A\tilde{A}}}{m^2 - 1}, \tag{7.48}$$

where $m := |A|$. Let $\mathcal{N}^{A\tilde{A}B \to A\tilde{A}}$ be a quantum channel defined by

$$\mathcal{N}^{A\tilde{A}B \to A\tilde{A}} := \mathcal{E}^{AB \to A\tilde{A}} \circ \mathrm{Tr}_{\tilde{A}}, \tag{7.49}$$

where for all $\omega \in \mathfrak{L}(AB)$

$$\mathcal{E}^{AB \to A\tilde{A}}(\omega^{AB}) := \mathrm{Tr}[\Phi^{AB}\omega^{AB}]\psi^{A\tilde{A}} + \mathrm{Tr}[(I^{AB} - \Phi^{AB})\omega^{AB}]\tau^{A\tilde{A}}. \tag{7.50}$$

By definition, the channel above satisfies

$$\mathcal{N}^{A\tilde{A}B \to A\tilde{A}}\left(\Phi^{AB} \otimes \mathbf{u}^{\tilde{A}}\right) = \psi^{A\tilde{A}}. \tag{7.51}$$

Therefore, it is left to show that the channel $\mathcal{N}$ is CMO. Since the channel $\mathcal{N}^{A\tilde{A}B \to A\tilde{A}}$ does not have an output on Bob's side, it is trivially $A \not\to B'$ semi-causal ($|B'| = 1$). To show that it is conditionally unital observe that for any $\sigma \in \mathfrak{D}(B)$ we have

$$\mathcal{N}^{A\tilde{A}B \to A\tilde{A}}\left(I^{A\tilde{A}} \otimes \sigma^B\right) = \mathcal{E}^{AB \to A\tilde{A}}\left(mI^A \otimes \sigma^B\right)$$

$$= \psi^{A\tilde{A}} + m\mathrm{Tr}\left[\left(mI^B - \mathbf{u}^B\right)\sigma^B\right]\tau^{A\tilde{A}} \tag{7.52}$$

$$\mathbf{(7.48)} \to\ = I^{A\tilde{A}}.$$

Hence, $\mathcal{N} \in \mathrm{CMO}(A\tilde{A}B \to A\tilde{A})$. This completes the proof that $\Phi^{AB} \otimes \mathbf{u}^{\tilde{A}} \succ_{A\tilde{A}} \psi^{A\tilde{A}}$.

To prove that $\psi^{A\tilde{A}} \succ_{A\tilde{A}} \Phi^{AB} \otimes \mathbf{u}^{\tilde{A}}$, we set $\tau^{AB} := (I^{AB} - \psi^{AB})/(m^2 - 1)$ and denote by $\mathcal{N}^{A\tilde{A} \to A\tilde{A}B}$ a quantum channel defined for all $\omega \in \mathfrak{L}(A\tilde{A})$ as

$$\mathcal{N}^{A\tilde{A} \to A\tilde{A}B}(\omega^{A\tilde{A}}) := \left(\mathcal{N}^{A\tilde{A} \to AB}(\omega^{A\tilde{A}})\right) \otimes \mathbf{u}^{\tilde{A}}, \tag{7.53}$$

where

$$\mathcal{N}^{A\tilde{A} \to AB}(\omega^{A\tilde{A}}) := \mathrm{Tr}[\psi^{A\tilde{A}}\omega^{A\tilde{A}}]\Phi^{AB} + \mathrm{Tr}[(I^{A\tilde{A}} - \psi^{A\tilde{A}})\omega^{A\tilde{A}}]\tau^{AB}. \tag{7.54}$$

By definition, the channel above satisfies

$$\mathcal{N}^{A\tilde{A} \to A\tilde{A}B}\left(\psi^{A\tilde{A}}\right) = \Phi^{AB} \otimes \mathbf{u}^{\tilde{A}}. \tag{7.55}$$

Therefore, it is left to show that the channel $\mathcal{N}$ is CMO. To show that it is conditionally unital observe that

$$\mathcal{N}^{A\tilde{A} \to A\tilde{A}B}\left(I^{A\tilde{A}}\right) = \left(\Phi^{AB} + (I^{AB} - \Phi^{AB})\right) \otimes \mathbf{u}^{\tilde{A}}$$

$$= I^{AB} \otimes \mathbf{u}^{\tilde{A}}. \tag{7.56}$$

To show that it is $A \not\rightarrow B$ semi-causal observe that for all $\omega \in \mathfrak{D}(A\tilde{A})$ the marginal channel $\mathcal{N}^{A\tilde{A} \to B}$ satisfies

$$\mathcal{N}^{A\tilde{A} \to B}\left(\omega^{A\tilde{A}}\right) = \mathrm{Tr}[\psi^{A\tilde{A}}\omega^{A\tilde{A}}]\mathrm{u}^{B} + \mathrm{Tr}[(I^{A\tilde{A}} - \psi^{A\tilde{A}})\omega^{A\tilde{A}}]\tau^{B}$$

$$\textbf{Exercise (7.6)}\rightarrow = \mathrm{u}^{B},$$

(7.57)

which is independent of $\omega^{A\tilde{A}}$, so that $\mathcal{N}^{A\tilde{A} \to A\tilde{A}B}$ is $A \not\rightarrow B$ semi-causal. Hence, $\mathcal{N} \in$ CMO$(A\tilde{A} \to A\tilde{A}B)$. This completes the proof. $\blacksquare$

**Exercise 7.6.** *Use the definition of $\tau^{AB}$ to verify the first equality in (7.56) and the second equality in (7.57).*

## 7.2 Definition of Conditional Entropy

In Section 7.1, we introduced the concept of conditional majorization between bipartite states $\rho \in \mathfrak{D}(AB)$ and $\sigma \in \mathfrak{D}(AB')$. This relationship is established if a conditionally unital and $A \not\rightarrow B'$ signaling channel in CPTP$(AB \to AB')$ can transform $\rho^{AB}$ into $\sigma^{AB'}$. Monotonic functions under this preorder quantify the conditional uncertainty Bob has about Alice's system $A$. Furthermore, if these functions are additive under tensor products, we refer to them as conditional entropies (see the forthcoming definition).

In the classical entropy's definition (see Definition 6.1), we require $\mathbb{H}$ to behave monotonically under majorization. Majorization typically involves comparing two probability vectors of equal dimensions, but this can be extended to vectors of differing dimensions by padding the shorter vector with zeros. In the quantum realm, we expand the concept of conditional majorization by introducing additional isometries to account for varying dimensions of Alice's input and output systems.

**Extension of Conditional Majorization**

Consider $\rho \in \mathfrak{D}(AB)$ and $\sigma \in \mathfrak{D}(A'B')$, where $|A|$ may differ from $|A'|$. We state that $\rho^{AB}$ conditionally majorizes $\sigma^{A'B'}$, denoted as $\rho^{AB} \succ_{A} \sigma^{A'B'}$, if for $|A| \geqslant |A'|$ an isometry $\mathcal{V} \in$ CPTP$(A' \to A)$ exists such that $\rho^{AB} \succ_{A} \mathcal{V}^{A' \to A}\left(\sigma^{A'B'}\right)$. Conversely, if $|A'| > |A|$, an isometry $\mathcal{U} \in$ CPTP$(A \to A')$ exists so that $\mathcal{U}^{A \to A'}\left(\rho^{AB}\right) \succ_{A'} \sigma^{A'B'}$.

With this extension, we can now define conditional entropy. Specifically, we consider

$$\mathbb{H}: \bigcup_{A,B} \mathfrak{D}(AB) \to \mathbb{R}$$

(7.58)

as a function mapping the set of all bipartite states across finite dimensions to the real line. The function $\mathbb{H}$ assigns to each bipartite density matrix $\rho^{AB}$ a real number, denoted by $\mathbb{H}(A|B)_\rho$. This notation distinguishes conditional entropy from the entropy of the marginal state $\rho^A$.

Our objective is to identify when $\mathbb{H}$ constitutes a conditional entropy. For systems where $|B| = 1$, we denote $\mathbb{H}(A|B)_\rho$ as $\mathbb{H}(A)_\rho := \mathbb{H}(\rho^A)$, aligning with the notation of conditional entropy. This notation proves useful when examining composite systems with multiple subsystems. Since we define entropy functions as nonconstant zero functions, we assume (implicitly throughout this book, and in the definition below) the existence of a quantum system $A$ and a state $\rho \in \mathfrak{D}(A)$ such that $\mathbb{H}(A)_\rho \neq 0$.

---

**Quantum Conditional Entropy**

**Definition 7.3.** The function $\mathbb{H}$ in (7.58) is termed a conditional entropy if for every $\rho \in \mathfrak{D}(AB)$ and $\sigma \in \mathfrak{D}(A'B')$ it satisfies the following two properties:

1. Monotonicity: If $\rho^{AB} \succ_A \sigma^{A'B'}$, then $\mathbb{H}(A|B)_\rho \leqslant \mathbb{H}(A'|B')_\sigma$.
2. Additivity: $\mathbb{H}(AA'|BB')_{\rho \otimes \sigma} = \mathbb{H}(A|B)_\rho + \mathbb{H}(A'|B')_\sigma$.

---

There are several properties of conditional entropy that follow from the definition above. First, observe that the case that system $B$ is trivial, that is, $|B| = 1$, a conditional entropy function reduces to an entropy function. Moreover, if $\rho^{AB} = \omega^A \otimes \tau^B$ is a product state, then it can be converted reversibly to the product state $\omega^A \otimes \mathbf{u}^B$ by a product channel of the form $\mathrm{id}^{A \to A} \otimes \mathcal{E}^{B \to B}$, which is in $\mathrm{CMO}(AB \to AB)$. Therefore, from the monotonicity property above it follows that

$$\mathbb{H}(A|B)_{\omega^A \otimes \iota^B} = \mathbb{H}(A|B)_{\omega^A \otimes \mathbf{u}^B}, \tag{7.59}$$

so that $\mathbb{H}(A|B)_\rho$ depends only on $\omega^A$. Moreover, the function $\omega^A \mapsto \mathbb{H}(A|B)_{\omega^A \otimes \mathbf{u}^B}$ satisfies the two axioms of entropy and therefore can be considered itself as an entropy of $\omega^A$. In other words, conditional entropy reduces to entropy on product states as intuitively expected.

Next, conditional entropy is invariant under the action of local isometric channels. That is, for a bipartite state $\rho^{AB}$ and isometric channels $\mathcal{U}^{A \to A'}$ and $\mathcal{V}^{B \to B'}$,

$$\mathbb{H}(A|B)_\rho = \mathbb{H}(A|B')_{\mathcal{V}(\rho)} = \mathbb{H}(A'|B)_{\mathcal{U}(\rho)}. \tag{7.60}$$

To see the first equality, observe that $\mathrm{id}^A \otimes \mathcal{V}^{B \to B'} \in \mathrm{CMO}(AB \to AB')$ so that $\mathbb{H}(A|B)_\rho \leqslant \mathbb{H}(A|B')_{\mathcal{V}(\rho)}$. For the converse, let $\mathcal{V}^{-1} \in \mathrm{CPTP}(B' \to B)$ be one of the left inverses of the isometry $\mathcal{V}^{B \to B'}$ (see, for example, (3.201)). Again, since $\mathrm{id}^A \otimes \mathcal{V}^{-1} \in \mathrm{CMO}(AB' \to AB)$, we conclude that

$$\mathbb{H}(A|B')_{\mathcal{V}(\rho)} \leqslant \mathbb{H}(A|B)_{\mathcal{V}^{-1} \circ \mathcal{V}(\rho)} = \mathbb{H}(A|B)_\rho. \tag{7.61}$$

Finally, the second equality in (7.60) (i.e. the invariance under an isometry on system $A$) follows directly from the monotonicity of conditional entropy under the extended version of conditional majorization.

**Exercise 7.7.** *Prove the second equality in (7.60). Hint: Use twice the monotonicity of conditional entropy under the extended version of conditional majorization.*

Consider the product state $\mathbf{u}^A \otimes \rho^B$, where $|A|=2$. Similar to what we found previously for unconditional entropy, the following inequality holds:

$$\mathbb{H}(A|B)_{\mathbf{u}\otimes\rho} = \mathbb{H}(A)_{\mathbf{u}} > 0. \tag{7.62}$$

This strict inequality allows us to set a normalization factor for conditional entropy. To be consistent with the normalization convention for unconditional entropy, we set for the case $|A| = 2$ that $\mathbb{H}(A|B)_{\mathbf{u}\otimes\rho} = 1$, which in turn implies that for $|A| > 2$

$$\mathbb{H}(A|B)_{\mathbf{u}\otimes\rho} = \log_2 |A|. \tag{7.63}$$

In the rest of this book, we will always assume that $\mathbb{H}$ is normalized in this way.

**Exercise 7.8.** *Let $\mathbb{H}$ be conditional entropy. Show that for every Hilbert spaces $A$ and $B$*

$$\mathbb{H}(A|B)_\rho \leqslant \log|A| \qquad \forall\, \rho \in \mathfrak{D}(AB) \tag{7.64}$$

*with equality if $\rho^{AB} = \mathbf{u}^A \otimes \tau^B$ for some $\tau \in \mathfrak{D}(B)$. Hint: Find a channel in $\mathrm{CMO}(AB \to AB)$ that takes $\rho^{AB}$ to $\mathbf{u}^A \otimes \rho^B$.*

## 7.3 Inevitability of Negative Quantum Conditional Entropy

In Theorem (7.2), we saw that the maximally entangled state $\Phi^{AB}$ conditionally majorizes all the states in $\mathfrak{D}(AB)$. Therefore, from the monotonicity property of the conditional entropy we get that for every $\rho \in \mathfrak{D}(AB)$, with $|A| = |B|$, and every conditional entropy function $\mathbb{H}$,

$$\mathbb{H}(A|B)_\rho \geqslant \mathbb{H}(A|B)_\Phi. \tag{7.65}$$

That is, the maximally entangled state has the least amount of conditional entropy. We will see next that $\mathbb{H}(A|B)_\Phi$ is negative.

Unlike entropy, quantum conditional entropy can be negative. This unintuitive phenomena puzzled the community for quite some time until an operational interpretation for the quantum conditional entropy was found. This operational interpretation is given in terms of a protocol known as quantum state merging (which we study in volume 2 of this book). In the following theorem we show that certain entangled states *must* have negative conditional entropy, while classical conditional entropy is always nonnegative.

The lower bound in the theorem below is given in terms of the conditional min-entropy defined on every bipartite state $\rho \in \mathfrak{D}(AB)$ as

$$H_{\min}(A|B)_\rho := -\inf_{\lambda\geqslant 0} \log_2\{\lambda : \rho_{AB} \leqslant \lambda I_A \otimes \rho_B\}. \tag{7.66}$$

In the next subsection, we will see that the conditional min-entropy is indeed a conditional entropy. Originally, this quantity was given the name conditional min-entropy because it was known to be the least among all Rényi conditional entropies. The theorem below strengthens this observation by proving that all plausible quantum conditional entropies are not smaller than the conditional min-entropy.

**Exercise 7.9.** *Consider the conditional min-entropy as defined in (7.66).*

1. *Show that for the maximally entangled state $\Phi^{AB}$ (with $|A| = |B|$) we have*

$$H_{\min}(A|B)_\Phi = -\log|A|. \tag{7.67}$$

2. *Show that if a density matrix $\rho \in \mathfrak{D}(AB)$ satisfies $H_{\min}(A|B)_\rho = \log|A|$, then $\rho^{AB} = \mathbf{u}^A \otimes \rho^B$.*
3. *Show that a state $\rho \in \mathfrak{D}(AB)$ has nonnegative conditional min-entropy if and only if $I^A \otimes \rho^B \geqslant \rho^{AB}$.*
4. *Show that if $\rho^{AB}$ is separable, then its conditional min-entropy is nonnegative.*

---

**Theorem 7.5.** Let $\mathbb{H}$ be a quantum conditional entropy. For all $\rho \in \mathfrak{D}(AB)$,

$$\mathbb{H}(A|B)_\rho \geqslant H_{\min}(A|B)_\rho, \tag{7.68}$$

with equality if $\rho^{AB}$ is the maximally entangled state $\Phi^{AB}$.

---

*Remark.* This theorem states that for the maximally entangled state $\Phi^{AB}$ we have $\mathbb{H}(A|B)_\Phi = H_{\min}(A|B)_\Phi$. Combining this with Exercise 7.9 we conclude that

$$\mathbb{H}(A|B)_\Phi = -\log|A|. \tag{7.69}$$

That is, *all* conditional entropies are negative on the maximally entangled state and equal to $-\log|A|$. Moreover, in conjunction with the third part of Exercise 7.9, the theorem above implies that all conditional entropies are nonnegative on separable states (and therefore also on classical states).

**Proof** Let's start by examining the scenario where $H_{\min}(A|B)\rho \geqslant 0$, and denote $m := |A|$. The proof strategy in this case revolves around identifying the largest integer $k \in [m]$ such that a classical system $X$, with dimension $|X| = k$, fulfills the condition $\mathbf{u}^X \succ_A \rho^{AB}$. Once we establish this optimal value of $k$, we can infer that every conditional entropy $\mathbb{H}$ must satisfy

$$\mathbb{H}(A|B)_\rho \geqslant H(X)_\mathbf{u} = \log k. \tag{7.70}$$

We begin by establishing that it is feasible to set $k := \left\lfloor 2^{H_{\min}(A|B)_\rho} \right\rfloor$.

Let $X$ be a classical system with dimension $k := \left\lfloor 2^{H_{\min}(A|B)_\rho} \right\rfloor$. By the assumption that $H_{\min}(A|B)_\rho \geqslant 0$, and given the dimension bound $H_{\min}(A|B)_\rho \leqslant \log_2 m$, it follows that $k \in [m]$. Observe that the case $k = m$ implies that $H_{\min}(A|B)_\rho = \log|A|$. In this case, according to the second part of Exercise 7.9 we must have $\rho^{AB} = \mathbf{u}^A \otimes \rho^B$ so that $\mathbb{H}(A|B)_\rho = \log|A| = H_{\min}(A|B)_\rho$.

We therefore assume now that $k < m$. We look for a channel $\mathcal{N} \in \mathrm{CMO}(A \to AB)$ that satisfies

$$\mathcal{N}^{A \to AB}\left(\frac{1}{k}\Pi^A\right) = \rho^{AB}, \tag{7.71}$$

where $\Pi^A$ is a projection onto a $k$-dimensional subspace of $A$ (i.e. set $\Pi^A :=$ $\sum_{x \in [k]} |x\rangle\langle x|^A$, where $\{|x\rangle\}_{x \in [m]}$ is some orthonormal basis of $A$). The existence of such a channel will prove that $\mathbf{u}^X \succ_A \rho^{AB}$ since $\mathbf{u}^X$ (with $|X| = k$) is equivalent to $\frac{1}{k}\Pi^A$ under conditional majorization. We choose $\mathcal{N}^{A \to AB}$ to be a measure-and-prepare channel of the form

$$\mathcal{N}^{A \to AB}(\omega^A) := \mathrm{Tr}\,[\Pi^A \omega^A]\rho^{AB} + \mathrm{Tr}\,[\left(I^A - \Pi^A\right)\omega^A]\tau^{AB} \qquad \forall\, \omega \in \mathfrak{L}(A), \quad (7.72)$$

where $\tau^{AB}$ is some density matrix that is chosen (see equation (7.76)) such that $\mathcal{N}^{A \to AB}$ is CMO. Indeed, the action of this channel is to perform a measurement according to the POVM $\{\Pi^X, I^X - \Pi^X\}$ and prepare the state $\rho^{AB}$ if the first outcome is obtained and the state $\tau^{AB}$ if the second outcome is obtained. By definition, this channel satisfies (7.71).

The channel $\mathcal{N}^{A \to AB}$ is $A \not\to B$ signaling if and only if the marginal channel $\mathcal{N}^{A \to B} := \mathrm{Tr}_A \circ \mathcal{N}^{A \to AB}$ satisfies $\mathcal{N}^{A \to B} \circ \mathcal{M}^{A \to A} = \mathcal{N}^{A \to B}$ for all $\mathcal{M} \in \mathrm{CPTP}(A \to A)$.

In other words, $\mathcal{N}^{A \to AB}$ is $A \not\to B$ signaling if and only if the marginal channel $\mathcal{N}^{A \to B}$ is a replacement channel. Now, for every $\omega \in \mathfrak{D}(A)$ we have that

$$\mathcal{N}^{A \to B}(\omega^A) = \mathrm{Tr}\,[\Pi^A \omega^A]\rho^B + \mathrm{Tr}\,[\left(I^A - \Pi^A\right)\omega^A]\tau^B. \qquad (7.73)$$

Therefore, by taking $\tau^{AB}$ to have the property that its marginal $\tau^B = \rho^B$, we get that the right-hand side does not depend on $\omega^A$, so that $\mathcal{N}^{A \to AB}$ is $A \not\to B$ semi-causal.

The channel $\mathcal{N}^{A \to AB}$ is conditionally unital if and only if the state

$$\mathcal{N}^{A \to AB}(I^A) = \mathrm{Tr}\,[\Pi^A I^A]\rho^{AB} + \mathrm{Tr}\,[\left(I^A - \Pi^A\right)I^A]\tau^{AB}$$
$$= k\rho^{AB} + (m - k)\,\tau^{AB} \qquad\qquad\qquad (7.74)$$

is equal to the state

$$I^A \otimes \mathcal{N}^{A \to B}(\mathbf{u}^A) = I^A \otimes \rho^B, \qquad (7.75)$$

where we used (7.73) with $\tau^B = \rho^B$. The equality between the two states above forces $\tau^{AB}$ to be

$$\tau^{AB} := \frac{I^A \otimes \rho^B - k\rho^{AB}}{m - k}. \qquad (7.76)$$

The operator $\tau^{AB}$ is positive semi-definite because $m - k > 0$ and

$$\frac{1}{k}I^A \otimes \rho^B - \rho^{AB} \geqslant 2^{-H_{\min}(A|B)_\rho} I^A \otimes \rho^B - \rho^{AB} \qquad (7.77)$$
$$(7.66) \to\ \geqslant 0.$$

Also, $\tau^{AB}$ has trace equal to 1 (so it is a density matrix) with marginal $\tau^B = \rho^B$. We therefore proved that

$$\mathbb{H}(A|B)_\rho \geqslant \log k = \log \left\lfloor 2^{H_{\min}(A|B)_\rho} \right\rfloor. \qquad (7.78)$$

Finally, since all conditional entropies are additive for tensor-product states, we conclude that

$$
\mathbb{H}(A|B)_\rho = \lim_{n\to\infty} \frac{1}{n}\mathbb{H}(A^n|B^n)_{\rho^{\otimes n}}
$$

$$
\textbf{(7.78)}\to \;\geqslant\; \lim_{n\to\infty}\frac{1}{n}\log\left\lfloor 2^{H_{\min}(A^n|B^n)_{\rho^{\otimes n}}} \right\rfloor
$$

$$
= \lim_{n\to\infty}\frac{1}{n}\log\left\lfloor 2^{n\,H_{\min}(A|B)_\rho} \right\rfloor
$$

$$
= H_{\min}(A|B)_\rho. \tag{7.79}
$$

Next, consider the case $H_{\min}(A|B) < 0$. The idea of the proof in this case is to find the largest possible $k \in \mathbb{N}$ such that the system $X$ (can be taken to be a classical system) with dimension $|X| = k$ satisfies

$$
\psi^{XA} \succ_{AX} \rho^A \otimes \mathbf{u}^X, \tag{7.80}
$$

where $\psi \in \mathrm{Pure}(XA)$ is some pure state. Due to the monotonicity property of every conditional entropy $\mathbb{H}$ the above relation implies that

$$
0 = \mathbb{H}(XA)_\psi
$$

$$
\textbf{(7.80)}\to \;\leqslant\; \mathbb{H}(XA|B)_{\rho\otimes u}
$$

$$
\textbf{Additivity}\to \;=\; \mathbb{H}(X)_\mathbf{u} + \mathbb{H}(A|B)_\rho. \tag{7.81}
$$

Finally, since $\mathbb{H}(X)_\mathbf{u} = \log k$ we get that $\mathbb{H}(A|B)_\rho \geqslant -\log(k)$. We first show that (7.80) holds with $k := \left\lceil 2^{-H_{\min}(A|B)} \right\rceil$.

To prove the relation (7.80), consider the measure-and-prepare channel $\mathcal{N} \in \mathrm{CPTP}$ $(AX \to AXB)$ defined on all $\omega \in \mathfrak{L}(XA)$ as

$$
\mathcal{N}^{XA \to XAB}\left(\omega^{XA}\right) := \mathrm{Tr}\left[\psi^{XA}\omega^{XA}\right]\mathbf{u}^X \otimes \rho^{AB} + \mathrm{Tr}\left[\Pi^{XA}\omega^{XA}\right]\tau^{XAB}, \tag{7.82}
$$

where $\Pi^{AX} := I^{AX} - \psi^{XA}$ and $\tau^{XAB}$ are chosen (see equation (7.87)) such that $\mathcal{N}$ is CMO. Observe that by definition we have

$$
\mathcal{N}^{XA \to XAB}\left(\psi^{XA}\right) = \mathbf{u}^X \otimes \rho^{AB}. \tag{7.83}
$$

We next show that there exists $\tau \in \mathfrak{D}(XAB)$ such that $\mathcal{N}$ as defined above is indeed CMO. If $\mathcal{N}$ is $XA \nrightarrow B$ semi-causal, then we must have that the marginal channel $\mathcal{N}^{XA \to B} := \mathrm{Tr}_{XA}\circ\mathcal{N}^{XA \to XAB}$ is a replacement channel (i.e. a constant channel). Now, for all $\omega \in \mathfrak{D}(XA)$,

$$
\mathcal{N}^{XA \to B}\left(\omega^{XA}\right) = \mathrm{Tr}\left[\psi^{XA}\omega^{XA}\right]\rho^B + \mathrm{Tr}\left[\Pi^{XA}\omega^{XA}\right]\tau^B. \tag{7.84}
$$

Thus, by choosing $\tau^{XAB}$ to have the property $\tau^B = \rho^B$, we get that the right-hand side of the equation above does not depend on $\omega$, so that $\mathcal{N}^{XA \to XAB}$ is $XA \nrightarrow B$ semi-causal.

Next, the channel $\mathcal{N}^{XA \to XAB}$ is conditionally unital if the operator

$$\mathcal{N}^{XA \to XAB}\left(I^{XA}\right) = \mathbf{u}^{X} \otimes \rho^{AB} + km\tau^{XAB} \tag{7.85}$$

equals the operator

$$I^{XA} \otimes \mathcal{N}^{XA \to B}\left(\mathbf{u}^{XA}\right) = I^{XA} \otimes \tau^{B}, \tag{7.86}$$

where the last equality follows from (7.84) with $\tau^{B} = \rho^{B}$. We therefore conclude that $\mathcal{N} \in \mathrm{CPTP}(XA \to XAB)$ as defined above is CMO if and only if $\tau^{XAB}$ equals

$$\tau^{XAB} := \mathbf{u}^{X} \otimes \frac{kI^{A} \otimes \rho^{B} - \rho^{AB}}{km - 1}. \tag{7.87}$$

Observe that this $\tau^{XAB}$ is indeed a density matrix since by definition of $k$ we have

$$kI^{A} \otimes \rho^{B} \geqslant 2^{-H_{\min}(A|B)}I^{A} \otimes \rho^{B}$$
$$(7.66)\to \;\geqslant \rho^{AB}. \tag{7.88}$$

Moreover, from its definition in (7.87) we have $\tau^{B} = \rho^{B}$. Hence, with this $\tau^{XAB}$ the channel $\mathcal{N}^{XA \to XAB}$ is CMO that maps the pure states $\psi^{XA}$ to the state $\mathbf{u}^{X} \otimes \rho^{AB}$. We therefore conclude that

$$\mathbb{H}(A|B)_{\rho} \geqslant -\log k = -\log\left[2^{-H_{\min}(A|B)}\right]. \tag{7.89}$$

Finally, from the additivity property of conditional entropies we get

$$\begin{aligned}
\mathbb{H}(A|B)_{\rho} &= \lim_{n \to \infty} \frac{1}{n}\mathbb{H}(A^{n}|B^{n})_{\rho^{\otimes n}} \\
(7.89)\to \;&\geqslant -\lim_{n \to \infty} \frac{1}{n}\log\left[2^{-H_{\min}(A^{n}|B^{n})_{\rho^{\otimes n}}}\right] \\
&= -\lim_{n \to \infty} \frac{1}{n}\log\left[2^{-nH_{\min}(A|B)_{\rho}}\right] \\
&= H_{\min}(A|B)_{\rho}.
\end{aligned} \tag{7.90}$$

It is left to prove the equality on maximally entangled states. Since conditional entropy is invariant under local isometries we can assume without loss of generality that $m := |A| = |B|$. From Theorem 7.4 we know that the state $\Phi^{AB} \otimes \mathbf{u}^{\tilde{A}}$ is equivalent under conditional majorization to any pure state in $\mathrm{Pure}(A\tilde{A})$. Since the entropy of every pure state in $\mathrm{Pure}(A\tilde{A})$ is zero, we conclude that

$$0 = \mathbb{H}(A\tilde{A}|B)_{\Phi \otimes \mathbf{u}}$$
$$\mathbf{Additivity}\to \; = \mathbb{H}(A|B)_{\Phi} + \mathbb{H}(\tilde{A})_{\mathbf{u}} \tag{7.91}$$
$$= \mathbb{H}(A|B)_{\Phi} + \log m,$$

where we used the fact that the entropy of the uniform state $\mathbf{u}^{\tilde{A}}$ is $\log_{2} m$. Hence,

$$\mathbb{H}(A|B)_{\Phi} = -\log m = H_{\min}(A|B)_{\Phi}. \tag{7.92}$$

This completes the proof.     ∎

We saw in Theorem 7.5 that the conditional entropy is positive for all separable states. This does not mean that the conditional entropy is positive just for separable states. In fact, some entangled states (i.e. states that are not separable) have positive conditional entropies. The following corollary provides a simple criterion to determine if a bipartite state has a positive conditional entropy.

**Corollary 7.1.** Let $\rho \in \mathfrak{D}(AB)$. Then, the following statements are equivalent:

1. For any choice of conditional entropy $\mathbb{H}$ we have $\mathbb{H}(A|B)_\rho \geqslant 0$.
2. The state $\rho^{AB}$ satisfies

$$I^A \otimes \rho^B \geqslant \rho^{AB}. \tag{7.93}$$

3. For every pure state $\psi \in \mathrm{Pure}(A)$

$$\psi^A \succ_A \rho^{AB}. \tag{7.94}$$

4. The state $\rho^{AB}$ can be obtained by CMO channel from a classical distribution; that is, there exists $\omega \in \mathfrak{D}(XY)$ such that

$$\omega^{XY} \succ_A \rho^{AB}. \tag{7.95}$$

**Proof** The proof follows directly from Theorem 7.5 in conjunction with Theorem 7.3. ∎

As an example, consider the Werner state (cf. (3.247))

$$\rho_t^{AB} = t\frac{2}{d(d+1)}\Pi_{\mathrm{Sym}}^{AB} + (1-t)\frac{2}{d(d-1)}\Pi_{\mathrm{Asy}}^{AB} \qquad \forall\, t \in [0,1], \tag{7.96}$$

where $\Pi_{\mathrm{Sym}}^{AB}$ and $\Pi_{\mathrm{Asy}}^{AB}$ are the projections, respectively, to the symmetric and anti-symmetric subspaces of $AB$. We will see in Chapter 12 that these states are entangled if and only if $t < \frac{1}{2}$. Moreover, observe that the Werner states have uniform marginals, particularly, $\rho_t^B = u^B$ for all $t \in [0,1]$. In addition, the Werner states have only two distinct eigenvalues given by $\frac{2t}{d(d+1)}$ and $\frac{2(1-t)}{d(d-1)}$. Combining all this information we get that

$$I^A \otimes \rho_t^B \geqslant \rho_t^{AB} \quad \Longleftrightarrow \quad \frac{3-d}{2} \leqslant t \leqslant \frac{d+1}{2}. \tag{7.97}$$

For $d = 2$ this condition holds only for $t \in [\frac{1}{2}, 1]$, in which case $\rho_t^{AB}$ is separable. On the other hand, for $d \geqslant 3$ this condition holds for all $t \in [0,1]$. We therefore conclude that for $d \geqslant 3$ all Werner states (including the entangled ones) have positive conditional entropy.

**Exercise 7.10.** *Use Exercise 4.70 to show that in the classical domain every entropy function $\mathbb{H}$ is also a conditional entropy; that is, show that function*

$$H(X|Y)_\rho := H(X)_\rho \qquad \forall\, \rho \in \mathfrak{D}(XY) \tag{7.98}$$

*is a conditional entropy of classical states. Moreover, give a counterexample to the same statement in the quantum domain.*

## 7.3.1 Negativity Precludes Extensions from the Classical Domain

In Sections 5.3 and 6.4, we examined the expansion of quantum divergences and relative entropies from the classical to the quantum domain. This process involved a systematic approach to their extensions, incorporating both minimal and maximal forms. This subsection focuses on exploring optimal extensions of classical conditional entropy to the quantum domain. However, we will show that such extensions are not feasible, primarily because quantum conditional entropies exhibit negativity when applied to the maximally entangled state.

Consider a classical conditional entropy $\mathbb{H}$. Following the methodologies in Sections 5.3 and 6.4, for any $\rho \in \mathfrak{D}(AB)$, we can define the minimal and maximal extensions of $\mathbb{H}$ as

$$\underline{\mathbb{H}}(A|B)_\rho := \sup \left\{ \mathbb{H}(X|Y)_\omega : \omega^{XY} \succ_X \rho^{AB}, \ \omega \in \mathfrak{D}(XY) \right\}$$
$$\overline{\mathbb{H}}(A|B)_\rho := \inf \left\{ \mathbb{H}(X|Y)_\omega : \rho^{AB} \succ_A \omega^{XY}, \ \omega \in \mathfrak{D}(XY) \right\}. \tag{7.99}$$

At first glance, these functions appear quite reasonable. For example, for any two states $\rho, \sigma \in \mathfrak{D}(AB)$ with $\rho^{AB} \succ_A \sigma^{AB}$, if $\omega^{XY} \succ_X \rho^{AB}$, then it necessarily follows that $\omega^{XY} \succ_X \sigma^{AB}$. Consequently,

$$\underline{\mathbb{H}}(A|B)_\rho \leqslant \sup \left\{ \mathbb{H}(X|Y)_\omega : \omega^{XY} \succ_X \sigma^{AB}, \ \omega \in \mathfrak{D}(XY) \right\}$$
$$= \underline{\mathbb{H}}(A|B)_\sigma. \tag{7.100}$$

This means $\underline{\mathbb{H}}(A|B)_\rho$ exhibits monotonic behavior under conditional majorization, aligning with expectations for a measure of conditional uncertainty.

However, in general, $\underline{\mathbb{H}}(A|B)_\rho$ is not well defined! This is because CMO channels form a subset of one-way LOCC and thus cannot generate entanglement. Since $\omega \in \mathfrak{D}(XY)$ is classical and hence separable, any state $\mathcal{N}(\omega)$ resulting from a one-way LOCC channel $\mathcal{N} \in \text{CMO}(XY \to AB)$ lies within SEP$(AB)$. Therefore, $\underline{\mathbb{H}}(A|B)_\rho$ is undefined if $\rho^{AB}$ is entangled.

**Exercise 7.11.** *Show that $\underline{\mathbb{H}}(A|B)_\rho$ is well defined if and only if $I^A \otimes \rho^B \geqslant \rho^{AB}$. Hint: Recall Corollary 7.1.*

In contrast to $\underline{\mathbb{H}}(A|B)\rho$, the quantity $\overline{\mathbb{H}}(A|B)\rho$ is well-defined for all $\rho \in \mathfrak{D}(AB)$. It also exhibits monotonic behavior under conditional majorization. Specifically, consider two states $\rho, \sigma \in \mathfrak{D}(AB)$, where $\rho^{AB} \succ_A \sigma^{AB}$. Then, if $\sigma^{AB} \succ_X \omega^{XY}$ for some $\omega \in \mathfrak{D}(XY)$, then $\rho^{AB} \succ_X \omega^{XY}$ is necessarily true. Hence,

$$\overline{\mathbb{H}}(A|B)_\sigma \geqslant \inf \left\{ \mathbb{H}(X|Y)_\omega : \rho^{AB} \succ_X \omega^{XY}, \ \omega \in \mathfrak{D}(XY) \right\}$$
$$= \overline{\mathbb{H}}(A|B)_\rho. \tag{7.101}$$

Therefore, $\underline{\mathbb{H}}$ qualifies as a measure of conditional uncertainty. However, since it is defined via an optimization problem, it exhibits only sub-additivity. Consequently, we introduce its regularized version:

$$\underline{\mathbb{H}}^{\mathrm{reg}}(A|B)_\rho := \lim_{n\to\infty} \frac{1}{n}\underline{\mathbb{H}}\left(A^n\big|B^n\right)_{\rho^{\otimes n}}. \tag{7.102}$$

From its definition, $\underline{\mathbb{H}}^{\mathrm{reg}}$ satisfies

$$\underline{\mathbb{H}}^{\mathrm{reg}}\left(A^n\big|B^n\right)_{\rho^{\otimes n}} = n\underline{\mathbb{H}}^{\mathrm{reg}}(A|B)_\rho. \tag{7.103}$$

Furthermore, this measure exhibits monotonic behavior under conditional majorization and is inherently nonnegative by definition. Does this not contradict equation (7.69). The answer is no. The key lies in understanding that $\underline{\mathbb{H}}^{\mathrm{reg}}$ is only weakly additive, rather than strongly additive, in general.

**Exercise 7.12.** *Calculate* $\underline{\mathbb{H}}^{\mathrm{reg}}(A\tilde{A}|B)_\rho$, *where* $\rho$ *is defined as* $\Phi^{AB} \otimes \mathbf{u}^{\tilde{A}}$. *Question: Does this result in a value of zero?*

Interestingly, the function $\underline{\mathbb{H}}^{\mathrm{reg}}$ serves as an example of a function meeting all the criteria expected of a quantum conditional entropy, except that it is only weakly additive. Its nonnegativity teaches us an important lesson: The tendency of quantum conditional entropies to assume negative values on certain entangled states is intrinsically connected to their property of full additivity.

## 7.4 Conditional Entropies from Relative Entropies

Any quantum relative entropy can be used to define conditional entropy. In fact, for a given quantum relative entropy $\mathbb{D}$, there are two candidates for conditional entropy:

$$\begin{aligned}
\mathbb{H}(A|B)_\rho &:= \log|A| - \mathbb{D}\left(\rho^{AB}\big\|\mathbf{u}^A \otimes \rho^B\right), \\
\mathbb{H}^\uparrow(A|B)_\rho &:= \log|A| - \min_{\sigma\in\mathfrak{D}(B)} \mathbb{D}\left(\rho^{AB}\big\|\mathbf{u}^A \otimes \sigma^B\right).
\end{aligned} \tag{7.104}$$

The up arrow in the notation above indicates the optimization over $\sigma^B$. By definition, $\mathbb{H}^\uparrow(A|B)_\rho \geqslant \mathbb{H}(A|B)_\rho$ for all $\rho \in \mathfrak{D}(AB)$.

> **Theorem 7.6.** Let $\mathbb{D}$ be a quantum relative entropy. Then, the function $\mathbb{H}$ as defined in (7.104) is a quantum conditional entropy.

**Proof**   First, we demonstrate that $\mathbb{H}$ satisfies the monotonicity property of conditional entropy. Let $\mathcal{N}^{AB\to A'B'}$ be a CMO, and consider the bipartite density matrix $\rho^{AB}$. We begin by considering the case where $A = A'$ so that

$$\mathbb{H}(A|B)_{\mathcal{N}(\rho)} = \log|A| - \mathbb{D}\left(\mathcal{N}(\rho^{AB})\big\|\mathbf{u}^A \otimes \mathrm{Tr}_A\left[\mathcal{N}\left(\rho^{AB}\right)\right]\right). \tag{7.105}$$

Since $\mathcal{N}$ is $A \nrightarrow B'$ semi-causal, the marginal channel $\mathcal{N}^{AB \to B'} := \mathrm{Tr}_A \circ \mathcal{N}^{AB \to AB'}$ satisfies

$$\mathcal{N}^{AB \to B'}\left(\rho^{AB}\right) = \mathcal{N}^{AB \to B'}\left(\mathbf{u}^A \otimes \rho^B\right). \tag{7.106}$$

To see this, take $\mathcal{M}^{A \to A}$ in (7.13) to be the completely randomizing channel. With this at hand, we get

$$\mathbf{u}^A \otimes \mathrm{Tr}_A\left[\mathcal{N}\left(\rho^{AB}\right)\right] = \mathbf{u}^A \otimes \mathrm{Tr}_A\left[\mathcal{N}\left(\mathbf{u}^A \otimes \rho^B\right)\right] \tag{7.107}$$

$$\mathcal{N}\text{ is conditionally unital} \to \; = \mathcal{N}\left(\mathbf{u}^A \otimes \rho^B\right). \tag{7.108}$$

Substituting this into (7.105) we obtain

$$\mathbb{H}(A|B)_{\mathcal{N}(\rho)} = \log |A| - \mathbb{D}\left(\mathcal{N}\left(\rho^{AB}\right) \middle\| \mathcal{N}\left(\mathbf{u}^A \otimes \rho^B\right)\right)$$

$$\mathbf{DPI} \to \; \geqslant \log |A| - \mathbb{D}\left(\rho^{AB} \middle\| \mathbf{u}^A \otimes \rho^B\right) \tag{7.109}$$

$$= \mathbb{H}(A|B)_\rho. \tag{7.110}$$

We also need to prove that $\mathbb{H}$ is invariant under the action of a local isometric channel acting on system $A$. Let $\mathcal{V} \in \mathrm{CPTP}(A \to A')$ be an isometry channel, $\rho \in \mathfrak{D}(AB)$, and $C$ be a Hilbert space of dimension $|C| = |A'| - |A|$. Observe that

$$\mathbb{H}(A|B)_{\mathcal{V}(\rho)} = \log |A'| - \mathbb{D}\left(\mathcal{V}^{A \to A'}\left(\rho^{AB}\right) \middle\| \mathbf{u}^{A'} \otimes \rho^B\right). \tag{7.111}$$

Clearly, if $\mathcal{V}$ is a unitary channel (i.e. $A \cong A'$), we have $\mathbb{H}(A'|B)_{\mathcal{V}(\rho)} = \mathbb{H}(A|B)_\rho$ since in this case $|A| = |A'|$ and $\mathcal{V}(\mathbf{u}^A) = \mathbf{u}^{A'}$. We can therefore assume without loss of generality that

$$\mathcal{V}^{A \to A'}\left(\rho^{AB}\right) = \rho^{AB} \oplus \mathbf{0}^{CB} := \begin{pmatrix} \rho^{AB} & \mathbf{0} \\ \mathbf{0} & \mathbf{0}^{CB} \end{pmatrix} \tag{7.112}$$

since the conditional entropy of $\mathcal{V}^{A \to A'}\left(\rho^{AB}\right)$ does not change by a unitary channel on $A'$. Moreover, denote by $t := \frac{|A|}{|A'|}$ and observe that $\mathbf{u}^{A'}$ can be expressed as

$$\mathbf{u}^{A'} = t\mathbf{u}^A \oplus (1 - t)\mathbf{u}^C. \tag{7.113}$$

Hence, substituting (7.112) and (7.113) into (7.111) gives

$$\mathbb{H}(A|B)_{\mathcal{V}(\rho)} = \log |A'| - \mathbb{D}\left(\rho^{AB} \oplus \mathbf{0}^{CB} \middle\| \left(t\mathbf{u}^A \otimes \rho^B\right) \oplus \left((1 - t)\mathbf{u}^C \otimes \rho^B\right)\right)$$

$$\mathbf{Exercise\ 6.24} \to \; = \log |A'| - \mathbb{D}\left(\rho^{AB} \middle\| \mathbf{u}^A \otimes \rho^B\right) + \log t$$

$$t = \frac{|A|}{|A'|} \quad\longrightarrow\quad = \mathbb{H}(A|B)_\rho.$$

$$\tag{7.114}$$

To prove the additivity property, let $\rho \in \mathfrak{D}(A_1 B_1)$ and $\sigma \in \mathfrak{D}(A_2 B_2)$ and observe that since $\mathbf{u}^{A_1 A_2} = \mathbf{u}^{A_1} \otimes \mathbf{u}^{A_2}$ we have

$$\mathbb{H}(A_1 A_2 | B_1 B_2)_{\rho \otimes \sigma} = \log |A_1 A_2| - \mathbb{D}\left( \rho^{A_1 B_1} \otimes \sigma^{A_2 B_2} \middle\| \mathbf{u}^{A_1} \otimes \rho^{A_1 B_1} \otimes \mathbf{u}^{A_2} \otimes \sigma^{A_2 B_2} \right)$$

$$\mathbb{D} \text{ is additive} \rightarrow = \mathbb{H}(A_1 | B_1)_\rho + \mathbb{H}(A_2 | B_2)_\sigma .$$

$$(7.115)$$

Finally, the normalization property follows from the fact that when $|B| = 1$ and $|A| = 2$, we have by definition

$$\mathbb{H}(A|B)_{\mathbf{u}} = \log 2 - \mathbb{D}\left( \mathbf{u}^A \middle\| \mathbf{u}^A \right) = 1. \tag{7.116}$$

This completes the proof. $\blacksquare$

**Exercise 7.13.** *Show that if $\mathbb{D} = D_{\max}$, then its corresponding conditional entropy $\mathbb{H}$ is the conditional min-entropy.*

In the following exercise we will show that $H^\uparrow$ behaves monotonically under conditionally unital channels and consequently behaves monotonically under conditional majorization. However, in general, $H^\uparrow$ does not necessarily satisfy the additivity property (at least a general proof of additivity of $\mathbb{H}^\uparrow$ is unknown to the author). Still, this expression has been used extensively by the community, particularly since it can be shown that for $\mathbb{D} = D_\alpha$ or $\mathbb{D} = \tilde{D}_\alpha$ it is additive (here $\tilde{D}_\alpha$ is the sandwiched Rényi divergence; see Definition 6.7). This includes the Umegaki relative entropy; however, as we will see shortly, in this case $H(A|B)_\rho = H^\uparrow(A|B)_\rho = H(A|B)_\rho$ for all $\rho \in \mathfrak{D}(AB)$.

**Exercise 7.14.** *Consider the function $\mathbb{H}^\uparrow$ as defined (7.104).*

1. *Show that $\mathbb{H}^\uparrow$ does not increase under conditionally unital channels, and use it to conclude that it satisfies the monotonicity property of conditional entropy.*
2. *Prove that $\mathbb{H}^\uparrow$ satisfies the invariance and normalization property of conditional entropy.*

## 7.5 Examples of Quantum Conditional Entropies

### The von-Neumann Conditional Entropy

Let $D$ be the Umegaki relative entropy. The conditional entropy with respect to this divergence is defined by

$$H(A|B)_\rho := \log |A| - D\left( \rho^{AB} \middle\| \mathbf{u}^A \otimes \rho^B \right). \tag{7.117}$$

To simplify the expression above, observe that

$$D\left(\rho^{AB}\|\mathbf{u}^A \otimes \rho^B\right) = \text{Tr}\left[\rho^{AB}\log\rho^{AB}\right] - \text{Tr}\left[\rho^{AB}\log\left(\mathbf{u}^A \otimes \rho^B\right)\right]$$

$$= -H(\rho^{AB}) + \log|A| - \text{Tr}\left[\rho^{AB}\log\left(I^A \otimes \rho^B\right)\right]$$

$$\boxed{\log\left(I^A \otimes \rho^B\right) = I^A \otimes \log\rho^B} \longrightarrow \quad = -H(\rho^{AB}) + \log|A| - \text{Tr}\left[\rho^B\log\rho^B\right]$$

$$= H(\rho^B) - H(\rho^{AB}) + \log|A|.$$

$$(7.118)$$

Therefore, the conditional von-Neumann entropy can be expressed simply as

$$H(A|B)_\rho = H\left(\rho^{AB}\right) - H\left(\rho^B\right). \tag{7.119}$$

This formula is consistent with the intuition of conditional entropy as depicted in Figure 7.1.

The finding from the previous section, which establishes that conditional entropy is nonnegative for separable states, leads to an intriguing implication regarding the von-Neumann entropy.

> **Corollary 7.2.** Let $\{p_x, \rho_x\}_{x\in[m]}$ be an ensemble of quantum states in $\mathfrak{D}(A)$. The von-Neumann entropy satisfies
>
> $$\sum_{x\in[m]} p_x H(\rho_x) \leqslant H\left(\sum_{x\in[m]} p_x\rho_x\right) \leqslant \sum_{x\in[m]} p_x H(\rho_x) + H(\mathbf{p}). \tag{7.120}$$

**Proof**  The lower bound follows from the concavity of $H$ (see Exercise 6.25). To get the upper bound, let $\rho^{XA} := \sum_{x\in[m]} p_x|x\rangle\langle x|^X \otimes \rho_x^A$. Since $\rho^{XA}$ is a cq-state, and in particular separable, it follows that

$$0 \leqslant H(X|A)_\rho = H(\rho^{AX}) - H(\rho^A), \tag{7.121}$$

where $\rho^A = \sum_{x\in[m]} p_x\rho_x^A$ is the marginal state. Therefore,

$$H\left(\sum_{x\in[m]} p_x\rho_x\right) \leqslant H(\rho^{AX})$$

$$\text{Exercise (7.15)}\rightarrow \quad = \sum_{x\in[m]} p_x H(\rho_x) + H(\mathbf{p}). \tag{7.122}$$

This completes the proof.  ∎

**Exercise 7.15.**  *Using the same notations as in the proof above, show that*

$$H(\rho^{AX}) = \sum_x p_x H(\rho_x) + H(\mathbf{p}). \tag{7.123}$$

**Exercise 7.16.**  *The quantum mutual information is a quantity defined for any* $\rho \in \mathfrak{D}(AB)$ *as*

$$I(A:B)_\rho := D\left(\rho^{AB}\|\rho^A \otimes \rho^B\right). \tag{7.124}$$

1. *Express the quantum mutual information in terms of $H(A|B)_\rho$ and $H(A)_\rho$.*
2. *Show that the von-Neumann entropy is subadditive; that is, prove that for all $\rho \in \mathfrak{D}(AB)$*

$$H(\rho^{AB}) \leqslant H(\rho^A) + H(\rho^B). \tag{7.125}$$

*Hint: Use the fact that the mutual information is nonnegative.*

The von-Neumann conditional entropy behaves monotonically even under conditionally unital channels that are not $A \not\rightarrow B$ semi-causal. To see this, we first prove the following type of triangle equality of the Umegaki relative entropy.

> **Triangle Equality**
>
> **Lemma 7.3.** Let $D$ be the Umegaki relative entropy. Then, for any $\rho \in \mathfrak{D}(AB)$, $\sigma, \tau \in \mathfrak{D}(B)$, and $\omega \in \mathfrak{D}(A)$, we have
>
> $$D\left(\rho^{AB} \,\|\, \omega^A \otimes \sigma^B\right) = D\left(\rho^{AB} \,\|\, \omega^A \otimes \rho^B\right) + D\left(\rho^B \,\|\, \sigma^B\right). \tag{7.126}$$

*Remark.* The relation in this lemma is equivalent to

$$D\left(\rho^{AB} \,\|\, \omega^A \otimes \sigma^B\right) = D\left(\rho^{AB} \,\|\, \omega^A \otimes \rho^B\right) + D\left(\omega^A \otimes \rho^B \,\|\, \omega^A \otimes \sigma^B\right), \tag{7.127}$$

which explains why we view this relation as a type of triangle equality.

**Proof** By definition

$$D\left(\rho^{AB} \,\|\, \omega^A \otimes \sigma^B\right) = -H\left(\rho^{AB}\right) - \mathrm{Tr}\left[\rho^{AB} \log\left(\omega^A \otimes \sigma^B\right)\right]. \tag{7.128}$$

Using the property that

$$\log\left(\omega^A \otimes \sigma^B\right) = \log \omega^A \otimes I^B + I^A \otimes \log \sigma^B, \tag{7.129}$$

we get by direct calculation

$$\begin{aligned}
D\left(\rho^{AB} \,\|\, \omega^A \otimes \sigma^B\right) &= -H\left(\rho^{AB}\right) - \mathrm{Tr}\left[\rho^A \log \omega^A\right] - \mathrm{Tr}\left[\rho^B \log \sigma^B\right] \\
&= -H\left(\rho^{AB}\right) - \mathrm{Tr}\left[\rho^A \log \omega^A\right] - \mathrm{Tr}\left[\rho^B \log \rho^B\right] + D\left(\rho^B \,\|\, \sigma^B\right) \\
&= -H\left(\rho^{AB}\right) - \mathrm{Tr}\left[\rho^{AB} \log\left(\omega^A \otimes \rho^B\right)\right] + D\left(\rho^B \,\|\, \sigma^B\right) \\
&= D\left(\rho^{AB} \,\|\, \omega^A \otimes \rho^B\right) + D\left(\rho^B \,\|\, \sigma^B\right).
\end{aligned} \tag{7.130}$$

This completes the proof. ∎

Note that by taking in the lemma above $\omega^A = \mathbf{u}^A$ we get that

$$D\left(\rho^{AB} \,\|\, \mathbf{u}^A \otimes \sigma^B\right) = D\left(\rho^{AB} \,\|\, \mathbf{u}^A \otimes \rho^B\right) + D\left(\rho^B \,\|\, \sigma^B\right). \tag{7.131}$$

Therefore,

$$H^{\uparrow}(A|B)_\rho := \log|A| - \min_{\sigma \in \mathfrak{D}(B)} D\left(\rho^{AB} \| \mathbf{u}^A \otimes \sigma^B\right)$$

$$\textbf{(7.131)}\rightarrow = \log|A| - D\left(\rho^{AB} \| \mathbf{u}^A \otimes \rho^B\right) - \min_{\sigma \in \mathfrak{D}(B)} D\left(\rho^B \| \sigma^B\right) \tag{7.132}$$

$$= \log|A| - D\left(\rho^{AB} \| \mathbf{u}^A \otimes \rho^B\right)$$

$$= H(A|B)_\rho.$$

The equality $H(A|B)_\rho = H^{\uparrow}(A|B)_\rho$ reveals that the von-Neumann conditional entropy is monotonic under conditionally unital channels that are not necessarily $A \not\to B$ semi-causal (see part 1 of Exercise 7.14).

Another useful property satisfied by the von-Neumann entropy is known as the strong subadditivity property. Recall from Exercise 7.16 that the von-Neumann entropy is subadditive, that is, for any $\rho \in \mathfrak{D}(AB)$,

$$H(AB)_\rho \leqslant H(A)_\rho + H(B)_\rho, \tag{7.133}$$

where $H(AB)_\rho$ denotes $H(\rho^{AB})$. A stronger version of this inequality, known as the "strong subadditivity of the von-Neumann entropy" states that for any $\rho \in \mathfrak{D}(ABC)$ we have

$$H(ABC)_\rho + H(B)_\rho \leqslant H(AB)_\rho + H(BC)_\rho. \tag{7.134}$$

Note that this is a stronger version of the previous inequality since for $|B| = 1$ it reduces to subadditivity. The above inequality is unique to the von-Neumann entropy and in general is not satisfied by other entropy functions (at least not in this form).

We can express the strong subadditivity in terms of conditional entropies. Observe that since $H(A|BC)_\rho = H(ABC)_\rho - H(BC)_\rho$ and $H(A|B)_\rho = H(AB)_\rho - H(B)_\rho$, the strong subadditivity can be expressed as

$$H(A|BC)_\rho \leqslant H(A|B)_\rho. \tag{7.135}$$

This version of the strong subadditivity is perhaps more intuitive than (7.134) since it can be interpreted as the statement that by removing the access to system $C$, one can only increase the uncertainty about system $A$. Note also that the above form of the strong subadditivity is satisfied by any conditional entropy function. That is, for any conditional entropy $\mathbb{H}$ and $\rho \in \mathfrak{D}(ABC)$ we have

$$\mathbb{H}(A|BC)_\rho \leqslant \mathbb{H}(A|B)_\rho. \tag{7.136}$$

The above inequality is simply a consequence of the monotonicity property of conditional entropy, since tracing out system $C$ is a map belonging to $\mathrm{CMO}(ABC \to AB)$. In terms of conditional majorization, we can express it as

$$\rho^{ABC} \succ_A \rho^{AB}. \tag{7.137}$$

**Exercise 7.17.** *Let $\rho \in \mathfrak{D}(ABC)$. Show that*

$$H(A|B)_\rho + H(A|C)_\rho \geqslant 0 \tag{7.138}$$

*with equality if $\rho^{ABC}$ is a pure state. Hint: If $\rho^{ABC}$ is a mixed state, let $\psi^{ABCD}$ be its purification, and express $H(A|C)_\rho$ in terms of systems $A, B, D$ (e.g., $H(AC)_\psi = H(BD)_\psi$). Finally, use (7.134) with $D$ replacing $C$.*

## The Conditional Rényi Entropies

In Chapter 6, we encountered three types of quantum relative entropies that generalized the classical Rényi divergences:

1. The Petz quantum Rényi divergence, $D_\alpha$, as defined in Definition 6.6 for $\alpha \in [0, 2]$.

2. The sandwiched Rényi relative entropy, $\tilde{D}_\alpha$, as defined in Definition 6.7 for all $\alpha \in [0, \infty]$.

3. The geometric relative entropy, $\widehat{D}_\alpha$, as defined in Definition 6.8 for all $\alpha \in [0, 2]$.

Each type of the quantum Rényi relative entropy above gives rise to two types of conditional entropies as given in (7.104). We denote the corresponding six conditional entropies by $H_\alpha$, $H_\alpha^\uparrow$, $\tilde{H}_\alpha$, $\tilde{H}_\alpha^\uparrow$, $\widehat{H}_\alpha$, and $\widehat{H}_\alpha^\uparrow$. The functions $H_\alpha$, $\tilde{H}_\alpha$, and $\widehat{H}_\alpha$ are all quantum conditional entropies since they are additive. We next show that also $H_\alpha^\uparrow$ is additive, and later we will see that this also implies that $\tilde{H}_\alpha^\uparrow$ is additive. Therefore, $H_\alpha^\uparrow$ and $\tilde{H}_\alpha^\uparrow$ are conditional entropies as well. However, to the author's knowledge, the additivity of $\widehat{H}_\alpha^\uparrow$ has not been explored.

In the following theorem, we provide a closed form for the function

$$H_\alpha^\uparrow (A|B)_\rho := \log |A| - \min_{\sigma \in \mathfrak{D}(B)} D_\alpha \left( \rho^{AB} \| \mathbf{u}^A \otimes \sigma^B \right) \qquad \forall \, \rho \in \mathfrak{D}(AB). \qquad (7.139)$$

**Closed Formula**

**Theorem 7.7.** Let $\rho \in \mathfrak{D}(AB)$ and $\alpha \in [0, 2]$. Then,

$$H_\alpha^\uparrow (A|B)_\rho = \frac{\alpha}{1-\alpha} \log \mathrm{Tr} \left[ \left( \eta_\alpha^B \right)^{1/\alpha} \right], \qquad \text{where} \quad \eta_\alpha^B := \mathrm{Tr}_A \left[ \left( \rho^{AB} \right)^\alpha \right]. \qquad (7.140)$$

**Proof**   First, observe that

$$\begin{aligned} H_\alpha^\uparrow (A|B)_\rho &= \max_{\sigma \in \mathfrak{D}(B)} \frac{1}{1-\alpha} \log \mathrm{Tr} \left[ \left( \rho^{AB} \right)^\alpha \left( I^A \otimes \sigma^B \right)^{1-\alpha} \right] \\ &= \max_{\sigma \in \mathfrak{D}(B)} \frac{1}{1-\alpha} \log \mathrm{Tr} \left[ \eta_\alpha^B \left( \sigma^B \right)^{1-\alpha} \right]. \end{aligned} \qquad (7.141)$$

Set $t := \mathrm{Tr} \left[ \left( \eta_\alpha^B \right)^{1/\alpha} \right]$ and denote by $\tau^B := \left( \eta_\alpha^B \right)^{1/\alpha} / t$ so that $\eta_\alpha^B = t^\alpha \left( \tau^B \right)^\alpha$. Substituting this into the previous equation we conclude

$$\begin{aligned} H_\alpha^\uparrow(A|B)_\rho &= \frac{\alpha}{1-\alpha}\log t + \max_{\sigma\in\mathfrak{D}(B)}\frac{1}{1-\alpha}\log\mathrm{Tr}\left[\left(\tau^B\right)^\alpha\left(\sigma^B\right)^{1-\alpha}\right]\\ &= \frac{\alpha}{1-\alpha}\log t - \min_{\sigma\in\mathfrak{D}(B)} D_\alpha\left(\tau^B\|\sigma^B\right)\\ &= \frac{\alpha}{1-\alpha}\log t. \end{aligned}$$ (7.142)

This completes the proof. ∎

**Exercise 7.18.** *Use the closed formula* (7.140) *in Theorem 7.7 to show that $H_\alpha^\uparrow$ is additive and therefore a conditional entropy.*

## The Optimized Conditional Min-Entropy

Among the plethora of families of quantum conditional entropies we saw above, there exists one conditional entropy that appears often in applications, known as the optimized conditional min-entropy. The optimized conditional min-entropy is defined as the quantum conditional entropy $\tilde{H}_\alpha^\uparrow(A|B)_\rho$ with $\alpha=\infty$. That is,

$$H_{\min}^\uparrow(A|B)_\rho := \log|A| - \min_{\sigma\in\mathfrak{D}(B)} D_{\max}\left(\rho^{AB}\|\mathbf{u}^A\otimes\sigma^B\right).$$ (7.143)

The conditional min-entropy is defined in terms of an SDP. To see this, first observe that from the above formula and from the definition of $D_{\max}$ we get

$$\begin{aligned} 2^{-H_{\min}^\uparrow(A|B)_\rho} &= \min\left\{t: tI^A\otimes\sigma^B \geqslant \rho^{AB},\ \sigma\in\mathfrak{D}(B),\ t\in\mathbb{R}\right\}\\ \boxed{\Lambda^B := t\sigma^B}\longrightarrow\ &= \min\left\{\mathrm{Tr}\left[\Lambda^B\right]: I^A\otimes\Lambda^B\geqslant\rho^{AB},\ \Lambda\in\mathrm{Pos}(B)\right\}. \end{aligned}$$ (7.144)

Then, using the notations $\mathfrak{K}_1:=\mathrm{Pos}(B)$, $\mathfrak{K}_2:=\mathrm{Pos}(AB)$, $H_1:=I^B$, $H_2:=\rho^{AB}$, and $\mathcal{N}\in\mathfrak{L}(A\to AB)$ given by

$$\mathcal{N}(\omega^B):=I^A\otimes\omega^B\qquad \forall\,\omega\in\mathfrak{L}(B),$$ (7.145)

we conclude that

$$2^{-H_{\min}^\uparrow(A|B)_\rho} = \min\left\{\mathrm{Tr}[\Lambda H_1]: \Lambda\in\mathfrak{K}_1,\mathcal{N}(\Lambda)-H_2\in\mathfrak{K}_2\right\}.$$ (7.146)

The above optimization problem has precisely the same form as the conic linear programming given in (A.52 of online version). Since the cones $\mathfrak{K}_1$ and $\mathfrak{K}_2$ are the sets of positive semi-definite matrices, this conic program is an SDP program.

The expression in (7.146) has a dual given by (A.57 of online version). Therefore, the conditional min-entropy can be expressed in terms of the following optimization problem

$$\begin{aligned} 2^{-H_{\min}^\uparrow(A|B)_\rho} &= \max\left\{\mathrm{Tr}[\eta H_2]: \eta\in\mathfrak{K}_2^*,\ H_1-\mathcal{N}^*(\eta)\in\mathfrak{K}_1^*\right\}\\ \text{\textbf{Exercise 7.19}}\rightarrow\ &= \max\left\{\mathrm{Tr}\left[\eta^{AB}\rho^{AB}\right]: \eta\in\mathrm{Pos}(AB),\ \eta^B=I^B\right\}. \end{aligned}$$ (7.147)

Any $\eta^{AB}$ as (7.147) is a Choi matrix; hence, it can be expressed as $\eta^{AB} = \mathcal{E}^{*\tilde{A} \to B}\left(\Omega^{A\tilde{A}}\right)$ for some channel $\mathcal{E} \in \mathrm{CPTP}(B \to A)$. We therefore get that

$$
\begin{aligned}
2^{-H_{\min}^{\uparrow}(A|B)_\rho} &= \max_{\mathcal{E} \in \mathrm{CPTP}(B \to \tilde{A})} \mathrm{Tr}\left[\rho^{AB} \mathcal{E}^{*\tilde{A} \to B}\left(\Omega^{A\tilde{A}}\right)\right] \\
&= |A| \max_{\mathcal{E} \in \mathrm{CPTP}(B \to \tilde{A})} \left\langle \Phi^{A\tilde{A}} \left| \mathcal{E}^{B \to \tilde{A}}\left(\rho^{AB}\right) \right| \Phi^{A\tilde{A}}\right\rangle \qquad (7.148) \\
&= |A| \max_{\mathcal{E} \in \mathrm{CPTP}(B \to \tilde{A})} F^2\left(\mathcal{E}^{B \to \tilde{A}}\left(\rho^{AB}\right), \Phi^{A\tilde{A}}\right),
\end{aligned}
$$

where $F$ is the fidelity. That is, the conditional min-entropy can be expressed in terms of the maximal overlap of $\mathcal{E}^{B \to \tilde{A}}\left(\rho^{AB}\right)$ with the maximally entangled state. We now use the expression of (7.148) to prove that the optimized conditional min-entropy is additive under tensor products, and thereby prove that the optimized conditional min-entropy is indeed a quantum conditional entropy as defined in Definition 7.3.

> **Lemma 7.4.** The optimized conditional min-entropy is a quantum conditional entropy satisfying both properties of Definition 7.3.

**Proof** Since the optimized conditional min-entropy equals $\tilde{H}_\alpha^{\uparrow}$ with $\alpha = \infty$, it is left to prove that it is additive. Let $\rho \in \mathfrak{D}(AB)$, $\tau \in \mathfrak{D}(A'B')$, and denote by $Q_{\min}(A|B)_\rho := 2^{-H_{\min}^{\uparrow}(A|B)_\rho}$. Therefore, the additivity of $H_{\min}$ would follow from the multiplicativity of $Q_{\min}$. On the one hand, from the primal problem (7.144) we have

$$
\begin{aligned}
Q_{\min}(AA'|BB')_{\rho \otimes \tau} &= \min\left\{\mathrm{Tr}[\Lambda^{BB'}] : I^{AA'} \otimes \Lambda^{BB'} \geqslant \rho^{AB} \otimes \tau^{A'B'}, \ \Lambda \in \mathrm{Pos}(BB')\right\} \\
&\leqslant \min\left\{\mathrm{Tr}[\Lambda_1^B]\mathrm{Tr}[\Lambda_2^B] : I^{AA'} \otimes \Lambda_1^B \otimes \Lambda_2^{B'} \geqslant \rho^{AB} \otimes \tau^{A'B'}, \ \Lambda_1 \in \mathrm{Pos}(B), \ \Lambda_2 \in \mathrm{Pos}(B')\right\} \\
&\leqslant Q_{\min}(A|B)_\rho Q_{\min}(A'|B')_\tau, \qquad (7.149)
\end{aligned}
$$

where in the first inequality we restricted $\Lambda^{BB'}$ to have the form $\Lambda_1^B \otimes \Lambda_2^B$, and in the last inequality we replaced the condition $I^{AA'} \otimes \Lambda_1^B \otimes \Lambda_2^{B'} \geqslant \rho^{AB} \otimes \tau^{A'B'}$ with the two conditions $I^A \otimes \Lambda_1^B \geqslant \rho^{AB}$ and $I^{A'} \otimes \Lambda_2^{B'} \geqslant \tau^{A'B'}$.

To get the opposite inequality we use the dual expression of the conditional min-entropy as given in (7.148). Specifically,

$$
\begin{aligned}
Q_{\min}(AA'|BB')_{\rho \otimes \tau} &= |AA'| \max_{\mathcal{E} \in \mathrm{CPTP}(BB' \to \tilde{A}\tilde{A}')} F\left(\mathcal{E}^{BB' \to \tilde{A}\tilde{A}'}\left(\rho^{AB} \otimes \tau^{BB'}\right), \Phi^{A\tilde{A}} \otimes \Phi^{A'\tilde{A}'}\right)^2 \\
\boxed{\mathcal{E} = \mathcal{E}_1 \otimes \mathcal{E}_2} \longrightarrow \quad &\geqslant |AA'| \max_{\substack{\mathcal{E}_1 \in \mathrm{CPTP}(B \to \tilde{A}) \\ \mathcal{E}_2 \in \mathrm{CPTP}(B' \to \tilde{A}')}} F\left(\mathcal{E}_1^{B \to \tilde{A}}\left(\rho^{AB}\right) \otimes \mathcal{E}_2^{B' \to \tilde{A}'}\left(\tau^{BB'}\right), \Phi^{A\tilde{A}} \otimes \Phi^{A'\tilde{A}'}\right)^2 \\
&= Q_{\min}(A|B)_\rho Q_{\min}(A'|B')_\tau. \qquad (7.150)
\end{aligned}
$$

Combining the two equations above we conclude that

$$
Q_{\min}(AA'|BB')_{\rho \otimes \tau} = Q_{\min}(A|B)_\rho Q_{\min}(A'|B')_\tau, \qquad (7.151)
$$

so that $H_{\min}^{\uparrow}(AA'|BB')_{\rho \otimes \tau} = H_{\min}^{\uparrow}(A|B)_\rho + H_{\min}^{\uparrow}(A'|B')_\tau$. This completes the proof. $\blacksquare$

When system $A$ is classical, the right-hand side of (7.148) has a simple interpretation as a guessing probability. Indeed, suppose that $A = X$ is a classical system with $m := |X|$. Then the state $\rho^{AB}$, which we denote as $\rho^{XB}$, takes the form of a classical-quantum (cq) state:

$$\rho^{XB} = \sum_{x \in [m]} p_x |x\rangle\langle x|^X \otimes \rho_x^B, \tag{7.152}$$

where $\{p_x\}_{x \in [m]}$ is a probability distribution, and each $\rho_x \in \mathfrak{D}(B)$. Furthermore, it's important to note that $\mathrm{CPTP}(B \to \tilde{A})$, which is the same as $\mathrm{CPTP}(B \to \tilde{X})$, comprises POVM channels that were initially introduced in Section 3.5.4. Under these circumstances, the aforementioned equation simplifies to the following (Exercise 7.20):

$$2^{-H_{\min}^{\uparrow}(X|B)_\rho} = \max_{\{\Lambda_x\}} \sum_{x \in [m]} p_x \mathrm{Tr}\left[\Lambda_x^B \rho_x^B\right], \tag{7.153}$$

where the maximum is over all POVMs $\{\Lambda_x^B\}_{x \in [m]}$ on system $B$. The expression in Equation (7.153) can be interpreted as the maximum probability for Bob to guess correctly the value of $X$. Specifically, given the cq-state $\rho^{XB}$, Bob can try to learn the classical value of $X$ by performing a quantum measurement/POVM, $\{\Lambda_x^B\}_{x \in [m]}$, on his system with $m := |X|$ possible outcomes. The probability that $X = x$ is $p_x$, and the probability that Bob gets the outcome $y$ given that $X = x$ is given by $\mathrm{Tr}\left[\Lambda_y^B \rho_x^B\right]$. If Bob takes $y$ to be his guess for the value of $X$, then $\mathrm{Tr}\left[\Lambda_x^B \rho_x^B\right]$ is the probability that Bob guesses correctly the value of $X$. Given that $X = x$ with probability $p_x$, we get that

$$\mathrm{Pr}_g(X|B)_\rho := \max_{\{\Lambda_x\}} \sum_{x \in [m]} p_x \mathrm{Tr}\left[\Lambda_x^B \rho_x^B\right] \tag{7.154}$$

is the maximal overall probability that Bob's guess of $X$ is correct. With this notation, the conditional entropy of $\rho^{XB}$ can be expressed as

$$H_{\min}^{\uparrow}(X|B)_\rho = -\log \mathrm{Pr}_g(X|B)_\rho. \tag{7.155}$$

Note that $H_{\min}(X|B)_\rho \geq 0$ as expected.

**Exercise 7.19.** *Prove the second equality in (7.147).*

**Exercise 7.20.** *Prove the reduction of (7.148) to (7.153) when $A = X$ is classical.*

## The Conditional Max-Entropy

In Theorem 7.5, we demonstrated that the conditional min-entropy $H_{\min}$ represents the lowest possible conditional entropy for a bipartite quantum state. This naturally leads to the question: What is the highest possible conditional entropy? To address this, we explore the upper limit of all conditional entropies.

> **Definition 7.4.** The conditional max-entropy of a quantum state $\rho \in \mathfrak{D}(AB)$ is defined as follows:
>
> $$H_{\max}(A|B)_\rho := \log |A| - \min_{\sigma \in \mathfrak{D}(B)} D_{\min}\left(\rho^{AB} \big\| \mathbf{u}^A \otimes \sigma^B\right). \qquad (7.156)$$

According to Theorem 7.7 and Exercise 7.18, the function $H_\alpha^\uparrow$ is additive for all $\alpha \in [0,2]$. Hence, the additivity of the conditional max-entropy (i.e. $H_{\alpha=0}^\uparrow$) is established. Therefore, we can affirm that the conditional max-entropy qualifies as a legitimate conditional entropy measure. Moreover, given that the min-relative entropy $D_{\min}$ is the smallest relative entropy, the following inequality holds for all conditional entropies $\mathbb{H}^\uparrow$ derived from a relative entropy $\mathbb{D}$ (as specified in (7.104)), for any quantum state $\rho \in \mathfrak{D}(AB)$:

$$\mathbb{H}^\uparrow(A|B)_\rho \leqslant H_{\max}(A|B)_\rho. \qquad (7.157)$$

This inequality signifies that the conditional max-entropy establishes an upper limit for all conditional entropies defined in relation to a relative entropy. Additionally, as will be explored in subsequent discussions, the conditional max-entropy is essentially the counterpart, or the dual, of the conditional min-entropy.

## 7.6 Duality Relations

We saw in Exercise 7.17 that for a pure state $\varphi \in \text{Pure}(ABC)$ the conditional von-Neumann entropy satisfies

$$H(A|B)_\varphi + H(A|C)_\varphi = 0. \qquad (7.158)$$

Such a relation is called a duality relation, and motivates us to define a dual for any conditional entropy.

> **Definition 7.5.** Let $\mathbb{H}$ be a conditional entropy. For any $\rho \in \mathfrak{D}(AB)$ with a purification $\varphi \in \text{Pure}(ABC)$ (where $C$ is the purifying system; that is, $\rho^{AB} = \text{Tr}_C \varphi^{ABC}$), we define the dual of $\mathbb{H}$ as
>
> $$\mathbb{H}^{\text{dual}}(A|B)_\rho := -\mathbb{H}(A|C)_\varphi. \qquad (7.159)$$

*Remark.* Since the conditional entropy is invariant under local isometries (specifically, $\mathbb{H}(A|C)_\varphi$ remains invariant under isometries on system $C$) the dual to a conditional entropy is well defined as it does not depend on the choice of the purifying system $C$.

By definition, the dual to a conditional entropy satisfies the invariance and additivity properties of conditional entropy (see Exercise 7.21). To see that it satisfies also the normalization property of a conditional entropy, let $\rho^{AB} = \mathbf{u}^A$ with $|A| = 2$ and $|B| = 1$. A purification of $\rho^{AB}$ can be expressed as the maximally entangled state $\Phi^{AC}$ with $C = \tilde{A}$. Therefore,

$$\mathbb{H}^{\mathrm{dual}}(A)_{\mathbf{u}} = \mathbb{H}^{\mathrm{dual}}(A|B)_{\rho}$$

$$\textbf{by definition} \rightarrow \; = -\mathbb{H}(A|C)_{\Phi} \tag{7.160}$$

$$\textbf{(7.69)} \rightarrow \; = \log 2 = 1.$$

Therefore, the dual to a conditional entropy would be itself a conditional entropy if it satisfies the monotonicity property. We will see shortly that this is indeed the case for all the conditional entropies studied in literature, although a general proof for all conditional entropy functions is unknown to the author.

**Exercise 7.21.** *Show that the dual to a conditional entropy satisfies the invariance and additivity properties of a conditional entropy.*

The relation (7.158) implies that the conditional von-Neumann entropy is self dual; that is, $H^{\mathrm{dual}}(A|B)_{\rho} = H(A|B)_{\rho}$ for all $\rho \in \mathfrak{D}(AB)$. Consider the Petz conditional Rényi entropy of order $\alpha \in [0, 2]$ given for all $\rho \in \mathfrak{D}(AB)$ by

$$H_{\alpha}(A|B)_{\rho} = \frac{1}{1-\alpha} \log \mathrm{Tr}\left[ \left(\rho^{AB}\right)^{\alpha} \left(I^A \otimes \rho^B\right)^{1-\alpha} \right]. \tag{7.161}$$

In the following lemma, we compute it's dual.

---

**Lemma 7.5.** For any $\alpha \in [0, 2]$, the dual of $H_{\alpha}$ is given by

$$H_{\alpha}^{\mathrm{dual}}(A|B)_{\rho} = H_{2-\alpha}(A|B)_{\rho}. \tag{7.162}$$

---

**Proof** Let

$$\rho^{AB} = \sum_{x\in[n]} p_x |\varphi_x\rangle\langle\varphi_x|^{AB} \tag{7.163}$$

be the spectral decomposition of $\rho^{AB}$, and let $\rho^{ABC} = |\varphi\rangle\langle\varphi|^{ABC}$, with $C \cong AB$, be the purification of $\rho^{AB}$ given by

$$|\varphi^{ABC}\rangle = \sum_{x\in[n]} \sqrt{p_x} |\varphi_x\rangle^{AB} |\varphi_x\rangle^C = \left(\rho^{AB}\right)^{\frac{1}{2}} \otimes I^C |\Omega^{(AB)C}\rangle, \tag{7.164}$$

where $|\Omega^{(AB)C}\rangle = \sum_{x\in[n]} |\varphi_j\rangle^{AB} |\varphi_j\rangle^C$ is the maximally entangled operator between system $AB$ and $C$. Now, observe that (see Exercise 7.22)

$$\left(\rho^{AB}\right)^{\alpha} \otimes I^C |\Omega^{(AB)C}\rangle = I^{AB} \otimes \left(\rho^C\right)^{\alpha} |\Omega^{(AB)C}\rangle, \tag{7.165}$$

where $\rho^C := \mathrm{Tr}_{AB}\varphi^{ABC}$. Therefore, from part 1 of Exercise 2.38 we get

$$\text{Tr}\left[\left(\rho^{AB}\right)^{\alpha}\left(I^{A}\otimes\rho^{B}\right)^{1-\alpha}\right] = \left\langle\Omega^{ABC}\left|\left(\rho^{AB}\otimes I^{C}\right)^{\alpha}\left(I^{A}\otimes\rho^{B}\otimes I^{C}\right)^{1-\alpha}\right|\Omega^{ABC}\right\rangle$$

$$(7.165)\rightarrow\ = \left\langle\Omega^{ABC}\left|I^{A}\otimes\left(\rho^{B}\right)^{1-\alpha}\otimes\left(\rho^{C}\right)^{\alpha}\right|\Omega^{ABC}\right\rangle$$

$$= \left\langle\varphi^{ABC}\left|I^{A}\otimes\left(\rho^{B}\right)^{1-\alpha}\otimes\left(\rho^{C}\right)^{\alpha-1}\right|\varphi^{ABC}\right\rangle, \qquad (7.166)$$

where in the last equality we used the fact that $|\varphi^{ABC}\rangle = I^{AB}\otimes\left(\rho^{C}\right)^{1/2}|\Omega^{(AB)C}\rangle$. Next, let

$$\rho^{B} = \sum_{y\in[m]} q_{y}|y\rangle\langle y|^{B} \quad\text{and}\quad \rho^{AC} = \sum_{y\in[m]} q_{y}|\chi_{y}\rangle\langle\chi_{y}|^{B} \qquad (7.167)$$

be the spectral decompositions of $\rho^{B}$ and $\rho^{AC}$, respectively, and consider the following Schmidt decomposition between system $B$ and system $AC$

$$|\varphi^{ABC}\rangle = \sum_{y\in[m]} \sqrt{q_{y}}|y\rangle^{B}|\chi_{y}\rangle^{AC} = \left(\rho^{B}\right)^{1/2}\otimes I^{AC}|\Omega^{B(AC)}\rangle, \qquad (7.168)$$

where $|\Omega^{B(AC)}\rangle = \sum_{y\in[m]}|y\rangle^{B}|\chi_{y}\rangle^{AC}$. Substituting the above expression for $|\varphi^{ABC}\rangle$ into (7.166) gives

$$\text{Tr}\left[\left(\rho^{AB}\right)^{\alpha}\left(I^{A}\otimes\rho^{B}\right)^{1-\alpha}\right] = \langle\Omega^{B(AC)}|I^{A}\otimes\left(\rho^{B}\right)^{2-\alpha}\otimes\left(\rho^{C}\right)^{\alpha-1}|\Omega^{B(AC)}\rangle. \ (7.169)$$

Finally, using the relation

$$\left(\rho^{B}\right)^{2-\alpha}\otimes I^{AC}|\Omega^{B(AC)}\rangle = I^{B}\otimes\left(\rho^{AC}\right)^{2-\alpha}|\Omega^{B(AC)}\rangle \qquad (7.170)$$

we conclude that

$$\text{Tr}\left[\left(\rho^{AB}\right)^{\alpha}\left(I^{A}\otimes\rho^{B}\right)^{1-\alpha}\right] = \langle\Omega^{B(AC)}|I^{B}\otimes\left(\left(\rho^{AC}\right)^{2-\alpha}\left(I^{A}\otimes\rho^{C}\right)^{\alpha-1}\right)|\Omega^{B(AC)}\rangle$$

$$= \text{Tr}\left[\left(\rho^{AC}\right)^{2-\alpha}\left(I^{A}\otimes\rho^{C}\right)^{\alpha-1}\right].$$

$$(7.171)$$

Therefore,

$$H_{\alpha}(A|B)_{\rho} = \frac{1}{1-\alpha}\log\text{Tr}\left[\left(\rho^{AC}\right)^{2-\alpha}\left(I^{A}\otimes\rho^{C}\right)^{\alpha-1}\right]$$

$$= -H_{2-\alpha}(A|C)_{\rho}. \qquad (7.172)$$

Note that the above equality is equivalent to

$$H_{\alpha}^{\text{dual}}(A|B)_{\rho} := -H_{\alpha}(A|C)_{\rho} = H_{2-\alpha}(A|B)_{\rho}. \qquad (7.173)$$

This completes the proof. ∎

**Exercise 7.22.** *Prove the relations (7.165) and (7.170).*

## The Dual of the Optimized Conditional Min-Entropy

**Lemma 7.6.** Let $\rho \in \text{Pure}(ABE)$. Then,

$$H^{\uparrow}_{\min}(A|B)_\rho = -\tilde{H}^{\uparrow}_{1/2}(A|E)_\rho = -\max_{\tau \in \mathfrak{D}(E)} \log F^2\left(\rho^{AE}, I^A \otimes \tau^E\right), \quad (7.174)$$

where $F$ is the fidelity.

*Remark.* It is noteworthy that the lemma above establishes $\tilde{H}^{\uparrow}_{1/2}$ as the dual of $H^{\uparrow}_{\min}$. Consequently, $\tilde{H}^{\uparrow}_{1/2}(A|B)\rho$ is sometimes referred to as the conditional max-entropy. However, we choose not to use this terminology here because, generally speaking, $\tilde{H}^{\uparrow}_{1/2}(A|B)_\rho \neq H_{\max}(A)_\rho$, particularly when $\rho^{AB} = \rho^A \otimes \rho^B$. In fact, as we will show later, the true dual of $H_{\min}$ (as opposed to $H^{\uparrow}_{\min}$) aligns with the conditional max-entropy as defined in (7.156). Additionally, when integrating the above lemma with (7.148), we derive the following relationship:

$$\max_{\mathcal{E} \in \text{CPTP}(B \to \tilde{A})} F\left(\mathcal{E}^{B \to \tilde{A}}\left(\rho^{AB}\right), \Omega^{A\tilde{A}}\right) = \max_{\tau \in \mathfrak{D}(E)} F\left(\rho^{AE}, I^A \otimes \tau^E\right). \quad (7.175)$$

**Proof** We start with the expression for $Q_{\min}(A|B)_\rho = 2^{-H^{\uparrow}_{\min}(A|B)_\rho}$, given in (7.148) as

$$Q_{\min}(A|B)_\rho = \max_{\mathcal{E} \in \text{CPTP}(B \to \tilde{A})} F^2\left(\mathcal{E}^{B \to \tilde{A}}\left(\rho^{AB}\right), \Omega^{A\tilde{A}}\right). \quad (7.176)$$

For any $\mathcal{E} \in \text{CPTP}(B \to \tilde{A})$ let $V_{\mathcal{E}} \in \text{CPTP}(B \to \tilde{A}R)$ be its Stinespring's isometry. Observe that $V_{\mathcal{F}}^{B \to \tilde{A}R}\left(\rho^{ABE}\right)$ is a purification of $\mathcal{F}^{B \to \tilde{A}}\left(\rho^{AB}\right)$. Moreover, since $\Omega^{A\tilde{A}}$ is already pure, any purification of $\Omega^{A\tilde{A}}$ in $A\tilde{A}RE$ must be of the form $\Omega^{A\tilde{A}} \otimes \chi^{RE}$, where $\chi \in \text{Pure}(RE)$. Hence, from the Uhlmann's theorem we get that

$$F^2\left(\Omega^{A\tilde{A}}, \mathcal{E}^{B \to \tilde{A}}\left(\rho^{AB}\right)\right) = \max_{\chi \in \text{Pure}(RE)} F^2\left(\Omega^{A\tilde{A}} \otimes \chi^{RE}, V_{\mathcal{E}}^{B \to \tilde{A}R}\left(\psi^{ABE}\right)\right). \quad (7.177)$$

Now, observe that *any* purification of the state $\rho^{AE} := \text{Tr}_B\left[\rho^{ABE}\right]$ in $\text{Pure}(A\tilde{A}ER)$ has the form $V_{\mathcal{E}}^{B \to \tilde{A}R}\left(\rho^{ABE}\right)$ for some $\mathcal{E} \in \text{CPTP}(B \to \tilde{A})$. Therefore, when we add the maximization over all $\mathcal{E} \in \text{CPTP}(B \to \tilde{A})$ to both sides of the equation above we get

$$Q_{\min}(A|B)_\rho = \max_{\substack{\psi \in \text{Pure}(A\tilde{A}RE) \\ \psi^{AE} = \rho^{AE},\, \chi \in \text{Pure}(RE)}} F^2\left(\Omega^{A\tilde{A}} \otimes \chi^{RE}, \psi^{A\tilde{A}RE}\right), \quad (7.178)$$

where on the right-hand side we replaced that maximum over all $\mathcal{E} \in \text{CPTP}(B \to \tilde{A})$ with a maximum over all pure states $\psi \in \text{Pure}(A\tilde{A}RE)$ with marginal $\psi^{AE} = \rho^{AE}$. Finally, applying Uhlmann's theorem to the expression above we conclude that

$$Q_{\min}(A|B)_\rho = \max_{\chi \in \mathfrak{D}(E)} F^2\left(I^A \otimes \chi^E, \rho^{AE}\right). \quad (7.179)$$

This completes the proof. ∎

**Exercise 7.23.** *Show that $Q_{\min}$ is a convex function. That is, show that for every set of n bipartite quantum states $\{\rho_x^{AB}\}_{x\in[n]}$ and every $\mathbf{p} \in \mathrm{Prob}(n)$ we have*

$$Q_{\min}(A|B)_\rho \leqslant \sum_{x\in[n]} p_x\, Q_{\min}(A|B)_{\rho_x}, \quad \text{where} \quad \rho^{AB} = \sum_{x\in[n]} p_x\rho_x^{AB}. \tag{7.180}$$

More generally, the duals of $H_\alpha^\uparrow$ and $\tilde{H}_\alpha^\uparrow$ can also be computed and they are given by (see the section "Notes and References" for more details)

$$\tilde{H}_\alpha^{\uparrow\,\mathrm{dual}}(A|B)_\rho = \tilde{H}_\beta^\uparrow(A|B)_\rho \quad \text{for} \quad \frac{1}{\alpha} + \frac{1}{\beta} = 2, \ \alpha, \beta \in [1/2, \infty]$$

$$\tilde{H}_\alpha^{\mathrm{dual}}(A|B)_\rho = H_\beta^\uparrow(A|B)_\rho \quad \text{for} \quad \alpha\beta = 1, \ \alpha, \beta \in [0, \infty]. \tag{7.181}$$

Observe that from the first equality above, by taking $\alpha = \infty$ (and hence $\beta = 1/2$) we get the statement given in the lemma above that the dual to the optimized conditional min-entropy is $\tilde{H}_{1/2}^\uparrow$. On the other hand, from the second equality we see that the dual of $H_{\min}$ is $H_0^\uparrow = H_{\max}$ (see Definition (7.4)). That is, for all $\rho \in \mathfrak{D}(AB)$

$$H_{\min}^{\mathrm{dual}}(A|B)_\rho = H_{\max}(A|B)_\rho. \tag{7.182}$$

We therefore get the following corollary.

> **Corollary 7.3.** Let $\mathbb{H}$ be a quantum conditional entropy and suppose its dual $\mathbb{H}^{\mathrm{dual}}$ is also a quantum conditional entropy. Then, for all $\rho \in \mathfrak{D}(AB)$,
>
> $$\mathbb{H}(A|B)_\rho \leqslant H_{\max}(A|B)_\rho. \tag{7.183}$$

*Remark.* Previously, we established that conditional entropies defined as in (7.104) are upper bounded by the conditional max-entropy. However, the corollary above does not require the conditional entropy $\mathbb{H}$ to be defined with respect to a relative entropy. Instead, it assumes that the dual entropy $\mathbb{H}^{\mathrm{dual}}$ is also a valid conditional entropy. It is worth noting that it remains an open problem whether this additional assumption can be removed, that is, whether the upper bound provided by the conditional max-entropy applies to all conditional entropies or just to those whose dual is also a conditional entropy.

**Proof**  Let $\varphi \in \mathrm{Pure}(ABC)$ be a purification of $\rho^{AB}$. From the definition of $\mathbb{H}^{\mathrm{dual}}$ we get

$$\mathbb{H}(A|B)_\rho = -\mathbb{H}^{\mathrm{dual}}(A|C)_\varphi$$

$$\textbf{Theorem 7.5 applied to } \mathbb{H}^{\mathrm{dual}} \rightarrow \ \leqslant -H_{\min}(A|C)_\varphi \tag{7.184}$$

$$\textbf{(7.182)} \rightarrow \ = H_{\max}(A|B)_\rho.$$

This completes the proof.  ∎

**Exercise 7.24.** *Use the Lemma 7.5 and the relations* (7.181) *to show that for any* $\varphi \in$ Pure$(ABC)$ *we have*

$$H_\alpha(A|B)_\varphi + H_\beta(A|C)_\varphi = 0 \quad \text{for} \quad \alpha + \beta = 2, \ \alpha, \beta \in [0,2]$$

$$\tilde{H}_\alpha^\uparrow(A|B)_\varphi + \tilde{H}_\beta^\uparrow(A|C)_\varphi = 0 \quad \text{for} \quad \frac{1}{\alpha} + \frac{1}{\beta} = 2, \ \alpha, \beta \in [1/2, \infty] \tag{7.185}$$

$$H_\alpha^\uparrow(A|B)_\varphi + \tilde{H}_\beta(A|C)_\varphi = 0 \quad \text{for} \quad \alpha\beta = 1, \ \alpha, \beta \in [0, \infty].$$

**Exercise 7.25.** *In the following, use the above duality relations.*

1. *Show that the dual of $H_\alpha$ is itself a quantum conditional entropy for all $\alpha \in [0,2]$ (i.e. you need to show the monotonicity property).*
2. *Show that $\tilde{H}_\alpha^\uparrow$ is a quantum conditional entropy for all $\alpha \in [0, \infty]$ (i.e. you need to show the additivity property).*
3. *Show that the dual of $\tilde{H}_\alpha^\uparrow$ is itself a quantum conditional entropy for all $\alpha \in [0, \infty]$.*
4. *Use part 2 to provide an alternative proof for the additivity of the optimized conditional min-entropy.*

## 7.7 The Decoupling Theorem

The decoupling theorem identifies the conditions under which a system, initially correlated with another system (such as the environment), becomes decoupled from that environment following a physical evolution. The conditional entropy, as examined in the previous sections, can be employed to measure the extent of this decoupling. The decoupling theorem serves as a pivotal instrument in both quantum Shannon theory and quantum resource theories. A "smoothed" version of this theorem will also be discussed in Section 10.4.2.

In Section C.4 of online version, we introduced the concept of the **G**-twirling operation over a compact Lie group. When the group $\mathbf{G} = \mathfrak{U}(A)$ encompasses all unitary matrices, the resulting **G**-twirling map, denoted as $\mathcal{G} \in \mathrm{CPTP}(A \to A)$, is given by the channel in (3.242). Therefore, the channel $\mathcal{G}^{A \to A}$ transforms a state $\rho \in \mathfrak{D}(AE)$ – representing a system $A$ correlated with an environment $E$ – into:

$$\mathcal{G}^{A \to A}\left(\rho^{AE}\right) := \int_{\mathfrak{U}(A)} dU^A \, U^A \rho^{AE} U^{*A} = \mathbf{u}^A \otimes \rho^E. \tag{7.186}$$

In this context, the **G**-twirling map acts as a completely randomizing channel, also known as the completely depolarizing channel.

When a quantum channel $\mathcal{N} \in \mathrm{CPTP}(A \to B)$ is applied to both sides of the equation above, we obtain:

$$\int_{\mathfrak{U}(A)} dU^A \, \mathcal{N}^{A \to B}\left(U^A \rho^{AE} U^{*A}\right) = \tau^B \otimes \rho^E, \tag{7.187}$$

where $\tau^B := \mathcal{N}^{A \to B}(\mathbf{u}^A)$. The decoupling theorem estimates how closely $\mathcal{N}^{A \to B}\left(U^A \rho^{AE} U^{*A}\right)$ (i.e. removing the integral and considering one specific unitary matrix) can approximate the decoupled state $\tau^A \otimes \rho^E$. Our discussion begins with a lemma using the square of the Frobenius norm for this estimation. In this lemma, we utilize the function

$$f\left(\omega^{AB}\right) := \frac{m}{\sqrt{m^2 - 1}}\left(\operatorname{Tr}\left(\omega^{AB}\right)^2 - \operatorname{Tr}\left(\mathbf{u}^A \otimes \omega^B\right)^2\right) \qquad \forall\, \omega \in \mathfrak{L}(AB), \quad (7.188)$$

where $m := |A|$. To simplify the notation in this section, we will omit the square brackets in certain expressions. For instance, in the above formula, we used $\operatorname{Tr}\left(\omega^{AB}\right)^2$ instead of $\operatorname{Tr}\left[\left(\omega^{AB}\right)^2\right]$. It's important to note that with this revised notation, all powers are included within the trace operation.

**Exercise 7.26.** *Let $\rho \in \operatorname{Pos}(AB)$ (we also assume $\rho^{AB}$ is not the zero matrix) and set $m := |A|$.*

1. *Show that*

$$\frac{1}{m} \leqslant \frac{\operatorname{Tr}(\rho^{AB})^2}{\operatorname{Tr}(\rho^A)^2} \leqslant m. \qquad (7.189)$$

*Hint: Start by showing $\operatorname{Tr}(\rho^B)^2 = \operatorname{Tr}\left[\left(\rho^{AB} \otimes I^{\tilde{A}}\right)\left(I^A \otimes \rho^{\tilde{A}B}\right)\right]$ and then use the Cauchy–Schwarz inequality. For the other side, show first that $\rho^{AB} \leqslant m I^A \otimes \rho^B$.*

2. *Consider the function $f$ as defined above.*

   *(a) Show that $f\left(\rho^{AB}\right) \geqslant 0$ with equality if and only if $\rho^{AB} = \mathbf{u}^A \otimes \rho^B$.*
   *(b) Show that*

$$f(\rho^{AB}) \leqslant \operatorname{Tr}\left(\rho^{AB}\right)^2. \qquad (7.190)$$

---

**Lemma 7.7.** Let $\rho \in \mathfrak{L}(AE)$, $m := |A|$, $\mathcal{N} \in \mathfrak{L}(A \to B)$, and $\tau^{AB} := \frac{1}{m} J_{\mathcal{N}}^{AB}$, where $J_{\mathcal{N}}^{AB}$ is the Choi matrix of $\mathcal{N}^{A \to B}$.

$$\int_{\mathfrak{U}(A)} dU^A \operatorname{Tr}\left(\mathcal{N}^{A \to B}\left(U^A \rho^{AE}\left(U^A\right)^*\right) - \tau^B \otimes \rho^E\right)^2 = f\left(\rho^{AE}\right) f\left(\tau^{AB}\right),$$

$$\tag{7.191}$$

where $f$ is defined in (7.188).

---

*Remark.* Observe that we do not assume that $\rho^{AE}$ is a density matrix (not even Hermitian) nor that $\mathcal{N}^{A \to B}$ is a quantum channel (just a linear map). However, if $\rho^{AE} \geqslant 0$ we can use the bound (7.190) in conjunction with the lemma above to get the relatively simple upper bound

$$\int_{\mathfrak{U}(A)} dU^A \left\|\mathcal{N}_U^{A \to B}\left(\rho^{AE}\right) - \tau^B \otimes \rho^E\right\|_2^2 \leqslant \operatorname{Tr}\left(\rho^{AE}\right)^2 \operatorname{Tr}\left(\tau^{AB}\right)^2, \qquad (7.192)$$

where $\mathcal{N}_U^{A\to B} := \mathcal{N}^{A\to B} \circ \mathcal{U}^{A\to A}$, with $\mathcal{U}^{A\to A}(\cdot) := U^A(\cdot)U^{*A}$, and we used the fact that any Hermitian matrix $\eta \in \mathrm{Herm}(BE)$ satisfies $\|\eta\|_2^2 = \mathrm{Tr}[\eta^2]$. Moreover, taking the square root on both sides of the equation above and using Jensen's inequality (see Section B.4 of online version) we obtain that

$$
\sqrt{\mathrm{Tr}\left(\rho^{AE}\right)^2 \mathrm{Tr}\left(\tau^{AB}\right)^2} \geqslant \sqrt{\int_{\mathfrak{U}(A)} dU^A \left\|\mathcal{N}_U^{A\to B}\left(\rho^{AE}\right) - \tau^B \otimes \rho^E\right\|_2^2}
$$
$$
\textbf{Jensen's Inequality} \to \; \geqslant \int_{\mathfrak{U}(A)} dU^A \left\|\mathcal{N}_U^{A\to B}\left(\rho^{AE}\right) - \tau^B \otimes \rho^E\right\|_2 .
$$
(7.193)

Finally, it's important to recognize that since the average of the integrand in the above equation is less than the expression on the left-hand side, it implies the existence of at least one unitary $U^A$ for which $\left\|\mathcal{N}_U^{A\to B}\left(\rho^{AE}\right) - \tau^B \otimes \rho^E\right\|_2$ is smaller than $\mathrm{Tr}\left[\rho^{AE}\right]\mathrm{Tr}\left[\tau^{AB}\right]$.

**Proof**  For simplicity of the exposition we will omit the superscript from $\mathcal{N}_U^{A\to B}$ and simply write it as $\mathcal{N}_U$. With these notations, the integrand of (7.191) can be decomposed into three terms:

$$
\mathrm{Tr}\left(\mathcal{N}_U\left(\rho^{AE}\right) - \tau^B \otimes \rho^E\right)^2
$$
$$
= \mathrm{Tr}\left(\mathcal{N}_U\left(\rho^{AE}\right)\right)^2 - 2\mathrm{Tr}\left[\left(\tau^B \otimes \rho^E\right)\mathcal{N}_U\left(\rho^{AE}\right)\right] + \mathrm{Tr}\left(\tau^B \otimes \rho^E\right)^2 .
$$
(7.194)

From (7.187), the integral of the second term above can be simplified as

$$
\int_{\mathfrak{U}(A)} dU \, \mathrm{Tr}\left[\left(\tau^B \otimes \rho^E\right)\mathcal{N}_U\left(\rho^{AE}\right)\right] = \mathrm{Tr}\left(\tau^B \otimes \rho^E\right)^2 .
$$
(7.195)

Therefore, taking the integral over $\mathfrak{U}(A)$ on both sides of (7.194) gives

$$
\int_{\mathfrak{U}(A)} dU \, \mathrm{Tr}\left(\mathcal{N}_U\left(\rho^{AE}\right) - \tau^B \otimes \rho^E\right)^2
$$
$$
= \int_{\mathfrak{U}(A)} dU \, \mathrm{Tr}\left(\mathcal{N}_U\left(\rho^{AE}\right)\right)^2 - \mathrm{Tr}\left(\tau^B\right)^2 \mathrm{Tr}\left(\rho^E\right)^2 .
$$
(7.196)

To compute the remaining integral we use a linearization technique that is based on Exercise 3.70. That is, we linearize the square in the integrand by using Exercise 3.70 with the flip operator $F^{B\tilde{B}E\tilde{E}} = F^{B\tilde{B}} \otimes F^{E\tilde{E}}$. Explicitly,

$$
\mathrm{Tr}\left(\mathcal{N}_U\left(\rho^{AE}\right)\right)^2 = \mathrm{Tr}\left[\left(\mathcal{N}_U\left(\rho^{AE}\right) \otimes \mathcal{N}_U\left(\rho^{AE}\right)\right) F^{B\tilde{B}E\tilde{E}}\right]
$$
$$
= \mathrm{Tr}\left[\mathcal{N}_U^{\otimes 2}\left(\left(\rho^{AE}\right)^{\otimes 2}\right) F^{B\tilde{B}E\tilde{E}}\right]
$$
$$
= \mathrm{Tr}\left[\left(\rho^{AE}\right)^{\otimes 2}\left(\mathcal{N}_U^{*\otimes 2}(F^{B\tilde{B}}) \otimes F^{E\tilde{E}}\right)\right] .
$$
(7.197)

Taking the integral over $\mathfrak{U}(A)$ on both sides and using the fact that $\mathcal{N}_U^* = \mathcal{U}^* \circ \mathcal{N}^*$ we obtain

$$\int_{\mathfrak{U}(A)} dU \, \mathrm{Tr}\left(\mathcal{N}_U\left(\rho^{AE}\right)\right)^2 = \mathrm{Tr}\left[\left(\rho^{AE}\right)^{\otimes 2}\left(\mathcal{G}(\mathcal{N}^{*\otimes 2}(F^{B\tilde{B}})) \otimes F^{E\tilde{E}}\right)\right], \quad (7.198)$$

where $\mathcal{G} \in \mathrm{CPTP}(A\tilde{A} \to A\tilde{A})$ denotes the twirling channel

$$\mathcal{G}(\cdot) := \int_{\mathfrak{U}(A)} dU \, U^* \otimes U^*(\cdot)U \otimes U. \quad (7.199)$$

Next, we make use of the fact that the twirling channel turns states to symmetric ones (see (3.247)). Specifically, observe that from (3.247) we get

$$\mathcal{G}\left(\mathcal{N}^{*\otimes 2}(F^{B\tilde{B}})\right) = aI^{A\tilde{A}} + bF^{A\tilde{A}}, \quad (7.200)$$

where the coefficients $a, b \in \mathbb{R}$ will be computed shortly using (3.247). Substituting (7.200) into (7.198) gives

$$\int_{\mathfrak{U}(A)} dU \, \mathrm{Tr}\left(\mathcal{N}_U\left(\rho^{AE}\right)\right)^2 = \mathrm{Tr}\left[\left(\rho^{AE}\right)^{\otimes 2}\left((aI^{A\tilde{A}} + bF^{A\tilde{A}}) \otimes F^{E\tilde{E}}\right)\right]$$

$$= a\mathrm{Tr}\left[\left(\rho^{E}\right)^{\otimes 2} F^{E\tilde{E}}\right] + b\mathrm{Tr}\left[\left(\rho^{AE}\right)^{\otimes 2} F^{A\tilde{A}E\tilde{E}}\right] \quad (7.201)$$

$$(3.248)\to = a\mathrm{Tr}\left(\rho^{E}\right)^2 + b\mathrm{Tr}\left(\rho^{AE}\right)^2.$$

Combining this with (7.196) we obtain

$$\int_{\mathfrak{U}(A)} dU \, \mathrm{Tr}\left(\mathcal{N}_U\left(\rho^{AE}\right) - \tau^B \otimes \rho^E\right)^2 = \left(a - \mathrm{Tr}\left(\tau^B\right)^2\right)\mathrm{Tr}\left(\rho^E\right)^2 + b\mathrm{Tr}\left(\rho^{AE}\right)^2. \quad (7.202)$$

It is therefore left to compute the coefficients $a$ and $b$. From (3.247) they can be expressed as

$$a := \frac{m\mathrm{Tr}\left[\mathcal{N}^{*\otimes 2}(F^{B\tilde{B}})\right] - \mathrm{Tr}\left[\mathcal{N}^{*\otimes 2}(F^{B\tilde{B}})F^{A\tilde{A}}\right]}{m(m^2 - 1)} \quad (7.203)$$

and

$$b := \frac{m\mathrm{Tr}\left[\mathcal{N}^{*\otimes 2}(F^{B\tilde{B}})F^{A\tilde{A}}\right] - \mathrm{Tr}\left[\mathcal{N}^{*\otimes 2}(F^{B\tilde{B}})\right]}{m(m^2 - 1)}. \quad (7.204)$$

To simplify the expressions above we use the definition of the adjoint map to get

$$\mathrm{Tr}\left[\mathcal{N}^{*\otimes 2}(F^{B\tilde{B}})\right] = \mathrm{Tr}\left[\left(\mathcal{N}(I^A)\right)^{\otimes 2} F^{B\tilde{B}}\right]$$

$$\boxed{\mathcal{N}\left(I^A\right) = J_{\mathcal{N}}^B = m\tau^B} \longrightarrow = m^2\mathrm{Tr}\left[\left(\tau^B\right)^{\otimes 2} F^{B\tilde{B}}\right] \quad (7.205)$$

$$\boxed{F^B := \mathrm{Tr}_{\tilde{B}}\left[F^{B\tilde{B}}\right] = I^B} \longrightarrow = m^2\mathrm{Tr}\left[\left(\tau^B\right)^2\right],$$

and

$$\mathrm{Tr}\left[F^{A\tilde{A}}\mathcal{N}^{*\otimes 2}(F^{B\tilde{B}})\right] = \mathrm{Tr}\left[F^{B\tilde{B}}\mathcal{N}^{\otimes 2}(F^{A\tilde{A}})\right]. \tag{7.206}$$

Moreover, since the Choi matrix of $\mathcal{N}^{\otimes 2}$ is given by $m^2\left(\tau^{AB}\right)^{\otimes 2}$ we get

$$\mathrm{Tr}\left[F^{B\tilde{B}}\mathcal{N}^{\otimes 2}(F^{A\tilde{A}})\right] = m^2\mathrm{Tr}\left[F^{B\tilde{B}}\mathrm{Tr}_{A\tilde{A}}\left[\left(\tau^{AB}\right)^{\otimes 2}\left(F^{A\tilde{A}}\otimes I^{B\tilde{B}}\right)\right]\right]$$

$$= m^2\mathrm{Tr}\left[\left(\tau^{AB}\right)^{\otimes 2}\left(F^{A\tilde{A}}\otimes F^{B\tilde{B}}\right)\right] \tag{7.207}$$

$$(\mathbf{3.248})\rightarrow = m^2\mathrm{Tr}\left[(\tau^{AB})^2\right].$$

We therefore conclude that $a$ and $b$ can be expressed as

$$a = \frac{m^2\mathrm{Tr}\left(\tau^B\right)^2 - m\mathrm{Tr}\left(\tau^{AB}\right)^2}{m^2 - 1} \quad \text{and}$$

$$b = \frac{m^2\mathrm{Tr}\left(\tau^{AB}\right)^2 - m\mathrm{Tr}\left(\tau^B\right)^2}{m^2 - 1} = \frac{m}{\sqrt{m^2 - 1}}f\left(\tau^{AB}\right). \tag{7.208}$$

Finally, substituting these expressions into (7.202), and observing that

$$a - \mathrm{Tr}\left(\tau^B\right)^2 = \frac{\mathrm{Tr}\left(\tau^B\right)^2 - m\mathrm{Tr}\left(\tau^{AB}\right)^2}{m^2 - 1} = -\frac{1}{\sqrt{m^2 - 1}}f\left(\tau^{AB}\right), \tag{7.209}$$

we get that the right-hand side of (7.202) equals the right-hand side of (7.191). This completes the proof. ∎

**Exercise 7.27.** *Demonstrate clearly that substituting the expressions in the proof above for $a - \mathrm{Tr}\left(\tau^B\right)^2$ and $b$ into (7.202) results in the equality (7.191).*

**Exercise 7.28.** *Using the same notations as in the lemma above, with $\rho \in \mathfrak{D}(AE)$ and $\mathcal{N} \in \mathrm{CP}(A \to B)$, show that for all $\sigma \in \mathfrak{D}(E)$,*

$$\int_{\mathfrak{U}(A)} dU^A\mathrm{Tr}\left(\mathcal{N}^{A\to B}\left(U^A\rho^{AE}\left(U^A\right)^*\right) - \tau^B\otimes\sigma^E\right)^2 \geqslant f\left(\rho^{AE}\right)f\left(\tau^{AB}\right), \tag{7.210}$$

*with equality if and only if $\sigma^E = \rho^E$.*

In the proof of Lemma 7.7, we used the twirling operation $\mathcal{G}$ as defined in (7.199). In Section 15.2.1 we will see that this channel belongs to a family of channels that we call the **G-twirling** operations. One of the properties of all **G** twirling operations, particularly the channel $\mathcal{G}$ as defined in (7.199), is that they can be expressed as a finite convex combination of unitary channels. In our context here, it means that there exists $k \in \mathbb{N}$, $\mathbf{p} \in \mathrm{Prob}(k)$, and a set of unitary matrices $\{U_x\}_{x\in[k]} \subset \mathfrak{U}(A)$ such that $\mathcal{G} \in \mathrm{CPTP}(A\tilde{A} \to A\tilde{A})$ as defined in (7.199) can be expressed as

$$\mathcal{G}(\omega^{A\tilde{A}}) = \sum_{x\in[k]} p_x\,(U_x \otimes U_x)\,\omega^{A\tilde{A}}\,(U_x \otimes U_x)^* \qquad \forall\omega \in \mathcal{L}(A\tilde{A}). \tag{7.211}$$

Therefore, working with this expression for the twirling map, we obtain the following corollary.

---

**Corollary 7.4.** Let $k \in \mathbb{N}$, $\mathbf{p} \in \text{Prob}(k)$, and $\{U_x\}_{x \in [k]} \subset \mathfrak{U}(A)$ be as in (7.211). Then, using the same notations as in Lemma 7.7, we have

$$\sum_{x \in [k]} p_x \text{Tr}\left(\mathcal{N}^{A \to B}\left(U_x^A \rho^{AE}\left(U_x^A\right)^*\right) - \tau^B \otimes \rho^E\right)^2 = f\left(\rho^{AE}\right) f\left(\tau^{AB}\right).$$

$$(7.212)$$

---

**Exercise 7.29.** *Prove the corollary above.*

In the following theorem, we make use of the conditional entropy $\tilde{H}_2^{\uparrow}(A|B)_\omega$, which is defined on every $\omega \in \text{Pos}(AB)$ terms of the divergence $\tilde{D}_2$ as (c.f. (7.104))

$$\tilde{H}_2^{\uparrow}(A|B)_\omega = - \min_{\eta \in \mathfrak{D}(B)} \tilde{D}_2\left(\omega^{AB} \| I^A \otimes \eta^B\right)$$

$$(7.213)$$

$$\textbf{Definition 6.7} \to \; = - \min_{\eta \in \mathfrak{D}(B)} \log \text{Tr}\left(\left(I^A \otimes \eta^{-1/4}\right) \omega^{AB}\left(I^A \otimes \eta^{-1/4}\right)\right)^2.$$

---

**Decoupling Theorem**

**Theorem 7.8.** Let $\rho \in \mathfrak{D}_{\leqslant}(AE)$, $\mathcal{N} \in \text{CP}(A \to B)$, and $\tau^{AB} := \frac{1}{|A|} J_{\mathcal{N}}^{AB}$, where $J_{\mathcal{N}}^{AB}$ is the Choi matrix of $\mathcal{N}^{A \to B}$. Then,

$$\int_{\mathfrak{U}(A)} dU^A \left\| \mathcal{N}^{A \to B}\left(U^A \rho^{AE}\left(U^A\right)^*\right) - \tau^B \otimes \rho^E \right\|_1 \leqslant 2^{-\frac{1}{2}\left(\tilde{H}_2^{\uparrow}(A|E)_\rho + \tilde{H}_2^{\uparrow}(A|B)_\tau\right)}.$$

$$(7.214)$$

---

**Proof**   In the first step of the proof we upper bound the trace norm with the Hilbert–Schmidt norm. Working with the Frobenius norm we will be able to use (7.193). From the third part of Exercise 5.24 it follows that for any matrix $M \in \text{Herm}(A)$ and $\sigma \in \text{Pos}(A)$

$$\|M\|_1 \leqslant \sqrt{\text{Tr}[\sigma]} \left\| \sigma^{-1/4} M \sigma^{-1/4} \right\|_2.$$

$$(7.215)$$

Taking $\sigma = \eta^B \otimes \zeta^E \in \mathfrak{D}(BE)$ and

$$M = \mathcal{N}_U\left(\rho^{AE}\right) - \tau^B \otimes \rho^E$$

$$(7.216)$$

gives

$$\left\| \mathcal{N}_U\left(\rho^{AE}\right) - \tau^B \otimes \rho^E \right\|_1$$

$$\leqslant \left\| \left(\eta^B \otimes \zeta^E\right)^{-\frac{1}{4}} \left(\mathcal{N}_U\left(\rho^{AE}\right) - \tau^B \otimes \rho^E\right) \left(\eta^B \otimes \zeta^E\right)^{-\frac{1}{4}} \right\|_2$$

$$(7.217)$$

The choice of $\eta^B$ and $\zeta^E$ will be made later. Denoting by $\tilde{\mathcal{N}}_U^{A \to B}(\cdot) := (\eta^B)^{-\frac{1}{4}} \mathcal{N}_U^{A \to B}$ $(\cdot)(\eta^B)^{-\frac{1}{4}}$, $\tilde{\rho}^{AE} := (\zeta^E)^{-\frac{1}{4}} \rho^{AE} (\zeta^E)^{-\frac{1}{4}}$, and $\tilde{\tau}^{AB} := \frac{1}{m} J_{\mathcal{N}}^{AB}$, we get

$$\left\| \mathcal{N}_U \left( \rho^{AE} \right) - \tau^B \otimes \rho^E \right\|_1 \leqslant \left\| \tilde{\mathcal{N}}_U \left( \tilde{\rho}^{AE} \right) - \tilde{\tau}^B \otimes \tilde{\rho}^E \right\|_2. \tag{7.218}$$

Taking the integral over $U \in \mathfrak{U}(m)$ on both sides we obtain

$$\int_{\mathfrak{U}(A)} dU \left\| \mathcal{N}_U \left( \rho^{AE} \right) - \tau^B \otimes \rho^E \right\|_1 \leqslant \int_{\mathfrak{U}(A)} dU \left\| \tilde{\mathcal{N}}_U \left( \tilde{\rho}^{AE} \right) - \tilde{\tau}^B \otimes \tilde{\rho}^E \right\|_2$$
$$\mathbf{(7.193)} \to \ \leqslant \sqrt{\mathrm{Tr}\left[ (\tilde{\tau}^{AB})^2 \right] \mathrm{Tr}\left[ (\tilde{\rho}^{AE})^2 \right]} \tag{7.219}$$

Finally, choosing $\eta^E$ and $\zeta^B$ such that

$$\mathrm{Tr}\left[ (\tilde{\rho}^{AE})^2 \right] = \mathrm{Tr}\left[ \left( (I^A \otimes (\eta^E)^{-\frac{1}{4}}) \rho^{AE} (I^A \otimes (\eta^E)^{-\frac{1}{4}}) \right)^2 \right] = 2^{-\tilde{H}_2^{\uparrow}(A|E)_\rho}$$
$$\mathrm{Tr}\left[ (\tilde{\tau}^{AB})^2 \right] = \mathrm{Tr}\left[ \left( (I^A \otimes (\zeta^B)^{-\frac{1}{4}}) \tau^{AB} (I^A \otimes (\zeta^B)^{-\frac{1}{4}}) \right)^2 \right] = 2^{-\tilde{H}_2^{\uparrow}(A|B)_\tau} \tag{7.220}$$

completes the proof. ∎

---

**Corollary 7.5.** Let $k \in \mathbb{N}$, $\mathbf{p} \in \mathrm{Prob}(k)$, and $\{U_x\}_{x \in [k]} \subset \mathfrak{U}(A)$ be as in (7.211). Then, using the same notations as in Theorem 7.8, we have

$$\sum_{x \in [k]} p_x \left\| \mathcal{N}^{A \to B} \left( U_x^A \rho^{AE} \left( U_x^A \right)^* \right) - \tau^B \otimes \rho^E \right\|_1 \leqslant 2^{-\frac{1}{2}\left( \tilde{H}_2^{\uparrow}(A|E)_\rho + \tilde{H}_2^{\uparrow}(A|B)_\tau \right)}.$$

$$\tag{7.221}$$

---

**Exercise 7.30.** *Use Theorem (7.8) and Corollary (7.4) to prove the corollary above.*

**Exercise 7.31.** *Show that if $\omega^{AB} := \frac{1}{t} J_{\mathcal{N}}^{AB}$, where $t := \mathrm{Tr}\left[ J_{\mathcal{N}}^{AB} \right]$ (i.e. $\omega^{AB} = \frac{|A|}{t} \tau^{AB}$ is a density matrix), and similarly, $\sigma^{AB} := \frac{1}{r} \rho^{AB}$ with $r := \mathrm{Tr}\left[ \rho^{AB} \right]$, then the decoupling theorem above can be expressed as*

$$\int_{\mathfrak{U}(A)} dU^A \left\| \mathcal{N}^{A \to B} \left( U^A \rho^{AE} \left( U^A \right)^* \right) - \tau^B \otimes \rho^E \right\|_1 \leqslant \frac{rt}{|A|} 2^{-\frac{1}{2}\left( \tilde{H}_2^{\uparrow}(A|E)_\sigma + \tilde{H}_2^{\uparrow}(A|B)_\omega \right)}.$$

$$\tag{7.222}$$

## 7.8 Notes and References

Conditional majorization was first introduced in Ref. [90] in the context of the quantum uncertainty principle, but its quantum version was fully defined in Ref. [103].

Many textbooks on quantum information includes a section on quantum conditional entropies and their properties. Specifically, the books in Refs. [211] and [209] provide a comprehensive review on the various quantum Rényi conditional entropies and their properties. The axiomatic approach for the quantum conditional entropy, as presented in this chapter, is due to Ref. [103]. Semi-causal maps were first introduced in Ref. [14], and was conjectured to have the characterization given in Theorem 7.1. Shortly after this conjecture was proved in Ref. [72]. The relatively simplified proof provided here for Theorem 7.1 is due to Ref. [182]. The duality relation given in Lemma 7.5 was first introduced in Ref. [213]. The proofs for the two other relations in (7.181) were given for the first relation in Ref. [170] and independently in Ref. [15], and for the second relation of (7.181) in Ref. [212].

# 8

# The Asymptotic Regime

As the dimension of a physical system grows, one can employ several tools from probability theory and statistics (e.g. the law of large numbers) to study its behavior and properties. Specifically, one of the main goals of quantum resource theories is to determine the rate at which many copies of one resource can be converted into many copies of another. The methods and tools developed here provide the foundations for several topics in this asymptotic domain. We start by reviewing some of these concepts and their generalizations to the quantum world.

## 8.1 Classical Typicality

A central theme in information theory is finding efficient methods for transmitting information from one party (Alice) to another (Bob). Consider a scenario where Alice wishes to send Bob a message $x$ from a set of $m$ possibilities. This transmission would require $\log_2(m)$ classical bits, achievable through $\log_2(m)$ uses of a perfectly noiseless classical bit-channel. At first glance, it seems the resource cost for sending a message of size $m$ is

$$\log_2(m)[c \to c], \tag{8.1}$$

where $[c \to c]$ denotes one usage of a noiseless cbit-channel.

However, if Alice's message to Bob consists of English alphabet letters (where $m = 26$), additional information about the message emerges. For instance, the letter "E" appears with a frequency of about 12%, while "Z" occurs only about 0.07% of the time. Thus, it's more probable for Bob to receive "E" than "Z." Shannon's groundbreaking 1948 paper illustrated how to leverage this additional information to significantly reduce classical communication costs. This approach is based on the crucial concept of typicality, which plays a major role in information science and, by extension, in quantum resource theories.

### 8.1.1 i.i.d. Information Source

Building on this, we model the messages that Alice sends to Bob as being drawn from a *classical information source*. This source can be conceptualized as a sequence of random variables $X_1$, $X_2$, ..., each representing an output. For instance, the word "PEACE" sent by Alice translates into a five-letter sequence: $X_1 = \text{P}$, $X_2 = \text{E}$, $X_3 = \text{A}$, $X_4 = \text{C}$, and $X_5 = \text{E}$, with each letter having a specific probability of occurrence. To simplify the mathematics and as a first approximation, we make two assumptions about this source.

First, we assume that the source's various uses are *independent*, meaning that each letter is emitted without being influenced by the previous ones. However, it's clear that this assumption does *not* strictly apply to the English alphabet. Take, for instance, the letter "T," which is the second most frequent letter in English, with an occurrence frequency of about 9%. If Alice sends the letter "M" to Bob, the likelihood of the next letter being "T" is significantly lower than 9%, given the relative rarity of the "MT" combination in English words. Thus, for many information sources, including English, the assumption of independence should be considered more as a first-order approximation than a definitive rule.

Second, we assume the source's uses are *identically distributed*, implying each use of the source can be represented by a random variable, sharing the same alphabet and possessing an identical probability distribution. That is, for all $k \in \mathbb{N}$, the probability, $\Pr(X_k = x) := p_x$, that the random variable $X_k$ is equal to some $x$ in the alphabet will be independent of $k$. We will therefore consider here information sources that are both *independent and identically distributed*, or in short, i.i.d. information sources. An i.i.d. source is thus represented by a single random variable $X$, with an alphabet $x \in \mathcal{X}$ and a corresponding probability distribution $\{p_x\}_{x \in \mathcal{X}}$. For simplicity, we consider finite alphabets, taking $\mathcal{X} = \{1, \ldots, m\}$, and denote the distribution as i.i.d. $\sim \mathbf{p}$, emphasizing the probability vector $\mathbf{p} = (p_1, \ldots, p_m)^T$ of $X$.

Consider, for example, $n$ uses of a binary i.i.d. source, producing a sequence of $n$ bits $X^n = (X_1, \ldots, X_n)$, with $p$ being the probability of outcome "0" and $1-p$ for "1" (hence $X_k \in \{0, 1\}$). For large $n$, it's highly likely that the sequence $X^n$ will contain roughly $np$ zeros and $n(1 - p)$ ones. The occurrence probability of such a typical sequence is approximately

$$\Pr\left(X^n = x^n\right) = p_{x_1} p_{x_2} \cdots p_{x_n} \approx p^{np}(1 - p)^{n(1-p)}. \qquad (8.2)$$

This probability can be further simplified to

$$\Pr\left(X^n = x^n\right) \approx 2^{-nH(X)}, \qquad (8.3)$$

where $H(X)$ is the binary Shannon entropy

$$H(X) := - p \log_2 (p) - (1 - p) \log_2 (1 - p). \qquad (8.4)$$

Despite the variety of typical sequences $x^n$, they all share approximately the same probability of occurrence. This effect is known as *the asymptotic equipartition property*, a direct result of the (weak) *law of large numbers*.

## 8.1.2 The Law of Large Numbers

Let $X_1, X_2, \ldots$ be an i.i.d.$\sim \mathbf{p}$ source with the same distribution as of $X$. Suppose that

$$\mathbb{E}(X) := \sum_{x \in \mathcal{X}} p_x x < \infty \quad \text{and} \quad \mathbb{E}(X^2) := \sum_{x \in \mathcal{X}} p_x x^2 < \infty. \tag{8.5}$$

Since we only consider in this book sets with finite cardinality, these conditions will trivially hold.

> **The Law of Large Numbers**
>
> **Theorem 8.1.** With the notations as above, for all $\varepsilon > 0$,
>
> $$\lim_{n \to \infty} \Pr\left( \left| \frac{1}{n} \sum_{j \in [n]} X_j - \mathbb{E}(X) \right| > \varepsilon \right) = 0. \tag{8.6}$$

This law is very intuitive as it shows that for very large $n$, the probability that $\frac{1}{n} \sum_{j \in [n]} X_j$ is close to $\mathbb{E}(X)$ is almost 1. In particular, (8.6) is equivalent to the statement that

$$\lim_{n \to \infty} \frac{1}{n} \sum_{j \in [n]} X_j = \mathbb{E}(X) \quad \text{in probability.} \tag{8.7}$$

**Proof**  We first prove the theorem for the case that the expectation value of $X$ is zero; that is, $\mathbb{E}(X) = 0$. Denote by $S_n := \frac{1}{n} \sum_{j \in [n]} X_j$. Then,

$$\mathbb{E}(S_n^2) = \frac{1}{n^2} \sum_{j,k=1}^{n} \mathbb{E}(X_j X_k). \tag{8.8}$$

A key observation is that for $j \neq k$, the two random variables $X_j$ and $X_k$ are independent, and consequently

$$\mathbb{E}(X_j X_k) = \sum_{X_j, X_k \in \mathcal{X}} X_j x_k p_{X_j} p_{x_k} = \mathbb{E}(X_j) \mathbb{E}(X_k) = 0. \tag{8.9}$$

Therefore, the only contributing terms in (8.8) are those with $j = k$. Hence,

$$\mathbb{E}(S_n^2) = \frac{1}{n^2} \sum_{j \in [n]} \mathbb{E}(X_j^2) = \frac{1}{n} \mathbb{E}(X^2). \tag{8.10}$$

Equation (8.10) demonstrates that for very large $n$ the *variance* of $S_n$ is very small, indicating that it will reach a single value in the limit $n \to \infty$. On the other hand, $\mathbb{E}(S_n^2)$ can be split into two terms, those for which the value of $S_n$ is close to zero and those for which it is at least $\varepsilon$-distance away from zero:

$$\mathbb{E}(S_n^2) = \sum_{s_n^2} s_n^2 \Pr(S_n^2 = s_n^2) = \sum_{|s_n| \leqslant \varepsilon} s_n^2 \Pr(S_n^2 = s_n^2) + \sum_{|s_n| > \varepsilon} s_n^2 \Pr(S_n^2 = s_n^2)$$

$$\geqslant \sum_{|s_n| > \varepsilon} s_n^2 \Pr(S_n^2 = s_n^2)$$

$$\geqslant \varepsilon^2 \sum_{|s_n| > \varepsilon} \Pr(S_n^2 = s_n^2) = \varepsilon^2 \Pr(|S_n| > \varepsilon).$$

We therefore conclude that

$$\Pr(|S_n| > \varepsilon) \leqslant \frac{\mathbb{E}(S_n^2)}{\varepsilon^2} = \frac{\mathbb{E}(X^2)}{n\varepsilon^2} \xrightarrow[n \to \infty]{} 0. \tag{8.11}$$

This completes the proof for the case $\mathbb{E}(X) = 0$. If $\mathbb{E}(X) \neq 0$, then define a sequence of i.i.d.$\sim$ **p** random variables $Y_j := X_j - \mathbb{E}(X)$. With this definition we get the $\mathbb{E}(Y_j) = 0$ so that

$$\Pr\left(\left|\frac{1}{n} \sum_{j \in [n]} X_j - \mathbb{E}(X)\right| > \varepsilon\right) = \Pr\left(\left|\frac{1}{n} \sum_{j \in [n]} Y_j\right| > \varepsilon\right) \xrightarrow{n \to \infty} 0. \tag{8.12}$$

This completes the proof. ∎

**Exercise 8.1.** (Markov Inequality). *Prove that for any $t > 0$ and any nonnegative random variable $X$*

$$\Pr(X \geqslant t) \leqslant \frac{\mathbb{E}(X)}{t}, \tag{8.13}$$

The law of large numbers above does not tell us much how fast the probability in (8.6) goes to zero. In the proof we saw in (8.11) that it goes to zero at least as $1/n$. If instead of requiring that the expectation value of $X$ and $X^2$ are finite, we require a stronger condition that the alphabets of $X$ themselves are all bounded, then it is possible to show that in this case the probability in (8.6) goes to zero exponentially fast with $n$.

**Hoeffding's Inequality**

**Theorem 8.2.** Let $X_1, \ldots, X_n$ be $n$ independent random variable satisfying $a_j \leqslant X_j \leqslant b_j$ for all $j = 1, \ldots, n$. Then,

$$\Pr\left(\left|\frac{1}{n} \sum_{j \in [n]} \left(X_j - \mathbb{E}(X_j)\right)\right| > \varepsilon\right) \leqslant \exp\left(-\frac{2n^2\varepsilon^2}{\sum_{j \in [n]} (b_j - a_j)^2}\right). \tag{8.14}$$

Note that Hoeffding's inequality above does not assume that the random variables $X_1, \ldots, X_n$ are identically distributed. If we add this assumption (so that $X_1, \ldots, X_n$ are i.i.d.), then we get a simplified version of Hoeffding's inequality given by

$$\Pr\left(\left|\frac{1}{n} \sum_{j \in [n]} X_j - \mathbb{E}(X)\right| > \varepsilon\right) \leqslant \exp\left(-\frac{2n\varepsilon^2}{(b - a)^2}\right), \tag{8.15}$$

where we assumed that $a \leqslant X \leqslant b$.

To prove Hoeffding's inequality we will need the following lemma.

> **Hoeffding's Lemma**
>
> **Lemma 8.1.** Let $X$ be a real-valued bounded random variable with expected value $\mathbb{E}(X) = \mu$ and $a \leqslant X \leqslant b$ for some $a, b \in \mathbb{R}$ with $b > a$. Then, for all $t \in \mathbb{R}$ we have
>
> $$\mathbb{E}\left(e^{tX}\right) \leqslant \exp\left(t\mu + \frac{t^2(b-a)^2}{8}\right). \tag{8.16}$$

**Proof**   Consider first the case $\mu = 0$. We therefore must have $a \leqslant 0 \leqslant b$. Also, it's important to note that if $a = 0$, then the condition $\mathbb{E}(X) = 0$ leads to $\mathbb{E}\left(e^{tX}\right) = 1$ (can you see why?). As a result, the inequality (8.16) is valid under these circumstances. Therefore, we will proceed with the assumption that $a < 0$. The convexity of the function $f(x) := e^{tx}$ implies that for any $a \leqslant x \leqslant b$ we have

$$e^{tx} \leqslant \frac{b-x}{b-a}e^{ta} + \frac{x-a}{b-a}e^{tb}, \tag{8.17}$$

where we wrote $x$ as the convex combination $x = a\frac{b-x}{b-a} + b\frac{x-a}{b-a}$. The key idea of the inequality above is that the right-hand side depends linearly on $x$. Applying this inequality to the random variable $X$ we get

$$\mathbb{E}\left(e^{tX}\right) \leqslant \frac{b - \mathbb{E}(X)}{b-a}e^{ta} + \frac{\mathbb{E}(X) - a}{b-a}e^{tb}$$

$$\boxed{\mu = 0} \longrightarrow \quad = \frac{b}{b-a}e^{ta} - \frac{a}{b-a}e^{tb} \tag{8.18}$$

$$\boxed{c := -\frac{a}{b-a}} \longrightarrow \quad = (1 - c + ce^{t(b-a)})e^{ta}.$$

Note that $c > 0$ since $a < 0$. Finally, denote by $s := t(b-a)$ the right-hand side of the equation above, which becomes equal to

$$(1 - c + ce^s)e^{-cs} = e^{f(s)}, \quad \text{where} \quad f(s) := -cs + \log\left(1 - c + ce^s\right), \tag{8.19}$$

and we use the equality $e^{ta} = e^{-cs}$. Consider the Tylor expansion of $f(s)$ up to its second order

$$f(s) = f(0) + sf'(0) + \frac{1}{2}s^2 f''(q), \tag{8.20}$$

where $q$ is some real number between zero and $s$. By straightforward calculation, we get that $f(0) = f'(0) = 0$ and $f''(q) \leqslant \frac{1}{4}$ (see Exercise 8.2 for more details). Combining everything we conclude that

$$\mathbb{E}\left(e^{tX}\right) \leqslant e^{f(s)} = e^{\frac{1}{2}s^2 f''(q)} \leqslant e^{\frac{1}{8}s^2} = e^{\frac{1}{8}t^2(b-a)^2}. \tag{8.21}$$

This completes the proof for the case $\mu = 0$. The proof for the case $\mu \neq 0$ is obtained immediately by defining $\tilde{X} := X - \mu$ and applying the theorem for $\tilde{X}$ (see Exercise 8.3). $\blacksquare$

**Exercise 8.2.** *Show that for $f''(q) \leqslant \frac{1}{4}$. Hint: Calculate the second derivative $f''(q)$ and show that it can be expressed as $p(1 - p)$ for some number $p > 0$ (that depends on $q$) and use the fact that $p(1 - p) \leqslant \frac{1}{4}$.*

**Exercise 8.3.** *Show that the proof for the case $\mu \neq 0$ in the lemma above follows immediately by defining $\tilde{X} := X - \mu$ and applying the theorem for $\tilde{X}$.*

We are now ready to prove Hoeffding's inequality.

**Proof of Theorem 8.2.** Denote by $S_n := X_1 + \cdots + X_n$ and observe that for any $r, t > 0$ we have

$$
\Pr\left(S_n - \mathbb{E}(S_n) \geqslant t\right) = \Pr\left(e^{r(S_n - \mathbb{E}(S_n))} \geqslant e^{rt}\right)
$$

$$
\textbf{Markov's Inequality (8.13)} \rightarrow \; \leqslant e^{-rt}\,\mathbb{E}\left(e^{r(S_n - \mathbb{E}(S_n))}\right)
$$

$$
= e^{-rt}\,\mathbb{E}\left(\prod_{j=1}^{n} e^{r(X_j - \mathbb{E}(X_j))}\right)
$$

$$
\{X_j\} \text{ are independent} \rightarrow \; = e^{-rt}\prod_{j=1}^{n}\mathbb{E}\left(e^{r(X_j - \mathbb{E}(X_j))}\right)
$$

$$
\textbf{Hoeffding's Lemma} \rightarrow \; \leqslant e^{-rt}\prod_{j=1}^{n} e^{\frac{1}{8}r^2(b_j - a_j)^2}
$$

$$
= e^{g(r)},
$$

(8.22)

where $g(r) := -rt + \frac{1}{8}r^2 \sum_{j \in [n]}(b_j - a_j)^2$ is a quadratic function whose minimum is given by the right-hand side of (8.14). This completes the proof. ∎

**Exercise 8.4.** *Show that the minimum of the function $g(r)$ above is given by the right-hand side of (8.14).*

## 8.1.3 Typical Sequences

As we have already seen, a typical sequence $x^n = (x_1, \ldots, x_n)$ that is drawn from an i.i.d.$\sim$ p source has a probability to occur that is close to $2^{-nH(\mathrm{p})}$. In this section, we apply the law of large numbers to make the notion of typical sequences rigorous.

> **Typical Sequence**
>
> **Definition 8.1.** Let $\varepsilon > 0$ and let $X$ be a random variable with cardinality $|\mathcal{X}| = m$, corresponding to an i.i.d. source. A sequence of $n$ source outputs $x^n := (x_1, \ldots, x_n)$ is called $\varepsilon$-typical if
>
> $$
> 2^{-n(H(X)+\varepsilon)} \leqslant \Pr\left(X^n = x^n\right) \leqslant 2^{-n(H(X)-\varepsilon)}, \tag{8.23}
> $$
>
> where $H(X) := -\sum_{x \in [m]} p_x \log p_x$ is the Shannon entropy.

By taking the log on all sides of (8.23), the condition in (8.23) can be reexpressed as

$$\left| \frac{1}{n} \log_2 \left( \frac{1}{\Pr(X^n = x^n)} \right) - H(X) \right| \leq \varepsilon. \tag{8.24}$$

We will denote the set of all $\varepsilon$-typical sequences by

$$\mathfrak{T}_\varepsilon(X^n) := \left\{ x^n = (x_1, \ldots, x_n) : x^n \text{ is } \varepsilon\text{-typical} \right\}. \tag{8.25}$$

Therefore, the probability that a sequence is $\varepsilon$-typical is given by

$$\Pr(\mathfrak{T}_\varepsilon(X^n)) := \sum_{x^n \in \mathfrak{T}_\varepsilon(X^n)} \Pr(X^n = x^n). \tag{8.26}$$

More generally, for any set of sequences $\mathfrak{K}_n \subseteq [m]^n$ we will use the notation $\Pr(\mathfrak{K}_n)$ to denote the probability that a sequence belongs to $\mathfrak{K}_n$. That is,

$$\Pr(\mathfrak{K}_n) := \sum_{x^n \in \mathfrak{T}_\varepsilon(X^n)} \Pr(X^n = x^n). \tag{8.27}$$

In the following theorem, we denote the constant

$$c := \frac{2}{\left( \log(p_{\max}/p_{\min}) \right)^2} > 0, \tag{8.28}$$

where $p_{\min} > 0$ and $p_{\max}$ are the smallest and largest positive (i.e. nonzero) components of $\mathbf{p} := (p_1, \ldots, p_m)^T$.

---

**Theorem 8.3.** Let $\mathbf{p} \in \mathrm{Prob}(m)$, $\varepsilon \in (0, 1)$, $\delta_n := e^{-c\varepsilon^2 n}$, where $c$ is defined in (8.28), $X$ be a random variable associated with an i.i.d.$\sim \mathbf{p}$ source, and for each $n \in \mathbb{N}$, let $\mathfrak{K}_n \subseteq [m]^n$ be a set of sequences with cardinality $|\mathfrak{K}_n| \leq 2^{nr}$ for some $r < H(X)$. Then, for all $n \in \mathbb{N}$ the following three inequalities hold:

1. $\Pr(\mathfrak{T}_\varepsilon(X^n)) > 1 - \delta_n$.
2. $(1 - \delta_n) 2^{n(H(X) - \varepsilon)} \leq |\mathfrak{T}_\varepsilon(X^n)| \leq 2^{n(H(X) + \varepsilon)}$.
3. $\Pr(\mathfrak{K}_n) \leq e^{-c'n}$, for some $c' > 0$.

---

**Proof**  For the first inequality, we assume without loss of generality that $\mathbf{p} > 0$, since any $x$ with $p_x = 0$ never occurs and can be removed from the alphabet of $X$. Let $Y := -\log_2(X)$ be the random variable whose alphabet symbols are given by $\mathcal{Y} := \left\{ -\log_2 \Pr(X = x) \right\}_{x \in [m]}$, with corresponding probabilities $p_x := \Pr(X = x)$. Let $Y_1, Y_2, \ldots$ be an i.i.d. sequence of random variables where each $Y_j$ corresponds to $X_j$. By definition, each $Y_j$ satisfies $-\log p_{\max} \leq Y_j \leq -\log p_{\min}$. Therefore, from Hoeffding's inequality, particularly (8.15), we get that

$$\Pr \left( \left| \frac{1}{n} \sum_{j \in [n]} Y_j - \mathbb{E}(Y) \right| > \varepsilon \right) \leq e^{-c\varepsilon^2 n}. \tag{8.29}$$

Observe that

$$\mathbb{E}(Y) = -\sum_{x\in[m]} p_x \log_2 (\Pr(X=x)) = -\sum_{x\in[m]} p_x \log_2 p_x = H(X). \tag{8.30}$$

Moreover,

$$\frac{1}{n}\sum_{j\in[n]} Y_j = -\frac{1}{n}\sum_{j\in[n]} \log_2 \Pr(X_j) = -\frac{1}{n}\log_2 \Pr(X^n) = \frac{1}{n}\log_2\left(\frac{1}{\Pr(X^n)}\right). \tag{8.31}$$

We therefore conclude that

$$\Pr\left(\left|\frac{1}{n}\log_2\left(\frac{1}{\Pr(X^n)}\right) - H(X)\right| > \varepsilon\right) \leqslant e^{-c\varepsilon^2 n}. \tag{8.32}$$

Equation (8.32) states that the probability that the random variable $X^n = (X_1,\ldots,X_n)$ is not an $\varepsilon$-typical sequence is no greater than $e^{-c\varepsilon^2 n}$. This completes the proof of the first part of the theorem.

For the second inequality, we get from the definition of $\varepsilon$-typical sequences that

$$1 \geqslant \sum_{x^n\in\mathcal{T}_\varepsilon(X^n)} \Pr(X^n = x^n) \geqslant \sum_{x^n\in\mathcal{T}_\varepsilon(X^n)} 2^{-n(H(X)+\varepsilon)} = \left|\mathcal{T}_\varepsilon(X^n)\right| 2^{-n(H(X)+\varepsilon)}. \tag{8.33}$$

Therefore,

$$\left|\mathcal{T}_\varepsilon(X^n)\right| \leqslant 2^{n(H(X)+\varepsilon)}. \tag{8.34}$$

On the other hand, observe that from the first part and the definition of $\varepsilon$-typical sequences we get

$$1 - \delta_n \leqslant \sum_{x^n\in\mathcal{T}_\varepsilon(X^n)} \Pr(X^n = x^n) \leqslant \sum_{x^n\in\mathcal{T}_\varepsilon(X^n)} 2^{-n(H(X)-\varepsilon)} = \left|\mathcal{T}_\varepsilon(X^n)\right| 2^{-n(H(X)-\varepsilon)}. \tag{8.35}$$

Hence,

$$(1 - \delta_n)2^{n(H(X)-\varepsilon)} \leqslant \left|\mathcal{T}_\varepsilon(X^n)\right|. \tag{8.36}$$

For the last part of the proof (i.e. third inequality), let $0 < \varepsilon' < \frac{1}{2}(H(X) - r)$. The probability of $\mathcal{R}_n$ can be expressed as

$$\sum_{x^n\in\mathcal{R}_n} \Pr(X^n = x^n) = \sum_{x^n\in\mathcal{R}_n\cap\mathcal{T}_{\varepsilon'}(X^n)} \Pr(X^n = x^n) + \sum_{\substack{x^n\in\mathcal{R}_n \\ x^n\notin\mathcal{T}_{\varepsilon'}(X^n)}} \Pr(X^n = x^n). \tag{8.37}$$

From the first part of the theorem, the last term cannot exceed $\delta_n' := e^{-c\varepsilon'^2 n}$ so that

$$\sum_{x^n\in\mathcal{R}_n} \Pr(X^n = x^n) \leqslant \sum_{x^n\in\mathcal{R}_n\cap\mathcal{T}_{\varepsilon'}(X^n)} \Pr(X^n = x^n) + \delta_n'$$

$$\boxed{x^n \text{ is } \varepsilon'\text{-typical}} \rightarrow \leqslant 2^{-n(H(X)-\varepsilon')}|\mathcal{R}_n| + \delta_n'$$

$$\boxed{|\mathcal{R}_n| \leqslant 2^{nr}} \qquad\quad \leqslant 2^{-n(H(X)-\varepsilon'-r)} + \delta_n' \tag{8.38}$$

$$\boxed{\varepsilon' < \frac{1}{2}(H(X) - r)} \longrightarrow \leqslant 2^{-n\frac{1}{2}(H(X)-r)} + \delta_n'.$$

Since both $\delta'_n$ and $2^{-n\frac{1}{2}(H(X)-r)}$ decrease exponentially fast with zero, there exists $c' > 0$ sufficiently small such that $\Pr(\mathfrak{K}_n) \leqslant e^{-c'n}$. This completes the proof. ∎

It's also pertinent to mention that (8.32) can be expressed equivalently as

$$\frac{1}{n} \log_2 \left( \frac{1}{\Pr(X^n)} \right) \xrightarrow[n\to\infty]{} H(X) \quad \text{in probability.} \tag{8.39}$$

This expression is the exact formulation of the asymptotic equipartition property, which will be examined in greater detail later in the book.

**Exercise 8.5.** *Using the same notations as in Theorem 8.3, show that if instead of Hoeffding's inequality we use the law of large numbers (i.e. Theorem 8.1), then we can still show that for any $\delta > 0$ and sufficiently large $n \in \mathbb{N}$*

$$\Pr\left(\mathfrak{T}_\varepsilon(X^n)\right) > 1 - \delta. \tag{8.40}$$

**Exercise 8.6.** (Variant of Part 3 of Theorem 8.3). *Prove the following variant of part 3 of the theorem above: Let $r < H(X)$ and let $\{\mathfrak{K}_n\}_{n\in\mathbb{N}}$ be sets of sequences of size $n$, and suppose for each $a \in \mathbb{N}$ there exists $n > a$ such that $|\mathfrak{K}_n| \leqslant 2^{nr}$. Then, for any $\delta > 0$ and every $b \in \mathbb{N}$ there exists $n > b$ such that $\Pr(\mathfrak{K}_n) \leqslant \delta$.*

## 8.1.4 Application: Data Compression

A data compression scheme is a process by which a sender (Alice) transmits to a receiver (Bob) a message of size $2^n$ by communicating less than $n$ cbits of communication (See Fig. 8.1). This is possible because Alice draws the message from an i.i.d. information source so that nontypical messages are highly unlikely to occur. Specifically, in a compression scheme of *rate $r$*, Alice encodes (compresses) a message $x^n = (x_1, \ldots, x_n)$ (drawn from an i.i.d. source; particularly, $x_j \in \mathcal{X}$, where $\mathcal{X}$ is the alphabet set of the source) into a bit string of size $y^m = (y_1, \ldots, y_m) \in \{0, 1\}^n$, with $m = \lfloor rn \rfloor$, and transmits the sequence $y^m$ to Bob. Bob then decompresses $y^m$ into a sequence $z^n = (z_1, \ldots, z_n)$ with $z_j \in \mathcal{X}$. The goal is that $z^n$ will be almost identical to $x^n$; see Figure 8.1. We will denote the compression map by $\mathcal{C}$ and the decompression map by $\mathcal{D}$ so that $y^m = \mathcal{C}(x^n)$ and $z^n = \mathcal{D}(y^m)$.

**Definition 8.2.** A compression–decompression scheme, $(\mathcal{C}, \mathcal{D})$ of rate $r$ is said to be *reliable* if

$$\lim_{n\to\infty} \Pr(Z^n = X^n) = \lim_{n\to\infty} \Pr\left(\mathcal{D}\left(\mathcal{C}(X^n)\right) = X^n\right) = 1. \tag{8.41}$$

**Shannon Theorem (1948)**

**Theorem 8.4.** Given an i.i.d. source with entropy $H(X)$, a reliable compression–decompression scheme of rate $r$ exists if and only if $r > H(X)$.

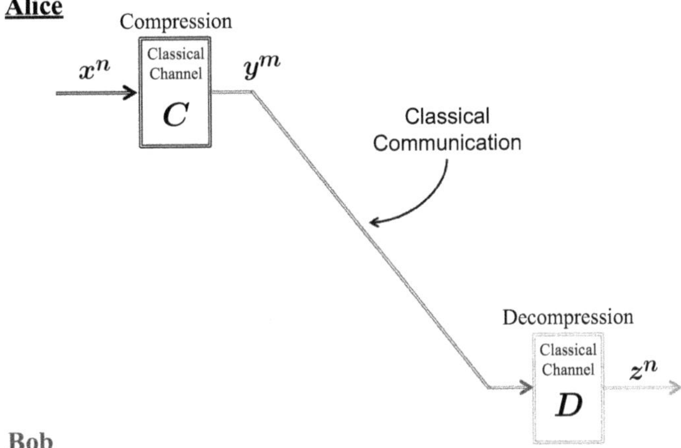


Figure 8.1    A compression–decompression scheme of rate $r$.

**Proof** Suppose $r > H(X)$. We need to show that there exists a reliable compression scheme of rate $r$. Let $\delta > 0$ and let $\varepsilon > 0$ be such that $r > H(X) + \varepsilon$. Then, from the first part of Theorem 8.3 for all $n \in \mathbb{N}$ we have $\Pr\left(\mathfrak{T}_\varepsilon(X^n)\right) \geqslant 1 - e^{-c\varepsilon^2 n}$, where $c > 0$ is a constant defined in (8.28). Let $k \in \left\{1, 2, \ldots, |\mathfrak{T}_\varepsilon(X^n)|\right\}$ be the index labeling all the $\varepsilon$-typical sequences in $\mathfrak{T}_\varepsilon(X^n)$. We assume that Alice and Bob agreed on the order before hand. Define the compression map $C \colon \mathcal{X}^n \to \{0, 1\}^m$, with $m := \lceil \log_2 |\mathfrak{T}_\varepsilon(X^n)| \rceil$ as follows. If $x^n$ is the $k$th sequence of $\mathfrak{T}_\varepsilon(X^n)$, then $C(x^n)$ is the binary representation of $k$. If $x^n \notin \mathfrak{T}_\varepsilon(X^n)$, then $C(x^n) = (0, \ldots, 0)$; that is, if Bob receives the zero sequence he knows there is an error.

Now, from the second part of the theorem of typical sequences we know that

$$\left|\mathfrak{T}_\varepsilon(X^n)\right| \leqslant 2^{n(H(X)+\varepsilon)} < 2^{nr}. \tag{8.42}$$

Therefore, for large enough $n$, the sequence $y^m = C(x^n)$ is of size $m = \lceil \log_2 |\mathfrak{T}_\varepsilon(X^n)| \rceil \leqslant nr$.

The decoding scheme $\mathcal{D} \colon \{0, 1\}^m \to \mathcal{X}^n$ is defined as follows. If $y^m$ is the zero sequence Bob declares an error. Otherwise, if $y^m$ is the binary representation of $k$, then $\mathcal{D}(y^m) = x^n$ with $x^n$ being the $k^{\text{th}}$ sequence of $\mathfrak{T}_\varepsilon(X^n)$. It is left to show that the success probability goes to one in the asymptotic limit $n \to \infty$. Indeed, by construction,

$$\Pr\left(Z^n = x^n \middle| X^n = x^n\right) = \begin{cases} 0 & \text{if } x^n \text{ is not } \varepsilon\text{-typical} \\ 1 & \text{if } x^n \text{ is } \varepsilon\text{-typical} \end{cases}. \tag{8.43}$$

Therefore,

$$\Pr\left(Z^n = X^n\right) = \sum_{x^n \in \mathcal{X}^n} \Pr\left(X^n = x^n\right) \Pr\left(Z^n = x^n \middle| X^n = x^n\right)$$

$$(8.43) \to = \sum_{x^n \in \mathfrak{T}_\varepsilon(X^n)} \Pr\left(X^n = x^n\right)$$

$$= \Pr\left(\mathfrak{T}_\varepsilon(X^n)\right) \geqslant 1 - \delta_n.$$

Since $\lim_{n\to\infty} \delta_n = 0$ we conclude that $\lim_{n\to\infty} \Pr(Z^n = X^n) = 1$. Hence, the compression–decompression scheme above is reliable.

Conversely, suppose there exists compression–decompression scheme of rate $r < H(X)$. Then, there are at most $2^{nr}$ outputs for $\mathcal{D}(y^m)$. Consequently, the set

$$\mathfrak{K}_n := \left\{ x^n : \mathcal{D}\left(\mathcal{C}(x^n)\right) = x^n \right\} \tag{8.44}$$

satisfies $|\mathfrak{K}_n| \leqslant 2^{nr}$ for all $n$. From the third part of Theorem 8.3 we get that $\lim_{n\to\infty} \Pr(\mathfrak{K}_n) = 0$. Hence, a compression–decompression scheme of rate $r < H(X)$ cannot be reliable. This completes the proof. $\blacksquare$

Note that in the proof above we showed that if $r < H(X)$, then not only the scheme is not reliable but also in fact the probability that $Z^n = X^n$ goes to zero; that is, $\lim_{n\to\infty} \Pr(\mathfrak{K}_n) = 0$. In other words, the error probability goes to 1. This type of behavior is known in classical and quantum Shannon theories as the *strong converse*, whereas the *weak converse* corresponds to a proof in which the error probability is shown to be bounded away from zero as $n$ goes to infinity (but not necessarily goes to 1).

## 8.2 Quantum Typicality

In this section, we extend the concept of typicality to the quantum realm. We begin by introducing the definition of an i.i.d. quantum information source.

### 8.2.1 i.i.d. Quantum Source

An i.i.d. quantum source can be considered as an ensemble of states $\left\{q_y, \phi_y\right\}_{y\in[k]}$ with each $\phi_y \in \text{Pure}(A)$, from which states are independently selected. Consequently, after utilizing the source $n$ times, we obtain a sequence of $n$ quantum states:

$$|\phi_{y^n}\rangle := |\phi_{y_1}\rangle|\phi_{y_2}\rangle \cdots |\phi_{y_n}\rangle. \tag{8.45}$$

In contrast to classical sequences $y^n := (y_1, \ldots, y_n)$, the quantum sequence above may not be distinguishable, as the states $\{|\phi_y\rangle\}_{y\in[k]}$ of the source are not orthogonal in general.

Imagine that Alice wishes to transmit the aforementioned state $|\phi_{y^n}\rangle$ to Bob. If Alice is aware of the value of $y^n$, she can employ Shannon's compression coding to send $y^n$ to Bob over a *classical* channel at a rate of $H(Y)$ (meaning the transmission of each source symbol incurs a cost of $H(Y)[c \to c]$). Upon receiving $y^n$, Bob can recreate the state $|\phi_{y^n}\rangle$ (assuming Bob knows the quantum source). However, as we will discuss, if Alice and Bob have access to noiseless quantum channels, not only is it unnecessary for Alice to know $y^n$ but she can also transmit the state $|\phi_{y^n}\rangle$ to Bob more efficiently! Specifically, each source state transmission costs $H(A)_\rho[q \to q]$, where $H(A)_\rho := -\text{Tr}[\rho^A \log \rho^A]$ denotes the von-Neumann entropy of the state

$$\rho^A := \sum_{y\in[k]} q_y |\phi_y\rangle\langle\phi_y|^A. \tag{8.46}$$

Observe that from the upper bound in (7.120) with $\rho_x$ replaced by the pure state $\phi_y$ and $\mathbf{p}$ replaced by $\mathbf{q}$, we get that $H(A)_\rho \leqslant H(\mathbf{q}) = H(Y)$.

Without the classical knowledge of $y$, the quantum source generates the state $\rho$ as mentioned above at each usage. Thus, we will denote by i.i.d.$\sim \rho$, an i.i.d. quantum source drawn from an ensemble of states whose average state, as described in (8.46), is $\rho$. Additionally, after $n$ uses of the source, the produced state is

$$\rho \otimes \rho \otimes \cdots \otimes \rho := \rho^{\otimes n}. \tag{8.47}$$

While we assume Alice lacks access to the classical register $y$ of the source, it's plausible that this value is recorded in some register system $R$. If $R$ is classical, each source use generates the cq-state:

$$\rho^{RA} := \sum_{y\in[k]} q_y |y\rangle\langle y|^R \otimes \phi_y^A. \tag{8.48}$$

Alternatively, if the registrar system $R$ is quantum, then each use of the source produces the state

$$|\psi^{RA}\rangle = \sum_{y\in[k]} \sqrt{q_y} |y\rangle^R |\phi_y\rangle^A. \tag{8.49}$$

Both the classical and quantum register systems record the value of $y$. However, without access to $R$, Alice and Bob cannot distinguish the source $\{q_y, \phi_y\}_{y\in[k]}$ from another source $\{r_z, \psi_z\}_{z\in[\ell]}$, whose average state is $\sum_{z\in[\ell]} r_z \psi_z = \rho$.

Although we assume that Alice and Bob do not have access to system $R$, it is necessary to think about the quantum source with a recording system. Otherwise, without the knowledge of $y^n$, the states $|\phi_{y^n}\rangle$ become equivalent to $\rho^{\otimes n}$ so that Bob, in principle, can prepare any number of copies of $\rho$ without any communication from Alice. However, if other parties have access to $y^n$, then they can verify that the state $\rho^{\otimes n}$ that Bob prepared is not the original state $|\phi_{y^n}\rangle$ that Alice intended to send.

Note that by applying the completely dephasing map $\Delta^R$ on system $R$, the entangled state $|\psi^{RA}\rangle$ in (8.49) becomes the cq-state in (8.48). This demonstrates that taking the registrar to be quantum is more general and we therefore adopt the entangled description $|\psi^{RA}\rangle$ of a quantum source. Note that in this picture, after $n$ uses of the source, Alice shares with the registrar the state

$$|\psi^{R^n A^n}\rangle := |\psi^{RA}\rangle^{\otimes n}. \tag{8.50}$$

In the quantum version of the compression scheme discussed above, the task of Alice is to transfer her system $A^n$ to Bob using the smallest possible number of noiseless qubit channels $[q \to q]$. We postpone the full details of this task to volume 2 of this book where we study quantum Shannon theory in more detail.

## 8.2.2 Typical Subspaces

Consider an i.i.d. quantum source generated from the ensemble $\{q_y, \phi_y^A\}_{y\in[k]}$. Let

$$\rho^A := \sum_{y\in[k]} q_y |\phi_y\rangle\langle\phi_y|^A = \sum_{x\in[m]} p_x |x\rangle\langle x|^A, \tag{8.51}$$

where $\{p_x\}$ are the eigenvalues of $\rho$ and $\{|x\rangle\}$ are the corresponding eigenvectors. We define the classical system (random variable) $X$ to have alphabet symbols $x \in [m]$ corresponding to a probability distribution $\{p_x\}$. We point out that the alphabet symbols of the system $Y$ that we discussed above corresponds to a different probability distribution $\{q_y\}$.

Now, observe that the state

$$\begin{aligned}
\rho^{\otimes n} &= \sum_{x_1\in[m]} \sum_{x_2\in[m]} \cdots \sum_{x_n\in[m]} p_{x_1} p_{x_2} \cdots p_{x_n} |x_1 \cdots x_n\rangle\langle x_1 \cdots x_n|^{A^n} \\
&:= \sum_{x^n\in[m]^n} p_{x^n} |x^n\rangle\langle x^n|^{A^n},
\end{aligned} \tag{8.52}$$

where $|x^n\rangle := |x_1 \cdots x_n\rangle$ and $p_{x^n} := p_{x_1} p_{x_2} \cdots p_{x_n}$. We use the notation $\mathcal{T}_\varepsilon(X^n)$ to denote the set of all $\varepsilon$-typical sequences $x^n$ with respect to a *classical* system $X$ corresponding to an i.i.d.$\sim$ **p** source. It is important to note that the components of the vector $\mathbf{p} \in \mathrm{Prob}(m)$ are the eigenvalues of $\rho$. With this notation, for every such i.i.d.$\sim \rho^A$ source, we define a corresponding typical subspace

$$\mathcal{T}_\varepsilon(A^n) := \mathrm{span}\left\{ |x_1 \cdots x_n\rangle : \ x^n \in \mathcal{T}_\varepsilon(X^n)\right\} \subseteq A^n, \tag{8.53}$$

and a typical projection

$$\Pi_\varepsilon^n := \sum_{x^n\in\mathcal{T}_\varepsilon(X^n)} |x^n\rangle\langle x^n|. \tag{8.54}$$

---

**Theorem 8.5.** Let $\rho \in \mathfrak{D}(A)$, $\varepsilon \in (0, 1)$, $\delta_n := e^{-c\varepsilon^2 n}$, where $c$ is defined in (8.28), $\mathcal{T}_\varepsilon(A^n)$ and $\Pi_\varepsilon^n$ be the typical subspaces and projections associated with a quantum i.i.d.$\sim \rho$ source. Furthermore, for each $n \in \mathbb{N}$, let $P_n \in \mathrm{Pos}(A^n)$ be an orthogonal projection to a subspace with dimension $\mathrm{Tr}\,[P_n] \leqslant 2^{nr}$ for some $r < H(A)_\rho$ ($r$ is independent on $n$). Then, for all $n \in \mathbb{N}$ the following three inequalities hold:

$$\begin{aligned}
&1. &&\mathrm{Tr}\left[\Pi_\varepsilon^n \rho^{\otimes n}\right] \geqslant 1 - \delta_n. &&(8.55)\\
&2. &&(1 - \delta_n)2^{n(H(A)_\rho-\varepsilon)} \leqslant \mathrm{Tr}\left[\Pi_\varepsilon^n\right] \leqslant 2^{n(H(A)_\rho+\varepsilon)}. &&(8.56)\\
&3. &&\mathrm{Tr}\left[P_n \rho^{\otimes n}\right] \leqslant e^{-c'n} \text{ for some } c' > 0. &&(8.57)
\end{aligned}$$

**Proof**  The proofs of the first two parts of the theorem follow from their classical counterparts. Particularly, for the first part

$$\mathrm{Tr}\left[\Pi_\varepsilon^n \rho^{\otimes n}\right] = \sum_{x^n \in \mathfrak{T}_\varepsilon(X^n)} p_{x^n} \tag{8.58}$$

$$= \Pr\left(\mathfrak{T}_\varepsilon(X^n)\right) \geqslant 1 - \delta_n.$$

For the second part

$$\mathrm{Tr}\left[\Pi_\varepsilon^n\right] = \dim\left(\mathfrak{T}_\varepsilon(A^n)\right) = \left|\mathfrak{T}_\varepsilon(X^n)\right|. \tag{8.59}$$

Therefore, this part follows as well from its classical counterpart. It is therefore left to prove the third part.

We first split the trace into two parts:

$$\mathrm{Tr}\left[P_n \rho^{\otimes n}\right] = \mathrm{Tr}\left[P_n \rho^{\otimes n} \Pi_\varepsilon^n\right] + \mathrm{Tr}\left[P_n \rho^{\otimes n}(I - \Pi_\varepsilon^n)\right]. \tag{8.60}$$

Our objective is to show that both of these terms are going to zero as $n \to \infty$. For the first one

$$\mathrm{Tr}\left[P_n \rho^{\otimes n} \Pi_\varepsilon^n\right] = \sum_{x^n \in \mathfrak{T}_\varepsilon(X^n)} p_{x^n} \left\langle x^n \left| P_n \right| x^n \right\rangle$$

$$\boxed{p_{x^n} \leqslant 2^{-n(H(A)-\varepsilon)}} \longrightarrow \quad \leqslant 2^{-n(H(A)-\varepsilon)} \sum_{x^n \in \mathfrak{T}_\varepsilon(X^n)} \left\langle x^n \left| P_n \right| x^n \right\rangle \tag{8.61}$$

$$\leqslant 2^{-n(H(A)-\varepsilon)} \mathrm{Tr}\left[P_n\right]$$

$$\boxed{\mathrm{Tr}\left[P_n\right] \leqslant 2^{nr}} \longrightarrow \quad \leqslant 2^{-n(H(A)-\varepsilon-r)} \xrightarrow[n\to\infty]{} 0,$$

where we assume that $\varepsilon > 0$ is small enough so that $H(A) - r > \varepsilon$. For the second term,

$$\mathrm{Tr}\left[P_n \rho^{\otimes n}(I - \Pi_\varepsilon^n)\right] = \sum_{x^n \notin \mathfrak{T}_\varepsilon(X^n)} p_{x^n} \left\langle x^n \left| P_n \right| x^n \right\rangle \leqslant \sum_{x^n \notin \mathfrak{T}_\varepsilon(X^n)} p_{x^n} \xrightarrow[n\to\infty]{} 0. \tag{8.62}$$

This completes the proof. $\blacksquare$

## 8.3 Relative Typicality

As already discussed, in the realm of information theory and quantum information, we often encounter scenarios where large sequences or states, denoted as $x^n$ or $\rho^{\otimes n}$, respectively, are drawn from i.i.d. sources. We saw that for sufficiently large values of $n$, these sequences or states are highly likely to be typical, adhering to a certain expected pattern. Building on this foundation, we now extend this concept of typicality to pairs of sequences and pairs of quantum states. This extension is analogous to how majorization, a mathematical concept used to compare probability vectors based on their spread or dispersion, is extended to relative majorization. In the case of relative majorization, we transitioned from comparing individual probability vectors to comparing pairs of vectors. Similarly, our extension involves analyzing pairs of sequences and quantum states, assessing their typicality not just individually but also in relation to each other.

## 8.3.1 The Classical Domain

We start with the following definition of a relative $\varepsilon$-typical sequence.

---

**Definition 8.3.** Let $n, m \in \mathbb{N}$, $\mathbf{p}, \mathbf{q} \in \mathrm{Prob}(m)$, and $\varepsilon \in (0, 1)$. A sequence $x^n \in [m]^n$ is said to be relative $\varepsilon$-typical with respect to an i.i.d.$\sim \mathbf{p}$ source, and a probability vector $\mathbf{q} \in \mathrm{Prob}(m)$ if

$$\left| D(\mathbf{p}\|\mathbf{q}) - \frac{1}{n} \log \frac{p_{x^n}}{q_{x^n}} \right| \leqslant \varepsilon. \tag{8.63}$$

The set of all relative $\varepsilon$-typical sequences is denoted by $\mathfrak{T}_{\varepsilon}^{\mathrm{rel}}(X^n)$.

---

**Exercise 8.7.** *Show that if* $x^n \in \mathfrak{T}_{\varepsilon}^{\mathrm{rel}}(X^n)$, *then*

$$2^{-n(D(\mathbf{p}\|\mathbf{q})+\varepsilon)} p_{x^n} \leqslant q_{x^n} \leqslant 2^{-n(D(\mathbf{p}\|\mathbf{q})-\varepsilon)} p_{x^n}. \tag{8.64}$$

We saw in Section 8.1.1 that given an i.i.d.$\sim \mathbf{p}$ source, all typical sequences with large size $n$, that are generated by the source, have approximately the same probability to occur given by $\approx 2^{-nH(\mathbf{p})}$. This phenomenon was dubbed as the asymptotic equipartition property (AEP). Here we study a variant of this property as described in the following theorem.

---

**Theorem 8.6.** Let $\mathbf{p}, \mathbf{q} \in \mathrm{Prob}(m)$ with $\mathrm{supp}(\mathbf{p}) \subseteq \mathrm{supp}(\mathbf{q})$, $\varepsilon \in (0, 1)$, $\delta_n := e^{-c\varepsilon^2 n}$, where $c$ is defined in (8.28), $X$ be a random variable associated with an i.i.d.$\sim \mathbf{p}$ source, and for each $n \in \mathbb{N}$, let $\mathfrak{K}_n \subseteq [m]^n$ be a set of sequences with cardinality $|\mathfrak{K}_n| \leqslant 2^{nr}$ for some $r < D(\mathbf{p}\|\mathbf{q})$. Then, for all $n \in \mathbb{N}$ the following three inequalities hold:

1. $\mathrm{Pr}(\mathfrak{T}_{\varepsilon}^{\mathrm{rel}}(X^n)) > 1 - \delta_n$.
2. $(1 - \delta_n)2^{n(D(\mathbf{p}\|\mathbf{q})-\varepsilon)} \leqslant \left| \mathfrak{T}_{\varepsilon}^{\mathrm{rel}}(X^n) \right| \leqslant 2^{n(D(\mathbf{p}\|\mathbf{q})+\varepsilon)}$.
3. $\mathrm{Pr}(\mathfrak{K}_n) \leqslant e^{-c'n}$, for some $c' > 0$.

---

*Remark.* Roughly speaking, the above theorem indicates that almost all sequences $x^n \in [m]^n$ have the same ratio $\frac{p_{x^n}}{q_{x^n}} \approx 2^{-nD(\mathbf{p}\|\mathbf{q})}$. Observe also that although we consider two probability distributions $\mathbf{p}$ and $\mathbf{q}$, the sequences $X_1, X_2, \ldots$ are drawn from a *single* $\mathbf{p}$-source.

**Proof**  Due to the similarity of this theorem to Theorem 8.3, we provide here only the proof of the first inequality, leaving the remaining two inequalities as an exercise for the reader. Without loss of generality suppose $\mathbf{q} > 0$ and let $Y := \log \frac{p_X}{q_X}$

be the random variable with alphabet $\{\log(p_x/q_x)\}_{x\in[m]}$ and corresponding probability $\mathbf{p} = (p_1, \ldots, p_m)^T$. Consider the sequence $X^n = (X_1, X_2, \ldots, X_n)$ drawn from an i.i.d. $\sim \mathbf{p}$ source, and for each $j \in [n]$ let $Y_j = \log \frac{p_{X_j}}{q_{X_j}}$. Observe that for each $j \in \mathbb{N}$

$$\mathbb{E}(Y_j) = \mathbb{E}(Y) = \sum_{x\in[m]} p_x \log(p_x/q_x) = D(\mathbf{p}\|\mathbf{q}). \tag{8.65}$$

Therefore, since

$$\begin{aligned}
\frac{1}{n} \log \frac{p_{X^n}}{q_{X^n}} &= \frac{1}{n} \log \frac{p_{X_1} \cdots p_{X_n}}{q_{X_1} \cdots q_{X_n}} \\
&= \frac{1}{n} \sum_{j\in[n]} \log \frac{p_{X_j}}{q_{X_j}}
\end{aligned} \tag{8.66}$$

we conclude that

$$\begin{aligned}
\Pr\left(\mathfrak{T}_\varepsilon^{\mathrm{rel}}(X^n)\right) &= \Pr\left(\left|\frac{1}{n} \log \frac{p_{X^n}}{q_{X^n}} - D(\mathbf{p}\|\mathbf{q})\right| < \varepsilon\right) \\
&= \Pr\left(\left|\frac{1}{n} \sum_{j\in[n]} Y_j - \mathbb{E}(Y)\right| < \varepsilon\right)
\end{aligned} \tag{8.67}$$

**Hoeffding's Inequality** $\rightarrow\ > 1 - e^{-nc\varepsilon^2}.$

This completes the proof of the first inequality. ∎

**Exercise 8.8.** *Prove the outstanding inequalities of the aforementioned theorem. Hint: Utilize a method similar to that applied in deriving the analogous inequalities in Theorem 8.3.*

The theorem above demonstrates that the probability that a sequence is relative $\varepsilon$-typical is very high. Note however that the probability $\Pr\left(\mathfrak{T}_\varepsilon^{\mathrm{rel}}(X^n)\right)$ is computed with respect to an i.i.d. $\sim \mathbf{p}$ source. If on the other hand we change $\mathbf{p}$ with $\mathbf{q}$, we would get the probability

$$\Pr\left(\mathfrak{T}_\varepsilon^{\mathrm{rel}}(X^n)\right)_\mathbf{q} := \sum_{x^n \in \mathfrak{T}_\varepsilon^{\mathrm{rel}}(X^n)} q_{x^n}. \tag{8.68}$$

In the following exercise you show that this probability goes to zero exponentially fast with $n$.

**Exercise 8.9.** *Let $\mathbf{p}, \mathbf{q} \in \mathrm{Prob}(m)$ with $\mathrm{supp}(\mathbf{p}) \subseteq \mathrm{supp}(\mathbf{q})$, and $\varepsilon \in (0, 1)$. Show that for all $n \in \mathbb{N}$*

$$(1 - \delta_n) 2^{-n(D(\mathbf{p}\|\mathbf{q})+\varepsilon)} \leqslant \Pr\left(\mathfrak{T}_\varepsilon^{\mathrm{rel}}(X^n)\right)_\mathbf{q} \leqslant 2^{-n(D(\mathbf{p}\|\mathbf{q})-\varepsilon)}. \tag{8.69}$$

*Hint: Take the sum over all $x^n \in \mathfrak{T}_\varepsilon(X^n)$ in all sides of (8.64), and use the inequalities $1 \geqslant \Pr\left(\mathfrak{T}_\varepsilon^{\mathrm{rel}}(X^n)\right) > 1 - \delta_n$.*

The pair of probability vectors $(\mathbf{p}^{\otimes n}, \mathbf{q}^{\otimes n})$ becomes more distinguishable as we increase $n$. In the following corollary, we use the theorem above to characterize this distinguishability with relative majorization.

> **Corollary 8.1.** Let $m \in \mathbb{N}$, $\mathbf{s}, \mathbf{t} \in \mathrm{Prob}_{>0}(2)$, and $\mathbf{p}, \mathbf{q} \in \mathrm{Prob}(m)$ with $\mathrm{supp}\,(\mathbf{p}) \subseteq \mathrm{supp}\,(\mathbf{q})$ and $\mathbf{p} \neq \mathbf{q}$. Then, for large enough $n \in \mathbb{N}$,
>
> $$(\mathbf{p}^{\otimes n}, \mathbf{q}^{\otimes n}) \succ (\mathbf{s}, \mathbf{t}). \tag{8.70}$$

**Proof**  Let $\varepsilon > 0$ be a small number and define the stochastic evolution matrix $E \in \mathrm{STOC}(2, 2^n)$ by its action on the standard basis $\{\mathbf{e}_{x^n} := \mathbf{e}_{x_1} \otimes \cdots \otimes \mathbf{e}_{x_n}\}_{x^n \in [m]^n}$ as

$$E\mathbf{e}_{x^n} := \begin{cases} \mathbf{e}_1^{(2)} & \text{if } x^n \in \mathfrak{T}_\varepsilon^{\mathrm{rel}}(X^n) \\ \mathbf{e}_2^{(2)} & \text{if } x^n \notin \mathfrak{T}_\varepsilon^{\mathrm{rel}}(X^n), \end{cases} \tag{8.71}$$

where $\mathbf{e}_1^{(2)} := (1,0)^T$ and $\mathbf{e}_2^{(2)} = (0,1)^T$ form the standard basis of $\mathbb{R}^2$. We then get

$$\left(\mathbf{p}^{\otimes n}, \mathbf{q}^{\otimes n}\right) \succ \left(E\mathbf{p}^{\otimes n}, E\mathbf{q}^{\otimes n}\right)$$
$$= \left(\begin{bmatrix} \Pr\left(\mathfrak{T}_\varepsilon^{\mathrm{rel}}(X^n)\right) \\ 1 - \Pr\left(\mathfrak{T}_\varepsilon^{\mathrm{rel}}(X^n)\right) \end{bmatrix}, \begin{bmatrix} \Pr\left(\mathfrak{T}_\varepsilon^{\mathrm{rel}}(X^n)\right)_{\mathbf{q}} \\ 1 - \Pr\left(\mathfrak{T}_\varepsilon^{\mathrm{rel}}(X^n)\right)_{\mathbf{q}} \end{bmatrix}\right). \tag{8.72}$$

Observe that due to the bounds $\Pr\left(\mathfrak{T}_\varepsilon^{\mathrm{rel}}(X^n)\right) \geqslant 1 - \delta_n$ and the bound $\Pr\left(\mathfrak{T}_\varepsilon^{\mathrm{rel}}(X^n)\right) \leqslant 2^{-n(D(\mathbf{p}\|\mathbf{q})-\varepsilon)}$ (see (8.69)) the pair of vectors on the right-hand side approaches the pair $(\mathbf{e}_1, \mathbf{e}_2)$ as $n \to \infty$, where $\{\mathbf{e}_1, \mathbf{e}_2\}$ is the standard basis of $\mathbb{R}^2$. Therefore, combining this with Exercise 4.47 we conclude that for any $\mathbf{s}, \mathbf{t} \in \mathrm{Prob}_{>0}(2)$, and sufficiently large $n \in \mathbb{N}$,

$$(\mathbf{p}^{\otimes n}, \mathbf{q}^{\otimes n}) \succ (\mathbf{s}, \mathbf{t}). \tag{8.73}$$

This concludes the proof. ∎

## 8.3.2 Relative Typical Subspaces

In this subsection, we aim to expand the concept of relative typical sequences into the quantum domain. However, we soon encounter a significant challenge: Unlike classical states, quantum states generally do not commute. This noncommutativity makes a straightforward extension from the classical framework, as shown in (8.63), nontrivial. Consequently, the definition of a relative typical subspace in the quantum context will significantly diverge from its classical counterpart.

Let $\rho, \sigma \in \mathfrak{D}(A)$ be two density matrices (which we can be viewed here as two i.i.d. quantum sources). Suppose also that $\mathrm{supp}\,(\rho) \subseteq \mathrm{supp}\,(\sigma)$. Let the spectral decomposition of $\rho$ and $\sigma$ be

$$\rho = \sum_{x \in [m]} p_x |\psi_x\rangle\langle\psi_x| \quad \text{and} \quad \sigma = \sum_{y \in [m]} q_y |\phi_y\rangle\langle\phi_y|. \tag{8.74}$$

Similar to the notations in the previous section, for any integer $n$ we will denote by $y^n := (y_1, \ldots, y_n)$, $q_{y^n} = q_{y_1} \cdots q_{y_n}$, and $|\phi_{y^n}\rangle := |\phi_{y_1}\rangle \otimes \cdots \otimes |\phi_{y_n}\rangle$. Then, for any $\varepsilon > 0$ and $n \in \mathbb{N}$ the *relative typical subspace*, $\mathfrak{T}_\varepsilon^{\mathrm{rel}}(A^n)$, is defined as

$$\mathfrak{T}_\varepsilon^{\mathrm{rel}}(A^n) := \mathrm{span}\left\{ |\phi_{y^n}\rangle \ : \ \left| \mathrm{Tr}\left[\rho \log \sigma\right] - \frac{1}{n} \log\left(q_{y^n}\right) \right| \leqslant \varepsilon \right\}. \tag{8.75}$$

We also denote by $\Pi_\varepsilon^{\mathrm{rel},n}$ the projection to the relative typical subspace $\mathfrak{T}_\varepsilon^{\mathrm{rel}}(A^n)$. Note that $\mathrm{Tr}\left[\rho \log \sigma\right]$ is well defined since $\mathrm{supp}\,(\rho) \subseteq \mathrm{supp}\,(\sigma)$.

The definition provided in Equation (8.75) clearly does not revert to the classical definition of a relative typical sequence when $\rho$ and $\sigma$ commute. Nevertheless, as we will explore in Theorem 8.7 and the subsequent sections, this definition proves to be an effective tool for examining the distinguishability of quantum states. Furthermore, as demonstrated in the upcoming exercise, relative typical subspaces do indeed converge to typical subspaces in the case where $\rho = \sigma$.

**Exercise 8.10.** *Show that if $\rho = \sigma$, then the relative typical subspace reduces to the typical subspace of $\rho$ as defined in the previous section. That is, show that in this case* $\mathfrak{T}_\varepsilon^{\mathrm{rel}}(A^n) = \mathfrak{T}_\varepsilon(A^n)$.

---

**Theorem 8.7.** Let $\varepsilon > 0$, $\rho, \sigma \in \mathfrak{D}(A)$ with $\mathrm{supp}\,(\rho) \subseteq \mathrm{supp}\,(\sigma)$, and $c > 0$ as defined in (8.82). Then, for all $n \in \mathbb{N}$

$$\mathrm{Tr}\left[\Pi_\varepsilon^{\mathrm{rel},n} \rho^{\otimes n}\right] \geqslant 1 - e^{-nc\varepsilon^2}. \tag{8.76}$$

---

**Proof** Note that the function $\mathrm{Tr}[\rho \log \sigma]$ can be expressed as

$$\mathrm{Tr}[\rho \log \sigma] = \sum_{y \in [m]} \langle \phi_y | \rho \log \sigma | \phi_y \rangle = \sum_{y \in [m]} \langle \phi_y | \rho | \phi_y \rangle \log q_y. \tag{8.77}$$

Therefore, denoting the relative distribution $r_y := \langle \phi_y | \rho | \phi_y \rangle$, and by $Y$ the random variable whose alphabet is $[m]$, and its corresponding distribution is $\{r_y\}_{y \in [m]}$, we get that

$$\mathrm{Tr}[\rho \log \sigma] = \mathbb{E}\left(\log q_Y\right). \tag{8.78}$$

Furthermore, we denote by $\mathcal{C}_\varepsilon^n$ the classical typical set

$$\mathcal{C}_\varepsilon^n := \left\{ y^n \ : \ \left| \mathrm{Tr}\left[\rho \log \sigma\right] - \frac{1}{n} \log\left(q_{y^n}\right) \right| \leqslant \varepsilon \right\}. \tag{8.79}$$

Then, by definition,

$$\begin{aligned} \mathrm{Tr}\left[\Pi_\varepsilon^{\mathrm{rel},n} \rho^{\otimes n}\right] &= \sum_{y^n \in \mathcal{C}_\varepsilon^n} \langle \phi_{y^n} | \rho^{\otimes n} | \phi_{y^n} \rangle \\ &= \sum_{y^n \in \mathcal{C}_\varepsilon^n} r_{y^n}, \end{aligned} \tag{8.80}$$

where the last term is the probability that a sequence $Y^n$ belongs to $\mathcal{C}^n_\varepsilon$. Therefore,

$$\mathrm{Tr}\left[\Pi^{\mathrm{rel},n}_\varepsilon \rho^{\otimes n}\right] = \mathrm{Pr}\left\{\mathcal{C}^n_\varepsilon\right\}$$

$$= \mathrm{Pr}\left\{\left|\mathbb{E}\left(\log q_Y\right) - \frac{1}{n}\sum_{i\in[n]}\log q_{Y_i}\right| \leqslant \varepsilon\right\} \tag{8.81}$$

**Hoeffding's Inequality** $\rightarrow$ $\geqslant 1 - e^{-cn\varepsilon^2}$.

This completes the proof. ∎

**Exercise 8.11.** *Using the same notations as above, show that:*

1. *The constant $c$ can be taken to be (cf. (8.28))*

$$c = \frac{2}{\left(\log\left(q_{\max}/q_{\min}\right)\right)^2}, \tag{8.82}$$

*where $q_{\min}$ is the smallest nonzero eigenvalue of $\sigma$, and $q_{\max}$ is the largest eigenvalue of $\sigma$.*

2. *The relative typical projector $\Pi^{\mathrm{rel},n}_\varepsilon$ satisfies*

$$2^{n(\mathrm{Tr}[\rho\log\sigma]-\varepsilon)}\Pi^{\mathrm{rel},n}_\varepsilon \leqslant \Pi^{\mathrm{rel},n}_\varepsilon \sigma^{\otimes n}\Pi^{\mathrm{rel},n}_\varepsilon \leqslant 2^{n(\mathrm{Tr}[\rho\log\sigma]+\varepsilon)}\Pi^{\mathrm{rel},n}_\varepsilon. \tag{8.83}$$

## 8.4 The Method of Types

The method of types is a fundamental concept in information theory that provides a powerful framework for analyzing the statistical properties of sequences of symbols. Originating from the work of Claude Shannon, this method revolves around the idea of categorizing sequences into types based on the frequency of each symbol's occurrence. By treating sequences with similar compositions as a single type, the method simplifies the analysis of large sets of data. This approach is particularly effective in understanding the behavior of random processes, quantifying the efficiency of coding schemes, and in the study of large deviations and typicality. The method of types has become a cornerstone in both classical and quantum information theory, underpinning many key theorems and applications in areas such as data compression, communication theory, and statistical inference.

---

**Type of a Sequence**

**Definition 8.4.** Let $n, m \in \mathbb{N}$. For every $x^n := (x_1, \ldots, x_n) \in [m]^n$ and $z \in [m]$, let $N(z|x^n)$ be the number of elements in the sequence $x^n$ that are equal to $z$. The *type* of the sequence $x^n$ is a probability vector in $\mathrm{Prob}(m)$ given by

$$\mathbf{t}(x^n) := \left(t_1(x^n), \ldots, t_m(x^n)\right)^T, \quad \text{where} \quad t_z(x^n) := \frac{1}{n}N(z|x^n) \quad \forall z \in [m]. \tag{8.84}$$

For example, for $m = 3$, the type of the sequence $x^6 = (2, 1, 1, 3, 2, 2)$ is the probability vector $t(x^6) = (1/3, 1/2, 1/6)$.

The significance of types comes into play when considering an i.i.d$\sim$ $\mathbf{p}$ source. In this case, the probability of a sequence $x^n \in [m]^n$ drawn from the source is given by

$$
\begin{aligned}
p_{x^n} := p_{x_1} \cdots p_{x_n} &= p_1^{N(1|x^n)} \cdots p_m^{N(m|x^n)} \\
&\quad \xrightarrow{\forall r > 0\, r = 2^{\log r}} = 2^{\sum_{z \in [m]} N(z|x^n) \log_2 p_z} \\
&\quad \xrightarrow{N(z|x^n) = nt_z(x^n)} = 2^{n \sum_{z \in [m]} t_z(x^n) \log_2 p_z} \\
&= 2^{-n\left(H(t(x^n)) + D(t(x^n)\|\mathbf{p})\right)},
\end{aligned}
$$
(8.85)

where $H(t(x^n))$ is the Shannon entropy of the type of the sequence $x^n$, and $D\left(t(x^n)\|\mathbf{p}\right)$ is the KL-divergence between $t(x^n)$ and $\mathbf{p}$ (see (5.24)). The above formula manifests that the probability distributions of sequences drawn from an i.i.d. source only depend on the type of the sequence. As we will see next, this property can lead to a significant simplification in some applications.

We denote by $\mathrm{Type}(n, m) \subseteq \mathrm{Prob}(m)$ the set of all types of sequences in $[m]^n$. For example, for sequences of bits (i.e. $m = 2$),

$$
\mathrm{Type}(n, 2) = \left\{ (1, 0)^T, \left(\frac{n-1}{n}, \frac{1}{n}\right)^T, \left(\frac{n-2}{n}, \frac{2}{n}\right)^T, \ldots, (0, 1)^T \right\}.
$$
(8.86)

Note that any type $t \in \mathrm{Type}(n, m)$ has $m$ components of the form $\frac{k}{n}$, where $k \in \{0, \ldots, n\}$. Therefore, the number of types in $\mathrm{Type}(n, m)$ cannot exceed $(n + 1)^m$, which is polynomial in $n$. The exact number of types can be computed using the "stars and bars" method in combinatorics. It is given by

$$
|\mathrm{Type}(n, m)| = \binom{n + m - 1}{n} \leqslant (n + 1)^m.
$$
(8.87)

On the other hand, the number all sequences of size $n$ is $m^n$, which is exponential in $n$.

The set of all sequences $x^n$ of a given type $t = (t_1, \ldots, t_m)$ will be denoted as $X^n(t)$. We emphasize that $X^n(t)$ denotes a *set* of all sequences in $[m]^n$ whose type is $t$, whereas $t(x^n)$ denotes a *single* probability vector (i.e. the type of a specific sequence $x^n$). The number of sequences in the set $X^n(t)$ is given by the combinatorial formula of arranging $nt_1, \ldots, nt_m$ objects in a sequence:

$$
|X^n(t)| = \binom{n}{nt_1, \ldots, nt_m} := \frac{n!}{\prod_{x=1}^{m} (nt_x)!}.
$$
(8.88)

The above formula is somewhat cumbersome, but by using Stirling's approximation we can find simpler lower and upper bounds.

> **Lemma 8.2.** Let $t \in \mathrm{Type}(n, m)$. Then,
>
> $$\frac{1}{(n+1)^m} 2^{nH(t)} \leqslant |X^n(t)| \leqslant 2^{nH(t)}. \tag{8.89}$$

**Proof**  Let $x^n$ be a sequence of size $n$ drawn from an i.i.d. source according to the distribution $t$. Then,

$$
\begin{aligned}
1 = \sum_{x^n \in [m]^n} t_{x^n} &\geqslant \sum_{x^n \in X^n(t)} t_{x^n} \\
\xrightarrow{\;(8.85)\text{ with } p = t\;} &= \sum_{x^n \in X^n(t)} 2^{-nH(t)} \\
&= |X^n(t)| 2^{-nH(t)}.
\end{aligned}
\tag{8.90}
$$

This proves that $|X^n(t)| \leqslant 2^{nH(t)}$. For the other inequality, we make use of the Stirling's bounds

$$\sqrt{2\pi}\, n^{n+\frac{1}{2}} e^{-n} \leqslant n! \leqslant e n^{n+\frac{1}{2}} e^{-n}. \tag{8.91}$$

By using the lower bound for $n!$ and the upper bound for each $(nt_x)!$ of (8.88) we get that

$$
\begin{aligned}
|X^n(t)| = \frac{n!}{\prod_{x=1}^m (nt_x)!} &\geqslant \frac{\sqrt{2\pi}\, n^{n+\frac{1}{2}} e^{-n}}{\prod_{x=1}^m e(nt_x)^{nt_x+\frac{1}{2}} e^{-nt_x}} \\
&= \frac{\sqrt{2\pi}\, n^{\frac{1}{2}}}{e^m n^{\frac{m}{2}} \prod_{x=1}^m t_x^{nt_x+\frac{1}{2}}} \\
&= \frac{\sqrt{2\pi}\,\sqrt{n}}{(e\sqrt{n})^m \sqrt{t_1 \cdots t_m}} 2^{-nH(t)}.
\end{aligned}
\tag{8.92}
$$

It is left as an exercise (see Exercise 8.12) to show that

$$\frac{\sqrt{2\pi}\,\sqrt{n}}{(e\sqrt{n})^m \sqrt{t_1 \cdots t_m}} \geqslant \frac{1}{(n+1)^m} \tag{8.93}$$

for all $n$, $m$, and $t_1, \ldots, t_m$.  ∎

**Exercise 8.12.** *Prove the inequality in (8.93). Hint: Use the fact that the product $t_1 \cdots t_n$ is Schur concave and achieves its maximum when $t_1 = t_2 = \cdots = t_m = \frac{1}{m}$.*

**Exercise 8.13.** *Let $\mathfrak{K} \subset \mathrm{Type}(n, m)$ be a set of probability distributions (that are types), and define $\mathfrak{C}_n := \{x^n \in [m]^n : t(x^n) \in \mathfrak{K}\}$. Fix $q \in \mathfrak{K}$. Show that*

$$\Pr(\mathfrak{C}_n)_q := \sum_{x^n \in \mathfrak{C}_n} q_{x^n} \tag{8.94}$$

*approaches 1 in the limit $n \to \infty$. Hint: Denote by $\mathfrak{C}_n^c$ the complement of $\mathfrak{C}_n$ in $[m]^n$ and show that $\Pr(\mathfrak{C}_n^c)_q$ approaches zero in the limit $n \to \infty$. Use (8.85), (8.89), and the fact that $D(t\|q) > 0$ for any type $t \neq q$.*

**Exercise 8.14.** *Let $n \in \mathbb{N}$. Use the strong Stirling's approximation, which states that*

$$\sqrt{2\pi n}\left(\frac{n}{e}\right)^n \leqslant n! \leqslant \sqrt{2\pi n}\left(\frac{n}{e}\right)^n e^{\frac{1}{12n}} \tag{8.95}$$

*to show that:*

1. *For any $p \in (0,1)$ such that $np \in \mathbb{N}$, we have*

$$\binom{n}{np} \leqslant \frac{2^{nh(p)}}{\sqrt{\pi np(1-p)}}, \tag{8.96}$$

    *where $h(p) = -p\log p - (1-p)\log(1-p)$ is the binary Shannon entropy.*
2. *For any integer $k \leqslant n/2$,*

$$\binom{n}{k} \leqslant 2^{nh\left(\frac{k}{n}\right)}. \tag{8.97}$$

## 8.4.1 Many Copies of a Quantum State

In this subsection, we show that the method of types can be used to simplify the expression of $n$ copies of some quantum state $\sigma \in \mathcal{D}(A)$. Let $m$ be the number of distinct eigenvalues of $\sigma$, so that

$$\sigma = \sum_{x \in [m]} q_x P_x, \tag{8.98}$$

where $\operatorname{spec}(\sigma) = \{q_1, \ldots, q_m\}$, and $\{P_x\}_{x \subset [m]}$ are orthogonal projectors. For $n$ copies of sigma

$$\sigma^{\otimes n} = \sum_{x^n \in [m]^n} q_{x^n} P_{x^n}, \tag{8.99}$$

where $q_{x^n} := q_{x_1} \cdots q_{x_n}$ and $P_{x^n} = P_{x_1} \otimes \cdots \otimes P_{x_n}$. From (8.85) the probability $q_{x^n} = 2^{-n\left(H(\mathsf{t}(x^n)) + D(\mathsf{t}(x^n)\|\mathsf{q})\right)}$ depends only on the type of $x^n$. Therefore, we can express $\sigma^{\otimes n}$ as

$$\sigma^{\otimes n} = \sum_{\mathsf{t} \in \mathrm{Type}(n,m)} 2^{-n\left(H(\mathsf{t}) + D(\mathsf{t}\|\mathsf{q})\right)} P_{\mathsf{t}}, \tag{8.100}$$

where for any $\mathsf{t} \in \mathrm{Type}(n,m)$,

$$P_{\mathsf{t}} := \sum_{x^n \in X^n(\mathsf{t})} P_{x^n}. \tag{8.101}$$

Note that the set $\{P_{\mathsf{t}}\}_{\mathsf{t} \in \mathrm{Type}(n,m)}$ is itself a set of orthogonal projectors. The significance of the formula above is that the number of terms in the sum is given by

$$\left|\operatorname{spec}\left(\sigma^{\otimes n}\right)\right| = |\mathrm{Type}(n,m)| \leqslant (n+1)^m, \tag{8.102}$$

which is polynomial in $n$. Therefore, the original sum in (8.99) that consists of $m^n$ terms has been reduced to a sum with a polynomial number of terms.

This exponential reduction in the number of terms can be applied to the pinching map $\mathcal{P}_H$ given in (3.227,3.233). For the case that $H = \sigma^{\otimes n}$ for some $\sigma \in \mathfrak{D}(A)$ we have $|\mathrm{spec}(H)| = \left|\mathrm{spec}\left(\sigma^{\otimes n}\right)\right| \leqslant (n+1)^m$. Therefore, when combined with the pinching inequality (3.235) we conclude that for all $\rho \in \mathfrak{D}(A^n)$,

$$\mathcal{P}_{\sigma^{\otimes n}}\left(\rho^{A^n}\right) \geqslant \frac{1}{(n+1)^m}\rho^{A^n}. \tag{8.103}$$

We will see later on that this inequality can be very useful as polynomial terms such as $(n+1)^m$ turns out to be "negligible" in some applications.

## 8.4.2 Sanov's Theorem

Let $\mathcal{C} \subseteq \mathrm{Prob}(m)$ be a set of probability vectors. Sanov's theorem provides an estimate on the probability that a given sequence $x^n$, that is drawn from an i.i.d.$\sim \mathbf{q}$ source, has a type belonging to $\mathcal{C}$. Clearly, this probability can be zero if the set $\mathcal{C}$ does not contain types. It is therefore necessary to assume that $\mathcal{C}$ is a "nice" set; particularly, we will assume that the set $\mathcal{C}$ is nonempty and is such that $\mathcal{C}$ is the closure of its interior. Thus, the interior of $\mathcal{C}$ is not empty (otherwise, $\mathcal{C}$ will also be empty) and consequently $\mathcal{C}$ contains a ball of nonzero radius. In particular, since the set $\bigcup_{n\in\mathbb{N}} \mathrm{Type}(n,m)$ is dense in $\mathrm{Prob}(m)$ it follows that for sufficiently large $n$ the set $\mathcal{C} \cap \mathrm{Type}(n,m)$ is nonempty.

The probability that a sequence of size $n$ has a type in $\mathcal{C}$ is given by

$$\mathrm{Pr}_n(\mathcal{C}) := \sum_{x^n \in \mathcal{R}_n} q_{x^n}, \tag{8.104}$$

where

$$\mathcal{R}_n := \left\{ x^n \in [m]^n : \mathsf{t}(x^n) \in \mathcal{C} \cap \mathrm{Type}(n,m) \right\}. \tag{8.105}$$

When $n$ is very large one can expect the type of $x^n$ to be relatively close to $\mathbf{q}$ (we will make this notion precise in the next subsection when we study strong typicality). Therefore, if $\mathbf{q} \notin \mathcal{C}$ one can expect that the probability $\mathrm{Pr}_n(\mathcal{C})$ decreases with $n$. Indeed, Sanov's theorem states that for large $n$ we have $\mathrm{Pr}_n(\mathcal{C}) \approx 2^{-nD(\mathbf{p}^\star \| \mathbf{q})}$, where the probability vector $\mathbf{p}^\star \in \mathrm{Prob}(m)$ is defined as

$$\mathbf{p}^\star := \arg\min_{\mathbf{p}\in\mathcal{C}} D(\mathbf{p}\|\mathbf{q}), \tag{8.106}$$

where $D$ is the KL-divergence. This result has a geometrical interpretation that the exponential decay of $\mathrm{Pr}_n(\mathcal{C})$ is increasing with the "distance" (as measured by the KL-divergence) of $\mathbf{q}$ from the set $\mathcal{C}$ (see Figure 8.2).

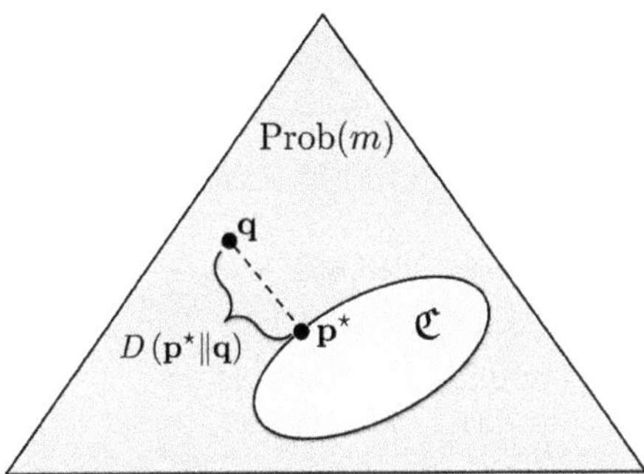


**Figure 8.2**   Sanov's theorem. The exponential decay factor is determined by the distance of **q** from $\mathcal{C}$ (as measured by the KL-divergence). The triangle represents the probability simplex Prob($m$), and the oval shape set $\mathcal{C}$.

---

**Sanov's Theorem**

**Theorem 8.8.** Let $\mathcal{C} \in \mathrm{Prob}(m)$ be a nonempty set of probability distributions such that $\mathcal{C}$ is the closure to its interior, and consider an i.i.d.$\sim$ **q** source. Using the same notations as above,

$$\lim_{n\to\infty} -\frac{1}{n}\log\mathrm{Pr}_n(\mathcal{C}) = \min_{\mathbf{p}\in\mathcal{C}} D(\mathbf{p}\|\mathbf{q}) := D(\mathbf{p}^{\star}\|\mathbf{q}). \tag{8.107}$$

*Remark.* Note that while the vector $\mathbf{p}^{\star}$ is *not* necessarily in $\bigcup_{n\in\mathbb{N}} \mathrm{Type}(n,m)$, there exists a sequence of types $\{\mathbf{p}_n\}_{n\in\mathbb{N}}$, with $\mathbf{p}_n \subset \mathcal{C} \cap \mathrm{Type}(n,m)$ for sufficiently large $n$, such that $\mathbf{p}_n \to \mathbf{p}^*$ as $n \to \infty$.

**Proof**   We will prove the theorem by finding upper and lower bounds for $\mathrm{Pr}_n(\mathcal{C})$. For the upper bound, using (8.85) we get

$$\mathrm{Pr}_n(\mathcal{C}) = \sum_{x^n\in\mathfrak{R}_n} 2^{-n\left(H(\mathsf{t}(x^n))+D(\mathsf{t}(x^n)\|\mathbf{q})\right)} = \sum_{\mathsf{t}\in\mathcal{C}\cap\mathrm{Type}(n,m)} |X^n(\mathsf{t})|2^{-n\left(H(\mathsf{t})+D(\mathsf{t}\|\mathbf{q})\right)}$$

$$(8.89)\to\ \leqslant \sum_{\mathsf{t}\in\mathcal{C}\cap\mathrm{Type}(n,m)} 2^{-nD(\mathsf{t}\|\mathbf{q})}$$

$$\text{By definition of } \mathbf{p}^{\star}\to\ \leqslant \sum_{\mathsf{t}\in\mathcal{C}\cap\mathrm{Type}(n,m)} 2^{-nD(\mathbf{p}^{\star}\|\mathbf{q})} \tag{8.108}$$

$$\leqslant |\mathrm{Type}(n,m)|2^{-nD(\mathbf{p}^{\star}\|\mathbf{q})}$$

$$(8.87)\to\ \leqslant (n+1)^m 2^{-nD(\mathbf{p}^{\star}\|\mathbf{q})}.$$

Note that we got this upper bound without assuming that $\mathcal{C}$ is the closure of its interior. We only assumed that $\mathcal{C}$ is nonempty so that $\mathbf{p}^{\star}$ exists.

For the lower bound, let $\{\mathbf{p}_n\}_{n\in\mathbb{N}}$, with $\mathbf{p}_n \in \mathcal{C} \cap \mathrm{Type}(n,m)$ for sufficiently large $n$, such that $\mathbf{p}_n \to \mathbf{p}^*$ as $n \to \infty$ (see the remark above). Then, for sufficiently large $n$ we have

$$\mathrm{Pr}_n(\mathcal{C}) = \sum_{t \in \mathcal{C} \cap \mathrm{Type}(n,m)} |X^n(t)| 2^{-n\left(H(t)+D(t\|q)\right)}$$

$$\overset{\substack{\textit{Taking only the term} \\ t=\mathbf{p}_n \textit{ in the sum}}}{\longrightarrow} \geqslant |x^n(\mathbf{p}_n)| 2^{-n\left(H(\mathbf{p}_n)+D(\mathbf{p}_n\|q)\right)} \tag{8.109}$$

$$\overset{(8.89)}{\longrightarrow} \geqslant \frac{1}{(n+1)^m} 2^{-nD(\mathbf{p}_n\|q)}.$$

From the two bounds above we get

$$D(\mathbf{p}^\star\|\mathbf{q}) - \frac{1}{n}\log(n+1)^m \leqslant -\frac{1}{n}\log \mathrm{Pr}_n(\mathcal{C}) \leqslant D(\mathbf{p}_n\|\mathbf{q}) + \frac{1}{n}\log(n+1)^m. \tag{8.110}$$

The proof is concluded by taking the limit $n \to \infty$. ∎

## 8.5 Strong Typicality

In this section, we employ the method of types to define the concepts of strong typical sequences and strong typical subspaces. While weak typicality serves as a valuable tool in various applications due to its simplicity, it often yields less robust results. As we will see in certain examples, there are instances where the application of strong typicality methods becomes necessary to achieve more robust outcomes.

### 8.5.1 Strong Typical Sequences

Previously, we examined what is often called *weak* typicality. To understand its limitations, consider an i.i.d.$\sim$ $\mathbf{p}$ source sequence that is $\varepsilon$-typical, satisfying (8.24). Simplifying (8.24) using (8.85), a sequence is (weakly) $\varepsilon$-typical if

$$\left| H\left(\mathbf{t}\left(x^n\right)\right) + D\left(\mathbf{t}(x^n)\|\mathbf{p}\right) - H(\mathbf{p}) \right| \leqslant \varepsilon. \tag{8.111}$$

In other words, $x^n$ is $\varepsilon$-typical if the entropy of its type approximates the entropy of $\mathbf{p}$, with a negligible correction term $D(\mathbf{t}(x^n)\|\mathbf{p})$ as $\mathbf{t}(x^n)$ converges to $\mathbf{p}$. To establish a more robust concept of typicality, one might require that the type of $x^n$ is $\varepsilon$-close to $\mathbf{p}$. This is a somewhat more natural requirement and the question is then: which metric to use? The fundamental requirement is that each element of $\mathbf{t}(x^n)$ should be $\varepsilon$-close to its counterpart in $\mathbf{p}$, necessitating

$$\left| p_z - t_z(x^n) \right| \leqslant \varepsilon \qquad \forall\, z \in [m]. \tag{8.112}$$

Note that any sequence $x^n$ that satisfies $\|\mathbf{p} - \mathbf{t}(x^n)\|_p \leqslant \varepsilon$ (for some $p \geqslant 1$) will satisfy the above equation, and therefore it imposes a slightly stronger condition on the sequence $x^n$. On the other hand, the condition $\|\mathbf{p} - \mathbf{t}(x^n)\|_\infty \leqslant \varepsilon$ is precisely equivalent to the above condition, and we will therefore use it to measure the distance between $\mathbf{t}(x^n)$ and $\mathbf{p}$.

**Strongly Typical Sequences**

**Definition 8.5.** Let $x^n \in [m]^n$ be a sequence of size $n$ drawn from an i.i.d. $\sim \mathbf{p}$ source. The sequence $x^n$ is said to be strongly $\varepsilon$-typical if $N(z|x^n) = 0$ for all $z \notin \mathrm{supp}\,(\mathbf{p})$ and

$$\left\| \mathbf{p} - \mathbf{t}(x^n) \right\|_\infty \leqslant \varepsilon. \tag{8.113}$$

The set of all strongly typical sequences of size $n$ is denoted by $\mathfrak{T}^{\mathrm{st}}_\varepsilon(X^n)$.

Note that, in accordance with our previous notation, the condition stipulated in the above definition can be equivalently expressed as

$$\left| p_z - \frac{N(z|x^n)}{n} \right| \leqslant \varepsilon \qquad \forall\, z \in [m] \text{ such that } p_z > 0, \tag{8.114}$$

and $N(z|x^n) = 0$ whenever $p_z = 0$. This latter condition intuitively implies that a typical sequence should not include any alphabet characters that have a zero probability of occurrence.

In the next theorem we prove several properties of typical sequences. We will use the notation

$$\Pr\left(\mathfrak{T}^{\mathrm{st}}_\varepsilon(X^n)\right) := \sum_{x^n \in \mathfrak{T}^{\mathrm{st}}_\varepsilon(X^n)} p_{x^n} \tag{8.115}$$

to denote the probability that a sequence is strongly $\varepsilon$-typical (with respect to an i.i.d. $\sim \mathbf{p}$ source). Moreover, we set the constant $a > 0$ to be

$$a := -\log \prod_{z \in \mathrm{supp}\,(\mathbf{p})} p_z. \tag{8.116}$$

**Properties of Strongly Typical Sequences**

**Theorem 8.9.** Let $X$ be the random variable of an i.i.d. $\sim \mathbf{p}$ source, and let $\varepsilon > 0$. For all $n \in \mathbb{N}$ the following inequalities hold:

1. $\Pr\left(\mathfrak{T}^{\mathrm{st}}_\varepsilon(X^n)\right) \geqslant 1 - e^{-2\varepsilon^2 n}$.
2. $2^{-n(H(X)+a\varepsilon)} \leqslant p_{x^n} \leqslant 2^{-n(H(X)-a\varepsilon)}$, for all $x^n \in \mathfrak{T}^{\mathrm{st}}_\varepsilon(X^n)$.
3. $\left(1 - e^{-2\varepsilon^2 n}\right) 2^{n(H(X)-\varepsilon a)} \leqslant \left| \mathfrak{T}^{\mathrm{st}}_\varepsilon(X^n) \right| \leqslant 2^{n(H(X)+\varepsilon a)}$.

*Remark.* Observe that the probability that a sequence is strongly $\varepsilon$-typical approach one exponentially fast with $n$. The second property highlights the equipartition property, indicating that the probability of every typical sequence approximates $2^{-nH(X)}$. The first and third properties bear resemblance to their counterparts related to weak typical sequences. However, due to the subtle distinctions between the two concepts of typicality, the upcoming proof will incorporate additional tools to address these differences.

**Proof** For any $z \in [m]$, let $1_z(X)$ be the *indicator* random variable that equals 1 if $X = z$ and 0 otherwise. Fix $z \in [m]$, and let $Z_1, Z_2, \ldots$ be an i.i.d. sequence of random variables where each $Z_j := 1_z(X_j)$ is an indicator random variable as above that corresponds to $X_j$. From the law of large numbers, particularly the application of Hoeffding's inequality (8.15) to the sequence $Z_1, \ldots, Z_n$, we get

$$\Pr\left( \left| \frac{1}{n} \sum_{j \in [n]} Z_j - \mathbb{E}(Z) \right| > \varepsilon \right) \leqslant e^{-2n\varepsilon^2}, \tag{8.117}$$

where we used the fact that $0 \leqslant Z_j \leqslant 1$ so that the constants $a$ and $b$ in (8.15) are given by $a = 0$ and $b = 1$. By definition,

$$\mathbb{E}(Z) = \sum_{x \in [m]} p_x \delta_{xz} = p_z, \tag{8.118}$$

and

$$\frac{1}{n} \sum_{j \in [n]} Z_j = \frac{1}{n} \sum_{j \in [n]} \delta_{x_j z} = \frac{1}{n} N(z|x^n) = t_z(x^n). \tag{8.119}$$

We therefore conclude that

$$\Pr\left( \left| t_z(x^n) - p_z \right| > \varepsilon \right) \leqslant e^{-2n\varepsilon^2}. \tag{8.120}$$

The equation above holds for all $z \in [m]$ and all $n \in \mathbb{N}$. Hence, it states that the probability that the random variable $X^n = (X_1, \ldots, X_n)$ is not an $\varepsilon$-typical sequence, is no greater than $e^{-2n\varepsilon^2}$. This completes the proof of first part of the theorem.

Suppose now that $x^n \in \mathcal{T}_\varepsilon^{\mathrm{st}}(X^n)$ and observe that

$$p_{x^n} = p_{x_1} p_{x_2} \cdots p_{x_n} = \prod_{z \in \mathrm{supp}\,(\mathbf{p})} p_z^{N(z|x^n)}. \tag{8.121}$$

Taking the log on both sides and dividing by $n$ gives

$$\frac{1}{n} \log p_{x^n} = \sum_{z \in \mathrm{supp}\,(\mathbf{p})} t_z(x^n) \log p_z. \tag{8.122}$$

Now, since $x^n$ is strongly $\varepsilon$-typical sequence, for every $z \in [m]$,

$$p_z - \varepsilon \leqslant t_z(x^n) \leqslant p_z + \varepsilon. \tag{8.123}$$

Combining this with the previous equation gives

$$\sum_{z \in \mathrm{supp}\,(\mathbf{p})} (p_z + \varepsilon) \log p_z \leqslant \frac{1}{n} \log p_{x^n} \leqslant \sum_{z \in \mathrm{supp}\,(\mathbf{p})} (p_z - \varepsilon) \log p_z. \tag{8.124}$$

That is,

$$-\varepsilon a - H(X) \leqslant \frac{1}{n} \log p_{x^n} \leqslant \varepsilon a - H(X). \tag{8.125}$$

Multiplying both sides by $n$ and raising all sides to the power of 2 completes the proof of the second part.

To prove the third part, observe that from the lower bound of Part 2 we get

$$1 \geqslant \sum_{x^n \in \mathfrak{T}^{\mathrm{st}}_\varepsilon(X^n)} p_{x^n} \geqslant \sum_{x^n \in \mathfrak{T}^{\mathrm{st}}_\varepsilon(X^n)} 2^{-n(H(X)+\varepsilon a)}$$
$$= \left| \mathfrak{T}^{\mathrm{st}}_\varepsilon(X^n) \right| 2^{-n(H(X)+\varepsilon a)}. \tag{8.126}$$

Therefore, $\left| \mathfrak{T}^{\mathrm{st}}_\varepsilon(X^n) \right| \leqslant 2^{n(H(X)+\varepsilon a)}$. On the other hand, from Part 1 we get

$$1 - e^{-2n\varepsilon^2} \leqslant \sum_{x^n \in \mathfrak{T}^{\mathrm{st}}_\varepsilon(X^n)} p_{x^n}$$

$$\textbf{Upper bound of Part 2} \rightarrow \; \leqslant \sum_{x^n \in \mathfrak{T}^{\mathrm{st}}_\varepsilon(X^n)} 2^{-n(H(X)-\varepsilon a)} \tag{8.127}$$

$$= \left| \mathfrak{T}^{\mathrm{st}}_\varepsilon(X^n) \right| 2^{-n(H(X)-\varepsilon a)}.$$

Hence, $\left(1 - e^{-2n\varepsilon^2}\right) 2^{n(H(X)-\varepsilon a)} \leqslant \left| \mathfrak{T}^{\mathrm{st}}_\varepsilon(X^n) \right|$. This completes the proof of the third part.

∎

**Exercise 8.15.** *Let $x^n \in [m]^n$ and $y^k \in [m]^k$ be two sequences of size $n$ and $k$, respectively, drawn from the same i.i.d. $\sim \mathbf{p}$ source. Show that if both $x^n$ and $y^k$ are strongly $\varepsilon$-typical, then also the joint sequence $(x^n, y^k) \in [m]^{n+k}$ is strongly $\varepsilon$-typical with respect to the i.i.d. $\sim \mathbf{p}$ source.*

**Exercise 8.16.** *Let $1 < n \in \mathbb{N}$, $\varepsilon \in (\frac{1}{n}, 1)$, and consider an i.i.d. $\sim \mathbf{p}$ source, where $\mathbf{p} \in \mathrm{Prob}(m)$.*

1. *Show that if a sequence $x^{n-1}$ is strongly $\varepsilon$-typical, then for any $y \in [m]$ the sequence $x^n := (y, x^{n-1})$ is strongly $(\varepsilon + \frac{1}{n})$-typical.*
2. *Let $\varepsilon' := \varepsilon - \frac{1}{n}$. Show that $\mathfrak{T}' \subset \mathfrak{T}^{\mathrm{st}}_\varepsilon(X^n)$, where*

$$\mathfrak{T}' := \left\{ (y, x^{n-1}) : \; y \in [m], \quad x^{n-1} \in [m]^{n-1}, \quad \left\| \mathbf{t}(x^{n-1}) - \mathbf{p} \right\|_\infty \leqslant \varepsilon' \right\}. \tag{8.128}$$

3. *Show that for any $\mathbf{q} \in \mathrm{Prob}(m)$ we have*

$$\sum_{x^n \in \mathfrak{T}^{\mathrm{st}}_\varepsilon(X^n)} q_{x_1} p_{x_2} \cdots p_{x_n} \geqslant 1 - e^{-2(n-1)\varepsilon'^2}. \tag{8.129}$$

*Hint: Use the first part of Theorem 8.9.*

## 8.5.2 Strong Typical Subspace

The extension of strong typicality from classical sequences to quantum states follows similar lines to the quantum extension of weak classical typicality. Let $\rho$ be an i.i.d. quantum source with spectral decomposition

$$\rho^A = \sum_{x \in [m]} p_x |x\rangle\langle x|^A, \tag{8.130}$$

where $\{p_x\}_{x\in[m]}$ are the eigenvalues of $\rho$ and $\{|x\rangle\}_{x\in[m]}$ are the corresponding eigenvectors. As before, we denote

$$\rho^{\otimes n} = \sum_{x^n\in[m]^n} p_{x^n}|x^n\rangle\langle x^n|^{A^n}, \tag{8.131}$$

where $|x^n\rangle := |x_1,\ldots,x_n\rangle$ and $p_{x^n} := p_{x_1}p_{x_2}\cdots p_{x_n}$. For any i.i.d. quantum source $\rho$ we define a corresponding strongly typical subspace

$$\mathcal{T}_\varepsilon^{\mathrm{st}}(A^n) := \mathrm{span}\left\{|x^n\rangle \in A^n : x^n \in \mathcal{T}_\varepsilon^{\mathrm{st}}(X^n)\right\}, \tag{8.132}$$

where $\mathcal{T}_\varepsilon^{\mathrm{st}}(X^n)$ is the set of (classical) sequences of size $n$, drawn from an i.i.d.$\sim \mathbf{p}$ source (with $\mathbf{p}$ being the probability vector whose components are the eigenvalues of $\rho$). The strongly typical projection to this subspace is given by

$$\Pi_\varepsilon^{n,\mathrm{st}} := \sum_{x^n\in\mathcal{T}_\varepsilon^{\mathrm{st}}(X^n)} |x^n\rangle\langle x^n|. \tag{8.133}$$

---

**Theorem 8.10.** Let $\rho \in \mathfrak{D}(A)$, $\varepsilon > 0$, and for each $n \in \mathbb{N}$ let $\mathcal{T}_n^\varepsilon(\rho)$ and $\Pi_\varepsilon^{n,\mathrm{st}}$ be the strongly typical subspace and projection associated with a quantum i.i.d.$\sim \rho$ source. The following inequalities hold for all $n \in \mathbb{N}$:

1. $\mathrm{Tr}\left[\Pi_\varepsilon^{n,\mathrm{st}}\rho^{\otimes n}\right] \geq 1 - e^{-2\varepsilon^2 n}$.
2. $2^{-n(H(A)_\rho+a\varepsilon)}\Pi_\varepsilon^{n,\mathrm{st}} \leq \Pi_\varepsilon^{n,\mathrm{st}}\rho^{\otimes n}\Pi_\varepsilon^{n,\mathrm{st}} \leq 2^{-n(H(A)_\rho-a\varepsilon)}\Pi_\varepsilon^{n,\mathrm{st}}$.
3. $(1-\delta)2^{n(H(\rho)-\varepsilon a)} \leq \left|\mathcal{T}_\varepsilon^{\mathrm{st}}(A^n)\right| \leq 2^{n(H(\rho)+\varepsilon a)}$.

---

The proof follows directly from the classical version of this theorem and is left as an exercise.

**Exercise 8.17.** *Prove Theorem 8.10.*

**Exercise 8.18.** *Let $\rho \in \mathfrak{D}(A)$, $\varepsilon > 0$, integer $m = o(n)$ (e.g. $m = \lfloor n^s \rfloor$ for some $0 < s < 1$), and $\sigma_m \in \mathfrak{D}(A^m)$. Let also $\Pi_\varepsilon^{n,\mathrm{st}}$ be the strongly typical projection associated with the quantum i.i.d.$\sim \rho$ source. Show that*

$$\lim_{n\to\infty} \mathrm{Tr}\left[\Pi_\varepsilon^{n,\mathrm{st}}\left(\rho^{\otimes(n-m)}\otimes\sigma_m\right)\right] = 1. \tag{8.134}$$

## 8.6 Classical Hypothesis Testing

Imagine a game where a player, named Alice, is handed one of two biased dice. These dice are characterized by probability vectors, denoted as $\mathbf{p}$ and $\mathbf{q}$. Alice's challenge is to determine if she holds the $\mathbf{p}$ dice or the $\mathbf{q}$ dice. To make an informed decision, she's allowed to roll the dice $n$ times. With a single roll (i.e. $n = 1$), the chances of Alice making an incorrect assumption can be high. However, with an increase in the number of rolls, her probability of making a mistake significantly reduces. This prompts a natural question: How rapidly does the error probability decrease as the number of rolls, $n$, gets larger and larger? This situation encapsulates the essence of a classical hypothesis testing problem.

In the realm of hypothesis testing, an observer or player aims to decide between two hypotheses related to two i.i.d. sources. These hypotheses are represented as the p-source and the q-source. Upon $n$ independent interactions with this source, the observer receives a sequence denoted as $x^n = (x_1, \ldots, x_n)$ that belongs to the set $[m]^n$. The challenge is to ascertain the correct hypothesis based on this sequence.

The observer's decision-making process can be represented by a function $g_n \colon [m]^n \to \{0, 1\}$. This function divides all potential sequences into two distinct groups:

1. The set $\{x^n \in [m]^n : g_n(x^n) = 0\}$ corresponds to the first hypothesis. Here, the observer believes the sequences in this set are from the p-source.
2. The set $\{x^n \in [m]^n : g_n(x^n) = 1\}$ pertains to the second hypothesis, indicating that the observer surmises the sequences are from the q-source.

Given this decision-making framework, two potential errors can emerge:

1. **Type I Error.** The observer incorrectly concludes that the sequence is from the q-source when, in reality, it is from the p-source. The probability of this error occurring is

$$\alpha(g_n) := \sum_{\substack{x^n \in [m]^n \\ g_n(x^n) = 1}} p_{x^n}. \tag{8.135}$$

2. **Type II Error.** The observer mistakenly assumes the sequence is from the p-source when it actually originates from the q-source. The likelihood of this error is

$$\beta(g_n) := \sum_{\substack{x^n \in [m]^n \\ g_n(x^n) = 0}} q_{x^n}. \tag{8.136}$$

In these two errors, we considered a deterministic hypothesis test, where the function $g_n \colon [m]^n \to \{0, 1\}$ remains fixed. A more general approach introduces an element of randomness to the problem. Here, the observer randomly selects the function $g_n$ based on a specific probability distribution.

To illustrate, consider a set of $\ell$ functions denoted as $\{g_{n,k}\}_{k \in [\ell]}$, where each $g_{n,k} \colon [m]^n \to \{0, 1\}$. Accompanying these functions is a probability vector $s \in \mathrm{Prob}(\ell)$. In this probabilistic framework, when given the sequence $x^n$, the observer first samples a values $k$ according to the distribution $s$. The observer then attributes the sequence to the p-source if $g_{n,k}(x^n) = 0$, and to the q-source if $g_{n,k}(x^n) = 1$. It's crucial to note that for each $k \in \lfloor \ell \rfloor$,

$$\beta(g_{n,k}) = \sum_{\substack{x^n \in [m]^n \\ g_{n,k}(x^n) = 0}} q_{x^n} = \mathbf{q}^{\otimes n} \cdot \mathbf{b}_k, \tag{8.137}$$

where the $x^n$-component of the bit vector $\mathbf{b}_k \in \{0, 1\}^{m^n}$ is one if $g_{n,k}(x^n) = 0$ and zero otherwise. The vector

$$\mathbf{t} := \sum_{k \in [\ell]} s_k \mathbf{b}_k \tag{8.138}$$

is termed the *probabilistic hypothesis test*. With the aforementioned notations and considering this broader context, the two types of errors can be described as follows:

1. **Type I Error.** This pertains to the likelihood of the observer incorrectly attributing the sequence to the **q**-source when it originates from the **p**-source:

$$\alpha(\mathbf{t}) := \sum_{k \in [\ell]} s_k \sum_{\substack{x^n \in [m]^n \\ g_{n,k}(x^n) = 1}} p_{x^n} = 1 - \mathbf{p}^{\otimes n} \cdot \mathbf{t}. \tag{8.139}$$

2. **Type II Error.** This represents the chance of the observer mistakenly deducing the sequence belongs to the **p**-source when it is from the **q**-source:

$$\beta(\mathbf{t}) := \sum_{k \in [\ell]} s_k \sum_{\substack{x^n \in [m]^n \\ g_{n,k}(x^n) = 0}} q_{x^n} = \mathbf{q}^{\otimes n} \cdot \mathbf{t}. \tag{8.140}$$

From its definition (8.138), all the components of the probabilistic hypothesis test vector $\mathbf{t}$ are between zero and 1 (i.e. $\mathbf{t} \in [0, 1]^{m^n}$). Conversely, any vector in $[0, 1]^{m^n}$ can be expressed as a convex combination of bit vectors in $\{0, 1\}^{m^n}$. Hence, $\mathbf{t}$ uniquely characterizes the probabilistic hypothesis test performed by the observer.

The goal of the observer is therefore to choose a probabilistic test vector $\mathbf{t}$ such that both types of error are very small. There are two common ways to do that, and we discuss both now. The first one is the *asymmetric* method in which the observer minimizes the type II error, $\beta(\mathbf{t})$, while at the same time keep the type I error, $\alpha(\mathbf{t})$, below a certain threshold $\varepsilon > 0$. The optimal way to do it is characterized by the Stein's lemma. The second method is known as the *symmetric* way, in which one assumes a prior $\{s_0, s_1\}$ known to the observer in which the **p**-source occurs with probability $s_0$, and the **q**-source with probability $s_1$. In this case, the goal is to minimize the error probability that is given by $s_0\alpha(\mathbf{t}) + s_1\beta(\mathbf{t})$. The optimal value of this probability of error is characterized by the Chernoff information. A fundamental instrument in these methods is the divergence used in hypothesis testing.

## 8.6.1 The Classical Hypothesis Testing Divergence

For a single application of the source, specifically when $n = 1$ as presented in (8.139,8.140), we define the two error types as $\alpha(\mathbf{t}) := 1 - \mathbf{p} \cdot \mathbf{t}$ and $\beta(\mathbf{t}) := \mathbf{q} \cdot \mathbf{t}$. By minimizing $\beta(\mathbf{t})$ while constraining $\alpha(\mathbf{t})$ to a specific threshold, we obtain the following divergence.

> **Definition 8.6.** For any $\mathbf{p}, \mathbf{q} \in \mathrm{Prob}(m)$ and $\varepsilon \in [0, 1)$, the *classical* hypothesis testing divergence is defined as
>
> $$D_{\min}^{\varepsilon}(\mathbf{p}\|\mathbf{q}) := -\log \min\left\{\mathbf{q} \cdot \mathbf{t} : \mathbf{p} \cdot \mathbf{t} \geqslant 1 - \varepsilon \quad , \quad \mathbf{t} \in [0, 1]^m\right\}, \qquad (8.141)$$
>
> where the minimization is over all probabilistic hypothesis test vectors $\mathbf{t}$ whose components are in the interval $[0, 1]$.

The hypothesis testing divergence is always nonnegative and equals infinity if $\mathbf{p} \cdot \mathbf{q} = 0$. The reason for the notation $D_{\min}^{\varepsilon}$ is that for $\varepsilon = 0$, it reduces to the min relative entropy:

$$D_{\min}^{0}(\mathbf{p}\|\mathbf{q}) = -\log \min\left\{\mathbf{q} \cdot \mathbf{t} : \mathbf{p} \cdot \mathbf{t} \geqslant 1 \quad , \quad \mathbf{t} \in [0, 1]^m\right\}$$

$$\boxed{\begin{array}{c} \forall\, x \in [m] \\ t_x = 1 \text{ if } p_x \neq 0 \end{array}} \longrightarrow \quad = -\log \sum_{x \in \mathrm{supp}\,(\mathbf{p})} q_x \qquad (8.142)$$

$$= D_{\min}(\mathbf{p}\|\mathbf{q}).$$

To see that $D_{\min}^{\varepsilon}$ in the definition above is indeed an (unnormalized) divergence, let $\mathbf{p}, \mathbf{q} \in \mathrm{Prob}(m)$, $E \in \mathrm{STOC}(n, m)$, and observe that

$$D_{\min}^{\varepsilon}(E\mathbf{p}\|E\mathbf{q}) = -\log \min\left\{(E\mathbf{q})^T \mathbf{s} : (E\mathbf{p})^T \mathbf{s} \geqslant 1 - \varepsilon \quad , \quad \mathbf{s} \in [0, 1]^n\right\}$$

$$\boxed{\begin{array}{l} \text{Replacing } E^T\mathbf{s} \in [0,1]^m \\ \text{with arbitrary } \mathbf{t} \in [0,1]^m \end{array}} \longrightarrow \quad \leqslant -\log \min\left\{\mathbf{q} \cdot \mathbf{t} : \mathbf{p} \cdot \mathbf{t} \geqslant 1 - \varepsilon \quad , \quad \mathbf{t} \in [0, 1]^m\right\} \qquad (8.143)$$

$$= D_{\min}^{\varepsilon}(\mathbf{p}\|\mathbf{q}).$$

**Exercise 8.19.** *Show that the constraint $\mathbf{p} \cdot \mathbf{t} \geqslant 1 - \varepsilon$ in (8.141) can be replaced with $\mathbf{p} \cdot \mathbf{t} = 1 - \varepsilon$ (i.e. both constraints leads to the same value of $D_{\min}^{\varepsilon}(\mathbf{p}\|\mathbf{q})$).*

**Exercise 8.20.** *Show that for all $\mathbf{p}, \mathbf{q} \in \mathrm{Prob}(m)$, $D_{\min}^{\varepsilon}(\mathbf{p}\|\mathbf{q})$ is nondecreasing in $\varepsilon$, and*

$$D_{\min}^{\varepsilon}(\mathbf{p}\|\mathbf{q}) \geqslant -\log (1 - \varepsilon), \qquad (8.144)$$

*with equality if $\mathbf{p} = \mathbf{q}$.*

The classical hypothesis testing divergence is closely related to the testing region defined in (4.139). To see the connection, first observe that we can replace the condition $\mathbf{p} \cdot \mathbf{t} \geqslant 1 - \varepsilon$ in (8.141) with the equality $\mathbf{p} \cdot \mathbf{t} = 1 - \varepsilon$ (since any $\mathbf{t}$ that satisfies $\mathbf{p} \cdot \mathbf{t} > 1 - \varepsilon$ is not optimal). With this change, the optimal $\mathbf{q} \cdot \mathbf{t}$ in (8.141) can be interpreted as the lowest point of the intersection of the testing region $\mathfrak{T}(\mathbf{p}, \mathbf{q})$ with the vertical line $x = 1 - \varepsilon$ (see Figure 8.3). That is, the optimal $\mathbf{q} \cdot \mathbf{t}$ is the $y$ component of the lower Lorenz curve $\mathrm{LC}(\mathbf{p}, \mathbf{q})$ at $x = 1 - \varepsilon$.

We can use the above geometrical interpretation of the hypothesis testing divergence to obtain a closed formula for $D_{\min}^{\varepsilon}$. Without loss of generality, suppose that the components of $\mathbf{p}$ and $\mathbf{q}$ are ordered as in (4.116). Then, from Theorem 4.9 we know that the vertices of the lower Lorenz curve of $(\mathbf{p}, \mathbf{q})$ are given by $\{(a_k, b_k)\}_{k=0}^{m}$ as defined in (4.142). Let $\ell$ be an integer such that $a_\ell < 1 - \varepsilon \leqslant a_{\ell+1}$. Then, the optimal point on

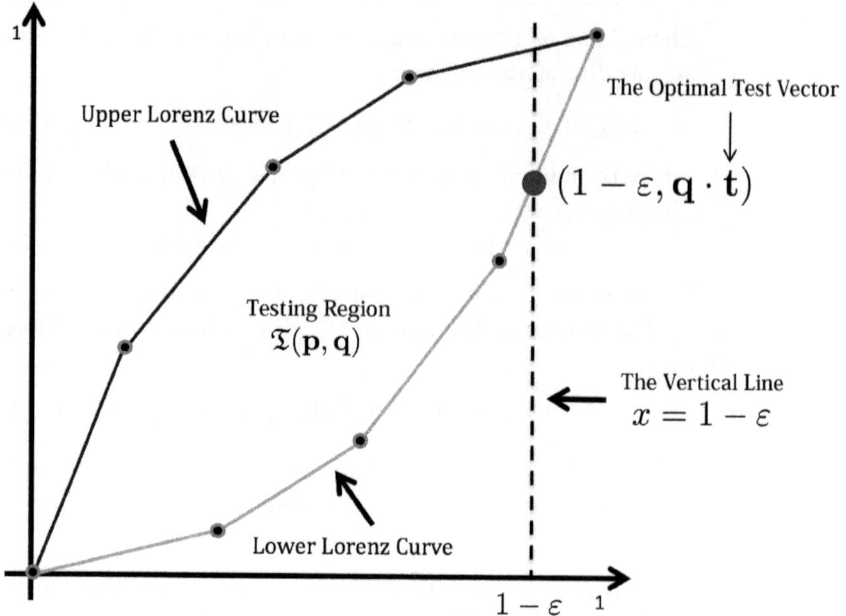


**Figure 8.3**    The location of the point in the testing region $\mathfrak{T}(\mathbf{p}, \mathbf{q})$ with the optimal testing vector $\mathbf{t}$ that minimizes (8.141).

the lower Lorenz curve is located between the $\ell$ and the $\ell + 1$ vertices. The line between these two vertices has a slope

$$\frac{b_{\ell+1} - b_\ell}{a_{\ell+1} - a_\ell} = \frac{q_{\ell+1}}{p_{\ell+1}}. \tag{8.145}$$

Hence, the $y$ component of the optimal point is given by

$$\mathbf{q} \cdot \mathbf{t} = b_\ell + \frac{q_{\ell+1}}{p_{\ell+1}}(1 - \varepsilon - a_\ell). \tag{8.146}$$

To summarize, we can express the hypothesis testing divergence as

$$D_{\min}^\varepsilon \left(\mathbf{p} \| \mathbf{q}\right) = -\log\left(b_\ell + \frac{q_{\ell+1}}{p_{\ell+1}}(1 - \varepsilon - a_\ell)\right), \tag{8.147}$$

where $\ell \in \{0, \ldots, m - 1\}$ is the integer satisfying $a_\ell < 1 - \varepsilon \leqslant a_{\ell+1}$. Recall that $D_{\min}^\varepsilon$ is nondecreasing with $\varepsilon$ (see Exercise 8.29), as it is also evident from the equation above. Therefore, we can bound the hypothesis testing divergences by taking above the two extreme cases $1 - \varepsilon = a_\ell$ and $1 - \varepsilon = a_{\ell+1}$ to get the simpler bounds

$$-\log b_{\ell+1} \leqslant D_{\min}^\varepsilon \left(\mathbf{p} \| \mathbf{q}\right) \leqslant -\log b_\ell, \tag{8.148}$$

where $\ell$, as before, is the integer satisfying $a_\ell < 1 - \varepsilon \leqslant a_{\ell+1}$.

## 8.6.2 The Stein's Lemma

Here we study the minimization of the type II error $\beta_n(t)$ while at the same time keeping the type I error $\alpha_n(t)$ below a certain threshold $\varepsilon \geqslant 0$.

> **The Stein's Lemma**
>
> **Theorem 8.11.** For any $0 < \varepsilon < 1$, $m \in \mathbb{N}$, and $\mathbf{p}, \mathbf{q} \in \mathrm{Prob}(m)$
>
> $$\lim_{n\to\infty} \frac{1}{n} D^{\varepsilon}_{\min}\left(\mathbf{p}^{\otimes n} \,\big\|\, \mathbf{q}^{\otimes n}\right) = D(\mathbf{p}\|\mathbf{q}), \tag{8.149}$$
>
> where $D$ is the KL-divergence.

*Remark.* The theorem above states that the type II error can be made as small as $\approx 2^{-nD(\mathbf{p}\|\mathbf{q})}$ while at the same time keeping the type I error below the threshold $\varepsilon$. The rate of this exponential decay is given by the KL-divergence. We postpone the proof of this theorem to the next section in which we prove the more general theorem known as the quantum Stein's lemma. For the interested reader, we also provide two more direct proofs of this theorem (that applicable only to the classical case) in Appendix D.4 of online version.

## 8.6.3 The Chernoff Information

In the second method of optimizing $\alpha(t)$ and $\beta(t)$, also known as the *symmetric* method, there exists a prior $\{s_0, s_1\}$ known to the observer in which the $\mathbf{p}$-source occurs with probability $s_0$, and the $\mathbf{q}$-source with probability $s_1$. In this case, for a single use of the source (i.e. $n = 1$) the error probability in which the observer guesses incorrectly the type of the source is given for any $\mathbf{p}, \mathbf{q} \in \mathrm{Prob}(m)$ by

$$\mathrm{Pr}_{\mathrm{error}}(\mathbf{p}, \mathbf{q}, s_0) := \min\left\{ s_0 \alpha(t) + s_1 \beta(t) : t \in [0,1]^m \right\}. \tag{8.150}$$

Note that this probability of error can be expressed as

$$\begin{aligned}
\mathrm{Pr}_{\mathrm{error}}(\mathbf{p}, \mathbf{q}, s_0) &= \min_{t \in [0,1]^m} \left\{ s_0(1 - \mathbf{p} \cdot t) + s_1 \mathbf{q} \cdot t \right\} \\
&= s_0 - \max_{t \in [0,1]^m} (s_0 \mathbf{p} - s_1 \mathbf{q}) \cdot t \\
&= s_0 - \sum_{x \in [m]} (s_0 p_x - s_1 q_x)_+ \\
\lambda := \frac{s_1}{s_0} \rightarrow &= s_0\left( 1 - \sum_{x \in [m]} (p_x - \lambda q_x)_+ \right),
\end{aligned} \tag{8.151}$$

where $(s - t)_+ := s - t$ if $s \geqslant t$ and is zero otherwise.

**Exercise 8.21.** *Show that the probability of error above can be expressed also as*

$$\mathrm{Pr}_{\mathrm{error}}(\mathbf{p}, \mathbf{q}, s_0) = \frac{1}{2}\left( 1 - \|s_0 \mathbf{p} - s_1 \mathbf{q}\|_1 \right). \tag{8.152}$$

**Exercise 8.22.** *Let* $\mathbf{p}, \mathbf{q} \in \mathrm{Prob}(m)$ *and* $\mathbf{p}', \mathbf{q}' \in \mathrm{Prob}(m')$. *Show that* $(\mathbf{p}, \mathbf{q}) \succ (\mathbf{p}', \mathbf{q}')$ *if and only if*

$$\mathrm{Pr}_{\mathrm{error}}(\mathbf{p}, \mathbf{q}, s_0) \geqslant \mathrm{Pr}_{\mathrm{error}}(\mathbf{p}', \mathbf{q}', s_0) \qquad \forall \, s_0 \in [0, 1]. \tag{8.153}$$

**Exercise 8.23.** *Let* $\mathbf{p}, \mathbf{q} \in \mathrm{Prob}(m)$ *and* $s_0, \alpha \in (0, 1)$.

1. *Show that*

$$\mathrm{Pr}_{\mathrm{error}}(\mathbf{p}, \mathbf{q}, s_0) = \sum_{x \in [m]} \min\{s_0 p_x, s_1 q_x\}. \tag{8.154}$$

    *Hint: Use the formula* $\min\{a, b\} = \frac{1}{2}(a + b) - \frac{1}{2}|a - b|$.

2. *Use the inequality* $\min\{a, b\} \leqslant a^\alpha b^{1-\alpha}$ *to show that for all* $\alpha \in [0, 1]$,

$$\mathrm{Pr}_{\mathrm{error}}(\mathbf{p}, \mathbf{q}, s_0) \leqslant s_0^\alpha s_1^{1-\alpha} \sum_{x \in [m]} p_x^\alpha q_x^{1-\alpha}. \tag{8.155}$$

3. *Show that*

$$\liminf_{n \to \infty} -\frac{1}{n} \log \mathrm{Pr}_{\mathrm{error}}(\mathbf{p}^{\otimes n}, \mathbf{q}^{\otimes n}, s_0) \geqslant -\log \sum_{x \in [m]} p_x^\alpha q_x^{1-\alpha}. \tag{8.156}$$

The exercise above demonstrates that in the asymptotic limit the optimal probability of error is bounded by

$$\liminf_{n \to \infty} -\frac{1}{n} \log \mathrm{Pr}_{\mathrm{error}}(\mathbf{p}^{\otimes n}, \mathbf{q}^{\otimes n}, s_0) \geqslant \xi(\mathbf{p}, \mathbf{q}), \tag{8.157}$$

where

$$\xi(\mathbf{p}, \mathbf{q}) := -\log \min_{\alpha \in [0, 1]} \sum_{x \in [m]} p_x^\alpha q_x^{1-\alpha}$$

$$\textbf{Definition 6.4} \rightarrow \; = \max_{\alpha \in [0, 1]} \left\{ (1 - \alpha) D_\alpha(\mathbf{p} \| \mathbf{q}) \right\}. \tag{8.158}$$

In the following theorem we show that the inequality in (8.157) is in fact an equality.

**The Chernoff Bound**

**Theorem 8.12.** Let $\mathbf{p}, \mathbf{q} \in \mathrm{Prob}(m)$ and $s_0 \in (0, 1)$. Then,

$$\lim_{n \to \infty} -\frac{1}{n} \log \mathrm{Pr}_{\mathrm{error}}(\mathbf{p}^{\otimes n}, \mathbf{q}^{\otimes n}, s_0) = \xi(\mathbf{p}, \mathbf{q}). \tag{8.159}$$

*Remark.* Note that the Chernoff bound $\xi(\mathbf{p}, \mathbf{q})$ does not depend on $s_0$. Moreover, the theorem above also states that for very large $n$ we have that the probability of error, $\mathrm{Pr}_{\mathrm{error}}(\mathbf{p}^{\otimes n}, \mathbf{q}^{\otimes n}, s_0) \approx 2^{-n\xi(\mathbf{p}, \mathbf{q})}$, decays exponentially fast with $n$ with an exponential factor given by the Chernoff bound.

**Proof** We need to prove the opposite inequality of (8.157). We will establish it by finding a lower bound on the probability of error $\mathrm{Pr}_{\mathrm{error}}(\mathbf{p}^{\otimes n}, \mathbf{q}^{\otimes n}, s_0)$. For this purpose, set $\lambda := s_1/s_0$ and $\mathfrak{K} := \{x \in [m] : p_x \geqslant \lambda q_x\}$. From (8.151) we have

$$\mathrm{Pr}_{\mathrm{error}}(\mathbf{p}, \mathbf{q}, s_0) = s_0 \sum_{x \in \mathfrak{K}^c} p_x + s_1 \sum_{x \in \mathfrak{K}} q_x \geqslant \min\left\{ \sum_{x \in \mathfrak{K}^c} p_x, \sum_{x \in \mathfrak{K}} q_x \right\}, \tag{8.160}$$

where $\mathfrak{K}^c$ is the complement of $\mathfrak{K}$ in $[m]$. Similarly, for any $n \in \mathbb{N}$ we denote $\mathfrak{K}_n := \{x^n \in [m]^n : p_{x^n} \geqslant \lambda q_{x^n}\}$ so that

$$\mathrm{Pr}_{\mathrm{error}}(\mathbf{p}^{\otimes n}, \mathbf{q}^{\otimes n}, s_0) \geqslant \min\left\{ \sum_{x^n \in \mathfrak{K}_n^c} p_{x^n}, \sum_{x^n \in \mathfrak{K}_n} q_{x^n} \right\}. \tag{8.161}$$

Next, we characterize the set $\mathfrak{K}_n$. From (8.85) the inequality $p_{x^n} \geqslant \lambda q_{x^n}$ holds if and only if

$$2^{-n\left(H(\mathbf{t}(x^n)) + D(\mathbf{t}(x^n)\|\mathbf{p})\right)} \geqslant \lambda 2^{-n\left(H(\mathbf{t}(x^n)) + D(\mathbf{t}(x^n)\|\mathbf{q})\right)}, \tag{8.162}$$

which is equivalent to

$$D\left(\mathbf{t}(x^n)\|\mathbf{q}\right) - D\left(\mathbf{t}(x^n)\|\mathbf{p}\right) \geqslant \frac{1}{n} \log \lambda. \tag{8.163}$$

Therefore, denoting by

$$\mathfrak{C}_n := \left\{ \mathbf{t} \in \mathrm{Prob}(m) : D\left(\mathbf{t}\|\mathbf{q}\right) - D\left(\mathbf{t}\|\mathbf{p}\right) \geqslant \frac{1}{n} \log \lambda \right\} \tag{8.164}$$

we get that the error probability can be bounded by

$$\mathrm{Pr}_{\mathrm{error}}(\rho^{\otimes n}, \sigma^{\otimes n}, s_0) \geqslant \min\left\{ \sum_{\substack{x^n \in [m]^n \\ \mathbf{t}(x^n) \in \mathfrak{C}_n^c}} p_{x^n}, \sum_{\substack{x^n \in [m]^n \\ \mathbf{t}(x^n) \in \mathfrak{C}_n}} q_{x^n} \right\}. \tag{8.165}$$

The first sum corresponds to the probability (with respect to an i.i.d.$\sim \mathbf{p}$ source) that a sequence of size $n$ has a type belonging to $\mathfrak{C}_n^c$, whereas the second sum corresponds to the probability (with respect to an i.i.d.$\sim \mathbf{q}$ source) that a sequence of size $n$ has a type belonging to $\mathfrak{C}_n$. These probabilities are very similar to the one appearing in Sanov's theorem (see Theorem 8.8) except that here the set $\mathfrak{C}_n$ depends on $n$. To remove this dependancy on $n$, observe that in the limit $n \to \infty$ the set $\mathfrak{C}_n$ approaches the set

$$\mathfrak{C} := \{\mathbf{t} \in \mathrm{Prob}(m) : \ D\left(\mathbf{t}\|\mathbf{q}\right) \geqslant D\left(\mathbf{t}\|\mathbf{p}\right)\}, \tag{8.166}$$

and similarly, the set $\mathfrak{C}_n^c$ (in $\mathrm{Prob}(m)$) approaches the set

$$\mathfrak{S} := \{\mathbf{t} \in \mathrm{Prob}(m) : \ D\left(\mathbf{t}\|\mathbf{q}\right) \leqslant D\left(\mathbf{t}\|\mathbf{p}\right)\}. \tag{8.167}$$

Let $\mathbf{p}^\star, \mathbf{q}^\star \in \mathrm{Prob}(m)$ be the optimizers

$$D(\mathbf{p}^\star\|\mathbf{p}) := \min_{\mathbf{r} \in \mathfrak{S}} D(\mathbf{r}\|\mathbf{p}) \quad \text{and} \quad D(\mathbf{q}^\star\|\mathbf{q}) := \min_{\mathbf{r} \in \mathfrak{C}} D(\mathbf{r}\|\mathbf{q}). \tag{8.168}$$

From definitions and the continuity of the relative entropy, there exists two sequences of vectors $\{\mathbf{p}_n\}_{n \in \mathbb{N}}$ and $\{\mathbf{q}_n\}_{n \in \mathbb{N}}$ with limits $\mathbf{p}_n \to \mathbf{p}^*$ and $\mathbf{q}_n \to \mathbf{q}^*$ as $n \to \infty$ such that for each $n \in \mathbb{N}$ the vector $\mathbf{p}_n \in \mathfrak{C}_n^c \cap \mathrm{Type}(n, m)$ and the vector $\mathbf{q}_n \in \mathfrak{C}_n \cap \mathrm{Type}(n, m)$. Therefore, following similar lines as given in Sanov's theorem we get for the first sum

$$\sum_{\substack{x^n \in [m]^n \\ t(x^n) \in \mathfrak{C}_n^c}} p_{x^n} = \sum_{t \in \mathfrak{C}_n^c \cap \mathrm{Type}(n,m)} |X^n(\mathbf{t})| 2^{-n\left(H(\mathbf{t})+D(\mathbf{t}\|\mathbf{p})\right)}$$

$$\underset{t=\mathbf{p}_n \text{ in the sum}}{\overset{\textit{Taking only the term}}{\longrightarrow}} \geqslant |X^n(\mathbf{p}_n)| 2^{-n\left(H(\mathbf{p}_n)+D(\mathbf{p}_n\|\mathbf{p})\right)} \tag{8.169}$$

$$\mathbf{(8.89)} \rightarrow \; \geqslant \frac{1}{(n+1)^m} 2^{-nD(\mathbf{p}_n\|\mathbf{p})}.$$

Similarly, the second sum is bounded from below by

$$\sum_{\substack{x^n \in [m]^n \\ t(x^n) \in \mathfrak{C}_n}} q_{x^n} \geqslant \frac{1}{(n+1)^m} 2^{-nD(\mathbf{q}_n\|\mathbf{q})}. \tag{8.170}$$

Substituting the two lower bounds above into (8.161) yields

$$\limsup_{n\to\infty} -\frac{1}{n}\log \mathrm{Pr}_{\mathrm{error}}(\rho^{\otimes n}, \sigma^{\otimes n}, s_0) \leqslant \limsup_{n\to\infty} -\frac{1}{n}\log \frac{\min\left\{2^{-nD(\mathbf{p}_n\|\mathbf{p})}, 2^{-nD(\mathbf{q}_n\|\mathbf{q})}\right\}}{(n+1)^m}$$

$$= \min\{D(\mathbf{p}^\star\|\mathbf{p}), D(\mathbf{q}^\star\|\mathbf{q})\}. \tag{8.171}$$

To compute $\mathbf{p}^\star$, observe that

$$\min_{\mathbf{r}\in\mathfrak{R}} D(\mathbf{r}\|\mathbf{p}) = \min_{\mathbf{r}\in\mathrm{Prob}(m)} \left\{D(\mathbf{r}\|\mathbf{p}) : D(\mathbf{r}\|\mathbf{q}) \leqslant D(\mathbf{r}\|\mathbf{p})\right\}$$

$$\mathbf{Exercise\ 8.24} \rightarrow = \min_{\mathbf{r}\in\mathrm{Prob}(m)} \left\{D(\mathbf{r}\|\mathbf{p}) : D(\mathbf{r}\|\mathbf{q}) = D(\mathbf{r}\|\mathbf{p})\right\} \tag{8.172}$$

$$= \min_{\mathbf{r}\in\mathrm{Prob}(m)} \left\{\sum_{x\in[m]} r_x \log \frac{r_x}{p_x} : \sum_{x\in[m]} r_x \log \frac{q_x}{p_x} = 0\right\}.$$

Using the method of Lagrange multipliers, we define the Lagrangian

$$\mathcal{L}(\mathbf{r}) := \sum_{x\in[m]} r_x \log \frac{r_x}{p_x} + \mu \sum_{x\in[m]} r_x \log \frac{q_x}{p_x} + \nu \sum_{x\in[m]} r_x, \tag{8.173}$$

where $\mu$ and $\nu$ are some coefficients (Lagrange multipliers). Hence,

$$\frac{\partial\mathcal{L}(\mathbf{r})}{\partial r_x} = \log(e) + \log \frac{r_x}{p_x} + \mu \log \frac{q_x}{p_x} + \nu = 0, \tag{8.174}$$

which gives after isolating $r_x$

$$r_x = a p_x \left(\frac{q_x}{p_x}\right)^\mu = a p_x^{1-\mu} q_x^\mu, \tag{8.175}$$

where $a = e^{-1} 2^{-\nu}$ is some constant determined by the condition $\sum_x r_x = 1$. Hence,

$$r_x = \frac{p_x^{1-\mu} q_x^\mu}{\sum_{x'\in[m]} p_{x'}^{1-\mu} q_{x'}^\mu}. \tag{8.176}$$

Hence, denoting by $\mathbf{r}_\mu$ the probability vector whose components are as above we conclude that

$$\min_{\mathbf{r}\in\mathfrak{R}} D(\mathbf{r}\|\mathbf{p}) = D(\mathbf{r}_\mu\|\mathbf{p}), \tag{8.177}$$

where $\mu$ is the number satisfies

$$D\left(\mathbf{r}_\mu\|\mathbf{p}\right) = D\left(\mathbf{r}_\mu\|\mathbf{q}\right). \tag{8.178}$$

Moreover, note that for $\mu = 0$, $\mathbf{r}_0 = \mathbf{p}$, and for $\mu = 1$, $\mathbf{r}_1 = \mathbf{q}$. Therefore, from the continuity of the KL-divergence it follows that the equality above is achieved for some $\mu \in [0, 1]$.

From the symmetry of the expression for $\mathbf{r}_\mu$ it is clear that by the repetition of the same computation as above we get $\mathbf{q}^\star = \mathbf{p}^\star = \mathbf{r}_\mu$. We therefore conclude that

$$\lim_{n\to\infty} -\frac{1}{n} \log \mathrm{Pr}_{\mathrm{error}}(\rho^{\otimes n}, \sigma^{\otimes n}, t_0) \leqslant D\left(\mathbf{r}_\mu\|\mathbf{p}\right), \tag{8.179}$$

where $\mu$ is determined by $D\left(\mathbf{r}_\mu\|\mathbf{q}\right) = D\left(\mathbf{r}_\mu\|\mathbf{p}\right)$. In the exercise below you will show that for this $\mu$

$$D\left(\mathbf{r}_\mu\|\mathbf{q}\right) = -\log \min_{\alpha\in[0,1]} \sum_{x\in[m]} p_x^{1-\alpha} q_x^\alpha. \tag{8.180}$$

This completes the proof. ∎

**Exercise 8.24.** *Prove the second equality of (8.172). Hint: Suppose that the minimum is obtained for some $\mathbf{r} \in \mathrm{Prob}(m)$ that satisfies $D\left(\mathbf{r}\|\mathbf{q}\right) < D\left(\mathbf{r}\|\mathbf{p}\right)$ and get a contradiction by showing that the vector $\mathbf{t} - (1-\varepsilon)\mathbf{r} + \varepsilon\mathbf{p}$ (with small $\varepsilon > 0$) satisfies $D(\mathbf{t}\|\mathbf{p}) < D(\mathbf{r}\|\mathbf{p})$.*

**Exercise 8.25.** *Prove (8.180). Hint: Show that the condition $D(\mathbf{r}_\mu\|\mathbf{p}) = D(\mathbf{r}_\mu\|\mathbf{q})$ is equivalent to $\sum_{x\in[m]} p_x^{1-\mu} q_x^\mu \log \frac{q_x}{p_x} = 0$ and compare it with the derivative of the function $f(s) := \sum_{x\in[m]} p_x^{1-s} q_x^s$.*

# 8.7 Quantum Hypothesis Testing

One of the foundational aspects of quantum mechanics is the inability to perfectly distinguish between quantum systems. To elucidate this concept of distinguishability, let's build upon the ideas presented in the previous section. Consider a scenario where an experimenter, named Alice, possesses a quantum system in her lab (e.g. an electron in a certain spin state) that could be in one of two potential states, denoted as $\rho \in \mathfrak{D}(A)$ and $\sigma \in \mathfrak{D}(A)$. Alice can carry out a POVM denoted by $\{\Lambda_x\}_{x\in[m]}$ on her system to deduce its state. Depending on the measurement outcome $x$ she may infer the state to be $\rho$ or $\sigma$. In this section, we delve into the best strategy Alice can employ to accurately determine the state of her quantum system.

We note that it's adequate to contemplate POVMs composed of only two elements. We can define $\Lambda \in \mathrm{Eff}(A)$ to be the sum of all effects $\{\Lambda_x\}$, where $x$ leads Alice to infer

$\rho$. Conversely, $I - \Lambda$ is the sum of the remaining POVM elements, corresponding to $x$ values that result in Alice inferring $\sigma$. Two types of errors might arise:

1. **Type I Error:** Alice possesses the state $\rho$ but incorrectly infers it as $\sigma$. The associated probability is

$$\alpha(\Lambda) := \mathrm{Tr}\left[\rho(I - \Lambda)\right]. \tag{8.181}$$

2. **Type II Error:** Alice has the state $\sigma$ but mistakenly deduces it as $\rho$. The corresponding probability is

$$\beta(\Lambda) := \mathrm{Tr}\left[\sigma\Lambda\right]. \tag{8.182}$$

As in the classical scenario, we explore strategies to minimize the error probabilities $\alpha(\Lambda)$ and $\beta(\Lambda)$. With the asymmetric approach, the objective is to minimize the Type II error, $\beta(\Lambda)$, while ensuring that the Type I error, $\alpha(\Lambda)$, stays below a specific threshold $\varepsilon > 0$. The optimal approach is encapsulated by the quantum Stein's lemma. In the symmetric strategy, the observer is aware of a prior $\{s_0, s_1\}$, where $\rho$ occurs with probability $s_0$, and $\sigma$ with probability $s_1$. The aim here is to reduce the overall error probability represented by $s_0\alpha(\Lambda) + s_1\beta(\Lambda)$.

It's worth noting that any pair of quantum states $\rho, \sigma \in \mathfrak{D}(A)$ that aren't identical satisfy $\mathrm{Tr}[\rho\sigma] < 1$. This implies:

$$\lim_{n\to\infty} \mathrm{Tr}\left[\rho^{\otimes n}\sigma^{\otimes n}\right] = \lim_{n\to\infty} (\mathrm{Tr}[\rho\sigma])^n = 0. \tag{8.183}$$

In essence, as $n$ approaches infinity, the states $\rho^{\otimes n}$ and $\sigma^{\otimes n}$ become orthogonal with respect to the Hilbert–Schmidt inner product. Naturally, we might question the rate at which these states turn distinguishable. Both the quantum Stein's lemma and the quantum Chernoff bound address this question, in the asymmetric and symmetric contexts, respectively.

## 8.7.1 The Quantum Hypothesis Testing Divergence

In asymmetric hypothesis testing, the goal is to minimize the Type II error while simultaneously ensuring that the Type I error remains bounded by a small parameter $\varepsilon > 0$. Under this scenario, the pertinent error probability is defined as

$$\beta^*(\varepsilon) := \min\left\{\beta(\Lambda) : \alpha(\Lambda) \leqslant \varepsilon, \ 0 \leqslant \Lambda \leqslant I\right\}. \tag{8.184}$$

**Exercise 8.26.** *Show that the optimal probability $\beta^*(\varepsilon) = 0$ for all $\varepsilon \geqslant 0$ if and only if $\rho$ and $\sigma$ satisfies $\rho\sigma = 0$.*

By taking the $-\log$ of the above error probability one obtains a quantity known as the *quantum hypothesis testing divergence*.

> **Definition 8.7.** The quantum hypothesis testing divergence is defined for all $\rho, \sigma \in \mathfrak{D}(A)$ and $\varepsilon \in [0, 1)$ as
>
> $$D_{\min}^{\varepsilon}(\rho\|\sigma) := -\log \min_{\Lambda \in \mathrm{Eff}(A)} \{ \mathrm{Tr}\,[\sigma\Lambda] : \mathrm{Tr}\,[\rho\Lambda] \geqslant 1 - \varepsilon \}. \tag{8.185}$$

The hypothesis testing divergence is invariably nonnegative and reaches infinity when $\rho$ and $\sigma$ are orthogonal. Furthermore, Exercise 8.27 guides you to demonstrate that when $\rho$ and $\sigma$ are diagonal in the same basis, the quantum hypothesis testing divergence, denoted as $D_{\min}^{\varepsilon}(\rho\|\sigma)$, simplifies to its classical equivalent, $D_{\min}^{\varepsilon}(\mathbf{p}\|\mathbf{q})$. Here, $\mathbf{p}$ and $\mathbf{q}$ represent the diagonal elements of $\rho$ and $\sigma$, respectively. In the classical context, we observed that for $\varepsilon = 0$, the divergence $D_{\min}^{\varepsilon}$ reduces to the min relative entropy. This observation is consistent in the quantum scenario as well (refer to Exercise 8.27). Such parallels justify the use of the "min" subscript in naming the quantum hypothesis testing divergence. Next, we aim to establish that this function indeed qualifies as an (unnormalized) divergence.

> **Data Processing Inequality**
>
> **Theorem 8.13.** Let $\varepsilon \in [0, 1)$, $\rho, \sigma \in \mathfrak{D}(A)$, and $\mathcal{E} \in \mathrm{CPTP}(A \to B)$. Then,
>
> $$D_{\min}^{\varepsilon}\big(\mathcal{E}(\rho)\big\|\mathcal{E}(\sigma)\big) \leqslant D_{\min}^{\varepsilon}(\rho\|\sigma). \tag{8.186}$$

**Proof**   Let $0 \leqslant \Gamma \leqslant I^B$ be an optimal effect such that

$$2^{-D_{\min}^{\varepsilon}(\mathcal{E}(\rho)\|\mathcal{E}(\sigma))} = \mathrm{Tr}\,[\mathcal{E}(\sigma)\Gamma] \tag{8.187}$$

and $\mathrm{Tr}\,[\mathcal{E}(\rho)\Gamma] \geqslant 1 - \varepsilon$. Note that $\Lambda := \mathcal{E}^*(\Gamma)$ satisfies $0 \leqslant \Lambda \leqslant I^A$ and $\mathrm{Tr}\,[\rho\Lambda] = \mathrm{Tr}[\mathcal{E}(\rho)\Gamma] \geqslant 1 - \varepsilon$. Hence,

$$
\begin{aligned}
2^{-D_{\min}^{\varepsilon}(\mathcal{E}(\rho)\|\mathcal{E}(\sigma))} &= \mathrm{Tr}\,[\mathcal{E}(\sigma)\Gamma] \\
&= \mathrm{Tr}\,[\sigma\Lambda] \\
&\geqslant \min\big\{ \mathrm{Tr}\,[\sigma\Lambda'] : \mathrm{Tr}\,[\rho\Lambda'] \geqslant 1 - \varepsilon, \ \Lambda' \in \mathrm{Eff}(A) \big\} \\
&= 2^{-D_{\min}^{\varepsilon}(\rho\|\sigma)}.
\end{aligned}
\tag{8.188}
$$

This completes the proof.      ∎

**Exercise 8.27.** *Consider Definition 8.7 of the quantum hypothesis testing.*

1. *Show that if $\rho, \sigma \in \mathfrak{D}(A)$ are diagonal in the same basis of $A$, then $D_{\min}^{\varepsilon}(\rho\|\sigma)$ reduces to its classical counterpart $D_{\min}^{\varepsilon}(\mathbf{p}\|\mathbf{q})$, where $\mathbf{p}$ and $\mathbf{q}$ are the diagonals of $\rho$ and $\sigma$, respectively.*

2. *Show that for $\varepsilon = 0$, the quantum hypothesis testing divergence simplifies to the quantum min relative entropy. That is, show that for all $\rho, \sigma \in \mathfrak{D}(A)$,*

$$D_{\min}^{\varepsilon=0}(\rho\|\sigma) = D_{\min}(\rho\|\sigma) = -\log \mathrm{Tr}\,\lfloor \sigma \Pi_{\rho} \rfloor. \tag{8.189}$$

**Exercise 8.28.** *Show that the constraint* $\mathrm{Tr}\,[\rho\Lambda] \geq 1 - \varepsilon$ *in* (8.185) *can be replaced with* $\mathrm{Tr}\,[\rho\Lambda] = 1 - \varepsilon$ *(i.e. both constraints leads to the same value of* $D^{\varepsilon}_{\min}(\rho\|\sigma)$*).*

**Exercise 8.29.**

1. *Show that for all* $\rho, \sigma \in \mathfrak{D}(A)$ *we have*

$$D^{\varepsilon}_{\min}(\rho\|\sigma) \geq -\log\,(1-\varepsilon), \tag{8.190}$$

    *with equality if* $\rho = \sigma$.
2. *Show that* $D^{\varepsilon}_{\min}(\rho\|\sigma)$ *is nondecreasing in* $\varepsilon$.

**Exercise 8.30.** *Show that the quantum hypothesis testing divergence equals its minimal extension from classical states. That is, show that for all* $\rho, \sigma \in \mathfrak{D}(A)$,

$$D^{\varepsilon}_{\min}\left(\rho^{A}\|\sigma^{A}\right) = \sup_{\mathcal{E}\in\mathrm{CPTP}(A\to X)} D^{\varepsilon}_{\min}\left(\mathcal{E}^{A\to X}(\rho^{A})\|\mathcal{E}^{A\to X}(\sigma^{A})\right), \tag{8.191}$$

*where the supremum is over all classical systems* $X$ *and POVM channels* $\mathcal{E} \in \mathrm{CPTP}$ $(A \to X)$ *that take* $\rho$ *and* $\sigma$ *to diagonal density matrices (i.e. probability vectors).*

## Computation of the Quantum Hypothesis Testing Divergence

The quantum hypothesis testing divergence can be computed using semidefinite programming (SDP) techniques. In particular, we can express it for $\rho, \sigma \in \mathfrak{D}(A)$ as in (A.52 of online version) via

$$2^{-D^{\varepsilon}_{\min}(\rho\|\sigma)} = \min_{\substack{\mathcal{N}(\eta)-H_2\geq 0 \\ \eta\geq 0}} \mathrm{Tr}[\eta H_1] \tag{8.192}$$

with the identifications $H_1 := \sigma$, $H_2 := (1 - \varepsilon) \oplus ( - I^A)$, and $\mathcal{N}: \mathrm{Herm}(A) \to \mathbb{R} \oplus \mathrm{Herm}(A)$ defined for all $\eta \in \mathrm{Herm}(A)$ as

$$\mathcal{N}(\eta) := \mathrm{Tr}[\rho\eta] \oplus ( - \eta). \tag{8.193}$$

Note that $\mathcal{N}$ is a linear map and its dual map $\mathcal{N}^*: \mathbb{R} \oplus \mathrm{Herm}(A) \to \mathrm{Herm}(A)$ is given by (see the following exercise)

$$\mathcal{N}^*(t \oplus \omega) = t\rho - \omega. \tag{8.194}$$

From (A.57 of online version) it then follows that the dual to the above SDP optimization problem is given by

$$2^{-D^{\varepsilon}_{\min}(\rho\|\sigma)} = \max_{\substack{H_1-\mathcal{N}^*(t\oplus\omega)\geq 0 \\ t\in\mathbb{R}_+,\, \omega\in\mathrm{Pos}(A)}} \mathrm{Tr}[(t \oplus \omega)H_2]. \tag{8.195}$$

Substituting the expressions above for $H_1$, $H_2$, and $\mathcal{N}^*$ gives

$$2^{-D^{\varepsilon}_{\min}(\rho\|\sigma)} = \max\left\{t(1 - \varepsilon) - \mathrm{Tr}[\omega] : t\rho \leq \sigma + \omega,\ t \in \mathbb{R}_+,\ \omega \in \mathrm{Pos}(A)\right\}. \tag{8.196}$$

The maximization above is over all $t \in \mathbb{R}_+$ and $\omega \in \mathrm{Pos}(A)$. For a fixed $t$, we want to minimize $\mathrm{Tr}[\omega]$ such that $\omega \geq 0$ and $\omega \geq t\rho - \sigma$. Under these constraints, it follows from Exercise 8.32 that the choice $\omega := (t\rho - \sigma)_+$ has the minimal trace. We therefore conclude that

$$D_{\min}^{\varepsilon}(\rho\|\sigma) = -\log\max_{t\in\mathbb{R}_+} f(t), \tag{8.197}$$

where $f: \mathbb{R}_+ \to \mathbb{R}_+$ is the function

$$f(t) := t(1 - \varepsilon) - \mathrm{Tr}(t\rho - \sigma)_+. \tag{8.198}$$

**Exercise 8.31.**

1. *Verify that indeed (8.192) is equivalent to (8.185).*
2. *Prove that the dual map $\mathcal{N}^*$ is given by (8.194), and use it to derive (8.196) from (8.195).*

**Exercise 8.32.**

1. *Let $\eta \in \mathrm{Herm}(A)$. Show that*

$$\mathrm{Tr}[\eta_+] = \inf\left\{\mathrm{Tr}[\zeta] : \zeta \geqslant \eta, \ \zeta \in \mathrm{Pos}(A)\right\}, \tag{8.199}$$

*where $\eta_+$ is the positive part of $\eta$ (i.e. $\eta = \eta_+ - \eta_-$ with $\eta_+, \eta_- \in \mathrm{Pos}(A)$ and $\eta_+\eta_- = 0$).*
2. *Show that the function in (8.198) can be expressed as*

$$f(t) = \frac{1}{2}\left(1 + t(1 - 2\varepsilon) - \|t\rho - \sigma\|_1\right). \tag{8.200}$$

*Hint: Recall that $\eta_+ = \frac{1}{2}(|\eta| + \eta)$.*

## The Relationship between the Hypothesis Testing and Rényi Divergences

In the subsequent theorem, we establish a connection between the hypothesis testing divergence and the R'enyi relative entropies. For this purpose, we use the notation $D_\alpha$ to represent the Petz quantum $\alpha$-R'enyi divergence (as defined in Definition 6.6), and $\tilde{D}_\alpha$ to denote the quantum sandwiched (minimal) $\alpha$-R'enyi divergence (as per Definition 6.7). Additionally, we use $h(\alpha) = -\alpha\log\alpha - (1 - \alpha)\log(1 - \alpha)$ to denote the binary Shannon entropy.

---

**Theorem 8.14.** Let $\varepsilon \in (0, 1)$ and $\rho, \sigma \in \mathfrak{D}(A)$.

1. For all $\alpha > 1$:
$$D_{\min}^{\varepsilon}(\rho\|\sigma) \leqslant \tilde{D}_\alpha(\rho\|\sigma) + \frac{\alpha}{\alpha - 1}\log\left(\frac{1}{1 - \varepsilon}\right) \tag{8.201}$$

2. For all $\alpha \in (0, 1)$:
$$D_{\min}^{\varepsilon}(\rho\|\sigma) \geqslant D_\alpha(\rho\|\sigma) + \frac{\alpha}{1 - \alpha}\left(\frac{h(\alpha)}{\alpha} - \log\left(\frac{1}{\varepsilon}\right)\right) \tag{8.202}$$

---

*Remark.* Given that $\tilde{D}_\alpha$ represents the minimal quantum extension of the classical $\alpha$-Rényi relative entropy, it follows that $\tilde{D}_\alpha(\rho\|\sigma) \leqslant D_\alpha(\rho\|\sigma)$. Consequently, in the upper bound of (8.201), we can substitute $\tilde{D}_\alpha(\rho\|\sigma)$ with $D_\alpha(\rho\|\sigma)$.

**Proof**  Let $\Lambda \in \mathrm{Eff}(A)$ be such that $2^{-D_{\min}^\varepsilon(\rho\|\sigma)} = \mathrm{Tr}[\Lambda\sigma]$ and $\mathrm{Tr}[\Lambda\rho] = 1 - \varepsilon$. Set $p := \mathrm{Tr}[\Lambda\sigma]$, and define the binary POVM Channel $\mathcal{E} \in \mathrm{CPTP}(A \to X)$ via

$$\mathcal{E}(\omega) := \mathrm{Tr}[\omega\Lambda]|0\rangle\langle 0| + \mathrm{Tr}[\omega(I - \Lambda)]|1\rangle\langle 1| \qquad \forall\, \omega \in \mathcal{L}(A). \tag{8.203}$$

From the DPI of $\tilde{D}_\alpha$ we get that

$$\tilde{D}_\alpha(\rho\|\sigma) \geqslant \tilde{D}_\alpha\big(\mathcal{E}(\rho)\big\|\mathcal{E}(\sigma)\big) = D_\alpha\left((1 - \varepsilon, \varepsilon)^T \big\| (p, 1 - p)^T\right)$$

$$\text{By definition} \to = \frac{1}{\alpha - 1}\log\left((1 - \varepsilon)^\alpha p^{1-\alpha} + \varepsilon^\alpha(1 - p)^{1-\alpha}\right)$$

$$\boxed{\text{Removing } \varepsilon^\alpha(1 - p)^{1-\alpha}} \longrightarrow \;\geqslant \frac{1}{\alpha - 1}\log\left((1 - \varepsilon)^\alpha p^{1-\alpha}\right) \tag{8.204}$$

$$= \frac{\alpha}{\alpha - 1}\log(1 - \varepsilon) - \log p$$

$$\text{By definition of } p \to = \frac{\alpha}{\alpha - 1}\log(1 - \varepsilon) + D_{\min}^\varepsilon(\rho\|\sigma).$$

This concludes the proof of (8.201).

To prove (8.202), let $\alpha \in (0, 1)$. We will use the expression for $D_{\min}^\varepsilon(\rho\|\sigma)$ as given in (8.197) and (8.198). To bound the expression $\mathrm{Tr}(t\rho - \sigma)_+$ in equation (8.198), we employ the quantum weighted geometric-mean inequality given by (B.68). This inequality asserts that for any pair of matrices $M, N \in \mathrm{Pos}(A)$ and any value of $\alpha$ within the range $[0,1]$,

$$\frac{1}{2}\mathrm{Tr}\Big[M + N - |M - N|\Big] \leqslant \mathrm{Tr}\big[M^\alpha N^{1-\alpha}\big]. \tag{8.205}$$

Since the term $|M - N|$ can be expressed as $|M - N| = 2(M - N)_+ - (M - N)$, the above inequality is equivalent to

$$\mathrm{Tr}(M - N)_+ \geqslant \mathrm{Tr}[M] - \mathrm{Tr}\big[M^\alpha N^{1-\alpha}\big]. \tag{8.206}$$

Taking $M = t\rho$ and $N = \sigma$ we have

$$\mathrm{Tr}(t\rho - \sigma)_+ \geqslant t - t^\alpha \mathrm{Tr}\big[\rho^\alpha \sigma^{1-\alpha}\big]$$
$$= t - t^\alpha 2^{(\alpha-1)D_\alpha(\rho\|\sigma)}. \tag{8.207}$$

Substituting this into (8.197) and (8.198) we get

$$2^{-D_{\min}^\varepsilon(\rho\|\sigma)} = \max_{t\in\mathbb{R}_+}\left\{t(1 - \varepsilon) - \mathrm{Tr}(t\rho - \sigma)_+\right\}$$

$$(\textbf{8.207}) \to \;\leqslant \max_{t\in\mathbb{R}_+}\left\{-t\varepsilon + t^\alpha 2^{(\alpha-1)D_\alpha(\rho\|\sigma)}\right\}. \tag{8.208}$$

It is straightforward to check that for fixed $\alpha, \rho, \sigma, \varepsilon$, the function $t \mapsto -t\varepsilon + t^\alpha 2^{(\alpha-1)D_\alpha(\rho\|\sigma)}$ obtains its maximal value at

$$t = \left(\frac{\alpha}{\varepsilon}\right)^{\frac{1}{1-\alpha}} 2^{-D_\alpha(\rho\|\sigma)}. \tag{8.209}$$

Substituting this value into the optimization in (8.208) gives

$$2^{-D_{\min}^{\varepsilon}(\rho\|\sigma)} \leqslant (1-\alpha)\left(\frac{\alpha}{\varepsilon}\right)^{\frac{\alpha}{1-\alpha}} 2^{-D_\alpha(\rho\|\sigma)}. \tag{8.210}$$

By taking $-\log$ on both sides we get (8.202). This concludes the proof. ∎

## 8.7.2 Asymmetric Discrimination of Quantum States

The subsequent theorem extends Theorem 8.11 to encompass the quantum domain. Consequently, the proof we present for the quantum scenario also substantiates the classical Stein's lemma, as given in Theorem 8.11.

> **The Quantum Stein's Lemma**
>
> **Theorem 8.15.** Let $A$ be a finite dimensional system, $0 < \varepsilon < 1$, and $\rho, \sigma \in \mathfrak{D}(A)$ with $\mathrm{supp}\,(\rho) \subseteq \mathrm{supp}\,(\sigma)$. Then,
>
> $$\lim_{n\to\infty} \frac{1}{n} D_{\min}^{\varepsilon}\left(\rho^{\otimes n}\|\sigma^{\otimes n}\right) = D(\rho\|\sigma), \tag{8.211}$$
>
> where $D(\rho\|\sigma) := \mathrm{Tr}[\rho\log\rho] - \mathrm{Tr}[\rho\log\sigma]$ is known as *the Umegaki relative entropy*.

*Remark.* The quantum Stein's lemma indicates that the optimal type II error approximately follows the behavior of $\approx 2^{-nD(\rho\|\sigma)}$ with respect to the number of copies, $n$, of $\rho$ and $\sigma$. Specifically, the lemma offers an operational interpretation of the Umegaki divergence, $D(\rho\|\sigma)$, defining it as the maximal rate at which the type II error diminishes to zero in an exponential manner with increasing $n$. Additionally, it's worth noting that the theorem above implies that the limit on the right-hand side of (8.211) exists and is independent on $\varepsilon$.

**Proof** The proof follows from the bounds in Theorem 8.14. Specifically, from (8.201) we get for any $\varepsilon \in (0, 1)$ and any $\alpha > 1$

$$\limsup_{n\to\infty} \frac{1}{n} D_{\min}^{\varepsilon}\left(\rho^{\otimes n}\|\sigma^{\otimes n}\right) \leqslant \limsup_{n\to\infty} \frac{1}{n}\left(\tilde{D}_\alpha\left(\rho^{\otimes n}\|\sigma^{\otimes n}\right) + \frac{\alpha}{\alpha-1}\log\left(\frac{1}{1-\varepsilon}\right)\right)$$
$$= \tilde{D}_\alpha\left(\rho\|\sigma\right), \tag{8.212}$$

where in the last equality we used the additivity (under tensor products) of $\tilde{D}_\alpha$. Since the equation above holds for all $\alpha > 1$ we conclude that

$$\limsup_{n\to\infty} \frac{1}{n} D_{\min}^{\varepsilon}\left(\rho^{\otimes n}\|\sigma^{\otimes n}\right) \leqslant \lim_{\alpha\to 1^+} \tilde{D}_\alpha\left(\rho\|\sigma\right)$$
$$= D(\rho\|\sigma), \tag{8.213}$$

where the equality above follows from continuity in $\alpha$ of the function $\alpha \mapsto \tilde{D}_\alpha(\rho\|\sigma)$.

For the opposite inequality, we use the bound (8.202) to get for all $\alpha \in (0, 1)$

$$\liminf_{n \to \infty} \frac{1}{n} D_{\min}^{\varepsilon} \left( \rho^{\otimes n} \big\| \sigma^{\otimes n} \right) \geqslant \liminf_{n \to \infty} \frac{1}{n} \left( D_\alpha \left( \rho^{\otimes n} \big\| \sigma^{\otimes n} \right) + \frac{\alpha}{1 - \alpha} \left( \frac{h(\alpha)}{\alpha} + \log \varepsilon \right) \right)$$

$$= D_\alpha \left( \rho \| \sigma \right), \tag{8.214}$$

where we used the additivity of $D_\alpha$. Since $D_\alpha$ is continuous in $\alpha$, and since the equation above holds for all $\alpha \in (0, 1)$, it must also hold for $\alpha = 1$; that is,

$$\liminf_{n \to \infty} \frac{1}{n} D_{\min}^{\varepsilon} \left( \rho^{\otimes n} \big\| \sigma^{\otimes n} \right) \geqslant D(\rho \| \sigma). \tag{8.215}$$

Combining this with the inequality (8.213), we conclude that the limit

$$\lim_{n \to \infty} \frac{1}{n} D_{\min}^{\varepsilon} \left( \rho^{\otimes n} \big\| \sigma^{\otimes n} \right) \tag{8.216}$$

exists and equals to $D(\rho \| \sigma)$.     ∎

**Exercise 8.33.** [The Umegaki Relative Entropy] *Let $D$ be the Umegaki relative entropy.*

1. *Show that $D$ satisfies the DPI. Hint: Use (8.211) and the fact that $D_{\min}^{\varepsilon}$ satisfies the DPI.*

2. *Show by direct calculation that for any two cq-states in $\mathfrak{D}(AX)$, $\rho^{AX} := \sum_{x \in [n]} p_x \rho_x^A \otimes |x\rangle\langle x|^X$ and $\sigma^{AX} := \sum_{x \in [n]} q_x \sigma_x^A \otimes |x\rangle\langle x|^X$ we have*

$$D \left( \rho^{AX} \big\| \sigma^{AX} \right) = \sum_{x \in [n]} p_x D \left( \rho_x^A \big\| \sigma_x^A \right) + D(\mathbf{p} \| \mathbf{q}), \tag{8.217}$$

*where the components of the probability vectors $\mathbf{p}$ and $\mathbf{q}$ are $\{p_x\}_{x \in [n]}$ and $\{q_x\}_{x \in [n]}$, respectively.*

3. *Use the above two properties to show that for any two ensembles of states $\{p_x, \rho_x\}_{x \in [n]}$ and $\{q_x, \sigma_x\}_{x \in [n]}$ we have*

$$D \left( \sum_{x \in [n]} p_x \rho_x \, \Big\| \, \sum_{x \in [n]} q_x \sigma_x \right) \leqslant \sum_{x \in [n]} p_x D \left( \rho_x \| \sigma_x \right) + D(\mathbf{p} \| \mathbf{q}). \tag{8.218}$$

*In particular, show that the Umegaki relative entropy is jointly convex.*

## 8.7.3 Symmetric Discrimination of Quantum States

Consider the following setup, in which an observer (say Alice) is given a quantum state $\rho$ with probability $t_0 := t \in [0, 1]$, and a quantum state $\sigma \in \mathfrak{D}(A)$ with probability $t_1 := 1 - t$. As before, the goal is for Alice to guess correctly which state she was given. For this purpose Alice performs a binary outcome POVM consisting of two POVM elements $\Lambda_0 := \Lambda$ and $\Lambda_1 := I - \Lambda$, where $0 \leqslant \Lambda \leqslant I$. If Alice gets outcome 0 she declares that the state is $\rho$, and if the outcome is 1 she declares that the state is $\sigma$. For any such POVM the probability of error is given by

$$\begin{aligned}
\mathrm{Pr}_{\mathrm{error}}(\Lambda,\rho,\sigma,t) &:= t_0\alpha(\Lambda) + t_1\beta(\Lambda) \\
&= t_0\mathrm{Tr}[\rho\Lambda_1] + t_1\mathrm{Tr}[\sigma\Lambda_0] \\
&= t_0 + \mathrm{Tr}\big[(t_1\sigma - t_0\rho)\Lambda\big].
\end{aligned} \tag{8.219}$$

Minimizing the above expression over all $0 \leqslant \Lambda \leqslant I$ gives

$$\begin{aligned}
\mathrm{Pr}_{\mathrm{error}}(\rho,\sigma,t) &:= \min_{0\leqslant\Lambda\leqslant I} \mathrm{Pr}_{\mathrm{error}}(\Lambda,\rho,\sigma,t) \\
\textbf{(8.199)} \rightarrow &= t_0 - \mathrm{Tr}\big(t_1\sigma - t_0\rho\big)_-
\end{aligned}$$

$$\boxed{\forall\eta \in \mathrm{Herm}(A),\ \eta_- = \tfrac{1}{2}(|\eta| - \eta)} \longrightarrow \begin{aligned} &= t_0 - \frac{1}{2}\big(\big\|t_1\sigma - t_0\rho\big\|_1 - 1 + 2t_0\big) \\[4pt] &= \frac{1}{2}\big(1 - \big\|t_0\rho - t_1\sigma\big\|_1\big). \end{aligned} \tag{8.220}$$

**Exercise 8.34.** *Show that with $t := t_0$ and $r := t_1/t_0$ we can express the probability of error as:*

$$\mathrm{Pr}_{\mathrm{error}}(\rho,\sigma,t) = t\big(1 - \mathrm{Tr}(\rho - r\sigma)_+\big). \tag{8.221}$$

**Exercise 8.35.** *Let $\rho,\sigma \in \mathfrak{D}(A)$.*

*1. Let $\varepsilon \in (0,1)$. Show that*

$$2^{-D^{\varepsilon}_{\min}(\rho\|\sigma)} = \sup_{t\in(0,1)} \frac{\mathrm{Pr}_{\mathrm{error}}(\rho,\sigma,t) - t\varepsilon}{1 - t}. \tag{8.222}$$

*Hint: Use (8.221) and (8.198).*

*2. Let $t$ and $r$ be as in (8.221). Show that*

$$\mathrm{Pr}_{\mathrm{error}}(\rho,\sigma,t) = t \inf_{\varepsilon\in(0,1)} \Big\{\varepsilon + r2^{-D^{\varepsilon}_{\min}(\rho\|\sigma)}\Big\}. \tag{8.223}$$

*Hint: Recall that $\mathrm{Tr}(\rho - r\sigma)_+ = \sup_{\Lambda\in\mathrm{Eff}(A)} \mathrm{Tr}[\Pi(\rho - r\sigma)]$ and split the supremum over all $\varepsilon \in (0,1)$ and all $\Lambda \in \mathrm{Eff}(A)$ such that $\mathrm{Tr}[\Lambda\rho] = 1 - \varepsilon$.*

As previously discussed, with increasing $n$ copies of $\rho$ and $\sigma$, the states $\rho^{\otimes n}$ and $\sigma^{\otimes n}$ become more distinguishable. We will demonstrate in the upcoming theorem that the error probability, $\mathrm{Pr}_{\mathrm{error}}(\rho^{\otimes n},\sigma^{\otimes n},t)$, diminishes at an exponential rate as $n$ approaches infinity. This rate is characterized by what is known as the quantum Chernoff bound. The classical counterpart of the subsequent theorem, along with its proof, can be found in Section 8.6 (see Theorem 8.12). We will use the notation $\xi_Q(\rho,\sigma)$ to denote the quantum extension of the classical Chernoff bound $\xi(\mathbf{p},\mathbf{q})$ as given in (8.158). In the quantum domain it is defined as

$$\begin{aligned}
\xi_Q(\rho,\sigma) &:= -\log \min_{0\leqslant\alpha\leqslant 1} \mathrm{Tr}[\rho^{\alpha}\sigma^{1-\alpha}] \\
\textbf{Definition 6.6} \ \rightarrow &= \max_{\alpha\in[0,1]} \big\{(1 - \alpha)D_{\alpha}(\rho\|\sigma)\big\}.
\end{aligned} \tag{8.224}$$

> **The Quantum Chernoff Bound**
>
> **Theorem 8.16.** Let $\rho, \sigma \in \mathfrak{D}(A)$. For any probability distribution $\{t_0 := t, t_1 := 1 - t\}$ with $0 < t < 1$,
>
> $$\lim_{n \to \infty} -\frac{1}{n} \log \mathrm{Pr}_{\mathrm{error}}(\rho^{\otimes n}, \sigma^{\otimes n}, t) = \xi_Q(\rho, \sigma). \tag{8.225}$$

**Proof**　In the proof of Theorem 8.14 we used (8.207) to bound $\mathrm{Tr}(t\rho - \sigma)_+$. Dividing both sides of (8.207) by $t$ and denoting $r := 1/t$ we get that (8.207) is equivalent to

$$\mathrm{Tr}(\rho - r\sigma)_+ \geqslant 1 - r^{1-\alpha} 2^{(\alpha-1)D_\alpha(\rho\|\sigma)}. \tag{8.226}$$

Combining this with (8.221) we get that for all $\alpha \in (0, 1)$

$$\mathrm{Pr}_{\mathrm{error}}(\rho, \sigma, t) \leqslant tr^{1-\alpha} 2^{(\alpha-1)D_\alpha(\rho\|\sigma)}$$

$$= t_0^\alpha t_1^{1-\alpha} \mathrm{Tr}\left[\rho^\alpha \sigma^{1-\alpha}\right]. \tag{8.227}$$

Hence,

$$\mathrm{Pr}_{\mathrm{error}}(\rho^{\otimes n}, \sigma^{\otimes n}, t) \leqslant t_0^\alpha t_1^{1-\alpha} \left(\mathrm{Tr}\left[\rho^\alpha \sigma^{1-\alpha}\right]\right)^n \tag{8.228}$$

so that

$$\liminf_{n \to \infty} -\frac{1}{n} \log \mathrm{Pr}_{\mathrm{error}}(\rho^{\otimes n}, \sigma^{\otimes n}, t) \geqslant -\log \mathrm{Tr}\left[\rho^\alpha \sigma^{1-\alpha}\right]. \tag{8.229}$$

Since the above equation holds for all $0 \leqslant \alpha \leqslant 1$ we have

$$\lim_{n \to \infty} -\frac{1}{n} \log \mathrm{Pr}_{\mathrm{error}}(\rho^{\otimes n}, \sigma^{\otimes n}, t) \geqslant \max_{\alpha \in [0,1]} \left\{ -\log \mathrm{Tr}\left[\rho^\alpha \sigma^{1-\alpha}\right] \right\}$$

$$= -\log \min_{\alpha \in [0,1]} \mathrm{Tr}\left[\rho^\alpha \sigma^{1-\alpha}\right]. \tag{8.230}$$

To prove the opposite inequality, let

$$\rho = \sum_{x \in [m]} p_x \psi_x \quad \text{and} \quad \sigma = \sum_{y \in [m]} q_y \phi_y \tag{8.231}$$

be the spectral decomposition of $\rho$ and $\sigma$ (here $m := |A|$), where $\psi_x, \phi_y \in \mathrm{Pure}(A)$ for all $x, y \in [m]$. Then, for any projection $\Pi \in \mathrm{Pos}(A)$ (i.e. $\Pi^2 = \Pi$) we have

$$\mathrm{Tr}\left[\Pi\rho\right] = \sum_{x \in [m]} p_x \langle \psi_x | \Pi | \psi_x \rangle = \sum_{x \in [m]} p_x \langle \psi_x | \Pi^2 | \psi_x \rangle$$

$$\boxed{\sum_{y \in [m]} \phi_y = I^A} \longrightarrow \quad = \sum_{x \in [m]} p_x \langle \psi_x | \Pi \sum_{y \in [m]} |\phi_y\rangle\langle\phi_y| \Pi | \psi_x \rangle \tag{8.232}$$

$$= \sum_{x,y \in [m]} p_x |\langle \psi_x | \Pi | \phi_y \rangle|^2.$$

Similarly, since $I - \Pi$ is also a projection we get

$$\mathrm{Tr}\left[(I - \Pi)\sigma\right] = \sum_{x,y \in [m]} q_y |\langle \psi_x | (I - \Pi) | \phi_y \rangle|^2. \tag{8.233}$$

Therefore, taking $\Lambda = \Pi$ in (8.219) gives

$$\mathrm{Pr}_{\mathrm{error}}(\Pi, \rho, \sigma, t) = t_0 \mathrm{Tr}[\rho \Pi] + t_1 \mathrm{Tr}[\sigma(I - \Pi)]$$

$$\textbf{(8.232), (8.233)} \rightarrow = \sum_{x,y \in [m]} \left( t_0 p_x |\langle \psi_x | I - \Pi | \phi_y \rangle|^2 + t_1 q_y |\langle \psi_x | \Pi | \phi_y \rangle|^2 \right)$$

$$\geqslant \sum_{x,y \in [m]} \min\{t_0 p_x, t_1 q_y\} \left( |\langle \psi_x | I - \Pi | \phi_y \rangle|^2 + |\langle \psi_x | \Pi | \phi_y \rangle|^2 \right).$$

$$(8.234)$$

Moreover, since for any two complex numbers $c_1$ and $c_2$ satisfies $|c_1|^2 + |c_2|^2 \geqslant \frac{1}{2}|c_1 + c_2|^2$, we get that

$$\mathrm{Pr}_{\mathrm{error}}(\Pi, \rho, \sigma, t_0) \geqslant \frac{1}{2} \sum_{x,y \in [m]} \min\{t_0 p_x, t_1 q_y\} \left| \langle \psi_x | I - \Pi | \phi_y \rangle + \langle \psi_x | \Pi | \phi_y \rangle \right|^2$$

$$= \frac{1}{2} \sum_{x,y \in [m]} \min\{t_0 p_x, t_1 q_y\} |\langle \psi_x | \phi_y \rangle|^2$$

$$(8.235)$$

$$= \frac{1}{2} \sum_{x,y \in [m]} \min\{t_0 p_{xy}, t_1 q_{xy}\}$$

$$\textbf{(8.154)} \rightarrow = \frac{1}{2} \mathrm{Pr}_{\mathrm{error}}(\mathbf{p}, \mathbf{q}, t),$$

where $\mathbf{p} = (p_{xy}) \in \mathrm{Prob}(m^2)$ and $\mathbf{q} = (q_{xy}) \in \mathrm{Prob}(m^2)$ are probability vectors with components

$$p_{xy} := p_x |\langle \psi_x | \phi_y \rangle|^2 \quad \text{and} \quad q_{xy} := q_y |\langle \psi_x | \phi_y \rangle|^2. \tag{8.236}$$

Moreover, note that the relation (8.236) respects tensor products. That is, for $\rho^{\otimes n}$ and $\sigma^{\otimes n}$ the corresponding probability vectors are $\mathbf{p}^{\otimes n}$ and $\mathbf{q}^{\otimes n}$, respectively. Hence,

$$\liminf_{n \to \infty} -\frac{1}{n} \log \mathrm{Pr}_{\mathrm{error}}(\rho^{\otimes n}, \sigma^{\otimes n}, t) \leqslant \liminf_{n \to \infty} -\frac{1}{n} \log \frac{1}{2} \mathrm{Pr}_{\mathrm{error}}(\mathbf{p}^{\otimes n}, \mathbf{q}^{\otimes n}, t)$$

$$\textbf{Theorem 8.12} \rightarrow = \max_{0 \leqslant \alpha \leqslant 1} \left\{ (1 - \alpha) D_\alpha(\mathbf{p} \| \mathbf{q}) \right\}$$

$$\textbf{(6.103)} \rightarrow = \max_{0 \leqslant \alpha \leqslant 1} \left\{ (1 - \alpha) D_\alpha(\rho \| \sigma) \right\} \tag{8.237}$$

$$= \xi_Q(\rho, \sigma).$$

This completes the proof.      ∎

## 8.8 Notes and References

Many of the classical concepts in this chapter such as typicality of sequences, the method of types, and classical hypothesis testing can be found in standard textbooks on information theory and statistics, for example, Ref. [54]. The topic of quantum typicality is covered by many books on quantum information, including Refs. [173, 235, 233]. The concept of relative typical subspace was first introduced in Ref. [27].

The expression of the hypothesis testing divergence in terms of the function in (8.198) is due to Ref. [40], whereas other variants can be found in Ref. [67]. The direct part of the quantum Stein's lemma was first proved in Ref. [122], while the strong converse part was proved almost 10 years later in Ref. [176]. A shorter version for both the direct and strong converse parts was found later in Ref. [27]. However, our extremely shorter version presented in this chapter is based on the moderm approach that involves the inequalities given in Theorem 8.14.

We followed Ref. [175] for the proof of the optimality of the quantum Chenoff bound, and Ref. [9] for its achievability.

# PART III

# THE GENERAL FRAMEWORK OF RESOURCE THEORIES

# 9

# Static Quantum Resource Theories

In this chapter, we present a precise definition of a quantum resource theory (QRT) and explore its general characteristics. As mentioned in the introduction, any set of natural constraints on a physical system results in a QRT. A prime example is the spatial separation between two individuals, Alice and Bob, which naturally leads to the LOCC (Local Operations and Classical Communication) constraint, forming the basis of entanglement theory. In this theory, every physical system is analyzed in the context of spatial separation. This implies that any physical system, for instance, system $A$, is considered a bipartite composite system, denoted as $A = (A_A, A_B)$. Here, $A_A$ represents a subsystem on Alice's side, and $A_B$ is a subsystem on Bob's side. It's important to note that even if $A$ is not inherently a composite system and is solely located on Alice's side, it can still be regarded in this framework with $A_A := A$ and $A_B$ being a trivial subsystem (i.e. $|A_B| = 1$). For simplicity, in entanglement theory, the notations $A$ for $A_A$ and $B$ for $A_B$ are often used. However, in the context of general resource theories, it is crucial to remember that physical systems, symbolized as $A, B, C$, and so on, are interpreted in relation to the constraints applied to them.

## 9.1 The Structure of Quantum Resource Theories

According to the first axiom of quantum mechanics, each physical system is uniquely associated with a corresponding Hilbert space. However, the reverse of this statement is not necessarily true. For instance, mathematically, the Hilbert space $\mathbb{C}^4$ is isomorphic to the Hilbert space $\mathbb{C}^2 \otimes \mathbb{C}^2$. Yet, physically, $\mathbb{C}^4$ may represent two entirely distinct physical systems. The space $\mathbb{C}^4$ could describe a single atom with four energy levels, or it might represent a composite system of two spatially separated electrons (spins). Therefore, it is important to clarify that while we use the notations $A, B, C$, and so on to denote both physical systems and their corresponding Hilbert spaces, in the forthcoming discussion, these notations will primarily refer to specific physical systems.

In the rest of this book, we will use the symbol 1 to represent the trivial system. In this context, the only element of $\mathrm{CPTP}(A \to 1)$ is the trace operation. Additionally, this notation allows us to equate quantum channels in $\mathrm{CPTP}(1 \to A)$ with density matrices in $\mathfrak{D}(A)$, and conversely. By adopting this identification, we can interpret all entities in quantum mechanics – such as states, POVMs, quantum instruments, and others – as specific forms of quantum channels. This integrative perspective aligns with the methodologies utilized in resource theories. We will embrace this approach in our discussions throughout the book.

> **Quantum Resource Theory**
>
> **Definition 9.1.** Let $\mathfrak{F}$ be a mapping that takes any two physical systems $A$ and $B$ to a set of quantum channels $\mathfrak{F}(A \to B) \subset \mathrm{CPTP}(A \to B)$. The mapping $\mathfrak{F}$ is called a *quantum resource theory* if it satisfies the following conditions:
>
> 1. *Doing nothing is free.* For any physical system $A$, the identity channel $\mathrm{id}^A \in \mathfrak{F}(A \to A)$.
> 2. *Concatenation is free.* For any three physical systems, $A$, $B$, and $C$, if $\mathcal{E} \in \mathfrak{F}(A \to B)$ and $\mathcal{N} \in \mathfrak{F}(B \to C)$, then $\mathcal{N} \circ \mathcal{E} \in \mathfrak{F}(A \to C)$.
> 3. *Discarding a system is free.* For any system $A$, the set $\mathfrak{F}(A \to 1) \neq \varnothing$; that is, $\mathfrak{F}(A \to 1) = \mathrm{CPTP}(A \to 1) = \{\mathrm{Tr}\}$.
>
> Moreover, the set $\mathfrak{F}(A \to B)$ is called the set of *free operations* from system $A$ to system $B$, and the set $\mathfrak{F}(A) := \mathfrak{F}(1 \to A)$ is identified as the set of *free states*.

The physical interpretation of Definition 9.1 is as follows. Consider a (possibly composite) quantum system held by one agent or distributed to a group of parties. A QRT models what the parties can physically accomplish given some restrictions or constraints that result from technical or experimental limitations, the rules of some game, or simply the laws of physics. What operations the agents can still perform given these restrictions is mathematically described by $\mathfrak{F}(A \to B)$, which is typically much smaller than the set of all quantum channels in $\mathrm{CPTP}(A \to B)$.

The first condition in Definition 9.1 simply says that the identity map (i.e. doing nothing) is free, an obvious requirement for any meaningful QRT. We point out, however, that in some QRTs "doing nothing" can be considered resourceful, particularly, if the systems involved decohere with the environment and resources degrade in time, so that the preservation or storage of a resource is itself a resource. Nonetheless, in such resource theories, the identity channel $\mathrm{id}^A$ in the definition above corresponds to a channel with zero time delay (i.e. instantaneous) so that it is indeed free. In general, the time delay of the channels in such resource theories needs to be incorporated into the formalism, and we will discuss it in more detail in volume 2 of this book when we introduce dynamical resource theories. For all the static QRTs that we study in this book, these considerations will not affect the formalism.

The second property in Definition 9.1 can be viewed as the defining property of a QRT. It essentially states that free operations cannot generate a resource. A resource in our model is a quantum state that is not free (i.e. $\rho \in \mathfrak{D}(A)$ but $\rho \notin \mathfrak{F}(A)$) or a quantum channel that is not free (i.e. $\mathcal{E} \in \mathrm{CPTP}(A \to B)$ but $\mathcal{E} \notin \mathfrak{F}(A \to B)$). In particular, the second property implies the following rule, known as the "golden" rule of QRTs.

> **The Golden Rule of QRTs**
>
> For any two physical systems $A$ and $B$, if a free channel $\mathcal{E} \in \mathfrak{F}(A \to B)$ acts on a free state $\rho \in \mathfrak{F}(A)$, the resulting state $\mathcal{E}(\rho)$ is a free quantum state in $\mathfrak{F}(B)$.

We included in the definition above the property that the trace is a free operation. In all QRTs studied in literature, this is indeed the case although one can consider a QRT in which "waste" or "trash" is considered a resource. In this book, we will not consider such resource theories, and will always consider the trace as a free operation. This assumption also leads to the following very useful property of QRTs.

Suppose $\sigma \in \mathfrak{F}(B)$ is a free state, and define the replacement channel

$$\mathcal{N}_\sigma^{A \to B}(\rho^A) := \mathrm{Tr}[\rho^A]\sigma^B \qquad \forall\, \rho \in \mathfrak{L}(A). \tag{9.1}$$

Then, if we view $\sigma^B$ as a channel, $\sigma^{1 \to B}$, from the trivial system 1 to $B$, the channel $\mathcal{N}_\sigma^{A \to B}$ can be expressed as a combination of the trace channel and the channel $\sigma^{1 \to B}$; specifically,

$$\mathcal{N}_\sigma^{A \to B} = \sigma^{1 \to B} \circ \mathrm{Tr}, \tag{9.2}$$

and since both $\mathrm{Tr}$ and $\sigma^{1 \to B}$ are free, it follows that $\mathcal{N}_\sigma^{A \to B}$ is free. Note that this means that we can convert any state $\rho \in \mathfrak{D}(A)$ to any free state $\sigma \in \mathfrak{F}(B)$ by free operations (as intuitively expected).

Quantum resource theories emerge from a specific set of limitations or constraints applied to the entire spectrum of quantum operations. The mapping $\mathfrak{F}$ exemplifies this, as the set $\mathfrak{F}(A \to B)$ generally forms a strict subset of all channels in $\mathrm{CPTP}(A \to B)$. While every QRT is linked to a unique set of restrictions, these restrictions frequently share common characteristics that contribute to extra structural complexity. These characteristics are so prevalent that some researchers have integrated them into the foundational definition of a QRT.

## 9.1.1 The Axiom of Free Instruments

In some resource theories, like the QRT of athermality, quantum measurements are not considered free. However, in most QRTs, certain measurements are free. Mathematically, this implies the existence of systems $A, B, X$ – with $X$ being a classical system – such that the set $\mathfrak{F}(A \to BX)$ is nonempty. Consider a quantum instrument in $\mathfrak{F}(A \to BX)$, denoted as

$$\mathcal{E}^{A \to BX} = \sum_{x \in [m]} \mathcal{E}_x^{A \to B} \otimes |x\rangle\langle x|^X. \tag{9.3}$$

According to the fundamental principle of QRTs, if $\rho \in \mathfrak{F}(A)$ is a free state, then the state $\mathcal{E}^{A \to BX}(\rho^A)$ must also be a free state in $\mathfrak{F}(BX)$. This state, expressed as

$$\mathcal{E}^{A \to BX}(\rho^A) = \sum_{x \in [m]} p_x \sigma_x^B \otimes |x\rangle\langle x|^X \tag{9.4}$$

is a classical-quantum (cq) state, where for each $x \in [m]$, $p_x := \mathrm{Tr}[\mathcal{E}_x(\rho)]$ and $\sigma_x^B := \frac{1}{p_x}\mathcal{E}_x^{A \to B}(\rho^A)$. If there existed an $x \in [m]$ for which $p_x \neq 0$ and $\sigma_x \notin \mathfrak{F}(B)$, then the quantum instrument $\mathcal{E}^{A \to BX}$ would create a resource $\sigma_x^B$ from the free state $\rho \in \mathfrak{F}(A)$ with a nonzero probability. To prevent such scenarios in QRTs, in this book we always assume the axiom of free instruments.

> **The Axiom of Free Instruments**
>
> Let $A$ and $B$ be two quantum systems and $X$ be a classical system. If $\mathfrak{F}(A \to BX)$ is nonempty and $\mathcal{E} := \sum_{x \in [m]} \mathcal{E}_x \otimes |x\rangle\langle x| \in \mathfrak{F}(A \to BX)$, then for every $\rho \in \mathfrak{F}(A)$ and $x \in [m]$, the state $\mathcal{E}_x^{A \to B}(\rho^A)/\mathrm{Tr}[\mathcal{E}_x^{A \to B}(\rho^A)]$ belongs to $\mathfrak{F}(B)$.

Note that the axiom of free instruments (AFI) reduces to the Chernoff of QRTs when $|X| = 1$, thus serving as an extension of this rule to encompass quantum measurements. Additionally, when $|X| > 1$, the golden rule of QRTs only ensures that $\mathcal{E}^{A \to BX}(\rho^A)$ is a free cq-state. Without further assumptions like the AFI, we cannot infer that each $\frac{\mathcal{E}_x^{A \to B}(\rho^A)}{\mathrm{Tr}[\mathcal{E}_x(\rho)]}$ is a free state. Since physical QRTs comply with the AFI (as do all QRTs studied in the literature), the rest of this book will proceed under the assumption that QRTs adhere to the AFI, without explicitly stating it each time. We will use the notation $\mathfrak{F}_{\leqslant}(A \to B) \subset \mathrm{CP}_{\leqslant}(A \to B)$ for the set of trace nonincreasing CP maps that are part of free quantum instruments. Specifically, $\mathcal{E} \in \mathfrak{F}_{\leqslant}(A \to B)$ if there exists a classical system $X$ with dimension $m \in \mathbb{N}$ and maps $\mathcal{E}_1, \ldots, \mathcal{E}_m \in \mathrm{CP}_{\leqslant}(A \to B)$, with the properties that (1) $\mathcal{E}_x = \mathcal{E}$ for some $x \in [m]$, and (2) $\sum_{x \in [m]} \mathcal{E}_x \otimes |x\rangle\langle x| \in \mathfrak{F}(A \to BX)$.

## 9.1.2 QRTs with a Tensor Product Structure

Since the free operations arise from certain physical constraints, it is natural to assume that they can act on a subsystem of a composite system. That is, a free operation $\mathcal{E} \in \mathfrak{F}(A \to B)$ can act on the state $\rho^{AC}$ as $\mathcal{E}^{A \to B}(\rho^{AC}) := \mathcal{E}^{A \to B} \otimes \mathrm{id}^C(\rho^{AC})$. This leads us to the following definition.

> **Tensor Product Structure**
>
> **Definition 9.2.** A QRT, $\mathfrak{F}$, is said to *admit a tensor-product structure* if it fulfills the following additional criteria:
>
> 4. *Completely free operations:* For any three systems $A$, $B$, and $C$, and a channel $\mathcal{E} \in \mathfrak{F}(A \to B)$, it holds that $\mathcal{E}^{A \to B} \otimes \mathrm{id}^C \in \mathfrak{F}(AC \to BC)$.
> 5. *Freedom of relabeling:* For any integer $n$, a free channel $\mathcal{N} \in \mathfrak{F}(A^n \to B^n)$, and permutation channels $\mathcal{P}_\pi^{A^n}$ and $\mathcal{P}_{\pi^{-1}}^{B^n}$ corresponding to a permutation $\pi$ on $n$ elements, the composition $\mathcal{P}_\pi^{B^n} \circ \mathcal{N}^{A^n \to B^n} \circ \mathcal{P}_{\pi^{-1}}^{A^n}$ is in $\mathfrak{F}(A^n \to B^n)$.

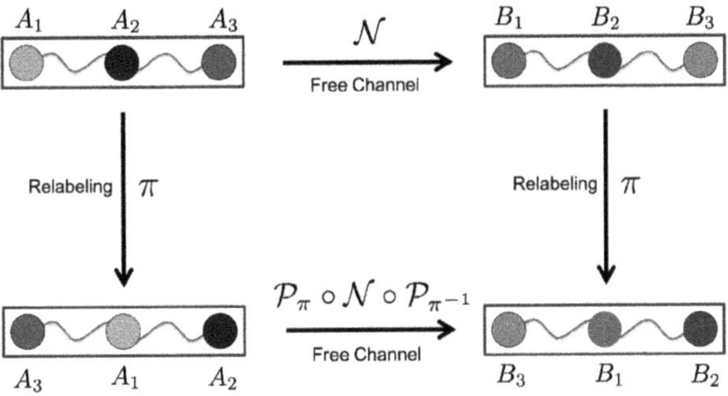


**Figure 9.1** Illustration of the fifth condition. Relabeling maintains the "freeness" of $\mathcal{N}$.

The condition 5 is very intuitive as it just states that the relabeling of $A^n = (A_1, \ldots, A_n)$ and $B^n = (B_1, \ldots, B_n)$, respectively, as $(A_{\pi(1)}, \ldots, A_{\pi(n)})$ and $(B_{\pi(1)}, \ldots, B_{\pi(n)})$ keeps the channel $\mathcal{N}$ free. Note that we do *not* require that $\mathcal{P}_\pi \circ \mathcal{N} \circ \mathcal{P}_{\pi^{-1}} = \mathcal{N}$, but only that $\mathcal{P}_\pi \circ \mathcal{N} \circ \mathcal{P}_{\pi^{-1}}$ is itself a free channel. See Figure 9.1 for an illustration.

Conditions 1–5 in Definitions 9.1 and 9.2 have several additional implications. First, note that since the trace is a free channel (property 3), the partial trace is also free (since $\mathrm{id} \otimes \mathrm{Tr}$ is free). Second, note that if $\mathcal{N} \in \mathfrak{F}(A \to B)$ and $\mathcal{M} \in \mathfrak{F}(A' \to B')$, then

$$\mathcal{N} \otimes \mathcal{M} = (\mathcal{N} \otimes \mathrm{id}) \circ (\mathrm{id} \otimes \mathcal{M}) \in \mathfrak{F}(AA' \to BB'). \tag{9.5}$$

In particular, this means that if two states are free, then their tensor product is also free. Finally, appending a free state is also a free channel. Specifically, let $\sigma \in \mathfrak{F}(B)$ and define

$$\mathcal{N}_\sigma^{A \to AB}(\rho^A) := \rho^A \otimes \sigma^B \qquad \forall \, \rho \in \mathcal{L}(A). \tag{9.6}$$

This channel can be viewed as a tensor product of two free channels, namely, $\mathcal{N}^{A \to AB} = \mathrm{id}^A \otimes \sigma^{1 \to B}$, and therefore is free.

## 9.1.3 Two Additional Properties: Closedness and Convexity

Resource theories can vary a lot as the set of free operations is unique to each resource theory. For example, LOCC in entanglement theory (Chapter 12) seems to be a very different set of operations than the set of thermal operations in thermodynamics (Chapter 17). Yet, in addition to the tensor product structure, both of these sets of operations have additional structure that is common to both of them. Here we discuss two additional common properties that are satisfied by almost all QRTs studied in literature.

> 6. For any physical system $A$ the set of free states $\mathfrak{F}(A)$ is closed.
> 7. For any physical system $A$ the set of free states $\mathfrak{F}(A)$ is convex.

Property 6 states that if for a sequence of states $\{\rho_n\}_{n\in\mathbb{N}} \subset \mathfrak{F}(A)$ the limit $\rho :=$ $\lim_{n\to\epsilon fty} \rho_n$ exists, then that limit is in $\mathfrak{F}(A)$ as well. Equivalently, if $\{\rho_n\}_{n\in\mathbb{N}} \subset \mathfrak{F}(A)$ and there exists $\rho \in \mathfrak{D}(A)$ such that $\lim_{n\to\epsilon fty} T(\rho,\rho_n) = 0$, where $T$ is the trace distance (or any other distance measure), then $\rho \in \mathfrak{F}(A)$. Note that if this property does not hold, then it would mean that there exists a sequence of free states that is approaching a resource $\rho$. However, if $T(\rho,\rho_n)$ is extremely small, say $10^{-100}$, for all practical purposes it is not possible to distinguish between $\rho$ and $\rho_n$. Therefore, the assumption that $\mathfrak{F}(A)$ is closed is very practical and consequently satisfied by all the QRTs studied in literature so far.

Property 7 is not satisfied by all QRTs, for example, non-Gaussianity in quantum optics, although many resource theories like entanglement do satisfy it and it is quite common. Besides being a convenient mathematical property, we can develop some intuition for this property. Consider a QRT in which an agent, say Alice, has access to an unbiased coin. She can flip the unbiased coin and prepare the state $\rho \in \mathfrak{F}(A)$ if she gets a head and otherwise prepare the state $\sigma \in \mathfrak{F}(A)$. Since $\rho$ and $\sigma$ are free, she can prepare them at no cost. Suppose now that Alice forgets which state she prepared. We will assume here that this "forgetting" is itself a free operation. Then, her state of the system is now $\frac{1}{2}\rho + \frac{1}{2}\sigma$. Therefore, we can assume that the convex combination $\frac{1}{2}\rho + \frac{1}{2}\sigma := \tau$ is also free since Alice prepared it at no cost. Moreover, if $\tau$ is free, Alice can repeat the same process with $\rho$ and $\tau$ to get that $\frac{3}{4}\rho + \frac{1}{4}\sigma$ is also free. Repeating this process, Alice can prepare any combination $\frac{k}{2^n}\rho + \left(1 - \frac{k}{2^n}\right)\sigma$, with $n \in \mathbb{N}$ and $k \in [2^n]$. Therefore, such convex combinations must also be free. Finally, since the set $\left\{\frac{k}{2^n}\right\}$ is dense in $[0,1]$, Property 6 implies that for any $t \in [0,1]$ the convex combination $t\rho + (1-t)\sigma$ is free.

**Exercise 9.1.** *Consider the process just described.*

1. *Show that for any $n \in \mathbb{N}$ and $k \in [2^n]$ Alice can prepare the state $\frac{k}{2^n}\rho + \left(1 - \frac{k}{2^n}\right)\sigma$. How many times Alice has to flip the coin.*
2. *Suppose that Alice has access to a biased coin, with probability $0 < p < 1$ to get a head (and $1 - p$ to get a tail). Show that Alice can use the coin to prepare any convex combination of free states.*

The two properties of closedness and convexity can also be applied to quantum channels. That is, we can require that for any two physical systems $A$ and $B$ the set $\mathfrak{F}(A \to B)$ is both closed and convex. However, we postpone the discussion of them to the second volume of this book where we study dynamical QRTs.

## 9.2 State-Based Resource Theories

Certain types of quantum phenomena can be identified directly on the level of states without involving constraints on quantum processes. This is particularly true for phenomena such as coherence and certain forms of Bell nonlocality. Within the framework of QRTs, the challenge becomes identifying sets of free operations that align with a predefined set of free states. It's interesting to note that these free states, represented as $\mathfrak{F}(A) = \mathfrak{F}(1 \to A)$, can themselves be considered a unique kind of free operations, specifically as preparation channels. This approach affords substantial flexibility in choosing a consistent set of free operations for any given set of free states, even within QRTs that admit a tensor-product structure.

Consider, for example, the phenomenon of quantum coherence. Quantum coherence epitomizes a key aspect of quantum mechanics, illustrating the principle that particles, such as electrons or photons, can simultaneously exist in multiple states. This phenomenon stems from the principle of superposition, enabling particles to exist in a mixture of states, or in *coherent* superposition, thus allowing them to interfere with one another in predictable manners. However, coherence is a fragile state, easily disturbed by external influences in a process known as decoherence, where quantum systems relinquish their superposition and adopt more classical behaviors. In recent developments, the capability to control and preserve quantum coherence has become crucial for the advancement of cutting-edge quantum technologies, including quantum computing and quantum cryptography, empowering the execution of tasks that surpass the capabilities of classical physics.

Considering the significance of this pivotal phenomenon, extensive efforts have been dedicated to characterizing it within the realm of QRTs. How is this achieved? We start by identifying the set of free states in $\mathfrak{D}(A)$. This is accomplished as follows: For any system $A$, a *classical* basis of the system is identified, denoted as $\{|x\rangle\}_{x \in [m]} \subset A$. Subsequently, the set of free states, or incoherent states, is defined as all diagonal density matrices in $\mathfrak{D}(A)$ with respect to the classical basis. Thus, in the QRT of coherence, the set of free states is clearly defined and is specified for any system $A$ with dimension $m := |A|$ as

$$\mathfrak{F}(A) = \left\{ \sum_{x \in [m]} p_x |x\rangle\langle x| \ : \ \mathbf{p} \in \text{Prob}(m) \right\}. \tag{9.7}$$

Consequently, the primary challenge in the resource theory of quantum coherence lies in identifying a set of free operations that align consistently with the above set of free states.

**Exercise 9.2.** *Let $\mathfrak{F}(A)$ be the set of free states defined in (9.7), and let $\Delta \in$ CPTP$(A \to A)$ be the completely dephasing map defined with respect to the classical basis. Show that for all $\rho \in \mathfrak{D}(A)$ we have that $\rho \in \mathfrak{F}(A)$ if and only if $\Delta(\rho) = \rho$.*

Physical factors often play a pivotal role in determining the choice of free operations within the realm of quantum mechanics. Nonetheless, even when these free operations are well-defined and grounded in physical principles, it is advantageous to investigate other classes of free operations that correspond to the same set of free states. This exploration can provide valuable insights and potentially reveal alternative mathematical or theoretical frameworks, as alternate classes might offer simpler or more elegant solutions that are not immediately apparent in operations primarily motivated by physical factors. A pertinent example is found in entanglement theory, where characterizing the class of LOCC is notably complex. To circumvent these complexities, considerable research has focused on entanglement theory within broader and more mathematically accessible sets of operations, such as separable operations and nonentangling operations (refer to Chapter 12). A commonality among these resource theories of entanglement is the identification of the set of separable states as the free states. Exploring more advanced operations can result in demonstrating no-go theorems for the less powerful but physically motivated free operations. In essence, if a quantum information task is unachievable with a more capable class of operations, it will certainly be infeasible with a weaker set. This section delves into various consistent sets of free operations in general QRTs, emphasizing their physical justifications and unique properties. To illustrate these abstract concepts, the QRTs of coherence and entanglement will frequently be used as examples.

Often some physical consideration will motivate a certain choice of free operations. But even in this case, it is valuable to study different classes of free operations for the same set of free states. This is because different classes may have an easier or more elegant mathematical structure than the physically motivated class of operations. This is the case, for example, in entanglement theory where LOCC is a notoriously difficult class of operations to characterize. To avoid the technical difficulties that arise when using these operations, much work has been devoted to the study of entanglement theory under larger and more analytically friendly sets of operations such as separable operations, nonentangling operations, and more (see Chapter 12). In all these resource theories of entanglement, a shared characteristic is the designation of separable states as the set of free states. Investigating more advanced operations can facilitate the proof of no-go theorems for less powerful, albeit more intuitive, free operations. This is based on the principle that if a quantum information task is unachievable using a more capable class of operations, it will inevitably be impossible with a weaker set. In this section, we study different consistent sets of free operations in general QRTs, highlighting their various physical motivations and properties. As examples to illustrate abstract ideas, we will often use the QRTs of coherence and entanglement to demonstrate them.

## 9.2.1 Resource Nongenerating Operations

Let $\mathfrak{F}(A)$ be a set of free density matrices. Any conceivable set of free operations $\mathfrak{F}(A \to B)$ must satisfy the properties given in Definition 9.1. In particular, any channel $\mathcal{N} \in \mathfrak{F}(A \to B)$ must satisfy the Chernoff of QRTs. We use this property in the following definition.

RNG Operations

**Definition 9.3.** Let $\mathfrak{F}(A) \subset \mathfrak{D}(A)$ be the set of free states on any physical system $A$. The set of *resource nongenerating operations* (RNG) between two physical systems $A$ and $B$ is defined as

$$\mathrm{RNG}(A \to B) := \left\{ \mathcal{N} \in \mathrm{CPTP}(A \to B) : \mathcal{N}(\rho) \in \mathfrak{F}(B), \ \forall \, \rho \in \mathfrak{F}(A) \right\}. \quad (9.8)$$

Resource nongenerating operations form the maximal set of free operations. That is, every other QRT $\mathfrak{F}$ with the same set of free state $\mathfrak{F}(A)$ must satisfy

$$\mathfrak{F}(A \to B) \subseteq \mathrm{RNG}(A \to B). \tag{9.9}$$

In the QRT of coherence this set of RNG operations is denoted by $\mathrm{MIO}(A \to B)$, where the acronym MIO stands for *maximally incoherent operations*. Denoting the $\Delta^A \in \mathrm{CPTP}(A \to A)$ and $\Delta^B \in \mathrm{CPTP}(B \to B)$ the completely dephasing channels with respect to the classical systems $A$ and $B$, respectively, we get from the definition above in conjunction with Exercise 9.3 that

$$\mathrm{MIO}(A \to B) = \left\{ \mathcal{N} \in \mathrm{CPTP}(A \to B) \; : \; \Delta^B \circ \mathcal{N}^{A \to B} \circ \Delta^A = \mathcal{N}^{A \to B} \circ \Delta^A \right\}. \tag{9.10}$$

**Exercise 9.3.** *Prove* (9.10).

**Exercise 9.4.** *Consider the QRT of coherence where $\mathfrak{F}(A)$ and $\mathfrak{F}(B)$ are sets of diagonal density matrices with respect to some fixed bases $\left\{ |x\rangle^A \right\}_{x \in [m]}$ and $\left\{ |y\rangle^B \right\}_{y \in [n]}$ of $A$ and $B$, respectively. Show that a quantum channel $\mathcal{N} \in MIO(A \to B)$ if and only if there exists conditional probability distribution $\{ p_{y|x} \}$ such that for all $x \in [m]$ the state*

$$\mathcal{N}^{A \to B} \left( |x\rangle\langle x|^A \right) = \sum_{y \in [n]} p_{y|x} |y\rangle\langle y|^B. \tag{9.11}$$

In entanglement theory, the set of RNG operations is called *nonentangling operations*. They consists of all bipartite quantum channels that cannot generate entanglement from separable states. Nonentangling operations form a strict superset of LOCC. For example, the (global) swap operator, that swap all the subsystems of Alice with those of Bob is nonentangling. Such a swap operation is highly nonlocal and clearly cannot be simulated with LOCC.

In general, RNG operations do not qualify as completely free. This means there are instances where an operation $\mathcal{E} \in \mathrm{RNG}(A \to B)$, when combined with an identity operation on some system $C$, that is, $\mathcal{E}^{A \to B} \otimes \mathrm{id}^C$, does not remain a free operation. To illustrate this, consider the context of entanglement theory and a product state $\Phi^{A\tilde{A}} \otimes |0\rangle\langle 0|^B$ shared between Alice and Bob, where the sizes of $A$ and $B$ are equal. Previously, we noted that a global swap operation is nonentangling. Specifically, applying a global swap to the aforementioned state results in $|0\rangle\langle 0|^A \otimes \Phi^{B\tilde{B}}$, maintaining its product state nature. However, if a partial swap is applied between subsystems $A$ and $B$ (keeping $\tilde{A}$ intact), the resulting state becomes $|0\rangle\langle 0|^A \otimes \Phi^{\tilde{A}B}$, which is entangled. This demonstrates that while a swap operation does not generate entanglement when applied to the entire system, it can induce entanglement when acting on individual subsystems.

## 9.2.2 Completely Resource Nongenerating Operations

The realization that RNG operations are generally not completely free (as outlined in condition 4 of Definition 9.2) underscores their somewhat nonphysical nature. This leads us to a new definition that not only adheres to the Chernoff of QRTs but also integrates the tensor product structure.

**Completely RNG Operations**

**Definition 9.4.** Given $k \in \mathbb{N}$, a quantum channel $\mathcal{N} \in \mathrm{RNG}(A \to B)$ is defined as $k$-RNG if for any reference system $R$ of dimension $k$, the channel $\mathrm{id}^R \otimes \mathcal{N}$ belongs to $\mathrm{RNG}(RA \to RB)$. Furthermore, the channel $\mathcal{N}$ is termed *completely RNG* (CRNG) if it is $k$-RNG for all $k \in \mathbb{N}$. The set of all CRNG channels in $\mathrm{CPTP}(A \to B)$ is denoted by $\mathrm{CRNG}(A \to B)$.

The definition above generalizes the concepts of $k$-positivity and complete-positivity (see Definition 3.3) to QRTs. Specifically, if we take the free set $\mathfrak{F}(A) = \mathfrak{D}(A)$ to be the set of *all* density matrices acting on $A$, then maps that are $k$-RNG and completely-RNG are equivalent to maps that are $k$-positive and completely positive, respectively. Moreover, the set of $k$-RNG maps with $k = 1$ is simply the set of RNG maps.

As an example, let $\mathfrak{F}(AB) := \mathrm{SEP}(AB) \subset \mathfrak{D}(AB)$ be the set of all separable states, and let $\mathcal{N} \in \mathrm{CRNG}(AB \to A'B')$ be a (bipartite) quantum channel that takes separable states to separable states even when acting on subsystems. Specifically, for any composite reference system $R = R_A R_B$ the channel $\mathrm{id}^R \otimes \mathcal{N}^{AB \to A'B'}$ is nonentangling (i.e. RNG). Recall the discussion at the beginning on this chapter that every system $R$ in entanglement theory is viewed as a bipartite system $R_A R_B$ with $R_A$ on Alice's side and $R_B$ on Bob's side. Taking $R_A \cong A$ and $R_B \cong B$ we get that the state $\Phi^{R_A A} \otimes \Phi^{R_B B}$ is a product state between Alice composite system $R_A A$ and Bob's composite system $R_B B$ (see Figure 9.2a). Therefore, since product states are in particular separable, and since $\mathrm{id}^R \otimes \mathcal{N}^{AB \to A'B'}$ is nonentangling we get that

$$\mathcal{N}^{AB \to A'B'} \left( \Phi^{R_A A} \otimes \Phi^{R_B B} \right) \in \mathrm{SEP}(R_A A' R_B B') \tag{9.12}$$

is a separable state between Alice's system $R_A A'$ and Bob's system $R_B B'$.

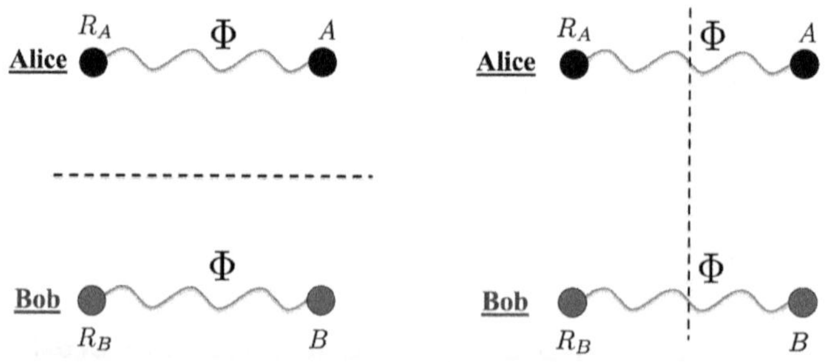


Figure 9.2    The state $\Phi^{R_A A} \otimes \Phi^{R_B B}$ in the lens of two bipartite cuts.

A key observation in this example is that the state $\Phi^{R_A A} \otimes \Phi^{R_B B}$ can be viewed as a maximally entangled state between system $R = R_A R_B \cong AB$ and system $AB$ (see Figure 9.2b); that is,

$$\Phi^{(R_A R_B)(AB)} = \Phi^{R_A A} \otimes \Phi^{R_B B}. \tag{9.13}$$

Therefore, the state in (9.12) is proportional to the Choi matrix of $\mathcal{N}$. Since $R_A \cong A$ and $R_B \cong B$ we conclude that the Choi matrix

$$J_{\mathcal{N}}^{ABA'B'} := \mathcal{N}^{\tilde{A}\tilde{B} \to A'B'} \left( \Omega^{A\tilde{A}} \otimes \Omega^{B\tilde{B}} \right) \tag{9.14}$$

is an unnormalized separable state between Alice's system $AA'$ and Bob's system $BB'$. From Exercise 3.8 it follows that the Choi matrix can be expressed as.

$$J_{\mathcal{N}}^{ABA'B'} = \sum_{j \in [k]} \psi_j^{AA'} \otimes \phi_j^{BB'}, \tag{9.15}$$

where the sets $\{\psi_j^{AA'}\}_{j \in [k]}$ and $\{\phi_j^{BB'}\}_{j \in [k]}$ are sets of (possibly unnormalized) pure states (i.e. rank one operators) in $\mathrm{Pos}(AA')$ and $\mathrm{Pos}(BB')$, respectively. For each $j \in [k]$, we can write

$$|\psi_j^{AA'}\rangle = \left( I^A \otimes M_j \right) |\Omega^{A\tilde{A}}\rangle \quad \text{and} \quad |\phi_j^{BB'}\rangle = \left( I^B \otimes N_j \right) |\Omega^{B\tilde{B}}\rangle, \tag{9.16}$$

for some complex matrices $M_j \in \mathfrak{L}(A, A')$ and $N_j \in \mathfrak{L}(B, B')$. Using this notation in (9.15) and comparing it with (9.14) we conclude that the channel $\mathcal{N}$ has the following operator sum representation

$$\mathcal{N}^{AB \to A'R'} \left( \rho^{AR} \right) = \sum_{j \in [k]} (M_j \otimes N_j) \rho^{AB} (M_j \otimes N_j)^* \quad \forall \, \rho \in \mathfrak{L}(AB). \tag{9.17}$$

A channel is classified as separable if it possesses at least one operator sum representation where each Kraus operator can be expressed as a tensor product in the form of $M_j \otimes N_j$. The collection of all such separable channels is denoted by $\mathrm{SEP}(AB \to A'B')$. This leads to an important conclusion in entanglement theory: The set of completely nonentangling operations, that is, CRNG, aligns exactly with the set of separable channels. In other words, we have $\mathrm{CRNG}(AB \to A'B') = \mathrm{SEP}(AB \to A'B')$.

**Exercise 9.5.** *Let* $\mathrm{RNG}(AB \to A'B')$ *be the set of nonentangling operations (i.e. RNG with respect to the set* $\mathfrak{F}(AB) = \mathrm{SEP}(AB)$*).*

1. *Show that if* $\mathcal{N} \in \mathrm{RNG}(AB \to A'B')$ *is k-RNG with* $k \geqslant |A|$, *then it is completely RNG.*
2. *Show that*

$$\mathrm{RNG} = 1\text{-RNG} \supseteq 2\text{-RNG} \supseteq \cdots \supseteq |AB|\text{-RNG} = CRNG = \mathrm{SEP}. \tag{9.18}$$

We saw above that for $\mathfrak{F}(AB) = \mathrm{SEP}(AB)$ we have $\mathrm{CRNG} \subset \mathrm{RNG}$ where the inclusion is strict since the global swap operator is nonentangling but also not a separable channel. Moreover, we will see in Chapter 12 that also the inclusion $\mathrm{RNG} \supseteq 2\text{-RNG}$

in the exercise above can be strict. However, in other resources theories some of these inclusions can be equalities. For example, in the QRT of coherence, in which $\mathfrak{F}(A) \subset \mathfrak{D}(A)$ consists of diagonal states with respect to a fixed basis $\{|x\rangle^A\}_{x \in [m]}$, we have that $\mathrm{RNG} = \mathrm{CRNG}$. To see this, let $\mathcal{N} \in \mathrm{MIO}(A \to B)$ (recall that in the QRT of coherence we denote all RNG operations from system $A$ to $B$ by $\mathrm{MIO}(A \to B)$). According to (9.10), $\Delta^B \circ \mathcal{N} \circ \Delta^A = \mathcal{N} \circ \Delta^A$. We need to show that for any system $C$, we have $\mathcal{N} \otimes \mathrm{id}^C \in \mathrm{MIO}(AC \to BC)$. Let $\Delta^C$ be the completely dephasing channel with respect to the classical basis of system $C$. In the following exercise we show that $\Delta^{AC} = \Delta^A \otimes \Delta^C$. Therefore,

$$\Delta^{CB} \circ \left( \mathrm{id}^C \otimes \mathcal{N} \right) \circ \Delta^{CA} = \left( \Delta^C \otimes \Delta^B \right) \circ \left( \mathrm{id}^C \otimes \mathcal{N} \right) \circ \left( \Delta^C \otimes \Delta^A \right)$$

$$\boxed{\Delta^C \circ \Delta^C = \Delta^C} \quad \longrightarrow \quad = \left( \mathrm{id}^C \otimes \Delta^B \circ \mathcal{N} \circ \Delta^A \right) \circ \left( \Delta^C \otimes \mathrm{id}^A \right)$$

$$\boxed{\mathcal{N} \in \mathrm{MIO}(A \to B)} \quad \longrightarrow \quad = \left( \mathrm{id}^C \otimes \mathcal{N} \circ \Delta^A \right) \circ \left( \Delta^C \otimes \mathrm{id}^A \right)$$

$$\boxed{\Delta^{CA} = \Delta^C \otimes \Delta^A} \quad \longrightarrow \quad = \left( \mathrm{id}^C \otimes \mathcal{N} \right) \circ \Delta^{CA}. \tag{9.19}$$

Thus, $\mathcal{N} \otimes \mathrm{id}^C \in \mathrm{MIO}(AC \to BC)$.

**Exercise 9.6.** *Let $\{|x\rangle^A\}_{x \in [m]}$ and $\{|y\rangle^B\}_{y \in [n]}$ be, respectively, two orthonormal bases of $A$ and $B$. Furthermore, let $\Delta^{AB}$ be the completely dephasing channel with respect to the basis $\{|xy\rangle^{AB}\}_{x,y}$. Show that $\Delta^{AB} = \Delta^A \otimes \Delta^B$, where $\Delta^A$ and $\Delta^B$ are the completely dephasing channels with respect to the bases $\{|x\rangle^A\}_{x \in [m]}$ and $\{|y\rangle^B\}_{y \in [n]}$, respectively.*

## 9.2.3 Physically Implementable Operations

The use of CPTP maps and generalized measurements in quantum information science is so common that their physical implementations are often taken for granted. Specifically, in Section 3.4.5 we saw that the Stinespring dilation theorem ensures that every quantum channel in $\mathrm{CPTP}(A \to A)$ can be implemented with a unitary evolution on the joint system $AE$ followed by tracing out the environment degrees of freedom (i.e. tracing out system $E$). Similarly, generalized measurements can be implemented with a joint unitary operation followed by a projective measurement as described in Figure 3.1. In the context of QRTs, such implementations of free channels (or free generalized measurements) may not be free since the joint unitary (or projective measurement) identified in such implementation is itself not necessarily free.

As an example, consider the QRT of coherence, with $\mathfrak{F} = \mathrm{MIO}$ being the set of free operations. If MIO were physically implementable, one would expect that for any channel $\mathcal{N} \in \mathrm{MIO}(A \to A)$ there exists a system $E$, a free unitary channel $\mathcal{U} \in \mathrm{MIO}(AE \to AE)$, and a diagonal state $\gamma \in \mathfrak{F}(E)$ such that

$$\mathcal{N}^{A \to A}(\omega^A) = \mathrm{Tr}_E \left[ U^{AE} \left( \omega^A \otimes \gamma^E \right) U^{*AE} \right]. \tag{9.20}$$

Now, from Exercise 9.4 we know that a channel $\mathcal{V} \in \text{CPTP}(A \to A)$ is MIO if and only if $\mathcal{V}(|x\rangle\langle x|^A)$ is a diagonal state in $\mathfrak{D}(A)$ for all $x \in [m]$ (here $m := |A|$). Therefore, if $\mathcal{V}$ is a unitary channel, then we must have for all $x \in [m]$

$$\mathcal{V}^{A \to A}(|x\rangle\langle x|^A) := V|x\rangle\langle x|^A V^* = |\pi(x)\rangle\langle\pi(x)|^A, \tag{9.21}$$

where $\pi$ is some permutation on $m$ elements. This relation implies that the unitary matrix $V$ satisfies $V|x\rangle^A = e^{i\theta_x}|\pi(x)\rangle^A$. In other words, up to phases, all the free unitary operations in the QRT of coherence are permutations. Given that permutation matrices form an extremely small set of operations relative to the set of all unitary channels, it is not too hard to show (see the relevant references at the end of this chapter) that there exists channels in $\text{MIO}(A \to A)$ that do not have the form (9.20) with free (i.e. incoherent) $U^{AE}$ and free (i.e. diagonal) $\gamma^E$. In other words, it costs coherence (i.e. resources) to implement some free channels in $\text{MIO}(A \to A)$.

This problem does not occur in the QRT of entanglement in which the set of free operations is LOCC. This is always the case whenever a QRT is defined in terms of a *physical* restriction (e.g. distant labs in entanglement theory) that is imposed on the set of free operations. On the other hand, any QRT such as quantum coherence, in which first the free states are identified, and only then consistent free operations are proposed, may face such an implementation problem. Aside from the QRT of coherence, all the free operations of the QRTs studied in this book will have a physically implementable set of free operations.

> **Definition 9.5.** Let $\mathfrak{F}$ be a QRT, and $A$ and $B$ two physical systems. We say that $\mathfrak{F}(A \to B)$ is *physically implementable* if any channel in $\mathfrak{F}(A \to B)$ can be generated by a sequence of unitary channels (possibly on composite systems), projective measurements, appending of free states, and processing of the classical outcomes, where each element in the sequence is itself a free action (see Figure 9.3 for an illustration).

*Remark.* In the definition above we added classical processing as a possible free physically implementable operation. This includes, for example, classical communication between subsystems, if these were allowed in the QRT (e.g. entanglement theory). Note also that if the free operations in a QRT are not physically implementable (according to the Definition 9.5), then the QRT would identify certain maps as being free with no way to physically implement these processes using free operations.

For a given designation of free states $\mathfrak{F}(A)$, it is possible to construct a unique physically implementable QRT that admits a tensor-product structure. Simply define the free operations to be any composition of (i) appending arbitrary free states, (ii) CRNG unitaries and projective measurements, (iii) discarding subsystems, and (iv) all free classical-processing maps. For a given two subsystems $A$ and $B$ we denote this set of *physically implementable operations* (PIO) as $\text{PIO}(A \to B)$. By design, $\text{PIO}(A \to B)$ is

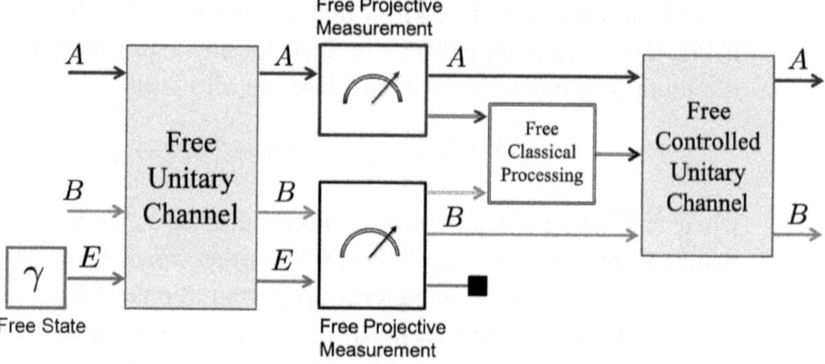



**Figure 9.3**    Example of a physically implementable free operation on a composite system $AB$.

physically implementable and has tensor-product structure. Most QRTs that were studied in literature have the property that all the isometries in RNG are completely free. In such QRTs, PIO is the minimal set of free operations that is consistent with the set of free states $\mathfrak{F}(A)$. The class $\mathrm{PIO}(A \to B)$ fits into the hierarchy of operations as

$$\mathrm{PIO} \subseteq \mathrm{CRNG} \subseteq \mathrm{RNG}. \tag{9.22}$$

## 9.2.4 Dually RNG Operations

Let $\mathcal{E} \in \mathfrak{F}(A \to B) \subseteq \mathrm{CPTP}(A \to B)$ represent a free operation within a specific resource theory $\mathfrak{F}$. As $\mathcal{E}$ is a completely positive map, its dual map $\mathcal{E}^*$ is also completely positive. However, it is crucial to note that the dual map $\mathcal{E}^*$ may not necessarily preserve trace. Despite this, in many resource theories, although $\mathcal{E}^*$ is not trace-preserving, it satisfies the following condition:

$$\frac{\mathcal{E}^*(\sigma)}{\mathrm{Tr}\left[\mathcal{E}^*(\sigma)\right]} \in \mathfrak{F}(A) \qquad \forall \, \sigma \in \mathfrak{F}(B). \tag{9.23}$$

This means that $\mathcal{E}^*$ is an RNG operation up to normalization.

**Exercise 9.7.** *Consider the resource theory of quantum entanglement. Show that if* $\mathcal{E} \in \mathrm{SEP}(AB \to A'B')$, *then*

$$\frac{\mathcal{E}^*\left(\sigma^{A'B'}\right)}{\mathrm{Tr}\left[\mathcal{E}^*\left(\sigma^{A'B'}\right)\right]} \in \mathrm{SEP}(AB) \qquad \forall \, \sigma \in \mathrm{SEP}(A'B'). \tag{9.24}$$

The need for normalization in (9.23) can be eliminated by broadening the definition of RNG operations to include cone-preserving operations. Specifically, for each system $A$, let us define $\mathfrak{K}(A) \subseteq \mathrm{Pos}(A)$ as the cone represented by

$$\mathfrak{K}(A) := \left\{ t\sigma \ : \ \sigma \in \mathfrak{F}(A), \ t \in \mathbb{R}_+ \right\}. \tag{9.25}$$

We then classify a map $\mathcal{E} \in \mathrm{CP}(A \to B)$ as a $\mathfrak{K}$-preserving operation if, for every $\eta \in \mathfrak{K}(A)$, it holds that $\mathcal{E}(\eta) \in \mathfrak{K}(B)$. By extending the scope of RNG operations to maps that are not necessarily trace-preserving, the condition in (9.23) essentially signifies that $\mathcal{E}^*$ is a $\mathfrak{K}$-preserving operation.

> **Definition 9.6.** Using the same notations as above, we say that a quantum channel $\mathcal{E} \in \mathrm{CPTP}(A \to B)$ is *dually resource nongenerating* if both $\mathcal{E}$ and its dual map $\mathcal{E}^*$ are $\mathfrak{K}$-preserving operations.

*Remark.* Note that since $\mathcal{E}$ is a trace-preserving map, the requirement for it to be a $\mathfrak{K}$-preserving operation is essentially equivalent to the condition of it being a RNG operation. Accordingly, we will use $\mathrm{dRNG}(A \to B)$ to represent the collection of all dually RNG channels within $\mathrm{CPTP}(A \to B)$.

In a resource theory $\mathfrak{F}$, if every operation $\mathcal{E} \in \mathfrak{F}(A \to B)$ satisfies (9.23), then it follows that

$$\mathfrak{F}(A \to B) \subseteq \mathrm{dRNG}(A \to B) \subseteq \mathrm{RNG}(A \to B). \tag{9.26}$$

This implies that dually RNG channels form a subset of the RNG channels, which includes all the free channels. For instance, in the theory of entanglement, the dual of any separable map is clearly separable, making separable maps dually nonentangling (and thus, LOCC channels as well). Protocols involving dually RNG operations are more restricted compared to other operations, such as nonentangling maps. This can provide a more precise approximation to LOCC, potentially leading to improved bounds on their operational power.

Take, for example, the resource theory of coherence. Let $\mathcal{E} \in \mathrm{CPTP}(A \to B)$ be an RNG operation; that is, $\mathcal{E} \in \mathrm{MIO}(A \to B)$. The condition in (9.23) for this theory is equivalent to:

$$\Delta^A \circ \mathcal{E}^* \left( \sigma^B \right) = \mathcal{E}^* \left( \sigma^B \right) \qquad \forall\, \sigma \in \mathfrak{F}(B). \tag{9.27}$$

For any $\rho \in \mathfrak{D}(A)$, the above equation leads to

$$\mathrm{Tr}\left[ \sigma^B \mathcal{E} \left( \rho^A \right) \right] = \mathrm{Tr}\left[ \sigma^B \mathcal{E} \circ \Delta^A \left( \rho^A \right) \right]. \tag{9.28}$$

Since $\sigma \in \mathfrak{F}(B)$ if and only if $\sigma = \Delta^B(\tau)$ for some $\tau \in \mathfrak{D}(B)$, we can rewrite the left-hand side of the equation as

$$\mathrm{Tr}\left[ \sigma^B \mathcal{E} \left( \rho^A \right) \right] = \mathrm{Tr}\left[ \tau^B \Delta^B \circ \mathcal{E} \left( \rho^A \right) \right], \tag{9.29}$$

utilizing the self-adjoint nature of $\Delta^B$. Similarly, applying the self-adjoint property of $\Delta^B$ to the right-hand side of (9.28) results in

$$\mathrm{Tr}\left[ \sigma^B \mathcal{E} \circ \Delta^A \left( \rho^A \right) \right] = \mathrm{Tr}\left[ \tau^B \Delta^B \circ \mathcal{E} \circ \Delta^A \left( \rho^A \right) \right]$$
$$\underset{\mathcal{E} \in \mathrm{MIO}(A \to B)}{\phantom{xxxxx}} = \mathrm{Tr}\left[ \tau^B \mathcal{E} \circ \Delta^A \left( \rho^A \right) \right]. \tag{9.30}$$

Combining these equations, it is concluded that for all $\rho \in \mathfrak{D}(A)$ and $\tau \in \mathfrak{D}(B)$, the following holds:

$$\text{Tr}\left[\tau^B \mathcal{E} \circ \Delta^A\left(\rho^A\right)\right] = \text{Tr}\left[\tau^B \Delta^B \circ \mathcal{E}\left(\rho^A\right)\right]. \tag{9.31}$$

Hence, the condition becomes

$$\mathcal{E}^{A \to B} \circ \Delta^A = \Delta^B \circ \mathcal{E}^{A \to B}. \tag{9.32}$$

This means that if $\mathcal{E} \in \text{dRNG}(A \to B)$, it must satisfy the above condition. The exercise below demonstrates that the converse is also true, leading to the conclusion that a quantum channel $\mathcal{E} \in \text{CPTP}(A \to B)$ is in $\text{dRNG}(A \to B)$ if and only if it satisfies the condition in (9.32). Notably, for the case where $A = B$, this condition simplifies to the commutation relation $[\mathcal{E}, \Delta] = 0$.

We note that in the QRT of coherence, quantum channels that satisfy (9.32) are identified as Dephasing-covariant Incoherent Operations, abbreviated as DIO. Therefore, we have demonstrated that within the resource theory of coherence, the set $\text{dRNG}(A \to B)$, is equivalent to the set of DIO channels, denoted as $\text{DIO}(A \to B)$.

**Exercise 9.8.** *Show that $DIO(A \to B) \subseteq dRNG(A \to B)$; that is, let $\mathcal{E} \in \text{CPTP}(A \to B)$ be a quantum channel satisfying (9.32), and show that $\mathcal{E} \in dRNG(A \to B)$.*

**Exercise 9.9.** *Show that the inclusion $DIO(A \to B) \subset MIO(A \to B)$ is strict.*

## 9.3 Affine Resource Theories

Some QRTs satisfy a stronger condition than convexity, and we call them *affine resource theories* (ARTs). We will see later on that ARTs have many desired properties, and in particular, their conversion rates can be computed efficiently and algorithmically with an SDP.

**An Affine Set**

**Definition 9.7.** Let $A$ be a physical system, and $\mathfrak{F}(A) \subseteq \mathfrak{D}(A)$ be a set of density matrices. The set $\mathfrak{F}(A)$ is called an *affine* set, if every affine combination of $n$ free states

$$\sigma := \sum_{x \in [n]} t_x \rho_x \qquad \rho_1, \ldots, \rho_n \in \mathfrak{F}(A), \quad t_1, \ldots, t_n \in \mathbb{R}, \tag{9.33}$$

such that $\sigma \in \mathfrak{D}(A)$, satisfies $\sigma \in \mathfrak{F}(A)$.

As an example of an affine set, let $\mathfrak{F}$ be the QRT of coherence in which $\mathfrak{F}(A)$ is the set of all diagonal states in $\mathfrak{D}(A)$ with respect to a fixed basis of $A$. The set of diagonal states $\mathfrak{F}(A)$ is affine since for any affine combination of diagonal states is diagonal; that is, if $\sum_{x \in [n]} t_x \rho_x \geq 0$, where each state $\rho_x \in \mathfrak{F}(A)$ (i.e. each $\rho_x$ is diagonal) and $\sum_{x \in [n]} t_x = 1$, then also $\sum_{x \in [n]} t_x \rho_x$ is a diagonal state. Therefore, the set of free states in the QRT of coherence is affine.

**Exercise 9.10.** *Let $\mathfrak{F}(A)$ be the set of all density matrices in $\mathfrak{D}(A)$ with real components with respect to a fixed basis $\{|x\rangle\}_{x \in [m]}$ of A. That is, $\rho \in \mathfrak{F}(A) \subset \mathfrak{D}(A)$ if and only if the number $\langle x | \rho | x' \rangle$ is real for all $x, x' \in [m]$. Show that $\mathfrak{F}(A)$ is affine.*

The following exercise provides another characterization of affine sets.

**Exercise 9.11.** *Let $\mathfrak{F}(A) \subseteq \mathfrak{D}(A)$ be a set of density matrices, and let $\mathfrak{K}(A) := \mathrm{span}_{\mathbb{R}} \{\mathfrak{F}(A)\}$ be the subspace of $\mathrm{Herm}(A)$ consisting of all linear combinations of the elements of $\mathfrak{F}(A)$. Show that $\mathfrak{F}(A)$ is affine if and only if*

$$\mathfrak{F}(A) = \mathfrak{K}(A) \cap \mathfrak{D}(A) \tag{9.34}$$

Not all convex sets are affine. For example, the set of separable states in entanglement theory is not affine. To see why, recall that product states of the form $\psi^A \otimes \phi^B$ are free states. Let $m := |A|$, $n := |B|$, $\{\psi_x^A\}_{x \in [m^2]}$ be a rank 1 basis of $\mathrm{Herm}(A)$, and $\{\phi_y^B\}_{y \in [n^2]}$ be a rank 1 basis of $\mathrm{Herm}(B)$. Then, the $m^2 n^2$ states $\{\psi_x^A \otimes \phi_y^B\}_{x,y}$ form a basis of $\mathrm{Herm}(AB)$. This means that *any* density matrix $\rho \in \mathfrak{D}(AB)$ can be expressed as an affine combination of product states; that is,

$$\rho^{AB} = \sum_{x \in [m^2]} \sum_{y \in [n^2]} t_{xy} \psi_x^A \otimes \phi_y^B, \tag{9.35}$$

where $\{t_{xy}\}$ is a set of real numbers. Hence, the set of separable states is not affine since even entangled states can be expressed as affine combination of product states. In this sense, the set of separable states is maximally nonaffine. More generally, we say that a set $\mathfrak{F}(A)$ is maximally nonaffine if

$$\mathrm{Herm}(A) = \mathrm{span}_{\mathbb{R}} \{\mathfrak{F}(A)\}. \tag{9.36}$$

**Exercise 9.12.** *Show that the set $\{t_{xy}\}_{x,y}$ in (9.35) must satisfy $\sum_{x,y} t_{xy} = 1$.*

**Exercise 9.13.** *Let $|\Phi_+^{AB}\rangle = \frac{1}{\sqrt{2}}(|00\rangle + |11\rangle)$ be the maximally entangled states.*

*1. Express (explicitly) $\Phi^{AB}$ as an affine combination of product states.*

*2. Show that $\Phi^{AB}$ cannot be expressed as a convex combination of product states.*

It is straightforward to extend the definition of an affine set in $\mathfrak{D}(A)$ to sets in $\mathrm{CPTP}(A \to B)$.

> **Affine Resource Theories**
>
> **Definition 9.8.** A QRT, $\mathfrak{F}$, is called an *affine resource theory* (AFT) if for any two systems $A$ and $B$, and any affine combination of $n$ free channels
>
> $$\mathcal{N}^{A \to B} := \sum_{x \in [n]} t_x \mathcal{E}_x^{A \to B}, \qquad \mathcal{E}_1, \ldots, \mathcal{E}_n \in \mathfrak{F}(A \to B), \quad t_1, \ldots, t_n \in \mathbb{R}, \quad (9.37)$$
>
> such that $\mathcal{N} \in \mathrm{CPTP}(A \to B)$, the channel $\mathcal{N}$ is free (i.e. $\mathcal{N} \in \mathfrak{F}(A \to B)$).

We will see in the next two chapters that ARTs have several properties that make them much easier to study. Particularly, many problems in ARTs can be solved with semidefinite programming, unlike certain convex QRTs, such as entanglement theory, in which even the determination of whether a state is free or not is very hard (more precisely, belongs to a complexity class known as NP-hard). In the following theorem we show that if the set of free operations is RNG or CRNG, then the QRT is affine if and only if the set of free states is affine.

> **Theorem 9.1.** Let $\mathfrak{F}(A) \subseteq \mathfrak{D}(A)$ be a set of free states on any physical system $A$, and for any two physical systems $A$ and $B$, let $\mathrm{RNG}(A \to B)$ and $\mathrm{CRNG}(A \to B)$ be as defined in Definitions 9.3 and 9.4 (with respect to the sets $\mathfrak{F}(A)$ and $\mathfrak{F}(B)$). Then, the following statements are equivalent:
>
> 1. For all systems $A$, the set $\mathfrak{F}(A)$ is affine.
> 2. For all systems $A$ and $B$, the set $\mathrm{RNG}(A \to B)$ is affine.
> 3. For all systems $A$ and $B$, the set $\mathrm{CRNG}(A \to B)$ is affine.

**Proof**   The implication $1 \Rightarrow 2$: Let $\mathcal{E}_1, \ldots, \mathcal{E}_n \in \mathrm{RNG}(A \to B)$ and $t_1, \ldots, t_n \in \mathbb{R}$ be such that

$$\mathcal{N}^{A \to B} := \sum_{x \in [n]} t_x \mathcal{E}_x^{A \to B} \in \mathrm{CPTP}(A \to B). \qquad (9.38)$$

We need to show that $\mathcal{N} \in \mathrm{RNG}(A \to B)$. Let $\sigma \in \mathfrak{F}(A)$ be a free state. Since $\mathcal{N}$ is a quantum channel $\mathcal{N}^{A \to B}(\sigma^A) \in \mathfrak{D}(B)$. On the other hand, from the definition of $\mathcal{N}$, we can write this relation as

$$\mathcal{N}^{A \to B}(\sigma^A) = \sum_{x \in [n]} t_x \omega_x^B \in \mathfrak{D}(B), \quad \text{where} \quad \omega_x^B := \mathcal{E}_x^{A \to B}\left(\sigma^A\right). \qquad (9.39)$$

Since $\sigma \in \mathfrak{F}(A)$ and $\mathcal{E}_x \in \mathrm{RNG}(A \to B)$ it follows that each $\omega_x \in \mathfrak{F}(B)$. Finally, using the assumption that $\mathfrak{F}(B)$ is affine we get that $\mathcal{N}^{A \to B}(\sigma^A) \in \mathfrak{F}(B)$. Since $\sigma$ was arbitrary state in $\mathfrak{F}(A)$ we conclude that $\mathcal{N} \in \mathrm{RNG}(A \to B)$.

**The implication** $2 \Rightarrow 3$: Let $\mathcal{E}_1, \ldots, \mathcal{E}_n \in \mathrm{CRNG}(A \to B)$ be a set of $n$ CRNG channels, and let $t_1, \ldots, t_n \in \mathbb{R}$ be such that (9.38) holds. We need to show that $\mathcal{N} \in \mathrm{CRNG}(A \to B)$ or equivalently, that for any reference system $R$, $\mathrm{id}^R \otimes \mathcal{N} \in \mathrm{RNG}(RA \to RB)$. Since each $\mathrm{id}^R \otimes \mathcal{E}_x \in \mathrm{RNG}(RA \to RB)$ and since we assume that $\mathrm{RNG}(RA \to RB)$ is affine it follows that

$$\mathrm{id}^R \otimes \mathcal{N} = \sum_{x \in [n]} t_x \, \mathrm{id}^R \otimes \mathcal{E}_x^{A \to B} \in \mathrm{RNG}(RA \to RB). \tag{9.40}$$

Hence, $\mathrm{CRNG}(A \to B)$ is affine.

**The implications** $3 \Rightarrow 1$: By taking $A$ to be the trivial system (i.e. $|A| = 1$) in $\mathrm{CRNG}(A \to B)$ we get that $\mathfrak{F}(B)$ is affine for all systems $B$. This completes the proof. ∎

## 9.3.1 Resource Destroying Maps

In specific QRTs, notably those concerning coherence and asymmetry, a distinct transformation exists. This transformation has the ability to map any density matrix into a free state, essentially "destroying" the resource. At the same time, it functions as the identity channel when applied to free states. This dual capability highlights its unique role in these particular QRTs. We point out that such maps do not necessarily have to be channels, and they might not even be linear. However, in the context of this book, where our exploration is confined to theories within the realm of quantum mechanics, we will consistently assume that these resource-destroying maps are at least linear.

> **Resource Destroying Map**
>
> **Definition 9.9.** Let $\mathfrak{F}$ be a QRT. A *resource destroying map* (RDM) is a linear map $\Delta \in \mathfrak{L}(A \to A)$ with the following two properties:
>
> 1. For all $\rho \in \mathfrak{D}(A)$, $\Delta(\rho) \in \mathfrak{F}(A)$.
> 2. For all $\sigma \in \mathfrak{F}(A)$, $\Delta(\sigma) = \sigma$.

*Remark.* It is important to note that the definition of a RDM is relative to the set of free states $\mathfrak{F}(A)$. Consequently, different QRTs with an identical set of free states $\mathfrak{F}(A)$ may share the same RDM. Conversely, there could exist multiple RDMs corresponding to the same set $\mathfrak{F}(A)$. Furthermore, it is evident that $\Delta \in \mathrm{Pos}(A \to A)$. This is because for any positive operator $\Lambda \in \mathrm{Pos}(A)$, the transformed state $\Delta(\Lambda/\mathrm{Tr}[\Lambda])$ is a free state and therefore positive semidefinite. However, it is crucial to understand that a RDM is not necessarily completely positive.

As an example of a RDM, consider the QRT of coherence in which $\mathfrak{F}(A) \subset \mathfrak{D}(A)$ is the set of diagonal states with respect to the basis $\{|x\rangle\}_{x \in [m]}$ (here $m := |A|$). Relative to this basis, define the completely dephasing channel

$$\Delta(\rho) := \sum_{x \in [n]} \langle x|\rho|x\rangle |x\rangle\langle x|. \tag{9.41}$$

It is simple to check that $\Delta$ as defined above is a RDM with respect to the set of diagonal matrices. In this example, $\Delta$ is a quantum channel.

**Exercise 9.14.** *Show that $\Delta$ as defined in (9.41) is a RDM. Moreover, show that it is self-adjoint; that is, $\Delta^* = \Delta$.*

**Exercise 9.15.** *Show that any RDM $\Delta \in \mathfrak{L}(A \to A)$ is idempotent; that is, $\Delta \circ \Delta = \Delta$.*

In the following lemma we show that not all QRTs have a RDM.

> **Lemma 9.1.** Let $\mathfrak{F}$ be a QRT and let $A$ be a quantum system. If $\mathfrak{F}(A)$ is not affine, then the QRT $\mathfrak{F}$ does not have a RDM.

**Proof**  Suppose by contradiction that $\mathfrak{F}(A)$ is not affine, and yet, there exists a RDM $\Delta \in \mathfrak{L}(A \to A)$. Then, by definition, there exists $t_1, \ldots, t_n \in \mathbb{R}$ and $\sigma_1, \ldots, \sigma_n \in \mathfrak{F}(A)$ such that

$$\sigma := \sum_{x \in [n]} t_x \sigma_x \in \mathfrak{D}(A) \quad \text{but} \quad \sigma \notin \mathfrak{F}(A) \tag{9.42}$$

Moreover, since $\Delta$ is a RDM we must have $\Delta(\sigma) \subset \mathfrak{F}(A)$. On the other hand,

$$\Delta(\sigma) = \sum_{x \in [n]} t_x \Delta(\sigma_x)$$

$$= \sum_{x \in [n]} t_x \sigma_x \tag{9.43}$$

$$= \sigma \notin \mathfrak{F}(A)$$

in contradiction with the fact that $\Delta(\sigma) \in \mathfrak{F}(A)$. Therefore, if a QRT admits a RDM, then it's set of states must be affine. $\blacksquare$

## Characterization of Self-adjoint RDM

All the RDMs that play a role in well-studied QRTs, such as the QRTs of coherence and asymmetry, have the property that they are self-adjoint. Self-adjoint RDMs have a relatively simple characterization, and they always exists if the set of free states is affine. In the following definition, we consider an affine set $\mathfrak{F}(A) \subseteq \mathfrak{D}(A)$, and denote by $\mathfrak{K}(A) := \text{span}_{\mathbb{R}}\{\mathfrak{F}(A)\}$ the subspace of $\text{Herm}(A)$ consisting of all linear combinations of the elements of $\mathfrak{F}(A)$. Moreover, we denote by $\mathfrak{K}(A)^{\perp}$ the orthogonal complement of $\mathfrak{K}(A)$ in $\text{Herm}(A)$ that satisfies $\text{Herm}(A) = \mathfrak{K}(A) \oplus \mathfrak{K}(A)^{\perp}$.

> **self-adjoint RDM**
>
> **Definition 9.10.** Let $\mathfrak{F}(A)$, $\mathfrak{K}(A)$, and $\mathfrak{K}(A)^{\perp}$ be as defined above. The linear map $\Delta : \text{Herm}(A) \to \text{Herm}(A)$
>
> $$\Delta(\eta + \zeta) := \eta \qquad \forall \, \eta \in \mathfrak{K}(A) \text{ and } \forall \, \zeta \in \mathfrak{K}(A)^{\perp}, \tag{9.44}$$
>
> is called the *self-adjoint RDM*.

*Remark.* Note that in this context, the map $\Delta$ is referred to as *the* self-adjoint RDM. This designation is justified by demonstrating that for every ART, there exists a unique self-adjoint RDM (see the following theorem). This uniqueness within each ART underscores the specific role that such a map plays in the theory.

**Exercise 9.16.** *Verify that $\Delta$ in the definition above is indeed a RDM which is self-adjoint (with respect to the Hilbert–Schmidt inner product).*

> **Theorem 9.2.** Let $\mathfrak{F}(A) \subseteq \mathfrak{D}(A)$ be an affine set. Then, there exists a unique self-adjoint RDM associated with $\mathfrak{F}(A)$ which is given by $\Delta$ as defined in (9.44).

**Proof** The existence of $\Delta$ follows from Definition 9.10 and Exercise 9.16. To prove uniqueness, let $\Delta \colon \mathrm{Herm}(A) \to \mathrm{Herm}(A)$ be a self-adjoint RDM. We would like to show that it coincides with the RDM given in Definition 9.10. Indeed, from the linearity of $\Delta$, and the fact that $\Delta$ is a RDM, we must have $\Delta(\eta) = \eta$ for all $\eta \in \mathfrak{K}(A)$. Moreover, since $\Delta$ is self-adjoint for every $\zeta \in \mathfrak{K}(A)^{\perp}$ and $\eta \in \mathfrak{K}(A)$, we have

$$0 = \mathrm{Tr}[\eta\zeta] = \mathrm{Tr}[\Delta(\eta)\zeta] = \mathrm{Tr}[\eta\Delta(\zeta)]. \tag{9.45}$$

Since the above equation holds for all $\eta \in \mathfrak{K}(A)$ we conclude that $\Delta(\zeta) \in \mathfrak{K}(A)^{\perp}$. However, since $\Delta$ is a RDM we must have $\Delta(\zeta) \in \mathfrak{K}(A)$. Both conditions hold only if $\Delta(\zeta) = 0$. To summarize, we got that for all $\eta \in \mathfrak{K}(A)$ and all $\zeta \in \mathfrak{K}(A)^{\perp}$ we have

$$\begin{aligned}
\Delta(\eta + \zeta) &= \Delta(\eta) + \Delta(\zeta) \\
&\xrightarrow{\boxed{\Delta(\zeta) = 0}} = \Delta(\eta) \\
&= \eta.
\end{aligned} \tag{9.46}$$

Hence, $\Delta$ coincides with the map defined in (9.44). This completes the proof. ∎

**Exercise 9.17.** *Let $\mathfrak{F}(A) \subseteq \mathfrak{D}(A)$ be an affine set, and let $\{\eta_x\}_{x \in [m]}$ be an orthonormal basis of $\mathrm{span}_{\mathbb{R}}\{\mathfrak{F}(A)\}$ (w.r.t. to Hilbert–Schmidt inner product). Show that the linear map*

$$\Delta(\omega) := \sum_{x \in [m]} \mathrm{Tr}\,[\eta_x \omega]\, \eta_x \qquad \forall \omega \in \mathfrak{L}(A), \tag{9.47}$$

*is a self-adjoint resource destroying map.*

We have seen that if a $\mathfrak{F}(A) \subseteq \mathfrak{D}(A)$ has a RDM, then it must be affine. On the other hand, any affine set $\mathfrak{F}(A)$ has a unique self-adjoint RDM. Therefore, if $\mathfrak{F}(A)$ has a RDM, then it also has a *self-adjoint* RDM. Note however that the self-adjoint RDM, $\Delta$, as defined in Definition 9.10, is not necessarily a quantum channel. To be a quantum channel, the Choi matrix of $\Delta$, denoted as $J_{\Delta}^{A\tilde{A}}$, must satisfy the two conditions of a quantum channel, namely, $J_{\Delta}^{A\tilde{A}} \geqslant 0$ and $J_{\Delta}^{A} = I^{A}$. In the QRT of coherence, the completely decohering map as defined in (9.41) is indeed a self-adjoint RDM that is also a quantum channel.

However, there exists affine sets for which the self-adjoint RDM is not a quantum channel. For example, let $\mathfrak{F}(A)$ be the set of all density matrices in $\mathfrak{D}(A)$ whose components are real with respect to a fixed basis $\{|x\rangle\}_{x \in [m]}$ (with $m := |A|$). A QRT with this set of states have been used in literature to quantify the *imaginarity* of quantum mechanics. Now, in the following exercise we will show that the self-adjoint RDM of this resource theory is given by

$$\Delta(\eta) = \frac{1}{2}\left(\eta + \eta^T\right) \qquad \forall\, \eta \in \mathrm{Herm}(A), \tag{9.48}$$

where the transpose is taken with respect to the fixed basis $\{|x\rangle\}_{x \in [m]}$. Therefore, since the transpose is not completely positive it follows that $\Delta$ above is not completely positive (see the following exercise).

**Exercise 9.18.** *Consider the affine set, $\mathfrak{F}(A)$, of all density matrices in $\mathfrak{D}(A)$ whose components are real with respect to a fixed basis $\{|x\rangle\}_{x \in [m]}$ (with $m := |A|$).*

*1. Show that the self-adjoint RDM associated with $\mathfrak{F}(A)$ is given by (9.48).*
*2. Show that this self-adjoint RDM is not completely positive.*

## 9.4 Resource Witnesses

In certain resource theories, the complexity of the set of free states is such that determining whether a given state is part of this set can be challenging. This is particularly evident in the theory of mixed bipartite entanglement. To address this challenge, a valuable tool from convex analysis can be employed to ascertain whether a quantum state qualifies as a free state. This approach provides a practical method for navigating the intricacies of these theories and identifying free states within complex sets.

---

**Resource Witness**

**Definition 9.11.** Let $\mathfrak{F}$ be a QRT and let $A$ be a physical system. An operator $W \in \mathrm{Herm}(A)$ is called a resource witness if the following two conditions hold:

1. For any $\sigma \in \mathfrak{F}(A)$,

$$\mathrm{Tr}[W\sigma] \geqslant 0. \tag{9.49}$$

2. There exists $\rho \in \mathfrak{D}(A)$ such that

$$\mathrm{Tr}[W\rho] < 0. \tag{9.50}$$

---

Since a resource-witness is a Hermitian matrix, it corresponds to an observable and the expressions $\mathrm{Tr}[W\rho]$ and $\mathrm{Tr}[W\sigma]$ corresponds to *expectation values* of $W$ and in principle can be measured in a laboratory. Therefore, resource witnesses provide a practical method to test if a given quantum system is a resource.

The condition (9.50) in the definition above is equivalent to the statement that $W$ is not a positive semidefinite matrix. Clearly, every positive semidefinite matrix satisfies (9.49). Therefore, resource witnesses can be viewed as the elements of $\mathfrak{F}(A)^*$ that are not positive semidefinite. Recall from Sec. A.8 of online version that $\mathfrak{F}(A)^*$ denotes the dual cone of $\mathfrak{F}(A)$ defined by

$$\mathfrak{F}(A)^* := \left\{ W \in \mathrm{Herm}(A) : \ \mathrm{Tr}\,[W\sigma] \geqslant 0 \ \ \forall\, \sigma \in \mathfrak{F}(A) \right\}. \qquad (9.51)$$

Since $\mathrm{Pos}(A) \subseteq \mathfrak{F}(A)^*$ we conclude that the set of all resource-witnesses can be viewed as the nonpositive semidefinite matrices in $\mathfrak{F}(A)^*$. If $\mathfrak{F}(A)$ is closed and convex, then the set of all witnesses completely determines the set of free states.

> **Theorem 9.3.** Let $A$ be a physical system, $\mathfrak{F}(A) \subseteq \mathfrak{D}(A)$ be a closed and convex subset of density matrices, and $\sigma \in \mathfrak{D}(A)$. Then, $\sigma \in \mathfrak{F}(A)$ if and only if
>
> $$\mathrm{Tr}\,[W\sigma] \geqslant 0 \qquad (9.52)$$
>
> for *all* resource witnesses $W \in \mathrm{Herm}(A)$.

**Proof** This theorem follows from the property that any closed and convex set $\mathfrak{K} \subset \mathrm{Herm}(A)$ satisfies $\mathfrak{K}^{**} = \mathfrak{K}$ (see Theorem A.11 of online version). Hence, in particular, $\mathfrak{F}(A)^{**} = \mathfrak{F}(A)$. The latter means that $\sigma \in \mathfrak{F}(A)$ if and only if $\sigma \in \mathfrak{F}(A)^{**}$; that is, if and only if

$$\mathrm{Tr}[W\sigma] \geqslant 0 \ \ \forall\, W \in \mathfrak{F}(A)^*. \qquad (9.53)$$

Note that the inequality above holds trivially for all $W \geqslant 0$. Therefore, it is sufficient to check it for all $0 \not\leqslant W \in \mathfrak{F}(A)^*$; that is, for all resource witnesses. $\blacksquare$

We point out that for affine QRTs, determining whether a quantum state is free or not is relatively an easy task. Since any affine set $\mathfrak{F}(A)$ has a self-adjoint resource destroying map (see Definition 9.10), to determine if a state $\rho \in \mathfrak{D}(A)$ is free or not, all we have to do is to check if $\Delta(\rho) = \rho$. Such a simplification does not occur in certain important convex QRTs (e.g. entanglement theory).

# 9.5 Notes and References

The term "resource theory" has a bit of a history, starting with the early recognition that quantum information theory, particularly quantum Shannon theory (which we cover in volume 2 of this book), is a theory of interconversions among different resources; see Refs. [19] and [62]. Originally coined by Schumacher in 2003 (unpublished), the term "resource theory" first appeared in a paper on the QRT of quantum reference frames in Ref. [97], although the framework for a QRT of information had already been investigated in a series of earlier papers in Refs. [177, 127, 131]. The definition given here for a QRT is due to Refs. [30] and [48]. Other definitions involving symmetric monoidal categories can be found in papers in Refs. [52, 80]. However, these definitions involve terms from category theory and goes beyond the scope of this book.

Affine resource theories were introduced in Ref. [88] and resource destroying maps in Ref. [151].

One of the most useful aspects of a QRT is that it generates precise and operationally meaningful ways to quantify a given physical resource. Here we study a variety of resource measures that can be introduced in any QRT. We start with the definition of a resource measure, and discuss some additional desirable properties that any resource measure should satisfy. After that, we study different families of specific resource measures applicable to any QRT. We put emphasis on the Umegaki relative entropy of a resource, as it turns out that this resource measure has several operational interpretations, and it plays a major role throughout this book.

## 10.1 Definitions and Properties of Resource Measures

In their definition, the set of free states, $\mathfrak{F}(A)$, and the set of free operations, $\mathfrak{F}(A \to B)$, are defined on *any* Hilbert spaces $A$ and $B$. Consequently, a resource measure is defined as a function whose domain is the set of all density matrices in every finite dimension. In the following definition we consider a function

$$\mathbf{M} \colon \bigcup_A \mathfrak{D}(A) \to \mathbb{R}, \tag{10.1}$$

where the union is over all Hilbert spaces $A$. This union also includes the trivial system $A = \mathbb{C}$ (i.e. $|A| = 1$) in which case $\mathfrak{D}(A) = \{1\}$ consists of only one element, namely, the number 1.

---

**Definition 10.1.** The function $\mathbf{M}$ in (10.1) is called a *resource measure* if it satisfies the following:

1. Monotonicity: $\mathbf{M}\big(\mathcal{E}(\rho)\big) \leqslant \mathbf{M}(\rho)$ for all $\mathcal{E} \in \mathfrak{F}(A \to B)$ and all $\rho \in \mathfrak{D}(A)$.
2. Normalization: $\mathbf{M}(1) = 0$.

---

The first property is fundamental in resource theories. It asserts that the value of any resource measure cannot be increased through the use of free operations. This principle, known as monotonicity, is consistent with the "Chernoff" of QRTs that free operations cannot generate resources. The normalization condition, in conjunction with monotonicity, leads to the positivity of every resource measure $M$, which can be expressed as

$$\mathrm{M}(\rho) \geqslant 0 \qquad \forall \, \rho \in \mathfrak{D}(A). \tag{10.2}$$

The positivity follows from the fact that the trace is a free operation in any QRT, and therefore, from the monotonicity of $M$ under free operations we have

$$\mathrm{M}(\rho) \geqslant \mathrm{M}(\mathrm{Tr}[\rho]) = \mathrm{M}(1) = 0. \tag{10.3}$$

Similarly, the two conditions of normalization and monotonicity implies that for any finite dimensional system $A$,

$$\sigma \in \mathfrak{F}(A) \quad \Rightarrow \quad \mathrm{M}(\sigma) = 0. \tag{10.4}$$

Indeed, any state $\sigma \in \mathfrak{F}(A)$ can be viewed as a free channel $\sigma^{1 \to A}$, where $1$ represents the trivial system corresponding to the Hilbert space $\mathbb{C}$. Hence,

$$\mathrm{M}\left(\sigma^{A}\right) = \mathrm{M}\left(\sigma^{1 \to A}(1)\right) \leqslant \mathrm{M}(1) = 0, \tag{10.5}$$

where the inequality follows from the monotonicity property of M. Therefore, since M is nonnegative we must have $\mathrm{M}(\sigma) = 0$ for all $\sigma \in \mathfrak{F}(A)$ and all finite dimensional systems $A$. We discuss now several properties that are satisfied by some, but not all, resource measures.

## Faithfulness

The condition expressed in (10.4) quantitatively defines the notion of "no resource." Intuitively, one might be inclined to consider that the reverse of (10.4) should also hold true. This concept is referred to as faithfulness. A general resource measure $M$ is deemed *faithful* if $M(\rho) = 0$ necessarily implies that $\rho$ is a free state.

However, it's important to recognize that for certain tasks, some resource states may not offer any operational advantage over free states. In such scenarios, these states should be assigned a zero value by any measure that quantifies their utility for performing the specified task. For instance, as we will explore later, the measure of distillable entanglement, which is a significant measure of entanglement, is zero for all bound entangled states. Therefore, although faithfulness is an intuitively attractive property, it is not an essential requirement for a resource measure. This perspective allows for a more nuanced understanding of resource measures and their application in various contexts within quantum resource theories.

## Resource Monotones

In certain QRTs, quantum measurements do not belong to the set of free operations. One such example is quantum thermodynamics as we will see in Chapter 17. However, in many other QRTs, like entanglement, quantum measurements can be free, and they represent an important component of the theory. In such QRTs, the set of quantum instruments $\mathfrak{F}(A \to BX)$, where $X$ is a classical "flag" system, is not empty. In particular, any such channel in $\mathcal{E} \in \mathfrak{F}(A \to BX)$ can be express as $\mathcal{E}^{A \to BX} = \sum_{x \in [m]} \mathcal{E}_x^{A \to B} \otimes |x\rangle\langle x|^X$, and consequently, any resource measure M satisfies for all $\rho \in \mathfrak{D}(A)$

$$\mathrm{M}\left(\rho^{A}\right) \geqslant \mathrm{M}\left(\mathcal{E}^{A \to BX}(\rho^{A})\right) = \mathrm{M}\left(\sum_{x \in [m]} p_x \sigma_x^{B} \otimes |x\rangle\langle x|^X\right), \tag{10.6}$$

where

$$p_x := \mathrm{Tr}[\mathcal{E}_x^{A \to B}(\rho^A)] \quad \text{and} \quad \sigma_x^B := \frac{1}{p_x}\mathcal{E}_x^{A \to B}(\rho^A). \tag{10.7}$$

We say that M is convex linear on qc-states if for all $\sigma \in \mathfrak{D}(BX)$ as above

$$\mathrm{M}\Big( \sum_{x \in [m]} p_x \sigma_x^B \otimes |x\rangle\langle x|^X \Big) = \sum_{x \in [m]} p_x \mathrm{M}(\sigma_x^B \otimes |x\rangle\langle x|^X). \tag{10.8}$$

Almost all resource measures studied in literature are convex linear on QC states. One reason for that is that the equality above is satisfied by many functions, like the von-Neuman entropy, Rényi entropies, all the Schatten $p$-norms, and so on. If a QRT admits a tensor product structure, then the partial trace is considered free so that for every $x \in [m]$

$$\mathrm{M}(\sigma_x^B \otimes |x\rangle\langle x|^X) \geqslant \mathrm{M}(\sigma_x^B). \tag{10.9}$$

Combining this with (10.6) and (10.8) we get that in such QRTs

$$\mathrm{M}\big(\rho^A\big) \geqslant \sum_{x \in [m]} p_x \mathrm{M}(\sigma_x^B). \tag{10.10}$$

This property is sometimes referred to as *strong monotonicity*. An intuitive justification for requiring strong monotonicity is to prevent $M$ from increasing on average when the experimenter can post-select or "flag" the multiple outcomes of a quantum measurement.

We point out that in many QRTs, the classical flag states $\{|x\rangle\langle x|^X\}_{x \in [m]}$ are themselves considered as free states in $\mathfrak{F}(X)$. As physical QRTs admit a tensor product structure, appending or discarding flags is considered a free operation (recall that the partial trace is free and appending free states as in (9.6) is a free operation). For such QRTs we have in fact equality in (10.9).

<hr>

**Resource Monotone**

**Definition 10.2.** A resource measure M is called a resource monotone if it satisfies the following:

1. *Strong monotonicity.* For any resource $\rho \in \mathfrak{D}(A)$, and any free quantum instrument $\mathcal{E} = \sum_{x \in [m]} \mathcal{E}_x \otimes |x\rangle\langle x| \in \mathfrak{F}(A \to BX)$

$$\mathrm{M}\big(\rho^A\big) \geqslant \sum_{x \in [m]} p_x \mathrm{M}\big(\sigma_x^B\big), \tag{10.11}$$

   where $\{p_x\}_{x \in [m]}$ and $\{\sigma_x^B\}_{x \in [m]}$ are defined in (10.7).
2. *Convexity.* For any ensemble of states $\{p_x, \rho_x\}_{x \in [m]}$ in $\mathfrak{D}(A)$

$$\mathrm{M}\Big( \sum_{x \in [m]} p_x \rho_x \Big) \leqslant \sum_{x \in [m]} p_x \mathrm{M}(\rho_x). \tag{10.12}$$

As we delve further, we will see that the convexity property above is extremely useful from a mathematical perspective in calculating the resource monotone for a specific state. Concurrently, a common physical interpretation of convex measures is that the process of mixing states does not result in an increase in the resource quantity. However, it's important to be cautious in drawing parallels between this mathematical notion of convexity and the physical process of mixing states, as the latter typically involves discarding information. We have previously discussed this important distinction in Section 9.1.3. Additionally, it's worth noting that in QRTs where a freely available classical (flag) basis does not exist, and thus strong monotonicity is not a relevant concept, convex resource measures will be referred to as resource monotones.

## Subadditivity

Some resource measures have additional properties that are mathematically convenient. One of such properties is *subadditivity*. A resource measure $M$ is said to be subadditive if for any $\rho \in \mathfrak{D}(A)$ and $\sigma \in \mathfrak{D}(B)$,

$$\mathbf{M}(\rho \otimes \sigma) \leqslant \mathbf{M}(\rho) + \mathbf{M}(\sigma). \tag{10.13}$$

While subadditivity is a natural property to expect from a resource measure, it will not hold for all measures in a general QRT. In particular, we will see examples of that when we discuss superactivation.

## Additivity

An even stronger property of a resource measure is *additivity*. That is, $\mathbf{M}$ is said to be additive when equality holds in (10.13) for all states. While most resource measures do not satisfy this property, there exists a procedure known as *regularization* that allows for the general construction of measures that are additive on multiple copies of the same state. We have already encountered this procedure implicitly in a few places of previous chapters. The regularization of a resource measure $\mathbf{M}$ is defined for all $\rho \in \mathfrak{D}(A)$ as

$$\mathbf{M}^{\text{reg}}(\rho) := \lim_{n \to \infty} \frac{1}{n} \mathbf{M}\left(\rho^{\otimes n}\right), \tag{10.14}$$

provided the limit exists. In the following exercise we will show that the limit above exists if $\mathbf{M}$ satisfies a weaker form of subadditivity.

**Exercise 10.1.** *Show that the limit in (10.14) exists if $M$ satisfies for all $n, m \in \mathbb{N}$ and any density matrix $\rho$*

$$\mathbf{M}\left(\rho^{\otimes(m+n)}\right) \leqslant \mathbf{M}\left(\rho^{\otimes m}\right) + \mathbf{M}\left(\rho^{\otimes n}\right). \tag{10.15}$$

*Hint: Use Exercise 6.37.*

## Asymptotic Continuity

It's a reasonable expectation for any resource measure with physical significance to exhibit continuity. This expectation stems from the idea that if one quantum state is a slight perturbation of another, their resource contents should be very similar. However, it's important to note that a function $f : \bigcup_A \mathfrak{D}(A) \to \mathbb{R}_+$ satisfying the following condition

$$|A|\|\rho - \sigma\|_1 \leqslant |f(\rho) - f(\sigma)| \leqslant |A|^2 \|\rho - \sigma\|_1 \qquad \forall\, \rho, \sigma \in \mathfrak{D}(A) \tag{10.16}$$

is indeed continuous. Yet, in the context of very large dimensions (i.e. $|A| \gg 1$), this type of continuity may not be practically useful. This is because, for the difference $|f(\rho) - f(\sigma)|$ to be small, $\rho$ and $\sigma$ need to be so closely aligned that they are virtually identical for all practical purposes. Therefore, a more robust notion of continuity, known as asymptotic continuity, is often considered. Asymptotic continuity is especially pertinent in the realm of large dimensions. It limits the dependence on dimension to a logarithmic scale, thereby providing a more practical and realistic measure of continuity when dealing with high-dimensional quantum states. This concept is particularly useful in assessing the continuity of resource measures in quantum systems where the dimensionality plays a significant role.

> **Asymptotic Continuity**
>
> **Definition 10.3.** A resource measure $\mathbf{M}$ is said to be asymptotically continuous if for any $\rho, \sigma \in \mathfrak{D}(A)$, and $\varepsilon := \frac{1}{2}\|\rho - \sigma\|_1$,
>
> $$\left|\mathbf{M}(\rho) - \mathbf{M}(\sigma)\right| \leqslant f(\varepsilon) \log |A|, \tag{10.17}$$
>
> where $f : \mathbb{R} \to \mathbb{R}$ is some continuous function, independent on the dimensions, and satisfies $\lim_{\varepsilon \to 0^+} f(\varepsilon) = 0$.

Note that the above notion of continuity is stronger than regular notion of continuity in the sense that the right-hand side of (10.17) depends on the dimension through a log function. This in particular implies that the regularization of $\mathbf{M}$, if it exists, is bounded. To see why, note that if $\mathbf{M}$ is an asymptotically continuous resource measure, then for any $n \in \mathbb{N}$

$$\left|\frac{1}{n}\mathbf{M}\left(\rho^{\otimes n}\right) - \frac{1}{n}\mathbf{M}\left(\sigma^{\otimes n}\right)\right| \leqslant \frac{1}{n}\log\left(|A|^n\right) f\left(\frac{1}{2}\|\rho^{\otimes n} - \sigma^{\otimes n}\|_1\right)$$

$$= f\left(\frac{1}{2}\|\rho^{\otimes n} - \sigma^{\otimes n}\|_1\right) \log |A|$$

$$\boxed{\tfrac{1}{2}\|\rho^{\otimes n} - \sigma^{\otimes n}\|_1 \leqslant 1} \longrightarrow \quad \leqslant \max_{0 \leqslant \varepsilon \leqslant 1} f(\varepsilon) \log |A|.$$

Moreover, since $f$ is continuous, $\max_{0 \leqslant \varepsilon \leqslant 1} f(\varepsilon) := c < \infty$. Hence, taking $\sigma \in \mathfrak{F}(A)$ to be free we get from the above equation that for all $n \in \mathbb{N}$

$$\frac{1}{n} \mathbf{M}\left(\rho^{\otimes n}\right) \leqslant c \log(|A|). \tag{10.18}$$

Hence, taking the limit $n \to \infty$ we get $\mathbf{M}^{\mathrm{reg}}(\rho) < \infty$.

**Exercise 10.2.** *Let* $f: \cup_A \mathfrak{D}(A) \to \mathbb{R}_+$ *be a function that satisfies* (10.16)*, and suppose there exists a state* $\sigma \in \mathfrak{D}(A)$ *such that* $\lim_{n \to \infty} \frac{1}{n} f(\sigma^{\otimes n}) < \infty$. *Show that for all other* $\rho \in \mathfrak{D}(A)$ *we must have*

$$\lim_{n \to \infty} \frac{1}{n} f\left(\rho^{\otimes n}\right) = \infty. \tag{10.19}$$

*Hint: Prove first that* $\left\| \rho^{\otimes n} - \sigma^{\otimes n} \right\|_1 \geqslant \|\rho - \sigma\|_1$ *for all* $n \in \mathbb{N}$.

If the set of free states, $\mathfrak{F}(A)$, contains a full rank state for any system $A$, then one can define a slightly weaker version of asymptotic continuity that will be very useful for our study, since most QRTs have this property.

> **Asymptotic Continuity (Alternative Definition)**
>
> **Definition 10.4.** A resource measure $\mathbf{M}$ is said to be asymptotically continuous if for all $\rho, \sigma \in \mathfrak{D}(A)$, and $\varepsilon := \frac{1}{2}\|\rho - \sigma\|_1$,
>
> $$\left| \mathbf{M}(\rho) - \mathbf{M}(\sigma) \right| \leqslant f(\varepsilon) \log \min_{\eta \in \mathfrak{F}(A)} \left\| \eta^{-1} \right\|_\infty \tag{10.20}$$
>
> where $f: \mathbb{R} \to \mathbb{R}$ is some continuous function, independent on the dimensions, and satisfies $\lim_{\varepsilon \to 0^+} f(\varepsilon) = 0$.

Observe that any density matrix $\eta \in \mathfrak{D}(A)$ satisfies $\left\|\eta^{-1}\right\|_\infty \geqslant |A|$. Therefore, the above notion of asymptotic continuity is a weaker one than the version given in Definition 10.3. On the other hand, if the QRT $\mathfrak{F}$ has the property that there exists a constant $0 < c < \infty$, independent of the dimensions, such that

$$\min_{\eta \in \mathfrak{F}(A)} \|\eta^{-1}\|_\infty \leqslant c|A| \tag{10.21}$$

for any choice of system $A$ (and $c$ is independent on $|A|$), then the two notions of asymptotic continuity become equivalent. Since all the QRTs studied in this book satisfy the above condition, we will use these two notions of asymptotic continuity interchangeably.

**Exercise 10.3.** *Let* $\mathfrak{F}$ *be a QRT in which the maximally mixed state is free. Show that the two notions of asymptotic continuity coincide in this case.*

Asymptotic continuity is a property that is extensively utilized in QRTs, especially in the asymptotic regime. Functions that are asymptotically continuous often incorporate the von-Neumann entropy or the Umegaki relative entropy. This reliance is partly because the Umegaki relative entropy is the only asymptotically continuous relative entropy, making it a unique and pivotal tool in QRTs. The proof of this uniqueness theorem, which establishes the singular nature of the Umegaki relative entropy in terms of asymptotic continuity, is an important aspect of these theories. However, we will delve into the details of this proof later in Section 11.4. For now, our focus will shift to introducing key examples of resource measures. These examples will provide a practical illustration of how the theoretical concepts discussed above applied in QRTs.

## 10.2 Distance-Based Resource Measures

In this section, we introduce a general distance-based recipe for constructing resource measures in QRTs. The idea is to quantify the amount of a resource in a quantum state by "how far" it is from the set of free states. In Chapter 5, we saw several examples of well-defined measures that satisfy the mathematical requirements of distance between two density matrices. We also saw that from an operational perspective, the property of monotonicity under quantum channels (i.e. data processing inequality) offers a more useful foundation for quantifying distance than standard metric space approaches. We will therefore employ quantum divergences to define resource measures.

> ### Divergence-Based Resource Measure
>
> **Definition 10.5.** Let $\mathfrak{F}$ be a QRT, and let $\mathbb{D}$ be a quantum divergence. The $\mathbb{D}$-*divergence of a resource* is the function
>
> $$\mathbb{D}(\rho\|\mathfrak{F}) := \inf_{\sigma \in \mathfrak{F}(A)} \mathbb{D}(\rho\|\sigma). \tag{10.22}$$
>
> Moreover, if $\mathbb{D}(\,\cdot\,\|\,\cdot\,)$ is a quantum relative entropy, then $\mathbb{D}(\,\cdot\,\|\mathfrak{F})$ is called a *relative entropy of a resource*.

*Remark.* In this book, we consistently regard the set of free states, $\mathfrak{F}(A)$, as a closed and compact set. Consequently, the infimum in (10.22) can be substituted with a minimum. This means that there exists an optimal state $\sigma^\star \in \mathfrak{F}(A)$ which fulfills the following equation:

$$\mathbb{D}(\rho\|\mathfrak{F}) = \mathbb{D}(\rho\|\sigma^\star). \tag{10.23}$$

The state $\sigma^\star$ is called a *closest free state* (CFS); see Figure 10.1 for an illustration.

Remarkably, the two conditions that a quantum divergence has to satisfy, namely, DPI and normalization, are sufficient to guarantee that $\mathbb{D}(\,\cdot\,\|\mathfrak{F})$ is a resource measure. Indeed, $\mathbb{D}(\,\cdot\,\|\mathfrak{F})$ is nonnegative since $\mathbb{D}$ is nonnegative, and if $\rho \in \mathfrak{F}(A)$, then $\mathbb{D}(\rho\|\mathfrak{F}) = 0$. To see the monotonicity of $\mathbb{D}(\,\cdot\,\|\mathfrak{F})$ under free operations observe that for any $\mathcal{E} \in \mathfrak{F}(A \to B)$ and any $\rho \in \mathfrak{D}(A)$ we have

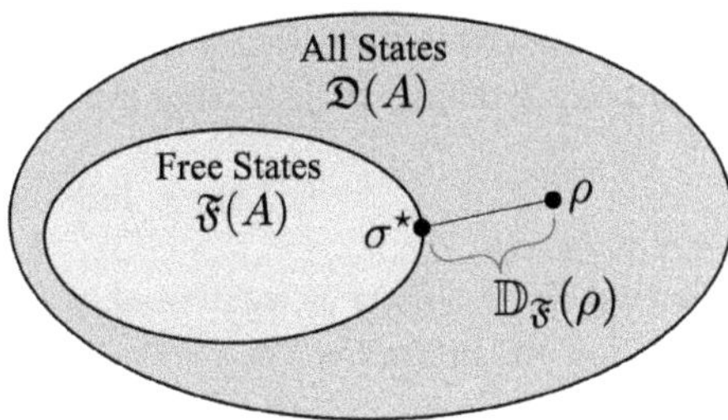



**Figure 10.1**   The closest free state (CFS).

$$
\begin{aligned}
\mathbb{D}\left(\mathcal{E}^{A\to B}(\rho^A)\,\big\|\,\mathfrak{F}\right) &:= \inf_{\omega\in\mathfrak{F}(B)} \mathbb{D}\left(\mathcal{E}^{A\to B}(\rho^A)\,\big\|\,\omega^B\right) \\
\text{restricting } \omega=\mathcal{E}(\sigma) \quad\longrightarrow\quad &\leqslant \inf_{\sigma\in\mathfrak{F}(A)} \mathbb{D}\left(\mathcal{E}^{A\to B}(\rho^A)\,\big\|\,\mathcal{E}^{A\to B}(\sigma^A)\right) \\
\mathbf{DPI}\to \quad &\leqslant \inf_{\sigma\in\mathfrak{F}(A)} \mathbb{D}\left(\rho^A\,\big\|\,\sigma^A\right) \\
&= \mathbb{D}\left(\rho^A\,\big\|\,\mathfrak{F}\right).
\end{aligned}
\tag{10.24}
$$

Hence, $\mathbb{D}(\,\cdot\,\|\mathfrak{F})$ is indeed a resource measure.

Several of the properties of the divergence $\mathbb{D}$ carry over to $\mathbb{D}(\,\cdot\,\|\mathfrak{F})$. For example, suppose that $\mathbb{D}$ is faithful, which is the case for almost all quantum relative entropies. Then, also $\mathbb{D}(\,\cdot\,\|\mathfrak{F})$ is faithful; that is, for any $\rho \in \mathfrak{D}(A)$,

$$
\mathbb{D}(\rho\|\mathfrak{F})=0 \iff \rho\in\mathfrak{F}(A). \tag{10.25}
$$

Indeed, if $\mathbb{D}(\rho\|\mathfrak{F})=0$, it means that there exists a $\sigma \in \mathfrak{F}(A)$ such that $\mathbb{D}(\rho\|\sigma)=0$, but this is possible only if $\rho=\sigma$, which means that $\rho \in \mathfrak{F}(A)$.

Also subadditivity carries over from $\mathbb{D}$ to $\mathbb{D}(\,\cdot\,\|\mathfrak{F})$. To see this, suppose that $\mathbb{D}$ is subadditive so that for any $\rho_1,\sigma_1 \in \mathfrak{D}(A)$ and $\sigma_1,\sigma_2 \in \mathfrak{D}(B)$

$$
\mathbb{D}\left(\rho_1\otimes\rho_2\,\big\|\,\sigma_1\otimes\sigma_2\right) \leqslant \mathbb{D}(\rho_1\|\sigma_1)+\mathbb{D}(\rho_2\|\sigma_2). \tag{10.26}
$$

Then,

$$
\begin{aligned}
\mathbb{D}\left(\rho_1^A\otimes\rho_2^B\,\big\|\,\mathfrak{F}\right) &= \inf_{\sigma\in\mathfrak{F}(AB)} \mathbb{D}\left(\rho_1^A\otimes\rho_2^A\,\big\|\,\sigma^{AB}\right) \\
\text{restricting } \sigma=\sigma_1\otimes\sigma_2 \quad\longrightarrow\quad &\leqslant \inf_{\substack{\sigma_1\in\mathfrak{F}(A)\\ \sigma_2\in\mathfrak{F}(B)}} \mathbb{D}\left(\rho_1^A\otimes\rho_2^B\,\big\|\,\sigma_1^A\otimes\sigma_2^B\right) \\
\text{subadditivity of } \mathbb{D} \quad\longrightarrow\quad &\leqslant \inf_{\sigma_1\in\mathfrak{F}(A)} \mathbb{D}\left(\rho_1^A\,\big\|\,\sigma_1^A\right) + \inf_{\sigma_2\in\mathfrak{F}(B)} \mathbb{D}\left(\rho_2^B\,\big\|\,\sigma_2^B\right) \\
&= \mathbb{D}\left(\rho_1^A\,\big\|\,\mathfrak{F}\right) + \mathbb{D}\left(\rho_2^B\,\big\|\,\mathfrak{F}\right).
\end{aligned}
\tag{10.27}
$$

If $\mathbb{D}$ is a relative entropy (in particular, it is additive), then $\mathbb{D}(\,\cdot\,\|\mathfrak{F})$ is not necessarily additive since the restriction $\sigma^{AB}=\sigma_1^A\otimes\sigma_2^B$ in the equation above can lead to a strict inequality $\mathbb{D}(\rho_1 \otimes \rho_2 \| \mathfrak{F}) < \mathbb{D}(\rho_1 \| \mathfrak{F}) + \mathbb{D}(\rho_2 \| \mathfrak{F})$. Still, subadditive resource measures can be regularized (see Exercise 10.1) so that the limit in the function

$$\mathbb{D}^{\mathrm{reg}}(\rho \| \mathfrak{F}) := \lim_{n \to \infty} \frac{1}{n} \mathbb{D}\left(\rho^{\otimes n} \| \mathfrak{F}\right) \tag{10.28}$$

exists. Note that $\mathbb{D}^{\mathrm{reg}}(\cdot \| \mathfrak{F})$ is at least weakly additive in the sense that $\mathbb{D}^{\mathrm{reg}}\left(\rho^{\otimes m} \| \mathfrak{F}\right) = m \mathbb{D}^{\mathrm{reg}}(\rho \| \mathfrak{F})$ for any integer $m$.

If the set $\mathfrak{F}(A)$ is convex, and the quantum divergence $\mathbb{D}$ is jointly convex, then the resulting $\mathbb{D}$-divergence of a resource is convex as well. To see this, let $\{p_x, \rho_x\}_{x \in [m]}$ be an ensemble of quantum states in $\mathfrak{D}(A)$, and for each $x \in [m]$ let $\sigma_x$ be the corresponding CFS of $\rho_x$. We then have

$$\mathbb{D}\left(\sum_{x \in [m]} p_x \rho_x \Big\| \mathfrak{F}\right) = \inf_{\sigma \in \mathfrak{F}(A)} \mathbb{D}\left(\sum_{x \in [m]} p_x \rho_x \Big\| \sigma\right)$$

Taking $\sigma = \sum_{x \in [m]} p_x \sigma_x \longrightarrow \leqslant \mathbb{D}\left(\sum_{x \in [m]} p_x \rho_x \Big\| \sum_{x \in [m]} p_x \sigma_x\right)$

$$\text{Joint convexity} \rightarrow \leqslant \sum_{x \in [m]} p_x \mathbb{D}(\rho_x \| \sigma_x) \tag{10.29}$$

Each $\sigma_x$ is CFS $\longrightarrow = \sum_{x \in [m]} p_x \mathbb{D}(\rho_x \| \mathfrak{F}).$

## 10.2.1 The Umegaki Relative Entropy of a Resource

The Umegaki relative entropy of a resource (sometimes we will call it in short *the* relative entropy of a resource) is a resource measure that plays a major role in QRTs. It demonstrates that many seemingly unrelated properties of physical systems can be characterized with the same mechanism of resource theories. For instance, the entropy of entanglement that we will encounter in Chapter 12, the free energy in quantum thermodynamics (Chapter 17), and the various communication capacities of a quantum channel (to appear in volume 2), all can be characterized as relative entropies of a resource.

The relative entropy of a resource is a $D$-divergence of a resource, where $D$ is the Umegaki relative entropy. It is defined as

$$D(\rho \| \mathfrak{F}) := \inf_{\sigma \in \mathfrak{F}(A)} D(\rho \| \sigma) \qquad \forall \rho \in \mathfrak{D}(A), \tag{10.30}$$

where the Umegaki relative entropy $D(\rho \| \sigma) := \mathrm{Tr}[\rho \log \rho] - \mathrm{Tr}[\rho \log \sigma]$, and the infimum is taken over all free states $\sigma \in \mathfrak{F}(A)$. Under the construction of $\mathbb{D}$-divergence of a resource, strong monotonicity (10.11) is not guaranteed to be satisfied, but for the relative entropy of a resource this is indeed the case.

> **Theorem 10.1.** Let $D$ be the Umegaki relative entropy and $\mathfrak{F}$ be a convex QRT. Then, the relative entropy of a resource, $D(\cdot \| \mathfrak{F})$, is a resource monotone.

**Proof** Since $\mathfrak{F}$ is convex, and since the Umegaki relative entropy is jointly convex (see Exercise 8.33), it follows from (10.29) that $D(\,\cdot\,\|\mathfrak{F})$ is convex. It is therefore left to show that $D(\,\cdot\,\|\mathfrak{F})$ satisfies the strong monotonicity property. Consider a free quantum instrument $\mathcal{E} = \sum_{x\in[m]} \mathcal{E}_x \otimes |x\rangle\langle x|^X \in \mathfrak{F}(A \to BX)$, with each $\mathcal{E}_x \in \mathrm{CP}(A \to B)$, and $\sum_{x\in[m]} \mathcal{E}_x$ is trace-preserving. For any resource state $\rho \in \mathfrak{D}(A)$, denote $\sigma^{BX} := \sum_{x\in[m]} p_x \sigma_x^B \otimes |x\rangle\langle x|^X$, where $p_x := \mathrm{Tr}[\mathcal{E}_x^{A \to B}(\rho^A)]$ and $\sigma_x^B := \frac{1}{p_x}\mathcal{E}_x^{A \to B}(\rho^A)$. Then,

$$D\left(\rho^A\big\|\mathfrak{F}\right) \geqslant D\left(\mathcal{E}^{A\to BX}\left(\rho^A\right)\big\|\mathfrak{F}\right) = D\left(\sigma^{BX}\big\|\mathfrak{F}\right) = \min_{\omega\in\mathfrak{F}(BX)} D\left(\sigma^{BX}\big\|\omega^{BX}\right),$$

$$(10.31)$$

where we used the monotonicity under free operations of the relative entropy of a resource. Denoting $\omega^{BX} := \sum_{x\in[m]} q_x \omega_x^B \otimes |x\rangle\langle x|^X$ we continue

$$D\left(\rho^A\big\|\mathfrak{F}\right) \geqslant \min_{\omega\in\mathfrak{F}(BX)} D\Big(\sum_{x\in[m]} p_x \sigma_x^B \otimes |x\rangle\langle x|^X \Big\| \sum_{x\in[m]} q_x \omega_x^B \otimes |x\rangle\langle x|^X\Big)$$

$$\textbf{Exercise 8.33}\rightarrow\ =\ \min_{\omega\in\mathfrak{F}(BX)} \sum_{x\in[m]} p_x D\left(\sigma_x^B\big\|\omega_x^B\right) + D(\mathbf{p}\|\mathbf{q})$$

$$\boxed{D(\mathbf{p}\|\mathbf{q}) \geqslant 0}\ \longrightarrow\ \geqslant\ \min_{\{\omega_x\}\subset\mathfrak{F}(B)} \sum_{x\in[m]} p_x D\left(\sigma_x^B\big\|\omega_x^B\right)$$

$$=\ \sum_{x\in[m]} p_x \min_{\omega\in\mathfrak{F}(B)} D\left(\sigma_x^B\big\|\omega^B\right) = \sum_{x\in[m]} p_x D\left(\sigma_x^B\big\|\mathfrak{F}\right).$$

This completes the proof. ∎

So far we have learned that the relative entropy of a resource is a faithful, subadditive resource monotone assuming $\mathfrak{F}$ is a QRT whose set of free states $\mathfrak{F}(A)$ is convex. The final property that we prove is that the Umegaki relative entropy of a resource is also asymptotically continuous. Later on, we will see that this measure is the *only* asymptotically continuous relative entropy of a resource, making the Umegaki relative entropy of a resource very unique. Moreover, from the following chapters it will follow that asymptotic continuity has several important applications in QRTs.

The relative entropy of a resource is not always bounded. As a very simple example, suppose the set of free states $\mathfrak{F}(A)$ consists of only one pure state $|0\rangle\langle 0|$. In this case, we get that

$$\mathbb{D}(|1\rangle\langle 1|\|\mathfrak{F}) = \infty. \qquad (10.32)$$

Hence, for such a (pathological) example the relative entropy of a resource is not bounded, and in particular, cannot be asymptotically continuous. However, in most QRTs, the set of free states $\mathfrak{F}(A)$ contains a full rank state. If such a state exists, say $\eta$, then since $\eta^{-1}$ exists and satisfies $\eta^{-1} \leqslant \|\eta^{-1}\|_\infty I^A$ it follows that

$$D(\rho\|\mathfrak{F}) \leqslant D(\rho\|\eta) = \mathrm{Tr}[\rho\log\rho] - \mathrm{Tr}[\rho\log\eta]$$

$$\mathbf{Tr[\rho\log\rho] < 0}\rightarrow\ \leqslant -\mathrm{Tr}[\rho\log\eta] = \mathrm{Tr}\left[\rho\log\left(\eta^{-1}\right)\right]$$

$$\textbf{log is an operator monotone}\rightarrow\ \leqslant \mathrm{Tr}\left[\rho\log\left(\|\eta^{-1}\|_\infty I^A\right)\right]$$

$$= \log\|\eta^{-1}\|_\infty.$$

For example, if the set of free states contains the maximally mixed state, then $D\left(\rho^A \| \mathfrak{F}\right) \leqslant \log |A|$.

> ### Asymptotic Continuity of the Relative Entropy of a Resource
>
> **Theorem 10.2.** Let $\mathfrak{F}(A)$ be a closed and convex set, $\rho, \sigma \in \mathfrak{D}(A)$, and set $\varepsilon := \frac{1}{2}\|\rho - \sigma\|_1$. If
>
> $$\kappa := \max_{\omega \in \mathfrak{D}(A)} D(\omega \| \mathfrak{F}) < \infty, \qquad (10.33)$$
>
> then
>
> $$\left| D(\rho \| \mathfrak{F}) - D(\sigma \| \mathfrak{F}) \right| \leqslant \varepsilon \kappa + (1 + \varepsilon) h\left(\frac{\varepsilon}{1 + \varepsilon}\right), \qquad (10.34)$$
>
> where $h(x) = -x \log x - (1 - x)\log(1 - x)$ is the binary Shannon entropy.

**Proof**  Let $\rho, \sigma \in \mathfrak{D}(A)$ and write $\rho - \sigma = \varepsilon(\omega_+ - \omega_-)$ as in (5.170) with $\omega_\pm := \frac{1}{\varepsilon}(\rho - \sigma)_\pm$, and with $\varepsilon := \frac{1}{2}\|\rho - \sigma\|_1$ being the trace distance between $\rho$ and $\sigma$. Set $t := 1/(1 + \varepsilon)$ so that

$$\gamma := t\rho + (1 - t)\omega_- = t\sigma + (1 - t)\omega_+, \qquad (10.35)$$

where for the second equality we use (5.170). The key idea of the proof is to find lower and upper bounds for $D(\gamma \| \mathfrak{F})$ in terms of $D(\gamma \| \mathfrak{F})$, $D(\sigma \| \mathfrak{F})$, and $\kappa$.

Since $\mathfrak{F}(A)$ is convex, we saw above that the relative entropy of a resource is convex, so that

$$\begin{aligned} D(\gamma \| \mathfrak{F}) &\leqslant t D(\sigma \| \mathfrak{F}) + (1 - t) D(\omega_+ \| \mathfrak{F}) \\ &\leqslant t D(\sigma \| \mathfrak{F}) + (1 - t)\kappa. \end{aligned} \qquad (10.36)$$

To get a lower bound, let $\eta$ be such that $D(\gamma \| \mathfrak{F}) = D(\gamma \| \eta)$. Then, from the definition of the Umegaki relative entropy we have

$$\begin{aligned} D(\gamma \| \mathfrak{F}) &= -H(\gamma) - \mathrm{Tr}[\gamma \log \eta] \\ \mathbf{(10.35)} \rightarrow &= -H\big(t\rho + (1 - t)\omega_-\big) - t\mathrm{Tr}[\rho \log \eta] - (1 - t)\mathrm{Tr}[\omega_- \log \eta]. \end{aligned} \qquad (10.37)$$

From Corollary 7.2 we have that the first term on the right-hand side above satisfies

$$-H\left(t\rho + (1 - t)\omega_-\right) \geqslant -tH(\rho) - (1 - t)H\left(\omega_-\right) - h\left(t\right). \qquad (10.38)$$

Substituting this into the previous equation gives

$$\begin{aligned} D(\gamma \| \mathfrak{F}) &\geqslant -t\big(H\left(\rho\right) + \mathrm{Tr}[\rho \log \eta]\big) - (1 - t)\big(H\left(\omega_-\right) + \mathrm{Tr}[\omega_- \log \eta]\big) - h\left(t\right) \\ &= tD(\rho \| \eta) + (1 - t)D(\omega_- \| \eta) - h\left(t\right) \\ &\geqslant tD(\rho \| \mathfrak{F}) - h\left(t\right), \end{aligned} \qquad (10.39)$$

where in the last line we removed the term $(1 - t)D(\omega_- \| \eta) \geqslant 0$, and replaced $D(\rho \| \eta)$ with $D(\rho \| \mathfrak{F})$. Combining the lower bound above with the upper bound in (10.36) we conclude that

$$D(\rho\|\mathfrak{F}) - D(\sigma\|\mathfrak{F}) \leqslant t^{-1}h(t) + \frac{1-t}{t}\kappa$$
$$= (1+\varepsilon)h\left(\frac{\varepsilon}{1+\varepsilon}\right) + \varepsilon\kappa. \tag{10.40}$$

The same upper bound holds for $D(\sigma\|\mathfrak{F}) - D(\rho\|\mathfrak{F})$ by exchanging between $\rho$ and $\sigma$ everywhere above. Hence, this completes the proof. ∎

**Exercise 10.4.** *Let $\mathfrak{F}$ be as in Theorem 10.2 and suppose further that there exists a free state $\eta \in \mathfrak{F}(A)$ that is full rank; that is, $\eta > 0$. Show that there exists a continuous function $f: [0,1] \to \mathbb{R}_+$, independent on the dimension of $A$, such that $f(0) = 0$ and*

$$\left|D(\rho\|\mathfrak{F}) - D(\sigma\|\mathfrak{F})\right| \leqslant f(\varepsilon)\log\|\eta^{-1}\|_\infty. \tag{10.41}$$

## Asymptotic Continuity of the Umegaki Relative Entropy

In Definition 10.3, we defined asymptotic continuity on resource measures. This definition can be extended to divergences and relative entropies. Since a relative entropy $\mathbb{D}$ takes two density matrices as its input, we will consider its continuity on the first argument. Moreover, we saw that relative entropies can be unbounded, so this property needs to be accommodated into the definition.

---

**Definition 10.6.** A relative entropy $\mathbb{D}$ is said to be *asymptotically continuous* if there exists a continuous function $f: [0,1] \to \mathbb{R}_+$ such that $f(0) = 0$, and for all $\rho, \rho', \sigma \in \mathfrak{D}(A)$, with $\operatorname{supp}(\rho) \subseteq \operatorname{supp}(\sigma)$ and $\operatorname{supp}(\rho') \subseteq \operatorname{supp}(\sigma)$

$$\left|\mathbb{D}(\rho\|\sigma) - \mathbb{D}(\rho'\|\sigma)\right| \leqslant f(\varepsilon)\log\|\sigma^{-1}\|_\infty, \tag{10.42}$$

where $\varepsilon := \frac{1}{2}\|\rho - \rho'\|_1$. We emphasize that $f$ is independent of $|A|$.

---

We now argue that the Umegaki relative entropy is asymptotically continuous. This is a simple consequence of Theorem 10.2.

---

**Corollary 10.1.** The Umegaki relative entropy is asymptotically continuous.

---

**Proof** Let $\rho, \rho', \sigma \in \mathfrak{D}(A)$ and set $\varepsilon := \frac{1}{2}\|\rho - \rho'\|_1$. Since $\operatorname{supp}(\rho) \subseteq \operatorname{supp}(\sigma)$ and $\operatorname{supp}(\rho') \subseteq \operatorname{supp}(\sigma)$ we can assume without loss of generality that $\sigma > 0$. Let $\mathfrak{F}(A) := \{\sigma\}$ be the set consisting of $\sigma$ (i.e. $\mathfrak{F}(A)$ contains only one density matrix). The set $\mathfrak{F}(A)$ is trivially closed and convex. Moreover, note that for this $\mathfrak{F}$ we get $D(\rho\|\mathfrak{F}) = D(\rho\|\sigma)$ and similarly $D(\rho'\|\mathfrak{F}) = D(\rho'\|\sigma)$. Therefore, applying Theorem 10.2 gives

$$\left|D(\rho\|\sigma) - D(\rho'\|\sigma)\right| \leqslant \varepsilon\kappa + (1+\varepsilon)h\left(\frac{\varepsilon}{1+\varepsilon}\right), \tag{10.43}$$

with

$$\kappa = \max_{\omega\in\mathfrak{D}(A)} D(\omega\|\sigma) \leqslant \max_{\omega\in\mathfrak{D}(A)} -\operatorname{Tr}[\omega\log\sigma], \tag{10.44}$$

where we droped the term $\mathrm{Tr}[\omega \log \omega] = -H(\omega)$ as it is negative. Moreover, note that $-\mathrm{Tr}[\omega \log \sigma] = \mathrm{Tr}[\omega \log(\sigma^{-1})]$, and since $\sigma^{-1} \leqslant \|\sigma^{-1}\|_\infty I^A$ we get that

$$\kappa \leqslant \max_{\omega \in \mathfrak{D}(A)} \mathrm{Tr}[\omega \log(\sigma^{-1})] \leqslant \log \|\sigma^{-1}\|_\infty, \tag{10.45}$$

where we used the operator monotonicity of the log function. This completes the proof (see Exercise 10.5). $\blacksquare$

**Exercise 10.5.** *Show that there exists function $f : [0, 1] \to \mathbb{R}_+$, independent on the dimension of $A$, such that $f(0) = 0$ and*

$$\varepsilon \log \|\sigma^{-1}\|_\infty + (1+\varepsilon)h\left(\frac{\varepsilon}{1+\varepsilon}\right) \leqslant f(\varepsilon) \log \|\sigma^{-1}\|_\infty. \tag{10.46}$$

## Asymptotic Continuity of the von-Neumann Conditional Entropy

The conditional entropy (see Chapter 7) can be viewed (at least mathematically) as a resource measure in a QRT in which the free operations are conditionally mixing operation (CMO) channels and the free states are given by

$$\mathfrak{F}(AB) = \left\{ \mathbf{u}^A \otimes \sigma^B : \sigma \in \mathfrak{D}(B) \right\}. \tag{10.47}$$

**Exercise 10.6.** *Show that any conditional entropy $\mathbb{H}$ (see Definition 7.3) is a resource measure in the QRT in which $\mathfrak{F}(AB \to AB') = \mathrm{CMO}(AB \to AB')$.*

For the case $\mathfrak{F} := \mathrm{CMO}$ we have

$$D\left(\rho^{AB} \,\|\, \mathfrak{F}\right) = \min_{\sigma \in \mathfrak{D}(B)} D\left(\rho^{AB} \,\|\, \mathbf{u}^A \otimes \sigma^B\right)$$

$$\textit{cf. (7.132)} \to = D\left(\rho^{AB} \,\|\, \mathbf{u}^A \otimes \rho^B\right) \tag{10.48}$$

$$\textbf{(7.117)} \to = \log|A| - H(A|B)_\rho.$$

Moreover, observe that $\kappa$ as defined in (10.33) satisfies

$$\kappa := \max_{\omega \in \mathfrak{F}(AB)} D\left(\omega^{AB} \,\|\, \mathfrak{F}\right)$$

$$= \log|A| - \max_{\omega \in \mathfrak{F}(AB)} H(A|B)_\rho \tag{10.49}$$

$$\textbf{Theorem 7.5} \to \leqslant 2\log|A|.$$

Therefore, taking this choice of $\mathfrak{F}$ in Theorem 10.2 yields the following corollary.

> **Corollary 10.2.** Let $\rho, \sigma \in \mathfrak{D}(AB)$ and $\varepsilon := \frac{1}{2}\|\rho^{AB} - \sigma^{AB}\|_1$. Then,
>
> $$\left|H(A|B)_\rho - H(A|B)_\sigma\right| \leqslant 2\varepsilon \log|A| + (1+\varepsilon)h\left(\frac{\varepsilon}{1+\varepsilon}\right). \tag{10.50}$$

**Exercise 10.7.** *Use Theorem 10.2 and the expressions above to prove the corollary.*

**Exercise 10.8.** *Using the same notations as in the corollary above, show that there exists a continuous function* $f : [0, 1] \to \mathbb{R}_+$ *satisfying* $\lim_{\delta \to 0^+} f(\delta) = 0$ *and*

$$H(A|B)_\rho \geqslant H(A|B)_\sigma - f(\varepsilon) \log |A|. \tag{10.51}$$

**Exercise 10.9.** (Asymptotic Continuity of the von-Neumann Entropy) *Let* $\rho, \sigma \in \mathfrak{D}(A)$ *and* $\varepsilon := \frac{1}{2} \| \rho - \sigma \|_1$.

1. *Show that the von-Neumann entropy* $H$ *satisfies*

$$\left| H(\rho) - H(\sigma) \right| \leqslant \varepsilon \log |A| + (1 + \varepsilon) h\left( \frac{\varepsilon}{1 + \varepsilon} \right). \tag{10.52}$$

   *Hint: Consider the QRT in which* $\mathfrak{F}(A \to A)$ *is the set of all unital channels.*

2. *Show that there exists a continuous function* $f : [0, 1] \to \mathbb{R}_+$ *satisfying* $\lim_{\delta \to 0^+} f(\delta) = 0$ *and*

$$H(\rho) \geqslant H(\sigma) - f(\varepsilon) \log |A|. \tag{10.53}$$

## 10.2.2 The Robustness of a Resource

Given a QRT $\mathfrak{F}$, the robustness and global robustness of a resource $\rho \in \mathfrak{D}(A)$ are defined as

$$\mathbf{R}(\rho) := \inf \left\{ s \geqslant 0 : \frac{\rho + s\omega}{1 + s} \in \mathfrak{F}(A), \ \omega \in \mathfrak{F}(A) \right\}$$

$$\mathbf{R}_g(\rho) := \inf \left\{ s \geqslant 0 : \frac{\rho + s\omega}{1 + s} \in \mathfrak{F}(A), \ \omega \in \mathfrak{D}(A) \right\}. \tag{10.54}$$

Moreover, if there is no such $s \geqslant 0$ and $\omega \in \mathfrak{F}(A)$ such that $\frac{\rho + s\omega}{1+s} \in \mathfrak{F}(A)$, then $\mathbf{R}(\rho) := \infty$. The robustness and global robustness of a resource measure of how robust a resource $\rho$ is to mixing with noise. In other words, they quantify the ability of the resource to maintain its usefulness in the presence of disturbances. The term "global" refers to the fact that the noise can be any density matrix $\omega \in \mathfrak{D}(A)$, which represents a wide range of possible disturbances. However, if we limit the density matrix $\omega$ to only represent free states, then the resulting quantity is called the robustness.

By definition, $\mathbf{R}_g(\rho) \leqslant \mathbf{R}(\rho)$ (see the following exercise) and if $\rho \in \mathfrak{F}(A)$, then $\mathbf{R}(\rho) = \mathbf{R}_g(\rho) = 0$ since $s$ above can be taken to be zero. Furthermore, from the following exercise, the converse of this statement is also true; that is, $\mathbf{R}_g$ is faithful.

**Exercise 10.10.** *Consider the robustness and global robustness as defined above.*

1. *Show that* $\mathbf{R}_g(\rho) \leqslant \mathbf{R}(\rho)$ *for all* $\rho \in \mathfrak{D}(A)$.
2. *Show that if* $\mathfrak{F}$ *is affine, then* $\mathbf{R}(\rho) = \infty$ *for all* $\rho \in \mathfrak{D}(A)$ *that is not free.*
3. *Show that* $\mathbf{R}_g(\rho) = 0$ *if and only if* $\rho \in \mathfrak{F}(A)$.

**Exercise 10.11.** *Let* $\rho \in \mathfrak{D}(A)$ *and suppose* $\mathbf{R}(\rho) < \infty$.

1. *Show that there exist* $\tau, \omega \in \mathfrak{F}(A)$ *such that*

$$\rho = \left(1 + \mathbf{R}(\rho)\right)\tau - \mathbf{R}(\rho)\omega. \tag{10.55}$$

2. *Show that there exist $\tau \in \mathfrak{F}(A)$ and $\omega \in \mathfrak{D}(A)$ such that*

$$\rho = \big(1 + \mathbf{R}_g(\rho)\big)\tau - \mathbf{R}_g(\rho)\omega. \tag{10.56}$$

*The above decompositions of $\rho$ are sometimes referred to as pseudo-mixtures of states.*

Note that from the exercise above it follows that $\mathbf{R}_g(\rho)$ can also be expressed as

$$\mathbf{R}_g(\rho) := \min \big\{s \geqslant 0 : \ \rho = (1+s)\tau - s\omega, \ \ \tau \in \mathfrak{F}(A), \ \ \omega \in \mathfrak{D}(A)\big\}, \tag{10.57}$$

so that we can think of the pseudo-mixture in (10.56) as the optimal one achieved with $s = \mathbf{R}_g(\rho)$.

> **Theorem 10.3.** The global robustness of a resource, $\mathbf{R}_g$, is a resource measure satisfying the strong monotonicity property. Moreover, if the set of free states is convex, then $\mathbf{R}_g$ is a resource monotone.

**Proof**   Since we have already seen that $\mathbf{R}_g(\rho) = 0$ for all $\rho \in \mathfrak{F}(A)$, we prove now the strong monotonicity property. Let $\mathcal{E} := \sum_{x \in [m]} \mathcal{E}_x \otimes |x\rangle\langle x| \in \mathfrak{F}(A \to BX)$ be a free quantum instrument (if $\mathfrak{F}(A \to BX)$ is an empty set, then strong monotonicity holds trivially). Let $\rho \in \mathfrak{D}(A)$ be as in (10.56) and for all $x \in [m]$ denote by

$$
\begin{aligned}
\sigma_x &:= \frac{1}{\mathrm{Tr}[\mathcal{E}_x(\rho)]} \mathcal{E}_x(\rho) = \frac{1}{\mathrm{Tr}[\mathcal{E}_x(\rho)]}\Big(\big(1 + \mathbf{R}_g(\rho)\big)\mathcal{E}_x(\tau) - \mathbf{R}_g(\rho)\mathcal{E}_x(\omega)\Big) \\
&= (1+s)\frac{\mathcal{E}_x(\tau)}{\mathrm{Tr}[\mathcal{E}_x(\tau)]} - s\frac{\mathcal{E}_x(\omega)}{\mathrm{Tr}[\mathcal{E}_x(\omega)]},
\end{aligned} \tag{10.58}
$$

where $s := \frac{\mathrm{Tr}[\mathcal{E}_x(\omega)]}{\mathrm{Tr}[\mathcal{E}_x(\rho)]} \mathbf{R}_g(\rho)$. On the other hand, the state $\sigma_x$ can also be expressed in its optimal pseudo-mixture as in (10.56) via

$$\sigma_x = \big(1 + \mathbf{R}_g(\sigma_x)\big)\tau_x - \mathbf{R}_g(\sigma_x)\omega_x \tag{10.59}$$

for some $\omega_x \in \mathfrak{D}(B)$ and $\tau_x \in \mathfrak{F}(B)$. Therefore, from the two expressions above for $\sigma_x$, and the optimality of the pseudo-mixture in (10.59), we get that $\mathbf{R}_g(\sigma_x)$ is no greater than $s$; that is,

$$\mathbf{R}_g(\sigma_x) \leqslant \frac{\mathrm{Tr}[\mathcal{E}_x(\omega)]}{\mathrm{Tr}[\mathcal{E}_x(\rho)]}\mathbf{R}_g(\rho). \tag{10.60}$$

From the above inequality we conclude that

$$\sum_{x \in [m]} \mathrm{Tr}[\mathcal{E}_x(\rho)]\mathbf{R}_g(\sigma_x) \leqslant \sum_{x \in [m]} \mathrm{Tr}[\mathcal{E}_x(\omega)]\mathbf{R}_g(\rho) = \mathbf{R}_g(\rho), \tag{10.61}$$

where in the last equality we use the fact that $\sum_{x \in [m]} \mathcal{E}_x$ is trace-preserving. This completes the proof of strong monotonicity.

Next, suppose that $\mathfrak{F}(A)$ is convex, and let $\{p_x, \rho_x\}_{x \in [m]}$ be an ensemble of quantum states in $\mathfrak{D}(A)$. Express each $\rho_x$ as a pseudo-mixture

$$\rho_x = \big(1 + \mathbf{R}_g(\rho_x)\big)\tau_x - \mathbf{R}_g(\rho_x)\omega_x \tag{10.62}$$

for some $\tau_x \in \mathfrak{F}(A)$ and $\omega_x \in \mathfrak{D}(A)$. Denote by $\bar{\rho} := \sum_{x \in [m]} p_x \rho_x$. Then, from the equation above we have

$$\bar{\rho} = \sum_{x \in [m]} p_x \left( \left(1 + \mathbf{R}_g(\rho_x)\right)\tau_x - \mathbf{R}_g(\rho_x)\omega_x \right) = (1 + r)\tau - r\omega, \tag{10.63}$$

where

$$r := \sum_{x \in [m]} p_x \mathbf{R}_g(\rho_x), \quad \tau := \frac{1}{1 + r} \sum_{x \in [m]} p_x \left(1 + \mathbf{R}_g(\rho_x)\right)\tau_x, \quad \omega := \frac{1}{r} \sum_{x \in [m]} p_x \mathbf{R}_g(\rho_x)\omega_x. \tag{10.64}$$

Note that $\tau \in \mathfrak{F}(A)$ since each $\tau_x \in \mathfrak{F}(A)$ and $\mathfrak{F}(A)$ is convex. Since the pseudo-mixture in (10.63) is not necessarily the optimal one, we conclude that

$$\mathbf{R}_g(\bar{\rho}) \leqslant r = \sum_{x \in [m]} p_x \mathbf{R}_g(\rho_x). \tag{10.65}$$

That is, $\mathbf{R}_g$ is a convex function. This completes the proof.  ∎

**Exercise 10.12.** *Let $\mathfrak{F}$ be a convex QRT, and let $\mathbf{R}$ be the corresponding robustness measure. Prove that $\mathbf{R}$ is a resource monotone. Hint: Follow similar steps as in the proof of Theorem 10.3.*

## The Logarithmic Global Robustness of a Resource

The logarithmic global robustness is another important distance-based measure given by replacing $D$ in (10.30) with the max relative entropy $D_{\max}$. It is given by

$$D_{\max}(\rho \| \mathfrak{F}) := \min_{\sigma \in \mathfrak{F}(A)} D_{\max}(\rho \| \sigma). \tag{10.66}$$

The terminology of $D_{\max}(\rho \| \mathfrak{F})$ is due to the following connection between $D_{\max}(\rho \| \mathfrak{F})$ and $\mathbf{R}_g$.

> **Lemma 10.1.** Let $\mathfrak{F}$ be a QRT. Then, for any $\rho \in \mathfrak{D}(A)$,
>
> $$D_{\max}(\rho \| \mathfrak{F}) = \log \left(1 + \mathbf{R}_g(\rho)\right). \tag{10.67}$$

**Proof**  By definition of $D_{\max}$ and the logarithmic global robustness of a resource, we have for all $\rho \in \mathfrak{D}(A)$

$$D_{\max}(\rho \| \mathfrak{F}) = \min \left\{ \log t \ : \ t\sigma \geqslant \rho, \ \sigma \in \mathfrak{F}(A) \right\}$$

$$= \min \left\{ \log t \ : \ t\sigma - \rho = (t - 1)\omega, \ \sigma \in \mathfrak{F}(A), \ \omega \in \mathfrak{D}(A), \ t \geqslant 1 \right\}, \tag{10.68}$$

since $t\sigma - \rho \geqslant 0$ implies that $t\sigma - \rho = (t - 1)\omega$ for some density matrix $\omega$. Denoting by $s = t - 1$ we continue

$$D_{\max}(\rho\|\mathfrak{F}) = \min\left\{\log(1 + s) : (1 + s)\sigma - \rho = s\omega, \ \sigma \in \mathfrak{F}(A), \ \omega \in \mathfrak{D}(A), \ s \geqslant 0\right\}$$

**Isolating** $\sigma \rightarrow = \min\left\{\log(1 + s) : \dfrac{\rho + s\omega}{1 + s} \in \mathfrak{F}(A), \ \omega \in \mathfrak{D}(A), \ s \geqslant 0\right\}$

$$= \log\left(1 + \mathbf{R}_g(\rho)\right).$$

$$(10.69)$$

This completes the proof. ∎

## 10.2.3 The Hypothesis Testing and $\alpha$-Relative Entropy of a Resource

Let $\mathfrak{F}$ be a quantum resource theory such that the set of free states $\mathfrak{F}(A) \subseteq \mathfrak{D}(A)$ is closed and convex. We define the hypothesis testing measure of a resource as

$$D_{\min}^{\varepsilon}(\rho\|\mathfrak{F}) := \min_{\omega \in \mathfrak{F}(A)} D_{\min}^{\varepsilon}(\rho\|\omega) \qquad \forall \, \rho \in \mathfrak{D}(A). \qquad (10.70)$$

For any $\alpha \in [0, 2]$, we also define the $\alpha$-Rényi relative entropy of a resource as

$$D_{\alpha}(\rho\|\mathfrak{F}) := \min_{\omega \in \mathfrak{F}(A)} D_{\alpha}(\rho\|\omega) \qquad \forall \, \rho \in \mathfrak{D}(A). \qquad (10.71)$$

Note that the case $\alpha = 0$ corresponds to $D_{\min}(\rho\|\mathfrak{F})$. The special case of $\alpha = 1$ is $D_{\alpha=1}(\rho\|\mathfrak{F}) = D(\rho\|\mathfrak{F})$ (the relative entropy of a resource). Since the Petz quantum Rényi divergence, $D_{\alpha}(\cdot\|\cdot)$, is nondecreasing with $\alpha$, also $D_{\alpha}(\cdot\|\mathfrak{F})$ is not decreasing in $\alpha$. The continuity of $D_{\alpha}(\cdot\|\cdot)$ in $\alpha$ also carries over to $D_{\alpha}(\cdot\|\mathfrak{F})$, including the continuity at $\alpha = 1$. This result is a simple consequence of Sion's minimax theorem.

**Lemma 10.2 (Sion's Minimax Theorem).** Let $X$ be a compact convex subset of a linear topological space, and let $Y$ be a convex subset of a topological space. Let $f : X \times Y \to \mathbb{R} \cup \{-\infty, +\infty\}$ be a real valued function satisfying the following:

1. For every fixed $y \in Y$, the function $x \mapsto f(x, y)$ is lower semicontinuous and quasi-convex on X.
2. For every fixed $x \in X$, the function $y \mapsto f(x, y)$ is upper semicontinuous and quasi-concave on Y.

Then,

$$\min_{x \in X} \sup_{y \in Y} f(x, y) = \sup_{y \in Y} \min_{x \in X} f(x, y). \qquad (10.72)$$

**Lemma 10.3.** Let $\rho \in \mathfrak{D}(A)$ be a fixed density matrix, and define $g : [0, 2] \to \mathbb{R}_+$ as $g(\alpha) := D_{\alpha}(\rho\|\mathfrak{F})$ for all $\alpha \in [0, 2]$. Then, $g(\alpha)$ is a continuous function.

**Proof**  Let $\beta \in (0,2)$. Since $D_\alpha$ is monotonically nondecreasing in $\alpha$ we have that

$$
\begin{aligned}
\lim_{\alpha \to \beta^+} g(\alpha) &= \lim_{\alpha \to \beta^+} \min_{\omega \in \mathfrak{F}(A)} D_\alpha(\rho \| \omega) \\
&= \inf_{\alpha \in (\beta,2)} \min_{\omega \in \mathfrak{F}(A)} D_\alpha(\rho \| \omega) \\
&= \min_{\omega \in \mathfrak{F}(A)} \inf_{\alpha \in (\beta,2)} D_\alpha(\rho \| \omega) \\
&= \min_{\omega \in \mathfrak{F}(A)} D_\beta(\rho \| \omega) = g(\beta).
\end{aligned}
\tag{10.73}
$$

When approaching $\beta$ from below observe that

$$
\begin{aligned}
\lim_{\alpha \to \beta^-} g(\alpha) &= \lim_{\alpha \to \beta^-} \min_{\omega \in \mathfrak{F}(A)} D_\alpha(\rho \| \omega) \\
&= \sup_{\alpha \in (0,\beta)} \min_{\omega \in \mathfrak{F}(A)} D_\alpha(\rho \| \omega).
\end{aligned}
\tag{10.74}
$$

In order to switch the order between the sup and min above we need to verify that all the conditions in Sion's minimax theorem are satisfied. Indeed, the function $f(\omega,\alpha) := D_\alpha(\rho \| \omega)$ has the property that it is continuous in $\omega$ (and therefore lower semi-continuous). Moreover, note that for a fixed $\alpha \in [0,2]$, the function $\omega \mapsto f(\omega,\alpha)$ is a quasi-convex function since for any $t \in [0,1]$ and $\omega_0, \omega_1 \in \mathfrak{F}(A)$ we have

$$
\begin{aligned}
f(t\omega_0 + (1-t)\omega_1, \alpha) &= D_\alpha \left( \rho \| t\omega_0 + (1-t)\omega_1 \right) \\
\mathbf{(6.111)} \to \quad &\leqslant \max \left\{ D_\alpha(\rho \| \omega_0), D_\alpha(\rho \| \omega_1) \right\} \\
&= \max \left\{ f(\omega_0, \alpha), f(\omega_1, \alpha) \right\}.
\end{aligned}
\tag{10.75}
$$

On the other hand, for a fixed $\omega \in \mathfrak{F}(A)$ the function $\alpha \mapsto f(\omega,\alpha)$ is a continuous function (and therefore upper semi-continuous) and quasi-concave since for any $t \in [0,1]$ and $\alpha_0, \alpha_1 \in [0,2]$ we have

$$
\begin{aligned}
f(\omega, t\alpha_0 + (1-t)\alpha_1) &= D_{t\alpha_0 + (1-t)\alpha_1}(\rho \| \omega) \\
\mathbf{monotonicity\ of\ } D_\alpha \mathbf{\ in\ } \alpha \to \quad &\geqslant D_{\min\{\alpha_0,\alpha_1\}}(\rho \| \omega) \\
&= \min \left\{ f(\omega, \alpha_0), f(\omega, \alpha_1) \right\}.
\end{aligned}
\tag{10.76}
$$

Therefore, $f(\omega,\alpha)$ satisfies all the requirements of Sion's minimax theorem. This means that we can switch the order of the sup and min in (10.74) to get

$$
\begin{aligned}
\lim_{\alpha \to \beta^-} g(\alpha) &= \min_{\omega \in \mathfrak{F}(A)} \sup_{\alpha \in (0,\beta)} D_\alpha(\rho \| \omega) \\
\mathbf{continuity\ of\ } D_\alpha \mathbf{\ in\ } \alpha \to \quad &= \min_{\omega \in \mathfrak{F}(A)} D_\beta(\rho \| \omega) = g(\beta).
\end{aligned}
\tag{10.77}
$$

This completes the proof of the lemma. $\blacksquare$

Using the bounds (8.201) and (8.202) we get Corollary 10.3

> **Corollary 10.3.** Let $\varepsilon \in (0,1)$ and $\rho, \sigma \in \mathfrak{D}(A)$.
>
> 1. For all $\alpha \in (1,2]$
>
> $$D_{\min}^{\varepsilon}(\rho\|\mathfrak{F}) \leqslant D_{\alpha}(\rho\|\mathfrak{F}) + \frac{\alpha}{\alpha-1}\log\left(\frac{1}{1-\varepsilon}\right). \qquad (10.78)$$
>
> 2. For all $\alpha \in (0,1)$
>
> $$D_{\min}^{\varepsilon}(\rho\|\mathfrak{F}) \geqslant D_{\alpha}(\rho\|\mathfrak{F}) + \frac{\alpha}{1-\alpha}\left(\frac{h(\alpha)}{\alpha} - \log\left(\frac{1}{\varepsilon}\right)\right). \qquad (10.79)$$

The regularized $\alpha$-Rényi relative entropy of a resource is defined as

$$D_{\alpha}^{\mathrm{reg}}(\rho\|\mathfrak{F}) := \lim_{n\to\infty} \frac{1}{n} D_{\alpha}\left(\rho^{\otimes n}\|\mathfrak{F}\right) \qquad \forall\, \rho \in \mathfrak{D}(A). \qquad (10.80)$$

The limit above exists since the $\alpha$-Rényi relative entropy of a resource is subadditive (see Exercise 10.1). From Corollary 10.3 it follows that

$$\limsup_{n\to\infty} \frac{1}{n} D_{\min}^{\varepsilon}\left(\rho^{\otimes n}\|\mathfrak{F}\right) \leqslant \lim_{\alpha\to 1^+} D_{\alpha}^{\mathrm{reg}}(\rho\|\mathfrak{F}), \qquad (10.81)$$

and similarly

$$\liminf_{n\to\infty} \frac{1}{n} D_{\min}^{\varepsilon}\left(\rho^{\otimes n}\|\mathfrak{F}\right) \geqslant \lim_{\alpha\to 1^-} D_{\alpha}^{\mathrm{reg}}(\rho\|\mathfrak{F}). \qquad (10.82)$$

Observe that in general we do not know if $D_{\alpha}^{\mathrm{reg}}(\rho\|\mathfrak{F})$ is continuous at $\alpha = 1$, but we can show continuity from the right.

> **Lemma 10.4.** Let $\rho \in \mathfrak{D}(A)$ and let $\mathfrak{F}$ be a quantum resource theory admitting a tensor product structure, and has the property that $\mathfrak{F}(A) \subseteq \mathfrak{D}(A)$ is closed and convex. Then,
>
> $$\lim_{\alpha\to 1^+} D_{\alpha}^{\mathrm{reg}}(\rho\|\mathfrak{F}) = D^{\mathrm{reg}}(\rho\|\mathfrak{F}). \qquad (10.83)$$

**Proof**   Observe that

$$\begin{aligned}
\lim_{\alpha\to 1^+} D_{\alpha}^{\mathrm{reg}}(\rho\|\mathfrak{F}) &= \inf_{\alpha\in(1,\infty)}\inf_{n\in\mathbb{N}} \frac{1}{n} D_{\alpha}\left(\rho^{\otimes n}\|\mathfrak{F}\right) \\
&= \inf_{n\in\mathbb{N}}\inf_{\alpha\in(1,\infty)} \frac{1}{n} D_{\alpha}\left(\rho^{\otimes n}\|\mathfrak{F}\right) \\
\textbf{Lemma 10.3}\to &= \sup_{n\in\mathbb{N}} \frac{1}{n} D\left(\rho^{\otimes n}\|\mathfrak{F}\right) \\
&= D^{\mathrm{reg}}(\rho\|\mathfrak{F}).
\end{aligned} \qquad (10.84)$$

We therefore conclude that

$$\limsup_{n\to\infty} \frac{1}{n} D_{\min}^{\varepsilon}\left(\rho^{\otimes n}\|\mathfrak{F}\right) \leqslant D^{\mathrm{reg}}(\rho\|\mathfrak{F}). \qquad (10.85)$$

In several resource theories the opposite inequality also holds, but in general we do not know if the limit $\lim_{\alpha \to 1^-} D_\alpha^{\mathrm{reg}}(\rho\|\mathfrak{F})$ equals to $D^{\mathrm{reg}}(\rho\|\mathfrak{F})$. At the time of writing this book it is a big open problem in the field to determine under what conditions the inequality in the equation above can be replaced with an equality.

# 10.3 Computation of the Relative Entropy of a Resource

The computation of the relative entropy of a resource can be hard, depending of course on the complexity of the set $\mathfrak{F}(A)$. As we will see in the next chapter, in entanglement theory its computation belongs to a class of problems known as NP hard. If $\mathfrak{F}(A)$ is closed and convex, some techniques from convex analysis can be employed to compute the relative entropy. Particularly, in this case the converse problem can be computed efficiently, as we discuss now.

## 10.3.1 The Converse Problem

Let $\sigma \in \mathfrak{F}(A)$ be the free state that optimizes (10.30) for a given resource state $\rho \in \mathfrak{D}(A)$. We will see shortly that, as it suggests intuitively, $\sigma$ must be on the *boundary* of the set $\mathfrak{F}(A)$. The boundary of $\mathfrak{F}(A)$ is defined as

$$\partial\mathfrak{F}(A) := \left\{\omega \in \mathfrak{D}(A) \;:\; \forall \varepsilon > 0 \;\; \exists \eta, \zeta \in \mathfrak{B}_\varepsilon(\omega) \text{ s.t. } \eta \in \mathfrak{F}(A) \text{ and } \zeta \notin \mathfrak{F}(A)\right\}, \quad (10.86)$$

where $\mathfrak{B}_\varepsilon(\omega)$ is the set of all density matrices that are $\varepsilon$-close (in trace distance) to $\omega$. That is, in any neighborhood of a state on the boundary of $\mathfrak{F}(A)$ there exists at least one state in $\mathfrak{F}(A)$ and at least one state not in $\mathfrak{F}(A)$.

The state $\sigma$ can be thought of as the closest free state (CFS) to $\rho$, when we measure the "distance" with the relative entropy. As we have already mentioned, the computation of $\sigma$ can be very hard. However, for a given state $\omega \in \partial\mathfrak{F}(A)$ we can compute all the resource states in $\mathfrak{D}(A)$ for which $\omega$ is the CFS. This converse problem has several applications and can be used to produce examples of resource states for which one knows the value of the relative entropy of a resource.

Note that if $0 < \rho \notin \mathfrak{F}(A)$ and $D(\rho\|\mathfrak{F}) = D(\rho\|\sigma)$ (i.e. $\sigma$ is a CFS), then $\sigma > 0$ or otherwise $D(\rho\|\mathfrak{F}) = \infty$. For simplicity of the exposition here, we will always assume that $\sigma$ has full rank, and refer the interested reader to the end of this chapter for more details and references on the singular case. We start by showing that if $0 < \sigma \in \mathfrak{F}(A)$ is a CFS, then $\sigma \in \partial\mathfrak{F}(A)$.

> **Theorem 10.4.** Let $0 < \sigma \in \mathfrak{F}(A)$ be a closest free state of a resource state $\rho \in \mathfrak{D}(A)$. Then, $\sigma \in \partial\mathfrak{F}(A)$.

**Proof**  Consider the following Taylor expansion of the logarithmic function. This expansion is based on the divided difference approach discussed in Appendix D.1 of online version. For any $t > 0$, $0 < \sigma \in \mathfrak{D}(A)$, and $\eta \in \mathrm{Herm}(A)$ we have

$$\log{(\sigma + t\eta)} = \log{\sigma} + t\mathcal{L}_\sigma(\eta) + O(t^2), \tag{10.87}$$

where $\mathcal{L}_\sigma : \mathrm{Herm}(A) \to \mathrm{Herm}(A)$ is a linear operator defined as follows. Let $\{p_x\}_{x \in [m]}$ (with $m := |A|$) be the eigenvalues of $\sigma$, and let $\{\eta_{xy}\}_{x,y \in [m]}$ be the matrix components of a matrix $\eta \in \mathrm{Herm}(A)$ in the eigenbasis of $\sigma$. Then, the matrix components of $\mathcal{L}_\sigma(\eta)$ are given by

$$\left[\mathcal{L}_\sigma(\eta)\right]_{xy} := \eta_{xy} \frac{\log p_x - \log p_y}{p_x - p_y} \quad \forall x, y \in [m], \tag{10.88}$$

where the case $p_x = p_y$ is understood in terms of the limit

$$\frac{\log p_x - \log p_y}{p_x - p_y} := \lim_{p_y \to p_x} \frac{\log p_x - \log p_y}{p_x - p_y} = \frac{1}{p_x}. \tag{10.89}$$

In Exercise 10.13, we will show that $\mathcal{L}_\sigma$ is a linear self-adjoint map that satisfies $\mathcal{L}_\sigma(\sigma) = I$.

Now, suppose by contradiction that $0 < \sigma \in \mathfrak{F}(A)$ is a CFS of $\rho \in \mathfrak{D}(A)$, and $\sigma \notin \partial\mathfrak{F}(A)$. This means that $\sigma$ is in the interior of $\mathfrak{F}(A)$, and, in particular, there exists $\varepsilon > 0$ such that $\mathfrak{B}_\varepsilon(\sigma)$ does not contain any resource state (i.e. $\mathfrak{B}_\varepsilon(\sigma) \subset \mathfrak{F}(A)$). Moreover, since $\sigma > 0$ it follows that for any $\sigma' \in \mathfrak{D}(A)$ (i.e. not necessarily free) and small enough $|t|$, where $t \in \mathbb{R}$ can be negative, the state $\omega := (1 - t)\sigma + t\sigma' \in \mathfrak{B}_\varepsilon(A) \subset \mathfrak{F}(A)$. Hence, for small enough $|t|$,

$$D(\rho\|\mathfrak{F}) = D(\rho\|\sigma) \leqslant D(\rho\|\omega), \tag{10.90}$$

since $\sigma$ is a CFS of $\rho$. The above expression is equivalent to $f(t) \leqslant \mathrm{Tr}[\rho \log \sigma]$, where

$$f(t) := \mathrm{Tr}\left[\rho \log\left(\sigma + t(\sigma' - \sigma)\right)\right]. \tag{10.91}$$

Since $f(0) = \mathrm{Tr}[\rho \log \sigma]$ achieves the maximum value, we must have $f'(0) = 0$. Using (10.87) we get

$$\begin{aligned} f'(0) &= \mathrm{Tr}\left[\rho\mathcal{L}_\sigma(\sigma' - \sigma)\right] \\ &= \mathrm{Tr}\left[\rho\left(\mathcal{L}_\sigma(\sigma') - I\right)\right] \\ &= \mathrm{Tr}\left[\mathcal{L}_\sigma(\rho)\sigma'\right] - 1. \end{aligned} \tag{10.92}$$

Therefore, the condition that $f'(0) = 0$ implies that $\mathrm{Tr}\left[\mathcal{L}_\sigma(\rho)\sigma'\right] = 1$ for all $\sigma' \in \mathfrak{D}(A)$. This means that $\mathcal{L}_\sigma(\rho) = I$, which is possible only if $\rho = \sigma$. But since we assume that $\rho \notin \mathfrak{F}(A)$ we get a contradiction. This completes the proof. $\blacksquare$

**Exercise 10.13.** *Let $0 < \sigma \in \mathfrak{D}(A)$ and $m := |A|$.*

1. *Show that $\mathcal{L}_\sigma(\sigma) = I^A$.*
2. *Show that $\mathcal{L}_\sigma$ is a linear self-adjoint map. That is, show that for any $\eta, \zeta \in \mathrm{Herm}(A)$*

$$\mathrm{Tr}\left[\eta\mathcal{L}_\sigma(\zeta)\right] = \mathrm{Tr}\left[\mathcal{L}_\sigma(\eta)\zeta\right]. \tag{10.93}$$

3. *Show that $\mathcal{L}_\sigma$ is invertible, and its inverse, $\mathcal{L}_\sigma^{-1}$, is also self-adjoint and is given by*

$$\left[\mathcal{L}_\sigma^{-1}(\zeta)\right]_{xy} := \zeta_{xy} \frac{p_x - p_y}{\log p_x - \log p_y} \quad \forall \, x, y \in [m], \quad \forall \, \zeta \in \mathrm{Herm}(A). \tag{10.94}$$

The following theorem provides a formula for all the resource states that have the same CFS. We will use the notation $\mathrm{WIT}_{\mathfrak{F}}(A)$ to denote the subset of $\mathrm{Herm}(A)$ that consists of all the normalized resource witnesses of the QRT $\mathfrak{F}$. Explicitly,

$$\mathrm{WIT}_{\mathfrak{F}}(A) := \left\{ \eta \in \mathfrak{F}(A)^* \ : \ \eta \not\geqslant 0, \ \|\eta\|_1 = 1 \right\}. \tag{10.95}$$

Note that we normalized the resource witnesses to have a unit trace norm since if $\eta \in \mathrm{Herm}(A)$ is a resource witness also $a\eta$ with $0 < a \in \mathbb{R}$ is a resource witness, and for our purposes it will be sufficient to consider only one representative of the set $\{a\eta\}_{a > 0}$.

> **Closed Formula for the Relative Entropy of a Resource**
>
> **Theorem 10.5.** Let $\mathfrak{F}(A)$ be convex and $0 < \sigma \in \partial\mathfrak{F}(A)$. Then, the set $\mathfrak{R}(\sigma)$ of all resource states in $\mathfrak{D}(A)$ for which $\sigma$ is the closest free state is given by
>
> $$\mathfrak{R}(\sigma) = \left\{ \sigma - a\mathcal{L}_{\sigma}^{-1}(\eta) \ : \ \eta \in \mathrm{WIT}_{\mathfrak{F}}(A), \ \mathrm{Tr}[\sigma\eta] = 0, \ 0 < a \leqslant a_{\max} \right\}, \tag{10.96}$$
>
> where $a_{\max}$ is the largest positive number that satisfies $a_{\max}\mathcal{L}_{\sigma}^{-1}(\eta) \leqslant \sigma$.

*Remark.* The conditions $\mathrm{Tr}[\sigma\eta] = 0$ and $a \leqslant a_{\max}$ ensure that the state $\sigma - a\mathcal{L}_{\sigma}^{-1}(\eta)$ is a density matrix. Indeed, the condition $a \leqslant a_{\max}$ ensures that it is positive semidefinite, and its trace is 1 since the self-adjointness of $\mathcal{L}_{\sigma}^{-1}$ gives

$$\mathrm{Tr}\left[\mathcal{L}_{\sigma}^{-1}(\eta)\right] = \mathrm{Tr}\left[\mathcal{L}_{\sigma}^{-1}(I)\eta\right] = \mathrm{Tr}[\sigma\eta] = 0. \tag{10.97}$$

**Proof** From the supporting hyperplane theorem, (see Theorem A.9 of online version) it follows that for any (fixed) $\sigma \in \partial\mathfrak{F}(A)$ there exists a Hermitian matrix $\eta \in \mathrm{Herm}(A)$ such that

$$\mathrm{Tr}[\sigma'\eta] \geqslant \mathrm{Tr}[\sigma\eta] \qquad \forall \, \sigma' \in \mathfrak{F}(A). \tag{10.98}$$

Moreover, since both $\sigma$ and $\sigma'$ are normalized, if $\eta$ satisfies the equation above, also $\eta + aI$ satisfies it for any $a \in \mathbb{R}$. We will therefore assume without loss of generality that $\mathrm{Tr}[\sigma\eta] = 0$, which means that $\mathrm{Tr}[\sigma'\eta] \geqslant 0$ for all $\sigma' \in \mathfrak{F}(A)$; that is, $\eta$ is a resource witness (observe that the condition $\mathrm{Tr}[\sigma\eta] = 0$ implies that $\eta \not\geqslant 0$ since $\sigma > 0$). Note also that we can always normalize $\eta$ such that $\|\eta\|_1 = 1$. Quite often, such a resource witness that satisfies these three conditions (i.e. $\mathrm{Tr}[\sigma\eta] = 0$, $\mathrm{Tr}[\sigma'\eta] \geqslant 0$ for all $\sigma' \in \mathfrak{F}(A)$, and $\|\eta\|_1 = 1$) is unique, although for some special boundary points $\sigma \in \partial\mathfrak{F}(A)$, there is a cone of such witnesses of dimension greater than 1 (see Figure 10.2).

Let $\rho$ be a resource state in $\mathfrak{D}(A)$ for which $\sigma$ is the closest free state. The main idea of the proof is the observation that $\eta' := I^A - \mathcal{L}_{\sigma}(\rho)$ is a affine. To see that, first observe that

$$\mathrm{Tr}[\eta'\sigma] = 1 - \mathrm{Tr}[\sigma\mathcal{L}_{\sigma}(\rho)]$$

$$\mathbf{\mathcal{L}_{\sigma} \text{ is self adjoint}} \rightarrow \ = 1 - \mathrm{Tr}[\mathcal{L}_{\sigma}(\sigma)\rho] \tag{10.99}$$

$$\boxed{\mathcal{L}_{\sigma}(\sigma) = I^A} \longrightarrow \ = 1 - \mathrm{Tr}[\rho] = 0.$$

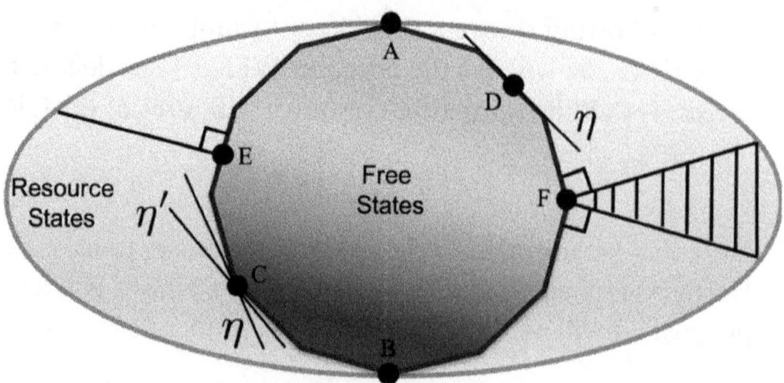



**Figure 10.2**  A schematic diagram of free states (dodecagon) and resource states (oval). Most points on the boundary, like the points D and E, have a unique supporting hyperplane (which is also the tangent plane). The point E is the closest free state of all the points on the vertical line from it. Some of the points, like the points C and F, have more than one supporting hyperplane. The point F is the closest free state of all the points in the shaded grid area. Some points on the boundary, like the points A and B, cannot be a closest free state; for example, separable states of rank 1 (i.e. product states) are on the boundary of separable states, but can never be the closest separable states of some entangled state.

Moreover, for every $\sigma' \in \mathfrak{F}(A)$, define $f(t)$ as in (10.91), but with nonnegative $t \in |, [0, 1]$ (recall that here $\sigma$ is a boundary point not in the interior of $\mathfrak{F}(A)$, so that we can only conclude that $\omega := (1 - t)\sigma + t\sigma'$ is a free state for nonnegative $t \in [0, 1]$). Since $\sigma$ is the closest free state to $\rho$, we must have that $f'(0) \leqslant 0$ (we cannot conclude that the derivative is zero since $t$ cannot be negative). From (10.92) we get for all $\sigma' \in \mathfrak{F}(A)$

$$
\begin{aligned}
0 \leqslant -f'(0) &= 1 - \operatorname{Tr}\left[\mathcal{L}_\sigma(\rho)\sigma'\right] \\
&= \operatorname{Tr}\left[\left(I - \mathcal{L}_\sigma(\rho)\right)\sigma'\right] \\
&= \operatorname{Tr}[\eta'\sigma'].
\end{aligned}
\tag{10.100}
$$

Hence, $\eta'$ is an affine. We can then normalize it $\eta := \frac{1}{a}\eta'$ with $a > 0$ such that $\|\eta\|_1 = 1$. We then conclude from the definition $\eta' := I^A - \mathcal{L}_\sigma(\rho)$ that

$$
\begin{aligned}
\rho &= \mathcal{L}_\sigma^{-1}\left(I^A - \eta'\right) \\
\overset{\eta' = a\eta}{=}\; &= \mathcal{L}_\sigma^{-1}\left(I^A\right) - a\mathcal{L}_\sigma^{-1}(\eta) \\
\overset{\mathcal{L}_\sigma^{-1}(I) = \sigma}{=}\; &= \sigma - a\mathcal{L}_\sigma^{-1}(\eta).
\end{aligned}
\tag{10.101}
$$

Conversely, suppose $\rho = \sigma - a\mathcal{L}_\sigma^{-1}(\eta)$ for some $\eta \in \operatorname{WIT}_{\mathfrak{F}}(A)$ and $a > 0$. We need to show that $D(\rho\|\mathfrak{F}) = D(\rho\|\sigma)$. For this purpose, let $\sigma' \in \mathfrak{F}(A)$ be any free state, and observe that $D(\rho\|\sigma) \leqslant D(\rho\|\sigma')$ if and only if $f(0) \geqslant f(1)$, where $f(t)$ is defined in (10.91). From the joint convexity of the relative entropy (and particularly its convexity in the second argument) it follows that the function $f(t)$ is concave (see Exercise 10.14). This means that if $f'(0) \leqslant 0$, then we must have $f(0) \geqslant f(1)$ (Exercise 10.14). Now, note that from (10.92)

$$f'(0) = \text{Tr}\left[\mathcal{L}_\sigma(\rho)\sigma'\right] - 1$$

$$\boxed{\rho = \sigma - a\mathcal{L}_\sigma^{-1}(\eta)} \longrightarrow = \text{Tr}\left[\mathcal{L}_\sigma\left(\sigma - a\mathcal{L}_\sigma^{-1}(\eta)\right)\sigma'\right] - 1 \tag{10.102}$$

$$= -a\,\text{Tr}\left[\eta\sigma'\right]$$

**$\eta$ is a resource witness**$\rightarrow \leqslant 0$.

Hence, $f(0) \geqslant f(1)$, which is equivalent to $D(\rho\|\sigma) \leqslant D(\rho\|\sigma')$. Since $\sigma'$ was arbitrary state in $\mathfrak{F}(A)$, this completes the proof. ∎

The significance of the theorem above is that if for a given resource state $\rho$ we have a candidate $\sigma$ that we believe to be a closest free state, then we can check it with the formula in (10.96). Specifically, what needs to be checked is whether the matrix $I - \mathcal{L}_\sigma(\rho)$ is a affine. We will see how this can be done when we compute the relative entropy of entanglement on pure bipartite states. We also point out that the techniques used above are not limited to the Umegaki relative entropy, and similar results can be obtained for the $\alpha$-Rényi relative entropy of a resource as defined in (10.71).

**Exercise 10.14.** *Let $f(t)$ be the function defined in* (10.91).

1. *Show that $f(t)$ is concave. Hint: Use the convexity of $D(\rho\|\sigma)$ in $\sigma$ (with fixed $\rho$).*
2. *Show that if $f'(0) \leqslant 0$, then $f(0) \geqslant f(1)$.*

**Exercise 10.15.** *Let $0 < \rho \in \mathfrak{D}(A)$ be a full rank resource state (i.e. $\rho \notin \mathfrak{F}(A)$). Show that the closest free state to $\rho$ is unique. Hint: Let $\sigma \neq \sigma'$ be two closest free states, define $t\sigma + (1 - t)\sigma'$, and use the strict concavity of the function $f(\sigma) - \text{Tr}[\rho \log \sigma]$.*

## 10.3.2 Exact Formula for Special Cases

In (10.71), for any $\alpha \in [0, 2]$, we defined the $\alpha$-Rényi relative entropy of a resource as

$$D_\alpha(\rho\|\mathfrak{F}) := \min_{\omega \in \mathfrak{F}(A)} D_\alpha(\rho\|\omega) \qquad \forall\, \rho \in \mathfrak{D}(A). \tag{10.103}$$

This quantity has a simple closed formula if the set of free states is affine (see Section 9.3) and in addition satisfy for any $\alpha \in [0, 2]$

$$\frac{\sigma^\alpha}{\text{Tr}[\sigma^\alpha]} \in \mathfrak{F}(A) \qquad \forall\, \sigma \in \mathfrak{F}(A). \tag{10.104}$$

---

**Theorem 10.6.** Let $\alpha \in [0, 2]$ and $\mathfrak{F}(A) \subseteq \mathfrak{D}(A)$ be an affine set satisfying (10.104). Then, for all $\rho \in \mathfrak{D}(A)$ we have

$$D_\alpha(\rho\|\mathfrak{F}) = \frac{1}{\alpha - 1} \log \left\| \Delta\left(\rho^\alpha\right)\right\|_{1/\alpha}, \tag{10.105}$$

where $\Delta \in \text{Pos}(A \to A)$ is the self-adjoint RDM associated with the affine set $\mathfrak{F}(A)$ (see Definition 9.10).

**Proof**   Observe that for all $\sigma \in \mathfrak{F}(A)$ we have

$$
\begin{aligned}
\mathrm{Tr}\left[\rho^\alpha \sigma^{1-\alpha}\right] &= \mathrm{Tr}\left[\rho^\alpha \Delta\left(\sigma^{1-\alpha}\right)\right] \\
&= \mathrm{Tr}\left[\Delta\left(\rho^\alpha\right)\sigma^{1-\alpha}\right] \\
&= \left\|\Delta\left(\rho^\alpha\right)\right\|_{1/\alpha}\mathrm{Tr}\left[\gamma^\alpha \sigma^{1-\alpha}\right]
\end{aligned}
\tag{10.106}
$$

where

$$
\gamma := \frac{\left(\Delta\left(\rho^\alpha\right)\right)^{1/\alpha}}{\mathrm{Tr}\left[\left(\Delta\left(\rho^\alpha\right)\right)^{1/\alpha}\right]}.
\tag{10.107}
$$

The proof is concluded with the observation that for $\alpha \leqslant 1$ we have $\mathrm{Tr}\left[\gamma^\alpha \sigma^{1-\alpha}\right] \leqslant 1$ (Hölder's inequality), and for $\alpha > 1$ we have $\mathrm{Tr}\left[\gamma^\alpha \sigma^{1-\alpha}\right] \geqslant 1$ (reverse Hölder's inequality), where equality holds in both cases for $\sigma = \gamma$. Therefore, $\sigma = \gamma$ is the optimizer. $\blacksquare$

**Exercise 10.16.** *Let $\alpha \in [0,2]$. Give a closed expression for $D_\alpha(\rho\|\mathfrak{F})$ for the following cases:*

1. *$\rho \in \mathfrak{D}(A)$ and $\mathfrak{F}(A)$ consists of a set of diagonal density matrices in some fixed basis.*
2. *$\rho \in \mathfrak{D}(AB)$ and $\mathfrak{F}(AB) = \{\sigma^A \otimes \mathbf{u}^B : \sigma \in \mathfrak{D}(A)\}$.*
3. *$\rho \in \mathfrak{D}(A)$ and $\mathfrak{F}(A)$ consists of a set of symmetric density matrices (i.e. $\sigma \in \mathfrak{F}(A)$ if and only if $\sigma = \sigma^T$ where the transpose is taken in some fixed basis).*
4. *$G$ is a unitary group, $\rho \in \mathfrak{D}(A)$, and $\mathfrak{F}(A)$ consists of the set of $G$-invariant states (i.e. $\sigma \in \mathfrak{F}(A)$ if and only if $\sigma = U\sigma U^*$ for all $U \in G$).*

**Exercise 10.17.** *Show that the expression given in (10.105) for the $\alpha$-relative entropy of a resource can be rewritten as*

$$
D_\alpha(\rho\|\mathfrak{F}) = H_{1/\alpha}\left(\Delta(\rho_\alpha)\right) - H_\alpha(\rho),
\tag{10.108}
$$

*where $H_\alpha$ is the Rényi entropy and $\rho_\alpha := \rho^\alpha/\mathrm{Tr}[\rho^\alpha]$. Conclude that for $\alpha = 1$,*

$$
D(\rho\|\mathfrak{F}) = H\left(\Delta(\rho)\right) - H(\rho).
\tag{10.109}
$$

## 10.4 Smoothing of Resource Measures

Resource measures, by definition, do not necessarily have to be smooth. An illustrative example is the logarithmic robustness, expressed as

$$
D_{\max}(\rho\|\mathfrak{F}) = \min_{\sigma \in \mathfrak{F}(A)} D_{\max}(\rho\|\sigma).
\tag{10.110}
$$

This lack of smoothness in the logarithmic robustness can be traced back to the discontinuity present in the max relative entropy. Contrasting with the Umegaki relative entropy, $D_{\max}$, does not exhibit asymptotic continuity. In fact, more broadly, it is not continuous with respect to its first argument. For example, take

$$\sigma = \frac{1}{2}|0\rangle\langle 0| + \frac{1}{2}|1\rangle\langle 1| \quad \text{and} \quad \rho_\varepsilon = \left(\frac{1}{2} - \varepsilon\right)|0\rangle\langle 0| + \frac{1}{2}|1\rangle\langle 1| + \varepsilon|2\rangle\langle 2|. \tag{10.111}$$

For these choices we have $D(\rho_\varepsilon\|\sigma) = \infty$ for all $\varepsilon \in (0, 1/2]$, whereas $D(\rho_{\varepsilon=0}\|\sigma) = 0$. On the other hand, in the laboratory, the preparation of a physical system in a state $\rho$ always results in some error such that the intended state $\rho$ is different (in trace distance) from the prepared state by some small $\varepsilon > 0$. Therefore, discontinuous resource measures are unlikely to have practical physical significance unless some smoothing process has been applied to them. In the following definition, we provide a simple method to smooth a resource measure.

---

**Definition 10.7.** Let $\mathfrak{F}$ be a QRT, $\mathbf{M}$ be a resource measure, and $\varepsilon > 0$. The $\varepsilon$-smoothed version of $\mathbf{M}$ is the function

$$\mathbf{M}^\varepsilon(\rho) := \min_{\rho' \in \mathfrak{B}_\varepsilon(\rho)} \mathbf{M}(\rho'). \tag{10.112}$$

---

*Remark.* In the definition we employed the notation $\mathfrak{B}_\varepsilon(\rho) := \{\rho' \in \mathfrak{D}(A): \frac{1}{2}\|\rho' - \rho\|_1 \leqslant \varepsilon\}$ to denote the "ball" of states $\rho'$ that are $\varepsilon$-close to $\rho$ in trace distance. The rationale for choosing the minimum over the ball $\mathfrak{B}_\varepsilon(\rho)$ stems from the intention to identify the minimum amount of resource present within this ball. This approach ensures that the value $\mathbf{M}^\varepsilon(\rho)$ represents the minimum guaranteed resource level in the system, even when our knowledge is limited to the state of the system being $\varepsilon$-close to $\rho$. Essentially, this method accounts for uncertainty in the system's state by considering the least amount of resource that can be confidently ascribed to states within an $\varepsilon$-radius of $\rho$. This approach is both cautious and practical, as it provides a conservative estimate of the resource quantity in the practical situations where exact state information is not available.

---

**Lemma 10.5.** Let $\mathbf{M}$ be a resource measure. Then, its smoothed version, $\mathbf{M}^\varepsilon$, is also a resource measure for all $\varepsilon > 0$.

---

**Proof**  By definition, $\mathbf{M}^\varepsilon$ is nonnegative since $\mathbf{M}$ is nonnegative. On the other hand, for any $\rho \in \mathfrak{F}(A)$ we have $\mathbf{M}^\varepsilon(\rho) \leqslant \mathbf{M}(\rho) = 0$, where we took $\rho' = \rho$ in (10.112). Thus, we must have $\mathbf{M}^\varepsilon(\rho) = 0$ for all $\rho \in \mathfrak{F}(A)$. It is therefore left to prove the monotonicity of $\mathbf{M}^\varepsilon$.

Let $\mathcal{N} \in \mathfrak{F}(A \to B)$ and $\rho \in \mathfrak{D}(A)$. Then,

$$\mathbf{M}^\varepsilon\left(\mathcal{N}(\rho)\right) = \min_{\sigma \in \mathfrak{B}_\varepsilon(\mathcal{N}(\rho))} \mathbf{M}(\sigma)$$

$$= \min\left\{\mathbf{M}(\sigma) : \frac{1}{2}\|\sigma - \mathcal{N}(\rho)\|_1 \leqslant \varepsilon, \ \sigma \in \mathfrak{D}(B)\right\}$$

$$\textbf{Restricting } \sigma = \mathcal{N}(\rho') \to \ \leqslant \min\left\{\mathbf{M}(\mathcal{N}(\rho')) : \frac{1}{2}\|\mathcal{N}(\rho') - \mathcal{N}(\rho)\|_1 \leqslant \varepsilon, \ \rho' \in \mathfrak{D}(A)\right\}$$

$$\textbf{Monotonicity of } \mathbf{M} \to \ \leqslant \min\left\{\mathbf{M}(\rho') : \frac{1}{2}\|\mathcal{N}(\rho') - \mathcal{N}(\rho)\|_1 \leqslant \varepsilon, \ \rho' \in \mathfrak{D}(A)\right\}$$

$$\textbf{DPI of trace distance} \to \ \leqslant \min\left\{\mathbf{M}(\rho') : \frac{1}{2}\|\rho' - \rho\|_1 \leqslant \varepsilon, \ \rho' \in \mathfrak{D}(A)\right\}$$

$$= \mathbf{M}^\varepsilon(\rho), \tag{10.113}$$

where in the last inequality we used the DPI of the trace distance, particularly, the property that if $\frac{1}{2}\|\rho' - \rho\|_1 \leqslant \varepsilon$, then $\frac{1}{2}\|\mathcal{N}(\rho') - \mathcal{N}(\rho)\|_1 \leqslant \varepsilon$. Therefore, the former impose a stronger constraint than the latter, which give rise to the last inequality. This completes the proof. $\blacksquare$

## 10.4.1 Smoothing of Divergence-Based Resource Measures

Let $\varepsilon > 0$, $\mathbb{D}$ be a quantum divergence, and $\mathbb{D}(\,\cdot\,\|\mathfrak{F})$ its associated resource measure as defined in (10.22). Then, the $\varepsilon$-smoothed version of $\mathbb{D}(\,\cdot\,\|\mathfrak{F})$ is given for all $\rho \in \mathfrak{D}(A)$ by

$$
\begin{aligned}
\mathbb{D}^{\varepsilon}(\rho\|\mathfrak{F}) &:= \min_{\rho' \in \mathcal{B}_{\varepsilon}(\rho)} \mathbb{D}(\rho'\|\mathfrak{F}) \\
&= \min_{\rho' \in \mathcal{B}_{\varepsilon}(\rho)} \inf_{\sigma \in \mathfrak{F}(A)} \mathbb{D}\left(\rho'\|\sigma\right) \\
&= \inf_{\sigma \in \mathfrak{F}(A)} \mathbb{D}^{\varepsilon}\left(\rho\|\sigma\right),
\end{aligned}
\tag{10.114}
$$

where $\mathbb{D}^{\varepsilon}$ is defined as

$$
\mathbb{D}^{\varepsilon}\left(\rho\|\sigma\right) := \min_{\rho' \in \mathcal{B}_{\varepsilon}(\rho)} \mathbb{D}\left(\rho'\|\sigma\right).
\tag{10.115}
$$

The quantity $\mathbb{D}^{\varepsilon}$ is called the $\varepsilon$-smoothed version of the quantum divergence $\mathbb{D}$. Smoothed divergences play key roles in QRTs and in the next theorem we prove some useful relationships among some of them.

**Exercise 10.18.** *Let $\mathbb{D}$ be a quantum divergence and $\varepsilon > 0$. Show that $\mathbb{D}^{\varepsilon}$ is itself a quantum divergence.*

We denote the $\varepsilon$-smoothed version of $D_{\max}$ as $D_{\max}^{\varepsilon}$. For $D_{\min}$, the notation $D_{\min}^{\varepsilon}$ already signifies the quantum hypothesis testing divergence (see Section 8.7.1). This aligns with the notion that the quantum hypothesis testing divergence is a smoothed version of $D_{\min}$ (refer to Exercise 8.29). Additionally, when smoothing $D_{\min}$ in the form $\min_{\rho' \in \mathcal{B}_{\varepsilon}(\rho)} D_{\min}\left(\rho'|\sigma\right)$, the result is always zero. This is because for any $\varepsilon > 0$ and $\rho \in \mathfrak{D}(A)$, there's a $\rho' \in \mathfrak{D}(A)$ that's $\varepsilon$-close to $\rho$ with $\rho' > 0$, making $D_{\min}(\rho'\|\sigma) = 0$. Henceforth, $D_{\min}^{\varepsilon}$ will exclusively represent the quantum hypothesis testing divergence in this book.

In the forthcoming theorem, we will establish specific inequalities that involve $D_{\min}^{\varepsilon}$ and $D_{\max}^{\varepsilon}$. These inequalities are crucial and will play a significant role in the subsequent discussions and analyses. The relationships between $D_{\min}^{\varepsilon}$ and $D_{\max}^{\varepsilon}$ are fundamental in understanding various aspects of quantum resources both in the single-shot regime as well as in the asymptotic domain. Furthermore, later on we will use some of these relationships to provide operational interpretation to both $D_{\min}^{\varepsilon}$ and $D_{\max}^{\varepsilon}$.

> **Theorem 10.7.** Let $\rho, \sigma \in \mathfrak{D}(A)$ and $\varepsilon, \varepsilon_1, \varepsilon_2 \in (0,1)$ such that $\varepsilon_1 + \varepsilon_2 \leqslant 1$. Then, the following relations hold:
>
> $$D_{\max}^{\varepsilon_1}(\rho\|\sigma) \geqslant D_{\min}^{\varepsilon_2}(\rho\|\sigma) + \log(1 - \varepsilon_1 - \varepsilon_2) \tag{10.116}$$
>
> $$D_{\min}^{\varepsilon}(\rho\|\sigma) \leqslant \frac{D(\rho\|\sigma) + h(\varepsilon)}{1 - \varepsilon} \tag{10.117}$$
>
> $$D_{\min}^{\varepsilon}(\rho\|\sigma) \geqslant D_{\max}^{\sqrt{1-\varepsilon^2}}(\rho\|\sigma) + \log\left(\frac{\varepsilon}{1-\varepsilon}\right), \tag{10.118}$$
>
> where $h(\varepsilon) := -\varepsilon \log \varepsilon - (1 - \varepsilon)\log(1 - \varepsilon)$ is the binary Shannon entropy.

We prove each of the inequalities separately:

**Proof of inequality** (10.116)   Let $\tilde{\rho} \in \mathfrak{D}(A)$ be such that $D_{\max}^{\varepsilon_1}(\rho\|\sigma) = D_{\max}(\tilde{\rho}\|\sigma)$ and $\frac{1}{2}\|\tilde{\rho} - \rho\|_1 \leqslant \varepsilon_1$. From (5.170) it follows that there exists $\omega_\pm \in \mathfrak{D}(A)$ such that

$$\tilde{\rho} = \rho + \varepsilon_1(\omega_+ - \omega_-), \tag{10.119}$$

where without loss of generality we assume the equality $\frac{1}{2}\|\tilde{\rho} - \rho\|_1 = \varepsilon_1$. Denote $r := D_{\max}^{\varepsilon_1}(\rho\|\sigma)$ so that $\tilde{\rho} \leqslant 2^r \sigma$. Combining this with the inequality $\rho \leqslant \tilde{\rho} + \varepsilon_1 \omega_-$ (that follows from the equation above) gives

$$\rho \leqslant 2^r \sigma + \varepsilon_1 \omega_-. \tag{10.120}$$

Finally, let $\Lambda \in \mathrm{Eff}(A)$ be the optimal operator satisfying

$$D_{\min}^{\varepsilon_2}(\rho\|\sigma) = -\log \mathrm{Tr}\,[\sigma\Lambda] \quad \text{and} \quad \mathrm{Tr}[\rho\Lambda] \geqslant 1 - \varepsilon_2. \tag{10.121}$$

Then,

$$1 - \varepsilon_2 \leqslant \mathrm{Tr}[\rho\Lambda]$$

$$\textbf{(10.120)}\rightarrow\ \leqslant 2^r \mathrm{Tr}[\sigma\Lambda] + \varepsilon_1 \mathrm{Tr}[\omega_-\Lambda] \tag{10.122}$$

$$\boxed{\textbf{(10.121) and } \Lambda \leqslant I} \longrightarrow\ \leqslant 2^{r - D_{\min}^{\varepsilon_2}(\rho\|\sigma)} + \varepsilon_1.$$

Recalling the definition of $r$, we get from the equation above that

$$D_{\max}^{\varepsilon_1}(\rho\|\sigma) - D_{\min}^{\varepsilon_2}(\rho\|\sigma) \geqslant \log(1 - \varepsilon_1 - \varepsilon_2). \tag{10.123}$$

This completes the proof. ∎

**Proof of inequality** (10.117)   Let $\Lambda \in \mathrm{Eff}(A)$ be the optimal effect such that $D_{\min}^{\varepsilon}(\rho\|\sigma) = -\log \mathrm{Tr}[\Lambda\sigma]$ and $\mathrm{Tr}[\Lambda\rho] = 1 - \varepsilon$. Define the channel $\mathcal{E} \in \mathrm{CPTP}(A \to X)$, with $|X| = 2$, as

$$\mathcal{E}(\omega) := \mathrm{Tr}[\Lambda\omega]|0\rangle\langle 0|^X + \mathrm{Tr}[(I - \Lambda)\omega]|1\rangle\langle 1|^X \qquad \forall\, \omega \in \mathfrak{L}(A). \tag{10.124}$$

By definition, $\mathcal{E}(\rho) = \mathrm{Diag}(1 - \varepsilon, \varepsilon)$ and $\mathcal{E}(\sigma) = \mathrm{Diag}(t, 1 - t)$, where $t := 2^{-D_{\min}^{\varepsilon}(\rho\|\sigma)}$ and $\mathrm{Diag}(\cdot, \cdot)$ denotes $2 \times 2$ diagonal matrix. With these definitions we get from the DPI that

$$D(\rho\|\sigma) \geqslant D\left(\mathcal{E}(\rho)\|\mathcal{E}(\sigma)\right) = D\left(\mathrm{Diag}(1-\varepsilon,\varepsilon)\|\mathrm{Diag}(t,1-t)\right)$$
$$= -h(\varepsilon) - (1-\varepsilon)\log t - \varepsilon\log(1-t)$$

$$\boxed{-\varepsilon\log(1-t)\geqslant 0,\ \log t = -D_{\min}^{\varepsilon}(\rho\|\sigma)}\longrightarrow \geqslant -h(\varepsilon) + (1-\varepsilon)D_{\min}^{\varepsilon}(\rho\|\sigma).$$

$$(10.125)$$

This completes the proof. ∎

**Proof of inequality** (10.118)  Recall from (8.196) that there exists $t \in \mathbb{R}_+$ and $\omega \in \mathrm{Pos}(A)$ such that $t\rho \leqslant \sigma + \omega$ and

$$2^{-D_{\min}^{\varepsilon}(\rho\|\sigma)} = (1-\varepsilon)t - \mathrm{Tr}[\omega]. \tag{10.126}$$

Observe that since $2^{-D_{\min}^{\varepsilon}(\rho\|\sigma)} \geqslant 0$ the equation above implies in particular that

$$\mathrm{Tr}[\omega] \leqslant (1-\varepsilon)t. \tag{10.127}$$

Since we would like to relate between $D_{\min}^{\varepsilon}(\rho\|\sigma)$ and a smoothing of $D_{\max}(\rho\|\sigma)$, we rewrite $t\rho \leqslant \sigma + \omega$ in a way that involves $D_{\max}(\rho'\|\sigma)$ for some $\rho' \in \mathfrak{D}(A)$ and then estimate how close $\rho'$ to $\rho$. For this purpose, we define the operator $G := \sigma^{\frac{1}{2}}(\sigma+\omega)^{-\frac{1}{2}}$ and observe that

$$tG\rho G^* \leqslant G(\sigma+\omega)G^*$$
$$= \sigma. \tag{10.128}$$

Moreover, denoting by $\Lambda := G^*G$ we get that

$$\Lambda = (\sigma+\omega)^{-\frac{1}{2}}\sigma(\sigma+\omega)^{-\frac{1}{2}}$$
$$\boxed{\sigma=\sigma+\omega-\omega}\rightarrow\rightarrow \leqslant I^A - (\sigma+\omega)^{-\frac{1}{2}}\omega(\sigma+\omega)^{-\frac{1}{2}} \tag{10.129}$$
$$\leqslant I^A.$$

Hence, $\Lambda \in \mathrm{Eff}(A)$. Finally, denoting by $\rho' := \frac{G\rho G^*}{\mathrm{Tr}[\Lambda\rho]}$ (so that $\rho' \in \mathfrak{D}(A)$), the condition $t\rho \leqslant \sigma + \omega$ implies that $t\mathrm{Tr}[\Lambda\rho]\rho' \leqslant \sigma$, so that

$$D_{\max}(\rho'\|\sigma) \leqslant -\log\left(t\mathrm{Tr}[\Lambda\rho]\right). \tag{10.130}$$

Next, we estimate $\mathrm{Tr}[\Lambda\rho]$:

$$\mathrm{Tr}[\Lambda\rho] = 1 - \mathrm{Tr}\left[(I-\Lambda)\rho\right]$$
$$\boxed{t\rho\leqslant\sigma+\omega}\longrightarrow \geqslant 1 - \frac{1}{t}\mathrm{Tr}\left[(I-\Lambda)(\sigma+\omega)\right]$$
$$\boxed{\mathrm{Tr}[\Lambda(\sigma+\omega)]=1}\longrightarrow = 1 - \frac{\mathrm{Tr}[\omega]}{t} \tag{10.131}$$
$$\mathbf{(10.127)}\rightarrow \geqslant \varepsilon,$$

where we used the expression for $\Lambda$ in (10.129) to get that $\mathrm{Tr}[\Lambda(\sigma+\omega)] = 1$. Combining the two equations above gives $D_{\max}(\rho'\|\sigma) \leqslant -\log(t\varepsilon)$. To estimate $t$, we use the fact that $\mathrm{Tr}[\omega] \geqslant 0$ to get from the relation in (10.126) that $2^{-D^\varepsilon_{\min}(\rho\|\sigma)} \leqslant (1-\varepsilon)t$. Substituting this lower bound on $t$ into the inequality $D_{\max}(\rho'\|\sigma) \leqslant -\log(t\varepsilon)$ gives

$$D_{\max}(\rho'\|\sigma) \leqslant -\log\left(\frac{\varepsilon}{1-\varepsilon} 2^{-D^\varepsilon_{\min}(\rho\|\sigma)}\right)$$

$$= D^\varepsilon_{\min}(\rho\|\sigma) - \log\left(\frac{\varepsilon}{1-\varepsilon}\right). \tag{10.132}$$

It is therefore left to show that $\rho'$ is $\sqrt{1-\varepsilon^2}$-close to $\rho$.

Let $|\psi^{A\tilde{A}}\rangle := (\sqrt{\rho} \otimes I)|\Omega^{A\tilde{A}}\rangle$ and $|\tilde{\psi}^{A\tilde{A}}\rangle := (G \otimes I)|\psi^{A\tilde{A}}\rangle$. Observe that $\psi^{A\tilde{A}}$ and $\tilde{\psi}^{A\tilde{A}}$ are purifications of $\rho$ and $\tilde{\rho} := G\rho G^*$, respectively. Moreover, observe that $|\psi'^{A\tilde{A}}\rangle := \frac{1}{\sqrt{\mathrm{Tr}[\tilde{\rho}]}}|\tilde{\psi}^{A\tilde{A}}\rangle$ is a purification of $\rho'$. From Uhlmann's theorem the fidelity between $\rho$ and $\rho'$ satisfies:

$$F(\rho,\rho') \geqslant \left|\langle\psi'^{A\tilde{A}}|\psi^{A\tilde{A}}\rangle\right|$$

$$\boxed{\mathrm{Tr}[\tilde{\rho}] \leqslant 1} \longrightarrow \geqslant \left|\langle\tilde{\psi}^{A\tilde{A}}|\psi^{A\tilde{A}}\rangle\right|$$

$$\textbf{Real Part}\rightarrow \geqslant \frac{1}{2}\left(\langle\tilde{\psi}^{A\tilde{A}}|\psi^{A\tilde{A}}\rangle + \langle\psi^{A\tilde{A}}|\tilde{\psi}^{A\tilde{A}}\rangle\right) \tag{10.133}$$

$$\boxed{P := \frac{1}{2}(G+G^*)} \longrightarrow = \langle\psi^{A\tilde{A}}|P \otimes I^{\tilde{A}}|\psi^{A\tilde{A}}\rangle$$

$$= \mathrm{Tr}[\rho P].$$

Combining this with the fact that $P \leqslant I^A$ (see Exercise 2.26) we obtain

$$F(\rho,\rho') \geqslant \mathrm{Tr}[\rho P] = 1 - \mathrm{Tr}[\rho(I-P)]$$

$$\boxed{t\rho \leqslant \sigma + \omega} \longrightarrow \geqslant 1 - \frac{1}{t}\mathrm{Tr}[(\sigma+\omega)(I-P)]$$

$$\textbf{By definition of P}\rightarrow = 1 - \frac{1}{t} - \frac{\mathrm{Tr}[\omega]}{t} + \frac{1}{t}\mathrm{Tr}\left[\sigma^{\frac{1}{2}}(\sigma+\omega)^{\frac{1}{2}}\right] \tag{10.134}$$

$$(\sigma+\omega)^{\frac{1}{2}} \geqslant \sigma^{\frac{1}{2}}\rightarrow \geqslant 1 - \frac{\mathrm{Tr}[\omega]}{t}$$

$$\textbf{(10.127)}\rightarrow \geqslant \varepsilon.$$

Therefore, from the relation (5.202) between the trace distance and the fidelity we get $\frac{1}{2}\|\rho - \rho'\|_1 \leqslant \sqrt{1-\varepsilon^2}$. This completes the proof. ∎

The relation between the smoothed max and min entropies can be used to obtain a generalized version of the AEP property.

> **Corollary 10.4.** For any $\rho,\sigma \in \mathfrak{D}(A)$ with $\mathrm{supp}(\rho) \subseteq \mathrm{supp}(\sigma)$, and any $\varepsilon \in (0,1)$
>
> $$\lim_{n\to\infty} \frac{1}{n} D^\varepsilon_{\max}\left(\rho^{\otimes n}\|\sigma^{\otimes n}\right) = D(\rho\|\sigma). \tag{10.135}$$

**Proof**    From (10.116) we get that for any $\varepsilon_1, \varepsilon_2 \in (0,1)$ with $\varepsilon_1 + \varepsilon_2 < 1$

$$\liminf_{n \to \infty} \frac{1}{n} D_{\max}^{\varepsilon_1}\left(\rho^{\otimes n} \| \sigma^{\otimes n}\right) \geq \liminf_{n \to \infty} \frac{1}{n}\left(D_{\min}^{\varepsilon_2}\left(\rho^{\otimes n} \| \sigma^{\otimes n}\right) + \log\left(1 - \varepsilon_1 - \varepsilon_2\right)\right)$$

$$= \liminf_{n \to \infty} \frac{1}{n} D_{\min}^{\varepsilon_2}\left(\rho^{\otimes n} \| \sigma^{\otimes n}\right) \tag{10.136}$$

$$(8.211) \to\ = D(\rho \| \sigma).$$

Conversely, from (10.118), after replacing the roles between $\delta := \sqrt{1 - \varepsilon^2}$ and $\varepsilon$, we get for every $\varepsilon \in (0,1)$

$$\limsup_{n \to \infty} \frac{1}{n} D_{\max}^{\varepsilon}\left(\rho^{\otimes n} \| \sigma^{\otimes n}\right) \leq \limsup_{n \to \infty} \frac{1}{n}\left(D_{\min}^{\delta}\left(\rho^{\otimes n} \| \sigma^{\otimes n}\right) - \log\left(\frac{\delta}{1 - \delta}\right)\right)$$

$$= \limsup_{n \to \infty} \frac{1}{n} D_{\min}^{\delta}\left(\rho^{\otimes n} \| \sigma^{\otimes n}\right) \tag{10.137}$$

$$(8.211) \to\ = D(\rho \| \sigma).$$

From the two equations above it follows that (10.135) must hold. ∎

The technique applied in the aforementioned theorem, especially in the proof of (10.118), is also applicable for upper-bounding the smoothed max relative entropy. This can be achieved by smoothing the second argument of $D_{\max}$. For further details, please refer to Appendix D.3 of the online version.

**Exercise 10.19.** *Let $\rho \in \mathfrak{D}(AB)$ and $\varepsilon \in (0,1)$. The smoothed version of $H_{\min}$ and $H_{\max}$ (see Definition 7.4) are defined, respectively, as*

$$H_{\min}^{\varepsilon}(A|B)_\rho := \max_{\rho' \in \mathfrak{B}_\varepsilon(\rho)} H_{\min}(A|B)_{\rho'} \quad and \quad H_{\max}^{\varepsilon}(A|B)_\rho := \min_{\rho' \in \mathfrak{B}_\varepsilon(\rho)} H_{\max}(A|B)_{\rho'}. \tag{10.138}$$

*1. Show that*

$$\lim_{n \to \infty} \frac{1}{n} H_{\min}^{\varepsilon}\left(A^n | B^n\right)_{\rho^{\otimes n}} = H(A|B)_\rho. \tag{10.139}$$

*Hint: Use Corollary 10.4.*

*2. Show that*

$$\lim_{n \to \infty} \frac{1}{n} H_{\max}^{\varepsilon}\left(A^n | B^n\right)_{\rho^{\otimes n}} = H(A|B)_\rho. \tag{10.140}$$

*Hint: Use the duality relation between $H_{\min}(A|B)$ and $H_{\max}(A|B)$.*

## Closed Formula for the Smoothing of an Entropy Function

Generally, the task of smoothing resource measures or relative entropies tends not to result in functions with closed forms. Yet, this scenario is notably different when it comes to the smoothing of entropy functions. Here, the utilization of the concepts of standard and flattest approximations of a probability vector, as delineated in Section 4.2 on approximate majorization, becomes particularly relevant. Intriguingly, these approximations reveal that a smoothed entropy function can indeed be articulated in a closed form. This insight introduces an element of simplicity and exactness into a domain typically characterized by its complexity.

Let $\mathbb{H}$ be a quantum entropy and $\varepsilon \in [0, 1)$. Then, the $\varepsilon$-smoothed version of $\mathbb{H}$ is defined as

$$\mathbb{H}^{\varepsilon}(\rho) := \max_{\rho' \in \mathfrak{B}_{\varepsilon}(\rho)} \mathbb{H}\left(\rho'\right). \tag{10.141}$$

Observe that this definition is consistent with the definition of a smoothed relative entropy. Specifically, if $\mathbb{H}$ is related to a quantum divergence $\mathbb{D}$ as $\mathbb{H}(\rho) = \log|A| - \mathbb{D}(\rho\|\mathbf{u})$, then $\mathbb{H}^{\varepsilon}$ as defined above is related to the smooth relative entropy $\mathbb{D}^{\varepsilon}$ as $\mathbb{H}^{\varepsilon}(\rho) = \log|A| - \mathbb{D}^{\varepsilon}(\rho\|\mathbf{u})$.

---

**Theorem 10.8.** Let $\mathbb{H}$ be a quantum entropy and let $\varepsilon \in [0, 1)$. Then, the $\varepsilon$-smoothed version of $\mathbb{H}$ is given by

$$\mathbb{H}^{\varepsilon}(\rho) = \mathbb{H}\left(\underline{\mathbf{p}}^{(\varepsilon)}\right) \qquad \forall\, \rho \in \mathfrak{D}(A), \tag{10.142}$$

where $\underline{\mathbf{p}}^{(\varepsilon)}$ is the probability vector defined in (4.76).

---

**Proof**   Let $\rho'$ be an optimal quantum state in $\mathfrak{B}_{\varepsilon}(\rho)$ such that $\mathbb{H}^{\varepsilon}(\rho) = \mathbb{H}(\rho')$. We first argue that without loss of generality we can assume that $\rho'$ commutes with $\rho$. To see this, let $\Delta \in \mathrm{CPTP}(A \to A)$ be the completely dephasing channel in the eigenbasis of $\rho$. Then, since $\Delta$ is a doubly stochastic channel we have $\mathbb{H}\left(\Delta(\rho')\right) \geqslant \mathbb{H}\left(\rho'\right)$. Moreover, since $\Delta(\rho) = \rho$ we have

$$\frac{1}{2}\left\|\Delta\left(\rho'\right) - \rho\right\|_1 = \frac{1}{2}\left\|\Delta\left(\rho'\right) - \Delta(\rho)\right\|_1$$

$$\mathbf{DPI} \!\to \; \leqslant \frac{1}{2}\left\|\rho' - \rho\right\|_1 \leqslant \varepsilon. \tag{10.143}$$

Hence, $\Delta\left(\rho'\right)$ is also $\varepsilon$-close to $\rho$, so that $\Delta(\rho')$ is also an optimizer of (10.141). Hence, without loss of generality we can assume that $\rho'$ is diagonal in the same eigenbasis of $\rho$.

Let $\mathbf{p} = \mathbf{p}^{\downarrow}$ be the vector consisting of the eigenvalues of $\rho$. From the argument above we can express the smoothed entropy in (10.141) as

$$\mathbb{H}^{\varepsilon}(\rho) = \max_{\mathbf{p}' \in \mathfrak{B}_{\varepsilon}(\mathbf{p})} \mathbb{H}\left(\mathbf{p}'\right). \tag{10.144}$$

Now, since $\mathfrak{B}_{\varepsilon}(\mathbf{p})$ has the property that for every $\mathbf{p}' \in \mathfrak{B}_{\varepsilon}(\mathbf{p})$ the vector $\underline{\mathbf{p}}^{(\varepsilon)}$ as defined in (4.76) satisfies $\mathbf{p}' \succ \underline{\mathbf{p}}^{(\varepsilon)}$ so that $\mathbb{H}(\mathbf{p}') \leqslant \mathbb{H}(\underline{\mathbf{p}}^{(\varepsilon)})$. Therefore, the choice $\mathbf{p}' = \underline{\mathbf{p}}^{(\varepsilon)}$ gives the maximum value.      $\blacksquare$

As a simple example, consider the min-entropy as defined in (6.22); that is, $H_{\min}(\rho) = -\log\|\rho\|_{\infty}$ for all $\rho \in \mathfrak{D}(A)$. Note that this entropy is related to the max relative entropy via $H_{\min}(\rho) = \log|A| - D_{\max}(\rho\|\mathbf{u})$. From the theorem above,

$H_{\min}^{\varepsilon}(\rho) = - \log \left\| \underline{\mathbf{p}}^{(\varepsilon)} \right\|_{\infty}$. Using the definition of $\underline{\mathbf{p}}^{(\varepsilon)}$ in (4.76) we get from (10.142) that

$$H_{\min}^{\varepsilon}(\rho) = -\log(a)$$

$$(\mathbf{4.81}) \rightarrow = \log\left( \frac{k}{\|\mathbf{p}\|_{(k)} - \varepsilon} \right), \tag{10.145}$$

where $k$ is the integer satisfying (4.82), which is equivalent to

$$p_{k+1} < \frac{\|\mathbf{p}\|_{(k)} - \varepsilon}{k} \leqslant p_k. \tag{10.146}$$

Alternatively, observe that from (4.87) we can also express $H_{\min}^{\varepsilon}(\rho)$ as

$$H_{\min}^{\varepsilon}(\rho) = -\log \max_{\ell \in [n]} \left\{ \frac{\|\mathbf{p}\|_{(\ell)} - \varepsilon}{\ell} \right\}. \tag{10.147}$$

It is worth noting that for the case that $\mathbb{H} = H_{\max}$ (recall that $H_{\max}(A)_\rho := \log \mathrm{Rank}(\rho)$ for all $\rho \in \mathfrak{D}(A)$), the definition in (10.141) results with a quantity that always equals $\log|A|$ since for any $\varepsilon \in (0, 1)$ and any $\rho \in \mathfrak{D}(A)$, there always exists a full rank state $\rho' \in \mathfrak{D}(A)$ that is $\varepsilon$-close to $\rho$. Therefore, in this case instead of taking the maximum in (10.141) we take the minimum, so that the smoothed version of $H_{\max}$ is defined as

$$H_{\max}^{\varepsilon}(A)_\rho := \min_{\rho' \in \mathfrak{B}_\varepsilon(\rho)} H_{\max}(A)_{\rho'}. \tag{10.148}$$

Also this formula has a closed form.

---

**Lemma 10.6.** Let $\varepsilon \in [0, 1)$ and $\rho \in \mathfrak{D}(A)$. Then, the $\varepsilon$-smoothed max-entropy is given by

$$H_{\max}^{\varepsilon}(A)_\rho = \log(m), \tag{10.149}$$

where $m$ is the integer satisfying $\|\rho\|_{(m-1)} < 1 - \varepsilon \leqslant \|\rho\|_{(m)}$.

---

**Proof** From its definition we get

$$H_{\max}^{\varepsilon}\left( \rho^A \right) = \min \left\{ \log \mathrm{Rank}(\rho') : \ T(\rho, \rho') \leqslant \varepsilon \right\}$$

$$\boxed{m := \mathrm{Rank}(\rho')} \longrightarrow = \min \left\{ \log m : \ T\left( \rho, \mathfrak{D}_m(A) \right) \leqslant \varepsilon \right\}, \tag{10.150}$$

where we used the notation $\mathfrak{D}_m(A)$ to denote the set of all density matrices in $\mathfrak{D}(A)$, whose rank is not greater than $m$. In Theorem 5.11 we showed that $T\left( \rho, \mathfrak{D}_m(A) \right) = 1 - \|\rho\|_{(m)}$. Substituting this to the equation above we conclude that

$$H_{\max}^{\varepsilon}(A)_\rho = \min \left\{ \log m : \ \|\rho\|_{(m)} \geqslant 1 - \varepsilon \right\}. \tag{10.151}$$

This completes the proof.     $\blacksquare$

**Exercise 10.20.** *Using the same notations as above:*

1. *Show that if we replace the maximum in (10.144) with a minimum we get*

$$\max_{\mathbf{p}' \in \mathfrak{B}_\varepsilon(\mathbf{p})} \mathbb{H}(\mathbf{p}') = \mathbb{H}\left(\overline{\mathbf{p}}^{(\varepsilon)}\right), \tag{10.152}$$

   *where $\overline{\mathbf{p}}^{(\varepsilon)}$ is the steepest $\varepsilon$-approximation of $\mathbf{p}$ as defined in (4.57).*
2. *Show that if $\rho, \sigma \in \mathfrak{D}(A)$ are such that $\rho \succ \sigma$, then*

$$H_{\max}^\varepsilon(A)_\rho \leqslant H_{\max}^\varepsilon(A)_\sigma. \tag{10.153}$$

## Another Smoothed Version of $D_{\min}$

As previously mentioned, the smoothed version of $D_{\min}$ given by $\min_{\rho' \in \mathfrak{B}_\varepsilon(\rho)} D_{\min}\left(\rho' \| \sigma\right)$ is always zero. Hence, we've recognized the hypothesis testing divergence as its smoothed variant. Yet, one might seek to define the smoothed version of the min relative entropy using a maximum instead of a minimum. Specifically, for any $\rho, \sigma \in \mathfrak{D}(A)$ and $\varepsilon \in [0, 1)$, we introduce another smoothed version of $D_{\min}$ as

$$D_{\min}^{(\varepsilon)}(\rho \| \sigma) := \max_{\rho' \in \mathfrak{B}_\varepsilon(\rho)} D_{\min}\left(\rho' \| \sigma\right). \tag{10.154}$$

It's crucial to emphasize, however, that this quantity is not necessarily a divergence, given the optimization uses a maximum instead of a minimum (thus, Lemma 10.5 isn't applicable). Nevertheless, in this book, we will encounter this function in some applications.

**Exercise 10.21.** *Show that for every $\rho \in \mathfrak{D}(A)$ we have*

$$H_{\max}^\varepsilon(A)_\rho = \log |A| - D_{\min}^{(\varepsilon)}(\rho \| \mathbf{u}). \tag{10.155}$$

---

> **Theorem 10.9.** Let $\rho, \sigma \in \mathfrak{D}(A)$ and $\varepsilon \in [0, 1)$. Then,
>
> $$D_{\min}^{(\varepsilon)}(\rho \| \sigma) \leqslant D_{\min}^\varepsilon(\rho \| \sigma). \tag{10.156}$$

---

**Proof** Let $\tilde{\rho} \in \mathfrak{D}(A)$ be a state that satisfies $D_{\min}^{(\varepsilon)}(\rho \| \sigma) = D_{\min}(\tilde{\rho} \| \sigma)$. By definition, $\tilde{\rho}$ is $\varepsilon$ close to $\rho$ and from the relation (5.152) we get that

$$\max_{0 \leqslant \Lambda \leqslant I} \mathrm{Tr}[(\tilde{\rho} - \rho)\Lambda] \leqslant \varepsilon. \tag{10.157}$$

In particular, taking $\Lambda = \Pi_{\tilde{\rho}}$ to be the projection to the support of $\tilde{\rho}$ we obtain

$$\mathrm{Tr}\left[\rho \Pi_{\tilde{\rho}}\right] \geqslant 1 - \varepsilon. \tag{10.158}$$

Therefore, taking $\Lambda = \Pi_{\tilde{\rho}}$ in the definition (8.185) we conclude

$$D_{\min}^\varepsilon(\rho) \geqslant -\log \mathrm{Tr}\left[\sigma \Pi_{\tilde{\rho}}\right]$$
$$= D_{\min}(\tilde{\rho} \| \sigma) = D_{\min}^{(\varepsilon)}(\rho \| \sigma). \tag{10.159}$$

This completes the proof. ∎

**Exercise 10.22.** *Let $\rho, \sigma \in \mathfrak{D}(A)$ and $\varepsilon \in (0, 1)$. Show that*

$$D_{\min}^{(\varepsilon)}(\rho\|\sigma) \geqslant D_{\min}\left(\frac{\sqrt{\Lambda}\rho\sqrt{\Lambda}}{\mathrm{Tr}[\Lambda\rho]}\bigg\|\sigma\right), \tag{10.160}$$

*for any $\Lambda \in \mathrm{Eff}(A)$ that satisfies $\mathrm{Tr}[\Lambda\rho] \geqslant 1 - \varepsilon^2$. Hint: Use the gentle measurement lemma (Lemma 5.8).*

In the quantum Stein's lemma (Theorem 8.15) we saw that the regularization of $D_{\min}^{\varepsilon}$ yields the Umegaki relative entropy. We show now that the same holds also for $D_{\min}^{(\varepsilon)}$. We will use this result later on when we discuss the uniqueness of the Umegaki relative entropy.

> **Theorem 10.10.** Let $\varepsilon \in (0, 1)$, and $\rho, \sigma \in \mathfrak{D}(A)$ with $\mathrm{supp}\,(\rho) \subseteq \mathrm{supp}\,(\sigma)$. Then,
>
> $$\lim_{n\to\infty} \frac{1}{n} D_{\min}^{(\varepsilon)}(\rho^{\otimes n}\|\sigma^{\otimes n}) = D(\rho\|\sigma). \tag{10.161}$$

**Proof**　From Theorem 10.9 we get that

$$\limsup_{n\to\infty} \frac{1}{n} D_{\min}^{(\varepsilon)}(\rho^{\otimes n}\|\sigma^{\otimes n}) \leqslant \limsup_{n\to\infty} \frac{1}{n} D_{\min}^{\varepsilon}(\rho^{\otimes n}\|\sigma^{\otimes n}) \tag{10.162}$$
$$\mathbf{Theorem\ 8.15}\!\rightarrow\; = D(\rho\|\sigma).$$

In order to prove the opposite inequality, we make use of the method of relative typical subspace introduced in Section 8.3.2. Set $\varepsilon, \delta \in (0, 1)$ and let $\Pi_{\delta}^{\mathrm{rel},n}$ be the projection to the relative typical subspace given in (8.75), $P_{\delta}^n$ be the projection to the $\delta$-typical subspace associated with $\rho$, and define

$$\rho_n := \frac{\Pi_{\delta}^{\mathrm{rel},n} P_{\delta}^n \rho^{\otimes n} P_{\delta}^n \Pi_{\delta}^{\mathrm{rel},n}}{\mathrm{Tr}\left[\Pi_{\delta}^{\mathrm{rel},n} P_{\delta}^n \rho^{\otimes n}\right]}. \tag{10.163}$$

Observe that for any $\delta_1, \delta_2 > 0$ we have for large enough $n$

$$\mathrm{Tr}\left[\Pi_{\delta}^{\mathrm{rel},n} P_{\delta}^n \rho^{\otimes n}\right] = \mathrm{Tr}\left[\Pi_{\delta}^{\mathrm{rel},n} \rho^{\otimes n}\right] - \mathrm{Tr}\left[\Pi_{\delta}^{\mathrm{rel},n} \left(I^{A^n} - P_{\delta}^n\right)\rho^{\otimes n}\right]$$
$$\geqslant \mathrm{Tr}\left[\Pi_{\delta}^{\mathrm{rel},n} \rho^{\otimes n}\right] - \mathrm{Tr}\left[\left(I^{A^n} - P_{\delta}^n\right)\rho^{\otimes n}\right] \tag{10.164}$$
$$\mathbf{(8.76),\ (8.55)}\!\rightarrow\; \geqslant 1 - \delta_1 - \delta_2,$$

where we use the properties of typical and relative-typical projectors. Taking $\delta_1 + \delta_2 = \varepsilon^2$ and using the gentle measurement lemma (see Lemma 5.8) we get that $\frac{1}{2}\|\rho_n - \rho^{\otimes n}\|_1 \leqslant \varepsilon$. Hence,

$$D_{\min}^{(\varepsilon)}(\rho^{\otimes n}\|\sigma^{\otimes n}) \geqslant D_{\min}(\rho_n\|\sigma^{\otimes n}) = -\log \mathrm{Tr}\left[\Pi_{\rho_n}\sigma^{\otimes n}\right]. \tag{10.165}$$

Now, we make two observations about $\Pi_{\rho_n}$: (1) It projects into a subspace of $\mathfrak{T}_\varepsilon^{\mathrm{rel}}(A^n)$, and (2) its trace is no greater than

$$\mathrm{Rank}(\rho_n) \leqslant \mathrm{Tr}\left[P_\delta^n\right] \leqslant 2^{n(H(A)_\rho+\delta)}. \tag{10.166}$$

Due to the first property, it follows by definition (8.75) of the relative-typical subspace that $\mathrm{Tr}\left[\Pi_{\rho_n}\sigma^{\otimes n}\right] \leqslant 2^{n(\mathrm{Tr}[\rho\log\sigma]+\delta)}\mathrm{Tr}\left[\Pi_{\rho_n}\right]$. Combining this with the above equation we conclude that for sufficiently large $n$

$$\begin{aligned}
D_{\min}^{(\varepsilon)}(\rho^{\otimes n}\|\sigma^{\otimes n}) &\geqslant -\log\left(2^{n(\mathrm{Tr}[\rho\log\sigma]+\delta)}\mathrm{Tr}\left[\Pi_{\rho_n}\right]\right) \\
&= -n\,(\mathrm{Tr}[\rho\log\sigma]+\delta) - \log\mathrm{Tr}\left[\Pi_{\rho_n}\right] \\
\mathbf{(10.166)}\rightarrow\ &\geqslant -n\,(\mathrm{Tr}[\rho\log\sigma]+\delta) - n\,(H(A)_\rho+\delta) \\
&= n\,(D(\rho\|\sigma)-2\delta)
\end{aligned} \tag{10.167}$$

Dividing both sides by $n$ and taking the limit $n\to\infty$ we conclude that

$$\liminf_{n\to\infty}\frac{1}{n}D_{\min}^{(\varepsilon)}(\rho^{\otimes n}\|\sigma^{\otimes n}) \geqslant D(\rho\|\sigma)-2\delta. \tag{10.168}$$

Since the above inequality holds for all $\delta > 0$, it must also hold for $\delta = 0$. This completes the proof. ∎

As a simple application of the result above, consider the smoothed max-entropy as defined in (10.148). Then, from Exercise 10.21 and the theorem above we get the following version of the AEP: For all $\varepsilon \in (0,1)$ and all $\rho \in \mathfrak{D}(A)$

$$\lim_{n\to\infty}\frac{1}{n}H_{\max}^\varepsilon(A^n)_{\rho^{\otimes n}} = H(A)_\rho, \tag{10.169}$$

where $H(A)_\rho$ is the von-Neumann entropy of $\rho$.

**Exercise 10.23.** *Prove this AEP version, and compare it with* (10.140) *for the case* $|B|=1$.

## 10.4.2 Smoothed Decoupling Theorem

In this subsection we show that the decoupling theorem as given in Theorem 7.8 can be expressed with smoothed quantities. In particular, we will replace the optimized conditional entropies $\tilde{H}_2^\uparrow(A|E)$ and $\tilde{H}_2^\uparrow(A|B)$ with the smoothed conditional min-entropies.

> **Decoupling Theorem (Smoothed Version)**
>
> **Corollary 10.5.** Let $\rho \in \mathfrak{D}_{\leqslant}(AE)$, $\mathcal{E} \in \mathrm{CP}_{\leqslant}(A \to B)$, and $\tau^{AB} := \frac{1}{|A|} J_{\mathcal{E}}^{AB}$, where $J_{\mathcal{E}}^{AB}$ is the Choi matrix of $\mathcal{E}^{A \to B}$. Then, for any $\varepsilon > 0$
>
> $$\int_{\mathfrak{U}(A)} dU^A \left\| \mathcal{E}^{A \to B}\left(U^A \rho^{AE} \left(U^A\right)^*\right) - \tau^B \otimes \rho^E \right\|_1 \leqslant 2^{-\frac{1}{2}\left(H_{\min}^{\varepsilon}(A|E)_\rho + H_{\min}^{\varepsilon}(A|B)_\tau\right)} + 8\varepsilon$$
>
> $$(10.170)$$
>
> where $\mathfrak{U}(A)$ is the group of all unitary matrices acting on $A$, and $\int_{\mathfrak{U}(A)} dU^A$ denotes the integral over the Haar measure on $\mathfrak{U}(A)$.

**Proof** This corollary concerns with the replacement of the terms involving $H_2$ in the decoupling theorem with the smoothed min-entropy. For this purpose, let $\tilde{\rho}^{AE}$ and $\tilde{\tau}^{AB}$ be such that $H_{\min}^{\varepsilon}(A|E)_\rho = H_{\min}(A|E)_{\tilde{\rho}}$ and $H_{\min}^{\varepsilon}(A|B)_\tau = H_{\min}(A|B)_{\tilde{\tau}}$. Note also that by definition $\|\rho^{AE} - \tilde{\rho}^{AE}\|_1 \leqslant 2\varepsilon$ and $\|\tau^{AB} - \tilde{\tau}^{AB}\|_1 \leqslant 2\varepsilon$. Denoting by $\tilde{\mathcal{E}}$ the CP map whose Choi matrix is $\tilde{\tau}^{AB}$ we get

$$2^{-\frac{1}{2}\left(H_{\min}^{\varepsilon}(A|E)_\rho + H_{\min}^{\varepsilon}(A|B)_\tau\right)} = 2^{-\frac{1}{2}\left(H_{\min}(A|E)_{\tilde{\rho}} + H_{\min}(A|B)_{\tilde{\tau}}\right)}$$

$$\boxed{H_{\min} \leqslant H_2} \longrightarrow \geqslant 2^{-\frac{1}{2}\left(H_2(A|E)_{\tilde{\rho}} + H_2(A|B)_{\tilde{\tau}}\right)}$$

$$\textbf{Theorem 7.8} \to \geqslant \int_{\mathfrak{U}(A)} dU^A \left\| \tilde{\mathcal{E}}^{A \to B}\left(U^A \tilde{\rho}^{AE} \left(U^A\right)^*\right) - \tilde{\tau}^B \otimes \tilde{\rho}^E \right\|_1$$

$$\textbf{See (10.172)} \to \geqslant \int_{\mathfrak{U}(A)} dU^A \left\| \tilde{\mathcal{E}}^{A \to B}\left(U^A \tilde{\rho}^{AE} \left(U^A\right)^*\right) - \tau^B \otimes \rho^E \right\|_1 - 4\varepsilon,$$

$$(10.171)$$

where in the last inequality we used the fact that $\eta := \tilde{\mathcal{E}}\left(U\tilde{\rho}U^*\right) \in \mathrm{Pos}(BE)$ satisfies

$$\|\eta - \tau \otimes \rho\|_1 = \|\eta - (\tilde{\tau} + \tau - \tilde{\tau}) \otimes \rho\|_1$$

$$\textbf{Triangle inequality} \to \leqslant \|\eta - \tilde{\tau} \otimes \rho\|_1 + \|(\tau - \tilde{\tau}) \otimes \rho\|_1$$

$$\boxed{\tau \approx_\varepsilon \tilde{\tau}} \longrightarrow \leqslant \|\eta - \tilde{\tau} \otimes (\tilde{\rho} + \rho - \tilde{\rho})\|_1 + 2\varepsilon \qquad (10.172)$$

$$\textbf{Triangle inequality} \to \leqslant \|\eta - \tilde{\tau} \otimes \tilde{\rho}\|_1 + \|\tilde{\tau} \otimes (\rho - \tilde{\rho})\|_1 + 2\varepsilon$$

$$\boxed{\rho \approx_\varepsilon \tilde{\rho}} \longrightarrow \leqslant \|\eta - \tilde{\tau} \otimes \tilde{\rho}\|_1 + 4\varepsilon.$$

Next, observe that

$$\tilde{\mathcal{E}}\left(U\tilde{\rho}U^*\right) = \mathcal{E}\left(U\rho U^*\right) + \mathcal{E}\left(U(\tilde{\rho} - \rho)U^*\right) + (\tilde{\mathcal{E}} - \mathcal{E})\left(U\tilde{\rho}U^*\right) \qquad (10.173)$$

so that from the triangle inequality of the trace norm we get

$$\left\| \tilde{\mathcal{E}}\left(U\tilde{\rho}U^*\right) - \tau \otimes \rho \right\|_1$$

$$\geqslant \left\| \mathcal{E}\left(U\rho U^*\right) - \tau \otimes \rho \right\|_1 - \left\| \mathcal{E}\left(U(\tilde{\rho} - \rho)U^*\right) \right\|_1 - \left\| (\tilde{\mathcal{E}} - \mathcal{E})\left(U\tilde{\rho}U^*\right) \right\|_1.$$

$$(10.174)$$

It is therefore left to bound the average of the last two terms over the group $\mathfrak{U}(A)$. Denote by $\eta_\pm := (\tilde{\rho}^{AE} -, \rho^{AE})_\pm$ and $\zeta_\pm := (\tilde{\tau}^{AB} -, \tau^{AB})_\pm$. Since $\tilde{\rho}$ and $\tilde{\tau}$ are $\varepsilon$-,close to $\rho$ and $\tau$, respectively, we have $\mathrm{Tr}[\eta_+ +, \eta_-,] \leqslant 2\varepsilon$ and $\mathrm{Tr}[\zeta_+ +, \zeta_-,] \leqslant 2\varepsilon$. Now, denote by $\mathcal{N}_\pm$ the CP maps whose Choi matrices are $\zeta_\pm$, respectively. We then have $\tilde{\mathcal{E}} - \mathcal{E} = \mathcal{N}_+ - \mathcal{N}_-$ and $\tilde{\rho} - \rho = \eta_+ - \eta_-$, so that

$$\int_{\mathfrak{U}(A)} dU \,\left\| \mathcal{E}\Big(U(\tilde{\rho} - \rho)U^*\Big) \right\|_1 = \int_{\mathfrak{U}(A)} dU \,\left\| \mathcal{E}\Big(U(\eta_+ - \eta_-)U^*\Big) \right\|_1$$

$$\textbf{Triangle inequality} \rightarrow \,\leqslant \int_{\mathfrak{U}(A)} dU \,\mathrm{Tr}\Big[\mathcal{E}\Big(U\eta_+U^*\Big)\Big]$$

$$+ \int_{\mathfrak{U}(A)} dU \,\mathrm{Tr}\Big[\mathcal{E}\Big(U\eta_-U^*\Big)\Big] \tag{10.175}$$

$$\boxed{\int_{\mathfrak{U}(A)} dU \, U^A \eta_\pm^{AE} U^{A*} = \mathbf{u}^A \otimes \eta_\pm^E} \longrightarrow \; = \mathrm{Tr}\,[\mathcal{E}(\mathbf{u})]\,\mathrm{Tr}\,[\eta_+] + \mathrm{Tr}\,[\mathcal{E}(\mathbf{u})]\,\mathrm{Tr}\,[\eta_-]$$

$$= \mathrm{Tr}[\tau]\big(\mathrm{Tr}\,[\eta_+] + \mathrm{Tr}[\eta_-]\big)$$

$$\leqslant 2\varepsilon.$$

Similarly,

$$\int_{\mathfrak{U}(A)} dU \,\left\| (\tilde{\mathcal{E}} - \mathcal{E})\Big(U\rho U^*\Big) \right\|_1 = \int_{\mathfrak{U}(A)} dU \,\left\| (\mathcal{N}_+ - \mathcal{N}_-)\Big(U\rho U^*\Big) \right\|_1$$

$$\leqslant \int_{\mathfrak{U}(A)} dU \,\mathrm{Tr}\Big[\mathcal{N}_+\Big(U\rho U^*\Big)\Big]$$

$$+ \int_{\mathfrak{U}(A)} dU \,\mathrm{Tr}\Big[\mathcal{N}_-\Big(U\rho U^*\Big)\Big] \tag{10.176}$$

$$= \mathrm{Tr}\Big[\mathcal{N}_+(\mathbf{u}^A) \otimes \rho^E\Big] + \mathrm{Tr}\Big[\mathcal{N}_-(\mathbf{u}^A) \otimes \rho^E\Big]$$

$$= \mathrm{Tr}[\zeta_+ + \zeta_-]\mathrm{Tr}[\rho^E]$$

$$\leqslant 2\varepsilon.$$

Combining everything we get

$$2^{-\frac{1}{2}\left(H_{\min}^\varepsilon(A|E)_\rho + H_{\min}^\varepsilon(A|B)_\tau\right)} \geqslant \int_{\mathfrak{U}(A)} dU \,\left\| \mathcal{E}\Big(U\rho U^*\Big) - \tau \otimes \rho \right\|_1 - 8\varepsilon. \tag{10.177}$$

This completes the proof. ∎

## 10.5 Resource Monotones and Support Functions

Let $\mathfrak{F}$ be a convex QRT and for any fixed $\eta \in \mathrm{Herm}(B)$ define the function $G_\eta : \bigcup_A \mathfrak{D}(A) \to \mathbb{R}$ as

$$G_\eta(\rho^A) := \sup_{\mathcal{E} \in \mathfrak{F}(A \to B)} \mathrm{Tr}\Big[\eta^B \mathcal{E}^{A \to B}(\rho^A)\Big] - \sup_{\omega \in \mathfrak{F}(B)} \mathrm{Tr}\Big[\eta^B \omega^B\Big] \qquad \forall \rho \in \mathfrak{D}(A). \tag{10.178}$$

> **Theorem 10.11.** For any $\eta \in \mathrm{Herm}(A)$ the function $G_\eta$ is a resource monotone.

**Proof**  We prove the following properties:

1. *Monotonicity.* Let $\mathcal{N} \in \mathfrak{F}(A \to A')$ and denote $c_\eta := \sup_{\omega \in \mathfrak{F}(B)} \mathrm{Tr}\left[\eta^B \omega^B\right]$. Then,

$$G_\eta\left(\mathcal{N}^{A \to A'}(\rho^A)\right) = \sup_{\mathcal{M} \in \mathfrak{F}(A' \to B)} \mathrm{Tr}\left[\eta^B \mathcal{M}^{A' \to B}\left(\mathcal{N}^{A \to A'}(\rho^A)\right)\right] - c_\eta$$

$$\boxed{\text{Replacing } \mathcal{M} \circ \mathcal{N} \text{ with } \mathcal{E}} \longrightarrow \leq \sup_{\mathcal{E} \in \mathfrak{F}(A \to B)} \mathrm{Tr}\left[\eta^B \mathcal{E}^{A \to B}\left(\rho^A\right)\right] - c_\eta$$

$$= G_\eta\left(\rho^A\right).$$

$$(10.179)$$

2. *Normalization.* Let $\sigma \in \mathfrak{F}(A)$. Then,

$$\sup_{\mathcal{E} \in \mathfrak{F}(A \to B)} \mathrm{Tr}\left[\eta^B \mathcal{E}^{A \to B}\left(\sigma^A\right)\right] = \sup_{\omega \in \mathfrak{F}(B)} \mathrm{Tr}\left[\eta^B \omega^B\right], \tag{10.180}$$

where $\omega^B = \mathcal{E}^{A \to B}\left(\sigma^A\right) \in \mathfrak{F}(B)$ can be taken to be any free state (by choosing $\mathcal{E}$ to be a replacement channel in $\mathfrak{F}(A \to B)$ that outputs $\omega^B$). Hence, $G_\eta\left(\sigma^A\right) = 0$ for all $\sigma \in \mathfrak{F}(A)$.

3. *Strong monotonicity.* Let $\mathcal{N} = \sum_{x \in [m]} \mathcal{N}_x \otimes |x\rangle\langle x| \in \mathfrak{F}(A \to A'X)$ be a free quantum instrument. Then, observe that for any free channel $\mathcal{M} \in \mathfrak{F}(A'X \to B)$ we have

$$\mathcal{M}^{A'X \to B} \circ \mathcal{N}^{A \to A'X} = \sum_{x \in [m]} \mathcal{M}_x^{A' \to B} \circ \mathcal{N}_x^{A \to A'}, \tag{10.181}$$

where $\mathcal{M}_x \in \mathfrak{F}(A' \to B)$ is defined by

$$\mathcal{M}_x^{A' \to B}(\omega^{A'}) := \mathcal{M}^{A'X \to B}\left(\omega^{A'} \otimes |x\rangle\langle x|^X\right) \qquad \forall\, \omega \in \mathfrak{L}(A'). \tag{10.182}$$

We therefore get that

$$G_\eta\left(\mathcal{N}^{A \to A'X}(\rho^A)\right) = \sup_{\mathcal{M} \in \mathfrak{F}(A'X \to B)} \mathrm{Tr}\left[\eta^B \mathcal{M}^{A'X \to B}\left(\mathcal{N}^{A \to A'X}(\rho^A)\right)\right] - c_\eta$$

$$= \sum_{x \in [m]} \sup_{\mathcal{M}_x \in \mathfrak{F}(A' \to B)} \mathrm{Tr}\left[\eta^B \mathcal{M}_x^{A' \to B}\left(\mathcal{N}_x^{A \to A'}(\rho^A)\right)\right] - c_\eta$$

$$= \sum_{x \in [m]} p_x G_\eta\left(\sigma_x^{A'}\right),$$

$$(10.183)$$

where $\sigma_x^{A'} := \frac{1}{p_x} \mathcal{N}_x^{A \to A'}(\rho^A)$ and $p_x := \mathrm{Tr}\left[\mathcal{N}_x^{A \to A'}(\rho^A)\right]$. Combining this with the monotonicity property of $G_\eta$ we conclude that

$$G_\eta\left(\rho^A\right) \geq \sum_{x \in [m]} p_x G_\eta\left(\sigma_x^{A'}\right). \tag{10.184}$$

4. *Convexity.* Let $\rho^{AX} := \sum_{x\in[m]} p_x \rho_x^A \otimes |x\rangle\langle x|^X$ be a cq-state in $\mathfrak{D}(AX)$. Using a similar argument as above we have

$$G_\eta\left(\rho^{AX}\right) = \sum_{x\in[m]} p_x G_\eta\left(\rho_x^A\right). \tag{10.185}$$

On the other hand, since the partial trace is a free operation we get that

$$G_\eta\left(\rho^A\right) \leqslant G_\eta\left(\rho^{AX}\right). \tag{10.186}$$

This equation is equivalent to

$$G_\eta\left(\sum_{x\in[m]} p_x \rho_x^A\right) \leqslant \sum_{x\in[m]} p_x G_\eta\left(\rho_x^A\right). \tag{10.187}$$

$\blacksquare$

Recall that the combination of monotonicity and normalization properties ensures that $G_\eta(\rho) \geqslant 0$ for all density matrices. Additionally, if we define $\mathfrak{C}_\rho := \mathcal{E}(\rho) : \mathcal{E} \in \mathfrak{F}(A \to B)$, then the *support function* of $\mathfrak{C}_\rho$ in the space of Hermitian matrices $\mathrm{Herm}(B)$ is described by

$$f_\rho(\eta) := \sup_{\omega\in\mathfrak{C}_\rho} \langle \omega^B, \eta^B \rangle = \sup_{\mathcal{E}\in\mathfrak{F}(A \to B)} \mathrm{Tr}\left[\eta^B \mathcal{E}^{A\to B}\left(\rho^A\right)\right]. \tag{10.188}$$

As we will explore later, this family of resource monotones is *complete*, meaning that it can be utilized to fully determine exact interconversions among resources. Furthermore, these monotones are formulated as conic linear programming problems, and in some QRTs, they reduce to semidefinite programming, which are comparatively simpler to compute.

## Example: Measures of Conditional Uncertainty

Consider a QRT in which the free operations are CMO channels (see Section 7.1.3). Given that conditional majorization can be determined by an SDP feasibility problem, it follows that also the corresponding support functions can be computed with an SDP. Recall that in this QRT the set of free states is given by

$$\mathfrak{F}(AB) = \left\{ \mathbf{u}^A \otimes \sigma^B \ : \ \sigma \in \mathfrak{D}(B) \right\}. \tag{10.189}$$

Therefore, for this QRT, for any $\eta \in \mathfrak{D}(AB')$ the coefficient $c_\eta$ is given by

$$c_\eta = \sup_{\sigma\in\mathfrak{D}(B')} \mathrm{Tr}\left[\eta^{AB'}\left(\mathbf{u}^A \otimes \sigma^{B'}\right)\right] = \frac{1}{|A|}\left\|\eta^{B'}\right\|_\infty. \tag{10.190}$$

Hence, for every $\eta \in \mathfrak{D}(AB')$ the function $G_\eta$ as defined above can be expressed as

$$G_\eta(A|B)_\rho := \sup_{\mathcal{N}\in\mathrm{CMO}(AB \to AB')} \mathrm{Tr}\left[\eta^{AB'}\mathcal{N}^{AB\to AB'}\left(\rho^{AB}\right)\right] - \frac{1}{|A|}\left\|\eta^{B'}\right\|_\infty. \tag{10.191}$$

We denote by $f_\eta$ the first term on the right-hand side above. In terms of the Choi matrix of $\mathcal{N}$, this function can be expressed as

$$f_\eta(A|B)_\rho = \sup_{J^{AB\tilde{A}B'}} \mathrm{Tr}\left[ J^{AB\tilde{A}B'} \left( \rho^{AB} \otimes \eta^{\tilde{A}B'} \right) \right], \tag{10.192}$$

where the maximization is over all Choi matrices of a CMO.

Recall from Exercise 7.4 that a Choi matrix $J^{AB\tilde{A}B'}$ is a Choi matrix of a CMO if and only if $\Upsilon\left( J^{AB\tilde{A}B'} \right) = 0$. Moreover, observe that the trace-preserving condition $J^{AB} = I^{AB}$ follows from $J^B = |A| I^B$ and $J^{ABB'} = \mathbf{u}^A \otimes J^{BB'}$. Hence,

$$f_\eta(A|B)_\rho = |A| \sup_{J^{AB\tilde{A}B'} \geq 0} \left\{ \mathrm{Tr}\left[ J^{AB\tilde{A}B'} \left( \rho^{AB} \otimes \eta^{\tilde{A}B'} \right) \right] : \Upsilon\left( J^{AB\tilde{A}B'} \right) = 0, \ J^B = I^B \right\}. \tag{10.193}$$

The above optimization problem can be solved with an SDP.

**Exercise 10.24.** *Use the strong duality relation of an SDP to show that the function $f_\eta$ can also be expressed as*

$$f_\eta(A|B)_\rho = |A| \inf_{\xi^{AB\tilde{A}B'} \geq 0} \left\{ \mathrm{Tr}\left[ \xi^B \right] : \mathbf{u}^{A\tilde{A}B'} \otimes \xi^B + \Upsilon\left( \xi^{AB\tilde{A}B'} \right) \geq \rho^{AB} \otimes \eta^{\tilde{A}B'} \right\}. \tag{10.194}$$

**Exercise 10.25.** *Show that if $|A| = |B|$, then for the maximally entangled state $\rho^{AB} = \Phi^{AB}$ $G_\eta$ as defined in (10.191) is given by*

$$G_\eta(A|B)_\Phi = \left\| \eta^{AB'} \right\|_\infty - \frac{1}{|A|} \left\| \eta^{B'} \right\|_\infty. \tag{10.195}$$

*Hint: Recall that for all states $\rho \in \mathfrak{D}(AB)$ we have $\Phi^{AB} \succ_A \rho^{AB}$.*

## 10.6 Notes and References

Measures of quantum resources in general QRTs where introduced in Ref. [30] and studied intensively Refs. [37] and [190]. The asymptotic continuity of the relative entropy of a resource as given in Ref. (10.34) is due to Ref. [237]. The robustness of a resource as defined here is sometimes called the *global* robustness and it was first introduced Ref. [114] for entanglement theory.

The converse problem introduced here for the relative entropy of a resource was introduced in Ref. [165] for the case of two-qubit entangled states, in Ref. [79] for all finite dimensions in entanglement theory, and finally in Ref. [84] for arbitrary QRTs. This technique was also used in Ref. [47] to compute the $\alpha$-relative entropy of entanglement for pure bipartite states. More additivity properties of the relative entropy of a resource can be found in Ref. [195].

The inequality (10.116) is due to Ref. [67], the inequality (10.117) is due to Ref. [231], and the inequality (10.118) is due to Ref. [67]. The decoupling theorem presented here is due to Ref. [68]. The relations between resource monotones and support functions as presented here is due to Ref. [95]. The streamlined proof of Theorem 10.9, as outlined here, was graciously shared with the author by Ryuji Takagi (private communication). An earlier version of the book had featured a substantially more complicated proof.

# 11

# Manipulation of Resources

One of the central goals of QRTs is to understand optimal and efficient ways to convert one resource to another. A resource in this context corresponds to a class of equivalent resource states. We say that two resource states $\rho, \sigma \in \mathfrak{D}(A)$ are equivalent if both $\rho \xrightarrow{\mathfrak{F}} \sigma$ (i.e. $\rho$ can be converted to $\sigma$ by free operations) and $\sigma \xrightarrow{\mathfrak{F}} \rho$. In this chapter, we study the conversion of resources in two regimes: the single-shot regime and the asymptotic regime.

## 11.1 Single-Shot Interconversions

The single-shot regime encompasses exact, probabilistic, and approximate interconversions. By "exact" interconversions, we refer to the transformation of a resource, such as $\rho$, into another target resource, like $\sigma$, with both 100% success probability and accuracy. However, in practical scenarios, it is frequently unfeasible to achieve such perfect conversion of $\rho$ to $\sigma$, necessitating a tolerance for slight errors. Additionally, we will discover that allowing for a small margin of error not only accommodates practical limitations but also provides theoretical insight. This flexibility facilitates a smoother transition from the single-shot regime to the asymptotic regime, highlighting the interconnectedness and practicality of these concepts in resource theory.

### 11.1.1 Exact Interconversions

Different QRTs have different sets of free operations. Consequently, one cannot expect to find a set of simple necessary and sufficient conditions that can be used in any QRT to determine the conversion of one resource $\rho$ to another resource $\sigma$. Still, in the following theorem, we show that there exists a common dual characterization for the problem of exact state conversion that is given in terms of the resource monotones discussed in Section 10.5. We will assume that $\mathfrak{F}$ is a closed convex QRT, meaning that for every two Hilbert spaces $A$ and $B$ the set $\mathfrak{F}(A \to B)$ is a closed convex set in the real vector space $\mathrm{Herm}(A \to B)$. In particular, $\mathfrak{F}(A)$ is a closed convex set in $\mathrm{Herm}(A)$.

> **Theorem 11.1.** Let $\mathfrak{F}$ be a closed convex QRT, $\rho \in \mathfrak{D}(A)$, and $\sigma \in \mathfrak{D}(B)$. The following are equivalent:
>
> 1. There exists $\mathcal{N} \in \mathfrak{F}(A \to B)$ such that $\sigma^B = \mathcal{N}^{A \to B}(\rho^A)$.
> 2. For all $\eta \in \mathfrak{D}(B)$,
>
> $$G_\eta\left(\rho^A\right) \geqslant G_\eta\left(\sigma^B\right), \tag{11.1}$$
>
> where $G_\eta$ are the resource monotones defined in (10.178).

**Proof** Let $\mathfrak{C}_\rho := \left\{\mathcal{E}^{A \to B}\left(\rho^A\right) : \mathcal{E} \in \mathfrak{F}(A \to B)\right\}$. Observe that $\mathfrak{C}_\rho$ is a convex set in $\mathrm{Herm}(B)$. From the hyperplane separation theorem (see Theorem A.1 of online version), $\sigma \notin \mathfrak{C}_\rho$ if and only if there exists a hyperplane $\eta \in \mathrm{Herm}(B)$ that separates them; that is,

$$\mathrm{Tr}\left[\eta^B \sigma^B\right] > \max_{\omega \in \mathfrak{C}_\rho} \mathrm{Tr}\left[\eta^B \omega^B\right]. \tag{11.2}$$

Alternatively, $\sigma \in \mathfrak{C}_\rho$ if and only if for *all* $\eta \in \mathrm{Herm}(B)$ we have

$$\begin{aligned}
\mathrm{Tr}\left[\eta^B \sigma^B\right] &\leqslant \max_{\omega \in \mathfrak{C}_\rho} \mathrm{Tr}\left[\eta^B \omega^B\right] \\
&= \max_{\mathcal{E} \in \mathfrak{F}(A \to B)} \mathrm{Tr}\left[\eta^B \mathcal{E}^{A \to B}\left(\rho^A\right)\right].
\end{aligned} \tag{11.3}$$

Note that if (11.3) holds for some $\eta \in \mathrm{Herm}(B)$, then it also holds if we replace $\eta^B$ with $\eta^B + cI^B$ and vice versa (here $c$ is any real number). Therefore, the equation holds for all $\eta \in \mathrm{Herm}(B)$ if and only if it holds for all $\eta \in \mathrm{Pos}(B)$. Similarly, by dividing both sides of the equation by $\mathrm{Tr}[\eta]$ we conclude that $\sigma \in \mathfrak{C}_\rho$ if and only if (11.3) holds for all density matrices $\eta \in \mathfrak{D}(B)$.

Now, observe that $\sigma \in \mathfrak{C}_\rho$ if and only if for all $\mathcal{M} \in \mathfrak{F}(B \to B)$ we have that $\mathcal{M}(\sigma) \in \mathfrak{C}_\rho$. To see this, suppose $\sigma \in \mathfrak{C}_\rho$ so that $\sigma^B = \mathcal{E}^{A \to B}\left(\rho^A\right)$ for some $\mathcal{E} \in \mathfrak{F}(A \to B)$. Then, $\mathcal{M}^{B \to B}\left(\sigma^B\right) = \mathcal{M}^{B \to B} \circ \mathcal{E}^{A \to B}\left(\rho^A\right)$ and since $\mathcal{M} \circ \mathcal{E} \in \mathfrak{F}(A \to B)$ we conclude that $\mathcal{M}(\sigma) \in \mathfrak{C}_\rho$. Conversely, if $\mathcal{M}(\sigma) \in \mathfrak{C}_\rho$ for all $\mathcal{M} \in \mathfrak{F}(B \to B)$, by taking the identity channel $\mathcal{M} = \mathrm{id}^B \in \mathfrak{F}(B \to B)$ we get immediately that $\sigma \in \mathfrak{C}_\rho$.

Finally, from (11.3) we get that for any $\mathcal{M} \in \mathfrak{F}(B \to B)$ we have $\mathcal{M}(\sigma) \in \mathfrak{C}_\rho$ if and only if for all $\eta \in \mathfrak{D}(B)$

$$\mathrm{Tr}\left[\eta^B \mathcal{M}^{B \to B}\left(\sigma^B\right)\right] \leqslant \max_{\mathcal{E} \in \mathfrak{F}(A \to B)} \mathrm{Tr}\left[\eta^B \mathcal{E}^{A \to B}\left(\rho^A\right)\right]. \tag{11.4}$$

Hence, $\sigma \in \mathfrak{C}_\rho$ if and only if Equation (11.4) holds for all $\mathcal{M} \in \mathfrak{F}(B \to B)$. Taking the maximum over all such $\mathcal{M} \in \mathfrak{F}(B \to B)$ we conclude that $\sigma \in \mathfrak{C}_\rho$ if and only if for all $\eta \in \mathfrak{D}(B)$

$$\max_{\mathcal{M} \in \mathfrak{F}(B \to B)} \mathrm{Tr}\left[\eta^B \mathcal{M}^{B \to B}\left(\sigma^B\right)\right] \leqslant \max_{\mathcal{E} \in \mathfrak{F}(A \to B)} \mathrm{Tr}\left[\eta^B \mathcal{E}^{A \to B}\left(\rho^A\right)\right]. \tag{11.5}$$

The proof is concluded by recognizing that the above inequality is equivalent to (11.1). $\blacksquare$

In general, the theorem above does not provide an efficient way to determine if one resource can be converted to another by free operations. This is the case even if the resource monotones $G_\eta$ themselves can be computed efficiently, as we need to check the conditions for all $\eta \in \mathfrak{D}(B)$. Therefore, instead, we can use (11.3) to conclude that $\rho^A$ can be converted to $\sigma^B$ by free operations if and only if

$$\min_{\eta \in \mathfrak{D}(B)} \max_{\mathcal{E} \in \mathfrak{F}(A \to B)} \mathrm{Tr}\left[\eta^B \left(\mathcal{E}^{A \to B}\left(\rho^A\right) - \sigma^B\right)\right] \geqslant 0. \tag{11.6}$$

For some QRTs, the optimization problem (11.6) is an SDP and therefore can be solved efficiently.

## 11.1.2 Stochastic (Probabilistic) Interconversions

Probabilistic interconversion of one resource to another can also be considered as an *exact* interconversion except that the conversion does not occur with a 100% success. Specifically, let $\rho \in \mathfrak{D}(A)$ and $\sigma \in \mathfrak{D}(B)$ be two resource states, and let $\mathrm{Pr}(\rho \xrightarrow{\mathfrak{F}} \sigma)$ be the maximum probability that $\rho$ can be converted to $\sigma$ by free operations. For some states it may be that $\mathrm{Pr}(\rho \xrightarrow{\mathfrak{F}} \sigma) = 0$, in which case $\rho$ cannot be converted to sigma even with small probability. In another extreme case, $\mathrm{Pr}(\rho \xrightarrow{\mathfrak{F}} \sigma) = 1$, meaning that $\rho$ can be converted to $\sigma$ deterministically. If $0 < \mathrm{Pr}(\rho \xrightarrow{\mathfrak{F}} \sigma) < 1$ we say that $\rho$ can be converted to $\sigma$ *stochastically*.

Any free probabilistic transformation can be characterized with a channel/instrument $\mathcal{E} \in \mathfrak{F}(A \to BX)$ of the form

$$\mathcal{E}^{A \to BX} = \sum_{x \in [n]} \mathcal{E}_x^{A \to B} \otimes |x\rangle\langle x|^X, \tag{11.7}$$

where each $\mathcal{E}_x \in \mathrm{CP}(A \to B)$ is trace nonincreasing such that $\sum_{x \in [m]} \mathcal{E}_x$ is trace preserving. From the axiom of free instruments we require that $\mathcal{E}_x^{A \to B}(\omega^A)/\mathrm{Tr}[\mathcal{E}_x^{A \to B}(\omega^A)]$ is a free state in $\mathfrak{F}(B)$ whenever $\omega \in \mathfrak{F}(A)$. If there exists such a free quantum instrument with the property that

$$\sigma^B = \frac{\mathcal{E}_x^{A \to B}(\rho^A)}{\mathrm{Tr}[\mathcal{E}_x^{A \to B}(\rho^A)]} \tag{11.8}$$

for some $x \in [n]$, then we say that $\rho$ can be converted to $\sigma$ by free operations with probability $p_x := \mathrm{Tr}[\mathcal{E}_x^{A \to B}(\rho^A)]$.

Recall that we used the notation $\mathfrak{F}_\leqslant(A \to B)$ to denote the set of trace nonincreasing CP maps in $\mathrm{CP}_\leqslant(A \to B)$ that are part of free quantum instruments. With this notation, $\rho$ can be converted to $\sigma$ by free operations with probability $p$ if and only if there exists $\mathcal{E} \in \mathfrak{F}_\leqslant(A \to B)$ such that $p = \mathrm{Tr}[\mathcal{E}(\rho)]$ and $\sigma = \mathcal{E}(\rho)/p$. Therefore, $\mathrm{Pr}(\rho \xrightarrow{\mathfrak{F}} \sigma)$ can be defined as

$$\mathrm{Pr}(\rho \xrightarrow{\mathfrak{F}} \sigma) := \max_{\mathcal{E} \in \mathfrak{F}_\leqslant(A \to B)} \left\{ \mathrm{Tr}[\mathcal{E}(\rho)] \ : \ \sigma = \frac{\mathcal{E}(\rho)}{\mathrm{Tr}[\mathcal{E}(\rho)]} \right\}. \tag{11.9}$$

We will now demonstrate that this probability is, in fact, a resource monotone.

> **Theorem 11.2.** Let $\sigma \in \mathfrak{D}(B)$ be such that $\sigma \notin \mathfrak{F}(B)$ (i.e. $\sigma$ is a resource state). Then, the function $f_\sigma : \mathfrak{D}(A) \to [0,1]$, defined via
>
> $$f_\sigma(\rho) := \Pr(\rho \xrightarrow{\mathfrak{F}} \sigma) \qquad \forall\, \rho \in \mathfrak{D}(A) \tag{11.10}$$
>
> is a resource measure satisfying the strong monotonicity property.

**Proof**  First observe that from the axiom of free instruments and the fact that $\sigma$ is a resource state, we must have $f_\sigma(\rho) = 0$ for all $\rho \in \mathfrak{F}(A)$. Next, we show that $f_\sigma$ is a resource measure. Let $\mathcal{N} \in \mathfrak{F}(A \to C)$ be a free channel. Let $\mathcal{M} \in \mathfrak{F}_{\leqslant}(C \to B)$ be an optimal free instrument satisfying

$$\Pr(\mathcal{N}(\rho) \xrightarrow{\mathfrak{F}} \sigma) = \operatorname{Tr}\left[\mathcal{M}(\mathcal{N}(\rho))\right]. \tag{11.11}$$

Define $\mathcal{E} := \mathcal{M} \circ \mathcal{N}$, and observe that $\mathcal{E} \in \mathfrak{F}_{\leqslant}(A \to B)$ and $\Pr(\mathcal{N}(\rho) \xrightarrow{\mathfrak{F}} \sigma) = \operatorname{Tr}[\mathcal{E}(\rho)]$. Hence, from the definition of $\Pr(\rho \xrightarrow{\mathfrak{F}} \sigma)$ in (11.9) we get

$$\Pr(\rho \xrightarrow{\mathfrak{F}} \sigma) \geqslant \Pr(\mathcal{N}(\rho) \xrightarrow{\mathfrak{F}} \sigma). \tag{11.12}$$

By definition, this is equivalent to $f_\sigma(\rho) \geqslant f_\sigma(\mathcal{N}(\rho))$. We therefore establish that $f_\sigma$ is a resource measure.

To prove strong monotonicity, let $\mathcal{N} \in \mathfrak{F}(A \to BY)$ and denote by

$$\tau^{BY} := \mathcal{N}^{A \to CY}(\rho^A) = \sum_{y \in [n]} t_y \tau_y^C \otimes |y\rangle\langle y|^Y, \tag{11.13}$$

where each $\tau_y \in \mathfrak{D}(C)$ and $\{t_y\}_{y \in [n]}$ is a probability distribution. From the monotonicity of $f_\sigma$ under free channels (in particular, under $\mathcal{N}$) we get

$$\begin{aligned}
f_\sigma\left(\rho^A\right) &\geqslant f_\sigma\left(\tau^{CY}\right) \\
&= \Pr\left(\sum_{y \in [n]} t_y \tau_y^C \otimes |y\rangle\langle y|^Y \xrightarrow{\mathfrak{F}} \sigma^B\right).
\end{aligned} \tag{11.14}$$

Let $\mathcal{E}^{(y)} \in \mathfrak{F}_{\leqslant}(C \to B)$ be an optimal trace nonincreasing CP map such that $\Pr\left(\tau_y^C \xrightarrow{\mathfrak{F}} \sigma^B\right) = \operatorname{Tr}\left[\mathcal{E}^{(y)}(\tau_y)\right]$ and $\mathcal{E}^{(y)}(\tau_y)$ is proportional to $\sigma$. We also define $\mathcal{M} \in \mathfrak{F}_{\leqslant}(CY \to B)$ as

$$\mathcal{M}^{CY \to B}\left(\omega^C \otimes |y\rangle\langle y|^Y\right) := \mathcal{E}^{(y)}\left(\omega^C\right) \qquad \forall\, \omega \in \mathfrak{L}(C) \quad \forall\, y \in [n]. \tag{11.15}$$

In Exercise 11.1 we will show that $\mathcal{M}^{CY \to B}$ is indeed an element of $\mathfrak{F}_{\leqslant}(CY \to B)$. By definition, the state $\mathcal{M}^{CY \to B}\left(\sum_{y \in [n]} t_y \tau_y^C \otimes |y\rangle\langle y|^Y\right)$ is proportional to $\sigma^B$.

Therefore,

$$
\Pr\!\left( \sum_{y\in[n]} t_y \tau_y^C \otimes |y\rangle\langle y|^Y \xrightarrow{\mathfrak{F}} \sigma^B \right) \geq \mathrm{Tr}\!\left[ \mathcal{M}^{CY\to B}\!\left( \sum_{y\in[n]} t_y \tau_y^C \otimes |y\rangle\langle y|^Y \right) \right]
$$

$$
= \sum_{y\in[n]} t_y \, \mathrm{Tr}\!\left[ \mathcal{E}^{(y)}(\tau_y) \right]
$$

$$
= \sum_{y\in[n]} t_y \Pr\!\left( \tau_y^C \xrightarrow{\mathfrak{F}} \sigma^B \right) \tag{11.16}
$$

$$
= \sum_{y\in[n]} t_y \, f_\sigma\!\left( \tau_y^C \right).
$$

Combining this with (11.14) we conclude that

$$
f_\sigma\!\left( \rho^A \right) \geq \sum_{y\in[n]} t_y \, f_\sigma\!\left( \tau_y^C \right). \tag{11.17}
$$

This completes the proof.        ∎

**Exercise 11.1.** *Show that* $\mathcal{M}^{CY\to B}$ *as defined in (11.15) belongs to* $\mathfrak{F}_{\leqslant}(CY\to B)$. *Hint: Define a free channel* $\mathcal{F}^{CY\to BX} = \sum_{x\in[m]} \mathcal{F}_x^{CY\to B} \otimes |x\rangle\langle x|^X$ *such that* $\mathcal{F}_1^{CY\to B} = \mathcal{M}^{CY\to B}$.

**Exercise 11.2.** *Let* $\rho \in \mathfrak{D}(A)$ *and* $\sigma \in \mathfrak{D}(B)$. *Show that for any* $\mathcal{M} \in \mathfrak{F}(B\to C)$

$$
\Pr\!\left( \rho \xrightarrow{\mathfrak{F}} \mathcal{M}(\sigma) \right) \geq \Pr\!\left( \rho \xrightarrow{\mathfrak{F}} \sigma \right). \tag{11.18}
$$

As already discussed, if $\rho^A$ can be converted by free operations to $\sigma^B$, with some nonzero probability, then there exists a free quantum instrument $\mathcal{E} = \sum_{x\in[m]} \mathcal{E}_x \otimes |x\rangle\langle x| \in \mathfrak{F}(A\to BX)$ such that $\mathcal{E}_1^{A\to B}(\rho^A)$ is proportional to $\sigma^B$. For every $x \in [m]$, let $p_x := \mathrm{Tr}\!\left[ \mathcal{E}_x^{A\to B}(\rho^A) \right]$, and for $x \geq 2$ let $\omega_x^B := \frac{1}{p_x}\mathcal{E}_x^{A\to B}(\rho^A)$. With this notation we have

$$
\mathcal{E}^{A\to BX}(\rho^A) = p_1 \sigma^B \otimes |1\rangle\langle 1|^X + \sum_{x=2}^{|X|} p_x \omega_x^B \otimes |x\rangle\langle x|^X. \tag{11.19}
$$

Now, let **M** be a resource measure that satisfies the strong monotonicity property. Then, by definition **M** satisfies

$$
\mathbf{M}(\rho^A) \geq p_1 \mathbf{M}(\sigma^B) + \sum_{x=2}^{|X|} p_x \mathbf{M}(\omega_x^B) \geq p_1 \mathbf{M}(\sigma^B). \tag{11.20}
$$

In other words, the probability $p_1$ to convert $\rho^A$ to $\sigma^B$ cannot exceed the ratio $\mathbf{M}(\rho^A)/\mathbf{M}(\sigma^B)$. Since this is true for all resource measures that satisfy the strong monotonicity property, we get that

$$
\Pr(\rho^A \xrightarrow{\mathfrak{F}} \sigma^B) \leqslant \inf_{\mathbf{M}} \frac{\mathbf{M}(\rho^A)}{\mathbf{M}(\sigma^B)}, \tag{11.21}
$$

where the infimum is over all resource measures, $\mathbf{M}$, that satisfy the strong monotonicity property. Moreover, from Theorem 11.2, for a fixed $\sigma$, the function $\mathbf{M}_\sigma(\omega^A) := \mathrm{Pr}(\omega^A \xrightarrow{\mathfrak{F}} \sigma^B)$ is itself a resource measure that satisfies the strong monotonicity property. Hence,

$$\inf_{\mathbf{M}} \frac{\mathbf{M}(\rho^A)}{\mathbf{M}(\sigma^B)} \leqslant \frac{\mathbf{M}_\sigma(\rho^A)}{\mathbf{M}_\sigma(\sigma^B)} = \mathbf{M}_\sigma(\rho^A) = \mathrm{Pr}(\rho^A \xrightarrow{\mathfrak{F}} \sigma^B). \tag{11.22}$$

We therefore arrive at the following corollary.

---

**Corollary 11.1.** Let $\rho \in \mathfrak{D}(A)$ and $\sigma \in \mathfrak{D}(B)$. Then,

$$\mathrm{Pr}(\rho^A \xrightarrow{\mathfrak{F}} \sigma^B) = \inf_{\mathbf{M}} \frac{\mathbf{M}(\rho^A)}{\mathbf{M}(\sigma^B)}, \tag{11.23}$$

where the infimum is over all resource measures, $\mathbf{M}$, that satisfy the strong monotonicity property.

---

## 11.1.3 Approximate Interconversion

In this section, we provide the precise definitions of cost and distillation of a resource in the single-shot regime. We start with the definitions of conversion distance and the "golden" unit of resource theories.

### The Conversion Distance

The conversion distance quantifies the proximity to which a resource state $\rho \in \mathfrak{D}(A)$ can be transformed into another resource state $\sigma \in \mathfrak{D}(B)$ using free operations (refer to Figure 11.1). Mathematically, it is defined as follows:

$$T\left(\rho \xrightarrow{\mathfrak{F}} \sigma\right) := \min_{\mathcal{E} \in \mathfrak{F}(A \to B)} \frac{1}{2} \left\| \sigma^B - \mathcal{E}^{A \to B}\left(\rho^A\right) \right\|_1. \tag{11.24}$$

It's evident that the conversion distance is zero if $\rho \xrightarrow{\mathfrak{F}} \sigma$ is achievable. However, deterministic conversion from $\rho$ to $\sigma$ is often not feasible, raising the question of how closely $\sigma$ can be approximated by applying free operations to $\rho$. As such, conversion distance not only provides a meaningful way to evaluate the efficiency of these conversions but, as the following lemma demonstrates, also serves as a resource measure in its own right.

---

**Lemma 11.1.** Let $\rho \in \mathfrak{D}(A)$, $\sigma \in \mathfrak{D}(B)$, $\mathcal{M} \in \mathfrak{F}(A \to A')$, and $\mathcal{N} \in \mathfrak{F}(B \to B')$. Then,

$$T\left(\mathcal{M}(\rho) \xrightarrow{\mathfrak{F}} \sigma\right) \geqslant T\left(\rho \xrightarrow{\mathfrak{F}} \sigma\right) \quad \text{and} \quad T\left(\rho \xrightarrow{\mathfrak{F}} \mathcal{N}(\sigma)\right) \leqslant T\left(\rho \xrightarrow{\mathfrak{F}} \sigma\right). \tag{11.25}$$

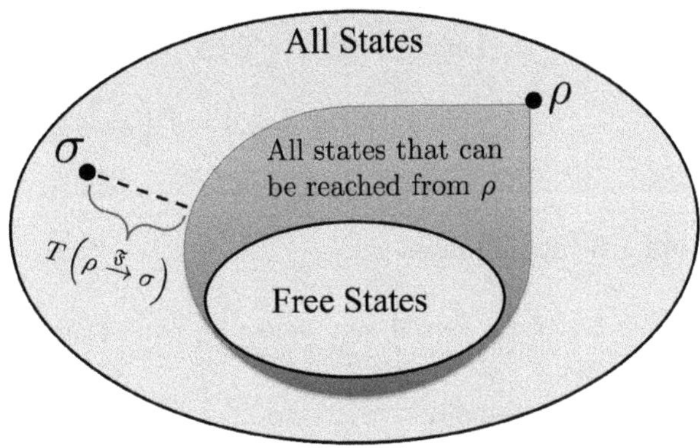


**Figure 11.1** The conversion distance from $\rho$ to $\sigma$.

**Proof**  For the first inequality, by definition, for any $\mathcal{M} \in \mathfrak{F}(A \to A')$

$$T\left(\mathcal{M}(\rho) \xrightarrow{\mathfrak{F}} \sigma\right) = \min_{\mathcal{E}' \in \mathfrak{F}(A' \to B)} \frac{1}{2} \left\| \sigma - \mathcal{E}' \circ \mathcal{M}(\rho) \right\|_1$$

**Replacing $\mathcal{E}' \circ \mathcal{M}$ with $\mathcal{E} \to$** $\geqslant \min_{\mathcal{E} \in \mathfrak{F}(A \to B)} \frac{1}{2} \left\| \sigma - \mathcal{E}(\rho) \right\|_1$     (11.26)

$$= T\left(\rho \xrightarrow{\mathfrak{F}} \sigma\right).$$

For the second inequality, we have for any $\mathcal{N} \in \mathfrak{F}(B \to B')$

$$T\left(\rho \xrightarrow{\mathfrak{F}} \mathcal{N}(\sigma)\right) = \min_{\mathcal{E}' \in \mathfrak{F}(A \to B')} \frac{1}{2} \left\| \mathcal{N}(\sigma) - \mathcal{E}'(\rho) \right\|_1$$

**Taking $\mathcal{E}' = \mathcal{N} \circ \mathcal{E} \to$** $\leqslant \min_{\mathcal{E} \in \mathfrak{F}(A \to B)} \frac{1}{2} \left\| \mathcal{N}(\sigma) - \mathcal{N} \circ \mathcal{E}(\rho) \right\|_1$

                     (11.27)

**DPI $\to$** $\leqslant \min_{\mathcal{E} \in \mathfrak{F}(A \to B)} \frac{1}{2} \left\| \sigma - \mathcal{E}(\rho) \right\|_1$

$$= T\left(\rho \xrightarrow{\mathfrak{F}} \sigma\right).$$

∎

**Exercise 11.3.** *Let $\mathfrak{F}$ be a QRT, $\rho, \sigma \in \mathfrak{D}(A)$, $\varepsilon \in [0, 1]$, and $k \in \mathbb{N}$. Show that if* $T(\rho \xrightarrow{\mathfrak{F}} \sigma) \leqslant \varepsilon$, *then*

$$T\left(\rho^{\otimes k} \xrightarrow{\mathfrak{F}} \sigma^{\otimes k}\right) \leqslant k\varepsilon. \tag{11.28}$$

The subsequent lemma highlights an additional property of the conversion distance: small changes in $\rho$ result in only minor variations in the conversion distance. This property underscores the stability of the conversion distance measure against slight perturbations in the resource state.

> **Lemma 11.2.** Let $\varepsilon \in (0, 1)$, $\rho \in \mathfrak{D}(A)$, $\sigma \in \mathfrak{D}(B)$, and $\tilde{\rho} \in \mathfrak{B}_\varepsilon(\rho)$. Then,
>
> $$\left| T\left(\rho \xrightarrow{\mathfrak{F}} \sigma\right) - T\left(\tilde{\rho} \xrightarrow{\mathfrak{F}} \sigma\right) \right| \leqslant \varepsilon. \tag{11.29}$$

**Proof**   By definition we get

$$
\begin{aligned}
T\left(\tilde{\rho} \xrightarrow{\mathfrak{F}} \sigma\right) &= \min_{\mathcal{E} \in \mathfrak{F}(A \to B)} \frac{1}{2} \|\sigma - \mathcal{E}(\tilde{\rho})\|_1 \\
\text{Triangle inequality} \to \;\geqslant\; & \min_{\mathcal{E} \in \mathfrak{F}(A \to B)} \left\{ \frac{1}{2} \|\sigma - \mathcal{E}(\rho)\|_1 - \frac{1}{2} \|\mathcal{E}(\tilde{\rho} - \rho)\|_1 \right\} \\
\text{DPI} \to \;\geqslant\; & \min_{\mathcal{E} \in \mathfrak{F}(A \to B)} \frac{1}{2} \|\sigma - \mathcal{E}(\rho)\|_1 - \frac{1}{2} \|\tilde{\rho} - \rho\|_1 \\
\geqslant\; & T\left(\rho \xrightarrow{\mathfrak{F}} \sigma\right) - \varepsilon.
\end{aligned}
\tag{11.30}
$$

The proof is concluded by repeating the same lines as above after exchanging between $\rho$ with $\tilde{\rho}$.  ∎

## The Golden Unit

Most resource theories contain a special type of a resource state, that we call here a "golden unit." For example, in entanglement theory, the maximally entangled state $\Phi^{AB} \in \mathfrak{D}(AB)$ with $|A| = |B|$ is a golden unit. Maximally entangled states are desirable since they can be used to accomplish quantum information processesing tasks such as quantum teleportation and superdense coding. Maximally entangled states have the property that they are closed under tensor products; that is, a tensor product of two maximally entangled states is itself a maximally entangled state. This motivates us to extend this property to all resource theories.

In the following definition, for any integer $m \in \mathbb{N}$, we denote $\Phi_m \in \mathfrak{D}(A)$, where $m$ is defined as $m \equiv |A|$, to represent a resource state, indicating that $\Phi_m$ does not belong to the set of free states $\mathfrak{F}(A)$. Importantly, we do not presuppose its specific form (such as assuming it to be the maximally entangled state, which is common in entanglement theory). Instead, we focus on outlining the necessary properties it must satisfy. Additionally, we use the equivalence notation $\rho \sim \sigma$ to signify that both conversions $\rho \xrightarrow{\mathfrak{F}} \sigma$ and $\sigma \xrightarrow{\mathfrak{F}} \rho$ are possible. This notation helps clarify the relationship between resource states in terms of their interconvertibility within the given resource theory.

> **Definition 11.1.** The sequence of resource states $\{\Phi_m\}_{m \in \mathbb{N}}$ is called *a golden unit* if for all $m, n \in \mathbb{N}$ the following two conditions hold:
>
> 1. $\Phi_n \otimes \Phi_m \sim \Phi_{nm}$.
> 2. If $n \geqslant m$ then, $\Phi_n \xrightarrow{\mathfrak{F}} \Phi_m$.

A golden unit can be used as a scale to measure the resourcefulness of a given state $\rho \in \mathfrak{D}(A)$. There are two distinct ways to do that.

---

**Definition 11.2.** Let $\rho \in \mathfrak{D}(A)$, $\{\Phi_m\}_{m \in \mathbb{N}}$ be *a golden unit*, and $\varepsilon \in (0, 1)$.

1. The $\varepsilon$-single-shot resource cost of $\rho$ is defined as

$$\mathrm{Cost}^\varepsilon(\rho) := \min \left\{ \log m \;:\; T\left( \Phi_m \xrightarrow{\mathfrak{F}} \rho \right) \leqslant \varepsilon. \quad m \in \mathbb{N} \right\}. \tag{11.31}$$

2. The $\varepsilon$-single-shot distillable resource of $\rho$ is defined as

$$\mathrm{Distill}^\varepsilon(\rho) := \max \left\{ \log m \;:\; T\left( \rho \xrightarrow{\mathfrak{F}} \Phi_m \right) \leqslant \varepsilon. \quad m \in \mathbb{N} \right\}. \tag{11.32}$$

---

*Remark.* Both the $\varepsilon$-resource cost and the $\varepsilon$-distillable resource are defined relative to the golden unit $\{\Phi_m\}_{m \in \mathbb{N}}$. If there is no $m \in \mathbb{N}$ such that $T\left( \Phi_m \xrightarrow{\mathfrak{F}} \rho \right) \leqslant \varepsilon$, then we define $\mathrm{Cost}^\varepsilon(\rho) := \infty$. On the other hand, for the trivial dimension $m = 1$ we must have $\Phi_m = 1$ so that in this case $T\left( \rho \xrightarrow{\mathfrak{F}} \Phi_m \right) = 0$. Hence, trivially, there is always $m \in \mathbb{N}$ such that $T\left( \rho \xrightarrow{\mathfrak{F}} \Phi_m \right) \leqslant \varepsilon$.

**Exercise 11.4.** *Let $\mathcal{E} \in \mathfrak{F}(A \to B)$ and $\rho \in \mathfrak{D}(A)$. Show that*

$$\mathrm{Cost}^\varepsilon(\mathcal{E}(\rho)) \leqslant \mathrm{Cost}^\varepsilon(\rho) \quad and \quad \mathrm{Distill}^\varepsilon(\mathcal{E}(\rho)) \leqslant \mathrm{Distill}^\varepsilon(\rho). \tag{11.33}$$

*Hint: Use Lemma 11.1.*

**Exercise 11.5.** *Let $\rho \in \mathfrak{D}(A)$. Show that*

$$\mathrm{Cost}^\varepsilon(\rho) = \min_{\rho' \in \mathfrak{B}_\varepsilon(\rho)} \mathrm{Cost}^{\varepsilon=0}(\rho'). \tag{11.34}$$

*That is, the $\varepsilon$-single-shot cost can be seen as the smoothed version of its respective zero-error counterpart. Why something similar does not hold for $\mathrm{Distill}^\varepsilon(\rho)$?*

In some resource theories there exists a golden unit $\{\Phi_m\}_{m \in \mathbb{N}}$ with a property that

$$\kappa := \max_{\omega \in \mathfrak{D}(A)} D(\omega \| \mathfrak{F}) = D(\Phi_m \| \mathfrak{F}) = \log(m). \tag{11.35}$$

In such QRTs, one can use the asymptotic continuity of the relative entropy of a resource to obtain an upper bound on the single-shot $\varepsilon$-distillable resource of some resource state $\rho \in \mathfrak{D}(A)$. Specifically, let $m \in \mathbb{N}$ be such that $T\left( \rho \xrightarrow{\mathfrak{F}} \Phi_m \right) \leqslant \varepsilon$. Then, for such $m$ there exists $\sigma \in \mathfrak{D}(A')$ with $m := |A'|$ such that $\rho \xrightarrow{\mathfrak{F}} \sigma$ and $\sigma \approx_\varepsilon \Phi_m$. Now, since the relative entropy of a resource is an entanglement monotone we get

$$D(\rho\|\mathfrak{F}) \geqslant D(\sigma\|\mathfrak{F})$$

$$\textbf{(10.34)} \rightarrow \; \geqslant D\left(\Phi_m\|\mathfrak{F}\right) - \varepsilon\kappa - (1+\varepsilon)h\left(\frac{\varepsilon}{1+\varepsilon}\right). \tag{11.36}$$

In many resource theories there exists a golden unit $\{\Phi_m\}_{m\in\mathbb{N}}$ with a property that $\kappa = D\left(\Phi_m\|\mathfrak{F}\right) = \log(m)$. Therefore, for such resource theories the inequality (11.36) takes the form

$$(1-\varepsilon)\log(m) \leqslant D(\rho\|\mathfrak{F}) + (1+\varepsilon)h\left(\frac{\varepsilon}{1+\varepsilon}\right). \tag{11.37}$$

Since $m$ was an arbitrary integer satisfying $T\left(\rho \xrightarrow{\mathfrak{F}} \Phi_m\right) \leqslant \varepsilon$, the inequality above implies that

$$\mathrm{Distill}^{\varepsilon}(\rho) \leqslant \frac{1}{1-\varepsilon}D(\rho\|\mathfrak{F}) + \frac{1+\varepsilon}{1-\varepsilon}h\left(\frac{\varepsilon}{1+\varepsilon}\right). \tag{11.38}$$

Note that for a small $\varepsilon > 0$ the upper bound is close to the relative entropy of a resource.

## 11.2 Generalized Asymptotic Equipartition Property

In Section 8.1.1, we saw that given an i.i.d.$\sim \mathbf{p}$ source, all typical sequences with large size $n$, that are generated by the source, have approximately the same probability to occur given by $\approx 2^{-nH(\mathbf{p})}$. This phenomenon is known as the asymptotic equipartition property (AEP). In Section 8.3.1, we saw a variant of this property that involved the relative entropy. In this subsection, we generalize further this property, and express it in terms of the relative entropy of a resource.

For any physical system $A$, let $\mathfrak{F}(A) \subseteq \mathfrak{D}(A)$ be a convex closed subset of density matrices. Recall that the relative entropy of a resource and the logarithmic robustness are defined, respectively, as

$$D(\rho\|\mathfrak{F}) := \min_{\sigma\in\mathfrak{F}(A)} D\left(\rho\|\sigma\right),$$

$$D_{\max}(\rho\|\mathfrak{F}) := \min_{\sigma\in\mathfrak{F}(A)} D_{\max}\left(\rho\|\sigma\right). \tag{11.39}$$

Recall also that for any $\varepsilon > 0$ the smoothed version of the logarithmic robustness is defined as

$$D_{\max}^{\varepsilon}(\rho\|\mathfrak{F}) := \min_{\rho'\in\mathfrak{B}_{\varepsilon}(\rho)} D_{\max}(\rho'\|\mathfrak{F}), \tag{11.40}$$

and the regularization of $D\left(\cdot\|\mathfrak{F}\right)$ as

$$D^{\mathrm{reg}}(\rho\|\mathfrak{F}) := \lim_{n\to\infty} \frac{1}{n}D\left(\rho^{\otimes n}\|\mathfrak{F}\right). \tag{11.41}$$

From the following exercise it follows that $D^{\mathrm{reg}}(\rho\|\mathfrak{F})$ is well defined since the limit on the right-hand side of the equation (11.41) exists.

**Exercise 11.6.**

1. *Show that for any sequence of real numbers, $\{a_j\}_{n=1}^\infty$, that is sub-additive, that is, $a_{n+m} \leqslant a_n + a_m$, the limit $\lim_{n\to\infty} a_n$ exists. Hint: See the hint given in Exercise 6.37.*
2. *Show that the limit of the sequence $\{a_n\}$, with $a_n := \frac{1}{n} D\left(\rho^{\otimes n} \| \mathfrak{F}\right)$, exists.*

> **The Generalized AEP**
>
> **Theorem 11.3.** Let $\mathfrak{F}$ be a QRT whose set of free states has the four properties listed below (11.57). Then, for all $\varepsilon \in (0,1)$ and all $\rho \in \mathfrak{D}(A)$,
>
> $$\lim_{n\to\infty} \frac{1}{n} D_{\max}^\varepsilon \left(\rho^{\otimes n} \| \mathfrak{F}\right) = D^{\mathrm{reg}}(\rho \| \mathfrak{F}). \tag{11.42}$$

*Remark.* At first glance it may not be very clear why Theorem 11.3 corresponds to the AEP property. Therefore, after the proof we will will give examples demonstrating that for different choices of $\mathfrak{F}(A)$, the theorem reduces to the various variants of the AEP studied in literature. In other words, it *unifies* all the variants of AEP into a single formula.

We divide the proof into two lemmas.

> **Lemma 11.3.** For any $\rho \in \mathfrak{D}(A)$ and $\varepsilon \in (0,1)$,
>
> $$D^{\mathrm{reg}}(\rho \| \mathfrak{F}) \geqslant \limsup_{n\to\infty} \frac{1}{n} D_{\max}^\varepsilon \left(\rho^{\otimes n} \| \mathfrak{F}\right). \tag{11.43}$$

**Proof**   From (10.118) we have for any $0 < \varepsilon < 1$

$$\limsup_{n\to\infty} \frac{1}{n} D_{\max}^\varepsilon \left(\rho^{\otimes n} \| \mathfrak{F}\right) = \limsup_{n\to\infty} \frac{1}{n} \min_{\sigma_n \in \mathfrak{F}(A^n)} D_{\max}^\varepsilon(\rho^{\otimes n} \| \sigma_n)$$

$$(10.118) \to \;\; \leqslant \limsup_{n\to\infty} \frac{1}{n} \min_{\sigma_n \in \mathfrak{F}(A^n)} D_{\min}^{\sqrt{1-\varepsilon^2}}(\rho^{\otimes n} \| \sigma_n) \tag{11.44}$$

$$(10.70) \to \;\; = \limsup_{n\to\infty} \frac{1}{n} D_{\min}^{\sqrt{1-\varepsilon^2}}(\rho^{\otimes n} \| \mathfrak{F}).$$

Combining this with the inequality (10.78), we then get for all $\alpha \in (1,2)$

$$\limsup_{n\to\infty} \frac{1}{n} D_{\max}^\varepsilon \left(\rho^{\otimes n} \| \mathfrak{F}\right) \leqslant \limsup_{n\to\infty} \frac{1}{n} D_\alpha \left(\rho^{\otimes n} \| \mathfrak{F}\right) = D_\alpha^{\mathrm{reg}}(\rho \| \mathfrak{F}). \tag{11.45}$$

where $D_\alpha(\,\cdot\, \| \mathfrak{F})$ is the $\alpha$-relative entropy of a resource as defined in (10.71). Finally, since the (11.45) inequality holds for all $\alpha \in (1,2)$ we conclude that for all $\varepsilon \in (0,1)$,

$$\limsup_{n\to\infty} \frac{1}{n} D_{\max}^\varepsilon \left(\rho^{\otimes n} \| \mathfrak{F}\right) \leqslant \lim_{\alpha \to 1^+} D_\alpha^{\mathrm{reg}}(\rho \| \mathfrak{F}) \tag{11.46}$$

$$\text{Lemma 10.4} \to \;\; = D^{\mathrm{reg}}(\rho \| \mathfrak{F}).$$

This completes the proof. $\blacksquare$

> **Lemma 11.4.** For any $\rho \in \mathfrak{D}(A)$,
> $$D^{\mathrm{reg}}(\rho\|\mathfrak{F}) \leqslant \liminf_{n\to\infty} \frac{1}{n} D^{\varepsilon}_{\max}\left(\rho^{\otimes n}\|\mathfrak{F}\right). \tag{11.47}$$

**Proof**  Recall the relation (10.116) from Lemma 10.7. The relation (10.116) implies that for every $\varepsilon_1, \varepsilon_2 \in (0,1)$ with $\varepsilon_1 + \varepsilon_2 < 1$ we have

$$D^{\varepsilon_1}_{\max}(\rho\|\mathfrak{F}) \geqslant D^{\varepsilon_2}_{\min}(\rho\|\mathfrak{F}) + \log\left(1 - \varepsilon_1 - \varepsilon_2\right). \tag{11.48}$$

This in turn gives

$$\liminf_{n\to\infty} \frac{1}{n} D^{\varepsilon_1}_{\max}\left(\rho^{\otimes n}\|\mathfrak{F}\right) \geqslant \liminf_{n\to\infty} \frac{1}{n} D^{\varepsilon_2}_{\min}\left(\rho^{\otimes n}\|\mathfrak{F}\right)$$
$$\text{Theorem } 11.4\to \; = D^{\mathrm{reg}}(\rho\|\mathfrak{F}), \tag{11.49}$$

where we get the generalized quantum Stein's lemma (Theorem 11.4) that will be proved in the next subsection. This completes the proof. ∎

## 11.2.1 Two Examples of AEP

In this subsection, we give two simple examples of QRTs that satisfy the AEP property. We start with the case that for each $n \in \mathbb{N}$ the set $\mathfrak{F}(A^n)$ consists of the single state $\sigma^{\otimes n}$, where $\sigma$ is some fixed state in $\mathfrak{D}(A)$. Note that in this case, $D^{\mathrm{reg}}(\rho\|\mathfrak{F}) = D(\rho\|\sigma)$ and $D^{\varepsilon}_{\max}(\rho\|\mathfrak{F}) = D^{\varepsilon}_{\max}(\rho\|\sigma)$. We therefore get the following corollary.

> **Corollary 11.2.** Let $\rho, \sigma \in \mathfrak{D}(A)$. Then, for any $\varepsilon \in (0,1)$,
> $$D(\rho\|\sigma) = \lim_{n\to\infty} \frac{1}{n} D^{\varepsilon}_{\max}\left(\rho^{\otimes n}\|\sigma^{\otimes n}\right). \tag{11.50}$$

To see how this corollary relates the AEP discussed in Section 8.1.1, take $\sigma^A = \mathbf{u}^A$ and observe that $D(\rho\|\mathbf{u}) = \log|A| - H(\rho)$ and $D^{\varepsilon}_{\max}(\rho\|\mathbf{u}) = \log|A| - H^{\varepsilon}_{\min}(\rho)$, where

$$H^{\varepsilon}_{\min}(\rho) := \max_{\rho' \in \mathfrak{B}_{\varepsilon}(\rho)} H_{\min}(\rho') \tag{11.51}$$

is known as the *smoothed min-entropy* of $\rho$. Therefore, the corollary above implies in particular that for any $\varepsilon \in (0,1)$,

$$\lim_{n\to\infty} \frac{1}{n} H^{\varepsilon}_{\min}\left(\rho^{\otimes n}\right) = H(\rho). \tag{11.52}$$

Recall that $H_{\min}(\rho) = -\log \lambda_{\max}(\rho)$ so that (11.52) states that for any $\varepsilon > 0$ and sufficiently large $n$, there exists a state $\rho'_n \in \mathfrak{D}(A^n)$ that is $\varepsilon$-close to $\rho^{\otimes n}$, and for which $\lambda_{\max}(\rho'_n) \approx 2^{-nH(\rho)}$. In other words, it only requires a small perturbation to make all of the eigenvalues of $\rho^{\otimes n}$ to be bounded from above by $2^{-nH(\rho)}$.

The second example we consider here is a variant of the AEP involving the conditional entropy. In this variant, we take the set of free states to be

$$\mathfrak{F}(AB) := \left\{ \mathbf{u}^A \otimes \rho^B \ : \ \rho \in \mathfrak{D}(A) \right\}. \tag{11.53}$$

With this set of free states we get

$$
\begin{aligned}
D\left(\rho^{AB} \,\middle\|\, \mathfrak{F}\right) &:= \min_{\sigma \in \mathfrak{F}(AB)} D\left(\rho^{AB} \,\middle\|\, \sigma^{AB}\right) \\
&= \min_{\sigma \in \mathfrak{D}(B)} D\left(\rho^{AB} \,\middle\|\, \mathbf{u}^A \otimes \sigma^B\right)
\end{aligned} \tag{11.54}
$$

$$\textbf{(7.132)}\rightarrow = \log |A| - H(A|B)_\rho$$

$$\textbf{Additivity of the conditional entropy}\rightarrow = D^{\mathrm{reg}}\left(\rho^{AB} \,\middle\|\, \mathfrak{F}\right).$$

Similarly,

$$
\begin{aligned}
D_{\max}\left(\rho^{AB} \,\middle\|\, \mathfrak{F}\right) &:= \min_{\substack{\sigma \in \mathfrak{F}(AB) \\ \tilde{\rho} \in \mathfrak{B}_\varepsilon(\rho)}} D_{\max}\left(\tilde{\rho}^{AB} \,\middle\|\, \sigma^{AB}\right) \\
&= \min_{\substack{\sigma \in \mathfrak{D}(B) \\ \tilde{\rho} \in \mathfrak{B}_\varepsilon(\rho)}} D_{\max}\left(\tilde{\rho}^{AB} \,\middle\|\, \mathbf{u}^A \otimes \sigma^B\right) \\
&= \min_{\tilde{\rho} \in \mathfrak{B}_\varepsilon(\rho)} \left\{ \log |A| - H_{\min}^{\uparrow}(A|B)_{\tilde{\rho}} \right\} \\
&= \log |A| - H_{\min}^{\uparrow\varepsilon}(A|B)_\rho.
\end{aligned} \tag{11.55}
$$

We can therefore apply Theorem 11.3 to get the following corollary.

---

**Corollary 11.3.** Let $\varepsilon \in (0,1)$ and $\rho \in \mathfrak{D}(AB)$. Then,

$$H(A|B)_\rho = \lim_{n \to \infty} \frac{1}{n} H_{\min}^{\uparrow\varepsilon}(A^n|B^n)_{\rho^{\otimes n}}. \tag{11.56}$$

---

**Exercise 11.7.** *Prove the corollary.*

## 11.3 The Generalized Quantum Stein's Lemma

In this section, we consider a generalization of the quantum Stein's lemma given in (8.211) by optimizing over $\sigma \in \mathfrak{F}(A)$. Specifically, recall the hypothesis-testing relative entropy of a resource defined in (10.70) as

$$D_{\min}^{\varepsilon}(\rho \| \mathfrak{F}) := \min_{\sigma \in \mathfrak{F}(A)} D_{\min}^{\varepsilon}(\rho \| \sigma). \tag{11.57}$$

The following theorem applies for a QRT whose set of free states $\mathfrak{F}_n := \mathfrak{F}(A^n)$ has the following properties for all $n, m \in \mathbb{N}$:

1. The set $\mathfrak{F}_n$ is closed and convex.
2. The set $\mathfrak{F}_n$ is closed under permutations of the subsystems of $A^n$.
3. If $\sigma \in \mathfrak{F}_n$ and $\omega \in \mathfrak{F}_m$, then $\sigma \otimes \omega \in \mathfrak{F}_{nm}$.
4. There exists $\gamma \in \mathfrak{F}_1$ with $\gamma > 0$.

**The Generalized Quantum Stein's Lemma**

**Theorem 11.4.** Let $\mathfrak{F}$ be a QRT whose sets of free states satisfy the properties above. Then, for all $\varepsilon \in (0,1)$ and all $\rho \in \mathfrak{D}(A)$

$$\lim_{n\to\infty} \frac{1}{n} D_{\min}^{\varepsilon}\left(\rho^{\otimes n} \big\| \mathfrak{F}\right) = D^{\mathrm{reg}}(\rho \| \mathfrak{F}). \tag{11.58}$$

Recall from Lemma 10.7 that for any $\varepsilon_1, \varepsilon_2 \in (0,1)$ with $\varepsilon_1 + \varepsilon_2 < 1$ we have (cf. (10.116))

$$D_{\min}^{\varepsilon_1}(\rho \| \sigma) \leqslant D_{\max}^{\varepsilon_2}(\rho \| \sigma) - \log(1 - \varepsilon_1 - \varepsilon_2). \tag{11.59}$$

Then, from the definitions it follows that for any such $\varepsilon_1, \varepsilon_2 \in (0,1)$ with $\varepsilon_1 + \varepsilon_2 < 1$ we have

$$D_{\min}^{\varepsilon_1}(\rho \| \mathfrak{F}) \leqslant D_{\max}^{\varepsilon_2}(\rho \| \mathfrak{F}) - \log(1 - \varepsilon_1 - \varepsilon_2), \tag{11.60}$$

so that

$$\limsup_{n\to\infty} \frac{1}{n} D_{\min}^{\varepsilon_1}\left(\rho^{\otimes n} \big\| \mathfrak{F}\right) \leqslant \limsup_{n\to\infty} \frac{1}{n} D_{\max}^{\varepsilon_2}\left(\rho^{\otimes n} \big\| \mathfrak{F}\right) \tag{11.61}$$

$$\text{\textbf{Lemma 11.43}} \rightarrow \ \leqslant D^{\mathrm{reg}}(\rho \| \mathfrak{F}).$$

This provides a proof for the strong converse of the theorem above (note that we already showed it in (10.85) using a different approach). We therefore focus now on the opposite inequality; specifically, we would like to prove that

$$s := \liminf_{n\to\infty} \frac{1}{n} D_{\min}^{\varepsilon}\left(\rho^{\otimes n} \big\| \mathfrak{F}\right) \geqslant D^{\mathrm{reg}}(\rho \| \mathfrak{F}). \tag{11.62}$$

To prove the inequality given in (11.62), we will develop several tools. We begin with a property of the hypothesis testing divergence. For every two operators $\Gamma, \Lambda \in \mathrm{Herm}(A)$ we use the notation $\{\Lambda \geqslant \Gamma\}$ to denote the orthogonal projection to the positive part of $\Lambda - \Gamma$; in other words, $\{\Lambda \geqslant \Gamma\}$ denotes the projection to the support of $(\Lambda - \Gamma)_+$.

**Lemma 11.5.** Let $\varepsilon, \delta \in (0,1)$, and let $\{\rho_n\}_{n\in\mathbb{N}}$ and $\{\sigma_n\}_{n\in\mathbb{N}}$ be two sets of density matrices, with $\rho_n, \sigma_n \in \mathfrak{D}(A_n)$, where $A_n$ is a finite dimensional Hilbert space whose dimension can depend on $n$. If

$$t := \liminf_{n\to\infty} \frac{1}{n} D_{\min}^{\varepsilon}(\rho_n \| \sigma_n), \tag{11.63}$$

then

$$\limsup_{n\to\infty} \mathrm{Tr}\left[\{2^{n(t+\delta)}\sigma_n \geqslant \rho_n\}\rho_n\right] \geqslant \varepsilon. \tag{11.64}$$

*Remark.* Note that if $\sigma_n$ is chosen as the optimal state such that $D_{\min}^{\varepsilon}\left(\rho^{\otimes n}\|\mathfrak{F}\right) = D_{\min}^{\varepsilon}\left(\rho^{\otimes n}\|\sigma_n\right)$, then the value of $t$, as defined in (11.63), is equal to $s$, as defined in (11.62). In the proof of the generalized quantum Stein's lemma, we will also consider other sequences $\{\sigma_n\}_{n\in\mathbb{N}}$ where, in those cases, $s \neq t$.

**Proof**    Set $\Pi_n := \{2^{n(t+\delta)}\sigma_n \geqslant \rho_n\}$ and suppose by contradiction that

$$\limsup_{n\to\infty} \operatorname{Tr}\left[\Pi_n\rho_n\right] < \varepsilon. \tag{11.65}$$

Let $\{n_k\}_{k\in\mathbb{N}}$ be a subsequence such that

$$t = \lim_{k\to\infty} \frac{1}{n_k} D_{\min}^{\varepsilon}(\rho_{n_k}\|\sigma_{n_k}). \tag{11.66}$$

From the assumption in (11.65) it follows that there exists an integer $k_0$ such that for all $k > k_0$

$$\operatorname{Tr}\left[\Pi_{n_k}\rho_{n_k}\right] < \varepsilon. \tag{11.67}$$

Denote by $\Lambda_{n_k} := I - \Pi_{n_k}$, so that $\operatorname{Tr}\left[\Lambda_{n_k}\rho_{n_k}\right] > 1 - \varepsilon$. On the other hand, from the definition of $\Pi_n$ we get that the projection $\Lambda_{n_k}$ projects to the negative part of $2^{n_k(t+\delta)}\sigma_{n_k} - \rho_{n_k}$. Thus,

$$\operatorname{Tr}\left[\Lambda_{n_k}\left(2^{-n(t+\delta)}\rho_{n_k} - \sigma_{n_k}\right)\right] \geqslant 0. \tag{11.68}$$

so that

$$\begin{aligned} \operatorname{Tr}\left[\Lambda_{n_k}\sigma_{n_k}\right] &\leqslant 2^{-n_k(t+\delta)}\operatorname{Tr}\left[\Lambda_{n_k}\rho_{n_k}\right] \\ &\leqslant 2^{-n_k(t+\delta)}. \end{aligned} \tag{11.69}$$

The above inequality implies that for all $k > k_0$

$$\begin{aligned} t + \delta &\leqslant -\frac{1}{n_k}\log\operatorname{Tr}\left[\Lambda_{n_k}\sigma_{n_k}\right] \\ &\leqslant \frac{1}{n_k}D_{\min}^{\varepsilon}\left(\rho_{n_k}\|\sigma_{n_k}\right), \end{aligned} \tag{11.70}$$

where in the last inequality we used the definition of $D_{\min}^{\varepsilon}$. However, since the above inequality holds for all $k > k_0$ it also hold for the limit $k \to \infty$, which according to (11.66) exists and equal to $t$. We therefore get the contradiction $t + \delta \leqslant t$. This completes the proof.    ∎

Next we show that $\mathfrak{F}$ can be replaced with its symmetric counterpart.

---

**Lemma 11.6.** Let $\{\mathfrak{F}_n\}_{n\in\mathbb{N}}$ be sets of free states satisfying the conditions outlined above Theorem 11.4, and for every $n \in \mathbb{N}$ let $\operatorname{Sym}(\mathfrak{F}_n)$ be the set of all states in $\mathfrak{F}_n$ that are symmetric under permutations of the $n$ subsystems of $A^n$. Then, for all $\rho \in \mathfrak{D}(A)$, $n \in \mathbb{N}$, and a quantum divergence $\mathbb{D}$, we have

$$\mathbb{D}\left(\rho^{\otimes n}\|\mathfrak{F}_n\right) = \mathbb{D}\left(\rho^{\otimes n}\|\operatorname{Sym}(\mathfrak{F}_n)\right). \tag{11.71}$$

**Proof** Let $\mathcal{G}_n \in \mathrm{CPTP}(A^n \to A^n)$ be the $\mathbf{S}_n$-twirling operation with respect to $\mathbf{S}_n$, the group of permutations on $n$-elements. That is,

$$\mathcal{G}_n\left(\omega^{A^n}\right) := \frac{1}{n!} \sum_{\pi \in \mathbf{S}_n} P_\pi^{A^n} \omega^{A^n} P_{\pi^{-1}}^{A^n}. \tag{11.72}$$

where $\{P_\pi^{A^n}\}_{\pi \in \mathbf{S}_n}$ for the set of all $n!$ matrices that permutes the $n$ subsystems of $A^n$ (see (C.160 of online version) for more details). Since we assume that $\mathfrak{F}_n$ is both convex and closed under permutation, for every $\sigma_n \in \mathfrak{F}_n$ we have necessarily that $\mathcal{G}_n(\sigma_n) \in \mathfrak{F}_n$. Thus, since $\mathcal{G}_n(\rho^{\otimes n}) = \rho^{\otimes n}$ we get from the DPI that for all $\sigma_n \in \mathfrak{F}_n$ and every quantum divergence $\mathbb{D}$ (in particular for $D_{\min}^\varepsilon$ and $D$) we have

$$\begin{aligned} \mathbb{D}\left(\rho^{\otimes n} \big\| \sigma_n\right) &\geqslant \mathbb{D}\left(\mathcal{G}_n\left(\rho^{\otimes n}\right) \big\| \mathcal{G}_n(\sigma_n)\right) \\ &= \mathbb{D}\left(\rho^{\otimes n} \big\| \mathcal{G}_n(\sigma_n)\right) \\ &\geqslant \mathbb{D}\left(\rho^{\otimes n} \big\| \mathrm{Sym}(\mathfrak{F}_n)\right). \end{aligned} \tag{11.73}$$

This complets the proof since the above inequality holds for all $\sigma_n \in \mathfrak{F}_n$. ∎

The proof outlined here can be adapted for any compact Lie group $\mathbf{G}$, where the $\mathbf{G}$-twirling channel is well-defined. In our case, this implies that proving Theorem 11.4 is sufficient if all free states are symmetric under permutations. Consequently, although not explicitly stated, we now assume that every $\sigma_n \in \mathfrak{F}_n$ satisfies $\mathcal{G}_n(\sigma_n) = \sigma_n$.

Our goal is to find a sequence of free symmetric states, $\{\sigma_n\}_{n \in \mathbb{N}}$, with each $\sigma_n \in \mathrm{Sym}(\mathfrak{F}_n)$, such that

$$r := \liminf_{n \to \infty} \frac{1}{n} D\left(\rho^{\otimes n} \big\| \sigma_n\right) \leqslant s. \tag{11.74}$$

Note that in general $\rho^{\otimes n}$ does not commute with $\sigma_n$, but in the following lemma we show that we can replace $\rho^{\otimes n}$ with $\mathcal{P}_n\left(\rho^{\otimes n}\right)$, where $\mathcal{P}_n$ is the pinching channel (see Section 3.5.12) with respect to the state $\sigma_n$.

---

**Lemma 11.7.** Let $\{\sigma_n\}_{n \in \mathbb{N}}$ be a sequence of free symmetric states with each $\sigma_n \in \mathrm{Sym}(\mathfrak{F}_n)$, and for every $n \in \mathbb{N}$ let $\mathcal{P}_n \in \mathrm{CPTP}(A^n \to A^n)$ be the pinching channel associated with the state $\sigma_n$. Then,

$$\liminf_{n \to \infty} \frac{1}{n} D\left(\rho^{\otimes n} \big\| \sigma_n\right) = \liminf_{n \to \infty} \frac{1}{n} D\left(\mathcal{P}_n\left(\rho^{\otimes n}\right) \big\| \sigma_n\right). \tag{11.75}$$

---

**Proof** Observe that

$$\begin{aligned} D\left(\mathcal{P}_n\left(\rho^{\otimes n}\right) \big\| \sigma_n\right) &= -H\left(\mathcal{P}_n\left(\rho^{\otimes n}\right)\right) - \mathrm{Tr}\left[\mathcal{P}_n\left(\rho^{\otimes n}\right) \log \sigma_n\right] \\ \mathcal{P}_n \text{ is self-adjoint} \to &= -H\left(\mathcal{P}_n\left(\rho^{\otimes n}\right)\right) - \mathrm{Tr}\left[\rho^{\otimes n} \mathcal{P}_n\left(\log \sigma_n\right)\right] \\ \sigma_n \text{ is symmetric} \to &= -H\left(\mathcal{P}_n\left(\rho^{\otimes n}\right)\right) - \mathrm{Tr}\left[\rho^{\otimes n} \log \sigma_n\right] \\ &= D\left(\rho^{\otimes n} \big\| \sigma_n\right) + H\left(\rho^{\otimes n}\right) - H\left(\mathcal{P}_n\left(\rho^{\otimes n}\right)\right). \end{aligned} \tag{11.76}$$

Now, from (3.233) it follows that the pinching channel can be written as a uniform mixture of unitary channels with $|\mathrm{spec}(\sigma_n)|$ terms. Since $\sigma_n$ is symmetric, its support is a subspace of the symmetric subspace $\mathrm{Sym}_n(A)$ (see Appendix C.10 of online version for many properties of the symmetric subspace). Thus,

$$|\mathrm{spec}(\sigma_n)| \leqslant \dim\left(\mathrm{Sym}_n(A)\right)$$
$$\textit{cf. } \textbf{(8.102)} \rightarrow \leqslant (n+1)^{|A|}. \tag{11.77}$$

Combining this with the upper bound in Corollary 7.2 we obtain

$$H\left(\mathcal{P}_n\left(\rho^{\otimes n}\right)\right) \leqslant H\left(\rho^{\otimes n}\right) + |A|\log(n+1). \tag{11.78}$$

Since $\mathcal{P}_n$ is a doubly stochastic channel, we also have $H\left(\mathcal{P}_n\left(\rho^{\otimes n}\right)\right) \geqslant H\left(\rho^{\otimes n}\right)$. Combining this with (11.78) we obtain

$$\lim_{n\to\infty}\frac{1}{n}\left(H\left(\mathcal{P}_n\left(\rho^{\otimes n}\right)\right) - H\left(\rho^{\otimes n}\right)\right) = 0. \tag{11.79}$$

Thus, dividing both sides of (11.76) by $n$ and taking the $\liminf_{n\to\infty}$ we obtain (11.75). This completes the proof. ∎

From Lemma 11.7 we get that in order to prove Theorem 11.4 it is sufficient to find a sequence of free symmetric states, $\{\sigma_n\}_{n\in\mathbb{N}}$, with each $\sigma_n \in \mathrm{Sym}(\mathfrak{F}_n)$, such that

$$r = \liminf_{n\to\infty}\frac{1}{n}D\left(\mathcal{P}_n\left(\rho^{\otimes n}\right)\,\middle\|\,\sigma_n\right) \leqslant s. \tag{11.80}$$

The idea of the proof is to pick an arbitrary sequence of symmetric states $\{\sigma_n\}_{n\in\mathbb{N}}$, and look at the difference $r - s$. Clearly, if this difference is nonpositive then we are done. Thus, we assume that $r - s > 0$ and then construct another sequence $\{\sigma_{n,1}\}_{n\in\mathbb{N}}$ with the property that

$$r_1 := \liminf_{n\to\infty}\frac{1}{n}D\left(\mathcal{P}_n\left(\rho^{\otimes n}\right)\,\middle\|\,\sigma_{n,1}\right) \tag{11.81}$$

satisfies

$$\frac{r_1 - s}{r - s} \leqslant 1 - \varepsilon. \tag{11.82}$$

Before proving the existence of the sequence $\{\sigma_{n,1}\}_{n\in\mathbb{N}}$, let us first demonstrate how the inequality in (11.82) can be applied to prove the generalized quantum Stein's lemma.

Observe that unless $s > r_1$ (in which case the proof is done) we get that $r_1$ is closer to $s$ than $r$ is. Repeating the process by starting from the sequence $\{\sigma_{n,1}\}_{n\in\mathbb{N}}$ and constructing the sequence $\{\sigma_{n,2}\}_{n\in\mathbb{N}}$ we can obtain that $r_2 := \liminf_{n\to\infty}\frac{1}{n}D\left(\mathcal{P}_n\left(\rho^{\otimes n}\right)\,\middle\|\,\sigma_{n,1}\right)$ satisfies

$$\frac{r_2 - s}{r_1 - s} \leqslant 1 - \varepsilon. \tag{11.83}$$

The combination of (11.82) and (11.83) results with

$$\frac{r_2 - s}{r - s} \leqslant (1 - \varepsilon)^2. \tag{11.84}$$

Similarly, by repetition, for every $k \in \mathbb{N}$ there exists a sequence of symmetric free states $\{\sigma_{n,k}\}_{n\in\mathbb{N}}$ such that

$$r_k - s \leqslant (1 - \varepsilon)^k(r - s), \quad \text{where} \quad r_k := \liminf_{n\to\infty}\frac{1}{n}D\left(\mathcal{P}_n\left(\rho^{\otimes n}\right)\,\middle\|\,\sigma_{n,k}\right). \tag{11.85}$$

Thus, from the definition of $r_k$ it follows that for every sequence $\{\delta_k\}_{k \in \mathbb{N}} \subset (0, 1)$ with limit $\lim_{k \to \infty} \delta_k = 0$, there exists a sequence of integers $\{n_k\}_{k \in \mathbb{N}}$ such that for all $k \in \mathbb{N}$

$$\frac{1}{n_k} D \left( \mathcal{P}_{n_k} \left( \rho^{\otimes n_k} \right) \, \middle\| \, \sigma_{n_k} \right) - s \leqslant (1 - \varepsilon)^k \, (r - s) + \delta_k. \tag{11.86}$$

Since the right-hand side goes to zero as $k$ goes to infinity, we obtain the desired result that

$$D^{\mathrm{reg}} \left( \rho \, \middle\| \, \mathrm{Sym}(\mathfrak{F}) \right) - s \leqslant \liminf_{k \to \infty} \frac{1}{n_k} D \left( \mathcal{P}_{n_k} \left( \rho^{\otimes n_k} \right) \, \middle\| \, \sigma_{n_k} \right) - s$$

$$(11.86) \to \; \leqslant 0. \tag{11.87}$$

We therefore conclude that in order to prove the theorem, starting from some sequence of symmetric states $\{\sigma_n\}_{n \in \mathbb{N}}$, it is sufficient to construct another sequence of symmetric states $\{\sigma_{n,1}\}_{n \in \mathbb{N}}$ that satisfy the inequality in (11.81).

## Construction of the Sequence $\{\sigma_{n,1}\}_{n \in \mathbb{N}}$

If the sequence $\{\sigma_n\}_{n \in \mathbb{N}}$ satisfies (11.80), we can take $\sigma_{n,1} = \sigma_n$ for all $n \in \mathbb{N}$. We therefore assume that $r > s$. Let $\varepsilon_0 > 0$ and observe that from the definition of $r$ there exists integer $m \in \mathbb{N}$ such that

$$\frac{1}{m} D \left( \rho^{\otimes m} \, \middle\| \, \sigma_m \right) < r + \varepsilon_0. \tag{11.88}$$

For all $n \in \mathbb{N}$ denote by $k_n := \left\lfloor \frac{n}{m} \right\rfloor$ and by

$$\sigma_{n,1} := \frac{1}{3} \left( \sigma_n^\star + \mathcal{G}_n \left( \sigma_m^{\otimes k_n} \otimes \gamma^{\otimes (n - k_n m)} \right) + \gamma^{\otimes n} \right), \tag{11.89}$$

where $\sigma_n^\star \in \mathrm{Sym}(\mathfrak{F}_n)$ is the state satisfying

$$D_{\min}^\varepsilon \left( \rho^{\otimes n} \, \middle\| \, \mathfrak{F} \right) = D_{\min}^\varepsilon \left( \rho^{\otimes n} \, \middle\| \, \sigma_n^\star \right). \tag{11.90}$$

Observe that from its definition, for every $n \in \mathbb{N}$ we have that $\sigma_{n,1} \in \mathrm{Sym}(\mathfrak{F}_n)$.

## Properties of $\{\sigma_{n,1}\}_{n \in \mathbb{N}}$

Let $\delta > 0$ to be sufficiently small such that $r > s + \delta$ (recall that we assume that $r > s$), and denoted by

$$P_n := \{ 2^{n(r + \varepsilon_0)} \sigma_{n,1} \geqslant \mathcal{P}_n \left( \rho^{\otimes n} \right) \} \quad \text{and} \quad Q_n := \{ 2^{n(s + \delta)} \sigma_{n,1} \geqslant \mathcal{P}_n \left( \rho^{\otimes n} \right) \}. \tag{11.91}$$

---

**Lemma 11.8.** Let $\varepsilon_0, \varepsilon_1, \delta \in (0, 1)$ with $\delta$ sufficiently small such that $r > s + \delta$. Then, the operators in (11.91) satisfy the following:

$$\limsup_{n \to \infty} \mathrm{Tr} \left[ Q_n P_n \left( \rho^{\otimes n} \right) \right] \geqslant \varepsilon.$$

$$\limsup_{n \to \infty} \mathrm{Tr} \left[ P_n P_n \left( \rho^{\otimes n} \right) \right] \geqslant 1 - \varepsilon_1. \tag{11.92}$$

**Proof** First, observe that from Lemma 11.5, it is sufficient to show that

$$\liminf_{n\to\infty} \frac{1}{n} D_{\min}^{\varepsilon}\left(\mathcal{P}_n\left(\rho^{\otimes n}\right)\,\big\|\,\sigma_{n,1}\right) \leqslant s \tag{11.93}$$

$$\limsup_{n\to\infty} \frac{1}{n} D_{\min}^{1-\varepsilon_1}\left(\mathcal{P}_n\left(\rho^{\otimes n}\right)\,\big\|\,\sigma_{n,1}\right) \leqslant r + \varepsilon_0. \tag{11.94}$$

To show these two inequalities we use the property that if $\sigma' \geqslant t\sigma$, then $D_{\min}^{\varepsilon}(\rho\|\sigma') \leqslant D_{\min}^{\varepsilon}(\rho\|\sigma) - \log t$ (see Exercise 11.8). Indeed, to prove (11.93), we use the inequality $\sigma_{n,1} \geqslant \frac{1}{3}\sigma_n^{\star}$ in conjunction with the DPI to get

$$
\begin{aligned}
D_{\min}^{\varepsilon}\left(\mathcal{P}_n\left(\rho^{\otimes n}\right)\,\big\|\,\sigma_{n,1}\right) &= D_{\min}^{\varepsilon}\left(\mathcal{P}_n\left(\rho^{\otimes n}\right)\,\big\|\,\mathcal{P}_n\left(\sigma_{n,1}\right)\right)\\
\mathbf{DPI}\to\ &\leqslant D_{\min}^{\varepsilon}\left(\rho^{\otimes n}\,\big\|\,\sigma_{n,1}\right)\\
\boxed{\sigma_{n,1} \geqslant \tfrac{1}{3}\sigma_n^{\star}} \longrightarrow\ &\leqslant D_{\min}^{\varepsilon}\left(\rho^{\otimes n}\,\big\|\,\sigma_n^{\star}\right) + \log 3\\
\mathbf{By\ definition\ of\ }\sigma_n^{\star}\to\ &= D_{\min}^{\varepsilon}\left(\rho^{\otimes n}\,\big\|\,\mathfrak{F}\right) + \log 3.
\end{aligned}
\tag{11.95}
$$

so that

$$\liminf_{n\to\infty} \frac{1}{n} D_{\min}^{\varepsilon}\left(\mathcal{P}_n\left(\rho^{\otimes n}\right)\,\big\|\,\sigma_{n,1}\right) \leqslant \liminf_{n\to\infty} \frac{1}{n} D_{\min}^{\varepsilon}\left(\rho^{\otimes n}\,\big\|\,\mathfrak{F}\right)$$
$$= s. \tag{11.96}$$

Similarly, to prove (11.94), we use the inequality $\sigma_{n,1} \geqslant \frac{1}{3}\mathcal{G}_n\left(\sigma_m^{\otimes k_n} \otimes \gamma^{\otimes(n-k_n m)}\right)$ in conjunction with the DPI to get

$$
\begin{aligned}
D_{\min}^{1-\varepsilon_1}\left(\mathcal{P}_n\left(\rho^{\otimes n}\right)\,\big\|\,\sigma_{n,1}\right) &= D_{\min}^{1-\varepsilon_1}\left(\mathcal{P}_n\left(\rho^{\otimes n}\right)\,\big\|\,\mathcal{P}_n\left(\sigma_{n,1}\right)\right)\\
\mathbf{DPI}\to\ &\leqslant D_{\min}^{1-\varepsilon_1}\left(\rho^{\otimes n}\,\big\|\,\sigma_{n,1}\right)\\
\boxed{\sigma_{n,1} \geqslant \tfrac{1}{3}\mathcal{G}_n\left(\sigma_m^{\otimes k_n} \otimes \gamma^{\otimes(n-k_n m)}\right)} \longrightarrow\ &\leqslant D_{\min}^{1-\varepsilon_1}\left(\rho^{\otimes n}\,\Big\|\,\mathcal{G}_n\left(\sigma_m^{\otimes k_n} \otimes \gamma^{\otimes(n-k_n m)}\right)\right) + \log 3.
\end{aligned}
\tag{11.97}
$$

Substituting $\rho^{\otimes n} = \mathcal{G}_n(\rho^{\otimes n})$ into the equation above and applying the DPI gives

$$
\begin{aligned}
D_{\min}^{1-\varepsilon_1}\left(\mathcal{P}_n\left(\rho^{\otimes n}\right)\,\big\|\,\sigma_{n,1}\right) &\leqslant D_{\min}^{1-\varepsilon_1}\left(\rho^{\otimes n}\,\big\|\,\sigma_m^{\otimes k_n} \otimes \gamma^{\otimes(n-k_n m)}\right) + \log 3\\
\mathbf{(8.201)}\to\ &\leqslant \tilde{D}_{\alpha}\left(\rho^{\otimes n}\,\big\|\,\sigma_m^{\otimes k_n} \otimes \gamma^{\otimes(n-k_n m)}\right) - \frac{\alpha}{\alpha-1}\log\left(1-\varepsilon\right) + \log 3\\
&= k_n \tilde{D}_{\alpha}\left(\rho^{\otimes m}\,\big\|\,\sigma_m\right) + (n - k_n m)\tilde{D}_{\alpha}\left(\rho\,\big\|\,\gamma\right)\\
&\quad - \frac{\alpha}{\alpha-1}\log\left(1-\varepsilon\right) + \log 3.
\end{aligned}
\tag{11.98}
$$

By definition, $k_n/n$ goes to $1/m$ as $n$ goes to infinity. Thus, by dividing both sides of Equation (11.98) by $n$ and taking the limit $n \to \infty$ we obtain

$$\limsup_{n\to\infty} \frac{1}{n} D_{\min}^{1-\varepsilon_1}\left(\mathcal{P}_n\left(\rho^{\otimes n}\right)\,\big\|\,\sigma_{n,1}\right) \leqslant \frac{1}{m}\tilde{D}_{\alpha}\left(\rho^{\otimes m}\,\big\|\,\sigma_m\right). \tag{11.99}$$

Since the equation above holds for all $\alpha > 1$, taking the limit $\alpha \to 1^+$ we get from the continuity of $\tilde{D}_\alpha$ at $\alpha = 1$ that

$$\limsup_{n \to \infty} \frac{1}{n} D^{1-\varepsilon_1}_{\min}\left(\mathcal{P}_n\left(\rho^{\otimes n}\right) \big\| \sigma_{n,1}\right) \leqslant \frac{1}{m} D\left(\rho^{\otimes m} \big\| \sigma_m\right) \tag{11.100}$$

$$\mathbf{(11.88)} \to\ < r + \varepsilon_0.$$

This completes the proof of the lemma. ∎

**Exercise 11.8.** *Show that if $\sigma' \geqslant t\sigma$, then $D^\varepsilon_{\min}(\rho \| \sigma') \leqslant D^\varepsilon_{\min}(\rho \| \sigma) - \log t$.*

We are finally ready to prove the generalized quantum Stein's lemma.

## Proof of Theorem 11.4

Since $\mathcal{P}_n\left(\rho^{\otimes n}\right)$ commutes with $\sigma_{n,1}$, the three projections $\Pi_n^{(1)} := I - P_n$, $\Pi_n^{(2)} := P_n - Q_n$, and $\Pi_n^{(3)} := Q_n$, form a three-outcome orthogonal von-Neumann projective measurement of system $A^n$. The key idea of the proof is to split the computation of $D\left(\mathcal{P}_n\left(\rho^{\otimes n}\right) \big\| \sigma_{n,1}\right)$ into three parts based on these projections. Explicitly, we write $D\left(\mathcal{P}_n\left(\rho^{\otimes n}\right) \big\| \sigma_{n,1}\right) = a_n^{(1)} + a_n^{(2)} + a_n^{(3)}$, where for all $j \in \{1, 2, 3\}$,

$$a_n^{(j)} := \mathrm{Tr}\left[\Pi_n^{(j)}\mathcal{P}_n\left(\rho^{\otimes n}\right)\left(\log \mathcal{P}_n\left(\rho^{\otimes n}\right) - \log \sigma_{n,1}\right)\right]. \tag{11.101}$$

*Upper bound for $a_n^{(1)}$:* Recall that $\sigma_{n,1} \geqslant \frac{1}{3}\gamma^{\otimes n} > 0$, and let $\lambda$ be such that $\gamma^{-1} \leqslant 2^\lambda I^A$. For reasons to be clear later on, we choose $\lambda$ to be sufficiently large so that we also have $\lambda \geqslant r + \varepsilon_0$. Then,

$$\begin{aligned}
\sigma_{n,1}^{-1} &\leqslant 3(\gamma^{-1})^{\otimes n} \\
&\leqslant 2^{n\lambda + \log 3} I^{A^n}.
\end{aligned} \tag{11.102}$$

We therefore bound the first term as

$$a_n^{(1)} \leqslant \mathrm{Tr}\left[\Pi_n^{(1)}\mathcal{P}_n\left(\rho^{\otimes n}\right) \log \sigma_{n,1}^{-1}\right]$$
$$\mathbf{(11.102)} \to\ \leqslant (n\lambda + \log 3)\,\mathrm{Tr}\left[\Pi_1 \mathcal{P}_n\left(\rho^{\otimes n}\right)\right]. \tag{11.103}$$

*Upper bound for $a_n^{(2)}$:* When restricted to the support of $\Pi_n^{(2)} := P_n - Q_n$, we get from the definitions of $P_n$ and $Q_n$ that $\mathcal{P}_n(\rho^{\otimes n}) \leqslant 2^{n(r+\varepsilon_0)}\sigma_{n,1}$ or equivalently $\mathcal{P}_n(\rho^{\otimes n})\sigma_{n,1}^{-1} \leqslant 2^{n(r+\varepsilon_0)} I^{A^n}$ (recall that $\mathcal{P}_n(\rho^{\otimes n})$ and $\sigma_{n,1}$ commutes). Thus,

$$a_n^{(2)} \leqslant n(r + \varepsilon_0)\mathrm{Tr}\left[\Pi_2 \mathcal{P}_n\left(\rho^{\otimes n}\right)\right]. \tag{11.104}$$

*Upper bound for $a_n^{(3)}$:* When restricted to the support of $\Pi_n^{(3)} := Q_n$, we get that $\mathcal{P}_n(\rho^{\otimes n}) \leqslant 2^{n(s+\delta)}\sigma_{n,1}$ or equivalently $\mathcal{P}_n(\rho^{\otimes n})\sigma_{n,1}^{-1} \leqslant 2^{n(s+\delta)} I^{A^n}$. Thus,

$$a_n^{(3)} \leqslant n(s + \delta)\mathrm{Tr}\left[\Pi_n^{(3)}\mathcal{P}_n\left(\rho^{\otimes n}\right)\right]. \tag{11.105}$$

*Upper bound for $r_1$:* From the upper bounds for $a_n^{(1)}$, $a_n^{(2)}$, and $a_n^{(3)}$, we get that

$$
\begin{aligned}
r_1 &= \liminf_{n\to\infty} \frac{1}{n}(a_n^{(1)} + a_n^{(2)} + a_n^{(3)}) \\
&\leqslant \liminf_{n\to\infty} \left\{ \lambda \operatorname{Tr}\left[ \Pi_n^{(1)} \mathcal{P}_n\left(\rho^{\otimes n}\right)\right] + (r+\varepsilon_0)\operatorname{Tr}\left[ \Pi_n^{(2)} \mathcal{P}_n\left(\rho^{\otimes n}\right)\right] \right. \\
&\qquad \left. + (s+\delta)\operatorname{Tr}\left[ \Pi_n^{(3)} \mathcal{P}_n\left(\rho^{\otimes n}\right)\right] \right\} \\
&= \lambda - \limsup_{n\to\infty} \left\{ (\lambda - r - \varepsilon_0)\operatorname{Tr}\left[ P_n \mathcal{P}_n\left(\rho^{\otimes n}\right)\right] + (r+\varepsilon_0 - s - \delta)\operatorname{Tr}\left[ Q_n \mathcal{P}_n\left(\rho^{\otimes n}\right)\right] \right\}.
\end{aligned}
$$
$$(11.106)$$

Finally, since both $\lambda - r - \varepsilon_0 \geqslant 0$ and $r + \varepsilon_0 - s - \delta \geqslant 0$, we apply Lemma 11.5 to the equation above to get the upper bound

$$
r_1 \leqslant \lambda - (\lambda - r - \varepsilon_0)(1 - \varepsilon_1) - (r + \varepsilon_0 - s - \delta)\varepsilon. \tag{11.107}
$$

Note that the inequality in (11.107) holds for all $\varepsilon_0, \varepsilon_1, \delta \in (0,1)$ where $\delta$ is sufficiently small such that $s + \delta < r$. Thus, by taking the limits $\delta, \varepsilon_0, \varepsilon_1 \to 0$ we obtain $r_1 \leqslant r - (r - s)\varepsilon$, which is equivalent to (11.82). This completes the proof of the generalized Stein's lemma.

**Exercise 11.9.** *Without using Theorem 11.4, show that all $\rho \in \mathfrak{D}(A)$ we have*

$$
\lim_{\varepsilon\to 1^-} \liminf_{n\to\infty} \frac{1}{n} D_{\min}^{\varepsilon}\left(\rho^{\otimes n}\|\mathfrak{F}\right) = \lim_{\varepsilon\to 1^-} \limsup_{n\to\infty} \frac{1}{n} D_{\min}^{\varepsilon}\left(\rho^{\otimes n}\|\mathfrak{F}\right) = D^{\mathrm{reg}}(\rho\|\mathfrak{F}). \tag{11.108}
$$

**Exercise 11.10.** *Provide an alternative proof of Theorem 11.4 for the special case that the set of free states is the one defined in (11.53); that is,*

$$
\mathfrak{F}(AB) := \left\{ \mathbf{u}^A \otimes \rho^B : \rho \in \mathfrak{D}(A) \right\}. \tag{11.109}
$$

*Follow the following steps:*

1. *First, show that*

$$
D_\alpha\left(\rho^{AB}\|\mathfrak{F}\right) = \log|A| - H_\alpha^{\uparrow}(A|B)_\rho. \tag{11.110}
$$

2. *Use the additivity of $H_\alpha^{\uparrow}$ (see Theorem 7.7 and Exercise 7.18) to show that $D_\alpha(\rho\|\mathfrak{F}) = D_\alpha^{\mathrm{reg}}(\rho\|\mathfrak{F})$.*

3. *Use (10.82) to conclude that*

$$
\liminf_{n\to\infty} \frac{1}{n} D_{\min}^{\varepsilon}\left(\rho^{\otimes n}\|\mathfrak{F}\right) \geqslant D(\rho\|\mathfrak{F}). \tag{11.111}
$$

4. *Complete the proof of the theorem.*

# 11.4 The Uniqueness of the Umegaki Relative Entropy

In this section, we use Stein's lemma and the AEP property proved in the previous sections, to show the uniqueness of the Umegaki relative entropy. Recall that the Umegaki relative entropy is defined for any $\rho, \sigma \in \mathfrak{D}(A)$ with $\operatorname{supp}(\rho) \subseteq \operatorname{supp}(\sigma)$ as

$$D(\rho\|\sigma) := \operatorname{Tr}[\rho \log \rho] - \operatorname{Tr}[\rho \log \sigma]. \tag{11.112}$$

This quantum relative entropy plays a key role in numerous applications in quantum information theory and beyond. We have already seen in the quantum Stein's lemma that it can be interpreted as the optimal decay rate of the type-II error exponent. Among all relative entropies, it is the most well known, and in this section we show that the Umegaki relative entropy can be singled out as the only quantum relative entropy that is asymptotically continuous. We will see later on in the book that this is the key reason of its "popularity."

Following Definition 10.6, we say that a relative entropy $\mathbb{D}$ is asymptotically continuous if there exists a continuous function $f : [0, 1] \to \mathbb{R}_+$ such that $f(0) = 0$ and for all $\rho, \rho', \sigma \in \mathfrak{D}(A)$, with $\operatorname{supp}(\rho) \subseteq \operatorname{supp}(\sigma)$ and $\operatorname{supp}(\rho') \subseteq \operatorname{supp}(\sigma)$,

$$\left|\mathbb{D}(\rho\|\sigma) - \mathbb{D}(\rho'\|\sigma)\right| \leqslant f(\varepsilon) \log \|\sigma^{-1}\|_\infty, \tag{11.113}$$

where $\varepsilon := \frac{1}{2}\|\rho - \rho'\|_1$, and $\sigma^{-1}$ is the *generalized* inverse of $\sigma$. We emphasize that $f$ is independent of $|A|$.

**Uniqueness of the Umegaki Relative Entropy**

**Theorem 11.5.** Let $\mathbb{D}$ be a relative entropy that is asymptotically continuous. Then, $\mathbb{D} = D$, where $D$ is the Umegaki relative entropy.

Recall from Corollary 10.1 that the Umegaki relative entropy is shown to be asymptotically continuous. The theorem we are discussing asserts that no other relative entropy possesses this property of asymptotic continuity. To substantiate this claim, we will utilize the following lemma, which introduces a notation for any relative entropy $\mathbb{D}$:

$$\mathbb{D}^{(\varepsilon)}(\rho\|\sigma) := \max_{\rho' \in \mathfrak{B}_\varepsilon(\rho)} \mathbb{D}(\rho'\|\sigma). \tag{11.114}$$

In other words, $\mathbb{D}^{(\varepsilon)}$ represents a form of smoothing, albeit using the maximum rather than the minimum over all states that are $\varepsilon$-close to $\rho$. Consequently, unlike $\mathbb{D}^\varepsilon$, the function $\mathbb{D}^{(\varepsilon)}$ does not qualify as a divergence. It's also worth noting that this specific notation was previously used in the context of the min relative entropy in Theorem 10.10. Both the lemma and this notation are crucial for our proof, as they facilitate the examination of how relative entropies respond to minor perturbations in the state $\rho$.

**Lemma 11.9.** Let $\rho, \sigma \in \mathfrak{D}(A)$ with $\mathrm{supp}(\rho) \subseteq \mathrm{supp}(\sigma)$, and $\mathbb{D}$ be a quantum relative entropy satisfying (11.113) (i.e. $\mathbb{D}$ is asymptotically continuous). Then,

$$\mathbb{D}(\rho\|\sigma) = \lim_{\varepsilon \to 0^+} \liminf_{n \to \infty} \frac{1}{n}\mathbb{D}^\varepsilon\left(\rho^{\otimes n}\|\sigma^{\otimes n}\right) = \lim_{\varepsilon \to 0^+} \limsup_{n \to \infty} \frac{1}{n}\mathbb{D}^{(\varepsilon)}\left(\rho^{\otimes n}\|\sigma^{\otimes n}\right).$$

$$(11.115)$$

**Proof** Let $\varepsilon \in (0, 1)$ and for each $n \in \mathbb{N}$, let $\rho_n' \in \mathfrak{D}(A^n)$ be such that $\frac{1}{2}\|\rho_n' - \rho^{\otimes n}\|_1 \leqslant \varepsilon$. Then, applying (11.113) to $n$ copies of $\rho$ and $\sigma$ gives

$$\left|\mathbb{D}(\rho\|\sigma) - \frac{1}{n}\mathbb{D}(\rho_n'\|\sigma^{\otimes n})\right| \leqslant f(\varepsilon)\log\|\sigma^{-1}\|_\infty. \qquad (11.116)$$

Therefore, by taking the $\liminf_{n\to\infty}$ or $\limsup_{n\to\infty}$ on both sides of the equation above followed by $\lim_{\varepsilon \to 0^+}$ completes the proof. ∎

We are now ready to prove Theorem 11.5.

**Proof of Theorem 11.5** Since the lemma above states that (11.113) implies (11.115), it is sufficient to prove that the Umegaki relative entropy is the only relative entropy that satisfies (11.115). Let $\mathbb{D}(\rho\|\sigma)$ be a relative entropy satisfying (11.115). Therefore,

$$\mathbb{D}(\rho\|\sigma) = \lim_{\varepsilon \to 0^+} \liminf_{n \to \infty} \frac{1}{n}\mathbb{D}^\varepsilon\left(\rho^{\otimes n}\|\sigma^{\otimes n}\right)$$

$$\textbf{(6.113)} \to \; \leqslant \lim_{\varepsilon \to 0^+} \liminf_{n \to \infty} \frac{1}{n}D_{\max}^\varepsilon\left(\rho^{\otimes n}\|\sigma^{\otimes n}\right) \qquad (11.117)$$

$$\textbf{AEP; see (11.50)} \to \; = D(\rho\|\sigma).$$

Conversely,

$$\mathbb{D}(\rho\|\sigma) = \lim_{\varepsilon \to 0^+} \limsup_{n \to \infty} \frac{1}{n}\mathbb{D}^{(\varepsilon)}\left(\rho^{\otimes n}\|\sigma^{\otimes n}\right)$$

$$\textbf{(6.113)} \to \; \geqslant \lim_{\varepsilon \to 0^+} \limsup_{n \to \infty} \frac{1}{n}D_{\min}^{(\varepsilon)}\left(\rho^{\otimes n}\|\sigma^{\otimes n}\right) \qquad (11.118)$$

$$\textbf{Theorem 10.10} \to \; = D(\rho\|\sigma).$$

This completes the proof. ∎

## 11.5 Asymptotic Interconversions

In the first section of this chapter, we studied interconversions of a single copy of a resource into another resource state by free operations. In this section, we study the asymptotic rate at which many copies of a given state can be converted, by free operations, to many copies of another state. We will consider two types of rates. In one type, the goal is to distill (by free operations) as many copies as possible of a desirable state

$\sigma$ from $n$ copies of a less desirable resource state $\rho$. That is, this type of rate is defined as the maximization of the ratio $\frac{m}{n}$ given that the conversion $\rho^{\otimes n} \xrightarrow{\mathfrak{F}} \sigma^{\otimes m}$ is possible. We will therefore call this rate the distillation rate of converting $\rho$ into $\sigma$. In the second type of rate, the goal is to find the smallest number $n$ for which the conversion $\rho^{\otimes n} \xrightarrow{\mathfrak{F}} \sigma^{\otimes m}$ is possible. We therefore call this rate the cost rate of converting $\rho$ into $\sigma$, and define it as the minimization of the ration $\frac{n}{m}$ such that $\rho^{\otimes n} \xrightarrow{\mathfrak{F}} \sigma^{\otimes n}$ is possible (see the following precise definition).

As we have already discussed, in many resource theories, the set of free operations is not so large, so that the exact conversion of one resource state to another is typically not possible. Therefore, instead of considering the exact conversion of $\rho^{\otimes n}$ to $\sigma^{\otimes m}$ we will allow the output state to be $\varepsilon$-close to $\sigma^{\otimes m}$ as long as the error $\varepsilon$ goes to zero in the asymptotic limit $n, m \to \infty$. This idea is made rigorous in the following definition.

---

**Definition 11.3.** Let $\rho \in \mathfrak{D}(A)$ and $\sigma \in \mathfrak{D}(B)$ be two resource states.

1. The *asymptotic distillable rate* of $\rho$ into $\sigma$ is defined as

$$\text{Distill}(\rho \to \sigma) := \lim_{\varepsilon \to 0^+} \sup_{n,m \in \mathbb{N}} \left\{ \frac{m}{n} : T\left( \rho^{\otimes n} \xrightarrow{\mathfrak{F}} \sigma^{\otimes m} \right) \leqslant \varepsilon \right\}. \quad (11.119)$$

2. The *asymptotic cost rate* of $\rho$ into $\sigma$ is defined as

$$\text{Cost}(\rho \to \sigma) := \lim_{\varepsilon \to 0^+} \inf_{n,m \in \mathbb{N}} \left\{ \frac{n}{m} : T\left( \rho^{\otimes n} \xrightarrow{\mathfrak{F}} \sigma^{\otimes m} \right) \leqslant \varepsilon \right\}. \quad (11.120)$$

---

*Remark.* These two definitions are not independent of each other. Specifically, observe that

$$\text{Distill}(\rho \to \sigma) = \frac{1}{\text{Cost}(\rho \to \sigma)}. \quad (11.121)$$

This relationship is consistent with the intuition that if $\rho$ is a free state and $\sigma$ is a resource state, then $\text{Cost}(\rho \to \sigma)$ is equal to infinity, while $\text{Distill}(\rho \to \sigma)$ equals zero. This is because, in the former case, no matter how many copies of $\rho$ you have, they are insufficient to prepare even a single copy of $\sigma$. In the latter case, it is impossible to distill or extract a resource state $\sigma$ from a free state $\rho$.

In the above definitions of cost and distillation, we did not impose any constraints on the integers $m$ and $n$. However, as intuition suggests, it is typically the case that $m$ and $n$ are both very large. In fact, for any natural number $a$, we can include the condition $n, m \geqslant a$ in the aforementioned definitions without altering their value. Specifically, we argue that

$$\text{Cost}(\rho \to \sigma) = \lim_{\varepsilon \to 0^+} \inf_{\substack{n,m \in \mathbb{N} \\ n,m \geqslant a}} \left\{ \frac{n}{m} : T\left( \rho^{\otimes n} \xrightarrow{\mathfrak{F}} \sigma^{\otimes m} \right) \leqslant \varepsilon \right\}, \quad (11.122)$$

(and similarly we can add $n, m \geqslant a$ to $\text{Distill}(\rho \to \sigma)$). To see why, observe first that the left-hand side of the equation above cannot be greater than the right-hand side since by adding the restriction $n, m \geqslant a$ one can only increase the infimum. To prove that we must have equality, recall from Exercise 11.3 that for any such $a \in \mathbb{N}$, if $T\left(\rho^{\otimes n} \xrightarrow{\mathfrak{F}} \sigma^{\otimes m}\right) \leqslant \varepsilon$, then $T\left(\rho^{\otimes na} \xrightarrow{\mathfrak{F}} \sigma^{\otimes ma}\right) \leqslant a\varepsilon$. Therefore,

$$
\mathrm{Cost}(\rho \to \sigma) \geqslant \lim_{\varepsilon \to 0^+} \inf_{n,m \in \mathbb{N}} \left\{ \frac{n}{m} : T\left(\rho^{\otimes na} \xrightarrow{\mathfrak{F}} \sigma^{\otimes ma}\right) \leqslant a\varepsilon \right\}
$$

$$
\textbf{replacing } \textit{na, ma} \textbf{ with } \textit{n', m'} \to \; \geqslant \lim_{\varepsilon \to 0^+} \inf_{\substack{n',m' \in \mathbb{N} \\ m',n' \geqslant a}} \left\{ \frac{n'}{m'} : T\left(\rho^{\otimes n'} \xrightarrow{\mathfrak{F}} \sigma^{\otimes m'}\right) \leqslant a\varepsilon \right\}
$$

$$
\boxed{\varepsilon' := a\varepsilon} \longrightarrow \; = \lim_{\varepsilon' \to 0^+} \inf_{\substack{n',m' \in \mathbb{N} \\ m',n' \geqslant a}} \left\{ \frac{n'}{m'} : T\left(\rho^{\otimes n'} \xrightarrow{\mathfrak{F}} \sigma^{\otimes m'}\right) \leqslant \varepsilon' \right\}.
$$

$$(11.123)$$

Hence, the equality in 11.122.

In the following exercise we will show that the asymptotic cost and distillable rates are themselves resource measures.

**Exercise 11.11.** *Consider the asymptotic cost and distillable rates defined in Definition 11.3.*

1. *Show that for a fixed resource state $\sigma \in \mathfrak{D}(B)$, the function $f_\sigma(\rho) := \mathrm{Distill}(\rho \to \sigma)$ is a resource measure.*
2. *Show that for a fixed resource state $\rho \in \mathfrak{D}(B)$, the function $g_\rho(\sigma) := \mathrm{Cost}(\rho \to \sigma)$ is a resource measure.*

**Exercise 11.12.** *Let $T'$ be another metric that is topologically equivalent to the trace distance $T$ (i.e. there exists $a, b > 0$ such that $aT \leqslant T' \leqslant bT$). Furthermore, for every $\rho \in \mathfrak{D}(A)$ and $\sigma \in \mathfrak{D}(B)$, let $\mathrm{Distill}'(\rho \to \sigma)$ and $\mathrm{Cost}'(\rho \to \sigma)$ be the distillation and cost rates obtained by replacing the trace distance in (11.119) and (11.120) with the metric $T'$. Show that for all $\rho \in \mathfrak{D}(A)$ and all $\sigma \in \mathfrak{D}(B)$,*

$$
\mathrm{Distill}'(\rho \to \sigma) = \mathrm{Distill}(\rho \to \sigma) \quad \textit{and} \quad \mathrm{Cost}'(\rho \to \sigma) = \mathrm{Cost}(\rho \to \sigma). \quad (11.124)
$$

Typically, in the process of converting $n$ copies of $\rho$ into $m$ copies of $\sigma$, some resource is consumed. This is reflected by the fact that the target state $\sigma^{\otimes m}$ (up to a small error) is less resourceful than the source state $\rho^{\otimes n}$. If this loss of a resource is not too high (e.g. sublinear in $n$), then typically one can use the $m$ copies of $\sigma$ to recover the $n$ copies of $\rho$. If the $n$ copies of $\rho$ can always be recovered (up to an error that goes to zero asymptotically), we say that the resource theory is asymptotically reversible.

> **Definition 11.4.** A QRT $\mathfrak{F}$ is called *asymptotically reversible* if for all $\rho \in \mathfrak{D}(A)$ and $\sigma \in \mathfrak{D}(B)$
>
> $$\mathrm{Distill}(\rho \to \sigma) = \mathrm{Cost}(\sigma \to \rho). \quad (11.125)$$

Note that due to (11.121) the condition that a QRT is reversible can also be expressed as

$$\text{Distill}(\rho \to \sigma)\text{Distill}(\sigma \to \rho) = 1, \tag{11.126}$$

or as

$$\text{Cost}(\rho \to \sigma)\text{Cost}(\sigma \to \rho) = 1. \tag{11.127}$$

> **Theorem 11.6.** Let $\mathfrak{F}$ be a QRT, $\rho \in \mathfrak{D}(A)$ and $\sigma \in \mathfrak{D}(B)$. Then,
>
> $$\text{Distill}(\rho \to \sigma) \leqslant \frac{D^{\text{reg}}(\rho\|\mathfrak{F})}{D^{\text{reg}}(\sigma\|\mathfrak{F})} \tag{11.128}$$
>
> and equality holds if the QRT $\mathfrak{F}$ is reversible.

*Remark.* The theorem above can also be expressed in terms of the asymptotic cost rate. Specifically, we have the bound

$$\text{Cost}(\rho \to \sigma) \geqslant \frac{D^{\text{reg}}(\sigma\|\mathfrak{F})}{D^{\text{reg}}(\rho\|\mathfrak{F})}, \tag{11.129}$$

where we used (11.121) in (11.128).

**Proof**  Let $\{\varepsilon_n\}_{n\in\mathbb{N}}$ be a sequence of positive numbers with zero limit, let $\{m_n\}_{n\in\mathbb{N}}$ be a sequence of integers, and $\{\mathcal{E}_n\}_{n\in\mathbb{N}}$ be a sequence of free channels with $\mathcal{E}_n \in \mathfrak{F}(A^n \to B^{m_n})$, such that

$$\left|\text{Distill}(\rho \to \sigma) - \frac{m_n}{n}\right| \leqslant \varepsilon_n \quad \text{and} \quad \frac{1}{2}\left\|\sigma^{\otimes m_n} - \mathcal{E}_n\left(\rho^{\otimes n}\right)\right\|_1 \leqslant \varepsilon_n. \tag{11.130}$$

Recall that the relative entropy of a resource is asymptotically continuous. Therefore, for all $n \in \mathbb{N}$

$$\begin{aligned}
D\left(\rho^{\otimes n}\big\|\mathfrak{F}\right) &\geqslant D\left(\mathcal{E}_n\left(\rho^{\otimes n}\right)\big\|\mathfrak{F}\right) \\
\mathbf{(10.34)} \to &\geqslant D\left(\sigma^{\otimes m_n}\big\|\mathfrak{F}\right) - \varepsilon_n \kappa_n - (1+\varepsilon_n)h\left(\frac{\varepsilon_n}{1+\varepsilon_n}\right),
\end{aligned} \tag{11.131}$$

where $\kappa_n := \max_{\omega\in\mathfrak{D}(B^{m_n})} D(\omega\|\mathfrak{F})$. Dividing both sides by $n$ and taking the limit $n \to \infty$ yields

$$D^{\text{reg}}(\rho\|\mathfrak{F}) \geqslant \lim_{n\to\infty} \frac{m_n}{n}\frac{1}{m_n} D\left(\sigma^{\otimes m_n}\big\|\mathfrak{F}\right) = \text{Distill}(\rho \to \sigma)D^{\text{reg}}(\sigma\|\mathfrak{F}), \tag{11.132}$$

where we used the assumption that

$$\limsup_{n\to\infty} \frac{\kappa_n}{n} < \infty. \tag{11.133}$$

This completes the proof of the inequality (11.128). For the equality, observe first that both bounds (11.128) and (11.129) can be written as

$$\text{Distill}(\rho \to \sigma) \leqslant \frac{D^{\text{reg}}(\rho\|\mathfrak{F})}{D^{\text{reg}}(\sigma\|\mathfrak{F})} \leqslant \text{Cost}(\sigma \to \rho). \tag{11.134}$$

Hence, if $\mathfrak{F}$ is reversible, then both the inequalities in (11.134) must be equalities. This completes the proof. ∎

## 11.5.1 Asymptotic Cost and Distillation of a Resource

Some QRTs contain a golden unit (see Definition 11.1) like the maximally entangled states in entanglement theory. In such cases, quite often one is interested in computing asymptotic conversion rates when $\rho$ or $\sigma$ is taken to be the element of the golden unit. Specifically, let $\{\Phi_k\}_{k\in\mathbb{N}}$ be a golden unit, and take $\sigma = \Phi_2$ be the two-dimensional element of the golden unit. For this choice, the rate $\mathrm{Distill}(\rho \to \Phi_2)$ quantifies the number of resource units (i.e. copies of $\Phi_2$) that can be *distilled* from each copy of $\rho$. For this reason the quantity $\mathrm{Distill}(\rho \to \Phi_2)$ is called the distillable resource of $\rho$, and denoted by

$$\mathrm{Distill}(\rho) := \mathrm{Distill}(\rho \to \Phi_2). \tag{11.135}$$

Conversely, one can use the asymptotic cost rate to quantify the cost (in resource units $\Phi_2$) of a resource state $\rho$. Specifically, the quantity

$$\mathrm{Cost}(\rho) := \mathrm{Cost}(\Phi_2 \to \rho) \tag{11.136}$$

quantifies the cost in resource units (i.e. copies of $\Phi_2$) that are needed to prepare each copy of $\rho$. The asymptotic cost and distillation of a resource are related to their single-shot versions as follows.

> **Lemma 11.10.** Let $\mathfrak{F}$ be a QRT and $\rho \in \mathfrak{D}(A)$. Then,
>
> $$\mathrm{Cost}(\rho) = \lim_{\varepsilon \to 0^+} \liminf_{n \to \infty} \frac{1}{n}\mathrm{Cost}^\varepsilon\left(\rho^{\otimes n}\right) \tag{11.137}$$
>
> $$\mathrm{Distill}(\rho) = \lim_{\varepsilon \to 0^+} \limsup_{n \to \infty} \frac{1}{n}\mathrm{Distill}^\varepsilon\left(\rho^{\otimes n}\right), \tag{11.138}$$
>
> where $\mathrm{Distill}^\varepsilon$ and $\mathrm{Cost}^\varepsilon$ have been defined in Definition 11.2.

**Proof** We prove the first equality and leave the second one to Exercise 11.15. By definition,

$$\inf_{n\in\mathbb{N}} \frac{1}{n}\mathrm{Cost}^\varepsilon\left(\rho^{\otimes n}\right) = \inf\left\{\frac{\log m}{n} : T\left(\Phi_m \xrightarrow{\mathfrak{F}} \rho^{\otimes n}\right) \leqslant \varepsilon. \quad n,m \in \mathbb{N}\right\}$$

$$\textbf{restricting } m = 2^k \to \; \leqslant \inf\left\{\frac{k}{n} : T\left(\Phi_{2^k} \xrightarrow{\mathfrak{F}} \rho^{\otimes n}\right) \leqslant \varepsilon. \quad n,k \in \mathbb{N}\right\}$$

$$\textbf{property of a golden unit} \to \; = \inf\left\{\frac{k}{n} : T\left(\Phi_2^{\otimes k} \xrightarrow{\mathfrak{F}} \rho^{\otimes n}\right) \leqslant \varepsilon. \quad n,k \in \mathbb{N}\right\}.$$

$$\tag{11.139}$$

Hence,

$$\mathrm{Cost}(\rho) \geqslant \lim_{\varepsilon \to 0^+} \inf_{n \in \mathbb{N}} \frac{1}{n} \mathrm{Cost}^\varepsilon \left( \rho^{\otimes n} \right)$$

$$\textbf{See Exercise 11.14 below} \to \quad = \lim_{\varepsilon \to 0^+} \inf_{a \leqslant n \in \mathbb{N}} \frac{1}{n} \mathrm{Cost}^\varepsilon \left( \rho^{\otimes n} \right) \qquad \forall\, a \in \mathbb{N}. \tag{11.140}$$

Since the above inequality holds for all $a \in \mathbb{N}$ we must have

$$\mathrm{Cost}(\rho) \geqslant \lim_{\varepsilon \to 0^+} \liminf_{n \to \infty} \frac{1}{n} \mathrm{Cost}^\varepsilon \left( \rho^{\otimes n} \right). \tag{11.141}$$

Conversely, let $\varepsilon, \delta \in (0,1)$ and let $a \in \mathbb{N}$ be large enough such that $\frac{1}{a} < \delta$. Then,

$$\liminf_{n \to \infty} \frac{1}{n} \mathrm{Cost}^\varepsilon \left( \rho^{\otimes n} \right) \geqslant \inf_{a \leqslant n \in \mathbb{N}} \frac{1}{n} \mathrm{Cost}^\varepsilon \left( \rho^{\otimes n} \right)$$

$$\textbf{by definition} \to \quad = \inf_{\substack{n, m \in \mathbb{N} \\ n \geqslant a}} \left\{ \frac{\log(m)}{n} : T\left( \Phi_m \xrightarrow{\mathfrak{F}} \rho^{\otimes n} \right) \leqslant \varepsilon \right\}. \tag{11.142}$$

Combining this with the inequality $\log(m) \geqslant \lceil \log m \rceil - 1$ gives

$$\liminf_{n \to \infty} \frac{1}{n} \mathrm{Cost}^\varepsilon \left( \rho^{\otimes n} \right) \geqslant \inf_{\substack{n, m \in \mathbb{N} \\ n \geqslant a}} \left\{ \frac{\lceil \log m \rceil - 1}{n} : T\left( \Phi_m \xrightarrow{\mathfrak{F}} \rho^{\otimes n} \right) \leqslant \varepsilon \right\}$$

$$\boxed{k := \lceil \log m \rceil} \quad \longrightarrow \quad \geqslant \inf_{\substack{n, k \in \mathbb{N} \\ n \geqslant a}} \left\{ \frac{k - 1}{n} : T\left( \Phi_{2^k} \xrightarrow{\mathfrak{F}} \rho^{\otimes n} \right) \leqslant \varepsilon \right\}, \tag{11.143}$$

where we used the fact that $m \leqslant 2^k$ so that $\Phi_{2^k} \xrightarrow{\mathfrak{F}} \Phi_m$ and consequently $T\left( \Phi_{2^k} \xrightarrow{\mathfrak{F}} \rho^{\otimes n} \right) \leqslant T\left( \Phi_m \xrightarrow{\mathfrak{F}} \rho^{\otimes n} \right)$. Now, observe that for $n \geqslant a$ we have $(k - 1)/n \geqslant k/n - \delta$ so that

$$\lim_{\varepsilon \to 0^+} \liminf_{n \to \infty} \frac{1}{n} \mathrm{Cost}^\varepsilon \left( \rho^{\otimes n} \right) \geqslant \lim_{\varepsilon \to 0^+} \inf_{n, k \in \mathbb{N}} \left\{ \frac{k}{n} : T\left( \Phi_{2^k} \xrightarrow{\mathfrak{F}} \rho^{\otimes n} \right) \leqslant \varepsilon \right\} - \delta$$

$$= \mathrm{Cost}(\rho) - \delta. \tag{11.144}$$

Since the above inequality holds for all $\delta \in (0,1)$ we conclude that

$$\lim_{\varepsilon \to 0^+} \liminf_{n \to \infty} \frac{1}{n} \mathrm{Cost}^\varepsilon \left( \rho^{\otimes n} \right) \geqslant \mathrm{Cost}(\rho). \tag{11.145}$$

The two inequalities (11.141) and (11.145) then give the desired equality (11.137). ∎

**Exercise 11.13.** *Show that for any $m \in \mathbb{N}$ and $\rho \in \mathfrak{D}(A)$ we have*

$$\frac{1}{m} \mathrm{Cost}\left( \rho^{\otimes m} \right) \geqslant \mathrm{Cost}(\rho) \quad \textit{and} \quad \frac{1}{m} \mathrm{Distill}\left( \rho^{\otimes m} \right) \leqslant \mathrm{Distill}(\rho). \tag{11.146}$$

**Exercise 11.14.** *Let $\rho \in \mathfrak{D}(A)$. Show that for any $a \in \mathbb{N}$ we have*

$$\lim_{\varepsilon \to 0^+} \inf_{n \in \mathbb{N}} \frac{1}{n} \mathrm{Cost}^\varepsilon \left( \rho^{\otimes n} \right) = \lim_{\varepsilon \to 0^+} \inf_{a \leqslant n \in \mathbb{N}} \frac{1}{n} \mathrm{Cost}^\varepsilon \left( \rho^{\otimes n} \right). \tag{11.147}$$

*Hint: Use similar arguments that were used to prove the equality in* (11.122).

**Exercise 11.15.** *Prove the equality in* (11.138).

In many QRTs it is possible to choose the golden unit such that $D_{\mathfrak{F}}^{\mathrm{reg}}(\Phi_2) = 1$. With this normalization we get from Theorem 11.6 that

$$\mathrm{Distill}(\rho) \leqslant D^{\mathrm{reg}}(\rho \| \mathfrak{F}) \leqslant \mathrm{Cost}(\rho). \tag{11.148}$$

Particularly, if the QRT $\mathfrak{F}$ is reversible, then both the asymptotic cost and the asymptotic distillation of the resource $\rho$ equals $D^{\mathrm{reg}}(\rho \| \mathfrak{F})$. Therefore, for reversible QRTs, the regularized relative entropy of a resource is the unique measure of a resource in the asymptotic domain. We make this statement rigorous in the following corollary.

---

**Corollary 11.4.** Let $\mathfrak{F}$ be a reversible QRT with a golden unit $\{\Phi_k\}_{k \in \mathbb{N}}$ such that $D(\Phi_2 \| \mathfrak{F}) = 1$, and let $\mathrm{M}$ be a resource measure that is asymptotically continuous and normalized such that $\mathrm{M}(\Phi_2) = 1$. Then,

$$\mathrm{M}^{\mathrm{reg}}(\rho) = D^{\mathrm{reg}}(\rho \| \mathfrak{F}) \qquad \forall\, \rho \in \mathfrak{D}(A). \tag{11.149}$$

---

**Exercise 11.16.** *Prove the corollary above. Hint: Follow all the lines leading to* (11.148), *but with* $\mathrm{M}$ *replacing everywhere* $D(\cdot \| \mathfrak{F})$.

## 11.5.2 Achieving Reversibility

The reversibility property of a QRT is extremely desirable, as quantum resources are expensive and reversibility ensures resources are not wasted during quantum information processing tasks. Moreover, in the previous subsections we saw that if a QRT is reversible, then the relative entropy of a resource characterizes uniquely all asymptotic interconversions. However, many QRTs are not asymptotically reversible. This typically happens when the set of free operations is not large enough to enable efficient interconversion of resources. It is therefore natural to ask if a QRT is reversible under the maximal set of free operations; that is, under RNG operations.

### Asymptotically RNG Operations

In Section 9.2.1, we discussed RNG operations and argued that they form the largest possible set of free operations. However, in the asymptotic regime, when one considers many copies of resources, one can define a set of operations that are RNG only in the asymptotic limit. That is, the operations become closer to RNG operations when we take the number of copies of the resources involved to infinity. To make this idea rigorous, we first define a set of operations that are approximately RNG.

> **Definition 11.5.** Let $\delta \in [0, 1]$ and $\mathfrak{F}$ be a QRT. We say that a quantum channel $\mathcal{N} \in \mathrm{CPTP}(A \to B)$ is $\mathrm{RNG}_\delta$ if it belong to the set
>
> $$\mathrm{RNG}_\delta(A \to B) := \left\{ \mathcal{E} \in \mathrm{CPTP}(A \to B) \; : \; \mathbf{R}_g\big(\mathcal{E}(\sigma)\big) \leqslant \delta \quad \forall \, \sigma \in \mathfrak{F}(A) \right\}. \tag{11.150}$$
>
> where $\mathbf{R}_g$ is the global robustness of a resource as defined in (10.54).

We have used the global robustness in the definition above since it is a resource monotone that is faithful (see Exercise 10.10) so that the inequality $\mathbf{R}_g\big(\mathcal{E}(\sigma)\big) \leqslant \delta$ implies that $\mathcal{E}(\sigma)$ is close to a free state. Specifically, suppose $\mu := \mathbf{R}_g\big(\mathcal{E}(\sigma)\big) \leqslant \delta$. Then, from (10.56) it follows that

$$\mathcal{E}(\sigma) = (1 + \mu)\tau - \mu\omega \tag{11.151}$$

for some $\tau \in \mathfrak{F}(B)$ and $\omega \in \mathfrak{D}(B)$. Hence, from the above equality we get

$$\frac{1}{2}\big\|\mathcal{E}(\sigma) - \tau\big\|_1 = \frac{1}{2}\big\|\mu(\tau - \omega)\big\|_1 \leqslant \mu \leqslant \delta. \tag{11.152}$$

In other words, if $\mathbf{R}_g\big(\mathcal{E}(\sigma)\big) \leqslant \delta$, then $\mathcal{E}(\sigma)$ is $\delta$-close to a free state.

> **Definition 11.6.** Let $\mathfrak{F}$ be a QRT, and for each $n \in \mathbb{N}$ let $A_n$ and $B_n$ be two physical systems. A sequence of quantum channel $\{\mathcal{E}_n\}_{n \in \mathbb{N}}$, with $\mathcal{E}_n \in \mathrm{CPTP}(A_n \to B_n)$, is said to be asymptotically RNG if there exists a sequence of nonnegative real numbers $\{\delta_n\}_{n \in \mathbb{N}}$ with $\lim_{n \to \infty} \delta_n = 0$ such that for each $n \in \mathbb{N}$, $\mathcal{E}_n \in \mathrm{RNG}_{\delta_n}(A_n \to B_n)$.

Note that in the definition above we do not specify how quickly $\delta_n$ goes to zero. The main result of this section will not be effected even if we require in addition that $\delta_n$ goes to zero exponentially fast with $n$. However, to keep the notion of asymptotically RNG in its most generality we did not include such a condition in the definition above.

## Cost and Distillation

The asymptotic distillable rate of $\rho$ into $\sigma$ under asymptotically RNG operations is defined slightly different than the definitions given in Definition 11.3. Recall that the distillable rate is defined as the supremum of the ratio $\frac{m}{n}$ under the constraints that the conversion distance between $\rho^{\otimes n}$ and $\sigma^{\otimes m}$ is not too high. In the context of *asymptotically* RNG, instead of taking the *supremum* of $\frac{m}{n}$ we take the *limit* when $n$ goes to infinity since we are only interested in the optimal *asymptotic* behavior of this ratio.

For any $\varepsilon \in (0, 1)$ we will use the notation $\mathfrak{R}_\varepsilon(\rho \to \sigma)$ to denote the set of all $r \in \mathbb{R}_+$ such that there exists a sequence $\{m_n\}_{n \in \mathbb{N}} \subset \mathbb{N}$ fulfilling the following two criteria:

1. $\lim_{n\to\infty} \frac{m_n}{n} = r$.
2. There exists another sequence $\{\delta_n\}_{n\in\mathbb{N}} \subset \mathbb{R}_+$ with a limit of zero, such that for every $n \in \mathbb{N}$

$$T\left(\rho^{\otimes n} \xrightarrow{\mathrm{RNG}_{\delta_n}} \sigma^{\otimes m_n}\right) \leqslant \varepsilon. \tag{11.153}$$

That is, the sequence $\{m_n\}_{n\in\mathbb{N}}$ is such that $\rho^{\otimes n}$ can be converted by $\mathrm{RNG}_{\delta_n}$ to a state that is $\varepsilon$-close to $\sigma^{\otimes m_n}$. Hence, the set $\mathfrak{R}_\varepsilon(\rho \to \sigma) \subset \mathbb{R}_+$ consists of all achievable conversion rates under asymptotically RNG that tolerate an $\varepsilon$-error. To get the optimal distillable rate we will have to take the limit $\varepsilon \to 0^+$.

**Exercise 11.17.** *Let $\rho \in \mathfrak{D}(A)$, $\sigma \in \mathfrak{D}(B)$, and*

$$r_\varepsilon := \sup\left\{r : r \in \mathfrak{R}_\varepsilon(\rho \to \sigma)\right\}. \tag{11.154}$$

1. *Show that $\mathfrak{R}_\varepsilon(\rho \to \sigma) = [0, r_\varepsilon]$; in particular, show that the supremum in the definition of $r_\varepsilon$ can be replaced with a maximum.*
2. *Show that $r_\varepsilon$ is nonincreasing in $\varepsilon$.*

> **Definition 11.7.** Let $\mathfrak{F}$ be a QRT, $\rho \in \mathfrak{D}(A)$, and $\sigma \in \mathfrak{D}(B)$. Using the notation given in (11.154) of the exercise above, the asymptotically RNG distillable rate is defined as
>
> $$\mathrm{Distill}(\rho \to \sigma) := \lim_{\varepsilon \to 0^+} r_\varepsilon. \tag{11.155}$$

The condition (11.153) implies that there exists $\mathcal{E}_n \in \mathrm{RNG}_{\delta_n}(A^n \to B^{m_n})$ such that $\mathcal{E}_n(\rho^{\otimes n}) \approx_\varepsilon \sigma^{\otimes m_n}$. Moreover, the condition that $\lim_{n\to\infty} \delta_n = 0$ implies that this sequence of channels $\{\mathcal{E}_n\}$ is asymptotically RNG. Therefore, for a given $\varepsilon \in (0,1)$, we get that $\rho^{\otimes n}$ can be converted by $\mathrm{RNG}_{\delta_n}$ to $\sigma^{\otimes m_n}$ up to an $\varepsilon$-error. The reason that we require $\lim_{n\to\infty} \frac{m_n}{n} = r$ instead of just $\sup\{\frac{m_n}{n}\} = r$ is that the supremum can be achieved with a finite $n$ in which case $\delta_n$ may not be very small. Taking the limit $n \to \infty$ ensures that the conversion $\rho^{\otimes n} \xrightarrow{\mathrm{RNG}_{\delta_n}} \sigma^{\otimes m_n}$ (up to an $\varepsilon$-error) is achieved with a very small $\delta_n$.

**Exercise 11.18.** *Show that if in the definition of $\mathfrak{R}_\varepsilon(\rho \to \sigma)$ we require $\sup\{\frac{m_n}{n}\} = r$ instead of $\lim_{n\to\infty} \frac{m_n}{n} = r$, then we will get that $\mathrm{Distill}(\rho \to \sigma) = \infty$. Hint: Let $n_0$ be a large integer and take $\delta_n = n_0$ for $n \leqslant n_0$ and $\delta_n = 0$ if $n > n_0$.*

For simplicity of the notation, we did not include a subscript in $\mathrm{Distill}(\rho \to \sigma)$ to indicate that the asymptotic distillable rate is calculated with respect to asymptotically RNG operations. Similarly, we denote by $\mathrm{Cost}(\rho \to \sigma) = 1/\mathrm{Distill}(\rho \to \sigma)$ the asymptotic cost rate of $\rho$ into $\sigma$ under asymptotic RNG operations.

## Toward Reversibility

In this book, we will restrict our attention to QRTs that meet the following condition:

$$\kappa(A) := \limsup_{n \to \infty} \frac{1}{n} \max_{\omega \in \mathfrak{D}(A^n)} D(\omega \| \mathfrak{F}) < \infty. \tag{11.156}$$

It's worth noting that this assumption is extremely lenient and is fulfilled by the majority, if not all, of the QRTs discussed in the existing literature. In fact, for many QRTs $\kappa(A) = 0$.

> **Theorem 11.7.** For any $\rho \in \mathfrak{D}(A)$ and $\sigma \in \mathfrak{D}(B)$, the asymptotic distillable rate of $\rho$ into $\sigma$ under asymptotic RNG is bounded by
>
> $$\mathrm{Distill}(\rho \to \sigma) \leqslant \frac{D^{\mathrm{reg}}(\rho \| \mathfrak{F})}{D^{\mathrm{reg}}(\sigma \| \mathfrak{F})}. \tag{11.157}$$

*Remark.* Theorem 11.7 does *not* follow from Theorem 11.6 since $\mathrm{Distill}(\rho \to \sigma)$ is calculated with respect to asymptotically RNG operations. Since these operations allow for the generation of a resource (although small amount that vanishes asymptotically), the proof of Theorem 11.6 cannot be applied directly, and a revised version is necessary to accommodate this case.

**Proof**  Suppose by contradiction that

$$\mathrm{Distill}(\rho \to \sigma) > \frac{D^{\mathrm{reg}}(\rho \| \mathfrak{F})}{D^{\mathrm{reg}}(\sigma \| \mathfrak{F})} + 2\delta \tag{11.158}$$

for some small positive $\delta$. By definition, this means in particular that for sufficiently small $\varepsilon \in (0, 1)$ there exists $r \in \mathfrak{R}_\varepsilon(\rho \to \sigma)$ such that

$$r > \frac{D^{\mathrm{reg}}(\rho \| \mathfrak{F})}{D^{\mathrm{reg}}(\sigma \| \mathfrak{F})} + \delta. \tag{11.159}$$

Since $r \in \mathfrak{R}_\varepsilon(\rho \to \sigma)$ there exists a sequence $\{m_n\}_{n \in \mathbb{N}} \subset \mathbb{N}$ satisfying both $r = \lim_{n \to \infty} \frac{m_n}{n}$ and (11.153). From (11.153) it follows that there exists $\mathcal{E}_n \in \mathrm{RNG}_{\delta_n}(A^n \to B^{m_n})$ such that

$$\mathcal{E}_n\left(\rho^{\otimes n}\right) \approx_\varepsilon \sigma^{\otimes m_n}. \tag{11.160}$$

Now, since $D(\cdot \| \mathfrak{F})$ is asymptotically continuous it follows that (cf. (10.34))

$$\left| D\left(\mathcal{E}_n\left(\rho^{\otimes n}\right) \| \mathfrak{F}\right) - D\left(\sigma^{\otimes m_n} \| \mathfrak{F}\right) \right| \leqslant c_n \varepsilon + (1 + \varepsilon) h\left(\frac{\varepsilon}{1 + \varepsilon}\right), \tag{11.161}$$

where

$$c_n := \max_{\omega \in \mathfrak{D}(B^{m_n})} D(\omega \| \mathfrak{F}). \tag{11.162}$$

Dividing both sides by $m_n$ and taking the limit $n \to \infty$ gives

$$D^{\mathrm{reg}}(\sigma \| \mathfrak{F}) \leqslant \lim_{n \to \infty} \frac{1}{m_n} D\left(\mathcal{E}_n\left(\rho^{\otimes n}\right) \| \mathfrak{F}\right) + \kappa(B)\varepsilon. \tag{11.163}$$

For each $n \in \mathbb{N}$, let $\omega_n \in \mathfrak{F}\left(A^n\right)$ be an optimizer state satisfying

$$D\left(\rho^{\otimes n} \,\middle\|\, \omega_n\right) = D\left(\rho^{\otimes n} \,\middle\|\, \mathfrak{F}\right). \tag{11.164}$$

Then, for each $n \in \mathbb{N}$

$$D\left(\mathcal{E}_n\left(\rho^{\otimes n}\right) \,\middle\|\, \mathfrak{F}\right) = \min_{\tau_n \in \mathfrak{F}(B^{m_n})} D\left(\mathcal{E}_n\left(\rho^{\otimes n}\right) \,\middle\|\, \tau_n\right)$$

$$\mathbf{(6.114)} \to \; \leqslant D\left(\mathcal{E}_n\left(\rho^{\otimes n}\right) \,\middle\|\, \mathcal{E}_n(\omega_n)\right) + \min_{\tau_n \in \mathfrak{F}(B^{m_n})} D_{\max}\left(\mathcal{E}_n\left(\omega_n\right) \,\middle\|\, \tau_n\right) \tag{11.165}$$

$$\mathbf{DPI} \to \; \leqslant D\left(\rho^{\otimes n} \,\middle\|\, \mathfrak{F}\right) + D_{\max}\left(\mathcal{E}_n\left(\omega_n\right) \,\middle\|\, \mathfrak{F}\right).$$

Since $\omega_n$ is a free state and since each $\mathcal{E}_n$ is $\mathrm{RNG}_{\delta_n}$, the global robustness of $\mathcal{E}_n\left(\omega_n\right)$ cannot exceed $\delta_n$, and in particular $D_{\max}\left(\mathcal{E}_n\left(\omega_n\right) \,\middle\|\, \mathfrak{F}\right) \leqslant \log\left(1 + \delta_n\right)$. Therefore,

$$D\left(\mathcal{E}_n\left(\rho^{\otimes n}\right) \,\middle\|\, \mathfrak{F}\right) \leqslant D\left(\rho^{\otimes n} \,\middle\|\, \mathfrak{F}\right) + \log\left(1 + \delta_n\right). \tag{11.166}$$

Substituting this into (11.163) gives

$$D^{\mathrm{reg}}(\sigma \,\|\, \mathfrak{F}) \leqslant \lim_{n \to \infty} \frac{1}{m_n}\left(D\left(\rho^{\otimes n} \,\middle\|\, \mathfrak{F}\right) + \log\left(1 + \delta_n\right)\right) + \kappa(B)\varepsilon$$

$$= \lim_{n \to \infty} \frac{n}{m_n} \frac{1}{n} D\left(\rho^{\otimes n} \,\middle\|\, \mathfrak{F}\right) + \kappa(B)\varepsilon \tag{11.167}$$

$$\boxed{\lim_{n \to \infty} \frac{m_n}{n} = r} \longrightarrow \quad = \frac{1}{r} D^{\mathrm{reg}}(\rho \,\|\, \mathfrak{F}) + \kappa(B)\varepsilon.$$

However, since $r > \frac{D^{\mathrm{reg}}(\rho\|\mathfrak{F})}{D^{\mathrm{reg}}(\sigma\|\mathfrak{F})} + \delta$ for sufficiently small $\varepsilon \in (0, 1)$, we get the contradiction

$$D^{\mathrm{reg}}(\sigma \,\|\, \mathfrak{F}) \leqslant \frac{1}{r} D^{\mathrm{reg}}(\rho \,\|\, \mathfrak{F}) + \kappa(B)\varepsilon \tag{11.168}$$

$$\mathbf{Exercise\ 11.19} \to \; < D^{\mathrm{reg}}(\sigma \,\|\, \mathfrak{F}).$$

This completes the proof. $\blacksquare$

**Exercise 11.19.** *Set $a := D^{\mathrm{reg}}(\rho\|\mathfrak{F})$ and $b := D^{\mathrm{reg}}(\sigma\|\mathfrak{F})$. Show that if $\varepsilon \in (0, 1)$ is chosen small enough such that*

$$\kappa(B)\varepsilon < \frac{a^2}{b}\delta. \tag{11.169}$$

*then for $r > \frac{a}{b} + \delta$*

$$\frac{a}{r} + \kappa(B)\varepsilon < b. \tag{11.170}$$

If the channels $\{\mathcal{E}_n\}_{n \in \mathbb{N}}$ in the proof above where generating a sublinear amount of a resource (instead of being asymptotically RNG) the result would still not change. That is, in the proof above we could replace the condition $\mathcal{E}_n \in \mathrm{RNG}_{\delta_n}(A^n \to B^{m_n})$ with the weaker condition that

$$\lim_{n \to \infty} \max_{\omega_n \in \mathfrak{F}(A^n)} \frac{D_{\max}\left(\mathcal{E}_n(\omega_n) \,\middle\|\, \mathfrak{F}\right)}{m_n} - 0. \tag{11.171}$$

To see why, observe that the only change in the proof above would be to replace the term $\frac{\log(1+\varepsilon)}{n}$ in the first line of (11.167) with the ratio $D_{\max}\left(\mathcal{E}_n(\omega_n)\|\mathfrak{F}\right)/m_n$ which also goes to zero in the limit $n \to \infty$. Hence, if the logarithmic robustness of $\mathcal{E}_n(\omega_n)$ grows sublinearly with $m_n$ the bound on the distillable rate would still hold. This observation is consistent with the intuition that a sublinear amount of a resource becomes negligible in the asymptotic limit and therefore cannot increase the distillable rate.

> **Theorem 11.8.** For any $\rho \in \mathfrak{D}(A)$, $\sigma \in \mathfrak{D}(B)$, and $\varepsilon \in (0, 1)$, the asymptotic distillable rate of $\rho$ into $\sigma$ under asymptotic RNG is given by
>
> $$\text{Distill}(\rho \to \sigma) = \frac{D^{\text{reg}}(\rho\|\mathfrak{F})}{D^{\text{reg}}(\sigma\|\mathfrak{F})}. \qquad (11.172)$$

**Proof** Due to Theorem 11.7, it is sufficient to prove that

$$\text{Distill}(\rho \to \sigma) \geqslant \frac{D^{\text{reg}}(\rho\|\mathfrak{F})}{D^{\text{reg}}(\sigma\|\mathfrak{F})}. \qquad (11.173)$$

For this purpose, let $r$ be a positive number satisfying $r < D^{\text{reg}}(\rho\|\mathfrak{F})/D^{\text{reg}}(\sigma\|\mathfrak{F})$, fix $\varepsilon \in (0, 1)$, and denote by $m_n := \lceil nr \rceil$ so that $\lim_{n\to\infty} \frac{m_n}{n} = r$. We will construct a sequence of channels $\{\mathcal{E}_n\}_{n\in\mathbb{N}}$ with the following two properties:

1. For sufficiently large $n \in \mathbb{N}$, the channel $\mathcal{E}_n \in \text{RNG}_{\delta_n}(A^n \to B^{m_n})$ with $\delta_n := 2^{-n\delta}$ (for some $\delta > 0$). Hence, the sequence $\{\mathcal{E}_n\}_{n\in\mathbb{N}}$ is asymptotically RNG.
2. For sufficiently large $n \in \mathbb{N}$, we have $\mathcal{E}_n(\rho^{\otimes n}) \approx_\varepsilon \sigma^{\otimes m_n}$.

Note that from the definition of $\text{Distill}(\rho \to \sigma)$, if for any choice of $\varepsilon \in (0, 1)$ there exists a sequence $\{\mathcal{E}_n\}_{n\in\mathbb{N}}$ that satisfies the above two conditions then we must have $\text{Distill}(\rho \to \sigma) \geqslant r$.

The idea behind the construction of the channels $\{\mathcal{E}_n\}_{n\in\mathbb{N}}$ is to try to achieve the rate $r$ with a (two-outcome) measurement-prepare channel of the form

$$\mathcal{E}_n(\eta) := \text{Tr}\left[\Lambda_n\eta\right]\sigma_n + \text{Tr}\left[\left(I^{A^n} - \Lambda_n\right)\eta\right]\omega_n \qquad \forall\, \eta \in \mathfrak{L}(A^n). \qquad (11.174)$$

for some $\sigma_n, \omega_n \in \mathfrak{D}(B^{m_n})$ and some $\Lambda_n \in \text{Eff}(A^n)$. We therefore need to check if there exist $\sigma_n$, $\omega_n$, and $\Lambda_n$ that satisfy both $\mathcal{E}_n(\rho^{\otimes n}) \approx_\varepsilon \sigma^{\otimes m_n}$ and $\mathcal{E}_n \in \text{RNG}_{\delta_n}(A^n \to B^{m_n})$. Note that if we choose $\Lambda_n$ such that $\text{Tr}[\Lambda_n\rho^{\otimes n}]$ is close to 1, then $\mathcal{E}_n(\rho^{\otimes n})$ will be close to $\sigma_n$. Therefore, if $\sigma_n$ is close to $\sigma^{\otimes m_n}$ we will get in this case that $\mathcal{E}_n(\rho^{\otimes n})$ is also close to $\sigma^{\otimes m_n}$. We take $\omega_n \in \mathfrak{F}(B^{m_n})$ to be any free density matrix, and define now $\Lambda_n$ and $\sigma_n$.

1. **Definition of $\Lambda_n$:** Denote by $a := D^{\text{reg}}(\rho\|\mathfrak{F})$ and observe that from the generalized quantum Stein's lemma (see Theorem 11.4) for every $\delta > 0$ and sufficiently large $n \in \mathbb{N}$ we get

$$\min_{\tau_n \in \mathfrak{F}(A^n)} D_{\min}^{\varepsilon/2}\left(\rho^{\otimes n}\|\tau_n\right) \geqslant n\,(a - \delta). \qquad (11.175)$$

(the choice $\varepsilon/2$ instead of $\varepsilon$ will be clear shortly). The left-hand side of the inequality above can be expressed as

$$
\min_{\tau_n \in \mathfrak{F}(A^n)} D_{\min}^{\varepsilon/2}\left(\rho^{\otimes n} \middle\| \tau_n\right) = -\log \max_{\substack{\tau_n \in \mathfrak{F}(A^n)}} \min_{\substack{\Lambda_n' \in \mathrm{Eff}(A^n) \\ \mathrm{Tr}[\rho^{\otimes n}\Lambda_n'] \geqslant 1-\varepsilon/2}} \mathrm{Tr}\left[\Lambda_n' \tau_n\right]
$$

$$
\overset{\substack{\textbf{Minimax Theorem} \\ \textbf{See Lemma 10.2}}}{\longrightarrow} = -\log \min_{\substack{\Lambda_n' \in \mathrm{Eff}(A^n) \\ \mathrm{Tr}[\rho^{\otimes n}\Lambda_n'] \geqslant 1-\varepsilon/2}} \max_{\tau_n \in \mathfrak{F}(A^n)} \mathrm{Tr}\left[\Lambda_n' \tau_n\right] \qquad (11.176)
$$

$$
\overset{\textbf{Definition of } \Lambda_n}{\longrightarrow} := -\log \max_{\tau_n \in \mathfrak{F}(A^n)} \mathrm{Tr}\left[\Lambda_n \tau_n\right].
$$

Combining the two equations above implies that the optimal effect $\Lambda_n$ satisfies (for and $\delta > 0$ and sufficiently large $n \in \mathbb{N}$)

$$
\max_{\tau_n \in \mathfrak{F}(A^n)} \mathrm{Tr}\left[\Lambda_n \tau_n\right] \leqslant 2^{-n(a-\delta)} \quad \text{and} \quad \mathrm{Tr}[\rho^{\otimes n}\Lambda_n] = 1 - \frac{\varepsilon}{2}. \qquad (11.177)
$$

2. **Definition of $\sigma_n$:** First, observe that if we choose $\sigma_n$ to be $\varepsilon/2$-close to $\sigma^{\otimes m_n}$ we get from the triangle inequality

$$
\frac{1}{2}\left\|\mathcal{E}_n\left(\rho^{\otimes n}\right) - \sigma^{\otimes m_n}\right\|_1 \leqslant \frac{1}{2}\left\|\mathcal{E}_n\left(\rho^{\otimes n}\right) - \sigma_n\right\|_1 + \frac{1}{2}\left\|\sigma_n - \sigma^{\otimes m_n}\right\|_1
$$

$$
\overset{(11.174)}{\longrightarrow} = \frac{1}{2}\left\|\frac{\varepsilon}{2}(\omega_n - \sigma_n)\right\|_1 + \frac{1}{2}\left\|\sigma_n - \sigma^{\otimes m_n}\right\|_1 \qquad (11.178)
$$

$$
\leqslant \frac{\varepsilon}{2} + \frac{\varepsilon}{2} = \varepsilon.
$$

That is, $\mathcal{E}_n\left(\rho^{\otimes n}\right) \approx_\varepsilon \sigma^{\otimes m_n}$. Therefore, we would like to define $\sigma_n$ that is $\varepsilon/2$-close to $\sigma^{\otimes m_n}$ such that $\mathcal{E}_n \in \mathrm{RNG}_{\delta_n}(A^n \to B^{m_n})$.

We take $\sigma_n \in \mathfrak{D}(B^{m_n})$ to be a density matrix that satisfies $D_{\max}(\sigma_n\|\mathfrak{F}) = D_{\max}^{\varepsilon/2}\left(\sigma^{\otimes m_n}\|\mathfrak{F}\right)$. The intuition behind this choice is that besides of being $\varepsilon/2$-close to $\sigma^{\otimes m_n}$, the density matrix $\sigma_n$ does not have "too much" robustness. To see why, recall first that from Lemma 11.3 it follows that

$$
D^{\mathrm{reg}}(\sigma\|\mathfrak{F}) \geqslant \limsup_{n\to\infty} \frac{1}{m_n} D_{\max}^{\varepsilon/2}\left(\sigma^{\otimes m_n}\|\mathfrak{F}\right)
$$

$$
= \limsup_{n\to\infty} \frac{1}{m_n} D_{\max}(\sigma_n\|\mathfrak{F}) \qquad (11.179)
$$

$$
\boxed{\lim_{n\to\infty}\frac{m_n}{n} = r} \longrightarrow = \frac{1}{r}\limsup_{n\to\infty}\frac{1}{n} D_{\max}(\sigma_n\|\mathfrak{F}).
$$

Now, since the inequality $r < a/D^{\mathrm{reg}}(\sigma\|\mathfrak{F})$ is strict, there exists $\delta > 0$ sufficiently small such that $r < (a - 2\delta)/D^{\mathrm{reg}}(\sigma\|\mathfrak{F})$, or equivalently

$$
r D^{\mathrm{reg}}(\sigma\|\mathfrak{F}) < a - 2\delta. \qquad (11.180)
$$

Hence, by combining the two Equations (11.179) and (11.180) we get that for sufficiently large $n$

$$
D_{\max}(\sigma_n\|\mathfrak{F}) \leqslant n\left(a - 2\delta\right). \qquad (11.181)
$$

That is, the global robustness (as defined in (10.54)) of $\sigma_n$ satisfies

$$\mathbf{R}_g(\sigma_n) \leqslant 2^{n(a-2\delta)} - 1. \tag{11.182}$$

To show that for these choices the channel $\mathcal{E}_n$ is $\mathrm{RNG}_{\delta_n}$, let $\eta \in \mathfrak{F}(A^n)$ be a free state, and denote by $t_n := \mathrm{Tr}\,[\Lambda_n \eta]$ and $r_n := \mathbf{R}_g(\sigma_n)$. Then, from the convexity of the global robustness we get

$$\mathbf{R}_g\big(\mathcal{E}_n\,(\eta)\big) \leqslant t_n \mathbf{R}_g(\sigma_n) + (1 - t_n)\mathbf{R}_g(\omega_n)$$

$$\boxed{\omega_n \in \mathfrak{F}\left(B^{m_n}\right)} \longrightarrow \; = t_n r_n \leqslant t_n(1 + r_n). \tag{11.183}$$

Now, from (11.182) we have $r_n + 1 \leqslant 2^{n(a-2\delta)}$ and from (11.177)

$$t_n := \mathrm{Tr}\,[\Lambda_n \eta] \leqslant \max_{\tau_n \in \mathfrak{F}(A^n)} \mathrm{Tr}\,[\Lambda_n \tau_n] \leqslant 2^{-n(a-\delta)}. \tag{11.184}$$

Combining everything we get

$$\mathbf{R}_g\,(\mathcal{E}_n\,(\eta)) \leqslant 2^{-n\delta} = \delta_n. \tag{11.185}$$

That is, $\mathcal{E}_n \in \mathrm{RNG}_{\delta_n}\,(A^n \to B^{m_n})$. This completes the proof. ∎

**Exercise 11.20.** *Let $\sigma_n$ and $\sigma^{\otimes m_n}$ be as in the proof above.*

*1. Show that the robustness of $\sigma^{\otimes m_n}$ is bounded by*

$$\mathbf{R}_g\left(\sigma^{\otimes m_n}\right) \leqslant 2^{m_n D_{\max}(\sigma\|\mathfrak{F})} - 1. \tag{11.186}$$

*2. Show that the right-hand side of (11.186) is larger than the bound on $R(\sigma_n)$ given in (11.182).*

## Examples

As an example, consider the resource theory of quantum coherence. In this example, the set of free states are diagonal with respect to a fixed basis. In this example, the $\alpha$-Rényi relative entropy of a resource is additive. To see this recall that (see (10.105))

$$D_\alpha(\rho\|\mathfrak{F}) = \frac{1}{\alpha - 1} \log \left\|\Delta\left(\rho^\alpha\right)\right\|_{1/\alpha}, \tag{11.187}$$

where $\Delta \in \mathrm{CPTP}(A \to A)$ is the completely dephasing map. Moreover, on $n$ copies of $A$, the completely dephasing map $\Delta_n \in \mathrm{CPTP}(A^n \to A^n)$ satisfy $\Delta_n = \Delta^{\otimes n}$, where $\Delta$ is the completely dephasing map on a single copy of $A$. We therefore get that

$$\begin{aligned}
D_\alpha\left(\rho^{\otimes n}\|\mathfrak{F}\right) &= \frac{1}{\alpha - 1} \log \left\|\Delta_n\left((\rho^\alpha)^{\otimes n}\right)\right\|_{1/\alpha} \\
&= \frac{1}{\alpha - 1} \log \left\|(\Delta\left(\rho^\alpha\right))^{\otimes n}\right\|_{1/\alpha} \\
&= n\frac{1}{\alpha - 1} \log \left\|\Delta\left(\rho^\alpha\right)\right\|_{1/\alpha} = n D_\alpha(\rho\|\mathfrak{F}).
\end{aligned} \tag{11.188}$$

Therefore, in the case,

$$D_\alpha^{\mathrm{reg}}(\rho\|\mathfrak{F}) = D_\alpha(\rho\|\mathfrak{F}). \tag{11.189}$$

Since $D_\alpha(\,\cdot\,\|\mathfrak{F})$ is continuous at $\alpha = 1$ (see Lemma 10.3), we conclude that the QRT of quantum coherence is reversible with

$$\mathrm{Distill}(\rho \to \sigma) = \frac{H\big(\Delta(\rho)\big) - H(\rho)}{H\big(\Delta(\sigma)\big) - H(\sigma)}. \tag{11.190}$$

**Exercise 11.21.** *Prove the equality above.*

As a second example, consider the QRT consisting of conditionally unital channels. In this QRT the free states are given by

$$\mathfrak{F}(AB) = \Big\{ \mathbf{u}^A \otimes \sigma^B : \sigma \in \mathfrak{D}(B) \Big\}. \tag{11.191}$$

Since this QRT is also an affine QRT it has a self-adjoint resource destroying channel given by

$$\Delta^{AB \to AB}\left(\omega^{AB}\right) := \mathbf{u}^A \otimes \omega^B \qquad \forall\, \omega \in \mathfrak{L}(AB). \tag{11.192}$$

Also in this QRT the $\alpha$-relative entropy of a resource is additive so that we get the distillable rate to be

$$\mathrm{Distill}\left(\rho^{AB} \to \sigma^{AB}\right) = \frac{\log|A| - H(A|B)_\rho}{\log|A| - H(A|B)_\sigma}. \tag{11.193}$$

**Exercise 11.22.** *Prove the equality above.*

## 11.6 Notes and References

Theorem 11.1 is credited in Ref. [95]. In deriving Corollary 11.1, we adopted the approach in Ref. [227], which utilized similar methodologies to establish the same corollary within entanglement theory. The generalized Asymptotic Equipartition Property (AEP) presented in Theorem 11.3 is a novel addition to this book, although a less robust version was initially proven in Ref. [32]. In the same publication, the authors also supported the validity of the generalized Stein's lemma (Theorem 11.4); however, subsequent analyses identified a flaw in the proof, as discussed in Refs. [75] and [24]. Fortunately, the gap was recently addressed by two independent corrections found in Refs. [116] and [146], with this book adopting the proof strategy in Ref. [116]. The proof for the uniqueness of the Umegaki relative entropy, as outlined in Ref. [98], traces its origins back to the foundational work in Ref. [162].

PART IV

# ENTANGLEMENT THEORY

# 12

# Pure-State Entanglement

Entanglement theory is the poster child of quantum resource theories. As we explored in Chapter 1.4, entanglement not only piques our curiosity from a fundamental perspective but also represents a valuable resource that can facilitate specific quantum information processing tasks. In the chapters ahead, we embark on a journey to rigorously define entanglement and formulate its corresponding resource theory. Our focus encompasses the classification, detection, quantification, and manipulation of entanglement.

## 12.1 Definition of Quantum Entanglement

As already discussed, entanglement can be regarded as a resource with practical utility in specific quantum information processing tasks. Take, for instance, quantum teleportation, a process where entanglement is harnessed to simulate a quantum channel. Naturally, if Alice and Bob already possess a noiseless quantum channel that they can freely employ an unlimited number of times, entanglement holds no value for them since they can generate it without constraints. In this context, entanglement serves as the means by which parties surmount the limitations imposed by their apparatuses, enabling them to perform operations that extend beyond local quantum operations (e.g. quantum measurements) aided by classical communication. Thus, we can operationally define entanglement as follows:

> **Quantum Entanglement**
>
> **Definition 12.1.** Entanglement is a characteristic of a composite physical system that cannot be created or enhanced through local (quantum) operations and classical communication (LOCC).

This definition precisely captures the intuition that entanglement is a quantum property of a composite system that corresponds to correlations that are not classical. Historically, this intuition led many researchers to associate entanglement with the nonlocal correlations exhibited by composite physical systems. These correlations find expression in the probability distribution $p(ab|xy)$ observed in a Bell-type scenario. However, as we will explore later, while the above definition of entanglement relates to Bell's nonlocal correlations, it is not an exact replica of the same property. Consequently, in general, entanglement and Bell nonlocality represent subtly different concepts.

To better understand the properties of entanglement, it is essential first to grasp the structure of LOCC. We first encountered LOCC in the context of quantum teleportation. In this process, Alice performs a quantum measurement on her two systems (a local quantum operation) and then transmits the measurement's outcome to Bob (using classical communication). At the end of the protocol, Bob executes a local (unitary) operation on his system. The LOCC in the teleportation protocol is particularly unique because it involves only a one-way classical communication channel from Alice to Bob. We refer to this restricted set of LOCC as $\text{LOCC}_1$.

Generally, LOCC allows for unlimited rounds of classical communication. Local (quantum) operations and classical communication limited to $n$ rounds of classical communication is denoted as $\text{LOCC}_n$. We also use the notation $\text{LO} = \text{LOCC}_0$ for local operations that occur without any communication. An example of an LO map is the local unitary operation $U^{AB} = U^A \otimes U^B$. More broadly, an LO operation can be defined as a quantum channel $\mathcal{E} \in \text{LOCC}(AB \to A'B')$ of the form $\mathcal{E} = \mathcal{M} \otimes \mathcal{N}$, where $\mathcal{M} \in \text{CPTP}(A \to A')$ and $\mathcal{N} \in \text{CPTP}(B \to B')$ are channels on Alice's and Bob's sides, respectively.

The most general quantum operation involving Alice performing a local operation (such as a generalized measurement) and then sending classical information to Bob can be characterized by a quantum instrument, $\mathcal{E} \in \text{CPTP}(A \to A'Y)$. Here, $Y$ represents the classical system that Bob receives from Alice. Upon receiving $Y$, Bob can implement a local operation, $\mathcal{F} \in \text{CPTP}(BY \to B')$. Consequently, the set of all $\text{LOCC}_1$ operations is defined mathematically as follows:

$$\text{LOCC}_1(AB \to A'B') := \left\{ \mathcal{F}^{BY \to B'} \circ \mathcal{E}^{A \to A'Y} : \begin{matrix} \mathcal{E} \in \text{CPTP}(A \to A'Y) \\ \mathcal{F} \in \text{CPTP}(BY \to B') \end{matrix}, |Y| < \infty \right\}. \quad (12.1)$$

Note that we do not impose any constraint on the classical system $Y$, only that it is finite dimensional. Setting $n := |Y|$, an $\text{LOCC}_1$ channel can be expressed as

$$\mathcal{F}^{BY \to B'} \circ \mathcal{E}^{A \to A'Y} = \sum_{y \in [n]} \mathcal{F}_{(y)}^{B \to B'} \otimes \mathcal{E}_y^{A \to A'}. \quad (12.2)$$

Here, for every $y \in [n]$, the operation $\mathcal{F}_{(y)} \in \text{CPTP}(B \to B')$, and $\mathcal{E}_y \in \text{CP}(A \to A')$. Furthermore, the sum $\sum_{y \in [n]} \mathcal{E}_y$ is trace preserving.

By incorporating an additional round of communication from Bob to Alice, we obtain channels in $\text{LOCC}_2$. Specifically, a channel $\mathcal{N} \in \text{LOCC}_2(AB \to A'B')$ can be expressed as follows:

$$\mathcal{N}^{AB \to A'B'} = \mathcal{E}_1^{A_1 X_1 \to A'} \circ \mathcal{F}^{BY_1 \to B_1 X_1} \circ \mathcal{E}_0^{A \to A_1 Y_1}, \quad (12.3)$$

where $A_1$ represents an additional system on Alice's side, $\mathcal{E}_0$ and $\mathcal{E}_1$ are channels on Alice's side, and $\mathcal{F}$ is a channel on Bob's side. It's important to note that without the second round of communication, which corresponds to the case when $|X_1| = 1$, the description reverts to a channel in $\text{LOCC}_1$.

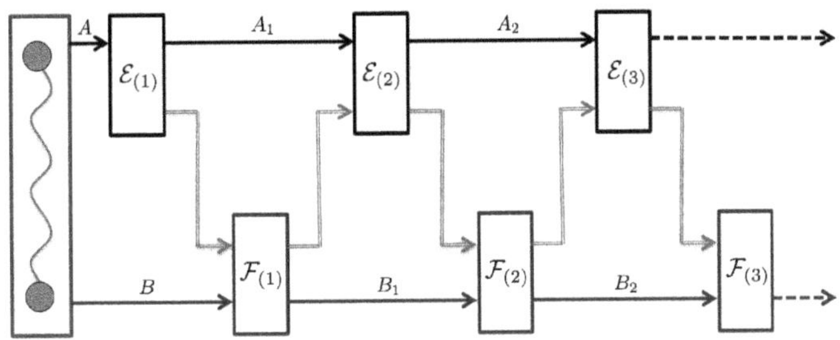



**Figure 12.1** An LOCC operation. The double lines represents classical communication between the parties.

**Exercise 12.1.** *Show that if $|X_1| = 1$, then the channel in (12.3) belongs to* $\text{LOCC}_1$.

In the same fashion, one can continue and express the most general protocol in $\text{LOCC}_n$. Clearly, from the construction above it is obvious that the expression of LOCC protocols can be very complicated particularly if it involves a large number of classical communication rounds (see Figure 12.1). Moreover, it is also known that $\text{LOCC}_n$ is a strict subset of $\text{LOCC}_{n+1}$ for all $n \in \mathbb{N}$. Due to this notorious complexity of LOCC, and despite the enormous body of work in recent years on the study of LOCC, there are still many open problems in entanglement theory. For this reason, it is sometimes convenient to consider a slightly larger class of operations that contains LOCC and have a simpler characterization. We will consider in the next chapter two such sets of operations known as the *separable set* and the PPT set. However, as we will see in this chapter, the complexity of LOCC is reduced dramatically when the bipartite system is initially in a *pure* state.

In entanglement theory we consider all systems to be composite systems consisting of at least two subsystems that we will denote by $A$ (operated by Alice) and $B$ (operated by Bob). This means that even a single system on Alice's side that is described by a density matrix $\rho \in \mathfrak{D}(A)$ can be treated mathematically as a bipartite system with $\rho$ acting on $AB$, with $B$ being the trivial system (i.e. $|B| = 1$) in this case. The set of free operations of this theory is $\mathfrak{F} = \text{LOCC}$, and we will typically use the notation $\text{LOCC}(AB \to A'B')$ to indicated bipartite channels in $\text{CPTP}(AB \to A'B')$ that take bipartite states in $\mathfrak{D}(AB)$ to bipartite states in $\mathfrak{D}(A'B')$ (i.e. $A$ and $A'$ represent Alice's subsystems, and $B$ and $B'$ Bob's subsystems).

**Exercise 12.2 (Separable Operations).** *Let* $\text{SEP}(AB \to A'B') \subset \text{CPTP}(AB \to A'B')$ *be the set of all quantum channels that has a separable operator sum representation; that is, $\mathcal{E} \in \text{SEP}(AB \to A'B')$ if and only if there exist sets of operators $M_x : A \to A'$ and $N_x : B \to B'$ such that*

$$\mathcal{E}^{AB \to A'B'}\left(\rho^{AB}\right) := \sum_x (M_x \otimes N_x)\, \rho^{AB}\, (M_x \otimes N_x)^* \qquad \forall\, \rho \in \mathfrak{L}(AB). \qquad (12.4)$$

*1. Show that*

$$\text{LOCC}(AB \to A'B') \subseteq \text{SEP}(AB \to A'B'). \qquad (12.5)$$

2. *Show that if $\mathcal{E} \in \mathrm{SEP}(AB \to A'B')$, then its normalized Choi matrix is a separable density matrix in $\mathfrak{D}(AA'BB')$ ( the seperability is between Alice system $AA'$ and Bob's system $BB'$).*

## 12.2 Exact Manipulations of Entanglement

Since entanglement is a resource, it is typically not available in its purest form as described for example by the pure (bipartite) singlet state. Instead, entangled systems are quite often only partially entangled, and given in a form that is mixed with noise. Consequently, such systems are not ideal resources for certain quantum information processing tasks (e.g. quantum teleportation). To identify which entangled states are more resourceful than others, we study here manipulations of entanglement; that is, the conversion of one (partially) entangled state to another by LOCC.

The most general LOCC protocol can be characterized by a quantum instrument $\{\mathcal{E}_x\}$, with each $\mathcal{E}_x \in \mathrm{CP}(AB \to A'B')$ being a trace nonincreasing map. The parameter $x$ corresponds to all the measurement outcomes that were involved in the protocol (and that were not discarded). Therefore, if the initial state of the system was $\rho^{AB}$, the final state of the system after outcome $x$ occurred is given by $\mathcal{E}_x\left(\rho^{AB}\right)/p_x$, where

$$p_x = \mathrm{Tr}\left[\mathcal{E}_x\left(\rho^{AB}\right)\right]. \tag{12.6}$$

If for all $x$, the post-measurement state is the same, that is, $\mathcal{E}_x\left(\rho^{AB}\right)/p_x = \sigma^{AB}$ for *all* $x$, then $\sigma^{AB} = \sum_x \mathcal{E}_x\left(\rho^{AB}\right)$, and the map $\sum_x \mathcal{E}_x := \mathcal{E}$ is an LOCC CPTP map. In particular, we say that a bipartite state $\rho^{AB}$ is *more entangled* (i.e. more resourceful) than another state $\sigma^{AB}$ if there exists an LOCC CPTP map $\mathcal{E}$ such that $\sigma^{AB} = \mathcal{E}(\rho^{AB})$.

In general, given two entangled mixed states $\rho^{AB}$ and $\sigma^{AB}$, it can be very difficult to determine if they are related by an LOCC CPTP map. There are two main reasons for that: as we have already discussed, LOCC is a very complex set to characterize, and in addition the states $\rho^{AB}$ and $\sigma^{AB}$ are mixed. It turns out that the entanglement properties of mixed states are very hard to characterize in general. For example, it is known to be NP-hard to determine if a mixed bipartite state $\rho^{AB}$ is entangled or separable. We therefore first treat in this chapter the pure bipartite entanglement manipulations, and postpone the treatment of mixed-state entanglement manipulations to the next chapter.

### 12.2.1 LOCC on Pure Bipartite States

Here we consider the effect of LOCC operations on a pure bipartite state. The set of all pure bipartite states will be denoted by Pure($AB$), and without loss of generality, we will assume in this section that $|A| = |B| := d$, since we can always embed any bipartite system in a larger Hilbert space with both local dimensions equal $\max\{|A|, |B|\}$.

Recall that any such state $\psi \in \text{Pure}(AB)$ has a Schmidt decomposition of the form (see Exercise 2.38)

$$|\psi^{AB}\rangle = \sum_{x\in[d]} \sqrt{p_x}|\psi_x^A\rangle \otimes |\phi_x^B\rangle := \left(\Lambda^A \otimes I^B\right)|\Omega^{AB}\rangle, \tag{12.7}$$

where $\{|\psi_x\rangle^A\}_{x\in[d]}$ and $\{|\phi_x\rangle^B\}_{x\in[d]}$ are orthonormal bases of $A$ and $B$, respectively, $\Lambda^A = \text{Diag}(\sqrt{p_1}, \ldots, \sqrt{p_d})$ is a diagonal matrix in the basis $\{|\psi_x\rangle^A\}_{x\in[d]}$, and $|\Omega^{AB}\rangle = \sum_{x\in[d]} |\psi_x^A\rangle \otimes |\phi_x^B\rangle$ is an (unnormalized) maximally entangled state.

We consider now the effect of a local measurement on Bob side. For this purpose, let $N$ be some $d \times d$ complex matrix, and note that

$$\left(I^A \otimes N\right)|\psi^{AB}\rangle = (\Lambda \otimes N)|\Omega^{AB}\rangle = \left(\Lambda N^T \otimes I^B\right)|\Omega^{AB}\rangle. \tag{12.8}$$

**Exercise 12.3.** *Show that the matrix $\Lambda N^T$ has same singular values as $N\Lambda$. Hint: Show that for any square matrix $C$, the matrices $C$ and $C^T$ have the same singular values.*

Since $\Lambda N^T$ and $N\Lambda$ are two square matrices with the same singular values, it follows from the singular value decomposition that there exists two unitary matrices $U$ and $V$ such that

$$\Lambda N^T = U N \Lambda V^T, \tag{12.9}$$

where the transpose on $V$ is for convenience. Substituting this into (12.8) gives

$$\begin{aligned}
\left(I^A \otimes N\right)|\psi^{AB}\rangle &= \left(UN\Lambda V^T \otimes I^B\right)|\Omega^{AB}\rangle \\
&- (UN\Lambda \otimes V)|\Omega^{AB}\rangle \\
&= (UN \otimes V)|\psi^{AB}\rangle.
\end{aligned} \tag{12.10}$$

That is, the vectors $\left(I^A \otimes N\right)|\psi^{AB}\rangle$ and $\left(N \otimes I^B\right)|\psi^{AB}\rangle$ are equivalent up to the local unitary map $U \otimes V$. This observation leads to the following result.

> **Lo-Popescu's Theorem**
>
> **Theorem 12.1.** The effect of any LOCC map on a pure bipartite state can be simulated by the following protocol: Alice performs a generalized quantum measurement $\{M_{yx}\}_{x,y}$, sends the result $(x, y)$ to Bob, who then performs a local unitary map $V_{yx}$ on his system, and in the final step, Alice and Bob discard the value of $y$.

**Proof** Consider first a single local instrument, $\{\mathcal{E}_x\}_{x\in[m]}$, performed by Bob. Any CP map $\mathcal{E}_x$ can be expressed as $\mathcal{E}_x(\,\cdot\,) = \sum_{y\in[n]} N_{yx}(\,\cdot\,)N_{yx}^*$, so that

$$\begin{aligned}
\mathcal{E}_x^{B\to B}(\psi^{AB}) &= \sum_{y\in[n]} \left(I^A \otimes N_{yx}\right)\psi^{AB}\left(I^A \otimes N_{yx}^*\right) \\
(12.10)\to &= \sum_{y\in[n]} \left(U_{yx}N_{yx} \otimes V_{yx}\right)\psi^{AB}\left(N_{yx}^* U_{yx}^* \otimes V_{yx}^*\right),
\end{aligned} \tag{12.11}$$

where we used (12.10) for each $x \in [m]$ and $y \in [n]$, with $U_{xy}$ and $V_{xy}$ being unitary matrices. Denoting by $M_{yx} := U_{yx} N_{yx}$ the above equation becomes

$$\mathcal{E}_x^{B \to B}(\psi^{AB}) = \sum_{y \in [n]} \left( M_{yx} \otimes V_{yx} \right) \psi^{AB} \left( M_{yx}^* \otimes V_{yx}^* \right). \tag{12.12}$$

Moreover, since $\sum_{x \in [m]} \sum_{y \in [n]} M_{yx}^* M_{yx} = I^A$, we conclude that any quantum instrument that is performed by Bob can be simulated with the following protocol: Alice performs a generalized quantum measurement $\{M_{xy}\}_{x \in [m], y \in [n]}$, sends the outcome $(x, y)$ to Bob, who then performs a unitary matrix $V_{xy}$. At the end of the protocol, Alice and Bob discard or forget the value of $y$. Therefore, in any LOCC protocol, all the local quantum instruments on Bob's side can be simulated with unitaries and measurements on Alice's side. Since a sequence of quantum instruments (generalized measurements) on Alice's side can be combined into a single generalized measurement (followed by coarse graining, that is, discarding of information), we conclude that the most general LOCC protocol on a pure bipartite state can be simulated with a single generalized measurement by Alice's side followed by a unitary on Bob's side that depends on Alice's measurement outcome, and ends with the discarding of partial information of the measurement outcome. ∎

**Exercise 12.4.** *Show that a sequence of two generalized measurements can be viewed as a single generalized measurement. That is, given two generalized measurements $\{M_x\}_{x \in [m]}$ and $\{N_y\}_{y \in [n]}$ show that the set of matrices $\{L_{xy} := M_x N_y\}_{x \in [m], y \in [n]}$ is also a generalized measurement.*

**Exercise 12.5.** *Let $\psi \in \mathrm{Pure}(AB)$ and $\sigma \in \mathfrak{D}(AB)$. Show that if there exists a deterministic LOCC protocol that converts $\psi^{AB}$ to $\sigma^{AB}$, that is, $\psi^{AB} \xrightarrow{LOCC} \sigma^{AB}$, then there exists a set $\{M_x\}_{x \in [m]}$ of complex matrices in $\mathfrak{L}(A)$, and a set $\{U_x\}_{x \in [m]}$ of unitary matrices in $\mathfrak{L}(B)$ such that*

$$\sigma^{AB} = \sum_{x \in [m]} \left( M_x \otimes U_x \right) \psi^{AB} \left( M_x \otimes U_x \right)^*. \tag{12.13}$$

*Show further that the above relation can be expressed as*

$$\sigma^{AB} = \mathcal{U}^{BX \to B} \circ \mathcal{E}^{A \to AX} \left( \psi^{AB} \right), \tag{12.14}$$

*where $\mathcal{E} \in \mathrm{CPTP}(A \to AX)$ is a quantum instrument and $\mathcal{U} \in \mathrm{CPTP}(BX \to B)$ is a controlled unitary channel. Note that in particular this implies that $\psi^{AB} \xrightarrow{LOCC_1} \sigma^{AB}$.*

Theorem 12.1 can be simplified further if we consider only LOCC protocols that take pure bipartite states to pure bipartite states. In this case, any LOCC transformation can be simulated by the following simple protocol: Alice performs a generalized measurement $\{M_x\}$, sends the outcome $x$ to Bob, who then performs a local unitary operation $V_x$. This simplification of LOCC will be crucial for the study of pure-state entanglement theory.

**Exercise 12.6.** *The Schmidt rank of a pure bipartite state is defined as the number of non-zero Schmidt coefficients; for example, the Schmidt rank of the state given in (12.7) is the rank of the matrix $\Lambda^A$. We denote the Schmidt rank of a bipartite state $\psi \in \mathrm{Pure}(AB)$ by $\mathrm{SR}(\psi)$. Show that for two bipartite states $\psi, \phi \in \mathrm{Pure}(AB)$ with $\mathrm{SR}(\phi) > \mathrm{SR}(\psi)$ it is impossible to convert $\psi$ to $\phi$ by LOCC (not even with probability less than one).*

## 12.2.2 Exact Deterministic Interconversions

In this section, we provide the precise conditions that determine if one quantum state can be converted to another by LOCC. We will use the notation $|\psi^{AB}\rangle \xrightarrow{LOCC} |\phi^{AB}\rangle$ whenever it is possible to convert a bipartite pure state $\psi \in \mathrm{Pure}(AB)$ into the state $\phi \in \mathrm{Pure}(AB)$. Recall that any bipartite quantum state $\psi \in \mathrm{Pure}(AB)$ (with $|A| = |B| := d$) has a Schmidt decomposition of the form

$$|\psi^{AB}\rangle = \sum_{x \in [d]} \sqrt{p_x} |\psi_x\rangle^A |\phi_x\rangle^B, \tag{12.15}$$

where $\{|\psi_x\rangle^A\}_{x \in [d]}$ and $\{|\phi_x\rangle^B\}_{x \in [d]}$ are orthonormal bases of $A$ and $B$, respectively. Let $U$ and $V$ be unitary matrices such that $U|\psi_x\rangle^A = |x\rangle^A$ and $V|\phi_x\rangle^B = |x\rangle^B$, where $\{|x\rangle^A\}$ and $\{|x\rangle^B\}$ are the standard bases of $A$ and $B$, respectively. Hence,

$$U \otimes V |\psi^{AB}\rangle = \sum_{x \in [d]} \sqrt{p_x} |xx\rangle^{AB}. \tag{12.16}$$

Note also that by applying additional local permutations (which are unitaries) to the state above we can rearrange the order that the Schmidt coefficients. Therefore, there exist unitary matrices $U' \in \mathfrak{L}(A)$ and $V' \in \mathfrak{L}(B)$ such that

$$|\tilde{\psi}^{AB}\rangle := U' \otimes V' |\psi^{AB}\rangle = \sum_{x \in [d]} \sqrt{p_x} |xx\rangle^{AB} \quad \text{and} \quad p_1 \geqslant p_2 \geqslant \cdots \geqslant p_d. \tag{12.17}$$

The above form is called the *standard form* of $|\psi^{AB}\rangle$. Note that $|\psi^{AB}\rangle$ can be converted by LOCC to another state $|\phi^{AB}\rangle$ if and only if the standard form of $|\psi^{AB}\rangle$ can be converted by LOCC to the standard form of $|\phi^{AB}\rangle$ (see Figure 12.2). Therefore, without loss of generality we will assume here that both $|\psi^{AB}\rangle$ and $|\phi^{AB}\rangle$ are given in their standard forms.

The next theorem provides a connection between LOCC conversions and majorization. For any two density matrices $\rho, \sigma \in \mathfrak{D}(A)$, we will say that $\rho$ majorizes $\sigma$,

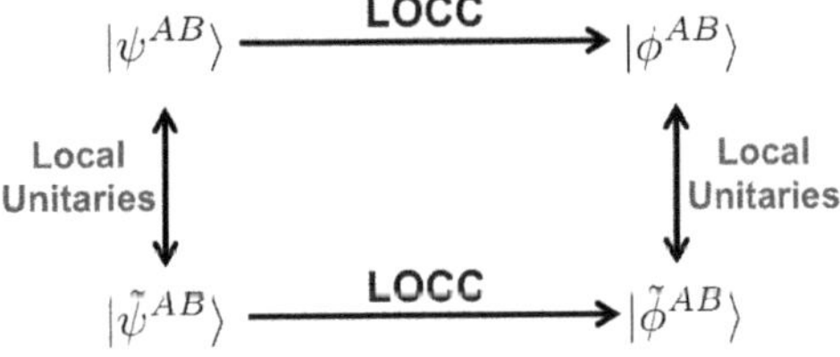



Figure 12.2   LOCC maps between two pure bipartite states and their standard forms.

and write $\rho \succ \sigma$, if the probability vectors $\mathbf{p}$ and $\mathbf{q}$, consisting, respectively, of the eigenvalues of $\rho$ and $\sigma$, satisfy $\mathbf{p} \succ \mathbf{q}$.

**Nielsen's Majorization Theorem**

**Theorem 12.2.** Let $\psi, \phi \in \in \mathrm{Pure}(AB)$ be two bipartite quantum states, and let $\rho^A := \mathrm{Tr}_B\left[\psi^{AB}\right]$ and $\sigma^A := \mathrm{Tr}_B\left[\phi^{AB}\right]$ be their corresponding reduced density matrices. Then,

$$\psi^{AB} \xrightarrow{LOCC} \phi^{AB} \qquad \Longleftrightarrow \qquad \sigma^A \succ \rho^A. \tag{12.18}$$

**Proof**   From the argument above we can assume without loss of generality that both $\psi^{AB}$ and $\phi^{AB}$ are given in their standard forms. Furthermore, Exercise 12.6 implies that we can assume without loss of generality that $\mathrm{supp}\,(\sigma^A) \subseteq \mathrm{supp}\,(\rho^A)$. This in turn implies that we can assume without loss of generality that $\rho^A > 0$ since otherwise we embed both $|\psi^{AB}\rangle$ and $|\phi^{AB}\rangle$ in $\mathrm{supp}\,(\rho^A) \otimes B$.

From Theorem 12.1 any LOCC map that takes pure bipartite state $|\psi^{AB}\rangle$ to another pure bipartite state $|\phi^{AB}\rangle$ can be simulated by one-way LOCC (i.e. $\mathrm{LOCC}_1$) of the following form: Alice performs a single generalized measurement $\{M_z\}_{z \in [m]}$ on her system, sends the measurement outcome $z$ to Bob, who then performs the unitary $V_z$. Therefore, after outcome $z$ occurred, the post-measurement state is given by

$$\frac{1}{\sqrt{t_z}} M_z \otimes V_z |\psi^{AB}\rangle, \tag{12.19}$$

where $t_z := \langle \psi^{AB} | M_z^* M_z \otimes I^B | \psi^{AB}\rangle$ is the probability that Alice's measurement outcome is $z$. Hence, if $|\psi^{AB}\rangle$ can be converted by LOCC to $|\phi^{AB}\rangle$ with 100% success rate, then there must exist a generalized measurement $\{M_z\}_{z \in [m]}$ and a collection of unitary matrices $\{V_z\}_{z \in [m]}$ such that

$$\frac{1}{\sqrt{t_z}} M_z \otimes V_z |\psi^{AB}\rangle = |\phi^{AB}\rangle \qquad \forall z \in [m]. \tag{12.20}$$

Since we assume without loss of generality that both $|\psi^{AB}\rangle$ and $|\phi^{AB}\rangle$ are given in their standard form, we have

$$|\psi^{AB}\rangle = \sum_{x \in [d]} \sqrt{p_x} |xx\rangle^{AB} = \sqrt{\rho} \otimes I^B |\Omega^{AB}\rangle$$

$$|\phi^{AB}\rangle = \sum_{x \in [d]} \sqrt{q_x} |xx\rangle^{AB} = \sqrt{\sigma} \otimes I^B |\Omega^{AB}\rangle, \tag{12.21}$$

where $\rho$ and $\sigma$ are, respectively, the reduced density matrices of $|\psi^{AB}\rangle$ and $|\phi^{AB}\rangle$. Explicitly, $\rho = \sum_{x \in [d]} p_x |x\rangle\langle x|^A$ and $\sigma = \sum_{x \in [d]} q_x |x\rangle\langle x|^A$. Substituting this into (12.20) gives

$$\frac{1}{\sqrt{t_z}} \left(M_z \sqrt{\rho} \otimes V_z\right) |\Omega^{AB}\rangle = \left(\sqrt{\sigma} \otimes I^B\right) |\Omega^{AB}\rangle. \tag{12.22}$$

From Exercise 2.38 it follows that the above equation holds if and only if

$$\frac{1}{\sqrt{t_z}} M_z \sqrt{\rho} V_z^T = \sqrt{\sigma}. \tag{12.23}$$

Note that the matrix $U_z := \left(V_z^{-1}\right)^T$ is unitary. With this notation, the above equation is equivalent to

$$M_z = \sqrt{t_z} \sqrt{\sigma} U_z \rho^{-1/2}. \tag{12.24}$$

The only constraint on $M_z$ is that $\sum_{z\in[m]} M_z^* M_z = I^A$. We therefore conclude that $|\psi^{AB}\rangle$ can be converted to $|\phi^{AB}\rangle$ by LOCC if and only if there exists $m \in \mathbb{N}$, unitary matrices $\{U_z\}_{z\in[m]}$, and probabilities $\{t_z\}_{z\in[m]}$ such that

$$\sum_{z\in[m]} t_z \rho^{-1/2} U_z^* \sigma U_z \rho^{-1/2} = I^A, \tag{12.25}$$

or equivalently,

$$\rho = \sum_{z\in[m]} t_z U_z^* \sigma U_z. \tag{12.26}$$

In other words, $|\psi^{AB}\rangle$ can be converted to $|\phi^{AB}\rangle$ by LOCC if and only if there exists a mixture of unitaries that transforms the reduced density matrix of $|\phi^{AB}\rangle$ to the reduced density matrix of $|\psi^{AB}\rangle$. Observe that such a random unitary channel is a unital channel. In Section 3.5.9, we showed that if $\rho = \mathcal{E}(\sigma)$, with $\mathcal{E}$ being a unital channel, then there exists a doubly stochastic matrix $D$ such that $\mathbf{p} = D\mathbf{q}$ (recall that $\mathbf{p}$ and $\mathbf{q}$ are the probability vectors whose components consist of the eigenvalues of $\rho$ and $\sigma$, respectively). From Theorem 4.1 it then follows that $\mathbf{q} \succ \mathbf{p}$ or equivalently, $\sigma \succ \rho$. We therefore conclude that if $\psi^{AB}$ can be converted to $\phi^{AB}$ by LOCC, then we must have $\sigma \succ \rho$.

Conversely, if $\sigma \succ \rho$, then from Theorem 4.1 we have $\mathbf{p} = D\mathbf{q}$ for some doubly stochastic matrix $D$. From Birkhoff/von-Neumann theorem (see Theorem A.5) every doubly stochastic matrix can be written as a convex combination of permutation matrices. Therefore, there there exists $m \in \mathbb{N}$, permutation matrices $\{\Pi_z\}_{z\in[m]}$, and probabilities $\{t_z\}_{z\in[m]}$, such that $\mathbf{p} = \sum_z t_z \Pi_z \mathbf{q}$. In Exercise 12.7, you will show that this relation can be expressed as

$$\rho = \sum_{z\in[m]} t_z \Pi_z \sigma \Pi_z^T. \tag{12.27}$$

The above equality is equivalent to (12.26) by taking $U_z = \Pi_z^T$. Therefore, $\psi^{AB}$ can be converted to $\phi^{AB}$ by LOCC. This completes the proof. ■

**Exercise 12.7.** *Show that if $\rho$ is a diagonal matrix, and $\mathbf{p}$ is the vector consisting of its diagonal elements, then $\Pi_z \rho \Pi_z^T$ is also a diagonal matrix, with the diagonal elements given by the components of $\Pi_z \mathbf{p}$. Use this to show that $\rho = \sum_{z\in[m]} t_z \Pi_z \sigma \Pi_z^T$ if and only if $\mathbf{p} = \sum_{z\in[m]} t_z \Pi_z \mathbf{q}$.*

**Exercise 12.8.** *Show that the maximally entangled state $|\Phi^{AB}\rangle := \frac{1}{\sqrt{d}}\sum_{x\in[d]}|xx\rangle$ can be converted by LOCC to any other state in $\mathrm{Pure}(AB)$. Moreover, show that any state $|\psi\rangle \in AB$ can be converted by LOCC to any product state of the form $|\phi\rangle|\chi\rangle \in AB$.*

**Exercise 12.9.** *For any bipartite state $\psi \in \mathrm{Pure}(AB)$, and for any $k = 1,\ldots,d$, define*

$$E_{(k)}(\psi^{AB}) := 1 - \sum_{x\in[k]} p_x^{\downarrow} = 1 - \|\mathbf{p}\|_{(k)}, \tag{12.28}$$

*where $\mathbf{p}$ is the Schmidt vector of $|\psi^{AB}\rangle$, and $\|\mathbf{p}\|_{(k)}$ is the Ky Fan norm of $\mathbf{p}$ (cf. Definition 2.4). Show that Nielsen Majorization Theorem can be expressed as*

$$\psi^{AB} \xrightarrow{\;LOCC\;} \phi^{AB} \qquad \Longleftrightarrow \qquad E_{(k)}(\psi^{AB}) \geqslant E_{(k)}(\phi^{AB}) \quad \forall\, k \in [d]. \tag{12.29}$$

**Exercise 12.10.** *Let $\psi \in \mathrm{Pure}(AB)$, with $m := |A| = |B|$.*

1. *Prove the following theorem, assuming that Alice and Bob share the state $\psi^{AB}$ (but no other entangled systems).* **Theorem.** *Faithful teleportation of a d-dimensional qudit is possible if, and only if,*

$$E_t(\psi^{AB}) := -\log_2 p_{\max} \geqslant \log_2 d, \tag{12.30}$$

*where $p_{\max}$ is the largest Schmidt coefficient of $\psi^{AB}$. That is, teleportation is possible if and only if none of the Schmidt coefficients are greater than $1/d$. This also implies that the Schmidt rank $m$ is greater than or equal to $d$. Hint: Use Nielsen majorization theorem.*

2. *Find a protocol for faithful teleportation of a* qubit *from Alice's lab to Bob's lab assuming Alice and Bob share the partially entangled state*

$$|\psi^{AB}\rangle = \frac{1}{\sqrt{2}}|0\rangle_A|0\rangle_B + \frac{1}{2}|1\rangle_A|1\rangle_B + \frac{1}{2}|2\rangle_A|2\rangle_B. \tag{12.31}$$

*In particular, determine the projective measurement performed by Alice and the unitary operators performed by Bob. What is the optimal classical communication cost? That is, how many classical bits Alice has to send to Bob?*

**Exercise 12.11.** *Consider $m$ bipartite states $\{\psi_z^{AB}\}_{z\in[m]}$ in $\mathrm{Pure}(AB)$. Find an optimal state $\phi^{AB} \in \mathrm{Pure}(AB)$ such that:*

1. *The state $\phi^{AB}$ can be converted by LOCC to $\psi_z^{AB}$, for all $z \in [m]$.*
2. *If another state, $\chi \in \mathrm{Pure}(AB)$, can be converted by LOCC to $\psi_z^{AB}$, for all $z \in [m]$, then $\chi^{AB}$ can also be converted to $\phi^{AB}$. That is, $\phi^{AB}$ is optimal.*

## 12.2.3 Entanglement Catalysis

As we have already observed, some LOCC conversions of pure bipartite states cannot be realized with a 100% success rate. This limitation arises because LOCC constitutes a restricted subset of quantum operations. To expand the capabilities of LOCC, one could allow Alice and Bob to temporarily use an entangled state during their LOCC protocols. The condition here is that they must return the entangled systems in their original state at the end of the protocols. At first glance, it might seem that borrowing an entangled system wouldn't provide any advantage for tasks that cannot be accomplished with standard LOCC. However, as we will now demonstrate, Nielsen's majorization theorem reveals that this entanglement-assisted LOCC (eLOCC) actually represents a significantly broader set of operations compared to LOCC alone.

We start with the following example. Consider the two entangled states

$$|\psi^{AB}\rangle = \sqrt{2/5}|00\rangle + \sqrt{2/5}|11\rangle + \sqrt{1/10}|22\rangle + \sqrt{1/10}|33\rangle,$$
$$|\phi^{AB}\rangle = \sqrt{1/2}|00\rangle + \sqrt{1/4}|11\rangle + \sqrt{1/4}|22\rangle. \tag{12.32}$$

The Schmidt probability vectors associated with the two states above are given by

$$\mathbf{p} := \left(\frac{2}{5}, \frac{2}{5}, \frac{1}{10}, \frac{1}{10}\right)^T \quad \text{and} \quad \mathbf{q} = \left(\frac{1}{2}, \frac{1}{4}, \frac{1}{4}, 0\right)^T, \tag{12.33}$$

respectively. In Exercise 4.52, you confirmed that neither $\mathbf{p}$ majorizes $\mathbf{q}$ nor $\mathbf{q}$ majorizes $\mathbf{p}$, symbolized as $\mathbf{p} \not\succ \mathbf{q}$ and $\mathbf{q} \not\succ \mathbf{p}$. Consequently, Nielsen's majorization theorem implies that neither $\psi^{AB}$ can be converted to $\phi^{AB}$ nor can $\phi^{AB}$ be converted to $\psi^{AB}$ using LOCC. Now, consider the state

$$|\chi^{A'B'}\rangle = \sqrt{3/5}|00\rangle + \sqrt{2/5}|11\rangle, \tag{12.34}$$

and let its Schmidt vector be denoted by $\mathbf{r} := (3/5, 2/5)^T$. Interestingly, it is easy to verify that

$$\mathbf{q} \otimes \mathbf{r} \succ \mathbf{p} \otimes \mathbf{r}. \tag{12.35}$$

Therefore, according to Nielsen's theorem, the transformation

$$|\psi^{AB}\rangle \otimes |\chi^{A'B'}\rangle \xrightarrow{\text{LOCC}} |\phi^{AB}\rangle \otimes |\chi^{A'B'}\rangle \tag{12.36}$$

is achievable with a 100% success rate. In this context, the state $\chi^{AB}$ functions as a catalyst for the conversion of $\psi^{AB}$ into $\phi^{AB}$, and thus is referred to as an *entanglement catalyst*.

**Exercise 12.12.** *Show that there is no entanglement catalyst if both $\psi^{AB}$ and $\phi^{AB}$ have Schmidt rank 3.*

**Exercise 12.13.** *Show that the maximally entangled state cannot act as a catalyst in any eLOCC conversions that are not possible by LOCC.*

Entanglement catalysis motivates the definition of a new partial order between probability vectors that we studied in Section 4.4 and is called the *trumping relation*. Recall that for any $\mathbf{p}, \mathbf{q} \in \text{Prob}(n)$ we say that $\mathbf{q}$ trumps $\mathbf{p}$ and write

$$\mathbf{q} \succ_* \mathbf{p}, \tag{12.37}$$

if there exists an integer $m \in \mathbb{N}$ and a vector $\mathbf{r} \in \text{Prob}(m)$ such that $\mathbf{q} \otimes \mathbf{r} \succ \mathbf{p} \otimes \mathbf{r}$.

**Exercise 12.14.** *Show that any Schur concave function $f : \mathrm{Prob}(n) \to \mathbb{R}$ that is additive under tensor product, behaves monotonically under the trumping relation. That is, for any $\mathbf{p}, \mathbf{q} \in \mathrm{Prob}(n)$ we have*

$$\mathbf{q} \succ_* \mathbf{p} \quad \Rightarrow \quad f(\mathbf{p}) \geqslant f(\mathbf{q}). \tag{12.38}$$

A well-known family of functions that behaves monotonically under the trumping relation are the Rényi entropies. The set of functions

$$f_\alpha(\mathbf{p}) := \begin{cases} \frac{\mathrm{sign}(\alpha)}{1-\alpha} \log \sum_{x \in [m]} p_x^\alpha & \text{if } 0 \neq \alpha \in [-\infty, \infty] \\ -\log(p_1 \cdots p_m) & \text{if } \alpha = 0 \end{cases} \tag{12.39}$$

satisfies the monotonicity under the tramping relation and additivity. For $\alpha \geqslant 0$ (i.e. $f_\alpha = H_\alpha$ is the Rényi entropy) these functions are entropy functions as they also satisfy the normalization condition that they are zero for $\mathbf{p} = (1, 0, \ldots, 0)$. For $\alpha \leqslant 0$ they are defined to be $-\infty$ if $\mathbf{p} \not> 0$. Such functions are not entropy functions since they do not satisfies the normalization condition; however, they are useful in the characterization of the tramping relation. Note that any convex combination of the functions above is also additive and monotonic under the trumping relation.

**Exercise 12.15.** *Show that the condition (4.181) of Theorem 4.11 is equivalent to the condition that*

$$f_\alpha(\mathbf{p}) > f_\alpha(\mathbf{q}) \qquad \forall\, \alpha \in [-\infty, \infty], \tag{12.40}$$

*where $f_\alpha$ is defined in (12.39)*

For each $\alpha \in [-\infty, \infty]$ and $\psi \in \mathrm{Pure}(AB)$ define

$$E_\alpha\left(\psi^{AB}\right) := f_\alpha(\mathbf{p}), \tag{12.41}$$

where $\mathbf{p}$ is the Schmidt probability vector of $\psi^{AB}$, and $f_\alpha$ is defined in (12.39). We will see later on that the functions $\{E_\alpha\}_\alpha$ are measures of entanglement on pure states. From Theorem 4.11 it follows that these functions can be used to characterize eLOCC transformations.

---

**Corollary 12.1.** Let $\psi, \phi \in \mathrm{Pure}(AB)$ be two bipartite quantum states. Then,

$$\psi^{AB} \xrightarrow{\;e\mathrm{LOCC}\;} \phi^{AB} \quad \Longleftrightarrow \quad E_\alpha\left(\psi^{AB}\right) > E_\alpha\left(\phi^{AB}\right) \qquad \forall\, \alpha \in [-\infty, \infty],$$

$$\tag{12.42}$$

where $\{E_\alpha\}_\alpha$ are defined in (12.41).

---

**Exercise 12.16.** *Prove the corollary above using Theorem 4.11.*

## 12.3 Quantification of Pure Bipartite Entanglement

How can entanglement be quantified? According to its definition (see Definition 12.1), entanglement cannot be created or increased by LOCC. Consequently, entanglement must be quantified using functions that are monotonic under LOCC. In this section, we focus on pure state entanglement and consider a measure of entanglement to be a function

$$\mathbf{E} \colon \bigcup_{A,B} \mathrm{Pure}(AB) \to \mathbb{R} \tag{12.43}$$

that exhibits monotonic behavior under pure-state LOCC transformations.

Any pure bipartite state $\psi \in \mathrm{Pure}(AB)$ can be transformed by a local unitary operation (i.e. a reversible LOCC) into the state

$$|\tilde{\psi}^{AB}\rangle := U \otimes V |\psi^{AB}\rangle = \sum_{x \in [d]} \sqrt{p_x} |xx\rangle^{AB}, \tag{12.44}$$

where $U$ and $V$ are $d \times d$ unitary matrices, $\{p_x\}_{x \in [d]}$ are the Schmidt coefficients of $\psi^{AB}$, and $\{|x\rangle^A\}$ and $\{|x\rangle^B\}$ are fixed bases of $A$ and $B$. By definition, since the LOCC map $U \otimes V$ is reversible (having an LOCC inverse $U^* \otimes V^*$), we must have for any measure of entanglement on pure states:

$$\mathbf{E}\left(\psi^{AB}\right) = \mathbf{E}(\tilde{\psi}^{AB}) = f(\mathbf{p}), \tag{12.45}$$

where $\mathbf{p} := (p_1, \ldots, p_d)^T$ is the probability Schmidt vector of $\psi^{AB}$ and $f \colon \mathrm{Prob}(d) \to \mathbb{R}_+$ is some nonnegative function of the Schmidt coefficients of $|\psi^{AB}\rangle$. That is, any measure of entanglement on pure states depends only on the Schmidt coefficients of the entangled state.

From Nielsen theorem we have that $|\psi^{AB}\rangle \xrightarrow{LOCC} |\phi^{AB}\rangle$ if and only if the corresponding Schmidt probability vectors $\mathbf{p}$ and $\mathbf{q}$ of $\psi^{AB}$ and $\phi^{AB}$, respectively, satisfy $\mathbf{p} \prec \mathbf{q}$. On the other hand, by definition, if $\psi^{AB} \xrightarrow{LOCC} \phi^{AB}$, then any measure of entanglement must satisfy $\mathbf{E}\left(\psi^{AB}\right) \geqslant \mathbf{E}\left(\phi^{AB}\right)$. Hence, the function $f$ above must be monotonic under majorization; that is, $f$ must be a Schur concave function. Specifically,

$$\mathbf{p} \prec \mathbf{q} \implies f(\mathbf{p}) \geqslant f(\mathbf{q}). \tag{12.46}$$

### Example 1: Entropy of Entanglement

The entropy of entanglement is arguably the most important measure of pure-state entanglement with several operational interpretations. It is detonated by $E$ and defined for any $\psi \subset \mathrm{Pure}(AB)$ by

$$E\left(\psi^{AB}\right) := H(\mathbf{p}), \tag{12.47}$$

where $H$ is the Shannon entropy and $\mathbf{p}$ is the Schmidt probability vector of $\psi^{AB}$. We will see in the next sections that the entropy of entanglement equals both the entanglement cost and the distillable entanglement in the asymptotic regime.

## Example 2: $\alpha$-Entropy of Entanglement

Similar to the entropy of entanglement, for any $\alpha \in [0, \infty]$ the $\alpha$-entropy of entanglement is defined for any $\psi \in \text{Pure}(AB)$ as

$$E_\alpha\left(\psi^{AB}\right) := H_\alpha(\mathbf{p}), \tag{12.48}$$

where $H_\alpha$ is the $\alpha$-Rényi entropy and $\mathbf{p}$ is the Schmidt probability vector of $\psi^{AB}$. In addition to being monotonic under LOCC the $\alpha$-entropy of entanglement is additive under tensor products. We saw in Corollary 12.1 that this in turn implies that the $E_\alpha$ is also monotonic under eLOCC and therefore can be used to characterize entanglement catalysis.

## Example 3: The Concurrence Monotones

In (4.49) we saw that the elementary symmetric functions are Schur concave. Therefore, they can be used to define a family of measures of entanglement known as the concurrence monotones. Specifically, they are defined for any $\psi \in \text{Pure}(AB)$ and any $k \in [d]$ as

$$C_k\left(\psi^{AB}\right) := \left(\frac{f_k(\mathbf{p})}{f_k(\mathbf{u})}\right)^{1/k}, \quad \text{where} \quad f_k(\mathbf{p}) := \sum_{\substack{x_1 < \cdots < x_k \\ x_1,\ldots,x_k \in [n]}} p_{x_1} \cdots p_{x_k}, \tag{12.49}$$

$\mathbf{p}$ is the Schmidt probability vector of $\psi^{AB}$, and $\mathbf{u}$ is the $d$-dimensional uniform probability vector. Note that the $k$-concurrence is normalized such that it equals one of maximally entangled states. Thus, the concurrence monotones take values between 0 and 1. The power $1/k$ above is necessary to make these functions entanglement monotones; that is, it can be shown that these functions not only behave monotonically under LOCC but also non-increasing on average under nondeterministic LOCC (see the following sections).

For the two extreme cases $k = 2$ and $k = d$, we get for any $\psi \in \text{Pure}(AB)$ with reduced density matrix $\rho := \text{Tr}_B\left[\psi^{AB}\right]$ that

$$\begin{aligned}
C\left(\psi^{AB}\right) &:= C_2\left(\psi^{AB}\right) = \sqrt{\frac{d}{d-1}\left(1 - \text{Tr}\left[\rho^2\right]\right)}, \\
G\left(\psi^{AB}\right) &:= C_d\left(\psi^{AB}\right) = d\left(\det\left(\rho\right)\right)^{1/d}.
\end{aligned} \tag{12.50}$$

In literature, $C$ is called simply the concurrence, and $G$ is called the $G$-concurrence since it can be interpreted as the geometric mean of the Schmidt coefficients. Note that for $d = 2$ we have $C = G = C_k$.

**Exercise 12.17.** *Prove the equalities in* (12.50).

**Exercise 12.18.** *Let $\{|0\rangle, |1\rangle)\}$ be a basis in which the second Pauli matrix, $\sigma_y$, has the form $\sigma_y = \begin{bmatrix} 0 & i \\ -i & 0 \end{bmatrix}$. Show that for any $\psi \in \text{Pure}(AB)$ with $|A| = |B| = 2$,*

$$C\left(\psi^{AB}\right) = \left|\langle \bar{\psi}^{AB} | \sigma_y \otimes \sigma_y | \psi^{AB} \rangle\right|, \tag{12.51}$$

*where $\bar{\psi}^{AB}$ is defined such that if $|\psi^{AB}\rangle = \sum_{x,y \in \{0,1\}} c_{xy} |x\rangle |y\rangle$, then $|\bar{\psi}^{AB}\rangle = \sum_{x,y \in \{0,1\}} \bar{c}_{xy} |x\rangle |y\rangle$.*

## 12.4 Stochastic Interconversions

Quantum mechanics is inherently nondeterministic. This is manifested by quantum measurements that transform a quantum state into several possible post-measurement states. Typically, an LOCC protocol does not convert a quantum state into another quantum state. Instead, it converts it to *ensemble of states* with each state occurring with different probability. Here we consider such probabilistic LOCC transformations among pure bipartite states.

Suppose Alice and Bob share the pure bipartite state $\psi \in \text{Pure}(AB)$. After they apply LOCC to this state they end up sharing one out of $n$ states $\{|\phi_z^{AB}\rangle\}_{z \in [n]}$ each with corresponding probability $t_z$. The resulting ensemble of states $\{t_z, |\phi_z^{AB}\rangle\}_{z \in [n]}$ can be described with a cq-state

$$\sigma^{ZAB} := \sum_{z \in [n]} t_z |z\rangle \langle z|^Z \otimes |\phi_z^{AB}\rangle \langle \phi_z^{AB}|, \tag{12.52}$$

where $Z$ is a "flag" system registering the value $z$. In this view, the LOCC protocol converted the state $\psi^{AB}$ to the cq-state $\sigma^{ZAB}$. The question we study here is to which cq-states the pure state $\psi^{AB}$ can be transformed into by LOCC. Since the output state $\sigma^{ZAB}$ is not a pure state, we cannot apply Nielsen majorization theorem. Yet, as we show now, Nielsen theorem is imperative to answer such a question.

---

**Theorem 12.3.** A pure bipartite state $\psi \in \text{Pure}(AB)$ can be converted by LOCC to the ensemble $\left\{\phi_z^{AB}, t_z\right\}_{z \in [n]}$ of pure states in $\text{Pure}(AB)$ if and only if for all $k \in [d]$ $(d := |A| = |B|)$

$$E_{(k)}\left(\psi^{AB}\right) \geqslant \sum_{z \in [n]} t_z E_{(k)}\left(\phi_z^{AB}\right), \tag{12.53}$$

where the functions $E_{(k)}$ are defined in (12.28).

---

**Proof** Let $\mathbf{p}$ be the Schmidt probability vector associated with $\psi^{AB}$. For each $z \in [n]$, define $\mathbf{q}_z := (q_{1|z}, \ldots, q_{d|z})^T$ as the Schmidt probability vector associated with $\phi_z^{AB}$. We can assume without loss of generality that $\mathbf{p} = \mathbf{p}^{\downarrow}$ and $\mathbf{q}_z = \mathbf{q}_z^{\downarrow}$, since the order of the Schmidt vectors can always be rearranged by applying local unitary (permutation) maps to $|\psi^{AB}\rangle$ and $|\phi_z^{AB}\rangle$.

Also, denote by

$$\mathbf{q} := \sum_{z\in[n]} t_z \mathbf{q}_z \quad \text{and by} \quad |\phi^{AB}\rangle := \sum_{x\in[d]} \sqrt{q_x}|xx\rangle^{AB} \tag{12.54}$$

the bipartite state whose Schmidt vector is $\mathbf{q}$. Since $\mathbf{q}_z = \mathbf{q}_z^\downarrow$ for all $z \in [n]$, it follows that $\mathbf{q} = \mathbf{q}^\downarrow$ as well. The components of $\mathbf{q}$ are thus given by

$$q_x = \sum_{z\in[n]} t_z q_{x|z} \qquad \forall\, x \in [d]. \tag{12.55}$$

Consequently, for each $k \in [d]$, the entanglement measure $E_{(k)}$ for $\phi^{AB}$ is given by

$$E_{(k)}(\phi^{AB}) = \sum_{x=k+1}^{d} q_x$$

$$(12.55)\rightarrow = \sum_{z\in[n]} t_z \sum_{x=k+1}^{d} q_{x|z} \tag{12.56}$$

$$= \sum_{z\in[n]} t_z E_{(k)}(\phi_z^{AB}).$$

Hence, from Nielsen majorization theorem as expressed in Exercise 12.9, it follows that (12.53) holds if and only if $\psi^{AB}$ can be converted to $\phi^{AB}$ by LOCC. Therefore, to complete the proof we now show that $\phi^{AB}$ can be converted by LOCC to the ensemble $\left\{\phi_z^{AB},\, t_z\right\}_{z\in[n]}$. The conversion is achieved by the following single measurement performed by Alice

$$M_z := \sum_{x\in[d]} \sqrt{\frac{t_z q_{x|z}}{q_x}}|x\rangle\langle x|^A. \tag{12.57}$$

Observe that

$$\sum_{z\in[n]} M_z^* M_z = \sum_{x\in[d]}\sum_{z\in[n]} \frac{t_z q_{x|z}}{q_x}|x\rangle\langle x| = \sum_{x\in[d]} |x\rangle\langle x|^A = I^A, \tag{12.58}$$

where we used (12.55). Furthermore, note that

$$\left(M_z \otimes I^B\right)|\phi^{AB}\rangle = \sum_{x\in[d]} \sqrt{t_z q_{x|z}}|xx\rangle^{AB} = \sqrt{t_z}|\phi_z^{AB}\rangle. \tag{12.59}$$

That is, Alice's measurement produces the state $|\phi_z^{AB}\rangle$ with probability $t_z$. This completes the proof.  ∎

This theorem generalizes Nielsen majorization theorem to probabilistic transformations, and in addition leads to the following corollary.

> **Corollary 12.2.** Let $\psi, \phi \in \mathrm{Pure}(AB)$ be two pure bipartite entangled states. Then, the maximal probability with which $\psi^{AB}$ can be converted to $\phi^{AB}$ by LOCC is given by
>
> $$\mathrm{Pr}\left(\psi^{AB} \xrightarrow{\text{LOCC}} \phi^{AB}\right) = \min_{k \in [d]} \frac{E_{(k)}(\psi^{AB})}{E_{(k)}(\phi^{AB})}. \qquad (12.60)$$

*Remark.* Note that for $k = 1$, $E_1(\psi^{AB}) = 1$ for all $\psi \in \mathrm{Pure}(AB)$. Therefore, the expression on the right-hand side of the equation above can never exceed 1. Note further that the corollary above is a simplification of the formal result given in Corollary 11.1. That is, for pure bipartite states it is sufficient to check the ratios of only $d$ resource measures in order to compute the maximum probability of conversion.

**Proof** Consider an optimal LOCC that converts $\psi^{AB}$ to $\phi^{AB}$ with the maximum possible probability. Such LOCC protocol yields $\phi^{AB}$ with probability $p :=$ $\mathrm{Pr}\left(\psi^{AB} \xrightarrow{\text{LOCC}} \phi^{AB}\right)$ and other states with probability $1 - p$. All such other states can always be converted deterministically (by LOCC) to the product state $|0\rangle\langle 0|^A \otimes |0\rangle\langle 0|^B$. Therefore, without loss of generality we can assume that $\psi^{AB}$ is converted to $\phi^{AB}$ with probability $p$, and to $|0\rangle\langle 0|^A \otimes |0\rangle\langle 0|^B$ with probability $1 - p$. Since $E_k(|0\rangle\langle 0|^A \otimes |0\rangle\langle 0|^B) = 0$ for all $k = 2, \ldots, d$, Theorem 12.3 implies that such an LOCC protocol is possible if and only if

$$E_{(k)}(\psi^{AB}) \geqslant p E_{(k)}(\phi^{AB}) \quad \forall k \in \{2, \ldots, d\}. \qquad (12.61)$$

The proof is concluded by recognizing that the equation above is equivalent to

$$p \leqslant \min_{k \in \{2,\ldots,d\}} \frac{E_{(k)}(\psi^{AB})}{E_{(k)}(\phi^{AB})}. \qquad (12.62)$$

Therefore, the maximum probability is the one given in (12.60). $\blacksquare$

**Exercise 12.19.** *Let $|\psi^{AB}\rangle = \sqrt{p_1}|11\rangle + \ldots + \sqrt{p_d}|dd\rangle \in \mathbb{C}^d \otimes \mathbb{C}^d$ be a 2-qudit entangled state.*

1. *What is the maximum probability to convert by LOCC $|\psi^{AB}\rangle$ to the maximally entangled state $|\Phi^{AB}\rangle$.*
2. *For $d = 2$ find the LOCC protocol that achieves the maximum probability you found in Part 1.*

**Exercise 12.20.** *Show that a maximally entangled state $\Phi^{AB}$ can be converted to any mixed bipartite quantum state $\rho \in \mathfrak{D}(AB)$.*

Theorem 12.3 can also be use to provide necessary and sufficient conditions to determine the conversion of a pure entangled state to a mixed entangled state by LOCC.

> **Corollary 12.3.** Let $\psi \in \mathrm{Pure}(AB)$ and $\sigma \in \mathfrak{D}(AB)$ be two bipartite entangled states with $d := |A| = |B|$. Then, $\psi^{AB}$ can be converted to $\sigma^{AB}$ by LOCC if and only if
>
> $$\max_{\{t_x,\,\phi_x\}_{x\in[m]}} \min_{k\in[d]} \left\{ E_{(k)}(\psi^{AB}) - \sum_{x\in[m]} p_x E_{(k)}(\phi_x^{AB}) \right\} \geq 0, \tag{12.63}$$
>
> where the maximum is over all pure-state decompositions of $\sigma$ (i.e. over all pure-state ensembles $\{p_x,\,\phi_x\}_{x\in[m]}$ that satisfy $\sigma = \sum_{x\in[m]} p_x \phi_x$).

**Proof**  Suppose the condition in (12.63) holds. Then, there exists an ensemble of states $\{p_x,\,\phi_x^{AB}\}_{x\in[m]}$ such that $E_{(k)}(\psi^{AB}) \geq \sum_{x\in[m]} p_x E_{(k)}(\phi_x^{AB})$ for all $k \in [d]$. From Theorem 12.3 it follows that $\psi^{AB}$ can be converted by LOCC to the cq-state

$$\sigma^{XAB} = \sum_{x\in[m]} p_x |x\rangle\langle x|^X \otimes \phi_x^{AB}. \tag{12.64}$$

Since tracing out the classical system $X$ is an LOCC operation, we conclude that $\psi^{AB} \xrightarrow{\mathrm{LOCC}} \sigma^{AB}$.

Conversely, suppose $\psi^{AB} \xrightarrow{\mathrm{LOCC}} \sigma^{AB}$. From Theorem 12.1 it follows that there exists a generalized measurement on Alice's system $\{M_x\}_{x\in[m]}$ and a set of unitary matrices on Bob's system $\{U_x\}_{x\in[m]}$ such that

$$\sigma^{AB} = \sum_{x\in[m]} (M_x \otimes U_x)\, \psi^{AB}\, (M_x \otimes U_x)^*. \tag{12.65}$$

Denote by $|\phi_x^{AB}\rangle := \frac{1}{\sqrt{p_x}} (M_x \otimes U_x) |\psi^{AB}\rangle$, where $p_x := \langle \psi^{AB} | M_x^* M_x \otimes I^B | \psi^{AB} \rangle$. Then, from the equation above we get that the ensemble $\{p_x,\,\phi_x^{AB}\}_{x\in[m]}$ form a pure state decomposition of $\sigma^{AB}$. Moreover, by definition, $\psi^{AB}$ can be converted by LOCC to $\phi_x^{AB}$ with probability $p_x$. Therefore, from Theorem 12.3 it follows that

$$\min_{k\in[d]} \left\{ E_{(k)}(\psi^{AB}) - \sum_{x\in[n]} p_x E_{(k)}(\phi_x^{AB}) \right\} \geq 0. \tag{12.66}$$

Hence, the condition in (12.63) holds. This completes the proof.  ∎

## 12.5 Approximate Single-Shot Conversions

In practical scenarios, state transformations are often imperfect. Rather than seeking perfect LOCC conversion of one state into another, it is more realistic to allow the final state to be $\varepsilon$-close to the target state, where $\varepsilon > 0$ is a small threshold typically reflective of the inaccuracy inherent in the apparatus used. Allowing for the final state to be any state within $\varepsilon$-proximity to the desired state opens the door to additional state transformations that are not encompassed by Nielsen's majorization condition.

## 12.5.1 The Conversion Distance

Consider two pure bipartite states $\psi, \phi \in \text{Pure}(AB)$ (with $d := |A| = |B|$). The conversion distance of $\psi^{AB}$ into $\phi^{AB}$ is given by (refer to (11.24) with $\mathfrak{F} = \text{LOCC}$)

$$T\left(\psi^{AB} \xrightarrow{\text{LOCC}} \phi^{AB}\right) := \min_{\sigma \in \mathfrak{D}(AB)} \left\{ \frac{1}{2} \left\| \phi^{AB} - \sigma^{AB} \right\|_1 : \psi^{AB} \xrightarrow{\text{LOCC}} \sigma^{AB} \right\}. \tag{12.67}$$

One might wonder whether this conversion distance changes if we restrict the optimization over $\mathfrak{D}(AB)$ to states over $\text{Pure}(AB)$. Given that the trace distance between pure states equals the purified distance, constraining the above optimization to pure states yields

$$P_\star\left(\psi^{AB} \xrightarrow{\text{LOCC}} \phi^{AB}\right) := \min_{\varphi \in \text{Pure}(AB)} \left\{ P\left(\phi^{AB}, \varphi^{AB}\right) : \psi^{AB} \xrightarrow{\text{LOCC}} \varphi^{AB} \right\}. \tag{12.68}$$

It's important to note that the value of $T(\psi^{AB} \xrightarrow{\text{LOCC}} \phi^{AB})$ is not greater than that of $P_\star(\psi^{AB} \xrightarrow{\text{LOCC}} \phi^{AB})$. This is because restricting $\sigma^{AB}$ to be a pure state in the calculation of $T(\psi^{AB} \xrightarrow{\text{LOCC}} \phi^{AB})$ can only increase the minimum value obtained in the optimization process. Furthermore, as we will explore in the next chapter (specifically, in Lemma 13.3), it will be shown that $T(\psi^{AB} \xrightarrow{\text{LOCC}} \phi^{AB})$ can actually be strictly smaller than $P_\star(\psi^{AB} \xrightarrow{\text{LOCC}} \phi^{AB})$. However, we will soon discover that the value of $P_\star$ remains unchanged even when the optimization is extended from $\text{Pure}(AB)$ to the full set of density matrices $\mathfrak{D}(AB)$.

The optimization problem in (12.68) can be simplified as follows. Initially, observe that if $\mathbf{p}, \mathbf{r} \in \text{Prob}(d)$ are the Schmidt vectors of $\psi^{AB}$ and $\varphi^{AB}$, respectively, then according to Nielsen's theorem, the condition $\psi^{AB} \xrightarrow{\text{LOCC}} \varphi^{AB}$ is equivalent to $\mathbf{r} \succ \mathbf{p}$. Thus, for any given Schmidt vector $\mathbf{r}$ of $\varphi^{AB}$, we first perform the optimization over all states $\varphi \in \text{Pure}(AB)$ with the same Schmidt vector $\mathbf{r}$. Denoting by $|\tilde{\varphi}^{AB}\rangle = \sum_{x \in [d]} \sqrt{r_x}|xx\rangle$, this is equivalent to optimization over the local unitaries $U \otimes V$, such that $|\varphi^{AB}\rangle = U \otimes V |\tilde{\varphi}^{AB}\rangle$. Due to the relationship between trace distance and fidelity, we have

$$\min_{U,V \in \mathfrak{U}(d)} P\left(\phi^{AB}, \varphi^{AB}\right) = \sqrt{1 - \max_{U,V \in \mathfrak{U}(d)} |\langle \phi^{AB}|\varphi^{AB}\rangle|^2}. \tag{12.69}$$

Denoting by $|\phi^{AB}\rangle = N \otimes I^B |\Omega^{AB}\rangle$ and by $D$ the diagonal matrix with diagonal $\mathbf{r}$, we obtain

$$\max_{U,V \in \mathfrak{U}(d)} \left| \langle \phi^{AB}|\varphi^{AB}\rangle \right| = \max_{U,V \in \mathfrak{U}(d)} \left| \text{Tr}[N^* U D V^T] \right|$$

$$\xrightarrow[\substack{\text{von-Neumann trace inequality} \\ \text{(see (B.25))}}]{} = \sum_{x \in [d]} \sqrt{r_x^\downarrow q_x^\downarrow}, \tag{12.70}$$

where $\mathbf{q} \in \text{Prob}(d)$ is the Schmidt vector of $\phi^{AB}$. Taking everything into consideration we obtain the following simplification:

$$P_\star\left(\psi^{AB} \xrightarrow{\text{LOCC}} \phi^{AB}\right) = \min_{\mathbf{r} \in \text{Prob}(d)} \left\{ P(\mathbf{q}, \mathbf{r}) : \mathbf{r} \succ \mathbf{p} \right\}, \tag{12.71}$$

where $\mathbf{p}$ is the Schmidt vector of $\psi^{AB}$, $\mathbf{q}$ the Schmidt vector of $\phi^{AB}$, and $P(\mathbf{q}, \mathbf{r}) := \sqrt{1 - F^2(\mathbf{q}, \mathbf{r})}$ is the purified distance between probability vectors.

Observe that we added the subscript $\star$ to $P_\star$ since the conversion distance between two pure states $\psi, \phi \in \mathrm{Pure}(AB)$ as measured by the purified distance is defined as

$$P\left(\psi^{AB} \xrightarrow{\mathrm{LOCC}} \phi^{AB}\right) := \min_{\sigma \in \mathfrak{D}(AB)} \left\{ P\left(\phi^{AB}, \sigma^{AB}\right) : \psi^{AB} \xrightarrow{\mathrm{LOCC}} \sigma^{AB} \right\}. \tag{12.72}$$

At first glance, this conversion distance may seem to be different than $P_\star$; however, the following theorem demonstrates the two are equal.

> **Theorem 12.4.** Let $\psi, \phi \in \mathrm{Pure}(AB)$. Then,
> $$P\left(\psi^{AB} \xrightarrow{\mathrm{LOCC}} \phi^{AB}\right) = P_\star\left(\psi^{AB} \xrightarrow{\mathrm{LOCC}} \phi^{AB}\right). \tag{12.73}$$

**Proof** The inequality

$$P\left(\psi^{AB} \xrightarrow{\mathrm{LOCC}} \phi^{AB}\right) \leqslant P_\star\left(\psi^{AB} \xrightarrow{\mathrm{LOCC}} \phi^{AB}\right) \tag{12.74}$$

follows trivially since restricting $\sigma^{AB}$ in (12.72) to be a pure state $\varphi^{AB}$ can only increase the quantity. To prove the opposite inequality let $\sigma^{AB}$ be an optimizer of (12.72). Since $\psi^{AB} \xrightarrow{\mathrm{LOCC}} \sigma^{AB}$ it follows from Corollary 12.3 and its proof that there exists an ensemble $\{t_z, \varphi_z^{AB}\}_{z \in [k]}$ such that $\sigma^{AB} = \sum_{z \in [k]} t_z \varphi_z^{AB}$ and $\psi^{AB} \xrightarrow{\mathrm{LOCC}} \{t_z, \varphi_z^{AB}\}_{z \in [k]}$. For each $z \in [k]$ let $\mathbf{r}_z$ be the Schmidt vector of $\varphi_z^{AB}$ and define

$$\mathbf{r} := \sum_{z \in [k]} t_z \mathbf{r}_z^{\downarrow}. \tag{12.75}$$

Let $\varphi^{AB}$ be a pure state with a Schmidt vector $\mathbf{r}$, so that $\psi^{AB} \xrightarrow{\mathrm{LOCC}} \varphi^{AB}$. Observe that the square fidelity is given by

$$F\left(\phi^{AB}, \sigma^{AB}\right)^2 = \langle \phi^{AB} | \sigma^{AB} | \phi^{AB} \rangle = \sum_{z \in [k]} t_z \left| \langle \phi^{AB} | \varphi_z^{AB} \rangle \right|^2$$

$$cf. \ (12.70) \rightarrow \ \leqslant \sum_{z \in [k]} t_z F\left(\mathbf{q}^{\downarrow}, \mathbf{r}_z^{\downarrow}\right)^2 \tag{12.76}$$

$$(5.188) \rightarrow \ \leqslant F\left(\mathbf{q}^{\downarrow}, \mathbf{r}^{\downarrow}\right)^2.$$

Hence,

$$P\left(\psi^{AB} \xrightarrow{\mathrm{LOCC}} \phi^{AB}\right) = \sqrt{1 - F\left(\phi^{AB}, \sigma^{AB}\right)^2}$$

$$(12.76) \rightarrow \ \geqslant \sqrt{1 - F\left(\mathbf{q}^{\downarrow}, \mathbf{r}^{\downarrow}\right)^2} \tag{12.77}$$

$$\geqslant P_\star\left(\psi^{AB} \xrightarrow{\mathrm{LOCC}} \phi^{AB}\right).$$

Comparing the above inequality with (12.74) we get the equality of (12.73). $\blacksquare$

> **Corollary 12.4.** Let $\psi \in \mathrm{Pure}(AB)$ and $\Phi_m \in \mathrm{Pure}(A'B')$ be the maximally entangled state with $m := |A'| = |B'|$. Then,
>
> $$P\left(\Phi_m \xrightarrow{\ \mathrm{LOCC}\ } \psi^{AB}\right) = \sqrt{E_{(m)}\left(\psi^{AB}\right)}, \qquad (12.78)$$
>
> where $E_{(m)}$ is the measure of entanglement defined in (12.28) for $k = m$.

*Remark.* Observe that the corollary above provides an operational meaning to the entanglement monotones $E_{(m)}$. That is, $E_{(m)}(\psi^{AB})$ measures how close (in terms of the square of the purified distance) $\Phi_m$ can reach to $\psi^{AB}$ by LOCC. Furthermore, it's noteworthy that when $m \geqslant |A|$, the conversion distance $P\left(\Phi_m \xrightarrow{\ \mathrm{LOCC}\ } \psi^{AB}\right)$ equals zero. This outcome arises because, in this scenario, the majorization theorem by Nielsen guarantees that the conversion $\Phi_m \xrightarrow{\ \mathrm{LOCC}\ } \psi^{AB}$ can be accomplished exactly.

**Proof** Let $\mathbf{p} \in \mathrm{Prob}^{\downarrow}(n)$, with $n := |A|$, be the Schmidt vector corresponding to $\psi^{AB}$. From Theorem 12.4 it follows that

$$P\left(\Phi_m \xrightarrow{\ \mathrm{LOCC}\ } \psi^{AB}\right) = \min_{\mathbf{r} \in \mathrm{Prob}^{\downarrow}(n)} \left\{ P(\mathbf{p}, \mathbf{r}) : \mathbf{r} \succ \mathbf{u}^{(m)} \right\}, \qquad (12.79)$$

where $\mathbf{u}^{(m)}$ is the uniform density matrix in $\mathrm{Prob}(m)$. Now, observe that the condition $\mathbf{r} \succ \mathbf{u}^{(m)}$ holds if and only if $\mathbf{r}$ has at most $m$ nonzero components. Denoting by $|\mathbf{r}|$ the number of nonzero components in $\mathbf{r}$, and using the fact that the square of the purified distance equals 1 minus the square of the fidelity, we get

$$P^2\left(\Phi_m \xrightarrow{\ \mathrm{LOCC}\ } \psi^{AB}\right) = 1 - \max_{\mathbf{r} \in \mathrm{Prob}(n), |\mathbf{r}| = m} \left( \sum_{x \in [n]} \sqrt{r_x p_x} \right)^2$$

$$= 1 - \max_{\mathbf{r} \in \mathrm{Prob}(m)} \left( \sum_{x \in [m]} \sqrt{r_x p_x} \right)^2 \qquad (12.80)$$

$$\text{Exercise } 12.21 \rightarrow \ = 1 - \sum_{x \in [m]} p_x$$

$$= E_{(m)}\left(\psi^{AB}\right).$$

This completes the proof. ∎

**Exercise 12.21.** *Let $\{s_x\}_{x \in [n]}$ be a set of nonnegative real numbers. Show that*

$$\max_{\mathbf{r} \in \mathrm{Prob}(n)} \sum_{x \in [n]} \sqrt{r_x} s_x = \|\mathbf{s}\|_2 := \sqrt{\sum_{x \in [n]} s_x^2}. \qquad (12.81)$$

Drawing on the more straightforward formula in (12.71), a different kind of conversion distance can be defined, one that is based on the trace distance between Schmidt probability vectors. Specifically, let

$$T_\star\left(\psi^{AB} \xrightarrow{\ \mathrm{LOCC}\ } \phi^{AB}\right) := \min_{\mathbf{r} \in \mathrm{Prob}(d)} \left\{ \frac{1}{2}\|\mathbf{q} - \mathbf{r}\|_1 : \mathbf{r} \succ \mathbf{p} \right\}, \qquad (12.82)$$

where $\mathbf{p}$ is the Schmidt vector of $\psi^{AB}$, and $\mathbf{q}$ is the Schmidt vector of $\phi^{AB}$.

This metric can be viewed as the conversion distance of $\psi^{AB}$ into $\phi^{AB}$, in a framework where all bipartite pure states are identified with their corresponding Schmidt vectors. It quantifies the proximity to which the Schmidt vector of $\psi^{AB}$ can approach the Schmidt vector of $\phi^{AB}$ via doubly stochastic matrices. Given that the resource content of pure bipartite entangled states is fully encapsulated in their Schmidt vectors, this conversion distance arguably becomes the most natural way to gauge the effectiveness of interconversion between two pure resources.

In fact, the two conversion distances defined in (12.67) and (12.82) are topologically equivalent. This equivalence becomes apparent when considering that the inequalities in (5.258) lead to the following relation:

$$T\left(\psi^{AB} \xrightarrow{\text{LOCC}} \phi^{AB}\right) \leqslant P\left(\psi^{AB} \xrightarrow{\text{LOCC}} \phi^{AB}\right) \leqslant \sqrt{2T\left(\psi^{AB} \xrightarrow{\text{LOCC}} \phi^{AB}\right)} \tag{12.83}$$

$$T_\star\left(\psi^{AB} \xrightarrow{\text{LOCC}} \phi^{AB}\right) \leqslant P_\star\left(\psi^{AB} \xrightarrow{\text{LOCC}} \phi^{AB}\right) \leqslant \sqrt{2T_\star\left(\psi^{AB} \xrightarrow{\text{LOCC}} \phi^{AB}\right)}. \tag{12.84}$$

Therefore, combining these inequalities with Theorem 12.4 gives

$$T\left(\psi^{AB} \xrightarrow{\text{LOCC}} \phi^{AB}\right) \leqslant \sqrt{2T_\star\left(\psi^{AB} \xrightarrow{\text{LOCC}} \phi^{AB}\right)} \tag{12.85}$$

$$T_\star\left(\psi^{AB} \xrightarrow{\text{LOCC}} \phi^{AB}\right) \leqslant \sqrt{2T\left(\psi^{AB} \xrightarrow{\text{LOCC}} \phi^{AB}\right)}. \tag{12.86}$$

Considering the established equivalence between the two conversion distances, it makes sense to primarily use $T_\star$ for further analysis, owing to its following closed-form expression.

**Closed Formula**

**Theorem 12.5.** Let $\psi, \phi \in \text{Pure}(AB)$ be two bipartite states with $d := |A| = |B|$, and let $\mathbf{p}, \mathbf{q} \in \text{Prob}(d)$ be the corresponding Schmidt probability vectors of $\psi^{AB}$ and $\phi^{AB}$, respectively. Then,

$$T_\star\left(\psi^{AB} \xrightarrow{\text{LOCC}} \phi^{AB}\right) = \max_{k \in [d]} \left\{ \|\mathbf{p}\|_{(k)} - \|\mathbf{q}\|_{(k)} \right\}. \tag{12.87}$$

**Proof**   The proof follows directly from Theorem 4.4. To see this, observe that

$$T_\star\left(\psi^{AB} \xrightarrow{\text{LOCC}} \phi^{AB}\right) := \min_{\mathbf{r} \in \text{Majo}(\mathbf{p})} \frac{1}{2}\|\mathbf{q} - \mathbf{r}\|_1$$

$$= T\left(\mathbf{q}, \text{Majo}(\mathbf{p})\right) \tag{12.88}$$

$$\text{Theorem 4.4} \rightarrow = \max_{k \in [d]} \left\{ \|\mathbf{p}\|_{(k)} - \|\mathbf{q}\|_{(k)} \right\}.$$

This concludes the proof.   ∎

**Exercise 12.22.** *Show that the maximization (12.87) over all $k \in [d]$ can be replaced with maximization over all $k \in [r]$, where $r = \text{SR}\left(\psi^{AB}\right)$ is the Schmidt rank of $\psi^{AB}$.*

## 12.5.2 Distillable Entanglement

In this subsection we calculate the single-shot distillable entanglement. Consider first the zero-error case in which $\varepsilon = 0$, and let $\mathbf{p} = \mathbf{p}^\downarrow \in \mathrm{Prob}(d)$ be the Schmidt vector of an entangled pure state $\psi \in \mathrm{Pure}(AB)$, with $d := |A| = |B|$. For the zero-error case, the single-shot distillable entanglement is defined as

$$\mathrm{Distill}^{\varepsilon=0}\left(\psi^{AB}\right) := \max_{m \in \mathbb{N}} \left\{ \log m \ : \ \psi^{AB} \xrightarrow{\mathrm{LOCC}} \Phi_m \right\}. \tag{12.89}$$

From Nielsen majorization theorem, $\psi^{AB} \xrightarrow{\mathrm{LOCC}} \Phi_m$ if and only if $\frac{1}{m} \geqslant p_1$, where $p_1$ is the first component of $\mathbf{p} = \mathbf{p}^\downarrow$. Hence,

$$\mathrm{Distill}^{\varepsilon=0}\left(\psi^{AB}\right) = \max_{m \in \mathbb{N}} \left\{ \log m \ : \ m \leqslant \frac{1}{p_1} \right\}$$

$$m \text{ is an integer} \to \ = \log \left\lfloor \frac{1}{p_1} \right\rfloor. \tag{12.90}$$

The quantity $1/p_1$ can be expressed in terms of the min-entropy of $\mathbf{p}$. Specifically,

$$\mathrm{Distill}^{\varepsilon=0}\left(\psi^{AB}\right) = \log \left\lfloor 2^{H_{\min}(A)_\rho} \right\rfloor, \tag{12.91}$$

where $H_{\min}(A)_\rho$ is the min-entropy (see (6.22)) of the reduced density matrix $\rho^A := \mathrm{Tr}_B\left[\psi^{AB}\right]$.

To extend the formula above for the case that $\varepsilon > 0$, we use the computable conversion distance, to calculate the single-shot distillable entanglement. Specifically, for any $\varepsilon \in (0, 1)$ and $\psi \in \mathrm{Pure}(AB)$, we define the $\varepsilon$-single-shot distillable entanglement as

$$\mathrm{Distill}^{\varepsilon}\left(\psi^{AB}\right) := \max_{m \in \mathbb{N}} \left\{ \log m \ : \ T_\star\left(\psi^{AB} \xrightarrow{\mathrm{LOCC}} \Phi_m\right) \leqslant \varepsilon \right\}. \tag{12.92}$$

From the closed formula for $T_\star$ we get the following result.

---

**Theorem 12.6.** Using the same notations as above, for every $\varepsilon \in [0, 1)$ and $\psi \in \mathrm{Pure}(AB)$ we have

$$\mathrm{Distill}^{\varepsilon}\left(\psi^{AB}\right) = \log \left\lfloor 2^{H^{\varepsilon}_{\min}(A)_\rho} \right\rfloor, \tag{12.93}$$

where $H^{\varepsilon}_{\min}(A)_\rho$ is the smoothed min-entropy as given in (10.147).

---

*Remark.* In (10.147) we found a closed form to the smoothed min-entropy. Using this form we can express the $\varepsilon$-single-shot distillable entanglement of $\psi^{AB}$ as

$$\mathrm{Distill}^{\varepsilon}\left(\psi^{AB}\right) = \min_{k \in \{\ell,\dots,d\}} \log \left\lfloor \frac{k}{\|\mathbf{p}\|_{(k)} - \varepsilon} \right\rfloor, \tag{12.94}$$

where $\ell \in [d]$ is the integer satisfying $\|\mathbf{p}\|_{(\ell-1)} \leqslant \varepsilon < \|\mathbf{p}\|_{(\ell)}$.

**Proof**  From Theorem 12.5 and Exercise 12.22 we have

$$T_\star\left(\psi^{AB} \xrightarrow{\text{LOCC}} \Phi_m\right) = \max_{k\in[d]}\left\{\|\mathbf{p}\|_{(k)} - \|\mathbf{u}^{(m)}\|_{(k)}\right\}$$

$$= \max_{k\in[d]}\left\{\|\mathbf{p}\|_{(k)} - \frac{k}{m}\right\}. \tag{12.95}$$

Combining this with the definition in (12.92) we obtain

$$\text{Distill}^\varepsilon\left(\psi^{AB}\right) = \max_{m\in\mathbb{N}}\left\{\log m \,:\, \|\mathbf{p}\|_{(k)} - \frac{k}{m} \leqslant \varepsilon \quad \forall\, k \in [d]\right\}$$

$$= \max_{m\in\mathbb{N}}\left\{\log m \,:\, \|\mathbf{p}\|_{(k)} - \frac{k}{m} \leqslant \varepsilon \quad \forall\, k \in \{\ell,\dots,d\}\right\}, \tag{12.96}$$

since from the definition of $\ell$, if $k < \ell$, then the inequality $\|\mathbf{p}\|_{(k)} - \frac{k}{m} \leqslant \varepsilon$ holds trivially. Finally, observe that for each $k \in \{\ell,\dots,d\}$, the condition $\|\mathbf{p}\|_{(k)} - \frac{k}{m} \leqslant \varepsilon$ can be expressed as $m \leqslant \frac{k}{\|\mathbf{p}\|_{(k)}-\varepsilon}$, and since $m$ is an integer, this condition is equivalent to $m \leqslant a_k$, where

$$a_k := \left\lfloor \frac{k}{\|\mathbf{p}\|_{(k)} - \varepsilon} \right\rfloor. \tag{12.97}$$

Therefore, with this notation (12.96) gives

$$\text{Distill}^\varepsilon\left(\psi^{AB}\right) = \max_{m\in\mathbb{N}}\left\{\log m \,:\, m \leqslant a_k \;\forall\, k \in \{\ell,\dots,d\}\right\}$$

$$= \log \min_{k\in\{\ell,\dots,d\}}\{a_k\}. \tag{12.98}$$

This completes the proof.  ∎

**Exercise 12.23.**  *Use the formula in (12.94) to compute* $\text{Distill}^\varepsilon(\psi^{AB})$ *for the two extreme cases: (1)* $\psi^{AB}$ *is a maximally entangled state; (2)* $\psi^{AB}$ *is a product state. Give a physical interpretation to the results.*

**Exercise 12.24.**  *Let* $\psi \in \text{Pure}(AB)$. *Show that*

$$\lim_{\varepsilon \to 1^-} \text{Distill}^\varepsilon\left(\psi^{AB}\right) = \infty. \tag{12.99}$$

## 12.5.3 Entanglement Cost

In this subsection, we present a closed-form expression for the single-shot entanglement cost. Intriguingly, as we will explore in the next chapter, for the entanglement cost, both conversion distances (referenced in equations (12.67) and (12.82)) yield the same entanglement cost. This is a result we could not demonstrate for the distillable entanglement of a pure bipartite state, though we suspect it to be true.

For any $\varepsilon \in (0, 1)$ and $\psi \in \text{Pure}(AB)$, the $\varepsilon$-single-shot entanglement cost is defined as

$$\text{Cost}^\varepsilon\left(\psi^{AB}\right) := \min_{m\in\mathbb{N}}\left\{\log m \,:\, T_\star\left(\Phi_m \xrightarrow{\text{LOCC}} \psi^{AB}\right) \leqslant \varepsilon\right\}. \tag{12.100}$$

From the closed formula for $T_\star$ we get the following result.

> **Theorem 12.7.** Let $\varepsilon \in [0, 1)$, $\psi \in \mathrm{Pure}(AB)$, $d := \mathrm{SR}(\psi^{AB})$, and $\mathbf{p} \in \mathrm{Prob}(d)$ be the Schmidt probability vector of $\psi^{AB}$. Then, the $\varepsilon$-single-shot entanglement cost of $\psi^{AB}$ is given by
>
> $$\mathrm{Cost}^\varepsilon\left(\psi^{AB}\right) = \log m, \qquad (12.101)$$
>
> where $m \in [d]$ is the integer satisfying $\|\mathbf{p}\|_{(m-1)} < 1 - \varepsilon \leqslant \|\mathbf{p}\|_{(m)}$.

**Proof** From Theorem 12.5 and Exercise 12.22 we have for any $m \in \mathbb{N}$

$$T_\star\left(\Phi_m \xrightarrow{\text{LOCC}} \psi^{AB}\right) = \max_{k \in [m]} \sum_{x \in [k]} \left(\frac{1}{m} - p_x\right) = \max_{k \in [m]} \left\{\frac{k}{m} - \|\mathbf{p}\|_{(k)}\right\}. \qquad (12.102)$$

We therefore have

$$\begin{aligned}
\mathrm{Cost}^\varepsilon\left(\psi^{AB}\right) &= \min_{m \in \mathbb{N}}\left\{\log m : \frac{k}{m} - \|\mathbf{p}\|_{(k)} \leqslant \varepsilon \;\; \forall\, k \in [m]\right\} \\
&= \min_{m \in \mathbb{N}}\left\{\log m : m \geqslant \frac{k}{\|\mathbf{p}\|_{(k)} + \varepsilon} \;\; \forall\, k \in [m]\right\} \\
\text{Exercise 12.25} \rightarrow &= \min_{m \in \mathbb{N}}\left\{\log m : m \geqslant \frac{m}{\|\mathbf{p}\|_{(m)} + \varepsilon}\right\} \\
&= \min_{m \in \mathbb{N}}\left\{\log m : \|\mathbf{p}\|_{(m)} \geqslant 1 - \varepsilon\right\}.
\end{aligned} \qquad (12.103)$$

This completes the proof. ∎

**Exercise 12.25.** *Let*

$$b_k := \frac{k}{\|\mathbf{p}\|_{(k)} + \varepsilon} \qquad \forall\, k \in [d]. \qquad (12.104)$$

*Show that*

$$b_1 \leqslant b_2 \leqslant \cdots \leqslant b_d. \qquad (12.105)$$

As a simple example of this formula, consider the case that $\varepsilon = 0$. In this case, the smallest $m$ for which $\|\mathbf{p}\|_{(m)} \geqslant 1$ is $m = \mathrm{SR}(\psi^{AB})$. Therefore, as expected, the exact single-shot cost of $\psi^{AB}$ is given by

$$\mathrm{Cost}^{\varepsilon=0}\left(\psi^{AB}\right) = \log \mathrm{SR}\left(\psi^{AB}\right). \qquad (12.106)$$

**Exercise 12.26.** *Use the formula in (12.101) to compute $\mathrm{Cost}^\varepsilon(\psi^{AB})$ for the two extreme cases: (1) $\psi^{AB}$ is a maximally entangled state; (2) $\psi^{AB}$ is a product state. Give a physical interpretation for your results.*

In the following corollary we provide an operational interpretation for the smoothed max-entropy as the single-shot entanglement cost. Recall that the max-entropy of $\rho \in \mathfrak{D}(A)$ is defined as (cf. (6.20))

$$H_{\max}(A)_\rho := \log \operatorname{Rank}(\rho), \tag{12.107}$$

and its $\varepsilon$-smoothed version as

$$H_{\max}^\varepsilon(A)_\rho := \min_{\rho' \in \mathfrak{B}_\varepsilon(\rho)} H_{\max}(A)_{\rho'}. \tag{12.108}$$

**Corollary 12.5.** Let $\varepsilon \in [0, 1)$, $\psi \in \operatorname{Pure}(AB)$, and denote by $\rho^A :=_B \left[ \psi^{AB} \right]$ its reduced density matrix. Then, the $\varepsilon$-single-shot entanglement cost of $\psi^{AB}$ is given by

$$\operatorname{Cost}^\varepsilon \left( \psi^{AB} \right) = H_{\max}^\varepsilon (A)_\rho . \tag{12.109}$$

**Proof**   The proof follows trivially from a combination of the theorem above and the expression for $H_{\max}^\varepsilon$ as given in Lemma 10.6. ∎

## 12.5.4 Embezzlement of Entanglement

In the single-shot regime, we explore the approximate interconversion of resources where the final states are $\varepsilon$-close to the intended states. As already discussed, this flexibility introduces additional state transformations that go beyond the scope of Nielsen's majorization theorem. Notably, this includes the phenomenon known as *embezzlement of entanglement*.

Embezzlement of entanglement can be regarded as an extreme form of entanglement catalysis. Consider a family of bipartite states $|\chi_n\rangle \in \mathbb{C}^n \otimes \mathbb{C}^n$, defined by

$$|\chi_n\rangle := \frac{1}{\sqrt{H_n}} \sum_{y \in [n]} \frac{1}{\sqrt{x}} |x\rangle |x\rangle, \tag{12.110}$$

where the normalization factor $H_n := \sum_{x \in [n]} \frac{1}{x}$ is known as the *harmonic number*. We will demonstrate that $|\chi_n\rangle$ can serve as a catalyst for the *generation* of any arbitrary bipartite state $|\psi\rangle \in \mathbb{C}^m \otimes \mathbb{C}^m$, with $|\chi_n\rangle$ undergoing minimal change. More precisely, for any $\varepsilon > 0$, there exists an $n \in \mathbb{N}$ such that

$$T_\star \left( \chi_n \xrightarrow{\text{LOCC}} \psi \otimes \chi_n \right) \leqslant \varepsilon. \tag{12.111}$$

This remarkable result implies that it's feasible to "embezzle" a copy of $|\psi\rangle$ from the catalyst $|\chi_n\rangle$, effectively borrowing some of its entanglement while leaving it largely unchanged.

To see how it works, recall from Lemma 11.1 that

$$T_\star \left( \chi_n \xrightarrow{\text{LOCC}} \psi \otimes \chi_n \right) \leqslant T_\star \left( \chi_n \xrightarrow{\text{LOCC}} \Phi_m \otimes \chi_n \right), \tag{12.112}$$

since the maximally entangled state $|\Phi_m\rangle = \frac{1}{\sqrt{m}} \sum_{x \in [m]} |xx\rangle$ can be converted by LOCC to the state $\psi$. It is therefore sufficient to show that the right-hand side of the equation above can be made arbitrarily small as we increase the dimension $n$. Let $\mathbf{p}$ be the Schmidt vector of $\chi_n$ and $\mathbf{q}$ be the Schmidt vector of $\Phi_m \otimes \chi_n$. Observe that $\mathbf{p} \in \mathrm{Prob}(n)$ and $\mathbf{q} \in \mathrm{Prob}(nm)$. From Theorem 12.5 we know that

$$T_\star \left( \chi_n \xrightarrow{\text{LOCC}} \Phi_m \otimes \chi_n \right) = \max_{k \in [n]} \left\{ \|\mathbf{p}\|_{(k)} - \|\mathbf{q}\|_{(k)} \right\}. \tag{12.113}$$

Now, observe that the components of $\mathbf{q}$ have the form $\frac{p_x}{m}$. Therefore, for any decomposition $k = am + b$, with $a := \lfloor \frac{k}{m} \rfloor$ and some $b \in \{0, 1, \ldots, m-1\}$, we have

$$\|\mathbf{q}\|_{(k)} = \|\mathbf{p}\|_{(a)} + \frac{b}{m} p_{a+1}. \tag{12.114}$$

Substituting this into (12.113) gives

$$T_\star \left( \chi_n \xrightarrow{\text{LOCC}} \Phi_m \otimes \chi_n \right) = \max_{\substack{k \in [n] \\ b \in \{0, \ldots, m-1\}}} \left\{ \|\mathbf{p}\|_{(k)} - \|\mathbf{p}\|_{(\lfloor k/m \rfloor)} - \frac{b}{m} p_{\lfloor k/m \rfloor + 1} \right\} \tag{12.115}$$
$$= \max_{k \in [n]} \left\{ \|\mathbf{p}\|_{(k)} - \|\mathbf{p}\|_{(\lfloor k/m \rfloor)} \right\}.$$

Now, from the specific form of $\chi_n$ in (12.110) we have $\|\mathbf{p}\|_{(k)} = H_k/H_n$ so that the above equality is equivalent to

$$T_\star \left( \chi_n \xrightarrow{\text{LOCC}} \Phi_m \otimes \chi_n \right) = \max_{k \in [n]} \left\{ \frac{H_k - H_{\lfloor k/m \rfloor}}{H_n} \right\}. \tag{12.116}$$

Finally, we use the well-known bounds on the harmonic number $H_n$ given by

$$\ln(n) + \frac{1}{n} \leqslant H_n \leqslant \ln(n) + 1. \tag{12.117}$$

Using these bounds we estimate

$$H_k - H_{\lfloor k/m \rfloor} \leqslant \ln(k) + 1 - \ln \left\lfloor \frac{k}{m} \right\rfloor - \frac{1}{\left\lfloor \frac{k}{m} \right\rfloor} \tag{12.118}$$
$$\text{Exercise 12.27} \rightarrow \leqslant 1 + \ln(2m).$$

We therefore conclude that

$$T_\star \left( \chi_n \xrightarrow{\text{LOCC}} \phi_m^+ \otimes \chi_n \right) \leqslant \frac{1 + \ln(2m)}{H_n} \xrightarrow{n \to \infty} 0, \tag{12.119}$$

since $H_n$ goes to infinity as $n$ goes to infinity.

**Exercise 12.27.** *Prove the second inequality of (12.118).*

**Exercise 12.28.** *Fix $\alpha \in \mathbb{R}$, and consider the bipartite entangled state*

$$|\varphi_n\rangle := \frac{1}{\sqrt{N_n}} \sum_{x \in [n]} \sqrt{x^\alpha} |xx\rangle, \tag{12.120}$$

*where $N_n = \sum_{x\in[n]} x^\alpha$ is the normalization factor. Show that only for $\alpha = -1$ the state $|\varphi_n\rangle$ can be used to embezzle entanglement.*

In the general case of arbitrary family of pure bipartite states, $\{\chi_n\}_{n\in\mathbb{N}}$, observe that for any integer $\ell \leqslant a := \lfloor n/m \rfloor + 1$,

$$\max_{k\in[m(\ell-1),\ldots,m\ell-1]} \left\{ \|\mathbf{p}\|_{(k)} - \|\mathbf{p}\|_{(\lfloor k/m \rfloor)} \right\} = \max_{k\in[m(\ell-1),\ldots,m\ell-1]} \left\{ \|\mathbf{p}\|_{(k)} - \|\mathbf{p}\|_{(\ell-1)} \right\}$$

$$= \|\mathbf{p}\|_{(m\ell-1)} - \|\mathbf{p}\|_{(\ell-1)}$$

$$(12.121)$$

where we used the convention that $\|\mathbf{p}\|_{(k)} := 1$ for an integer $k > n$. With this convention, we conclude that

$$T_\star\left( \chi_n \xrightarrow{\text{LOCC}} \Phi_m \otimes \chi_n \right) = \max_{\ell\in[a]} \left\{ \|\mathbf{p}\|_{(m\ell-1)} - \|\mathbf{p}\|_{(\ell-1)} \right\}. \qquad (12.122)$$

Therefore, if $\{\chi_n\}_{n\in\mathbb{N}}$ is an embezzling family, then in particular it satisfies

$$\lim_{n\to\infty} \sum_{x=\ell}^{m\ell-1} p_x^{(n)} = 0 \qquad \forall\, \ell \in \mathbb{N}. \qquad (12.123)$$

The above equation holds if and only if

$$\lim_{n\to\infty} p_x^{(n)} = 0 \qquad \forall\, x \in \mathbb{N}. \qquad (12.124)$$

However, observe that the condition above is in general insufficient to determine if the states $\{\chi_n\}_{n\in\mathbb{N}}$ form an embezzling family, since the maximizer $\ell$ in (12.122) can depend on $n$.

## 12.6 Asymptotic Entanglement Theory of Pure States

In Section 11.5.1, we defined the asymptotic resource cost and the asymptotic distillation of a resource. Applying these general definitions to the case of pure bipartite entanglement results in the following definitions of the cost and distillable rates of converting $\psi \in \text{Pure}(AB)$ to $\phi \in \text{Pure}(AB)$:

$$\text{Cost}\left( \psi^{AB} \to \phi^{AB} \right) := \lim_{\varepsilon\to 0^+} \inf_{n,m\in\mathbb{N}} \left\{ \frac{n}{m} : T_\star\left( \psi^{\otimes n} \xrightarrow{\text{LOCC}} \phi^{\otimes m} \right) \leqslant \varepsilon \right\},$$

$$\text{Distill}\left( \psi^{AB} \to \phi^{AB} \right) := \lim_{\varepsilon\to 0^+} \sup_{n,m\in\mathbb{N}} \left\{ \frac{m}{n} : T_\star\left( \psi^{\otimes n} \xrightarrow{\text{LOCC}} \phi^{\otimes m} \right) \leqslant \varepsilon \right\}. \qquad (12.125)$$

These definitions can then be used to define the entanglement cost and distillable entanglement as follows:

$$\text{Cost}\left( \psi^{AB} \right) = \text{Cost}\left( \Phi_2 \to \psi^{AB} \right) \quad \text{and} \quad \text{Distill}\left( \psi^{AB} \right) = \text{Distill}\left( \psi^{AB} \to \Phi_2 \right), \qquad (12.126)$$

where $|\Phi_2\rangle := \frac{1}{\sqrt{2}}(|00\rangle + |11\rangle)$ is the $2 \times 2$ dimensional maximally entangled state (i.e. a Bell state). In the following subsections we provide closed formulas for these measures of entanglement and discuss their relations to the single-shot quantities $\text{Cost}^\varepsilon$ and $\text{Distill}^\varepsilon$ that we studied in the previous section.

## 12.6.1 Entanglement Cost

From Lemma 11.10, particularly Exercise 11.14, it follows that the entanglement cost as already defined can be expressed for any $\psi \in \text{Pure}(AB)$ as

$$\text{Cost}\left(\psi^{AB}\right) = \lim_{\varepsilon \to 0^+} \liminf_{n \to \infty} \frac{1}{n} \text{Cost}^\varepsilon\left(\psi^{\otimes n}\right). \tag{12.127}$$

In the following theorem we compute the entanglement cost and prove a stronger version of the above relation.

> **Theorem 12.8.** Let $\psi \in \text{Pure}(AB)$. Then, for any $\varepsilon \in (0,1)$
>
> $$\text{Cost}\left(\psi^{AB}\right) = \lim_{n \to \infty} \frac{1}{n} \text{Cost}^\varepsilon\left(\psi^{\otimes n}\right) = E\left(\psi^{AB}\right), \tag{12.128}$$
>
> where $E$ is the entropy of entanglement defined in (12.47).

*Remark.* Observe that from the theorem above it follows that there is no need to take the limit $\varepsilon \to 0^+$ in (12.127); that is, by taking the limit $n \to \infty$ the dependance on $\varepsilon$ is eliminated (as long as $\varepsilon \in (0,1)$).

**Proof**  The proof follows directly from a combination of Corollary 12.5 and the variant of the AEP property given in (10.169). Specifically, denoting by $\rho^A := \text{Tr}_B\left[\psi^{AB}\right]$, we get from Corollary 12.5 that

$$\lim_{n \to \infty} \frac{1}{n} \text{Cost}^\varepsilon\left(\psi^{\otimes n}\right) = \lim_{n \to \infty} \frac{1}{n} H_{\max}^\varepsilon\left(A^n\right)_{\rho^{\otimes n}} \tag{12.129}$$
$$(10.169) \to = H(A)_\rho.$$

This completes the proof.  ∎

**Exercise 12.29.** *Provide a more direct proof of the theorem above using the concept of typicality. Hint: The proof is a bit involved and can be found in Appendix D.5 of online version.*

## 12.6.2 Distillable Entanglement

From Lemma 11.10, particularly Exercise 11.14, it follows that the distillable entanglement as already defined can be expressed for any $\psi \in \text{Pure}(AB)$ as

$$\text{Distill}\left(\psi^{AB}\right) = \lim_{\varepsilon \to 0^+} \limsup_{n \to \infty} \frac{1}{n} \text{Distill}^\varepsilon\left(\psi^{\otimes n}\right). \tag{12.130}$$

In the following theorem we compute the distillable entanglement and prove a stronger version of the above relation.

> **Theorem 12.9.** Let $\psi \in \text{Pure}(AB)$. Then, for any $\varepsilon \in (0, 1)$
>
> $$\text{Distill}\left(\psi^{AB}\right) = \lim_{n\to\infty} \frac{1}{n}\text{Distill}^{\varepsilon}\left(\psi^{\otimes n}\right) = E\left(\psi^{AB}\right), \qquad (12.131)$$
>
> where $E$ is the entropy of entanglement defined in (12.47).

**Proof**   The proof follows directly from a combination of Theorem 12.6 and the variant of the AEP property given in (11.50). Specifically, denoting by $\rho^A := \text{Tr}_B\left[\psi^{AB}\right]$, we get from Corollary 12.5 that

$$\lim_{n\to\infty} \frac{1}{n}\text{Cost}^{\varepsilon}\left(\psi^{\otimes n}\right) = \lim_{n\to\infty} \frac{1}{n}H_{\min}^{\varepsilon}\left(A^n\right)_{\rho^{\otimes n}}. \qquad (12.132)$$

Moreover, taking $\sigma^A = \mathbf{u}^A$ in (11.50) yields

$$\lim_{n\to\infty} \frac{1}{n}H_{\min}^{\varepsilon}\left(A^n\right)_{\rho^{\otimes n}} = H(A)_{\rho}. \qquad (12.133)$$

This completes the proof.   ∎

For those readers seeking additional insights, an alternative proof utilizing the concept of typicality is provided in Appendix D.5 of online version. This proof offers a different perspective and leverages the principles of typicality, which may be of interest to readers who are keen on exploring diverse approaches and methodologies within the field.

## 12.6.3 Reversibility of Pure Bipartite Entanglement Theory

We saw in the last two theorems that both the entanglement cost and the distillable entanglement of a pure bipartite state are equal to the entropy of entanglement. This equality between the entanglement cost and the distillable entanglement implies that the QRT of pure bipartite entanglement is reversible. Note that we get reversibility under LOCC, which is a strict subset of RNG (i.e. "nonentangling") operations. In the next chapter, we will discuss in more detail the relationship between nonentangling operations and LOCC.

**Exercise 12.30.** *Use Theorems 12.8 and 12.9 above to show that for any $\psi, \phi \in \text{Pure}(AB)$*

$$\text{Distill}\left(\psi^{AB} \to \phi^{AB}\right) = \frac{E(\psi)}{E(\phi)} \quad \text{and} \quad \text{Cost}\left(\psi^{AB} \to \phi^{AB}\right) = \frac{E(\phi)}{E(\psi)}. \qquad (12.134)$$

## 12.7 Notes and References

Comprehensive reviews on entanglement theory can be found in Refs. [135] and [184]. Additional information on LOCC operations can be found in Refs. [49, 47], and references therein. Theorem 12.1 is a slight modified version of the one given in Ref. [152]. The relation between entanglement and majorization, specifically Theorem 12.2, was first established in Ref. [172]. The entanglement monotones defined in (12.28) were first introduced in Ref. [227]. Entanglement catalysis was introduced in Ref. [143]. The concurrence measure of entanglement was first introduced in Ref. [238] for the purpose of computing the entanglement of formation of a mixed bipartite state. Later on it was generalized in Ref. [86] to the family of entanglement monotones given in (12.49). Theorem 12.3 is due to Ref. [144], while the formula for the maximum probability to convert one state to another by LOCC, that is, Corollary 12.2, was first proved independently in Ref. [226] without the use of Theorem 12.3 as we do here. The second corollary of Theorem 12.3, that is, Corollary 12.3, as well as the closed formula for $T_\star$, where first introduced in Ref. [242]. Embezzlement of entanglement was first introduced in Ref. [217].

# 13

# Mixed-State Entanglement

To gain a better understanding of bipartite entanglement theory, we will delve into its most general form in this chapter. The free states of the theory consists of separable states, denoted for any composite system $AB$ by

$$\mathrm{SEP}(AB) := \left\{ \sum_{x \in [n]} p_x \sigma_x^A \otimes \omega_x^B : \sigma_x \in \mathfrak{D}(A), \omega_x \in \mathfrak{D}(B), \mathbf{p} \in \mathrm{Prob}(n), n \in \mathbb{N} \right\}, \quad (13.1)$$

where we used the notation $\mathbf{p} := (p_1, \ldots, p_n)^T$. Observe that $\mathrm{SEP}(AB)$ is a closed convex set. Any quantum state $\rho \in \mathfrak{D}(AB)$ that does not belong to $\mathrm{SEP}(AB)$ is referred to as an entangled state. This chapter will reveal the intricate structure of entangled states, highlighting the complexity of mixed-state entanglement theory.

## 13.1 Detection of Entanglement

Unlike pure-state entanglement, detecting mixed state entanglement is a challenging task even from a theoretical standpoint. In particular, density matrices in $\mathfrak{D}(AB)$ are usually expressed as positive semidefinite matrices with a size of $|AB| \times |AB|$ and a trace of 1. Establishing whether these matrices belong to $\mathrm{SEP}(AB)$ is a complex undertaking, and in most cases, it falls under the category of NP-hard problems.

### 13.1.1 Entanglement Witnesses

One technique for detecting entanglement involves the concept of resource witnesses, which is discussed in Section 9.4. Specifically, Definition 9.11 of a resource witness can be adapted to apply to entanglement theory.



**Definition 13.1.** An operator $\Gamma \in \mathrm{Herm}(AB)$ is called an entanglement witness if the following two conditions hold:

1. For any $\sigma \in \mathrm{SEP}(AB)$,

$$\mathrm{Tr}\left[\Gamma^{AB} \sigma^{AB}\right] \geqslant 0. \tag{13.2}$$

2. There exists $\rho \in \mathfrak{D}(AB)$ such that

$$\mathrm{Tr}\left[\Gamma^{AB} \rho^{AB}\right] < 0. \tag{13.3}$$

From the condition in (13.2), and the fact that SEP($AB$) is the convex hull of product states, it follows that if $\Gamma \in \mathrm{Herm}(AB)$ is an entanglement witness, then for any product state $\psi \otimes \phi \in \mathrm{Pure}(AB)$ we must have

$$\left\langle \psi^A \otimes \phi^B \left| \Gamma^{AB} \right| \psi^A \otimes \phi^B \right\rangle \geqslant 0. \tag{13.4}$$

On the other hand, the condition 13.3 also implies that there exists a state $\chi \in \mathrm{Pure}(AB)$ such that

$$\left\langle \chi^{AB} \left| \Gamma^{AB} \right| \chi^{AB} \right\rangle < 0. \tag{13.5}$$

In other words, the condition (13.3) implies that $\Gamma^{AB}$ is not positive semidefinite so that we can take $\chi^{AB}$, for example, to be an eigenstate corresponding to a negative eigenvalue of $\Gamma^{AB}$.

Based on Theorem 9.3, we can conclude that entanglement witnesses are an effective tool for detecting entanglement. Specifically, $\rho \in \mathfrak{D}(AB)$ is an entangled state if and only if there exists an entanglement witness $\Gamma \in \mathrm{Herm}(AB)$ such that

$$\mathrm{Tr}\left[ \Gamma^{AB} \rho^{AB} \right] < 0. \tag{13.6}$$

This characteristic can be employed to demonstrate that the set of separable states occupies a nonzero volume.

> **Theorem 13.1.** The set SEP($AB$) has a nonzero volume in $\mathfrak{D}(AB)$. Specifically, there exists $\varepsilon > 0$ such that $\mathfrak{B}_\varepsilon(\mathbf{u}^{AB}) \subset \mathrm{SEP}(AB)$, where $\mathfrak{B}_\varepsilon(\mathbf{u}^{AB})$ is the "ball" of all states in $\mathfrak{D}(AB)$ that are $\varepsilon$-close to the maximally mixed state $\mathbf{u}^{AB} = \mathbf{u}^A \otimes \mathbf{u}^B$.

**Proof**   Suppose by contradiction that the statement in the theorem is false. Then, there exists a sequence of bipartite entangled states $\{\tau_n^{AB}\}_{n\in\mathbb{N}}$ such that

$$\lim_{n\to\infty} \frac{1}{2} \left\| \tau_n^{AB} - \mathbf{u}^{AB} \right\|_1 = 0. \tag{13.7}$$

Since we assume that $\tau_n^{AB}$ is entangled, we have $\tau_n \notin \mathrm{SEP}(AB)$. Therefore, there exists an entanglement witness $\Gamma_n^{AB}$ such that

$$\mathrm{Tr}\left[ \tau_n^{AB} \Gamma_n^{AB} \right] < 0. \tag{13.8}$$

Without loss of generality we can assume that for each $n \in \mathbb{N}$ the witness $\Gamma_n^{AB}$ is normalized with respect to the Hilbert–Schmidt inner product; that is,

$$\mathrm{Tr}\left[ \left( \Gamma_n^{AB} \right)^2 \right] = 1. \tag{13.9}$$

Therefore, the sequence $\{\Gamma_n^{AB}\}$ is a sequence of Hermitian operators in the unit sphere of $\mathrm{Herm}(AB)$. Since the unit sphere is compact, there exists a subsequence $\{n_k\}_{k\in\mathbb{N}}$ of integers such that the limit $\lim_{k\to\infty} \Gamma_{n_k}^{AB}$ exists and equal to some normalized operator $\Gamma_\star^{AB} \in \mathrm{Herm}(AB)$. Since each $\Gamma_{n_k}^{AB}$ is an entanglement witness, the limit $\Gamma_\star^{AB}$ must satisfy

$$\mathrm{Tr}\left[\Gamma_\star^{AB}\sigma^{AB}\right] \geqslant 0 \qquad \forall\, \sigma \in \mathrm{SEP}(AB). \tag{13.10}$$

On the other hand, taking the limit $k \to \infty$ on both sides of the inequality $\mathrm{Tr}\left[\tau_{n_k}^{AB}\Gamma_{n_k}^{AB}\right] < 0$ gives

$$\mathrm{Tr}\left[\Gamma_\star^{AB}\mathbf{u}^{AB}\right] \leqslant 0, \tag{13.11}$$

so that $\mathrm{Tr}\left[\Gamma_\star^{AB}\right] \leqslant 0$. Now, let $\{|\psi_x\rangle^A\}_{x\in[m]}$ be an orthonormal basis of $A$, and $\{|\phi_y\rangle^B\}_{y\in[\ell]}$ be an orthonormal basis of $B$. Then,

$$0 \geqslant \mathrm{Tr}\left[\Gamma_\star^{AB}\right] = \sum_{x\in[m]}\sum_{y\in[\ell]} \mathrm{Tr}\left[\Gamma_\star^{AB}\left(\psi_x^A \otimes \phi_y^B\right)\right]. \tag{13.12}$$

From (13.10) it follows that for each $x \in [m]$ and $y \in [\ell]$ we have $\mathrm{Tr}\left[\Gamma_\star^{AB}\left(\psi_x^A \otimes \phi_y^B\right)\right] \geqslant 0$. We therefore get that for any $x \in [m]$ and $y \in [\ell]$, $\mathrm{Tr}\left[\Gamma_\star^{AB}\left(\psi_x^A \otimes \phi_y^B\right)\right] = 0$. Finally, since the orthonormal bases $\{|\psi_x\rangle^A\}_{x\in[m]}$ and $\{|\phi_y\rangle^B\}_{y\in[\ell]}$ were arbitrary, we conclude that

$$\mathrm{Tr}\left[\Gamma_\star^{AB}\left(\psi^A \otimes \phi^B\right)\right] = 0 \qquad \forall\, \psi \in \mathrm{Pure}(A) \quad \forall\, \phi \in \mathrm{Pure}(B). \tag{13.13}$$

However, from Exercise 3.14 it follows that (13.13) holds if and only if $\Gamma_\star^{AB} = 0$ in contradiction with the fact that $\Gamma_\star^{AB}$ is normalized, so in particular, cannot be the zero matrix. This completes the proof. ∎

**Exercise 13.1.** *Let $A_1, \ldots, A_m$ be $m$ physical systems and let $\mathrm{SEP}(A_1 \cdots A_m)$ be the set of multipartite separable states; that is, $\mathrm{SEP}(A_1 \cdots A_m)$ is the convex hull of the set of all $m$-fold product states of the form $\rho_1 \otimes \cdots \otimes \rho_m$, with $\rho_x \in \mathfrak{D}(A_x)$ for all $x \in [m]$. Show that $\mathrm{SEP}(A_1 \cdots A_m)$ has a nonzero volume in $\mathfrak{D}(A_1 \cdots A_m)$.*

The following theorem shows a close connection between entanglement witnesses and positive maps.

> **Theorem 13.2.** Any entanglement witness is the Choi matrix of a positive map that is not completely positive. Explicitly, $\Gamma \in \mathrm{Herm}(AB)$ is an entanglement witness if and only if $\Gamma^{AB} = J_{\mathcal{E}}^{AB}$ for some positive map $\mathcal{E} \in \mathrm{Pos}(A \to B)$ that is not completely positive (i.e. $\mathcal{E} \notin \mathrm{CP}(A \to B)$).

**Proof** Suppose first that $\Gamma^{AB} = J_{\mathcal{E}}^{AB}$ for some positive map $\mathcal{E} \in \mathrm{Pos}(A \to B)$ and suppose $\mathcal{E} \notin \mathrm{CP}(A \to B)$. Then, for any product state $\rho \otimes \sigma \in \mathrm{Pure}(AB)$ we have

$$\mathrm{Tr}\left[\Gamma^{AB}\left(\rho^A \otimes \sigma^B\right)\right] = \mathrm{Tr}\left[J_{\mathcal{E}}^{AB}\left(\rho^A \otimes \sigma^B\right)\right]$$

$$\boxed{\mathrm{Tr}_A\left[J_{\mathcal{E}}^{AB}\left(\rho^A \otimes I^B\right)\right] = \mathcal{E}^{A\to B}\left((\rho^A)^T\right)} \longrightarrow = \mathrm{Tr}\left[\sigma^B \mathcal{E}^{A\to B}\left((\rho^A)^T\right)\right] \tag{13.14}$$

$$\geqslant 0,$$

where the last inequality follows from the fact that $\mathcal{E} \in \mathrm{Pos}(A \to B)$ so that $\mathcal{E}\left(\rho^T\right) \geq 0$. Finally, the existence of a state $\chi \in \mathrm{Pure}(AB)$ that satisfies (13.5) follows from the fact that $\mathcal{E}$ is not completely positive so its Choi matrix $\Gamma^{AB}$ is not positive semidefinite.

Conversely, suppose $\Gamma^{AB}$ is an entanglement witness and let $\mathcal{E} \in \mathfrak{L}(A \to B)$ be such that $\Gamma^{AB} = J_{\mathcal{E}}^{AB}$ (but we do not assume that $\mathcal{E}$ is positive). Then, for any $\rho \in \mathfrak{D}(A)$ and $\sigma \in \mathfrak{D}(B)$ we have

$$
\begin{aligned}
\mathrm{Tr}\left[\sigma^B \mathcal{E}^{A \to B}\left(\rho^A\right)\right] &= \mathrm{Tr}\left[J_{\mathcal{E}}^{AB}\left(\left(\rho^A\right)^T \otimes \sigma^B\right)\right] \\
&= \mathrm{Tr}\left[\Gamma^{AB}\left(\left(\rho^A\right)^T \otimes \sigma^B\right)\right] \qquad (13.15) \\
&\geq 0,
\end{aligned}
$$

where the last inequality follows from the fact that $\Gamma^{AB}$ is an entanglement witness and $\left(\rho^A\right)^T \otimes \sigma^B \in \mathrm{SEP}(AB)$. Since $\rho^A$ and $\sigma^B$ were arbitrary states, (13.15) implies that $\mathcal{E} \in \mathrm{Pos}(A \to B)$. The map $\mathcal{E}$ is not completely positive since its Choi matrix, $\Gamma^{AB}$, is not positive semidefinite (as $\Gamma^{AB}$ is an entanglement witness). This completes the proof. $\blacksquare$

Observe that the set of all entanglement witnesses consist of all the nonpositive semidefinite matrices that are in the dual cone of the set of separable states. Specifically, for the composite system $AB$, the set of all entanglement witnesses, denoted by $\mathrm{WIT}(AB)$, is given by

$$
\mathrm{WIT}(AB) = \left\{\Gamma \in \mathrm{SEP}(AB)^* : \Gamma^{AB} \not\geq 0\right\}. \qquad (13.16)
$$

We will now provide two examples of how Theorem 9.3, adapted to entanglement theory with the set $\mathrm{WIT}(AB)$ already mentioned, can be used to determine whether a quantum state is entangled.

## Example 1: The Isotropic State

The isotropic state in $\mathfrak{D}(AB)$, with $m := |A| = |B|$, is defined as

$$
\rho_t^{AB} = t\Phi_m^{AB} + (1-t)\tau^{AB}, \qquad (13.17)
$$

where $t \in (0, 1)$, and

$$
\tau^{AB} := \frac{I^{AB} - \Phi_m}{m^2 - 1}. \qquad (13.18)
$$

Observe that $\Phi_m \tau = \tau \Phi_m = 0$, and furthermore, since $\Phi_m^{AB}$ is invariant under the action of the twirling channel $\mathcal{G}$ defined in (3.251) also $\rho_t^{AB}$ has this property. In fact, the state $\left\{\rho_t^{AB}\right\}_{t \in [0,1]}$ can be viewed as the set of *all* quantum states that are invariant under $\mathcal{G}$ (see (3.255)). In the following, we will utilize this property to make the argument that the isotropic state $\rho_t^{AB}$ satisfies

$$
\rho_t^{AB} \in \mathrm{SEP}(AB) \quad \Longleftrightarrow \quad t \leq \frac{1}{m}. \qquad (13.19)
$$

To prove this statement we follow Theorem 9.3. Specifically, $\rho_t^{AB}$ is separable if and only if $\mathrm{Tr}[\Gamma^{AB}\rho_t^{AB}] \geqslant 0$ for all entanglement witnesses $\Gamma \in \mathrm{WIT}(AB)$. The key idea is to use the invariance of $\rho_t^{AB}$ under $\mathcal{G}$ to get that

$$\mathrm{Tr}\left[\Gamma^{AB}\rho_t^{AB}\right] = \mathrm{Tr}\left[\Gamma^{AB}\mathcal{G}\left(\rho_t^{AB}\right)\right]$$

$$\mathcal{G} \text{ is self-adjoint} \rightarrow = \mathrm{Tr}\left[\mathcal{G}\left(\Gamma^{AB}\right)\rho_t^{AB}\right]. \tag{13.20}$$

Now, observe that if $\sigma \in \mathrm{SEP}(AB)$, then also $\mathcal{G}(\sigma) \in \mathrm{SEP}(AB)$ so that

$$\mathrm{Tr}\left[\mathcal{G}\left(\Gamma^{AB}\right)\sigma^{AB}\right] = \mathrm{Tr}\left[\Gamma^{AB}\mathcal{G}\left(\sigma^{AB}\right)\right] \geqslant 0. \tag{13.21}$$

Combining this with (13.20) we conclude that $\rho_t^{AB}$ is separable if and only if $\mathrm{Tr}[\Gamma^{AB}\rho_t^{AB}] \geqslant 0$ for all entanglement witnesses of the form

$$\Gamma^{AB} = \mathcal{G}\left(\Gamma^{AB}\right) = aI^{AB} + b\Phi_m^{AB}, \tag{13.22}$$

where $a, b \in \mathbb{R}$. In the final equality, we made use of the fact that $I^{AB}$ and $\Phi_m^{AB}$ spans the subspace of $\mathcal{G}$-invariant operators in $\mathrm{Herm}(AB)$.

To ensure that the matrix $\Gamma = aI + b\Phi_m$ is an entanglement witness, we need to appropriately specify the coefficients $a$ and $b$. Let's start by noting that $a$ must be nonnegative, as evidenced by

$$a = \mathrm{Tr}\left[\Gamma\left(|1\rangle\langle 1| \otimes |2\rangle\langle 2|\right)\right] \geqslant 0. \tag{13.23}$$

Furthermore, it's important to recognize that $\Gamma$ exhibits two distinct eigenvalues: $a$ with multiplicity $|AB| - 1$, and $a + b$ with multiplicity 1. Given that an entanglement witness has at least one negative eigenvalue, and considering that $a \geqslant 0$, it is necessary for $b$ to satisfy $b < -a$. It's also worth mentioning that the scenario where $a = 0$ does not yield an entanglement witness (this is an interesting point to ponder – why this is the case?).

Consequently, after rescaling $\Gamma^{AB}$ by a positive factor $a > 0$, we can, without loss of generality, assume that $\Gamma^{AB}$ takes the form

$$W = I^{AB} - r\Phi_m^{AB}, \tag{13.24}$$

where $r > 1$. From (13.4) the matrix $\Gamma^{AB}$ is an entanglement witness if and only if for any product state $\psi \otimes \phi \in \mathrm{Pure}(AB)$

$$0 \leqslant \mathrm{Tr}\left[\Gamma^{AB}\left(\psi^A \otimes \phi^B\right)\right] = 1 - r\mathrm{Tr}\left[\Phi_m^{AB}\left(\psi^A \otimes \phi^B\right)\right]. \tag{13.25}$$

In Exercise 13.2 we will show that

$$\max_{\substack{\psi \in \mathrm{Pure}(A) \\ \phi \in \mathrm{Pure}(B)}} \mathrm{Tr}\left[\Phi_m^{AB}\left(\psi^A \otimes \phi^B\right)\right] = \frac{1}{m}. \tag{13.26}$$

We therefore conclude that $\Gamma = I - r\Phi_m$ is an entanglement witness if and only if $1 < r \leqslant m$. Hence, $\rho_t^{AB}$ is separable if and only if for all $1 < r \leqslant m$ we have

$$0 \leqslant \mathrm{Tr}\left[\rho_t^{AB}\left(I^{AB} - r\Phi_m^{AB}\right)\right] = 1 - rt. \tag{13.27}$$

This inequality holds for all $1 < r \leqslant m$ if and only if $t \leqslant 1/m$. This completes the proof of (13.19).

**Exercise 13.2.** *Prove* (13.26).

**Exercise 13.3.** *Show that the isotropic state can be expressed as*

$$\rho_t^{AB} = \frac{m^2}{m^2 - 1}\left((1-t)\mathbf{u}^{AB} + \left(t - \frac{1}{m^2}\right)\Phi_m^{AB}\right). \tag{13.28}$$

## Example 2: Werner States

The Werner state is a density matrix in $\mathfrak{D}(AB)$ that remains invariant under the action of the $\mathcal{G}$-twirling map, where $\mathcal{G}$ in this case is defined as the map given in (3.243). The Werner state $\rho_{\mathrm{w}}^{AB}$ is defined for all $p \in [0, 1]$ by (cf. (3.246))

$$\rho_{\mathrm{w}}^{AB} := p\frac{2}{m(m+1)}\Pi_{\mathrm{Sym}}^{AB} + (1-p)\frac{2}{m(m-1)}\Pi_{\mathrm{Asy}}^{AB}. \tag{13.29}$$

Recall that $\Pi_{\mathrm{Sym}}^{AB} = \frac{1}{2}\left(I^{AB} + F^{AB}\right)$ and $\Pi_{\mathrm{Asy}}^{AB} = \frac{1}{2}\left(I^{AB} - F^{AB}\right)$, where $F^{AB}$ is the flip operator $F^{AB}$ defined in (C.187 of online version). Therefore, the Werner state can also be expressed more compactly as (see Exercise 13.4)

$$\rho_{\mathrm{w}}^{AB} = \frac{1}{m(m-\alpha)}\left(I^{AB} - \alpha F^{AB}\right), \tag{13.30}$$

where the new parameter $\alpha \in [-1, 1]$ is related to $p$ via

$$\alpha := \frac{1 + m(1 - 2p)}{1 - 2p + m}. \tag{13.31}$$

**Exercise 13.4.** *Prove* (13.30) *by substituting the expressions* $\Pi_{\mathrm{Sym}}^{AB} = \frac{1}{2}\left(I^{AB} + F^{AB}\right)$ *and* $\Pi_{\mathrm{Asy}}^{AB} = \frac{1}{2}\left(I^{AB} - F^{AB}\right)$ *into* (13.29).

Similar to the analysis of the isotropic state, we can determine for which values of $p$ (or $\alpha$) the Werner state is entangled. We find that $\rho_{\mathrm{w}}^{AB} \in \mathrm{SEP}(AB)$ if and only if $p \leqslant \frac{1}{2}$ or equivalently if and only if $\alpha \leqslant \frac{1}{m}$. We leave the proof as an exercise.

**Exercise 13.5.** *Prove that* $\rho_w^{AB} \in \mathrm{SEP}(AB)$ *if and only if* $\alpha \leqslant \frac{1}{m}$. *Hint: Show first that the Werner state* $\rho_w^{AB}$ *is entangled if and only if* $\mathrm{Tr}\left[\Gamma^{AB}\rho_w^{AB}\right] \geqslant 0$ *for all* $\Gamma \in \mathrm{WIT}(AB)$ *of the form* $\Gamma^{AB} = aI^{AB} + bF^{AB}$. *Then find the values of a and b for which* $aI^{AB} + bF^{AB} \in \mathrm{WIT}(AB)$ *and continue from there.*

## Entanglement Witnesses in Small Dimensions

For small dimensions, $\mathrm{WIT}(AB)$ has the following simple characterization.

> **Theorem 13.3.** Let $\Gamma \in \mathrm{WIT}(AB)$ with $|AB| \leqslant 6$. Then, there exists $\eta_1, \eta_2 \in \mathrm{Pos}(AB)$ such that
>
> $$\Gamma^{AB} = \eta_1^{AB} + \mathcal{T}^{B \to B}\left(\eta_2^{AB}\right), \tag{13.32}$$
>
> where $\mathcal{T} \in \mathrm{Pos}(B \to B)$ is the transpose map.

**Proof**  Without loss of generality suppose $|A| = 2$ and $|B| \leqslant 3$. From Theorem 13.2 there exists $\mathcal{E} \in \mathrm{Pos}(A \to B)$ such that

$$\Gamma^{AB} = \mathcal{E}^{\tilde{A} \to B}\left(\Omega^{A\tilde{A}}\right). \tag{13.33}$$

Furthermore, from Theorem 3.3 it follows that $\mathcal{E} = \mathcal{N}_1 + \mathcal{T} \circ \mathcal{N}_2$ for some $\mathcal{N}_1, \mathcal{N}_2 \in \mathrm{CP}(A \to B)$. Substituting this into (13.33), and denoting by $\eta_j := \mathcal{N}_j^{\tilde{A} \to B}\left(\Omega^{A\tilde{A}}\right)$ for $j = 1, 2$, we get that $\Gamma^{AB}$ has the form (13.32). Finally, observe that $\eta_1, \eta_2 \geqslant 0$ since $\mathcal{N}_1$ and $\mathcal{N}_2$ are completely positive maps. This concludes the proof. $\blacksquare$

### 13.1.2 The PPT Criterion

In this subsection we consider a simple criterion to detect entanglement. The criterion is known as the Peres–Horodecki criterion or the PPT criterion since it is based on the partial transpose. The criterion states that if $\rho^{AB} = \sum_{x \in [m]} p_x \rho_x^A \otimes \rho_x^B$ is a separable density matrix, then its partial transpose

$$\mathcal{T}^{B \to B}\left(\rho^{AB}\right) = \sum_{x \in [m]} p_x \rho_x^A \otimes \left(\rho_x^B\right)^T \geqslant 0 \tag{13.34}$$

is a positive semidefinite matrix. We will say that $\rho^{AB}$ has positive partial transpose (PPT) if this property hold, and otherwise, we will say that it has a negative partial transpose (NPT) or simply that the state is an NPT state.[1]

We have seen before that the 2-qubit maximally entangled state is an NPT state, and in Exercise 3.21 you showed that all pure entangled states are NPT. Therefore, it is natural to ask if *all* entangled states are NPT. In low dimensions, the following theorem states that this is indeed the case.

> **Theorem 13.4.** Let $\rho \in \mathfrak{D}(AB)$ be a bipartite density matrix with dimensions of the underlying Hilbert spaces satisfy $|AB| \leqslant 6$. Then, $\rho^{AB}$ is entangled if and only if it is an NPT state.

**Proof**  If $\rho^{AB}$ is an NPT state, then from (13.34) it cannot be separable. Conversely, suppose $\rho^{AB}$ is a PPT state, and recall from Theorem 9.3 (when applied to entanglement theory) that $\rho^{AB}$ is separable if and only if

---

[1] Note that the partial transpose of NPT states can have positive eigenvalues. We use the term NPT only to indicate that the partial transpose of the state has at least one negative eigenvalue.

$$\mathrm{Tr}\left[\rho^{AB}\Gamma^{AB}\right]\geqslant 0 \qquad \forall\,\Gamma\in\mathrm{WIT}(AB). \tag{13.35}$$

Now, fix $\Gamma\in\mathrm{WIT}(AB)$. From Theorem 13.3 $\Gamma^{AB}$ have the form (13.32) for some $\eta_1,\eta_2\in\mathrm{Pos}(AB)$. Hence,

$$\begin{aligned}
\mathrm{Tr}\left[\rho^{AB}\Gamma^{AB}\right] &= \mathrm{Tr}\left[\rho^{AB}\left(\eta_1^{AB}+\mathcal{T}^{B\to B}\left(\eta_2^{AB}\right)\right)\right]\\
&\geqslant \mathrm{Tr}\left[\rho^{AB}\mathcal{T}^{B\to B}\left(\eta_2^{AB}\right)\right]\\
\mathcal{T}=\mathcal{T}^{*}\;\to\; &= \mathrm{Tr}\left[\mathcal{T}^{B\to B}\left(\rho^{AB}\right)\eta_2^{AB}\right]\\
\rho^{AB}\text{ is PPT}\;\to\; &\geqslant 0.
\end{aligned} \tag{13.36}$$

Since $\Gamma^{AB}$ was an arbitrary entanglement witness in $\mathrm{WIT}(AB)$ we conclude that (13.36) holds for all $\Gamma\in\mathrm{WIT}(AB)$ so that $\rho^{AB}$ must be a separable state. This concludes the proof. ∎

The condition $|AB|\leqslant 6$ in the theorem above is optimal. Indeed, there are examples of PPT entangled states in higher dimensions, including the case $|A|=2$ and $|B|=4$, as well as the case $|A|=|B|=3$.

## Example of a PPT Entangled State

Consider the five product states in $A\otimes B:=\mathbb{C}^3\otimes\mathbb{C}^3$ given by

$$|\psi_1\rangle:=\frac{1}{\sqrt{2}}|0\rangle\big(|0\rangle-|1\rangle\big)\quad,\quad |\psi_2\rangle:=\frac{1}{\sqrt{2}}\big(|0\rangle-|1\rangle\big)|2\rangle$$

$$|\psi_3\rangle:=\frac{1}{\sqrt{2}}|2\rangle\big(|1\rangle-|2\rangle\big)\quad,\quad |\psi_4\rangle:=\frac{1}{\sqrt{2}}\big(|1\rangle-|2\rangle\big)|0\rangle \tag{13.37}$$

$$|\psi_5\rangle:=\frac{1}{3}\big(|0\rangle+|1\rangle+|2\rangle\big)\big(|0\rangle+|1\rangle+|2\rangle\big).$$

Observe that these five states are orthonormal and they are invariant under the partial transpose (i.e. $\mathcal{T}^{B\to B}(\psi_x^{AB})=\psi_x^{AB}$ for all $x\in[5]$). They also have the property that any pure state in $\mathbb{C}^3\otimes\mathbb{C}^3$ that is orthogonal to all the five states must be entangled (see Exercise 13.6). The set $\{|\psi_x\rangle\}_{x\in[5]}$ is consequently called an *unextendible product basis* (UPB) of the subspace $\mathcal{H}:=\mathrm{span}\{|\psi_x\rangle\}_{x\in[5]}$. It therefore follows that the orthogonal complement of $\mathcal{H}$ in $\mathbb{C}^3\otimes\mathbb{C}^3$, denoted by $\mathcal{H}^{\perp}$, contains only entangled states. Let $\Pi$ be the projection to $\mathcal{H}^{\perp}$; that is,

$$\Pi^{AB}=I_9-\sum_{x\in[5]}\psi_x^{AB}, \tag{13.38}$$

where $I_9$ is the $9\times 9$ identity matrix. It then follows that the bipartite density matrix $\rho:=\frac{1}{3}\Pi$ is entangled (see Exercise 13.6). However, the state $\rho$ is also PPT since

$$\mathcal{T}^{B\to B}(\rho^{AB})=\frac{1}{3}\left(I^{AB}-\sum_{x\in[5]}\mathcal{T}^{B\to B}\left(\psi_x^{AB}\right)\right)=\frac{1}{3}\left(I^{AB}-\sum_{x\in[5]}\psi_x^{AB}\right)=\rho^{AB}\geqslant 0. \tag{13.39}$$

Therefore, the two-qutrit state $\rho^{AB}$ is a PPT entangled state.

**Exercise 13.6.** *Consider the five states defined in* (13.37).

1. *Show that any nonzero state in* $\mathbb{C}^3 \otimes \mathbb{C}^3$ *that is orthogonal to all the five states in* (13.37) *must be entangled.*
2. *Show that the state* $\frac{1}{3}\left(I_9 - \sum_{x \in [5]} \psi_x\right)$ *is entangled. Hint: Use the fact that* $\Pi$ *projects into a subspace consisting of only entangled states.*

**Exercise 13.7.**

1. *Show that the number elements of every UPB in* $\mathbb{C}^m \otimes \mathbb{C}^n$ *cannot be less than* $m + n - 1$.
2. *Let* $\mathcal{W} \subset \mathbb{C}^m \otimes \mathbb{C}^n$ *be a subspace containing no product states (i.e. containing no normalized product vectors). Show that*

$$\dim\left(\mathcal{W}\right) \leqslant (m-1)(n-1). \tag{13.40}$$

**Exercise 13.8.** *Let* $\mathfrak{K}$ *be the set of PPT operators in* $\mathrm{Pos}(AB)$. *Show that* $\mathfrak{K}^*$ *consists of all operators* $\Gamma \in \mathrm{Herm}(AB)$ *of the form* (13.32).

## 13.1.3 The Reduction Criterion

In Section 7.3, we saw that separable states have nonnegative conditional entropy. Moreover, in Corollary 7.1 we saw that if a quantum state $\rho^{AB}$ satisfies $\mathbb{H}(A|B)_\rho \geqslant 0$ for every measure $\mathbb{H}$ of conditional entropy, then $\rho^{AB}$ must satisfy

$$I^A \otimes \rho^B \geqslant \rho^{AB}. \tag{13.41}$$

In particular, if $\rho^{AB}$ is separable then it must satisfy the condition above. This criterion for separability is known as *the reduction criterion*.

The reduction criterion can be expressed in terms of the positive map $\mathcal{P} \in \mathrm{Pos}(A \to A)$:

$$\mathcal{P}(\omega^A) := \mathrm{Tr}\left[\omega^A\right] I^A - \omega^A \qquad \forall\, \omega \in \mathfrak{L}(A). \tag{13.42}$$

Recall from Exercise 3.34 that the map already described is positive but not 2-positive, and therefore not completely positive. Utilizing this map, the reduction criterion can be expressed as:

$$\mathcal{P}^{A \to A}\left(\rho^{AB}\right) \geqslant 0. \tag{13.43}$$

Exercise 13.9 provides an alternative expression for the positive map $\mathcal{P}^{A \to A}$ when $|A| = 2$; specifically, for this case

$$\mathcal{P}^{A \to A}\left(\omega^A\right) = \left(\sigma_y \omega^A \sigma_y\right)^T \qquad \forall\, \omega \in \mathfrak{L}(A). \tag{13.44}$$

Therefore, in this case the reduction criterion is equivalent to the PPT criterion.

In general, however, the PPT criterion is a more powerful criterion for detecting entanglement than the reduction criterion, since there exist entangled states that violate the PPT criterion, yet cannot be detected by the reduction criterion, whereas the converse is not true. This means that the PPT criterion can detect a larger class of entangled states than the reduction criterion. The reason for that is that for $|A| > 2$ the map $\mathcal{P}$ has the form (see Exercise 13.9)

$$\mathcal{P} = \mathcal{P}_1 + \mathcal{T} \circ \mathcal{P}_2, \tag{13.45}$$

where $\mathcal{P}_1, \mathcal{P}_2 \in \mathrm{CP}(A \to A)$ and $\mathcal{T} \in \mathrm{Pos}(A \to A)$ is the transpose map.

Although the reduction criterion may not be as effective as the PPT criterion in detecting entanglement, further investigation reveals that it still holds importance in the field of quantum resource theories. In fact, quantum states that do not satisfy the reduction criterion possess nonzero distillable entanglement, which emphasizes the usefulness of the criterion in other aspects of quantum information. In the upcoming sections, we will explore some of these implications in greater detail.

**Exercise 13.9.** *Let* $\mathcal{P}^{A \to A}$ *be as in* (13.42).

1. *Prove the relation* (13.44) *for the case that* $|A| = 2$.
2. *Prove the relation* (13.45) *for the case* $|A| \geqslant 2$. *Hint: Show that the partial transpose of the Choi matrix of* $\mathcal{P}^{A \to A}$ *is positive semidefinite.*

## 13.1.4 The Realignment Criterion

The realignment criterion is another powerful tool used to detect entanglement in quantum systems. One of the advantages of the realignment criterion is that it can be used in conjunction with other criteria to strengthen the detection of entanglement. For example, if a state satisfies the PPT criterion but fails the realignment criterion, then the state is entangled. Similarly, if a state fails the PPT criterion but satisfies the realignment criterion, then it is entangled.

The realignment criterion is based on the operator Schmidt decomposition introduced in Exercise 2.45. Specifically, let $A$ and $B$ be two Hilbert spaces of dimensions $m := |A|$ and $n := |B|$ and denote by $k := \min\{m^2, n^2\}$. Every $\rho \in \mathfrak{D}(AB)$ can be expressed in terms of $k$ nonnegative real numbers $\{\lambda_x\}_{x \in [k]}$, and two orthonormal sets of Hermitian matrices (w.r.t. the Hilbert–Schmidt inner product) $\{\eta_x\}_{x \in [k]} \subset \mathrm{Herm}(A)$ and $\{\zeta_y\}_{y \in [k]} \subset \mathrm{Herm}(B)$:

$$\rho^{AB} = \sum_{x \in [k]} \lambda_x \eta_x^A \otimes \zeta_x^B. \tag{13.46}$$

---

**Theorem 13.5.** Using the same notations as in (13.46), a quantum state $\rho \in \mathfrak{D}(AB)$ is entangled if it satisfies

$$\sum_{x \in [k]} \lambda_x > 1. \tag{13.47}$$

---

**Proof** Since the condition given in (13.47) can be expressed as

$$\mathrm{Tr}\left[\Lambda^{AB} \rho^{AB}\right] < 0, \quad \text{where} \quad \Lambda := I^{AB} - \sum_{x \in [k]} \eta_x^A \otimes \zeta_x^B, \tag{13.48}$$

it is sufficient to show that $\Lambda^{AB}$ is an entanglement witness. Indeed, let $\psi \in \text{Pure}(A)$ and $\phi \in \text{Pure}(B)$. Then,

$$\text{Tr}\left[\Lambda^{AB}\left(\psi^A \otimes \phi^B\right)\right] = 1 - \sum_{x \in [k]} \text{Tr}[\psi \eta_x]\text{Tr}[\phi \zeta_x]. \qquad (13.49)$$

Now, let $\mathbf{v}, \mathbf{u} \in \mathbb{R}^k$ be the vectors whose components are $\{\text{Tr}[\psi \eta_x]\}_{x \in [k]}$ and $\{\text{Tr}[\phi \zeta_x]\}_{x \in [k]}$, respectively. Then, we need to show that $\mathbf{v} \cdot \mathbf{u} \leqslant 1$. Since $\{\eta_x\}_{x \in [k]}$ is an orthonormal set in $\text{Herm}(A)$ (which can be completed to a full orthonormal basis of $\text{Herm}(A)$), in terms of the Frobenius norm (i.e. the norm induced by the Hilbert–Schmidt inner product)

$$1 = \|\psi\|_2^2 \geqslant \left\| \sum_{x \in [k]} \text{Tr}[\psi \eta_x]\eta_x \right\|_2^2$$

$$\boxed{\{\eta_x\}_{x \in [k]} \text{ is orthonormal}} \longrightarrow = \sum_{x \in [k]} |\text{Tr}[\psi \eta_x]|^2 = \mathbf{v} \cdot \mathbf{v}. \qquad (13.50)$$

Similarly, we get that $\mathbf{u} \cdot \mathbf{u} \leqslant 1$. Hence, we must have $\mathbf{v} \cdot \mathbf{u} \leqslant 1$. This completes the proof. ∎

The term "realignment" comes from the fact that the sum of the singular values $\{\lambda_x\}_{x \in [k]}$ can be expressed in terms of the trace norm of a *realigned* version of the state $\rho^{AB}$. For simplicity, suppose $m = |A| = |B|$, let $\rho \in \mathfrak{D}(AB)$, and expend $\rho^{AB}$ in the standard basis as

$$\rho^{AB} = \sum_{x,x',y,y' \in [m]} r_{xx'yy'}|x\rangle\langle x'|^A \otimes |y\rangle\langle y'|^B, \qquad (13.51)$$

where $r_{xx'yy'} \in \mathbb{C}$. We then define the *realigned* state $\tilde{\rho}^{AB}$ as

$$\tilde{\rho}^{AB} := \sum_{x,x',y,y' \in [m]} r_{xx'yy'}|x\rangle\langle y|^A \otimes |x'\rangle\langle y'|^B. \qquad (13.52)$$

We then argue (see Exercise 13.10) that the sum appearing in (13.47) can be expressed as

$$\sum_{x \in [m^2]} \lambda_x = \left\| \tilde{\rho}^{AB} \right\|_1. \qquad (13.53)$$

In other words, the realignment criterion can be stated as follows: *If the trace-norm of the realigned matrix $\tilde{\rho}^{AB}$ is greater than 1 (i.e. $\|\tilde{\rho}^{AB}\|_1 \geqslant 1$), then the state $\rho^{AB}$ is entangled.*

**Exercise 13.10.** *Prove the relation* (13.53).

**Exercise 13.11.** *Consider the case $|A| = 2$ and $|B| = 4$. Let $p \in [0, 1]$ and $\rho \in \text{Herm}(AB)$ be the matrix*

$$\rho^{AB} = \frac{1}{7p + 1}\begin{pmatrix} pI_4 & p\xi \\ p\xi^T & \eta \end{pmatrix}, \qquad (13.54)$$

*where $I_4$ is the $4 \times 4$ identity matrix,*

$$\xi := \begin{pmatrix} 0 & 1 & 0 & 0 \\ 0 & 0 & 1 & 0 \\ 0 & 0 & 0 & 1 \\ 0 & 0 & 0 & 0 \end{pmatrix}, \quad \text{and} \quad \eta := \begin{pmatrix} \frac{1+p}{2} & 0 & 0 & \frac{\sqrt{1-p^2}}{2} \\ 0 & p & 0 & 0 \\ 0 & 0 & p & 0 \\ \frac{\sqrt{1-p^2}}{2} & 0 & 0 & \frac{1+p}{2} \end{pmatrix}. \tag{13.55}$$

1. *Show that $\rho \in \mathfrak{D}(AB)$ for all $p \in [0,1]$.*
2. *Show that $\rho^{AB}$ is PPT for all $p \in [0,1]$.*
3. *Show that for some $p \in [0,1]$ the state $\rho^{AB}$ is entangled.*

*Hint: Use the Schur complement (see Section B.8 of online version).*

## 13.1.5 The $k$-Extendability Criterion

The concept of symmetric extensions of quantum states provides the most powerful criterion for separability currently known. Consider a state $\sigma^{AB}$ that can be expressed as a convex combination of product states:

$$\sigma^{AB} = \sum_{x \in [m]} p_x \, \psi_x^A \otimes \phi_x^B \in \mathrm{SEP}(AB). \tag{13.56}$$

We can construct a symmetric extension $\sigma \in \mathrm{SEP}(AB\tilde{B})$ of this state by introducing an additional system $\tilde{B}$ and defining the extension as

$$\sigma^{AB\tilde{B}} = \sum_{x \in [m]} p_x \, \psi_x^A \otimes \phi_x^B \otimes \phi_x^{\tilde{B}}. \tag{13.57}$$

This extension is considered symmetric since the original state can be obtained by tracing out either the $B$ or the $\tilde{B}$ systems; that is, the marginals of $\sigma^{AB\tilde{B}}$ satisfy $\sigma^{AB} = \sigma^{A\tilde{B}}$. On the other hand, when dealing with entangled states, it is not immediately clear whether a symmetric extension, $\rho^{AB\tilde{B}}$, with the property $\rho^{AB} = \rho^{A\tilde{B}}$ exists for an entangled state $\rho \in \mathfrak{D}(AB)$. While this property holds trivially for separable states, it doesn't hold for all entangled states.

Note that the extension of the separable state in (13.56) can also be extended to $k$-copies of $B$ via

$$\rho^{AB^k} = \sum_{x \in [m]} p_x \, \psi_x^A \otimes \phi_x^{B_1} \otimes \cdots \otimes \phi_x^{B_k}, \tag{13.58}$$

where $B \cong B_1 \cong \cdots \cong B_k$. We say that $\rho^{AB^k}$ has a symmetric $k$-extension of $\rho^{AB}$.

> **Definition 13.2.** We say that $\rho \in \mathfrak{D}(AB)$ is $k$-extendible if there exists $\rho \in \mathfrak{D}(AB^k)$, with $B^k = B_1 \cdots B_k$ and $B \cong B_1 \cong \cdots \cong B_k$, such that its marginals satisfy
>
> $$\rho^{AB_m} = \rho^{AB_{m'}} = \rho^{AB} \qquad \forall \, m, m' \in [k]. \tag{13.59}$$
>
> If such a state exists, then $\rho^{AB^k}$ is called a symmetric $k$-extension of $\rho^{AB}$.

We have already seen that every separable quantum state is $k$-extendible for all $k \in \mathbb{N}$. Surprisingly, the converse of this statement is also true! This means that if a quantum state $\rho \in \mathfrak{D}(AB)$ is $k$-extendible for all $k \in \mathbb{N}$, then it must be separable. However, proving this statement requires certain techniques that are beyond the scope of this book. Specifically, it involves the use of the quantum de Finetti theorem. Interested readers can find more information in the "notes and references" section at the end of this chapter.

Given a quantum state $\rho \in \mathfrak{D}(AB)$, how can we determine if it is $k$-extendible? Observe that the conditions $\rho^{AB_1} = \rho^{AB_j}$, for all $j \in [k]$, can be expressed as

$$\mathrm{Tr}\left[\rho^{AB^k} \Lambda_j^{AB^k}\right] = 0, \tag{13.60}$$

for all $\Lambda_j \in \mathrm{Herm}(AB^k)$ of the form

$$\Lambda_j^{AB^k} = \eta^{AB_1} \otimes I^{B_2 \cdots B_k} - \xi^{AB_j} \otimes I^{B_1 \cdots B_{j-1} B_{j+1} \cdots B_k}, \tag{13.61}$$

where $\eta \in \mathrm{Herm}(AB_1)$ and $\xi \in \mathrm{Herm}(AB_j)$. Note that the linearity of the condition above implies that we can restrict $\eta$ and $\xi$ to belong to orthonormal bases of Herm $(AB_1)$ and $\mathrm{Herm}(AB_j)$, respectively. Thus, we conclude that there exists a finite number of operators $\{\Lambda_{j\ell}\}_{j\in[k],\ell\in[n]}$ such that $\rho^{AB_1} = \rho^{AB_j}$, for all $j \in [k]$, if and only if

$$\mathrm{Tr}\left[\rho^{AB^k} \Lambda_{j\ell}^{AB^k}\right] = 0 \qquad \forall\, j \in [k],\ \ell \in [n]. \tag{13.62}$$

The conditions specified here indicate that the determination of whether $\rho^{AB}$ is $k$-extendible requires the solution of an SDP feasibility problem. Therefore, the criterion for $k$-extendibility can be computed algorithmically and efficiently.

**Exercise 13.12.** *Using the same notations as above:*

1. *Find an upper bound on n.*
2. *Use Farkas lemma of Exercise 4.79 to express the dual form of* (13.62).

## 13.2 Quantification of Entanglement

Entanglement is quantified by functions that behave monotonically under LOCC. More precisely, adapting Definition 10.1 to entanglement theory we get that a *measure of entanglement* is a function

$$E: \bigcup_{A,B} \mathfrak{D}(AB) \to \mathbb{R} \tag{13.63}$$

that satisfies the following two conditions:

1. For any LOCC map $\mathcal{E} \in \mathrm{LOCC}(AB \to A'B')$, and any bipartite state $\rho \in \mathfrak{D}(AB)$

$$E\left(\mathcal{E}\left(\rho^{AB}\right)\right) \leqslant E\left(\rho^{AB}\right). \tag{13.64}$$

2. $E(1) = 0$, where 1 corresponds to the only element of $\mathfrak{D}(AB)$ when $|A| = |B| = 1$.

**Exercise 13.13.** *Show that any measure of entanglement $E$ as already defined satisfies the following two conditions: (1) It is always nonnegative, that is, $E(\rho^{AB}) \geq 0$ for all $\rho \in \mathfrak{D}(AB)$, and (2) it satisfies $E(\sigma^{AB}) = 0$ for all $\sigma \in \mathrm{SEP}(AB)$.*

In general, LOCC can be stochastic, in the sense that $\rho^{AB}$ can be converted to $\sigma_x^{AB}$ with some probability $p_x$. In this case, the map from $\rho^{AB}$ to $\sigma_x^{AB}$ cannot be described by a CPTP map. However, by introducing a classical "flag" system $X$, we can view the ensemble $\{\sigma_x^{AB}, p_x\}_{x \in [m]}$ as a classical quantum state $\sigma^{XAB} := \sum_{x \in [m]} p_x |x\rangle\langle x|^X \otimes \sigma_x^{AB}$. Hence, if $\rho^{AB}$ can be converted by LOCC to $\sigma_x^{AB}$ with probability $p_x$, then there exists a map $\mathcal{E} \in \mathrm{LOCC}(AB \to XAB)$ such that $\mathcal{E}(\rho^{AB}) = \sigma^{XAB}$. Since the "flag" system $X$ is classical, both Alice and Bob have access to it since if Alice holds it she can communicate it to Bob, and vice versa. Therefore, the definition as in (13.63) and (13.64) of a measure of entanglement capture also probabilistic transformations. Particularly, $E$ must satisfy $E\left(\sigma^{XAB}\right) \leq E\left(\rho^{AB}\right)$.

Almost all measures of entanglement studied in literature (although not all) satisfy

$$E\left(\sigma^{XAB}\right) = \sum_{x \in [m]} p_x E(\sigma_x^{AB}), \tag{13.65}$$

which is very intuitive since $X$ is just a classical system encoding the value of $x$. We call this relation in (13.65) the *direct sum property* since, mathematically, $\sigma^{XAB}$ can also be viewed as $\bigoplus_{x \in [m]} p_x \sigma_x^{AB}$. If the direct sum property holds, then the condition $E\left(\sigma^{XAB}\right) \leq E\left(\rho^{AB}\right)$ becomes $\sum_x p_x E(\sigma_x^{AB}) \leq E\left(\rho^{AB}\right)$, meaning that LOCC cannot increase entanglement on average. Therefore, the direct sum property is in general stronger than the strong monotonicity property (10.10) of a resource measure $\mathbf{M}$. In fact, the condition (13.65) also implies that $E$ is convex (see Exercise 13.14). We therefore conclude that any measure of entanglement that satisfies the direct sum property is an entanglement monotone.

**Exercise 13.14.** *Let $E$ be a measure of entanglement satisfying the direct sum property (13.65). Show that $E$ is convex; that is, for any ensemble of states $\{p_x, \sigma_x^{AB}\}_{x \in [m]}$ we have*

$$E\left(\sum_{x \in [m]} p_x \sigma_x^{AB}\right) \leq \sum_{x \in [m]} p_x E\left(\sigma_x^{AB}\right). \tag{13.66}$$

## 13.2.1 Extension of Entanglement from Pure to Mixed States

Quantifying entanglement in mixed states is considerably more complex than quantifying entanglement in pure states. This is partly because there is no equivalent to Nielsen's majorization theorem, and as we have already discussed, LOCC on mixed bipartite states is much more intricate and difficult to characterize. Nevertheless, several approaches have been developed to help characterize entanglement.

## The Convex Roof Extension

Let

$$E: \bigcup_{A,B} \text{Pure}(AB) \to \mathbb{R} \tag{13.67}$$

be a measure of pure-state entanglement. We can extend the domain of $E$ to mixed states using a method known as the convex roof extension. The method is based on the fact that any bipartite mixed state $\rho^{AB}$ has many pure-state decompositions. Recall that a pure-state decomposition of $\rho^{AB}$ is an ensemble of pure states, $\{p_x, \psi_x^{AB}\}_{x\in[m]}$ (where $\psi_x \in \text{Pure}(AB)$ and $\{p_x\}_{x\in[m]}$ is a probability distribution) that satisfies $\rho^{AB} = \sum_{x\in[m]} p_x \psi_x^{AB}$. In Exercise 2.27, we saw that every unitary matrix in $\mathfrak{U}(m)$ can be used to define a particular pure-state decomposition.

> **Entanglement of Formation**
>
> **Definition 13.3.** Let $E$ be a measure of pure-state entanglement. The convex roof extension of $E$ is a function $E_F: \bigcup_{A,B} \mathfrak{D}(AB) \to \mathbb{R}$ defined on any $\rho \in \mathfrak{D}(AB)$ via
>
> $$E_F\left(\rho^{AB}\right) = \inf \sum_{x\in[m]} p_x E\left(\psi_x^{AB}\right), \tag{13.68}$$
>
> where the infimum is over all pure-state decompositions $\{p_x, \psi_x^{AB}\}_{x\in[m]}$ of $\rho^{AB}$. $E_F$ is called the *entanglement of formation* associated with the pure-state measure $E$.

*Remark.* The term "entanglement of formation" originated from historical reasons, as it was originally believed that $E_F(\rho^{AB})$, with $E$ taken as the entropy of entanglement, represented the entanglement cost required to create the state $\rho^{AB}$. However, we will discover later on that it is actually the regularized entanglement of formation that can be interpreted as the entanglement cost of $\rho^{AB}$.

**Exercise 13.15.** *Show that if $E$ is the entropy of entanglement, then its corresponding entanglement of formation satisfies for all $\rho \in \mathfrak{D}(AB)$:*

$$E_F\left(\rho^{AB}\right) \leqslant \min\left\{H\left(\rho^A\right), H\left(\rho^B\right)\right\}. \tag{13.69}$$

To show that the entanglement of formation is indeed a measure of entanglement, recall that any measure of pure-state entanglement, $E$, can be expressed for all $\psi \in \text{Pure}(AB)$ as

$$E\left(\psi^{AB}\right) = g\left(\rho^A\right) \quad \text{with} \quad \rho^A := \text{Tr}_B\left[\psi^{AB}\right], \tag{13.70}$$

for some Schur concave function $g$. A slightly stronger condition than Schur concavity is the condition that $g$ is both symmetric and concave. In this case, the resulting entanglement of formation is an entanglement monotone.

> **Theorem 13.6.** Let $E$ be a measure of entanglement on pure states given as in (13.70) with $g$ being a concave symmetric function. Then, its convex roof extension, $E_F$, as defined in Definition 13.3, is an entanglement monotone.

We first prove the following auxiliary lemma. In this lemma we only consider a quantum instrument on Bob's system and consider a pure initial bipartite state. We show that for this simpler case, the convex roof extension of $E$ satisfies strong monotonicity.

> **Lemma 13.1.** Let $\psi \in \mathrm{Pure}(AB)$, $\mathcal{E}^{B \to B'X} := \sum_{x \in [m]} \mathcal{E}_x^{B \to B'} \otimes |x\rangle\langle x|^X$ be a quantum instrument on Bob's subsystem, and for all $x \in [m]$,
>
> $$\sigma_x^{AB'} := \frac{1}{p_x} \mathcal{E}_x^{B \to B'}\left(\psi^{AB}\right), \quad \text{where} \quad p_x := \mathrm{Tr}\left[\mathcal{E}_x^{B \to B'}\left(\psi^{AB}\right)\right]. \qquad (13.71)$$
>
> Then,
>
> $$\sum_{x \in [m]} p_x E_F\left(\sigma_x^{AB'}\right) \leqslant E_F\left(\psi^{AB}\right). \qquad (13.72)$$

*Remark.* In the proof next, we adopt the notations like $\phi^A := \mathrm{Tr}_B\left[\phi^{AB}\right]$ to denote the reduced density matrix of a pure bipartite state. This notation is instrumental in reducing the number of symbols used, enhancing clarity and conciseness. However, it's crucial to remember that $\phi^A$ in this context represents a mixed state, despite the notation resembling that typically used for pure states. This distinction is important for a correct understanding of the concepts and calculations involved in the proof.

**Proof**  For every $x \in [m]$, let

$$\sigma_x^{AB'} = \sum_{y \in [n]} r_{y|x} \phi_{xy}^{AB'}, \qquad (13.73)$$

where $\{r_{y|x}, \phi_{xy}^{AB'}\}_{y \in [n]}$ is the optimal pure-state decomposition of $\sigma_x^{AB'}$. Therefore, for each $x \in [m]$ we have

$$E_F\left(\sigma_x^{AB'}\right) = \sum_{y \in [n]} r_{y|x} E\left(\phi_{xy}^{AB'}\right)$$

$$\textbf{cf. (13.70)} \to = \sum_{y \in [n]} r_{y|x} g\left(\phi_{xy}^A\right)$$

$$\textbf{\textit{g} is concave} \to \leqslant g\left(\sum_{y \in [n]} r_{y|x} \phi_{xy}^A\right) \qquad (13.74)$$

$$\textbf{cf. (13.73)} \to = g\left(\sigma_x^A\right).$$

Hence,

$$\sum_{x \in [m]} p_x E_F\left(\sigma_x^{AB'}\right) \leqslant \sum_{x \in [m]} p_x g\left(\sigma_x^A\right)$$

$$\textbf{\textit{g} is concave} \to \leqslant g\left(\sum_{x \in [m]} p_x \sigma_x^A\right). \qquad (13.75)$$

Finally, observe that the reduced density matrix of $\psi^{AB}$ satisfies $\psi^A = \sum_{x\in[m]} p_x \sigma_x^A$. Substituting this into the equation (13.75) gives

$$\sum_{x\in[m]} p_x E_F\left(\sigma_x^{AB'}\right) \leqslant g\left(\psi^A\right) = E\left(\psi^{AB}\right) = E_F\left(\psi^{AB}\right). \qquad (13.76)$$

This completes the proof of the lemma. ∎

We are now ready to prove the theorem.

**Proof of Theorem 13.6**   Let $\rho \in \mathfrak{D}(AB)$ and let $\{p_x, \psi_x^{AB}\}_{x\in[m]}$ be an optimal pure-state decomposition of $\rho^{AB}$ satisfying

$$E_F\left(\rho^{AB}\right) = \sum_{x\in[m]} p_x E\left(\psi_x^{AB}\right). \qquad (13.77)$$

We first prove the strong monotonicity of $E_F$. The proof strategy aims to show that $E_F$ does not increase on average under a general quantum instrument on Bob's subsystem. We can then apply a similar argument to Alice's side, demonstrating that $E_F$ remains non-increasing under quantum instruments on either subsystem. The significance of this lies in the fact that LOCC consists of such local quantum instruments, coupled with rounds of classical communication, which do not affect the monotonicity property. Therefore, by demonstrating the nonincreasing nature of $E_F$ under a quantum instrument on both Bob's side (and, by symmetry arguments, also on Alice's side), it can be concluded that $E_F$ satisfies the strong monotonicity property under LOCC.

Let $\{\mathcal{E}_z^{B\to B'}\}_{z\in[k]}$ be a quantum instrument on Bob's subsystem, and for each $z \in [k]$, let $\rho_z^{AB'}$ be the post-measurement state after outcome $z$ occurred. Moreover, for every $z \in [k]$ and $x \in [m]$ we denote by $r_z := \mathrm{Tr}\left[\mathcal{E}_z^{B\to B'}\left(\rho^{AB}\right)\right]$, $t_{z|x} := \mathrm{Tr}\left[\mathcal{E}_z^{B\to B'}\left(\psi_x^{AB}\right)\right]$, and

$$\sigma_{xz}^{AB'} := \frac{1}{t_{z|x}}\mathcal{E}_z^{B\to B'}\left(\psi_x^{AB}\right). \qquad (13.78)$$

With these notations, the post-measurement state can be expressed as

$$\rho_z^{AB'} := \frac{1}{r_z}\mathcal{E}_z^{B\to B'}\left(\rho^{AB}\right) = \frac{1}{r_z}\sum_{x\in[m]} p_x \mathcal{E}_z^{B\to B'}\left(\psi_x^{AB}\right)$$

$$= \sum_{x\in[m]} \frac{p_x t_{z|x}}{r_z}\sigma_{xz}^{AB'}. \qquad (13.79)$$

We need to show that the average entanglement $\sum_{z\in[k]} r_z E_F\left(\rho_z^{AB'}\right)$ cannot exceed $E_F(\rho^{AB})$. For this purpose, for each state $\rho_z^{AB}$ we need to find a suitable pure-state decomposition that can be related to $\rho^{AB}$. Observe that the equation (13.79) involves the mixed states $\{\sigma_{xz}^{AB'}\}_{x\in[m]}$. Therefore, for each $\sigma_{xz}^{AB'}$ we denote by $\{s_{y|xz}, \phi_{xyz}^{AB'}\}_{y\in[n]}$ its optimal pure-state decomposition, so that

$$E_F\left(\sigma_{xz}^{AB'}\right) = \sum_{y\in[n]} s_{y|xz} E\left(\phi_{xyz}^{AB'}\right). \qquad (13.80)$$

With this final notation, we get our desirable pure-state decomposition of $\rho_z^{AB'}$:

$$\rho_z^{AB'} = \sum_{x \in [m]} \sum_{y \in [n]} \frac{p_x t_{z|x} s_{y|xz}}{r_z} \phi_{xyz}^{AB'}. \tag{13.81}$$

Since this pure-state decomposition of $\rho_z^{AB'}$ is not necessarily optimal, we conclude

$$\sum_{z \in [k]} r_z E_F\left(\rho_z^{AB}\right) \leqslant \sum_{x,y,z} r_z \frac{p_x t_{z|x} s_{y|xz}}{r_z} E\left(\phi_{xyz}^{AB'}\right) = \sum_{x,y,z} p_x t_{z|x} s_{y|xz} E\left(\phi_{xyz}^{AB'}\right)$$

$$\mathbf{(13.80)} \rightarrow = \sum_{x,z} p_x t_{z|x} E_F\left(\sigma_{xz}^{AB'}\right) \tag{13.82}$$

$$\mathbf{Lemma\ 13.1} \rightarrow \leqslant \sum_{x \in [m]} p_x E_F\left(\psi_x^{AB}\right)$$

$$\mathbf{(13.77)} \rightarrow = E_F\left(\rho^{AB}\right).$$

We therefore conclude that $E_F$ does not increase on average under a general quantum instrument on Bob's subsystem. This completes the proof of strong monotonicity.

It is therefore left to show that $E_F$ is convex. Indeed, let $\{p_x, \rho_x^{AB}\}_{x \in [m]}$ be an ensemble of bipartite entangled states, and for each $x \in [m]$ let $\{q_{y|x}, \psi_{xy}^{AB}\}_{y \in [n]}$ be an optimal pure-state decomposition of $\rho_x^{AB}$ such that

$$E_F\left(\rho_x^{AB}\right) = \sum_{y \in [n]} q_{y|x} E\left(\psi_{xy}^{AB}\right). \tag{13.83}$$

Now, observe that $\{p_x q_{y|x}, \psi_{xy}^{AB}\}_{x,y}$ is a pure-state decomposition of $\sum_x p_x \rho_x^{AB}$. Thus,

$$E_F\left(\sum_x p_x \rho_x^{AB}\right) \leqslant \sum_{x,y} p_x q_{y|x} E\left(\psi_{xy}^{AB}\right)$$

$$\mathbf{(13.83)} \rightarrow = \sum_{x \in [m]} p_x E_F\left(\rho_x^{AB}\right). \tag{13.84}$$

This completes the proof.      ∎

The task of computing convex roof extensions is notably complex, particularly when determining the entanglement of formation for a bipartite quantum state. If there were a straightforward, closed formula for the entanglement of formation, identifying whether a bipartite quantum state is entangled would be relatively easy. However, given that this identification task is known to be hard (specifically, NP-hard), it's unrealistic to expect a simple formula for the entanglement of formation. Nevertheless, for two-qubit systems, such a formula does exist, as outlined in the following theorem.

Recall the concurrence monotones defined in (12.49) for pure bipartite states. In the case of two-qubit states, all concurrences are equivalent, so we denote them simply by $C$. The concurrence of formation, which is the convex roof extension of $C$, is then denoted as $C_F$.

**Exercise 13.16.** *Consider $E$ and $C$ as the entropy of entanglement and concurrence for pure states, respectively. Let $A$ and $B$ be two-qubit systems (i.e. $|A| = |B| = 2$), and define the function $g : [0, 1] \to [0, 1]$ as*

$$g(x) = h_2 \left( \frac{1 + \sqrt{1 - x^2}}{2} \right), \tag{13.85}$$

*where $h_2(x) := -x \log x - (1 - x) \log (1 - x)$ is the binary Shannon entropy.*

1. *Show that for any $\psi \in \mathrm{Pure}(AB)$ we have*

$$E\left(\psi^{AB}\right) = g\left(C\left(\psi^{AB}\right)\right). \tag{13.86}$$

2. *Show that for any $\rho \in \mathfrak{D}(AB)$*

$$E_F\left(\rho^{AB}\right) = g\left(C_F\left(\rho^{AB}\right)\right). \tag{13.87}$$

   *Hint: Show first that the function $g$ is concave.*

This exercise shows that in order to compute the entanglement of formation of a two-qubit state $\rho^{AB}$ it is sufficient to compute its concurrence of formation. In the following theorem we give a closed formula for the concurrence of formation. The closed formula is given in terms of the density matrix

$$\rho_\star^{AB} := (\sigma_2 \otimes \sigma_2)\bar{\rho}^{AB}(\sigma_2 \otimes \sigma_2), \tag{13.88}$$

where $\bar{\rho}^{AB}$ is the density matrix whose components

$$\langle xy|\bar{\rho}^{AB}|x'y'\rangle := \overline{\langle xy|\rho^{AB}|x'y'\rangle} \qquad \forall, x, y, x', y' \in \{0, 1\}, \tag{13.89}$$

where the orthonormal basis $\{|x\rangle\}_{x \in \{0,1\}}$ (and similarly $\{|y\rangle\}_{y \in \{0,1\}}$) is such that $\sigma_2$ has the form $-i|0\rangle\langle 1| + i|1\rangle\langle 0|$.

**Closed Formula**

**Theorem 13.7.** Let $\rho \in \mathfrak{D}(AB)$ be a two-qubit mixed state (i.e. $|A| = |B| = 2$). Then, the concurrence of formation of $\rho^{AB}$ is given by

$$C_F\left(\rho^{AB}\right) = \max\{0, \lambda_1 - \lambda_2 - \lambda_3 - \lambda_4\}, \tag{13.90}$$

where $\{\lambda_1, \dots, \lambda_4\}$ are the four eigenvalues of the matrix $\left|\sqrt{\rho}\sqrt{\rho_\star}\right|$, arranged in a nonincreasing order.

In the derivation of the formula (13.90), we will use a bilinear form denoted as $(\cdot, \cdot) : \mathbb{C}^2 \otimes \mathbb{C}^2 \to \mathbb{C}$. This bilinear form will not only be instrumental in proving the formula but will also be valuable in the analysis of multipartite entanglement. It is defined for any two vectors $|\psi\rangle, |\phi\rangle \in AB$ as

$$\left(|\psi\rangle, |\phi\rangle\right) := \langle\bar{\psi}^{AB}|\sigma_2 \otimes \sigma_2|\phi^{AB}\rangle, \tag{13.91}$$

where $\bar{\psi}^{AB}$ is defined such that if $|\psi^{AB}\rangle = \sum_{x,y\in\{0,1\}} c_{xy}|x\rangle|y\rangle$ then $|\bar{\psi}^{AB}\rangle = \sum_{x,y\in\{0,1\}} \bar{c}_{xy}|x\rangle|y\rangle$. Note that $\bar{\psi}^{AB}$ is well define only with respect to some fixed bases $\{|0\rangle^A, |1\rangle^A\} \subset A$ and $\{|0\rangle^B, |1\rangle^B\} \subset B$ of $A$ and $B$, respectively. These orthonormal bases are chosen such that $\sigma_2$ has the form $-i|0\rangle\langle 1| + i|1\rangle\langle 0|$. The relation of this bilinear form to our study here can be found in Exercise 12.18, in which you had to show that the concurrence of a two-qubit pure state $\psi \in \mathrm{Pure}(AB)$ can be expressed as $C\left(\psi^{AB}\right) = |\langle|\psi\rangle, |\psi\rangle\rangle|$. In the following exercise you prove several additional properties of this bilinear form.

**Exercise 13.17.** *Consider the bilinear form defined in* (13.91).

1. *Show the linearity of the bilinear form; that is, show that for any vectors* $|\psi\rangle$, $|\psi_1\rangle$, $|\psi_2\rangle$, $|\phi\rangle$, $|\phi_1\rangle$, *and* $|\phi_2\rangle$ *in* $\mathbb{C}^2 \otimes \mathbb{C}^2$,

$$\langle|\psi_1\rangle + |\psi_2\rangle, |\phi\rangle\rangle = \langle|\psi_1\rangle, |\phi\rangle\rangle + \langle|\psi_2\rangle, |\phi\rangle\rangle$$
$$\langle|\psi\rangle, |\phi_1\rangle + |\phi_2\rangle\rangle = \langle|\psi\rangle, |\phi_1\rangle\rangle + \langle|\psi\rangle, |\phi_2\rangle\rangle. \tag{13.92}$$

2. *Show that the bilinear form is symmetric; that is, for any two vectors* $|\psi\rangle$, $|\phi\rangle \in \mathbb{C}^2 \otimes \mathbb{C}^2$,

$$\langle|\psi\rangle, |\phi\rangle\rangle = \langle|\phi\rangle, |\psi\rangle\rangle. \tag{13.93}$$

3. *Invariance property. Let* $M, N \in SL(2, \mathbb{C})$, $\psi, \phi \in \mathrm{Pure}(AB)$, *and denote* $|\tilde{\psi}\rangle := M \otimes N|\psi\rangle$ *and* $|\tilde{\phi}\rangle := M \otimes N|\phi\rangle$. *Show that*

$$\langle|\tilde{\psi}\rangle, |\tilde{\phi}\rangle\rangle = \langle|\psi\rangle, |\phi\rangle\rangle. \tag{13.94}$$

*Hint: Use the relation* (C.13 *of online version*).

**Proof of Theorem 13.7**  Consider first the case that $\lambda_1 > \lambda_2 + \lambda_3 + \lambda_4$. Let $\{p_x, \psi_x\}_{x\in[4]}$ and $\{q_y, \phi_y^{AB}\}_{y\in[n]}$ be two pure-state decompositions of $\rho^{AB}$, and for each $x \in [4]$ and $y \in [n]$ (here $n \geqslant 4$), let $|\tilde{\psi}_x\rangle := \sqrt{p_x}|\psi_x\rangle$ and $|\tilde{\phi}_y^{AB}\rangle := \sqrt{q_y}|\phi_y^{AB}\rangle$. Recall from Exercise 2.27 that there exists an $n \times 4$ isometry $V = (v_{yx})$ such that for all $y \in [n]$

$$|\tilde{\phi}_y\rangle = \sum_{x\in[4]} v_{yx}|\tilde{\psi}_x\rangle. \tag{13.95}$$

We can therefore relate the bilinear forms of the two decompositions as

$$\langle|\tilde{\phi}_y\rangle, |\tilde{\phi}_{y'}\rangle\rangle = \sum_{x,x'} v_{yx} v_{y'x'}\langle|\tilde{\psi}_x\rangle, |\tilde{\psi}_{x'}\rangle\rangle \qquad \forall\, y, y' \in [n]. \tag{13.96}$$

Denoting by $M_\psi$ and $M_\phi$ the matrices whose components are $\langle\tilde{\psi}_x, \tilde{\psi}_{x'}\rangle$ and $\langle\tilde{\phi}_y, \tilde{\phi}_{y'}\rangle$, respectively, we get that (13.96) can be written as

$$M_\phi = V M_\psi V^T. \tag{13.97}$$

Since $M_\psi$ is symmetric (see second part of Exercise 13.17), there exists a $4 \times 4$ unitary matrix $U$ such that $U M_\psi U^T$ is diagonal. Moreover, by appropriate choice of $U$, the diagonal elements of $U M_\psi U^T$ can always be made real and positive (i.e. they are the singular values of $M_\psi$), and arranged on the diagonal of $U M_\psi U^T$ in a nonincreasing order. We therefore conclude that there exists a pure-state decomposition that is diagonal with respect to the bilinear form. For simplicity of the exposition, we take it to be $\{p_x, \psi_x^{AB}\}_{x\in[4]}$ itself; that is,

$$\left(|\tilde{\psi}_x\rangle, |\tilde{\psi}_{x'}\rangle\right) = \lambda_x \delta_{xx'} \qquad \forall\, x, x' \in [4] \tag{13.98}$$

with real nonnegative numbers $\lambda_1 \geqslant \lambda_2 \geqslant \lambda_3 \geqslant \lambda_4$. From Exercise 13.18 it follows that $\lambda_1, \ldots, \lambda_4$ are precisely the eigenvalues of $|\sqrt{\rho}\sqrt{\rho_*}|$. With this specific choice of $\{\tilde{\psi}_x\}_{x=1}^{4}$, let $\{q_y, \phi_y\}_{y\in[n]}$ be another pure-state decomposition of $\rho^{AB}$ (and as before we set $|\tilde{\phi}_y\rangle := \sqrt{q_y}|\phi_y\rangle$). Then, the relation (13.96) gives

$$\left(|\tilde{\phi}_y\rangle, |\tilde{\phi}_{y'}\rangle\right) = \sum_{x\in[4]} v_{yx} v_{y'x} \lambda_x. \tag{13.99}$$

We therefore get that the average concurrence of $\{q_y, \phi_y^{AB}\}_{y\in[n]}$ can be expressed as

$$\sum_{y\in[n]} q_y C\left(\phi_y^{AB}\right) = \sum_{y\in[n]} q_y \left|\langle|\bar{\phi}_y\rangle, |\phi_y\rangle\rangle\right| = \sum_{y\in[n]} \left|\left(|\tilde{\phi}_y\rangle, |\tilde{\phi}_y\rangle\right)\right|$$

$$(13.99)\rightarrow = \sum_{y\in[n]} \left|\sum_{x\in[4]} v_{yx}^2 \lambda_x\right| \tag{13.100}$$

$$\geqslant \sum_{y\in[n]} \left(|v_{y1}|^2 \lambda_1 - |v_{y2}|^2 \lambda_2 - |v_{y3}|^2 \lambda_3 - |v_{y4}|^2 \lambda_4\right)$$

$$= \lambda_1 - \lambda_2 - \lambda_3 - \lambda_4,$$

where we used the inequality $|a+b+c+d| \geqslant |a| - |b| - |c| - |d|$ for every $a, b, c, d \in \mathbb{C}$. Moreover, this inequality can be saturated by taking $V$ to be the unitary matrix (i.e. taking $n = 4$)

$$V = \frac{1}{2}\begin{pmatrix} -1 & i & i & i \\ 1 & -i & i & i \\ 1 & i & -i & i \\ 1 & i & i & -i \end{pmatrix}. \tag{13.101}$$

Indeed, observe that matrix $V$ is unitary and has the property that for all $y \in [4]$, $\sum_{x\in[4]} v_{yx}^2 \lambda_x = \frac{1}{4}(\lambda_1 - \lambda_2 - \lambda_3 - \lambda_4)$. Therefore, with this $V$ we get from (13.100) that the average concurrence of $\{q_y, \phi_y\}_{y\in[4]}$ is

$$\sum_{y\in[4]} \left|\sum_{x\in[4]} v_{yx}^2 \lambda_x\right| = \lambda_1 - \lambda_2 - \lambda_3 - \lambda_4. \tag{13.102}$$

It is therefore left to show that if $\lambda_1 \leqslant \lambda_2 + \lambda_3 + \lambda_4$, then $C_F(\rho^{AB}) = 0$. In this case we take

$$V = \frac{1}{2}\begin{pmatrix} -1 & e^{i\theta_2} & e^{i\theta_3} & e^{i\theta_4} \\ 1 & -e^{i\theta_2} & e^{i\theta_3} & e^{i\theta_4} \\ 1 & e^{i\theta_2} & -e^{i\theta_3} & e^{i\theta_4} \\ 1 & e^{i\theta_2} & e^{i\theta_3} & -e^{i\theta_4} \end{pmatrix}, \tag{13.103}$$

where $\theta_2, \theta_3$, and $\theta_4$ are some choices of phases to be determined shortly. With this $V$, the average concurrence of $\{q_y, \phi_y\}_{y\in[4]}$ is given by (see (13.100))

$$\sum_{y\in[4]} q_y C\left(\phi_y^{AB}\right) = \left|\lambda_1 + \lambda_2 e^{2i\theta_2} + \lambda_3 e^{2i\theta_3} + \lambda_4 e^{2i\theta_4}\right|. \tag{13.104}$$

Since we assume that $\lambda_1 \leqslant \lambda_2 + \lambda_3 + \lambda_4$ (as well as $\lambda_1 \geqslant \lambda_2 \geqslant \lambda_3 \geqslant \lambda_4$), we can always find three angles $\theta_1, \theta_2$, and $\theta_3$ such the right-hand side above is zero (see Exercise 13.19). This completes the proof. ∎

**Exercise 13.18.** *Let $\{\tilde\psi_x\}_{x=1}^4$ be a set of four two-qubit subnormalized states satisfying* (13.98) *with some nonnegative real numbers $\{\lambda_x\}_{x=1}^4$. Show that $\{\lambda_x\}_{x=1}^4$ are the eigenvalues of $\left|\sqrt{\rho}\sqrt{\rho_\star}\right|$, where $\rho := \sum_{x\in[4]} \tilde\psi_x$. Hint: Recall from Exercise 5.35 that if $\lambda$ is an eigenvalue of $\left|\sqrt{\rho}\sqrt{\rho_\star}\right|$, then $\lambda^2$ is an eigenvalue of $\rho\rho_\star$, and compute $\rho\rho_\star|\tilde\psi_x\rangle$.*

**Exercise 13.19.** *Show that for any four nonnegative real numbers $\lambda_1, \dots, \lambda_4$ that satisfy $\lambda_1 \leqslant \lambda_2 + \lambda_3 + \lambda_4$ and $\lambda_1 \geqslant \lambda_2 \geqslant \lambda_3 \geqslant \lambda_4$ there exist three angles $\theta_2, \theta_3$, and $\theta_4$ such that the right-hand side of* (13.104) *is zero. Hint: Use a continuity argument.*

**Exercise 13.20.** *Compute the concurrence of the following two-qubit bipartite mixed states:*

1. *The isotropic state*

$$\rho_p^{AB} := p\mathbf{u}^{AB} + (1-p)\Phi^{AB}, \tag{13.105}$$

   *with $p \in [0, 1]$.*
2. *The Werner state*

$$\rho_t^{AB} = \frac{1}{3}t\Pi_{\mathrm{Sym}}^{AB} + (1-t)\Pi_{\mathrm{Asy}}^{AB}, \tag{13.106}$$

   *with $t \in [0, 1]$.*

**Exercise 13.21.** *Let $\rho \in \mathfrak{D}(AB)$ with $|A| = |B| = 2$, and define the quantity*

$$C_a\left(\rho^{AB}\right) := \max \sum_{x\in[m]} p_x C\left(\psi_x^{AB}\right), \tag{13.107}$$

*where the maximum is over all pure-state decompositions of $\rho^{AB}$ (i.e. $C_a$ is defined similarly to $C_F$ but with a maximum instead of a minimum). Show that*

$$C_a\left(\rho^{AB}\right) = F\left(\rho^{AB}, \rho_\star^{AB}\right), \tag{13.108}$$

*where $F$ is the fidelity. Hint: Use similar lines as in the proof of Theorem 13.7. Show also that the square of the fidelity in* (13.108) *can be expressed as*

$$\left\|\sqrt{\rho^{AB}}\sqrt{\rho_\star^{AB}}\right\|_1^2 = \mathrm{Tr}\left[\rho^{AB}\rho_\star^{AB}\right]. \tag{13.109}$$

## Monotones Based on the Ky Fan Norms

Let us revisit the entanglement measures introduced in (12.28). According to Nielsen's majorization theorem, these functions are indicative of whether a pure bipartite state can be transformed into another. In the upcoming sections, we will demonstrate that some of the operational significance of these measures can also be extended to their convex roof extensions. Specifically, we can express the convex roof extension of the pure-state entanglement measures defined in (12.28) as follows:

$$E_{(k)}\left(\rho^{AB}\right) := \min \sum_{x \in [m]} p_x \left(1 - \left\|\rho_x^A\right\|_{(k)}\right),$$
(13.110)

where the minimum is over all pure-state decompositions $\rho^{AB} = \sum_x p_x \psi_x^{AB}$, with $\rho_x^A := \mathrm{Tr}_B[\psi_x^{AB}]$, and $\|\cdot\|_{(k)}$ is the Ky Fan norm.

**Exercise 13.22.** *Show that the functions $E_{(k)}$ as already defined are entanglement monotones.*

Note that for the two-qubit case, the only nontrivial measure $E_{(k)}$ is when $k = 1$. In this case

$$E_{(1)}\left(\rho^{AB}\right) := \min \sum_{x \in [m]} p_x \lambda_{\min}\left(\rho_x^A\right),$$
(13.111)

since each $\rho_x^A$ is a qubit so its the minimum eigenvalue $\lambda_{\min}\left(\rho_x^A\right) = 1 - \left\|\rho_x^A\right\|_{(1)}$. Moreover, observe that

$$\begin{aligned}
\lambda_{\min}\left(\rho_x^A\right) &= \frac{1}{2}\left(1 - \sqrt{1 - 4\det\left(\rho_x^A\right)}\right) \\
&= \frac{1}{2}\left(1 - \sqrt{1 - C^2\left(\psi_x^{AB}\right)}\right),
\end{aligned}$$
(13.112)

where $C^2\left(\psi_x^{AB}\right)$ is the square of the concurrence of $\psi_x^{AB}$. In the following exercise we will show that the (13.112) relation can be used to show that for any two-qubit state $\rho \in \mathfrak{D}(AB)$ with $|A| = |B| = 2$ we have

$$E_{(1)}\left(\rho^{AB}\right) = \frac{1}{2}\left(1 - \sqrt{1 - C_f^2\left(\rho^{AB}\right)}\right),$$
(13.113)

where $C_f^2\left(\rho^{AB}\right)$ is the square of the concurrence of formation of $\rho^{AB}$. Hence, the closed formula for the concurrence of formation can be used to compute $E_{(1)}$.

**Exercise 13.23.** *Let $\rho \in \mathfrak{D}(AB)$ be a two-qubit state with $|A| = |B| = 2$.*
1. *Prove the relation (13.112).*
2. *Prove the relation (13.113). Hint: The proof is similar to the proof of (13.87).*

In Corollary 12.3 we found necessary and sufficient conditions to convert by LOCC a pure bipartite state to a mixed bipartite state. Interestingly, for the case that $d := |A| = |B| = 2$, the minimization in (12.63) over $k \in \{1, 2\}$ becomes trivial since for all $\psi \in \mathrm{Pure}(AB)$ we have $E_{(2)}\left(\psi^{AB}\right) = 0$. We therefore arrive at the following corollary.

> **Corollary 13.1.** Let $\psi \in \mathrm{Pure}(AB)$ and $\sigma \in \mathfrak{D}(AB)$ be two bipartite entangled states with $d := |A| = |B| = 2$. Then, $\psi^{AB}$ can be converted to $\sigma^{AB}$ by LOCC if and only if
>
> $$C\left(\psi^{AB}\right) \geqslant C_F\left(\sigma^{AB}\right), \tag{13.114}$$
>
> where $C$ is the concurrence.

**Proof** As already discussed, taking $k = 1$ in (12.63) gives that $\psi^{AB}$ can be converted to $\sigma^{AB}$ by LOCC if and only if

$$E_{(1)}\left(\psi^{AB}\right) \geqslant E_{(1)}\left(\sigma^{AB}\right). \tag{13.115}$$

The proof is concluded by expressing $E_{(1)}$ on both sides of (13.115) in terms of the concurrence (see the relation (13.113) between $E_{(1)}$ and the concurrence). $\blacksquare$

**Exercise 13.24.** *Let $\psi \in \mathrm{Pure}(AB)$ with $d := |A| = |B| > 2$ and let $\sigma \in \mathfrak{D}(A'B')$ with $|A'| = |B'| = 2$. Show that $\psi^{AB}$ can be converted to $\sigma^{AB}$ by LOCC if and only if*

$$E_{(1)}\left(\psi^{AB}\right) \geqslant E_{(1)}\left(\sigma^{A'B'}\right). \tag{13.116}$$

## Optimal Extensions

In Chapter 5, we introduced a method to extend divergences from classical to quantum systems. This method is in fact quite general and can be slightly modified to incorporate extensions of measures of entanglement from pure to mixed states. Specifically, let $E$ be a measure on pure-state entanglement. For any $\rho \in \mathfrak{D}(AB)$ the maximal extension of $E$ is defined as

$$\overline{E}\left(\rho^{AB}\right) := \inf\left\{E\left(\psi^{A'B'}\right) \; : \; \psi^{A'B'} \xrightarrow{\text{LOCC}} \rho^{AB}\right\}, \tag{13.117}$$

where the infimum is over all systems $A'B'$ and all pure states $\psi \in \mathrm{Pure}(A'B')$ for which $\psi^{A'B'}$ can be converted by LOCC to $\rho^{AB}$. Similarly, the minimal extension is defined as

$$\underline{E}\left(\rho^{AB}\right) := \sup\left\{E\left(\psi^{A'B'}\right) \; : \; \rho^{AB} \xrightarrow{\text{LOCC}} \psi^{A'B'}\right\}. \tag{13.118}$$

The following exercise demonstrates the optimality of these definitions.

**Exercise 13.25.** *Let $E$ be a measure of pure-state entanglement, and let $\overline{E}$ and $\underline{E}$ be its maximal and minimal extensions.*

1. *Show that $\overline{E}$ and $\underline{E}$ are measures of mixed bipartite entanglement.*
2. *Show that if $E'$ is a measure of bipartite mixed-state entanglement that reduces to $E$ on pure states, then*

$$\underline{E}\left(\rho^{AB}\right) \leqslant E'\left(\rho^{AB}\right) \leqslant \overline{E}\left(\rho^{AB}\right). \tag{13.119}$$

3. *Show that if $E$ is additive under tensor products of pure bipartite states, then $\overline{E}$ is sub-additive and $\underline{E}$ super-additive under tensor products of mixed bipartite states.*

The minimal extension $\underline{E}$ is not a very useful measure of entanglement since typically a mixed bipartite state cannot be converted by LOCC to a pure entangled state. Therefore, for such mixed entangled states $\underline{E}$ takes the zero value. On the other hand, the maximal extension is a faithful measure of entanglement (i.e. takes the zero value *only* on separable states).

As an example, recall that the Schmidt rank is a measure of entanglement on pure states. Its maximal extension to mixed states is given by

$$\overline{\mathrm{SR}}\left(\rho^{AB}\right) := \inf\left\{\mathrm{SR}\left(\psi^{A'B'}\right) \;:\; \psi^{A'B'} \xrightarrow{\mathrm{LOCC}} \rho^{AB}\right\}. \tag{13.120}$$

In general, the condition $\psi^{A'B'} \xrightarrow{\mathrm{LOCC}} \rho^{AB}$ can be very complicated. However, we can replace $\psi^{A'B'}$ in (13.120) with the maximally entangled state $\Phi_k$, where $k :=$ $\mathrm{SR}\left(\psi^{A'B'}\right) = \mathrm{SR}\left(\Phi_k\right)$, since whenever $\psi^{A'B'} \xrightarrow{\mathrm{LOCC}} \rho^{AB}$ we also have $\Phi_k \xrightarrow{\mathrm{LOCC}} \rho^{AB}$. We therefore conclude that

$$\mathrm{SR}\left(\rho^{AB}\right) := \overline{\mathrm{SR}}\left(\rho^{AB}\right) = \min\left\{k \;:\; \Phi_k \xrightarrow{\mathrm{LOCC}} \rho^{AB}\right\}, \tag{13.121}$$

where, for simplicity of the exposition, we removed the over-line symbol from $\overline{\mathrm{SR}}\left(\rho^{AB}\right)$.

**Exercise 13.26.** *Let $\rho \in \mathfrak{D}(AB)$. Show that $\mathrm{SR}\left(\rho^{AB}\right) = k$ for some $k \in \mathbb{N}$ if and only if the following two conditions hold:*

1. *At least one of the states, in any pure-state decomposition of $\rho^{AB}$, has a Schmidt rank no smaller than $k$.*
2. *There exists a pure-state decomposition of $\rho^{AB}$ with all states having Schmidt rank at most $k$.*

**Exercise 13.27.** *Let $\rho \in \mathfrak{D}(AB)$. Show that*

$$\mathrm{SR}\left(\rho^{AB}\right) = \inf \max_{x \in [k]} \mathrm{SR}(\psi_x^{AB}), \tag{13.122}$$

*where the infimum is over all pure-state decompositions of $\rho^{AB} = \sum_{x \in [k]} p_x \psi_x^{AB}$.*

## 13.2.2 The Relative Entropy of Entanglement

In Section 10.2 we studied many properties of the relative entropy of a resource, and in Section 10.3 we developed a method to compute it. In entanglement theory, the relative entropy of entanglement is defined for any $\rho \in \mathfrak{D}(AB)$ as (see Figure 13.1)

$$E_R\left(\rho^{AB}\right) := \min_{\sigma \in \mathrm{SEP}(AB)} D\left(\rho^{AB} \,\|\, \sigma^{AB}\right). \tag{13.123}$$

As discussed in Section 10.3, computing the relative entropy of entanglement can generally be quite challenging. However, in certain special cases, such as with pure states

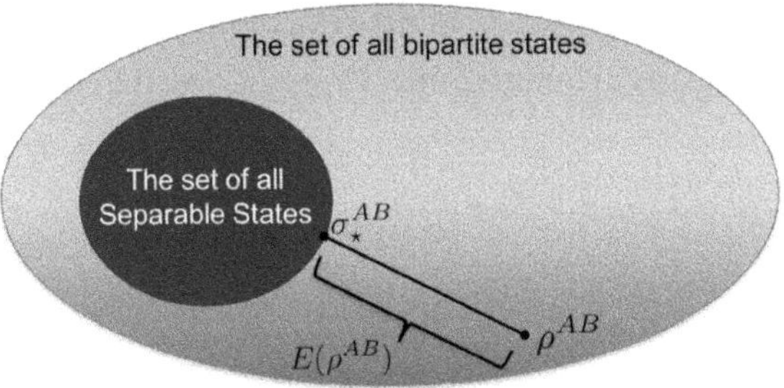


**Figure 13.1**    The closest separable state.

and symmetric states, it is feasible to compute this measure. The complexity in computing the relative entropy of entanglement typically arises from the need to optimize over the large set of separable states, which can be a demanding task for most mixed states. Yet, for pure states and certain states with specific symmetrical properties, this complexity is significantly reduced, making the calculation manageable.

## Relative Entropy of Entanglement for Pure States

In the next theorem, it's shown that for pure states, the relative entropy of entanglement simplifies to the entropy of entanglement. This finding connects two key entanglement measures, highlighting the elegant underlying structure of quantum entanglement in pure systems.

> **Theorem 13.8.** Let $\psi \in \mathrm{Pure}(AB)$. Then,
>
> $$E_R\left(\psi^{AB}\right) = E\left(\psi^{AB}\right) := H(A)_\rho, \qquad (13.124)$$
>
> where $H(A)_\rho$ is the von-Neumann entropy of the reduced density matrix $\rho^A := \mathrm{Tr}_B\left[\psi^{AB}\right]$.

**Proof**    The proof is based on the closed formula given in Theorem 10.5 for the relative entropy of a resource. Let

$$|\psi\rangle = \sum_{x\in[n]} \sqrt{p_x}|xx\rangle \qquad (13.125)$$

be given in its Schmidt form, where $n := \mathrm{SR}(\psi)$, and let

$$\sigma_\star := \sum_{x\in[n]} p_x |xx\rangle\langle xx|. \qquad (13.126)$$

We argue now that $\sigma_\star$ is the closest separable state (see Figure 13.1); that is, we argue that

$$\min_{\sigma\in\mathrm{SEP}(AB)} D\left(\psi^{AB}\big\|\sigma^{AB}\right) = D\left(\psi^{AB}\big\|\sigma_\star^{AB}\right). \qquad (13.127)$$

Indeed, from Theorem 10.5 it follows that $\sigma_\star^{AB}$ satisfies the above equality if and only if there exists an entanglement witness $\eta \in \mathrm{WIT}(AB)$ such that

$$\psi^{AB} = \sigma_\star^{AB} - a\mathcal{L}_{\sigma_\star}^{-1}(\eta). \tag{13.128}$$

The above equality can be expressed as $\mathcal{L}_{\sigma_\star}^{-1}(a\eta) = \sigma_\star^{AB} - \psi^{AB}$, which is equivalent to

$$a\eta = \mathcal{L}_{\sigma_\star}(\sigma_\star - \psi) = I - \mathcal{L}_{\sigma_\star}(\psi). \tag{13.129}$$

That is, $\sigma_\star$ satisfies (13.127) if and only if the right-hand side of (13.129) is an entanglement witness. Now, by direct computation we have (see Exercise 13.28)

$$\mathcal{L}_{\sigma_\star}(\psi) = \sum_{x,y\in[n]} \sqrt{p_x p_y}\frac{\log p_x - \log p_y}{p_x - p_y}|xx\rangle\langle yy|. \tag{13.130}$$

Denote by $r_{xy} := \frac{p_x}{p_y}$ and observe that for any $x, y \in [n]$

$$c_{xy} := \sqrt{p_x p_y}\frac{\log p_x - \log p_y}{p_x - p_y} = \frac{\sqrt{r_{xy}}\log(r_{xy})}{r_{xy} - 1} \leqslant 1, \tag{13.131}$$

where the last inequality follows from the fact that logarithm function satisfies $\log(r) \leqslant (r - 1)/\sqrt{r}$ for $r \geqslant 1$ and opposite inequality for $0 < r \leqslant 1$. We therefore get for any product state $\phi^A \otimes \varphi^B$

$$
\begin{aligned}
\mathrm{Tr}\left[\left(\phi^A \otimes \varphi^B\right)\mathcal{L}_{\sigma_\star}(\psi)\right] &= \sum_{x,y\in[n]} c_{xy}\langle\phi|x\rangle\langle\varphi|x\rangle\langle y|\phi\rangle\langle y|\varphi\rangle \\
&\leqslant \sum_{x,y\in[n]} c_{xy}\left|\langle\phi|x\rangle\langle\varphi|x\rangle\langle y|\phi\rangle\langle y|\varphi\rangle\right| \\
\boxed{c_{xy} \leqslant 1} \longrightarrow &\leqslant \sum_{x,y\in[n]} \left|\langle\phi|x\rangle\langle\varphi|x\rangle\langle y|\phi\rangle\langle y|\varphi\rangle\right| \\
\boxed{|ab| \leqslant \frac{1}{2}\left(|a|^2 + |b|^2\right)} \longrightarrow &\leqslant \frac{1}{4}\sum_{x,y\in[n]}\left(|\langle\phi|x\rangle|^2 + |\langle\varphi|x\rangle|^2\right)\left(|\langle y|\phi\rangle|^2 + |\langle y|\varphi\rangle|^2\right) \\
&= 1.
\end{aligned}
\tag{13.132}
$$

We therefore conclude that for any product state $\phi^A \otimes \varphi^B$

$$\mathrm{Tr}\left[\left(\phi^A \otimes \varphi^B\right)\left(I - \mathcal{L}_{\sigma_\star}(\psi)\right)\right] \geqslant 0, \tag{13.133}$$

so that $I - \mathcal{L}_{\sigma_\star}(\psi)$ is an entanglement witness. This completes the proof. ∎

**Exercise 13.28.** *Prove the equality in* (13.130).

## Relative Entropy of Entanglement for Symmetric States

Unlike the case of pure bipartite states, there isn't a general, straightforward formula for the relative entropy of entanglement of mixed bipartite states. However, for certain special cases like isotropic states or Werner states, such formulas do exist. The underlying reason for this is the symmetry inherent in these states. To understand why a formula exists for these specific states, consider that if there is a bipartite channel $\mathcal{G} \in \mathrm{LOCC}(AB \to AB)$ such that $\rho^{AB} = \mathcal{G}(\rho^{AB})$ (which is true for isotropic and Werner states under various twirling operations), then the relative entropy of entanglement of such a state can be expressed as follows:

$$
\begin{aligned}
E_R\left(\rho^{AB}\right) &:= \min_{\sigma \in \mathrm{SEP}(AB)} D\left(\rho^{AB} \,\middle\|\, \sigma^{AB}\right) \\
\boxed{\mathcal{G}(\mathrm{SEP}(AB)) \subseteq \mathrm{SEP}(AB)} \longrightarrow &\leqslant \min_{\sigma \in \mathrm{SEP}(AB)} D\left(\rho^{AB} \,\middle\|\, \mathcal{G}\left(\sigma^{AB}\right)\right) \\
\boxed{\rho^{AB} = \mathcal{G}(\rho^{AB})} \longrightarrow &= \min_{\sigma \in \mathrm{SEP}(AB)} D\left(\mathcal{G}\left(\rho^{AB}\right) \,\middle\|\, \mathcal{G}\left(\sigma^{AB}\right)\right) \\
\mathbf{DPI} \to &\leqslant \min_{\sigma \in \mathrm{SEP}(AB)} D\left(\rho^{AB} \,\middle\|\, \sigma^{AB}\right) = E_R\left(\rho^{AB}\right).
\end{aligned}
\tag{13.134}
$$

Therefore, all the inequalities must actually be equalities, and particularly:

$$
E_R\left(\rho^{AB}\right) = \min_{\sigma \in \mathrm{SEP}(AB)} D\left(\rho^{AB} \,\middle\|\, \mathcal{G}\left(\sigma^{AB}\right)\right).
\tag{13.135}
$$

The importance of (13.135) lies in the fact that the minimization over separable states of the form $\mathcal{G}\left(\sigma^{AB}\right)$ might involve significantly fewer parameters compared to the minimization over the entire set of separable states $\mathrm{SEP}(AB)$. This simplification is what makes the computation of the relative entropy of entanglement feasible for these symmetric states.

As an example, let's consider the isotropic state defined for all $t \in [0, 1]$ (with $m := |A| = |B|$) as

$$
\rho_t^{AB} = t\Phi_m^{AB} + (1-t)\tau^{AB}, \quad \text{where} \quad \tau^{AB} := \frac{I^{AB} - \Phi_m}{m^2 - 1}.
\tag{13.136}
$$

Previously, we observed that this state is invariant under the twirling channel described in (3.251) and is separable if and only if $t \leqslant 1/m$. At one extreme ($t = 0$), the isotropic state is the separable state $\tau^{AB}$. At the other extreme ($t = 1/m$), the isotropic state is the separable state

$$
\rho_{1/m}^{AB} = \frac{m}{m+1}\mathrm{u}^{AB} + \frac{1}{m+1}\Phi_m^{AB}.
\tag{13.137}
$$

For an entangled isotropic state $\rho_t^{AB}$ with $t > 1/m$, equation (13.135) yields

$$
E_R\left(\rho_t^{AB}\right) = \min_{t' \in [0, 1/m]} D\left(\rho_t^{AB} \,\middle\|\, \rho_{t'}^{AB}\right),
\tag{13.138}
$$

since for every separable state $\sigma \in \mathrm{SEP}(AB)$, the twirled state $\mathcal{G}\left(\sigma^{AB}\right) = \rho_{t'}^{AB}$ for some $t' \in [0, 1/m]$. This demonstrates how the symmetry of $\rho_t^{AB}$ simplifies the optimization problem. Furthermore, in Exercise 13.29, we will show that the optimal $t'$ is $t' = 1/m$, resulting in

$$
\begin{aligned}
E_R\left(\rho_t^{AB}\right) &= D\left(\rho_t^{AB} \,\middle\|\, \rho_{1/m}^{AB}\right) \\
&= \log m + t \log t + (1-t)\log \frac{1-t}{m-1}.
\end{aligned}
\tag{13.139}
$$

This result is intuitive as the separable state $\rho_{t'}^{AB}$ with $t' = 1/m$ is on the boundary of the set of separable states and, roughly speaking, the closest to $\rho_t^{AB}$.

**Exercise 13.29.** *Prove the equalities in* (13.139). *Hint. Observe that* $[\rho_t^{AB}, \rho_{t'}^{AB}] = 0$ *for all* $t, t' \in [0, 1]$ *and use it to show that* $D\left(\rho_t^{AB} \big\| \rho_{t'}^{AB}\right) = D(\mathbf{t} \| \mathbf{t}')$, *where* $\mathbf{t} := (t, 1 - t)^T$ *and* $\mathbf{t}' := (t', 1 - t')^T$.

**Exercise 13.30.** *Consider the Werner state* $\rho_W^{AB}$ *as given in* (13.29) *with* $p > 1/2$ *(i.e. an entangled Werner state). Show that*

$$E_R\left(\rho_W^{AB}\right) = D\left(\rho_W^{AB} \big\| \omega^{AB}\right), \tag{13.140}$$

*where*

$$\omega^{AB} := \frac{1}{m(m+1)} \Pi_{\text{Sym}}^{AB} + \frac{1}{m(m-1)} \Pi_{\text{Asy}}^{AB}. \tag{13.141}$$

## 13.2.3 The Robustness of Entanglement

Section 10.2.2 introduced a resource measure called the robustness. However, this measure is not very useful for affine resource theories, as noted in the Exercise 10.10. On the other hand, since the set of separable states is maximally nonaffine (as discussed around (9.35)), the robustness measure is highly relevant in the context of entanglement theory. Later on, we will explore its operational interpretation.

To apply the definition of the robustness measure to entanglement theory, we define the robustness of entanglement for a bipartite state $\rho \in \mathfrak{D}(AB)$ as

$$\mathbf{R}\left(\rho^{AB}\right) := \min\left\{s \geqslant 0 : \frac{\rho^{AB} + s\omega^{AB}}{1 + s} \in \text{SEP}(AB), \quad \omega \in \text{SEP}(AB)\right\}. \tag{13.142}$$

The global robustness, $\mathbf{R}_g$, is defined in a manner similar to the definition in (13.142), but with the key distinction that the state $\omega$ is chosen from the entire set of density matrices $\mathfrak{D}(AB)$, rather than being restricted to separable states $\text{SEP}(AB)$.

We have established that the logarithmic global robustness is connected to $\mathbf{R}_g$ through the relationship:

$$D_{\max}\left(\rho^{AB} \big\| \mathfrak{F}\right) = \log\left(1 + \mathbf{R}_g\left(\rho^{AB}\right)\right). \tag{13.143}$$

In a similar vein, the logarithmic robustness of a state $\rho^{AB}$ is defined as

$$\mathbf{LR}\left(\rho^{AB}\right) := \log\left(1 + \mathbf{R}\left(\rho^{AB}\right)\right). \tag{13.144}$$

**Exercise 13.31.** *Show that in entanglement theory, for all* $\rho \in \mathfrak{D}(AB)$ *there exists a finite* $s \geqslant 0$ *and a separable density matrix* $\omega^{AB}$ *such that the density matrix* $\left(\rho^{AB} + s\omega^{AB}\right)/(1 + s)$ *is separable.*

**Exercise 13.32.** *Show that the logarithmic robustness is subadditive; that is, for all* $\rho \in \mathfrak{D}(AB)$ *and* $\sigma \in \mathfrak{D}(A'B')$ *we have*

$$\mathbf{LR}\left(\rho^{AB} \otimes \sigma^{A'B'}\right) \leqslant \mathbf{LR}\left(\rho^{AB}\right) + \mathbf{LR}\left(\sigma^{A'B'}\right). \tag{13.145}$$

**Exercise 13.33.** *Let $\rho \in \mathcal{D}(AB)$. Show that $R\left(\rho^{AB}\right) = 0$ if and only if $\rho \in \mathrm{SEP}(AB)$.*

**Exercise 13.34.** *Show that all $\rho \in \mathcal{D}(AB)$ can be written as*

$$\rho^{AB} = \left(1 + \mathbf{R}\left(\rho^{AB}\right)\right)\tau^{AB} - \mathbf{R}\left(\rho^{AB}\right)\omega^{AB}, \tag{13.146}$$

*for some $\tau, \omega \in \mathrm{SEP}(AB)$. This decomposition of $\rho^{AB}$ is sometimes referred to as a pseudo-mixture of $\rho^{AB}$.*

Note that from this exercise it follows that $\mathbf{R}\left(\rho^{AB}\right)$ can also be expressed as

$$\mathbf{R}\left(\rho^{AB}\right) := \min\left\{s \geqslant 0 \; : \; \rho^{AB} = (1+s)\tau^{AB} - s\omega^{AB}, \; \tau, \omega \in \mathrm{SEP}(AB)\right\}, \tag{13.147}$$

so that we can think of the pseudo-mixture in (13.146) as the optimal one achieved with $s = \mathbf{R}\left(\rho^{AB}\right)$.

**Exercise 13.35.** *Show that the robustness of entanglement is an entanglement monotone.*

## 13.2.4 Partial Transpose-Based Entanglement Measures

In the previous sections, we learned that a bipartite quantum state is entangled if the partial transpose of the state has a negative eigenvalue. In this subsection, we will explore how this property can be leveraged to measure the extent of entanglement in a bipartite state. However, entanglement measures constructed in this manner have certain limitations. For instance, they assign a value of zero to all PPT states, including those that are entangled. Despite these shortcomings, these measures possess the advantage of being computationally tractable and can be efficiently computed using SDP algorithms, as we will demonstrate.

### The Negativity of Entanglement

The negativity of a bipartite state $\rho^{AB}$ is a measure of entanglement that is defined as the absolute value of the sum of all the negative eigenvalues of the partial transpose of $\rho^{AB}$. Unlike other measures of entanglement, the negativity of entanglement is relatively easy to compute even for nonqubit systems. Therefore, it is used quite extensively in literature.

In this section, for any $\rho \in \mathcal{D}(AB)$ we will use the notation $\rho^{\Gamma}$ to denote the partial transpose of the state $\rho^{AB}$; that is,

$$\rho^{\Gamma} := T^{B \to B}\left(\rho^{AB}\right), \tag{13.148}$$

where $T \in \mathrm{Pos}(B \to B)$ is the transpose map. The intuition behind this notation is that $\rho^{\Gamma}$ represents half of the full transposed state $\rho^{\mathsf{T}}$.

**Definition 13.4.** Let $\rho \in \mathfrak{D}(AB)$. The negativity of $\rho^{AB}$ is defined as

$$\mathbf{N}\left(\rho^{AB}\right) := \frac{\left\|\rho^{\Gamma}\right\|_1 - 1}{2}. \tag{13.149}$$

*Remark.* We chose as a convention for this book that $\rho^{\Gamma}$ represents partial transpose on Bob's side. This convention does not effect this definition since

$$\left\|\mathcal{T}^{B \to B}\left(\rho^{AB}\right)\right\|_1 = \left\|\mathcal{T}^{A \to A}\left(\rho^{AB}\right)\right\|_1. \tag{13.150}$$

Therefore, the definition of the negativity is independent of whether the partial transpose is taken on Bob's side or Alice's side.

Note that $\rho^{\Gamma}$ has trace 1 since the partial transpose does not effect the trace. Therefore, if $\lambda_1, \ldots, \lambda_n$ are the eigenvalues of $\rho^{\Gamma}$ then they sum to 1. Suppose, without loss of generality, that the first $k$ eigenvalues of $\rho^{\Gamma}$ are nonnegative, and the remaining $n - k$ are negative. We then get

$$\left\|\rho^{\Gamma}\right\|_1 = \sum_{x \in [n]} |\lambda_x| = \sum_{x \in [k]} \lambda_x - \sum_{x = k+1}^{n} \lambda_x$$

$$\boxed{\sum_{x \in [n]} \lambda_x = 1} \longrightarrow = 1 - 2 \sum_{x = k+1}^{n} \lambda_x. \tag{13.151}$$

Substituting this into (13.149) gives

$$\mathbf{N}\left(\rho^{AB}\right) = -\sum_{x = k+1}^{n} \lambda_x = \left| \sum_{x = k+1}^{n} \lambda_x \right|. \tag{13.152}$$

That is, the negativity of $\rho^{AB}$ is the absolute value of the sum of all the negative eigenvalues of $\rho^{\Gamma}$. We can therefore express it also as

$$\mathbf{N}\left(\rho^{AB}\right) = \mathrm{Tr}\left[\rho_-^{\Gamma}\right], \tag{13.153}$$

where $\rho_-^{\Gamma} := \left(\rho^{\Gamma}\right)_-$ is the negative part of $\rho^{\Gamma}$. Note that this also demonstrates that the negativity is zero on separable states.

**Exercise 13.36.** *Let $\rho \in \mathfrak{D}(AB)$. Show that there exists density matrices $\rho_+, \rho_- \in \mathfrak{D}(AB)$ such that $\rho_+ \rho_- = \rho_- \rho_+ = 0$ and*

$$\rho^{\Gamma} = \left(1 + \mathbf{N}\left(\rho^{AB}\right)\right)\rho_+^{AB} - \mathbf{N}\left(\rho^{AB}\right)\rho_-^{AB}. \tag{13.154}$$

The decomposition (13.154) of $\rho^{\Gamma}$ in the Exercise 13.36 is optimal in the following sense. Suppose there exists $\sigma, \tau \in \mathfrak{D}(AB)$ such that

$$\rho^{\Gamma} = (1 + a)\sigma^{AB} - a\tau^{AB}, \tag{13.155}$$

for some $a \in \mathbb{R}_+$. Then, from (13.154) we have

$$\left(1 + \mathbf{N}\left(\rho^{AB}\right)\right)\rho_+^{AB} - \mathbf{N}\left(\rho^{AB}\right)\rho_-^{AB} = (1+a)\sigma^{AB} - a\tau^{AB}. \tag{13.156}$$

Let $\Pi_-$ be the projector to the support of $\rho_-^{AB}$. Multiplying both sides of (13.156) by $\Pi_-$ and taking the trace gives

$$\begin{aligned}
-\mathbf{N}\left(\rho^{AB}\right) &= \mathrm{Tr}\left[\Pi_-\left((1+a)\sigma^{AB} - a\tau^{AB}\right)\right] \\
&\geqslant -a\,\mathrm{Tr}\left[\Pi_-\tau^{AB}\right] \\
&\geqslant -a.
\end{aligned} \tag{13.157}$$

Hence, we must have $a \geqslant \mathbf{N}(\rho^{AB})$. In other words, we can express the negativity of $\rho^{AB}$ as

$$\mathbf{N}\left(\rho^{AB}\right) = \inf\left\{a \in \mathbb{R} : \exists\, \sigma, \tau \in \mathfrak{D}(AB)\ \text{s.t.}\ \rho^\Gamma = (1+a)\sigma^{AB} - a\tau^{AB}\right\}. \tag{13.158}$$

> **Theorem 13.9.** The negativity measure as defined in (13.149) is an entanglement monotone.

**Proof**  To prove the strong monotonicity property, let $\{\mathcal{E}_x\}_{x\in[m]}$ be a quantum instrument on Alice's system, with each $\mathcal{E}_x \in \mathrm{CP}(A \to A')$ being trace nonincreasing and $\sum_{x\in[m]} \mathcal{E}_x \in \mathrm{CPTP}(A \to A')$. For each $x \in [m]$, denote by $\rho_x^{A'B} := \frac{1}{p_x}\mathcal{E}_x^{A\to A'}\left(\rho^{AB}\right)$, where $p_x := \mathrm{Tr}\left[\mathcal{E}_x^{A\to A'}\left(\rho^{AB}\right)\right]$. Finally, set $v := \mathbf{N}\left(\rho^{AB}\right)$. By definition we have

$$\begin{aligned}
\rho_x^\Gamma &= \frac{1}{p_x}\left(\mathcal{E}_x^{A\to A'}\left(\rho^{AB}\right)\right)^\Gamma \\
\xrightarrow[\textit{acts on Bob's side}]{\textit{Partial transpose}} &= \frac{1}{p_x}\mathcal{E}_x^{A\to A'}\left(\rho^\Gamma\right) \\
\overset{(13.154)}{\longrightarrow} &= \frac{1+v}{p_x}\mathcal{E}_x^{A\to A'}\left(\rho_+^{AB}\right) - \frac{v}{p_x}\mathcal{E}_x^{A\to A'}\left(\rho_-^{AB}\right).
\end{aligned} \tag{13.159}$$

Since the decomposition of $\rho_x^\Gamma$ is not necessarily optimal (in the sense of (13.158)) we must have

$$\mathbf{N}\left(\rho_x^{A'B}\right) \leqslant \frac{v}{p_x}\mathrm{Tr}\left[\mathcal{E}_x^{A\to A'}\left(\rho_-^{AB}\right)\right]. \tag{13.160}$$

We therefore get that

$$\begin{aligned}
\sum_{x\in[m]} p_x \mathbf{N}\left(\rho_x^{A'B}\right) &\leqslant v \sum_{x\in[m]} \mathrm{Tr}\left[\mathcal{E}_x^{A\to A'}\left(\rho_-^{AB}\right)\right] \\
&= v = \mathbf{N}\left(\rho^{AB}\right),
\end{aligned} \tag{13.161}$$

where we used the fact that $\sum_{x\in[m]} \mathcal{E}_x$ is trace preserving. That is, the negativity of entanglement cannot increase on average by a quantum instrument on Alice's side. Since the negativity is not affected if we take the partial transpose on Alice's system (instead of Bob's), using similar arguments, we get that the negativity cannot increase on average under any quantum instrument applied on Bob's side. We therefore conclude that the negativity satisfies the strong monotonicity condition of an entanglement monotone. It is left to show that the negativity is convex.

Let $\{p_x, \rho_x^{AB}\}_{x\in[m]}$ be an ensemble of density matrices in $\mathfrak{D}(AB)$. Then, by definition,

$$
\mathbf{N}\left(\sum_{x\in[m]} p_x \rho_x^{AB}\right) = \frac{1}{2}\left\|\left(\sum_{x\in[m]} p_x \rho_x\right)^{\Gamma}\right\|_1 - \frac{1}{2}
$$

$$
= \frac{1}{2}\left\|\sum_{x\in[m]} p_x \rho_x^{\Gamma}\right\|_1 - \frac{1}{2} \tag{13.162}
$$

$$
\leq \frac{1}{2}\sum_{x\in[m]} p_x \left\|\rho_x^{\Gamma}\right\|_1 - \frac{1}{2} = \sum_{x\in[m]} p_x \mathbf{N}\left(\rho_x^{AB}\right).
$$

This completes the proof. ∎

## The Logarithmic Negativity

The negativity has many nice properties, but it is not additive. It turns out that by a small tweak to its definition in (13.149), we can get an additive measure of entanglement.

**Exercise 13.37.** *Show that the negativity of a pure bipartite state $\psi \in \mathrm{Pure}(AB)$ with $m := |A| = |B|$ is given by*

$$
\mathbf{N}(\psi^{AB}) = \sum_{\substack{x < y \\ x,y\in[m]}} \sqrt{p_x p_y}, \tag{13.163}
$$

*where $\{p_x\}_{x\in[m]}$ are the Schmidt coefficients of $\psi^{AB}$.*

The Logarithmic Negativity

**Definition 13.5.** Let $\rho \in \mathfrak{D}(AB)$. The logarithmic negativity of $\rho^{AB}$ is defined as

$$
\mathbf{LN}\left(\rho^{AB}\right) = \log\left\|\rho^{\Gamma}\right\|_1. \tag{13.164}
$$

Note that the logarithmic negativity can be expressed as a function of the negativity, namely,

$$
\mathbf{LN}\left(\rho^{AB}\right) = \log\left(2\mathbf{N}\left(\rho^{AB}\right) + 1\right). \tag{13.165}
$$

Therefore, the logarithmic negativity is a measure of entanglement since the negativity is an entanglement monotone. On the other hand, the logarithmic negativity is not an entanglement monotone; in particular, it is in general not convex (Exercise 13.38).

The logarithmic negativity is additive under tensor products. To see why, let $\rho \in \mathfrak{D}(AB)$ and $\sigma \in \mathfrak{D}(A'B')$. Then,

$$
\begin{aligned}
\mathbf{LN}\left(\rho^{AB} \otimes \sigma^{A'B'}\right) &= \log \left\| (\rho \otimes \sigma)^{\Gamma} \right\|_1 \\
&= \log \left\| \rho^{\Gamma} \otimes \sigma^{\Gamma} \right\|_1 \\
&= \log \left\| \rho^{\Gamma} \right\|_1 \left\| \sigma^{\Gamma} \right\|_1 = \log \left\| \rho^{\Gamma} \right\|_1 + \log \left\| \sigma^{\Gamma} \right\|_1 \\
&= \mathbf{LN}\left(\rho^{AB}\right) + \mathbf{LN}(\sigma^{A'B'}).
\end{aligned}
\tag{13.166}
$$

We will see later on that the logarithmic negativity provides an upper bound to the distillable entanglement.

**Exercise 13.38.** *Show that the logarithmic negativity is not convex.*

## The $\kappa$-Entanglement

The $\kappa$-Entanglement is another measure of entanglement that is based on the partial transpose. In Section 13.9 we will see that the regularized version of this measure has an operational meaning as the zero-error entanglement cost under PPT operations. The $\kappa$-entanglement is defined for all $\rho \in \mathfrak{D}(AB)$ as

$$
E_\kappa\left(\rho^{AB}\right) = \min_{\Lambda \in \mathrm{Pos}(AB)} \left\{ \log \mathrm{Tr}[\Lambda] : -\Lambda^{\Gamma} \leqslant \rho^{\Gamma} \leqslant \Lambda^{\Gamma} \right\}.
\tag{13.167}
$$

In Section 13.9 we will see that $E_\kappa$ behaves monotonically under a set of operations that is larger than LOCC. Moreover, if $\rho \in \mathrm{PPT}(AB)$, then we can take in the equation above $\Lambda = \rho$, so that the $\kappa$-Entanglement vanishes on PPT states and in particular on separable states. Therefore, $E_\kappa$ is a measure of entanglement.

---

**Lemma 13.2.** Let $\rho \in \mathfrak{D}(AB)$. Then,

$$
\mathbf{LN}\left(\rho^{AB}\right) \leqslant E_\kappa\left(\rho^{AB}\right) \leqslant \min_{\sigma \in \mathrm{PPT}(AB)} D_{\max}\left(|\rho^{\Gamma}| \,\|\, \sigma\right).
\tag{13.168}
$$

---

*Remark.* From this lemma it follows that the $\kappa$-entanglement equals the logarithmic negativity if

$$
|\rho^{\Gamma}|^{\Gamma} \geqslant 0.
\tag{13.169}
$$

To see why, note that in this case we have that the state $\rho_\star := |\rho^{\Gamma}| / \| \rho^{\Gamma} \|_1 \in \mathrm{PPT}(AB)$, so by taking $\sigma = \rho_\star$ we get that the upper bound

$$
\min_{\sigma \in \mathrm{PPT}(AB)} D_{\max}\left(|\rho^{\Gamma}| \,\|\, \sigma\right) \leqslant D_{\max}\left(|\rho^{\Gamma}| \,\|\, \rho_\star\right)
\tag{13.170}
$$

$$
\textbf{Exercise 13.39} \rightarrow -\mathbf{LN}(\rho).
$$

**Proof** The condition $-\Lambda^{\Gamma} \leqslant \rho^{\Gamma} \leqslant \Lambda^{\Gamma}$ in (13.167) can also be expressed as

$$
\Lambda^{\Gamma} \geqslant \rho^{\Gamma} \quad \text{and} \quad \Lambda^{\Gamma} \geqslant -\rho^{\Gamma}.
\tag{13.171}
$$

Combining this with the decomposition (13.154) of $\rho^\Gamma$ gives the following two inequalities:

$$\Lambda^\Gamma \geqslant \left(1+\mathbf{N}\left(\rho\right)\right)\rho_+ - \mathbf{N}\left(\rho\right)\rho_- \quad \text{and} \quad \Lambda^\Gamma \geqslant \mathbf{N}\left(\rho\right)\rho_- - \left(1+\mathbf{N}\left(\rho\right)\right)\rho_+. \quad (13.172)$$

Let $\Pi_\pm$ be the projections to the supports of $\rho_\pm$. Then, from equations in (13.172) we get

$$\Pi_+\Lambda^\Gamma\Pi_+ \geqslant \left(1+\mathbf{N}\left(\rho\right)\right)\rho_+ \quad \text{and} \quad \Pi_-\Lambda^\Gamma\Pi_- \geqslant \mathbf{N}\left(\rho\right)\rho_-. \quad (13.173)$$

Since $\Lambda^\Gamma \geqslant 0$, we get

$$\begin{aligned}
\mathrm{Tr}[\Lambda] = \mathrm{Tr}\left[\Lambda^\Gamma\right] &\geqslant \mathrm{Tr}\left[\Lambda^\Gamma\left(\Pi_+ + \Pi_-\right)\right]\\
&= \mathrm{Tr}\left[\Pi_+\Lambda^\Gamma\Pi_+\right] + \mathrm{Tr}\left[\Pi_-\Lambda^\Gamma\Pi_-\right] \quad (13.174)\\
\mathbf{(13.173)}\rightarrow &\geqslant 1 + 2\mathbf{N}(\rho) = \|\rho^\Gamma\|_1.
\end{aligned}$$

Since this inequality holds for all $\Lambda \in \mathrm{Pos}(AB)$ that satisfies $-\Lambda^\Gamma \leqslant \rho^\Gamma \leqslant \Lambda^\Gamma$, we conclude that the lower bound in (13.168) must hold.

To get an upper bound observe that for all $\rho \in \mathfrak{D}(AB)$

$$\begin{aligned}
E_\kappa\left(\rho\right) &\leqslant \min_{\Lambda\in\mathrm{Pos}(AB)} \left\{\log\mathrm{Tr}[\Lambda] : \Lambda^\Gamma \geqslant |\rho^\Gamma|\right\}\\
\mathbf{\Lambda = t\sigma} \rightarrow &= \min_{\sigma\in\mathrm{PPT}(AB)} \left\{\log\left(t\right) : t\sigma \geqslant |\rho^\Gamma|\right\} \quad (13.175)\\
&= \min_{\sigma\in\mathrm{PPT}(AB)} D_{\max}\left(|\rho^\Gamma|\,\big\|\,\sigma\right).
\end{aligned}$$

This completes the proof. $\blacksquare$

**Exercise 13.39.** *Prove the equality in* (13.170).

**Exercise 13.40.** *Show that $E_\kappa$ is subadditive. Hint: Show that if $-\Lambda_j^\Gamma \leqslant \rho_j^\Gamma \leqslant \Lambda_j^\Gamma$ for $j = 1, 2$, then*

$$-\Lambda_1^\Gamma \otimes \Lambda_2^\Gamma \leqslant \rho_1^\Gamma \otimes \rho_2^\Gamma \leqslant \Lambda_1^\Gamma \otimes \Lambda_2^\Gamma. \quad (13.176)$$

## 13.2.5 The Squashed Entanglement

In Exercise 7.16 we defined the mutual information as

$$I(A:B)_\rho := D\left(\rho^{AB}\,\big\|\,\rho^A \otimes \rho^B\right) = H(A)_\rho + H(B)_\rho - H(AB)_\rho. \quad (13.177)$$

This function quantifies the total (i.e. both quantum and classical) amount of correlation between Alice and Bob (see Figure 13.2a). An extension of this quantity, known as the conditional mutual information (CMI), is a function on a tripartite density matrix defined by

$$I(A:B|R)_\rho := H(A|R)_\rho + H(B|R)_\rho - H(AB|R)_\rho \qquad \forall\, \rho \in \mathfrak{D}(ABR). \quad (13.178)$$

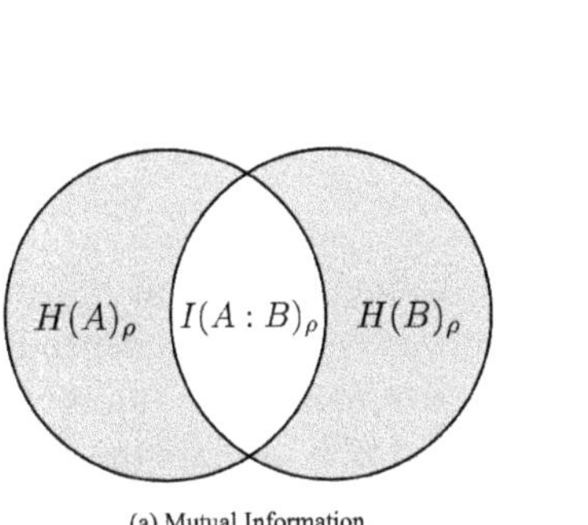


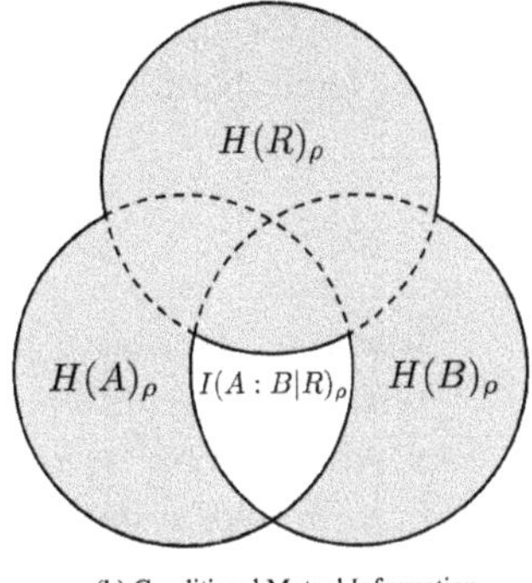


**Figure 13.2**   Venn diagrams. (a) The intersection area (white) illustrates the mutual information. (b) The white area illustrates the conditional mutual information.

Since the CMI is defined with respect to the conditional von-Neumann entropy, it can also be expressed for all $\rho \in \mathfrak{D}(ABR)$ as

$$I(A:B|R)_\rho = H(A|R)_\rho + H(BR)_\rho - H(ABR)_\rho$$
$$= H(A|R)_\rho - H(A|BR)_\rho \tag{13.179}$$
$$(\mathbf{7.136}) \to \; \geqslant 0.$$

Therefore, the CMI cannot be negative. The CMI quantifies the total correlations between Alice and Bob, given that one has access to a reference system $R$ (see Figure 13.2b for a heuristic description).

**Exercise 13.41 (Chain Rule).** *Let $\rho \in \mathfrak{D}(AA'BB'R)$.*

1. *Show that*
$$I(AA':B|R)_\rho = I(A':B|R)_\rho + I(A:B|RA')_\rho$$
$$I(A:BB'|R)_\rho = I(A:B'|R)_\rho + I(A:B|RB')_\rho. \tag{13.180}$$

2. *Show that*
$$I(AA':B|R)_\rho \geqslant I(A:B|R)_\rho. \tag{13.181}$$

*That is, tracing out a local subsystem cannot increase the CMI.*

**Exercise 13.42.** *Show that for any state of the form*
$$\sigma^{ABRA'} = \sum_{x\in[n]} p_x \sigma_x^{ABR} \otimes |x\rangle\langle x|^{A'}, \tag{13.182}$$

*we have that*
$$I(A:B|RA')_\sigma = \sum_{x\in[n]} p_x I(A:B|R)_{\sigma_x}. \tag{13.183}$$

The mutual information measures the overall correlation between Alice and Bob, and only takes on the value zero for product states. However, it does not necessarily vanish for separable states that are not products. In contrast, the CMI can be zero even for separable states. For instance, consider the state

$$\rho^{ABR} := \sum_{x \in [n]} p_x \rho_x^A \otimes \rho_x^B \otimes |x\rangle\langle x|^R, \tag{13.184}$$

where $p_x$ denotes a probability distribution, $\rho_x^A$ and $\rho_x^B$ are density matrices, and $|x\rangle\langle x|^R$ represents a pure state in a third system $R$ that depends on the discrete variable $x$. Although $\rho^{AB}$ is a separable state, its CMI is zero. This is because knowing the value of $x$ through system $R$ allows Alice and Bob to share the product state $\rho_x^A \otimes \rho_x^B$. The state $\rho^{ABR}$ belongs to a special type of state known as quantum Markov state.

Quantum Markov state is a type of quantum state that exhibits a special type of correlation structure between different subsystems. In a quantum Markov state, the correlation between two subsystems is entirely mediated by a third subsystem $R$, which serves as a kind of "bridge" or mediator between $A$ and $B$. This correlation structure is analogous to the Markov property in classical probability theory, where the future state of a system depends only on its present state and not on its past states. A quantum Markov state $\rho \in \mathfrak{D}(ABR)$ is defined as follows.

**Definition 13.6.** Let $A$, $B$, and $R$ be three quantum systems. A state $\rho \in \mathfrak{D}(ABR)$ is called a quantum Markov state if the following two conditions hold:

1. There exists $m \in \mathbb{N}$, and two sets of Hilbert spaces $\{R_x^{(1)}\}_{x\in[m]}$ and $\{R_x^{(2)}\}_{x\in[m]}$ such that $R = \bigoplus_{x\in[m]} R_x^{(1)} \otimes R_x^{(2)}$.

2. There exist two sets of density matrices $\left\{\rho_x^{AR_x^{(1)}}\right\}_{x\in[m]}$ and $\left\{\rho_x^{BR_x^{(2)}}\right\}_{x\in[m]}$ and a probability vector $\mathbf{p} \in \mathrm{Prob}(m)$ such that

$$\rho^{ABR} = \bigoplus_{x\in[m]} p_x \rho_x^{AR_x^{(1)}} \otimes \rho_x^{BR_x^{(2)}}. \tag{13.185}$$

**Exercise 13.43.**

1. *Show that the state in (13.184) is a quantum Markov state.*
2. *Show that for any quantum Markov state $\rho \in \mathfrak{D}(ABR)$ we have*

$$H(A : B|R)_\rho = 0. \tag{13.186}$$

*Remark: The converse of this statement is also true! That is, any density matrix $\rho \in \mathfrak{D}(ABR)$ with zero CMI is necessarily a quantum Markov state.*

This exercise demonstrates that if $\rho^{AB}$ is a separable state, then it has a tripartite extension $\rho^{ABR}$ as in (13.184), for which the conditional mutual information is zero. This observation motivates the following definition of a measure of entanglement known as the squashed entanglement.

> **The Squashed Entanglement**
>
> **Definition 13.7.** The squashed entanglement of a bipartite density matrix $\rho \in \mathfrak{D}(AB)$ is defined as
>
> $$E_{\text{sq}}\left(\rho^{AB}\right) := \frac{1}{2}\inf I(A{:}B|R)_\omega, \tag{13.187}$$
>
> where the infimum is over all finite dimensional systems $R$ and all $\omega \in \mathfrak{D}(ABR)$ with marginal $\omega^{AB} = \rho^{AB}$.

*Remark.* The term "squashed" entanglement is used because conditioning on a reference system R enables the removal (i.e. squash out) of all nonquantum correlations, similar to how conditioning on system R can eliminate classical correlations in separable states. Additionally, since squashed entanglement is based on the CMI, it is also known as the CMI entanglement.

**Exercise 13.44.** *Show that for $\psi \in \text{Pure}(AB)$, we have*

$$E_{sq}\left(\psi^{AB}\right) = H(A)_\psi, \tag{13.188}$$

*where $H(A)_\psi$ is the von-Neuman entropy of the reduced density matrix on system A.*

The squashed entanglement has many desirable properties of measures of entanglement. We start with the fact that it is an entanglement monotone.

> **Theorem 13.10.** The squashed entanglement is an entanglement monotone.

**Proof** Let $\rho \in \mathfrak{D}(AB)$ and let $\omega^{ABR}$ be an extension of $\rho^{AB}$; that is, $\omega^{AB} = \rho^{AB}$. Suppose Alice applies on her system $A$ a quantum instrument $\mathcal{E} \in \text{CPTP}(A \to A'A'')$ of the form

$$\mathcal{E}^{A \to A'A''} = \sum_{x \in [n]} \mathcal{E}_x^{A \to A'} \otimes |x\rangle\langle x|^{A''}. \tag{13.189}$$

Then, the state of the composite system $ABR$ after Alice's measurement is given by

$$\sigma^{A'A''BR} := \mathcal{E}^{A \to A'A''}\left(\omega^{ABR}\right). \tag{13.190}$$

From Stinespring's dilation theorem there exists an isometry $V : A \to A'A''E$ such that

$$\sigma^{A'A''BR} = \text{Tr}_E\left[V\omega^{ABR}V^*\right]. \tag{13.191}$$

Since conditional entropy is invariant under local isometry, we get that

$$\frac{1}{2}I(A{:}B|R)_\omega = \frac{1}{2}I(A'A''E{:}B|R)_{V\omega V^*}. \tag{13.192}$$

Combining the equality in (13.192) with the fact that by tracing out a local subsystem the CMI cannot increase (see (13.181)), we get by tracing out system $E$

$$\frac{1}{2}I(A:B|R)_\omega \geq \frac{1}{2}I(A'A'':B|R)_\sigma. \tag{13.193}$$

Combining this with the chain rule (13.180) gives

$$\frac{1}{2}I(A:B|R)_\omega \geq \frac{1}{2}I(A'':B|R)_\sigma + \frac{1}{2}I(A':B|RA'')_\sigma$$

$$\boxed{I(A'':B|R)_\sigma \geq 0} \longrightarrow \quad \geq \frac{1}{2}I(A':B|RA'')_\sigma \tag{13.194}$$

$$\textbf{Exercise 13.42}\rightarrow \; = \frac{1}{2}\sum_{x\in[n]} p_x I(A':B|R)_{\sigma_x},$$

where $p_x := \mathrm{Tr}\left[\mathcal{E}_x^{A\to A'}\left(\omega^{ABR}\right)\right]$ and $\sigma_x^{A'BR} := \frac{1}{p_x}\mathcal{E}_x^{A\to A'}\left(\omega^{ABR}\right)$. Finally, by definition, for any $x \in [n]$ we have $\frac{1}{2}I(A':B|R)_{\sigma_x} \geq E_{\mathrm{sq}}\left(\sigma_x^{A'B}\right)$ so that the inequality above gives

$$\frac{1}{2}I(A:B|R)_\omega \geq \sum_{x\in[n]} p_x E_{\mathrm{sq}}\left(\sigma_x^{A'B}\right). \tag{13.195}$$

Since $\omega^{ABR}$ was an arbitrary extension of $\rho^{AB}$ we conclude that

$$E_{\mathrm{sq}}\left(\rho^{AB}\right) \geq \sum_{x\in[n]} p_x E_{\mathrm{sq}}\left(\sigma_x^{A'B}\right). \tag{13.196}$$

In other words, the squashed entanglement does not increase on average under any local quantum instrument on system $A$. From symmetry, the same holds for any quantum instrument on Bob's system. Therefore, the squashed entanglement satisfies the strong monotonicity property of an entanglement monotone.

To prove the convexity of $E_{\mathrm{sq}}$, let $\rho_1, \rho_2 \in \mathfrak{D}(AB)$ and $t \in [0,1]$. Let $\omega_1^{ABR}$ and $\omega_2^{ABR}$ be extensions of $\rho_1^{AB}$ and $\rho_2^{AB}$, respectively. Note that in general we can always assume that these extensions have the same reference system $R$ as otherwise we embed the lower dimensional reference system in the higher dimensional one. Finally, let $R'$ be a qubit system and denote by

$$\omega^{ABRR'} := \omega_1^{ABR}\otimes|0\rangle\langle0|^{R'} + (1-t)\omega_2^{ABR}\otimes|1\rangle\langle1|^{R'}. \tag{13.197}$$

Since $\omega^{AB} = t\rho^{AB} + (1-t)\sigma^{AB}$, we get that

$$E_{\mathrm{sq}}\left(t\rho^{AB} + (1-t)\sigma^{AB}\right) \leq I\left(A:B|RR'\right)_\omega$$

$$\textbf{Exercise 13.42}\rightarrow \; = tI\left(A:B|R\right)_{\omega_1} + (1-t)I\left(A:B|R\right)_{\omega_2}. \tag{13.198}$$

Since the extensions $\omega_1^{ABR}$ and $\omega_2^{ABR}$ were arbitrary, we conclude that

$$E_{\mathrm{sq}}\left(t\rho^{AB} + (1-t)\sigma^{AB}\right) \leq tE_{\mathrm{sq}}\left(\rho^{AB}\right) + (1-t)E_{\mathrm{sq}}\left(\sigma^{AB}\right). \tag{13.199}$$

This completes the proof. ∎

Another interesting property of the squashed entanglement is that it is additive.

> **Theorem 13.11.** Let $\rho \in \mathfrak{D}(AA'BB')$. Then,
> $$E_{\text{sq}}\left(\rho^{AA'BB'}\right) \geqslant E_{\text{sq}}\left(\rho^{AB}\right) + E_{\text{sq}}\left(\rho^{A'B'}\right), \tag{13.200}$$
> with equality if $\rho^{AA'BB'} = \rho^{AB} \otimes \rho^{A'B'}$.

**Proof** Let $\omega \in \mathfrak{D}(AA'BB'R)$ be a quantum extension of $\rho^{AA'BB'}$. Then, by applying the chain rule in (13.180) one time with respect to Alice's systems and one time with respect to Bob's systems we get

$$I(AA' : BB'|R)_\omega = I(A' : BB'|R)_\omega + I(A : BB'|RA')_\omega$$

$$\mathbf{(13.180)}\rightarrow\; = I(A' : B'|R)_\omega + I(A' : B|RB')_\omega + I(A : B'|RA')_\omega + I(A : B|RA'B')_\omega$$

$$\mathbf{CMI \geqslant 0} \rightarrow\; \geqslant I(A' : B'|R)_\omega + I(A : B|RA'B')_\omega$$

$$\textbf{By definition} \rightarrow\; \geqslant 2E_{\text{sq}}\left(\rho^{AB}\right) + 2E_{\text{sq}}\left(\rho^{A'B'}\right). \tag{13.201}$$

Since $\omega^{AA'BB'R}$ was an arbitrary extension of $\rho^{AA'BB'}$, we conclude that the above inequality implies (13.200).

It is left to show that $\rho^{AA'BB'} = \rho^{AB} \otimes \rho^{A'B'}$, we have equality in (13.200). Let $\omega^{ABR_1}$ and $\omega^{A'B'R_2}$ be extensions of $\rho^{AB}$ and $\rho^{A'B'}$, respectively. Set $R = R_1 R_2$ and $\omega^{AA'BB'R} := \omega^{ABR_1} \otimes \omega^{A'B'R_2}$. In Exercise 13.45 we will show that for such $\omega^{AA'BB'R}$ we have

$$I(A' : B|RB')_\omega = I(A : B'|RA')_\omega = 0. \tag{13.202}$$

Combining this with the second equality in (13.201) we get

$$I(AA' : BB'|R)_\omega = I(A' : B'|R)_\omega + I(A : B|RA'B')_\omega. \tag{13.203}$$

Finally, since $\omega^{A'B'R} = \omega^{R_1} \otimes \omega^{A'B'R_2}$ we have $I(A' : B'|R)_\omega = I(A' : B'|R_2)_\omega$ as conditioning by an additional independent system $R_1$ does not change the CMI. Similarly,

$$I(A : B|RA'B')_\omega = I(A : B|R_1)_\omega. \tag{13.204}$$

We therefore get that for any extensions $\omega^{ABR_1}$ and $\omega^{A'B'R_2}$ of $\rho^{AB}$ and $\rho^{A'B'}$, we have

$$\begin{aligned} E_{\text{sq}}\left(\rho^{AB} \otimes \rho^{A'B'}\right) &\leqslant \frac{1}{2}I(AA' : BB'|R)_\omega \\ &= \frac{1}{2}I(A' : B'|R_1)_\omega + \frac{1}{2}I(A : B|R_2)_\omega. \end{aligned} \tag{13.205}$$

Since this inequality holds of any such extensions of $\rho^{AB}$ and $\rho^{A'B'}$, we must have

$$E_{\text{sq}}\left(\rho^{AB} \otimes \rho^{A'B'}\right) \leqslant E_{\text{sq}}\left(\rho^{AB}\right) + E_{\text{sq}}\left(\rho^{A'B'}\right). \tag{13.206}$$

Combining this with (13.200) gives an equality. This completes the proof. ∎

**Exercise 13.45.** *Prove* (13.202) *and* (13.204).

## 13.2.6 Coherent Information of Entanglement

In this section, we will explore a measure of entanglement that exhibits monotonic behavior under one-way LOCC. While this measure may not exhibit monotonicity under arbitrary LOCC, it can still be a valuable tool for providing bounds on the distillable entanglement of mixed bipartite states, as we will see in the next section.

The most general one-way LOCC operation that Alice and Bob can perform is for Alice to apply a quantum instrument $\{\mathcal{E}_x\}_{x \in [m]}$, with $\mathcal{E}_x \in \mathrm{CP}(A \to A')$ and $\sum_{x \in [m]} \mathcal{E}_x \in \mathrm{CPTP}(A \to A')$, and send the outcome $x$ to Bob, who then applies a quantum channel $\mathcal{F}_x \in \mathrm{CPTP}(B \to B')$ that depends on the outcome $x$ received from Alice. The overall operation can be described by the quantum channel

$$\mathcal{N}^{AB \to A'B'} := \sum_{x \in [m]} \mathcal{E}_x^{A \to A'} \otimes \mathcal{F}_x^{B \to B'}. \tag{13.207}$$

---

**Definition 13.8.** Let $\rho \in \mathrm{Pure}(ABE)$. The coherent information of the marginal state $\rho^{AB}$ is defined as

$$I(A \rangle B)_\rho := -H(A|B)_\rho = H(A|E)_\rho, \tag{13.208}$$

where the second equality is due to the duality relation of the conditional von-Neumann entropy given in (7.158). Moreover, the *coherent information of entanglement* of the state $\rho^{AB}$ is defined as

$$E_\to \left( \rho^{AB} \right) := \sup_{\mathcal{E} \in \mathrm{CPTP}(A \to AX)} I\left(A \rangle BX\right)_{\mathcal{E}(\rho)}, \tag{13.209}$$

where the supremum is also over all finite dimensions of the classical system $X$.

---

**Exercise 13.46.** Let $\rho^{ABX} := \sum_{x \in [m]} p_x \rho_x^{AB} \otimes |x\rangle\langle x|^X$ be a cq-state in $\mathfrak{D}(ABX)$.

*1. Show that*

$$I(A \rangle BX)_\rho = \sum_{x \in [m]} p_x I(A \rangle B)_{\rho_x}. \tag{13.210}$$

*2. Show that the coherent information is convex. That is, prove that*

$$I(A \rangle B)_\rho \leqslant \sum_{x \in [m]} p_x I(A \rangle B)_{\rho_x}. \tag{13.211}$$

*Hint: Use the joint convexity of the relative entropy.*

*3. Show that for every quantum channel $\mathcal{F} \in \mathrm{CPTP}(B \to B')$ we have*

$$I(A \rangle B')_{\mathcal{F}(\rho)} \leqslant I(A \rangle B)_\rho. \tag{13.212}$$

*Hint: Either use the DPI directly or recall that a single channel on Bob's system is a conditionally mixing operation.*

**Exercise 13.47.** *Show that the supremum in* (13.209) *can be restricted quantum channels of the form* $\mathcal{E}^{A \to AX} = \sum_{x \in [n]} \mathcal{E}_x^{A \to A} \otimes |x\rangle\langle x|^X$, *where each* $\mathcal{E}_x^{A \to A}$ *is a CP map with a single Kraus operator. Hint: Use the joint convexity of the relative entropy.*

**Exercise 13.48.** *Let* $\rho \in \mathfrak{D}(AB)$.

1. *Show that* $E_\to(\rho^{AB}) \geqslant 0$ *with equality if* $\rho^{AB}$ *is separable.*
2. *Show that* $E_\to(\rho^{AB}) \geqslant I(A\rangle B)_\rho$.

Note that while the supremum in Definition 13.8 is taken over all dimensions of $X$, we consider only quantum instruments from $A$ to $A$. However, this limitation is not necessary, as the coherent information of entanglement for the state $\rho^{AB}$ can be defined as

$$E_\to\left(\rho^{AB}\right) := \sup_{\mathcal{E} \in \mathrm{CPTP}(A \to A'X)} I\left(A'\rangle BX\right)_{\mathcal{E}(\rho)}, \tag{13.213}$$

with the supremum extending over all dimensions of system $A'$. To understand this, first consider that if $|A'| \leqslant |A|$, every channel $\mathcal{E} \in \mathrm{CPTP}(A \to A'B)$ can be embedded in $\mathrm{CPTP}(A \to AX)$, as the coherent information is invariant under local isometries (a property shared by all conditional entropies). Consequently, in this case, the supremum over $\mathrm{CPTP}(A \to AX)$ is at least as great as that over $\mathrm{CPTP}(A \to A'X)$.

Conversely, if $|A'| > |A|$, consider a quantum instrument $\mathcal{E}^{A \to A'X} = \sum_{x \in [n]} \mathcal{E}_x^{A \to A'} \otimes |x\rangle\langle x|^X$, where each $\mathcal{E}_x^{A \to A'}(\cdot) = M_x(\cdot)M_x^*$ is a CP map with a single Kraus operator $M_x : A \to A'$. Through polar decomposition, each $M_x$ can be written as $M_x = V_x N_x$, with each $N_x : A \to A$ being part of a generalized measurement, and each $V_x : A \to A'$ an isometry. Due to the invariant property of coherent information under isometries, the CP maps $\mathcal{E}_x^{A \to A'}(\cdot) = M_x(\cdot)M_x^*$ can be substituted with $\mathcal{N}_x^{A \to A}(\cdot) = N_x(\cdot)N_x^*$, allowing the optimization over all channels in $\mathrm{CPTP}(A \to A'X)$ to be replaced with optimization over all quantum instruments in $\mathrm{CPTP}(A \to AX)$.

This observation is significant as it can be used to prove that the coherent information of entanglement exhibits monotonic behavior under one-way LOCC.

> **Theorem 13.12.** Let $\rho \in \mathfrak{D}(AB)$ and $\mathcal{N} \in \mathrm{LOCC}_1(AB \to A'B')$. Then,
>
> $$E_\to\left(\mathcal{N}^{AB \to A'B'}\left(\rho^{AB}\right)\right) \leqslant E_\to\left(\rho^{AB}\right). \tag{13.214}$$

**Proof** For every quantum instrument $\mathcal{E} \in \mathrm{CPTP}(A' \to A'X)$, $\mathcal{E} \circ \mathcal{N} \in \mathrm{LOCC}_1(AB \to A'B'X)$. Therefore,

$$
\begin{aligned}
E_\to\left(\mathcal{N}^{AB \to A'B'}\left(\rho^{AB}\right)\right) &= \sup_{\mathcal{E} \in \mathrm{CPTP}(A' \to A'X)} I\left(A'\rangle B'X\right)_{\mathcal{E} \circ \mathcal{N}(\rho)} \\
&\leqslant \sup_{\mathcal{M} \in \mathrm{LOCC}_1(AB \to A'B'X)} I\left(A'\rangle B'X\right)_{\mathcal{M}(\rho)},
\end{aligned}
\tag{13.215}
$$

where we replaced $\mathcal{E} \circ \mathcal{N}$ with arbitrary $\mathcal{M} \in \mathrm{LOCC}_1(AB \to A'B'X)$. Now, recall that every element of $\mathrm{LOCC}_1(AB \to A'B'X)$ can be expressed as

$$\mathcal{M}^{AB \to A'B'X} := \sum_{y \in [n]} \mathcal{E}_y^{A \to A'X} \otimes \mathcal{F}_y^{B \to B'}, \tag{13.216}$$

with each $\mathcal{E}_y^{A \to A'X}$ being a CP map such that $\sum_{y \in [n]} \mathcal{E}_y \in \mathrm{CPTP}(A \to A'X)$, and each $\mathcal{F}_y \in \mathrm{CPTP}(B \to B')$. Thus,

$$\mathcal{M}^{AB \to A'B'X}\left(\rho^{AB}\right) = \sum_{y \in [n]} q_y \mathcal{F}_y^{B \to B'}\left(\sigma_y^{A'BX}\right), \tag{13.217}$$

where $\sigma_y^{A'BX} := \frac{1}{q_y} \mathcal{E}_y^{A \to A'X}\left(\rho^{AB}\right)$ and $q_y := \mathrm{Tr}\left[\mathcal{E}_y^{A \to A'}\left(\rho^{AB}\right)\right]$. Combining this with the convexity of the coherent information (see (13.211)) we get from (13.215) that

$$\begin{aligned}
E_\to\left(\mathcal{N}^{AB \to A'B'}\left(\rho^{AB}\right)\right) &\leqslant \sup_{\mathcal{M} \in \mathrm{LOCC}_1} \sum_{y \in [n]} q_y I\left(A' \rangle B'X\right)_{\mathcal{F}_y(\sigma_y)} \\
\mathbf{cf.\ (13.212)} \to\ &\leqslant \sup_{\mathcal{M} \in \mathrm{LOCC}_1} \sum_{y \in [n]} q_y I\left(A' \rangle B'X\right)_{\sigma_y}.
\end{aligned} \tag{13.218}$$

Finally, denoting by $Z := XY$, and by $\mathcal{E}^{A \to A'Z} := \sum_{y \in [n]} \mathcal{E}_y^{A' \to A'X} \otimes |y\rangle\langle y|^Y$ we conclude that

$$\begin{aligned}
E_\to\left(\mathcal{N}^{AB \to A'B'}\left(\rho^{AB}\right)\right) &\leqslant \sup_{\mathcal{E} \in \mathrm{CPTP}(A \to A'Z)} I\left(A' \rangle BZ\right)_{\mathcal{E}(\rho)} \\
\mathbf{(13.213)} \to\ &= E_\to\left(\rho^{AB}\right).
\end{aligned} \tag{13.219}$$

This completes the proof. $\blacksquare$

**Exercise 13.49.** *Show that $E_\to$ is an entanglement monotone under $\mathrm{LOCC}_1$. That is, prove the strong monotonicity property and convexity.*

**Exercise 13.50.** *Let $\Phi_m \in \mathfrak{D}(AB)$ be the maximally entangled state with $m := |A| = |B|$. Show that*

$$E_\to\left(\Phi_m^{AB}\right) = \log(m). \tag{13.220}$$

The coherent information of entanglement is superadditive. That is, for any $\rho \in \mathfrak{D}(AB)$ and $\sigma \in \mathfrak{D}(A'B')$ we have (see Exercise 13.51)

$$E_\to\left(\rho^{AB} \otimes \sigma^{A'B'}\right) \geqslant E_\to\left(\rho^{AB}\right) + E_\to\left(\sigma^{A'B'}\right). \tag{13.221}$$

From Exercise 10.1 it then follows that the limit in

$$E_\to^{\mathrm{reg}}\left(\rho^{AB}\right) := \lim_{n \to \infty} \frac{1}{n} E_\to\left(\rho^{\otimes n}\right) \tag{13.222}$$

exists. We will see in the next section that this regularized coherent information of entanglement has an operational meaning as the one-way distillable entanglement of $\rho^{AB}$.

**Exercise 13.51.** *Prove the superadditivity of the coherent information of entanglement as given in* (13.221).

## 13.3 The Conversion Distance

The conversion distance in entanglement theory is given by (cf. (11.24))

$$T\left(\rho^{AB} \xrightarrow{\text{LOCC}} \sigma^{A'B'}\right) := \frac{1}{2} \min_{\mathcal{N} \in \text{LOCC}} \left\| \mathcal{N}^{AB \to A'B'}\left(\rho^{AB}\right) - \sigma^{A'B'} \right\|_1. \tag{13.223}$$

Computing this quantity in general is a highly challenging task, so we often rely on establishing lower and upper bounds. In this section, we will narrow our focus to the special cases where either $\rho$ or $\sigma$ is maximally entangled. Recall that these cases are particularly relevant for calculating entanglement distillation and entanglement cost.

### 13.3.1 Conversion Distance to a Maximally Entangled State

We start with the following simplification of the expression given in (13.223) when $\sigma^{A'B'}$ is maximally entangled.

> **Lemma 13.3.** Let $\rho \in \mathfrak{D}(AB)$ and $\Phi_m \in \mathfrak{D}(A'B')$ be the maximally entangled state with $m := |A'| = |B'|$. Then,
>
> $$T\left(\rho \xrightarrow{\text{LOCC}} \Phi_m\right) = P^2\left(\rho \xrightarrow{\text{LOCC}} \Phi_m\right) = 1 - \sup_{\mathcal{N}} \text{Tr}\lfloor \Phi_m \mathcal{N}(\rho)\rfloor, \tag{13.224}$$
>
> where the supremum is over all $\mathcal{N} \in \text{LOCC}(AB \to A'B')$, and $P$ is the purified distance as given in (5.251).

*Remark.* In Section 12.5.1, we explored various conversion distances among pure bipartite states. It was established that the $P_\star$-conversion distance is equal to the $P$-conversion distance. Additionally, we speculated, albeit without formal proof, that the $T$-conversion distance might be strictly smaller than the $P$-conversion distance. The lemma above confirms this speculation by demonstrating that when the target state is $\Phi_m$, the $T$-conversion distance actually aligns with the $P^2$-conversion distance, which is indeed strictly smaller than the $P$-conversion distance. This outcome is based on the understanding that the purified distance is no greater than 1; thus, squaring it effectively reduces its magnitude.

**Proof**   Let $\mathcal{G} \in \text{LOCC}(A'B' \to A'B')$ be the twirling map given in (3.251). That is, for any $\omega \in \mathfrak{D}(A'B')$

$$\mathcal{G}(\omega) := \int_{\mathfrak{U}(m)} dU\,(U \otimes \overline{U})\omega(U \otimes \overline{U})^*$$

$$(3.255) \to\, = (1 - \text{Tr}\left[\Phi_m \omega\right])\,\tau + \text{Tr}\left[\Phi_m \omega\right] \Phi_m, \tag{13.225}$$

where $\tau \in \mathfrak{D}(A'B')$ is given by $\tau = (I - \Phi_m)/(m^2 - 1)$. In particular, observe that for all $\omega \in \mathfrak{D}(A'B')$

$$\frac{1}{2}\|\mathcal{G}(\omega) - \Phi_m\|_1 = \left(1 - \mathrm{Tr}[\Phi_m\omega]\right)\frac{1}{2}\|\tau - \Phi_m\|_1 = 1 - \mathrm{Tr}[\Phi_m\omega], \qquad (13.226)$$

where the last equality follows from the fact that $\tau\Phi_m = \Phi_m\tau = 0$. From the DPI of the trace distance, and the invariance of $\Phi_m$ under the twirling map $\mathcal{G}$, it follows that for all $\mathcal{N} \in \mathrm{LOCC}(AB \to A'B')$ and all $\rho \in \mathfrak{D}(AB)$,

$$\frac{1}{2}\|\mathcal{N}(\rho) - \Phi_m\|_1 \geqslant \frac{1}{2}\|\mathcal{G} \circ \mathcal{N}(\rho) - \Phi_m\|_1. \qquad (13.227)$$

Since $\mathcal{G} \circ \mathcal{N}$ is also an LOCC channel it follows from the inequality above that the conversion distance can be expressed as

$$T\left(\rho \xrightarrow{\text{LOCC}} \Phi_m\right) = \inf_{\mathcal{N}\in\mathrm{LOCC}} \frac{1}{2}\|\mathcal{G} \circ \mathcal{N}(\rho) - \Phi_m\|_1$$
$$\overset{(13.226)}{\longrightarrow} = 1 - \sup_{\mathcal{N}\in\mathrm{LOCC}} \mathrm{Tr}[\Phi_m\mathcal{N}(\rho)]. \qquad (13.228)$$

This completes the proof. ∎

**Exercise 13.52.** *Let $k := |A| = |B|$, $m = |A'| = |B'|$, and $\rho \in \mathfrak{D}(AB)$. Show that*

$$T\left(\rho^{AB} \xrightarrow{\text{LOCC}} \Phi_m^{A'B'}\right) \geqslant 1 - \frac{k}{m}. \qquad (13.229)$$

*Note that this bound is not trivial for $k < m$. Hint: Estimate $T\left(\Phi_k^{AB} \xrightarrow{\text{LOCC}} \Phi_m^{A'B'}\right)$.*

## Conversion Distance under One-Way LOCC

The lemma's expression for $T(\rho \xrightarrow{\text{LOCC}} \Phi_m)$ is complex, making its computation challenging. One contributing factor is the complexity inherent in LOCC. To simplify this, we begin by restricting the channel $\mathcal{N}$ in (13.224) to one-way LOCC, which yields an upper bound on the conversion distance $T(\rho \xrightarrow{\text{LOCC}} \Phi_m)$. More explicitly, the one-way conversion distance is defined as

$$T\left(\rho \xrightarrow{\text{LOCC}_1} \Phi_m\right) = 1 - \sup_{\mathcal{N}} \mathrm{Tr}[\Phi_m\mathcal{N}(\rho)], \qquad (13.230)$$

where the supremum is over all $\mathcal{N} \in \mathrm{LOCC}_1(AB \to A'B')$. It is evident that since $\mathrm{LOCC}_1$ is a subset of LOCC, we have the following inequality for all $\rho \in \mathfrak{D}(AB)$:

$$T\left(\rho \xrightarrow{\text{LOCC}} \Phi_m\right) \leqslant T\left(\rho \xrightarrow{\text{LOCC}_1} \Phi_m\right). \qquad (13.231)$$

Given that one-way LOCC is significantly easier to characterize than LOCC, we can represent the conversion distance $T(\rho \xrightarrow{\text{LOCC}_1} \Phi_m)$ as an optimization problem over quantum instruments (see the following lemma). This simplification is advantageous because it reduces the complexity involved in the calculation and allows for a more straightforward analysis of the conversion distance. By focusing on one-way LOCC, we limit the operations to a sequence where one party, say Alice, performs a quantum operation and communicates the outcome classically to the other party (Bob), who then performs a quantum operation based on that information. This constraint narrows down the set of operations to be considered in the optimization problem, making the task of determining the conversion distance more manageable and conceptually clearer.

In the following lemma, we relate between the conversion distance under one-way LOCC and the optimized conditional min-entropy $H_{\min}^{\uparrow}$ as defined in (7.143). The relation will be given in terms of the function

$$Q_{\min}(A'|BX)_\tau := 2^{-H_{\min}^{\uparrow}(A'|BX)_\tau} \qquad \forall \tau \in \mathfrak{D}(A'BX), \tag{13.232}$$

with $X$ being a classical system and $A'$ and $B$ being quantum systems.

**Exercise 13.53.** *Let $\rho \in \mathfrak{D}(AB)$ and $\mathcal{E}^{A \to A'X} := \sum_{x \in [k]} \mathcal{E}_x^{A \to A'} \otimes |x\rangle\langle x|^X$ be a quantum instrument. Show that*

$$Q_{\min}(A'|BX)_{\mathcal{E}(\rho)} = \sum_{x \in [k]} Q_{\min}(A'|B)_{\mathcal{E}_x(\rho)}, \tag{13.233}$$

*where we extended the definition of $Q_{\min}$ to subnormalized states such that for any $\sigma \in \mathfrak{D}_{\leqslant}(A'B)$*

$$Q_{\min}(A'|B)_\sigma := 2^{-H_{\min}^{\uparrow}(A'|B)_\sigma}$$
$$:= \min\left\{ \mathrm{Tr}\left[\Lambda^B\right] : I^{A'} \otimes \Lambda^B \geqslant \sigma^{A'B}, \ \Lambda \in \mathrm{Pos}(B) \right\}. \tag{13.234}$$

> **Lemma 13.4.** Let $\rho \in \mathfrak{D}(AB)$, and $\Phi_m \in \mathfrak{D}(A'B')$ be the maximally entangled state $(m := |A'| = |B'|)$. Then,
>
> $$T\left(\rho \xrightarrow{\text{LOCC}_1} \Phi_m\right) = 1 - \frac{1}{m} \sup_{\mathcal{E} \in \mathrm{CPTP}(A \to A'X)} Q_{\min}(A'|BX)_{\mathcal{E}(\rho)}, \tag{13.235}$$
>
> where $Q_{\min}$ is defined in (13.232).

*Remark.* Note that $Q_{\min}(A'|BX)_{\mathcal{E}(\rho)}$ depends on $m$ as $m := |A'|$. Replacing CPTP $(A \to A'X)$ in (13.235) with $\mathrm{CPTP}(A \to AX)$, we obtain a lower bound (see Exercise 13.54):

$$T\left(\rho \xrightarrow{\text{LOCC}_1} \Phi_m\right) \geqslant 1 - \frac{1}{m} \sup_{\mathcal{E} \in \mathrm{CPTP}(A \to AX)} Q_{\min}(A|BX)_{\mathcal{E}(\rho)}. \tag{13.236}$$

**Proof** Every $\mathcal{N} \in \mathrm{LOCC}_1(AB \to A'B')$ can be expressed as

$$\mathcal{N}^{AB \to A'B'} = \mathcal{F}^{BX \to B'} \circ \mathcal{E}^{A \to A'X}, \tag{13.237}$$

where $X$ is a classical system, and $\mathcal{E}^{A \to A'X}$ and $\mathcal{F}^{BX \to B'}$ are quantum channels. The channel $\mathcal{E}^{A \to A'X} = \sum_{x \in [k]} \mathcal{E}_x^{A \to A'} \otimes |x\rangle\langle x|^X$ can be viewed as a quantum instrument with $k = |X|$, where we can assume without loss of generality that each

$\mathcal{E}_x^{A \to A'}(\,\cdot\,) = M_x(\,\cdot\,)M_x^*$ has a single Kraus operator $M_x$ by increasing the dimension of $X$. Thus, the optimization in (13.230) over all one-way LOCC channels, $\mathcal{N}^{AB \to A'B'}$, can be decomposed into optimizations over all $X$, all $\mathcal{E}^{A \to A'X}$, and all $\mathcal{F}^{A \to A'X}$. Next, we optimize over $\mathcal{F}^{A \to A'X}$, while keeping $X$ and $\mathcal{E}^{A \to A'X}$ fixed. Denoting by $\sigma^{A'BX} := \mathcal{E}^{A \to A'X}(\rho^{AB})$, we get:

$$\max_{\mathcal{F} \in \text{CPTP}} \text{Tr}\left[\Phi_m^{A'B'} \mathcal{F}^{BX \to B'} \circ \mathcal{E}^{A \to A'X}(\rho^{AB})\right] = \max_{\mathcal{F} \in \text{CPTP}} \text{Tr}\left[\Phi_m^{A'B'} \mathcal{F}^{BX \to B'}(\sigma^{A'BX})\right]$$

$$(7.148) \to = \frac{1}{m} 2^{-H_{\min}^{\uparrow}(A'|BX)_{\sigma}},$$

$$\tag{13.238}$$

where we used (7.148) with $\mathcal{F}^{BX \to B'}$ replacing $\mathcal{E}^{B \to \tilde{A}}$. Substituting this into (13.230) we get that the one-way LOCC conversion distance is given as in (13.235). $\blacksquare$

**Exercise 13.54.** *Prove* (13.236). *Hint: Use the property that any conditional entropy is invariant under local isometries (particularly on Alice's system).*

Building on Lemma (13.4) and Exercise 13.53, the conversion distance can be re-expressed as follows:

$$T\left(\rho \xrightarrow{\text{LOCC}_1} \Phi_m\right) = 1 - \frac{1}{m} \sup_{\mathcal{E} \in \text{CPTP}(A \to A'X)} \sum_{x \in [k]} Q_{\min}(A'|B)_{\mathcal{E}_x(\rho)}. \tag{13.239}$$

In Lemma 7.6, we established a connection between the optimized conditional min-entropy and the square fidelity. Specifically, let $E$ as be a purifying system and $\rho^{ABE}$ be a purification of $\rho^{AB}$. The marginal of the (subnormalized) *pure*-state $\mathcal{E}_x^{A \to A'}(\rho^{ABE})$ is denoted as $\mathcal{E}_x^{A \to A'}(\rho^{AE})$. Recall that we can assume each $\mathcal{E}_x^{A \to A'}$ in the supremum comprises a single Kraus operator. Hence, the relation in Lemma 7.6 implies:

$$Q_{\min}(A'|B)_{\mathcal{E}_x(\rho)} = \max_{\tau \in \mathfrak{D}(E)} F^2\left(I^{A'} \otimes \tau^E, \mathcal{E}_x^{A \to A'}(\rho^{AE})\right). \tag{13.240}$$

Combining this with the relation $P^2 = 1 - F^2$ between the purified distance and the fidelity, we conclude that (13.235) can also be rewritten as

$$T\left(\rho^{AB} \xrightarrow{\text{LOCC}_1} \Phi_m\right) = \inf_{\{\mathcal{E}_x\}} \sum_{x \in [k]} \min_{\tau \in \mathfrak{D}(E)} P^2\left(u^{A'} \otimes \tau^E, \mathcal{E}_x^{A \to A'}(\rho^{AE})\right), \tag{13.241}$$

where the infimum is over all $k \in \mathbb{N}$ and all quantum instruments $\{\mathcal{E}_x^{A \to A'}\}_{x \in [k]}$. We will use this form of the conversion distance to get the following upper bound.

## Upper Bound

The main result of this subsection is the following upper bound on the right-hand side of (13.241).

**Upper Bound**

> **Theorem 13.13.** Let $\rho \in \mathrm{Pure}(ABE)$ and $m \in \mathbb{N}$. Then,
>
> $$T\left(\rho^{AB} \xrightarrow{\mathrm{LOCC_1}} \Phi_m\right) \leqslant \sqrt{m}\, 2^{-\frac{1}{2}\tilde{H}_2^{\uparrow}(A|E)_\rho}, \tag{13.242}$$
>
> where $\tilde{H}_2^{\uparrow}(A|E)_\rho := -\min_{\omega \in \mathfrak{D}(E)} \tilde{D}_2\left(\rho^{AE} \| I^A \otimes \rho^E\right)$ is the optimized conditional entropy as defined with respect to the quantum Sandwich divergence of order $\alpha = 2$.

*Remark.* When $m \geqslant |A|$, the upper bound in (13.242) is trivial, since in this case

$$\sqrt{m}\, 2^{-\frac{1}{2}\tilde{H}_2^{\uparrow}(A|E)_\rho} = 2^{\frac{1}{2}\left(\log(m) - \tilde{H}_2^{\uparrow}(A|E)_\rho\right)} \geqslant 1, \tag{13.243}$$

where we used the fact that $\tilde{H}_2^{\uparrow}(A|E)_\rho \leqslant \log|A| \leqslant \log(m)$. However, as we will soon see, this upper bound is very useful when $|A| > m$. Specifically, we will use it to derive a tight lower bound on the distillable entanglement.

**Proof** We get the upper bound on the conversion distance in two stages. First, by taking $\tau^E = \rho^E$ in (13.241) we get

$$T\left(\rho^{AB} \xrightarrow{\mathrm{LOCC_1}} \Phi_m^{A'B'}\right) \leqslant \inf_{\{\mathcal{E}_x\}} \sum_{x \in [k]} P^2\left(\mathbf{u}^{A'} \otimes \rho^E, \mathcal{E}_x^{A \to A'}(\rho^{AE})\right). \tag{13.244}$$

Second, we replace the maximization above over all quantum instruments $\{\mathcal{E}_x\}_{x \in [k]}$ with a specific choice of a quantum instrument to get a simpler upper bound. We will denote by $n := |A|$ and assume that $m \leqslant n$ (see the remark above).

Observe that the expression in (13.244) has a form that is somewhat similar to the decoupling theorem studied in Section 7.7. Therefore, our strategy is to choose $\{\mathcal{E}_x\}_{x \in [k]}$ in such a way that we will be able to use the upper bound given in the decoupling theorem. For this purpose, recall that the twirling operation $\mathcal{G} \in \mathrm{CPTP}(A\tilde{A} \to A\tilde{A})$ as defined in (7.199) can be expressed as a finite convex combination of product unitary channels as given in (7.211). With these $k \in \mathbb{N}$, $\mathbf{p} \in \mathrm{Prob}(k)$, and $\{U_x\}_{x \in [k]} \subset \mathfrak{U}(A)$, we define

$$\mathcal{E}_x^{A \to A'} := p_x\, \mathcal{N}^{A \to A'} \circ \mathcal{U}_x^{A \to A}, \tag{13.245}$$

where $\mathcal{U}_x(\cdot) = U_x(\cdot)U_x^*$, and $\mathcal{N}^{A \to A'}(\cdot) := \frac{n}{m} V^*(\cdot)V$, where $V : A' \to A$ is some isometry.

We now discuss the properties of the set $\{\mathcal{E}_x^{A \to A'}\}_{x \in [k]}$. First, observe that by definition, the channel

$$\mathcal{R}^{A \to A} := \sum_{x \in [k]} p_x\, \mathcal{U}_x^{A \to A} \tag{13.246}$$

correspond to the completely randomizing channel that outputs the maximally mixed state irrespective on the input state. This follows from the fact that both (7.199) and (7.211) correspond to the same twirling channel, so their marginal channels are also the same (see (3.242)). With this at hand, we get that

$$\mathcal{E}^{A \to A'} := \sum_{x \in [k]} \mathcal{E}_x^{A \to A'} = \mathcal{N}^{A \to A'} \circ \mathcal{R}^{A \to A}. \tag{13.247}$$

Next, we argue that $\mathcal{E}^{A \to A'}$ is trace preserving so that $\{\mathcal{E}_x^{A \to A'}\}_{x \in [k]}$ as already defined is indeed a quantum instrument. To see it, observe that for all $\omega \in \mathcal{L}(A)$ we have

$$\mathrm{Tr}\left[\mathcal{E}^{A \to A'}(\omega^A)\right] = \mathrm{Tr}\left[\mathcal{N}^{A \to A'} \circ \mathcal{R}^{A \to A}\left(\omega^A\right)\right]$$

$$\boxed{\mathcal{N}(\mathbf{u}^A) = \frac{1}{n}\mathcal{N}(I^A) = \frac{1}{m}VV^*} \longrightarrow = \mathrm{Tr}\left[\mathcal{N}^{A \to A'}\left(\mathrm{Tr}[\omega^A]\mathbf{u}^A\right)\right]$$

$$= \mathrm{Tr}[\omega^A]\frac{1}{m}\mathrm{Tr}[VV^*]$$

$$= \mathrm{Tr}[\omega^A], \tag{13.248}$$

where in the last line we used the fact that $VV^*$ is a projection of rank $m = |A'|$ since $V : A' \to A$ is an isometry.

Therefore, with this choice of quantum instrument, (13.244) becomes

$$T\left(\rho^{AB} \xrightarrow{\text{LOCC}_1} \Phi_m^{A'B'}\right) \le \sum_{x \in [k]} p_x P^2\left(\mathbf{u}^{A'} \otimes \rho^E, \mathcal{N}^{A \to A'} \circ \mathcal{U}_x^{A \to A}(\rho^{AE})\right)$$

$$\boxed{P^2(\rho,\sigma) \le \|\rho - \sigma\|_1} \longrightarrow \le \sum_{x \in [k]} p_x \left\|\mathcal{N}^{A \to A'}\left(U_x^A \rho^{AE}\left(U_x^A\right)^*\right) - \mathbf{u}^{A'} \otimes \rho^E\right\|_1, \tag{13.249}$$

where the last line follows from (5.202). Finally, to apply the decoupling theorem as given in (7.221), we define

$$\tau^{AA'} := \frac{1}{n}J_{\mathcal{N}}^{AA'} = \frac{1}{n}\mathcal{N}^{\tilde{A} \to A'}\left(\Omega^{A\tilde{A}}\right) = \frac{1}{m}\left(I^A \otimes V\right)^*\Omega^{A\tilde{A}}\left(I^A \otimes V\right). \tag{13.250}$$

Note in particular that the marginal $\tau^{A'} = \mathbf{u}^{A'}$ so that the right-hand side of (13.249) has the exact same form as given on the left-hand side of (7.221) (in the decoupling theorem). Hence, we can apply the decoupling theorem to get

$$T\left(\rho^{AB} \xrightarrow{\text{LOCC}_1} \Phi_m^{A'B'}\right) \le 2^{-\frac{1}{2}\left(\tilde{H}_2^{\uparrow}(A|E)_\rho + \tilde{H}_2^{\uparrow}(A|A')_\tau\right)}. \tag{13.251}$$

Moreover, since $\tau^{AA'}$ is maximally entangled we get that

$$\tilde{H}_2^{\uparrow}(A|A')_\tau = -\log(m). \tag{13.252}$$

Substituting this into the previous equation gives (13.242). This completes the proof. ∎

## 13.3.2 Conversion Distance from a Maximally Entangled State

The conversion distance in (13.223) is defined with respect to the trace distance. Since all metrics in finite dimensional Hilbert spaces are topologically equivalent, we can always select a different metric to simplify computations. In this section, we find that the square of the purified distance is a more convenient measure to work with. Recall from Lemma 13.3 that the $T$-conversion distance (measured with respect to the trace distance $T$) is equivalent to the $P^2$-conversion distance when computing the conversion distances to a maximally entangled state. However, it is important to note that this equivalence does not hold when the source/input state is maximally entangled.

The $P^2$-conversion distance from $\sigma \in \mathfrak{D}(A'B')$ to $\rho \in \mathfrak{D}(AB)$ is defined as

$$P^2\left(\sigma^{A'B'} \xrightarrow{\text{LOCC}} \rho^{AB}\right) := \max_{\mathcal{E} \in \text{LOCC}} P^2\left(\mathcal{E}^{A'B' \to AB}\left(\sigma^{A'B'}\right), \rho^{AB}\right). \tag{13.253}$$

In what follows, we take the source state $\sigma^{A'B'}$ to be the maximally entangled state $\Phi_m^{A'B'}$.

**Exercise 13.55.** *Show that for any $\rho \in \mathfrak{D}(AB)$ and any $\sigma \in \mathfrak{D}(A'B')$ we have*

$$T^2\left(\sigma \xrightarrow{\text{LOCC}} \rho\right) \leqslant P^2\left(\sigma \xrightarrow{\text{LOCC}} \rho\right) \leqslant 2T\left(\sigma \xrightarrow{\text{LOCC}} \rho\right). \tag{13.254}$$

*Hint: See Theorem 5.22.*

---

**Theorem 13.14.** Let $\rho \in \mathfrak{D}(AB)$ and $m \in \mathbb{N}$. Then,

$$P^2\left(\Phi_m \xrightarrow{\text{LOCC}} \rho^{AB}\right) = E_{(m)}\left(\rho^{AB}\right), \tag{13.255}$$

where $E_{(m)}\left(\rho^{AB}\right)$ is the entanglement monotone defined in (13.110) with $k = m$.

---

**Proof** Recall first that for the case that $\rho^{AB}$ is a pure pure state, the theorem follows from Corollary 12.4. We therefore need to generalize this result to the case that $\rho^{AB}$ is a mixed state. We start by showing that

$$\left\{\mathcal{E}\left(\Phi_m\right) : \mathcal{E} \in \text{LOCC}(A'B' \to AB)\right\} = \left\{\omega \in \mathfrak{D}(AB) : \text{SR}\left(\omega^{AB}\right) \leqslant m\right\}, \tag{13.256}$$

where $\text{SR}\left(\omega^{AB}\right)$ is the Schmidt rank as defined in (13.120). Indeed, since the Schmidt rank $\text{SR}$ as defined in (13.120) is a measure of entanglement, it follows that for $\omega^{AB} = \mathcal{E}\left(\Phi_m\right)$

$$\text{SR}\left(\omega^{AB}\right) \leqslant \text{SR}\left(\Phi_m\right) = m. \tag{13.257}$$

Therefore, the left-hand side of (13.256) is contained in the right-hand side. On the other hand, from Exercise 13.26 it follows that every state $\omega^{AB}$ with Schmidt rank no greater than $m$ has a pure-state decomposition with all states having Schmidt rank no greater than $m$. As a consequence of Nielsen's theorem, such a state $\omega^{AB}$ can be generated by LOCC from $\Phi_m$. That is, the right-hand side of (13.256) is contained in the left-hand side. This completes the proof of the equality in (13.256).

Let $E$ be a purifying system of dimension $n := |E| \leqslant |AB|$. From the equivalency of the two sets in (13.256) we get

$$\max_{\mathcal{E} \in \mathrm{LOCC}(A'B' \to AB)} F^2\left(\mathcal{E}(\Phi_m), \rho^{AB}\right) = \max_{\omega \in \mathfrak{D}(AB),\, \mathrm{SR}(\omega) \leqslant m} F^2\left(\omega^{AB}, \rho^{AB}\right)$$

$$\mathbf{Uhlmann's\ theorem} \to\ =\ \max_{\substack{\psi,\phi \in \mathrm{Pure}(ABE) \\ \mathrm{SR}(\phi^{AB}) \leqslant m,\, \psi^{AB} = \rho^{AB}}} \left|\langle \phi^{ABE} | \psi^{ABE}\rangle\right|^2.$$

$$(13.258)$$

Let $\{|x\rangle^E\}_{x \in [n]}$ be a fixed orthonormal basis of the purifying system $E$, and observe that every purification $\psi^{ABE}$ of $\rho^{AB}$ can be expressed as

$$|\psi^{ABE}\rangle := \sum_{x \in [n]} \sqrt{p_x} |\psi_x^{AB}\rangle |x\rangle^E, \tag{13.259}$$

where $\{p_x, \psi_x^{AB}\}_{x \in [n]}$ is a pure-state decomposition of $\rho^{AB}$. Specifically, there is a one-to-one correspondence between all purifications $\psi^{ABE}$ of $\rho^{AB}$ that have the form (13.259), and all pure-state decompositions $\{p_x, \psi_x^{AB}\}_{x \in [n]}$ of $\rho^{AB} = \sum_{x \in [n]} p_x \psi_x^{AB}$.

Similarly, for every $\omega^{AB}$ with Schmidt rank $\mathrm{SR}(\omega^{AB}) \leqslant m$ let

$$|\phi^{ABE}\rangle := \sum_{x \in [n]} \sqrt{q_x} |\phi_x\rangle^{AB} |x\rangle^E \tag{13.260}$$

be a purification of $\omega^{AB}$ with the property that each $|\phi_x\rangle^{AB}$ has a Schmidt rank no greater than $m$ (see Exercise 13.26). With this at hand, we get from (13.258) that

$$\max_{\mathcal{E} \in \mathrm{LOCC}(A'B' \to AB)} F^2\left(\mathcal{E}(\Phi_m), \rho^{AB}\right) = \max \left| \sum_{x \in [n]} \sqrt{q_x p_x} \langle \phi_x^{AB} | \psi_x^{AB}\rangle \right|^2, \tag{13.261}$$

where the maximum on the right-hand side is over all pure-state decompositions of $\rho^{AB} = \sum_x p_x \psi_x^{AB}$, all probability vectors $\mathbf{q} \in \mathrm{Prob}(n)$, and all pure states $\{\phi_x^{AB}\}_{x \in [n]}$ with Schmidt rank no greater than $m$. Now, from Corollary 12.4 it follows that (see Exercise 13.56) for every $\psi \in \mathrm{Pure}(AB)$

$$\max_{\substack{\phi \in \mathrm{Pure}(AB) \\ \mathrm{SR}(\phi) \leqslant m}} \left|\langle \phi^{AB} | \psi^{AB}\rangle\right|^2 = \|\psi^A\|_{(m)}. \tag{13.262}$$

Therefore, there exists $\{\phi_x^{AB}\}_{x \in [n]}$ such that

$$\langle \phi_x^{AB} | \psi_x^{AB}\rangle = \left|\langle \phi_x^{AB} | \psi_x^{AB}\rangle\right| = \sqrt{\|\psi_x^A\|_{(m)}}, \tag{13.263}$$

and that $\left|\langle \phi_x^{AB} | \psi_x^{AB}\rangle\right|$ cannot exceed this value. We therefore conclude that

$$\max_{\mathcal{E} \in \mathrm{LOCC}(A'B' \to AB)} F^2\left(\mathcal{E}(\Phi_m), \rho^{AB}\right) = \max_{\{p_x, \psi_x\}} \max_{\mathbf{q} \in \mathrm{Prob}(n)} \left( \sum_{x \in [n]} \sqrt{q_x p_x \|\psi_x^A\|_{(m)}} \right)^2$$

$$\mathbf{Exercise\ 12.21} \to\ =\ \max_{\{p_x, \psi_x\}} \sum_{x \in [n]} p_x \left\|\psi_x^A\right\|_{(m)},$$

$$(13.264)$$

where the maximum on the right-hand side stands for a maximum over all pure-state decompositions $\{p_x, \psi_x^{AB}\}_{x \in [n]}$ of $\rho^{AB} = \sum_{x \in [n]} p_x \psi_x^{AB}$. In terms of the purified distance we have

$$\max_{\mathcal{E} \in \mathrm{LOCC}(A'B' \to AB)} P^2 \left( \mathcal{E}(\Phi_m), \rho^{AB} \right) = 1 - \max \sum_{x \in [n]} p_x \left\| \psi_x^A \right\|_{(m)}$$

$$= \min \sum_{x \in [n]} p_x \left( 1 - \left\| \psi_x^A \right\|_{(m)} \right) \tag{13.265}$$

$$= E_{(m)} \left( \rho^{AB} \right),$$

where both the min and max above are over all pure-state decompositions $\{p_x, \psi_x^{AB}\}_{x \in [n]}$ of $\rho^{AB}$. This completes the proof. ∎

**Exercise 13.56.** *Use Corollary 12.4 to prove the equality in (13.262).*

## 13.4 Single-Shot Distillable Entanglement

The $\varepsilon$-single-shot distillable entanglement is then defined as (cf. (11.32))

$$\mathrm{Distill}^{\varepsilon} \left( \rho^{AB} \right) := \max \left\{ \log m \; : \; T \left( \rho^{AB} \xrightarrow{\mathrm{LOCC}} \Phi_m \right) \leqslant \varepsilon \right\}. \tag{13.266}$$

Since the computation of the distillable entanglement is hard, we start with the single-shot one-way distillable entanglement. As $\mathrm{LOCC}_1$ is a subset of LOCC, any lower bound on the one-way distillable entanglement will automatically provide a lower bound on the distillable entanglement already defined.

**Exercise 13.57.** *Let $k := |A| = |B|$, $m = |A'| = |B'|$, and $\rho \in \mathfrak{D}(AB)$. Show that*

$$\mathrm{Distill}^{\varepsilon} \left( \rho^{AB} \right) \leqslant \log(k) - \log(1 - \varepsilon). \tag{13.267}$$

### 13.4.1 One-Way Single-Shot Distillable Entanglement

Similar to (13.266), we define the $\varepsilon$-single-shot distillable entanglement under one-way LOCC as

$$\mathrm{Distill}^{\varepsilon}_{\to} \left( \rho^{AB} \right) := \max \left\{ \log m \; : \; T \left( \rho^{AB} \xrightarrow{\mathrm{LOCC}_1} \Phi_m^{A'B'} \right) \leqslant \varepsilon \right\}. \tag{13.268}$$

A simple formula for the above expression is not presently available. However, we can provide some useful lower and upper bounds.

In the following theorem, we present an upper bound on the single-shot one-way distillable entanglement. It is worth noting that the upper bound given in (11.38) will not be helpful in this case, as we are considering a subset of LOCC. Therefore, we can expect to obtain a tighter upper bound, particularly since the upper bound given in (11.38) remains valid even if we replace LOCC with nonentangling operations.

The upper bound presented in the following theorem is expressed in terms of the coherent information of entanglement, denoted as $E_\rightarrow$. This particular measure of entanglement has been defined and extensively examined in Section 13.2.6.

---

**Theorem 13.15.** Let $\rho \in \mathfrak{D}(AB)$ and $\varepsilon \in (0, 1/2)$. Then, the one-way $\varepsilon$-single-shot distillable entanglement is bounded by

$$\mathrm{Distill}^\varepsilon_\rightarrow \left(\rho^{AB}\right) \leqslant \frac{1}{1-2\varepsilon} E_\rightarrow \left(\rho^{AB}\right) + \frac{1+\varepsilon}{1-2\varepsilon} h\left(\frac{\varepsilon}{1+\varepsilon}\right), \qquad (13.269)$$

where $h(x) := -x \log x - (1-x) \log(1-x)$ is the binary Shannon entropy.

---

**Proof** Let $m \in \mathbb{N}$ be such that $\mathrm{Distill}^\varepsilon_\rightarrow \left(\rho^{AB}\right) = \log m$ so that $T\left(\rho^{AB} \xrightarrow{\mathrm{LOCC}_1} \Phi_m^{A'B'}\right) \leqslant \varepsilon$.
This means that $\rho^{AB} \xrightarrow{\mathrm{LOCC}_1} \sigma^{A'B'}$ for some state $\sigma \in \mathfrak{D}(A'B')$ that is $\varepsilon$-close to $\Phi_m^{A'B'}$.
Therefore, from the monotonicity of $E_\rightarrow$ under one-way LOCC we get that

$$E_\rightarrow \left(\rho^{AB}\right) \geqslant E_\rightarrow \left(\sigma^{A'B'}\right). \qquad (13.270)$$

Next, we use the fact that $\sigma^{A'B'}$ is $\varepsilon$-close the $\Phi_m^{A'B'}$ to show that the right-hand side of the equation (13.270) cannot be much smaller than $\log(m)$. Indeed, combining the continuity property of the function $I(A' \rangle B')_\rho := -H(A'|B')_\rho$ (see (10.50)), with the second part of Exercise 13.48, gives

$$E_\rightarrow \left(\sigma^{A'B'}\right) \geqslant I(A' \rangle B')_\sigma$$

$$(10.50) \rightarrow \; \geqslant I(A' \rangle B')_{\Phi_m} - 2\varepsilon \log m - (1+\varepsilon)h\left(\frac{\varepsilon}{1+\varepsilon}\right) \qquad (13.271)$$

$$= (1-2\varepsilon) \log(m) - (1+\varepsilon)h\left(\frac{\varepsilon}{1+\varepsilon}\right).$$

The proof is concluded by noting that the inequality above in conjunction with the inequality (13.270) yields the desired inequality (13.269). ∎

**Exercise 13.58.** *Use similar lines as in the proof above to prove the following bound on the $\varepsilon$-shot distillable entanglement*

$$\mathrm{Distill}^\varepsilon \left(\rho^{AB}\right) \leqslant \frac{1}{1-2\varepsilon} \sup_{\mathcal{E} \in \mathrm{LOCC}(AB \to A'B')} I(A' \rangle B')_{\mathcal{E}(\rho)} + \frac{1+\varepsilon}{1-2\varepsilon} h\left(\frac{\varepsilon}{1+\varepsilon}\right),$$
$$(13.272)$$

*where the supremum is also over all systems $A'$ and $B'$.*

In the context where the single-shot distillable entanglement (and the entanglement cost) is expressed as $\log(m)$ for some integer $m \in \mathbb{N}$, it is useful to introduce a specific notation, $\gtrsim$, to denote a particular type of inequality between two real numbers $a, b \in \mathbb{R}$. This notation is defined as follows:

$$a \gtrsim b \quad \Longleftrightarrow \quad a \geqslant \log\left\lfloor 2^b \right\rfloor. \qquad (13.273)$$

This definition provides a convenient way to express inequalities that are relevant in the quantification of entanglement, especially in scenarios involving logarithmic expressions and integer values. With this in mind, our next goal is a lower bound on the $\text{Distill}^{\varepsilon}_{\rightarrow}\left(\rho^{AB}\right)$. The lower bound given below is known as the single-shot *hashing bound*.

**Theorem 13.16.** Let $\rho \in \text{Pure}(ABE)$ and $\varepsilon \in (0,1)$. Then, for every $0 < \delta < \varepsilon$ we have

$$\text{Distill}^{\varepsilon}_{\rightarrow}\left(\rho^{AB}\right) \gtrsim H^{\delta}_{\min}(A|E)_{\rho} + \log(\varepsilon - \delta)^2. \tag{13.274}$$

*Remark.* Observe that unlike the upper bound, which is given in terms of the coherent information of entanglement, the lower bound does not involve an optimization over channels in $\text{CPTP}(A \to AX)$.

**Proof** The main strategy of the proof is to use the upper bound given in Theorem 13.13 for the one-way convcersion distance. Specifically, let $\delta \in (0,1)$ and let $\tilde{\rho} \in \mathfrak{B}_{\delta}(\rho^{AE})$ be such that $H^{\uparrow\delta}_{\min}(A|E)_{\rho} = H^{\uparrow}_{\min}(A|E)_{\tilde{\rho}}$. With these definitions we get

$$\sqrt{m}2^{-\frac{1}{2}H^{\uparrow\delta}_{\min}(A|E)_{\rho}} = \sqrt{m}2^{-\frac{1}{2}H^{\uparrow}_{\min}(A|E)_{\tilde{\rho}}}$$

$$\boxed{H^{\uparrow}_{\min} \leqslant H^{\uparrow}_2} \longrightarrow \;\geqslant \sqrt{m}2^{-\frac{1}{2}\tilde{H}^{\uparrow}_2(A|E)_{\tilde{\rho}}}$$

$$(\textbf{13.242}) \rightarrow \; \geqslant T\left(\tilde{\rho}^{AB} \xrightarrow{\text{LOCC}_1} \Phi^{A'B'}_m\right) \tag{13.275}$$

$$\textbf{Lemma 11.2} \rightarrow \; \geqslant T\left(\rho^{AB} \xrightarrow{\text{LOCC}_1} \Phi^{A'B'}_m\right) - \delta.$$

That is,

$$T\left(\rho^{AB} \xrightarrow{\text{LOCC}_1} \Phi^{A'B'}_m\right) \leqslant \sqrt{m}2^{-\frac{1}{2}H^{\uparrow\delta}_{\min}(A|E)_{\rho}} + \delta. \tag{13.276}$$

Therefore, for any $\varepsilon \in (0,1)$ and $0 < \delta < \varepsilon$ we get that the one-way $\varepsilon$-distillable entanglement satisfies

$$\begin{aligned}
\text{Distill}^{\varepsilon}_{\rightarrow}\left(\rho^{AB}\right) &:= \max\left\{\log m : T\left(\rho^{AB} \xrightarrow{\text{LOCC}_1} \Phi^{A'B'}_m\right) \leqslant \varepsilon\right\} \\
(\textbf{13.276}) \rightarrow &\geqslant \max\left\{\log m : \sqrt{m}2^{-\frac{1}{2}H^{\delta}_{\min}(A|E)_{\rho}} + \delta \leqslant \varepsilon\right\} \\
&= \max\left\{\log m : \log m \leqslant H^{\delta}_{\min}(A|E)_{\rho} + \log(\varepsilon - \delta)^2\right\} \\
&= \log\left\lfloor (\varepsilon - \delta)^2 2^{H^{\delta}_{\min}(A|E)_{\rho}}\right\rfloor.
\end{aligned} \tag{13.277}$$

This completes the proof. ∎

**Exercise 13.59.** *Show that for any $\rho \in \text{Pure}(ABE)$ and $\varepsilon \in (0,1)$ we have*

$$\text{Distill}^{\varepsilon}_{\rightarrow}\left(\rho^{AB}\right) \gtrsim \tilde{H}^{\uparrow}_2(A|E)_{\rho} + 2\log\varepsilon. \tag{13.278}$$

# 13.5 Asymptotic Distillable Entanglement

The asymptotic distillable entanglement is defined for any $\rho \in \mathfrak{D}(AB)$ as (cf. (11.119) and (11.135))

$$\text{Distill}\left(\rho^{AB}\right) := \lim_{\varepsilon \to 0^+} \sup_{n,m \in \mathbb{N}} \left\{ \frac{m}{n} : T\left(\rho^{\otimes n} \xrightarrow{\text{LOCC}} \Phi_2^{\otimes m}\right) \leqslant \varepsilon \right\}. \tag{13.279}$$

From (11.138) it follows that the asymptotic distillable entanglement can be expressed as

$$\text{Distill}\left(\rho^{AB}\right) = \lim_{\varepsilon \to 0^+} \limsup_{n \to \infty} \frac{1}{n} \text{Distill}^{\varepsilon}\left(\rho^{\otimes n}\right). \tag{13.280}$$

Similarly, the one-way $\varepsilon$-distillable entanglement is given by

$$\text{Distill}_{\to}\left(\rho^{AB}\right) = \lim_{\varepsilon \to 0^+} \limsup_{n \to \infty} \frac{1}{n} \text{Distill}_{\to}^{\varepsilon}\left(\rho^{\otimes n}\right). \tag{13.281}$$

**Exercise 13.60.** *Show that for any $n \in \mathbb{N}$ and any $\rho \in \mathfrak{D}(AB)$ we have*

$$\text{Distill}\,(\rho) \geqslant \frac{1}{n} \text{Distill}\left(\rho^{\otimes n}\right) \quad \text{and} \quad \text{Distill}_{\to}\,(\rho) \geqslant \frac{1}{n} \text{Distill}_{\to}\left(\rho^{\otimes n}\right). \tag{13.282}$$

## 13.5.1 Simple Lower Bound on the Distillable Entanglement

Our analysis begins with a simple lower bound.

**The Hashing Bound**

**Theorem 13.17.** Let $\rho \in \mathfrak{D}(AB)$. Then,

$$\text{Distill}_{\to}\left(\rho^{AB}\right) \geqslant I(A\rangle B)_\rho. \tag{13.283}$$

*Remark.* Since the distillable entanglement is always no smaller than the one-way distillable entanglement, we also have

$$\text{Distill}\left(\rho^{AB}\right) \geqslant I(A\rangle B)_\rho. \tag{13.284}$$

**Proof** Let $\rho^{ABE} \in \text{Pure}(ABE)$ be a purification of $\rho^{AB}$, and let $\varepsilon, \delta \in (0, 1)$ be such that $\delta < \varepsilon$. From the lower bound in (13.274) we get

$$\liminf_{n \to \infty} \frac{1}{n} \text{Distill}_{\to}^{\varepsilon}\left(\rho^{\otimes n}\right) \geqslant \liminf_{n \to \infty} \frac{1}{n}\left(H_{\min}^{\delta}(A^n|E^n)_{\rho^{\otimes n}} + 2\log\left(\varepsilon - \delta\right)\right)$$

$$= \liminf_{n \to \infty} \frac{1}{n} H_{\min}^{\delta}(A^n|E^n)_{\rho^{\otimes n}} \tag{13.285}$$

$$\textbf{Theorem 11.3} \to \; = H(A|E)_\rho$$

$$\textbf{Duality relation (7.158)} \to \; = -H(A|B)_\rho = I(A\rangle B)_\rho.$$

Since the equation above holds for all $\varepsilon \in (0, 1)$, it also holds if we take the limit $\varepsilon \to 0^+$. This completes the proof. ∎

It is worth noting that the hashing bound reveals that the distillable entanglement is nonzero whenever the conditional entropy of $\rho^{AB}$ is negative. As already discussed, the conditional entropy can only be negative for entangled states, which aligns with the fact that only entangled states can possess nonzero distillable entanglement. However, in the upcoming sections, we will discover that the converse statement is not true. Specifically, there exist entangled states with zero distillable entanglement.

## 13.5.2 One-Way Distillable Entanglement

Although the hashing bound is a useful and straightforward lower bound on the distillable entanglement, its tightness in general is not very clear. Interestingly, on pure states, the hashing bound is *equal* to the distillable entanglement. This is because, for any pure state $\psi \in \mathrm{Pure}(AB)$, the coherent information is given by

$$I(A\rangle B)_\psi = H(A)_\psi, \tag{13.286}$$

which is equal to the entropy of entanglement of the pure state $\psi^{AB}$. For mixed states, we have the following expression for the one-way distillable entanglement.

> **Theorem 13.18.** Let $\rho \in \mathfrak{D}(AB)$. Then, the one-way distillable entanglement of $\rho^{AB}$ is given by
>
> $$\mathrm{Distill}_\to \left(\rho^{AB}\right) = E_\to^{\mathrm{reg}}\left(\rho^{AB}\right), \tag{13.287}$$
>
> where $E_\to^{\mathrm{reg}}$ is the regularized coherent information of entanglement as defined in (13.222).

*Remark.* It is worth noting that the theorem provides an operational interpretation for the coherent information of entanglement as the one-way distillable entanglement.

**Proof** For the direct part (i.e. achievability), observe that any quantum instrument $\mathcal{E} \in \mathrm{CPTP}(A \to AX)$ can be considered as a special type of $\mathrm{LOCC}_1$ (i.e. Alice implies the instrument $\{\mathcal{E}_x\}_{x \in [m]}$ on her system and sends the outcome $x$ to Bob). Since the single-shot one-way distillable entanglement behaves monotonically under such $\mathrm{LOCC}_1$, we get that for all $\varepsilon \in (0, 1)$

$$\mathrm{Distill}_\to^\varepsilon \left(\rho^{AB}\right) \geqslant \mathrm{Distill}_\to^\varepsilon \left(\sigma^{ABX}\right), \tag{13.288}$$

where $\sigma^{ABX} := \mathcal{E}^{A \to AX}\left(\rho^{AB}\right)$. Since for every $n \in \mathbb{N}$ the equation above also holds with $n$ copies of $\rho$ and $\sigma$, we get

$$\liminf_{n \to \infty} \frac{1}{n} \mathrm{Distill}_\to^\varepsilon \left(\rho^{\otimes n}\right) \geqslant \liminf_{n \to \infty} \frac{1}{n} \mathrm{Distill}_\to^\varepsilon \left(\sigma^{\otimes n}\right) \tag{13.289}$$

$$\overset{(\mathbf{13.285})}{\to} \geqslant I(A\rangle BX)_\sigma.$$

Since this inequality holds for all $\mathcal{E} \in \mathrm{CPTP}(A \to AX)$, we conclude that

$$\liminf_{n\to\infty} \frac{1}{n} \mathrm{Distill}^{\mathcal{E}}_{\to} \left( \rho^{\otimes n} \right) \geq E_{\to} \left( \rho^{AB} \right). \tag{13.290}$$

We would like to replace the right-hand side with the regularized version of $E_{\to}$. For this purpose, fix $k \in \mathbb{N}$ and observe that by applying this inequality for $\rho^{\otimes k} \in \mathfrak{D}\left( A^k B^k \right)$ we get

$$\liminf_{n\to\infty} \frac{1}{kn} \mathrm{Distill}^{\mathcal{E}}_{\to} \left( \rho^{\otimes kn} \right) \geq \frac{1}{k} E_{\to} \left( \rho^{\otimes k} \right). \tag{13.291}$$

We next show that the left-hand side of the two equations above coincide. Indeed, by definition of the $\liminf$, the left-hand side of (13.290) is no greater than the left-hand side of (13.291). For the converse, let $\{n_j\}_{j\in\mathbb{N}}$ be a subsequence of integers such that

$$\liminf_{n\to\infty} \frac{1}{n} \mathrm{Distill}^{\mathcal{E}}_{\to} \left( \rho^{\otimes n} \right) = \lim_{j\to\infty} \frac{1}{n_j} \mathrm{Distill}^{\mathcal{E}}_{\to} \left( \rho^{\otimes n_j} \right). \tag{13.292}$$

Now, for any $j \in \mathbb{N}$, set $m_j := k \left\lfloor \frac{n_j}{k} \right\rfloor$; that is, $m_j$ is the largest multiple of $k$ that is no greater than $n_j$. In particular, note that $n_j - k < m_j \leq n_j$. Then,

$$
\begin{aligned}
\liminf_{n\to\infty} \frac{1}{kn} \mathrm{Distill}^{\mathcal{E}}_{\to} \left( \rho^{\otimes kn} \right) &\leq \liminf_{j\to\infty} \frac{1}{m_j} \mathrm{Distill}^{\mathcal{E}}_{\to} \left( \rho^{\otimes m_j} \right) \\
\boxed{m_j \leq n_j} \longrightarrow &\leq \liminf_{j\to\infty} \frac{1}{m_j} \mathrm{Distill}^{\mathcal{E}}_{\to} \left( \rho^{\otimes n_j} \right) \\
\boxed{n_j - k < m_j \leq n_j} \longrightarrow &= \lim_{j\to\infty} \frac{1}{n_j} \mathrm{Distill}^{\mathcal{E}}_{\to} \left( \rho^{\otimes n_j} \right) \\
\boxed{(13.292)} \longrightarrow &= \liminf_{n\to\infty} \frac{1}{n} \mathrm{Distill}^{\mathcal{E}}_{\to} \left( \rho^{\otimes n} \right),
\end{aligned}
\tag{13.293}
$$

where the first inequality follows from the fact that $\{m_j\}_{j\in\mathbb{N}}$ is a subset of $\{kn\}_{n\in\mathbb{N}}$, the second inequality from the fact that $\mathrm{Distill}^{\mathcal{E}}_{\to} \left( \rho^{\otimes m_j} \right) \geq \mathrm{Distill}^{\mathcal{E}}_{\to} \left( \rho^{\otimes n_j} \right)$ as $m_j \geq n_j$, and the third in equality from the fact that $\frac{1}{m_j} = \frac{n_j}{m_j} \frac{1}{n_j}$ and $\lim_{j\to\infty} n_j / m_j = 1$ (since $n_j - k < m_j \leq n_j$). This completes the proof that the left-hand side of (13.290) equals the left-hand side of (13.291) so that

$$\liminf_{n\to\infty} \frac{1}{n} \mathrm{Distill}^{\mathcal{E}}_{\to} \left( \rho^{\otimes n} \right) \geq \frac{1}{k} E_{\to} \left( \rho^{\otimes k} \right). \tag{13.294}$$

Since this equation holds for all $k \in \mathbb{N}$, it must also hold for the limit $k \to \infty$; that is, we conclude that

$$\liminf_{n\to\infty} \frac{1}{n} \mathrm{Distill}^{\mathcal{E}}_{\to} \left( \rho^{\otimes n} \right) \geq E^{\mathrm{reg}}_{\to} \left( \rho^{AB} \right). \tag{13.295}$$

For the converse inequality we apply the upper bound (13.269) with $\rho^{\otimes n}$ instead of $\rho$. Explicitly, observe that for any $\varepsilon \in (0, 1)$

$$\limsup_{n\to\infty} \frac{1}{n}\mathrm{Distill}^{\varepsilon}_{\to}\left(\rho^{\otimes n}\right) \leqslant \limsup_{n\to\infty} \frac{1}{n}\left\{\frac{1}{1-2\varepsilon}E_{\to}\left(\rho^{\otimes n}\right) + \frac{1+\varepsilon}{1-2\varepsilon}h\left(\frac{\varepsilon}{1+\varepsilon}\right)\right\}$$

$$= \frac{1}{1-2\varepsilon}E^{\mathrm{reg}}_{\to}\left(\rho^{AB}\right).$$

$$(13.296)$$

Taking the limit $\varepsilon \to 0^+$ and combining the resulting inequality with (13.295) we conclude that the equality in (13.287) holds. ∎

In the final step of the proof just discussed, we had to consider the limit as $\varepsilon$ approaches zero (from the positive side). It remains unclear to the author whether this step is essential, and whether the following equality holds true:

$$\lim_{n\to\infty} \frac{1}{n}\mathrm{Distill}^{\varepsilon}_{\to}\left(\rho^{\otimes n}\right) = E^{\mathrm{reg}}_{\to}\left(\rho^{AB}\right) \tag{13.297}$$

for all $\varepsilon$ within the interval $(0,1)$. This point raises an interesting question in the study of quantum information theory, particularly regarding the behavior of the distillable entanglement under asymptotic conditions. The uncertainty here revolves around whether the regularized entanglement measure, $E^{\mathrm{reg}}_{\to}$, aligns with the distillable entanglement rate, $\frac{1}{n}\mathrm{Distill}^{\varepsilon}_{\to}\left(\rho^{\otimes n}\right)$, for any nonzero $\varepsilon$. Resolving this would contribute to a deeper understanding of entanglement properties in quantum systems.

## 13.5.3 Bound Entanglement

The following theorem presents a fascinating discovery in quantum mechanics, demonstrating that certain entangled states have a distillable entanglement of zero. This means that their entanglement is bounded, and they cannot be transformed into a pure bipartite form, making them an example of *bound* entanglement. Despite its inability to be distilled, bound entanglement still has practical applications in quantum information theory.

One important application of bound entanglement is in quantum cryptography, where it can be used as a resource for various cryptographic protocols. For example, bound entangled states can be used to implement quantum key distribution (QKD) protocols, which allow two parties to establish a secure shared key even in the presence of an eavesdropper. Another potential application of bound entanglement is in quantum communication networks, where it can be used as a resource for distributing entanglement between multiple parties. This could be useful for tasks such as quantum teleportation, where entanglement is used to transmit quantum information between distant locations.

Overall, while bound entanglement cannot be used for some tasks that pure entanglement can, it still has practical uses in quantum information processing and communication. From a fundamental point of view, this remarkable result challenges our understanding of entanglement and highlights the complexity of quantum systems. It underscores the importance of exploring and understanding the different types of entanglement that can exist in quantum systems. It also presents a new avenue for research, as scientists continue to investigate the properties and applications of bound entanglement.

> **Theorem 13.19.** Let $\rho \in \mathfrak{D}(AB)$ be an entangled state with a positive partial transpose. Then,
> $$\text{Distill}\left(\rho^{AB}\right) = 0. \tag{13.298}$$

**Proof**   Suppose by contradiction that $\text{Distill}\left(\rho^{AB}\right) > 0$. This in particular means that there exists $n \in \mathbb{N}$ such that

$$\left(\rho^{AB}\right)^{\otimes n} \xrightarrow{\text{LOCC}} \sigma^{A'B'}, \tag{13.299}$$

for some two-qubit entangled state $\sigma \in \mathfrak{D}(A'B')$ with $|A'| = |B'| = 2$. However, since the logarithmic negativity is additive, we get

$$\text{LN}\left(\rho^{\otimes n}\right) = n\text{LN}\left(\rho^{AB}\right)$$
$$\rho^{AB} \text{ is PPT} \rightarrow = 0, \tag{13.300}$$

whereas

$$\text{LN}\left(\sigma^{A'B'}\right) > 0, \tag{13.301}$$

since $\sigma^{A'B'}$ is a two-qubit entangled state and from Theorem 13.4 it follows that it is NPT (recall that the logarithmic robustness is strictly positive on NPT states). We therefore get that

$$\text{LN}\left(\rho^{\otimes n}\right) < \text{LN}\left(\sigma^{A'B'}\right) \tag{13.302}$$

in contradiction with (13.299) and the fact that the logarithmic robustness is a measure of entanglement and therefore cannot increase by LOCC. This completes the proof. ∎

By applying the hashing bound, we can deduce that if $H(A|B)\rho < 0$, then the distillable entanglement of $\rho^{AB}$ is strictly positive. Combining this with the theorem mentioned above, we can conclude that if $\rho \in \mathfrak{D}(AB)$ is PPT, then its conditional entropy $H(A|B)\rho \geq 0$. This observation is consistent with the reduction criterion discussed in Section 13.1.3, as Corollary 7.1 and Theorem 7.5 show that states satisfying the reduction criterion (particularly PPT states) have nonnegative conditional entropy.

**Exercise 13.61.** *Let $\rho \in \text{Pure}(ABC)$ be a tripartite pure state. Show that if its marginals satisfy $I^A \otimes \rho^C > \rho^{AC}$, then* $\text{Distill}\left(\rho^{AB}\right) > 0$.

Theorem 13.19 states that PPT entangled states have zero distillable entanglement. This result is an important insight into the relationship between entanglement and the partial transpose operation. However, it raises the question of whether the converse of this property also holds. That is, are all entangled states with zero distillable entanglement (i.e. bound entangled states) necessarily PPT? This is one of the most challenging and long-standing open problems in quantum information theory, and despite significant efforts over the past two decades, the answer is still unknown. Despite the current lack of a definitive answer to this question, research in this area continues to progress, with new insights and techniques being developed to study the properties of entangled states and their relation to the partial transpose operation.

## 13.6 Single-Shot Entanglement Cost

In this section, we compute the single-shot entanglement cost of a bipartite mixed entangled state $\rho \in \mathfrak{D}(AB)$. For any $\varepsilon \in (0, 1)$, we define the $\varepsilon$-single-shot entanglement cost as

$$\text{Cost}^{\varepsilon}\left(\rho^{AB}\right) := \min\left\{\log m \; : \; P^2\left(\Phi_m \xrightarrow{\text{LOCC}} \rho^{AB}\right) \leqslant \varepsilon\right\}, \tag{13.303}$$

where $\Phi_m$ is the maximally entangled state in $\mathfrak{D}(A'B')$ with $m := |A'| = |B'|$. Since all metrics in finite dimensional Hilbert spaces are topologically equivalent, we chose the square of the purified distance as it is easier to work with, and in particular has the form given in Theorem 13.14.

### The Zero-Error Cost

Before providing a formula for this $\varepsilon$-single-shot entanglement cost, we first consider the case $\varepsilon = 0$. In this case, observe that

$$\text{Cost}^{\varepsilon=0}\left(\rho^{AB}\right) = \min\left\{\log m \; : \; \Phi_m \xrightarrow{\text{LOCC}} \rho^{AB}\right\}$$

$$\mathbf{(13.121)} \rightarrow = \log \text{SR}\left(\rho^{AB}\right). \tag{13.304}$$

That is, the logarithm of the Schmidt rank of $\rho^{AB}$ offers an operational interpretation as the zero-error entanglement cost of $\rho^{AB}$. Following this, we demonstrate that $\text{SR}(\rho^{AB})$ bears a close relationship to the conditional max-entropy. To elaborate further, let's first present an alternative method for describing the convex roof extension.

---

**Definition 13.9.** Let $\rho \in \mathfrak{D}(AB)$ be a bipartite density matrix, and let $X$ be a classical system of dimension $k$. We say that $\rho^{XAB}$ is a *regular extension* of $\rho^{AB}$ if there exists a pure-state decomposition of $\rho^{AB} = \sum_{x \in [k]} p_x \psi_x^{AB}$ such that

$$\rho^{XAB} := \sum_{x \in [k]} p_x |x\rangle\langle x|^X \otimes \psi_x^{AB}. \tag{13.305}$$

Note that if $\{\psi_x^{AB}\}_{x\in[k]}$ in the definition above were composed of mixed states instead of pure states, then $\rho^{XAB}$ would not necessarily qualify as a regular extension, even though it would still be an extension of $\rho^{AB}$. Moreover, we show now that the Schmidt rank of $\rho^{AB}$ can be expressed as an optimization problem over all marginal cq-states $\rho^{XA}$ that results from regular extensions. Explicitly, if $\rho^{XAB}$ is a regular extension of $\rho^{AB}$, as given in (13.305), then the marginal cq-state, $\rho^{XA}$, has the form

$$\rho^{XA} = \sum_{x\in[k]} p_x |x\rangle\langle x|^X \otimes \rho_x^A, \tag{13.306}$$

where $\rho_x^A := \mathrm{Tr}_B\left[\psi_x^{AB}\right]$. By definition, $\mathrm{SR}(\psi_x^{AB}) = \mathrm{Tr}\left[\Pi_{\rho_x}^A\right]$, where $\Pi_{\rho_x} \in \mathrm{Pos}(A)$ is the projection in $A$ to the support of $\rho_x^A$. Combining this with the relation (13.122) we can express the Schmidt rank of $\rho^{AB}$ as

$$\mathrm{SR}\left(\rho^{AB}\right) = \inf_{\rho^{ABX}} \max_{x\in[k]} \mathrm{Tr}\left[\Pi_{\rho_x}^A\right], \tag{13.307}$$

where the maximum is over all regular extensions $\rho^{ABX}$ of $\rho^{AB}$. The expression above can be rewritten in terms of the conditional max-entropy of the state $\rho^{AX}$. To see this, let $\Pi_\rho^{XA}$ be the projection to the support of $\rho^{XA}$ defined as

$$\Pi_\rho^{XA} = \sum_{x\in[k]} |x\rangle\langle x|^X \otimes \Pi_{\rho_x}^A, \tag{13.308}$$

and observe that from its definition

$$\begin{aligned}
H_{\max}(A|X)_\rho &= \max_{\tau\in\mathfrak{D}(X)} -D_{\min}\left(\rho^{XA}\big\|\tau^X \otimes I^A\right) \\
&= \max_{\tau\in\mathfrak{D}(X)} \log \mathrm{Tr}\left[\Pi_\rho^{XA}\left(\tau^X \otimes I^A\right)\right] \\
&= \log \max_{x\in[k]} \mathrm{Tr}\left[\Pi_{\rho_x}^A\right],
\end{aligned} \tag{13.309}$$

where in the last line we replaced the maximum over all $\tau \in \mathfrak{D}(X)$ with a maximum over all $x \in [k]$. Using this observation in conjunction with (13.307), we can express the logarithm of the Schmidt rank of $\rho^{AB}$ as

$$\log \mathrm{SR}\left(\rho^{AB}\right) = \inf_{\rho^{ABX}} H_{\max}(A|X)_\rho. \tag{13.310}$$

We therefore conclude that the zero-error entanglement cost of $\rho^{AB}$ is given by

$$\mathrm{Cost}^{\varepsilon=0}\left(\rho^{AB}\right) = \inf_{\rho^{ABX}} H_{\max}(A|X)_\rho, \tag{13.311}$$

where the infimum is over all classical systems $X$ and all regular extensions $\rho^{ABX}$ of $\rho^{AB}$. In the following exercise you show that we can remove the restriction to regular extensions.

**Exercise 13.62.** *Show that* (13.311) *still holds even if we take the infimum over all classical systems $X$ and all extensions $\rho^{ABX}$ of $\rho^{AB}$.*

**Exercise 13.63.** *Let $\rho \in \mathfrak{D}(AB)$. Show that the entanglement of formation of $\rho^{AB}$ can be expressed as*

$$E_F\left(\rho^{AB}\right) := \inf_{\rho^{ABX}} H(A|X)_\rho, \tag{13.312}$$

*where $H(A|X)_\rho$ is the von-Neumann conditional entropy, and the infimum is over all classical systems $X$ and all extensions $\rho^{ABX}$ of $\rho^{AB}$.*

**Exercise 13.64.** *Let $\rho \in \mathfrak{D}(AB)$ and $n \in \mathbb{N}$.*

*1 Show that if $\rho^{ABX}$ is a regular extension of $\rho^{AB}$, then $\left(\rho^{ABX}\right)^{\otimes n}$ is a regular extension of $\left(\rho^{AB}\right)^{\otimes n}$.*

*2. Show that there exists a density matrix $\omega^{A^n B^n X^n}$ that is a regular extension of $\left(\rho^{AB}\right)^{\otimes n}$ but it does not have the form $\left(\sigma^{ABX}\right)^{\otimes n}$ (for some $\sigma \in \mathfrak{D}(XAB)$).*

## The General Case

From the previous discussion, for every $\varepsilon \in (0, 1)$ the $\varepsilon$-single-shot entanglement cost can be expressed as

$$\begin{aligned}
\mathrm{Cost}^\varepsilon\left(\rho^{AB}\right) &= \min\left\{\log\mathrm{SR}\left(\omega^{AB}\right) \ : \ P^2\left(\omega^{AB}, \rho^{AB}\right) \leqslant \varepsilon, \ \omega \in \mathfrak{D}(AB)\right\} \\
&= \inf_{\omega^{ABX}}\left\{H_{\max}(A|X)_\omega \ : \ P^2\left(\omega^{AB}, \rho^{AB}\right) \leqslant \varepsilon\right\},
\end{aligned} \tag{13.313}$$

where the second infimum is over all density matrices $\omega^{AB}$, all classical systems $X$, *and* over all regular extensions $\omega^{ABX}$ of $\omega^{AB}$. Given the complexity of the above expression, we'll transition to an alternative approach, specifically employing the formula presented in Theorem 13.14 for the conversion distance.

---

**Theorem 13.20.** Let $\rho \in \mathfrak{D}(AB)$. Then, the $\varepsilon$-single-shot entanglement cost is given by

$$\mathrm{Cost}^\varepsilon\left(\rho^{AB}\right) = \inf_{\rho^{ABX}} H_{\max}^\varepsilon(A|X)_\rho, \tag{13.314}$$

where $H_{\max}^\varepsilon(A|X)_\rho$ is the smoothed conditional max-entropy (see remark below), and the infimum is over all classical systems $X$ and all extensions $\rho^{ABX}$ of $\rho^{AB}$.

---

*Remark.* Exercise 13.62 allows us to limit the infimum in (13.314) to *regular* extensions $\rho^{ABX}$ of $\rho^{AB}$. Additionally, from (13.309), the smoothed version of $H_{\max}(A|X)_\rho$ is expressed as

$$H_{\max}^\varepsilon(A|X)_\rho = \min_{\omega \in \mathfrak{D}_\varepsilon(\rho^{XA})} \max_{x \in [k]} \log\mathrm{Tr}\left[\Pi_{\omega_x}^A\right], \tag{13.315}$$

where $\omega_x^A := \frac{1}{q_x}\mathrm{Tr}_X\left[\left(|x\rangle\langle x|^X \otimes I^A\right)\omega^{XA}\right]$ with $q_x := \mathrm{Tr}\left[\left(|x\rangle\langle x|^X \otimes I^A\right)\omega^{XA}\right]$. Interestingly, the optimization's minimizer $\omega^{XA}$ can be an $m$-pruned version of $\rho^{XA}$ (as referenced in (5.156)). Therefore, before delving into the proof of Theorem 13.20, we first establish that the minimizer $\omega^{XA}$ in the optimization in (13.315) is the $m$-pruned state of the cq-state $\rho^{XA} = \sum_{x \in [k]} p_x |x\rangle\langle x| \otimes \rho_x^A$. This pruned state is given by

$$\rho^{(m)} = \sum_{x \in [k]} p_x |x\rangle\langle x| \otimes \rho_x^{(m)}, \tag{13.316}$$

where each $\rho_x^{(m)}$ is the $m$-pruned version of $\rho_x$ as defined in (5.156).

**Exercise 13.65.** *Use* (12.60) *to show that if* $\rho^{(m)} \neq \rho^{XA}$, *then* $H(A|X)_{\rho^{(m)}} = \log m$.

---

**Lemma 13.5.** Let $\varepsilon \in (0,1)$, $d := |A|$, $\rho \in \mathfrak{D}(XA)$, and for all $m \in [d]$ let $\rho^{(m)} \in \mathfrak{D}(XA)$ be the $m$-pruned version of $\rho^{XA}$ as defined in (13.316). Then,

$$H_{\max}^{\varepsilon}(A|X)_{\rho} = \min_{m \in [d]} \left\{ \log m \ : \ \rho^{(m)} \approx_{\varepsilon} \rho^{XA} \right\}. \tag{13.317}$$

---

*Remark.* Observe that the trace distance between $\rho^{XA}$ and its $m$-pruned version is given by

$$\frac{1}{2} \left\| \rho^{XA} - \rho^{(m)} \right\|_1 = \sum_{x \in [k]} p_x \frac{1}{2} \left\| \rho_x - \rho_x^{(m)} \right\|_1$$

$$(5.157)\rightarrow = \sum_{x \in [k]} p_x \left( 1 - \| \rho_x \|_{(m)} \right) \tag{13.318}$$

$$= 1 - \sum_{x \in [k]} p_x \| \rho_x \|_{(m)}.$$

Therefore, (13.317) can also be written as

$$H_{\max}^{\varepsilon}(A|X)_{\rho} = \min_{m \in [d]} \left\{ \log m \ : \ \sum_{x \in [k]} p_x \| \rho_x \|_{(m)} \geqslant 1 - \varepsilon \right\}. \tag{13.319}$$

**Proof** By definition,

$$H_{\max}^{\varepsilon}(A|X)_{\rho} = \min \left\{ H_{\max}(A|X)_{\omega} \ : \ \omega^{XA} \approx_{\varepsilon} \rho^{XA} \right\}$$

$$\boxed{\text{Restricting } \omega = \rho^{(m)}} \longrightarrow \ \leqslant \min_{m \in [d]} \left\{ H_{\max}(A|X)_{\rho^{(m)}} \ : \ \rho^{(m)} \approx_{\varepsilon} \rho^{XA} \right\} \tag{13.320}$$

$$\text{Exercise 13.65}\rightarrow = \min_{m \in [d]} \left\{ \log m \ : \ \rho^{(m)} \approx_{\varepsilon} \rho^{XA} \right\}.$$

For the converse direction, suppose by contradiction that (see (13.319))

$$H_{\max}^{\varepsilon}(A|X)_{\rho} < \min_{m \in [d]} \left\{ \log m \ : \ \sum_{x \in [k]} p_x \| \rho_x \|_{(m)} \geqslant 1 - \varepsilon \right\}. \tag{13.321}$$

Let $m$ be the minimizer of the right-hand side in (13.321), so that $\sum_{x \in [k]} p_x \| \rho_x \|_{(m)} \geqslant 1 - \varepsilon$. Furthermore, let $\omega \in \mathfrak{D}(XA)$ be such that $H_{\max}^{\varepsilon}(A|X)_{\rho} = H_{\max}(A|X)_{\omega}$. Then, by the assumption

$$\log m > H_{\max}(A|X)_\omega$$

$$\textbf{cf. (13.309)} \rightarrow = \log \max_{x \in [k]} \mathrm{Tr}\left[\Pi_{\omega_x}^A\right], \qquad (13.322)$$

so that $\mathrm{Tr}\left[\Pi_{\omega_x}^A\right] < m$ for all $x \in [k]$. Thus, since the rank of each $\omega_x$ is no greater than $m - 1$ we get that for any $x \in [k]$, $\left\|\rho_x^A\right\|_{(m-1)} \geq \mathrm{Tr}\left[\rho_x^A \Pi_{\omega_x}^A\right]$. Therefore, denoting $\Pi_\omega^{XA} := \sum_{x \in [k]} |x\rangle\langle x|^X \otimes \Pi_{\omega_x}^A$ we obtain (cf. (5.159))

$$\sum_{x \in [k]} p_x \left\|\rho_x^A\right\|_{(m-1)} \geq \mathrm{Tr}\left[\rho^{XA} \Pi_\omega^{XA}\right]$$

$$\boxed{\mathrm{Tr}\left[\omega^{XA} \Pi_\omega^{XA}\right] = 1} \longrightarrow = 1 + \mathrm{Tr}\left[\left(\rho^{XA} - \omega^{XA}\right) \Pi_\omega^{XA}\right]$$

$$\boxed{\eta \geq -(\eta)_- \quad \forall \eta \in \mathrm{Herm}(XA)} \longrightarrow \geq 1 - \mathrm{Tr}\left[\left(\rho^{XA} - \omega^{XA}\right)_- \Pi_\omega^{XA}\right] \qquad (13.323)$$

$$\boxed{\Pi_\omega^{XA} \leq I^{XA}} \longrightarrow \geq 1 - \mathrm{Tr}\left(\rho^{XA} - \omega^{XA}\right)_-$$

$$\geq 1 - \varepsilon,$$

where we used the fact that $\mathrm{Tr}\left(\rho^{XA} - \omega^{XA}\right)_- = \frac{1}{2}\|\omega^{XA} - \rho^{XA}\|_1 \leq \varepsilon$. Therefore, we get a contradiction with the assumption that $m$ was the minimizer of the right-hand side of (13.321). This completes the proof. ∎

We are now ready to prove Theorem 13.20.

**Proof of Theorem 13.20** From Theorem 13.14 it follows that the conversion distance that appears in (13.303) can be expressed as

$$P^2\left(\Phi_m \xrightarrow{\text{LOCC}} \rho^{AB}\right) = \min_{\rho^{XAB}} \sum_{x \in [k]} p_x \left(1 - \left\|\rho_x^A\right\|_{(m)}\right), \qquad (13.324)$$

where the minimum is over all regular extensions $\rho^{XAB}$ of $\rho^{AB}$, with the same notations as in (13.306). We therefore get that the $\varepsilon$-single-shot entanglement cost as defined in (13.303) can be expressed as

$$\mathrm{Cost}^\varepsilon\left(\rho^{AB}\right) = \min_{m \in [d]} \left\{\log m : \max_{\rho^{XAB}} \sum_{x \in [k]} p_x \left\|\rho_x^A\right\|_{(m)} \geq 1 - \varepsilon\right\}$$

$$\textbf{Exercise 13.66} \rightarrow = \min_{\rho^{XAB}} \min_{m \in [d]} \left\{\log m : \sum_{x \in [k]} p_x \left\|\rho_x^A\right\|_{(m)} \geq 1 - \varepsilon\right\} \qquad (13.325)$$

$$\textbf{(13.319)} \rightarrow = \min_{\rho^{XAB}} H_{\max}^\varepsilon(A|X)_\rho,$$

where the maximum is over all regular extensions $\rho^{XAB}$ of $\rho^{AB}$. This completes the proof. ∎

**Exercise 13.66.** *Let $\lambda \in \mathbb{R}_+$, $\mathfrak{K}_1, \mathfrak{K}_2 \subseteq \mathfrak{D}(A)$ be two subsets of density matrices, and $f : \mathfrak{D}(A) \to \mathbb{R}_+$ and $g : \mathfrak{D}(A) \times \mathfrak{D}(A) \to \mathbb{R}_+$ be two functions.*

1. *Show that*

$$\inf_{\rho \in \mathfrak{R}_1} \left\{ f(\rho) : \sup_{\sigma \in \mathfrak{R}_2} g(\rho, \sigma) \geqslant \lambda \right\} = \inf_{\sigma \in \mathfrak{R}_2} \inf_{\rho \in \mathfrak{R}_1} \left\{ f(\rho) : g(\rho, \sigma) \geqslant \lambda \right\}. \qquad (13.326)$$

2. *Use similar lines to prove the second equality in* (13.325).

**Exercise 13.67.** *Consider the two formulas given in* (12.101) *and* (13.314) *for the $\varepsilon$-single-shot entanglement cost of pure and mixed states, respectively.*

1. *Show that the two formulas coincide when $\rho^{AB} = \psi^{AB}$ is a pure state.*
2. *Without assuming* (13.314), *use Theorem 5.11 to show (by direct calculation) that for the pure state case, the formula in* (12.101) *can be expressed as*

$$\mathrm{Cost}^{\varepsilon}\left(\psi^{AB}\right) = H_{\max}^{\varepsilon}(\rho^A) := \inf_{\sigma \in \mathfrak{B}_{\varepsilon}(\rho)} H_{\max}(\sigma^A), \qquad (13.327)$$

*where $\rho^A := \mathrm{Tr}_B\left[\psi^{AB}\right]$.*

## 13.7 Asymptotic Entanglement Cost

The asymptotic entanglement cost is defined for any $\rho \in \mathfrak{D}(AB)$ as (cf. (11.120) and (11.136))

$$\mathrm{Cost}\left(\rho^{AB}\right) := \lim_{\varepsilon \to 0^+} \inf_{n,m \in \mathbb{N}} \left\{ \frac{n}{m} : T\left( \Phi_2^{\otimes n} \xrightarrow{\mathrm{LOCC}} \rho^{\otimes m} \right) \leqslant \varepsilon \right\}. \qquad (13.328)$$

Recall from Exercise 11.12 that the above cost does not change if we replace the trace distance with the square of the purified distance, as the two metrics are topologically equivalent. Thus, from (11.137) it follows that the asymptotic entanglement cost can be expressed as

$$\mathrm{Cost}\left(\rho^{AB}\right) := \lim_{\varepsilon \to 0^+} \liminf_{n \to \infty} \frac{1}{n} \mathrm{Cost}^{\varepsilon}\left(\rho^{\otimes n}\right). \qquad (13.329)$$

---

**Theorem 13.21.** Let $\rho \in \mathfrak{D}(AB)$. Then, the entanglement cost of $\rho^{AB}$ can be expressed as

$$\mathrm{Cost}\left(\rho^{AB}\right) = E_F^{\mathrm{reg}}\left(\rho^{AB}\right) := \lim_{n \to \infty} \frac{1}{n} E_F\left(\rho^{\otimes n}\right), \qquad (13.330)$$

where $E_F$ is the entanglement of formation (see Definition 13.3).

---

**Proof** We first prove that $\mathrm{Cost}\left(\rho^{AB}\right) \leqslant E_F^{\mathrm{reg}}\left(\rho^{AB}\right)$. From Theorem 13.20 we have

$$\mathrm{Cost}^{\varepsilon}\left(\left(\rho^{AB}\right)^{\otimes n}\right) = \inf_{\rho_n} H_{\max}^{\varepsilon}(A^n | Y_n)_{\rho_n}$$

$$\boxed{\text{taking } \rho_n = \left(\rho^{XAB}\right)^{\otimes n}} \longrightarrow \leqslant \inf_{\rho^{XAB}} H_{\max}^{\varepsilon}(A^n | X^n)_{\rho^{\otimes n}}, \qquad (13.331)$$

where the first infimum is over all classical systems $Y_n$ and all regular extensions $\rho_n \in \mathfrak{D}(Y_n A^n B^n)$ of $(\rho^{AB})^{\otimes n}$, and the second infimum is over all regular extensions $\rho \in \mathfrak{D}(XAB)$ of $\rho^{AB}$. For the inequality above we used the fact that if $\rho^{XAB}$ is a regular extension of $\rho^{AB}$, then $(\rho^{XAB})^{\otimes n}$ is a regular extension of $(\rho^{AB})^{\otimes n}$ (see Exercise 13.64). Therefore, the entanglement cost satisfies

$$\mathrm{Cost}\left(\rho^{AB}\right) \leqslant \lim_{\varepsilon \to 0^+} \inf_{\rho^{XAB}} \liminf_{n \to \infty} \frac{1}{n} H_{\max}^{\varepsilon}(A^n | X^n)_{\rho^{\otimes n}}$$

$$\boxed{\text{AEP of the form (10.140)}} \longrightarrow = \inf_{\rho^{XAB}} H(A|X)_{\rho} \qquad (13.332)$$

$$\textbf{Exercise 13.63} \to = E_F\left(\rho^{AB}\right).$$

Thus, $E_F\left(\rho^{AB}\right) \geqslant \mathrm{Cost}\left(\rho^{AB}\right)$. Repeating the same argument with $m \in \mathbb{N}$ copies of $\rho^{AB}$ gives $E_F\left(\rho^{\otimes m}\right) \geqslant \mathrm{Cost}\left(\rho^{\otimes m}\right)$. Combining this with (11.146) we get

$$\mathrm{Cost}\left(\rho^{AB}\right) \leqslant \frac{1}{m}\mathrm{Cost}\left(\rho^{\otimes m}\right) \leqslant \frac{1}{m}E_F\left(\rho^{\otimes m}\right). \qquad (13.333)$$

Since this inequality holds for all integers $m$, it also holds in the limit $m \to \infty$. Hence, $\mathrm{Cost}\left(\rho^{AB}\right) \leqslant E_F^{\mathrm{reg}}\left(\rho^{AB}\right)$.

For the converse inequality, observe that

$$\mathrm{Cost}^{\varepsilon}\left(\left(\rho^{AB}\right)^{\otimes n}\right) = \inf_{\rho_n} H_{\max}^{\varepsilon}(A^n | Y_n)_{\rho_n}$$

$$\textbf{By definition} \to = \inf_{\rho_n} \min_{\omega \subset \mathfrak{B}_\varepsilon\left(\rho_n^{Y_n A^n}\right)} H_{\max}(A^n | Y_n)_{\omega} \qquad (13.334)$$

$$\boxed{H_{\max}(A|X) \geqslant H(A|X)} \longrightarrow \geqslant \inf_{\rho_n} \min_{\omega \in \mathfrak{B}_\varepsilon\left(\rho_n^{Y_n A^n}\right)} H(A^n | Y_n)_{\omega}.$$

Now, the asymptotic continuity property (10.51) of the conditional entropy gives for any $\omega \in \mathfrak{B}_\varepsilon\left(\rho_n^{Y_n A^n}\right)$

$$H(A^n | Y_n)_{\omega} \geqslant H(A^n | Y_n)_{\rho_n} - n \log |A| f(\varepsilon). \qquad (13.335)$$

Substituting this into (13.334) gives

$$\mathrm{Cost}^{\varepsilon}\left(\rho^{\otimes n}\right) \geqslant \inf_{\rho_n} H(A^n | Y_n)_{\rho_n} - n \log |A| f(\varepsilon)$$

$$\textbf{Exercise 13.63} \to = E_F\left(\rho^{\otimes n}\right) - n \log |A| f(\varepsilon). \qquad (13.336)$$

Dividing both sides by $n$ and taking the limit $n \to \infty$ followed by $\varepsilon \to 0^+$ gives $\mathrm{Cost}\left(\rho^{AB}\right) \geqslant E_F^{\mathrm{reg}}\left(\rho^{AB}\right)$. This completes the proof. $\blacksquare$

If the entanglement of formation ($E_F$) were additive under tensor products, determining $E_F^{\mathrm{reg}}\left(\rho^{AB}\right)$ would be a more straightforward task. For quite some time, a prevalent belief among researchers in the field was that $E_F$ is indeed additive, implying its equivalence to the entanglement cost. However, in a pivotal development in 2008, Hastings refuted this additivity conjecture, demonstrating that the entanglement of formation is generally not additive. From its definition, for any states $\rho \in \mathfrak{D}(AB)$ and $\sigma \in \mathfrak{D}(A'B')$, the following inequality holds:

$$E_F\left(\rho^{AB} \otimes \sigma^{A'B'}\right) \leqslant E_F\left(\rho^{AB}\right) + E_F\left(\sigma^{A'B'}\right). \tag{13.337}$$

Hastings' result indicates that this inequality can be strict, even when $\rho$ is equal to $\sigma$. Notably, Hastings' proof is existential, meaning that it establishes the existence of such non-additivity without providing an explicit counterexample. To date, an explicit example where $E_F$ is not additive has not been identified, but Hastings' contribution significantly altered our understanding on this problem (further details can be found in the "Notes and References" section at the end of this chapter).

While the entanglement of formation (EoF) is not generally additive, it can be additive for certain specific states, allowing for efficient computation of their entanglement cost. An interesting concept relevant in this context is that of an "entanglement breaking subspace."

> **Definition 13.10.** A bipartite subspace $\mathcal{K} \subset AB$ is said to be entanglement breaking subspace (EBS) if for every other bipartite system $A'B'$ and any pure state in $\mathcal{K} \otimes A'B'$, its reduced density matrix that obtained after tracing out system $B$ can be written as a separable state between system $A$ and the joint system $A'B'$.

To be more precise, a subspace $\mathcal{K}$ is an EBS if, for every $\psi \in \mathrm{Pure}\left(\mathcal{K} \otimes A'B'\right)$, we have

$$\mathrm{Tr}_B\left[\psi^{ABA'B'}\right] = \sum_{x \in [m]} p_x \phi_x^A \otimes \varphi_x^{A'B'}, \tag{13.338}$$

where $m \in \mathbb{N}$, $\mathbf{p} = (p_1, \ldots, p_m)^T \in \mathrm{Prob}(m)$, and $\phi_x \in \mathrm{Pure}(A)$ and $\varphi_x \in \mathrm{Pure}(A'B')$ for each $x \in [m]$. Note that requiring $\mathcal{K}$ to be a subspace is a strong condition. Even if two states in $\mathrm{Pure}\left(\mathcal{K} \otimes A'B'\right)$ satisfy the above condition, it is not clear why all their linear combinations would satisfy it as well. This raises the question of whether EBSs exist. In the following discussion, we present a couple of examples of EBSs.

We start with an example in two qubits. Let $\mathcal{K}$ be the subspace spanned by the two product states $|00\rangle$ and $|11\rangle$. We argue now that this subspace of $\mathbb{C}^2 \otimes \mathbb{C}^2$ is indeed an EBS. Indeed, any state in $\psi \in \mathrm{Pure}\left(\mathcal{K} \otimes A'B'\right)$ can be expressed as

$$|\psi^{ABA'B'}\rangle = |00\rangle^{AB} \otimes |\phi_0^{A'B'}\rangle + |11\rangle^{AB} \otimes |\phi_1^{A'B'}\rangle, \tag{13.339}$$

where $\phi_0^{AB}$ and $\phi_1^{AB}$ are some subnormalized pure states. Clearly, the reduced density matrix $\psi^{AA'B'}$ has the form

$$\psi^{AA'B'} = |0\rangle\langle 0|^A \otimes \phi_0^{A'B'} + |1\rangle\langle 1|^A \otimes \phi_1^{A'B'}, \tag{13.340}$$

which is a separable state between $A$ and $A'B'$.

In the second example, we consider the two-qutrit space $\mathbb{C}^3 \otimes \mathbb{C}^3$. In Exercise 13.68 we show that its antisymmetric subspace that is spanned by the three states

$$|\chi_1^{AB}\rangle := \frac{1}{\sqrt{2}}(|01\rangle - |10\rangle),$$

$$|\chi_2^{AB}\rangle := \frac{1}{\sqrt{2}}(|12\rangle - |21\rangle), \tag{13.341}$$

$$|\chi_3^{AB}\rangle := \frac{1}{\sqrt{2}}(|20\rangle - |02\rangle), \text{ and}$$

is an EBS.

**Exercise 13.68.** *Let $\mathcal{K}$ be the subspace spanned by the three vectors above. Show that:*

1. *For all $x, y \in [3]$ we have $\mathrm{Tr}_B\left[|\chi_x^{AB}\rangle\langle\chi_y^{AB}|\right] = \left(\delta_{xy} I^A - |y\rangle\langle x|^A\right).$*
2. *The reduced density matrix of every $\phi^{AB} \in \mathrm{Pure}\,(\mathcal{K})$ of the form $|\phi^{AB}\rangle = \sum_{x\in[3]} \lambda_x |\chi_x^{AB}\rangle$ (with $\lambda_x \in \mathbb{C}$) can be expressed as*

$$\mathrm{Tr}_B[\phi^{AB}] = \frac{1}{2} I^A - \frac{1}{2}\varphi^T, \tag{13.342}$$

*where $|\varphi^A\rangle := \sum_{x\in[3]} \lambda_x |x\rangle^A.$*
3. *The subspace $\mathcal{K}$ is EBS.*

---

**Theorem 13.22.** Let $\rho \in \mathrm{Pure}(AB)$ be a bipartite quantum state whose support is an EBS. Then, for every bipartite system $A'B'$ and every $\sigma \in \mathfrak{D}(A'B')$ we have

$$E_F\left(\rho^{AB} \otimes \sigma^{A'B'}\right) = E_F\left(\rho^{AB}\right) + E_F\left(\sigma^{A'B'}\right). \tag{13.343}$$

In particular, the entanglement cost of $\rho^{AB}$ equals its entanglement of formation.

---

**Proof**   Due to (13.337) it is sufficient to prove that

$$E_F\left(\rho^{AB} \otimes \sigma^{A'B'}\right) \geqslant E_F\left(\rho^{AB}\right) + E_F\left(\sigma^{A'B'}\right). \tag{13.344}$$

Denote by $\mathcal{K} := \mathrm{supp}\left(\rho^{AB}\right)$. The key idea is to first show that for every $\psi \in \mathrm{Pure}\left(\mathcal{K} \otimes A'B'\right)$ we have

$$E\left(\psi^{ABA'B'}\right) \geqslant E_F\left(\psi^{AB}\right) + E_F\left(\psi^{A'B'}\right), \tag{13.345}$$

where $E$ on the left-hand side is the entropy of entanglement between systems $AA'$ and $BB'$, and for simplicity of notations we use $\psi^{AB}$ and $\psi^{A'B'}$ to denote the *mixed* marginal states of $\psi^{ABA'B'}$. To see why the above inequality holds, recall that $\psi^{ABA'B'}$ belongs to an EBS so that we can express it as

$$|\psi^{ABA'B'}\rangle := \sum_{x\in[m]} \sqrt{p_x} |x\rangle^B \otimes |\phi_x^A\rangle \otimes |\varphi_x^{A'B'}\rangle, \tag{13.346}$$

where $m \in \mathbb{N}$, $\mathbf{p} = (p_1, \ldots, p_m)^T \in \mathrm{Prob}(m)$, and $\phi_x \in \mathrm{Pure}(A)$ and $\varphi_x \in \mathrm{Pure}(A'B')$ for each $x \in [m]$. We therefore get from the definition of the entropy of entanglement that

$$E\left(\psi^{ABA'B'}\right) = H\left(AA'\right)_\psi, \tag{13.347}$$

where $H(AA')_\psi$ is the von-Neumann entropy of the marginal state

$$\psi^{AA'} := \sum_{x \in [m]} p_x \phi_x^A \otimes \varphi_x^{A'}, \quad \text{where} \quad \varphi_x^{A'} := \mathrm{Tr}_{B'}\left[\varphi_x^{A'B'}\right]. \tag{13.348}$$

Now, let $C$ be an $m$-dimensional system and let $\sigma \in \mathfrak{D}(AA'C)$ be the state

$$\sigma^{AA'C} := \sum_{x \in [m]} p_x \phi_x^A \otimes \varphi_x^{A'} \otimes |x\rangle\langle x|^C. \tag{13.349}$$

Then, from the strong subadditivity as given in (7.134) and the fact that the marginal $\sigma^{AA'} = \psi^{AA'}$ we get

$$\begin{aligned}
H(AA')_\psi &= H(AA')_\sigma \\
\textbf{Strong subadditivity (7.134)} &\to \geq H(A)_\sigma + H(AA'C)_\sigma - H(AC)_\sigma \\
\textbf{Exercise 13.69} &\to = H\left(\psi^A\right) + \sum_{x \in [m]} p_x H\left(\varphi_x^{A'}\right).
\end{aligned} \tag{13.350}$$

Now, from (13.69) we have $H\left(\psi^A\right) \geq E_F\left(\psi^{AB}\right)$. Moreover, by definition, for every $x \in [m]$ we have $H\left(\varphi_x^{A'}\right) = E\left(\varphi_x^{A'B'}\right)$. Combining this with the equation (13.350) and with (13.347) gives

$$\begin{aligned}
E\left(\psi^{ABA'B'}\right) &\geq E_F\left(\psi^{AB}\right) + \sum_{x \in [m]} p_x E\left(\varphi_x^{A'B'}\right) \\
&\geq E_F\left(\psi^{AB}\right) + E_F\left(\psi^{A'B'}\right).
\end{aligned} \tag{13.351}$$

This completes the proof. $\blacksquare$

**Exercise 13.69.** *Prove the last equality in (13.350).*

**Exercise 13.70.** *Compute the entanglement cost of the state*

$$\rho^{AB} := p\Phi_+^{AB} + (1-p)\Phi_-^{AB}, \tag{13.352}$$

*for all $p \in [0,1]$, where*

$$|\Phi_\pm^{AB}\rangle := \frac{1}{\sqrt{2}}(|00\rangle \pm |11\rangle). \tag{13.353}$$

## 13.8 Beyond LOCC: Nonentangling Operations

As already discussed, the study of mixed-state entanglement is significantly more complex than that of pure bipartite entanglement, leading to numerous unresolved issues.

The intricate nature of LOCC is the main contributing factor to this difficulty. To address this challenge, this section focuses on examining entanglement theory under different sets of operations that are comparatively simpler to handle. Specifically, we explore briefly in this section the theory of entanglement under RNG operations, also referred to in entanglement theory as *nonentangling operations*.

A quantum channel $\mathcal{E} \in \text{CPTP}(AB \to A'B')$ is considered nonentangling if, for any $\sigma \in \text{SEP}(AB)$, $\mathcal{E}(\sigma) \in \text{SEP}(A'B')$. It should be emphasized that the relative entropy of entanglement is a measure of entanglement that behaves monotonically under nonentangling operations. This subsection is dedicated to the calculation of the single-shot entanglement cost, as well as the distillable entanglement, under nonentangling operations. The calculation reveals that (1) the operational meaning of the relative entropy of a resource, as measured by hypothesis testing divergence, is the single-shot distillable entanglement and (2) the smoothed logarithmic robustness can be interpreted as the single-shot entanglement cost.

## 13.8.1 The Single-Shot Entanglement Cost

In order to compute the single-shot entanglement cost under nonentangling operations, we first need to simplify the expression for the conversion distance:

$$T\left(\Phi_m^{A'B'} \xrightarrow{\text{RNG}} \rho^{AB}\right) = \frac{1}{2} \min_{\mathcal{E} \in \text{RNG}} \left\| \rho^{AB} - \mathcal{E}^{A'B' \to AB}\left(\Phi_m^{A'B'}\right) \right\|_1. \tag{13.354}$$

Since $\Phi_m^{A'B'}$ is invariant under the action of the (self-adjoint) twirling map

$$\mathcal{G}\left(\omega^{A'B'}\right) = \int dU \ \left(U \otimes \overline{U}\right) \omega^{A'B'} \left(U \otimes \overline{U}\right)^* \qquad \forall \, \omega \in \mathfrak{L}(A'B'), \tag{13.355}$$

we can replace $\mathcal{E}$ in (13.354) with $\mathcal{E} \circ \mathcal{G}$, or in other words, we can assume without loss of generality that $\mathcal{E} = \mathcal{E} \circ \mathcal{G}$. Any such nonentangling (RNG) operation has the form (see (3.256))

$$\mathcal{E}(\omega) = \text{Tr}\left[(I - \Phi_m)\omega\right] \sigma^{AB} + \text{Tr}\left[\Phi_m \omega\right] \eta^{AB} \qquad \forall \, \omega \in \mathfrak{D}(A'B'), \tag{13.356}$$

where $\sigma, \eta \in \mathfrak{D}(AB)$. Note that the channel $\mathcal{E}$ is RNG (i.e. nonentangling) if and only if

$$\text{Tr}\left[(I - \Phi_m)\omega\right] \sigma + \text{Tr}\left[\Phi_m \omega\right] \eta \in \text{SEP}(AB) \qquad \forall \, \omega \in \text{SEP}(A'B'). \tag{13.357}$$

Now, recall that the density state $\tau := (I - \Phi_m)/(m^2 - 1)$ is a separable isotropic state (see (13.19)). Taking $\omega = \tau$ above we get $\mathcal{E}(\tau) = \sigma$. Therefore, since $\tau$ is a separable state we get that $\sigma$ must be separable as well. More generally, from (13.26) we have that $\text{Tr}\left[\Phi_m^{AB}\omega^{AB}\right] \leqslant \frac{1}{m}$ for all separable states $\omega \in \text{SEP}(AB)$. Therefore, the condition in (13.357) holds if and only if

$$\frac{1}{m}\eta^{AB} + \frac{m-1}{m}\sigma^{AB} \in \text{SEP}(AB). \tag{13.358}$$

To summarize, the channel $\mathcal{E}$ as given in (13.356) is RNG if and only if $\sigma \in \mathrm{SEP}(AB)$ and the equation above holds. Hence,

$$T\left(\Phi_m^{A'B'} \xrightarrow{\mathrm{RNG}} \rho^{AB}\right) = \min \frac{1}{2}\left\|\rho^{AB} - \eta^{AB}\right\|_1, \tag{13.359}$$

where the minimum is over all $\eta \in \mathfrak{D}(AB)$ that satisfies (13.358) with some $\sigma \in \mathrm{SEP}(AB)$.

The robustness measure of entanglement of the state $\eta^{AB}$ is defined as (see (10.2.2))

$$\mathbf{R}(\eta^{AB}) := \min\left\{s \geqslant 0 : \frac{\eta + s\sigma}{1 + s} \in \mathrm{SEP}(AB), \ \sigma \in \mathrm{SEP}(AB)\right\}. \tag{13.360}$$

Comparing this with the expression for the conversion distance above we get that conversion distance can be expressed compactly as

$$T\left(\Phi_m^{A'B'} \xrightarrow{\mathrm{RNG}} \rho^{AB}\right) = \frac{1}{2}\min_{\substack{\eta \in \mathfrak{D}(AB) \\ \mathbf{R}(\eta) \leqslant m-1}} \left\|\rho^{AB} - \eta^{AB}\right\|_1. \tag{13.361}$$

In other words, the conversion distance above can be interpreted as the distance of $\rho^{AB}$ to the set of states with robustness no greater than $m - 1$.

Using the compact expression for the conversion distance above, we get that for any $\varepsilon \in (0, 1)$, the $\varepsilon$-single-shot entanglement cost under nonentangling operations is given by

$$\mathrm{Cost}^\varepsilon(\rho^{AB}) = \min_{m \in \mathbb{N}}\left\{\log m : \frac{1}{2}\left\|\rho^{AB} - \eta^{AB}\right\|_1 \leqslant \varepsilon, \ \mathbf{R}\left(\eta^{AB}\right) \leqslant m - 1\right\}. \tag{13.362}$$

That is,

$$\mathrm{Cost}^\varepsilon\left(\rho^{AB}\right) = \log\left(1 + \mathbf{R}^\varepsilon\left(\rho^{AB}\right)\right) = \mathbf{LR}^\varepsilon\left(\rho^{AB}\right). \tag{13.363}$$

Equation (13.363) provides an operational interpretation for the smoothed logarithmic robustness as the single-shot entanglement cost under nonentangling operations.

## 13.8.2 Single-Shot Distillable Entanglement

The conversion distance is given by

$$T\left(\rho^{AB} \xrightarrow{\mathrm{RNG}} \Phi_m^{A'B'}\right) = 1 - \sup_{\mathcal{E} \in \mathrm{RNG}} \mathrm{Tr}\left[\Phi_m^{A'B'} \mathcal{E}^{AB \to A'B'}\left(\rho^{AB}\right)\right]. \tag{13.364}$$

Due to the symmetry of $\Phi_m$ we can assume without loss of generality that $\mathcal{E} = \mathcal{G} \circ \mathcal{E}$ so that the nonentangling operation $\mathcal{E}$ has the form (see (3.257))

$$\mathcal{E}(\omega) = (1 - \mathrm{Tr}[\Lambda\omega])\tau^{A'B'} + \mathrm{Tr}[\Lambda\omega]\Phi_m^{A'B'} \qquad \forall \omega \in \mathcal{L}(AB), \tag{13.365}$$

where $\Lambda \in \mathrm{Eff}(AB)$, and

$$\tau^{A'B'} := \frac{I^{A'B'} - \Phi_m}{m^2 - 1}. \tag{13.366}$$

Observe that for $\omega \in \mathfrak{D}(AB)$ the state $\mathcal{E}(\omega)$ is an isotropic state, and therefore is entangled if and only if $\mathrm{Tr}[\Lambda\omega] \leqslant \frac{1}{m}$ (we used again (13.19)). Since $\Lambda = \mathcal{E}^*(\Phi_m)$ we conclude that

$$T\left(\rho \xrightarrow{\text{RNG}} \Phi_m\right) = 1 - \sup_{\Lambda \in \mathrm{Eff}(AB)} \left\{ \mathrm{Tr}[\Lambda\rho] : \mathrm{Tr}[\Lambda\sigma] \leqslant \frac{1}{m}, \forall \sigma \in \mathrm{SEP}(AB) \right\}. \quad (13.367)$$

Therefore, the distillation is given by

$$\mathrm{Distill}^\varepsilon(\rho) := \sup_{m \in \mathbb{N}} \left\{ \log m \; : \; T\left(\rho \xrightarrow{\text{RNG}} \Phi_m\right) \leqslant \varepsilon \right\}$$

$$= \sup_{m \in \mathbb{N}} \left\{ \log m : \mathrm{Tr}[\Lambda\rho] \geqslant 1-\varepsilon, \; \max_{\sigma \in \mathrm{SEP}(AB)} \mathrm{Tr}[\Lambda\sigma] \leqslant \frac{1}{m}, , \Lambda \in \mathrm{Eff}(AB) \right\}. \quad (13.368)$$

The condition that $m$ is no greater than the reciprocal of $\max_{\sigma \in \mathrm{SEP}(AB)} \mathrm{Tr}[\Lambda\sigma]$ implies the following:

$$\mathrm{Distill}^\varepsilon(\rho) \leqslant - \log \min_{\substack{\Lambda \in \mathrm{Eff}(AB) \\ \mathrm{Tr}[\Lambda\rho] \geqslant 1-\varepsilon}} \max_{\sigma \in \mathrm{SEP}(AB)} \mathrm{Tr}[\Lambda\sigma]$$

$$= - \log \max_{\sigma \in \mathrm{SEP}(AB)} \min_{\substack{\Lambda \in \mathrm{Eff}(AB) \\ \mathrm{Tr}[\Lambda\rho] \geqslant 1-\varepsilon}} \mathrm{Tr}[\Lambda\sigma] \quad (13.369)$$

$$= \min_{\sigma \in \mathrm{SEP}(AB)} D_{\min}^\varepsilon(\rho\|\sigma)$$

$$= D_{\min}^\varepsilon(\rho\|\mathrm{SEP}).$$

**Exercise 13.71.** *Using the same notations as above, show that*

$$\mathrm{Distill}^\varepsilon(\rho) = \log\left\lfloor 2^{D_{\min}^\varepsilon(\rho\|\mathrm{SEP})} \right\rfloor. \quad (13.370)$$

*Hint: Observe that the optimal $m$ in (13.368) is the floor of the reciprocal of $\max_{\sigma \in \mathrm{SEP}(AB)} \mathrm{Tr}[\Lambda\sigma]$.*

## 13.8.3 The Asymptotic Regime

The results obtained in the single-shot regime lead directly to the following expressions for the cost and distillation of a bipartite state $\rho \in \mathfrak{D}(AB)$ under nonentangling operations:

$$\mathrm{Cost}\left(\rho^{AB}\right) = \lim_{\varepsilon \to 0} \liminf_{n \to \infty} \frac{1}{n} \mathbf{LR}^\varepsilon\left(\rho^{\otimes n}\right),$$

$$\mathrm{Distill}\left(\rho^{AB}\right) = \lim_{\varepsilon \to 0} \limsup_{n \to \infty} \frac{1}{n} D_{\min}^\varepsilon\left(\rho^{\otimes n}\|\mathrm{SEP}\right). \quad (13.371)$$

From these expressions, it becomes evident that if the generalized quantum Stein's lemma (as proposed in Conjecture 11.4) is valid, then the distillable entanglement under nonentangling operations would be equal to the regularized relative entropy of entanglement.

In contrast, regarding the entanglement cost, it is known (as detailed in the section on "Notes and References" at the end of this chapter) that there are states for which the entanglement cost is strictly greater than the distillable entanglement. This implies that even under a broad range of nonentangling operations, the reversibility of mixed state entanglement is not guaranteed. In essence, this reflects a fundamental asymmetry in the processes of creating and extracting entanglement from quantum systems.

## 13.9 Beyond LOCC: NPT-Entanglement Theory

In this subsection, we delve into the quantum resource theory where $\mathfrak{F}(AB)$ is defined as the set of PPT states within $\mathfrak{D}(AB)$, and $\mathfrak{F}(AB \to A'B')$ as the set of completely PPT preserving quantum channels. The exploration of this resource theory is not only intriguing from a theoretical standpoint but is also driven by the fact that the set of completely PPT preserving quantum channels encompasses LOCC. Consequently, the entanglement cost and distillation rates determined under these operations offer lower and upper bounds, respectively, on the corresponding rates under LOCC.

In the framework of completely PPT-preserving operations, entangled states that exhibit a positive partial transpose are regarded as free resources. Accordingly, in this resource theory, the focus is on *NPT-entanglement*, emphasizing interest in entangled states with a negative partial transpose (NPT), which are states whose partial transpose is not positive semidefinite. In the rest of this section, we will use the notation PPT$(AB)$ to refer to the set of all density matrices in $\mathfrak{D}(AB)$ that have a positive partial transpose. Thus, we obtain the following relation:

$$\mathrm{SEP}(AB) \subset \mathrm{PPT}(AB) \subset \mathfrak{D}(AB). \tag{13.372}$$

### 13.9.1 The Set of Completely PPT Preserving Operations

**Definition 13.11.** Let $\mathcal{E} \in \mathfrak{L}(AB \to A'B')$ be a linear map. We say that $\mathcal{E}$ is PPT preserving if

$$\mathcal{E}(\rho) \in \mathrm{PPT}\left(A'B'\right) \qquad \forall \, \rho \in \mathrm{PPT}(AB). \tag{13.373}$$

Moreover, we say the $\mathcal{E}^{AB \to A'B'}$ is completely PPT preserving if $\mathrm{id}^{A''B''} \otimes \mathcal{E}^{AB \to A'B'}$ is PPT preserving for every bipartite system $A''B''$.

To characterize the set of completely PPT preserving operations, we begin by defining the partial transpose of a linear map. Let $\mathcal{E} \in \mathfrak{L}(AB \to A'B')$ be a linear map. The partial transpose of $\mathcal{E}$ is defined as follows:

$$\mathcal{E}^{\Gamma}(\omega) := \left(\mathcal{E}\left(\omega^{\Gamma}\right)\right)^{\Gamma} \qquad \forall \, \omega \in \mathfrak{L}(AB), \tag{13.374}$$

where the superscript $\Gamma$ indicates partial transpose w.r.t. Bob's systems (see (13.148)). In the following exercise you prove some of the key properties of this extension of partial transpose to linear maps.

**Exercise 13.72.** *Let $\mathcal{E} \in \mathfrak{L}(AB \to A'B')$ be a bipartite linear map, $\mathcal{E}^\Gamma$ be its partial transpose, and $J_\mathcal{E} := J_\mathcal{E}^{ABA'B'}$ be its Choi matrix. Prove the following statements:*

1. *$\mathcal{E}$ is PPT preserving if and only if $\mathcal{E}^\Gamma$ is PPT preserving.*
2. *$\mathcal{E}$ is completely PPT preserving if and only if $\mathcal{E}^\Gamma$ is completely PPT preserving.*
3. *$\mathcal{E}$ satisfies*

$$J_{\mathcal{E}^\Gamma} = J_\mathcal{E}^\Gamma, \tag{13.375}$$

*where on the right-hand side the superscript $\Gamma$ denotes the partial transpose of $J_\mathcal{E}^{ABA'B'}$ with respect to both $B$ and $B'$.*

4. *$\mathcal{E}$ satisfies*

$$\left(\mathcal{E}^*\right)^\Gamma = \left(\mathcal{E}^\Gamma\right)^*. \tag{13.376}$$

---

**Theorem 13.23.** Let $\mathcal{E} \in \mathrm{CPTP}(AB \to A'B')$ be a bipartite quantum channel. Then, the following are equivalent:

1. $\mathcal{E}$ is completely PPT preserving.
2. $\mathcal{E}^\Gamma$ is a quantum channel.

---

**Proof** We can use 13.375 to observe that $\mathcal{E}^\Gamma$ is a quantum channel if and only if $J_\mathcal{E}^\Gamma \geqslant 0$. Thus, it suffices to show that $\mathcal{E}$ is completely PPT preserving if and only if its Choi matrix is PPT. To see this, note that the Choi matrix of $\mathcal{E}^{AB \to A'B'}$ can be expressed as

$$J_\mathcal{E}^{ABA'B'} = \mathcal{E}^{\tilde{A}\tilde{B} \to A'B'}\left(\Omega^{(AB)(\tilde{A}\tilde{B})}\right), \tag{13.377}$$

where $\Omega^{(AB)(\tilde{A}\tilde{B})}$ is an unnormalized maximally entangled state between system $\tilde{A}\tilde{B}$ and system $AB$. Furthermore, we can write

$$\Omega^{(AB)(\tilde{A}\tilde{B})} = \Omega^{A\tilde{A}} \otimes \Omega^{B\tilde{B}}, \tag{13.378}$$

where $\Omega^{A\tilde{A}}$ and $\Omega^{B\tilde{B}}$ are unnormalized maximally entangled states between the respective systems. Since we take the partial transpose with respect to system $B\tilde{B}$, we get from the equation (13.378) that $\Omega^{(AB)(\tilde{A}\tilde{B})}$ is PPT with respect to $B\tilde{B}$. Therefore, if $\mathcal{E}$ is completely PPT preserving, then its Choi matrix $J_\mathcal{E}^{ABA'B'}$ must be PPT, since $\Omega^{ABA'B'}$ is PPT.

Conversely, suppose $J_\mathcal{E}^\Gamma \geqslant 0$ (i.e. $\mathcal{E}^\Gamma$ is a quantum channel) and let $\rho^{A''B''AB}$ be a PPT state with respect to system $B''B$. For simplicity of the exposition here, we use the superscript $\Gamma$ to indicate partial transpose on all systems on Bob's side (i.e. for $\rho^{AB}$, the superscript $\Gamma$ in $\rho^\Gamma$ stands for partial transpose w.r.t. $B$, and for $\rho^{A''B''AB}$, the superscript in $\rho^\Gamma$ stands for partial transpose w.r.t. to system $B''B$). Then,

$$\left(\mathcal{E}^{AB \to A'B'}\left(\rho^{A''B''AB}\right)\right)^\Gamma = \mathcal{E}^\Gamma\left(\rho^\Gamma\right) \geqslant 0, \tag{13.379}$$

since $\rho^{\Gamma} \geqslant 0$ and $\mathcal{E}^{\Gamma}$ is completely positive. Therefore, the state $\mathcal{E}^{AB \to A'B'}\left(\rho^{A''B''AB}\right)$ is PPT, and since $\rho^{A''B''AB}$ was an arbitrary PPT state in $\mathfrak{D}(A''B''AB)$ we conclude that $\mathcal{E}$ is completely PPT preserving. This completes the proof. ∎

We denote by $\text{PPT}(AB \to A'B')$ the set of all completely PPT preserving channels in $\text{CPTP}(AB \to A'B')$.

**Exercise 13.73.** *Recall that $LOCC(AB \to A'B')$ is a subset of $\text{SEP}(AB \to A'B')$; see (12.5). Show that*

$$\text{SEP}(AB \to A'B') \subset \text{PPT}(AB \to A'B'). \tag{13.380}$$

**Exercise 13.74.** *Show that if $\mathcal{E} \in \text{PPT}(AB \to A'B')$, then there exists a channel $\mathcal{N} \in \text{PPT}(AB \to A'B')$, such that $\mathcal{E} = \mathcal{N}^{\Gamma}$.*

## Monotonicity of Entanglement Measures

Theorem 13.23 states that a quantum channel $\mathcal{E} \in \text{CPTP}(AB \to A'B')$ is completely PPT preserving if and only if its partial transpose $\mathcal{E}^{\Gamma}$ is also completely positive. Using this property, we can show that measures of entanglement based on the partial transpose behave monotonically under completely PPT preserving operations. Indeed, for $\rho \in \mathfrak{D}(AB)$ and $\mathcal{E} \in \text{PPT}(AB \to A'B')$, we have $(\mathcal{E}(\rho))^{\Gamma} = \mathcal{E}^{\Gamma}(\rho^{\Gamma})$, which gives $\left\|(\mathcal{E}(\rho))^{\Gamma}\right\|_1 = \|\mathcal{E}^{\Gamma}(\rho^{\Gamma})\|_1 \leqslant \|\rho^{\Gamma}\|_1$ by DPI and the fact that $\mathcal{E}^{\Gamma}$ is a quantum channel. Therefore, both the negativity and the logarithmic negativity behave monotonically under completely PPT preserving operations.

To see that $E_\kappa$ also behaves monotonically, consider an optimal operator $\Lambda \in \text{Pos}(AB)$ that satisfies $-\Lambda^{\Gamma} \leqslant \rho^{\Gamma} \leqslant \Lambda^{\Gamma}$ and $E_\kappa(\rho) = \log \text{Tr}[\Lambda]$. Then, for all $\mathcal{E} \in \text{PPT}(AB \to A'B')$, we have

$$-\mathcal{E}^{\Gamma}\left(\Lambda^{\Gamma}\right) \leqslant \mathcal{E}^{\Gamma}\left(\rho^{\Gamma}\right) \leqslant \mathcal{E}^{\Gamma}\left(\Lambda^{\Gamma}\right) \tag{13.381}$$

since $\mathcal{E}^{\Gamma}$ is CP. Since $(\mathcal{E}(\rho))^{\Gamma} = \mathcal{E}^{\Gamma}(\rho^{\Gamma})$, we get that the above equation is equivalent to

$$-\Lambda'^{\Gamma} \leqslant \left(\mathcal{E}(\rho)\right)^{\Gamma} \leqslant \Lambda'^{\Gamma}, \quad \text{where} \quad \Lambda' := \mathcal{E}(\Lambda). \tag{13.382}$$

Therefore, $\Lambda'$ is a feasible solution in the optimization problem of $E_\kappa\left(\mathcal{E}(\rho)\right)$ so that

$$E_\kappa\left(\mathcal{E}(\rho)\right) \geqslant \log \text{Tr}\left[\Lambda'\right] = \log \text{Tr}[\Lambda] = E_\kappa(\rho). \tag{13.383}$$

That is, $E_\kappa$ behaves monotonically under completely PPT preserving operations.

**Exercise 13.75.** *Consider the linear map $\mathcal{E} \in \text{CPTP}(AB \to AB)$ with $m := |A| = |B|$ defined for all $\omega \in \mathfrak{L}(AB)$ as*

$$\mathcal{E}(\omega) := \text{Tr}[\Lambda\omega]\mathbf{u}^{AB} + \text{Tr}[(I - \Lambda)\omega]\Phi_m^{AB}, \tag{13.384}$$

*where*

$$\Lambda := \frac{1}{m+1}\left(I^{AB} - \Phi_m^{AB}\right)^{\Gamma}. \tag{13.385}$$

1. *Show that $\Lambda \in \text{Eff}(AB)$.*
2. *Show that $\mathcal{E}$ is PPT preserving (but not necessarily completely PPT preserving).*
3. *Show that for $m > 3$ we have $N\left(\mathcal{E}\left(\rho_w^{AB}\right)\right) > N\left(\rho_w^{AB}\right)$, where $\rho_w^{AB}$ is the maximally entangled Werner state (see (13.30) with $\alpha = 1$)*

$$\rho_w^{AB} = \frac{1}{d(d-1)}\left(I^{AB} - F^{AB}\right). \tag{13.386}$$

This exercise demonstrates that the negativity measure, in general, does not exhibit monotonic behavior under PPT-preserving operations, but as we saw earlier, it does exhibit monotonic behavior under *completely* PPT-preserving operations.

## 13.9.2 The PPT Conversion Distance

In this section, we aim to simplify the calculation of the conversion distance under PPT operations. Since our focus is on the cost and distillation of entanglement under PPT operations, we will only consider the conversion distance $T(\rho \xrightarrow{\text{PPT}} \sigma)$ when either $\rho$ or $\sigma$ is maximally entangled. We begin with the case where $\sigma$ is maximally entangled.

### The PPT Conversion Distance to a Maximally Entangled State

Very similar to Lemma 13.3, the conversion distance under completely PPT preserving operations is given by

$$T\left(\rho^{AB} \xrightarrow{\text{PPT}} \Phi_m^{A'B'}\right) = 1 - \sup_{\mathcal{E} \in \text{PPT}} \text{Tr}\left[\Phi_m^{A'B'} \mathcal{E}^{AB \to A'B'}\left(\rho^{AB}\right)\right]. \tag{13.387}$$

In order to simplify this expression for the conversion distance, we will need the following lemma.

> **Lemma 13.6.** Let $A, B, A'$, and $B'$ be four quantum systems $m := |A'| = |B'|$, and let $\Lambda \in \text{Herm}(AB)$. The following are equivalent:
>
> 1. There exists $\mathcal{E} \in \text{PPT}(AB \to A'B')$ such that $\Lambda^{AB} = \mathcal{E}^*\left(\Phi_m^{A'B'}\right)$.
> 2. $\Lambda \in \text{Eff}(AB)$ and $\left\|\Lambda^{\Gamma}\right\|_\infty \leq \frac{1}{m}$.

**Proof**  Suppose first that $\Lambda = \mathcal{E}^*(\Phi_m)$ for some $\mathcal{E} \in \text{PPT}(AB \to A'B')$. Since $\mathcal{E}$ is a CPTP map it follows that $\mathcal{E}^*$ is a unital CP map. Therefore, since $0 \leq \Phi_m \leq I^{A'B'}$ we have $0 \leq \Lambda \leq I^{AB}$. Moreover, observe that

$$\Lambda^{\Gamma} = \left(\mathcal{E}^*(\Phi_m)\right)^{\Gamma} = \left(\mathcal{E}^*\right)^{\Gamma}\left(\Phi_m^{\Gamma}\right)$$
$$\textbf{(13.376)} \rightarrow = \left(\mathcal{E}^{\Gamma}\right)^*\left(\Phi_m^{\Gamma}\right) \tag{13.388}$$
$$= \frac{1}{m}\left(\mathcal{E}^{\Gamma}\right)^*\left(F^{A'B'}\right),$$

where $F^{AB}$ is the flip operator. Since $\mathcal{E}^\Gamma$ is also a CPTP map (see Theorem 13.23), it follows that $(\mathcal{E}^\Gamma)^*$ is a unital CP map. Combining this with the fact that $-I^{AB} \leqslant F^{AB} \leqslant I^{AB}$ (recall $F^2 = I^{AB}$) we conclude that

$$-\frac{1}{m}I^{AB} \leqslant \Lambda^\Gamma \leqslant \frac{1}{m}I^{AB}, \tag{13.389}$$

which is equivalent to $\left\| \Lambda^\Gamma \right\|_\infty \leqslant \frac{1}{m}$.

Conversely, suppose $\Lambda \in \mathrm{Eff}(AB)$ and $\left\| \Lambda^\Gamma \right\|_\infty \leqslant \frac{1}{m}$. Define the measurement-prepare channel $\mathcal{E} \in \mathrm{CPTP}(AB \to A'B')$ as

$$\mathcal{E}(\omega) := \mathrm{Tr}[\Lambda\omega]\, \Phi_m + \mathrm{Tr}[(I - \Lambda)\omega]\, \tau \qquad \forall\, \omega \in \mathfrak{L}(AB), \tag{13.390}$$

where $\tau := (I - \Phi_m)/(m^2 - 1) \in \mathfrak{D}(A'B')$. Observe that $\Lambda = \mathcal{E}^*(\Phi_m)$ (can you see why?). The intuition behind the definition in (13.390) comes from the observation that the optimization over the PPT channels in (13.387) can be further restricted to channels that satisfy $\mathcal{E} = \mathcal{G} \circ \mathcal{E}$, which according to (3.257) have the form of the channel in (13.390). It is therefore left to show that $\mathcal{E}$ as defined here is a PPT quantum channel.

Indeed, observe that by definition for all $\omega \in \mathfrak{L}(AB)$ we have

$$\begin{aligned}
\mathcal{E}^\Gamma(\omega) &= \left( \mathcal{E}\left(\omega^\Gamma\right) \right)^\Gamma \\
&= \mathrm{Tr}\left[\Lambda\omega^\Gamma\right] \Phi_m^\Gamma + \mathrm{Tr}\left[(I - \Lambda)\omega^\Gamma\right]\tau^\Gamma \\
\xrightarrow[\text{is self-adjoint}]{\textbf{The partial transpose}} &= \mathrm{Tr}\left[\Lambda^\Gamma\omega\right] \Phi_m^\Gamma + \mathrm{Tr}\left[(I - \Lambda^\Gamma)\omega\right]\tau^\Gamma.
\end{aligned} \tag{13.391}$$

Now, from (C.186 of online version) and (C.188 of online version) we get that

$$\Phi_m^\Gamma = \frac{1}{m}F^{A'B'} = \frac{1}{m}\left(\Pi_{\mathrm{Sym}} - \Pi_{\mathrm{Asy}}\right), \tag{13.392}$$

and (recall $I = \Pi_{\mathrm{Sym}} + \Pi_{\mathrm{Asy}}$)

$$\tau^\Gamma = \frac{I - \Phi_m^\Gamma}{m^2 - 1} = \frac{1}{m(m+1)}\Pi_{\mathrm{Sym}} + \frac{1}{m(m-1)}\Pi_{\mathrm{Asy}}. \tag{13.393}$$

Substituting these expressions for $\Phi_m^\Gamma$ and $\tau^\Gamma$ into (13.391) (and rearranging terms) gives (see Exercise 13.76)

$$\mathcal{E}^\Gamma(\omega) = \mathrm{Tr}\left[\Lambda'\omega\right]\sigma_{\mathrm{Sym}} + \mathrm{Tr}\left[(I - \Lambda')\omega\right]\sigma_{\mathrm{Asy}}, \tag{13.394}$$

where

$$\Lambda' := \frac{1}{2}(I + m\Lambda), \quad \sigma_{\mathrm{Sym}} := \frac{2}{m(m+1)}\Pi_{\mathrm{Sym}}, \quad \sigma_{\mathrm{Asy}} := \frac{2}{m(m-1)}\Pi_{\mathrm{Asy}}. \tag{13.395}$$

Hence, since $\|\Lambda\|_\infty \leqslant 1/m$ we get that $\Lambda' \in \mathrm{Eff}(A'B')$ so that $\mathcal{E}^\Gamma$ is itself a measurement-prepare quantum channel. That is, $\mathcal{E}$ is indeed a PPT channel. This completes the proof. $\blacksquare$

**Exercise 13.76.** *Give more details for the derivation of* (13.394).

We therefore get the following corollary.

> **Corollary 13.2.** The PPT conversion distance to a maximally entangled state is given by
>
> $$T\left(\rho^{AB} \xrightarrow{\text{PPT}} \Phi_m^{A'B'}\right) = 1 - \max_{\substack{\Lambda \in \text{Eff}(AB) \\ \|\Lambda^{\Gamma}\|_{\infty} \leqslant \frac{1}{m}}} \text{Tr}\left[\Lambda^{AB} \rho^{AB}\right]. \qquad (13.396)$$

**Exercise 13.77.** *Use Lemma 13.6 to prove the corollary.*

## The PPT Conversion Distance from a Maximally Entangled State

The conversion distance from a maximally entangled state is given by

$$T\left(\Phi_m^{A'B'} \xrightarrow{\text{PPT}} \rho^{AB}\right) = \frac{1}{2} \min_{\mathcal{E} \in \text{PPT}} \left\|\rho^{AB} - \mathcal{E}^{A'B' \to AB}\left(\Phi_m^{A'B'}\right)\right\|_1. \qquad (13.397)$$

In the following lemma we provide a characterization of the density matrix $\mathcal{E}^{A'B' \to AB}\left(\Phi_m^{A'B'}\right)$.

> **Lemma 13.7.** Let $A, B, A'$, and $B'$ be four quantum systems $m := |A'| = |B'|$, and let $\sigma \in \mathfrak{D}(AB)$. The following are equivalent:
>
> 1. There exists $\mathcal{E} \in \text{PPT}(AB \to A'B')$ such that $\sigma^{AB} = \mathcal{E}\left(\Phi_m^{A'B'}\right)$.
> 2. There exists $\omega \in \mathfrak{D}(AB)$ such that
>
> $$(1 - m)\omega^{\Gamma} \leqslant \sigma^{\Gamma} \leqslant (1 + m)\omega^{\Gamma}. \qquad (13.398)$$

*Remark.* The condition presented in equation (13.398) leads to the implication that $(m + 1)\omega^{\Gamma} \geqslant (1 - m)\omega^{\Gamma}$. This inequality can be simplified and equivalently restated as $m\omega^{\Gamma} \geqslant 0$. Thus, the partial transpose of $\omega$ is positive semidefinite, so that $\omega \in \text{PPT}(AB)$.

**Proof** Let $\mathcal{G} \in \text{CPTP}(A'B' \to A'B')$ be the twirling channel as defined in (3.251). Recall that $\mathcal{G}$ is an LOCC channel, and thus it is completely PPT preserving. If $\sigma = \mathcal{E}(\Phi_m)$ for some PPT channel $\mathcal{E}$, we can assume without loss of generality that $\mathcal{E} = \mathcal{E} \circ \mathcal{G}$. Otherwise, we can replace $\mathcal{E}$ with $\mathcal{E}' = \mathcal{E} \circ \mathcal{G}$, which has the desired property. From (3.256) it then follows that for all $\eta \in \mathcal{L}(AB)$

$$\mathcal{E}(\eta) = \text{Tr}\left[\eta \Phi_m\right]\sigma + \text{Tr}\left[(I - \Phi_m)\eta\right]\omega, \qquad (13.399)$$

where we replaced $\omega_1$ in (3.256) with $\mathcal{E}(\Phi_m) = \sigma$ and renamed the density matrix $\omega_2$ as $\omega \in \mathfrak{D}(AB)$. The partial transpose of $\mathcal{E}$ is given for all $\eta \in \mathcal{L}(AB)$ as

$$\mathcal{E}^{\Gamma}(\eta) = \text{Tr}\left[\eta^{\Gamma}\Phi_m\right]\sigma^{\Gamma} + \text{Tr}\left[(I - \Phi_m)\eta^{\Gamma}\right]\omega^{\Gamma}$$

$$\overset{\text{Partial transpose}}{\underset{\text{is self-adjoint}}{\longrightarrow}} = \text{Tr}\left[\eta\Phi_m^{\Gamma}\right]\sigma^{\Gamma} + \text{Tr}\left[(I - \Phi_m^{\Gamma})\eta\right]\omega^{\Gamma}. \qquad (13.400)$$

We next use (13.392) to express $\Phi_m^{\Gamma}$ in terms of the symmetric and antisymmetric projectors. Hence,

$$
\begin{aligned}
\mathcal{E}^{\Gamma}(\eta) &= \frac{1}{m}\mathrm{Tr}\left[\eta\left(\Pi_{\mathrm{Sym}} - \Pi_{\mathrm{Asy}}\right)\right]\sigma^{\Gamma} + \frac{1}{m}\mathrm{Tr}\left[\left((m-1)\Pi_{\mathrm{Sym}} + (m+1)\Pi_{\mathrm{Asy}}\right)\eta\right]\omega^{\Gamma} \\
&= \frac{1}{m}\mathrm{Tr}\left[\eta\Pi_{\mathrm{Sym}}\right]\left(\sigma^{\Gamma} + (m-1)\,\omega^{\Gamma}\right) + \frac{1}{m}\mathrm{Tr}\left[\eta\Pi_{\mathrm{Asy}}\right]\left((1+m)\,\omega^{\Gamma} - \sigma^{\Gamma}\right).
\end{aligned}
$$

$$(13.401)$$

Hence, $\mathcal{E}^{\Gamma}$ is completely positive if and only if the matrices $\sigma$ and $\omega$ satisfy (13.398). This completes the proof. ∎

From Lemma 13.7 it follows that the conversion distance can be expressed as

$$
T\left(\Phi_m^{A'B'} \xrightarrow{\mathrm{PPT}} \rho^{AB}\right) = \min_{\omega,\sigma\in\mathfrak{D}(AB)}\left\{\frac{1}{2}\|\rho - \sigma\|_1 : (1-m)\omega^{\Gamma} \leqslant \sigma^{\Gamma} \leqslant (1+m)\omega^{\Gamma}\right\}.
$$

$$(13.402)$$

Now that we have obtained formulas for the two types of PPT-conversion distances in this subsection, we can use them for the operational tasks of entanglement distillation and entanglement cost.

### 13.9.3 Distillation of NPT Entanglement

In this subsection, we express the single-shot distillable NPT-entanglement in terms of the hypothesis testing divergence and show that it can be computed using an SDP program. We then use it to derive a tight upper bound on the asymptotic distillable NPT-entanglement, known as Rains' bound. Since the distillable NPT-entanglement is not smaller than the LOCC distillable entanglement, we can conclude that this upper bound also applies to the distillable entanglement under LOCC.

### Single-Shot Distillable NPT-Entanglement

For every $\varepsilon \in (0, 1)$, the $\varepsilon$-single-shot distillable NPT-entanglement of a bipartite state $\rho \in \mathfrak{D}(AB)$ is defined as

$$
\mathrm{Distill}^{\varepsilon}\left(\rho^{AB}\right) = \sup_{m\in\mathbb{N}}\left\{\log m \;:\; T\left(\rho^{AB} \xrightarrow{\mathrm{PPT}} \Phi_m^{A'B'}\right) \leqslant \varepsilon\right\}. \tag{13.403}
$$

---

**Theorem 13.24.** Let $\rho \in \mathfrak{D}(AB)$ and $\varepsilon \in (0,1)$. Then, the $\varepsilon$-single-shot distillable NPT-entanglement is given by

$$
\mathrm{Distill}^{\varepsilon}\left(\rho^{AB}\right) = \min_{\substack{\|\eta^{\Gamma}\|_1 \leqslant 1 \\ \eta\in\mathrm{Herm}(AB)}} \log\left\lfloor 2^{D_{\min}^{\varepsilon}\left(\rho^{AB}\|\eta^{AB}\right)}\right\rfloor, \tag{13.404}
$$

where the definition of the hypothesis testing divergence in (13.404) has been extended to operators that are not necessarily density matrices.

**Proof** Using the expression for the conversion distance given in (13.396) we get

$$\text{Distill}^{\varepsilon}(\rho) = \sup_{m \in \mathbb{N}} \left\{ \log m : \text{Tr}[\Lambda\rho] \geq 1-\varepsilon, \left\|\Lambda^{\Gamma}\right\|_{\infty} \leq \frac{1}{m}, \Lambda \in \text{Eff}(AB) \right\}$$

$$\boxed{m = \left\lfloor \frac{1}{\left\|\Lambda^{\Gamma}\right\|_{\infty}} \right\rfloor} \longrightarrow = \max \left\{ \log \left\lfloor \frac{1}{\left\|\Lambda^{\Gamma}\right\|_{\infty}} \right\rfloor : \text{Tr}[\Lambda\rho] \geq 1 - \varepsilon, \Lambda \in \text{Eff}(AB) \right\}$$

$$(13.405)$$

Now, from Exercise 2.34 we have that

$$\left\|\Lambda^{\Gamma}\right\|_{\infty} = \max_{\substack{\|\eta\|_1 \leq 1 \\ \eta \in \text{Herm}(AB)}} \text{Tr}\left[\Lambda^{\Gamma}\eta\right]$$

$$\textbf{$\Gamma$ is self-adjoint} \rightarrow = \max_{\substack{\|\eta\|_1 \leq 1 \\ \eta \in \text{Herm}(AB)}} \text{Tr}\left[\Lambda\eta^{\Gamma}\right].$$

$$(13.406)$$

We therefore get that the minimal value that $\left\|\Lambda^{\Gamma}\right\|_{\infty}$ can take under the constraints given in (13.405) is given by

$$\min_{\substack{\text{Tr}[\Lambda\rho] \geq 1-\varepsilon \\ \Lambda \in \text{Eff}(AB)}} \left\|\Lambda^{\Gamma}\right\|_{\infty} = \min_{\substack{\text{Tr}[\Lambda\rho] \geq 1-\varepsilon \\ \Lambda \in \text{Eff}(AB)}} \max_{\substack{\|\eta\|_1 \leq 1 \\ \eta \in \text{Herm}(AB)}} \text{Tr}\left[\Lambda\eta^{\Gamma}\right]$$

$$\textbf{minmax theorem} \rightarrow = \max_{\substack{\|\eta\|_1 \leq 1 \\ \eta \in \text{Herm}(AB)}} \min_{\substack{\text{Tr}[\Lambda\rho] \geq 1-\varepsilon \\ \Lambda \in \text{Eff}(AB)}} \text{Tr}\left[\Lambda\eta^{\Gamma}\right]$$

$$= \max_{\substack{\|\eta\|_1 \leq 1 \\ \eta \in \text{Herm}(AB)}} 2^{-D_{\min}^{\varepsilon}\left(\rho\|\eta^{\Gamma}\right)}$$

$$(13.407)$$

$$= \max_{\substack{\|\eta^{\Gamma}\|_1 \leq 1 \\ \eta \in \text{Herm}(AB)}} 2^{-D_{\min}^{\varepsilon}(\rho\|\eta)}.$$

Substituting this into (13.405) we obtain (13.404). This completes the proof.　∎

**Exercise 13.78.** *Show that the $\varepsilon$-single shot distillable NPT-entanglement can be be computed with an SDP program and find the dual problem of (13.404).*

**Exercise 13.79.** *Let $\varepsilon \in (0,1)$, $\rho \in \mathfrak{D}(AB)$, and $\eta \in \text{Herm}(AB)$. Show that for all $\mathcal{E} \in \text{CPTP}(AB \to A'B')$ we have*

$$D_{\min}^{\varepsilon}\left(\mathcal{E}(\rho)\|\mathcal{E}(\eta)\right) \leq D_{\min}^{\varepsilon}(\rho\|\eta). \tag{13.408}$$

*That is, the DPI with $D_{\min}^{\varepsilon}$ still holds even if $\eta$ is not positive semidefinite.*

For states that have a certain symmetry, the optimization problem given in (13.404) can be performed analytically. For example, consider the Werner state, $\rho_{\text{w}}^{AB}$, as defined in (13.29). This state is invariant under the twirling map $\mathcal{G} \in \text{CPTP}(AB \to AB)$ as defined in (7.199). Let $\eta \in \text{Herm}(AB)$ be an optimal matrix such that

$$\text{Distill}^{\varepsilon}\left(\rho_{\text{w}}^{AB}\right) = D_{\min}^{\varepsilon}\left(\rho_{\text{w}}^{AB}\|\eta^{AB}\right). \tag{13.409}$$

Due to the invariance property of $\rho_{\text{w}}^{AB}$ we have

$$D_{\min}^{\varepsilon}(\rho_{\rm w}\|\eta) \geqslant D_{\min}^{\varepsilon}\big(\mathcal{G}(\rho_{\rm w})\|\mathcal{G}(\eta)\big)$$
$$= D_{\min}^{\varepsilon}\big(\rho_{\rm w}\|\mathcal{G}(\eta)\big). \tag{13.410}$$

Moreover, since $\mathcal{G} \in {\rm PPT}(AB \to AB)$ and $\|\eta^{\Gamma}\|_1 \leqslant 1$, we get that also $\zeta := \mathcal{G}(\eta)$ satisfies

$$\|\zeta^{\Gamma}\|_1 = \|\mathcal{G}^{\Gamma}\,(\eta^{\Gamma})\|_1$$
$$\overset{\textbf{DPI}}{\underset{}{\leqslant}} \|\eta^{\Gamma}\|_1 \leqslant 1. \tag{13.411}$$

Therefore, since $\eta$ was optimal, we must have $D_{\min}^{\varepsilon}(\rho_{\rm w}\|\eta) = D_{\min}^{\varepsilon}(\rho_{\rm w}\|\zeta)$, which means that $\zeta$ is also optimal. To summarize, without loss of generality, we can restrict the optimization in (13.404) to Hermitian matrices $\eta \in {\rm Herm}(AB)$ that satisfy both $\|\eta^{\Gamma}\|_1 \leqslant 1$ and $\mathcal{G}(\eta) = \eta$. This additional condition implies that $\eta$ can be written as a linear combination of $I^{AB}$ and $F^{AB}$, or equivalently, $\eta^{\Gamma}$ can be expressed as

$$\eta^{\Gamma} = a\Phi_m^{AB} + b\tau_m^{AB}, \tag{13.412}$$

for some $a, b \in \mathbb{R}$, and $\tau_m^{AB} := (I^{AB} - \Phi_m^{AB})/(m^2 - 1)$.

**Exercise 13.80.** *Let $\rho_w^{AB}$ be the Werner state with $\alpha > \frac{1}{m}$ (i.e. $\rho_w^{AB}$ is entangled). Show that for $m > 2$*

$$\mathrm{Distill}^{\varepsilon}\left(\rho_w^{AB}\right) = \log\left\lfloor \frac{m(m+\alpha)-2}{m(m-\alpha)(1-\varepsilon)} \right\rfloor. \tag{13.413}$$

*Hint: Optimize over $a, b \in \mathbb{R}$ that appear in (13.412) and note that since $\Phi_m^{AB}\tau_m^{AB} = 0$, we get $\|\eta^{\Gamma}\|_1 = |a| + |b|$, so the condition $\|\eta^{\Gamma}\|_1 \leqslant 1$ becomes equivalent to $|a| + |b| \leqslant 1$.*

In the optimization problem in (13.404), the operator $\eta \in {\rm Herm}(AB)$ does not have to be positive semidefinite. However, if we restrict it to be in ${\rm Pos}(AB)$ (instead of ${\rm Herm}(AB)$) we get the following upper bound.

---

**Corollary 13.3.** Let $\rho \in \mathfrak{D}(AB)$ and $\varepsilon \in (0, 1)$. Then, the $\varepsilon$-single-shot distillable NPT-entanglement is bounded from above by

$$\mathrm{Distill}^{\varepsilon}\left(\rho^{AB}\right) \leqslant \min_{\sigma \in \mathfrak{D}(AB)} \left\{ D_{\min}^{\varepsilon}(\rho\|\sigma) + \mathbf{LN}(\sigma) \right\}, \tag{13.414}$$

where **LN** stands for the logarithmic negativity as defined in (13.164).

---

**Proof** By removing the floor function, and replacing ${\rm Herm}(AB)$ in the right-hand side of (13.404) with the smaller set ${\rm Pos}(AB)$, we get

$$\mathrm{Distill}^{\varepsilon}(\rho) \leqslant \min_{\substack{\|\eta^{\Gamma}\|_1 \leqslant 1 \\ \eta \in \mathrm{Pos}(AB)}} D_{\min}^{\varepsilon}(\rho\|\eta)$$

$$\boxed{\eta = t\sigma} \longrightarrow \; = \min_{\substack{t\|\sigma^{\Gamma}\|_1 \leqslant 1 \\ \sigma \in \mathfrak{D}(AB),\, t \geqslant 0}} \left\{ D_{\min}^{\varepsilon}(\rho\|\sigma) - \log t \right\} \tag{13.415}$$

$$\substack{\textbf{Taking the largest} \\ \textbf{possible } t = 1/\|\sigma^{\Gamma}\|_1} \;\rightarrow\; = \min_{\sigma \in \mathfrak{D}(AB)} \left\{ D_{\min}^{\varepsilon}(\rho\|\sigma) + \log \|\sigma^{\Gamma}\|_1 \right\}.$$

Finally, observe that the second term on the right-hand side of the equation is the logarithmic negativity of $\sigma^{AB}$ as defined in (13.164). This completes the proof. $\blacksquare$

## Asymptotic Distillable NPT-Entanglement

Using Corollary 13.3, we can get the following upper bound on the distillable NPT-entanglement.

---

**Corollary 13.4.** Let $\rho \in \mathfrak{D}(AB)$. Then, the distillable NPT-entanglement is bounded from above by

$$\mathrm{Distill}\left(\rho^{AB}\right) \leqslant \min_{\sigma \in \mathfrak{D}(AB)} \left\{ D(\rho\|\sigma) + \mathbf{LN}(\sigma) \right\}. \tag{13.416}$$

The right-hand side is known as the Rains' bound.

---

**Proof**   The proof follows directly from a combination of Corollary 13.3 and the quantum Steins' lemma given in (8.211). Explicitly, let $\varepsilon \in (0,1)$ and $\sigma \in \mathfrak{D}(AB)$. Then, from Corollary 13.3 we get

$$\limsup_{n \to \infty} \frac{1}{n} \mathrm{Distill}^{\varepsilon}\left(\rho^{\otimes n}\right) \leqslant \limsup_{n \to \infty} \left\{ \frac{1}{n} D_{\min}^{\varepsilon}\left(\rho^{\otimes n} \| \sigma^{\otimes n}\right) + \frac{1}{n}\mathbf{LN}\left(\sigma^{\otimes n}\right) \right\}$$

$$\textbf{(8.211)} + \textbf{Additivity of LN} \rightarrow = D(\rho\|\sigma) + \mathbf{LN}(\sigma).$$

$$\tag{13.417}$$

Since the inequality above holds for all $\sigma \in \mathfrak{D}(AB)$ we can take the minimum over all density matrices so that

$$\limsup_{n \to \infty} \frac{1}{n} \mathrm{Distill}^{\varepsilon}\left(\rho^{\otimes n}\right) \leqslant \min_{\sigma \in \mathfrak{D}(AB)} \left\{ D(\rho\|\sigma) + \mathbf{LN}(\sigma) \right\}. \tag{13.418}$$

Note that this inequality is in fact stronger than (13.416) in the sense that it holds for all $\varepsilon \in (0,1)$. This completes the proof. $\blacksquare$

**Exercise 13.81.** *Show that the Rains' bound is a measure of entanglement and in particular does not increase under completely PPT preserving operations.*

**Exercise 13.82.** *Prove that the Rains' bound for a pure bipartite state $\psi \in \mathrm{Pure}(AB)$ is equal to the entropy of entanglement $E(\psi^{AB})$, which means that on pure bipartite states, the Rains' bound is equal to the distillable entanglement. Hint: Take a look at the proof of Theorem 13.8.*

## 13.9.4 Cost of NPT Entanglement

Using the conversion distance provided in (13.402), we can express the $\varepsilon$-single-shot NPT-entanglement cost as

$$\mathrm{Cost}^{\varepsilon}\left(\rho^{AB}\right) = \min_{m\in\mathbb{N}}\left\{\log\left(m\right) :\ (1-m)\omega^{\Gamma}\leqslant\sigma^{\Gamma}\leqslant(1+m)\omega^{\Gamma},\ \omega\in\mathfrak{D}(AB),\ \sigma\in\mathfrak{B}_{\varepsilon}(\rho)\right\}.$$
$$(13.419)$$

This expression can be simplified by replacing $1\pm m$ with $\pm m$. This change does not change much the entanglement cost.

> **Theorem 13.25.** For all $\varepsilon\in(0,1)$ and all $\rho\in\mathfrak{D}(AB)$ we have
>
> $$\log_2\left(2^{E_{\kappa}^{\varepsilon}(\rho^{AB})}-1\right)\leqslant\mathrm{Cost}^{\varepsilon}\left(\rho^{AB}\right)\leqslant\log_2\left(2^{E_{\kappa}^{\varepsilon}(\rho^{AB})}+1\right),\qquad(13.420)$$
>
> where
>
> $$E_{\kappa}^{\varepsilon}\left(\rho^{AB}\right)=\min_{\rho'\in\mathfrak{B}_{\varepsilon}(\rho)}E_{\kappa}^{\varepsilon}\left(\rho'^{AB}\right)\qquad(13.421)$$
>
> is the smoothed version of the $\kappa$-entanglement defined in (13.167).

**Proof**   Recall from Exercise 11.5 that

$$\mathrm{Cost}^{\varepsilon}\left(\rho\right)=\min_{\rho'\in\mathfrak{B}_{\varepsilon}(\rho)}\mathrm{Cost}_{\mathfrak{F}}^{\varepsilon=0}\left(\rho'\right)\qquad(13.422)$$

(this can also be verified directly from the expression in (13.419)). Therefore, it is sufficient to prove the lemma for the case $\varepsilon=0$. For $\varepsilon=0$ the entanglement cost given in (13.419) takes the form

$$\mathrm{Cost}^{\varepsilon=0}\left(\rho^{AB}\right)=\min_{m\in\mathbb{N}}\left\{\log\left(m\right) :\ (1-m)\omega^{\Gamma}\leqslant\rho^{\Gamma}\leqslant(1+m)\omega^{\Gamma},\ \omega\in\mathfrak{D}(AB)\right\}.$$
$$(13.423)$$

Now, observe that every $m\in\mathbb{N}$ and $\omega\in\mathrm{PPT}(AB)$ that satisfies $(1-m)\omega^{\Gamma}\leqslant\rho^{\Gamma}$ also satisfies $-(1+m)\omega^{\Gamma}\leqslant\rho^{\Gamma}$. Therefore, we get that

$$\mathrm{Cost}^{\varepsilon=0}\left(\rho^{AB}\right)\geqslant\min_{m\in\mathbb{N}}\left\{\log\left(m\right) :\ -(1+m)\omega^{\Gamma}\leqslant\rho^{\Gamma}\leqslant(1+m)\omega^{\Gamma},\ \omega\in\mathfrak{D}(AB)\right\}$$
$$\boxed{\Lambda:=(1+m)\omega}\ \longrightarrow\ =\min_{\Lambda\in\mathrm{Pos}(AB)}\left\{\log\left(\mathrm{Tr}\left[\Lambda\right]-1\right) :\ -\Lambda^{\Gamma}\leqslant\rho^{\Gamma}\leqslant\Lambda^{\Gamma}\right\}$$
$$(\mathbf{13.167})\rightarrow\ =\log_2\left(2^{E_{\kappa}(\rho^{AB})}-1\right).$$
$$(13.424)$$

Similarly, for the other inequality observe that every $m \in \mathbb{N}$ and $\omega \in \text{PPT}(AB)$ that satisfies $\rho^\Gamma \leqslant (m-1)\omega^\Gamma$ also satisfies $\rho^\Gamma \leqslant (m+1)\omega^\Gamma$. Therefore, we get that

$$
\begin{aligned}
\text{Cost}^{\varepsilon=0}\left(\rho^{AB}\right) &\leqslant \min_{m \in \mathbb{N}}\left\{\log(m): \ -(1-m)\omega^\Gamma \leqslant \rho^\Gamma \leqslant (1-m)\omega^\Gamma, \ \omega \in \mathfrak{D}(AB)\right\} \\
\boxed{\Lambda := (m-1)\omega} \longrightarrow &= \min_{\Lambda \in \text{Pos}(AB)}\left\{\log(\text{Tr}[\Lambda]+1): \ -\Lambda^\Gamma \leqslant \rho^\Gamma \leqslant \Lambda^\Gamma\right\} \\
\mathbf{(13.167)} \rightarrow &= \log_2\left(2^{E_\kappa(\rho^{AB})}+1\right).
\end{aligned}
$$

$$(13.425)$$

This completes the proof. $\blacksquare$

Theorem 13.25 demonstrates that the single-shot NPT-entanglement cost is essentially given by the smoothed version of the $\kappa$-entanglement. From the bounds in (13.168) it follows that

$$
\mathbf{LN}^\varepsilon\left(\rho^{AB}\right) \leqslant E_\kappa^\varepsilon\left(\rho^{AB}\right) \leqslant \min_{\substack{\sigma \in \text{PPT}(AB) \\ \omega \in \mathfrak{B}_\varepsilon(\rho)}} D_{\max}\left(|\omega^\Gamma| \,\|\, \sigma\right). \tag{13.426}
$$

While it is true that the lower and upper bounds here may appear to be simpler than computing the $E_\kappa^\varepsilon$ directly using an SDP program, it is important to note that this may not always be the case. In fact, in many instances, computing the bounds may require solving nontrivial optimization problems themselves, and as such may not necessarily be any easier to compute than the original quantity $E_\kappa^\varepsilon$. Therefore, while the bounds can be a useful tool for gaining insight into the behavior of $E_\kappa^\varepsilon$, they should not be relied upon exclusively as a substitute for computing the quantity directly using an SDP program. Moreover, it is not clear to the author how these bounds can be used in deriving computable bounds for the asymptotic NPT-entanglement cost.

## The Asymptotic NPT-Entanglement Cost

From the lower and upper bounds given in (13.420) we get the following expression for the NPT-entanglement cost.

> **Theorem 13.26.** Let $\rho \in \mathfrak{D}(AB)$. The NPT-entanglement cost of $\rho^{AB}$ is given by
>
> $$
> \text{Cost}\left(\rho^{AB}\right) = \lim_{\varepsilon \to 0^+} \lim_{n \to \infty} \frac{1}{n} E_\kappa^\epsilon\left(\rho^{\otimes n}\right). \tag{13.427}
> $$

**Exercise 13.83.** *Prove Theorem 13.26, and in particular show that the limit*

$$
\lim_{n \to \infty} \frac{1}{n} E_\kappa^\epsilon\left(\rho^{\otimes n}\right) \tag{13.428}
$$

*exists for all $\rho \in \mathfrak{D}(AB)$ and all $\varepsilon \in (0,1)$.*

From the theorem it follows that

$$\mathrm{Cost}\left(\rho^{AB}\right) \leqslant E_\kappa\left(\rho^{AB}\right), \tag{13.429}$$

since for all $n \in \mathbb{N}$ and all $\varepsilon \in (0, 1)$ we have

$$E_\kappa^\epsilon\left(\rho^{\otimes n}\right) \leqslant E_\kappa\left(\rho^{\otimes n}\right)$$
$$E_\kappa \text{ is subadditive} \rightarrow \leqslant n E_\kappa(\rho). \tag{13.430}$$

Therefore, (13.429) provides a computable upper bound on the NPT-entanglement cost.

## 13.10 Notes and References

For those interested to learn more about entanglement theory, several comprehensive reviews are available. Two recommended sources are Refs. [135] and [184]. If you're specifically interested in entanglement detection and entanglement witnesses, we suggest checking out Refs. [105] and [51]. For a study of the entanglement properties of isotropic states and reduction criteria, we recommend Ref. [128]. The Werner state, a key example in entanglement theory, was first introduced in Ref. [234].

In 1996, the positive partial transpose (PPT) criterion was proposed in Ref. [180]. In the same year, the proof that a state is separable if and only if it is PPT (as stated in Theorem 13.4) was given in Ref. [129]. Another useful technique for constructing PPT entangled states is to use unextendible product bases (UPB), as demonstrated in the example given in (13.37), which was first proposed in Ref. [22].

The realignment criterion, first introduced in Ref. [44], was developed further in Ref. [197]. If you're interested in the $k$-extendability criterion, take a look at the original paper in Ref. [66].

To learn more about entanglement monotones and convex roof extensions, please refer to Ref. [227]. The closed formulas for the entanglement of formation (13.87) and the concurrence of formation (13.90) were discovered in Ref. [238]. The characterization of pure-to-mixed state conversions, as given in Corollary 13.1 and (13.116), can be attributed to Ref. [242]. The Schmidt rank measure of mixed-state entanglement was initially introduced in Ref. [210], and its interpretation as the maximal extension from the Schmidt rank on pure states was introduced in Ref. [98].

The concept of relative entropy of entanglement was initially introduced in Ref. [220] (also see the review article in Ref. [219]). The robustness of entanglement was first introduced in Ref. [228]. The negativity was first introduced in Ref. [225], its logarithmic version in Refs. [8, 183], and the $\kappa$-entanglement in Ref. [232]. The squashed entanglement, also referred to as the CMI entanglement, was initially mentioned in Eq.(42) of Ref. [214]. However, it was later proven to be an additive measure of entanglement in Ref. [50]. More recently, in Ref. [33], it was shown that the squashed entanglement is a faithful measure of entanglement.

The study of single-shot distillable entanglement was first conducted in Ref. [39] using the quantum spectrum method, which we did not employ in this book. Instead, we utilized other techniques such as the decoupling theorem to obtain the bounds presented in Theorem 13.16 and Theorem 13.15.

The first protocols for asymptotic distillation of mixed states were introduced in the seminal paper in Ref. [21]. The (asymptotic) hashing bound (13.283) and the formula for one-way distillable entanglement, as presented in Theorem 13.18, were discovered in Ref. [63].

Bound entanglement was discovered in Ref. [130]. Since then, extensive research has been conducted on bound entanglement, with various applications in quantum cryptography, such as those presented in Ref. [126], and in distributing entanglement in quantum networks, such as those outlined in Ref. [166].

The study of single-shot entanglement cost was conducted in Ref. [34], where tight lower and upper bounds for single-shot entanglement cost were identified. Here, in Theorem 13.20, we have slightly improved upon those findings by providing the precise expression for the single-shot entanglement cost. The proof for the asymptotic entanglement cost, specifically in Theorem 13.21, was presented in Ref. [117].

The additivity of entanglement of formation under tensor products had been an unresolved problem since the mid-1990s, and it was discovered to be related to three other conjectures, namely, the classical capacity of a quantum channel, and the minimum entropy output of a quantum channel, strong superadditivity of the entanglement of formation, with the latter being the easiest to approach. In a remarkable development [115], it was established that the minimum entropy output of a quantum channel is not additive, disproving all conjectures, including the additivity of entanglement of formation. Nevertheless, an explicit example that demonstrates the nonadditivity of entanglement of formation is still missing.

Theorem 13.22 and the concept of an "entanglement breaking subspace" were first introduced in Ref. [224]. More recent developments on this topic can be found in Ref. [243].

The concept of nonentangling operations (i.e. operations beyond LOCC) was introduced in Ref. [35] for the asymptotic regime, and further developed for the single-shot regime in Ref. [34]. More recently, in Ref. [147], it was demonstrated that the entanglement cost can be strictly greater than the distillable entanglement under nonentangling operations, thereby illustrating that entanglement theory is not reversible even under the broad set of nonentangling operations.

Completely PPT preserving operations (sometimes referred to as PPT operations) were introduced in Ref. [188]. In the same work, among many other findings, the Rains' bound (13.416) on distillable NPT-entanglement was discovered. The monotonicity of negativity and logarithmic negativity under PPT operations was proven in Ref. [183]. The expression presented in (13.404) for the one-shot distillable NPT-entanglement was initially discovered in Ref. [76] (see also Ref. [191] for additional results on distillation beyond LOCC). Lastly, the NPT-entanglement cost was first studied in Ref. [8] and developed further in Ref. [232].

Thus far, our focus has been on entanglement that is shared solely between two parties. However, entanglement is not restricted to bipartite systems and can exist among any number of parties. In this section, we will examine the properties of multipartite entanglement, comparing and contrasting it with bipartite entanglement. It's important to note that the theory of multipartite entanglement can be quite complex. Therefore, in this chapter, we will restrict ourselves to pure multipartite states, and concentrate on simpler cases involving three and four qubits in greater detail.

## 14.1 Stochastic LOCC (SLOCC)

Consider a Hilbert space $A^n$ that is composed of $n$ subsystems, denoted as $A_1, A_2, \cdots, A_n$. In the previous chapter, we discussed the bipartite case where $n = 2$, and showed that the Nielsen majorization theorem can be utilized to determine state-conversion under LOCC. However, as we will see in the upcoming sections, this theorem cannot be applied for $n > 2$, and even for two pure states $\psi$ and $\phi$ that belong to Pure($A^n$), deterministic conversion from $\psi$ to $\phi$ under LOCC is almost always impossible, unless the states are related by local unitaries. Hence, the majority of the research on multipartite entanglement has been dedicated to studying stochastic interconversions.

For two pure states $\psi$ and $\phi$ belonging to Pure($A^n$), we say that $\psi$ can be converted to $\phi$ via stochastic LOCC, or SLOCC, denoted as $\psi \xrightarrow{\text{SLOCC}} \phi$, if there exists a matrix $M_x \in \mathfrak{L}(A_x)$ for each $x \in [n]$ such that $M_x^* M_x \leqslant I^{A_x}$ and

$$|\phi\rangle = M_1 \otimes \cdots \otimes M_n |\psi\rangle. \tag{14.1}$$

This relation implies that $\psi$ can be converted into $\phi$ through local measurements with some nonzero probability, since each $M_x$ can be considered as one Kraus element of a local generalized measurement on system $A_x$.

The set of all states $|\phi\rangle$ in $A^n$ that can be obtained from $|\psi\rangle$ as in (14.1) is called the SLOCC class of $\psi$. The SLOCC class of $\psi$ comprises two types of states: those that can be written in the form (14.1) with *invertible* matrices $M_1, \ldots, M_n$, and those in which at least one of the matrices $M_x$ is noninvertible. In the former case, $\psi$ can be converted to $\phi$ and vice versa using SLOCC, whereas in the latter, the resulting state $|\phi\rangle$ cannot be converted back to $\psi$ via SLOCC.

If $|A_x| = 2$ for some $x \in [n]$ (i.e. the $x$-th subsystem is a qubit) and $M_x$ is noninvertible, then the resulting state $|\phi\rangle$ is a product state between the qubit system $A_x$ and the remaining $n - 1$ subsystems. To demonstrate this, assume $x = 1$ for simplicity, so that $M_1$ is a $2 \times 2$ noninvertible matrix. Since $M_1$ is a rank 1 matrix, it can be written as $M_1 = |u\rangle\langle v|$, where $|u\rangle$ is an unnormalized vector in $A_1$, and $|v\rangle$ is a normalized vector in $A_1$. The state $|\psi\rangle$ can be expressed as

$$|\psi\rangle^{A^n} = a|v\rangle^{A_1}|\psi_1\rangle^{A_2\cdots A_n} + b|v^\perp\rangle^{A_1}|\psi_2\rangle^{A_2\cdots A_n}, \tag{14.2}$$

where $a, b \in \mathbb{C}$, $|v^\perp\rangle \in A_1$ is an orthogonal vector to $|v\rangle$, and $\psi_1$ and $\psi_2$ are some pure states in $A_2 \cdots A_n$. Thus,

$$M_1 \otimes M_2 \otimes \cdots \otimes M_n |\psi\rangle^{A^n} = a|u\rangle^{A_1} \otimes \left( M_2 \otimes \cdots \otimes M_n |\psi_1\rangle^{A_2\cdots A_n} \right), \tag{14.3}$$

which is a product state between system $A_1$ and system $A_2 \cdots A_n$.

**Exercise 14.1.** *Let $\psi, \phi \in \mathrm{Pure}(A^n)$ and suppose there exists matrices $M_1, \ldots, M_n$ such that (14.1) holds. Let $B := A_2 \cdots A_n$, $d := |A_1|$, and suppose $\det(M_1) = 0$. Show that*

$$\mathrm{SR}\left(\phi^{A_1 B}\right) \leqslant d - 1, \tag{14.4}$$

*where we view $\phi^{A_1 B}$ as a bipartite state between $A_1$ and $B$.*

As we observed in the previous exercise, if any of the matrices $M_1, \ldots, M_n$ is noninvertible, then the resulting state $|\phi\rangle$ belongs to a class of states with lower Schmidt rank. Hence, we can categorize $n$-partite entanglement in two stages:

1. Determine the $n$ Schmidt ranks between each subsystem and the other $n - 1$ subsystems.
2. Classify the $n$-partite entanglement based on a fixed set of $n$ Schmidt ranks obtained in the first step.

It is worth noting that for the second step, we only need to consider reversible SLOCC conversions where all matrices $M_1, \ldots, M_n$ in (14.1) are invertible. Hence, states $\psi, \phi \in \mathrm{Pure}(A^n)$ belong to the same reversible SLOCC class if and only if there exists a matrix

$$M \in GL^n := GL(A_1) \times \cdots \times GL(A_n), \tag{14.5}$$

such that $|\phi\rangle = M|\psi\rangle$. Here, for each $x \in [n]$, the set $GL(A_x)$ represents the group of invertible matrices in $\mathfrak{L}(A_x)$. It is also noteworthy that $M$ takes the form of $M_1 \otimes \cdots \otimes M_n$, where $M_x \in GL(A_x)$ for each $x \in [n]$. Please note that our notation $GL^n$ does not explicitly specify the system $A^n = A_1 \cdots A_n$. However, in the rest of this chapter we will assume that the context makes it clear which system we are referring to.

With these observations, we can use certain tools from representation theory to characterize the reversible SLOCC class of $|\psi\rangle$; that is, the set of states $M|\psi\rangle$. To achieve this, we begin by relaxing the normalization condition that $\|M|\psi\rangle\| = 1$ and allowing each $M_x$ to vary over any element of $SL(A_x)$. The group $SL(A_x)$ is a subgroup of $GL(A_x)$ with the property that the determinant of its elements is 1. The limitation to this group can only affect the normalization of the states in $M|\psi\rangle$. Therefore, we will consider the "orbit" of $\psi$ with respect to the group

$$SL^n := SL(A_1) \times \cdots \times SL(A_n). \tag{14.6}$$

Mathematically, the SL-orbit of a pure state $\psi \in \mathrm{Pure}(A^n)$ is defined as

$$SL^n|\psi\rangle := \left\{ M|\psi\rangle : M \in SL^n \right\}. \tag{14.7}$$

By working with SL-orbits rather than GL-orbits, we can classify multipartite entanglement using SL-invariant polynomials.

## 14.2 SL-Invariant Polynomials

Let us recall the bilinear form defined in (13.91). One of its properties (as shown in Exercise 14.5) is that for any two-qubit pure states $|\psi\rangle$ and $|\phi\rangle$

$$\langle M|\psi\rangle, M|\phi\rangle\rangle = \langle|\psi\rangle, |\phi\rangle\rangle \qquad \forall\, M \in SL^2. \tag{14.8}$$

The function $f : \mathbb{C}^2 \otimes \mathbb{C}^2 \to \mathbb{C}$ defined by

$$f(|\psi\rangle) := \langle|\psi\rangle, |\psi\rangle\rangle \qquad \forall\, \psi \in \mathbb{C}^2 \otimes \mathbb{C}^2 \tag{14.9}$$

is a polynomial in the coefficients of $|\psi\rangle := \sum_{x,y\in\{0,1\}} a_{xy}|xy\rangle$. Specifically, it is the polynomial

$$f(|\psi\rangle) = a_{00}a_{11} - a_{01}a_{10}. \tag{14.10}$$

This polynomial is invariant under the group $SL^2$; that is, $f(M|\psi\rangle) = f(|\psi\rangle)$ for any $M \in SL^2$. Hence, we refer to it as an SL-invariant polynomial. Note that if $f(|\psi\rangle) \neq 0$, then $\psi$ is not a product state, and furthermore, all states in the orbit $SL^2|\psi\rangle$ are not product states. On the other hand, if $|\phi\rangle \ /\in SL^2|\psi\rangle$, then $|\phi\rangle$ is necessarily a product state. Therefore, in the case of two qubits, there exist precisely two classes of states: $SL^2|00\rangle$ and $SL^2|\psi\rangle$ for some two-qubit nonproduct state $|\psi\rangle$ (i.e. $f(|\psi\rangle) \neq 0$). Similarly, as we will see shortly, for more general multipartite systems, SL-invariant polynomials can be used to classify all reversible SLOCC classes.

> **Definition 14.1.** A polynomial $f : A^n \to \mathbb{C}$ is called SL-invariant polynomial, or in short SLIP, if for any $M \in SL^n$ and any $\psi \in \mathrm{Pure}(A^n)$ we have $f(M|\psi\rangle) = f(|\psi\rangle)$. The set of all SLIPs on $A^n$ with $f(0) = 0$ is denoted by $\mathrm{SLIP}(A^n)$.

*Remark.* The condition that $f(0) = 0$ is a convention that we will adopt to eliminate trivial SLIPs that are constant for all vectors in $A^n$. Furthermore, we will see shortly that this convention implies that SLIPs vanish on product states.

The set of all SLIPs forms a vector space over $\mathbb{C}$. Additionally, the following exercise reveals that this vector space has a basis consisting of *homogeneous* SLIPs. Therefore, we will concentrate on homogeneous SLIPs of some fixed degree $k \in \mathbb{N}$. For instance, the SLIP in (14.10) is homogeneous of degree 2 since it satisfies $f(c|\psi\rangle) = c^2 f(|\psi\rangle)$ for any $c \in \mathbb{C}$. The dimension of the space of all homogeneous SLIPs of a fixed degree $k$ is finite, but, as we will see, it grows exponentially with $n$.

**Exercise 14.2.** *Show that the vector space of SLIPs has a basis consisting of homogeneous SLIPs.*

**Exercise 14.3.** *Let $AB$ be a bipartite system with $d := |A| = |B|$. Show that the function*

$$f(|\psi\rangle) = \det(M) \qquad \forall \ |\psi\rangle = M \otimes I^B |\Omega^{AB}\rangle \tag{14.11}$$

*is a homogeneous SLIP of degree $d$.*

The degrees of homogeneous SLIPs have a close relationship with the local dimensions of the subsystems. To understand this connection, consider a multipartite system $A^n = A_1 \cdots A_n$, where $m_x := |A_x|$ for each $x \in [n]$. Suppose $f_k : A^n \to \mathbb{C}$ is a homogeneous SLIP of degree $k \in \mathbb{N}$. Note that if $c \in \mathbb{C}$ satisfies $c^{m_x} = 1$ for some $x \in [n]$, then the matrix $cI^{A_x}$ has a determinant of 1, implying that $cI^{A^n} \in SL^n$. Hence, we obtain the following relationship: for every $|\psi\rangle \in A^n$

$$f_k(|\psi\rangle) = f_k\left(cI^{A^n}|\psi\rangle\right) = c^k f_k(|\psi\rangle), \tag{14.12}$$

where the first equality is due to the SL-invariance property of $f_k$, and the second equality is due to the homogeneity of $f_k$. Therefore, as long as $f_k$ is not the zero polynomial, it follows that $c^k = 1$ for any complex number $c$ that satisfies $c^{m_x} = 1$. This means that $m_x$ must divide $k$. Since this holds for all $x$ we can conclude that $k$ is divisible by the least common multiple $r := \mathrm{lcm}(m_1, \ldots, m_n)$.

**Exercise 14.4.** *Let $|\psi\rangle \in A^n$ be a product state; that is, $|\psi\rangle^{A^n} = |\psi_1\rangle^{A_1} \otimes \cdots \otimes |\psi_n\rangle^{A_n}$. Show that for any $f \in \mathrm{SLIP}(A^n)$ we have $f(|\psi\rangle) = 0$.*

Remember that for $\psi, \phi \in \mathrm{Pure}(A^n)$, the relation $\psi \xrightarrow{\text{SLOCC}} \phi$ holds if there exists a matrix $M_x \in \mathfrak{L}(A_x)$ for every $x \in [n]$ such that both $M_x^* M_x \leqslant I^{A_x}$ and $|\phi\rangle = M|\psi\rangle$ are satisfied, where $M := M_1 \otimes \cdots \otimes M_n$. The subsequent lemma establishes that if any of the matrices in the set $\{M_x\}_{x \in [n]}$ exhibits rank deficiency, then any SLIP will be nullified for $|\phi\rangle$.

> **Lemma 14.1.** Let $f : A^n \to \mathbb{C}$ be a SLIP and $|\psi\rangle$ and $|\phi\rangle$ be as above. If there exists $x \in [n]$ such that $\det(M_x) = 0$, then $f(|\phi\rangle) = 0$.

**Proof** Since every SLIP can be expressed as a linear combination of homogeneous SLIPs, we will assume without loss of generality that $f$ is a homogeneous SLIP of degree $k \in \mathbb{N}$. Denote by $d := |A^n|$, and for every $\varepsilon \in [0, 1)$, define $M_\varepsilon := M_1(\varepsilon) \otimes \cdots \otimes$

$M_n(\varepsilon)$, where each $M_x(\varepsilon)$ is a slight perturbation of $M_x$ ensuring that $\det(M_x(\varepsilon)) \neq 0$ for all $\varepsilon \in (0,1)$. As a consequence, $\mu_\varepsilon := \det(M_\varepsilon) \neq 0$ for all $\varepsilon \in (0,1)$, which implies $N_\varepsilon := M_\varepsilon / \mu_\varepsilon^{1/d}$ is an element of $SL^n$. By definition, we can express:

$$f\left(M_\varepsilon|\psi\rangle\right) = f\left(\mu_\varepsilon^{1/d} N_\varepsilon|\psi\rangle\right)$$

$$\boxed{f \text{ is homogeneous of degree } k} \longrightarrow = \mu_\varepsilon^{k/d} f\left(N_\varepsilon|\psi\rangle\right) \qquad (14.13)$$

$$\boxed{f \in \mathrm{SLIP}(A^n),\ N_\varepsilon \in SL^n} \longrightarrow = \mu_\varepsilon^{k/d} f\left(|\psi\rangle\right).$$

Upon taking the limit as $\varepsilon \to 0^+$ on both sides and noting that $M = \lim_{\varepsilon \to 0^+} M_\varepsilon$ and $\lim_{\varepsilon \to 0^+} \mu_\varepsilon = \det(M) = 0$, we infer that $f(M|\psi\rangle) = 0$, or equivalently, $f(|\phi\rangle) = 0$. This concludes the proof. ∎

## 14.2.1 SLIPs of $n$-Qubits

SLIPs, in general, involve cumbersome expressions. However, for systems of $n$-qubits, there exists a broad class of slips that can be expressed elegantly, due to the property (C.13 of online version) of the symplectic matrix $J_2 := -i\sigma_2 = |0\rangle\langle 1| - |1\rangle\langle 0|$. Recall the bilinear form $\langle \cdot, \cdot \rangle$ as defined in (13.91). This form can be extended to any number of qubits. Specifically, for any $n \in \mathbb{N}$, we define the bilinear form $\langle \cdot, \cdot \rangle_n : A^n \times A^n \to \mathbb{C}$ as follows:

$$\left\langle \psi^{A^n}, \phi^{A^n} \right\rangle_n := \left\langle \bar{\psi}^{A^n} \Big| \underbrace{J_2 \otimes \cdots \otimes J_2}_{n-\text{times}} \Big| \phi^{A^n} \right\rangle \qquad \forall\, |\psi\rangle, |\phi\rangle \in A^n, \qquad (14.14)$$

where for convenience we replaced the second Pauli matrix $\sigma_2$ that appears in (13.91) with $J_2 := -i\sigma_2 = |0\rangle\langle 1| - |1\rangle\langle 0|$.

**Exercise 14.5.** *Consider the bilinear form in (14.14).*

1. *Use the relation (C.13 of online version) to show that for any $M \in SL^n$, and any vectors $|\psi\rangle, |\phi\rangle \in A^n$*

$$\left\langle M|\psi\rangle, M|\phi\rangle \right\rangle_n = \left\langle |\psi\rangle, |\phi\rangle \right\rangle_n. \qquad (14.15)$$

2. *Show that for any odd integer $n \in \mathbb{N}$ we have*

$$\left\langle |\psi\rangle, |\psi\rangle \right\rangle_n = 0. \qquad (14.16)$$

The bilinear form already defined can be used to define SLIPs on systems of $n$ qubits. For an even number of qubits, it follows from this exercise that

$$f(|\psi\rangle) := \left\langle |\psi\rangle, |\psi\rangle \right\rangle_n \qquad (14.17)$$

is an homogeneous SLIP of degree two.

For an odd number of qubits, one can define a SLIP of degree 4 as follows: For any $|\psi\rangle \in A^n$ we write

$$|\psi^{A^n}\rangle = |0\rangle^{A_1}|\psi_0\rangle^{A_2 \cdots A_n} + |1\rangle^{A_1}|\psi_1\rangle^{A_2 \cdots A_n}, \qquad (14.18)$$

where $|\psi_0\rangle, |\psi_1\rangle \in A_2 \cdots A_n$. The basis $|0\rangle^{A_1}$ and $|1\rangle^{A_1}$ is chosen such that $J_2$ has the standard form. With these choices the determinant

$$g_4(|\psi\rangle) := \det \begin{pmatrix} \langle|\psi_0\rangle, |\psi_0\rangle\rangle_{n-1} & \langle|\psi_0\rangle, |\psi_1\rangle\rangle_{n-1} \\ \langle|\psi_1\rangle, |\psi_0\rangle\rangle_{n-1} & \langle|\psi_1\rangle, |\psi_1\rangle\rangle_{n-1} \end{pmatrix} \tag{14.19}$$

is a homogeneous SLIP of degree 4. To see that this function is a SLIP, observe that for any $M^{(n)} = M_1 \otimes M^{(n-1)} \in SL^n$, we have $g_4\left(M^{(n)}|\psi\rangle\right) = g_4\left(M_1 \otimes I^{A_2 \cdots A_n}|\psi\rangle\right)$ since $\langle \cdot, \cdot \rangle_{n-1}$ is invariant under the action of $M^{(n-1)}$. Now, let $M_1 = \begin{pmatrix} a & b \\ c & d \end{pmatrix}$ and observe that

$$\begin{aligned} M_1 \otimes I^{A_2 \cdots A_n}|\psi\rangle &= \left(a|0\rangle + c|1\rangle\right)|\psi_0\rangle + \left(b|0\rangle + d|1\rangle\right)|\psi_1\rangle \\ &= |0\rangle\left(a|\psi_0\rangle + b|\psi_1\rangle\right) + |1\rangle\left(c|\psi_0\rangle + d|\psi_1\rangle\right). \end{aligned} \tag{14.20}$$

For each $x, y \in \{0, 1\}$ we denote by $\mu_{xy} := \langle|\psi_x\rangle, |\psi_y\rangle\rangle_{n-1}$. With these notations

$$\begin{aligned} &g_4\left(M_1 \otimes I^{A_2 \cdots A_n}|\psi\rangle\right) \\ &= \det \begin{pmatrix} \langle a|\psi_0\rangle + b|\psi_1\rangle, a|\psi_0\rangle + b|\psi_1\rangle\rangle & \langle a|\psi_0\rangle + b|\psi_1\rangle, c|\psi_0\rangle + d|\psi_1\rangle\rangle \\ \langle c|\psi_0\rangle + d|\psi_1\rangle, a|\psi_0\rangle + b|\psi_1\rangle\rangle & \langle c|\psi_0\rangle + d|\psi_1\rangle, c|\psi_0\rangle + d|\psi_1\rangle\rangle \end{pmatrix} \\ &= (a^2\mu_{00} + b^2\mu_{11} + 2ab\mu_{01})(c^2\mu_{00} + d^2\mu_{11} + 2cd\mu_{10}) \\ &\quad - \left(ac\mu_{00} + bd\mu_{11} + (ad + cb)\mu_{01}\right)^2 \\ &= (ad - bc)^2 \left(\mu_{00}\mu_{11} - \mu_{01}^2\right), \end{aligned} \tag{14.21}$$

where the last line follows from direct algebraic simplification of all the terms involved. Since $M_1 \in SL(2, \mathbb{C})$ we have $ad - bc = 1$ so that

$$g_4\left(M_1 \otimes I^{A_2 \cdots A_n}|\psi\rangle\right) = \mu_{00}\mu_{11} - \mu_{01}^2 = g_4(|\psi\rangle). \tag{14.22}$$

To get other SLIPs, let $A^n$ be a system of $n$ qubits, and for any choice of $m < n$ of its qubits, we associate a *bipartite cut* denoted as $\mathbb{A}_m \otimes \mathbb{B}_{n-m}$, where $\mathbb{A}_m$ is a system of $m$ qubits of $A^n$, and $\mathbb{B}_{n-m}$ is the system comprising of the remaining $n - m$ qubits of $A^n$. With respect to this bipartite cut, any vector $|\psi\rangle \in A^n$ can be expressed as

$$\begin{aligned} |\psi^{A^n}\rangle &= \sum_{x \in [2^m]} \sum_{y \in [2^{n-m}]} \lambda_{xy} |u_x\rangle^{\mathbb{A}_m} |v_y\rangle^{\mathbb{B}_{n-m}} \\ &= \Lambda \otimes I^{\tilde{\mathbb{B}}_{n-m}}|\Omega^{\mathbb{B}_{n-m}\tilde{\mathbb{B}}_{n-m}}\rangle, \end{aligned} \tag{14.23}$$

where $\{|u_x\rangle^{\mathbb{A}_m}\}$ and $\{|v_y\rangle^{\mathbb{B}_{n-m}}\}$ are orthonormal bases of $\mathbb{A}_m$ and $\mathbb{B}_{n-m}$, respectively, and each coefficient $\lambda_{xy} \in \mathbb{C}$. Therefore, and vector $|\psi\rangle \in A^n$, and every bipartite cut $\mathbb{A}_m \otimes \mathbb{B}_{n-m}$ of $A^n$ defines a matrix $\Lambda := (\lambda_{xy})$ with $x \in [2^m]$ and $y \in [2^{n-m}]$.

> **Theorem 14.1.** Let $A^n$ be a system of $n$ qubits, $m \in [n-1]$, and $\mathbb{A}_m \otimes \mathbb{B}_{n-m}$ be a bipartite cut of $A^n$. For any $|\psi\rangle \in A^n$, let $\Lambda = (\lambda_{xy})$ be the $2^m \times 2^{n-m}$ matrix defined in (14.23), and set $J_2 := |0\rangle\langle 1| - |1\rangle\langle 0|$. Then, for any $\ell \in \mathbb{N}$ the function
>
> $$f_\ell(|\psi\rangle) := \mathrm{Tr}\left[\left(J_2^{\otimes m} \Lambda J_2^{\otimes(n-m)} \Lambda^T\right)^\ell\right] \tag{14.24}$$
>
> is a homogeneous SLIP of degree $2\ell$.

**Proof**  Let $M \in SL^n$. We need to show that $f(M|\psi\rangle) = f(|\psi\rangle)$. Let $N \in SL^m$ and $L \in SL^{n-m}$ be such that $M = N \otimes L$. Then,

$$\begin{aligned} M|\psi^{A^n}\rangle &= N\Lambda \otimes L\left|\Omega^{\mathbb{B}_{n-m}\tilde{\mathbb{B}}_{n-m}}\right\rangle \\ &= N\Lambda L^T \otimes I^{\tilde{\mathbb{B}}_{n-m}}\left|\Omega^{\mathbb{B}_{n-m}\tilde{\mathbb{B}}_{n-m}}\right\rangle. \end{aligned} \tag{14.25}$$

We therefore get that

$$\begin{aligned} f_\ell\left(M|\psi^{A^n}\rangle\right) &= \mathrm{Tr}\left[\left(J_2^{\otimes m} N\Lambda L^T J_2^{\otimes(n-m)} L\Lambda^T N^T\right)^\ell\right] \\ \mathbf{Cyclic\ permutation} \to\ &= \mathrm{Tr}\left[\left(N^T J_2^{\otimes m} N\Lambda L^T J_2^{\otimes(n-m)} L\Lambda^T\right)^\ell\right], \end{aligned} \tag{14.26}$$

where we have used the invariance of the trace under cyclic permutation (note that the power over $\ell$ does not effect this property). To complete the proof we now argue that $N^T J_2^{\otimes m} N = J_2^{\otimes m}$ and $L^T J_2^{\otimes(n-m)} L = J_2^{\otimes(n-m)}$ so that the right-hand side above equals $f\left(|\psi^{A^n}\rangle\right)$. Indeed, observe that $N = N_1 \otimes \cdots \otimes N_m$, where $N_x \in SL(2, \mathbb{C})$ so that

$$\begin{aligned} N^T J_2^{\otimes m} N &= N_1^T J_2 N_1 \otimes \cdots \otimes N_m J_2 N_m \\ \mathbf{(C.13)} \to\ &= J_2 \otimes \cdots \otimes J_2 = J_2^{\otimes m}. \end{aligned} \tag{14.27}$$

In the same way, one can prove that $L^T J_2^{\otimes(n-m)} L = J_2^{\otimes(n-m)}$. This completes the proof. $\blacksquare$

## Examples

1. The case $n = 2$. In this case the only nontrivial $m$ is $m = 1$. In this case, for any two-qubit state $|\psi\rangle = \sum_{x,y\in\{0,1\}} \lambda_{xy}|xy\rangle$ we get

$$\begin{aligned} f_\ell(|\psi\rangle) &= \mathrm{Tr}\left[\left(J_2 \Lambda J_2 \Lambda^T\right)^\ell\right] \\ \mathbf{(C.13)} \to\ &= \mathrm{Tr}\left[\left(J_2 \det(\Lambda) J_2\right)^\ell\right] \\ \boxed{J_2^2 = I} \longrightarrow\ &= 2\left(\det(\Lambda)\right)^\ell. \end{aligned} \tag{14.28}$$

Note that in this case $\det(\Lambda) = \langle|\psi\rangle, |\psi\rangle\rangle_2$ which is the same SLIP we have already discussed.

2. The case $n = 3$. Consider the bipartite cut $(A_1 A_2) \otimes A_3$ of the three-qubit system $A^3$. With respect to this bipartite cut, any state $|\psi\rangle \in A^3$ can be expressed as

$$|\psi^{A^3}\rangle = \Lambda \otimes I^{\tilde{A}_3} |\Omega^{A_3 \tilde{A}_3}\rangle, \tag{14.29}$$

where $\Lambda : A_3 \to A_1 A_2$ is a $4 \times 2$ matrix. It will be convenient to view the matrix $\Lambda$ in a block form

$$\Lambda = \begin{bmatrix} \Lambda_1 \\ \Lambda_2 \end{bmatrix}, \tag{14.30}$$

where $\Lambda_1$ and $\Lambda_2$ are both $2 \times 2$ matrices. For this choice of bipartite cut we have

$$f_\ell(|\psi\rangle) = \mathrm{Tr}\left[ \left( J_2^{\otimes 2} \Lambda J_2 \Lambda^T \right)^\ell \right]. \tag{14.31}$$

The term

$$\Lambda J_2 \Lambda^T = \begin{bmatrix} \Lambda_1 \\ \Lambda_2 \end{bmatrix} J_2 [\Lambda_1^T \ \Lambda_2^T] = \begin{bmatrix} \Lambda_1 J_2 \Lambda_1^T & \Lambda_1 J_2 \Lambda_2^T \\ \Lambda_2 J_2 \Lambda_1^T & \Lambda_2 J_2 \Lambda_2^T \end{bmatrix} \tag{14.32}$$

Hence, combining this with $J_2^{\otimes 2} = \begin{bmatrix} 0 & J_2 \\ -J_2 & \end{bmatrix}$ gives

$$J_2^{\otimes 2} \Lambda J_2 \Lambda^T = \begin{bmatrix} J_2 \Lambda_2 J_2 \Lambda_1^T & J_2 \Lambda_2 J_2 \Lambda_2^T \\ -J_2 \Lambda_1 J_2 \Lambda_1^T & -J_2 \Lambda_1 J_2 \Lambda_2^T \end{bmatrix}. \tag{14.33}$$

Since the trace of the matrix above is zero, the case $\ell = 1$ is trivial. For the case $\ell = 2$ we have

$$\begin{aligned}
\mathrm{Tr}\left[ \left( J_2^{\otimes 2} \Lambda J_2 \Lambda^T \right)^2 \right] = {} & \mathrm{Tr}\left[ J_2 \Lambda_2 J_2 \Lambda_1^T J_2 \Lambda_2 J_2 \Lambda_1^T \right] \\
& - \mathrm{Tr}\left[ J_2 \Lambda_2 J_2 \Lambda_2^T J_2 \Lambda_1 J_2 \Lambda_1^T \right] \\
& + \mathrm{Tr}\left[ J_2 \Lambda_1 J_2 \Lambda_2^T J_2 \Lambda_1 J_2 \Lambda_2^T \right] \\
& - \mathrm{Tr}\left[ J_2 \Lambda_1 J_2 \Lambda_1^T J_2 \Lambda_2 J_2 \Lambda_2^T \right],
\end{aligned} \tag{14.34}$$

and due to the cyclic permutation of the trace we have

$$f_{\ell=2}(|\psi\rangle) = 2\mathrm{Tr}\left[ J_2 \Lambda_2 J_2 \Lambda_1^T J_2 \Lambda_2 J_2 \Lambda_1^T - J_2 \Lambda_2 J_2 \Lambda_2^T J_2 \Lambda_1 J_2 \Lambda_1^T \right]. \tag{14.35}$$

This expression is a homogeneous SLIP of degree 4. Since for three qubits there are no homogeneous SLIPs of degree 2, any other SLIP must be proportional to some power of $f_{\ell=2}$. Hence, for three qubits, the SLIP is essentially the only one.

3. The case $n = 4$. Let $\mathbb{A}_2 \otimes \mathbb{B}_2$ be a bipartite cut with exactly two qubits on each side and let $|\psi\rangle \in A^4$ be given as $|\psi\rangle = \Lambda \otimes I^{\mathbb{B}_2} |\Omega^{\mathbb{A}_2 \mathbb{B}_2}\rangle$, where $\Lambda$ is a $4 \times 4$ matrix. Then, the function

$$f_\ell(|\psi\rangle) = \mathrm{Tr}\left[ \left( J_2^{\otimes 2} \Lambda J_2^{\otimes 2} \Lambda^T \right)^\ell \right] \tag{14.36}$$

is a homogeneous SLIP of degree $2\ell$. Specifically, consider the four-qubit state

$$|\psi\rangle = \lambda_1 |\Psi_+\rangle |\Psi_+\rangle + \lambda_2 |\Psi_-\rangle |\Psi_-\rangle + \lambda_3 |\Phi_+\rangle |\Phi_+\rangle + \lambda_4 |\Phi_-\rangle |\Phi_-\rangle, \tag{14.37}$$

where $\lambda_x \in \mathbb{C}$ for all $x \in [4]$ and the two-qubit states $|\Psi_\pm\rangle$ and $|\Phi_\pm\rangle$ form the Bell basis of $\mathbb{C}^2 \otimes \mathbb{C}^2$. For this states it follows that

$$f_\ell(|\psi\rangle) = \lambda_1^{2\ell} + \lambda_2^{2\ell} + \lambda_3^{2\ell} + \lambda_4^{2\ell}. \tag{14.38}$$

**Exercise 14.6.** *Prove the relation* (14.38).

## 14.2.2 The Set of All Homogeneous SLIPs

Homogeneous polynomials of degree 1 on $A^n$ can be defined using an inner product of the form $f_\chi(\psi) := \langle \chi | \psi \rangle$ for all $|\psi\rangle \in A^n$, where $\chi$ is a fixed coefficient vector in $A^n$. Similarly, homogeneous polynomials of degree $k \in \mathbb{N}$ can be expressed as

$$f_\chi(\psi) = \left\langle \chi | \psi^{\otimes k} \right\rangle \qquad \forall \, \psi \in A^n, \tag{14.39}$$

where the coefficient vector $|\chi\rangle \in (A^n)^{\otimes k}$. However, this relation is not one-to-one; $f_\chi$ is equal to $f_{\chi'}$ if the coefficient vectors $|\chi\rangle$ and $|\chi'\rangle$ are related by a permutation matrix. This permutation is with respect to the $k$ copies of $A^n$. As a result, there exists an isomorphism between the space of homogeneous polynomials of degree $k$ and the subspace $\mathrm{Sym}_k(A^n)$ of $(A^n)^{\otimes k}$ (see definition in (C.161 of online version)).

The polynomial $f_\chi$ above is SLIP if and only if for any $M \in SL^n$ for all $|\psi\rangle \in A^n$ we have $f_\chi(M|\psi\rangle) = f_\chi(|\psi\rangle)$, which is equivalent to

$$\left\langle \chi | M^{\otimes k} | \psi^{\otimes k} \right\rangle = \left\langle \chi | \psi^{\otimes k} \right\rangle. \tag{14.40}$$

Since (14.40) has to hold for all $|\psi\rangle$, and since if $M \in SL^n$, then $M^* \in SL^n$, we conclude that $f_\chi$ is SLIP if and only if (see Exercise 14.7)

$$M^{\otimes k} |\chi\rangle = |\chi\rangle \qquad \forall \, M \in SL^n. \tag{14.41}$$

In other words, $|\chi\rangle$ is a $SL^n$-fixed vector under the representation $\pi_k : SL^n \mapsto \mathfrak{L}\left((A^n)^{\otimes k}\right)$ given by $\pi_k(M) = M^{\otimes k}$. Denoting by $\mathrm{SLIP}_k(A^n)$ the set of all homogeneous SLIPs of degree $k$, and by $V := (A^n)^{\otimes k}$ we conclude that

$$\mathrm{SLIP}_k(A^n) = \left\{ f_\chi : |\chi\rangle \in V^{SL^n} \right\}, \tag{14.42}$$

where we used the same notations as in (C.175 of online version). Our task is therefore to characterize $V^{SL^n}$. For this purpose, observe that the vector space $V$ is isomorphic to

$$V = (A^n)^{\otimes k} \cong A_1^{\otimes k} \otimes \cdots \otimes A_n^{\otimes k}. \tag{14.43}$$

Let $P : V \to A_1^{\otimes k} \otimes \cdots \otimes A_n^{\otimes k}$ be this isomorphism (permutation) map. Under this isomorphism any matrix $M^{\otimes k}$, with $M := M_1 \otimes \cdots \otimes M_n \in SL^n$, goes to

$$P M^{\otimes k} P^{-1} = P\left(M_1 \otimes \cdots \otimes M_n\right)^{\otimes k} P^{-1} = M_1^{\otimes k} \otimes \cdots \otimes M_n^{\otimes k}. \tag{14.44}$$

Therefore, for any $|\chi\rangle \in V$ that satisfies (14.41) the vector $|\phi\rangle := P|\chi\rangle$ satisfies

$$M_1^{\otimes k} \otimes \cdots \otimes M_n^{\otimes k} |\phi\rangle = |\phi\rangle. \tag{14.45}$$

**Exercise 14.7.** *Let $|\chi\rangle \in (A^n)^{\otimes k}$. Show that $f_\chi$ as defined in (14.39) is a homogeneous SLIP of degree $k$ if and only if (14.41) holds.*

---

**Lemma 14.2.** Let $|\phi\rangle \in A_1^{\otimes k} \otimes \cdots \otimes A_n^{\otimes k}$. Then, $|\phi\rangle$ satisfies (14.45) if and only if

$$|\phi\rangle \in W := \left(A_1^{\otimes k}\right)^{SL(A_1)} \otimes \cdots \otimes \left(A_n^{\otimes k}\right)^{SL(A_n)}, \tag{14.46}$$

where we used the notations given in (C.175 of online version).

---

**Proof** Clearly, by definition, if $|\phi\rangle \in W$, then $|\phi\rangle$ satisfies (14.45). Conversely, suppose that $|\phi\rangle$ satisfies (14.45), and let $\{|v_x\rangle\}$ be an orthonormal basis of $\left(A_1^{\otimes k}\right)^{SL(A_1)}$, and $\{|u_y\rangle\}$ be an orthonormal basis of the orthogonal complement of $\left(A_1^{\otimes k}\right)^{SL(A_1)}$ in $A_1^{\otimes k}$. Finally, let $\{|\varphi_z\rangle\}$ be an orthonormal basis of $A_2^{\otimes k} \otimes \cdots \otimes A_n^{\otimes k}$. With these notations, since $|\phi\rangle \in A_1^{\otimes k} \otimes \cdots \otimes A_n^{\otimes k}$ it can be expressed as

$$|\phi\rangle = \sum_{x,z} \lambda_{xz}|v_x\rangle|\varphi_z\rangle + \sum_{y,z} \mu_{yz}|u_y\rangle|\varphi_z\rangle, \tag{14.47}$$

where $\lambda_{xz}, \mu_{yz} \in \mathbb{C}$. Using this expression in (14.45), and taking a special case in which $M_2 = I^{A_2}, \ldots, M_n = I^{A_n}$, we get

$$\sum_{x,z} \lambda_{xz} M_1^{\otimes k}|v_x\rangle|\varphi_z\rangle + \sum_{y,z} \mu_{yz} M_1^{\otimes k}|u_y\rangle|\varphi_z\rangle = \sum_{x,z} \lambda_{xz}|v_x\rangle|\varphi_z\rangle + \sum_{y,z} \mu_{yz}|u_y\rangle|\varphi_z\rangle. \tag{14.48}$$

Since $M_1 \in SL(A_1)$ we have for all $x$, $M_1^{\otimes k}|v_x\rangle = |v_x\rangle$ (by definition of $|v_x\rangle$). Hence, the above equation can be simplified to

$$\sum_{y,z} \mu_{yz} M_1^{\otimes k}|u_y\rangle|\varphi_z\rangle = \sum_{y,z} \mu_{yz}|u_y\rangle|\varphi_z\rangle. \tag{14.49}$$

Since the vectors $\{|\varphi_z\rangle\}$ are orthonormal, it follows that for all $z$ and all $M_1 \in SL(A_1)$ we have

$$\sum_{y} \mu_{yz} M_1^{\otimes k}|u_y\rangle = \sum_{y} \mu_{yz}|u_y\rangle. \tag{14.50}$$

The above equation implies that for each $z$ the vector $\sum_y \mu_{yz}|u_y\rangle$ belongs to the subspace $\left(A_1^{\otimes k}\right)^{SL(A_1)}$. However, by definition, the vectors $\{|u_y\rangle\}$ belong to the orthogonal complement of $\left(A_1^{\otimes k}\right)^{SL(A_1)}$. Therefore, the coefficients $\{\mu_{yz}\}$ must be zero, so that

$$|\phi\rangle = \sum_{x,z} \lambda_{xz}|v_x\rangle|\varphi_z\rangle. \tag{14.51}$$

Denoting by $C := A_2^{\otimes k} \otimes \cdots \otimes A_n^{\otimes k}$, the above equation can be expressed as $\Pi_1 \otimes I^C |\phi\rangle = |\phi\rangle$, where

$$\Pi_1 := \sum_x |v_x\rangle\langle v_x| \tag{14.52}$$

is the orthogonal projection to the subspace $\left(A_1^{\otimes k}\right)^{SL(A_1)}$. Denoting by the $\Pi_x$ the orthogonal projection to the subspace $\left(A_x^{\otimes k}\right)^{SL(A_x)}$, and repeating the same argument for any $x \in [n]$ we conclude that

$$\Pi_1 \otimes \cdots \otimes \Pi_n |\phi\rangle = |\phi\rangle. \tag{14.53}$$

That is, $|\phi\rangle \in W$. This completes the proof. ∎

Lemma 14.2 shows that characterizing $V^{SL^n}$ can be done by characterizing $\left(A_x^{\otimes k}\right)^{SL(A_x)}$ or the orthogonal projection $\Pi_x$. Therefore, the problem of characterizing all SLIPs of degree $k$ can be reduced to characterizing $\left(A^{\otimes k}\right)^{SL(A)}$, which is a classic representation theory problem that uses the Schur–Weyl duality. This duality connects the irreducible representations (irreps) of $SL(A)$ to the symmetric group on $k$ elements, with a natural action. Further information can be found in the "Notes and References" section at the end of this chapter.

## 14.2.3 SLIPs and Multipartite Entanglement Monotones

The following theorem illustrates the usefulness of SLIPs in quantifying entanglement.

**Theorem 14.2.** Let $f_k \colon A^n \to \mathbb{R}_+$ be an homogenous SLIP of degree $k \in \mathbb{N}$, and define for any $\psi \in \mathrm{Pure}(A^n)$

$$E\left(\psi^{A^n}\right) := \left| f_k\left(|\psi^{A^n}\rangle\right) \right|^{2/k}. \tag{14.54}$$

Then, $E$ is an entanglement monotone on pure multipartite states; that is, it is zero on product states and it does not increase on average under LOCC.

**Proof**　From Exercise 14.4, we deduce that $E$ vanishes on product states. Consider an arbitrary $m \in \mathbb{N}$ and $\psi \in \mathrm{Pure}(A^n)$. If there exists an LOCC protocol that transforms $\psi^{A^n}$ to $\phi_x^{A^n} \in \mathrm{Pure}(A^n)$ with a probability $p_x$, where $x \in [m]$, then each $\phi_x^{A^n}$ can be represented as

$$\left|\phi_x^{A^n}\right\rangle = \frac{1}{\sqrt{p_x}} M_x \left|\psi^{A^n}\right\rangle, \tag{14.55}$$

where each matrix $M_x$ is a tensor product of the form $M_x = \Lambda_{x1} \otimes \cdots \otimes \Lambda_{xn}$, and for every $y \in [n]$, $\Lambda_{xy} \in \mathcal{L}(A_y)$. Additionally, we have the relation $\sum_{x \in [m]} M_x^* M_x = I^{A^n}$.

Leveraging Lemma 14.1, we observe that $f(M_x|\psi\rangle) = 0$ when $\det(M_x) = 0$. We can then categorize the set $\{M_x\}_{x \in [m]}$ into two subsets: the matrices that are rank deficient and those that possess full rank. Without loss of generality, let's assume the first $r \in [m]$

matrices $\{M_x\}_{x\in[r]}$ are all of full rank, while the subsequent matrices, for all $x = r + 1, \ldots, m$, satisfy the condition $\det(M_x) = 0$.

Thus, for each $x \in [r]$, aside from a scalar coefficient, $M_x$ can be interpreted as a member of $SL^n$. More precisely, we can express $M_x$ as $M_x = \mu_x N_x$, where $\mu_x := (\det(M_x))^{1/d}$, $d := |A^n|$, and the normalized matrix $N_x := \frac{1}{\mu_x} M_x$ belongs to $SL^n$. With these notations, we can proceed as follows:

$$\sum_{x\in[m]} p_x E\left(\phi_x^{A^n}\right) = \sum_{x\in[r]} p_x \left| f_k\left(\frac{1}{\sqrt{p_x}} M_x | \psi^{A^n}\rangle\right)\right|^{2/k} = \sum_{x\in[r]} p_x \left| f_k\left(\frac{\mu_x}{\sqrt{p_x}} N_x | \psi^{A^n}\rangle\right)\right|^{2/k}$$

$$f_k \text{ is homogenous of degree } k \rightarrow = \sum_{x\in[r]} |\mu_x|^2 \left| f_k\left(N_x | \psi^{A^n}\rangle\right)\right|^{2/k}$$

$$SL^n \text{ invariance} \rightarrow = \sum_{x\in[r]} |\mu_x|^2 \left| f_k\left(|\psi^{A^n}\rangle\right)\right|^{2/k}$$

$$= E\left(\psi^{A^n}\right) \sum_{x\in[r]} |\mu_x|^2.$$

$$(14.56)$$

Now, observe that

$$\sum_{x\in[r]} |\mu_x|^2 \leqslant \sum_{x\in[m]} |\mu_x|^2 = \sum_{x\in[m]} \left|\det(M_x)\right|^{2/d} = \sum_{x\in[m]} \left(\det\left(M_x^* M_x\right)\right)^{1/d}, \quad (14.57)$$

where we removed the absolute value since $M_x^* M_x \geqslant 0$. From the geometric-arithmetic inequality we have that $\left(\det\left(M_x^* M_x\right)\right)^{1/d} \leqslant \frac{1}{d}\text{Tr}\left[M_x^* M_x\right]$. Hence, substituting this into the equation above gives

$$\sum_{x\in[r]} |\mu_x|^2 \leqslant \sum_{x\in[m]} \frac{1}{d}\text{Tr}\left[M_x^* M_x\right] = \frac{1}{d}\text{Tr}\left[I^{A_n}\right] = 1. \quad (14.58)$$

Combining this with (14.56) we conclude that

$$\sum_{x\in[m]} p_x E\left(\phi_x^{A^n}\right) \leqslant E\left(\psi^{A^n}\right). \quad (14.59)$$

This completes the proof. ∎

To extend the definition of $E$ as defined in (14.54) to mixed multipartite states, we can employ the convex roof extension. In particular, for a homogeneous SLIP of degree $k$, we can use the following approach:

$$E\left(\rho^{A^n}\right) := \min \sum_{x\in[m]} p_x \left| f_k\left(|\psi_x^{A^n}\rangle\right)\right|^{2/k}, \quad (14.60)$$

where the minimum is over all pure state decompositions of $\rho^{A^n} = \sum_{x\in[m]} p_x \psi^{A^n}$. The above theorem implies that $E$ is an entanglement monotone on mixed states.

**Exercise 14.8.** *Show that $E$ as defined in (14.60) is an entanglement monotone on multipartite mixed states.*

If we set $n = k = 2$ and choose $f_k$ to be the specific SLIP (14.9) in (14.60), we can see that $E$ corresponds to the concurrence. Although there is a simple closed formula for the concurrence of mixed bipartite states, it may not be immediately clear whether similar formulas exist for multipartite entanglement. Interestingly, there is one known example of such a formula for an even number of qubits.

**Exercise 14.9.** *Let $n \in \mathbb{N}$ be an even integer, $\rho \in \mathfrak{D}(A^n)$, $\langle \cdot, \cdot \rangle_n : A^n \times A^n \to \mathbb{C}$ be the bilinear form as defined in (14.14), and $f$ be the SLIP of degree 2 as defined in (14.17). Finally, let $E$ be the entanglement monotone as defined in (14.60) but with $f$ replacing $f_k$. Show that*

$$E\left(\rho^{A^n}\right) = \max\left\{0, \lambda_1 - \sum_{x=2}^{\ell} \lambda_x\right\}, \tag{14.61}$$

*where $\ell := 2^n$, and $\{\lambda_x\}_{x \in [\ell]}$ are the eigenvalues of the matrix $\left|\sqrt{\rho}\sqrt{\rho_\star}\right|$ arranged in non-decreasing order. The matrix $\rho_\star$ is defined similarly to (13.88) as*

$$\rho_\star^{A^n} := \sigma_2^{\otimes n} \bar{\rho}^{AB} \sigma_2^{\otimes n}. \tag{14.62}$$

*Hint: Follow the exact same lines as in the proof of Theorem 13.7 by with $\langle \cdot, \cdot \rangle_n$ replacing the bilinear form given in (13.91).*

## 14.3 Characteristics of Multipartite Entanglement

In this section, we will present several key results from representation theory and algebraic geometry that help characterize the structure of multipartite entangled states. While some of these results are presented without their full proofs, as they go beyond the scope of this book, interested readers can find more information in the book in Ref. [230]. By leveraging these powerful mathematical tools, we can gain deeper insight into the properties of multipartite entangled states.

### 14.3.1 Critical States

Let $\mathrm{Lie}(SL^n)$ be the Lie algebra of the group $SL^n$ defined in (14.6). Define the set of *critical* states in $A^n$ to be

$$\mathrm{Crit}(A^n) := \left\{|\psi\rangle \in A^n : \langle\psi|X|\psi\rangle = 0 \quad \forall\, X \in \mathrm{Lie}\left(SL^n\right)\right\}. \tag{14.63}$$

The reason for this terminology is that any state $|\psi\rangle \in \mathrm{Crit}(A^n)$ is a critical point of the function $f : SL^n|\psi\rangle \to \mathbb{R}_+$ defined by $f(|\phi\rangle) := \||\phi\rangle\|$. In fact, we have something that is a bit stronger.

> **Kempf–Ness Theorem (Part I)**
>
> **Theorem 14.3.** Let $|\psi\rangle \in A^n$. The following statements are equivalent:
>
> 1. The state $|\psi\rangle \in \mathrm{Crit}(A^n)$.
> 2. For any $M \in SL^n$ we have $\|M|\psi\rangle\| \geqslant \||\psi\rangle\|$.
> 3. For any $x \in [n]$, the reduced density matrix of $\psi^{A^n}$ on the $x^{\mathrm{th}}$ subsystem is proportional to the identity matrix $I^{A_x}$.

**Proof**  We start by proving that $1 \Rightarrow 2$. Suppose $|\psi\rangle \in \mathrm{Crit}(A^n)$ and observe that for any $M \in SL^n$ also $M^*M \in SL^n$. We can therefore write $M^*M = e^X$ for some $X \in \mathrm{Lie}\,(SL^n)$. Hence,

$$\|M|\psi\rangle\|^2 = \langle\psi|e^X|\psi\rangle$$

$$\boxed{e^X \geqslant I + X} \longrightarrow \ \geqslant \langle\psi|\,(I + X)\,|\psi\rangle \tag{14.64}$$

$$\boxed{|\psi\rangle \text{ is critical}} \rightarrow \ = \langle\psi|\psi\rangle = \||\psi\rangle\|^2.$$

To prove that $2 \Rightarrow 1$, suppose that for any $M \in SL^n$ we have $\|M|\psi\rangle\| \geqslant \||\psi\rangle\|$. Then, for any $X \in \mathrm{Lie}\,(SL^n)$ and $t \in \mathbb{R}$ we have

$$f(t) := \left\|e^{\frac{1}{2}tX}|\psi\rangle\right\|^2 = \langle\psi|e^{tX}|\psi\rangle \geqslant \langle\psi|\psi\rangle = f(0). \tag{14.65}$$

Hence, $t = 0$ must be a critical point of the function $f(t)$ so that $f'(0) = \langle\psi|X|\psi\rangle = 0$. Since this holds for any $X \in \mathrm{Lie}\,(SL^n)$ we conclude that $|\psi\rangle \in \mathrm{Crit}(A^n)$.

We next prove the equivalence of 1 and 3. Recall that any $X \in \mathrm{Lie}\,(SL^n)$ can be written as a linear combination of matrices that up to a permutation of the subsystems of $A^n$ have the form $X_1 \otimes I^{A_2} \otimes \cdots \otimes I^{A_n}$. Now, if $X = X_1 \otimes I^{A_2} \otimes \cdots \otimes I^{A_n}$, then the condition $\langle\psi|X|\psi\rangle = 0$ is equivalent to

$$\mathrm{Tr}\left[\rho^{A_1} X_1\right] = 0, \tag{14.66}$$

where $\rho^{A_1} := \mathrm{Tr}_{A_2 \cdots A_n}\left[\psi^{A^n}\right]$. Since the above condition has to hold for all $X_1 \in \mathrm{Lie}\big(SL(A_1)\big)$, we conclude that $\rho^{A_1}$ is proportional to the identity matrix. In other words, $|\psi\rangle \in \mathrm{Crit}(A^n)$ if and only if for any $x \in [n]$ the reduced density matrix of $\psi^{A^n}$ on the $x$th subsystem is proportional to the identity matrix $I^{A_x}$. This completes the proof. ∎

**Exercise 14.10.** *Let $|\psi\rangle \in Crit(A^n)$ and let $M \in SL^n$ be such that $\|M|\psi\rangle\| = \||\psi\rangle\|$.*

1. *Show that there exists a local unitary matrix; that is, $U \in SU(d_1) \times \cdots \times SU(d_n)$ such that $M|\psi\rangle = U|\psi\rangle$.*

2. *Show that if in addition, $M > 0$, then $M|\psi\rangle = |\psi\rangle$.*

## 14.3.2 The Null Cone

> **Definition 14.2.** Let $A^n$ be a multipartite system. The *null cone* of $A^n$, denoted by $\mathrm{Null}(A^n)$, is the set of all vectors in $A^n$ on which all SLIPs vanish. That is, $|\psi\rangle \in \mathrm{Null}(A^n)$ if and only if
>
> $$f(|\psi\rangle) = 0 \qquad \forall\, f \in \mathrm{SLIP}(A^n). \tag{14.67}$$

For two-qubit states the null cone consists only of product states. This can be easily verified by noting that the SLIP given by $f(|\psi\rangle) := \big(|\psi\rangle, |\psi\rangle\big)_2$ is zero if and only if $|\psi\rangle \in \mathbb{C}^2 \otimes \mathbb{C}^2$ is a product state. For higher number of qubits the null cone is not trivial. As an example, consider the three-qubit state, known as the W-state,

$$|W\rangle := \frac{1}{\sqrt{3}}\big(|100\rangle + |010\rangle + |001\rangle\big). \tag{14.68}$$

This state has the property that for any $0 \neq t \in \mathbb{C}$ the matrix $M_t := \begin{pmatrix} t & 0 \\ 0 & t^{-1} \end{pmatrix}^{\otimes 3}$ satisfies

$$M_t|W\rangle = t|W\rangle. \tag{14.69}$$

Since $M_t \in SL^3$, for any homogeneous SLIP $f_k$ of degree $k$ we have

$$f_k(|W\rangle) = f_k(M_t|W\rangle) = f_k(t|W\rangle) = t^k f_k(|W\rangle). \tag{14.70}$$

Since $t \neq 0$, this means $f_k(|W\rangle) = 0$. Since $f_k$ is an arbitrary homogenous SLIP, this implies that for any $f \in \mathrm{SLIP}(A^3)$ (where $A$ is a qubit; that is, $|A| = 2$) we have $f(|W\rangle) = 0$. Therefore, the W-state belongs to the null cone of three qubits.

From (14.69) it follows that

$$\lim_{t \to 0} M_t|W\rangle = \lim_{t \to 0} t|W\rangle = \mathbf{0}. \tag{14.71}$$

That is, the orbit $SL^3|W\rangle$ contains a sequence of vectors approaching the zero vector. This is precisely the key property of states in the null cone.

> **The Hilbert–Mumford Theorem**
>
> **Theorem 14.4.** Let $|\psi\rangle \in A^n$. The following statements are equivalent:
>
> 1. The vector $|\psi\rangle \in \mathrm{Null}(A^n)$.
> 2. There exists a sequence of vectors $\{|\psi_k\rangle\}_{k \in \mathbb{N}} \subset SL^n|\psi\rangle$ such that
>
> $$\lim_{k \to \infty} |\psi_k\rangle = \mathbf{0}. \tag{14.72}$$

The direction that $2 \Rightarrow 1$ is relatively simple to show. Indeed, suppose there is a sequence of vectors $|\psi_k\rangle_{k \in \mathbb{N}} \subset SL^n|\psi\rangle$ that approaches the zero vector in the limit $k \to \infty$. Let $f \in \mathrm{SLIP}(A^n)$. Then, for any $k \in \mathbb{N}$ we have $f(|\psi_k\rangle) = f(|\psi\rangle)$. Since this holds for any integer $k$, it must hold also for the limit $k \to \infty$. Combining this with the continuity of polynomial functions we get

$$f(|\psi\rangle) = \lim_{k \to \infty} f(|\psi_k\rangle) = f(\mathbf{0}) = 0. \tag{14.73}$$

As $f$ was an arbitrary SLIP we conclude that $|\psi\rangle \in \text{Null}(A^n)$. The other direction can be found in Theorem 43 in Ref. [230].

**Exercise 14.11.** *Let* $\lambda_1, \ldots, \lambda_n \in \mathbb{C}$, *and let*

$$|\psi\rangle := \lambda_1 |10\ldots0\rangle + \lambda_2 |01\ldots0\rangle + \cdots + \lambda_n |00\ldots1\rangle \in A^n. \tag{14.74}$$

*Show that* $|\psi\rangle \in Null(A^n)$.

## 14.3.3 Stable States

> **Definition 14.3.** Let $A^n$ be a multipartite system. A state $\psi \in \text{Pure}(A^n)$ is said to be *stable* if its orbit $SL^n|\psi\rangle$ is closed. The set of all stable states is denoted $\text{Stable}(A^n)$.

Note that the orbit $SL^n|\psi\rangle$ is closed if for any sequence of states $\{|\phi_k\rangle\}_{n\in\mathbb{N}} \subset SL^n|\psi\rangle$ with a limit $\lim_{k\to\infty} |\phi_k\rangle = |\phi\rangle$ we have that the limit $|\phi\rangle$ is also in $SL^n|\psi\rangle$. Therefore, states in the null cone are not stable since if $|\psi\rangle \in \text{Null}(A^n)$ is a nonzero vector, then $SL^n|\psi\rangle$ does not contain the zero vector. Still, $SL^n|\psi\rangle$ contains a sequence of vectors with zero limit since $|\psi\rangle$ is in the null cone. Hence, the null cone and the set of stable states forms two disjoint sets of states in $A^n$. The following theorem shows that any state in $A^n$ can be written as a linear combination of these two set of states.

> **Theorem 14.5.** Let $A^n := A_1 \cdots A_n$ be a multipartite system and $\psi \in A^n$. Then, there exists $|\phi\rangle \in \text{Stable}(A^n)$ and $|\chi\rangle \in \text{Null}(A^n)$ such that
>
> $$|\psi\rangle = |\phi\rangle + |\chi\rangle. \tag{14.75}$$

This result follows from a variant of the Hilbert–Mumford theorem given in Theorem 45 in Ref. [230]. Theorem 14.5 states that the vector space $A^n$ can be decomposed into the direct sum

$$A^n = \text{Stable}(A^n) \oplus \text{Null}(A^n). \tag{14.76}$$

In addition, it can be shown that almost all vectors in $A^n$ are stable in the sense that the closure of $\text{Stable}(A^n)$ is the whole space; that is,

$$A^n = \overline{\text{Stable}(A^n)}. \tag{14.77}$$

Therefore, much of the characterization in literature of multipartite entanglement is focused on stable states.

**The Kempf–Ness Theorem (Part II)**

> **Theorem 14.6.** Let $|\psi\rangle \in A^n$. Then, $|\psi\rangle \in \text{Stable}(A^n)$ if and only if $SL^n|\psi\rangle$ contains a critical state.

**Proof** Suppose $|\psi\rangle$ is stable so that $SL^n|\psi\rangle$ is closed. Then, there exists a state $|\phi\rangle \in SL^n|\psi\rangle$ with minimal norm; that is, for any $M \in SL^n$

$$\big\| M|\psi\rangle \big\| \geqslant \big\| |\phi\rangle \big\|. \tag{14.78}$$

But since $|\phi\rangle = N|\psi\rangle$ for some $N \in SL^n$ we can express the above equation as

$$\big\| MN^{-1}|\phi\rangle \big\| \geqslant \big\| |\phi\rangle \big\| \qquad \forall\, M \in SL^n. \tag{14.79}$$

Since any $M' \in SL^n$ can be expressed as $M' = MN^{-1}$ for some $M \in SL^n$ we conclude that $\big\| M'|\phi\rangle \big\| \geqslant \big\| |\phi\rangle \big\|$ for all $M' \in SL^n$. From Theorem 14.3 it then follows that $|\phi\rangle$ is a critical state. That is, the orbit $SL^n|\psi\rangle$ contains a critical state. The proof of the converse part can be found in Theorem 47 in Ref. [230]. $\blacksquare$

## 14.3.4 Characterization of SLOCC Classes

In this subsection we show that SLOCC classes of states can be characterized with SLIPs. Specifically, let $\psi, \phi \in \mathrm{Pure}(A^n)$. How can we determine if these two states belong to the same SLOCC class? According to the discussion above in (14.5), it is sufficient to consider reversible SLOCC classes. Hence, $\psi^{A^n}$ and $\phi^{A^n}$ belong to the same (reversible) SLOCC class if and only if there exists $\theta \in [0, 2\pi)$ and $M \in SL^n$ such that

$$\big| \phi^{A^n} \big\rangle = e^{i\theta} \frac{M \big| \psi^{A^n} \big\rangle}{\big\| M \big| \psi^{A^n} \big\rangle \big\|}. \tag{14.80}$$

Observe that if $f \in \mathrm{SLIP}_k(A^n)$ for some $k \in \mathbb{N}$, then the above equation gives

$$f\left( \big| \phi^{A^n} \big\rangle \right) = \frac{e^{i\theta k}}{\big\| M \big| \psi^{A^n} \big\rangle \big\|^k}\, f\left( \big| \psi^{A^n} \big\rangle \right). \tag{14.81}$$

Therefore, if $h \in \mathrm{SLIP}_k(A^n)$ is another homogenous SLIP of degree $k$ such that $h\left( \big| \psi^{A^n} \big\rangle \right) \neq 0$, then (14.81) is equivalent to

$$\frac{f\left( \big| \phi^{A^n} \big\rangle \right)}{h\left( \big| \phi^{A^n} \big\rangle \right)} = \frac{f\left( \big| \psi^{A^n} \big\rangle \right)}{h\left( \big| \psi^{A^n} \big\rangle \right)}. \tag{14.82}$$

That is, if $\psi^{A^n}$ and $\phi^{A^n}$ belong to the same reversible SLOCC, then (14.82) must hold. The following theorem demonstrates that the converse is also true for almost all states in $A^n$.

> **Theorem 14.7.** Let $|\psi\rangle, |\phi\rangle \in \mathrm{Stable}(A^n)$. Then, there exist $\theta \in [0, 2\pi)$ and $M \in SL^n$ such that (14.80) holds if and only if (14.82) holds for all $k \in \mathbb{N}$ and all $f, h \in \mathrm{SLIP}_k(A^n)$ with $h\left( \big| \psi^{A^n} \big\rangle \right) \neq 0$.

*Remark.* Note that in this theorem $k$ is unbounded. However, since it is known that the space of SLIPs has a finite dimension, it is possible to restrict $k$, although the best upper bound is unknown.

**Proof** We have already shown that (14.80) implies (14.82). It is therefore left to show the converse. If there exists $h \in \text{SLIP}_k(A^n)$ such that $h\left(|\psi^{A^n}\rangle\right) \neq 0$ but $h\left(|\phi^{A^n}\rangle\right) = 0$, then clearly $\psi^{A^n}$ and $\phi^{A^n}$ are not in the same invertible SLOCC class. We therefore assume without loss of generality that there exist $k \in \mathbb{N}$ and $h \in \text{SLIP}_k(A^n)$ such that both $h\left(|\psi^{A^n}\rangle\right) \neq 0$ and $h\left(|\phi^{A^n}\rangle\right) \neq 0$, and denote by

$$\lambda := \frac{h\left(|\phi^{A^n}\rangle\right)}{h\left(|\psi^{A^n}\rangle\right)} \neq 0. \tag{14.83}$$

With this notation, our assumption in (14.82) implies that for all $f \in \text{SLIP}_k(A^n)$,

$$f\left(|\phi^{A^n}\rangle\right) = \lambda f\left(|\psi^{A^n}\rangle\right) = f\left(\lambda^{1/k}|\psi^{A^n}\rangle\right). \tag{14.84}$$

Our first goal is to show that up to some phase factors, $f$ here can be replaced with *any* SLIP (even not homogeneous). For this purpose, consider the subgroup $\mathbf{G}_{n,k} \subset GL(A^n)$ defined by

$$\mathbf{G}_{n,k} := \left\{\mu M : \mu^k = 1, \ M \in SL^n, \ \mu \in \mathbb{C}\right\}, \tag{14.85}$$

and observe that in addition for being $SL^n$-invariant polynomial (i.e. SLIP), $h$ is also $\mathbf{G}_{n,k}$-invariant polynomial. Moreover, the degree of any homogeneous $\mathbf{G}_{n,k}$-invariant polynomial must be divisible by $k$. To see this, let $g$ be a $\mathbf{G}_{n,k}$-invariant polynomial of degree $m$. Then, since for any $\mu \in \mathbb{C}$ such that $\mu^k = 1$ we have $\mu I \in \mathbf{G}_{n,k}$, it follows that $g(|\psi\rangle) = g(\mu I |\psi\rangle) = \mu^m g(|\psi\rangle)$, so that $\mu^m = 1$. Since $m$ satisfies this property for any such $\mu$ (i.e. any $k$th root of unity), we conclude that $m = kr$ for some $r \in \mathbb{N}$.

Now, fix $k$, and let $g$ be a homogenous $\mathbf{G}_{n,k}$-invariant polynomial of degree $kr$ for some $r \in \mathbb{N}$. Since $g$ is a SLIP, from the assumption of the theorem

$$\frac{g\left(|\phi^{A^n}\rangle\right)}{h^r\left(|\phi^{A^n}\rangle\right)} = \frac{g\left(|\psi^{A^n}\rangle\right)}{h^r\left(|\psi^{A^n}\rangle\right)}. \tag{14.86}$$

Combining this with (14.83) yields

$$g\left(|\phi^{A^n}\rangle\right) = \lambda^r g\left(|\psi^{A^n}\rangle\right) = g\left(\lambda^{1/k}|\psi^{A^n}\rangle\right), \tag{14.87}$$

where in the last equality we used the fact that $g$ is homogeneous of degree $kr$. Since the above equation holds for any *homogenous* $SL^n$-invariant polynomial $g$ (recall that $r$ was arbitrary), it must also hold for all (possibly nonhomogeneous) $SL^n$-invariant polynomials. Hence, using a result from invariant theory that closed orbits of a reductive algebraic subgroup of $GL(A^n)$ are separated by their invariant polynomials, we conclude that there exists $\mu \in \mathbb{C}$ with $\mu^k = 1$ and $M \in SL^n$ such that $|\phi^{A^n}\rangle = \lambda^{1/k}\mu M|\psi^{A^n}\rangle$. The upshot is $|\phi^{A^n}\rangle = cM|\psi^{A^n}\rangle$ for some $c \in \mathbb{C}$, and the normalization $\||\phi^{A^n}\rangle\| = 1$ gives $c = e^{i\theta}/\|M|\psi^{A^n}\rangle\|$. This completes the proof. $\blacksquare$

This demonstrates that SLIPs can be used to classify multipartite entanglement. We give two examples of such classifications in three- and four-qubits systems.

# 14.4 Multipartite Entanglement of Three and Four Qubits

## 14.4.1 Classification of Three-Qubit Entanglement

### Canonical Form

In the previous chapters, we learned that pure bipartite states can always be represented in their Schmidt form. Specifically, for a two-qubit system $AB$, any state $\psi \in \mathrm{Pure}(AB)$ can be expressed, up to local unitaries, as

$$|\psi^{AB}\rangle = \sqrt{p}|00\rangle + \sqrt{1-p}|11\rangle, \tag{14.88}$$

where $p \in [0, 1]$. We refer to this representation as the canonical form of the state $\psi^{AB}$.

Now, our goal is to find a canonical form for any three-qubit state in $ABC$ where $|A| = |B| = |C| = 2$. To achieve this, we will utilize the following property presented in the following exercise.

**Exercise 14.12.** *Let $AB$ be a two-qubit system and let $|\psi_0\rangle, |\psi_1\rangle \in AB$ be two pure bipartite vectors. Show that if the vectors $|\psi_0^{AB}\rangle$ and $|\psi_1^{AB}\rangle$ are linearly independent, then there exists numbers $a, b \in \mathbb{C}$ such that $a|\psi_0^{AB}\rangle + b|\psi_1^{AB}\rangle$ is a product (i.e. nonentangled) state. Hint: Denote by $c := \frac{a}{b}$ and view the determinant of the reduced density matrix of the (nonnormalized) state $c|\psi_0^{AB}\rangle + |\psi_1^{AB}\rangle$ as a quadratic polynomial in $c$. Recall that over the complex field, all quadratic polynomials have roots.*

---

**Theorem 14.8.** Let $ABC$ be a three-qubit system, and let $\psi \in \mathrm{Pure}(ABC)$. Then, up to a local unitary matrix in $\mathfrak{L}(ABC)$, the state $|\psi^{ABC}\rangle$ can be expressed as

$$|\psi^{ABC}\rangle = \lambda_0|000\rangle + \lambda_1 e^{i\theta}|100\rangle + \lambda_2|101\rangle + \lambda_3|110\rangle + \lambda_4|111\rangle, \tag{14.89}$$

where $\lambda_0, \ldots, \lambda_4 \in \mathbb{R}_+$ and $\theta \in [0, \pi]$.

---

*Remark.* The normalization of $\psi^{ABC}$ implies that $\lambda_0^2 + \cdots + \lambda_4^2 = 1$. In the proof here, the fact that we can restrict $\theta$ to the domain $[0, \pi]$ will be left as an exercise.

**Proof** Every three-qubit state $|\psi\rangle \in ABC$ can be expressed as

$$|\psi^{ABC}\rangle = |0\rangle^A|\psi_0^{BC}\rangle + |1\rangle^A|\psi_1^{BC}\rangle, \tag{14.90}$$

where $|\psi_0\rangle, |\psi_1\rangle \in BC$ are two orthogonal (possibly unnormalized) vectors. From Exercise 14.12 it follows that there exist two complex numbers $a, b \in \mathbb{C}$ such that $a|\psi_0^{BC}\rangle + b|\psi_1^{BC}\rangle$ is a product state. Note that without loss of generality we can assume that $|a|^2 + |b|^2 = 1$. Therefore, the matrix $U = \begin{pmatrix} a & b \\ -\bar{b} & \bar{a} \end{pmatrix}$ is a unitary matrix, so that by applying $U$ to the first qubit of $|\psi^{ABC}\rangle$ we get

$$U^A \otimes I^{BC} |\psi^{ABC}\rangle = \left(a|0\rangle^A - \bar{b}|1\rangle^A\right)|\psi_0^{BC}\rangle + \left(b|0\rangle^A + \bar{a}|1\rangle^A\right)|\psi_1^{BC}\rangle$$
$$= |0\rangle^A \left(a|\psi_0^{BC}\rangle + b|\psi_1^{BC}\rangle\right) + |1\rangle^A \left(\bar{a}|\psi_1^{BC}\rangle - \bar{b}|\psi_0^{BC}\rangle\right). \tag{14.91}$$

Since $a|\psi_0^{BC}\rangle + b|\psi_1^{BC}\rangle$ is a (possibly unnormalized) product state, there exists a local unitary on $BC$ that transforms it to the state $\lambda_0|00\rangle^{BC}$, where $\lambda_0 \in \mathbb{C}$ is some normalization factor. We therefore conclude that, up to local unitaries, the state $|\psi^{ABC}\rangle$ can be expressed as

$$|\psi^{ABC}\rangle = \lambda_0|000\rangle + |1\rangle|\phi^{BC}\rangle, \tag{14.92}$$

where $|\phi^{BC}\rangle$ is some vector in $BC$. Let $\lambda_1, \ldots, \lambda_4 \in \mathbb{C}$ be such that

$$|\phi^{BC}\rangle = \lambda_1|00\rangle + \lambda_2|01\rangle + \lambda_3|10\rangle + \lambda_4|11\rangle. \tag{14.93}$$

Note that by applying to the state (14.93), the local unitary

$$U^{BC} := \begin{pmatrix} e^{i\theta_1} & 0 \\ 0 & e^{i\theta_2} \end{pmatrix} \otimes \begin{pmatrix} e^{i\theta_3} & 0 \\ 0 & e^{i\theta_4} \end{pmatrix}, \tag{14.94}$$

we get

$$U^{BC}|\phi^{BC}\rangle = \lambda_1 e^{i(\theta_1+\theta_3)}|00\rangle + \lambda_2 e^{i(\theta_1+\theta_4)}|01\rangle + \lambda_3 e^{i(\theta_2+\theta_3)}|10\rangle + \lambda_4 e^{i(\theta_2+\theta_4)}|11\rangle. \tag{14.95}$$

Therefore, by choosing appropriately the four phases $\theta_1, \theta_2, \theta_3$, and $\theta_4$ we can make three of the $\lambda$s nonnegative real numbers. We choose them to be $\lambda_2, \lambda_3, \lambda_4 \in \mathbb{R}_+$. Moreover, observe that by applying $e^{i\theta}|0\rangle\langle 0| + |1\rangle\langle 1|$ to system $A$ in (14.92) we can add a phase to $\lambda_0$. Therefore, we can assume without loss of generality that $\lambda_0$ is a real nonnegative number. ∎

**Exercise 14.13.** *Complete the proof above by showing that $\theta$ in (14.89) can be restricted to $[0, \pi]$.*

**Exercise 14.14.** *Let $|\psi^{ABC}\rangle$ be the three-qubit state given in (14.89). Show that its three local marginals (i.e. reduced density matrices) are given by*

$$\psi^A = \begin{pmatrix} \lambda_0^2 & \lambda_0\lambda_1 e^{-i\theta} \\ \lambda_0\lambda_1 e^{-i\theta} & 1 - \lambda_0^2 \end{pmatrix} \quad , \quad \psi^B = \begin{pmatrix} \lambda_0^2 + \lambda_1^2 + \lambda_2^2 & \lambda_1\lambda_3 e^{i\theta} + \lambda_2\lambda_4 \\ \lambda_1\lambda_3 e^{-i\theta} + \lambda_2\lambda_4 & \lambda_3^2 + \lambda_4^2 \end{pmatrix}, \tag{14.96}$$

*and*

$$\psi^C = \begin{pmatrix} \lambda_0^2 + \lambda_1^2 + \lambda_3^2 & \lambda_1\lambda_2 e^{i\theta} + \lambda_3\lambda_4 \\ \lambda_1\lambda_2 e^{-i\theta} + \lambda_3\lambda_4 & \lambda_2^2 + \lambda_4^2 \end{pmatrix}. \tag{14.97}$$

From Theorem 14.8 and Exercise 14.14 we get that up to local unitaries, there is only one normalized critical state given by the GHZ state

$$|GHZ\rangle := \frac{1}{\sqrt{2}}\left(|000\rangle + |111\rangle\right). \tag{14.98}$$

> **Corollary 14.1.** Let $ABC$ be a composite system of three qubits. Then, if $|\psi\rangle \in$ Crit$(ABC)$ is normalized, then there exists a local unitary matrix $U_1 \otimes U_2 \otimes U_3 \in \mathfrak{L}(ABC)$ such that
>
> $$|\psi^{ABC}\rangle = U_1 \otimes U_2 \otimes U_3|GHZ\rangle. \tag{14.99}$$

**Proof** From the properties of critical states (see Theorem 14.3) we know that if $|\psi\rangle \in$ Crit$(ABC)$ is normalized, then all three local marginals $\psi^A$, $\psi^B$, and $\psi^C$ must be maximally mixed. Now, from Theorem 14.8 we know that up to local unitaries the state $\psi^{ABC}$ can be expressed as in (14.89). Hence, using this form, we get from Exercise 14.14 that the condition $\psi^A = \frac{1}{2}I^A$ holds if and only if $\lambda_0^2 = \frac{1}{2}$ and $\lambda_1 = 0$. The condition $\psi^B = \frac{1}{2}I^B$ gives in particular $\lambda_0^2 + \lambda_1^2 + \lambda_2^2 = \frac{1}{2}$. Therefore, also $\lambda_2 = 0$. Finally, the condition $\psi^C = \frac{1}{2}I^C$ gives $\lambda_3 = 0$ and $\lambda_4^2 = \frac{1}{2}$. Hence, the state $\psi^{ABC}$ as given in (14.89) is critical if and only in it is the GHZ state. This concludes the proof. ∎

Recall the homogeneous SLIP of degree 4 as defined in (14.19) for odd number of qubits. For three qubit system $ABC$ (with $|A| = |B| = |C| = 2$), its absolute value is called *the 3-tangle*, and it is given for any vector

$$|\psi^{ABC}\rangle := |0\rangle^A|\psi_0^{BC}\rangle + |1\rangle^A|\psi_1^{BC}\rangle \in ABC \tag{14.100}$$

by

$$\text{Tangle}\left(|\psi^{ABC}\rangle\right) := \left| \det \begin{pmatrix} \langle|\psi_0\rangle, |\psi_0\rangle\rangle & \langle|\psi_0\rangle, |\psi_1\rangle\rangle \\ \langle|\psi_1\rangle, |\psi_0\rangle\rangle & \langle|\psi_1\rangle, |\psi_1\rangle\rangle \end{pmatrix} \right|, \tag{14.101}$$

where $\langle|\psi_x\rangle, |\psi_y\rangle\rangle := \langle\bar\psi_x^{BC}|J_2 \otimes J_2|\psi_y^{BC}\rangle$ for each $x, y \in \{0, 1\}$; recall that $J_2 := |0\rangle\langle1| - |1\rangle\langle0|$.

From the corollary discussed earlier, it follows that all stable vectors in $ABC$ are, up to normalization, contained in the $\mathbf{G}_3$ orbit of the GHZ state $|GHZ\rangle$. In other words, almost all three-qubit normalized states are in the SLOCC class of the GHZ state. This, in turn, implies that almost all three-qubit states have a nonzero 3-tangle, which is consistent with the formula for the 3-tangle given in the following exercise.

**Exercise 14.15.** *Show that the 3-tangle of the state $\psi^{ABC}$ in (14.89) is given by*

$$\text{Tangle}\left(|\psi^{ABC}\rangle\right) = \lambda_0\lambda_4. \tag{14.102}$$

The formula presented in this exercise shows that the 3-tangle is zero when $\lambda_0 = 0$, which makes sense because in this case, the state $\psi^{ABC}$ is a product state between system $A$ and system $BC$. This implies that the state is in the null cone, that is, it has no genuine tripartite entanglement. On the other hand, if $\lambda_4 = 0$, then the 3-tangle is also zero. In this case, the state $\psi^{ABC}$ can be expressed as

$$|\psi^{ABC}\rangle = \lambda_0|000\rangle + \lambda_1 e^{i\theta}|100\rangle + \lambda_2|101\rangle + \lambda_3|110\rangle. \tag{14.103}$$

If we apply the flip operator $|10| + |0\rangle\langle 1|$ to the first qubit of the state $\psi^{ABC}$, the resulting state takes the form:

$$|\psi^{ABC}\rangle = \lambda_1 e^{i\theta}|000\rangle + \lambda_0|100\rangle + \lambda_2|001\rangle + \lambda_3|010\rangle. \tag{14.104}$$

Moreover, observe that by applying the local unitary matrix $\left(e^{-i\theta/3}|0\rangle\langle 0| + e^{2i\theta/3}|1\rangle\langle 1|\right)^{\otimes 3}$ we can eliminate the phase attached to the $|000\rangle$ term. Therefore, after renaming the coefficients $\lambda_0, \ldots, \lambda_3$ we conclude that unless the state $\psi^{ABC}$ is a product state between $A$ and $BC$, its 3-tangle is zero if and only if, up to local unitaries, it can be expressed as

$$|\psi^{ABC}\rangle = \lambda_0|000\rangle + \lambda_1|100\rangle + \lambda_2|010\rangle + \lambda_3|001\rangle, \tag{14.105}$$

with $\lambda_0, \ldots, \lambda_3 \in \mathbb{R}_+$.

**Exercise 14.16.** *Show that for any three-qubit pure state $\psi^{ABC}$ of the form* (14.105), *there exist three matrices $M, N, L \in GL(2, \mathbb{C})$ such that*

$$|\psi^{ABC}\rangle = M \otimes N \otimes L |W\rangle, \tag{14.106}$$

*where $|W\rangle$ is the W-state as defined in* (14.68).

The preceding discussion and exercise demonstrate that the SLOCC class of the $W$-state consists of all states whose 3-tangle vanishes. Furthermore, since the $W$-state lies in the null cone (as shown in the discussion below equation (14.68)), we can conclude that the null cone precisely consists of the SLOCC class of the $W$-state. This, in turn, implies that a three-qubit vector lies in the null cone if and only if its 3-tangle vanishes. This also implies that any other SLIP must be proportional to a power of the 3-tangle. Therefore, the 3-tangle is essentially the absolute value of the only SLIP in three qubits.

In summary, we can divide the space of three qubits into six invertible SLOCC classes:

- The "genuine" tripartite entanglement classes: the GHZ class and the W-class.
- Three bipartite entanglement classes: the three $SL^3$-orbits generated by $|0\rangle^A|\Phi^{BC}\rangle$, $|0\rangle^B|\Phi^{AC}\rangle$, and $|\Phi^{AB}\rangle|0\rangle^C$.
- The unentangled class generated by $|000\rangle$.

## 14.4.2 Four-Qubit Entanglement

In this section, our goal is to analyze the classification of SLOCC classes of four-qubit states by seeking their canonical forms under both local unitaries and SLOCC. We will find that characterizing these forms is considerably more complex than in the three-qubit case. Therefore, we will focus our attention on the set of critical states, which is somewhat simpler to characterize. For interested readers, we refer to the references listed in the section "Notes and References" at the end of this chapter for more details on this topic.

# The Canonical Form under Local Unitaries

In this section, we consider the composite system ABCD consisting of four qubits, where $|A| = |B| = |C| = |D| = 2$. The primary technique employed in the study of four-qubit entanglement is the "accident" in Lie-group theory, which states an isomorphism between the special orthogonal group $SO(4)$ (consisting of $4 \times 4$ real orthogonal matrices with determinant one) and the group $SU(2) \otimes SU(2)$. We denote this isomorphism as

$$SU(2) \otimes SU(2) \cong SO(4). \tag{14.107}$$

In the following exercise, we will prove this isomorphism.

**Exercise 14.17.** *Consider the $4 \times 4$ complex matrix*

$$T := \frac{1}{\sqrt{2}} \begin{pmatrix} 1 & 0 & 0 & 1 \\ 0 & i & i & 0 \\ 0 & -1 & 1 & 0 \\ i & 0 & 0 & -i \end{pmatrix}. \tag{14.108}$$

1. *Show that $T$ is a unitary matrix.*
2. *Show that for all $U_1, U_2 \in SU(2)$ we have*

$$T(U_1 \otimes U_2)T^* \in SO(4). \tag{14.109}$$

*Hint: Show that $T^T T = J \otimes J$, where $J = |0\rangle\langle 1| - |1\rangle\langle 0|$ is the matrix that satisfies (C.13 of online version).*

We can use the isomorphism in (14.107) to get the canonical form of a four-qubit state. Let $|\psi^{ABCD}\rangle \in ABCD$ be a four qubit state, and let $M : AB \to AB$ be the $4 \times 4$ complex matrix defined via

$$|\psi^{ABCD}\rangle = M \otimes I^{CD} |\Omega^{(AB)(CD)}\rangle, \tag{14.110}$$

where

$$|\Omega^{(AB)(CD)}\rangle = \sum_{x,y \in \{0,1\}} |xy\rangle^{AB} |xy\rangle^{CD}. \tag{14.111}$$

In other words, we view four-qubit states as $4 \times 4$ complex matrices. Consider now a state

$$|\phi^{ABCD}\rangle := U_1 \otimes U_2 \otimes U_3 \otimes U_4 |\psi^{ABCD}\rangle \tag{14.112}$$

with each $U_x \in SU(2)$. That is, $|\psi\rangle$ and $|\phi\rangle$ are related by local unitaries. Let $N$ be the $4 \times 4$ matrix representing $|\phi^{ABCD}\rangle$ similarly to (14.110). Then, from the second part of Exercise 2.38 we get that $M$ and $N$ are related by

$$\begin{aligned} N &= (U_1 \otimes U_2)M(U_3 \otimes U_4)^T \\ &= T^* O_1 T M T^* O_2 T, \end{aligned} \tag{14.113}$$

where $T$ is the unitary matrix (14.108) and

$$O_1 := T(U_1 \otimes U_2)T^* \quad \text{and} \quad O_2 := T(U_3 \otimes U_4)^T T^*. \tag{14.114}$$

Observe that $O_1, O_2 \in SO(4)$.

Continuing, let $TMT^* = M_1 + iM_2$, where $M_1$ and $M_2$ are matrices with real coefficients. Then, $TNT^* = N_1 + iN_2$, where $N_1 := O_1 M_1 O_2$ and $N_2 := O_1 M_2 O_2$ are real matrices. Finally, observe that by appropriate choice of $O_1$ and $O_2$, the matrix $N_1$ (or $N_2$) can be made diagonal using the (real) singular value decomposition. The resulting $N = N_1 + iN_2$ and consequently $|\phi^{ABCD}\rangle$ can be viewed as the canonical form of $|\psi^{ABCD}\rangle$. However, this canonical form of four-qubit states is not very useful as it involves too many parameters. Specifically, even if $N_1$ is diagonal, the matrix $N_2$ is not.

**Exercise 14.18.** *Prove that any four-qubit state can be expressed, up to local unitary transformations, in the form given by (14.110), where the matrix M satisfies the condition*

$$M^* M = D + i\Lambda, \tag{14.115}$$

*and $D, \Lambda \in \mathbb{R}^{4\times4}$, with $D$ being a diagonal matrix with nonnegative diagonal elements and $\Lambda$ being a skew-symmetric matrix.*

**Exercise 14.19.** *Prove that if $\Lambda_1, \Lambda_2 \in SL(2, \mathbb{C})$, then*

$$T(\Lambda_1 \otimes \Lambda_2)T^* \in SO(4, \mathbb{C}), \tag{14.116}$$

*where $SO(4, \mathbb{C})$ is the (noncompact) special orthogonal group over $\mathbb{C}$; that is, $O \in SO(4, \mathbb{C})$ if and only if $O \in \mathbb{C}^{4\times4}$, $O^T O = I_4$, and $\det(O) = 1$. Here $T$ is the same matrix that was used in Exercise 14.17.*

## Critical States

In this subsection, we will characterize the set $\mathrm{Crit}(ABCD)$ of critical states in the four-qubit system by leveraging the isomorphism described in (14.116). Specifically, we begin by considering $\Lambda = \Lambda_1 \otimes \Lambda_2 \otimes \Lambda_3 \otimes \Lambda_4 \in \mathbf{G}_4$, $|\psi\rangle \in ABCD$, and the matrix $M$ defined in (14.110). We observe that

$$\Lambda|\psi^{ABCD}\rangle = N \otimes I^{CD}|\Omega^{(AB)(CD)}\rangle, \quad \text{where} \quad N = (\Lambda_1 \otimes \Lambda_2)M(\Lambda_3 \otimes \Lambda_4)^T. \tag{14.117}$$

Next, under the isomorphism in (14.116), the matrix $M$ is transformed into $\tilde{M} = TMT^*$ and $N$ into $\tilde{N} = TNT^*$. We can then express the relation between $\tilde{M}$ and $\tilde{N}$ as $\tilde{N} = O_1 \tilde{M} O_2$, where

$$O_1 := T(\Lambda_1 \otimes \Lambda_2)T^* \quad \text{and} \quad O_2 := T(\Lambda_3 \otimes \Lambda_4)^T T^*. \tag{14.118}$$

Note that if $O_1$ and $O_2$ were unitaries, we could have diagonalized $\tilde{M}$ using the singular value decomposition. However, since they are orthogonal, this is not always possible. Nevertheless, a somewhat cumbersome canonical form does exist (see, for example, [221]).

We now focus on four-qubit states in $ABCD$ whose corresponding $\tilde{M}$ matrix has the form $O_1' D O_2'$, where $D$ is a $4 \times 4$ complex diagonal matrix, and $O_1'$ and $O_2'$ are $4 \times 4$ orthogonal complex matrices. We will show that all critical states in four qubits belong to this class. Therefore, by the Kempf–Ness theorem (Theorem 14.6) in conjunction with (14.77), this class of states is dense in $ABCD$. In other words, almost all four-qubit pure states have this property.

We begin by noting that the diagonalizable property of $\tilde{M}$ remains invariant under the action of $\mathbf{G}_4$. This is because we have already shown that for every $|\psi\rangle \in ABCD$, the transformation $|\psi\rangle \to \Lambda|\psi\rangle$ translates, under the isomorphism, to the transformation of $\tilde{M}$ to $O_1 \tilde{M} O_2 = O_1 O_1' D O_2' O_2$, which is of the form $Q_1 D Q_2$, where $Q_1 = O_1 O_1'$ and $Q_2 = O_2 O_2'$ are two orthogonal matrices.

Next, for a fixed diagonal matrix $D = \mathrm{Diag}(\lambda_1, \lambda_2, \lambda_3, \lambda_4)$, where each $\lambda_x \in \mathbb{C}$ ($x \in$ [4]), we take the state corresponding to $\tilde{M} = D$ to represent this $\mathbf{G}_4$ orbit. Note that $M = T^*\tilde{M}T = T^*DT$, so the representative state has the form

$$\begin{aligned} |\psi^{ABCD}\rangle &= T^*DT \otimes I^{CD}|\Omega^{(AB)(CD)}\rangle \\ &= (T^*D \otimes T^T)|\Omega^{(AB)(CD)}\rangle. \end{aligned} \tag{14.119}$$

For any $j \in [4]$ with binary representation $(x, y)$ (with $x, y \in 0, 1$), we define $|u_j^{AB}\rangle := T^*|xy\rangle^{AB}$ and $|v_j^{CD}\rangle := T^T|xy\rangle^{CD}$. With these notations

$$|\psi^{ABCD}\rangle = \sum_{j\in[4]} \lambda_j |v_j^{AB}\rangle|u_j^{CD}\rangle. \tag{14.120}$$

It is simple to check that for each $j \in [4]$ the states $|u_j^{AB}\rangle$ and $|v_j^{CD}\rangle$ are maximally entangled. Hence, up to local unitary matrices, the state (14.120) can be expressed as

$$|\psi_\lambda^{ABCD}\rangle = \lambda_1|\Phi_+^{AB}\rangle|\Phi_+^{CD}\rangle + \lambda_2|\Phi_-^{AB}\rangle|\Phi_-^{CD}\rangle + \lambda_3|\Psi_+^{AB}\rangle|\Psi_+^{CD}\rangle + \lambda_4|\Psi_-^{AB}\rangle|\Psi_-^{CD}\rangle, \tag{14.121}$$

where $\{|\Phi_\pm\rangle, |\Psi_\pm\rangle\}$ denotes the Bell basis of maximally entangled states in four qubits.

Note that if there exists another diagonal matrix $D' = \mathrm{Diag}(\lambda_1', \lambda_2', \lambda_3', \lambda_4')$ such that $D' = O_1 D O_2$, we then must have

$$D'^2 = (D')^T D' = O_2^T D^2 O_2. \tag{14.122}$$

Therefore, the coefficients $\lambda_1, \ldots, \lambda_4$ must be equal to the coefficients $\lambda_1', \ldots, \lambda_4'$ up to a plus/minus sign. This means that the states $|\psi_\lambda\rangle$ and $|\psi_{\lambda'}\rangle$ belong to the same $\mathbf{G}_4$ orbit if and only if there exists a permutation $\pi$ on four elements such that for all $j \in [4]$, $\lambda_j' = \lambda_{\pi(j)}$ or $\lambda_j' = -\lambda_{\pi(j)}$.

Since local unitaries do not form a subgroup of $\mathbf{G}_4$, we cannot directly translate the above conclusion to the language of SLOCC classes. However, we can order the coefficients $\lambda_x$ such that $\lambda_1$ has the maximal absolute value and apply a global phase to the state $|\psi_\lambda\rangle$ to remove the phase of $\lambda_1$, so that $\lambda_1$ is a positive real number. With this convention, we arrive at the following result.

> **Theorem 14.9.** Let $|\psi_\lambda\rangle$ and $|\psi_{\lambda'}\rangle$ be two four qubit states as given in (14.121), with the coefficients $\lambda_1$ and $\lambda_1'$ being real positive such that for all $x = 2, 3, 4$ we have $\lambda_1 \geqslant |\lambda_x|$ and $\lambda_1' \geqslant |\lambda_x'|$. Then, the two states $|\psi_\lambda\rangle$ and $|\psi_{\lambda'}\rangle$ belong to the same SLOCC class if and only if $\lambda_1 = \lambda_1'$ and there exists a permutation, $\pi$, on three elements such that for each $x \in \{2, 3, 4\}$ we have $\lambda_x' = \lambda_{\pi(x)}$ or $\lambda_x' = -\lambda_{\pi(x)}$.

This theorem highlights a stark contrast between three-qubit systems and four-qubit systems. While three-qubit systems have a finite number of SLOCC classes, the same cannot be said for four-qubit systems. In fact, the theorem demonstrates that four-qubit systems have an uncountable number of SLOCC classes.

This has significant implications, as it means that converting $|\psi_\lambda\rangle$ to $|\psi_{\lambda'}\rangle$ by LOCC is impossible, even with a probability less than 1, unless $\lambda' = \lambda$ up to a permutation and a sign change of the components of $\lambda'$ and $\lambda$. In simpler terms, the components of $\lambda'$ and $\lambda$ must be identical except for a rearrangement and possibly a change in sign.

**Exercise 14.20.** *Show that the state* $|\psi_\lambda\rangle$ *in* (14.121) *is a critical state. Specifically, show that if* $|\psi_\lambda^{ABCD}\rangle$ *is normalized, then its four local marginals are maximally mixed; that is, show that*

$$\psi_\lambda^A = \psi_\lambda^B = \psi_\lambda^C = \psi_\lambda^D = \frac{1}{2}I_2.\tag{14.123}$$

It is worth noting that in Refs. [223, 221, 229] it has been shown that up to local unitaries, the set

$$\mathfrak{C} := \left\{ |\psi_\lambda^{ABCD}\rangle : \lambda_1, \lambda_2, \lambda_3, \lambda_4 \in \mathbb{C} \quad , \quad \sum_{x\in[4]} |\lambda_x|^2 = 1 \right\}\tag{14.124}$$

is the set of *all* critical states in four qubits.

## 14.5 Deterministic Interconversions of Multipartite Entanglement

As discussed in the previous section, four-qubit systems exhibit an uncountable number of SLOCC classes. This means that deterministic and nondeterministic LOCC conversions between two randomly selected states is typically not possible in multipartite systems. However, for two states that belong to the same SLOCC class, it may be possible to convert one state to another deterministically via LOCC. Recall, however, that LOCC operations can be very complex in multipartite systems. As such, it will be more convenient to consider the larger set of separable operations instead. These operations can be used to transform a state into any other state in the same SLOCC class, and are generally easier to handle than LOCC operations.

A quantum channel $\mathcal{E} \in \mathrm{CPTP}(A^n \to A^n)$ is said to be separable if it has an operator sum representation of the form

$$\mathcal{E}(\cdot) = \sum_{k\in[m]} M_k(\cdot)M_k^*, \quad \text{where} \quad M_k = N_1^{(k)} \otimes N_2^{(k)} \otimes \cdots \otimes N_n^{(k)},\tag{14.125}$$

and for each $j \in [n]$, $N_j^{(k)} \in \mathcal{L}(A_j)$. We denote the set of all such channels by $\mathrm{SEP}(A^n \to A^n)$. While the matrices $N_j^{(k)}$ (and, by extension, $M_k$) might not always be invertible, in this section, our attention is specifically on the conversion between one pure state to another under the assumption that all $\{M_k\}k \in [m]$ are nonsingular. We use the notation $\mathrm{SEP}_1(A^n \to A^n) \subset \mathrm{SEP}(A^n \to A^n)$ to represent all separable channels of this kind. In essence, our focus is restricted to separable operations as defined earlier, with each $M_k$ being an element of $GL^n$. For further insights and references into the relations between LOCC, $\mathrm{SEP}_1$, and SEP, readers interested are referred to the concluding section of this chapter, titled "Notes and References."

## 14.5.1 The Stabilizer Group

In order to fully characterize conversions among pure multipartite states, it is necessary to introduce the concept of the stabilizer group. This group plays a crucial role in determining the properties and symmetries of a given state, and can help us identify which states can be transformed into one another via separable operations.

It is worth noting that, unlike the stabilizer formalism commonly used in quantum error correction codes, the stabilizer group discussed here is not necessarily a subgroup of the Pauli group. Instead, it is a subgroup of $GL^n$, a much larger group that includes the Pauli group as a special case. This difference is important because the stabilizer group in quantum error correction codes is designed to protect against certain types of errors, whereas the stabilizer group in multipartite quantum systems reflects the underlying symmetries and properties of the system itself. By understanding the structure and properties of this group, we can develop new insights into the behavior of multipartite quantum systems and discover new ways to manipulate and control them.

---

**Definition 14.4.** Let $\psi \in \mathrm{Pure}(A^n)$. The stabilizer group of $\psi^{A^n}$ is a subgroup of $GL^n$ defined by

$$\mathrm{Stab}(\psi) := \left\{ \Lambda \in GL^n : \Lambda|\psi\rangle = |\psi\rangle \right\}. \tag{14.126}$$

---

Note that the set $\mathrm{Stab}(\psi)$ is not empty since the identity matrix belongs to it.

**Exercise 14.21.** *Let $\psi \in \mathrm{Pure}(A^n)$ and consider the stabilizer group* $\mathrm{Stab}(|\psi\rangle)$.

1. *Show that* $\mathrm{Stab}(|\psi\rangle)$ *is indeed a group.*
2. *Show that for any* $\Lambda \in GL^n$ *and* $|\phi\rangle := \Lambda|\psi\rangle$ *we have*

$$\mathrm{Stab}(\phi) = \Lambda\, \mathrm{Stab}(\psi)\, \Lambda^{-1}. \tag{14.127}$$

**Exercise 14.22.** *Let $AB$ be a bipartite system with $|A| = |B|$. Find the stabilizer group of the maximally entangled state $|\Phi^{AB}\rangle$.*

The stabilizer group for $\psi$ is a subgroup of $GL^n$. One may naturally wonder how this group is related to the same group, but with $GL^n$ replaced by $SL^n$. The following theorem demonstrates that, unless $\psi$ is in the null cone of $A^n$, every element in the stabilizer group of $\psi$ lies in $SL^n$ up to a factor given by a root of unity.

> **Theorem 14.10.** Let $\psi \in \mathrm{Pure}(A^n)$ and suppose that $|\psi\rangle \notin \mathrm{Null}(A^n)$. Then, there exists $m \in \mathbb{N}$ such that
> $$\mathrm{Stab}(\psi) \subset \mathbf{G}_m := \left\{ e^{i\frac{2\pi k}{m}} \Lambda : k \in [m], \ \Lambda \in SL^n \right\}. \tag{14.128}$$

**Proof** By definition, since $|\psi\rangle \notin \mathrm{Null}(A^n)$, there exists a homogeneous SLIP, $f$, with the property that $f(|\psi\rangle) \neq 0$. Let $m$ be the degree of $f$. Now, let $\Lambda' \in \mathrm{Stab}(\psi)$ and observe that since $\mathrm{Stab}(\psi) \subset GL^n$, there exists $a \in \mathbb{C}$ such that $\Lambda' = a\Lambda$, where $\Lambda \in SL^n$. Thus, the property $\Lambda'|\psi\rangle = |\psi\rangle$ gives

$$f(|\psi\rangle) = f\left(\Lambda'|\psi\rangle\right) = f(a\Lambda|\psi\rangle) = a^m f(\Lambda|\psi\rangle) = a^m f(|\psi\rangle). \tag{14.129}$$

Since $f(|\psi\rangle) \neq 0$ we must have $a^m = 1$ so that $\Lambda' = a\Lambda \in \mathbf{G}_m$. This completes the proof. ∎

> **Corollary 14.2.** Let $\psi \in \mathrm{Pure}(A^n)$ and suppose there exists a homogeneous SLIP of degree $m \in \mathbb{N}$ that is not vanishing on $\psi$. Let $\mathbf{G} := \mathrm{Stab}(\psi) \cap SL^n$. Then, the quotient group $\mathrm{Stab}(\psi)/\mathbf{G}$ is a group of order at most $m$.

**Proof** We first need to show that $\mathbf{G}$ is a normal subgroup of $\mathrm{Stab}(\psi)$. To see why, recall that for any $M \in \mathrm{Stab}(\psi)$ we have $M = e^{i\frac{2\pi k}{m}} \Lambda$, where $\Lambda \in SL^n$. Therefore, for any $\Gamma \in \mathbf{G}$ we get

$$M\Gamma = e^{i\frac{2\pi k}{m}} \Lambda\Gamma\Lambda^{-1}\Lambda = \Gamma'M, \tag{14.130}$$

where $\Gamma' := \Lambda\Gamma\Lambda^{-1}$. We therefore need to show that $\Gamma' \in \mathbf{G}$. Since $\Lambda'$ is a product of three matrices in $SL^n$, it is itself in $SL^n$. To show that $\Gamma' \in \mathrm{Stab}(\psi)$ observe that by definition $\Gamma'$ can also be expressed as $\Gamma' = M\Lambda M^{-1}$, which is a product of three matrices in $\mathrm{Stab}(\psi)$ and therefore also $\Lambda'$ is in $\mathrm{Stab}(\psi)$. Hence, $\mathbf{G}$ is a normal subgroup of $\mathrm{Stab}(\psi)$. Finally, note that any $M = e^{i\frac{2\pi k}{m}} \Lambda$ as above satisfies $M^m \in SL^n$ so that $M^m \in \mathbf{G}$. This completes the proof. ∎

In the second corollary of Theorem 14.10, we use the notation $d_x := |A_x|$ to represent the local dimension of subsystem $A_x$ for each $x \in [n]$. Additionally, we denote by

$$SU^n := SU(d_1) \times \cdots \times SU(d_n) \tag{14.131}$$

and for any $m \in \mathbb{N}$

$$\mathbf{K}_m := \left\{ e^{i\frac{2\pi k}{m}} U : k \in [m], \ U \in SU^n \right\}. \tag{14.132}$$

> **Corollary 14.3.** Let $\psi \in \mathrm{Crit}(A^n)$ be such that $\mathrm{Stab}(\psi)$ is a finite group. Then, there exists $m \in \mathbb{N}$ such that
> $$\mathrm{Stab}(\psi) \subset \mathbf{K}_m. \tag{14.133}$$

**Proof**  Let $M \in \mathrm{Stab}(\psi)$. Since $\psi$ is a critical state, it is not in the null cone of $A^n$, so that from Theorem 14.10 there exists $N \in SL^n$, and $a \in \mathbb{C}$ with $a^m = 1$, such that $M = aN$. Moreover, using the polar decomposition we can further express $N$ as $N = U\Lambda$, where $U \in SU^n$ and $\Lambda > 0$ is a positive matrix in $SL^n$. Hence,

$$\big\| |\psi\rangle \big\| = \big\| M |\psi\rangle \big\| = \big\| aU\Lambda |\psi\rangle \big\| = \big\| \Lambda |\psi\rangle \big\|. \tag{14.134}$$

As $\Lambda \in SL^n$ is positive definite, the Kempf–Ness theorem (as described in Exercise 14.10) implies that $\Lambda |\psi\rangle = |\psi\rangle$. In other words, $\Lambda$ belongs to the stabilizer group $\mathrm{Stab}(\psi)$. Since $\mathrm{Stab}(\psi)$ is a finite group, the sequence $\{\Lambda^k\}_{k \in \mathbb{N}}$ must contain elements that are equal to each other, and therefore, there exists $k \in \mathbb{N}$ such that $\Lambda^k = I^{A^n}$. Since $\Lambda > 0$ we must have $\Lambda = I^{A^n}$. Hence, $M = aU \in \mathbf{K}_m$. Since $M$ was an arbitrary element of $\mathrm{Stab}(\psi)$ we conclude that all the elements of $\mathrm{Stab}(\psi)$ belong to $\mathbf{K}_m$. This completes the proof.  ∎

## Example 1: The Stabilizer Group of the Three-Qubit GHZ State

Let $\phi := \frac{1}{\sqrt{2}}(|000\rangle + |111\rangle)$ be the GHZ state of three qubits, and let $\mathbf{G} := \mathrm{Stab}(\phi) \cap SL^3$. A straightforward calculation (see Exercise 14.23) shows that $\mathbf{G}$ is given by

$$\mathbf{G} = \left\{ \begin{pmatrix} s_1 & 0 \\ 0 & s_1^{-1} \end{pmatrix} \otimes \begin{pmatrix} s_2 & 0 \\ 0 & s_2^{-1} \end{pmatrix} \otimes \begin{pmatrix} s_3 & 0 \\ 0 & s_3^{-1} \end{pmatrix} : s_1 s_2 s_3 = 1, \; s_1, s_2, s_3 \in \mathbb{C} \right\}. \tag{14.135}$$

Now, recall that in three qubits, the 3-tangle is defined in terms of an homogeneous SLIP of degree 4. Therefore, from Corollary 14.2 we get that the quotient group $\mathrm{Stab}(\phi)/\mathbf{G}$ is a group of order at most 4. However, note that if $\Lambda \in SL^3$, then also $-\Lambda \in SL^3$ so that $\mathbf{G}_4 = \mathbf{G}_2$, where the groups $\mathbf{G}_2$ and $\mathbf{G}_4$ are defined in (14.128). We therefore conclude that $\mathrm{Stab}(\phi)/\mathbf{G}$ is a group of order at most 2. Since $X \otimes X \otimes X \in \mathrm{Stab}(\phi)/\mathbf{G}$ we conclude that $\mathrm{Stab}(\phi)/\mathbf{G}$ contains only the identity matrix and the flip matrix $X \otimes X \otimes X$ so that $\mathrm{Stab}(\phi)$ is the union of $\mathbf{G}$ and the coset $(X \otimes X \otimes X)\mathbf{G}$.

**Exercise 14.23.**  *Prove (14.135) by direct calculation.*

**Exercise 14.24.**  *Find the stabilizer group of the W-state of three qubits. Is it compact?*

## Example 2: The Stabilizer Group of a Generic State in Four Qubits

As a second example, let $\psi$ by the four-qubit state

$$|\psi\rangle = \lambda_1 |\Phi_+\rangle|\Phi_+\rangle + \lambda_2 |\Phi_-\rangle|\Phi_-\rangle + \lambda_3 |\Psi_+\rangle|\Psi_+\rangle + \lambda_4 |\Psi_-\rangle|\Psi_-\rangle, \tag{14.136}$$

where $\{|\Phi_\pm\rangle, |\Psi_\pm\rangle\}$ is the Bell basis of two qubits, $\lambda_1, \lambda_2, \lambda_3, \lambda_4 \in \mathbb{C}$, and $\lambda_x^2 \neq \lambda_{x'}^2$ for all $x \neq x' \in [4]$. For this four-qubit state, it can be shown (see [100] and [229]) that the group $\mathbf{G} = \mathrm{Stab}(\psi) \cap SL^4$ is the Klein group consisting of only four elements:

$$\mathbf{G} = \{ I, \; X \otimes X \otimes X \otimes X, \; Y \otimes Y \otimes Y \otimes Y, \; Z \otimes Z \otimes Z \otimes Z \}, \tag{14.137}$$

where $X, Y, Z$ are the three Pauli matrices.

Now, recall the homogeneous SLIP of degree 2 given in (14.38). This polynomial is nonzero if

$$\lambda_1^2 + \lambda_2^2 + \lambda_3^2 + \lambda_4^2 \neq 0. \tag{14.138}$$

Therefore, if the coefficients of $\psi$ in (14.136) satisfy the relation (14.138), then from Corollary 14.2 there is no difference between $\mathbf{G}$ and $\mathrm{Stab}(\psi)$ so we get that the stabilizer of $\psi$ is the Klein group given in (14.137).

**Exercise 14.25.** *Consider the four-qubit state*

$$|\psi\rangle := \frac{1}{\sqrt{3}}\Big(|\Phi_+\rangle|\Phi_+\rangle + \omega|\Phi_-\rangle|\Phi_-\rangle + \overline{\omega}|\Psi_+\rangle|\Psi_+\rangle\Big), \tag{14.139}$$

*where $\omega = e^{i\frac{2\pi}{3}}$.*
*1. Show that $\mathbf{G} \neq \mathrm{Stab}(\psi)$.*
*2. Find all the elements of $\mathrm{Stab}(\psi)$.*

## 14.5.2 Generalization of Nielsen Majorization Theorem

In this subsection, we extend Theorem 12.2 to the multipartite scenario by defining a positive semidefinite operator for each stable state in $A^n$. Recall that Theorem 14.6 establishes that any stable state $\psi \in \mathrm{Pure}(A^n)$ is related to some critical state $\chi \in \mathrm{Pure}(A^n)$ by an $SL^n$-orbit. Therefore, there exist $\theta \in [0, 2\pi)$ and $M \in SL^n$ such that

$$|\psi\rangle = e^{i\theta} \frac{M|\chi\rangle}{\|M|\chi\rangle\|}. \tag{14.140}$$

We say that $\Lambda \in \mathrm{Pos}(A^n)$ is an associated positive (semidefinite) operator (APO) of $\psi$ if $\Lambda$ can be expressed as

$$\Lambda := \frac{M^*M}{\langle\chi|M^*M|\chi\rangle}. \tag{14.141}$$

As demonstrated by the following exercise, the APO of the state $\psi$ in (14.140) is not unique.

**Exercise 14.26.** *Show that if $S \in \mathrm{Stab}(\chi)$, then both $\Lambda$ and $S^*\Lambda S$ are APOs of $\psi$.*

**Exercise 14.27.** *Consider a bipartite state $\psi \in \mathrm{Pure}(AB)$ with $|A| = |B|$. Show that one of its APOs is given by $\rho^A \otimes I^B$, where $\rho^A := \mathrm{Tr}_B\big[\psi^{AB}\big]$.*

---

**Theorem 14.11.** Let $\psi_1, \psi_2 \in \mathrm{Pure}(A^n)$ be two multipartite states in the same reversible SLOCC class of $\chi \in \mathrm{Pure}(A^n)$. Let $\Lambda_1$ and $\Lambda_2$ be any APOs corresponding to $\psi_1$ and $\psi_2$, respectively. Then, $\psi_1 \xrightarrow{\mathrm{SEP1}} \psi_2$ if and only if there exists a probability distribution $\{p_x\}_{x\in[m]}$ along with operators $\{S_x\}_{x\in[m]} \subset \mathrm{Stab}(\chi)$ such that

$$\Lambda_1 = \sum_{x\in[m]} p_x S_x^* \Lambda_2 S_x. \tag{14.142}$$

**Proof**   Let $N_1, N_2 \in SL^n$ be such that for $j = 1, 2$

$$|\psi_j\rangle = e^{i\theta_j} \frac{N_j|\chi\rangle}{\|N_j|\chi\rangle\|}, \tag{14.143}$$

and denote by

$$\Lambda_1 = \frac{N_1^* N_1}{\langle \chi|N_1^* N_1|\chi\rangle} \quad \text{and} \quad \Lambda_2 = \frac{N_2^* N_2}{\langle \chi|N_2^* N_2|\chi\rangle}. \tag{14.144}$$

Now, $\psi_1 \xrightarrow{\text{SEP1}} \psi_2$ if and only if there exists $M_x \in GL^n$ satisfying $\sum_x M_x^* M_x = I^{A^n}$ such that

$$M_x|\psi_1\rangle = c_x|\psi_2\rangle \tag{14.145}$$

for some $c_x \in \mathbb{C}$. After combining the relation of (14.145) with (14.143), and performing some algebra, we obtain

$$\frac{1}{c_x} \frac{\|N_2|\chi\rangle\|}{\|N_1|\chi\rangle\|} N_2^{-1} M_x N_1|\chi\rangle = |\chi\rangle. \tag{14.146}$$

Thus, for each $x$ we have

$$\frac{1}{c_x} \frac{\|N_2|\chi\rangle\|}{\|N_1|\chi\rangle\|} N_2^{-1} M_x N_1 = S_x, \tag{14.147}$$

where $S_x \in \text{Stab}(\chi)$. Isolating $M_x$ gives

$$M_x = c_x \frac{\|N_1|\chi\rangle\|}{\|N_2|\chi\rangle\|} N_2 S_x N_1^{-1}. \tag{14.148}$$

Using the notations (14.144), it can be verified easily that $\sum_x M_x^* M_x = I^{A^n}$ if and only if

$$\Lambda_1 = \sum_x p_x S_x^* \Lambda_2 S_x, \tag{14.149}$$

with $p_x := |c_x|^2$. Hence, $\psi_1 \xrightarrow{\text{SEP1}} \psi_2$ if and only if $\Lambda_1$ and $\Lambda_2$ satisfy the relation (14.149). This completes the proof. $\blacksquare$

Let $\chi \in \text{Crit}(A^n)$ have a finite stabilizer, that is, $\text{Stab}(\chi) = U_{x\,x\in[m]}$ is a finite set of unitaries, as established by Corollary 14.3. In this situation, we can define the $\text{Stab}(\chi)$-twirling operation as

$$\mathcal{G}\left(\omega^{A^n}\right) = \frac{1}{m} \sum_{x\in[m]} U_x \omega^{A^n} U_x^* \qquad \forall\, \omega \in \mathfrak{L}(A^n). \tag{14.150}$$

Now, according to Theorem 14.11 $\psi_1 \xrightarrow{\text{SEP1}} \psi_2$ if and only if there exists a probability distribution $\{p_x\}_{x\in[m]}$ such that

$$\Lambda_1 = \sum_{x\in[m]} p_x U_x^* \Lambda_2 U_x. \tag{14.151}$$

By taking the twirling map $\mathcal{G}$ on both sides of (14.151) we get that if $\psi_1 \xrightarrow{\text{SEP1}} \psi_2$, then

$$\mathcal{G}(\Lambda_1) = \mathcal{G}(\Lambda_2). \tag{14.152}$$

In other words, this condition is a necessary condition for the conversion $\psi_1 \xrightarrow{\text{SEP}_1} \psi_2$ (but not always sufficient). Moreover, if $\Lambda_1$ is symmetric, meaning $\mathcal{G}(\Lambda_1) = \Lambda_1$, then $\psi_1 \xrightarrow{\text{SEP}_1} \psi_2$ if and only if $\Lambda_1 = \mathcal{G}(\Lambda_2)$.

One special case that $\Lambda_1$ is symmetric is the case that $\psi_1 = \chi$. In this case, $\Lambda_1 = I^{A^n}$ and $\chi \xrightarrow{\text{SEP}_1} \psi_2$ if and only if $\mathcal{G}(\Lambda_2) = I^{A^n}$. Conversely, if $\psi_2 = \chi$, then $\Lambda_2 = I^{A^n}$, so the condition (14.151) becomes $\Lambda_1 = I^{A^n}$. In other words, $\psi_1 \xrightarrow{\text{SEP}_1} \chi$ if and only if up to local unitaries $\psi_1 = \chi$. This is consistent with the intuition that the critical state is the maximally entangled state of the SLOCC orbit.

To demonstrate how Theorem 14.11 generalizes Nielsen's majorization theorem, we now apply it to the bipartite case. Let us consider a bipartite system $AB$ with $m := |A| = |B|$. The only critical state of this system is the maximally entangled state $\Phi_m$, and its stabilizer is given by

$$\text{Stab}(\Phi_m) := \left\{ S^{-1} \otimes S^T : S \in GL(m) \right\}. \tag{14.153}$$

The stabilizer group already mentioned is clearly not compact. However, in the derivation of Nielsen's majorization theorem, we employed Lo-Popescu's Theorem (Theorem 12.1) which limits Bob's operations to be unitary operations. Thus, we can limit $S$ to be a unitary matrix without loss of generality.

Consider two bipartite states $\psi_1, \psi_2 \in \text{Pure}(AB)$. As per Exercise 14.27, an APO of $\psi_1$ has the form $\Lambda_1 = \rho_1^A \otimes I^B$, where $\rho_1^A$ is the reduced density matrix of $\psi_1^{AB}$. Similarly, an APO of $\psi_2$ has the form $\Lambda_2 = \rho_2^A \otimes I^B$, where $\rho_2^A$ is the reduced density matrix of $\psi_2^{AB}$. Therefore, from Theorem 14.11, we can infer that $\psi_1^{AB} \xrightarrow{\text{SEP}_1} \psi_2^{AB}$ if and only if there exists a probability distribution $\{p_x\}_{x \in [k]}$ along with $k$ unitary matrices $\{U_x\}_{x \in [k]} \subset U(m)$ such that

$$\rho_1^A \otimes I^B = \sum_{x \in [k]} p_x U_x^* \rho_2^A U_x \otimes I^B. \tag{14.154}$$

This condition is precisely the same as the condition we obtained in (12.26) for Nielsen's majorization criterion.

**Exercise 14.28.** *Let $\psi$ be the four-qubit state (14.136) and suppose it satisfies (14.138). Classify all the four-qubit states $\phi$ for which $\psi \xrightarrow{\text{SEP}_1} \phi$ holds.*

## 14.6 Entanglement of Assistance

In the previous sections, we have seen that LOCCs alone are limited in their ability to manipulate multipartite entanglement. For instance, the classification of four-qubit entanglement involves infinitely many SLOCC classes, highlighting the complexity of entanglement manipulation under LOCC for generic multipartite entangled pure states in $\text{Pure}(A^n)$, where $n \geq 4$. More generally, it has been shown that if two generic multipartite entangled pure states $\psi$ and $\phi$ cannot be locally converted into each other by a unitary operation, then the conversion cannot be achieved by LOCC as well. However, this situation can change significantly if the target state $\phi$ is not generic.

While bipartite entanglement is the most useful form of entanglement in quantum information, the concentration of multipartite systems into two parties has also been extensively studied. In this section, we focus on the conversion of a tripartite pure state into an ensemble of bipartite states. We begin by considering the following conversion of a tripartite state $\psi \in \mathrm{Pure}(ABR)$ into an ensemble of pure states, or equivalently, a cq-state of the form

$$\sigma^{ABX} = \sum_{x \in [m]} p_x \phi_x^{AB} \otimes |x\rangle\langle x|^X, \tag{14.155}$$

where $\mathbf{p} := (p_1, \ldots, p_m)^T \in \mathrm{Prob}(m)$, and each $\phi_x \in \mathrm{Pure}(AB)$.

**Lemma 14.3.** Let $\psi \in \mathrm{Pure}(AB)$ and let $\rho^B := \mathrm{Tr}_A\left[\psi^{AB}\right]$ be its reduced density matrix on system $B$. Then, for every pure state decomposition $\rho^B = \sum_{x \in [m]} p_x \phi_x^B$, there exists a POVM on Alices system, $\{\Lambda_x\}_{x \in [m]} \subset \mathrm{Eff}(A)$, such that for all $x \in [m]$

$$\phi_x^B = \frac{1}{p_x} \mathrm{Tr}_A\left[\left(\Lambda_x^A \otimes I^B\right) \psi^{AB}\right] \quad \text{and} \quad p_x = \mathrm{Tr}\left[\left(\Lambda_x^A \otimes I^B\right) \psi^{AB}\right]. \tag{14.156}$$

That is, every pure state decomposition of $\rho^B$ can be realized by a generalized measurement on Alice's system.

**Proof**   Consider the pure state

$$|\tilde{\psi}^{RB}\rangle := \sum_{x \in [n]} \sqrt{p_x} |x\rangle^R |\phi_x^B\rangle. \tag{14.157}$$

Since both $\psi^{AB}$ and $\tilde{\psi}^{RB}$ are purifications of $\rho^B$, there exists an isometry $V : A \to R$ such that $|\tilde{\psi}^{RB}\rangle = V \otimes I^B |\psi^{AB}\rangle$. Therefore, taking

$$\Lambda_x^A := V^* |x\rangle\langle x|^R V \qquad \forall\, x \in [m] \tag{14.158}$$

gives

$$\mathrm{Tr}_A\left[\left(\Lambda_x^A \otimes I^B\right) \psi^{AB}\right] = \mathrm{Tr}_R\left[\left(|x\rangle\langle x|^R \otimes I^B\right) \tilde{\psi}^{AB}\right] = \sqrt{p_x}\phi_x^B. \tag{14.159}$$

Therefore, with this choice of the POVM $\{\Lambda_x^A\}_{x \in [m]}$ we get (14.156). This completes the proof.   ∎

Let $\psi \in \mathrm{Pure}(ABR)$ be a tripartite pure state with marginal $\rho^{AB} := \mathrm{Tr}_R\left[\psi^{ABR}\right]$. From Lemma 14.3, we know that every pure state decomposition of $\rho^{AB}$ can be realized by a generalized measurement on the reference system $R$. Thus, for a given measure of bipartite entanglement $E$, we define the *entanglement of assistance* to be

$$E_a\left(\rho^{AB_1}\right) := \sup \sum_{x \in [m]} p_x E\left(\phi_x^{AB_1}\right), \tag{14.160}$$

where the supremum is taken over all pure state decompositions of $\rho^{AB} = \sum_{x\in[m]} p_x \phi_x^{AB}$. Note that this definition is similar to the definition of the entanglement of formation given in (13.68), except that we take the supremum instead of the infimum as taken in (13.68).

**Exercise 14.29.** *Compute the entanglement of assistance of the maximally mixed state $u^{AB}$ and conclude that the entanglement of assistance is not a measure of entanglement.*

**Exercise 14.30.** *Let $E$ be a measure of pure bipartite entanglement, and let $E_F$ and $E_a$ be its corresponding entanglement of formation and assistance, respectively. Let $\psi \in \mathrm{Pure}(AB_1 B_2)$ be a tripartite pure state with marginal $\rho^{AB_1} := \mathrm{Tr}_{B_2}\left[\psi^{AB_1 B_2}\right]$. Show that if $\psi^{AB_1 B_2}$ satisfies the disentangling condition (14.184), then*

$$E_F\left(\rho^{AB}\right) = E_a\left(\rho^{AB}\right). \tag{14.161}$$

Exercise 14.29 demonstrates that entanglement of assistance is not a measure of entanglement as a function of the bipartite state on system $AB$. However, one may wonder if the entanglement of assistance is a measure of tripartite pure entanglement. Specifically, for any $\psi \in \mathrm{Pure}(ABR)$, we define

$$\tilde{E}_a\left(\psi^{ABR}\right) := E_a\left(\rho^{AB}\right), \tag{14.162}$$

where $\rho^{AB} := \mathrm{Tr}_R\left[\psi^{ABR}\right]$. One can then ask whether it is impossible to increase $\tilde{E}_a$ by LOCC. In other words, can an LOCC prior to the measurement on the reference system increase the entanglement of assistance? Surprisingly, we now show that such a pre-LOCC can increase the entanglement of assistance, and therefore $\tilde{E}_a$ is also not an entanglement measure.

Let $A = A_1 A_2$ with $|A_1| = 2$, $|A_2| = |B| = 4$, and $|R| = 2$, and let $\psi \in \mathrm{Pure}(ABR)$ be the state

$$|\psi^{ABR}\rangle := \frac{1}{2}\left(|0\rangle^{A_1}|\Phi_0^{A_2 B}\rangle|0\rangle^R + |0\rangle^{A_1}|\Phi_1^{A_2 B}\rangle|1\rangle^R + |1\rangle^{A_1}|\Phi_2^{A_2 B}\rangle|+\rangle^R + |1\rangle^{A_1}|\Phi_3^{A_2 B}\rangle|-\rangle^R\right), \tag{14.163}$$

where $\{\Phi_x^{A_2 B}\}_{x\in\{0,1,2,3\}}$ are four maximally entangled states to be determined shortly. Consider a protocol where Alice measures system $A_1$ in the $|0\rangle, |1\rangle$ basis and sends the classical outcome to the referee. Based on the outcome, the referee performs either the measurement in the $|0\rangle^R, |1\rangle^R$ basis (if Alice's outcome is zero) or the measurement in the $|+\rangle^R, |-\rangle^R$ basis (if Alice's outcome is one). Regardless of the measurement outcomes of both Alice and the referee, Alice and Bob end up with one of the four maximally entangled states $\{\Phi_x^{A_2 B}\}_{x\in 0,1,2,3}$. In particular, by local unitary operation, Alice and Bob can transform any of these four states into the maximally entangled state $\Phi^{A_2 B}$. Thus, we conclude that

$$\psi^{ABR} \xrightarrow{\text{LOCC}} \Phi^{A_2 B}. \tag{14.164}$$

However, we now show that for certain choices of maximally entangled states $\{\Phi_x^{A_2 B}\}_{x\in 0,1,2,3}$, the transformation in (14.164) cannot be achieved (even with probability less than one) if we only allow system $R$ to perform a measurement.

The reduced density matrix $\rho^{AB} := \mathrm{Tr}_R\left[\psi^{ABR}\right]$ can be expressed as

$$\rho^{AB} = \varphi_0^{AB} + \varphi_1^{AB}, \tag{14.165}$$

where

$$|\varphi_0^{AB}\rangle := \frac{1}{2}|0\rangle^{A_1}|\Phi_0^{A_2B}\rangle + \frac{1}{2\sqrt{2}}|1\rangle^{A_1}\left(|\Phi_2^{A_2B}\rangle + |\Phi_3^{A_2B}\rangle\right) \tag{14.166}$$

and

$$|\varphi_1^{AB}\rangle := \frac{1}{2}|0\rangle^{A_1}|\Phi_1^{A_2B}\rangle + \frac{1}{2\sqrt{2}}|1\rangle^{A_1}\left(|\Phi_2^{A_2B}\rangle - |\Phi_3^{A_2B}\rangle\right). \tag{14.167}$$

We argue that there exists four maximally entangled states $\{\Phi_x^{A_2B}\}_{x\in\{0,1,2,3\}}$ such that any linear combination of $|\varphi_0^{AB}\rangle$ and $|\varphi_1^{AB}\rangle$ is not maximally entangled. Indeed, take

$$|\Phi_0^{A_2B}\rangle = |\Phi_2^{A_2B}\rangle = \frac{1}{2}(|00\rangle + |11\rangle + |22\rangle + |33\rangle) \tag{14.168}$$

and

$$|\Phi_1^{A_2B}\rangle = \frac{1}{2}(|00\rangle - i|11\rangle - |22\rangle + i|33\rangle)$$
$$|\Phi_3^{A_2B}\rangle = \frac{1}{2}(|00\rangle - i|11\rangle + |22\rangle - i|33\rangle). \tag{14.169}$$

With these choices we get by direct calculation that for any $a, b \in \mathbb{C}$ the linear combination

$$a|\varphi_0^{AB}\rangle + b|\varphi_1^{AB}\rangle \tag{14.170}$$

is not proportional to the maximally entangled state (see Exercise 14.31). Therefore, the state $\psi^{ABR}$ cannot be converted to $\Phi^{AB}$ (even with probability less than one) by a local measurement on system $R$. Alternatively, none of the pure-state decompositions of $\rho^{AB}$ contains a maximally entangled state (i.e. two-ebits).

**Exercise 14.31.** *Show that for any choice of $a, b \in \mathbb{C}$ with $|a|^2 + |b|^2 = 1$ the state in (14.170) is not maximally entangled. Hint: Write the state in (14.170) as a linear combination $\sum_{x=0}^{3}|\phi_x^A\rangle|x\rangle^B$ and show that the vectors $\{|\phi_x^A\rangle\}_x$ cannot all have the same norm and also orthogonal to each other.*

## 14.6.1 Entanglement of Collaboration

Since pre-LOCC operations performed by Alice and Bob can increase the entanglement of assistance, we can modify the definition of entanglement of assistance by including arbitrary LOCC operations performed by all parties. This modification results in a new measure called the "entanglement of collaboration," which more comprehensively quantifies the amount of bipartite entanglement available for collaborative tasks among all parties.

---

**Entanglement of Collaboration**

**Definition 14.5.** Let $E$ be a measure of bipartite entanglement for mixed states. Its corresponding measure of tripartite entanglement, known as the "entanglement of collaboration" and denoted by $E_c$, is defined as

$$E_c\left(\rho^{ABR}\right) := \sup_{\mathcal{N}\in\text{LOCC}} E\left(\mathcal{N}^{ABR\to A'B'}\left(\rho^{ABR}\right)\right) \qquad \forall\, \rho \in \mathfrak{D}(ABR),$$

$$(14.171)$$

where the supremum is over all quantum systems $A'$ and $B'$, and all LOCC channels $\mathcal{N}^{ABR\to A'B'}$.

In other words, the entanglement of collaboration is the maximal bipartite entanglement (as measured by $E$) that can be shared between Alice and Bob after all parties collaborate via LOCC. Observe that the entanglement of collaboration is never smaller than the entanglement of assistance. Indeed, suppose $\rho^{ABR} = \psi^{ABR}$ is a pure state and take $A' = A$, $B' = BX$, where $X$ is a classical system corresponding to the outcome of a POVM performed on system $R$. Then, taking the LOCC channel $\mathcal{N}^{ABR\to ABX} = \text{id}^{AB} \otimes \mathcal{E}^{R\to X}$, where $\mathcal{E} \in \text{CPTP}(R \to X)$ is a POVM Channel, we get that

$$\mathcal{N}^{ABR\to ABX}\left(\psi^{ABR}\right) = \sigma^{ABX} := \sum_{x\in[m]} p_x \phi_x^{AB} \otimes |x\rangle\langle x|^X \qquad (14.172)$$

is a cq-state, where $\{p_x, \phi_x^{AB}\}_{x\in[m]}$ is one of the pure-state decompositions of $\rho^{AB}$. Therefore, for this choice of $\mathcal{N}^{ABR\to A'B'}$ we get

$$E\left(\mathcal{N}^{ABR\to A'B'}\left(\rho^{ABR}\right)\right) = E\left(\sigma^{ABX}\right)$$

**Assuming $E$ is convex linear (10.8)** $\to$ $= \sum_{x\in[m]} p_x\, E(\phi_x^{AB}).$

$$(14.173)$$

Therefore, since entanglement of collaboration is defined as a supremum over all such LOCC channels $\mathcal{N}^{ABR\to A'B'}$, it must be no smaller than the entanglement of assistance. Furthermore, unlike entanglement of assistance, entanglement of collaboration is a measure of tripartite entanglement.

**Exercise 14.32.** *Show that the entanglement of collaboration is a measure of tripartite entanglement.*

---

**Lemma 14.4.** Let $E$ be a measure of bipartite entanglement. Then, for all $\rho \in \mathfrak{D}(ABR)$

$$E_c\left(\rho^{ABR}\right) \leqslant \min\left\{E\left(\rho^{A(BR)}\right), E\left(\rho^{B(AR)}\right)\right\}, \qquad (14.174)$$

where the parenthesis in $\rho^{A(BR)}$ indicates that the entanglement is computed between system $A$ and the composite system $BR$.

**Exercise 14.33.** *Prove Lemma 14.4.*

One of the most fundamental questions in entanglement theory is the distillation of Bell states from multiple copies of a bipartite entangled state. As we have already seen, for a given pure state $\psi^{AB}$, the distillable entanglement is determined by the von-Neumann entropy of the reduced density matrix $\psi^A$. A similar question arises in the multipartite regime: Given many copies of a tripartite pure entangled state $\psi^{ABR}$, how many Bell states can be distilled between Alice and Bob by LOCC of all three parties sharing the state?

Lemma 14.4 asserts that the optimal distillation rate cannot exceed the minimum between the entropies of system $A$ and system $B$. Remarkably, it has been shown that this upper bound can be attained. However, we will present the proof of this statement in volume 2 of this book, after introducing the quantum state merging protocol from quantum Shannon theory.

# 14.7 Monogamy of Entanglement

Monogamy of entanglement is a fundamental property of multipartite quantum systems, whereby the amount of entanglement between two subsystems limits the amount of entanglement that each subsystem can have with the rest of the system. Unlike classical correlation, entanglement cannot be freely shared among multiple parties, making this principle a fundamental concept in quantum information theory. For instance, if Alice's system is maximally entangled with Bob's system, it cannot be simultaneously entangled with another third system. That is, sharing entanglement is limited.

This principle has far-reaching implications in various areas, including quantum communication and cryptography. Overall, monogamy of entanglement is a fundamental property of quantum systems with significant implications in both theoretical and practical aspects of quantum information theory.

## 14.7.1 Quantification of Monogamy of Entanglement

We can quantify the phenomenon of monogamy of entanglement as follows. Let $B = B_1 B_2$ be a composite system, and let $E$ be a measure of entanglement. We say that $E$ is monogamous if for all $\rho \in \mathfrak{D}(AB)$

$$E\left(\rho^{AB}\right) \geqslant E\left(\rho^{AB_1}\right) + E\left(\rho^{AB_2}\right). \tag{14.175}$$

If $E$ is monogamous and $A$ and system $B_1$ are maximally entangled, then we must have $E\left(\rho^{AB}\right) = E\left(\rho^{AB_1}\right)$. From the inequality above, we can conclude that $E\left(\rho^{AB_2}\right) = 0$, meaning that $B_2$ has no entanglement with $A$.

It is worth noting that not all measures of entanglement satisfy the inequality in (14.175), and it is not immediately clear that monogamous measures of entanglement exist. Therefore, to demonstrate the monogamy of entanglement, we must first show that there are measures of entanglement that satisfy (14.175).

> **Theorem 14.12.** The squashed entanglement is a monogamous measure of entanglement satisfying (14.175).

**Proof** Let $\rho^{ABR}$ be an extension of the state $\rho^{AB}$. Using the chain rule of the conditional mutual information (see the second equality in (13.180)) we get

$$\frac{1}{2}I(A:B|R)_\rho = \frac{1}{2}I(A:B_1|R)_\rho + \frac{1}{2}I(A:B_2|RB_1)_\rho$$

$$\textbf{By definition} \rightarrow \; \geq E_{\text{sq}}\left(\rho^{AB_1}\right) + E_{\text{sq}}\left(\rho^{AB_2}\right). \tag{14.176}$$

Since the above inequality holds for all extensions $\rho^{ABR}$ of $\rho^{AB}$, we conclude that

$$E_{\text{sq}}\left(\rho^{AB}\right) \geq E_{\text{sq}}\left(\rho^{AB_1}\right) + E_{\text{sq}}\left(\rho^{AB_2}\right). \tag{14.177}$$

This completes the proof. ∎

It is important to recognize that not all measures of entanglement adhere to the monogamy condition specified by (14.175). Nevertheless, the squashed entanglement is currently the only known measure of entanglement that satisfies (14.175) in all finite dimensions, which is a remarkable property that highlights the unique nature of this measure. Other measures satisfy (14.175) on fixed dimensions. For example, on qubit systems, the square of the concurrence is also a monogamous measure of entanglement that satisfies (14.175) when $|A| = |B_1| = |B_2| = 2$.

## Qubit Monogamy Relations

For qubit systems one can use the concurrence to quantify entanglement. Let $\psi \in$ Pure($ABC$) be a three-qubit pure state; that is, $|A| = |B| = |C| = 2$. In this case, from (13.109) and the fact that $\rho^{AB}$ is at most rank 2, the concurrence of assistance (as defined in Exercise 13.21; see (13.107) and (13.108)) can be expressed as

$$C_a\left(\rho^{AB}\right)^2 = \text{Tr}\left[\rho^{AB}\rho_\star^{AB}\right]$$

$$\textbf{Exercise 14.34} \rightarrow \; = 2\left(\det\left(\rho^A\right) + \det\left(\rho^B\right) - \det\left(\rho^C\right)\right). \tag{14.178}$$

**Exercise 14.34.** *Use a direct calculation to prove the equality in (14.178).*

The equality above implies that for the pure state $\psi^{ABC}$ with marinals $\rho^{AB}$ and $\rho^{AC}$ we have

$$C_a\left(\rho^{AB}\right)^2 + C_a\left(\rho^{AC}\right)^2 = 4\det\left(\rho^A\right) = C\left(\psi^{A(BC)}\right)^2. \tag{14.179}$$

Denoting by $\tau\left(\rho^{AB}\right) := C\left(\rho^{AB}\right)^2$ (the square of the concurrence of formation, also know as the 2-tangle) and using the fact that $C\left(\rho^{AB}\right) \leq C_a\left(\rho^{AB}\right)$ we arrive at the following monogamy inequality

$$\tau\left(\psi^{A(BC)}\right) \geq \tau\left(\rho^{AB}\right) + \tau\left(\rho^{AC}\right), \tag{14.180}$$

where

$$\tau\left(\psi^{A(BC)}\right) := 2\left(1 - \mathrm{Tr}\left[\left(\rho^{A}\right)^{2}\right]\right) = 4\det\left(\rho^{A}\right). \tag{14.181}$$

The relation (14.180) is widely known in the literature as the Coffman–Kundu–Wootter (CKW) monogamy relation. This seminal result introduced the concept of monogamy of entanglement for the first time, demonstrating that the amount of entanglement between two subsystems is limited by the amount of entanglement between each subsystem and a third subsystem.

**Exercise 14.35.** *Show that*

$$\tau\left(\psi^{A(BC)}\right) - \tau\left(\rho^{AB}\right) - \tau\left(\rho^{AC}\right) = \mathrm{Tangle}\left(\psi^{ABC}\right), \tag{14.182}$$

*where the right-hand side is the 3-tangle defined in (14.101).*

## 14.7.2 Monogamy without Inequalities

The definition of a monogamous measure of entanglement, as provided in (14.175), only captures a partial aspect of monogamy of entanglement. This is due to the fact that many important measures of entanglement do not satisfy this relation, and some of them are not even additive under tensor product. Therefore, the summation in the right-hand side of (14.175) is only a convenient choice and not a necessity. For instance, it is well-known that if $E$ does not satisfy this relation, it is still possible to find a positive exponent $\alpha > 0$ such that the function $E^{\alpha}$ satisfies the relation. This is already evident in the CKW relation, where $E$ is taken to be the square of the concurrence instead of the concurrence itself.

Moreover, there exist measures of entanglement that are multiplicative under tensor product, which implies that they fail to satisfy (14.175) or any power of it. This highlights the necessity for a more nuanced and refined definition of monogamy of entanglement that encompasses these peculiarities. This becomes especially important when dealing with the range of measures of entanglement available, which exhibit varying properties.

One approach to address the issues with the current definition of a monogamous measure of entanglement is to replace the relation given in (14.175) with a family of monogamy relations of the form

$$E\left(\rho^{AB}\right) \geq f\left(E\left(\rho^{AB_1}\right), E\left(\rho^{AB_2}\right)\right), \tag{14.183}$$

where $f$ is a function of two variables that satisfies certain conditions. While this family of monogamy relations may be more flexible than the original definition, it still lacks a clear theoretical foundation. Thus, a more desirable solution would be to derive the monogamy relations from more basic principles, which would provide a deeper understanding of the nature of this phenomenon.

Recently, such approach to monogamy of entanglement has been proposed, which is more "fine-grained" in nature and avoids the need for introducing a function $f$. This approach does not involve monogamy relations such as (14.175) or (14.183). Instead, it defines a measure of entanglement $E$ to be monogamous if it satisfies a certain condition that does not involve inequalities. In particular, this approach takes into account the fact that different measures of entanglement have varying properties and limitations, rather than attempting to impose a one-size-fits-all definition. By adopting this more nuanced approach, we can gain a deeper understanding of the monogamy of entanglement and how it manifests itself across different measures.

---

**Definition 14.6.** Let $B = B_1 B_2$ denote a composite system, and let $E$ be a measure of bipartite entanglement. $E$ is said to be *monogamous* if for any bipartite state $\rho \in \mathcal{D}(AB)$ that satisfies

$$E\left(\rho^{AB}\right) = E\left(\rho^{AB_1}\right) \tag{14.184}$$

we have that $E\left(\rho^{AB_2}\right) = 0$.

---

The condition expressed in equation (14.184) is considered to be a strong one, and is usually not met by most states within $\mathcal{D}(AB)$. It is often referred to as the "disentangling condition." As we will explore further below, quantum Markov states (defined in Definition 13.6) are always guaranteed to satisfy this equality for any entanglement monotone $E$. Additionally, the condition presented in (14.175) is even stronger than the one defined in Definition 14.6. Specifically, if $E$ satisfies (14.175), then any $\rho^{AB}$ that satisfies (14.184) must have $E(\rho^{AB_2}) = 0$. Nonetheless, Definition 14.6 still captures the essence of monogamy, by stipulating that if system $A$ shares the maximum amount of entanglement with subsystem $B_1$, it is left with no entanglement to share with $B_2$.

In Definition 14.6, we do not invoke a particular monogamy relation such as (14.175). Instead, we propose a minimalist approach which is not quantitative, in which we only require what is essential from a measure of entanglement to be monogamous. Yet, this requirement is sufficient to generate a more quantitative monogamy relation as demonstrated in the following theorem.

---

**Theorem 14.13.** Let $E$ be a continuous measure of entanglement. Then, $E$ is monogamous according to Definition 14.6 if and only if for every $d \in \mathbb{N}$, and every systems $A$ and $B = B_1 B_2$ with $|AB| = d$, there exists $0 < \alpha < \infty$ such that

$$E^\alpha(\rho^{AB}) \geqslant E^\alpha(\rho^{AB_1}) + E^\alpha(\rho^{AB_2}) \qquad \forall \, \rho \in \mathcal{D}(AB). \tag{14.185}$$

---

**Proof**   We leave it as an exercise to show that if $E$ satisfies (14.185), then $E$ is a monogamous measure of entanglement (according to Definition 14.6). We therefore assume now that $E$ is monogamous and prove the relation (14.185). Since $E$ is a measure of entanglement, it is nonincreasing under partial traces, and therefore for all $\rho \in \mathcal{D}(AB)$

$$E(\rho^{AB}) \geqslant \max\{E(\rho^{AB_1}), E(\rho^{AB_2})\}. \tag{14.186}$$

Without loss of generality we assume that $E(\rho^{AB}) > 0$ and set

$$x_1 := \frac{E(\rho^{AB_1})}{E(\rho^{AB})} \quad \text{and} \quad x_2 := \frac{E(\rho^{AB_2})}{E(\rho^{AB})}. \tag{14.187}$$

From (14.186) we get that $x_1, x_2 \in [0, 1]$. Moreover, since $E$ is monogamous we get that if $x_1 = 1$, then we must have $x_2 = 0$ and vice versa. Therefore, there exists $\mu > 0$ such that

$$x_1^{\mu} + x_2^{\mu} \leqslant 1, \tag{14.188}$$

since either $x_j^{\mu} \Rightarrow 0$ when $\mu$ increases, or if $x_1 = 1$, then by assumption $x_2 = 0$ and similarly if $x_2 = 1$, then $x_1 = 0$. Let $f : \mathfrak{D}(AB) \to \mathbb{R}_+$ denote a function defined such that $f(\rho^{AB})$ represents the smallest value of $\mu$ that satisfies equality in (14.188). Since $E$ is continuous, so is $f$, and the compactness of $\mathfrak{D}(AB)$ gives

$$\alpha := \max_{\rho \in \mathfrak{D}(AB)} f(\rho^{AB}) < \infty. \tag{14.189}$$

By definition, $\alpha$ satisfies the condition in (14.185). $\blacksquare$

It is important to note that the relation given in (14.185) is *not* of the form given in (14.183), since the monogamy exponent $\alpha$ in (14.185) depends on the dimension $d$, whereas $f$ is considered universal in the sense that it does not depend on the dimension. Therefore, if a measure of entanglement such as the entanglement of formation is not monogamous according to the class of relations given in (14.183), it does not necessarily mean that it is not monogamous according to Definition 14.6.

In the next theorem we show that all quantum Markov states satisfy the disentangling condition. For this purpose, we will rename $B_1$ as $B$ and $B_2$ as $B'$, since the theorem involves further decomposition of system $B$ into subsystems. Specifically, an entangled Markov quantum state $\rho \in \mathfrak{D}(ABB')$ is a state of the form (cf. (13.185))

$$\rho^{ABB'} = \bigoplus_{x \in [m]} p_x \rho_x^{AB_x^{(1)}} \otimes \rho_x^{B_x^{(2)} B'}, \tag{14.190}$$

where

$$B = \bigoplus_{x \in [m]} B_x^{(1)} \otimes B_x^{(2)}, \tag{14.191}$$

and for each $x \in [m]$, $\rho_x^{AB_x^{(1)}}$ and $\rho_x^{B_x^{(2)} B'}$ are density matrices in $\mathfrak{D}(AB_x^{(1)})$ and $\mathfrak{D}(AB_x^{(2)})$, respectively.

> **Theorem 14.14.** Let $E$ be an entanglement monotone. Then, $E$ satisfies the disentangling condition $E\left(\rho^{ABB'}\right) = E\left(\rho^{AB}\right)$ (cf. (14.184)) for all quantum Markov states of the form (14.190).

**Proof** Since local ancillary systems are free in entanglement theory, one can append a classical ancillary system $X$ that encodes the orthogonality of the subspaces $B_x^{(1)} \otimes$

$B_x^{(2)}$. This can be done with an isometry that maps states in $B_x^{(1)} \otimes B_x^{(2)}$ to states in $B^{(1)} \otimes B^{(2)} \otimes |x\rangle\langle x|^X$, where systems $B^{(1)}$ and $B^{(2)}$ have dimensions $\max_x |B_x^{(1)}|$ and $\max_x |B_x^{(2)}|$, respectively. Therefore, without loss of generality we can write the above Markov state as

$$\sigma^{ABB'X} = \sum_{x \in [m]} p_x\, \rho_x^{AB^{(1)}} \otimes \rho_x^{B^{(2)}B'} \otimes |x\rangle\langle x|^X. \tag{14.192}$$

Now, note that with any entanglement monotone $E$, the entanglement between $A$ and $BB'$ is measured by

$$E\left(\rho^{ABB'}\right) = E\left(\sigma^{ABB'X}\right)$$
$$\mathbf{(13.65)}{\rightarrow} = \sum_{x \in [m]} p_x\, E\left(\rho_x^{AB^{(1)}} \otimes \rho_x^{B^{(2)}B'}\right) \tag{14.193}$$
$$= \sum_{x \in [m]} p_x\, E\left(\rho_x^{AB^{(1)}}\right).$$

Similarly, the entanglement between $A$ and $B$ is measured by

$$E\left(\rho^{AB}\right) = E\left(\sigma^{ABX}\right)$$
$$= \sum_{x \in [m]} p_x\, E\left(\rho_x^{AB^{(1)}} \otimes \rho_x^{B^{(2)}}\right) \tag{14.194}$$
$$= \sum_{x \in [m]} p_x\, E\left(\rho_x^{AB^{(1)}}\right).$$

We therefore obtain $E\left(\rho^{ABB'}\right) = E\left(\rho^{AB}\right)$. This completes the proof. ∎

The Markov state mentioned in the theorem has an important property: The marginal state $\rho^{AB'}$ is separable, which implies $E(\rho^{AB'}) = 0$. Therefore, Markov states always satisfy the condition given in Definition 14.6. However, one might question whether the converse of the statement in the theorem holds true. In other words, if a state $\rho^{ABB'}$ satisfies $E(\rho^{ABB'}) = E(\rho^{AB})$, is it necessarily a Markov state? For mixed tripartite states, the answer is obviously "no" because all separable states between systems $A$ and $BB'$ satisfy $E(\rho^{ABB'}) = E(\rho^{AB})$, but not all separable states are Markov states. Nonetheless, in the following theorem, we will see that under mild assumptions, the converse of the above theorem holds for pure tripartite states.

In Section 13.2.1, we observed that every entanglement monotone takes the form (13.70) when evaluated on pure states. Specifically, the entanglement monotone $E$ can be expressed as follows:

$$E\left(\psi^{AB}\right) = g\left(\rho^A\right) \quad \text{with} \quad \rho^A := \mathrm{Tr}_B\left[\psi^{AB}\right], \tag{14.195}$$

where the function $g \colon \mathfrak{D}(A) \to \mathbb{R}_+$ is Schur concave. Furthermore, we noted that if $g$ is symmetric (i.e. invariant under unitary channels) and concave, then the convex roof extension of $E$ corresponds to an entanglement monotone. As a reminder, given any measure of entanglement $E$ on mixed states, we can construct its convex roof extension as

$$E_F\left(\rho^{AB}\right) := \min \sum_{x\in[m]} p_x E\left(\psi_x^{AB}\right) \qquad \forall\, \rho \in \mathfrak{D}(AB), \tag{14.196}$$

where the minimum is over all pure state decompositions of $\rho^{AB} = \sum_{x\in[m]} p_x \psi_x^{AB}$. Moreover, if $E$ is convex (e.g. entanglement monotone), then $E\left(\rho^{AB}\right) \leq E_F\left(\rho^{AB}\right)$ for all $\rho \in \mathfrak{D}(AB)$.

---

**Theorem 14.15.** Let $E$ be an entanglement monotone and $\psi \in \mathrm{Pure}(ABB')$. Suppose $g$ as defined in (14.195) is strictly concave. The following statements are equivalent:

1. $E\left(\psi^{ABB'}\right) = E\left(\rho^{AB}\right)$, where $\rho^{AB} := \mathrm{Tr}_{B'}\left[\psi^{ABB'}\right]$.
2. There exists subsystems $B_1$ and $B_2$, $\chi \in \mathrm{Pure}(AB_1)$, $\phi \in \mathrm{Pure}(B_2B')$, and an isometry $U: B_1B_2 \to B$ such that $\psi^{ABB'} = U\left(\chi^{AB_1} \otimes \phi^{B_2B'}\right)U^*$.

---

*Remark.* This theorem states that if the pure state $\psi^{ABB'}$ satisfies the disentangling condition, then it is a Markov state (up to local unitary on system $B$). Additionally, keep in mind that since $E$ measures entanglement, the function $g$ defined in (14.195) is invariant under unitary channels. As $g$ is also strictly concave, the convex roof extension of $E$ yields an entanglement monotone, as stated in Theorem 13.6. However, we don't assume $E$ to be equal to its convex roof extension. Instead, we observe that since $E$ is convex, it is always lower than or equal to its convex roof extension.

**Proof**    We only prove the implication from the first statement to the second, as the converse is straightforward and left as an exercise for the reader. The first statement implies that

$$E\left(\psi^{ABB'}\right) = E\left(\rho^{AB}\right) \leq E_F\left(\rho^{AB}\right), \tag{14.197}$$

where we used the fact that $E$ is no greater than its convex roof extension. On the other hand, from Lemma 14.3 we get that every pure-state decomposition of $\rho^{AB} = \sum_{x\in[m]} p_x \psi_x^{AB}$ has a corresponding measurement on system $B'$ of $\psi^{ABB'}$, where the outcome $x$ occurs with probability $p_x$ and the post-measurement state on system $AB$ is $\psi_x^{AB}$. When combined with the fact that $E$ is an entanglement monotone, this implies that

$$E\left(\psi^{ABB'}\right) \geq \sum_{x\in[m]} p_x E\left(\psi_x^{AB}\right). \tag{14.198}$$

Equations (14.197) and (14.198) lead to the very strong conclusion that *all* pure-state decompositions of $\rho^{AB}$ have the *same* average entanglement, which equals $E\left(\psi^{ABB'}\right)$. In other words, the inequality in (14.198) is actually an equality, and it holds for every pure-state decomposition $\{p_x, \psi_x^{AB}\}_{x \in [m]}$ of $\rho^{AB}$. This equality can be expressed in terms of the function $g$ as follows:

$$g(\rho^A) = \sum_{x \in [m]} p_x g(\rho_x^A), \tag{14.199}$$

where $\rho_x^A := \mathrm{Tr}_B\left[\psi_x^{AB}\right]$. Since $g$ is strictly concave, (14.199) holds if and only if $\rho^A = \rho_x^A$ for all $x \in [m]$. Let $B_1$ be a system of dimension $r := |B_1| = \mathrm{Rank}(\rho^A)$, and let $\chi \in \mathrm{Pure}(AB_1)$ be a purification of $\rho^A$. Since each $|\psi_x^{AB}\rangle$ is also a purification of $\rho^A = \rho_x^A$, we can infer that there exists an isometry $V_x : B_1 \to B$ such that

$$|\psi_x^{AB}\rangle = I^A \otimes V_x^{B_1 \to B}|\chi^{AB_1}\rangle. \tag{14.200}$$

Our first goal is to show that if $\{|\psi_x^{AB}\rangle\}_{x \in [m]}$ are the eigenvectors of $\rho^{AB}$, then $V_{x'}^* V_x = \delta_{xx'} I^{B_1}$.

To prove it, let $\{q_y, \phi_y^{AB}\}_{x \in [m]}$ be another pure-state decomposition of $\rho^{AB}$, also with $m$ elements. Then, for the exact same reasons as already stated, for each $y \in [m]$ there exists an isometry $W_y : B_1 \to B$ such that

$$|\phi_y^{AB}\rangle = I^A \otimes W_y^{B_1 \to B}|\chi^{AB_1}\rangle. \tag{14.201}$$

Recall from Exercise 2.27 that the ensemble $\{q_y, \phi_y^{AB}\}_{x \in [m]}$ can be related to the spectral decomposition of $\rho^{AB}$ by a unitary matrix $U = (u_{yx})$ as follows:

$$\sqrt{q_y}|\phi_y^{AB}\rangle = \sum_{x \in [m]} u_{yx}\sqrt{p_x}|\psi_x^{AB}\rangle \qquad \forall\, y \in [m]. \tag{14.202}$$

Combining this with (14.200) and (14.201) gives

$$\left(I^A \otimes \sqrt{q_y}W_y\right)|\chi^{AB_1}\rangle = \left(I^A \otimes \sum_{y \in [m]} u_{yx}\sqrt{p_x}V_x\right)|\chi^{AB_1}\rangle. \tag{14.203}$$

By multiplying both sides by $\left(\rho^A\right)^{-1/2}$ we can replace $|\chi^{AB_1}\rangle$ on both sides of (14.203) with the (unnormalized) maximally entangled state $|\Omega^{\tilde{B}_1 B_1}\rangle$. Therefore, the equation gives

$$\sqrt{q_y}W_y = \sum_{x \in [m]} u_{yx}\sqrt{p_x}V_x. \tag{14.204}$$

Since $W_y$ is an isometry we get

$$q_y I^{B_1} = \sum_{x,x' \in [m]} \bar{u}_{yx'}u_{yx}\sqrt{p_{x'}p_x}V_{x'}^* V_x$$
$$= \sum_{x \in [m]} p_x|u_{yx}|^2 I^{B_1} + \sum_{\substack{x \neq x' \\ x,x' \in [m]}} \bar{u}_{yx'}u_{yx}\sqrt{p_{x'}p_x}V_{x'}^* V_x. \tag{14.205}$$

Now, using the fact that $\{|\psi_x^{AB}\rangle\}_{x \in [m]}$ forms an orthonormal set of vectors, we get from (14.202) that

$$q_y = \left\|\sqrt{q_y}|\phi_y^{AB}\rangle\right\|_2^2 = \sum_{x \in [m]} p_x|u_{yx}|^2. \tag{14.206}$$

Combining this with (14.205) gives

$$\sum_{\substack{x \neq x' \\ x,x' \in [m]}} \bar{u}_{yx'} u_{yx} \sqrt{p_{x'} p_x}\, V_{x'}^* V_x = 0. \tag{14.207}$$

This equation holds for all unitary matrices $U = (u_{yx})$ and all $y \in [m]$. Setting $y = 1$, and choosing $U$ to be a unitary matrix with its first row as $\frac{1}{\sqrt{2}}(1,1,0,\ldots,0)$ gives $V_2^* V_2 + V_1^* V_2 = 0$. Similarly, choosing the first row of $U$ to be $\frac{1}{\sqrt{2}}(1,i,0,\ldots,0)$ gives $V_2^* V_2 - V_1^* V_2 = 0$. Thus, we obtain $V_1^* V_2 = V_2^* V_1 = 0$. By repeating the same argument with permuted versions of $\frac{1}{\sqrt{2}}(1,1,0,\ldots,0)$ and $\frac{1}{\sqrt{2}}(1,i,0,\ldots,0)$, we conclude that for all $x, x' \in [m]$ such that $x \neq x'$, we have $V_{x'}^* V_x = 0$.

Let $\{|z\rangle\}_{z \in [r]}$ be an orthonormal basis of $B_1$, and define $|\varphi_{xz}^B\rangle := V_x |z\rangle$ for all $x \in [m]$ and $z \in [r]$. Using the fact that $V_{x'}^* V_x = \delta_{xx'} I^{B_1}$, we can derive that $\langle \varphi_{x'z'}^B | \varphi_{xz}^B \rangle = \delta_{xx'} \delta_{zz'}$ for all $x \in [m]$ and $z \in [r]$. Let $\mathfrak{K}$ be the subspace spanned by the orthonormal vectors $\{|\varphi_{xz}^B\rangle\}$, with $x \in [m]$ and $z \in [r]$, and note that the dimension of $\mathfrak{K}$ is $mr$. Thus, there exists a subspace $B_2$ of $B$ with $|B_2| = m$ such that $\mathfrak{K}$ is isomorphic to $B_1 \otimes B_2$. This isomorphism implies that there exists an isometry $U : B_1 B_2 \to B$ such that $|\varphi_{xz}^B\rangle = U^{B_1 B_2 \to B} |z\rangle^{B_1} |x\rangle^{B_2}$. Combining this with the definition $|\varphi_{xz}^B\rangle := V_x |z\rangle^{B_1}$ gives $V_x |z\rangle^{B_1} = U |z\rangle^{B_1} |x\rangle^{B_2}$ for all $x \in [m]$ and all $z \in [r]$. Hence, $V_x^{B_1 \to B} = U^{B_1 B_2 \to B} \left( I^{B_1} \otimes |x\rangle^{B_2} \right)$ so that $|\psi_x^{AB}\rangle = U^{B_1 B_2 \to B} |\chi^{AB_1}\rangle |x\rangle^{B_2}$ and

$$\rho^{AB} = U \left( \chi^{AB_1} \otimes \sigma^{B_2} \right) U^*, \quad \text{where} \quad \sigma^{B_2} := \sum_{x \in [m]} p_x |x\rangle\langle x|^{B_2}. \tag{14.208}$$

Observe that the state in (14.208) has a purification of the form $U^{B_1 B_2 \to B} |\chi^{AB_1}\rangle |\phi^{B_2 B'}\rangle$, where

$$|\phi^{B_2 B'}\rangle = \sum_{x \in [m]} \sqrt{p_x} |x\rangle^{B_2} |x\rangle^{B'}. \tag{14.209}$$

Therefore, since $|\psi^{ABB'}\rangle$ is also a purification of $\rho^{AB}$, we conclude that up to a local unitary on system $B$ and on system $B'$, the state $\psi^{ABB'}$ has the form $|\chi^{AB_1}\rangle |\phi^{B_2 B'}\rangle$. $\blacksquare$

**Exercise 14.36.** *Consider a monotonically increasing convex function $h : \mathbb{R}_+ \to \mathbb{R}_+$ with the property that $h(t) = 0$ if and only if $t = 0$. For any state $\rho \in \mathfrak{D}(AB)$, let*

$$E_h \left( \rho^{AB} \right) := \min \sum_{x \in [m]} p_x h \left( E \left( \psi_x^{AB} \right) \right), \tag{14.210}$$

*where the minimum is over all pure-state decompositions of $\rho^{AB} = \sum_{x \in [m]} p_x \psi_x^{AB}$. Show that if $E_h$ is monogamous on pure tripartite states, then $E$ is also monogamous on pure tripartite states.*

Many operational measures of entanglement, such as the relative entropy of entanglement, entanglement cost, and distillable entanglement, reduce to the entropy of entanglement for bipartite pure states. Recall that the entropy of entanglement is given in terms of the von-Neumann entropy of the reduced state, $H(\rho) = -\mathrm{Tr}[\rho \log \rho]$. Since the von-Neumann entropy is known to be strictly concave, these measures are monogamous on pure tripartite states. In the following theorem, we show that their convex roof extensions are also monogamous for mixed tripartite states.

> **Theorem 14.16.** Let $\rho \in \mathfrak{D}(ABB')$, and $E$ and $E_F$ be as in (14.195) and (14.196), respectively, where $g$ is both symmetric and strictly concave. Then, the convex roof extension $E_F$ is monogamous.

**Proof** Let $\{p_x, |\psi^{ABB'}\rangle\}_{x \in [m]}$ be the optimal pure state decomposition of $\rho^{ABB'}$ satisfying

$$E_F\left(\rho^{ABB'}\right) = \sum_{x \in [m]} p_x E\left(\psi_x^{ABB'}\right).\tag{14.211}$$

Denoting by $\rho_x^{AB} := \mathrm{Tr}_{B'}\left[\psi_x^{ABB'}\right]$ we have that

$$\rho^{AB} = \sum_{x \in [m]} p_x \rho_x^{AB}.\tag{14.212}$$

Suppose $E_F\left(\rho^{ABB'}\right) = E_F\left(\rho^{AB}\right)$. Combining this with Equations (14.211) and (14.212) above gives

$$\sum_{x \in [m]} p_x E\left(\psi_x^{ABB'}\right) = E_F\left(\rho^{AB}\right)$$
$$\boxed{\text{Convexity of } E_F} \longrightarrow \quad \leqslant \sum_{x \in [m]} p_x E_F\left(\rho_x^{AB}\right).\tag{14.213}$$

On the other hand, for every $x \in [m]$ we have

$$E\left(\psi_x^{ABB'}\right) \geqslant E_F\left(\rho_x^{AB}\right),\tag{14.214}$$

since $E_F$ is a measure of entanglement (in fact, an entanglement monotone) and does not increase under the tracing out of the local system $B'$. Hence, from these two inequalities we get that for all $x \in [m]$ we have

$$E\left(\psi_x^{ABB'}\right) = E_F\left(\rho_x^{AB}\right).\tag{14.215}$$

Now, from Theorem 14.15 we get that for each $x \in [m]$ there exists systems $B_x^{(1)}$ and $B_x^{(2)}$, and an isometry $V_x : B_x^{(1)} B_x^{(2)} \to B$ such that

$$|\psi_x^{ABB'}\rangle = V_x^{B_x^{(1)} B_x^{(2)} \to B} |\chi_x^{AB_x^{(1)}}\rangle |\phi_x^{B_x^{(2)} B'}\rangle,\tag{14.216}$$

for some $\chi_x \in \mathrm{Pure}(AB_x^{(1)})$ and $\phi_x \in \mathrm{Pure}\left(B_x^{(2)} B'\right)$. Tracing out system $B$ on both sides of the equation gives

$$\psi_x^{AB'} = \chi_x^A \otimes \phi_x^{B'}.\tag{14.217}$$

Therefore, the marginal state $\rho^{AB'} = \sum_{x \in [m]} p_x \psi_x^{AB'}$ is separable so that $E\left(\rho^{AB'}\right) = 0$. This completes the proof. ∎

## 14.8 Notes and References

A comprehensive review of multipartite entanglement can be found in Chapter 17 of the book in Ref. [16]. The classification of all homogeneous SL-invariant polynomials using the Schur-Weyl duality is presented in Ref. [102].

Critical states, also known as normal forms, are discussed in detail in Refs. [223] and [229] for mathematically inclined readers. The canonical form of three-qubit states as given in Theorem 14.8 is due to Ref. [1]. The classification of SLOCC classes in three qubits was done by in Ref. [69], while that for four qubits was done in Ref. [221]. A detailed analysis of all four-qubit maximally entangled states can be found in Refs. [100, 206]. The classification of maximally entangled sets of multipartite systems is presented in Ref. [59].

The generalization of Nielsen's majorization theorem to the multipartite case, as presented in Theorem 14.11, is due to Ref. [101]. The generalization of this theorem to the full set SEP can be found in Ref. [118]. Several other generalizations of this result, particularly deterministic interconversions of multipartite entanglement under various operations, including LOCC Refs. can be found in [207] and [60]. It is worth mentioning that in Refs. [92] and [199], it was shown that the stabilizer group of almost all multipartite entangled states is trivial, and consequently, LOCC conversion between two states in the *same* SLOCC class is almost never possible. Nevertheless, certain multipartite states that have symmetry (e.g. GHZ states, graph states, stabilizer states, etc.) have a nontrivial stabilizer group and consequently rich entanglement properties [150].

In Ref. [118], it was demonstrated that any pure-state transformation attainable by $LOCC$ using a finite number of communication rounds can also be accomplished using $SEP_1$. However, not every pure-state transformation possible with SEP is achievable with $SEP_1$. These findings underscore that $SEP_1$ serves as a robust outer approximation of LOCC, particularly given that infinite rounds of classical communication are less feasible in practice.

The concept of entanglement of assistance was first introduced in Ref. [65], and the example given in (14.163) that demonstrates that it is not a tripartite entanglement monotone was taken from Ref. [96]. The result that asymptotic entanglement of assistance is equal to the smaller of its two local entropies was discovered in Ref. [204]. The concept of localizable entanglement was first introduced in the context of spin chains in Ref. [222], and its comparison with entanglement of collaboration can be found in Ref. [87].

Monogamy of entanglement was first introduced in Ref. [53], in which the CKW monogamy relation was discovered. The monogamy of the squashed entanglement was discovered in Ref. [50]. The concept of "monogamy of entanglement without inequalities" was first introduced in Ref. [104] and developed further in Ref. [106]. Additional references on monogamy of entanglement can be found in those papers.

PART V

# ADDITIONAL EXAMPLES OF STATIC RESOURCE THEORIES

# The Resource Theory of Asymmetry

In Shannon's theory, information is considered to be "fungible," meaning that it can be encoded into any physical system's degree of freedom, and the information's content is independent of the encoding method. For instance, a simple yes/no message can be transmitted equally well by a 5/0-volt potential difference across a circuit element or by flipping a coin to heads/tails. This type of information is known as "speakable information," as it can be conveyed through speech or symbols.

However, there are also nonfungible types of information, such as a direction in space, the time of an event, or the relative phase between two quantum states in a superposition. Such information is referred to as "unspeakable information" because it cannot be conveyed verbally without a shared coordinate system, a synchronized clock, or a common phase reference. For example, directional information can only be transmitted between two parties through the exchange of a physical system whose state represents the direction itself, such as a classical gyroscope, in the absence of a common gravitational field or stellar background.

Unspeakable information can be communicated verbally when a reference frame is present, which is true for both classical and quantum information. However, despite speakable information being fungible, multiple parties must first agree on how to encode/decode this information in a physical system, which implicitly necessitates a common reference frame. As a result, quantum information processing tasks assume the existence of a shared reference frame, and the lack of this shared frame significantly restricts what can be accomplished.

The absence or deterioration of a common reference frame is a natural constraint that frequently occurs in the study of multiple physical systems. Consequently, this constraint gives rise to a resource theory of reference frames, which can be more broadly classified as a resource theory of asymmetry.

## 15.1 Free States and Free Operations

Consider two parties (Alice and Bob) who do not share a reference frame. Mathematically, we represent the information about the frame by an element $g$ of a compact group $\mathbf{G}$. For instance, $g \in \mathbf{G}$ could correspond to a particular orientation in space, clock synchronization, phase information, and so on.

In this chapter, it is assumed without explicit statement that **G** is either a finite group or a compact Lie group. Additionally, readers who are not well-versed in representation theory are advised to first read Appendix C of online version before proceeding with this chapter. The same notations as those in Appendix C of online version will be utilized, and frequent references to this online appendix will be made throughout the chapter.

## G-Invariant States

In the resource theory of asymmetry, each element $g \in$ **G** is denoted by a unitary matrix $U_g$. If $\rho \in \mathfrak{D}(A)$ represents the density matrix of a quantum system with respect to Alice's reference frame, then the state of the same physical system with respect to Bob's reference frame is given by

$$\mathcal{U}_g(\rho) := U_g \rho U_g^*. \tag{15.1}$$

If Alice and Bob are unaware of the element $g \in$ **G** that establishes the relation between their reference frames, then the states that Alice can prepare relative to Bob's reference frame are those satisfying $\rho = \mathcal{U}_g(\rho)$ for all $g \in$ **G**. Such states are referred to as **G**-invariant and satisfy $[\rho, U_g] = 0$, as indicated by Definition C.6 of online version.

The absence of a shared reference frame places a limitation on the types of states that Alice can generate relative to Bob's reference frame. She is only capable of creating **G**-invariant states, which comprise the free states in the QRT of reference frames, denoted as

$$\mathfrak{F}(A) = \mathrm{INV}_{\mathbf{G}}(A) := \left\{ \rho \in \mathfrak{D}(A) : \mathcal{U}_g(\rho) = \rho \quad \forall g \in \mathbf{G} \right\}. \tag{15.2}$$

For instance, suppose the group $\mathbf{G} = U(1)$ corresponds to an optical phase reference or to dynamics with rotational symmetry around a fixed axis (in which case the group is SO(2), which is known to be isomorphic to the group $U(1)$). In this scenario, a unitary representation of **G** is provided by $U_\theta = e^{i\hat{N}\theta}$, where $\theta \in U(1)$ and $\hat{N}$ is the total number operator (or in the case of rotational symmetry, $\hat{N}$ can be replaced with $L_{\mathbf{n}}$, the angular momentum operator in the **n** direction). In this instance, the free states are given by states of the form $\sum_n p_n |n\rangle\langle n|$, where $|n\rangle$ corresponds to the eigenvectors of $\hat{N}$.

More generally, it will be observed that the absence of a shared reference frame enforces a *superselection rule* regarding the types of states that Alice can generate. This superselection rule is characterized by the fact that coherent superpositions between states in specific subspaces are not feasible. For instance, coherent superpositions of $U(1)$ states among the eigenstates of the number operator are not free and cannot be prepared by Alice.

## G-Covariant Channels

The set of free operations in the QRT of reference frames can be defined similarly to the free states. Let $\sigma \in \mathfrak{D}(B)$ be an arbitrary density matrix of system $B$ described in

Bob's reference frame. Suppose Alice performs a quantum operation on this system described by the channel $\mathcal{E} \in \mathrm{CPTP}(A \to A)$ in her reference frame. How would this operation be described in Bob's reference frame? If Bob knows that their reference frames are linked by an element $g \in \mathbf{G}$, then $\mathcal{U}_g^*(\sigma)$ is Alice's description of the initial state, and $\mathcal{E}(\mathcal{U}_g^*(\sigma))$ is her description of the final state. Therefore, the final state in Bob's reference frame is given by $\mathcal{U}_g \circ \mathcal{E} \circ \mathcal{U}_g^*(\sigma)$, and his description of Alice's operation is $\mathcal{U}_g \circ \mathcal{E} \circ \mathcal{U}_g^*$.

Hence, if Alice and Bob are unaware of the value of $g \in \mathbf{G}$, they will have a similar description of the CPTP map $\mathcal{E}$ only if $\mathcal{E}$ satisfies

$$\mathcal{U}_g \circ \mathcal{E} \circ \mathcal{U}_g^* = \mathcal{E} \qquad \forall\, g \in \mathbf{G}. \tag{15.3}$$

Quantum channels of this kind are referred to as **G**-*covariant*, and they represent the free operations in the QRT of asymmetry. Similar to **G**-invariant states, a quantum channel is **G**-covariant if and only if it commutes with $\mathcal{U}_g$ for all $g \in \mathbf{G}$.

Therefore, the set of free operations in the QRT of reference frames can be expressed as

$$\mathfrak{F}(A \to A) = \mathrm{COV}_{\mathbf{G}}(A \to A) := \left\{ \mathcal{E} \in \mathrm{CPTP}(A \to A) : [\mathcal{E}, \mathcal{U}_g] = 0 \quad \forall\, g \in \mathbf{G} \right\}, \tag{15.4}$$

where $[\mathcal{E}, \mathcal{U}_g] := \mathcal{E} \circ \mathcal{U}_g - \mathcal{U}_g \circ \mathcal{E}$ (see Figure 15.1). For instance, for $\mathbf{G} = U(1)$, a **G**-covariant quantum channel, $\mathcal{E} \in \mathrm{CPTP}(A \to A)$, satisfies for all $\theta \in [0, 2\pi)$ and all $\rho \in \mathfrak{D}(A)$

$$\mathcal{E}\left( e^{i\theta\hat{N}} \rho e^{-i\theta\hat{N}} \right) = e^{i\theta\hat{N}} \mathcal{E}(\rho) e^{-i\theta\hat{N}}. \tag{15.5}$$

In subsequent sections, we'll delve into characterizations of **G**-covariant channels, employing covariant adaptations of the operator sum representation and the Stinespring representation of a quantum channel. Before we proceed to these broader characterizations, let's focus initially on the specific case of unitary channels. Consider a covariant unitary channel $\mathcal{E}(\,\cdot\,) = V(\,\cdot\,)V^*$, where $V : A \to A$ is a unitary matrix. According to condition (15.3), for every element $g \in \mathbf{G}$ and for any state $\rho \in \mathfrak{D}(A)$, the channel satisfies

$$\left( U_g V U_g^* \right) \rho^A \left( U_g V U_g^* \right)^* = V\rho^A V^*. \tag{15.6}$$

This implies that for every $g \in \mathbf{G}$, there exists a phase $\omega_g \in \mathbb{C}$ with $|\omega_g| = 1$ such that

$$U_g V U_g^* = \omega_g V. \tag{15.7}$$

Since this equation holds for all $g \in \mathbf{G}$, it follows that the map $g \mapsto \omega_g$ is a one-dimensional representation of $\mathbf{G}$. Specifically, when $g = e$ is the identity element, we have $U_g = I$, which gives $\omega_g = 1$. Furthermore, for $g, h \in \mathbf{G}$, we have

$$\begin{aligned}
\omega_{gh} V &= U_{gh} V U_{gh}^* \\
&= U_g U_h V U_h^* U_g^* \\
&= \omega_h U_g V U_g^* \\
&= \omega_h \omega_g V,
\end{aligned} \tag{15.8}$$

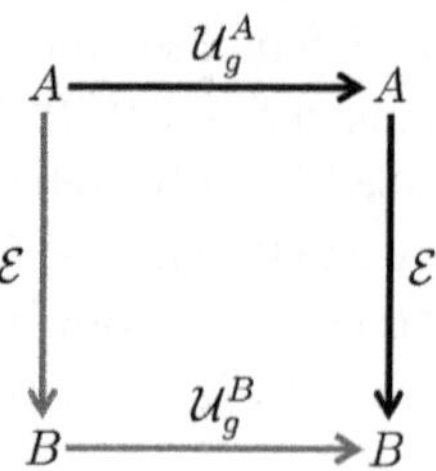



**Figure 15.1**   Heuristic description of **G**-covariant operations. The channel $\mathcal{E}$ is **G**-covariant if for every choice of group element $g \in \mathbf{G}$, the two different pathways yield the same outcome.

which implies that $\omega_{gh} = \omega_g \omega_h$. In other words, the set of all **G**-covariant unitary channels can be characterized by unitary matrices that are "almost" **G**-invariant, meaning that they commute with the elements of the group up to a phase, where this phase itself forms a one-dimensional representation of **G**.

In conclusion, we have observed that in the QRT of reference frames, the set of free states is the set of *symmetric* states (i.e. those states that commute with $U_g$ for all $g \in \mathbf{G}$), and the set of free operations is the set of *symmetric* operations (i.e. those operations that commute with $\mathcal{U}_g$ for all $g \in \mathbf{G}$). Symmetric evolutions are prevalent in physics and may arise in various contexts, not just from the absence of a shared reference frame. Therefore, the set of **G**-covariant operations defines a resource theory with applications extending beyond quantum reference frames. It may be referred to as a QRT of *asymmetry* because in any QRT in which $\mathfrak{F}$ specifies a set of **G**-covariant operations, asymmetric states and asymmetric operations are the resources of the theory.

Thus far, we have only examined **G**-covariant channels with the same input and output dimensions. More generally, a quantum channel $\mathcal{E}: \mathrm{CPTP}(A \to B)$ is **G**-covariant with respect to two (unitary) representations of **G**, $\{U_g^A\}_{g \in \mathbf{G}}$ and $\{U_g^B\}_{g \in \mathbf{G}}$, if

$$\mathcal{E}^{A \to B} \circ \mathcal{U}_g^A = \mathcal{U}_g^B \circ \mathcal{E}^{A \to B} \quad \forall g \in \mathbf{G}. \tag{15.9}$$

Refer to Figure 15.1 for an illustrative depiction of **G**-covariant operations. The set of all **G**-covariant quantum channels in $\mathrm{CPTP}(A \to B)$ will be denoted by $\mathrm{COV}_{\mathbf{G}}(A \to B)$. It is worth noting that this notation does not explicitly specify the two unitary representations of **G**, $\{U_g^A\}_{g \in \mathbf{G}}$ and $\{U_g^B\}_{g \in \mathbf{G}}$. The representations used will be clear from the context.

## $G$-Covariant Measurements

The result obtained from a quantum measurement, often referred to as the classical outcome, provides a form of information known as "speakable information." This type of information can be effectively communicated between parties who do not share a common reference frame. Let's consider an example where Alice and Bob do not have a shared Cartesian reference frame. Suppose Alice performs a measurement on the spin of an electron in the $z$-direction relative to her reference frame and obtains an outcome of "up" (indicating that the electron's spin is pointing in the positive $z$-direction).

Alice can then transmit this outcome to Bob, allowing him to determine that the electron's spin is aligned with the positive $z$-direction in relation to Alice's frame. Therefore, even though the specific information about the $z$-direction itself cannot be conveyed, the measurement outcome, that is, the "up"/"down" information, can be effectively communicated between the parties involved.

Consequently, we make the assumption that the group $\mathbf{G}$ associated with the resource theory of asymmetry has a trivial action on classical systems that represent measurement outcomes. Moving forward, in Section 3.5.10, we observed that a general quantum measurement can be characterized by a quantum instrument denoted as $\mathcal{E} \in \mathrm{CPTP}(A \to BX)$, where $X$ represents the classical outcome of the measurement. We refer to $\mathcal{E}$ as a $\mathbf{G}$-covariant quantum instrument if it satisfies the condition:

$$\mathcal{E}^{A \to BX} \circ \mathcal{U}_g^{A \to A} = \mathcal{U}_g^{B \to B} \circ \mathcal{E}^{A \to BX} \qquad \forall \, g \in \mathbf{G}. \tag{15.10}$$

The collection of all such $\mathbf{G}$-covariant quantum instruments is denoted by $\mathrm{COV}_{\mathbf{G}}$ $(A \to BX)$.

Every quantum instrument $\mathcal{E}^{A \to BX}$ as discussed above can be expressed as

$$\mathcal{E}^{A \to BX} = \sum_{x \in [m]} \mathcal{E}_x^{A \to B} \otimes |x\rangle\langle x|^X, \tag{15.11}$$

where $m \in \mathbb{N}$, and each $\mathcal{E}_x \in \mathrm{CP}(A \to B)$. If $\mathcal{E}^{A \to BX}$ is $\mathbf{G}$-covariant, the relation above in conjunction with (15.10) implies that for all $x \in [m]$ we have

$$\mathcal{E}_x^{A \to B} \circ \mathcal{U}_g^{A \to A} = \mathcal{U}_g^{B \to B} \circ \mathcal{E}_x^{A \to B} \qquad \forall \, g \in \mathbf{G}. \tag{15.12}$$

In other words, the quantum instrument $\mathcal{E}^{A \to BX}$ is $\mathbf{G}$-covariant if and only if each CP map $\mathcal{E}_x^{A \to B}$ is $\mathbf{G}$-covariant.

A special type of $\mathbf{G}$-covariant quantum instrument is a $\mathbf{G}$-covariant POVM. We get $\mathbf{G}$-covariant POVM by taking above $B$ to be the trivial system (i.e. $|B| = 1$) so that for each $x \in [m]$ and every $\rho \in \mathfrak{L}(A)$, $\mathcal{E}_x^{A \to B}(\rho^A) = \mathrm{Tr}[\Lambda_x^A \rho^A]$ for some $\Lambda_x \in \mathrm{Eff}(A)$ and the set $\{\Lambda_x^A\}_{x \in [m]}$ is a POVM. Now, for a trivial system $B$, the condition given in (15.12) becomes equivalent to

$$\mathrm{Tr}[\Lambda_x \rho] = \mathrm{Tr}[\Lambda_x \mathcal{U}_g(\rho)] = \mathrm{Tr}[\mathcal{U}_g^*(\Lambda_x)\rho] \qquad \forall \, g \in \mathbf{G}. \tag{15.13}$$

Since this condition holds for all $\rho \in \mathfrak{L}(A)$, we must have $\Lambda_x = \mathcal{U}_g^*(\Lambda_x)$ for all $g \in \mathbf{G}$ and $x \in [m]$. In other words, a POVM $\{\Lambda_x\}_{x \in [m]}$ is $\mathbf{G}$-covariant if and only if each element $\Lambda_x$ is $\mathbf{G}$-invariant; that is, each $\Lambda_x$ satisfies $[\Lambda_x, U_g] = 0$ for all $g \in \mathbf{G}$.

## 15.2 Distinctive Concepts in the QRT of Asymmetry

In this subsection, we introduce several mathematical tools and concepts that are distinctive to the resource theory of asymmetry and do not appear in other resource theories.

## 15.2.1 The G-Twirling Operation

If Alice lacks the information about $g$, her description of Bob's density matrix is obtained by averaging over all possible values of $g$. The uniform Haar measure over the group $\mathbf{G}$ is denoted by $dg$, and this average can be expressed as follows:

$$\mathcal{G}(\rho) := \int_{\mathbf{G}} dg\, \mathcal{U}_g(\rho). \tag{15.14}$$

The averaging CPTP map is known as the **G**-twirling map (see Section C.4 of online version). If the group $\mathbf{G}$ is finite, the integral is replaced by a discrete sum over the $|\mathbf{G}|$ elements of the group, that is, $\mathcal{G}(\rho) = \frac{1}{|\mathbf{G}|}\sum_{g\in\mathbf{G}}\mathcal{U}_g(\rho)$.

The free states in this QRT have a very particular structure. First, note that $\rho \in \mathfrak{F}(A)$ if and only if it is **G**-invariant, meaning that $\mathcal{U}_g(\rho) = \rho$ for all $g$. In particular, $\mathcal{G}(\rho) = \rho$ for all $\rho \in \mathfrak{F}(A)$. Combining this with the definition of $\mathfrak{F}(A)$ implies that **G**-twirling is a resource-destroying map (see Definition 9.9). Additionally, one can characterize the free states using techniques from representation theory. In particular, Theorem C.5 of online version states that $\rho \in \mathfrak{D}(A)$ is free if and only if $\rho^A$ has the following form:

$$\rho^A = \bigoplus_{\lambda\in\mathrm{Irr}(U)} \mathbf{u}^{B_\lambda} \otimes \rho_\lambda^{C_\lambda}, \quad \text{where} \quad \rho_\lambda^{C_\lambda} := \mathrm{Tr}_{B_\lambda}\left[\Pi^{A_\lambda}\rho^A\Pi^{A_\lambda}\right]. \tag{15.15}$$

Moreover, note that the expression in (15.15) implies the following corollary.

---

**Corollary 15.1.** The **G**-twirling map can be decomposed as

$$\mathcal{G} = \bigoplus_{\lambda\in\mathrm{Irr}(U)} \left(\mathcal{R}^{B_\lambda} \otimes \mathrm{id}^{C_\lambda}\right) \circ \mathcal{P}^{A_\lambda}, \tag{15.16}$$

where $\mathcal{P}^{A_\lambda}(\,\cdot\,) = \Pi^\lambda(\,\cdot\,)\Pi^\lambda$, $\Pi^{A_\lambda}: A \to A$ is a projection to subspace $A_\lambda$, $\mathcal{R}^{B_\lambda} \in \mathrm{CPTP}(B_\lambda \to B_\lambda)$ is the completely depolarizing (randomizing) channel, and $\mathrm{id}^{C_\lambda} \in \mathrm{CPTP}(C_\lambda \to C_\lambda)$ is the identity channel.

---

This corollary demonstrates that the **G**-twirling operation eliminates any correlations among distinct irreducible representations. For example, let us consider the case where $\mathbf{G} = \mathfrak{U}(1)$. As this group is Abelian, it has only one-dimensional irreps (i.e. $|B_\lambda| = 1$). The irreps of $\mathfrak{U}(1)$ are labeled by integers $\lambda = k \in \mathbb{Z}$, and the $k$-th irrep $u_k : \mathfrak{U}(1) \mapsto \mathbb{C}$ is of the form

$$u_k(\theta) = e^{ik\theta} \qquad \forall\theta \in \mathfrak{U}(1). \tag{15.17}$$

In this context, we will consider an infinite-dimensional (separable) Hilbert space denoted by $A$ with basis vectors $|n\rangle$, where $n$ belongs to the set of integers $\mathbb{Z}$. The "number" operator which generates the $\mathfrak{U}(1)$ symmetry can be defined as follows:

$$\hat{N} := \sum_{n\in\mathbb{Z}} n|n\rangle\langle n|. \tag{15.18}$$

Note that we allow negative values of $n$ and work with the representation $\theta \mapsto e^{i\hat{N}\theta}$.

For each irrep on a single copy of $A$, the multiplicity space is trivial (i.e. $|C_\lambda| = 1$), and the **G**-twirling operation can be easily represented as

$$\mathcal{G}(\,\cdot\,) = \sum_{k\in\mathbb{Z}} |k\rangle\langle k|(\,\cdot\,)|k\rangle\langle k|, \tag{15.19}$$

which means that $\mathcal{G}$ is the completely dephasing channel with respect to the basis $|k\rangle_{k\in\mathbb{Z}}$ in this case.

However, when considering $\ell$ copies of $A$, the multiplicity space of a given irrep is usually not trivial, and as a result, the **G**-twirling operation is not equivalent to the dephasing channel. Specifically, let $\hat{N}_x$ be the number operators associated with system $A_x$ for each $x \in [\ell]$. Consider the unitary representation on $A^n = (A_1, \ldots, A_\ell)$ defined by

$$\theta \mapsto \bigotimes_{x\in[\ell]} e^{i\hat{N}_x\theta} = e^{i\hat{N}_{\text{tot}}\theta}, \quad \text{where} \quad \hat{N}_{\text{tot}} := \sum_{x\in[\ell]} \hat{N}_x. \tag{15.20}$$

In this case, the irreps are denoted by the eigenvalues $n \in \mathbb{Z}$ of the total number operator $\hat{N}$tot. While the representation space $B_n$ (i.e. $B_\lambda$ with $\lambda = n$) is trivial (i.e. one dimensional) for every irrep $\lambda = n$, the multiplicity space $C_n$ is not. Let

$$\Pi_n^{(\ell)} := \sum_{\substack{k_1+\cdots+k_\ell=n \\ k_1,\ldots,k_\ell\in\mathbb{Z}}} |k_1\rangle\langle k_1| \otimes \cdots \otimes |k_\ell\rangle\langle k_\ell| \tag{15.21}$$

be the projection onto the eigenspace of $\hat{N}_{\text{tot}}$ corresponding to the eigenvalue $n$. Using this notation, the **G**-twirling operation can be expressed as

$$\mathcal{G}_\ell(\,\cdot\,) = \sum_{n\in\mathbb{Z}} \Pi_n^{(\ell)}(\,\cdot\,)\Pi_n^{(\ell)}. \tag{15.22}$$

Note that $\Pi_n^{(\ell)}$ is the projection onto the multiplicity space $C_n$. This space is often referred to as the *decoherence-free subspace*, as any pure state $\psi \in \text{Pure}(C_n)$ is $U(1)$-invariant, that is, $\mathcal{G}(\psi) = \psi$. For example, if $\ell = 3$, any linear combination of $|011\rangle$, $|101\rangle$, and $|110\rangle$ is an eigenvector of $\hat{N}_{\text{tot}}$ corresponding to an eigenvalue of 2. Therefore, the coherence of any state in the span of these three vectors remains unaffected by the **G**-twirling operation.

The **G**-twirling operation can be applied to quantum channels as well. Suppose Alice applies a quantum operation $\mathcal{E} \in \mathfrak{L}(A \to A)$ to her system. If Bob knows the relation between their reference frames, then he can describe the operation relative to his own system as $\mathcal{U}_g\circ\mathcal{E}\circ\mathcal{U}_g^*$, where $\mathcal{U}_g$ is a unitary operator that relates Alice's and Bob's frames. However, in the absence of a shared reference frame, Bob cannot use this description. Instead, the channel $\mathcal{E}$ appears to him as a mixture of the form $\int_G dg\, \mathcal{U}_g \circ \mathcal{E} \circ \mathcal{U}_g^\dagger$. In order for Alice and Bob to have the same description of the channel, the condition $\mathcal{E} = \int_G dg\, \mathcal{U}_g \circ \mathcal{E} \circ \mathcal{U}_g^\dagger$ must be satisfied. This integral is a type of twirling operation applied to the channel $\mathcal{E}$.

**Exercise 15.1.** *Show that the **G**-twirling map is unital and idempotent; that is, $\mathcal{G}\circ\mathcal{G} = \mathcal{G}$.*

For compact Lie groups the **G**-twirling is defined in terms of an integral over the group. From Carathéodory theorem (see Theorem A.3 of online version) it follows that the **G**-twirling can be expressed as a finite convex combination of unitary channels $\mathcal{U}_g$. To see why, for every $g \in \mathbf{G}$ let $|\psi_g^{A\tilde{A}}\rangle := U_g^A \otimes I^{\tilde{A}}|\Omega^{A\tilde{A}}\rangle$, and let $\mathfrak{C}$ be the convex hull of the set $\{\psi_g^{A\tilde{A}}\}_{g\in\mathbf{G}}$. Note that $\mathfrak{C} \subset \mathfrak{R}$, where $\mathfrak{R}$ is a subspace of $\mathrm{Herm}(A\tilde{A})$ given by

$$\mathfrak{R} := \left\{ \Lambda^{A\tilde{A}} \in \mathrm{Herm}(A\tilde{A}) : \ \Lambda^A \propto I^A, \quad \Lambda^{\tilde{A}} \propto I^{\tilde{A}} \right\}, \tag{15.23}$$

where the symbol $\propto$ stands for "proportional to."

Since we assume that **G** is a compact Lie group, the set $\{\psi_g^{A\tilde{A}}\}_{g\in\mathbf{G}}$ is also compact in $\mathfrak{R}$. Consequently, also its convex hull is compact (this follows from Carathéodory's theorem, see Exercise A.9 of online version). Therefore, the Choi matrix of the $\mathcal{G}$, $J_{\mathcal{G}}^{A\tilde{A}}$, also belongs to $\mathfrak{C}$, and from Carathéodory's theorem it can be expressed as a convex combination of at most $d$ elements of $\{\psi_g^{A\tilde{A}}\}_{g\in\mathbf{G}}$, where $d := \dim(\mathfrak{R})+1$. We therefore conclude that there exists $\mathbf{p} \in \mathrm{Prob}(d)$ and $d$ group elements $\{g_x\}_{x\in[d]} \subset \mathbf{G}$ such that

$$J_{\mathcal{G}}^{A\tilde{A}} = \sum_{x\in[d]} p_x \psi_{g_x}^{A\tilde{A}}. \tag{15.24}$$

In other words, $\mathcal{G}$ can be expressed as

$$\mathcal{G}(\cdot) = \sum_{x\in[d]} p_x U_{g_x}(\cdot)U_{g_x}^*. \tag{15.25}$$

**Exercise 15.2.** *Let* $m := |A|$. *Show that*

$$d := \dim(\mathfrak{R}) + 1 = (m^2 - 1)^2 + 2 \leqslant m^4. \tag{15.26}$$

**Exercise 15.3.** *Let* $\mathcal{G}_k \in \mathrm{CPTP}(A^k \to A^k)$ *be the* **G***-twirling, map as defined on k-copies of A. That is, for any* $\rho \in \mathfrak{D}(A^k)$ *we have*

$$\mathcal{G}_k(\rho^{A^k}) := \int_{\mathbf{G}} dg\, U_g^{\otimes k} \rho^{A^k} \left(U_g^{\otimes k}\right)^*. \tag{15.27}$$

*Show that*

$$\mathcal{G}_1^{\otimes k} \circ \mathcal{G}_k = \mathcal{G}_1^{\otimes k}. \tag{15.28}$$

**Exercise 15.4.** *Let* $\rho \in \mathfrak{D}(A)$ *and* $\alpha \in \mathbb{R}_+$. *Show that if* $\rho$ *is* **G***-invariant, then also* $\rho^\alpha/\mathrm{Tr}[\rho^\alpha]$. *Hint: Use* (15.15).

**Exercise 15.5.** *Let* $g \mapsto U_g$ *be a projective unitary representation of a finite or compact Lie group* **G**. *For each* $\lambda \in \mathrm{Irr}(U)$, *let* $U_g^{(\lambda)}$ *be the reduction of* $U_g$ *to the space* $B_\lambda$ *as given in* (C.45 of online version). *Show that for every* $\rho \in \mathfrak{L}(B_\lambda)$ *we have*

$$\int dg\, U_g^{(\lambda)} \rho^{B_\lambda} U_g^{*(\lambda)} = \mathrm{Tr}\left[\rho^{B_\lambda}\right] I^{B_\lambda}. \tag{15.29}$$

## The Weighted G-Twirling

The weighted **G**-twirling is a variant of the G-twirling that also plays an important role in the resource theory of asymmetry. It is defined as follows: Let $\mathbf{p}\colon \mathbf{G} \to \mathbb{R}_+\colon g \mapsto p(g)$ a probability distribution that is normalized such that $\int_\mathbf{G} dg \, p(g) = 1$. With respect to this distribution, the weighted **G**-twirling is defined as

$$\mathcal{G}_\mathbf{p}(\,\cdot\,) := \int_\mathbf{G} dg \, p(g) U_g(\,\cdot\,) U_g^*. \tag{15.30}$$

When $p$ is uniform, that is, $p(g) = 1$, we get that $\mathcal{G}_\mathbf{p} = \mathcal{G}$.

It is worth noting that by choosing different probability densities **p**, we can obtain any convex combination of the unitary channels $\mathcal{U}_g(\,\cdot\,) := U_g(\,\cdot\,)U_g^*$. Additionally, if $\rho \in \mathrm{INV}_\mathbf{G}(A)$, then $\mathcal{G}_\mathbf{p}(\rho) = \rho$. However, the converse is not true; that is, in general, there exists a density matrix $\rho \notin \mathrm{INV}_\mathbf{G}(A)$ such that $\mathcal{G}_\mathbf{p}(\rho) = \rho$. One such example is obtained by taking $p(g) = \delta(g)$ to be the Dirac delta function (i.e. $p(g) = 0$ for all $g \neq e$) so that $\mathcal{G}_\mathbf{p} = \mathrm{id}^A$ is the identity channel. In this extreme example, every resource state $\rho \notin \mathrm{INV}_\mathbf{G}(A)$ satisfies $\mathcal{G}_\mathbf{p}(\rho) = \rho$.

Unlike the **G**-twirling, the weighted **G**-twirling is not necessarily **G**-covariant. This depends on the group **G** and on the distribution $p(g)$. For example, for Abelian group the weighted **G**-twirling is covariant since $\mathcal{G}_\mathbf{p} \circ \mathcal{U}_g = \mathcal{U}_g \circ \mathcal{G}_\mathbf{p}$ for all $g \in \mathbf{G}$. For non-Abelian groups, the weighted **G**-twirling is **G**-covariant if $p(g)$ is a class function as introduced in Definition C.9 of online version.

> **Theorem 15.1.** Let $\mathbf{p}\colon \mathbf{G} \to \mathbb{R}_+\colon g \mapsto p(g)$ be a normalized probability distribution. If $p(g)$ is a class function, then $\mathcal{G}_\mathbf{p}$ is **G**-covariant.

**Proof**   Let $h \in \mathbf{G}$ and $\rho \in \mathfrak{L}(A)$. Then,

$$
\begin{aligned}
\mathcal{U}_h^* \circ \mathcal{G}_\mathbf{p} \circ \mathcal{U}_h(\rho) &= \int_\mathbf{G} dg \, p(g) U_{h^{-1}} U_g U_h \rho U_h^* U_g^* U_{h^{-1}}^* \\
&= \int_\mathbf{G} dg \, p(g) U_{h^{-1}gh} \rho U_{hgh^{-1}}^* \\
\boxed{g' := hgh^{-1}} \longrightarrow \quad &= \int_\mathbf{G} dg' \, p(hg'h^{-1}) U_{g'} \rho U_{g'}^* \\
\mathbf{\textit{p} \text{ is a class function}} \rightarrow \quad &= \int_\mathbf{G} dg' \, p(g') U_{g'} \rho U_{g'}^* = \mathcal{G}_\mathbf{p}(\rho).
\end{aligned}
\tag{15.31}
$$

Since $\rho \in \mathfrak{L}(A)$ was arbitrary we conclude that $\mathcal{U}_h^* \circ \mathcal{G}_\mathbf{p} \circ \mathcal{U}_h = \mathcal{G}_\mathbf{p}$ for all $h \in \mathbf{G}$. This completes the proof.   ∎

In the definition of the weighted **G**-twirling we assumed that $p(g)$ is an arbitrary probability density over **G**. However, as we saw earlier, thanks to Carathéodory's theorem, the **G**-twirling can be expressed as a finite convex combination of unitary channels of the form $U_g(\,\cdot\,)U_g^*$. The same arguments can be applied to $\mathcal{G}_\mathbf{p}$, so we can assume, without loss of generality, that $\mathbf{p}$ is a discrete probability distribution, that is, $\mathbf{p} \in \mathrm{Prob}(d)$ for some $d \leqslant m^4$, where $m := |A|$. Hence, the weighted $\mathbf{G}$-twirling takes the form

$$\mathcal{G}_{\mathbf{p}}(\,\cdot\,) = \sum_{x \in [d]} p_x U_{g_x}(\,\cdot\,)U_{g_x}^*. \tag{15.32}$$

**Exercise 15.6.** *Give an example of a group* $\mathbf{G}$, *a probability distribution* $p(g) \neq \delta(g)$, *and a state* $\rho \in \mathfrak{D}(A)$ *such that* $\mathcal{G}_{\mathbf{p}}(\rho) = \rho$ *but* $\rho \notin \mathrm{INV}_{\mathbf{G}}(A)$.

## 15.2.2 Three Representations of G-Covariant Maps

### Covariant Operator Sum Representation

In this subsection, we introduce the concept of an irreducible tensor operator and use it to characterize the operator sum representation of a covariant channel. The irreducible tensor operator is defined with respect to a projective representation, $g \mapsto U_g^E$, of the group $\mathbf{G}$. This projective representation induces a particular structure on the Hilbert space $E$ (see Theorem C.4 of online version). Specifically, $E$ can be decomposed as

$$E = \bigoplus_{\lambda \in \mathrm{Irr}(U^E)} B_\lambda \otimes C_\lambda, \tag{15.33}$$

where $B_\lambda$ is an irreducible $G$-invariant subspace of $A$, and $C_\lambda$ is the multiplicity subspace. According to Theorem C.4 of online version for all $g \in \mathbf{G}$ we can also decompose $U_g^E$ as

$$U_g^E \cong \bigoplus_{\lambda \in \mathrm{Irr}(U^E)} U_g^{(\lambda)} \otimes I^{C_\lambda}, \tag{15.34}$$

where each $U_g^{(\lambda)}$ acts irreducibly on $B_\lambda$. We will denote by $u_{m'm}^{(\lambda)}(g)$ the $m'm$-component of $U_g^{(\lambda)}$, and use the indices $\lambda$, $m$, and $x$ to label the basis of $E$ as given in (C.46 of online version). The index $x$ corresponds to the multiplicity index.

---

**Definition 15.1.** Let $g \mapsto U_g^E$ be as in (15.34), and let $g \mapsto U_g^A$ and $g \mapsto U_g^B$ be two additional projective unitary representations of $\mathbf{G}$. We say that a set of operators $\{K_{\lambda,m,x}\}_{\lambda,m,x} \subset \mathfrak{L}(A, B)$ is *an irreducible tensor operator* with respect to the three projective unitary representations on systems $A$, $B$, and $E$, if its elements are orthonormal and satisfy for all $\lambda$, $m$, and $x$,

$$U_g^B K_{\lambda,m,x} U_g^{*A} = \sum_{m'} u_{m'm}^{(\lambda)}(g) K_{\lambda,m',x} \qquad \forall\, g \in \mathbf{G}, \tag{15.35}$$

where $u_{m'm}^{(\lambda)}(g)$ is $m'm$ component of the matrix $U_g^{(\lambda)}$ that appears in (15.34).

---

*Remark.* The orthonormality condition for an irreducible tensor operator is defined in terms of the Hilbert–Schmidt inner product. Specifically, we make the assumption that

$$\mathrm{Tr}\left[K^{*}_{\lambda',m',x'}K_{\lambda,m,x}\right]=\delta_{\lambda\lambda'}\delta_{mm'}\delta_{xx'}. \tag{15.36}$$

The condition stated in (15.35) imposes constraints not only on the elements of the irreducible tensor operator but also on the three representations $U_g^A$, $U_g^B$, and $U_g^E$. This can be illustrated by the following exercise, where it is shown that the cocycle of the map $g \mapsto U_g^E$ is entirely determined by the cocycles of $g \mapsto U_g^A$ and $g \mapsto U_g^B$.

**Exercise 15.7.** *Suppose the representations $g \mapsto U_g^A$ and $g \mapsto U_g^B$ have cocycles*

$$\left\{e^{i\theta^A(g,h)}\right\}_{g,h\in G}\quad and \quad \left\{e^{i\theta^B(g,h)}\right\}_{g,h\in G}, \tag{15.37}$$

*respectively. Show that if (15.35) holds, then $g \mapsto U_g^E$ has a cocycle*

$$\left\{e^{i\left(\theta^B(g,h)-\theta^A(g,h)\right)}\right\}_{g,h\in G}. \tag{15.38}$$

It is important to note that if $A = B$ in Definition 15.1, then the exercise shows that the representation $g \mapsto U_g^E$ is nonprojective. Additionally, we emphasize that the irreps $\lambda \in \mathrm{Irr}(U^E)$ used in the definition of the irreducible tensor operator may not necessarily be the same irreps that appear in the decompositions of $U_g^A$ or $U_g^B$. Therefore, the dimension of the system $E = \mathrm{span}\{|\lambda,m,x\rangle^E\}$ depends on the irreps $\lambda \in \mathrm{Irr}(U^E)$ that appear in the decomposition of $U_g^E$. Specifically, the components $u_{mm'}^{(\lambda)}(g)$ of $U_g^{(\lambda)}$ appear in the decomposition of $U_g^E$ and not in the decompositions of $U_g^A$ or $U_g^B$.

---

**Theorem 15.2.** Let $\mathcal{E} \in \mathfrak{L}(A \to B)$ be trace preserving. The map $\mathcal{E} \in \mathrm{COV}_G(A \to B)$ if and only if there exists a projective unitary representation $g \mapsto U_g^E$ of $G$, and a corresponding orthonormal set of irreducible tensor operators $\{K_{\lambda,m,x}\}_{\lambda,m,x}$ in $\mathfrak{L}(A,B)$ such that

$$\mathcal{E}(\rho) = \sum_{\lambda,m,x} K_{\lambda,m,x}\rho K^{*}_{\lambda,m,x} \qquad \forall\, \rho \in \mathfrak{L}(A). \tag{15.39}$$

---

**Proof** We first prove that the channel given in (15.39) is G-covariant. Indeed, for any $\rho \in \mathfrak{L}(A)$ we have

$$\mathcal{U}_g^B \circ \mathcal{E} \circ \mathcal{U}_g^{*A}(\rho) = \sum_{\lambda,m,x} U_g^B K_{\lambda,m,x} U_g^{*A} \rho \left(U_g^B K_{\lambda,m,x} U_g^{*A}\right)^{*}$$

$$\mathbf{(15.35)}\!\to\; = \sum_{\lambda,m,x,k,k'} u_{km}^{(\lambda)}(g)\bar{u}_{k'm}^{(\lambda)}(g) K_{\lambda,k,x}\rho K^{*}_{\lambda,k',x} \tag{15.40}$$

$$\boldsymbol{U_g^{(\lambda)}}\textbf{ is a unitary matrix}\!\to\; = \sum_{\lambda,x,k,k'} \delta_{kk'} K_{\lambda,k,x}\rho K^{*}_{\lambda,k',x} = \mathcal{E}(\rho).$$

Hence, $\mathcal{E}$ is G-covariant.

Conversely, suppose $\mathcal{E}$ is a $G$-covariant quantum channel. Let $\{K_x\}_{x\in[n]} \in \mathfrak{L}(A, B)$ be a canonical Kraus decomposition of $\mathcal{E}$ (see Corollary 3.2). Since $\mathcal{E}$ is $G$-covariant it follows that for all $g \in G$ and $\rho \in \mathfrak{L}(A)$

$$\mathcal{E}(\rho) = \mathcal{U}_g^B \circ \mathcal{E} \circ \mathcal{U}_g^{A*}(\rho) = \sum_{x\in[n]} \left( U_g^B K_x U_g^{*A} \right) \rho \left( U_g^B K_x U_g^{*A} \right)^*. \tag{15.41}$$

Therefore, the set $\{U_g^B K_x U_g^{*A}\}_{x\in[n]}$ also form a canonical Kraus decomposition of $\mathcal{E}$. Now, recall from Section 3.4.4 that every two operator sum representations of $\mathcal{E}$ that have the same number of elements are related by a unitary matrix. Therefore, for any $g \in G$ there exists an $n \times n$ unitary matrix $U_g^E = \left(u_{xz}(g)\right) \in \mathfrak{L}(E)$ with $n := |E|$ such that for all $x \in [n]$, we have

$$U_g^B K_x U_g^{*A} = \sum_{z\in[n]} u_{zx}(g)K_z. \tag{15.42}$$

Furthermore, since $\{K_z\}_{z\in[n]}$ are linearly independent (as they are orthonormal in the Hilbert-Schmidt inner product), it follows that for every $g \in \mathbf{G}$, there is a unique $U_g^E$ that satisfies (15.42). Additionally, using the notation given in (15.37) for the cocycles, we get that for all $g, h \in \mathbf{G}$

$$U_{gh}^B K_x U_{gh}^{*A} = e^{i\left(\theta^B(g,h)-\theta^A(g,h)\right)} U_g^B U_h^B K_x U_h^{*A} U_g^{*A}$$

$$\mathbf{(15.42)} \rightarrow = e^{i\left(\theta^B(g,h)-\theta^A(g,h)\right)} \sum_{z\in[n]} u_{zx}(h)U_g^B K_z U_g^{*A}$$

$$\mathbf{Using\ (15.42)\ once\ more} \rightarrow = e^{i\left(\theta^B(g,h)-\theta^A(g,h)\right)} \sum_{z,z'\in[n]} u_{zx}(h)u_{z'z}(g)K_{z'} \tag{15.43}$$

$$= e^{i\left(\theta^B(g,h)-\theta^A(g,h)\right)} \sum_{z'\in[n]} (U_g^E U_h^E)_{z'x} K_{z'}.$$

We therefore conclude that

$$U_{gh}^E = e^{i\left(\theta^B(g,h)-\theta^A(g,h)\right)} U_g^E U_h^E. \tag{15.44}$$

That is, the mapping $g \mapsto U_g^E$ is a projective unitary representation of the group $G$. Finally, using the unitary freedom in the choice of the canonical Kraus decomposition $\{K_z\}_{z\in[n]}$ of $\mathcal{E}$ (see Exercise 3.40), we choose it in such a way that $U_g^E$ is block-diagonal with respect to the irreps of $G$. In this basis, $U_g^E = \bigoplus_\lambda U_g^{(\lambda)} \otimes I^{C_\lambda}$ so we can denote the Kraus operators by $\{K_{\lambda,m,x}\}$ (with $\lambda$ the irrep label and $x$ the multiplicity index). This completes the proof.  ∎

**Exercise 15.8.** *Extend Theorem 15.2 to CP maps that are not necessarily trace preserving. That is, show that $\mathcal{E} \in \mathrm{CP}(A \to B)$ is **G**-covariant if and only if it can be expressed as in (15.39).*

To illustrate this theorem, we will provide a few examples. Let's begin with the case of a covariant unitary channel. As we mentioned earlier, a unitary channel $\mathcal{E}(\cdot) = V(\cdot)V^*$, where $V: A \to A$ is a unitary matrix, is covariant if and only if (15.7) holds for all

$g \in \mathbf{G}$. Here, $g \mapsto \omega_g$ is a one-dimensional unitary representation of $\mathbf{G}$. As we will illustrate now, Theorem 15.2 can be used to derive the same conclusion.

Indeed, since a unitary channel has only one Kraus operator, the unitary representation $g \mapsto U_g^E$ must be an irreducible representation (therefore, a single $\lambda$), one dimensional (thus, a single $m$), and with no multiplicity (a single $x$). This implies that $|E| = 1$, so we have $U_g^E = \omega_g$ for some $\omega_g \in \mathbb{C}$, where $|\omega_g| = 1$. Let us denote the single Kraus operator of $\mathcal{E}$ by $V = K_{\lambda,m,x}$. In this case, the relation (15.35) can be expressed as follows:

$$U_g^A V U_g^{*A} = \omega_g V. \tag{15.45}$$

It's worth noting that if there exists $g \in \mathbf{G}$ such that $\omega_g \neq 1$ (i.e. $V$ is not $\mathbf{G}$-invariant), then $\mathcal{G}(V) = 0$. To see why, consider taking the integral over $\mathbf{G}$ (with respect to the Haar measure) on both sides of (15.45):

$$\mathcal{G}(V) = cV, \quad \text{where} \quad c := \int_{\mathbf{G}} dg\, \omega_g. \tag{15.46}$$

If $c \neq 0$, then we have $V = \frac{1}{c}\mathcal{G}(V) = \mathcal{G}\left(\frac{1}{c}V\right)$, which is $\mathbf{G}$-invariant. This implies that $\omega_g = 1$ for all $g \in \mathbf{G}$, and $c = 1$ in this case. Therefore, if $V$ is not $\mathbf{G}$-invariant, we must have $c = 0$, and consequently, $\mathcal{G}(V) = \mathbf{0}$.

As another example, let's consider the group $U(1)$. The Kraus operators of a $U(1)$ covariant channel can be labeled as $K_{k,\alpha} \in \mathcal{L}(A)$, where $\alpha$ is the multiplicity index. Then, from (15.35) we get that

$$e^{i\theta \hat{N}} K_{k,\alpha} e^{-i\theta \hat{N}} = e^{ik\theta} K_{k,\alpha} \qquad \forall\, \theta \in U(1). \tag{15.47}$$

Note that the irreducible representations of $U(1)$ are one dimensional. As a result, the Kraus operators are not mixed with one another under the action of $U(1)$. This provides a significant simplification compared to the non-Abelian case.

Any Kraus operator $K_{k,\alpha}$ that satisfies (15.47) must have the form (see Exercise 15.9)

$$K_{k,\alpha} = S_k D_{k,\alpha}, \tag{15.48}$$

where

$$S_k := \sum_{n \in \mathbb{Z}} |n + k\rangle\langle n| \tag{15.49}$$

is the "shift" operator, and $D_{k,\alpha}$ are diagonal operators in $\mathcal{L}(A)$; that is, $\langle n|D_{k,\alpha}|n'\rangle = 0$ for $n \neq n'$. Note that in the infinite-dimensional Hilbert space $A$, the shift operator $S_k$ is unitary. Therefore, in the QRT of $U(1)$-asymmetry, the set of free unitary operations consists of diagonal unitaries, shift operators, and combinations of the two.

**Exercise 15.9.** *Prove (15.48). Hint: Substitute $K_{k,\alpha} = \sum_{n,n'} c_{nn'}^{(k,\alpha)} |n\rangle\langle n'|$ in (15.47).*

**Exercise 15.10.** *Let $\mathcal{G}$ be the $U(1)$-twirling map, and $S_k$ the shift operator. Show that $S_k$ is not $U(1)$-invariant by showing that $\mathcal{G}(S_k) = 0$. Still, we emphasize that $\mathcal{E}(\cdot) = S_k(\cdot)S_k^*$ is $U(1)$-covariant.*

**Exercise 15.11.** *Let $A$ be an $n$-dimensional Hilbert space, $\mathbb{Z}_n$ be the cyclic group of $n$-elements, and for any $k \in \mathbb{Z}_n$ let*

$$G_k := \sum_{x \in [n]} |x + k \ (mod \ n)\rangle\langle x|. \tag{15.50}$$

*Find the general form of the Kraus operators constituting the operator sum representation of a $\mathbb{Z}_n$-covariant channel with respect to the representation $k \mapsto G_k$.*

Theorem 15.2 demonstrates that every $G$-covariant channel $\mathcal{E} \in \mathrm{COV}_G(A \to B)$ induces an auxiliary system $E$ and a projective unitary representation $g \mapsto U_g \in \mathcal{L}(E)$ that corresponds to its Kraus decomposition with irreducible tensor operators. The terms "induced space" and "induced representation" can be used to refer to $E$ and $g \mapsto U_g^E$, respectively. These concepts are important in the study of symmetry in physics, as they enable us to understand the structure of $G$-covariant channels in terms of irreducible tensor operators. Furthermore, the induced space $E$ appears in the covariant version of Stinespring's representation.

## Covariant Stinespring Representation

In Theorem 3.8, we learned that it is possible to represent every quantum channel as the action of an isometry followed by a partial trace. In this context, we now demonstrate that the Stinespring representation of a covariant channel involves an isometry that is itself $G$-invariant.

---

**Theorem 15.3.** Let $\mathcal{E} \in \mathcal{L}(A \to B)$ be a linear map. The map $\mathcal{E} \in \mathrm{COV}_G(A \to B)$ if and only if there exists a system $E$, a projective unitary representation $g \mapsto U_g^E$, and an (intertwiner) isometry $V : A \to BE$ such that $\mathcal{E}(\rho) = \mathrm{Tr}_E\left[V\rho V^*\right]$ and for all $g \in G$

$$\left(U_g^B \otimes \overline{U}_g^E\right) V = V U_g^A. \tag{15.51}$$

---

*Remark.* If $\mathcal{E}$ is $G$-covariant, then in the proof below we will see that the representation $g \mapsto U_g^E$ is given by the induced representation of $\mathcal{E}$. The components of the matrix $\overline{U}_g^E = \left(U_g^{*E}\right)^T$ equals the complex conjugate of the corresponding components of $U_g^E$. Recall that $g \mapsto \overline{U}_g^E$ is also a projective unitary representation (see Exercise C.5 of online version).

**Proof**  Suppose $\mathcal{E}$ has the form $\mathcal{E}(\rho) = \mathrm{Tr}_E\left[V\rho V^*\right]$, where the isometry $V$ satisfies (15.51). Using the standard Stinesprings dilation theorem, we know that $\mathcal{E} \in$ CPTP$(A \to B)$. Moreover, for all $g \in G$ and all $\rho \in \mathcal{L}(A)$ we have

$$\mathcal{E}^{A \to B} \circ \mathcal{U}_g^A(\rho) = \mathrm{Tr}_E\left[ V \mathcal{U}_g^A(\rho) V^* \right]$$

$$\textbf{(15.51)} \to \; = \mathrm{Tr}_E\left[ U_g^B \otimes \overline{U}_g^E \left( V\rho V^* \right) \right]$$

$$= U_g^B \left( \mathrm{Tr}_E\left[ (V\rho V^*) \right] \right)$$

$$= U_g^B \circ \mathcal{E}^{A \to B}(\rho), \tag{15.52}$$

where $\overline{\mathcal{U}}_g^E(\,\cdot\,) := \overline{U}_g^E(\,\cdot\,)(U_g^E)^T$.

Conversely, suppose $\mathcal{E} \in \mathrm{COV}_G(A \to B)$. Let $\{K_{\lambda,m,x}\}$ be its canonical covariant Kraus decomposition as given in Theorem 15.2, and define the isometry $V : A \to BE$ as

$$V := \sum_{\lambda,m,x} K_{\lambda,m,x} \otimes |\lambda, m, x\rangle^E, \tag{15.53}$$

where $E := \mathrm{span}\{|\lambda, m, x\rangle^E\}$ is the induced space of $\mathcal{E}$ decomposed according to the irreps of $G$ that appear in the induced representation of $\mathcal{E}$. By definition, for all $\rho \in \mathfrak{L}(A)$ we have

$$\mathrm{Tr}_E\left[ V\rho V^* \right] = \sum_{\lambda,m,x} K_{\lambda,m,x}\rho K^*_{\lambda,m,x}$$

$$\textbf{(15.39)} \to \; = \mathcal{E}(\rho). \tag{15.54}$$

Therefore, it is left to show that $V$ satisfies (15.51). Indeed, taking $g \mapsto U_g^E$ to be the induced representation of $\mathcal{E}$ we get

$$\left( U_g^B \otimes \overline{U}_g^E \right) V U_g^{*A} = \sum_{\lambda,m,x} U_g^B K_{\lambda,m,x} U_g^{*A} \otimes \overline{U}_g^E |\lambda, m, x\rangle^E$$

$$\textbf{(15.35)} \to \; = \sum_{\lambda,x} \sum_{m,m'} u_{m'm}^{(\lambda)}(g) K_{\lambda,m',x} \otimes \overline{U}_g^E |\lambda, m, x\rangle^E$$

$$= \sum_{\lambda,x} \sum_{m'} K_{\lambda,m',x} \otimes \overline{U}_g^E \left( U_g^E \right)^T |\lambda, m', x\rangle^E$$

$$= V. \tag{15.55}$$

where the first annotated step uses $\sum_m u_{m'm}^{(\lambda)}(g)|\lambda,m,x\rangle = (U_g^E)^T |\lambda,m',x\rangle$ and the second uses $\overline{U}_g^E (U_g^E)^T = I^E$.

Therefore, $V$ is an intertwiner isometry. ∎

When considering a channel $\mathcal{E} \in \mathrm{COV}_G(A \to A)$, a slightly different version of the covariant Stinespring dilation theorem is obtained. Note that in this case, not only is the output system $B$ replaced with the input system $A$, but the same projective unitary representation is also considered on both the input and output systems of $\mathcal{E}$. This enables us to obtain a covariant Stinespring dilation theorem that involves a $G$-invariant unitary matrix.

> **Theorem 15.4.** Let $\mathcal{E} \in \mathfrak{L}(A \to A)$ be a linear map. The map $\mathcal{E}$ is a $G$-covariant quantum channel, that is, $\mathcal{E} \in \mathrm{COV}_{\mathbf{G}}(A \to A)$, if and only if there exists a system $E$, **G**-invariat state $|0\rangle\langle 0| \in \mathrm{Pure}(E)$, and **G**-invariant unitary $W : AE \to AE$ such that
>
> $$\mathcal{E}^{A \to A}(\rho^A) = \mathrm{Tr}_E \left[ W^{AE} \left( \rho^A \otimes |0\rangle\langle 0|^E \right) W^{*AE} \right]. \tag{15.56}$$

*Remark.* The matrix $W^{AE}$ in Theorem 15.4 is **G**-invariant with respect to the projective unitary representation $g \mapsto U_g^A \otimes \overline{U}_g^E$, where the representation $g \mapsto U_g^E$ is the induced representation of $\mathcal{E}$. Moreover, from Theorem C.7 of online version it follows that a **G**-invariant pure state always exists. To see it, using the same notations as in Theorem C.7 of online version we get for all $\psi \in \mathrm{Pure}(E)$ the vector $\Pi|\psi\rangle$ is proportional to a **G**-invariant state.

**Proof**    The proof that (15.56) implies that $\mathcal{E}$ is **G**-covariant follows similar lines as the ones appear in the proof of Theorem 15.3 (we leave the details to Exercise 15.12). For the converse, if $\mathcal{E} \in \mathrm{COV}_{\mathbf{G}}(A \to A)$, Theorem 15.3 states that there exists an intertwiner isometry $V : A \to AE$ such that

$$\left( U_g^A \otimes \overline{U}_g^E \right) V = V U_g^A \quad \text{and} \quad \mathcal{E}(\rho) = \mathrm{Tr}_E \left[ V\rho V^* \right]. \tag{15.57}$$

Now, let $|0\rangle \in E$ be a **G**-invariant state and define $\tilde{A} := \{|\psi^A\rangle \otimes |0\rangle^E : |\psi\rangle \in A\}$. Clearly, $\tilde{A}$ is a subspace of $AE$ and we define the isometry $\tilde{V} : \tilde{A} \to AE$ via

$$\tilde{V}|\psi\rangle^A |0\rangle^E := V|\psi\rangle^A. \tag{15.58}$$

With this definition we have

$$\mathcal{E}(\rho) = \mathrm{Tr}_E \left[ \tilde{V} \left( \rho^A \otimes |0\rangle\langle 0|^E \right) \tilde{V}^* \right] \tag{15.59}$$

and from Exercise 15.13 it follows that $\tilde{V}$ is a **G**-invariant isometry. Denote by $\Pi := I^A \otimes |0\rangle\langle 0|^E$ the projection (in $AE$) to $\tilde{A}$ . Then, Theorem C.7 of online version states that there exists a **G**-invariant unitary $W \in \mathfrak{U}(AE)$ such that $\tilde{V}\Pi = W\Pi$. We therefore get from (15.59) that (15.56) must hold with this choice of $W$.     ∎

**Exercise 15.12.** *Show that if $W : AE \to AE$ is **G**-invariant unitary matrix and $|0\rangle\langle 0| \in \mathrm{Pure}(E)$ is **G** invariant state, then the map $\mathcal{E}$ as defined in (15.56) is **G**-covariant.*

**Exercise 15.13.** *Show that the operator $\tilde{V}$ as defined in (15.58) is a **G**-invariant isometry. Hint: Show that the operator $\tilde{V}\Pi : AE \to AE$ with $\Pi := I^A \otimes |0\rangle\langle 0|^E$ is **G**-invariant.*

## 15.2.3 Alignment of References Frames

As we have already discussed, when parties are separated, they often need to establish a common reference frame. For example, they may need to synchronize their clocks or align their Cartesian frames. While lacking a shared reference frame doesn't completely hinder tasks such as communication and computation, it does impose limitations, reducing their practical efficiency. This often requires more advanced encodings. Hence, parties may prioritize allocating communication resources to establish a shared reference frame initially. Later, they can utilize a standard encoding instead of continuously circumventing its absence with a relational encoding.

In tasks aimed at establishing a shared reference frame, parties can employ quantum particles to encode information regarding the relative orientation of their frames. For instance, spin-1/2 particles, like electrons, can encode the orientation of Cartesian frames, while exchanging quantum states of an optical mode can align phase references. Hence, in the realm of quantum reference frames, which involve quantum particles holding information about a shared reference frame, the usefulness of a quantum state is determined by the amount of information that can be extracted from it to establish such a reference frame.

This discussion illustrates that the resource theory of quantum reference frames introduces certain aspects that differ from what we have encountered thus far. Specifically, when Alice and Bob do not share a reference frame, Alice can gain at least partial information about Bob's reference frame by receiving a resource in the form of a quantum state that encodes it. As a result, the set of free operations (i.e. $G$-covariant operations) needs to be updated to incorporate this partial information. For instance, instead of using regular $G$-twirling operations, weighted $G$-twirling operations can be employed, taking into account that the parties have partial knowledge about the element $g \in G$ that relates their reference frames.

We now discuss the general approach to align reference frames making the notions discussed rigorous. Consider two parties, Alice and Bob, who do not share a reference frame, with $G$ being the corresponding group describing the reference frame. The goal is for Bob to learn the element $g \in G$ that relates between his reference frame and Alice's reference frame. To accomplish that, Alice sends Bob a quantum reference frame (e.g. spin-1/2 particles pointing in the $z$-direction of her Cartesian reference frame) in a form of a quantum state $\rho \in \mathfrak{D}(A)$. From Bob's perspective, he received one of the states $\{U_g \rho U_g^*\}_{g \in G}$, all occurring with uniform prior.

To determine the specific state he possesses, that is, to identify the group element $g \in G$, Bob conducts a POVM, $\{\Lambda_g\}_{g \in G}$, on his system. Consequently, the probability that Bob guesses the group element as $g'$, given the actual element is $g$, is denoted by

$$q(g'|g) := \mathrm{Tr}\left[\Lambda_{g'} U_g \rho U_g^*\right]. \tag{15.60}$$

In order to quantify how much information Bob gained after the measurement, consider first the case that $G$ is a finite group. In this senario, we can use the probability that Bob guesses $g$ correctly as our *figure of merit* and maximizes this function over all states and all POVMs. For a given state $\rho \in \mathfrak{D}(A)$ and a POVM $\{\Lambda_g\}_{g \in G}$, this probability is given by

$$\mathrm{Pr}_{\mathrm{guess}}\left(\rho, \{\Lambda_g\}_{g \in G}\right) := \frac{1}{|G|} \sum_{g \in G} p(g|g) = \frac{1}{|G|} \sum_{g \in G} \mathrm{Tr}\left[\Lambda_g U_g \rho U_g^*\right]. \tag{15.61}$$

Thus, Alice and Bob's objective is to maximize this guessing probability across all possible $\rho$ (referred to as the *fiducial* state) and all POVMs $\{\Lambda_g\}_{g\in\mathbf{G}}$.

Conversely, if $\mathbf{G}$ is a compact Lie group, the chance of Bob correctly inferring Alice's reference frame becomes infinitesimally small. In such instances, the direct likelihood or guessing probability cannot serve as an effective figure of merit. Instead, the *maximum likelihood* of a correct guess is adopted as the figure of merit. This maximum likelihood, akin to the formula above but integrated over the group, is defined as

$$\mu_{\max} := \max \int_{\mathbf{G}} dg \ p(g|g) = \max \int_{\mathbf{G}} dg \ \mathrm{Tr}\left[\Lambda_g U_g \rho U_g^*\right], \qquad (15.62)$$

with the maximization conducted over all fiducial states $\rho \in \mathfrak{D}(A)$ and all POVMs $\{\Lambda_g\}_{g\in\mathbf{G}}$. Given that the guessing probability in (15.62) is linear in $\rho$, the maximal value can always be achieved with a pure state, allowing us to assume, for simplification, that the fiducial state $\rho = \psi$ is pure.

From Theorem C.4 of online version it follows that the Hilbert space $A$ can be decomposed as $A = \bigoplus_{\lambda\in\mathrm{Irr}(U)} B_\lambda \otimes C_\lambda$, where for each irrep $\lambda$, $B_\lambda$ denotes the representation space, and $C_\lambda$ denotes the multiplicity space. We will denote by $d_\lambda := |B_\lambda|$ and $m_\lambda := |C_\lambda|$. With these notations we can write any pure state $\psi \in \mathrm{Pure}(A)$ as

$$|\psi\rangle = \bigoplus_{\lambda\in\mathrm{Irr}(U)} c_\lambda |\psi_\lambda\rangle, \qquad (15.63)$$

where each $\psi_\lambda \in \mathrm{Pure}(B_\lambda C_\lambda)$, and $c_\lambda \in \mathbb{C}$ with $\sum_{\lambda\in\mathrm{Irr}(U)} |c_\lambda|^2 = 1$. In the following theorem, we consider a projective unitary representation, $g \mapsto U_g^A$, and denote by $d_\lambda := |B_\lambda|$ and $m_\lambda := |C_\lambda|$ the dimensions of the representation and multiplicity spaces (respectively) associated with the irrep $\lambda \in \mathrm{Irr}(U)$. We also use the notation $n_\lambda := \min\{m_\lambda, d_\lambda\}$ for all $\lambda \in \mathrm{Irr}(U)$.

---

**Theorem 15.5.** Using the notations above, the maximum likelihood $\mu_{\max}$ as defined in (15.62) is given by

$$\mu_{\max} = \sum_{\lambda\in\mathrm{Irr}(U)} d_\lambda n_\lambda. \qquad (15.64)$$

---

**Proof**  Let $\psi$ and $\{\Lambda_g\}_{g\in\mathbf{G}}$ be the optimal state and POVM that maximizes the maximum likelihood. Expressing $\psi$ as in (15.63), observe that due to the Schmidt decomposition of $|\psi_\lambda\rangle$ there exists an orthogonal projector $\Pi_\lambda^{C_\lambda}$ such that $\mathrm{Tr}\left[\Pi_\lambda^{C_\lambda}\right] = n_\lambda$ and $I^{B_\lambda} \otimes \Pi_\lambda^{C_\lambda} |\psi_\lambda\rangle = |\psi_\lambda\rangle$. Therefore, the operator

$$\Pi := \bigoplus_{\lambda\in\mathrm{Irr}(U)} I^{B_\lambda} \otimes \Pi_\lambda^{C_\lambda} \qquad (15.65)$$

satisfies $\Pi|\psi\rangle = |\psi\rangle$ and $\mathrm{Tr}[\Pi] = \sum_{\lambda\in\mathrm{Irr}(U)} d_\lambda n_\lambda$. From (C.45 of online version) we also get that $[\Pi, U_g] = 0$ for all $g \in \mathbf{G}$. Therefore,

$$U_g \psi U_g^* = U_g \Pi \psi \Pi U_g^*$$

$$\boxed{[\Pi, U_g] = 0} \longrightarrow \quad = \Pi U_g \psi U_g^* \Pi \tag{15.66}$$

$$\boxed{U_g \psi U_g^* \leqslant I^A} \longrightarrow \quad \leqslant \Pi.$$

Substituting the above inequality into (15.62) we get that the maximum likelihood is bounded from above by

$$\mu_{\max} \leqslant \max \int_{\mathbf{G}} dg \, \mathrm{Tr}\left[\Lambda_g \Pi\right]$$

$$\boxed{\int_{\mathbf{G}} dg \, \Lambda_g = I} \longrightarrow \quad = \mathrm{Tr}[\Pi] = \sum_{\lambda\in\mathrm{Irr}(U)} d_\lambda n_\lambda. \tag{15.67}$$

For the converse, take $\rho = \psi$ to be the pure state

$$|\psi\rangle = \frac{1}{\sqrt{\nu}} \bigoplus_{\lambda\in\mathrm{Irr}(U)} \sqrt{d_\lambda n_\lambda}|\Phi_\lambda\rangle, \quad \text{where} \quad |\Phi_\lambda\rangle := \frac{1}{\sqrt{n_\lambda}} \sum_{x\in[n_\lambda]} |x\rangle^{B_\lambda}|x\rangle^{C_\lambda}, \tag{15.68}$$

and $\nu := \sum_{\lambda\in\mathrm{Irr}(U)} d_\lambda n_\lambda$. For the POVM we define $\Lambda_g := \nu U_g \psi U_g^*$, and observe that

$$\int_{\mathbf{G}} dg \, \Lambda_g = \nu \mathcal{G}(\psi)$$

$$\mathbf{Corollary\ 15.1} \rightarrow \quad = \bigoplus_{\lambda\in\mathrm{Irr}(U)} d_\lambda n_\lambda (\mathcal{R}^{B_\lambda} \otimes \mathrm{id}^{C_\lambda})(\Phi_\lambda) \tag{15.69}$$

$$= \bigoplus_{\lambda\in\mathrm{Irr}(U)} I^{B_\lambda} \otimes \Pi_\lambda^{C_\lambda},$$

where $\Pi_\lambda^{C_\lambda} := \sum_{x\in[n_\lambda]} |x\rangle\langle x|^{C_\lambda}$. Let $\Pi$ be the projector appearing on the right-hand side of the equation (15.69) and observe that it satisfies $\Pi|\psi\rangle = |\psi\rangle$ and $[U_g, \Pi] = 0$. Therefore, the set $\{I^A - \Pi\}\cup\{\Lambda_g\}_{g\in\mathbf{G}}$ is a POVM. Moreover, the measurement outcome corresponding to the element $I^A - \Pi$ occurs with probability

$$\mathrm{Tr}\left[\left(I^A - \Pi\right) U_g \psi U_g^*\right] = \mathrm{Tr}\left[U_g \left(I^A - \Pi\right) \psi U_g^*\right] = 0 \qquad \forall\, g \in \mathbf{G}. \tag{15.70}$$

In other words, the outcome corresponding to $I^A - \Pi$ never occurs. We therefore conclude that for this choice of POVM and fiducial state the maximum likelihood is given by

$$\int_{\mathbf{G}} dg\, \mathrm{Tr}\left[\Lambda_g U_g \psi U_g^*\right] = \nu \int_{\mathbf{G}} dg\, \mathrm{Tr}\left[\psi^2\right] = \nu. \tag{15.71}$$

Hence, $\mu_{\max} \geqslant \nu$ and since we already saw that $\mu_{\max} \leqslant \nu$ we conclude that $\mu_{\max} = \nu$. This completes the proof. ∎

## Example: Alignment of Cartesian Frame

As an example, we explore the alignment of an entire three-dimensional coordinate system using the communication of 1/2-spin particles. Suppose the Cartesian frames of Alice and Bob are related by a group element $g \in SO(3)$. Every group element of $SO(3)$ represents a rotation by some angle $\theta$ and along a unit vector $\mathbf{n} \in \mathbb{R}^3$. In Section 2.4.1, we studied the representation of $SO(3)$ in $\mathbb{C}^2$ and showed that $g := (\theta, \mathbf{n}) \mapsto U_g := e^{\frac{1}{2}\theta \mathbf{n} \cdot \boldsymbol{\sigma}}$ is a unitary representation of $SO(3)$ on $A := \mathbb{C}^2$.

Here we are interested in the representation of $SO(3)$ on the space $A^n := \left(\mathbb{C}^2\right)^{\otimes n}$ for some integer $n$. We extend here the group of rotations $SO(3)$ to the group $SU(2)$ to allow for spinor representations. For simplicity, we will assume that $n$ is even, and use some well-known results from representation theory. Specifically, the representation $g \mapsto U_g^{\otimes n}$ can be decomposed into a direct sum of $SU(2)$ irreps, labeled by the total angular momentum $j$ ranging from 0 to $n/2$. The decomposition (C.44 of online version), for the representation of $SU(2)$ on $A^n$, has been extensively studied in representation theory, and is given by

$$A^n = \bigoplus_{j=0}^{n/2} B_j \otimes C_j, \tag{15.72}$$

where $|B_j| = 2j + 1$ and

$$|C_j| = \binom{n}{n/2 - j} \frac{2j + 1}{n/2 + j + 1}. \tag{15.73}$$

From (15.73) we see that $|C_j| \geqslant |B_j|$ for all $j$ except the case $j = n/2$, for which $|B_j| \geqslant |C_j|$. Therefore, according to Theorem 15.5, the maximum likelihood of a correct guess is given by

$$\mu_{\max} = \sum_{j=0}^{n/2-1} (2j + 1)^2 + (n + 1) = \frac{1}{6}n^3 + \frac{5}{6}n + 1. \tag{15.74}$$

Moreover, in this case the optimal state (15.68) that achieves the maximum likelihood above is given by

$$|\psi\rangle = \frac{1}{\sqrt{\mu_{\max}}} \left( \sum_{j=0}^{n/2-1} (2j + 1)|\Phi_j\rangle + \sqrt{n + 1}|n/2, n/2\rangle \right), \tag{15.75}$$

where

$$|\Phi_j\rangle := \frac{1}{\sqrt{2j + 1}} \sum_{m=-j}^{j} |j, m\rangle^{B_j} |\phi_m^{C_j}\rangle \tag{15.76}$$

and $\{|\phi_m^{C_j}\rangle\}_{m \in \{-j,...,j\}}$ is an orthonormal set of vectors in $C_j$.

As a specific example, suppose $n = 2$. In this case we have two irreps, corresponding to total angular momentum $j = 0$ and $j = 1$. In this case, $|C_j| = 1$ for both $j = 0$ and

$j = 1$, and the representation space $B_0 = \mathrm{span}\{|0,0\rangle\}$ is one dimensional spanned by the singlet state

$$|0,0\rangle^{B_0} := |\Psi_-^{A\tilde{A}}\rangle := \frac{1}{\sqrt{2}}\left(|01\rangle - |10\rangle\right), \tag{15.77}$$

whereas $B_1$ is three dimensional spanned by the triplet states $|1,1\rangle^{B_1} := |0\rangle^A|0\rangle^A$, $|1,0\rangle^{B_1} = |\Psi_+^{A\tilde{A}}\rangle$, and $|1,-1\rangle := |1\rangle^A|1\rangle^A$. Therefore, the formula above implies that the two 1/2-spin particle state

$$|\psi\rangle = \frac{1}{2}\left(|0,0\rangle^{B_0} + \sqrt{3}|1,1\rangle^{B_1}\right) = \frac{1}{2}\left(|\Psi_-^{A\tilde{A}}\rangle + \sqrt{3}|11\rangle^{A\tilde{A}}\right) \tag{15.78}$$

achieves the largest maximum likelihood $\mu_{\max} = 4$, as defined in (15.62). It is worth pointing out that this state is not unique, and replacing $|1,1\rangle^{B_1}$ with any other normalized state in $B_1$ would still give the maximum likelihood $\mu_{\max} = 4$.

**Exercise 15.14.** *Let $\varphi \in \mathrm{Pure}(B_1)$ be a pure state in the triplet space of two spin-1/2 particles. Show that there exists a POVM $\{\Lambda_g\}_{g \in G}$ such that the state*

$$|\psi\rangle = \frac{1}{2}\left(|0,0\rangle^{B_0} + \sqrt{3}|\varphi^{B_1}\rangle\right) \quad \textit{satisfies} \quad \int_G dg\, \mathrm{Tr}\left[\Lambda_g U_g \rho U_g^*\right] = 4. \tag{15.79}$$

# 15.3 Quantification of Asymmetry

The reformulation of symmetric dynamics in the context of a resource theory has significant implications. Often, the dynamics of a system can be so complex that a complete characterization of its evolution becomes impractical. Instead, by understanding the symmetries of the Hamiltonian, partial information about its dynamics can be gained. Noether's theorem is one way to do this, as it states that a differentiable symmetry of the action of a physical system has a corresponding conservation law. However, Noether's theorem is not applicable to open systems, and as a result, it does not capture all the consequences of symmetric evolution of mixed states.

Recently, it was demonstrated that the QRT of asymmetry provides a systematic approach to capturing all the outcomes of symmetric evolution. The fundamental idea is that the conserved quantities of closed systems can be substituted with resource monotones in open systems. These resource monotones measure the degree of asymmetry in a quantum state, and they cannot increase under symmetric evolution.

In this section, we will explore different measures of asymmetry and their properties. We will focus on three different types of measures:

1. **Measures of Quantum Frameness:** Measures that quantify how well a resource state can be utilized for the alignment of quantum reference frames.
2. **Relative Entropies of Asymmetry:** Measures that are derived from the general framework of resource theories using different choices of relative entropies.

3. **Derivatives of Asymmetry:** Measures that are unique to the theory of asymmetry and involve taking derivatives of quantum divergences.

While some measures, like the relative entropy of asymmetry, are derived from the general framework of quantum resource theories, they have certain drawbacks, such as not being additive under tensor products and having zero regularized versions. To overcome these limitations, we introduce a new technique to construct measures of asymmetry that involve taking derivatives of quantum divergences. We refer to these measures as *derivatives of asymmetry*. The concept of derivatives of asymmetry encompasses significant measures like the quantum Fisher information and the Wigner–Yanase–Dyson skew information. These measures play pivotal roles in fields like quantum metrology, where precision and sensitivity are paramount. By exploring the derivatives of asymmetry, we can gain a deeper understanding of asymmetry in quantum systems and their applications in diverse fields beyond quantum information.

## 15.3.1 Measures of Quantum Frameness

In this section, we introduce a subset of measures of asymmetry that we call "measures of quantum frameness." These are measures that appear as "figure of merits" in the context of reference frame alignment, and they quantify the uncertainty (or more correctly, the certainty) that Bob has about Alice's reference frame. In other words, they quantify the distinguishability of the elements in the set $\mathfrak{S}(\rho) := \{U_g \rho U_g^*\}_{g \in \mathbf{G}}$. Note that since $\mathfrak{S}(\rho) = \mathfrak{S}(U_g \rho U_g^*)$, measures of quantum frameness must be invariant under the action $\rho \mapsto U_g \rho U_g^*$.

In any reference frame alignment scheme, Alice sends Bob a quantum state $\rho \in \mathfrak{D}(A)$ that is described relative to her reference frame. From Bob's perspective, he received one of the states $\{U_g \rho U_g^*\}_{g \in \mathbf{G}}$, all occurring with uniform prior. To learn $g \in \mathbf{G}$, Bob performs a POVM, $\{\Lambda_g\}_{g \in \mathbf{G}}$, on his system so that the probability that he guesses $g' \in \mathbf{G}$, given that the actual element relating the frames is $g$, is given by

$$q(g'|g) := \mathrm{Tr}\left[\Lambda_{g'} U_g \rho U_g^*\right]. \tag{15.80}$$

In order to quantify how much information Bob gained after the measurement, let $X$ be the random variable corresponding to the element $g \in \mathbf{G}$ that relates between Alice and Bob's reference, and let $Y$ be the random variable associated with Bob's measurement outcome $g' \in \mathbf{G}$. With these notations, any measure of conditional uncertainty can be used to quantify the uncertainty of $X$ given that Bob has access to $Y$. Let $S(X|Y)_{\mathbf{q}}$, with $\mathbf{q} := \{q(g'|g)\}_{g,g' \in \mathbf{G}}$ denoting the probability distribution, be some measure of conditional certainty such as the negative of the conditional entropy $H(X|Y)_{\mathbf{q}}$. Then, a measure of quantum frameness associated with $S(X|Y)_{\mathbf{q}}$ is defined for all $\rho \in \mathfrak{D}(A)$ as

$$\mathbf{F}(\rho) := \max_{\{\Lambda_g\}} S(X|Y)_{\mathbf{q}}, \tag{15.81}$$

where the maximum is over all POVMs that Bob can perform on his system. In other words, Bob chooses a POVM that maximizes his certainty about $X$. We say that $F$ is a measure of quantum reference frame only if it can be written in this way.

> **Theorem 15.6.** Every measure of quantum frameness is a measure of asymmetry.

**Proof**   Let $\rho \in \mathfrak{D}(A)$ and $\{\Gamma_g\}_{g \in \mathbf{G}} \subset \mathrm{Eff}(A)$. Let $\mathcal{N} \in \mathrm{COV}_{\mathbf{G}}(A \to B)$ and observe that since $\mathcal{N}$ is $\mathbf{G}$-covariant we get that

$$
\begin{aligned}
\mathrm{Tr}\left[\Gamma_{g'} U_g^B \mathcal{N}(\rho) U_g^{*B}\right] &= \mathrm{Tr}\left[\Gamma_{g'} \mathcal{N}\left(U_g^A \rho U_g^{*A}\right)\right] \\
&= \mathrm{Tr}\left[\mathcal{N}^*\left(\Gamma_{g'}\right) U_g \rho U_g^*\right].
\end{aligned}
\tag{15.82}
$$

Therefore,

$$
\mathbf{F}\big(\mathcal{N}(\rho)\big) = \max_{\{\mathcal{N}^*(\Gamma_g)\}} S(X|Y)_{\mathsf{q}} \leqslant \max_{\{\Lambda_g\}} S(X|Y)_{\mathsf{q}} = \mathbf{F}(\rho),
\tag{15.83}
$$

where the first maximum is over all POVMs of the form $\{\mathcal{N}^*(\Gamma_g)\}_{g \in \mathbf{G}}$, which is a subset of all possible POVMs $\{\Lambda_g\}_{g \in \mathbf{G}}$. This completes the proof.    ■

Note that in this proof we did not need to use any of properties of the function $S(X|Y)_{\mathsf{q}}$. However, since $S(X|Y)_{\mathsf{q}}$ measures the conditional certainty of $X$ given that Bob has access to $Y$, measures of quantum frameness has additional properties. In Chapter 7, we saw that all measures of conditional uncertainty have to behave monotonically under conditional majorization. However, in Chapter 7 we only considered finite-dimensional, discrete probability distributions. Therefore, for finite groups in which $X$ and $Y$ are discrete random variables with $|X| = |Y| = |\mathbf{G}|$, the function $S(X|Y)_{\mathsf{q}}$ must behave monotonically under conditional majorization. We called such functions in Section 4.6.5 conditionally Schur convex functions.

The extension of conditional majorization to continuous probability distributions is a complex and currently unresolved issue in the field. However, various functions, such as the family of conditional R'enyi entropies (including the conditional von-Neumann entropy), can be utilized to measure the conditional entropy of continuous distributions. For the purpose of our discussion here, we only need to focus on one common property shared by all such functions that quantify conditional uncertainty: their invariance under the action of the group, both from the left and from the right.

Let's recall that $S(X|Y)_{\mathsf{q}}$ represents a function of the conditional probability distribution $q(g'|g)$. Now, suppose Bob rotates his reference frame by an element $h \in \mathbf{G}$, causing the corresponding element $g \in \mathbf{G}$ relating his frame to Alice's frame to change to $h^{-1}g$. Consequently, the outcome of the measurement $g'$ transforms to $h^{-1}g'$ under this change in Bob's reference frame. Since such a transformation should not affect Bob's uncertainty about $g$, we deduce that both distributions $q(g'|g)$ and

$r(g'|g) := q(h^{-1}g'|h^{-1}g)$ represent the same conditional uncertainty. Hence, any function $S(X|Y)_{\mathsf{p}}$ that quantifies Bob's certainty about $X$ must be *left-invariant*, meaning that $S(X|Y)_{\mathsf{q}} = S(X|Y)_{\mathsf{r}}$ holds for all conditional distributions $\mathsf{q}$ and all $h \in \mathbf{G}$.

Similarly, let's consider the scenario where Bob changes his reference frame such that $g \mapsto gh$ and $g' \mapsto g'h$. As before, such a transformation should not affect Bob's uncertainty about $g$. Consequently, both distributions $q(g'|g)$ and $r(g'|g) := q(g'h|gh)$ represent the same conditional uncertainty. Thus, any function $S(X|Y)_{\mathsf{q}}$ that quantifies Bob's certainty about $X$ must also be *right-invariant*, meaning that $S(X|Y)_{\mathsf{q}} = S(X|Y)_{\mathsf{r}}$ holds for all conditional distributions $\mathsf{q}$ and all $h \in \mathbf{G}$.

Many of the functions $S(X|Y)_{\mathsf{q}}$ are not linear in $\mathsf{q}$, which makes the optimization in (15.81) very difficult for such choices. We therefore focus here on measure of conditional certainty that are linear in $\mathsf{q}$. We start with the maximum likelihood that we already encountered in Section 15.2.3.

## The Maximum Likelihood

In the previous section we introduced a figure of merit, called maximum likelihood, that characterizes how well a quantum state can be used to establish a shared reference frame. In particular, for a quantum state $\rho \in \mathfrak{D}(A)$ we define

$$\mu(\rho) := \max \int_G dg \; \mathrm{Tr}\left[ \Lambda_g U_g \rho U_g^* \right], \tag{15.84}$$

where the maximum is over all POVMs $\{\Lambda_g\}_{g \in \mathbf{G}} \subset \mathrm{Eff}(A)$. Observe that $\mu_{\max} := \max_{\rho \in \mathfrak{D}(A)} \mu(\rho)$ is the maximum likelihood for Bob's correct guess of Alice's reference frame.

**Exercise 15.15.** *Consider the maximum likelihood measure as just defined.*

1. *Show that $\mu(\rho)$ can be expressed as in (15.81) with*

$$S(X|Y)_{\mathsf{q}} := \int_G dg \; q(g|g). \tag{15.85}$$

2. *Show that the function in (15.85) is both left- and right-invariant.*

In the following theorem, we will see that the maximum likelihood $\mu(\rho)$ can be expressed in terms of the max relative entropy. Among other things, this result demonstrates that the function $\mu(\rho)$ behaves monotonically under **G**-covariant operations and therefore can be used to define a measure for asymmetry. To be more precise, since for $\rho \in \mathrm{INV}_{\mathbf{G}}(A)$ we have $\mu(\rho) = 1$, the function $\rho \mapsto \mu(\rho) - 1$ is a measure of asymmetry (in fact, it can be shown to be an asymmetry monotone, see the following exercise).

**Theorem 15.7.** Using the same notations as above, for any $\rho \in \mathfrak{D}(A)$

$$\log \mu(\rho) = \min_{\sigma \in \mathrm{INV}_{\mathbf{G}}(A)} D_{\max}(\rho \| \sigma). \tag{15.86}$$

**Proof**   For any POVM $\{\Lambda_g\}_{g\in\mathsf{G}}$ denote by

$$\Lambda := \int_{\mathsf{G}} dg\; U_g^* \Lambda_g U_g, \tag{15.87}$$

so that

$$\int_{\mathsf{G}} dg\; \mathrm{Tr}\left[\Lambda_g U_g \rho U_g^*\right] = \mathrm{Tr}\left[\Lambda\rho\right]. \tag{15.88}$$

Since $\{\Lambda_g\}_{g\in\mathsf{G}}$ is a POVM, we get that

$$\mathcal{G}(\Lambda) = \int_{\mathsf{G}} dg\; \mathcal{G}\left(U_g^* \Lambda_g U_g\right) = \int_{\mathsf{G}} dg\; \mathcal{G}(\Lambda_g) = \mathcal{G}\left(\int_{\mathsf{G}} dg\; \Lambda_g\right) = \mathcal{G}(I^A) = I^A. \tag{15.89}$$

Conversely, let $\Lambda \in \mathrm{Pos}(A)$ be such that $\mathcal{G}(\Lambda) = I^A$, and define, $\Lambda_g := U_g \Lambda U_g^*$ for every $g \in \mathsf{G}$, so that $\int_{\mathsf{G}} dg\; \Lambda_g = \mathcal{G}(\Lambda) = I^A$. By definition, this POVM $\{\Lambda_g\}_{g\in\mathsf{G}}$ satisfies (15.88).

We can therefore express $\mu(\rho)$ as the following SDP:

$$\mu(\rho) = \max_{\substack{\Lambda\in\mathrm{Pos}(A)\\ \mathcal{G}(\Lambda)=I}} \mathrm{Tr}\left[\Lambda\rho\right]. \tag{15.90}$$

The above optimization problem is an SDP. As such, it has a dual given by (see Section A.9 of online version)

$$\mu(\rho) = \min\left\{t \geqslant 0 : t\sigma \geqslant \rho,\; \sigma \in \mathrm{INV}_{\mathsf{G}}(A)\right\}. \tag{15.91}$$

That is,

$$\log \mu(\rho) = \min_{\sigma\in\mathrm{INV}_{\mathsf{G}}(A)} D_{\max}\left(\rho\|\sigma\right). \tag{15.92}$$

This completes the proof.                    ∎

**Exercise 15.16.**  *Use the duality relations discussed in Section A.9 of online version to show that the dual of (15.90) is given by the expression in (15.91).*

**Exercise 15.17.**  *Show that the maximum likelihood $\mu(\rho)$ is an asymmetry monotone.*

We now use the expression in Theorem 15.7 to compute the maximum likelihood of a pure state $\psi \in \mathfrak{D}(A)$. For this purpose, we will use the fact that any quantum state $\sigma \in \mathrm{INV}_{\mathsf{G}}(A)$ has the form

$$\sigma^A = \bigoplus_{\lambda\in\mathrm{Irr}(U)} I^{B_\lambda} \otimes \sigma_\lambda^{C_\lambda}, \tag{15.93}$$

for some $\sigma_\lambda \in \mathrm{Pos}(C_\lambda)$. Let $s_\lambda := \mathrm{Tr}\left[\sigma_\lambda\right]$ and observe that since $\sigma$ is normalized we must have

$$\sum_{\lambda\in\mathrm{Irr}(U)} d_\lambda s_\lambda = 1, \tag{15.94}$$

where $d_\lambda := |B_\lambda|$. We will also use the fact that any pure state $\psi \in \mathrm{Pure}(A)$ can be expressed as

$$|\psi\rangle = \bigoplus_{\lambda \in \mathrm{Irr}(U)} \sqrt{p_\lambda}|\psi_\lambda\rangle, \tag{15.95}$$

where $\{p_\lambda\}_{\lambda \in \mathrm{Irr}(U)}$ is a probability distribution, and each $\psi_\lambda \in \mathrm{Pure}(B_\lambda C_\lambda)$ is a pure state in the tensor product of the representation and multiplicity spaces.

> **Theorem 15.8.** The maximum likelihood of a pure state $\psi \in \mathrm{Pure}(A)$ is given by
>
> $$\log \mu(\psi) = H_{1/2}\big(\mathcal{G}(\psi)\big) := 2\log \mathrm{Tr}\left[\sqrt{\mathcal{G}(\psi)}\right], \tag{15.96}$$
>
> where $H_{1/2}$ is the Rényi entropy of order $\alpha = 1/2$.

**Proof**   Let $t \in \mathbb{R}_+$ and $\sigma \in \mathfrak{D}(A)$ be such that $t\sigma \geqslant \psi$. This condition holds if and only if

$$tI^A \geqslant \sigma^{-1/2}|\psi\rangle\langle\psi|\sigma^{-1/2} \tag{15.97}$$

(where all inverses are understood as generalized inverses). This condition holds if and only if $t \geqslant \langle\psi|\sigma^{-1}|\psi\rangle$. Therefore,

$$\mu(\psi) = \min_{\sigma \in \mathrm{INV}_\mathbf{G}(A)} \langle\psi|\sigma^{-1}|\psi\rangle. \tag{15.98}$$

Now, a density matrix $\sigma \in \mathrm{INV}_\mathbf{G}(A)$ if and only if it has the form (15.93). Therefore, the maximum likelihood of $\psi$ is given by

$$\mu(\psi) = \min \sum_{\lambda \in \mathrm{Irr}(U)} p_\lambda \langle\psi_\lambda|I^{B_\lambda} \otimes \sigma_\lambda^{-1}|\psi_\lambda\rangle, \tag{15.99}$$

where the minimum is over all $\sigma_\lambda \in \mathrm{Pos}(C_\lambda)$ whose traces satisfy (15.94). Using the notations $s_\lambda := \mathrm{Tr}[\sigma_\lambda]$ and $\eta_\lambda := \frac{1}{s_\lambda}\sigma_\lambda$, we split the minimization into two parts: first, we fix the numbers $\{s_\lambda\}_{\lambda \in \mathrm{Irr}(U)}$ and minimize the expression over all $\eta_\lambda \in \mathfrak{D}(C_\lambda)$, and then we minimize the resulting expression over all $\{s_\lambda\}_{\lambda \in \mathrm{Irr}(U)}$ that satisfy (15.94). That is,

$$\mu(\psi) = \min_{\{s_\lambda\}} \sum_{\lambda \in \mathrm{Irr}(U)} p_\lambda s_\lambda^{-1} \min_{\eta_\lambda \in \mathfrak{D}(C_\lambda)} \langle\psi_\lambda|I^{B_\lambda} \otimes \eta_\lambda^{-1}|\psi_\lambda\rangle. \tag{15.100}$$

Denote the reduced density matrix of $\psi_\lambda$ by $\rho_\lambda^{C_\lambda} := \mathrm{Tr}_{B_\lambda}\left[\psi_\lambda^{B_\lambda C_\lambda}\right]$, and observe that

$$\langle\psi_\lambda|I^{B_\lambda} \otimes \eta_\lambda^{-1}|\psi_\lambda\rangle = \mathrm{Tr}\left[\rho_\lambda \eta_\lambda^{-1}\right]$$

$$\boxed{\tau_\lambda := \frac{\sqrt{\rho_\lambda}}{\mathrm{Tr}\left[\sqrt{\rho_\lambda}\right]}} \longrightarrow = \left(\mathrm{Tr}\left[\sqrt{\rho_\lambda}\right]\right)^2 \mathrm{Tr}\left[\tau_\lambda^2 \eta_\lambda^{-1}\right] \tag{15.101}$$

$$\textbf{Definition 6.6 with } \boldsymbol{\alpha=2} \rightarrow = \left(\mathrm{Tr}\left[\sqrt{\rho_\lambda}\right]\right)^2 2^{D_2(\tau_\lambda \| \eta_\lambda)}.$$

Therefore, the minimum of the expression above over all $\eta \in \mathfrak{D}(C_\lambda)$ is obtained when $\eta = \tau_\lambda$. Hence,

$$\mu(\psi) = \min_{\{s_\lambda\}} \sum_{\lambda \in \mathrm{Irr}(U)} p_\lambda s_\lambda^{-1} \left(\mathrm{Tr}\left[\sqrt{\rho_\lambda}\right]\right)^2. \tag{15.102}$$

For the remaining of the optimization problem, for each $\lambda \in \mathrm{Irr}(U)$ we denote by $r_\lambda := p_\lambda \left(\mathrm{Tr}\left[\sqrt{\rho_\lambda}\right]\right)^2$ and $q_\lambda := \frac{1}{s}s_\lambda$, where $s := \sum_{\lambda \in \mathrm{Irr}(U)} s_\lambda$. Observe that from (15.94) we get that $s^{-1} = \sum_{\lambda \in \mathrm{Irr}(U)} d_\lambda q_\lambda$. Therefore, with these notations we get that

$$\mu(\psi) = \min \sum_{\lambda \in \mathrm{Irr}(U)} d_\lambda q_\lambda \sum_{\lambda \in \mathrm{Irr}(U)} r_\lambda q_\lambda^{-1}, \tag{15.103}$$

where the minimum is over all probability distributions $\{q_\lambda\}_{\lambda \in \mathrm{Irr}(U)}$. From the Cauchy–Schwarz inequality we get that the minimum is given by

$$\mu(\psi) = \left( \sum_{\lambda \in \mathrm{Irr}(U)} \sqrt{d_\lambda r_\lambda} \right)^2. \tag{15.104}$$

Taking the log on both sides and substituting the expression $p_\lambda \left(\mathrm{Tr}\left[\sqrt{\rho_\lambda}\right]\right)^2$ for $r_\lambda$ gives

$$\log \mu(\psi) = 2\log \sum_{\lambda \in \mathrm{Irr}(U)} \sqrt{d_\lambda p_\lambda}\, \mathrm{Tr}\left[\sqrt{\rho_\lambda}\right]. \tag{15.105}$$

Finally, observe the expression inside the log on the right-hand side of (15.105) is given by $\mathrm{Tr}\left[\sqrt{\mathcal{G}(\psi)}\right]$. Hence, $\log \mu(\psi)$ is given by (15.96). This completes the proof. ∎

**Exercise 15.18.** *Use the formula in* (15.96) *to prove Theorem 15.5.*

## The Weighted Maximum Likelihood

In Section 15.2.3, we highlighted the importance of maximizing the likelihood of making the correct guess. However, when it comes to practical considerations, relying solely on the maximum likelihood density as a measure of success is not particularly advantageous, as it only rewards a completely accurate guess. A more general approach is to employ a payoff function $f: \mathbf{G} \times \mathbf{G} \mapsto \mathbb{R}_+$, denoted as $f(g', g)$, which determines the reward or payoff associated with guessing group element $g'$ when the true group element that relates between the parties' reference frames is $g$. By assuming a uniform prior for the signal states, the figure of merit for the alignment scheme can be defined as the average payoff

$$\mu_f(\rho) := \max \int_{\mathbf{G}} dg \int_{\mathbf{G}} dg'\, f(g, g')q(g'|g) \qquad \forall\, \rho \in \mathfrak{D}(A), \tag{15.106}$$

where the maximum is over all POVMs $\{\Lambda_{g'}\}_{g' \in \mathbf{G}} \subset \mathrm{Eff}(A)$, and $q(g'|g) := \mathrm{Tr}\left[\Lambda_{g'} U_g \rho U_g^*\right]$ is the probability of guessing $g'$ given that the actual element that relates between the parties' reference frames is $g$. Note that by taking $f(g', g) = \delta(g'g^{-1})$ to be the Dirac delta function we can get back the maximum likelihood function. We therefore call the function above the weighted maximum likelihood.

Note that we can write $\mu_f(\rho) = \max S(X|Y)_{\mathbf{q}}$ as given in (15.81), where $\mathbf{q} := \{q(g'|g)\}_{g,g'\in\mathbf{G}}$, the maximum is over all POVM, and $S(X|Y)_{\mathbf{q}} = L_f(\mathbf{q})$ is taken to be the linear functional

$$L_f(\mathbf{q}) := \int_{\mathbf{G}} dg \int_{\mathbf{G}} dg' \, f(g,g')q(g'|g). \tag{15.107}$$

Since the function $L_f(\mathbf{q})$ represents the certainty that Bob has about $g$, it has to be both left and right invariant. Fix $h \in \mathbf{G}$ and denote by $r(g'|g) := q(hg'|hg)$. Since $L_f$ is left-invariant, we have $L_f(\mathbf{r}) = L_f(\mathbf{q})$ for all conditional distributions $\mathbf{p}$. Since

$$L_f(\mathbf{r}) := \int_{\mathbf{G}} dg \int_{\mathbf{G}} dg' \, f(g,g')q(h^{-1}g'|h^{-1}g) = \int_{\mathbf{G}} dg \int_{\mathbf{G}} dg' \, f(hg,hg')q(g'|g), \tag{15.108}$$

the condition $L_f(\mathbf{r}) = L_f(\mathbf{q})$ is equivalent to

$$\int_{\mathbf{G}} dg \int_{\mathbf{G}} dg' \, \Big( f(hg,hg') - f(g,g') \Big) q(g'|g) = 0. \tag{15.109}$$

As this condition holds for all conditional probability distributions $q(g'|g)$, we conclude that $f$ itself is *left-invariant*. That is,

$$f(hg,hg') = f(g,g') \qquad \forall \, g,g',h \in \mathbf{G}. \tag{15.110}$$

The left-invariance property of $f$ is consistent with the intuition that the payoff function should exclusively depend on the *relative* transformation between the transmitted state, characterized by the group element $g$, and the measurement outcome, represented by the group element $g'$.

Following similar arguments, the right-invariance property of $L_f$ implies that the function $f$ itself is also *right-invariant*, that is,

$$f(kh^{-1},k'h^{-1}) = f(k,k') \qquad \forall \, k,k',h \in \mathbf{G}. \tag{15.111}$$

The fact that $f$ is both right- and left-invariant has the following consequences.

First, by taking $h = g^{-1}$ in (15.110) we get that

$$f(g,g') = f(e,g^{-1}g'). \tag{15.112}$$

That is, $f(g,g')$ can be viewed as a function of $g^{-1}g'$. We will denote this function by $p$ so that $f(g,g') = p(g^{-1}g')$. Now, since $f(g,g')$ is also right-invariant we get that for all $h,g \in \mathbf{G}$ we have $p(hgh^{-1}) = p(g)$. That is, $p$ is a class function as introduced in Definition C.9 of online version.

Since $f$ is nonnegative so is $p$, and consequently, it is natural to normalize $p$ such that $\int_{\mathbf{G}} dg \, p(g) = 1$. That is, $\{p(g)\}_{g\in\mathbf{G}}$ is a probability distribution over the group. Moreover, for a function $f$ that is both left and right-invariant, we have

$$\mu_f(\rho) = \max \int_{\mathbf{G}} dg \int_{\mathbf{G}} dg' \, p(g^{-1}g')\mathrm{Tr}\Big[\Lambda_{g'} U_g \rho U_g^*\Big]$$

$$\boxed{h := g^{-1}g'} \longrightarrow \quad = \max \int_{\mathbf{G}} dg' \int_{\mathbf{G}} dh \, p(h)\mathrm{Tr}\Big[\Lambda_{g'} U_{g'} U_h^* \rho U_h U_{g'}^*\Big] \tag{15.113}$$

$$\textbf{renaming } g' \textbf{ as } g \rightarrow \quad = \max \int_{\mathbf{G}} dg \, \mathrm{Tr}\Big[U_g^* \Lambda_g U_g \, \mathcal{G}_{\mathbf{p}}(\rho)\Big],$$

where $\mathcal{G}_{\mathbf{p}}$ is the weighted **G**-twirling as defined in (15.30). We therefore conclude that

$$\mu_f(\rho) = \mu\big(\mathcal{G}_{\mathbf{p}}(\rho)\big) = \min_{\sigma \in \mathrm{INV}_{\mathbf{G}}(A)} D_{\max}\big(\mathcal{G}_{\mathbf{p}}(\rho)\big\|\sigma\big), \qquad (15.114)$$

where the last equality follows from Theorem 15.7.

**Exercise 15.19.** *Explain why for $f$ that is not left- and right-invariant, the function $\mu_f$ is not necessarily a measure of asymmetry.*

## 15.3.2 The Relative Entropy of Asymmetry

For every normalized quantum divergence $\mathbb{D}$ the "distance" (as measured by $\mathbb{D}$) of a state $\rho \in \mathfrak{D}(A)$ to the set $\mathrm{INV}_{\mathbf{G}}(A)$ is a measure of asymmetry. That is, the function

$$\mathrm{Asy}_{\mathbb{D}}(\rho) = \min_{\sigma \in \mathrm{INV}_{\mathbf{G}}(A)} \mathbb{D}(\rho\|\sigma) \qquad (15.115)$$

is a measure of asymmetry. For certain choices of the divergence $\mathbb{D}$, the function (15.115) can be hard to compute. However, for the relative entropy it has a very simple form.

For any $\alpha \in [0, 2]$, the $\alpha$-Rényi relative entropy of asymmetry is defined as

$$\mathrm{Asy}_\alpha(\rho) := \min_{\sigma \in \mathrm{INV}_{\mathbf{G}}(A)} D_\alpha(\rho\|\sigma) \qquad \forall\, \rho \in \mathfrak{D}(A). \qquad (15.116)$$

In Exercise 15.4, we have seen that if $\rho$ is **G**-invariant, then for all $\alpha \in [0, 2]$ the state $\rho_\alpha := \rho^\alpha/\mathrm{Tr}[\rho^\alpha]$ is also **G**-invariant. Therefore, from Theorem 10.105 it follows that the $\alpha$-Rényi relative entropy of asymmetry is given by

$$\mathrm{Asy}_\alpha(\rho) = \frac{1}{\alpha - 1} \log \big\|\mathcal{G}\left(\rho^\alpha\right)\big\|_{1/\alpha} \qquad (15.117)$$

$$\mathbf{(10.108)} \rightarrow\ = H_{1/\alpha}\big(\mathcal{G}(\rho_\alpha)\big) - H_\alpha(\rho),$$

where $H_\alpha$ is the $\alpha$-Rényi entropy, and $\mathcal{G}$ is the **G**-twirling map that is also the resource destroying map of the QRT of asymmetry. The special case of $\alpha = 1$ is also known as the **G**-asymmetry of the state $\rho \in \mathfrak{D}(A)$ and is given by

$$\mathrm{Asy}(\rho) = H\big(\mathcal{G}(\rho)\big) - H(\rho). \qquad (15.118)$$

From Theorem 15.7 it follows that the log of the maximum likelihood, $\log \mu(\rho)$, can be viewed as the max relative entropy of asymmetry, in which $D_\alpha$ in (15.116) is replaced by $D_{\max}$. Since $D_{\max}$ is the largest relative entropy, the formula in (15.117) with $\alpha = 2$ can be used to provide a lower bound for $\log \mu(\rho)$. Specifically, we have

$$\log \mu(\rho) \geqslant H_{1/2}\left(\mathcal{G}\left(\frac{\rho^2}{\mathrm{Tr}[\rho^2]}\right)\right) - H_2(\rho). \qquad (15.119)$$

Remarkably, due to Theorem 15.8, this inequality becomes an equality on *all* pure states.

Despite this elegant expression for the **G**-asymmetry, in general, the **G**-asymmetry is not additive under tensor products. In fact, in the following theorem we show that its regularization is zero!

> **Theorem 15.9.** Let **G** be a finite or compact Lie group and let $\rho \in \mathfrak{D}(A)$. Then,
>
> $$\lim_{n \to \infty} \frac{1}{n} \mathrm{Asy}\left(\rho^{\otimes n}\right) = 0. \tag{15.120}$$

*Remark.* Theorem 15.9 underscores a notable constraint associated with using the **G**-asymmetry as a measure of asymmetry in quantum systems. It signals the necessity to investigate other measures capable of surmounting this limitation, particularly in the asymptotic regime where numerous copies of asymmetric states are considered. We will see that venturing into alternative measures will pave the way for a broader and more nuanced comprehension of asymmetry's nature and characteristics when approached from the perspective of the asymptotic domain.

**Proof**  According to (15.25) the action of the **G**-twirling on the state $\rho$ is given by

$$\mathcal{G}(\rho) = \sum_{x \in [d]} p_x U_{g_x}(\rho) U_{g_x}^*, \tag{15.121}$$

where $d$ is an integer satisfying $d \leqslant m^4$ (see Exercise 15.2). Therefore, from the von-Neumman property (7.120) we get that

$$H\big(\mathcal{G}(\rho)\big) \leqslant H(\rho) + H(\mathbf{p}) \leqslant H(\rho) + \log(d). \tag{15.122}$$

Thus, combining this with the definition in (15.118) gives $\mathrm{Asy}(\rho) \leqslant \log(d)$.

Now, fix $n \in \mathbb{N}$ and consider the action of the **G**-twirling on $\rho^{\otimes n}$:

$$\mathcal{G}_n\left(\rho^{\otimes n}\right) := \int_{\mathbf{G}} dg\, U_g^{\otimes n} \rho^{\otimes n} U_g^{\otimes n}. \tag{15.123}$$

Observe that the support of $\rho^{\otimes n}$ is a subspace of the symmetric subspace $\mathrm{Sym}_n(A)$. Thus, we can view $\rho^{\otimes n}$ as a positive semidefinite operator acting on $\mathrm{Sym}_n(A)$. Moreover, the map $g \mapsto U_g^{\otimes n}$ can also be seen as a projective unitary representation of **G** on the space $\mathrm{Sym}_n(A)$. Therefore, if we repeat the same steps that led to the inequality $\mathrm{Asy}(\rho) \leqslant \log(d)$ but with $\rho^{\otimes n}$ instead of $\rho$, we obtain

$$\mathrm{Asy}\left(\rho^{\otimes n}\right) \leqslant \log(d_n), \tag{15.124}$$

where $d_n$ is an integer no greater than the dimension of $\mathrm{Sym}_n(A)$ to the power 4 (see (15.26)). Combining this with the formula (C.166 of online version) for the dimension of the symmetric subspace we arrive at

$$\mathrm{Asy}\left(\rho^{\otimes n}\right) \leqslant 4\log \binom{n+m-1}{n} \tag{15.125}$$

$$(\mathbf{8.87})\rightarrow \ \leqslant 4m\log(n+1).$$

Hence,

$$\lim_{n \to \infty} \frac{1}{n} \mathrm{Asy}\left(\rho^{\otimes n}\right) \leqslant 4m \lim_{n \to \infty} \frac{1}{n}\log(n+1) = 0. \tag{15.126}$$

This completes the proof.  ∎

To illustrate Theorem 15.9, let's consider the group $U(1)$ and the pure state $\psi$ belonging to $\mathrm{Pure}(A)$, given by

$$|\psi^A\rangle = \sum_{x \in [m]} c_x |x\rangle, \tag{15.127}$$

where $\{|x\rangle\}_{x \in [m]}$ are the eigenvectors of the number operator $\hat{N}$, and each $c_x \in \mathbb{C}$. We can express $n$ copies of $\psi$ as

$$|\psi^{\otimes n}\rangle = \sum_{j \in [mn]} a_j |\phi_j^{A^n}\rangle, \tag{15.128}$$

where $a_j \in \mathbb{C}$ and $|\phi_j^{A^n}\rangle$ is the eigenvector of the total number operator $\hat{N}_{\mathrm{tot}}$ corresponding to the eigenvalue $j$, for each $j \in [mn]$. By applying the G-twirling to $\psi^{\otimes n}$, we obtain (see (15.22))

$$\mathcal{G}_n\left(\psi^{\otimes n}\right) = \sum_{j \in [mn]} |a_j|^2 \phi_j^{A^n}. \tag{15.129}$$

Denoting by $\mathbf{p} \in \mathrm{Prob}(mn)$ with components $p_j := |a_j|^2$ for each $j \in [mn]$, we conclude that

$$H\left(\mathcal{G}_n\left(\psi^{\otimes n}\right)\right) = H(\mathbf{p}) \leqslant \log(mn). \tag{15.130}$$

Therefore,

$$\lim_{n \to \infty} \frac{1}{n} H\left(\mathcal{G}_n\left(\psi^{\otimes n}\right)\right) \leqslant \lim_{n \to \infty} \frac{1}{n} \log(nm) = 0. \tag{15.131}$$

The key observation in this example is that the rank of $\mathcal{G}\left(\psi^{\otimes n}\right)$ grows linearly with $n$.

## The Weighted G-Asymmetry

Using the weighted G-twirling, we define the weighted G-asymmetry as

$$\mathrm{Asy}_{\mathbf{p}}(\rho) := H\left(\mathcal{G}_{\mathbf{p}}(\rho)\right) - H(\rho). \tag{15.132}$$

**Theorem 15.10.** The weighted **G**-asymmetry as defined in (15.132) is a measure of asymmetry.

**Proof** We begin by expressing $\mathrm{Asy}_{\mathbf{p}}(\rho)$ as the mutual information of the state $\sigma^{XA}$, defined as

$$\sigma^{XA} := \sum_{x \in [d]} p_x |x\rangle\langle x|^X \otimes U_{g_x} \rho^A U_{g_x}^*, \tag{15.133}$$

where $X$ is a classical system of dimension $d$. By definition, the mutual information is given by (see (13.177))

$$D\left(\sigma^{XA} \| \sigma^A \otimes \sigma^X\right) = H\left(\sigma^X\right) + H\left(\sigma^A\right) - H\left(\sigma^{XA}\right), \tag{15.134}$$

where $\sigma^A = \mathcal{G}_{\mathbf{p}}\left(\rho^A\right)$ and $H\left(\sigma^X\right) = H(\mathbf{p})$. From Exercise 7.15, we have

$$H\left(\sigma^{XA}\right) = H(\mathbf{p}) + \sum_x p_x H\left(U_{g_x}\rho^A U_{g_x}^*\right)$$

$$\boxed{H\left(U_{g_x}\rho^A U_{g_x}^*\right) = H\left(\rho^A\right)} \quad\longrightarrow\quad = H(\mathbf{p}) + H\left(\rho^A\right). \tag{15.135}$$

Combining these results, we get

$$\mathrm{Asy}_{\mathbf{p}}\left(\rho^A\right) = D\left(\sigma^{XA}\big\|\sigma^X \otimes \sigma^A\right). \tag{15.136}$$

Next, let $\mathcal{N} \in \mathrm{COV}_{\mathbf{G}}(A \to B)$ and observe that since $\mathcal{N}^{A\to B} \circ \mathcal{U}_g^A = \mathcal{U}_g^B \circ \mathcal{N}^{A\to B}$ for all $g \in \mathbf{G}$, we have

$$\mathcal{N}^{A\to B}\left(\sigma^{XA}\right) = \mathcal{N}^{A\to B} \circ \mathcal{E}^{A\to XA}\left(\rho^A\right)$$

$$= \sum_{x\in[d]} p_x |x\rangle\langle x|^X \otimes \mathcal{U}_{g_x}^B \circ \mathcal{N}^{A\to B}\left(\rho^A\right). \tag{15.137}$$

Therefore,

$$\mathrm{Asy}_{\mathbf{p}}\left(\mathcal{N}^{A\to B}\left(\rho^A\right)\right) = D\left(\mathcal{N}^{A\to B}\left(\sigma^{XA}\right)\big\|\sigma^X \otimes \mathcal{N}^{A\to B}\left(\sigma^A\right)\right)$$

$$\mathbf{DPI} \to \;\leqslant\; D\left(\sigma^{XA}\big\|\sigma^X \otimes \sigma^A\right) \tag{15.138}$$

$$= \mathrm{Asy}_{\mathbf{p}}\left(\rho^A\right).$$

This completes the proof. ∎

## The Mutual Information of Asymmetry

The relation (15.136) can be generalized to an arbitrary divergence $\mathbb{D}$ in the following way. For any $\rho \in \mathfrak{D}(A)$ and $\mathbf{p} \in \mathrm{Prob}(d)$, we define the *mutual information of asymmetry* of $\rho^A$ with respect to $\mathbf{p}$ as

$$\mathbb{I}_{\mathbf{p}}\left(\rho^A\right) := \mathbb{D}\left(\sigma^{XA}\big\|\sigma^X \otimes \sigma^A\right), \tag{15.139}$$

where $\sigma^{XA}$ is defined as in (15.133). Note that the same argument used in (15.138) can be repeated with $\mathbb{D}$ replacing $D$. Therefore, $\mathbb{I}_{\mathbf{p}}$ is also a measure of asymmetry.

By tracing out system $X$ in (15.139), we obtain the function (recall that the marginal $\sigma^A = \mathcal{G}_{\mathbf{p}}\left(\rho^A\right)$)

$$\mathbb{A}_{\mathbf{p}}(\rho) := \mathbb{D}\left(\rho\big\|\mathcal{G}_{\mathbf{p}}(\rho)\right) \qquad \forall\, \rho \in \mathfrak{D}(A). \tag{15.140}$$

This function is also a measure of asymmetry, since if $\rho \in \mathrm{INV}_{\mathbf{G}}(A)$, then $\mathbb{A}_{\mathbf{p}}(\rho) = 0$, and for any $\mathcal{N} \in \mathrm{COV}_{\mathbf{G}}(A \to B)$ we have

$$\mathbb{A}_{\mathbf{p}}(\mathcal{N}(\rho)) = \mathbb{D}\left(\mathcal{N}(\rho)\big\|\mathcal{G}_{\mathbf{p}} \circ \mathcal{N}(\rho)\right)$$

$$\mathcal{N} \text{ is G-covariant} = \mathbb{D}\left(\mathcal{N}(\rho)\big\|\mathcal{N} \circ \mathcal{G}_{\mathbf{p}}(\rho)\right) \tag{15.141}$$

$$\mathbf{DPI} \to \;\leqslant\; \mathbb{D}\left(\rho\big\|\mathcal{G}_{\mathbf{p}}(\rho)\right) = \mathbb{A}_{\mathbf{p}}(\rho).$$

Therefore, $\mathbb{A}_{\mathbf{p}}$ is a valid measure of asymmetry.

**Exercise 15.20.** *Show that for every divergence* $\mathbb{D}$, *all* $\rho \in \mathfrak{D}(A)$, *and all* $\mathbf{p} \in \mathrm{Prob}(d)$ *we have*

$$\mathbb{A}_{\mathbf{p}}(\rho) \leqslant \mathbb{I}_{\mathbf{p}}(\rho). \tag{15.142}$$

### 15.3.3 Derivatives of Asymmetry

Consider the last measure of asymmetry $\mathbb{A}_{\mathbf{p}}$ that we discussed in the previous subsection, and take $\mathbf{p}$ to be the vector $\mathbf{p} = (1, 0, \ldots, 0)^T$. Furthermore, denote by $g := g_1$ so that $\mathbb{A}_{\mathbf{p}}$ can be denoted as $\mathbb{A}_g$ and is given by

$$\mathbb{A}_g(\rho) := \mathbb{D}\left(\rho \middle\| U_g \rho U_g^*\right) \qquad \forall \, \rho \in \mathfrak{D}(A). \tag{15.143}$$

Since a Lie group is a differentiable manifold, we can take $g$ to be infinitesimally close to the identity element. Specifically, we can always choose $U_g = e^{it\Lambda}$, where $\Lambda$ is some generator of the representation $g \mapsto U_g$, and $t \geqslant 0$ is some phase. The derivative of asymmetry with respect to the divergence $\mathbb{D}$ and generator $\Lambda$ is defined for all $\rho \in \mathfrak{D}(A)$ as

$$\mathbb{D}_\Lambda(\rho) := \frac{d}{dt}\mathbb{D}\left(\rho \middle\| e^{it\Lambda}\rho e^{-it\Lambda}\right)\bigg|_{t=0}. \tag{15.144}$$

As we will see shortly, quite often the derivative on the right-hand side yields the constant zero function. In such cases, $\mathbb{D}_\Lambda(\rho)$ is defined in terms of the second derivative as

$$\mathbb{D}_\Lambda(\rho) := \frac{1}{2}\frac{d^2}{dt^2}\mathbb{D}\left(\rho \middle\| e^{it\Lambda}\rho e^{-it\Lambda}\right)\bigg|_{t=0}. \tag{15.145}$$

**Exercise 15.21.** *Show that the derivative of asymmetry as already defined is a measure of asymmetry. Hint: Use the fact that for each* $g \in \mathbf{G}$ *the function* $\mathbb{A}_g$ *is a measure of asymmetry.*

In order to compute the derivatives in (15.144) and (15.145) we will use the expression

$$e^{it\Lambda}\rho e^{-it\Lambda} = \rho + it[\Lambda, \rho] - \frac{1}{2}t^2[\Lambda, [\Lambda, \rho]] + O(t^3). \tag{15.146}$$

It will be convenient to use the notations $\sigma := i[\Lambda, \rho]$ and $\eta := -\frac{1}{2}[\Lambda, [\Lambda, \rho]]$ so that

$$e^{it\Lambda}\rho e^{-it\Lambda} = \rho + t\sigma + t^2\eta + O(t^3). \tag{15.147}$$

We now use this expansion to compute the derivatives of asymmetry for several examples.

## Differential Trace Distance of Asymmetry

As our first example, let $\mathbb{D} = T$ be the trace distance. In this case, $\mathbb{D}_\Lambda = T_\Lambda$ is called the *differential trace distance of asymmetry* and is given by

$$
\begin{aligned}
T_\Lambda(\rho) &= \lim_{t\to 0^+} \frac{1}{t} T\left(\rho, \rho + t\sigma\right) \\
&= \frac{1}{2} \lim_{t\to 0^+} \frac{1}{t} \left\| t\sigma \right\|_1 \\
&= \frac{1}{2} \left\| [\rho, \Lambda] \right\|_1 .
\end{aligned}
\tag{15.148}
$$

The differential trace distance measures the asymmetry in a state $\rho$ relative to a subgroup of $\mathbf{G}$ associated with a generator $\Lambda$. This measure depends on the coherence of $\rho$ over the eigenspaces of $\Lambda$, which is indicated by the nonzero commutator $[\rho, \Lambda]$. The question then arises as to which norm should be used to measure the commutator $[\rho, \Lambda]$ and thus, the asymmetry of $\rho$. While the answer to this question may not be immediately apparent, the above discussion indicated that the trace norm is the most appropriate measure for this purpose.

**Exercise 15.22.** *Let $\psi \in \mathrm{Pure}(A)$ and $\Lambda \in \mathrm{Herm}(A)$. Show that*

$$
T_\Lambda(\psi) = \sqrt{\langle \psi | \Lambda^2 | \psi \rangle - \langle \psi | \Lambda | \psi \rangle^2}.
\tag{15.149}
$$

*That is, on pure states, the differential trace distance of asymmetry reduces to the variance of the observable $\Lambda$.*


## The Wigner–Yanase–Dyson Skew Information

For our second example, we take $\mathbb{D} = D_\alpha$ to be the Petz quantum Rényi divergence of over $\alpha \in [0, 2]$ (see Definition 6.6). In this case, we will see that the first derivative in (15.144) is zero, so we will have to expend the function $D_\alpha\left(\rho \| e^{it\Lambda} \rho e^{-it\Lambda}\right)$ up to second order in $t$. Up to second order in $t$ we have

$$
D_\alpha\left(\rho \| \rho + t\sigma + t^2\eta\right) = \frac{1}{\alpha - 1} \log \mathrm{Tr}\left[ \rho^\alpha \left(\rho + t\sigma + t^2\eta\right)^{1-\alpha} \right].
\tag{15.150}
$$

In what follows, we will consider the spectral decomposition of $\rho^A = \sum_{x\in[m]} p_x |x\rangle\langle x|^A$ (where $\{|x\rangle\}_{x\in[m]}$ is the basis of $A$ consisting of the eigenvectors of $\rho^A$), and make use of the divided difference approach discussed in Appendix D.1 of online version. Particularly, the trace here has the form given in Corollary D.1 of online version with $g(t) := t^\alpha$ and $f(t) := t^{1-\alpha}$. Therefore, the function $h(t)$ as defined in Corollary D.1 of online version is given by

$$
h(t) := g(t) f'(t) = 1 - \alpha,
\tag{15.151}
$$

so that $h(\rho) = (1-\alpha) I^A$ is a constant function. As such, $\mathrm{Tr}\left[h(\rho)\sigma\right] = (1-\alpha)\mathrm{Tr}[\sigma] = 0$ and similarly $\mathrm{Tr}\left[h(\rho)\eta\right] = (1 - \alpha)\mathrm{Tr}[\eta] = 0$. Observe further that since $h$ is a constant function, $\mathcal{L}_h(\sigma) = 0$. Substituting all this into Corollary D.1 of online version we conclude that

$$
\begin{aligned}
D_\alpha\left(\rho \| \rho + t\sigma + t^2\eta\right) &= \frac{1}{\alpha - 1} \log\left(1 - \frac{1}{2}t^2 \mathrm{Tr}\left[\mathcal{L}_f(\sigma)\mathcal{L}_g(\sigma)\right]\right) + O(t^3) \\
&= \frac{1}{2}\frac{t^2}{1 - \alpha} \mathrm{Tr}\left[\mathcal{L}_f(\sigma)\mathcal{L}_g(\sigma)\right] + O(t^3),
\end{aligned}
\tag{15.152}
$$

where the self-adjoint linear maps $\mathcal{L}_f, \mathcal{L}_g \in \mathrm{Herm}(A \to A)$ are defined by (see Appendix D.1 of online version for more details)

$$
\begin{aligned}
\langle x|\mathcal{L}_f(\sigma)|y\rangle &= \frac{p_x^{1-\alpha} - p_y^{1-\alpha}}{p_x - p_y}\langle x|\sigma|y\rangle \\
&= -i\left(p_x^{1-\alpha} - p_y^{1-\alpha}\right)\langle x|\Lambda|y\rangle,
\end{aligned}
\tag{15.153}
$$

and similarly

$$
\langle y|\mathcal{L}_g(\sigma)|x\rangle = -i\left(p_y^\alpha - p_x^\alpha\right)\langle y|\Lambda|x\rangle.
\tag{15.154}
$$

Therefore,

$$
\begin{aligned}
\mathrm{Tr}\left[\mathcal{L}_f(\sigma)\mathcal{L}_g(\sigma)\right] &= \sum_{x,y\in[m]}\left(p_y^{1-\alpha} - p_x^{1-\alpha}\right)\left(p_y^\alpha - p_x^\alpha\right)|\langle x|\Lambda|y\rangle|^2 \\
&= 2\sum_{x,y\in[m]} p_x|\langle x|\Lambda|y\rangle|^2 - 2\sum_{x,y\in[m]} p_x^{1-\alpha}p_y^\alpha|\langle x|\Lambda|y\rangle|^2 \\
&= 2\mathrm{Tr}\left[\rho\Lambda^2\right] - 2\mathrm{Tr}\left[\rho^{1-\alpha}\Lambda\rho^\alpha\Lambda\right].
\end{aligned}
\tag{15.155}
$$

Hence, for all $\alpha \in [0, 2]$, $\rho \in \mathfrak{D}(A)$ and $\Lambda \in \mathrm{Herm}(A)$, the differential $\alpha$-Rényi divergence of asymmetry is given by

$$
D_{\Lambda,\alpha}(\rho) = \frac{1}{1-\alpha}\left(\mathrm{Tr}\left[\rho\Lambda^2\right] - \mathrm{Tr}\left[\rho^{1-\alpha}\Lambda\rho^\alpha\Lambda\right]\right).
\tag{15.156}
$$

The expression in the parenthesis in (15.156) (i.e. without the factor $1/(1-\alpha)$) is known as the Wigner–Yanase–Dyson skew information. Note that as the previous example, on pure states the Wigner–Yanase–Dyson skew information reduces to the variance of $\Lambda$.

**Exercise 15.23.** *Let $\rho = \psi \in \mathrm{Pure}(A)$ and $\Lambda \in \mathrm{Herm}(A)$. Show that*

$$
D_{\Lambda,\alpha}(\psi) = \frac{1}{1-\alpha}\left(\langle\psi|\Lambda^2|\psi\rangle - \langle\psi|\Lambda|\psi\rangle^2\right).
\tag{15.157}
$$

**Exercise 15.24.** *Show that for $\alpha \in (0, 1)$ the function $D_{\Lambda,\alpha}(\rho)$ is concave in $\rho$. Hint: Use Lieb's Concavity Theorem (see Theorem B.7 of online version).*

**Exercise 15.25.** *Show that for $\alpha = 1$ and $\rho \in \mathfrak{D}(A)$ we have*

$$
D_\Lambda(\rho) := \lim_{\alpha\to 1} D_{\Lambda,\alpha}(\rho) = \mathrm{Tr}\left[\Lambda^2\rho\log\rho\right] - \mathrm{Tr}\left[\Lambda\rho\Lambda\log\rho\right].
\tag{15.158}
$$

## The Quantum Fisher Information

As our third example, we take $\mathbb{D} = \tilde{D}_\alpha$ to be the minimal quantum divergence. In this case, we use the invariance of every relative entropy under unitary operations to get

$$\tilde{D}_\alpha\left(\rho \| U_g \rho U_g^*\right) = \tilde{D}_\alpha\left(U_g^* \rho U_g \| \rho\right). \tag{15.159}$$

Since $U_g^* \rho U_g = \rho - t\sigma + t^2\eta + O(t^3)$, we get that

$$\tilde{D}_\alpha\left(\rho \| U_g \rho U_g^*\right) = \mathbb{D}_\alpha(\rho - t\sigma + t^2\eta \| \rho) + O(t^3). \tag{15.160}$$

By definition,

$$\begin{aligned}
\mathbb{D}_\alpha(\rho - t\sigma + t^2\eta \| \rho) &= \frac{1}{\alpha - 1} \log \operatorname{Tr}\left[\left(\rho^{\frac{1-\alpha}{2\alpha}}\left(\rho - t\sigma + t^2\eta\right)\rho^{\frac{1-\alpha}{2\alpha}}\right)^\alpha\right] \\
&= \frac{1}{\alpha - 1} \log \operatorname{Tr}\left[\left(\tilde{\rho} - t\tilde{\sigma} + t^2\tilde{\eta}\right)^\alpha\right],
\end{aligned} \tag{15.161}$$

where

$$\tilde{\rho} := \rho^{\frac{1-\alpha}{2\alpha}} \rho \rho^{\frac{1-\alpha}{2\alpha}} = \rho^{1/\alpha}, \quad \tilde{\sigma} := \rho^{\frac{1-\alpha}{2\alpha}} \sigma \rho^{\frac{1-\alpha}{2\alpha}}, \quad \text{and} \quad \tilde{\eta} := \rho^{\frac{1-\alpha}{2\alpha}} \eta \rho^{\frac{1-\alpha}{2\alpha}}. \tag{15.162}$$

The trace $\operatorname{Tr}\left[\left(\tilde{\rho} - t\tilde{\sigma} + t^2\tilde{\eta}\right)^\alpha\right]$ has the form given in Corollary D.1 of online version with $g(t) := 1$ and $f(t) := t^\alpha$. Therefore, $h(t) := g(t)f'(t) = \alpha t^{\alpha-1}$, so that

$$\operatorname{Tr}\left[h(\tilde{\rho})\tilde{\sigma}\right] = \alpha\operatorname{Tr}[\tilde{\rho}^{\alpha-1}\tilde{\sigma}] = \alpha\operatorname{Tr}[\sigma] = 0 \tag{15.163}$$

and similarly $\operatorname{Tr}\left[h(\tilde{\rho})\tilde{\eta}\right] = \alpha\operatorname{Tr}[\eta] = 0$. Observe further that since $g$ is a constant function, $\mathcal{L}_g(\sigma) = 0$. Substituting all this into Corollary D.1 of online version we conclude that

$$\operatorname{Tr}\left[\left(\tilde{\rho} - t\tilde{\sigma} + t^2\tilde{\eta}\right)^\alpha\right] = 1 + \frac{1}{2}t^2\operatorname{Tr}\left[\tilde{\sigma}\mathcal{L}_h(\tilde{\sigma})\right] + O(t^3). \tag{15.164}$$

It will be convenient to denote $s := \frac{1}{\alpha}$. Since we assume that $\alpha \in [1/2, \infty]$, we have that $s \in [0, 2]$. Working with the eigenbasis of $\rho$ we get for all $x, y \in [m]$

$$\begin{aligned}
\langle x|\tilde{\sigma}|y\rangle &= p_x^{\frac{1-\alpha}{2\alpha}} p_y^{\frac{1-\alpha}{2\alpha}} \langle x|\sigma|y\rangle \\
&= i p_x^{(s-1)/2} p_y^{(s-1)/2}(p_y - p_x)\langle x|\Lambda|y\rangle.
\end{aligned} \tag{15.165}$$

$$\boxed{s := 1/\alpha, \ \sigma := i[\Lambda, \rho]}$$

Furthermore, since the eigenvalues of $\tilde{\rho}$ are $\{p_x^s\}_{x \in [m]}$, we get by definition of $\mathcal{L}_h$ that

$$\langle y|\mathcal{L}_h(\tilde{\sigma})|x\rangle = \frac{h\left(p_y^s\right) - h\left(p_x^s\right)}{p_y^s - p_x^s}\langle y|\tilde{\sigma}|x\rangle = \frac{1}{s}\frac{p_y^{1-s} - p_x^{1-s}}{p_y^s - p_x^s}\langle y|\tilde{\sigma}|x\rangle, \tag{15.166}$$

where the case $p_x = p_y$ is understood in terms of the limit

$$\lim_{p_y \to p_x} \frac{p_y^{1-s} - p_x^{1-s}}{p_y^s - p_x^s} = \frac{1-s}{s}p_x^{1-2s}. \tag{15.167}$$

With these expressions we get

$$\mathrm{Tr}\left[\tilde{\sigma}\,\mathcal{L}_h(\tilde{\sigma})\right] = \sum_{x,y\in[m]} \langle x|\tilde{\sigma}|y\rangle\langle y|\mathcal{L}_h(\tilde{\sigma})|x\rangle$$

**(15.165), (15.166)**$\rightarrow$ $$= \frac{1}{s}\sum_{x,y\in[m]}\frac{p_x^{s-1}-p_y^{s-1}}{p_y^{s}-p_x^{s}}(p_y-p_x)^2|\langle x|\Lambda|y\rangle|^2. \tag{15.168}$$

Note that since the limit $p_y \to p_x$ of the components in the sum (15.168) is zero, we can restrict the sum to all $x,y \in [m]$ that satisfies $p_x \neq p_y$. Hence, we conclude that

$$\tilde{D}_{\Lambda,s}(\rho) := \tilde{D}_{\Lambda,\alpha}(\rho) = \frac{1}{2(s-1)}\sum_{\substack{x,y\in[m]\\ p_x\neq p_y}}\frac{p_x^{s-1}-p_y^{s-1}}{p_x^{s}-p_y^{s}}(p_x-p_y)^2|\langle x|\Lambda|y\rangle|^2. \tag{15.169}$$

If $\rho$ is given by the pure state $\psi = |1\rangle\langle 1| \in \mathrm{Pure}(A)$, then $p_x = \delta_{1x}$ for all $x \in [m]$. In this case, for all $s \in [0,2]$

$$\tilde{D}_{\Lambda,s}(\psi) = \frac{1}{s-1}\sum_{y=2}^{m}|\langle\psi|\Lambda|y\rangle|^2 + \frac{1}{s-1}\sum_{x=2}^{m}|\langle x|\Lambda|\psi\rangle|^2$$

$$= \frac{2}{s-1}\sum_{x=2}^{m}|\langle x|\Lambda|\psi\rangle|^2$$

$$\boxed{\sum_{x=2}^{m}|x\rangle\langle x|^A = I^A - \psi^A}\longrightarrow\quad = \frac{2}{s-1}\left\langle\psi^A\left|\Lambda\left(I^A-\psi^A\right)\Lambda\right|\psi^A\right\rangle$$

$$= \frac{2}{s-1}\left(\langle\psi|\Lambda^2|\psi\rangle - \langle\psi|\Lambda|\psi\rangle^2\right). \tag{15.170}$$

The Fisher information is a measure of asymmetry that is obtained by setting $s = 2$ in the family of asymmetry monotones given in (15.169), which yields

$$F_\Lambda(\rho) := 4\tilde{D}_{\Lambda,2}(\rho) = 2\sum_{x,y\in[m]}\frac{(p_x-p_y)^2}{p_x+p_y}|\langle x|\Lambda|y\rangle|^2. \tag{15.171}$$

The Fisher information is a fundamental concept in statistics and information theory with numerous applications in quantum metrology and quantum information. It plays a crucial role in studying the ultimate limits of precision in quantum measurements, commonly referred to as the quantum Cramér-Rao bound. Moreover, the Fisher information is employed to measure the distinguishability of quantum states, to characterize the entanglement properties of multipartite systems, and to devise optimal quantum measurement strategies. In the field of quantum thermodynamics, it has an operational interpretation as the coherence cost of preparing a system in a particular state without any restrictions on work consumption.

**Exercise 15.26.** *Show that for $s = \alpha = 1$*

$$\tilde{D}_\Lambda(\rho) := \lim_{\alpha\to 1}\tilde{D}_{\Lambda,\alpha}(\rho) = \sum_{x,y\in[m]}(\log p_x - \log p_y)(p_x - p_y)|\langle x|\Lambda|y\rangle|^2. \tag{15.172}$$

**Exercise 15.27.** *Show that $\tilde{D}_{\Lambda,s}$ can be expressed as*

$$\tilde{D}_{\Lambda,s}(\rho) = \operatorname{Tr}\left[\Lambda \mathcal{L}_\rho(\Lambda)\right] \qquad \forall\, \Lambda \in \operatorname{Herm}(A), \quad \forall\, \rho \in \mathfrak{D}(A), \tag{15.173}$$

*where $\mathcal{L}_\rho \in \operatorname{Herm}(A \to A)$ is a self-adjoint linear map.*

**Exercise 15.28.** *Show that the Fisher information upper bound the Wigner–Yanase skew information; that is, show that for all $\rho \in \mathfrak{D}(A)$ and $\Lambda \in \operatorname{Herm}(A)$ we have*

$$F_\Lambda(\rho) \geqslant 4\left(\operatorname{Tr}\left[\rho\Lambda^2\right] - \operatorname{Tr}\left[\sqrt{\rho}\Lambda\sqrt{\rho}\Lambda\right]\right). \tag{15.174}$$

*Hint: Recall that $\tilde{D}_\alpha(\rho\|\sigma) \leqslant D_\alpha(\rho\|\sigma)$ for all $\rho, \sigma \in \mathfrak{D}(A)$.*

## 15.4 Manipulation of Pure Asymmetry

In this section, we aim to investigate the circumstances in which it is possible to transform an asymmetric pure state $\psi$ into another pure state $\phi$, using **G**-covariant maps. To achieve this, we must first identify all the pure states that are equivalent under symmetric (i.e. **G**-covariant) operations. This characterization is crucial for understanding the limitations and possibilities of pure-state transformations in this resource theory.

### 15.4.1 Characterization of Asymmetry Equivalence Classes

> **Definition 15.2.** Consider a projective unitary representation $g \mapsto U_g \in \mathfrak{L}(A)$, where **G** is a finite or compact Lie group.
>
> 1. We say that two states $\rho, \sigma \in \mathfrak{D}(A)$ are **G**-equivalent if there exists $\mathcal{E}, \mathcal{F} \in \operatorname{COV}_{\mathbf{G}}(A \to A)$ such that $\rho = \mathcal{E}(\sigma)$ and $\sigma = \mathcal{F}(\rho)$.
> 2. Two pure states $\psi, \phi \in \operatorname{Pure}(A)$ are called unitarily **G**-equivalent if there exists a **G**-invariant unitary matrix $V : A \to A$ such that $V|\psi\rangle = |\phi\rangle$.

We will also refer to the set of all states $\sigma \in \mathfrak{D}(A)$ that are **G**-equivalent to $\rho$ as the **G**-*equivalence class of* $\rho$. In this subsection, our focus is on characterizing the **G**-equivalence class of a pure state. To achieve this goal, we begin by characterizing unitarily **G**-equivalent states. Note that if $[V, U_g] = 0$ holds for all $g$, then $[V^*, U_g] = 0$ holds for all $g$ as well. Therefore, we can replace the condition $V|\psi\rangle = |\phi\rangle$ in the definition of unitarily **G**-equivalent states with $|\psi\rangle = V|\phi\rangle$.

**Exercise 15.29.** *Let $g \mapsto \mathbf{G}$ be a projective unitary representation of **G** and for every $\rho \in \mathfrak{D}(A)$ let*

$$\operatorname{Sym}_{\mathbf{G}}(\rho) := \{g \in \mathbf{G} : U_g \rho U_g^* = \rho\}. \tag{15.175}$$

*1. Show that* $\text{Sym}_{\mathbf{G}}(\rho)$ *is a subgroup of* **G**.

*2. Show that if* $\rho \xrightarrow{\text{G–COV}} \sigma$ *for some* $\rho, \sigma \in \mathfrak{D}(A)$, *then* $\text{Sym}_{\mathbf{G}}(\rho)$ *is a subgroup of* $\text{Sym}_{\mathbf{G}}(\sigma)$.

*3. Show that if* $\rho$ *and* $\sigma$ *are* **G**-*equivalent, then* $\text{Sym}_{\mathbf{G}}(\rho) = \text{Sym}_{\mathbf{G}}(\sigma)$.

Every projective unitary representation, $g \mapsto U_g^A$, corresponds to a decomposition of the Hilbert space $A$ as given in (C.44 of online version). Specifically, $A = \bigoplus_{\lambda \in \text{Irr}(U)} A_\lambda$, where $A_\lambda = B_\lambda \otimes C_\lambda$. Accordingly, every two pure states $\psi, \phi \in \text{Pure}(A)$ can be expressed as

$$|\psi^A\rangle = \sum_{\lambda \in \text{Irr}(U)} |\psi_\lambda^{B_\lambda C_\lambda}\rangle \quad \text{and} \quad |\phi^A\rangle = \sum_{\lambda \in \text{Irr}(U)} |\phi_\lambda^{B_\lambda C_\lambda}\rangle, \tag{15.176}$$

where $|\psi_\lambda\rangle, |\phi_\lambda\rangle \in B_\lambda C_\lambda$ are subnormalized states in $A_\lambda$. In the following theorem, we show that $\psi^A$ and $\phi^A$ are unitarily **G**-equivalent if the marginals of $\phi_\lambda^{B_\lambda C_\lambda}$ and $\phi_\lambda^{B_\lambda C_\lambda}$ on $B_\lambda$ are the same. In addition, the theorem characterizes states that are unitarily **G**-equivalent in terms of their characteristic functions. In Section C.7 of online version, we discuss various properties of characteristic functions, and we encourage readers who are unfamiliar with this material to read Section C.7 of online version before proceeding to the following theorem.

---

**Theorem 15.11.** Let $\psi, \phi \in \text{Pure}(A)$. The following statements are equivalent:

1. The states $\psi$ and $\phi$ are unitarily **G**-equivalent.
2. There exists a unitary matrix $V \in \mathfrak{U}(A)$ such that $V U_g |\psi\rangle = U_g |\phi\rangle, \forall\, g \in \mathbf{G}$.
3. The characteristic functions of $\psi$ and $\phi$ are the same: $\chi_\psi(g) = \chi_\phi(g), \forall\, g \in \mathbf{G}$.
4. Using (15.176), for all $\lambda \in \text{Irr}(U)$, $\text{Tr}_{C_\lambda}\left[\psi_\lambda^{B_\lambda C_\lambda}\right] = \text{Tr}_{C_\lambda}\left[\phi_\lambda^{B_\lambda C_\lambda}\right]$.

---

**Proof**  *The implication* 1 $\Rightarrow$ 2: Suppose that $\psi$ and $\phi$ are unitarily **G**-equivalent. Then there exists a **G**-invariant unitary matrix $V : A \to A$ such that $|\phi\rangle = V|\psi\rangle$. Since $V$ is **G**-invariant, after multiplying both sides by $U_g$ from the left we get $U_g |\phi\rangle = U_g V |\psi\rangle = V U_g |\psi\rangle$.

*The implication* 2 $\Rightarrow$ 3: For all $g \in \mathbf{G}$ we have

$$\langle\psi|U_g|\psi\rangle = \langle\psi|V^* V U_g|\psi\rangle$$
$$\boxed{V|\psi\rangle = |\phi\rangle} \longrightarrow = \langle\phi|V U_g|\psi\rangle \tag{15.177}$$
$$\boxed{V U_g|\psi\rangle = U_g|\phi\rangle} \longrightarrow = \langle\phi|U_g|\phi\rangle.$$

*The implication* 3 $\Rightarrow$ 4: Since we assume that $\chi_\psi(g) = \chi_\phi(g)$ for all $g \in \mathbf{G}$, we get from Theorem C.11 of online version

$$\mathrm{Tr}_{C_\lambda}\left[\psi_\lambda^{B_\lambda C_\lambda}\right] = |B_\lambda| \int_G dg\, \chi_\psi(g^{-1}) U_g^{(\lambda)}$$

$$\boxed{\chi_\psi(g^{-1}) = \chi_\phi(g^{-1})} \longrightarrow \quad = |B_\lambda| \int_G dg\, \chi_\phi(g^{-1}) U_g^{(\lambda)} \qquad (15.178)$$

$$\mathbf{(C.129)} \rightarrow \quad = \mathrm{Tr}_{C_\lambda}\left[\phi_\lambda^{B_\lambda C_\lambda}\right].$$

*The implication* $4 \Rightarrow 1$: Since for each $\lambda \in \mathrm{Irr}(U)$ the states $\psi_\lambda^{B_\lambda C_\lambda}$ and $\phi_\lambda^{B_\lambda C_\lambda}$ have the same marginal on representation space $B_\lambda$, there exists a unitary matrix $V_\lambda : C_\lambda \to C_\lambda$ such that

$$I^{B_\lambda} \otimes V_\lambda^{C_\lambda} |\psi_\lambda^{B_\lambda C_\lambda}\rangle = |\phi_\lambda^{B_\lambda C_\lambda}\rangle. \qquad (15.179)$$

Let $V : A \to A$ be the unitary matrix

$$V := \bigoplus_{\lambda \in \mathrm{Irr}(U)} I^{B_\lambda} \otimes V_\lambda^{C_\lambda}. \qquad (15.180)$$

Then, by definition, $|\phi^A\rangle = V|\psi^A\rangle$, and since for each $g \in G$ the unitary matrix $U_g$ has the form (C.45 of online version) we get that $[V, U_g] = 0$. Hence, $\psi^A$ and $\phi^A$ are unitarily **G**-equivalent. This completes the proof. $\blacksquare$

**Exercise 15.30.** *Let* $\psi \in \mathrm{Pure}(A)$ *and* $\phi \in \mathrm{Pure}(B)$*, where* $|B| \neq |A|$*, and consider two projective unitary representations* $g \mapsto U_g^A$ *and* $g \mapsto U_g^B$ *in A and B, respectively. Show that if* $\chi_\psi(g) = \chi_\phi(g)$ *for all* $g \in G$*, then* $\psi$ *and* $\phi$ *are* **G**-*equivalent.*

Theorem 15.11 characterizes unitarily **G**-equivalent states. However, from a resource theory perspective, two states belong to the same resource equivalence class if they are **G**-equivalent (not necessarily unitarily). Our next theorem characterizes this **G**-equivalence class, assuming that the states involved are **G**-regular.

> **Definition 15.3.** Let $\psi, \phi \in \mathrm{Pure}(A)$ and **G** be a group. We say that $\psi$ and $\phi$ are **G**-regular with respect to a representation $g \mapsto U_g$ if one of the following two conditions holds:
>
> 1. The group **G** is finite and there is no $g \in \mathbf{G}$ such that $\chi_\psi(g) = \chi_\phi(g) = 0$; that is, the functions $\chi_\psi, \chi_\phi : \mathbf{G} \to \mathbb{C}$ cannot take the zero value simultaneously.
> 2. The group **G** is a compact Lie group and there is no open set (other than the trivial one) $\mathfrak{C}$ of **G** for which $\chi_\psi(g) = \chi_\phi(g) = 0$ for all $g \in \mathfrak{C}$.

It is worth pointing out that every *connected* compact Lie group **G** satisfies the second condition in Definition 15.3. In fact, if **G** is connected, for any state $\psi \in \mathrm{Pure}(A)$, there cannot be an open neighbourhood $\mathfrak{C}$ of **G** for which $\chi_\psi(g) = 0$ for all $g \in \mathfrak{C}$. To see why, by contradiction, suppose that $\chi_\psi(g) = 0$ for all $g \in \mathfrak{C}$. Since the function $\chi_\psi : \mathbf{G} \to \mathbb{C}$ is analytic, the identity theorem in complex analysis implies that $\chi_\psi$ is the zero function, which contradicts the fact that $\chi_\psi(e) = 1$ for the identity element $e \in \mathbf{G}$.

**Exercise 15.31.** *Let* **G** *be a compact Lie group and let* $\psi, \phi \in \mathrm{Pure}(A)$ *be such that for all* $g \in$ **G** *there exists elements* $h, h' \in$ **H** *of a connected subgroup* **H** *of* **G** *for which* $|\chi_\psi(hgh')| + |\chi_\phi(hgh')| \neq 0$. *Show that* $\psi$ *and* $\phi$ *are* **G**-*regular.*

To clarify the notion of **G**-regular states, let's consider the group $O(2)$ of $2 \times 2$ real orthogonal matrices. This group is a compact Lie group, but it is not connected because matrices with determinant 1 are not continuously connected to matrices with determinant $-1$. Let **H** $:= SO(2)$ be the subgroup of $O(2)$ consisting of all the elements of $O(2)$ with determinant 1. The question we want to answer is: Is there a state $\psi \in \mathrm{Pure}(\mathbb{C}^2)$ such that $\chi_\psi(g) = 0$ for all $g \notin$ **H**?

To answer this question, we first observe that all the matrices $g \in O(2)$ with $\det(g) = -1$ have the form $\begin{pmatrix} \cos\theta & \sin\theta \\ \sin\theta & -\cos\theta \end{pmatrix}$ for some $\theta \in [0, 2\pi]$. Therefore, $\chi_\psi(g) = 0$ for all $g \notin$ **H** if and only if

$$\langle\psi| \begin{pmatrix} \cos\theta & \sin\theta \\ \sin\theta & -\cos\theta \end{pmatrix} |\psi\rangle = 0 \qquad \forall\, \theta \in [0, 2\pi]. \tag{15.181}$$

The only pure state that satisfies (15.181) is $|\psi\rangle = \frac{1}{\sqrt{2}}(|0\rangle + i|1\rangle)$. Therefore, in this example, the second condition in Definition 15.3 is satisfied except in the case where $|\psi\rangle = |\phi\rangle = \frac{1}{\sqrt{2}}(|0\rangle + i|1\rangle)$.

---

**Theorem 15.12.** Let $g \mapsto U_g$ be a projective unitary representation of **G**, and let $\psi, \phi \in \mathrm{Pure}(A)$ be **G**-regular states. The states $\psi$ and $\phi$ are **G**-equivalent if and only if there exists a one-dimensional unitary representation of **G**, $\{e^{i\theta_g}\}_{g \in \mathbf{G}}$, such that for all $g \in$ **G**

$$\langle\psi|U_g|\psi\rangle = e^{i\theta_g}\langle\phi|U_g|\phi\rangle. \tag{15.182}$$

---

*Remark.* We will see in the following proof that for any finite or compact (not necessarily connected) Lie group **G**, if (15.182) holds, then $\phi$ and $\psi$ are **G**-equivalent. Therefore, we only need the assumption that $\psi$ and $\phi$ are **G**-regular for the converse part. In Section D.6 of online version we provide additional observations for the case that $\psi$ and $\phi$ are not **G**-regular. Moreover, it is worth noting that semi-simple compact Lie groups, such as $SU(2)$, do not have any nontrivial one-dimensional representation. Therefore, it follows from Theorem 15.12 and the preceding theorem that for such groups, the following statements are all equivalent:

1. $\psi$ and $\phi$ are **G**-equivalent.
2. $\psi$ and $\phi$ are unitarily **G**-equivalent.
3. $\psi$ and $\phi$ have the same characteristic function.

**Proof** We first prove that the condition (15.182) implies that $\psi$ and $\phi$ are **G**-equivalent. Let $E$ be a qubit system (i.e. $|E| = 2$) with an orthonormal basis $\{|\varphi_1\rangle, |\varphi_2\rangle\}$, and let $g \mapsto U_g^E$ be the nonprojective unitary representation of **G** with

$$U_g^E := \varphi_1^E + e^{i\theta_g}\varphi_2^E, \tag{15.183}$$

where $g \mapsto e^{i\theta_g}$ is the one-dimensional representation of $\mathbf{G}$ that appears in (15.182). Clearly, with respect to the representation $g \mapsto U_g^E$, the states $\varphi_1^E$ and $\varphi_2^E$ are $\mathbf{G}$-invariant, and for all $g \in \mathbf{G}$, $\chi_{\varphi_1}(g) = 1$ and $\chi_{\varphi_2}(g) = e^{i\theta_g}$.

Combining this with (15.182) we conclude that $|\psi\rangle^A|\varphi_1\rangle^E$ has the same characteristic function as $|\phi\rangle^A|\varphi_2\rangle^E$. From Theorem 15.11 it follows that there exists a $\mathbf{G}$-invariant unitary $V$ (with respect to the representation $g \mapsto U_g^A \otimes U_g^E$) such that

$$\psi^A \otimes \varphi_1^E = V\left(\phi^A \otimes \varphi_2^E\right)V^*. \tag{15.184}$$

Define a quantum channel $\mathcal{E}(\rho^A) := \mathrm{Tr}_E\left[V\left(\rho^A \otimes \varphi_2^E\right)V^*\right]$ for all $\rho \in \mathcal{L}(A)$. By definition $\mathcal{E}(\phi^A) = \psi^A$, and the $\mathbf{G}$-invariance of $V$ and $\varphi_2^E$ implies that $\mathcal{E} \in \mathrm{COV}_{\mathbf{G}}(A \to A)$. Therefore, $\phi$ can be converted to $\psi$ by symmetric (i.e. covariant) operations. Following similar lines we also get that $\psi$ can be converted to $\phi$ by $\mathbf{G}$-covariant operations. Hence, $\psi$ and $\phi$ are $\mathbf{G}$-equivalent.

For the converse part of the proof, suppose there exists a $\mathbf{G}$-covariant channel mapping $\psi$ to $\phi$ and another $\mathbf{G}$-covariant channel that maps $\phi$ to $\psi$. From the covariant version of Stinespring delation theorem (see Theorem 15.3) there exists two isometries $V_1 \colon A \to AE$ and $V_2 \colon A \to A\tilde{E}$, each satisfying (15.51) for all $g \in \mathbf{G}$, and with the property that

$$V_1|\psi\rangle^A = |\phi\rangle^A|\varphi_1\rangle^E \tag{15.185}$$

$$V_2|\phi\rangle^A = |\psi\rangle^A|\varphi_2\rangle^{\tilde{E}}, \tag{15.186}$$

for some $\varphi_1, \varphi_2 \in \mathrm{Pure}(E)$. Since, $V_1$ and $V_2$ satisfy (15.51) for all $g \in \mathbf{G}$, (15.185) and (15.186) imply that for all $g \in \mathbf{G}$

$$\begin{aligned} \chi_\psi(g) &= \chi_\phi(g)\chi_{\varphi_1}(g) \\ \chi_\phi(g) &= \chi_\psi(g)\chi_{\varphi_2}(g). \end{aligned} \tag{15.187}$$

Combining these two equations yields in particular that

$$\begin{aligned} \chi_\psi(g) &= \chi_\psi(g)\chi_{\varphi_1}(g)\chi_{\varphi_2}(g) \\ \chi_\phi(g) &= \chi_\phi(g)\chi_{\varphi_1}(g)\chi_{\varphi_2}(g). \end{aligned} \tag{15.188}$$

First, if for all $g \in \mathbf{G}$ $\chi_\psi(g) \neq 0$ and/or $\chi_\phi(g) \neq 0$, then $\chi_{\varphi_1}(g)\chi_{\varphi_2}(g) = 1$. Since the absolute value of characteristic functions cannot exceed 1, it follows that $|\chi_{\varphi_1}(g)| = |\chi_{\varphi_2}(g)| = 1$ for all $g \in \mathbf{G}$. Therefore, from Lemma C.1 of online version we get that the states $\varphi_1$ and $\varphi_2$ are $\mathbf{G}$-invariant in this case. Second, suppose $\mathbf{G}$ is a compact Lie group and suppose by contradiction that there exists $g \in \mathbf{G}$ such that $\chi_{\varphi_1}(g)\chi_{\varphi_2}(g) \neq 1$. Then, from the continuity of the characteristic function, there exists a neighborhood $\mathfrak{C} \subset \mathbf{G}$ of $g$ such that for all $g' \in \mathfrak{C}$ we have $\chi_{\varphi_1}(g')\chi_{\varphi_2}(g') \neq 1$. From (15.188) it then follows that $\chi_\psi(g') = \chi_\phi(g') = 0$ for all $g' \in \mathfrak{C}$ in contradiction with the assumption that $\psi$ and $\phi$ are $\mathbf{G}$-regular. Therefore, also in this case $\chi_{\varphi_1}(g)\chi_{\varphi_2}(g) = 1$ for all $g \in \mathbf{G}$, so that $\varphi_1$ and $\varphi_2$ are $\mathbf{G}$-invariant.

To summarize, in both cases we can express the characteristic functions of $\varphi_1$ and $\varphi_2$ as $\chi_{\varphi_1}(g) = e^{i\theta_g}$ and $\chi_{\varphi_2}(g) = e^{-i\theta_g}$, where $g \mapsto e^{i\theta_g}$ is a one-dimensional unitary representation of $\mathbf{G}$ (see Exercise C.13 of online version). Substituting this into (15.187) completes the proof. ∎

As an example, consider the group $U(1)$ and an arbitrary state

$$|\tilde{\psi}\rangle = \sum_{n\in[m]} \lambda_n |n\rangle, \tag{15.189}$$

where $\{|n\rangle\}_{n\in\mathbb{Z}}$ is the eigenbasis of the number operator. The characteristic function of $\tilde{\psi}$ is given by

$$\chi_{\tilde{\psi}}(\theta) = \langle\tilde{\psi}|e^{i\theta\hat{N}}|\tilde{\psi}\rangle = \sum_{n\in[m]} |\lambda_n|^2 e^{i\theta n}. \tag{15.190}$$

Since the characteristic function of $\tilde{\psi}$ depends only on the absolute values of the coefficients $\{\lambda_n\}_{n\in[m]}$, we get from Theorem 15.11 (particularly, the equivalence of the first and third statements of this theorem) that $\tilde{\psi}$ is unitarily $\mathbf{G}$-equivalent to the state

$$|\psi\rangle = \sum_{n\in[m]} \sqrt{p_\psi(n)}|n\rangle, \tag{15.191}$$

where $p_\psi(n) := |\lambda_n|^2$. Therefore, similar to the Schmidt decomposition in entanglement theory, in the QRT of $U(1)$ asymmetry, all pure states are unitarily $U(1)$-equivalent to a state of the form in (15.191). Therefore, in this resource theory, the resource is characterized by the probability distribution $p_\psi : \mathbb{Z} \to [0,1]$.

Every one-dimensional unitary representation of $U(1)$ is determined by some integer $k \in \mathbb{Z}$ and a mapping $\theta \mapsto e^{i\theta k}$. Therefore, according to Theorem 15.12 two pure states $\psi$ and $\phi$ are $\mathbf{G}$-equivalent if and only if there exists $k \in \mathbb{Z}$ such that $\chi_\psi(\theta) = e^{i\theta k}\chi_\phi(\theta)$ for all $\theta \in U(1)$. This condition can be written as

$$\sum_{n\in\mathbb{Z}} p_\psi(n)e^{i\theta n} = e^{i\theta k}\sum_{n\in\mathbb{Z}} p_\phi(n)e^{i\theta n} \qquad \forall\, \theta \in U(1), \tag{15.192}$$

where $p_\psi, p_\phi : \mathbb{Z} \to [0,1]$ are the probability distributions associated with $\psi$ and $\phi$, respectively. Using the Fourier transform (see Exercise 15.32) we get that the condition in (15.192) can be expressed as

$$p_\psi(n) = p_\phi(n+k). \tag{15.193}$$

As a specific example, observe that the states $|\psi\rangle = \frac{1}{\sqrt{2}}(|0\rangle+|1\rangle)$ and $|\phi\rangle = \frac{1}{\sqrt{2}}(|1\rangle+|2\rangle)$ are $\mathbf{G}$-equivalent since in this case $p_\psi(n) = p_\phi(n-1)$.

**Exercise 15.32.** *Show that the condition in* (15.192) *is equivalent to one in* (15.193). *Hint: Apply a Fourier transform on both sides of* (15.192).

## 15.4.2 Deterministic Transformations

In this subsection, we will explore the conditions under which a pure asymmetric state can be deterministically converted into another state by **G**-covariant operations. In entanglement theory, we learned that such conversions under LOCC are determined by Nielsen's majorization theorem. However, we will show that in the QRT of asymmetry convertibility is actually determined by the concept of a positive-definite function on a group. For readers who are not familiar with this topic, we provide a review in Section C.8 of online version.

> **Theorem 15.13.** Let $\psi, \phi \in \mathrm{Pure}(A)$ be pure states. There exists a **G**-covariant map $\mathcal{E} \in \mathrm{COV}_{\mathbf{G}}(A \to A)$ such that $\phi = \mathcal{E}(\psi)$ if and only if there exists a positive definite function $f : \mathbf{G} \to \mathbb{C}$ such that $\chi_\psi(g) = \chi_\phi(g)f(g)$ for all $g \in \mathbf{G}$.

*Remark.* If $\chi_\phi(g) \neq 0$ for all $g \in \mathbf{G}$, then this theorem states in this case that $\psi$ can be converted to $\phi$ by symmetric operations if and only if $\chi_\psi(g)/\chi_\phi(g)$ is a positive definite function over $\mathbf{G}$.

**Proof** Suppose first that $\psi \xrightarrow{\ \mathbf{G-COV}\ } \phi$. In the derivation of the relation in (15.187), using the covariant Stinespring dilation theorem we showed that the condition $\psi \xrightarrow{\ \mathbf{G-COV}\ } \phi$ implies that there exists a pure state $\varphi \in \mathrm{Pure}(A)$ such that

$$\chi_\psi(g) = \chi_\phi(g)\chi_\varphi(g). \tag{15.194}$$

Hence, taking $f(g) := \chi_\varphi(g)$ we get $\chi_\psi(g) = \chi_\phi(g)f(g)$. Finally, observe that the characteristic function $\chi_\varphi : \mathbf{G}\mathbb{C}$ is a normalized positive definite function over $\mathbf{G}$ (see Theorem C.12 of online version).

Conversely, suppose $\chi_\psi(g) = \chi_\phi(g)f(g)$ for some positive definite function $f$. Since for $g = e$ we get $f(e) = \chi_\psi(e)/\chi_\phi(e) = 1$, the function $f$ is normalized so that according to Theorem C12 of online version it corresponds to some characteristic function $f(g) = \langle\varphi|U_g^E|\varphi\rangle$, where $g \mapsto U_g^E$ is some unitary representation of $\mathbf{G}$ on some Hilbert space $E$. Moreover, there exists a **G**-invariant state $|0\rangle \in E$ whose characteristic function is constant and equal to 1 for all group elements. Therefore, from the relation $\chi_\psi(g) = \chi_\phi(g)f(g)$ we get that the states $|\psi\rangle^A|0\rangle^E$ and $|\phi\rangle^A|\varphi\rangle^E$ have the same characteristic function. Therefore, there exists a **G**-invariant unitary $V : AE \to AE$ such that $V\left(|\psi\rangle^A|0\rangle^E\right) = |\phi\rangle^A|\varphi\rangle^E$. Taking the trace over $E$ on both sides demonstrates that $\psi$ can be converted to $\phi$ by a **G**-covariant channel. $\blacksquare$

**Exercise 15.33.** *Consider two states $\psi, \phi \in \mathrm{Pure}(A)$ and suppose $\psi$ has the property that $\chi_\psi(g) = 0$ for all $g \in \mathbf{G}$ such that $g \neq e$. Show that $\psi \xrightarrow{\ \mathbf{G-COV}\ } \phi$. In other words, $\psi$ with such a property is a maximal resource state.*

## Example: The Cyclic Group $\mathbb{Z}_n$

Let $n \in \mathbb{N}$ be a fixed integer, and consider the group $\mathbb{Z}_n$, which represents a cyclic group of order $n$. The group $\mathbb{Z}_n$ is the group of integers $\{0, \ldots, n-1\}$, where the group operation is addition modulo $n$, and 0 is the identity element of the group. It is well known that every cyclic group or order $n$ is isomorphic to $\mathbb{Z}_n$. Since $\mathbb{Z}_n$ is an Abelian group, all of its irreps are one dimensional. Each irrep can be uniquely identified by an integer $y \in \mathbb{Z}_n$, under the action $x \mapsto e^{i\frac{2\pi yx}{n}}$ for all $x \in \mathbb{Z}_n$.

Consider a (nonprojective) unitary representation of $\mathbb{Z}_n$ in the space $A = \mathbb{C}^n$, given for all $x \in \mathbb{Z}_n$ by $x \mapsto U_x$, where

$$U_x := \sum_{y \in \mathbb{Z}_n} e^{i\frac{2\pi yx}{n}} |y\rangle\langle y|. \tag{15.195}$$

Note that this unitary representation of $\mathbb{Z}_n$ is composed of a direct sum of its irreps, each occurring with multiplicity 1.

We would like to find the conditions under which the quantum pure state $|\psi\rangle = \sum_{x=0}^{n-1} \sqrt{p_x}|x\rangle$ can be converted to another pure state $|\phi\rangle := \sum_{x=0}^{n-1} \sqrt{q_x}|x\rangle$ by $\mathbb{Z}_n$-covariant operations. Observe that the characteristic function of $|\psi\rangle$ is given for any $x \in \mathbb{Z}_n$ by

$$\chi_\psi(x) = \langle\psi|U_x|\psi\rangle = \sum_{y \in \mathbb{Z}_n} p_y e^{i\frac{2\pi yx}{n}}. \tag{15.196}$$

Similarly, $\chi_\phi(x)$ can be expressed as in (15.196) with $q_y$ replacing $p_y$. The above equation demonstrates that the characteristic function is nothing but the discrete Fourier transform of the sequence $\{p_0, \ldots, p_{n-1}\}$.

Theorem 15.13 implies that $\psi$ can be converted to $\phi$ by $\mathbb{Z}_n$-covariant operations if and only if the function $x \mapsto \chi_\psi(x)/\chi_\phi(x)$ is a positive definite function over $\mathbb{Z}_n$. From Exercise C.18 of online version we have that a function $f : \mathbb{Z}_n \to \mathbb{C}$ is positive definite if and only if its (discrete) Fourier transform is positive. We therefore conclude that $\psi \xrightarrow{\mathbb{Z}_n - \mathrm{COV}} \phi$ if and only if

$$\sum_{x \in \mathbb{Z}_n} \frac{\chi_\psi(x)}{\chi_\phi(x)} e^{i\frac{2\pi xy}{n}} \geqslant 0 \qquad \forall\, y \in \mathbb{Z}_n. \tag{15.197}$$

To illustrate the condition in (15.197), we consider now the case $n = 2$. For $n = 2$ the condition in (15.197) gives for $y \in \mathbb{Z}_2 = \{0, 1\}$

$$0 \leqslant \frac{\chi_\psi(0)}{\chi_\phi(0)} + (-1)^y \frac{\chi_\psi(1)}{\chi_\phi(1)} = 1 + (-1)^y \frac{p_0 - p_1}{q_0 - q_1}. \tag{15.198}$$

This condition can be expressed as

$$\frac{|p_0 - p_1|}{|q_0 - q_1|} \leqslant 1, \tag{15.199}$$

which is equivalent to

$$\max\{p_0, p_1\} \leqslant \max\{q_0, q_1\}. \tag{15.200}$$

The condition we obtained for the case $n = 2$ can be expressed also as $\psi \xrightarrow{\mathbb{Z}_2\text{-COV}} \phi$ if and only if $\mathbf{q} \succ \mathbf{p}$, where $\mathbf{p} := (p_0, p_1)^T$ and $\mathbf{q} := (q_0, q_1)^T$. More generally, for arbitrary integer $n \in \mathbb{N}$ we have that if $\psi \xrightarrow{\mathbb{Z}_2\text{-COV}} \phi$, then necessarily $\mathbf{q} \succ \mathbf{p}$. To see why, observe that the relation $\chi_\psi(x) = \chi_\phi(x) f(x)$ implies that $f(0) = 1$ and $f(x)$ itself can be expressed as a Fourier series

$$f(x) = \sum_{z \in \mathbb{Z}_n} r_z e^{i\frac{2\pi z x}{n}}, \tag{15.201}$$

where $r_z \in \mathbb{R}$. Since $f$ is positive definition over $\mathbb{Z}_n$, we must have that $r_z \geqslant 0$ for all $z \in \mathbb{Z}_n$. Since $f(0) = 1$ we conclude that $\{r_z\}_{z \in \mathbb{Z}_n}$ is a probability distribution. Substituting the expression of (15.201) for $f(x)$ into the relation $\chi_\psi(x) = \chi_\phi(x) f(x)$ gives

$$\sum_{y \in \mathbb{Z}_n} p_y e^{i\frac{2\pi y x}{n}} = \sum_{w, z \in \mathbb{Z}_n} q_w r_z e^{i\frac{2\pi (z+w)x}{n}}. \tag{15.202}$$

Considering all summations and subtractions to be modulus $n$, we change variables on the right-hand side of (15.202) by denoting $y = z + w$ so that

$$\sum_{y \in \mathbb{Z}_n} p_y e^{i\frac{2\pi y x}{n}} = \sum_{y, w \in \mathbb{Z}_n} q_w r_{y-w} e^{i\frac{2\pi y x}{n}}. \tag{15.203}$$

Hence, (15.203) implies that for all $y \in \mathbb{Z}_n$

$$p_y = \sum_{w \in \mathbb{Z}_n} q_w r_{y-w}. \tag{15.204}$$

Denoting by $\mathbf{p} := (p_0, \ldots, p_{n-1})^T$, $\mathbf{q} := (q_0, \ldots, q_{n-1})^T$, and by $R$ the $n \times n$ matrix whose $(y, w)$ component is $r_{y-w}$ we can express Equation (15.204) as $\mathbf{p} = R\mathbf{q}$. Observe that by definition $R$ is doubly stochastic so that $\mathbf{q} \succ \mathbf{p}$.

The relation given in (15.204) can be used to provide alternative necessary and sufficient conditions for $\psi \xrightarrow{\mathbb{Z}_n\text{-COV}} \phi$. First, note that by rewriting (15.204) with $z := y - w$ we get that $\psi \xrightarrow{\mathbb{Z}_n\text{-COV}} \phi$ if and only if there exists $\mathbf{r} = (r_0, \ldots, r_{n-1})^T \in \mathrm{Prob}(n)$ such that

$$p_y = \sum_{z \in \mathbb{Z}_n} q_{y-z} r_z. \tag{15.205}$$

Next, observe that this equation can be expressed simply as $\mathbf{p} = Q\mathbf{r}$, where $Q$ is an $n \times n$ matrix whose $(y, z)$ component is $q_{y-z}$. Hence, assuming $Q$ is invertible we conclude that $\psi \xrightarrow{\mathbb{Z}_n\text{-COV}} \phi$ if and only if $Q^{-1}\mathbf{p} \geqslant 0$, where the inequality is entry-wise. In order to avoid the computation of $Q^{-1}$ we can also use the Cramer's rule as we discuss now.

The matrix $Q$ as already defined is known as a *circulant* matrix. The eigenvalues of such matrices are given by the discrete Fourier transforms. Specifically, the $x$-th eigenvalue of $Q$ is given by

$$\lambda_x(Q) = \chi_\phi(x) = \sum_{y \in \mathbb{Z}_n} p_y e^{i\frac{2\pi y x}{n}} \qquad \forall\, x \in \mathbb{Z}_n. \tag{15.206}$$

The matrix $Q$ is also doubly stochastic so its determinant is in the interval $[0, 1]$. Therefore, as long as $\chi_\phi(x) \neq 0$ for all $x \in \mathbb{Z}_n$, we have $\det(Q) > 0$. Next, for any $x \in \mathbb{Z}_n$ let $Q_x$ be the matrix obtained from $Q$ by replacing the $x$-th column with the column $(p_0, p_1, \ldots, p_{n-1})^T$. Then, assuming $\det(Q) > 0$ we get from Cramer's rule that $\psi \xrightarrow{\mathbb{Z}_n-\mathrm{COV}} \phi$ if and only if $\det(Q_x) \geqslant 0$ for all $x \in \mathbb{Z}_n$.

As a specific example, consider the case $n = 3$. For this case the matrix $Q$ has the form

$$
Q = \begin{pmatrix} q_0 & q_2 & q_1 \\ q_1 & q_0 & q_2 \\ q_2 & q_1 & q_0 \end{pmatrix}. \tag{15.207}
$$

Observe that $\det(Q) \geqslant 0$ with equality if and only if $q_0 = q_1 = q_2$. That is, if $|\phi\rangle \neq \frac{1}{\sqrt{3}}(|0\rangle + |1\rangle + |2\rangle)$, then $\det(Q) > 0$. Hence, $\psi \xrightarrow{\mathbb{Z}_3-\mathrm{COV}} \phi$ if and only if the following three conditions hold:

$$
\begin{aligned}
\det(Q_0) &= p_0(q_0^2 - q_1 q_2) + p_1(q_1^2 - q_0 q_2) + p_2(q_2^2 - q_0 q_1) \geqslant 0 \\
\det(Q_1) &= p_0(q_2^2 - q_0 q_1) + p_1(q_0^2 - q_1 q_2) + p_2(q_1^2 - q_0 q_2) \geqslant 0 \\
\det(Q_2) &= p_0(q_1^2 - q_0 q_2) + p_1(q_2^2 - q_0 q_1) + p_2(q_0^2 - q_1 q_2) \geqslant 0.
\end{aligned} \tag{15.208}
$$

**Exercise 15.34.** *Show that for every $\psi \in \mathrm{Pure}(\mathbb{C}^n)$ we have $\Phi \xrightarrow{\mathbb{Z}_n-\mathrm{COV}} \psi$, where $|\Phi\rangle := \frac{1}{\sqrt{n}} \sum_{x=0}^{n-1} |x\rangle$. In other words, $\Phi$ is a state with maximal $\mathbb{Z}_n$-asymmetry.*

**Exercise 15.35.** *Consider the case $n = 3$, and let $|\psi\rangle = \sum_{x=0}^{3} \sqrt{p_x}|x\rangle$ and $|\phi\rangle = \sum_{x=0}^{3} \sqrt{q_x}|x\rangle$.*

1. *Show that if $q_1 = q_2$, then $\psi \xrightarrow{\mathbb{Z}_3-\mathrm{COV}} \phi$ if and only if $\mathbf{q} \succ \mathbf{p}$.*
2. *Show that for $\mathbf{p} = (5/12, 7/24, 7/24)^T$ and $\mathbf{q} = (5/12, 1/3, 1/4)^T$ it is not possible to convert $\psi$ to $\phi$ by $\mathbb{Z}_3$-covariant operations even though $\mathbf{q} \succ \mathbf{p}$.*

### 15.4.3 Catalysis

In every resource theory, if the state $\psi$ cannot be deterministically transformed into the state $\phi$ using the limited set of operations, the use of a catalyst provides a potential solution. As we explored in earlier chapters, a catalyst refers to an additional system that is initially prepared in a state not compatible with the constraints of the resource theory but must be restored to its original state at the conclusion of the process. An illustrative example can be found in the resource theory of entanglement, as discussed in Section 12.2.3, where we observed that certain conversions between states are prohibited under LOCC. However, by employing LOCC alongside a suitable catalyst, such conversions become achievable.

This notion of catalysis vividly demonstrates the significant variations encountered within the resource theory of asymmetry, contingent upon the choice of groups involved. Specifically, we will demonstrate that a catalyst holds no utility for a connected compact Lie group, whereas for a finite group, a catalyst always exists.

> **Theorem 15.14.** Let $\psi, \phi \in \mathrm{Pure}(A)$, and $\mathbf{G}$ be a group with a projective unitary representation $g \mapsto U_g^A$. Suppose further that $\psi$ cannot be converted to $\phi$ by $\mathbf{G}$-covariant operations. Then,
>
> 1. If $\mathbf{G}$ is a finite group, then there exists an ancillary system $C$ along with a projective unitary representation $g \mapsto U_g^C$, and a state $\varphi \in \mathrm{Pure}(C)$, such that
> $$\psi^A \otimes \varphi^C \xrightarrow{\;\mathrm{G-COV}\;} \phi^A \otimes \varphi^C. \tag{15.209}$$
>
> 2. If $\mathbf{G}$ is a connected compact Lie group, then (15.209) can never hold.

**Proof** Suppose first that $\mathbf{G}$ is a finite group, and let $g \mapsto U_g^C$ be the regular representation of $\mathbf{G}$ on the space $C := \mathbb{C}^{|\mathbf{G}|} = \mathrm{span}\{|g\rangle \,:\, g \in \mathbf{G}\}$ (see Section C.6 of online version). Fix an element $h \in \mathbf{G}$ and let $|\varphi^C\rangle := |h\rangle^C$. By the definition of the regular representation, we have that $\chi_\varphi(g) = \delta_{e,g}$, so that (15.210) holds trivially. Therefore, for any $h \in \mathbf{G}$, the state $|\varphi^C\rangle = |h\rangle$ satisfies (15.209).

We next prove that if $\mathbf{G}$ is a connected compact Lie group, then the relation (15.209) never holds. Suppose by contradiction that (15.209) does hold. Then, from Theorem 15.13 there exists a positive-definite function $f : \mathbf{G} \to \mathbb{C}$ such that $\chi_{\psi \otimes \varphi}(g) = \chi_{\phi \otimes \varphi}(g) f(g)$ for all $g \in \mathbf{G}$. Since the representation on system $AC$ is given by $g \mapsto U_g^A \otimes U_g^C$, we have $\chi_{\psi \otimes \varphi}(g) = \chi_\psi(g)\chi_\varphi(g)$ and similarly $\chi_{\phi \otimes \varphi}(g) = \chi_\phi(g)\chi_\varphi(g)$ so that

$$\chi_\psi(g)\chi_\varphi(g) = \chi_\phi(g)\chi_\varphi(g)f(g) \qquad \forall\, g \in \mathbf{G}. \tag{15.210}$$

As discussed in Definition 15.3, since $\mathbf{G}$ is a connected compact Lie group, there exists a neighborhood, $\mathfrak{C}$, around the identity element of the group such that $\chi_\varphi(g) \neq 0$ for all elements $g \in \mathfrak{C}$. Combining this with (15.210) gives

$$\chi_\psi(g) = \chi_\phi(g)f(g) \qquad \forall\, g \in \mathfrak{C}. \tag{15.211}$$

However, since the functions $\chi_\psi$, $\chi_\phi$, and $f$, are all analytic, the identity theorem in complex analysis implies that the equality in (15.211) holds for all $g \in \mathbf{G}$. Hence, from Theorem 15.13 we get that $\psi \xrightarrow{\;\mathrm{G-COV}\;} \phi$ in contradiction with the assumption of the theorem that $\psi$ cannot be converted to $\phi$ by $\mathbf{G}$-covariant operations. Hence, the relation (15.209) cannot hold if $\mathbf{G}$ is a connected compact Lie group. $\blacksquare$

The existence of a catalyst for finite groups is a consequence of the fact that for finite groups, it is possible to completely overcome the lack of a shared reference frame by sending a single resource from Alice to Bob. In the proof presented here, the state $|\varphi^C\rangle := |h\rangle^C$ serves as an "ultimate" resource that removes the restriction to $\mathbf{G}$-covariant operations. To understand why, let's revisit the guessing probability given in (15.61).

Taking $\rho = \varphi^C$ with the regular representation $U_g|h\rangle\langle h|U_g^* = |gh\rangle\langle gh|$ yields

$$\mathrm{Pr}_{\mathrm{guess}}\left(\rho, \{\Lambda_g\}_{g \in \mathbf{G}}\right) = \frac{1}{|\mathbf{G}|} \sum_{g \in \mathbf{G}} \mathrm{Tr}\left[\Lambda_g |gh\rangle\langle gh|\right]. \tag{15.212}$$

Therefore, by choosing $\Lambda_g := |gh\rangle\langle gh|$, we obtain $\mathrm{Pr}_{\mathrm{guess}}\left(\rho, \{\Lambda_g\}_{g\in G}\right) = 1$. Consequently, upon receiving the state $\varphi^C$, Bob can determine the group element $g$ that relates his reference frame to Alice's reference frame.

This discussion demonstrates that for finite groups, the notion of catalysis is not very useful as it merely reflects the existence of an ultimate resource for such groups, specifically in relation to the regular representation. On the other hand, for connected Lie groups, catalysis does not exist, and it remains an open problem whether catalysis can exist for arbitrary compact Lie groups.

## 15.4.4 Probabilistic Transformations

In this subsection, we study the conversion of a pure state to an ensemble of pure states by **G**-covariant operations. Recall that every ensemble of pure states $\{p_x, \phi_x^A\}_{x\in[m]}$ can be characterized with a cq-state in $\mathfrak{D}(AX)$ of the form

$$\sigma^{AX} := \sum_{x\in[m]} p_x \phi_x^A \otimes |x\rangle\langle x|^X. \tag{15.213}$$

Given a pure state $\psi \in \mathrm{Pure}(A)$, we want to find the conditions under which the conversion $\psi^A \xrightarrow{\text{G-COV}} \sigma^{AX}$ is possible.

> **Theorem 15.15.** Using the same notations as above, $\psi^A \xrightarrow{\text{G-COV}} \sigma^{AX}$ if and only if there exists normalized positive-definite and continuous (in the case of Lie group) functions $f_x : \mathbf{G} \to \mathbb{C}$ such that
>
> $$\chi_\psi(g) = \sum_{x\in[n]} p_x f_x(g) \chi_{\phi_x}(g). \tag{15.214}$$

**Proof** From the covariant version of Stinespring dilation theorem, $\mathcal{E} \in \mathrm{COV}_G(A \to AX)$ if and only if there exists a system $E$, a projective unitary representation $g \mapsto U_g^E$, and an intertwiner isometry $V : A \to AXE$ such that for all $\eta \in \mathfrak{L}(A)$ we have $\mathcal{E}(\eta) = \mathrm{Tr}_E(V\eta V^*)$. Therefore, $\psi^A \xrightarrow{\text{G-COV}} \sigma^{AX}$ if and only if there exists an intertwiner isometry $V : A \to AXE$ such that

$$\sigma^{AX} = \mathcal{E}^{A\to AX}\left(\psi^A\right) = \mathrm{Tr}_E\left[V\psi^A V^*\right]. \tag{15.215}$$

We first assume that such a covariant channel $\mathcal{E}^{A\to AX}$ exists, and prove the relation (15.214). Indeed, (15.215) implies that $V|\psi^A\rangle$ is a purification of $\sigma^{AX}$ and therefore has the form

$$V|\psi^A\rangle = \sum_{x\in[m]} \sqrt{p_x}|\phi_x^A\rangle|x\rangle^X|\varphi_x^E\rangle \tag{15.216}$$

for some orthonormal set $\{|\varphi_x^E\rangle\}_{x\in[m]}$ in $E$. Since $\mathbf{G}$ acts trivially on system $X$, and since $V$ is an intertwiner we get that

$$
\begin{aligned}
VU_g^A|\psi^A\rangle &= \left(U_g^B \otimes I^X \otimes U_g^E\right) V|\psi^A\rangle \\
&= \sum_{x\in[m]} \sqrt{p_x}\, U_g^A|\phi_x^A\rangle \otimes |x\rangle^X \otimes U_g^E|\varphi_x^E\rangle.
\end{aligned}
\tag{15.217}
$$

Finally, taking the inner product between the two states in (15.216) and (15.217) gives

$$
\chi_\psi(g) = \sum_{x\in[m]} p_x\, \chi_{\phi_x}(g)\chi_{\varphi_x}(g).
\tag{15.218}
$$

Since $f_x(g) := \chi_{\varphi_x}(g)$ is a positive-definitive function (see Theorem C.12 of online version) we get that (15.214) holds.

Conversely, suppose (15.214) holds. From Theorem C.12 of online version $f_x$ can be expressed as the characteristic function of some state $\varphi_x^E$. Without loss of generality we can assume that the states $\{|\varphi_x^E\rangle\}_{x\in[m]}$ are orthonormal since otherwise we can replace each $|\varphi_x^E\rangle$ with $|\varphi_x^E\rangle|x\rangle^{E'}$, where $E'$ is another ancillary system upon which the group $\mathbf{G}$ acts trivially (so that $|\varphi_x^E\rangle$ and $|\varphi_x^E\rangle|x\rangle^{E'}$ have the same characteristic function). With this in mind, let

$$
|\phi^{AXE}\rangle := \sum_{x\in[m]} \sqrt{p_x}|\phi_x^A\rangle|x\rangle^X|\varphi_x^E\rangle.
\tag{15.219}
$$

Then, from (15.214) we get that $\chi_\psi(g) = \chi_\phi(g)$ for all $g \in \mathbf{G}$. Moreover, there exists a $\mathbf{G}$-invariant state $|0\rangle \in XE$ whose characteristic function is constant and equal to one for all group elements. Therefore, from the relation $\chi_\psi(g) = \chi_\phi(g)$ we get that the states $|\psi^A\rangle|0\rangle^{XE}$ and $|\phi^{AXE}\rangle$ have the same characteristic function. Therefore, there exists a $\mathbf{G}$-invariant unitary $V: AXE \to AXE$ such that $V\left(|\psi^A\rangle|0\rangle^{XE}\right) = |\phi^{AXE}\rangle$. Taking the trace over $E$ on both sides demonstrates that $\psi^A$ can be converted to $\sigma^{AX}$ by a $\mathbf{G}$-covariant channel. This completes the proof. ∎

**Exercise 15.36.** *Prove the following corollary to the theorem above: The conversion $\psi^A \xrightarrow{\mathrm{G-COV}} \phi^A$ can be achieved with probability $q$ if and only if there exists a normalized positive definition function $f: \mathbf{G} \to \mathbb{C}$ such that $\chi_\psi(g) - qf(g)\chi_\phi(g)$ is positive definite.*

## 15.5 Manipulation of Mixed Asymmetry

In this section, we study interconversions among asymmetric mixed states in the single-shot regime. Unlike the pure-state case, for mixed state there is no simple criterion to determine if one state can be converted to another by $\mathbf{G}$-covariant operations. However, we will see that the problem can be casted as an SDP optimization problem.

We start with the following observation about the Choi matrix $J_{\mathcal{E}}^{AB}$ of a $\mathbf{G}$-covariant channel $\mathcal{E} \in \mathrm{COV}_\mathbf{G}(A \to B)$. The $\mathbf{G}$-covariance property implies that

$$\mathcal{E}^{A\to B} = \mathcal{U}_g^B \circ \mathcal{E}^{A\to B} \circ \mathcal{U}_g^{*A} \qquad \forall\, g \in \mathbf{G}. \tag{15.220}$$

Applying both sides of this equation to the maximally entangled state $|\Omega^{A\tilde{A}}\rangle$ gives

$$
\begin{aligned}
J_{\mathcal{E}}^{AB} &= \mathcal{U}_g^B \circ \mathcal{E}^{\tilde{A}\to B} \circ \mathcal{U}_g^{*\tilde{A}}\left(\Omega^{A\tilde{A}}\right) \\
\mathbf{(2.91)}\to\ &= \bar{\mathcal{U}}_g^A \otimes \mathcal{U}_g^B \circ \mathcal{E}^{\tilde{A}\to B}\left(\Omega^{A\tilde{A}}\right) \\
&= \bar{\mathcal{U}}_g^A \otimes \mathcal{U}_g^B\left(J_{\mathcal{E}}^{AB}\right).
\end{aligned}
\tag{15.221}
$$

Therefore, the matrix $J_{\mathcal{E}}^{AB}$ is a Choi matrix of a **G**-covariant channel if and only if

$$\left(\bar{U}_g^A \otimes U_g^B\right) J_{\mathcal{E}}^{AB} \left(\bar{U}_g^A \otimes U_g^B\right)^* = J_{\mathcal{E}}^{AB} \qquad \forall\, g \in \mathbf{G}. \tag{15.222}$$

That is, the Choi matrix $J_{\mathcal{E}}^{AB}$ is symmetric with respect to the projective unitary representation $g \mapsto \bar{U}_g^A \otimes U_g^B$. In this section, we will denote by $\mathcal{G} \in \mathrm{CPTP}(AB \to AB)$ the **G**-twirling operation with respect to this representation, so that $\mathcal{E}$ is **G**-covariant if and only if $\mathcal{G}\left(J_{\mathcal{E}}^{AB}\right) = J_{\mathcal{E}}^{AB}$.

With this property we can use Theorem 11.1 to get necessary and sufficient conditions for a conversion of one mixed state to another by **G**-covariant operations. To apply Theorem 11.1 for the case that $\mathfrak{F}(A \to B) = \mathrm{COV}_{\mathbf{G}}(A \to B)$, observe that

$$
\begin{aligned}
\sup_{\mathcal{E}\in\mathrm{COV}_{\mathbf{G}}(A\to B)} \mathrm{Tr}\left[\eta^B \mathcal{E}^{A\to B}\left(\rho^A\right)\right] &= \sup_{\mathcal{E}\in\mathrm{COV}_{\mathbf{G}}(A\to B)} \mathrm{Tr}\left[J_{\mathcal{E}}^{AB}\left(\rho^T \otimes \eta^B\right)\right] \\
&= \sup_{\substack{J\in\mathrm{Pos}(AB) \\ J^A = I^A}} \mathrm{Tr}\left[J^{AB} \mathcal{G}^{AB\to AB}\left(\rho^T \otimes \eta^B\right)\right] \\
\mathbf{(7.147)}\to\ &= 2^{-H_{\min}^{\uparrow}(B|A)_{\mathcal{G}(\rho^T\otimes\eta)}}.
\end{aligned}
\tag{15.223}
$$

Therefore, Theorem 11.1 implies the following characterization of $\rho^A \xrightarrow{\ \mathbf{G-COV}\ } \sigma^B$.

---

**Corollary 15.2.** Let $\rho \in \mathfrak{D}(A)$ and $\sigma \in \mathfrak{D}(B)$. The following are equivalent:

1. $\rho^A \xrightarrow{\ \mathbf{G-COV}\ } \sigma^B$.
2. For all $\eta \in \mathfrak{D}(B)$ we have $H_{\min}^{\uparrow}(B|A)_{\mathcal{G}(\rho^T\otimes\eta)} \leqslant H_{\min}^{\uparrow}(B|\tilde{B})_{\mathcal{G}(\sigma^T\otimes\eta)}$.

---

While the condition outlined in the corollary holds theoretical significance, it falls short of offering a practical methodology for assessing whether a quantum state $\rho^A$ can be transformed into another state $\sigma^B$ through **G**-covariant operations. To address this gap, a more applicable criterion is derived from the condition presented in (11.6). For the context at hand, this criterion is articulated in a specific format, which we encapsulate as a theorem for clarity and ease of application.

> **Theorem 15.16.** Let $\rho \in \mathfrak{D}(A)$, $\sigma \in \mathfrak{D}(B)$, and define
>
> $$f(\rho, \sigma) := \min_{\tau \in \mathrm{Pos}(A),\, \eta \in \mathfrak{D}(B)} \left\{ \mathrm{Tr}[\tau - \sigma\eta] \, : \, \tau^A \otimes I^B \geqslant \mathcal{G}\left( \rho^T \otimes \eta^B \right) \right\}. \tag{15.224}$$
>
> Then, $\rho^A \xrightarrow{\text{G--COV}} \sigma^B$ if and only if $f(\rho, \sigma) \geqslant 0$.

*Remark.* The optimization of the function $f(\rho, \sigma)$ can be solved efficiently and algorithmically with an SDP program.

The proof of this theorem is based on the fact that $\sigma = \mathcal{E}(\rho)$ if and only if for all $\Lambda \in \mathrm{Herm}(B)$ we have $\mathrm{Tr}[\Lambda\sigma] = \mathrm{Tr}[\Lambda\mathcal{E}(\rho)]$. This relation can be expressed as

$$\mathrm{Tr}[\Lambda\sigma] = \mathrm{Tr}\left[ J_{\mathcal{E}}^{AB} \left( \rho^T \otimes \Lambda^B \right) \right]. \tag{15.225}$$

In the following exercise we will use this to complete the proof.

**Exercise 15.37.** *Use* (11.6) *and* (15.223) *to prove Theorem 15.16.*

**Exercise 15.38.** *Let $\rho \in \mathfrak{D}(A)$ and $\sigma \in \mathfrak{D}(B)$. Show that $\rho^A \xrightarrow{\text{G--COV}} \sigma^B$ if and only if there exists $\mathcal{F} \in \mathrm{CPTP}(A \to B)$ such that for all $g \in \mathrm{G}$*

$$\mathcal{F}\big(\mathcal{U}_g(\rho)\big) = \mathcal{U}_g(\sigma). \tag{15.226}$$

**Exercise 15.39.** *The conversion distance from $\rho \in \mathfrak{D}(A)$ to $\sigma \in \mathfrak{D}(B)$ is defined as*

$$T\left( \rho \xrightarrow{\mathfrak{F}} \sigma \right) := \min_{\mathcal{E} \in \mathrm{COV}_G(A \to B)} \frac{1}{2} \left\| \sigma^B - \mathcal{E}^{A \to B}(\rho^A) \right\|_1. \tag{15.227}$$

*Use the trace distance property*

$$\frac{1}{2} \left\| \sigma^B - \mathcal{E}^{A \to B}(\rho^A) \right\|_1 = \min_{\substack{\Lambda \in \mathrm{Pos}(B) \\ \Lambda \geqslant \sigma^B - \mathcal{E}^{A \to B}(\rho^A)}} \mathrm{Tr}\,[\Lambda], \tag{15.228}$$

*to show that the conversion distance can be expressed as the following SDP:*

$$T\left( \rho \xrightarrow{\mathfrak{F}} \sigma \right) = \min \mathrm{Tr}\,[\Lambda] \tag{15.229}$$

*subject to the following:*

*1.* $\Lambda^B \geqslant \sigma^B - \mathrm{Tr}_A\left[ J^{A\tilde{A}} \left( \rho^T \otimes I^{\tilde{A}} \right) \right].$

*2.* $J^A = I^A.$

*3.* $\mathcal{G}(J^{A\tilde{A}}) = J^{A\tilde{A}}.$

*4.* $\Lambda \in \mathrm{Pos}(B),\, J \in \mathrm{Pos}(A\tilde{A}).$

# 15.6 Time Translation Symmetry

Time-translation symmetry, also known as time-translational invariance, is a fundamental concept in physics that relates to the behavior of physical systems under shifts or translations in time. It is a principle that states that the laws of physics remain unchanged or invariant over time. In simpler terms, time-translation symmetry implies that the fundamental laws of physics do not depend on the specific moment in time at which they are applied. This means that if a physical experiment or process is performed today or tomorrow, under the same conditions, the outcome should be the same.

The concept of time-translation symmetry is closely related (via Noether's theorem) to the conservation of energy, where the total energy of a closed system remains constant over time. It provides a foundation for many important principles and theories in physics, including the laws of motion, quantum mechanics, and relativity. Therefore, time-translation symmetry is a fundamental symmetry in the universe, and it plays a crucial role in our understanding of the laws governing the behavior of matter and energy over time. Given its importance, we devote this section to study the resource theory of asymmetry with respect to this time-translation symmetry.

## 15.6.1 Time-Translation Covariant Operations

We say that a quantum state $\rho \in \mathfrak{D}(A)$ is time-translation invariant with respect to a Hamiltonian $H^A \in \mathrm{Pos}(A)$ if, for all $t \in \mathbb{R}$, the following equation holds:

$$e^{-iH^A t} \rho^A e^{iH^A t} = \rho^A. \tag{15.230}$$

Note that $\rho^A$ is time-translation invariant if and only if it commutes with the Hamiltonian. This is why it is sometimes referred to in the literature as "quasi-classical," as both the state and the Hamiltonian are diagonal with respect to the same basis. Throughout this section, we will always work with the eigenbases of the Hamiltonians.

The Hamiltonian $H^A$ can be decomposed as

$$H^A = \sum_{x \in [m]} a_x \Pi_x^A, \tag{15.231}$$

where $\{a_x\}_{x \in [m]}$ is the set of distinct eigenvalues of $H^A$ and each $\Pi_x^A$ is a projection to the eigenspace of $a_x$. Without loss of generality, we will assume that $a_1 < a_2 < \cdots < a_m$ (noting that they are all distinct, allowing us to arrange $\{a_x\}_{x \in [m]}$ in increasing order). With the form of $H^A$ given in (15.231), the state $\rho^A$ is time-translation invariant, if and only if it takes the form

$$\rho = \sum_{x \in [m]} p_x \rho_x, \tag{15.232}$$

where $\mathbf{p} \in \mathrm{Prob}(m)$, $\rho_x \in \mathfrak{D}(A)$, and $\rho_x \rho_y = 0$ for every $x \neq y$. We will use the notation $\mathrm{INV}(A)$ to denote the set of states in $\mathfrak{D}(A)$ that are time-translation invariant.

**Exercise 15.40.** *Prove the above form of $\rho$. Hint:* $\mathrm{supp}(\rho_x) \subseteq \mathrm{supp}(\Pi_x)$.

Consider a quantum channel $\mathcal{N} \in \text{CPTP}(A \to B)$, where systems $A$ and $B$ have corresponding Hamiltonians $H^A \in \text{Pos}(A)$ and $H^B \in \text{Pos}(B)$. The channel $\mathcal{N}$ is said to be *time-translation covariant* if for all $t \in \mathbb{R}$

$$\mathcal{N}^{A \to B}\left(e^{-iH^A t} \rho^A e^{iH^A t}\right) = e^{-iH^B t}\mathcal{N}^{A \to B}(\rho^B)e^{iH^B t} \qquad \forall\, t \in \mathbb{R}. \tag{15.233}$$

We will use the notation $\text{COV}(A \to B)$ to denote the set of all time-translation covariant channels in $\text{CPTP}(A \to B)$.

In the Choi representation, the property given in (15.233) can be expressed as (see the relation (15.222))

$$\left(e^{-i\bar{H}^A t} \otimes e^{iH^B t}\right) J_{\mathcal{N}}^{AB} \left(e^{-i\bar{H}^A t} \otimes e^{iH^B t}\right)^* = J_{\mathcal{N}}^{AB} \qquad \forall\, t \in \mathbb{R}. \tag{15.234}$$

Note that $\bar{H}^A$ has the same eigenvalues as $H^A$. For our purposes, we can replace $\bar{H}^A$ (in the equation above) with $H^A$, since it will not make any difference in our analysis. Therefore, $\mathcal{N} \in \text{COV}(A \to B)$ if and only if $J_{\mathcal{N}}^{AB}$ commutes with the operator

$$\xi^{AB} := H^A \otimes I^B - I^A \otimes H^B. \tag{15.235}$$

Therefore, the degeneracy of the energy levels of the operator $\xi^{AB}$ will play a key role in the resource theory of time-translation asymmetry.

## The Pinching Channel

The set of channels $\text{COV}(A \to A)$ contains the pinching channel $\mathcal{P}_H$ as defined in Section 3.5.12. For $H^A$ as in (15.231), the pinching channel on system $A$ is given by

$$\mathcal{P}_H^{A \to A}\left(\rho^A\right) := \sum_{x \in [m]} \Pi_x^A \rho^A \Pi_x^A. \tag{15.236}$$

This pinching channel, also known as the "twirling channel" (as it is the **G**-twirling map with respect to the group $\mathbf{G} = \{e^{iH^A t}\}_{t \in \mathbb{R}}$), has the property that a state $\rho \in \mathfrak{D}(A)$ is quasi-classical if and only if $\mathcal{P}_H(\rho) = \rho$.

**Exercise 15.41.** *Show that the condition $\mathcal{P}_H(\rho) = \rho$ is equivalent to the condition that $\rho$ has the form given in (15.232).*

**Exercise 15.42.** *Let $\mathcal{P} \in \text{CPTP}(A \to A)$ and $\mathcal{P}' \in \text{CPTP}(A' \to A')$ be the pinching channel associated with the Hamiltonians $H^A$ and $H^{A'}$, respectively. Furthermore, let $\mathcal{N} \in \text{CPTP}(A \to A')$.*

*1. Show that $\mathcal{P}' \circ \mathcal{N} \circ \mathcal{P} \in \text{COV}(A \to A')$.*
*2. Show that if $\mathcal{N} \in \text{COV}(A \to A')$, then $\mathcal{P}' \circ \mathcal{N} = \mathcal{N} \circ \mathcal{P}$.*
*3. Show that if the Hamiltonian $H^A$ is nondegenerate, then $\mathcal{P}_H^{A \to A} = \Delta^{A \to A}$, where $\Delta^{A \to A}$ is the completely dephasing channel as defined in Section 3.5.2.*

Covariant channels can also be characterized in terms of the pinching channel. Consider $\mathcal{N} \in \text{CPTP}(A \to B)$ and let $\mathcal{P}_\xi \in \text{COV}(AB \to AB)$ be the pinching channel associated with the operator $\xi^{AB}$ given in (15.235). Then, the quantum channel $\mathcal{N}^{A \to B}$ is time-translation covariant if and only if its Choi matrix satisfies

$$\mathcal{P}_{\xi}^{AB \to AB}\left(J_{\mathcal{N}}^{AB}\right) = J_{\mathcal{N}}^{AB}. \tag{15.237}$$

This follows from Exercise 3.62 and our earlier observation that $\mathcal{N} \in \mathrm{COV}(A \to B)$ if and only if its Choi matrix commutes with $\xi^{AB}$.

The twirling channel can also be used to quantify time-translation asymmetry. For example, the relative entropy distance of a quantum state $\rho \in \mathfrak{D}(A)$ to its twirled state $\mathcal{P}(\rho)$ is a time-translation asymmetry (sometimes referred to as coherence) measure given by

$$C(\rho) := D\left(\rho \big\| \mathcal{P}_H(\rho)\right) = H\left(\mathcal{P}_H(\rho)\right) - H(\rho), \tag{15.238}$$

where $D(\rho\|\sigma) := \mathrm{Tr}[\rho \log \rho] - \mathrm{Tr}[\rho \log \sigma]$ is the Umegaki relative entropy and $H(\rho) := -\mathrm{Tr}[\rho \log \rho]$ is the von-Neumann entropy. This function is nonincreasing under time-translation covariant operations, and achieves its maximal value of $\log d$ (where $d := |A|$) on the maximally coherent state $|+\rangle := \frac{1}{\sqrt{d}} \sum_{x \in [d]} |x\rangle$, where $\{|x\rangle\}_{x \in [d]}$ is the energy eigenbasis. We next move to characterize the set $\mathrm{COV}(A \to B)$ in three different cases that depends on the level of degeneracy of the Hamiltonians involved.

## The Case of Relatively Non-degenerate Hamiltonians

Let $H^A$ and $H^B$ be the Hamiltonians of two systems $A$ and $B$, of dimensions $m := |A|$ and $n := |B|$. The Hamiltonians can be expressed in their spectral decomposition as

$$H^A = \sum_{x \in [m]} a_x |x\rangle\langle x|^A \quad \text{and} \quad H^B = \sum_{y \in [n]} b_y |y\rangle\langle y|^B, \tag{15.239}$$

where $\{a_x\}$ and $\{b_y\}$ are the energy eigenvalues of $H^A$ and $H^B$, respectively.

> **Definition 15.4.** We say that the Hamiltonians $H^A$ and $H^B$, as defined in (15.239), are *relatively nondegenerate* if for all $x, x' \in [m]$ and $y, y' \in [n]$ we have
>
> $$a_x - a_{x'} = b_y - b_{y'} \quad \Rightarrow \quad x = x' \text{ and } y = y'. \tag{15.240}$$
>
> If this condition does not hold we say that the Hamiltonians are relatively degenerate.

Note that if $H^A$ and $H^B$ are relatively nondegenerate, then each of them is also nondegenerate. For example, suppose $H^A$ is degenerate with $a_x = a_{x'}$ for some $x \neq x' \in [m]$. Then, for $y = y'$ we get $a_x - a_{x'} = 0 = b_y - b_{y'}$ even though $x \neq x'$. Therefore, relative nondegeneracy is a stronger notion than nondegeneracy. In fact, relative non-degeneracy of $H^A$ and $H^B$ is equivalent to the nondegeneracy of the operator $\xi^{AB}$ as defined in (15.235). Moreover, in the generic case in which $H^A$ and $H^B$ are arbitrary (chosen at random) the Hamiltonians are relatively nondegenerate. For this case, time-translation covariant channels have a very simple characterization.

---

**Theorem 15.17.** Let $A$ and $B$ be two physical systems with relatively nondegenerate Hamiltonians. Then, $\mathcal{N} \in \mathrm{CPTP}(A \to B)$ is a time-translation covariant channel if and only if

$$\mathcal{N}^{A \to B} = \Delta^{B \to B} \circ \mathcal{N}^{A \to B} \circ \Delta^{A \to A}, \qquad (15.241)$$

where $\Delta^{A \to A}$ and $\Delta^{B \to B}$ are the completely dephasing channels of systems $A$ and $B$, respectively. In other words, for physical systems with relatively nondegenerate Hamiltonians only classical channels are time-translation covariant.

---

**Proof**　Since we assume that the Hamiltonians $H^A$ and $H^B$ are relatively nondegenerate we get that the joint operator, $\xi^{AB}$, is nondegenerate. Hence, $J_{\mathcal{N}}^{AB}$ is diagonal in the same eigenbasis $\{|x\rangle^A |y\rangle^B\}_{x \in [m], y \in [n]}$ of $\xi^{AB}$, so that

$$\Delta^{A \to A} \otimes \Delta^{B \to B} \left( J_{\mathcal{N}}^{AB} \right) = J_{\mathcal{N}}^{AB}. \qquad (15.242)$$

This equation describes the same relation as the one given in (15.241). Hence, $\mathcal{N}^{A \to B}$ is a classical channel. This completes the proof. ∎

## The Case of Bohr Spectrum

We consider now the case in which $A = B$. Therefore, since $H^A = H^B$ we cannot apply the characterization Theorem 15.17 to this case. Instead, we will assume that $H^A$ has a nondegenerate Bohr spectrum. In its spectral decomposition $H^A$ has the form

$$H^A = \sum_{x \in [m]} a_x |x\rangle \langle x|^A, \qquad (15.243)$$

where $\{a_x\}_{x \in [m]}$ is the set of distinct eigenvalues of $H^A$.

---

**Definition 15.5.** We say that $H^A$ as given in (15.243) has a nondegenerate Bohr spectrum if it has the property that for any $x, y, x', y' \in [m]$

$$a_x - a_y = a_{x'} - a_{y'} \quad \Longleftrightarrow \quad x = x' \text{ and } y = y' \quad \text{or} \quad x = y \text{ and } x' = y' ;$$

that is, there are no degeneracies in the nonzero differences of the energy levels of $H^A$.

---

**Exercise 15.43.** *Show that $H^A$ has a nondegenerate Bohr spectrum if and only if all the nonzero eigenvalues of the operator*

$$\xi^{A\tilde{A}} := H^A \otimes I^{\tilde{A}} - I^A \otimes H^{\tilde{A}} \qquad (15.244)$$

*are distinct. In other words, $H^A$ has a nondegenerate Bohr spectrum if and only if the zero eigenvalue of $\xi^{A\tilde{A}}$ is the sole eigenvalue with a multiplicity greater than 1.*

It is noteworthy that the vast majority of Hamiltonians exhibit a nondegenerate Bohr spectrum, indicating that Hamiltonians lacking this feature are exceptionally rare, constituting a set of measure zero. This observation segues into a focused interest in time-translation covariant channels that are compatible with nondegenerate Bohr spectra, offering a unique area for characterization.

> **Theorem 15.18.** Consider a Hamiltonian, $H^A$, as outlined in (15.243), which is characterized by having a nondegenerate Bohr spectrum, and let $\mathcal{N} \in$ CPTP($A \to A$). The channel $\mathcal{N} \in$ COV($A \to A$) if and only if for all $x, x', y, y' \in [m]$, $\langle xx'|J_{\mathcal{N}}^{A\tilde{A}}|yy'\rangle = 0$ unless $x = x'$ and $y = y'$, or $x = y$ and $x' = y'$.

*Remark.* We will see next that even if the spectrum of the Hamiltonian $H^A$ has degeneracies, any quantum channel $\mathcal{N} \in$ CPTP($A \to A$) whose Choi matrix has the form (15.249) is necessarily time-translation covariant.

**Proof**   Following the same lines as in Theorem 15.17, by replacing $H^B$ with $H^A$ everywhere, we get that a quantum channel $\mathcal{N} \in$ CPTP($A \to A$) is time-translation covariant if and only if its Choi matrix $J_{\mathcal{N}}^{A\tilde{A}}$ commutes with the operator

$$\xi^{A\tilde{A}} := H^A \otimes I^{\tilde{A}} - I^A \otimes H^{\tilde{A}} = \sum_{x,y \in [m]} (a_x - a_y)|x\rangle\langle x|^A \otimes |y\rangle\langle y|^{\tilde{A}}. \tag{15.245}$$

Since $H^A$ has a nondegenerate Bohr spectrum, the set $\{a_x - a_y\}$ that appears in the sum in (15.245) consists of distinct eigenvalues, as we only consider indices $x, y \in [m]$ that satisfy $y \neq x$. We therefore conclude that the pinching channel $\mathcal{P}_\xi \in$ CPTP($A\tilde{A} \to A\tilde{A}$) associated with the operator $\xi^{A\tilde{A}}$ is given by

$$\mathcal{P}_\xi(\cdot) = \Pi(\cdot)\Pi + \sum_{\substack{x,y \in [m] \\ x \neq y}} P_{xy}(\cdot)P_{xy}, \tag{15.246}$$

where

$$P_{xy} := |xy\rangle\langle xy| \quad \text{and} \quad \Pi := \sum_{x \in [m]} |xx\rangle\langle xx|. \tag{15.247}$$

Observe that $\Pi$ is the projection to the zero eigenspace of $\xi^{A\tilde{A}}$. With these notations the condition $J_{\mathcal{N}}^{A\tilde{A}} = \mathcal{P}_\xi\left(J_{\mathcal{N}}^{A\tilde{A}}\right)$ is equivalent to

$$J_{\mathcal{N}}^{A\tilde{A}} = \Pi J_{\mathcal{N}}^{A\tilde{A}} \Pi + \sum_{\substack{x,y \in [m] \\ x \neq y}} P_{xy} J_{\mathcal{N}}^{A\tilde{A}} P_{xy}. \tag{15.248}$$

Observe that Choi matrix $J_{\mathcal{N}}^{A\tilde{A}}$ satisfies the condition above if and only if $\langle xx'|J_{\mathcal{N}}^{A\tilde{A}}|yy'\rangle = 0$ unless $x = x'$ and $y = y'$, or $x = y$ and $x' = y'$. This completes the proof. ∎

The condition in the theorem is equivalent to the statement that the Choi matrix has the form

$$J_{\mathcal{N}}^{A\tilde{A}} = \sum_{x,y\in[m]} \left( p_{y|x}|xy\rangle\langle xy|^{A\tilde{A}} + (1 - \delta_{xy})q_{xy}|xx\rangle\langle yy|^{A\tilde{A}} \right), \tag{15.249}$$

where $q_{xy} := \langle xx|J_{\mathcal{N}}^{A\tilde{A}}|yy\rangle$ and $p_{y|x} := \langle xy|J_{\mathcal{N}}^{A\tilde{A}}|xy\rangle$. Observe that by definition $p_{x|x} = q_{xx}$ for all $x \in [m]$. Given that $J_{\mathcal{N}}^{AB}$ is the Choi matrix of a quantum channel, it implies certain properties for the coefficients $\{p_{y|x}\}_{x,y\in[m]}$ and the matrix $Q_{\mathcal{N}}$, which consists of the components $q_{xy}$. Specifically, the first term on the right-hand side of (15.249) corresponds to $\Pi J_{\mathcal{N}}^{A\tilde{A}} \Pi$, and the second term is a sum over all $P_{xy} J_{\mathcal{N}}^{A\tilde{A}} P_{xy}$. Therefore, from the condition $\Pi J_{\mathcal{N}}^{A\tilde{A}} \Pi \geqslant 0$ we get that $Q_{\mathcal{N}} \geqslant 0$, where $Q_{\mathcal{N}}$ is the matrix whose components are $q_{xy}$. Similarly, the condition $P_{xy} J_{\mathcal{N}}^{A\tilde{A}} P_{xy} \geqslant 0$ implies that $p_{y|x} \geqslant 0$. Thus, we conclude that $J_{\mathcal{N}}^{A\tilde{A}} \geqslant 0$ if and only if $Q_{\mathcal{N}} \geqslant 0$ and each $p_{y|x} \geqslant 0$. Finally, the remaining condition $J_{\mathcal{N}}^{A} = I^{A}$ implies that for all $x \in [m]$ we have $\sum_{y\in[m]} p_{y|x} = 1$. To summarize, Theorem 15.18 implies that $\mathcal{N} \in \mathrm{COV}(A \to A)$ if its Choi matrix has the form (15.249), with $Q_{\mathcal{N}} \geqslant 0$ and $\{p_{y|x}\}_{x,y\in[m]}$ being a conditional probability distribution.

## 15.6.2 Exact State Conversion

In this section, we examine the precise state conversions for each of the two distinct degeneracies we have already discussed of the Hamiltonians involved. Similar to previous sections, we denote by $\rho \xrightarrow{\mathrm{COV}} \sigma$ the conversion of a quantum state $\rho$ to another quantum state $\sigma$ through covariant operations. It is worth mentioning that there are numerous other significant examples of Hamiltonians whose spectra are either degenerate or do not satisfy the condition stated in Definition 15.5. The QRT of time-translation asymmetry with such Hamiltonians is still in the process of being fully developed and remains an active area of research.

### The Case of Relatively Nondegenerate Hamiltonians

First, we consider the conversion of a state $\rho \in \mathfrak{D}(A)$ to a state $\sigma \in \mathfrak{D}(B)$ by covariant operation, with Hamiltonians $H^{A}$ and $H^{B}$ that are relatively nondegenerate. As we saw in Theorem 15.17 the set $\mathrm{COV}(A \to B)$ consists of all classical channels in $\mathrm{CPTP}(A \to B)$, with respect to the eigen-bases of the Hamiltonians $H^{A}$ and $H^{B}$. Therefore, in this case, $\rho^{A} \xrightarrow{\mathrm{COV}} \sigma$ if and only if $\sigma^{B}$ is classical.

**Exercise 15.44.** *Prove the statement above. That is, show that* $\rho^{A} \xrightarrow{\mathrm{COV}} \sigma^{B}$ *if and only if* $\sigma^{B} = \Delta^{B}(\sigma^{B})$.

## The Case of Bohr Spectrum

Next, we explore the exact single-shot interconversions of systems whose Hamiltonians possess a nondegenerate Bohr spectrum. This problem is more challenging compared to the similar problem-involving relatively nondegenerate Hamiltonians. We will use the notation $\{|x\rangle^A\}_{x\in[m]}$ to represent the energy eigenbasis of a Hamiltonian $H^A$, and consider two density matrices in $\mathfrak{D}(A)$:

$$\rho^A = \sum_{x,x'\in[m]} r_{xx'}|x\rangle\langle x'|^A \quad \text{and} \quad \sigma^A = \sum_{x,x'\in[m]} s_{xx'}|x\rangle\langle x'|^A \tag{15.250}$$

with components $\{r_{xx'}\}$ and $\{s_{xx'}\}$, respectively. In the following theorem, we assume that $r_{xx'} \neq 0$ for all $x, x' \in [m]$, and define the $m \times m$ matrix $Q$, with components

$$q_{xy} := \begin{cases} \min\left\{1, \frac{s_{xx}}{r_{xx}}\right\} & \text{if } x = y \\ \frac{s_{xy}}{r_{xy}} & \text{otherwise.} \end{cases} \tag{15.251}$$

---

**Theorem 15.19.** Let $\rho, \sigma \in \mathfrak{D}(A)$ be as in (15.250) with $r_{xx'} \neq 0$ for all $x, x' \in [m]$, and suppose the Hamiltonian $H^A$ has a nondegenerate Bohr spectrum. Then, the following statements are equivalent:

1. There exists $\mathcal{E} \in \mathrm{COV}(A \to A)$ such that $\sigma = \mathcal{E}(\rho)$.
2. The matrix $Q$ as defined in (15.251) is positive semidefinite.

---

*Remark.* We will see in the proof here that the second statement implies the first statement even if the Hamiltonian $H^A$ has a degenerate Bohr spectrum. Moreover, we will see that if $r_{xy} = 0$ for some off diagonal terms (i.e. $x \neq y$), then $s_{xy}$ must also be zero. However, in this case, for any $x \neq y \in [m]$ with $r_{xy} = 0$, the components of $q_{xy}$ can be arbitrary. This means that in this case the condition becomes cumbersome, as we will need to require that there *exists* $Q$ as already defined but with no restriction on the components $q_{xy}$ for which $r_{xy} = 0$.

**Proof**  From Theorem 15.18 and the preceeding discussion in (15.249), it follows that there exists $\mathcal{N} \in \mathrm{COV}(A \to A)$ such that $\sigma = \mathcal{N}(\rho)$ if and only if there exists a conditional probability distribution $\{p_{y|x}\}_{x,y\in[m]}$, and an $m \times m$ positive semidefinite matrix $Q$, such that

$$\sigma = \mathcal{N}(\rho) = \mathrm{Tr}_A\left[J_{\mathcal{N}}^{A\tilde{A}}(\rho^T \otimes I^{\tilde{A}})\right]$$
$$= \sum_{x,y\in[m]} p_{y|x}r_{xx}|y\rangle\langle y| + \sum_{\substack{x\neq y \\ x,y\in[m]}} q_{xy}r_{xy}|x\rangle\langle y|. \tag{15.252}$$

That is, $\sigma = \mathcal{N}(\rho)$ if and only if

$$s_{yy} = \sum_{x\in[m]} p_{y|x}r_{xx} \qquad \forall\, y \in [m] \quad \text{and}$$
$$s_{xy} = q_{xy}r_{xy} \qquad \forall\, x \neq y \in [m]. \tag{15.253}$$

Hence, for the off diagonal terms, $s_{xy} = 0$ whenever $r_{xy} = 0$. Since we assume that all the off-diagonal terms of $\rho$ are nonzero, that is, $r_{xy} \neq 0$ for $x \neq y$, there is no freedom left in the choice of the off diagonal terms of $Q_{\mathcal{N}}$ and we must have $q_{xy} = \frac{s_{xy}}{r_{xy}}$. Since $Q_{\mathcal{N}}$ must be positive semidefinite we will maximize its diagonal terms $\{p_{x|x}\}_{x \in [m]}$ given the constraint that $s_{yy} = \sum_{x \in [m]} p_{y|x} r_{xx}$. This constraint immediately gives $s_{yy} \geqslant p_{y|y} r_{yy}$ so that we must have $p_{y|y} \leqslant \frac{s_{yy}}{r_{yy}}$. Clearly, we also have $p_{y|y} \leqslant 1$ so we conclude that

$$p_{y|y} \leqslant \min\left\{1, \frac{s_{yy}}{r_{yy}}\right\}. \tag{15.254}$$

Remarkably, this condition is sufficient since there exists conditional probabilities $\{p_{y|x}\}$, with both $p_{y|y} = \min\left\{1, \frac{s_{yy}}{r_{yy}}\right\}$ and $s_{yy} = \sum_{x \in [m]} p_{y|x} r_{xx}$. Indeed, for simplicity set $r_x := r_{xx}$ and $s_x := s_{xx}$, and define

$$p_{y|x} := \begin{cases} \min\left\{1, \frac{s_x}{r_x}\right\} & \text{if } x = y \\ \frac{1}{\mu r_x}(s_y - r_y)_+(r_x - s_x)_+ & \text{otherwise,} \end{cases} \tag{15.255}$$

where

$$\mu := \sum_{y \in [m]} (s_y - r_y)_+ = \frac{1}{2}\|\mathbf{s} - \mathbf{r}\|_1, \tag{15.256}$$

and we used the notation $(s_y - r_y)_+ := s_y - r_y$ if $s_y \geqslant r_y$ and $(s_y - r_y)_+ := 0$ if $s_y < r_y$. Clearly, $p_{y|x} \geqslant 0$, and it is straightforward to check that $\sum_{y \in [m]} p_{y|x} = 1$ and $s_y = \sum_{x \in [m]} p_{y|x} r_x$; that is, this conditional probability distribution satisfies all the required conditions. This completes the proof. $\blacksquare$

Observe that if $H^A$ has degenerate Bohr spectrum and $Q \geqslant 0$, then we still get that the Choi matrix of the form (15.249) (with $p_{y|x}$ as in (15.255) and $q_{xy}$ as in (15.251)) corresponds to a quantum channel $\mathcal{N} \in \text{CPTP}(A \to A)$ with the property that $\sigma = \mathcal{N}(\rho)$. As discussed in the proof of Theorem 15.18, all channels with a Choi matrix of the form (15.249) are time-translation covariant. Hence, $\mathcal{N} \in \text{COV}(A \to A)$.

**Exercise 15.45.** *In the proof above we saw that if $r_{xy} = 0$ for some $x \neq y$, then $\sigma = \mathcal{E}(\rho)$ for some $\mathcal{E} \in \text{COV}(A \to A)$ only if $s_{xy} = 0$. Use this to show that if $\rho$ has a block diagonal form $\rho = \begin{pmatrix} \tilde{\rho} & 0 \\ 0 & 0 \end{pmatrix}$, and if it can be converted by a time-translation covariant channel to $\sigma$, then $\sigma$ must have the form $\sigma = \begin{pmatrix} \tilde{\sigma} & 0 \\ 0 & D \end{pmatrix}$, where $D$ is some diagonal matrix.*

**Exercise 15.46.** *Show that $J^{AB}$ as given in (15.249) is positive semidefinite if and only if both $p_{y|x} \geqslant 0$ for all $x$ and $y$, and $Q \geqslant 0$.*

**Exercise 15.47.** *Show that the coefficients $\{p_{y|x}\}$ as defined in (15.255) satisfy*

$$\sum_{y \in [m]} p_{y|x} = 1 \quad \text{and} \quad s_y = \sum_{x \in [m]} p_{y|x} r_x \quad \forall \, y \in [m]. \tag{15.257}$$

## Example: The Qubit Case

For the case that $|A| = 2$, all nondegenerate Hamiltonians (i.e. Hamiltonians with two distinct eigenvalues) have a Bohr spectrum. Let

$$\rho = \begin{pmatrix} a & z \\ \bar{z} & 1-a \end{pmatrix} \quad \text{and} \quad \sigma = \begin{pmatrix} b & w \\ \bar{w} & 1-b \end{pmatrix} \tag{15.258}$$

be two qubit states. Without loss of generality suppose that $a \geqslant b$. In this case the matrix $Q$ can be expressed as

$$Q = \begin{pmatrix} \frac{b}{a} & \frac{w}{z} \\ \frac{\bar{w}}{\bar{z}} & 1 \end{pmatrix}, \tag{15.259}$$

and $Q \geqslant 0$ if and only if

$$\frac{b}{a} \geqslant \left| \frac{w}{z} \right|^2. \tag{15.260}$$

Therefore, $\rho \xrightarrow{\text{COV}} \sigma$ if and only if $v(\rho) \geqslant v(\sigma)$, where $v : \mathfrak{D}(A) \to \mathbb{R}_+$ is a measure of qubit time-translation-asymmetry defined on every density matrix of the form (15.258) as

$$v(\rho) := \frac{|z|^2}{a}. \tag{15.261}$$

If $\rho$ is a pure state, so that $|z| = \sqrt{a(1-a)}$, then $v(\rho) \geqslant v(\sigma)$ holds if and only if $|w|^2 \leqslant b(1-a)$. Note that $|w|^2 \leqslant b(1-b)$ since $\sigma \geqslant 0$. Therefore, by taking

$$a \in \left[ b, 1 - \frac{|w|^2}{b} \right] \tag{15.262}$$

we get $|w|^2 \leqslant b(1-a)$ and also $a \geqslant b$. Hence, for any mixed state $\sigma$ there exists a pure state $\psi$ that can be converted to $\sigma$.

On the other hand, if $\sigma$ is pure (i.e. $|w|^2 = b(1-b)$) and $\rho$ arbitrary qubit, then the condition in (15.260) becomes

$$|z|^2 \geqslant a(1-b). \tag{15.263}$$

Since $\rho \geqslant 0$ we also have $|z|^2 \leqslant b(1-b)$. Combining both equation we find that the only way $\rho$ can be converted to a pure qubit state $\sigma$ is if $b = a$ (since $a \geqslant b$ was the initial assumption) and $|z|^2 = a(1-a)$. That is, $\rho$ is a pure state itself, and up to a diagonal unitary equals to $\sigma$. Hence, pure coherence cannot be obtained from mixed coherence, and deterministic interconversion among inequivalent pure resources is not possible.

The example here shows that there is no unique "golden unit" that can be used as the ultimate resource in two-dimensional systems. Instead, any pure resource (i.e. pure state that is not an energy eigenstate) is maximal in the sense that there is no other resource that can be converted into it. However, the set of all pure qubit resources is maximal (i.e. any mixed state can be reached from some pure state by translation covariant operations). We now show that this latter property holds in general.

> **Corollary 15.3.** Let $\sigma \in \mathfrak{D}(A)$ be an arbitrary state, and denote by $p_x :=$ $\langle x|\sigma|x\rangle$ the diagonal elements of $\sigma$ in the energy eigenbasis $\{|x\rangle\}_{x \in [m]}$ of system $A$. Then, the pure quantum state
>
> $$|\psi\rangle := \sum_{x \in [m]} \sqrt{p_x}|x\rangle \tag{15.264}$$
>
> can be converted to $\sigma$ by a time-translation covariant channel.

**Proof**   Observe that the diagonal elements of $Q$ are all 1, and the off-diagonal terms are given by

$$q_{xy} = \frac{\sigma_{xy}}{\sqrt{p_x p_y}} \qquad \forall\, x, y \in [m],\ x \neq y. \tag{15.265}$$

Therefore, we can express $Q = D_{\mathbf{p}}^{-1}\sigma D_{\mathbf{p}}^{-1}$, where $D_{\mathbf{p}}$ is the diagonal matrix whose diagonal is $(\sqrt{p_1}, \ldots, \sqrt{p_m})$. Since $D_{\mathbf{p}} > 0$ and $\sigma \geqslant 0$ it follows that $Q \geqslant 0$. This completes the proof. ∎

**Exercise 15.48.** *Show that if $\rho$ and $\sigma$ are two distinct pure states and both have non-zero off-diagonal terms (with respect to the energy eigenbasis), then the matrix $Q$ is not positive semidefinite.*

## 15.7 Notes and References

An outstanding review article on reference frames and superselection rules in quantum information can be found in Ref. [12]. The theory of quantum reference frames as a resource theory was initially introduced in Ref. [97] and further developed as the resource theory of asymmetry in Refs. [159, 161]. The Kraus representation of a **G**-covariant map was presented in Ref. [97], which was later utilized in Ref. [157] to derive the covariant version of Stinespring's dilation theorem.

The review article in Ref. [12] provides numerous references on the advancement of techniques for aligning reference frames. In particular, we adopted the group theoretical approach to frame alignment developed in Ref. [45].

The concept of relative entropy of asymmetry, initially referred to as **G**-asymmetry, was introduced in Ref. [216] and further developed in Ref. [93]. Other measures of asymmetry, including several derivatives of asymmetry, were investigated in Ref. [160].

The study of pure-state asymmetry originated in Ref. [97] for specific groups and was subsequently extended to all finite or compact Lie groups in Refs. [159, 161]. However, the manipulation of mixed-state asymmetry, particularly in the context of approximate state conversions, remains poorly understood. Nonetheless, the recognition that the exact state conversion problem can be solved using semidefinite programming (refer to Theorem 15.16) was first introduced in Ref. [91].

Exact conversions under time-translation covariant transformations were investigated in Ref. [89]. The positivity of the matrix $Q$ in Theorem 15.19 was discovered earlier in Ref. [171] through a slightly different approach. Lastly, the asymptotic regime under periodic Hamiltonians was recently studied in Ref. [158], where it was demonstrated that the quantum Fisher information can be interpreted as the cost of time-translation asymmetry.

This chapter introduces a specialized sub-theory of quantum thermodynamics, which will receive further attention in the subsequent chapter. This theory operates under the premise that the environment exhibits a high level of "noise," leading physical systems to inherently evolve toward a maximally mixed state. Within this framework, systems that have reached maximally mixed states are deemed readily accessible and free, whereas pure states are considered valuable resources. Consequently, this theory is often termed the resource theory of purity or nonuniformity, emphasizing that a state's value increases with its deviation from the uniform (maximally) mixed state.

In comparison to the broader field of quantum thermodynamics, the QRT of nonuniformity presents a more streamlined approach, facilitating simpler calculations of conversion rates and resource monotones. Notably, the profound link between majorization and relative majorization, explored in Section 4.3, allows numerous findings in the QRT of athermality to be directly inferred as corollaries from theorems within the QRT of nonuniformity. This interconnection suggests a logical progression: By initially laying down the principles and methodologies inherent to the QRT of nonuniformity, we can seamlessly apply these insights to thermodynamic systems. This approach not only streamlines our exploration but will also enrich our overall comprehension of quantum thermodynamics.

## 16.1 The Free Operations

As already discussed, for any system $A$, we consider the maximally mixed state $\mathbf{u}^A := \frac{1}{|A|} I^A$ to be the free state of the theory. That is, the set of free states $\mathfrak{F}(A) := \{\mathbf{u}^A\}$ consists of a single state. The set of free operations of the QRT of nonuniformity consists of physically implementable operations (see Section 9.2.3) relative to this set of states. They are called completely factorizable channels.

### 16.1.1 Completely Factorizable Channels

**Definition 16.1.** A quantum channel $\mathcal{N} \in \mathrm{CPTP}(A \to A')$ is said to be completely factorizable if there exist systems $B$ and $B'$ and a unitary channel $\mathcal{U} \in \mathrm{CPTP}(AB \to A'B')$ such that $|AB| = |A'B'|$ and

$$\mathcal{N}^{A \to A'}\left(\rho^A\right) = \mathrm{Tr}_{B'}\left[\mathcal{U}^{AB \to A'B'}\left(\rho^A \otimes \mathbf{u}^B\right)\right] \qquad \forall\, \rho \in \mathfrak{L}(A). \qquad (16.1)$$

Note that a completely factorizable channel $\mathcal{N}^{A \to A'}$ has the property that $\mathcal{N}^{A \to A'}(\mathbf{u}^A) = \mathbf{u}^{A'}$. That is, factorizable channels take maximally mixed states to maximally mixed states. In particular, if $|A| = |A'|$, then a completely factorizable channel is unital. However, as we will see shortly, not all unital channels are completely factorizable.

---

**Theorem 16.1.** Let $\mathcal{N} \in \mathrm{CPTP}(A \to A)$ be a mixture of unitaries of the form

$$\mathcal{N}^{A \to A} = \sum_{x \in [\ell]} p_x\, \mathcal{U}_x^{A \to A}, \tag{16.2}$$

where each $\mathcal{U}_x^{A \to A}$ is a unitary channel, $\ell \in \mathbb{N}$, and $\mathbf{p} := (p_1, \ldots, p_\ell)^T$ is a probability vector in $\mathbb{Q}^n$ (i.e. each $p_x$ is a nonnegative rational number). Then, $\mathcal{N}^{A \to A}$ is completely factorizable.

---

**Proof** Since all the $\{p_x\}_{x \in [\ell]}$ are rational, there exists a common denominator $m \in \mathbb{N}$ and $\ell$ integers $\{m_x\}_{x \in [\ell]}$ such that $p_x = \frac{m_x}{m}$, and in particular $\sum_{x \in [\ell]} m_x = m$ since $\sum_{x \in [\ell]} p_x = 1$. Set $n := |A|$, and let $B$ be a system with dimension $|B| = m$. Define a unitary matrix $U: AB \to AB$ via its action on a basis element $|xy\rangle \in AB$ with $x \in [n]$ and $y \in [m]$ as

$$U^{AB} |x\rangle^A |y\rangle^B = U_{k_y}^A |x\rangle^A |y\rangle^B, \tag{16.3}$$

where $k_y$ is the integer in $[\ell]$ satisfying

$$\sum_{j \in [k_y - 1]} m_j \leqslant y < \sum_{j \in [k_y]} m_j \tag{16.4}$$

(note that $k_y$ depends on $y$). That is, $U^{AB}$ is a controlled unitary that its action on $A$ depends on the input of system $B$. Using the notation $\mathcal{U}^{AB \to AB} := U^{AB}(\,\cdot\,)U^{*AB}$ get that for all $\omega \in \mathfrak{L}(A)$

$$\mathrm{Tr}_B\left[\mathcal{U}^{AB \to AB}\left(\omega^A \otimes \mathbf{u}^B\right)\right] = \frac{1}{m} \sum_{y \in [m]} \mathrm{Tr}_B\left[\mathcal{U}^{AB \to AB}\left(\omega^A \otimes |y\rangle\langle y|^B\right)\right]$$

$$= \frac{1}{m} \sum_{y \in [m]} \mathcal{U}_{k_y}^{A \to A}\left(\omega^A\right), \tag{16.5}$$

where we used the definition of $U^{AB}$ here. Now, observe that from the definition of $k_y$, for any $x \in [\ell]$ there exists $m_x$ values of $y \in [m]$ for which $k_y = x$. Therefore, continuing from the last line we get

$$\mathrm{Tr}_B\left[\mathcal{U}^{AB \to AB}\left(\omega^A \otimes \mathbf{u}^B\right)\right] = \sum_{x \in [\ell]} \frac{m_x}{m}\, \mathcal{U}_x^{A \to A}\left(\omega^A\right)$$

$$= \sum_{x \in [\ell]} p_x\, \mathcal{U}_x^{A \to A}(\omega^A). \tag{16.6}$$

Hence, $\sum_{x \in [\ell]} p_x \mathcal{U}_x^{A \to A}$ is a noisy operation. This completes the proof. ∎

## 16.1.2 Noisy Operations

In the proof of Theorem (16.1), we made the assumption that the coefficients $\{p_x\}_{x\in[\ell]}$ are rational numbers. Additionally, it should be noted that the dimension $m$ of system $B$ is determined by the common denominator of these rational coefficients. Consequently, we cannot employ a continuity argument to establish that any mixture of unitaries, potentially with irrational coefficients, qualifies as a completely factorizable channel. This limitation arises because the dimension of system $B$ tends to infinity when the rational coefficients $\{p_x\}_{x\in[\ell]}$ approach irrational numbers.

The aforementioned issue stems from the fact that the system $B$ appearing in the definition of completely factorizable channels, while finite, is yet unbounded. Consequently, it is possible, in principle, to define a sequence of completely factorizable channels $\mathcal{N}_j \in \mathrm{CPTP}(A \to A')$, which can only be realized using systems $B_j$ of increasing dimensions as $j$ grows. Specifically, we observe that $\lim_{j\to\infty} |B_j| = \infty$. In other words, the set encompassing all completely factorizable channels in $\mathrm{CPTP}(A \to A')$ is not closed. However, this unphysical attribute can be addressed by considering the closure of the set of completely factorizable channels.

> **Definition 16.2.** A quantum channel $\mathcal{N} \in \mathrm{CPTP}(A \to A')$ is called a *noisy operation* if there exists a sequence of completely factorizable channels $\{\mathcal{N}_k\}_{k\in\mathbb{N}} \subset \mathrm{CPTP}(A \to A')$ such that
>
> $$\lim_{k\to\infty} \mathcal{N}_k^{A \to A'} = \mathcal{N}^{A \to A'}. \tag{16.7}$$
>
> The set of all noisy operations in $\mathrm{CPTP}(A \to A')$ is denoted by $\mathrm{Noisy}(A \to A')$.

*Remark.* The limit (16.7) is understood in terms of the Choi matrices. That is, the relation (16.7) means that

$$\lim_{k\to\infty} \left\| J_{\mathcal{N}_k}^{AA'} - J_{\mathcal{N}}^{AA'} \right\|_1 = 0. \tag{16.8}$$

Note that by definition the set of noisy operations is closed. Moreover, the set of noisy operations in $\mathrm{CPTP}(A \to A)$ forms a subset of unital channels (see Exercise 16.1). However, it can be shown that not every unital channel is a noisy operation, so that noisy operations form a strict subset of unital channels.

**Exercise 16.1.** *Show that if $\mathcal{N} \in \mathrm{Noisy}(A \to A)$, then $\mathcal{N}^{A \to A}$ is a unital channel.*

> **Theorem 16.2.** Any mixture of unitary channels in $\mathrm{CPTP}(A \to A)$ is a noisy operation.

**Exercise 16.2.** *Use Theorem 16.1 and the definition of noisy operations to prove Theorem 16.2.*

### 16.1.3 The Structure of the QRT of Nonuniformity

We define the resource theory of nonuniformity as a framework where the set of free operations is noisy operations; that is, for any pair of systems $A$ and $A'$, $\mathfrak{F}(A \to A') = \mathrm{Noisy}(A \to A')$. In this resource theory, we can treat all states and operations as classical without any loss of generality. To see this, observe first that any quantum state $\rho^A$ can be converted into a diagonal state in the same basis by applying a unitary channel. Such a unitary channel constitutes a reversible noisy operation within the set of free operations. Hence, all resource states in $\mathfrak{D}(A)$ can be represented by diagonal density matrices in the same basis. We therefore fix a basis and denote the completely dephasing channel in this basis as $\Delta^A \in \mathrm{CPTP}(A \to A)$. To summarize, without loss of generality we can assume that all resources are characterized by states satisfying $\Delta^A(\rho^A) = \rho^A$.

Now, suppose we can convert $\rho \in \mathfrak{D}(A)$ to $\sigma \in \mathfrak{D}(B)$ using a free noisy operation $\mathcal{N} \in \mathrm{CPTP}(A \to B)$. Since we assume that $\rho^A = \Delta^A(\rho^A)$ and $\sigma^B = \Delta^B(\sigma^B)$, it follows that

$$\sigma^B = \mathcal{N}^{A \to B}\left(\rho^A\right)$$

$$\sigma \text{ is diagonal} \to \; = \Delta^B \circ \mathcal{N}^{A \to B}\left(\rho^A\right) \tag{16.9}$$

$$\rho \text{ is diagonal} \to \; = \Delta^B \circ \mathcal{N}^{A \to B} \circ \Delta^A\left(\rho^A\right).$$

Hence, if $\rho^A$ can be converted to $\sigma^B$ using a free quantum channel $\mathcal{N} \in \mathrm{Noisy}(A \to B)$, then $\rho^A$ can also be mapped to $\sigma^B$ using the *classical* channel $\Delta^B \circ \mathcal{N}^{A \to B} \circ \Delta^A$. The latter channel is also a noisy operation since $\Delta^A$ and $\Delta^B$ are themselves noisy operations; they can be expressed as random unitary channels (see Exercise 3.65). Therefore, all resources and free channels can be characterized using classical systems and classical channels.

Finally, considering that all pure states in a fixed dimension are equivalent, we select the qubit pure state $|0\rangle\langle0|$, where $|0\rangle \in \mathbb{C}^2$, as the chosen reference state for this resource theory, serving as the golden unit.

## 16.2  Measures of Nonuniformity

A measure of nonuniformity was defined earlier in Definition 5.3. On the other hand, according to the definition of a resource measure, a measure of nonuniformity is a function

$$g: \bigcup_A \mathfrak{D}(A) \to \mathbb{R} \cup \{\infty\} \tag{16.10}$$

that is nonincreasing under noisy operations and take the value zero on free states. To see that both definitions are equivalent, observe first that since we consider only diagonal states (in the same basis) we can replace $\mathfrak{D}(A)$ with the classical set $\mathrm{Prob}(d)$, where $d := |A|$.

Due to Corollary 16.3, the monotonicity of $g$ under noisy operation is equivalent to the Schur concavity of $g$ and to the third condition in Definition 5.3. The only additional assumption that we added in Definition 5.3 is that $g$ is continuous. This assumption is crucial for the bijection between divergences and measures of nonuniformity (see Theorem 5.3), and we will assume it also here.

The bijection given in Theorem 5.3 demonstrates that all measures of nonuniformity can be expressed as

$$g(\mathbf{p}) = \mathbb{D}\left(\mathbf{p}\,\big\|\,\mathbf{u}^{(d)}\right) \qquad \forall\, d \in \mathbb{N} \quad \forall\, \mathbf{p} \in \mathrm{Prob}(d), \tag{16.11}$$

where $\mathbb{D}$ is a classical divergence. Therefore, all the divergences and relative entropies that introduced in Chapters 5 and 6 can be used to quantify nonuniformity. A particular useful one is the nonuniformity measure obtained by taking $\mathbb{D}$ to be the KL-divergence. In this case, for all $\mathbf{p} \in \mathrm{Prob}(n)$ we have

$$g(\mathbf{p}) = D\left(\mathbf{p}\,\big\|\,\mathbf{u}^{(d)}\right) = \log(d) - H(\mathbf{p}), \tag{16.12}$$

where $H$ is the Shannon entropy. Similarly, for the Rényi divergences we have for all $\alpha \in [0, \infty]$

$$g_\alpha(\mathbf{p}) = D_\alpha\left(\mathbf{p}\,\big\|\,\mathbf{u}^{(d)}\right) = \log(d) - H_\alpha(\mathbf{p}). \tag{16.13}$$

It is worth mentioning that for pure states, specifically when taking $\mathbf{p} = (1, 0, \ldots, 0)^T$, we get that $g_\alpha(\mathbf{p}) = \log(d)$. This implies that the nonuniformity of pure states increases with the dimension $d$.

## 16.3 Interconversions in the Single-Shot Regime

In this section, we delve into the interconversions between nonuniformity states in the single-shot regime, considering both exact and approximate scenarios. We will demonstrate that, similar to pure bipartite entanglement, majorization plays a crucial role in determining these interconversions. We will use the notations $\rho \xrightarrow{\text{Noisy}} \sigma$ whenever $\sigma = \mathcal{N}(\rho)$ for some noisy operation $\mathcal{N} \in \mathrm{Noisy}(A \to A)$.

### 16.3.1 Exact Deterministic Conversions

**Theorem 16.3.** Let $\rho, \sigma \in \mathfrak{D}(A)$. Then, $\rho \xrightarrow{\text{Noisy}} \sigma$ if and only if $\rho \succ \sigma$.

**Proof** Suppose $\sigma = \mathcal{N}(\rho)$ for some noisy operation $\mathcal{N} \in \mathrm{Noisy}(A \to A)$. Since a noisy operation $\mathcal{N} \in \mathrm{Noisy}(A \to A)$ is also a unital channel, from Section 3.5.9 it follows that $\rho \succ \sigma$. In the same subsection we also proved that $\rho \succ \sigma$ if and only if there exists a random unitary channel that takes $\rho$ to $\sigma$. From the Theorem 16.2, this random unitary is also a noisy operation, so the proof is concluded. ∎

This theorem can be slightly modified to accommodate systems of different dimensions. In particular, if $\rho \in \mathfrak{D}(A)$ and $\sigma \in \mathfrak{D}(B)$, then $\sigma^B = \mathcal{N}^{A \to B}(\rho^A)$ for some noisy operation $\mathcal{N} \in \mathrm{Noisy}(A \to B)$ if and only if $\rho^A \otimes \mathbf{u}^B \succ \mathbf{u}^A \otimes \sigma^B$. This is because appending a maximally mixed state is a reversible free operation.

From here onward we consider the "states" of the QRT of nonuniformity to be probability vectors in $\mathrm{Prob}(d)$. Therefore, from the theorem and the discussion here it follows that for two given states $\mathbf{p} \in \mathrm{Prob}(d)$ and $\mathbf{q} \in \mathrm{Prob}(d')$ we have

$$\mathbf{p} \xrightarrow{\text{Noisy}} \mathbf{q} \quad \Longleftrightarrow \quad \left(\mathbf{p}, \mathbf{u}^{(d)}\right) \succ \left(\mathbf{q}, \mathbf{u}^{(d')}\right). \tag{16.14}$$

That is, conversion under noisy operations induce a pre-order that can be characterized with relative majorization. Note that if $d = d'$, this preorder reduces to the standard definition of majorization, however, for $d \neq d'$ it is not equivalent to majorization between $\mathbf{p}$ and $\mathbf{q}$. In particular, embedding a state, say $\mathbf{p} \in \mathrm{Prob}(d)$, in a higher dimensional space $\mathrm{Prob}(d')$ with $d' > d$ (by adding zero components) can increase the resourcefulness of $\mathbf{p}$. Therefore, such embeddings are not free.

**Exercise 16.3.** *Let* $\mathbf{q} = (1/2, 1/2, 0, 0)^T$ *be the vector obtained from the uniform state* $\mathbf{u}^{(2)}$ *by adding two zeros. Show that* $\mathbf{q}$ *can be converted by noisy operations to any state in* $\mathfrak{D}(2)$.

## 16.3.2 The Conversion Distance

Following the general definition given in (11.24), we define the conversion distance of nonuniformity, from a state $\mathbf{p} \in \mathrm{Prob}(d)$ into a state $\mathbf{q} \in \mathrm{Prob}(d')$ as

$$T\left(\mathbf{p} \xrightarrow{\text{Noisy}} \mathbf{q}\right) := \min_{\mathbf{r} \in \mathrm{Prob}(d')} \left\{ \frac{1}{2} \|\mathbf{q} - \mathbf{r}\|_1 : (\mathbf{p}, \mathbf{u}^{(d)}) \succ (\mathbf{r}, \mathbf{u}^{(d')}) \right\}. \tag{16.15}$$

From the properties of the conversion distance (see for example Lemma 11.1), it follows that $T(\mathbf{p} \xrightarrow{\text{Noisy}} \mathbf{q})$ remains invariant under any permutation of the components of $\mathbf{p}$ or $\mathbf{q}$. Therefore, in the rest of this chapter we will always assume without loss of generality that $\mathbf{p} = \mathbf{p}^{\downarrow}$ and $\mathbf{q} = \mathbf{q}^{\downarrow}$.

**Theorem 16.4.** *Let* $\mathbf{p}, \mathbf{q} \in \mathrm{Prob}(d)$ *be two probability vectors. Then,*

$$T\left(\mathbf{p} \xrightarrow{\text{Noisy}} \mathbf{q}\right) = \max_{\ell \in [d]} \left\{ \|\mathbf{q}\|_{(\ell)} - \|\mathbf{p}\|_{(\ell)} \right\}. \tag{16.16}$$

*Remark.* The case that $\mathbf{p} \in \mathrm{Prob}(d)$ and $\mathbf{q} \in \mathrm{Prob}(d')$ with $d \neq d'$ can be solved by applying this theorem to the vectors $\mathbf{p} \otimes \mathbf{u}^{(d')}$ and $\mathbf{u}^{(d)} \otimes \mathbf{q}$. Specifically,

$$T\left(\mathbf{p} \xrightarrow{\text{Noisy}} \mathbf{q}\right) = \max_{\ell \in [dd']} \left\{ \left\|\mathbf{u}^{(d)} \otimes \mathbf{q}\right\|_{(\ell)} - \left\|\mathbf{p} \otimes \mathbf{u}^{(d')}\right\|_{(\ell)} \right\}. \tag{16.17}$$

**Proof** Since we consider the case that both $\mathbf{p}$ and $\mathbf{q}$ are $d$-dimensional, the conversion distance can be expressed as

$$T\left(\mathbf{p} \xrightarrow{\text{Noisy}} \mathbf{q}\right) = \min_{\mathbf{r}\in\text{Prob}(d)}\left\{\frac{1}{2}\|\mathbf{q}-\mathbf{r}\|_1 : \mathbf{p}\succ\mathbf{r}\right\}. \tag{16.18}$$

This expression for the conversion distance represents the distance of $\mathbf{q}$ to the set majo($\mathbf{p}$) as defined in (4.96) (with $\mathbf{p}$ replacing $\mathbf{q}$). Hence,

$$T\left(\mathbf{p} \xrightarrow{\text{Noisy}} \mathbf{q}\right) = T\left(\mathbf{q}, \text{majo}(\mathbf{p})\right)$$

$$\textbf{Theorem 4.6}\!\rightarrow = \max_{\ell\in[n]}\left\{\|\mathbf{q}\|_{(\ell)} - \|\mathbf{p}\|_{(\ell)}\right\}. \tag{16.19}$$

This completes the proof. ∎

### 16.3.3 The Single-Shot Nonuniformity Cost

In order to define the single-shot nonuniformity cost of a resource state $\mathbf{p}\in\text{Prob}^{\downarrow}(d)$, we first recall that the vector $\mathbf{e}_1^{(m)}$, defined as $(1,0,\ldots,0)^T\in\text{Prob}^{\downarrow}(m)$, represents the maximal resource in dimension $m$. Moreover, the resourcefulness of $\mathbf{e}_1^{(m)}$ increases with $m$, and the set $\{\mathbf{e}_1^{(m)}\}_{m\in\mathbb{N}}$ forms a golden unit according to Definition 11.1. Therefore, we define the $\varepsilon$-nonuniformity cost of $\mathbf{p}$ as

$$\text{Cost}^{\varepsilon}(\mathbf{p}) := \min\left\{\log m : T\left(\mathbf{e}_1^{(m)} \xrightarrow{\text{Noisy}} \mathbf{p}\right)\leqslant\varepsilon\right\}. \tag{16.20}$$

Clearly, $\text{Cost}^{\varepsilon}(\mathbf{p})\leqslant\log d$ since $T\left(\mathbf{e}_1^{(d)} \xrightarrow{\text{Noisy}} \mathbf{p}\right) = 0$. We will therefore assume (implicitly) in the rest of this section that $m\leqslant d$. In the following theorem, we use the notation $H_{\min}^{\varepsilon}(\mathbf{p})$ for the smoothed min-entropy of $\mathbf{p}$. In (10.147) we found a closed form for this smoothed entropy given by

$$H_{\min}^{\varepsilon}(\mathbf{p}) = -\log\max_{\ell\in[d]}\left\{\frac{\|\mathbf{p}\|_{(\ell)}-\varepsilon}{\ell}\right\}. \tag{16.21}$$

---

**Theorem 16.5.** Let $\varepsilon\in(0,1)$ and $\mathbf{p}\in\text{Prob}^{\downarrow}(d)$. The $\varepsilon$-nonuniformity cost of $\mathbf{p}$ is given by

$$\text{Cost}^{\varepsilon}(\mathbf{p}) = \log\left\lceil d2^{-H_{\min}^{\varepsilon}(\mathbf{p})}\right\rceil. \tag{16.22}$$

---

**Proof** We first prove the theorem for the case $\varepsilon = 0$. In this case,

$$\text{Cost}^{\varepsilon=0}(\mathbf{p}) := \min\left\{\log m : \mathbf{e}_1^{(m)} \xrightarrow{\text{Noisy}} \mathbf{p}\right\}. \tag{16.23}$$

The condition $\mathbf{e}_1^{(m)} \xrightarrow{\text{Noisy}} \mathbf{p}$ is equivalent to $(\mathbf{e}_1^{(m)},\mathbf{u}^{(m)})\succ(\mathbf{p},\mathbf{u}^{(d)})$. Moreover, in Exercise 16.4 you show that the condition $(\mathbf{e}_1^{(m)},\mathbf{u}^{(m)})\succ(\mathbf{p},\mathbf{u}^{(d)})$ is equivalent to $p_1\leqslant\frac{m}{d}$. Since the smallest integer that satisfies this condition is $m = \lceil dp_1\rceil$, we conclude that

$$\text{Cost}^{\varepsilon=0}(\mathbf{p}) = \log \lceil dp_1 \rceil$$

$$\textbf{cf. (6.22)} \rightarrow = \log \left\lceil d2^{-H_{\min}(\mathbf{p})} \right\rceil. \tag{16.24}$$

This completes the proof for the case $\varepsilon = 0$. For $\varepsilon > 0$ we use (11.34) to get

$$\text{Cost}^{\varepsilon}(\mathbf{p}) = \min_{\mathbf{p}' \in \mathcal{B}_{\varepsilon}(\mathbf{p})} \text{Cost}^{\varepsilon=0}(\mathbf{p}')$$

$$\textbf{(16.24)} \rightarrow = \min_{\mathbf{p}' \in \mathcal{B}_{\varepsilon}(\mathbf{p})} \log \left\lceil d2^{-H_{\min}(\mathbf{p}')} \right\rceil \tag{16.25}$$

$$= \log \left\lceil d2^{-H^{\varepsilon}_{\min}(\mathbf{p})} \right\rceil,$$

where the last line follows from the definition of $H^{\varepsilon}_{\min}(\mathbf{p})$. This completes the proof. ∎

**Exercise 16.4.** *Show that the condition* $(\mathbf{e}_1^{(m)}, \mathbf{u}^{(m)}) \succ (\mathbf{p}, \mathbf{u}^{(d)})$ *is equivalent to* $p_1 \leqslant \frac{m}{d}$.

**Exercise 16.5.** *Let* $\mathbf{p} \in \text{Prob}^{\downarrow}(d)$ *and* $m \in [d]$.

1. *Show that*

$$T\left(\mathbf{e}_1^{(m)} \xrightarrow{\text{Noisy}} \mathbf{p}\right) = f_{\mathbf{p}}\left(\frac{m}{d}\right), \tag{16.26}$$

*where* $f_{\mathbf{p}}(t) := \sum_{x \in [d]} (p_x - t)_+$ *is the function studied at the end of Section 4.2.2.*
2. *Provide a direct proof of Theorem 16.5 using this conversion distance and the explicit expression given in (4.107) for* $f_{\mathbf{p}}^{-1}$.
3. *Show that the conversion distance here can also be expressed as*

$$T\left(\mathbf{e}_1^{(m)} \xrightarrow{\text{Noisy}} \mathbf{p}\right) = \frac{1}{2} \left\| \mathbf{p} - m\mathbf{u}^{(d)} \right\|_1 - \frac{m-1}{2}. \tag{16.27}$$

**Exercise 16.6.** *Show that the single-shot $\varepsilon$-nonuniformity cost of $\mathbf{p}$ is bounded by*

$$\log\left(\|\mathbf{p}\|_{(k)} - \varepsilon\right) \leqslant \text{Cost}^{\varepsilon}(\mathbf{p}) - \log(d/k) \leqslant \log\left(\|\mathbf{p}\|_{(k)} - \varepsilon + \frac{k}{d}\right), \tag{16.28}$$

*where $k \in [d]$ is the integer satisfying $\varepsilon \in (r_k, r_{k+1}]$, where $r_k$ is defined in (4.83).*

## 16.3.4 The Single-Shot Distillable Nonuniformity

For any $\varepsilon \in (0, 1)$ and $\mathbf{p} \in \text{Prob}^{\downarrow}(d)$, we define the $\varepsilon$-single-shot distillable nonuniformity of $\mathbf{p}$ as

$$\text{Distill}^{\varepsilon}(\mathbf{p}) := \max\left\{\log m : T\left(\mathbf{p} \xrightarrow{\text{Noisy}} \mathbf{e}_1^{(m)}\right) \leqslant \varepsilon\right\}. \tag{16.29}$$

Unlike the case for resource cost, an analogous formula to (11.34) does not exist for resource distillation. Therefore, the calculation of single-shot distillable nonuniformity necessitates a direct computation of the conversion distance $T\left(\mathbf{p} \xrightarrow{\text{Noisy}} \mathbf{e}_1^{(m)}\right)$. In the following lemma we provide a closed formula of this conversion distance in terms of the coefficient $\mu_m$, which is defined for all $m \in \mathbb{N}$ as

$$\mu_m := \|\mathbf{p}\|_{(\lfloor d/m \rfloor)} + (d/m - \lfloor d/m \rfloor)\, p_{\lfloor d/m \rfloor + 1}, \tag{16.30}$$

with the convention that $\|\mathbf{p}\|_{(0)} = 0$ so that $\mu_m = \frac{d}{m} p_1$ if $m > d$.

> **Lemma 16.1.** Let $d, m \in \mathbb{N}$, $\mathbf{p} \in \mathrm{Prob}^{\downarrow}(d)$, and $\mu_m$ as defined in (16.30). Then,
>
> $$T\left(\mathbf{p} \xrightarrow{\text{Noisy}} \mathbf{e}_1^{(m)}\right) = 1 - \mu_m. \tag{16.31}$$

**Proof** The case $m > d$ is left as an exercise, and we assume here that $m \leqslant d$. From the previous section, the conversion distance can be expressed as

$$T\left(\mathbf{p} \xrightarrow{\text{Noisy}} \mathbf{e}_1^{(m)}\right) = \max_{k \in [dm]} \sum_{j \in [k]} \left((\mathbf{e}_1^{(m)} \otimes \mathbf{u}^{(d)})_j^{\downarrow} - (\mathbf{u}^{(m)} \otimes \mathbf{p})_j^{\downarrow}\right). \tag{16.32}$$

Since the vector $\mathbf{e}_1^{(m)} \otimes \mathbf{u}^{(d)}$ has exactly $n$ nonzero components (all equal to $1/d$), we get that the optimizer $k$ above must satisfy $k \leqslant d$. Moreover, the $j$-th term in (16.32) has the form

$$(\mathbf{e}_1^{(m)} \otimes \mathbf{u}^{(d)})_j^{\downarrow} - (\mathbf{u}^{(m)} \otimes \mathbf{p})_j^{\downarrow} = \frac{1}{d} - \frac{p_x}{m}, \tag{16.33}$$

where $x = \left\lceil \frac{j}{m} \right\rceil$. Since $\mathbf{p} = \mathbf{p}^{\downarrow}$, the terms in (16.33) are nondecreasing with $j$. We therefore conclude that the optimal $k$ in (16.32) must be $k = d$. Denoting $a := \lfloor \frac{d}{m} \rfloor$ and $b := d - am$ (hence $d = am + b$) we get

$$T\left(\mathbf{p} \xrightarrow{\text{Noisy}} \mathbf{e}_1^{(m)}\right) = 1 - \left(\sum_{x \in [a]} p_x + b\frac{p_{a+1}}{m}\right)$$

$$\boxed{b = d - am} \longrightarrow \quad = 1 - \|\mathbf{p}\|_{(a)} - \left(\frac{d}{m} - a\right) p_{a+1} \tag{16.34}$$

$$= 1 - \mu_m.$$

This completes the proof. ∎

Combining the definition of the $\varepsilon$-single-shot distillable nonuniformity with Lemma 16.1 we obtain the following closed form for $\mathrm{Distill}^{\varepsilon}(\mathbf{p})$.

> **Theorem 16.6.** Let $\varepsilon \in (0, 1)$, $\mathbf{p} \in \mathrm{Prob}^{\downarrow}(d)$, $m \in [d]$, and $\mu_m$ as defined in (16.30). If $p_1 > 1 - \varepsilon$, then $\mathrm{Distill}^{\varepsilon}(\mathbf{p}) := \lfloor dp_1/(1 - \varepsilon) \rfloor$. Otherwise, the $\varepsilon$-single-shot distillable nonuniformity is given by
>
> $$\mathrm{Distill}^{\varepsilon}(\mathbf{p}) := \max_{m \in [d]} \{\log m : \mu_m \geqslant 1 - \varepsilon\}. \tag{16.35}$$

**Exercise 16.7.** *Use the closed form in* (16.31) *to prove Theorem 16.6.*

**Exercise 16.8.** *Show that for* $\varepsilon = 0$ *the single-shot distillable nonuniformity of* $\mathbf{p} \in$ Prob$(d)$ *is given by*

$$\text{Distill}^{\varepsilon = 0}(\mathbf{p}) = \log(d) - H_{\max}(\mathbf{p}), \tag{16.36}$$

*where* $H_{\max}$ *is the max-entropy given by* $H_{\max}(\mathbf{p}) := \log(k)$, *where* $k$ *is the number of nonzero components of* $\mathbf{p}$.

The formula in Theorem 16.6 is somewhat cumbersome. One can get somewhat simpler bounds on the single-shot distillable entanglement by removing the floor functions that appear in the definition of $\mu_m$. These simpler bounds can be expressed in terms of the formula for the smoothed max-entropy given in Lemma 10.6. Specifically, from Lemma 10.6 it follows that the smoothed max-entropy can be expressed as the logarithm of an integer $k$ satisfying

$$\|\mathbf{p}\|_{(k-1)} < 1 - \varepsilon \leqslant \|\mathbf{p}\|_{(k)}, \tag{16.37}$$

with the convention $\|\mathbf{p}\|_{(0)} := 0$.

> **Corollary 16.1.** Let $\varepsilon \in (0, 1)$, $\mathbf{p} \in \text{Prob}(d)$, and set $k := 2^{H_{\max}^{\varepsilon}(\mathbf{p})}$. Then,
>
> $$\log(d - 1 - k) - \log(1 + k) \leqslant \text{Distill}^{\varepsilon}(\mathbf{p}) \leqslant \log(d) - \log(k - 1). \tag{16.38}$$

**Proof**   First, observe that

$$\|\mathbf{p}\|_{(\lfloor d/m \rfloor)} \leqslant \mu_m \leqslant \|\mathbf{p}\|_{(\lfloor d/m \rfloor + 1)}. \tag{16.39}$$

Therefore,

$$\text{Distill}^{\varepsilon}(\mathbf{p}) \leqslant \max_{m \in [d]} \left\{ \log m : \|\mathbf{p}\|_{(\lfloor d/m \rfloor + 1)} \geqslant 1 - \varepsilon \right\}. \tag{16.40}$$

Now, observe that $\lfloor \frac{d}{m} \rfloor \leqslant \frac{d}{m}$ so that $m \leqslant \frac{d}{\lfloor d/m \rfloor}$. Hence,

$$\text{Distill}^{\varepsilon}(\mathbf{p}) \leqslant \max_{m \in [d]} \left\{ \log \frac{d}{\lfloor d/m \rfloor} : \|\mathbf{p}\|_{(\lfloor d/m \rfloor + 1)} \geqslant 1 - \varepsilon \right\}$$

$$\begin{array}{c}\text{\textit{Replacing} } \lfloor d/m \rfloor \\ \text{\textit{with arbitrary} } \ell \in [d]\end{array} \rightarrow \leqslant \max_{\ell \in [d]} \left\{ \log \frac{d}{\ell} : \|\mathbf{p}\|_{(\ell + 1)} \geqslant 1 - \varepsilon \right\} \tag{16.41}$$

$$(16.37) \rightarrow = \log \frac{d}{k - 1}.$$

For the lower bound, observe that (16.39) gives

$$\text{Distill}^{\varepsilon}(\mathbf{p}) \geqslant \max_{m \in [d]} \left\{ \log m : \|\mathbf{p}\|_{(\lfloor d/m \rfloor)} \geqslant 1 - \varepsilon \right\}. \tag{16.42}$$

Now, for the lower bound we cannot replace $\lfloor d/m \rfloor$ with arbitrary integer $\ell \in [d]$ since this will increase the right-hand side (16.42). Instead, we use the fact that for any $s \in [\frac{1}{d}, 1]$ there exists a unique $m \in [d]$ such that

$$s - \frac{1}{d} < \frac{m}{d} \leqslant s. \tag{16.43}$$

Observe further that for any such $s \in [\frac{1}{d}, 1]$ and $m \in [d]$, if in addition $\|\mathbf{p}\|_{(\lfloor s^{-1} \rfloor)} \geqslant 1 - \varepsilon$, then also $\|\mathbf{p}\|_{(\lfloor d/m \rfloor)} \geqslant 1 - \varepsilon$ since $s^{-1} \leqslant \frac{d}{m}$. Moreover, since such $m$ and $s$ also satisfy $\log m \geqslant \log (ds - 1)$ we get that

$$\mathrm{Distill}^{\varepsilon}(\mathbf{p}) \geqslant \max_{s \in [\frac{1}{d}, 1]} \left\{ \log (ds - 1) : \|\mathbf{p}\|_{(\lfloor s^{-1} \rfloor)} \geqslant 1 - \varepsilon \right\}. \tag{16.44}$$

In the last step, denote by $\ell := \lfloor s^{-1} \rfloor \in [d]$ and use the fact that $s^{-1} \leqslant \ell + 1$ to get $s \geqslant \frac{1}{\ell+1}$. Substituting this to the right-hand side of Equation (16.44) gives

$$\mathrm{Distill}^{\varepsilon}(\mathbf{p}) \geqslant \max_{\ell \in [d]} \left\{ \log \left( \frac{d}{1+\ell} - 1 \right) : \|\mathbf{p}\|_{(\ell)} \geqslant 1 - \varepsilon \right\}$$
$$= \log \left( \frac{d}{1+k} - 1 \right). \tag{16.45}$$

This completes the proof. $\blacksquare$

## 16.4 Asymptotic Conversions

The distillation rate of a nonuniformity state $\mathbf{q}$ from another nonuniformity state $\mathbf{p}$ is defined as

$$\mathrm{Distill}(\mathbf{p} \to \mathbf{q}) := \lim_{\varepsilon \to 0^+} \sup_{n,m \in \mathbb{N}} \left\{ \frac{m}{n} : T \left( \mathbf{p}^{\otimes n} \xrightarrow{\text{Noisy}} \mathbf{q}^{\otimes m} \right) \leqslant \varepsilon \right\}. \tag{16.46}$$

We will show in this section that the conversion rate has the following simple formula.

---

**Theorem 16.7.** Let $\mathbf{p} \in \mathrm{Prob}(d)$ and $\mathbf{q} \in \mathrm{Prob}(d')$ be two probability vectors. The conversion rate in (16.46) is given by

$$\mathrm{Distill}(\mathbf{p} \to \mathbf{q}) = \frac{D\left(\mathbf{p} \| \mathbf{u}^{(d)}\right)}{D\left(\mathbf{q} \| \mathbf{u}^{(d')}\right)} = \frac{\log (d) - H(\mathbf{p})}{\log (d') - H(\mathbf{q})}. \tag{16.47}$$

---

*Remark.* Note that the formula for the asymptotic conversion rate demonstrates that the resource theory of nonuniformity is reversible. Specifically, note that for any $\mathbf{p}$ and $\mathbf{q}$ as above, $\mathrm{Distill}(\mathbf{p} \to \mathbf{q})\mathrm{Distill}(\mathbf{q} \to \mathbf{p}) = 1$.

We prove Theorem 16.7 by computing separately the nonuniformity cost and the distillable nonuniformity. Recall from the discussion in Section 11.5.1, specifically (11.137), that the asymptotic cost of a nonuniformity state $\mathbf{p} \in \mathrm{Prob}(k)$ is given by

$$\mathrm{Cost}(\mathbf{p}) := \lim_{\varepsilon \to 0^+} \liminf_{n \to \infty} \frac{1}{n} \mathrm{Cost}^{\varepsilon} \left( \mathbf{p}^{\otimes n} \right). \tag{16.48}$$

Therefore, we can use the results from the single-shot case to compute this asymptotic rate.

> **Lemma 16.2.** Let $p \in \mathrm{Prob}(d)$ and $\varepsilon \in (0, 1)$. Then, the asymptotic nonuniformity cost of $p$ is given by
>
> $$\mathrm{Cost}(p) = \lim_{n \to \infty} \frac{1}{n} \mathrm{Cost}^{\varepsilon} \left( p^{\otimes n} \right) = \log(d) - H(p). \tag{16.49}$$

**Proof**   From the result in the single-shot case, specifically (16.22), we obtain

$$\lim_{n \to \infty} \frac{1}{n} \mathrm{Cost}^{\varepsilon} \left( p^{\otimes n} \right) = \lim_{n \to \infty} \frac{1}{n} \log \left\lceil d^n 2^{-H_{\min}^{\varepsilon}(p^{\otimes n})} \right\rceil \tag{16.50}$$
$$\mathbf{AEP\ (11.52)} \to \; = \log(d) - H(p).$$

This completes the proof.     $\blacksquare$

Similarly, from the discussion in Section 11.5.1, specifically (11.138), the asymptotic distillation rate of a nonuniformity state $p \in \mathrm{Prob}(d)$ is given by

$$\mathrm{Distill}(p) = \lim_{\varepsilon \to 0^+} \limsup_{n \to \infty} \frac{1}{n} \mathrm{Distill}^{\varepsilon} \left( p^{\otimes n} \right). \tag{16.51}$$

As before, we can use the results from the single-shot regime to compute this expression.

> **Lemma 16.3.** Let $p \in \mathrm{Prob}(d)$ and $\varepsilon \in (0, 1)$. Then, the asymptotic distillable nonuniformity is given by
>
> $$\mathrm{Distill}(p) = \lim_{n \to \infty} \frac{1}{n} \mathrm{Distill}^{\varepsilon}(p^{\otimes n}) = \log(d) - H(p). \tag{16.52}$$

**Proof**   From the upper bound in (16.38) we get

$$\limsup_{n \to \infty} \frac{1}{n} \mathrm{Distill}^{\varepsilon}(p^{\otimes n}) \leqslant \limsup_{n \to \infty} \frac{1}{n} \log \left( \frac{d^n}{2^{H_{\max}^{\varepsilon}(p^{\otimes n})} - 1} \right) \tag{16.53}$$
$$\mathbf{AEP\ (10.169)} \to \; = \log(d) - H(p).$$

Similarly, from the lower bound in (16.38) we get

$$\liminf_{n \to \infty} \frac{1}{n} \mathrm{Distill}^{\varepsilon}(p^{\otimes n}) \geqslant \liminf_{n \to \infty} \frac{1}{n} \log \left( \frac{d^n}{1 + 2^{H_{\max}^{\varepsilon}(p^{\otimes n})}} - 1 \right) \tag{16.54}$$
$$\mathbf{AEP\ (10.169)} \to \; = \log(d) - H(p).$$

Comparing the two inequalities in (16.53) and (16.54) we conclude that

$$\lim_{n \to \infty} \frac{1}{n} \mathrm{Distill}^{\varepsilon}(p^{\otimes n}) = \log(d) - H(p). \tag{16.55}$$

This completes the proof.     $\blacksquare$

**Exercise 16.9.** *Use the Lemmas 16.2 and 16.3 to prove Theorem 16.7.*

# 16.5 Notes and References

To the best of the author's knowledge, the inaugural paper on the resource theory of non-uniformity was presented in Ref. [132]. This paper introduced the term "Noisy Operations." The study of factorizable channels, in the context of von-Neumann algebra, was carried out independently, as can be seen in, for instance, in Ref. [108]. It was only much later that this resource theory was identified as a subset of the QRT of quantum athermality. For a comprehensive review and additional references, one can refer to the review article in Ref. [94].

# Quantum Thermodynamics

Thermodynamics stands as one of the most influential theories in physics, finding applications across a wide range of disciplines. Initially focused on steam engines, its relevance has expanded to encompass fields such as biochemistry, nanotechnology, and black hole physics, among others [83, 26, 61]. Despite its immense success, the foundational aspects of thermodynamics continue to be a subject of controversy. There persists a pervasive confusion regarding the relationship between macroscopic and microscopic laws, particularly concerning reversibility and time-symmetry. Furthermore, there is a lack of consensus on the optimal formulation of the second law. As early as 1941, Nobel laureate Percy Bridgman noted, "there are almost as many formulations of the Second Law as there have been discussions of it," and unfortunately, little progress has been made in resolving this situation since then. In recent years, researchers have taken a fresh perspective on these fundamental issues by approaching thermodynamics as a resource theory. This viewpoint considers a system that is not in equilibrium with its environment as a valuable resource known as "athermality." Athermality serves as the fuel utilized in work extraction, computational erasure operations, and other thermodynamic tasks.

The resource-theoretic approach to thermodynamics delves into the quantification of a state's deviation from equilibrium and explores its utility in quantum thermodynamics. It also investigates the necessary and sufficient conditions for transforming one state into another. Within this framework, different notions of state conversion can be examined, including exact and approximate conversions, single-copy and multiple-copy scenarios, and conversions with or without the aid of a catalyst.

These quantum-information techniques have brought forth numerous novel insights, particularly considering the historical importance of information in foundational topics such as Maxwell's demon [156], the thermodynamic reversibility of computation [17, 18], Landauer's principle regarding the work cost of erasure [148, 139], and Jaynes's utilization of maximum entropy principles in deriving statistical mechanics [140, 141].

Furthermore, the resource-theoretic approach to thermodynamics reveals that the conventional formulation of the second law of thermodynamics, which focuses on entropy non-decrease, is insufficient as a criterion for determining the feasibility of a given state conversion. However, we will discover that it is possible to identify a set of measures quantifying the degree of nonequilibrium (including entropy) such that a state conversion is feasible if and only if all of these measures do not increase.

# 17.1 Thermal States and Athermality States

We use the term "thermal bath" or "thermal reservoir" to indicate a thermodynamic system with the property that any amount of reasonable heat that is added or extracted from it does not change its temperature. In other words, its heat capacity is extremely large. For a heat bath that is held at a fixed inverse temperature $\beta := \frac{1}{k_B T}$, the state,

$$\gamma^B := \frac{e^{-\beta H^B}}{\text{Tr}\left[e^{-\beta H^B}\right]} \tag{17.1}$$

is the thermal equilibrium state known as the Gibbs state. The Gibbs state, $\gamma^B$, is also referred to as the thermal state of the system $B$, and the normalization factor

$$\mathcal{Z}^B := \text{Tr}\left[e^{-\beta H^B}\right] \tag{17.2}$$

is called the *partition function*.

The free states in the resource theory of athermality corresponds to physical systems that are in thermal equilibrium with their surrounding. We will therefore consider Gibbs states to be free. Note however that the Gibbs state of a system $B$ is defined with respect to the Hamiltonian $H^B$ of the system (or heat bath). As it follows from the following exercise, any density matrix is a Gibbs state with respect to some Hamiltonian.

**Exercise 17.1.** *Show that for any density matrix $\rho \in \mathfrak{D}_{>0}(A)$ there exists a Hamiltonian $H \in \text{Pos}(A)$ such that $\rho^A$ is the Gibbs state with respect to this Hamiltonian.*

## 17.1.1 Optimality of the Gibbs State

Energy and entropy are the two cornerstones of thermodynamics. Any quantum (or classical) system tends to evolve into equilibrium state with its environment. In such a spontaneous evolution, the second law of thermodynamics states that the entropy associated with the system cannot decrease, or alternatively, heat can never pass from a colder system to a warmer body without an external work put into the system. One can therefore think of the equilibrium state of a system with a given (fixed) entropy as the state that has the lowest amount of energy, so that no further heat exchange can occur with the environment.

Explicitly, let $A$ be a quantum system with Hamiltonian $H^A$ and a given entropy $S$. Among all states $\rho \in \mathfrak{D}(A)$ with entropy $H(\rho) = S$, we want to find the one that has the smallest amount of energy. That is, we want to minimize $\text{Tr}[\rho^A H^A]$ given the constraint that the entropy of $\rho^A$ equals $S$. This problem can be solved by minimizing the Lagrangian

$$\mathcal{L}(\rho, \lambda) := \text{Tr}[H\rho] + \lambda(\text{Tr}[\rho \log \rho] + S) \tag{17.3}$$

over all $\rho \in \mathfrak{D}(A)$, where $\lambda$ is a Lagrange multiplier. Let $\rho$ be the optimal density matrix that minimizes the Lagrangian in (17.3). Then, any other state in $\mathfrak{D}(A)$ can be written as $\rho + tY$ for some $t \in \mathbb{R}$ and $Y \in \text{Herm}(A)$ is a traceless matrix (we can also assume without loss of generality that $\|Y\|_\infty \leqslant 1$ although we will not need it). Since $\rho$ is optimal we must have for any such $Y$

$$0 = \frac{d}{dt}\mathcal{L}(\rho + tY, \lambda)\Big|_{t=0}$$

$$\textbf{Exercise 17.2} \rightarrow = \text{Tr}[HY] + \lambda\text{Tr}[Y\log\rho]. \tag{17.4}$$

That is, an optimal $\rho$ must satisfy

$$\text{Tr}[Y(H + \lambda\log\rho)] = 0 \tag{17.5}$$

for all traceless matrices $Y \in \text{Herm}(A)$, so that $H + \lambda\log\rho$ is orthogonal (in the Hilbert–Schmidt inner product) to the subspace of all traceless matrices in $\text{Herm}(A)$. Consequently, $H + \lambda\log\rho$ must be proportional to the identity matrix; that is, there exists $c \in \mathbb{R}$ such that $H + \lambda\log\rho = cI$. Hence, the optimal $\rho$ has the form

$$\rho = e^{\frac{c}{\lambda}}e^{-\frac{H}{\lambda}} = \frac{e^{-\beta H}}{\mathcal{Z}}, \tag{17.6}$$

where in the last equality we denoted by $\beta := \frac{1}{\lambda}$, and used the fact that $\text{Tr}[\rho] = 1$ so that $e^{\frac{c}{\lambda}} = 1/\text{Tr}[e^{-\beta H}]$.

**Exercise 17.2.** *Use Corollary D.1 of online version to prove the expression for the directional derivative given in* (17.4).

**Exercise 17.3.** *Let $\alpha \in [0, \infty]$. Find the state $\rho_u \in \mathfrak{D}(A)$ that minimizes $\text{Tr}[H^A\rho^A]$ while keeping the $\alpha$-Rényi entropy fixed.*

## 17.1.2 Passive States

In deriving the Gibbs state already mentioned, we sought a state that minimizes the average energy while maintaining a constant von-Neumann entropy. Although this method is mathematically sounds, it does not offer a convincing rationale for designating the Gibbs state as the only free state within the model. This raises the question of whether there exists a more operational method to derive the Gibbs state, one that does not rely on the concept of von-Neumann entropy. As we will explore here, such an approach does indeed exist.

Let $A$ be a physical system with a Hamiltonian $H^A$. Suppose that the system $A$ is described by the density matrix $\rho \in \mathfrak{D}(A)$. Therefore, the energy of the system is given by $\text{Tr}[H^A\rho^A]$. Under a closed (i.e. unitary) evolution/process, the system $A$ can be evolved into another state of the form $U\rho^A U^*$, where $U \in \mathfrak{U}(A)$ is some unitary matrix. The maximal amount of work that can be extracted from such a system cannot exceed the energy difference between the initial and final states. Therefore, we define the maximal extractable work from a system in a state $\rho^A$ as

$$W_{\max}\left(\rho^A\right) = \max_U \text{Tr}\left[H^A\left(\rho^A - U\rho^A U^*\right)\right], \tag{17.7}$$

where the maximum is over all unitary matrices $U \in \mathfrak{U}(A)$. Interestingly, this optimization problem can be solved analytically.

---

**Lemma 17.1.** Let $H^A = \sum_{x \in [m]} a_x |x\rangle\langle x|^A$ be the Hamiltonian of system $A$, with the energy eigenvalues arranged in nondecreasing order; that is, $a_1 \leqslant a_2 \leqslant \cdots \leqslant a_m$. Then,

$$W_{\max}\left(\rho^A\right) = \mathrm{Tr}\left[H^A \rho^A\right] - \mathrm{Tr}\left[H^A \sigma_\rho^A\right] \qquad \forall\, \rho \in \mathfrak{D}(A), \tag{17.8}$$

where $\sigma_\rho^A$ is defined with respect to the eigenvalues $\{p_x\}_{x \in [m]}$ of $\rho^A$ as

$$\sigma_\rho^A := \sum_{x \in [m]} p_x^{\downarrow} |x\rangle\langle x|^A. \tag{17.9}$$

---

**Proof** From a variant of the von-Neumann trace inequality, known as Ruhe's Trace Inequality, as given in Theorem B.3 of online version, it follows that

$$\mathrm{Tr}\left[H^A U \rho^A U^*\right] \geqslant \sum_{x \in [m]} a_x p_x^{\downarrow} \tag{17.10}$$
$$= \mathrm{Tr}\left[H^A \sigma_\rho^A\right],$$

where we used the lower bound in (B.26) of online version with $N := H^A$ and $M := U \rho^A U^*$. The proof is then concluded with the observation that there exists a unitary matrix $U$ satisfying $U \rho^A U^* = \sigma_\rho^A$. ∎

**Exercise 17.4.** *Let $n, m \in \mathbb{N}$ and $\rho, \sigma \in \mathfrak{D}(A)$.*

*1. Show that*

$$W_{\max}(\rho \otimes \sigma) \geqslant W_{\max}(\rho) + W_{\max}(\sigma). \tag{17.11}$$

*2. Show that if $n \geqslant m$, then*

$$\frac{1}{n} W_{\max}\left(\rho^{\otimes n}\right) \geqslant \frac{1}{m} W_{\max}\left(\rho^{\otimes m}\right). \tag{17.12}$$

In the following lemma, we demonstrate that the maximum extractable work can never exceed the difference between the energy of the system and the energy of the system at equilibrium.

---

**Lemma 17.2.** Let $\gamma^A$ be a Gibbs state of system $A$ with a temperature $\beta$ for which $H(\rho^A) = H(\gamma^A)$. Then,

$$W_{\max}\left(\rho^A\right) \leqslant \mathrm{Tr}\left[H^A \rho^A\right] - \mathrm{Tr}\left[H^A \gamma^A\right]. \tag{17.13}$$

**Proof** The Gibbs state $\gamma^A$ is the state with the smallest energy that has an entropy $H(\rho^A) = H(\sigma_\rho^A)$. Therefore, the state $\sigma_\rho^A$ has higher energy than $\gamma^A$ so that

$$\mathrm{Tr}\left[H^A\gamma^A\right] \leqslant \mathrm{Tr}\left[H^A\sigma_\rho^A\right]. \tag{17.14}$$

Combining this with (17.8) completes the proof. ∎

Lemma 17.2 establishes an additive upper bound. Consequently, it can be inferred from the lemma that

$$W_{\max}^{\mathrm{reg}}\left(\rho^A\right) := \lim_{n\to\infty} \frac{1}{n} W_{\max}\left(\rho^{\otimes n}\right)$$

**(17.13) applied to system $A^n$** $\to$ $\leqslant \lim_{n\to\infty} \frac{1}{n}\left(\mathrm{Tr}\left[H^{A^n}\rho^{\otimes n}\right] - \mathrm{Tr}\left[H^{A^n}\gamma^{A^n}\right]\right)$ (17.15)

$$= \mathrm{Tr}\left[H^A\rho^A\right] - \mathrm{Tr}\left[H^A\gamma^A\right],$$

where we used the relations $\mathrm{Tr}\left[H^{A^n}\rho^{\otimes n}\right] = n\mathrm{Tr}\left[H^A\rho^A\right]$ and $\mathrm{Tr}\left[H^{A^n}\gamma^{A^n}\right] = n\mathrm{Tr}\left[H^A\gamma^A\right]$. We now demonstrate that the aforementioned inequality actually holds as an equality.

---

**Theorem 17.1.** Let $\gamma^A$ be a Gibbs state of system $A$ with a temperature $\beta$ for which $H(\rho^A) = H(\gamma^A)$. Then,

$$W_{\max}^{\mathrm{reg}}\left(\rho^A\right) = \mathrm{Tr}\left[H^A\rho^A\right] - \mathrm{Tr}\left[H^A\gamma^A\right]. \tag{17.16}$$

---

**Proof** From (17.15) it is sufficient to prove that

$$W_{\max}^{\mathrm{reg}}\left(\rho^A\right) \geqslant \mathrm{Tr}\left[H^A\rho^A\right] - \mathrm{Tr}\left[H^A\gamma^A\right]. \tag{17.17}$$

Let

$$f(\rho) := \min_U \mathrm{Tr}[H^A U\rho^A U^*], \tag{17.18}$$

and observe that the inequality in (17.17) is equivalent to

$$f^{\mathrm{reg}}(\rho) := \lim_{n\to\infty} \frac{1}{n} f(\rho^{\otimes n}) \leqslant \mathrm{Tr}[H^A\gamma^A]. \tag{17.19}$$

Our goal is therefore to prove this inequality.

Since $f$ is invariant under unitaries we can assume without loss of generality that $\rho$ is diagonal and we denote its diagonal by $\mathbf{p} = (p_1, \ldots, p_m)^T$. We also denote by $X$ the random variable that takes the value $X = x \in [m]$ with probability $p_x$. Furthermore, let $\gamma' \in \mathfrak{D}(A)$ be another Gibbs state of system $A$, with a diagonal $\mathbf{g}' = (g_1', \ldots, g_m')^T$ and with inverse temperature $\beta' < \beta$ (i.e. $\gamma'$ corresponds to a Gibbs state with the same Hamiltonian $H^A$ and with a higher temperature). Furthermore, let $Y$ be the random variable that takes the value $Y = y \in [m]$ with probability $g_y'$. With these notations we have

$$H(Y) = H(\gamma') > H(\gamma) = H(\rho) = H(X). \tag{17.20}$$

Now, let $\varepsilon > 0$ and recall that the number of strongly $\varepsilon$-typical sequences drawn from an i.i.d.$\sim$ p source scales predominantly as $2^{nH(X)}$. Therefore, since $H(X) < H(Y)$ we get that for sufficiently small $\varepsilon > 0$ and sufficiently large $n$ we have $|\mathfrak{T}^{\mathrm{st}}_\varepsilon(X^n)| < |\mathfrak{T}^{\mathrm{st}}_\varepsilon(Y^n)|$. In particular, there exists a one-to-one function $\pi_n : [m]^n \to [m]^n$, with the property that for any $x^n \in \mathfrak{T}^{\mathrm{st}}_\varepsilon(X^n)$ we have $\pi_n(x^n) \in \mathfrak{T}^{\mathrm{st}}_\varepsilon(Y^n)$. Define the unitary $U_n \in \mathcal{L}(A^n)$ by its action on basis elements of $A^n$ as $U_n|x^n\rangle := |\pi_n(x^n)\rangle$ for all $x^n \in [m]^n$. Since $U_n$ is not necessarily optimal we get

$$
\begin{aligned}
f^{\mathrm{reg}}(\rho) &\leqslant \lim_{n\to\infty} \frac{1}{n} \mathrm{Tr}\left[ H^{A^n} U_n \rho^{\otimes n} U_n^* \right] \\
&= \lim_{n\to\infty} \frac{1}{n} \sum_{x^n \in [m]^n} p_{x^n} \mathrm{Tr}\left[ H^{A^n} |\pi(x^n)\rangle\langle\pi(x^n)| \right].
\end{aligned}
\tag{17.21}
$$

Next, observe that

$$
H^{A^n} = \sum_{x^n \in [m]^n} \left( a_{x_1} + \cdots + a_{x_n} \right) |x^n\rangle\langle x^n| = n \sum_{x^n \in [m]^n} \mathbf{t}(x^n) \cdot \mathbf{a} |x^n\rangle\langle x^n|,
\tag{17.22}
$$

where $\mathbf{t}(x^n) \in \mathrm{Type}(n,m)$ is the type of the sequence $x^n$ and $\mathbf{a} := (a_1, \ldots, a_m)^T$. Substituting this into the previous equation gives

$$
\begin{aligned}
f^{\mathrm{reg}}(\rho) &\leqslant \lim_{n\to\infty} \sum_{x^n \in [m]^n} p_{x^n} \mathbf{t}\left(\pi_n(x^n)\right) \cdot \mathbf{a} \\
\textbf{Exercise 17.5}\rightarrow \;&= \lim_{n\to\infty} \sum_{x^n \in \mathfrak{T}^{\mathrm{st}}_\varepsilon(X^n)} p_{x^n} \mathbf{t}\left(\pi_n(x^n)\right) \cdot \mathbf{a},
\end{aligned}
\tag{17.23}
$$

where in the second line we restricted $x^n$ to the set of strongly $\varepsilon$-typical sequences. The theorem of strongly typical sequences ensures that the contribution of nontypical sequences vanishes in the limit $n \to \infty$ (see Exercise 17.5). Since the inequality in (17.23) holds for all $\varepsilon \in (0,1)$, taking the limit $\varepsilon \to 0^+$ gives

$$
\begin{aligned}
f^{\mathrm{reg}}(\rho) &\leqslant \lim_{\varepsilon\to 0^+} \lim_{n\to\infty} \sum_{x^n \in \mathfrak{T}^{\mathrm{st}}_\varepsilon(X^n)} p_{x^n} \mathbf{t}\left(\pi_n(x^n)\right) \cdot \mathbf{a} \\
&= \mathbf{g}' \cdot \mathbf{a} = \mathrm{Tr}\left[ H^A \gamma'^A \right],
\end{aligned}
\tag{17.24}
$$

where we used the fact that $\pi_n(x^n) \in \mathfrak{T}^{\varepsilon,\mathrm{st}}_n(\mathbf{g}')$ so that $\mathbf{t}\left(\pi_n(x^n)\right) \to \mathbf{g}'$ as $n \to \infty$. Finally, since we have proved that $f^{\mathrm{reg}}(\rho) \leqslant \mathrm{Tr}\left[ H^A \gamma'^A \right]$ for any Gibbs state with inverse temperature $\beta' > \beta$, it follows that the inequality also hold for $\beta' = \beta$. This completes the proof. ∎

**Exercise 17.5.** *Prove the relation* (17.23). *Hint: Use Theorem 8.9 in conjunction with the fact that* $\mathbf{t}\left(\pi_n(x^n)\right) \cdot \mathbf{a}$ *is bounded from above; for example,* $\mathbf{t}\left(\pi_n(x^n)\right) \cdot \mathbf{a} \leqslant a_m$ *since* $\mathbf{t}\left(\pi_n(x^n)\right)$ *is a probability vector.*

A state $\rho \in \mathfrak{D}(A)$ characterized by $W^{\mathrm{reg}}\mathrm{max}(\rho) = 0$ is identified as a *completely passive state*. Such states are inherently unable to facilitate work extraction, irrespective of their quantity. As inferred from Exercise 17.4, if $W^{\text{reg}}\text{max}(\rho) = 0$, then it follows that

$$W_{\text{max}}\left(\rho^{\otimes n}\right) = 0 \qquad \forall\, n \in \mathbb{N}. \tag{17.25}$$

This insight, derived from Theorem 17.1, establishes the Gibbs state as the unique completely passive state. Consequently, this finding compellingly supports the designation of the Gibbs state, or thermal state, as the exclusive free state in the domain of quantum thermodynamics.

**Exercise 17.6.** *Give full details why the only state that is completely passive is the Gibbs state.*

## 17.1.3 Athermality States

An athermality state of system $A$ cannot be solely characterized by $\rho^A$ since the resourcefulness of the state also depends on the Hamiltonian. Therefore, in quantum thermodynamics, every thermodynamic state comprises a quantum state $\rho \in \mathfrak{D}(A)$ that acts on the Hilbert space $A$, and a time-independent Hamiltonian $H^A \in \text{Pos}(A)$ that governs the dynamics of system $A$. In other words, a state of athermality can be characterized by a pair $(\rho^A, H^A)$. This characterization is widely used in the literature.

However, from a resource-theoretic perspective, this characterization has several drawbacks. First, it is not invariant under an energy shift of the form $H^A \mapsto H^A + cI^A$, where $c \in \mathbb{R}$ is a constant. Indeed, the choice of setting the minimal energy of a system to be zero is somewhat arbitrary. Second, we will observe that the resourcefulness of the state $\rho^A$ is determined in relation to its deviation from the Gibbs state $\gamma^A$ of system $A$.

Therefore, it appears more natural to characterize athermality states (i.e. the "objects" of this theory) as pairs of the form $(\rho^A, \gamma^A)$. It is worth noting that all the relevant information about the Hamiltonian $H^A$ is contained in the Gibbs state $\gamma^A$, which is invariant under energy shifts. Using this notation, the Gibbs state can be represented as $(\gamma^A, \gamma^A)$. In the following lemma, we demonstrate that the Gibbs state can also be utilized to determine if a given unitary evolution exhibits time-translation symmetry.

**Lemma 17.3.** Let $U \in \mathfrak{U}(A)$ be a unitary matrix. Then $U^A$ commutes with a Hamiltonian $H \in \text{Pos}(A)$ if and only if $U^A$ commutes with the Gibbs state $\gamma^A$.

**Proof**  By definition, the Gibbs state $\gamma^A$ commutes with $U^A$ if and only if $e^{-\beta H^A}$ commutes with $U^A$. Therefore, if $[U^A, H^A] = 0$, then clearly $U^A$ commutes with $\gamma^A$. Conversely, suppose $U^A$ commutes with $e^{-\beta H^A}$ and express $H^A = \sum_x \lambda_x P_x$, where $\{P_x\}$ are orthogonal projections satisfying $P_x P_y = \delta_{xy} P_x$ and $\{\lambda_x\}$ are *distinct* eigenvalues of $H^A$. Then,

$$\sum_x e^{-\beta \lambda_x} U P_x = U e^{-\beta H^A} = e^{-\beta H^A} U = \sum_y e^{-\beta \lambda_y} P_y U. \tag{17.26}$$

Multiplying by $P_x$ from the right and $P_y$ from the left we get

$$e^{-\beta\lambda_x} P_y U P_x = e^{-\beta\lambda_y} P_y U P_x. \tag{17.27}$$

Since $\lambda_x \neq \lambda_y$, we conclude that $P_y U P_x = 0$ for all $x \neq y$. Hence, $U$ commutes with $H^A$. ∎

Note that in Lemma 17.3 the condition that $U^A$ commutes with $\gamma^A$ can be expressed as

$$\mathcal{U}^{A \to A}\left(\gamma^A\right) := U^A \gamma^A U^{*A} = \gamma^A. \tag{17.28}$$

That is, the unitary matrix $U^A$ commutes with the Hamiltonian if and only if the unitary channel $\mathcal{U}^{A \to A}$ preserves the Gibbs state.

Suppose now that system $B$ is comprised of two subsystems $B_1$ and $B_2$ and that the total Hamiltonian of system $B$ can be expressed as

$$H^B = H^{B_1} \otimes I^{B_2} + I^{B_1} \otimes H^{B_2}. \tag{17.29}$$

In this case, the Gibbs state $\gamma^B$ of the composite system can be expressed as a tensor product of the two Gibbs states of the subsystems. Indeed, we have

$$\gamma^B = \frac{e^{-\beta\left(H^{B_1}\otimes I^{B_2}+I^{B_1}\otimes H^{B_2}\right)}}{\mathrm{Tr}\left[e^{-\beta\left(H^{B_1}\otimes I^{B_2}+I^{B_1}\otimes H^{B_2}\right)}\right]}, \tag{17.30}$$

and since

$$e^{-\beta\left(H^{B_1}\otimes I^{B_2}+I^{B_1}\otimes H^{B_2}\right)} = e^{-\beta H^{B_1}} \otimes e^{-\beta H^{B_2}}, \tag{17.31}$$

we conclude that $\gamma^B = \gamma^{B_1} \otimes \gamma^{B_2}$, where for each $j = 1, 2$, $\gamma^{B_j} := e^{-\beta H^{B_j}}/\mathrm{Tr}\left[e^{-\beta H^{B_j}}\right]$ is the Gibbs state of subsystem $B_j$.

## 17.2 The Free Operations

There have been various formulations in the literature regarding the free operations of the QRT of thermodynamics, all of which involve operations conserving certain extensive quantities like energy, particle number, charge, and so on. In this chapter, we focus solely on free operations corresponding to energy conservation, giving rise to the resource theory of athermality. However, these operations can be analogously extended to conserve other extensive quantities. For further details, we direct the reader to the section titled "History and Further Reading."

Let us denote the Hilbert space associated with the thermal bath with a fixed temperature $T$ as $B$. Additionally, we will consider a quantum system $A$ that interacts with the heat bath (see Figure 17.1).

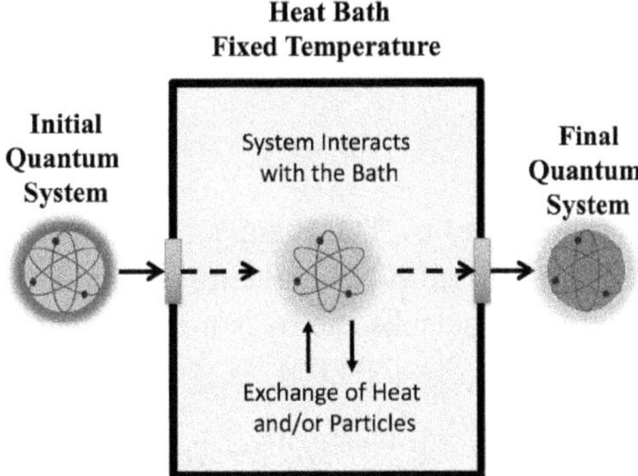


**Figure 17.1** A quantum system interacting with a heat bath.

## 17.2.1 Thermal Operations

The set of free operations relative to a background heat bath at temperature $T$ comprise of three basic steps:

1. Thermal equilibrium. Any subsystem $B$, with Hamiltonian $H^B \in \mathrm{Pos}(B)$, can be prepared in its thermal Gibbs state $\gamma^B$.
2. Conservation of energy. Unitary operation on a composite physical system that commutes with the total Hamiltonian can be implemented.
3. Discarding subsystems. It is possible to trace over any subsystem of a composite system.

*Remark.* For the second step of conservation of energy, it is assumed that the couplings among the subsystems of the composite system is controlled entirely by the experimenter. Therefore, in the absence of such intervention, it will be assumed that the total Hamiltonian is decoupled, and can be expressed as a sum of the free Hamiltonians of the subsystems.

Any CPTP map comprising of the above three steps is called *thermal operation* (TO). Thermal operation forms the set of free operations in the QRT of athermality. To investigate more closely these operations, let $(\rho^A, \gamma^A)$ be an athermality state with $\rho \in \mathfrak{D}(A)$ and corresponding Gibbs state $\gamma \in \mathfrak{D}(A)$ (or equivalently a Hamiltonian $H \in \mathrm{Pos}(A)$). From the first step above it follows that the transformation

$$\rho^A \to \rho^A \otimes \gamma^B \tag{17.32}$$

is a free operation, where $B$ is some ancillary system in the Gibbs state $\gamma^B$ and Hamiltonian $H^B$. The total Hamiltonian of system $AB$ is given by $H^{AB} := H^A \otimes I^B + I^A \otimes H^B$. Its corresponding Gibbs state is given by $\gamma^{AB} := \gamma^A \otimes \gamma^B$. According to the second step above, any unitary matrix $U : AB \to AB$ that commutes with the total Hamiltonian $H^{AB}$ yields a permissible evolution of the system $AB$. Combining this with Lemma 17.3 we conclude that a unitary evolution $\mathcal{U} \in \mathrm{CPTP}(AB \to AB)$ is free if and only if it preserves the Gibbs state $\gamma^{AB}$. For such a Gibbs preserving unitary channel $\mathcal{U}^{AB\to AB}$ we get that the transformation

$$\rho^A \otimes \gamma^B \to \mathcal{U}^{AB\to AB}\left(\rho^A \otimes \gamma^B\right) \tag{17.33}$$

can be implemented by TO. Finally, if systems $A = A_1 \cdots A_n$ and $B = B_1 \cdots B_m$ are themselves comprised of several subsystems $A_1, \ldots, A_n$ and $B_1, \ldots, B_m$, then tracing out several of these subsystems is a free operation. We therefore conclude that any thermal operation $\mathcal{E} \in \mathrm{CPTP}(A \to A')$ can be expressed as

$$\mathcal{E}^{A\to A'}(\rho^A) = \mathrm{Tr}_{B'}\left[\mathcal{U}^{AB\to AB}\left(\rho^A \otimes \gamma^B\right)\right], \tag{17.34}$$

where $A'$ and $B'$ are (possibly composite) subsystems of $AB$ such that $AB = A'B'$.

In the following lemma we show that the requirement $AB = A'B'$ is not necessary, only that $AB \cong A'B'$. This is not completely trivial since systems $A'$ and $B'$ can correspond to completely different physical systems.

---

**Lemma 17.4.** Let $AB$, $A'B'$ be two composite physical systems with corresponding Gibbs states $\gamma^{AB} := \gamma^A \otimes \gamma^B$ and $\gamma^{A'B'} := \gamma^{A'} \otimes \gamma^{B'}$. Suppose $|AB| = |A'B'|$ and let $\mathcal{U}^{AB\to A'B'} \in \mathrm{CPTP}(AB \to A'B')$ be a unitary channel satisfying

$$\mathcal{U}^{AB\to A'B'}\left(\gamma^{AB}\right) = \gamma^{A'B'}. \tag{17.35}$$

Then, the map

$$\mathcal{N}^{A\to A'}\left(\omega^A\right) := \mathrm{Tr}_{B'}\left[\mathcal{U}^{AB\to A'B'}\left(\omega^A \otimes \gamma^B\right)\right] \qquad \forall\, \omega \in \mathcal{L}(A) \tag{17.36}$$

is a thermal operation.

---

**Proof** Let $\gamma^{ABA'B'} := \gamma^{AB} \otimes \gamma^{A'B'}$ and let $\mathcal{V} \in \mathrm{CPTP}(ABA'B' \to ABA'B')$ be the unitary matrix given by

$$\mathcal{V} := \mathcal{U}^{AB\to A'B'} \otimes \mathcal{U}^{*A'B'\to AB}. \tag{17.37}$$

In the following exercise you show that $\mathcal{V}$ preserves the joint Gibbs state $\gamma^{ABA'B'}$. Hence, the channel

$$\mathrm{Tr}_{ABB'}\left[\mathcal{V}\left(\omega^A \otimes \gamma^{BA'B'}\right)\right] = \mathrm{Tr}_{ABB'}\left[\mathcal{U}^{AB\to A'B'}\left(\omega^A \otimes \gamma^B\right) \otimes \mathcal{U}^{*A'B'\to AB}\left(\gamma^{A'B'}\right)\right]$$

$$= \mathrm{Tr}_{B'}\left[\mathcal{U}^{AB\to A'B'}\left(\omega^A \otimes \gamma^B\right)\right] \tag{17.38}$$

is a thermal operation. ∎

**Exercise 17.7.** *Show that the matrix $\mathcal{V}$ as defined in the proof above is indeed Gibbs preserving. Hint: Apply $\mathcal{U}^*$ to both sides of (17.35) to show that $\mathcal{U}^*$ is Gibbs preserving.*

**Exercise 17.8.** *Consider the unitary matrix $U : AB \to A'B'$ associated with the unitary channel $\mathcal{U}^{AB \to A'B'}$ mentioned in Lemma 17.4 (i.e. $\mathcal{U}^{AB \to A'B'}(\cdot) := U(\cdot)U^*$). Demonstrate that the condition (17.35) is satisfied if and only if*

$$UH^{AB} = H^{A'B'}U. \tag{17.39}$$

Recall that a density matrix $\rho \in \mathfrak{D}(A)$ can be viewed as an athermality state only when the Hamiltonian or Gibbs state of system $A$ is specified. Similarly, a quantum channel $\mathcal{N} \in \mathrm{CPTP}(A \to A')$ on its own cannot be considered a thermal operation without specifying the Gibbs state associated with systems $A$ and $A'$. We will therefore view a thermal operation as a triple $(\mathcal{N}^{A \to A'}, \gamma^A, \gamma^{A'})$, where $\mathcal{N} \in \mathrm{CPTP}(A \to A')$, $\gamma^A$ is the input Gibbs state, and $\gamma^{A'}$ is the output Gibbs state. We use this perspective in the following formal definition of thermal operations.

---

**Definition 17.1.** Let $\mathcal{N} \in \mathrm{CPTP}(A \to A')$ and $\gamma^A$ and $\gamma^{A'}$ be two density matrices. The triple $(\mathcal{N}^{A \to A'}, \gamma^A, \gamma^{A'})$ is called a *thermal operation* if there exists a unitary channel $\mathcal{U} \in \mathrm{CPTP}(AB \to A'B')$ (with $|AB| = |A'B'|$), and density matrices $\gamma^B$ and $\gamma^{B'}$, such that both (17.35) and (17.36) hold.

---

In the following lemma we show that the triple $(\mathcal{N}^{A \to A'}, \gamma^A, \gamma^{A'})$ in this definition is not independent.

---

**Lemma 17.5.** Let $(\mathcal{N}^{A \to A'}, \gamma^A, \gamma^{A'})$ be a thermal operation. Then, $\mathcal{N}^{A \to A'}$ is Gibbs preserving; that is,

$$\mathcal{N}^{A \to A'}(\gamma^A) = \gamma^{A'}. \tag{17.40}$$

---

**Proof**  Observe that since $\mathcal{U}^{AB \to A'B'}$ in (17.36) is Gibbs preserving it follows that

$$\mathcal{N}^{A \to A'}(\gamma^A) = \mathrm{Tr}_{B'}\left[\mathcal{U}^{AB \to A'B'}\left(\gamma^{AB}\right)\right] = \mathrm{Tr}_{B'}\left[\gamma^{A'B'}\right] = \gamma^{A'}. \tag{17.41}$$

This completes the proof. ∎

We denote by $\mathrm{TO}(A \to A')$ the set of all quantum channels $\mathcal{N} \in \mathrm{CPTP}(A \to A')$ such that $(\mathcal{N}^{A \to A'}, \gamma^A, \gamma^{A'})$ is a thermal operation. For simplicity of the notation $\mathrm{TO}(A \to A')$, we do not explicitly specify the input and output Gibbs states. However, throughout this chapter, the physical systems $A$ and $A'$ will always possess well-defined Hamiltonians (and consequently Gibbs states). We also denote by $\mathrm{GPO}(A \to A')$ the set of all Gibbs preserving CPTP maps; that is,

$$\mathrm{GPO}(A \to A') := \left\{\mathcal{N} \in \mathrm{CPTP}(A \to A') : \mathcal{N}^{A \to A'}(\gamma^A) = \gamma^{A'}\right\}. \tag{17.42}$$

Clearly, from their definitions and the lemma above it follows that

$$\mathrm{TO}(A \to A') \subseteq \mathrm{GPO}(A \to A'). \tag{17.43}$$

> **Theorem 17.2.** The set TO($A \to A'$) is convex.

**Proof**   Let $\{\mathcal{N}_x\}_{x\in[m]}$ be a set of $m$ channels in TO($A \to A'$), and consider a convex combination of these $m$ channels:

$$\mathcal{N}^{A\to A'} := \sum_{x\in[m]} p_x \mathcal{N}_x^{A\to A'}, \tag{17.44}$$

where $\mathbf{p} := (p_1,\ldots,p_m)^T \in \mathrm{Prob}(m)$. Since each $\mathcal{N}_x$ is a thermal operation, it can be expressed as

$$\mathcal{N}_x^{A\to A'}\left(\omega^A\right) := \mathrm{Tr}_{B_x'}\left[\mathcal{U}_x^{AB_x\to A'B_x'}\left(\omega^A \otimes \gamma^{B_x}\right)\right] \qquad \forall\, \omega \in \mathfrak{L}(A), \tag{17.45}$$

where for each $x \in [m]$, $B_x$ and $B_x'$ are auxiliary thermal baths, and $\mathcal{U}_x$ is a Gibbs preserving unitary channel. Let

$$B := \bigoplus_{x\in[m]} B_x\,, \quad B' := \bigoplus_{x\in[m]} B_x', \quad \text{and} \quad \gamma^B := \bigoplus_{x\in[m]} p_x \gamma^{B_x}. \tag{17.46}$$

Finally, we define the unitary channel $\mathcal{U} \in \mathrm{CPTP}(AB \to A'B')$ as

$$\mathcal{U}^{AB\to A'B'}(\eta^{AB}) := \bigoplus_{x\in[m]} \mathcal{U}_x^{AB_x\to A'B_x'}(\eta^{AB_x}) \qquad \forall\, \eta^{AB} := \bigoplus_{x\in[m]} \eta^{AB_x} \in \mathfrak{D}(AB).$$

$$\tag{17.47}$$

With these definitions we get for all $\omega \in \mathfrak{L}(A)$

$$\mathrm{Tr}_{B'}\left[\mathcal{U}^{AB\to A'B'}\left(\omega^A \otimes \gamma^B\right)\right] = \sum_{x\in[m]} p_x \mathrm{Tr}_{B_x'}\left[\mathcal{U}^{AB_x\to A'B_x'}\left(\omega^A \otimes \gamma^{B_x}\right)\right]$$

$$\mathbf{(17.45)}\!\to \;\; = \sum_{x\in[m]} p_x \mathcal{N}_x^{A\to A'}(\omega^A) = \mathcal{N}^{A\to A'}(\omega^A). \tag{17.48}$$

Therefore, $\mathcal{N}^{A\to A'}$ is a thermal operation. This completes the proof.    ∎

## 17.2.2 Closed Thermal Operations

Thermal operations comprise of all quantum channels of the form given in (17.36). In general, these operations do not form a topologically closed set, as the dimension of system $B$ in (17.36) is unbounded (but finite since we will only consider finite dimensional systems). It is therefore possible that some given state cannot be converted to another by thermal operations, and yet, for *any* $\varepsilon > 0$ the conversion is possible up to an $\varepsilon$-error. This unphysical property can be tailored back to noisy operations which were defined as the closure of factorizable channels. We will therefore define *closed thermal operations* (CTO) to be the closure of thermal operations.

---

> **Definition 17.2.** Let $A$ and $A'$ be two physical systems. The set of *closed thermal operations*, denoted as $\text{CTO}(A \to A')$, is defined as
>
> $$\text{CTO}(A \to A') := \overline{\text{TO}(A \to A')}. \qquad (17.49)$$

*Remark.* By definition, $\mathcal{N} \in \text{CTO}(A \to A')$ if and only if there exists a sequence of thermal operations $\left\{ \mathcal{N}_n^{A \to A'} \right\}_{n \in \mathbb{N}} \subset \text{TO}(A \to A')$ such that

$$\lim_{n \to \infty} \mathcal{N}_n^{A \to A'} = \mathcal{N}^{A \to A'}. \qquad (17.50)$$

The limit in (17.50) is equivalent to

$$\lim_{n \to \infty} \left\| J_{\mathcal{N}_n}^{AA'} - J_{\mathcal{N}}^{AA'} \right\|_1 = 0, \qquad (17.51)$$

where $J_{\mathcal{N}_n}^{AA'}$ and $J_{\mathcal{N}}^{AA'}$ are the Choi matrices of $\mathcal{N}_n$ and $\mathcal{N}$, respectively.

The physical justification for CTO is obvious: for $\varepsilon := 10^{-100}$ being one over googol, it is impossible to discriminate between states or channels that are $\varepsilon$-close to each other and for all practical purposes the states or channels can be considered identical. We will see however that this assumption has significant consequences particularly in the quasi-classical regime.

**Exercise 17.9.** *Show that* CTO *is closed under concatenation. That is, show that if* $(\mathcal{N}, \gamma^A, \gamma^B)$ *and* $(\mathcal{M}, \gamma^B, \gamma^C)$ *are closed thermal operations, then* $(\mathcal{M} \circ \mathcal{N}, \gamma^A, \gamma^C)$ *is also a closed thermal operation.*

> **Lemma 17.6.** Let $(\rho^A, \gamma^A)$ and $(\sigma^{A'}, \gamma^{A'})$ be two athermality states. Then the following statements are equivalent (see Figure 17.2):
>
> 1. The state $\left(\rho^A, \gamma^A\right)$ can be converted to $\left(\sigma^{A'}, \gamma^{A'}\right)$ by CTO.
> 2. For any $\varepsilon > 0$ there exists states $\tilde{\rho}^A$ and $\tilde{\sigma}^{A'}$ that are $\varepsilon$-close to $\rho^A$ and $\sigma^{A'}$, respectively, such that
>
> $$\tilde{\sigma}^{A'} = \mathcal{N}^{A \to A'}\left(\tilde{\rho}^A\right) \text{ for some } \mathcal{N} \in \text{TO}(A \to A'). \qquad (17.52)$$

**Proof** The proof that $1 \Rightarrow 2$ is left as an exercise, and we prove that $2 \Rightarrow 1$. Let $\{\varepsilon_k\}_{k \in \mathbb{N}}$ be a sequence of positive numbers with zero limit, and for each $k \in \mathbb{N}$, let $(\rho_k^A, \gamma^A)$ be an athermality state that can be converted to a state that is $\varepsilon_k$-close to $(\sigma^{A'}, \gamma^{A'})$. That is, for each $k$ there exists a thermal operation $\mathcal{N}_k \in \text{TO}(A \to A')$ with the property that

$$\mathcal{N}_k^{A \to A'}\left(\rho_k^A\right) \approx_{\varepsilon_k} \sigma^{A'}. \qquad (17.53)$$

Since the set $\text{CPTP}(A \to A')$ is compact, there exists a converging subsequence of $\{\mathcal{N}_k\}_{k \in \mathbb{N}}$. For simplicity of the exposition here, we assume without loss of generality

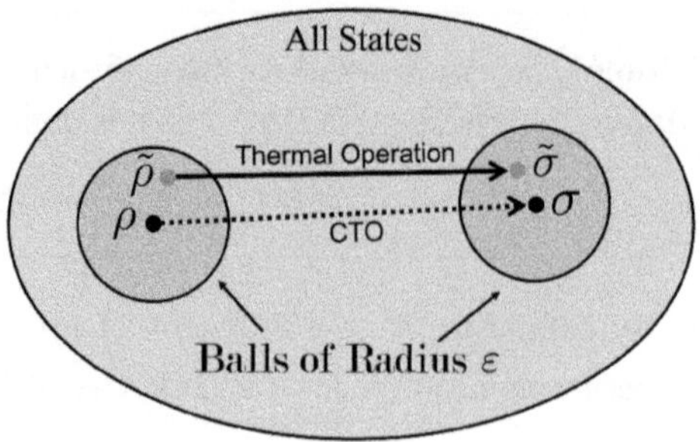


**Figure 17.2**  The conversion of $\rho$ to $\sigma$ by CTO. For any $\varepsilon > 0$ there exists states $\tilde{\rho}$ and $\tilde{\sigma}$ that are $\varepsilon$-close to $\rho$ and $\sigma$, respectively, such that $\tilde{\rho}$ can be converted to $\tilde{\sigma}$ by thermal operation.

that the sequence $\{\mathcal{N}_k\}_{k\in\mathbb{N}}$ itself is converging (otherwise, we have to replace $k$ with a subsequence $\{n_k\}_{k\in\mathbb{N}}$) and set $\mathcal{N} := \lim_{k\to\infty}\mathcal{N}_k$. By definition, $\mathcal{N} \in \mathrm{CTO}(A \to A')$ since each $\left(\mathcal{N}_k^{A\to A'}, \gamma^A, \gamma^{A'}\right)$ is a thermal operation. Moreover, observe that

$$\mathcal{N}^{A\to A'}\left(\rho^A\right) = \lim_{k\to\infty}\mathcal{N}_k^{A\to A'}(\rho_k) = \sigma^{A'}, \tag{17.54}$$

where we used (17.53). Hence, $\left(\rho^A, \gamma^A\right)$ can be converted to $\left(\sigma^{A'}, \gamma^{A'}\right)$ by CTO. This completes the proof.  ∎

**Exercise 17.10.** *Prove that* $1 \Rightarrow 2$ *in Lemma 17.6.*

**Exercise 17.11.** *Show that Lemma 17.6 still holds even if we replace in (17.52)* $\mathcal{N} \in \mathrm{TO}(A \to A')$ *with* $\mathcal{N} \in \mathrm{CTO}(A \to A')$.

## 17.2.3 Gibbs-Preserving Covariant Operations

In Section 15.6 we studied time-translation covariant channels. Such channels are defined with respect to the input and output Hamiltonians associated with the channel. Specifically, let $H^A \in \mathrm{Pos}(A)$ and $H^{A'} \in \mathrm{Pos}(A')$ be two Hamiltonians, and let $\mathcal{E} \in \mathrm{CPTP}(A \to A')$. Then, we say that $\mathcal{E}^{A\to A'}$ is time-translation covariant with respect to the Hamiltonians $H^A$ and $H^{A'}$ if for all $t \in \mathbb{R}$

$$\mathcal{U}_t^{A'\to A'} \circ \mathcal{E}^{A\to A'} = \mathcal{E}^{A\to A'} \circ \mathcal{U}_t^{A\to A} \qquad \forall t \in \mathbb{R}, \tag{17.55}$$

where $\mathcal{U}_t^{A\to A}(\cdot) := U_t^A(\cdot)U_t^{*A}$ is the unitary channel defined with $U_t^A := e^{-itH^A}$ and similarly $\mathcal{U}_t^{A'\to A'}(\cdot) := U_t^{A'}(\cdot)U_t^{*A'}$, where $U_t^{A'} := e^{-itH^{A'}}$. We show here that thermal operations are time-translation covariant.

> **Theorem 17.3.** Let $\gamma \in \mathfrak{D}(A)$ be a Gibbs state, and let $\mathcal{E} \in \mathrm{CTO}(A \to A')$. Then, $\mathcal{E}^{A \to A'}$ is time-translation covariant with respect to the Hamiltonians of systems $A$ and $A'$.

**Proof** Suppose first that $\mathcal{E} \in \mathrm{TO}(A \to A')$, and it has the form (17.36) with $AB \cong A'B'$. To see that $\mathcal{E}^{A \to A'}$ is time-translation covariant, observe that

$$
\begin{aligned}
\mathcal{E}^{A \to A'}\left(e^{-itH^A} \rho^A e^{itH^A}\right) &= \mathrm{Tr}_{B'}\left[\mathcal{U}^{AB \to A'B'}\left(e^{-itH^A} \rho^A e^{itH^A} \otimes \gamma^B\right)\right] \\
\boxed{[\gamma^B, H^B] = 0} \longrightarrow \quad &= \mathrm{Tr}_{B'}\left[\mathcal{U}^{AB \to A'B'}\left(e^{-itH^A} \rho^A e^{itH^A} \otimes e^{-itH^B} \gamma^B e^{itH^B}\right)\right] \\
&= \mathrm{Tr}_{B'}\left[\mathcal{U}^{AB \to A'B'} \circ \mathcal{V}_t^{AB \to AB}\left(\rho^A \otimes \gamma^B\right)\right],
\end{aligned}
\tag{17.56}
$$

where $V_t^{AB} := e^{-itH^A} \otimes e^{-itH^B} = e^{-itH^{AB}}$, $H^{AB} := H^A \otimes I^B + I^A \otimes H^B$ is the total Hamiltonian, and $\mathcal{V}_t^{AB \to AB} := V_t^{AB}(\cdot)V_t^{*AB}$. From (17.39) we get that the unitary channel $\mathcal{U}^{AB \to AB} := U^{AB}(\cdot)U^{*AB}$ satisfies

$$
\mathcal{U}^{AB \to A'B'} \circ \mathcal{V}_t^{AB \to AB} = \mathcal{V}_t^{A'B' \to A'B'} \circ \mathcal{U}^{AB \to A'B'},
\tag{17.57}
$$

where $\mathcal{V}_t^{A'B' \to A'B'}(\cdot) := e^{-itH^{A'B'}}(\cdot)e^{itH^{A'B'}}$. Combining this with (17.56) we get for any $\rho \in \mathfrak{D}(A)$

$$
\mathcal{E}^{A \to A'}\left(e^{-itH^A} \rho^A e^{itH^A}\right) = \mathrm{Tr}_{B'}\left[\mathcal{V}_t^{A'B' \to A'B'} \circ \mathcal{U}^{AB \to A'B'}\left(\rho^A \otimes \gamma^B\right)\right].
\tag{17.58}
$$

Finally, observe that $\mathcal{V}_t^{A'B' \to A'B'} = \mathcal{U}_t^{A' \to A'} \otimes \mathcal{U}_t^{B' \to B'}$, where $\mathcal{U}_t^{A' \to A'}(\cdot) := e^{-itH^{A'}}(\cdot)e^{itH^{A'}}$ and $\mathcal{U}_t^{B' \to B'}(\cdot) := e^{-itH^{B'}}(\cdot)e^{itH^{B'}}$. Substituting this into the equation above gives

$$
\begin{aligned}
\mathcal{E}^{A \to A'}\left(e^{-itH^A} \rho^A e^{itH^A}\right) &= \mathcal{U}_t^{A' \to A'}\left(\mathrm{Tr}_{B'}\left[\mathcal{U}^{AB \to A'B'}\left(\rho^A \otimes \gamma^B\right)\right]\right) \\
&= e^{-itH^{A'}} \mathcal{E}^{A \to A'}\left(\rho^A\right) e^{itH^{A'}}.
\end{aligned}
\tag{17.59}
$$

This completes the proof for $\mathcal{E} \in \mathrm{TO}(A \to A')$. The case $\mathcal{E} \in \mathrm{CTO}(A \to A')$ follows from the fact that the limit of time-translation covariant channels is itself time-translation covariant (see Exercise 17.12). $\blacksquare$

**Exercise 17.12.** *Let $G$ be a group, and let $\{\mathcal{E}_n\}_{n \in \mathbb{N}}$ be a sequence of channels in $\mathrm{COV}_G(A \to A')$ (with respect to some unitary representations of $G$ on $A$ and $A'$). Show that if the limit $\mathcal{E} := \lim_{n \to \infty} \mathcal{E}_n$ exists, then also $\mathcal{E} \in \mathrm{COV}_G(A \to A')$.*

**Definition 17.3.** Let $\gamma^A$ and $\gamma^{A'}$ be two Gibbs states. A channel $\mathcal{N} \in \mathrm{GPO}(A \to A')$ is called a *Gibbs-preserving covariant operation* (in short, GPC operation) if in addition of being Gibbs preserving it is also time-translation covariant satisfying (17.55). We denote by $\mathrm{GPC}(A \to A')$ the set of all such GPC channels in $\mathrm{GPO}(A \to A')$.

From its definition $\mathrm{GPC}(A \to A')$ forms a subset of $\mathrm{GPO}(A \to A')$. Furthermore, from Theorem 17.3 we have

$$\mathrm{CTO}(A \to A') \subseteq \mathrm{GPC}(A \to A'). \tag{17.60}$$

**Exercise 17.13.** *Show that* $\mathrm{GPC}(A \to A')$ *is convex.*

The following exercise shows another (possibly strictly) subclass of GPC operations. We say that isometry channel $\mathcal{V} \in \mathrm{CPTP}(A \to A')$ is *time-translation covariant* if it satisfies

$$V H^A = H^{A'} V, \tag{17.61}$$

where $H^A$ and $H^{A'}$ are the Hamiltonians of systems $A$ and $A'$.

**Exercise 17.14.** *Let $A$, $B$, $A'$, and $B'$ be four physical systems with corresponding Hamiltonians $H^A$, $H^B$, $H^{A'}$, and $H^{B'}$, and let $\mathcal{V}^{AB \to A'B'}(\cdot) = V(\cdot)V^*$ be a time-translation covariant isometry channel. Denote by*

$$\mathcal{E}^{A \to A'}\left(\omega^A\right) := \mathrm{Tr}_{B'}\left[\mathcal{V}^{AB \to A'B'}\left(\omega^A \otimes \gamma^B\right)\right] \qquad \forall\, \omega \in \mathfrak{L}(B), \tag{17.62}$$

*and set $t := \frac{Z^{AB}}{Z^{A'B'}}$. Show that the map*

$$\mathcal{N}^{A \to A'}\left(\omega^A\right) = t\mathcal{E}^{A \to A'}\left(\omega^A\right) + \left(\gamma^{A'} - t\mathcal{E}^{A \to A'}\left(\gamma^A\right)\right)\mathrm{Tr}\left[\omega^A\right] \tag{17.63}$$

*is a thermal operation (and in particular a quantum channel). Hint: Start with the covariance property $e^{-\beta H^{AB}} = V^* e^{-\beta H^{A'B'}} V$ to get*

$$Z^{AB} = \mathrm{Tr}\left[V V^* e^{-\beta H^{A'B'}}\right] = Z^{A'B'} \mathrm{Tr}\left[V V^* \gamma^{A'B'}\right] \leqslant Z^{A'B'}, \tag{17.64}$$

*with equality if and only if $|AB| = |A'B'|$ (in which case $V$ is a unitary matrix), and conclude that*

$$\tau^{A'} := \frac{\gamma^{A'} - t\mathcal{E}^{A \to A'}\left(\gamma^A\right)}{1 - t} \tag{17.65}$$

*is a density matrix.*

# 17.3 Quasi-Classical Athermality

In this section we examine a scenario in which every resource $(\rho^A, \gamma^A)$ is diagonal in the eigenbasis of $H^A$; that is, $[\rho^A, \gamma^A] = 0$. This implies that we are considering physical systems that lack quantum coherence between different energy levels. In particular, the Gibbs state, being commutative with the Hamiltonian, also lacks coherence between energy levels. Therefore, we refer to this scenario as the quasi-classical case. We start by showing that in the quasi-classical case thermal operations has the same capability for interconversions as Gibbs-preserving operations.

## 17.3.1 CTO versus GPO

In the semi-classical regime, it is convenient to denote an athermality state $(\rho^A, \gamma^A)$ (where we assume that $[\rho, \gamma] = 0$) by $(\mathbf{p}, \mathbf{g})$, where $\mathbf{p}, \mathbf{g} \in \mathrm{Prob}(m)$ are probability vectors of dimension $m := |A|$, and their components comprise the eigenvalues of $\rho$ and $\gamma$, respectively. It then follows that in the quasi-classical regime, a state $(\mathbf{p}, \mathbf{g})$ can be converted to $(\mathbf{p}', \mathbf{g}')$ by GPO if and only if there exists an $m' \times m$ column stochastic matrix, $E$, where $m := |A|$ and $m' := |A'|$, such that

$$\mathbf{p}' = E\mathbf{p} \quad \text{and} \quad \mathbf{g}' = E\mathbf{g}. \tag{17.66}$$

Note that $E$ corresponds to a Gibbs preserving channel. The relation of (17.66) corresponds precisely to the definition of relative majorization (see Section 4.3). Therefore, we conclude that

$$(\mathbf{p}, \mathbf{g}) \xrightarrow{\text{GPO}} (\mathbf{p}', \mathbf{g}') \quad \Longleftrightarrow \quad (\mathbf{p}, \mathbf{g}) \succ (\mathbf{p}', \mathbf{g}'). \tag{17.67}$$

Remarkably, the relation above remains unchanged even if we replace the set GPO with CTO.

---

**Theorem 17.4.** Let $(\rho, \gamma)$ and $(\rho', \gamma')$ be two quasi-classical states of systems $A$ and $A'$, respectively. The following statements are equivalent:

1. $(\rho, \gamma)$ can be converted to $(\rho', \gamma')$ by CTO.
2. $(\rho, \gamma)$ can be converted to $(\rho', \gamma')$ by GPO.

---

*Remark.* Note that this theorem does not state that CTO=GPO, only that they have the same conversion power. In general, we have CTO⊆GPO since GPO is a closed set of operations containing thermal operations. Therefore, the implication $1 \Rightarrow 2$ is trivial, and we only need to prove the direction $2 \Rightarrow 1$.

The proof of this theorem is technically involved and extensive; it has been deferred to Appendix D.7 of online version. It remains a compelling open challenge to discover a more concise and straightforward proof for this theorem.

## The Church of the Trivialized Hamiltonian

Theorem 17.4, in conjunction with (17.67), implies that interconversions under CTO can be characterized with relative majorization.

> **Corollary 17.1.** Let $(\mathbf{p}, \mathbf{g})$ and $(\mathbf{p}', \mathbf{g}')$ be two athermality states (in the quasi-classical regime) of systems $A$ and $A'$, respectively. Then,
>
> $$(\mathbf{p}, \mathbf{g}) \xrightarrow{\text{CTO}} (\mathbf{p}', \mathbf{g}') \quad \Longleftrightarrow \quad (\mathbf{p}, \mathbf{g}) \succ (\mathbf{p}', \mathbf{g}'). \tag{17.68}$$

We can therefore apply all the machinery of the theory of (relative) majorization to the theory of athermality. In particular, one of the immediate consequences of this corollary is that in the quasi-classical regime there exists a bijection between the resource theory of athermality and the resource theory of nonuniformity. This remarkable connection between the two theories essentially states that in the quasi-classical regime athermality *is* nonuniformity. This equivalence follows from Theorem 4.8.

Specifically, suppose $(\mathbf{p}, \mathbf{g})$ is an athermality state in the quasi-classical regime, and suppose that $\mathbf{g}$ has only rational components. Then, we can write the components of $\mathbf{g}$ as $g_x = \frac{n_x}{n}$ with $x \in [m]$, $n_x \in \mathbb{N}$, and $n := \sum_{x \in [m]} n_x$, and we have

$$(\mathbf{p}, \mathbf{g}) \sim \left( \mathbf{r}, \mathbf{u}^{(n)} \right), \quad \text{where} \quad \mathbf{r} := \bigoplus_{x=1}^{m} p_x \mathbf{u}^{(n_x)}. \tag{17.69}$$

That is, there exists an $n$-dimensional system $R$ in some state $\mathbf{r} \in \mathrm{Prob}(n)$, with trivial Hamiltonian (i.e. uniform Gibbs states), such that $(\mathbf{p}, \mathbf{g}) \sim (\mathbf{r}, \mathbf{u}^{(n)})$. Combining this with Theorem 17.4 we conclude that

$$(\mathbf{p}, \mathbf{g}) \xrightarrow{\text{CTO}} (\mathbf{r}, \mathbf{u}^{(n)}) \quad \text{and} \quad (\mathbf{r}, \mathbf{u}^{(n)}) \xrightarrow{\text{CTO}} (\mathbf{p}, \mathbf{g}). \tag{17.70}$$

In other words, $(\mathbf{p}, \mathbf{g})$ and $(\mathbf{r}, \mathbf{u}^{(n)})$ corresponds to the same resource, so that the athermality of $(\mathbf{p}, \mathbf{g})$ can be interpreted as the nonuniformity of $(\mathbf{r}, \mathbf{u}^{(n)})$.

**Exercise 17.15.** *Let $\varepsilon > 0$ and $(\mathbf{p}, \mathbf{g})$ be an athermality state (in the quasi-classical regime). We do not assume that $\mathbf{g}$ has rational components. Show that there exists an $n$-dimensional system $R$ with trivial Hamiltonian, and two states $\mathbf{r}_1, \mathbf{r}_2 \in \mathrm{Prob}(n)$ that satisfies $\frac{1}{2}\|\mathbf{r}_1 - \mathbf{r}_2\|_1 \leqslant \varepsilon$ and*

$$(\mathbf{r}_1, \mathbf{u}^{(n)}) \succ (\mathbf{p}, \mathbf{g}) \succ (\mathbf{r}_2, \mathbf{u}^{(n)}). \tag{17.71}$$

*Hint. Use Section 4.3.5.*

**Exercise 17.16.** *Prove that the relation (17.69) implies that for any $k \in \mathbb{N}$ we also have*

$$\left( \mathbf{p}^{\otimes k}, \mathbf{g}^{\otimes k} \right) \sim \left( \mathbf{r}^{\otimes k}, \left( \mathbf{u}^{(n)} \right)^{\otimes k} \right). \tag{17.72}$$

The equivalence between athermality and nonuniformity give rise to the following property.

> **The Many Second Laws of Thermodynamics**
>
> **Theorem 17.5.** Let $(\mathbf{p},\mathbf{g})$ and $(\mathbf{p}',\mathbf{g}')$ be two thermal states. The following statements are equivalent.
>
> 1. For every $\varepsilon > 0$ there exists a thermal catalyst $\kappa := (\mathbf{r},\tilde{\mathbf{g}})$ such that
>
> $$(\mathbf{p}_\varepsilon,\mathbf{g}) \otimes \kappa \xrightarrow{\text{CTO}} (\mathbf{p}'_\varepsilon,\mathbf{g}') \otimes \kappa \tag{17.73}$$
>
> for some $\mathbf{p}_\varepsilon \in \mathfrak{B}_\varepsilon(\mathbf{p})$ and $\mathbf{p}'_\varepsilon \in \mathfrak{B}_\varepsilon(\mathbf{p}')$.
> 2. For all $\alpha \geqslant \frac{1}{2}$
>
> $$D_\alpha(\mathbf{p}\|\mathbf{g}) \geqslant D_\alpha(\mathbf{p}'\|\mathbf{g}') \quad \text{and} \quad D_\alpha(\mathbf{g}\|\mathbf{p}) \geqslant D_\alpha(\mathbf{g}'\|\mathbf{p}'). \tag{17.74}$$

*Remark.* Very recently (see the notes and references at the end of this section) it was shown that Theorem 17.5 can be strengthened by replacing $\mathbf{p}_\varepsilon^A$ with $\mathbf{p}^A$ so that (17.73) becomes

$$(\mathbf{p},\mathbf{g}) \otimes \kappa \xrightarrow{\text{CTO}} (\mathbf{p}'_\varepsilon,\mathbf{g}') \otimes \kappa. \tag{17.75}$$

This improvement makes the result somewhat more physical, and furthermore provides simple characterization for catalytic majorization (cf. Lemma 4.6): $(\mathbf{p},\mathbf{g}) \succ_c (\mathbf{p}',\mathbf{g}')$ if and only if for every $\varepsilon > 0$ there exists $\mathbf{p}'_\varepsilon \in \mathfrak{B}_\varepsilon(\mathbf{p}')$ such that $(\mathbf{p},\mathbf{g}) \succ_* (\mathbf{p}'_\varepsilon,\mathbf{g}')$. The proof of this improvement involves techniques not covered in this book and the interested reader can find the relevant references at the last section of this chapter.

**Proof** From Theorem (17.4) the condition in (17.73) is equivalent to

$$(\mathbf{p}_\varepsilon,\mathbf{g}) \succ_* (\mathbf{p}'_\varepsilon,\mathbf{g}'). \tag{17.76}$$

Therefore, from Lemma 4.6 this condition is equivalent to

$$(\mathbf{p},\mathbf{g}) \succ_c (\mathbf{p}',\mathbf{g}'). \tag{17.77}$$

Hence, the equivalence of the two conditions in the theorem follows from Theorem 4.12. This completes the proof. $\blacksquare$

## 17.4 Quantification of Athermality

Measures of athermality are functions that take athermality states to the real numbers and behave monotonically under CTO. Recall that in the theory of athermality, any physical system is described by pair of states of the form $(\rho^A,\gamma^A)$, and consequently measures of athermality are functions of such pair of states. Since we consider Hamiltonians with bounded energy, all Gibbs states are positive-definite (i.e. we assume $\gamma^A > 0$).

> **Definition 17.4.** A measure of athermality is a function
>
> $$\mathbf{D}: \bigcup_A \mathfrak{D}(A) \times \mathfrak{D}_{>0}(A) \to \mathbb{R} \qquad (17.78)$$
>
> that satisfies the following two conditions:
>
> 1. Monotonicity: For any two athermality states $(\rho^A, \gamma^A)$ and $(\sigma^{A'}, \gamma^{A'})$
>
> $$(\rho^A, \gamma^A) \xrightarrow{\text{CTO}} (\sigma^{A'}, \gamma^{A'}) \quad \Rightarrow \quad \mathbf{D}(\rho^A, \gamma^A) \geqslant \mathbf{D}(\sigma^{A'}, \gamma^{A'}). \qquad (17.79)$$
>
> 2. Normalization: On the trivial system $|A| = 1$, $\mathbf{D}(1, 1) = 1$.

We have chosen the symbol $\mathbf{D}$ to denote a measure of athermality, given that every normalized quantum divergence $\mathbf{D}$ also serves as a measure of athermality. Such measures of athermality behave monotonically under the larger set of Gibbs-preserving operations. However, it's worth noting that not all measures of athermality are quantum divergences, as they only need to exhibit monotonic behavior under CTO.

**Exercise 17.17.** *Show that every normalized quantum divergence is a measure of athermality.*

In the quasi-classical regime, athermality measures are applied to pairs of probability vectors, with the stipulation that the second vector remains strictly positive due to the Gibbs states' inability to contain zero components (assuming finite energies). Furthermore, as previously discussed, two athermality states $(\mathbf{p}, \mathbf{g})$ and $(\mathbf{p}', \mathbf{g}')$ satisfy $(\mathbf{p}, \mathbf{g}) \xrightarrow{\text{CTO}} (\mathbf{p}', \mathbf{g}')$ if and only if $(\mathbf{p}, \mathbf{g}) \succ (\mathbf{p}', \mathbf{g}')$. Thus, within the quasi-classical framework, the earlier definition of an athermality measure essentially transforms into the definition of a divergence. This implies that, in the quasi-classical domain, athermality measures are indeed divergences. Additionally, the direct correlation between classical divergences and nonuniformity measures extends to form a bijection between nonuniformity measures and athermality measures, further intertwining these concepts.

## 17.4.1 A Complete Family of Monotones

In section Section 10.5, we introduced a complete family of resource monotones. Taking $\mathfrak{F} = \text{GPC}$ we compute these monotones for the theory of athermality. In order to apply Theorem 11.1 into our case here, we will view each state as a pair $(\rho^A, \gamma^A)$ and the free operations as channels of the form $\mathcal{E} \oplus \mathcal{E}$, with $\mathcal{E} \in \text{COV}(A \to A')$. With these identifications, the state $\eta$ in (11.1) is replaced by $\eta := (\eta_0, \eta_1)$ with $\eta_0, \eta_1 \in \text{Pos}(A)$ so that (10.178) becomes

$$G_\eta(\rho, \gamma) := \max \text{Tr}\left[ J^{AA'} \left( \rho^T \otimes \eta_0^{A'} + \gamma^A \otimes \eta_1^{A'} \right) \right] - \left\| \eta_0^{A'} + \eta_1^{A'} \right\|_\infty, \qquad (17.80)$$

where the maximum is over all $J \in \text{Pos}(AA')$ subject to the following:

1. $\mathcal{P}_\xi(J^{AA'}) = J^{AA'}$, where $\mathcal{P}_\xi$ is the pinching channel associated with the operator (cf. (15.235))

$$\xi^{AA'} := H^A \otimes I^{A'} - I^A \otimes H^{A'}. \tag{17.81}$$

2. $J^A = I^A$.

The above optimization problem is an SDP, and consequently has a dual given by (see Exercise 17.18)

$$G_\eta(\rho, \gamma) = \min \operatorname{Tr}\left[\sigma^A\right] - \left\|\eta_0^{A'} + \eta_1^{A'}\right\|_\infty, \tag{17.82}$$

where the minimum is over all $\sigma \in \operatorname{Pos}(A)$ subject to

$$\sigma^A \otimes I^{A'} \geqslant \omega^{AA'} := \mathcal{P}_\xi\left(\rho^T \otimes \eta_0^{A'} + \gamma^A \otimes \eta_1^{A'}\right). \tag{17.83}$$

Hence, $G_\eta(\rho, \gamma)$ can be expressed as

$$G_\eta(\rho, \gamma) = 2^{-H_{\min}^\uparrow(A'|A)_\omega} - \left\|\eta_0^{A'} + \eta_1^{A'}\right\|_\infty. \tag{17.84}$$

If the Hamiltonians $H^A$ and $H^{A'}$ are nondegenerate, then the operator $\xi^{AA'}$ is nondegenerate so that $\mathcal{P}_\xi$ is the completely dephasing channel in the energy eigenbasis. In this case, $\omega^{AA'}$ is diagonal and therefore we can assume without loss of generality that also $\eta_0$ and $\eta_1$ are diagonal. For this case, for every choice of $\eta$ we have

$$G_\eta(\rho, \gamma) = G_\eta(\Delta(\rho), \gamma), \tag{17.85}$$

where $\Delta \in \operatorname{CPTP}(A \to A)$ is the energy dephasing channel. Therefore, for such a choice of system $A'$, $G_\eta$ depends only on the diagonal elements of $\rho$.

**Exercise 17.18.** *Express the optimization problem in (17.80) as a conic linear programming of the form (A.57) (i.e. as a dual problem) and then use the primal problem A.52 of online version to obtain (17.82).*

**Exercise 17.19.** *Show that if $A = A'$ and $H^A$ has a nondegenerate Bohr spectrum, then without loss of generality we can assume that $\eta_1$ is diagonal in the energy eigenbasis (i.e. $G_\eta$ depends only on the diagonal elements of $\eta_1$).*

**Exercise 17.20.** *Let $\eta_0, \eta_1 \in \operatorname{Pos}(A')$, $\rho, \gamma \in \mathfrak{D}(A)$, and $\tilde{\omega}^{AA'} := \gamma^A \otimes \left(\eta_0^{A'} + \eta_1^{A'}\right)$.*

*1. Show that*

$$H_{\min}^\uparrow(A'|A)_{\tilde{\omega}} = H_{\min}(A')_\omega = -\log\left\|\eta_0^{A'} + \eta_1^{A'}\right\|_\infty. \tag{17.86}$$

*2. Show that the function*

$$f_\eta(\rho, \gamma) := H_{\min}(A')_\omega - H_{\min}^\uparrow(A'|A)_\omega \tag{17.87}$$

*is a measure of athermality.*

**Exercise 17.21.** *Show that for the case* $\mathfrak{F} = \text{GPO}$, *the athermality monotones* $G_\eta(\rho, \gamma)$ *are given as in* (17.84), *but with*

$$\omega^{AA'} := \rho^T \otimes \eta_0^{A'} + \gamma^A \otimes \eta_1^{A'}. \tag{17.88}$$

## 17.4.2 The Free Energy

We have seen in this book that the Umegaki relative entropy has several operational interpretations and play a key role in quantum resource theories. In particular, we will see in the following sections that under Gibbs preserving operations the relative entropy is the unique measure of athermality in the asymptotic setting, as in this setting both the distillable rate of athermality and the athermality cost are given in terms of the relative entropy. Therefore, it is not a surprise that the relative entropy to the Gibbs state is related to an important quantity in thermodynamics known as the *free energy*.

In thermodynamics, the free energy is a fundamental concept that represents the potential energy available in a system to do useful work. It is a state function, meaning its value depends only on the current state of the system and not on how the system reached that state. Free energy is denoted by the symbol "F" and for an athermality state $(\rho, \gamma)$ of system $A$ the free energy is defined as the energy available to do useful work and is given by

$$F(\rho) := \text{Tr}\left[\rho \hat{H}\right] - T H(\rho), \tag{17.89}$$

where $T$ is the temperature, and we add here the "hat" symbol to the Hamiltonian $\hat{H}$ of the system, in order to distinguish it from the entropy symbol $H(\rho)$, which stands for the von-Neumann entropy of $\rho$.

**Exercise 17.22.** *Show that the free energy of the Gibbs state* $\gamma := \frac{1}{\mathcal{Z}} e^{-\beta \hat{H}}$ *is given by*

$$F(\gamma) = -T \log \mathcal{Z}. \tag{17.90}$$

To see the relation of the free energy to the relative entropy, observe that the relative entropy of athermality is given by

$$
\begin{aligned}
D(\rho\|\gamma) &= -H(\rho) - \text{Tr}\left[\rho \log\left(\frac{1}{\mathcal{Z}} e^{-\beta \hat{H}}\right)\right] \\
&= \log \mathcal{Z} - H(\rho) + \beta \text{Tr}\left[\rho \hat{H}\right] \\
&= \beta F(\rho) + \log \mathcal{Z} \\
(\textbf{17.90})\!\rightarrow\ &= \beta\big(F(\rho) - F(\gamma)\big).
\end{aligned}
\tag{17.91}
$$

Hence, the free energy is the key factor that directly governs the optimal rate of interconversions of athermality.

The Umegaki relative entropy of athermality has another interesting representation. For a quantum athermality state $(\rho, \gamma)$, with Hamiltonian $\hat{H}$, we can express $D(\rho\|\gamma)$ as:

$$
\begin{aligned}
D(\rho\|\gamma) &= -H(\rho) - \mathrm{Tr}\left[\rho\log\gamma\right] \\
&= -H(\rho) - \mathrm{Tr}\left[\mathcal{P}_{\hat{H}}(\rho)\log\gamma\right] \\
&= D\left(\mathcal{P}_{\hat{H}}(\rho)\|\gamma\right) + H\left(\mathcal{P}_{\hat{H}}(\rho)\right) - H(\rho) \\
&= D\left(\mathcal{P}_{\hat{H}}(\rho)\|\gamma\right) + C(\rho),
\end{aligned}
\tag{17.92}
$$

where $\mathcal{P}_{\hat{H}}$ is the pinching channel associated with the Hamiltonian $\hat{H}$, and $C(\rho)$ is the coherence measure defined in (15.238) ($C(\rho)$ is also known as the **G**-asymmetry of the state $\rho$ as defined in 15.118, where **G** stands for the group of time-translation symmetry). That is, the athermality of the state $(\rho, \gamma)$ can be decomposed into two components:

1. Its nonuniformity that is quantified by $D\left(\mathcal{P}_{\hat{H}}(\rho)\|\gamma\right)$.
2. Its asymmetry (or coherence between energy eigenspaces) that is quantified by the coherence measure $C(\rho)$.

We will see later on that this decomposition has an operational meaning, in which (roughly speaking) $D\left(\mathcal{P}_{\hat{H}}(\rho)\|\gamma\right)$ is the cost to prepare the athermality state $(\mathcal{P}_{\hat{H}}(\rho), \gamma)$ and $C(\rho)$ is the cost to "rotate" $\mathcal{P}_{\hat{H}}(\rho)$ to $\rho$. Moreover, since the regularization of the $C(\rho)$ vanishes (see Theorem 15.9), we conclude that

$$
\lim_{n\to\infty} \frac{1}{n} D\left(\mathcal{P}_n\left(\rho^{\otimes n}\right)\|\gamma^{\otimes n}\right) = D(\rho\|\gamma),
\tag{17.93}
$$

where $\mathcal{P}_n$ is the pinching channel associated with the total Hamiltonian of system $A^n$.

## 17.5 Single-Shot Exact Interconversions

In this section, we study the condition under which an athermality state $(\rho^A, \gamma^A)$ can be converted to another state $(\sigma^A, \gamma^A)$ by either CTO, GPC, or GPO. We will work in the basis $\{|x\rangle\}_{x\in[m]}$ in which the Hamiltonian is diagonalized, and denote by $\hat{H} = \sum_{x\in[m]} a_x|x\rangle\langle x|$ the Hamiltonian of system $A$, where $\{a_x\}_{x\in[m]}$ are the energy eigenvalues of $\hat{H}$.

### 17.5.1 Exact Conversions under GPO

The state $(\rho, \gamma)$ can be converted to $(\rho', \gamma')$ by GPO if and only if there exists a channel $\mathcal{N} \in \mathrm{CPTP}(A \to A')$ such that $\mathcal{N}(\rho) = \rho'$ and $\mathcal{N}(\gamma) = \gamma'$. In the previous chapter, we saw that in the quasi-classical case these conditions are equivalent to relative majorization. Here we study its quantum version.

> **Definition 17.5.** Let $\rho, \gamma \in \mathfrak{D}(A)$ and $\rho', \gamma' \in \mathfrak{D}(A')$. We say that the pair $(\rho, \gamma)$ relatively majorizes the pair $(\rho', \gamma')$, and write
>
> $$(\rho, \gamma) \succ (\rho', \gamma') \tag{17.94}$$
>
> if there exists a channel $\mathcal{N} \in \mathrm{CPTP}(A \to A')$ such that
>
> $$\mathcal{N}(\rho) = \rho' \quad \text{and} \quad \mathcal{N}(\gamma) = \gamma'. \tag{17.95}$$

The two conditions in (17.95) are equivalent to the existence of a Choi matrix $J \in \mathrm{Pos}(AA')$ that satisfies

$$\mathrm{Tr}_{A'}\left[J^{AA'}\left(\rho^T \otimes I^{A'}\right)\right] = \rho' \quad \text{and} \quad \mathrm{Tr}_{A'}\left[J^{AA'}\left(\gamma^T \otimes I^{A'}\right)\right] = \gamma'. \tag{17.96}$$

This problem, of determining whether or not such a Choi matrix $J^{AA'}$ exists, is an SDP feasibility problem that can be solved efficiently and algorithmically using techniques from semi-definite programming. However, unlike the classical case, where relative majorization can be characterized with Lorenz curves, it is not known in the fully quantum case whether a similar geometrical characterization exists.

Observe that any quantum divergence behaves monotonically under quantum relative majorization. Specifically, if $\mathbb{D}$ is a quantum divergence, then

$$(\rho, \gamma) \succ (\rho', \gamma') \quad \Rightarrow \quad \mathbb{D}(\rho \| \gamma) \geqslant \mathbb{D}\left(\rho' \| \gamma'\right). \tag{17.97}$$

The converse to the above property also holds. That is, if for any choice of a quantum divergence $\mathbb{D}$ we have $\mathbb{D}(\rho \| \gamma) \geqslant \mathbb{D}\left(\rho' \| \gamma'\right)$ then we must have $(\rho, \gamma) \succ (\rho', \gamma')$. In fact, we show now that this assertion still holds even if we restrict $\mathbb{D}$ to have a very specific form.

Recall the complete family of monotones given in (17.84) with $\omega^{AA'} := \rho^T \otimes \eta_0^{A'} + \gamma^A \otimes \eta_1^{A'}$ given as in Exercise 17.21. From the completeness of the family of monotones, it follows that $(\rho, \gamma) \succ (\rho', \gamma')$ if and only if $G_\eta(\rho, \gamma) \geqslant G_\eta(\rho', \gamma')$ for all $\eta_0, \eta_1 \in \mathrm{Pos}(A')$. Similar to (17.87), for every $\eta_0, \eta_1 \in \mathrm{Pos}(A')$ we define

$$D_\eta(\rho \| \gamma) := H_{\min}(A')_\omega - H_{\min}(A'|A)_\omega. \tag{17.98}$$

The above functions form a family of normalized quantum divergences that can be used to characterize quantum relative majorization.

**Exercise 17.23.** *Show that for every $\eta_0, \eta_1 \in \mathrm{Pos}(A')$, the function $D_\eta$ as defined above is a quantum divergence.*

> **Theorem 17.6.** Let $\rho, \gamma \in \mathfrak{D}(A)$ and $\rho', \gamma' \in \mathfrak{D}(A')$. Then, the following are equivalent:
>
> 1. $(\rho, \gamma) \succ (\rho', \gamma')$.
> 2. For any $\eta_0, \eta_1 \in \mathrm{Pos}(A')$ we have $D_\eta(\rho \| \gamma) \geqslant D_\eta(\rho' \| \gamma')$.

**Exercise 17.24.** *Consider Theorem 17.6.*

1. *Prove the theorem.*
2. *Show that the theorem holds even if we restrict $\eta_0$ and $\eta_1$ to satisfy $\text{Tr}[\eta_0 + \eta_1] = 1$ (hence, we can assume without loss of generality that $\omega^{AA'}$ is a density matrix).*
3. *Show that the theorem holds even if we restrict $\eta_0$ and $\eta_1$ to satisfy $\text{Tr}[\eta_0] = \text{Tr}[\eta_1] = 1/2$.*

While the aforementioned theorem offers a characterization of quantum relative majorization, it doesn't provide a straightforward method for determining whether one pair of states relatively majorizes another due to the necessity of verifying an infinite number of conditions. Instead, as already mentioned, one can employ Semidefinite Programming (SDP) feasibility algorithms from convex analysis to address this challenge.

Nevertheless, when dealing with qubit states $\rho$ and $\gamma$, a far simpler method exists to characterize quantum relative majorization. In the upcoming theorem, we leverage the fidelity function $F(\eta, \zeta) := |\eta^{1/2}\zeta^{1/2}|_1$ for all $\eta, \zeta \in \text{Pos}(A)$, including unnormalized states, to provide a more accessible approach to understanding quantum relative majorization.

***

**Theorem 17.7.** Let $\rho, \gamma \in \mathfrak{D}(A)$ and $\rho', \gamma' \in \mathfrak{D}(A')$, and suppose $|A| = 2$. Furthermore, let $a := 2^{D_{\max}(\rho\|\gamma)}$ and $b := 2^{D_{\max}(\gamma\|\rho)}$. Then, $(\rho, \gamma) \succ (\rho', \gamma')$ if and only if the following conditions hold:

1. $D_{\max}(\rho\|\gamma) \geq D_{\max}(\rho'\|\gamma')$.
2. $D_{\max}(\gamma\|\rho) \geq D_{\max}(\gamma'\|\rho')$.
3. $F\left(a\gamma' - \rho', b\rho' - \gamma'\right) \geq F\left(a\gamma - \rho, b\rho - \gamma\right)$.

***

*Remark.* By definition of $a$ and $b$ we have $a\gamma - \rho \geq 0$ and $b\rho - \gamma \geq 0$. Moreover, since $\rho$ and $\gamma$ are qubits, $a\gamma - \rho$ and $b\rho - \gamma$ are rank 1. Finally, note that from the first two conditions above we also get $a\gamma' - \rho' \geq 0$ and $b\rho' - \gamma' \geq 0$.

**Proof**   The case $\rho = \gamma$ is left as an exercise, and we assume now that $\rho \neq \gamma$ so that both $a$ and $b$ are strictly greater than 1. The necessity of the first two conditions in the theorem follows from the fact that $D_{\max}$ satisfies the DPI. Similarly, the necessity of the third condition follows from the DPI of the fidelity. It is therefore left to show the sufficiency of the conditions. Define

$$\psi := \frac{a\gamma - \rho}{a - 1} \quad \text{and} \quad \phi := \frac{b\rho - \gamma}{b - 1}. \tag{17.99}$$

As discussed in the remark above, $\psi$ and $\phi$ are *pure* states. Denoting by

$$\eta := \frac{a\gamma' - \rho'}{a - 1} \quad \text{and} \quad \zeta := \frac{b\rho' - \gamma'}{b - 1}, \tag{17.100}$$

we can express the third condition as $F(\eta, \zeta) \geqslant F(\psi, \phi)$. From Uhlmann's theorem there exists purifications of $\eta$ and $\zeta$, denoted by $\psi', \phi' \in \mathrm{Pure}(A'\tilde{A}')$ such that $F(\eta, \zeta) = |\langle \psi'|\phi'\rangle|$, so that we get

$$|\langle \psi'|\phi'\rangle| \geqslant |\langle \psi|\phi\rangle|. \tag{17.101}$$

The main trick of the proof is to introduce two states $\varphi_1, \varphi_2 \in \mathrm{Pure}(R)$, where $R$ is some Hilbert space (which we can choose to be two dimensional), that satisfies

$$\langle \psi|\phi\rangle = \langle \psi'|\phi'\rangle\langle \varphi_1|\varphi_2\rangle. \tag{17.102}$$

The above invariant overlap implies that the matrix $V: A \to A'\tilde{A}'R$ defined by

$$V|\psi\rangle := |\psi'\rangle|\varphi_1\rangle \quad \text{and} \quad V|\phi\rangle := |\phi'\rangle|\varphi_2\rangle, \tag{17.103}$$

is an isometry. Finally, let $\mathcal{N} \in \mathrm{CPTP}(A \to A')$ be the channel defined via

$$\mathcal{N}^{A \to A'}\left(\omega^A\right) := \mathrm{Tr}_{\tilde{A}'R}\left[V\omega^A V^*\right] \qquad \forall\, \omega \in \mathcal{L}(A). \tag{17.104}$$

To see that the channel above satisfies the desired properties, first observe that by isolating $\rho$ and $\gamma$ from (17.99) we get

$$\rho = \frac{(a-1)\psi + a(b-1)\phi}{ab-1} \quad \text{and} \quad \gamma = \frac{b(a-1)\psi + (b-1)\phi}{ab-1}. \tag{17.105}$$

Therefore,

$$\begin{aligned}
\mathcal{N}(\rho) &= \frac{(a-1)\mathcal{N}(\psi) + a(b-1)\mathcal{N}(\phi)}{ab-1} \\
\textbf{By definition} \to &= \frac{(a-1)\eta + a(b-1)\zeta}{ab-1} \\
\textbf{(17.100)} \to &= \rho'.
\end{aligned} \tag{17.106}$$

Using similar lines it can be shown that $\mathcal{N}(\gamma) = \gamma'$ (see Exercise 17.25). This completes the proof. ∎

**Exercise 17.25.** *Prove the assertion in the proof above that $\mathcal{N}(\gamma) = \gamma'$.*

**Exercise 17.26.** *Let $\rho, \sigma, \gamma \in \mathfrak{D}(A)$ be three qubit states (i.e. $|A| = 2$). Show that*

$$(\rho, \gamma) \xrightarrow{\mathrm{GPO}} (\sigma, \gamma) \tag{17.107}$$

*if and only if*

$$D_{\max}\left(\rho\|\gamma\right) \geqslant D_{\max}\left(\sigma\|\gamma\right) \quad \text{and} \quad D_{\max}\left(\gamma\|\rho\right) \geqslant D_{\max}\left(\gamma\|\sigma\right). \tag{17.108}$$

*That is, the third fidelity condition of Theorem 17.7 is unnecessary in this case.*

## 17.5.2 Exact Conversions under CTO and GPC

Given two athermality states $(\rho, \gamma)$ of system $A$, and $(\rho', \gamma')$ of system $A'$, the condition $(\rho, \gamma) \xrightarrow{\text{GPC}} (\rho', \gamma')$ is equivalent to the existence of a Choi matrix $J \in \text{Pos}(AA')$ that satisfies the following four conditions:

1. $\text{Tr}_{A'}\left[ J^{AA'} \left( \rho^T \otimes I^{A'} \right) \right] = \rho'$
2. $\text{Tr}_{A'}\left[ J^{AA'} \left( \gamma^T \otimes I^{A'} \right) \right] = \gamma'$
3. $\mathcal{P}_\xi \left( J^{AA'} \right) = 0$, where $\mathcal{P}_\xi$ is the pinching channel of $\xi^{AA'} := H^A \otimes I^{A'} - I^A \otimes H^{A'}$.
4. $J^A = I^A$.

Similar to the GPO case, this problem, of determining whether or not such a Choi matrix $J^{AA'}$ exists, is an SDP feasibility problem that can be solved efficiently and algorithmically using techniques from semi-definite programming. However, for certain choices of Hamiltonians, there exists a much simpler way to characterize the conversion $(\rho, \gamma) \xrightarrow{\text{GPC}} (\rho', \gamma')$.

### The Case of Relatively Nondegenerate Hamiltonians

Theorem 15.17 of Section 15.6 has the following implication in thermodynamics.

> **Corollary 17.2.** Let $A$ and $A'$ be two physical systems with relatively nondegenerate Hamiltonians. Let $(\rho, \gamma)$ and $(\rho', \gamma')$ be two athermality states on system $A$ and $A'$, respectively. Furthermore, ler $\mathbf{r}$, $\mathbf{r}'$, $\mathbf{g}$, and $\mathbf{g}'$ be the probability vectors whose components are the elements on the diagonals of $\rho$, $\sigma'$, $\gamma$, and $\gamma'$, respectively. Then, for $\mathfrak{F}$ being CTO or GPC, the following are equivalent:
>
> 1. $(\rho, \gamma) \xrightarrow{\mathfrak{F}} (\rho', \gamma')$.
> 2. $\rho'$ is diagonal and $(\mathbf{r}, \mathbf{g}) \succ (\mathbf{r}', \mathbf{g}')$.

**Exercise 17.27.** *Use Theorem 15.17 to prove Corollary 17.2.*

Note that in the general case of relatively nondegenerate Hamiltonians, CTO and GPC operations can only disrupt the coherence between the energy levels of the input state $\rho^A$. In such scenarios, coherence cannot be manipulated, but only destroyed. Therefore, for the remainder of this chapter, we will focus on Hamiltonians that exhibit relative degeneracy.

When considering the conversion of one athermality state $(\rho, \gamma)$ to another athermality state $(\rho', \gamma')$ we will use the properties

$$(\rho, \gamma) \xleftrightarrow{\mathfrak{F}} (\rho \otimes \gamma', \gamma \otimes \gamma') \quad \text{and} \quad (\rho', \gamma') \xleftrightarrow{\mathfrak{F}} (\gamma \otimes \rho', \gamma \otimes \gamma'). \tag{17.109}$$

The equivalence relations above follow from the fact that appending or removing a Gibbs state is a free operation in the theory of athermality. Therefore, the conversion

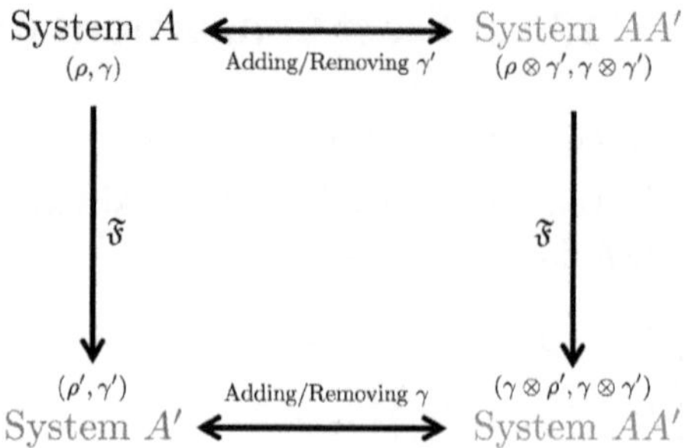



**Figure 17.3** Equivalence of a conversion from $A$ to $A'$ and a conversion from $AA'$ to itself.

$(\rho,\gamma)\xrightarrow{\mathfrak{F}}(\rho',\gamma')$ between a state of system $A$ and a state of system $A'$ is equivalent to the conversion $(\rho\otimes\gamma',\gamma\otimes\gamma')\xrightarrow{\mathfrak{F}}(\gamma\otimes\rho',\gamma\otimes\gamma')$ between two states of system $AA'$; see Figure 17.3. In other words, interconversions among states with the same dimensions (i.e. states with $|A|=|A'|$) is general enough to capture also interconversions with $|A'|\neq|A|$ (as long as we do not impose nondegeneracy constraints). We will therefore focus here on interconversions among states that are all in $\mathfrak{D}(A)$.

## The Case of Bohr Spectrum

Consider a physical system $A$ whose Hamiltonian, $\hat{H}=\sum_{x\in[m]}a_x|x\rangle\langle x|$, has a nondegenerate Bohr spectrum; that is, $a_x-a_{x'}=a_y-a_{y'}$ if and only if $x=x'$ and $y=y'$, or $x=y$ and $x'=y'$. Furthermore, consider a conversion of the form $(\rho,\gamma)\xrightarrow{\text{GPC}}(\sigma,\gamma)$, where *all* the off-diagonal terms of $\rho$ are nonzero. In this case, Theorem 15.19 states that $\rho$ can be converted to $\sigma$ by a time-translation covariant channel if and only if the matrix $Q$ as defined in (15.251) is positive semidefinite. Since GPC channels are in particular covariant under the time-translation group, the condition $Q\geqslant 0$ is a necessary (but not sufficient) condition for $(\rho,\gamma)\xrightarrow{\text{GPC}}(\sigma,\gamma)$.

To establish the full set of necessary and sufficient conditions, let $J^{AB}$ be the Choi matrix of a time-translation covariant channel $\mathcal{E}\in\text{COV}(A\to A)$ that satisfies $\mathcal{E}(\rho)=\sigma$ and $\mathcal{E}(\gamma)=\gamma$. Denoting by $\{r_{xy}\}_{x,y\in[m]}$ and $\{s_{xy}\}_{x,y\in[m]}$ the components of $\rho$ and $\sigma$, respectively, we get that the Choi matrix of $\mathcal{E}$ has the form (cf. (15.249))

$$J^{A\tilde{A}}=\sum_{x,y}p_{y|x}|x\rangle\langle x|^A\otimes|y\rangle\langle y|^{\tilde{A}}+\sum_{x\neq y}\frac{s_{xy}}{r_{xy}}|x\rangle\langle y|^A\otimes|x\rangle\langle y|^{\tilde{A}},\qquad(17.110)$$

where $P=(p_{y|x})$ is some column stochastic matrix, and we assumed that the off diagonal terms of $\rho^A$ are nonzero. Let $\mathbf{r}$ and $\mathbf{s}$ be the probability vectors consisting of the diagonals of $\rho$ and $\sigma$, and identify the diagonal matrix $\gamma$ with the Gibbs vector $\mathbf{g}$ consisting of its diagonal. Then, the Choi matrix above corresponds to such a GPC channel $\mathcal{E}$ if and only if it is positive semidefinite *and*

$$P\mathbf{r}=\mathbf{s}\quad\text{and}\quad P\mathbf{g}=\mathbf{g}.\qquad(17.111)$$

The above condition implies that $(\mathbf{r}, \mathbf{g}) \succ (\mathbf{s}, \mathbf{g})$; however, it is not sufficient since we also require that $J^{A\tilde{A}} \geqslant 0$. This latter condition is equivalent to the requirement that the matrix obtained by replacing the diagonal elements of $Q$ (as defined in (15.251)) with $\{p_{x|x}\}_{x \in [m]}$ is positive semidefinite. We summarize these considerations in the following exercise.

**Exercise 17.28.** *Let $(\rho, \gamma)$ and $(\sigma, \gamma)$ be two athermality states of a system A, whose Hamiltonian $\hat{H}$ has a nondegenerate Bohr spectrum. Suppose also that the off diagonal terms of $\rho$ are nonzero. Show that*

$$(\rho, \gamma) \xrightarrow{\;GPC\;} (\sigma, \gamma) \tag{17.112}$$

*if and only if there exists a column stochastic matrix P that satisfies both (17.111) and the matrix*

$$\sum_{x \in [m]} p_{x|x}|x\rangle\langle x| + \sum_{x \neq y \in [m]} \frac{s_{xy}}{r_{xy}}|x\rangle\langle y| \geqslant 0. \tag{17.113}$$

Exercise 17.28 does not offer significant computational simplification compared to the SDP feasibility problem discussed at the beginning of this section. This is because determining the existence of a column stochastic matrix $P$ itself constitutes an SDP problem. However, the exercise's significance lies in its ability to highlight the role of quantum coherence in converting athermality, as demonstrated by the following theorem. Furthermore, we will observe later that in the qubit case, Exercise 17.28 provides a straightforward criterion for exact inter-conversions under GPC.

---

**Theorem 17.8.** Let $(\rho, \gamma)$ and $(\sigma, \gamma)$ be two quantum athermality states of dimension $m := |A|$. For any $x, y \in [m]$ let $r_{xy} := \langle x|\rho|y\rangle$ and $s_{xy} := \langle x|\sigma|y\rangle$ be the $xy$-component of $\rho$ and $\sigma$, respectively. Suppose that $r_{xy} \neq 0$ for all $x, y \in [m]$ and that $r_{xx} = s_{xx}$ for all $x \in [m]$. Then,

$$(\rho, \gamma) \xrightarrow{\;GPC\;} (\sigma, \gamma) \quad \Longleftrightarrow \quad Q := I + \sum_{x \neq y \in [m]} \frac{s_{xy}}{r_{xy}}|x\rangle\langle y| \geqslant 0. \tag{17.114}$$

---

**Proof** Since the diagonals of $\rho$ and $\sigma$ are the same, we get that if $Q \geqslant 0$, then by taking the stochastic matrix $P$ to be the identity matrix, all the conditions in Exercise 17.28 are satisfied so that $(\rho, \gamma) \xrightarrow{\;GPC\;} (\sigma, \gamma)$. Conversely, if $(\rho, \gamma) \xrightarrow{\;GPC\;} (\sigma, \gamma)$, then by Exercise 17.28 there exists a stochastic matrix $P$ with a diagonal $\{p_{x|x}\}$ that satisfies (17.113). By adding the positive semidefinite matrix $\sum_{x \in [m]} (1 - p_{x|x})|x\rangle\langle x|$ to the matrix in (17.113) we get that also $Q \geqslant 0$. This completes the proof. $\blacksquare$

In simple terms, the condition stated in Theorem 17.8, that $\rho$ and $\sigma$ share the same diagonals, implies that they have the same nonuniformity and only differ in their coherence (asymmetry) properties. Interestingly, the condition $Q \geqslant 0$ turns out to be identical to the condition given in Theorem 15.19 when $\rho$ and $\sigma$ have the same diagonal elements. Thus, in this case, we can state that $(\rho, \gamma) \xrightarrow{\text{GPC}} (\sigma, \gamma)$ if and only if $\rho$ can be transformed into $\sigma$ through time-translation covariant operations. It is noteworthy that the Gibbs state, $\gamma$, does not play a role in such conversions because $\rho$ and $\sigma$ share the same nonuniformity (i.e. same diagonal elements).

> **Corollary 17.3.** Let $\sigma \in \mathfrak{D}(A)$ be an arbitrary state, and denote by $p_x :=$ $\langle x | \sigma | x \rangle$ the diagonal elements of $\sigma$ in the energy eigenbasis $\{|x\rangle\}_{x \in [m]}$ of system $A$. Then, the pure quantum state
>
> $$|\psi\rangle := \sum_{x \in [m]} \sqrt{p_x} |x\rangle \qquad (17.115)$$
>
> can be converted to $\sigma$ by GPC. That is, $(\psi, \gamma) \xrightarrow{\text{GPC}} (\sigma, \gamma)$.

**Exercise 17.29.** *Prove Corollary 17.3. Hint: See the proof of Corollary 15.3.*

**Exercise 17.30.** *Show that Corollary 17.3 still holds even if we replace GPC with CTO. Hint: Use the fact that $\psi$ can be converted to $\sigma$ by time-translation covariant channel, and then use Theorem 15.4.*

## The Qubit Case

In this subsection, we use the considerations above to provide the analytical conditions for inter-conversions of athermality when the systems invloved are qubits. Specifically, let $\rho, \sigma, \gamma \in \mathfrak{D}(A)$ with $|A| = 2$. Denote

$$\rho = \begin{pmatrix} r & a \\ \bar{a} & 1 - r \end{pmatrix} \quad , \quad \sigma = \begin{pmatrix} s & b \\ \bar{b} & 1 - s \end{pmatrix}, \quad \text{and} \quad \gamma = \begin{pmatrix} g & 0 \\ 0 & 1 - g \end{pmatrix}. \qquad (17.116)$$

We also denote the diagonals of the matrices above by $\mathbf{r} := (r, 1 - r)^T$, $\mathbf{s} := (s, 1 - s)^T$, and $\mathbf{g} = (g, 1 - g)^T$, respectively. We would like to find the conditions under which $(\rho, \gamma) \xrightarrow{\text{GPC}} (\sigma, \gamma)$. Recall that if $a = 0$, then we must have $b = 0$ since GPC cannot generate coherence between energy levels. Therefore, the case $a = 0$ has already been covered by the quasi-classical regime. We will therefore assume in the rest of this subsection that $a \neq 0$.

> **Theorem 17.9.** Let $\rho, \sigma, \gamma \in \mathfrak{D}(A)$ be three qubit states as before and suppose $a \neq 0$ and $\gamma \neq u$. Then, for $\mathbf{r} \neq \mathbf{g}$, $(\rho, \gamma) \xrightarrow{\text{GPC}} (\sigma, \gamma)$ if and only if $(\mathbf{r}, \mathbf{g}) \succ (\mathbf{s}, \mathbf{g})$ and
>
> $$\frac{|b|^2}{|a|^2} \leqslant \frac{s - g}{r - g} + g(1 - g)\left(\frac{r - s}{r - g}\right)^2. \qquad (17.117)$$
>
> For $\mathbf{r} = \mathbf{g}$, $(\rho, \gamma) \xrightarrow{\text{GPC}} (\sigma, \gamma)$ if and only if $s = g$ and $|a| \geqslant |b|$.

**Proof** From Exercise 17.28 it follows that $(\rho, \gamma)$ can be converted to $(\sigma, \gamma)$ by GPC if and only if there exists a $2 \times 2$ column stochastic matrix $P = \{p_{y|x}\}_{x,y \in \{0,1\}}$ that satisfies $P\mathbf{r} = \mathbf{s}$, $P\mathbf{g} = \mathbf{g}$, and

$$\begin{pmatrix} p_{0|0} & b/a \\ \bar{b}/\bar{a} & p_{1|1} \end{pmatrix} \geqslant 0. \tag{17.118}$$

Note that this last condition is equivalent to

$$\frac{|b|^2}{|a|^2} \leqslant p_{0|0}\, p_{1|1}. \tag{17.119}$$

The conditions $P\mathbf{r} = \mathbf{s}$ and $P\mathbf{g} = \mathbf{g}$ can be expressed as the following linear systems of equations

$$\begin{bmatrix} r & 1-r \\ g & 1-g \end{bmatrix}\begin{bmatrix} p_{0|0} \\ p_{0|1} \end{bmatrix} = \begin{bmatrix} s \\ g \end{bmatrix} \quad \text{and} \quad \begin{bmatrix} r & 1-r \\ g & 1-g \end{bmatrix}\begin{bmatrix} p_{1|0} \\ p_{1|1} \end{bmatrix} = \begin{bmatrix} 1-s \\ 1-g \end{bmatrix}. \tag{17.120}$$

Note that the equations involving $p_{1|0}$ and $p_{1|1}$ follow trivially from the ones involving $p_{0|0}$ and $p_{0|1}$ since $P$ is column stochastic. From Cramer's rule it then follows that for the case that $r \neq g$

$$p_{0|0} = \frac{\det\begin{pmatrix} s & 1-r \\ g & 1-g \end{pmatrix}}{\det\begin{pmatrix} r & 1-r \\ g & 1-g \end{pmatrix}} \quad \text{and} \quad p_{1|1} = \frac{\det\begin{pmatrix} r & 1-s \\ g & 1-g \end{pmatrix}}{\det\begin{pmatrix} r & 1-r \\ g & 1-g \end{pmatrix}}. \tag{17.121}$$

Finally, substituting the above expression in (17.119) gives (after some simple algebra) the inequality (17.117).

For the case that $r = g$ we also have $s = g$ (otherwise, $(\mathbf{r}, \mathbf{g}) \not\succ (\mathbf{s}, \mathbf{g})$) and the linear system of equations in (17.120) has a unique solution given by $p_{0|0} = p_{1|1} = 1$. Therefore, in this case, (17.119) gives $|b| \leqslant |a|$. This completes the proof. ∎

**Exercise 17.31.** *Show that if $s = g$ in (17.116), then $(\rho, \gamma)$ can be converted to $(\sigma, \gamma)$ by GPC if and only if*

$$\frac{|b|^2}{|a|^2} \leqslant \det(\gamma). \tag{17.122}$$

From Exercise 17.31 it follows that already in the qubit case, conversions under GPC have a certain type of discontinuity. To see this, consider the case $s = g$, and observe that the condition $|a|^2 \det(\gamma) \geqslant |b|^2$ is stronger than the condition $|a| \geqslant |b|$ that one obtains if also $r = g$. In particular, observe that $\det(\gamma) \leqslant \frac{1}{4}$. Hence, there exists an $\varepsilon > 0$ and $\rho, \sigma, \gamma \in \mathfrak{D}(A)$ such that for any $\rho \in \mathcal{B}_\varepsilon(o)$ the state $(\rho, \gamma)$ cannot be converted by GPC to $(\sigma, \gamma)$ unless $\rho = \sigma$.

**Exercise 17.32.** *Find explicit example of three qubit states $\rho, \sigma, \gamma$, and $\varepsilon > 0$ such that for any $\rho \in \mathcal{B}_\varepsilon(\sigma)$, $(\rho, \gamma) \xrightarrow{\text{GPC}} (\sigma, \gamma)$ unless $\rho = \sigma$.*

## 17.6 The Conversion Distance of Athermality

We consider first the case that $\mathfrak{F} = \text{GPO}$, and define the conversion distance as

$$T\left((\rho,\gamma) \xrightarrow{\text{GPO}} (\rho',\gamma')\right) := \min_{\mathcal{E}\in\text{CPTP}(A\to A')} \left\{\frac{1}{2}\left\|\rho' - \mathcal{E}(\rho)\right\|_1 : \gamma' = \mathcal{E}(\gamma)\right\}. \quad (17.123)$$

Using the fact that the trace distance between two density matrices can be expressed as

$$\frac{1}{2}\left\|\rho' - \mathcal{E}(\rho)\right\|_1 = \min_{\substack{\Lambda\in\text{Pos}(A') \\ \Lambda\geqslant\rho'-\mathcal{E}(\rho)}} \text{Tr}\,[\Lambda], \quad (17.124)$$

we can express the conversion distance as the following SDP:

$$T\left((\rho,\gamma) \xrightarrow{\text{GPO}} (\rho',\gamma')\right) = \min \text{Tr}\,[\Lambda], \quad (17.125)$$

where the minimum is over all $\Lambda \in \text{Pos}(A')$ that satisfy the following conditions:

1. $\Lambda \geqslant \rho' - \text{Tr}_A\left[J^{AA'}\left(\rho^T \otimes I^{A'}\right)\right]$.
2. $\gamma' = \text{Tr}_A\left[J^{AA'}\left(\gamma^A \otimes I^{A'}\right)\right]$.
3. $J \in \text{Pos}(AA')$ and $J^A = I^A$.

For the case that $\mathfrak{F} = \text{GPC}$ the conversion distance is evaluated exactly as above with the additional constraint on $J^{AA'}$ that

$$\mathcal{P}_\xi\left(J^{AA'}\right) = J^{AA'}, \quad \text{where } \xi^{AA'} := H^A \otimes I^{A'} - I^A \otimes H^{A'}. \quad (17.126)$$

Note that this additional condition is still in a form suitable for SDP.

The preceding discussion demonstrates that the conversion distance of athermality can be computed numerically. However, the formulation presented here for the conversion distance lacks insight and does not offer any practical means to calculate the distillable athermality or the athermality cost of a state $(\rho, \gamma)$. Therefore, we now turn our attention to the case where the target state is quasi-classical, and show that for this case there exists an analytical formula for the conversion distance.

> **Theorem 17.10.** Let $(\rho,\gamma)$ be an arbitrary state of system $A$ and $(\rho',\gamma')$ be a quasi-classical state of system $A'$. Then,
>
> $$T\left((\rho,\gamma) \xrightarrow{\text{GPC}} (\rho',\gamma')\right) = T\left((\mathcal{P}(\rho),\gamma) \xrightarrow{\text{GPC}} (\rho',\gamma')\right), \quad (17.127)$$
>
> where $\mathcal{P}$ denotes the pinching channel associated with the Hamiltonian of system $A$.

*Remark.* Note that on the right-hand side, we have a conversion distance between two quasi-classical states. In the next subsection, we will demonstrate that for such cases, an analytical formula exists.

**Proof**   Let $\mathcal{P}$ and $\mathcal{P}'$ be the pinching channels associated with the Hamiltonians of systems $A$ and $A'$, respectively, and observe that for any $\mathcal{E} \in \mathrm{COV}(A \to A')$ that satisfies $\gamma' = \mathcal{E}(\gamma)$ we have

$$\gamma' = \mathcal{P}'(\gamma') = \mathcal{P}' \circ \mathcal{E}(\gamma)$$
$$\textbf{Part 2 of Exercise 15.42} \to \; = \mathcal{E} \circ \mathcal{P}(\gamma). \tag{17.128}$$

Therefore,

$$T\left((\mathcal{P}(\rho),\gamma) \xrightarrow{\mathrm{GPC}} (\rho',\gamma')\right) = \min_{\mathcal{E} \in \mathrm{COV}(A \to A')} \left\{ \frac{1}{2} \left\| \rho' - \mathcal{E} \circ \mathcal{P}(\rho) \right\|_1 \; : \; \gamma' = \mathcal{E}(\gamma) \right\}$$

$$\textbf{17.128} \to \; = \min_{\mathcal{E} \in \mathrm{COV}(A \to A')} \left\{ \frac{1}{2} \left\| \rho' - \mathcal{E} \circ \mathcal{P}(\rho) \right\|_1 \; : \; \gamma' = \mathcal{E} \circ \mathcal{P}(\gamma) \right\}$$

$$\mathcal{N} := \mathcal{E} \circ \mathcal{P} \to \; \geqslant \min_{\mathcal{N} \in \mathrm{COV}(A \to A')} \left\{ \frac{1}{2} \left\| \rho' - \mathcal{N}(\rho) \right\|_1 \; : \; \gamma' = \mathcal{N}(\gamma) \right\}$$

$$= T\left((\rho,\gamma) \xrightarrow{\mathrm{GPC}} (\rho',\gamma')\right). \tag{17.129}$$

For the converse inequality, observe that by using Part 2 of Exercise 15.42 we get that for every $\mathcal{E} \in \mathrm{COV}(A \to A')$

$$\left\| \rho' - \mathcal{E} \circ \mathcal{P}(\rho) \right\|_1 = \left\| \rho' - \mathcal{P}' \circ \mathcal{E}(\rho) \right\|_1$$
$$\boxed{\mathcal{P}'(\rho') = \rho'} \longrightarrow \; = \left\| \mathcal{P}'\big(\rho' - \mathcal{E}(\rho)\big) \right\|_1 \tag{17.130}$$
$$\textbf{DPI} \to \; \leqslant \left\| \rho' - \mathcal{E}(\rho) \right\|_1 .$$

Combining this inequality with the definition of the conversion distance, specifically with the first equality in (17.129), gives

$$T\left((\mathcal{P}(\rho),\gamma) \xrightarrow{\mathrm{GPC}} (\rho',\gamma')\right) \leqslant \min_{\mathcal{E} \in \mathrm{COV}(A \to A')} \left\{ \frac{1}{2} \left\| \rho' - \mathcal{E}(\rho) \right\|_1 : \gamma' = \mathcal{E}(\gamma) \right\}$$

$$= T\left((\rho,\gamma) \xrightarrow{\mathrm{GPC}} (\rho',\gamma')\right). \tag{17.131}$$

Combining the two inequalities in (17.129) and (17.129) gives the equality in (17.127). This completes the proof.   ∎

**Exercise 17.33.** *Use Lemma 11.1 to provide a shorter proof of the inequality in* (17.129).

**Exercise 17.34.** *Let $\mathcal{P}$ and $\mathcal{P}'$ be the pinching channel associated with the Hamiltonians of systems $A$ and $A'$, respectively, and let $\mathcal{N} := \mathcal{P}' \circ \mathcal{E}$, where $\mathcal{E} \in \mathrm{CPTP}(A \to A')$. Show that $\mathcal{N} \in \mathrm{COV}(A \to A')$ if and only if*

$$\mathcal{N} = \mathcal{N} \circ \mathcal{P}. \tag{17.132}$$

## 17.6.1 The Conversion Distance between Quasi-Classical States

Consider two quasi-classical athermality states $(\mathbf{p}, \mathbf{g})$ of system $A$ and $(\mathbf{p}', \mathbf{g}')$ of system $A'$. Since applying a permutation on both components of $\mathbf{p}$ and $\mathbf{g}$ is a reversible thermal operation, without loss of generality we assume that the components of the probability vectors are ordered as

$$\frac{p_1}{g_1} \geqslant \frac{p_2}{g_2} \geqslant \cdots \geqslant \frac{p_m}{g_m} \quad \text{and} \quad \frac{p_1'}{g_1'} \geqslant \frac{p_2'}{g_2'} \geqslant \cdots \geqslant \frac{p_n'}{g_n'}, \tag{17.133}$$

where $m := |A|$ and $n := |A'|$. Suppose first that the Gibbs states $\mathbf{g}$ and $\mathbf{g}'$ have rational coefficients, and denote by $k \in \mathbb{N}$ the common denominator of all the components of $\mathbf{g}$ and $\mathbf{g}'$. That is, for each $x \in [m]$ and $y \in [n]$, we have $g_x = \frac{a_x}{k}$ and $g_y' = \frac{b_y}{k}$, where $a_x, b_y \in \mathbb{N}$ are integers satisfying

$$\sum_{x\in[m]} a_x = \sum_{y\in[n]} b_y = k. \tag{17.134}$$

Recall that from Theorem 4.8, there exists $\mathbf{r}, \mathbf{s} \in \mathrm{Prob}(k)$ such that $(\mathbf{p}, \mathbf{g}) \sim (\mathbf{r}, \mathbf{u}^{(k)})$ and $(\mathbf{p}', \mathbf{g}') \sim (\mathbf{r}', \mathbf{u}^{(k)})$. Specifically,

$$\mathbf{r} := \bigoplus_{x\in[m]} p_x \mathbf{u}^{(a_x)} \quad \text{and} \quad \mathbf{r}' := \bigoplus_{y\in[n]} p_y' \mathbf{u}^{(b_y)}. \tag{17.135}$$

From (17.133) we have $\mathbf{r} = \mathbf{r}^{\downarrow}$ and $\mathbf{r}' = \mathbf{r}'^{\downarrow}$, and moreover,

$$(\mathbf{p}, \mathbf{g}) \succ (\mathbf{p}', \mathbf{g}') \iff \mathbf{r} \succ \mathbf{r}'. \tag{17.136}$$

With these notations and the assumption that the Gibbs vectors have rational components, we have the following closed formula for the conversion distance.

---

**Theorem 17.11.** Let $(\mathbf{p}, \mathbf{g})$, $(\mathbf{p}', \mathbf{g}')$, and $\mathbf{r}, \mathbf{r}' \in \mathrm{Prob}(k)$ be as before. Then,

$$T\left((\mathbf{p}, \mathbf{g}) \xrightarrow{\mathrm{CTO}} (\mathbf{p}', \mathbf{g}')\right) = \max_{\ell\in[k]} \left\{\|\mathbf{r}'\|_{(\ell)} - \|\mathbf{r}\|_{(\ell)}\right\}. \tag{17.137}$$

---

**Proof**   Let $E$ be a $k \times n$ column stochastic matrix defined on every $\mathbf{q} \in \mathrm{Prob}(n)$ as (cf. (4.136))

$$E\mathbf{q} := \bigoplus_{y\in[n]} q_y \mathbf{u}^{(b_y)}. \tag{17.138}$$

Observe that $\mathbf{r}' = E\mathbf{p}'$ and that $\|\mathbf{p}' - \mathbf{q}\|_1 = \|\mathbf{r}' - E\mathbf{q}\|_1$ (see Exercise 17.35). Thus,

$$T\left((\mathbf{p}, \mathbf{g}) \xrightarrow{\mathrm{CTO}} (\mathbf{p}', \mathbf{g}')\right) = \min_{\mathbf{q}\in\mathrm{Prob}(n)} \left\{\frac{1}{2}\|\mathbf{r}' - E\mathbf{q}\|_1 \; : \; \mathbf{r} \succ E\mathbf{q}\right\}$$

$$\boxed{\mathbf{s} := E\mathbf{q}} \longrightarrow \quad \geqslant \min_{\mathbf{s}\in\mathrm{Prob}(k)} \left\{\frac{1}{2}\|\mathbf{r}' - \mathbf{s}\|_1 \; : \; \mathbf{r} \succ \mathbf{s}\right\} \tag{17.139}$$

$$\boxed{cf.\ (16.15)} \longrightarrow \quad := T\left(\mathbf{r} \xrightarrow{\mathrm{Noisy}} \mathbf{r}'\right).$$

We next show that the above inequality is in fact an equality.

Let

$$D := \bigoplus_{y \in [n]} D^{(b_y)} \quad \text{with} \quad D^{(b_y)} := \frac{1}{b_y} \begin{bmatrix} 1 & \cdots & 1 \\ \vdots & \ddots & \vdots \\ 1 & \cdots & 1 \end{bmatrix}, \tag{17.140}$$

and observe that $D$ is a $k \times k$ doubly stochastic matrix satisfying $D\mathbf{r}' = \mathbf{r}'$. Thus, using the data processing inequality with the matrix $D$ in the definition of $T\left(\mathbf{r} \xrightarrow{\text{Noisy}} \mathbf{r}'\right)$ above gives

$$T\left(\mathbf{r} \xrightarrow{\text{Noisy}} \mathbf{r}'\right) \geq \min_{\mathbf{s} \in \text{Prob}(k)} \left\{ \frac{1}{2} \left\| D\mathbf{r}' - D\mathbf{s} \right\|_1 \; : \; \mathbf{r} \succ \mathbf{s} \right\}$$

$$\boxed{D\mathbf{r}' = \mathbf{r}' \text{ and } \mathbf{s} \succ D\mathbf{s}} \longrightarrow \geq \min_{\mathbf{s} \in \text{Prob}(k)} \left\{ \frac{1}{2} \left\| \mathbf{r}' - D\mathbf{s} \right\|_1 \; : \; \mathbf{r} \succ D\mathbf{s} \right\} \tag{17.141}$$

since $\mathbf{r} \succ D\mathbf{s}$ is a weaker constraint than $\mathbf{r} \succ \mathbf{s}$. By definition $D\mathbf{s}$ has the form $E\mathbf{q}$ for some $\mathbf{q} \in \text{Prob}(n)$ (see Exercise 17.35). Therefore,

$$T\left(\mathbf{r} \xrightarrow{\text{Noisy}} \mathbf{r}'\right) \geq \min_{\mathbf{q} \in \text{Prob}(n)} \left\{ \frac{1}{2} \left\| \mathbf{r}' - E\mathbf{q} \right\|_1 \; : \; \mathbf{r} \succ E\mathbf{q} \right\} = T\left((\mathbf{p}, \mathbf{g}) \xrightarrow{\text{CTO}} (\mathbf{p}', \mathbf{g}')\right). \tag{17.142}$$

Combining this with (17.139) we conclude

$$T\left((\mathbf{p}, \mathbf{g}) \xrightarrow{\text{CTO}} (\mathbf{p}', \mathbf{g}')\right) = T\left(\mathbf{r} \xrightarrow{\text{Noisy}} \mathbf{r}'\right)$$

$$\text{Theorem 16.4} \rightarrow = \max_{\ell \in [k]} \left\{ \left\| \mathbf{r}' \right\|_{(\ell)} - \left\| \mathbf{r} \right\|_{(\ell)} \right\}. \tag{17.143}$$

This completes the proof. $\blacksquare$

**Exercise 17.35.** *Using the same symbols as in the proof above, show that $\|\mathbf{p}' - \mathbf{q}\|_1 = \|\mathbf{r}' - E\mathbf{q}\|_1$, and that for every $\mathbf{s} \in \text{Prob}(k)$ there exists $\mathbf{q} \in \text{Prob}(n)$ such that $D\mathbf{s} = E\mathbf{q}$.*

In Theorem 17.11 we assumed that the Gibbs states $\mathbf{g}$ and $\mathbf{g}'$ have rational components. In Appendix D.8 of online version we show that the conversion distance is continuous in $\mathbf{g}$ and $\mathbf{g}'$. This in turns implies that one can use Theorem 17.11 to estimate the conversion distance up to an arbitrary precision even for the case that $\mathbf{g}$ and $\mathbf{g}'$ have irrational components.

## 17.6.2 The Golden Unit of Athermality

We start by discussing the golden unit of athermality in the quasi-classical regime, and use the notation $\{\mathbf{e}_1, \ldots, \mathbf{e}_m\}$ for the standard basis of $\mathbb{R}^m$. From Exercise 4.48, the maximal resource of a system $A$ of dimension $|A| = m$, with a fixed Hamiltonian $H^A := \sum_{x \in [m]} a_x^\uparrow |x\rangle\langle x|^A$ (or equivalently with a fixed Gibbs state $\mathbf{g}$) is given by $(\mathbf{e}_m, \mathbf{g})$, where $\mathbf{e}_m$ corresponds to the maximal eigenvalue $a_m$ of $H^A$. In general, we cannot call $(\mathbf{e}_m, \mathbf{g})$ the "golden unit" of system $A$ since it depends on the Hamiltonian $H^A$. That is, without specifying the Hamiltonian of system $A$, and as long as the maximal energy $a_m < \infty$, we cannot specify a resource that is maximal on all systems with the same dimension $|A| = m$.

On the other hand, if we do take $a_m = \infty$, so that the Gibbs state has a zero $m$-component, then the resource $(\mathbf{e}_m, \mathbf{g})$ is an infinite resource in the sense that all other systems $A'$ in *any* dimension and any quasi-classical athermality state $(\mathbf{p}', \mathbf{g}')$ will satisfy $(\mathbf{e}_m, \mathbf{g}) \succ (\mathbf{p}', \mathbf{g}')$. Hence, such a system cannot serve as the golden unit since it is an infinite resource. In other words, even in dimension $m = 2$, by maximizing over all two-dimensional Hamiltonians, we get an infinite resource, by taking $a_2 = \infty$ so that $\mathbf{g}^A = \mathbf{e}_1$ and the thermal state is the infinite resource $(\mathbf{e}_2, \mathbf{e}_1)$.

We therefore choose the golden unit for a fixed dimension $|A| = m$ to be of the form $(\mathbf{e}_m, \mathbf{g})$, but instead of choosing the Hamiltonian that maximizes the resource in that dimension we choose the one that minimizes it. That is, we are looking for a Gibbs state $\mathbf{g}$ that satisfies

$$(|m\rangle\langle m|, \tilde{\mathbf{g}}) \succ (|m\rangle\langle m|, \mathbf{g}) \qquad \forall\, \tilde{\mathbf{g}} \in \mathrm{Prob}(m). \tag{17.144}$$

The only vector $\mathbf{g}$ that satisfies the above relation is the uniform vector $\mathbf{u}^{(m)}$. Hence, the golden unit for any system $A$ will be chosen as

$$(\mathbf{e}_m, \mathbf{u}^{(m)}) \sim (|0\rangle\langle 0|^A, \mathbf{u}^A), \tag{17.145}$$

where on the right-hand side we use the quantum notations, with $|0\rangle$ denoting any pure state of system $A$. Recall that when the Gibbs state is uniform, all pure states are equivalent under thermal operations (which are equivalent to noisy operations in this case).

**Exercise 17.36.** *Show that a vector* $\mathbf{g} \in \mathrm{Prob}(m)$ *satisfies* (17.144) *for all* $\tilde{\mathbf{g}} \in \mathrm{Prob}(m)$ *if and only if* $\mathbf{g} = \mathbf{u}^{(m)}$.

In the fully quantum case, under GPC and CTO, coherence among energy level is a resource that cannot be measured by the golden unit $(|0\rangle\langle 0|^A, \mathbf{u}^A)$. The reason is that this golden unit is quasi-classical, and it cannot be converted by GPC (or CTO) to any athermality state that is not quasi-classical (even if we take $m := |A| = \infty$). This means that in the QRT of quantum athermality, there exists another type of resource, namely, time-translation asymmetry, that cannot be quantified by the golden unit $(|0\rangle\langle 0|^A, \mathbf{u}^A)$. We conclude that quantum athermality can be viewed as a resource comprising of two types:

1. Nonuniformity (since in the quasi-classical regime athermality can be viewed as nonuniformity)
2. Time-translation asymmetry (also referred to as quantum coherence).

In contrast to GPC and CTO, GPO has the capability to induce coherence between energy levels. Consequently, as demonstrated in the subsequent exercise, we can retain the state $(|0\rangle\langle 0|^A, \mathbf{u}^A)$ as the golden unit of the resource theory.

**Exercise 17.37.** *Let $m := |A|$ and $(\rho', \gamma')$ be an athermality state of system $A'$. Show that for sufficiently large $m$*

$$(|0\rangle\langle0|^A, \mathbf{u}^A) \xrightarrow{\text{GPO}} (\rho', \gamma'). \tag{17.146}$$

**Exercise 17.38.** *Show that under GPO operations, the resource $(|0\rangle\langle0|^A, \mathbf{u}^A)$ is equivalent to the resource $(|0\rangle\langle0|^X, \mathbf{u}_m^X)$, where $X$ is a two-dimensional classical system, $m := |A|$, and*

$$\mathbf{u}_m^X := \frac{1}{m}|0\rangle\langle0|^X + \frac{m-1}{m}|1\rangle\langle1|^X. \tag{17.147}$$

This exercise demonstrates that we can always consider the golden unit to be a qubit. Moreover, note that $\mathbf{u}_m^X$ is well defined even if $m$ is not an integer. This can help simplifying certain expressions, and we will therefore consider also the states $(|0\rangle\langle0|^X, \mathbf{u}_m^X)$ with $m \in \mathbb{R}_+$. We will use the notation

$$\Upsilon_m := \left(|0\rangle\langle0|^X, \mathbf{u}_m^X\right) \tag{17.148}$$

to denote this golden unit.

**Exercise 17.39.** *Show that $\{\Upsilon_m\}_{m\in\mathbb{N}}$ satisfies the conditions of a golden unit outlined in Definition 11.1.*

### 17.6.3 The Conversion Distance to and from the Golden Unit

The conversion distance (under GPO) from an arbitrary state $(\rho, \gamma)$ of system $A$ to the golden unit $\Upsilon_m$ is given by

$$T\left((\rho,\gamma) \xrightarrow{\text{GPO}} \Upsilon_m\right) = \min_{\mathcal{E}\in\text{CPTP}(A\to X)}\left\{\frac{1}{2}\left\||0\rangle\langle0|^X - \mathcal{E}(\rho^A)\right\|_1 \ : \ \mathbf{u}_m^X = \mathcal{E}\left(\gamma^A\right)\right\}. \tag{17.149}$$

Observe that since any $\mathcal{E} \in \text{CPTP}(A \to X)$ is a binary POVM Channel, it can be expressed as

$$\mathcal{E}(\omega) = \text{Tr}\left[\omega\Lambda\right]|0\rangle\langle0|^X + \text{Tr}\left[\omega\,(I - \Lambda)\right]|1\rangle\langle1|^X \qquad \forall\,\omega \in \mathcal{L}(A), \tag{17.150}$$

for some effect $0 \leqslant \Lambda \leqslant I^A$. We therefore get the simplification (see Exercise 17.40)

$$\frac{1}{2}\left\||0\rangle\langle0|^X - \mathcal{E}(\rho^A)\right\|_1 = 1 - \text{Tr}[\rho\Lambda] \tag{17.151}$$

so that

$$T\left((\rho,\gamma) \xrightarrow{\text{GPO}} \Upsilon_m\right) = \min_{\Lambda\in\text{Eff}(A)}\left\{1 - \text{Tr}[\rho\Lambda] \ : \ \text{Tr}[\Lambda\gamma] = \frac{1}{m}\right\}. \tag{17.152}$$

Note that the expression in (17.152) is somewhat similar to the hypothesis testing divergence (see Exercise 17.41). Moreover, since the golden unit $\Upsilon_m$ is quasi-classical we get from (17.127) that under GPC we have

$$T\left((\rho,\gamma) \xrightarrow{\text{GPC}} \Upsilon_m\right) = \min_{\Lambda \in \text{Eff}(A)} \left\{ 1 - \text{Tr}[\mathcal{P}(\rho)\Lambda] \; : \; \text{Tr}[\Lambda\gamma] = \frac{1}{m} \right\}, \tag{17.153}$$

where $\mathcal{P}$ is the pinching channel associated with the Hamiltonian of system $A$.

**Exercise 17.40.** *Prove the equality* (17.151).

**Exercise 17.41.** *Set* $\varepsilon := 1 - \frac{1}{m}$. *Show that*

$$T\left((\rho,\gamma) \xrightarrow{\text{GPO}} \Upsilon_m\right) = 2^{D^{\varepsilon}_{\min}(\gamma\|\rho)}. \tag{17.154}$$

We next consider the conversion distance from the golden unit $\Upsilon_m$ to an arbitrary state $(\rho,\gamma)$ of system $A$. Here we only consider GPO since GPC cannot generate coherence. By definition,

$$T\left(\Upsilon_m \xrightarrow{\text{GPO}} (\rho,\gamma)\right) = \min_{\mathcal{E}\in\text{CPTP}(X\to A)} \left\{ \frac{1}{2} \left\| \rho^A - \mathcal{E}(|0\rangle\langle 0|^X) \right\|_1 \; : \; \gamma^A = \mathcal{E}\left(\mathsf{u}^X_m\right) \right\}. \tag{17.155}$$

Denoting by $\omega = \mathcal{E}(|0\rangle\langle 0|)$ and $\tau = \mathcal{E}(|1\rangle\langle 1|)$, the conversion distance can be simplified as

$$
\begin{aligned}
T\left(\Upsilon_m \xrightarrow{\text{GPO}} (\rho,\gamma)\right) &= \min_{\omega,\tau\in\mathfrak{D}(A)} \left\{ \frac{1}{2}\|\rho - \omega\|_1 \; : \; \gamma = \frac{1}{m}\omega + \frac{m-1}{m}\tau \right\} \\
&= \min_{\omega\in\mathfrak{D}(A)} \left\{ \frac{1}{2}\|\rho - \omega\|_1 \; : \; m\gamma \geqslant \omega \right\} \\
&= \min_{\omega\in\mathfrak{D}(A)} \left\{ \frac{1}{2}\|\rho - \omega\|_1 \; : \; D_{\max}(\omega\|\gamma) \leqslant \log m \right\}.
\end{aligned}
\tag{17.156}
$$

This expression will be instrumental in our calculations regarding the cost of athermality under GPO.

**Exercise 17.42.** *Let* $(\rho,\gamma)$ *be an athermality state of system* $A$. *Show that under GPC for any* $m \in \mathbb{N}$

$$T\left(\Upsilon_m \xrightarrow{\text{GPC}} (\rho,\gamma)\right) \geqslant \min_{\sigma\in\mathfrak{D}(A)} \frac{1}{2}\|\rho - \Delta(\sigma)\|_1\,, \tag{17.157}$$

*where* $\Delta \in \text{CPTP}(A \to A)$ *is the completely dephasing channel (with respect to the basis of the Hamiltonian of system* $A$).

## 17.7 Distillation and Cost in the Single-Shot Regime

### 17.7.1 Distillation of Athermality

As already discussed, for any $\varepsilon \in [0, 1]$ the $\varepsilon$-approximate single-shot distillation of an athermality state $(\rho, \gamma)$ of system $A$ is defined by

$$\text{Distill}^{\varepsilon}(\rho, \gamma) := \log \sup_{0 < m \in \mathbb{R}} \left\{ m \ : \ T\left((\rho, \gamma) \xrightarrow{\text{GPO}} \Upsilon_m\right) \leqslant \varepsilon \right\}. \tag{17.158}$$

Integrating this with the formulas from the preceding subsection that pertain to the conversion distance, we arrive at the subsequent outcome. We denote by $\mathcal{P} \in \text{CPTP}(A \to A)$ the pinching channel associated with the Hamiltonian of system $A$, and by $D_{\min}^{\varepsilon}$ the quantum hypothesis testing divergence as defined in (8.185).

> **Theorem 17.12.** Let $\varepsilon \in [0, 1]$. For any athermality state $(\rho, \gamma)$ of a quantum system $A$, the $\varepsilon$-approximate single-shot distillation of athermality is given by:
>
> 1. Under GPO: $\text{Distill}^{\varepsilon}(\rho, \gamma) = D_{\min}^{\varepsilon}(\rho \| \gamma)$.
> 2. Under GPC and CTO: $\text{Distill}^{\varepsilon}(\rho, \gamma) = D_{\min}^{\varepsilon}(\mathcal{P}(\rho) \| \gamma)$.

**Proof** From (17.158) we get

$$\text{Distill}^{\varepsilon}(\rho, \gamma) = -\log \inf_{0 < m \in \mathbb{R}} \left\{ \frac{1}{m} \ : \ T\left((\rho, \gamma) \xrightarrow{\text{GPO}} \Upsilon_m\right) \leqslant \varepsilon \right\}$$

$$\textbf{(17.152)} \to \ = -\log \inf_{0 < m \in \mathbb{R}} \left\{ \frac{1}{m} \ : \ 1 - \text{Tr}[\rho \Lambda] \leqslant \varepsilon \ , \ \text{Tr}[\Lambda \gamma] = \frac{1}{m} \ , \ \Lambda \in \text{Eff}(A) \right\}$$

$$= -\log \inf \left\{ \text{Tr}[\Lambda \gamma] \ : \ 1 - \text{Tr}[\rho \Lambda] \leqslant \varepsilon \ , \ \Lambda \in \text{Eff}(A) \right\}$$

$$= D_{\min}^{\varepsilon}(\rho \| \gamma). \tag{17.159}$$

This completes the proof of the first part. The second part of the proof follows from the first part in conjunction with (17.127). This concludes the proof. ∎

Observe that when we take $\varepsilon = 0$ we get that the exact single-shot distillation is given by

$$\text{Distill}^{0}\left(\rho^A, \gamma^A\right) = D_{\min}\left(\rho^A \| \gamma^A\right). \tag{17.160}$$

This result gives a physical meaning to the min relative entropy as the exact single-shot distillation rate under GPO.

## 17.7.2 Athermality Cost under GPO

For any $\varepsilon \in [0, 1]$, the $\varepsilon$-single-shot cost of an athermality state $(\rho, \gamma)$ is defined as

$$\mathrm{Cost}^{\varepsilon}(\rho, \gamma) := \log \inf_{0 < m \in \mathbb{R}} \left\{ m \ : \ T\left( \Upsilon_m \xrightarrow{\mathrm{GPO}} (\rho, \gamma) \right) \leqslant \varepsilon \right\}. \tag{17.161}$$

**Theorem 17.13.** Let $\varepsilon \in [0, 1]$. For any athermality state $(\rho, \gamma)$ of system $A$, the $\varepsilon$-single-shot distillation (under GPO) is given by

$$\mathrm{Cost}^{\varepsilon}(\rho, \gamma) = D_{\max}^{\varepsilon}(\rho \| \gamma). \tag{17.162}$$

**Proof**   Combining the expression (17.156) for the conversion distance together with the definition (17.161) gives

$$\begin{aligned}
\mathrm{Cost}^{\varepsilon}(\rho, \gamma) &= \inf_{0 < m \in \mathbb{R}} \left\{ \log m \ : \ \frac{1}{2}\|\rho - \omega\|_1 \leqslant \varepsilon \ , \ \ D_{\max}(\omega\|\gamma) \leqslant \log m \ , \ \ \omega \in \mathfrak{D}(A) \right\} \\
&= \inf \left\{ D_{\max}(\omega\|\gamma) \ : \ \frac{1}{2}\|\rho - \omega\|_1 \leqslant \varepsilon \ , \ \ \omega \in \mathfrak{D}(A) \right\} \\
&= D_{\max}^{\varepsilon}(\rho\|\gamma).
\end{aligned}$$
$$\tag{17.163}$$

This completes the proof.  ∎

Observe that for $\varepsilon = 0$ we get the exact single-shot athermality cost

$$\mathrm{Cost}^{0}(\rho, \gamma) = D_{\max}(\rho\|\gamma). \tag{17.164}$$

This result provides a physical meaning to the max relative entropy as the exact single-shot cost under GPO.

**Exercise 17.43.**  *Let $\gamma \in \mathfrak{D}_{>0}(A)$ be the Gibbs state of system $A$ with eigenvalues $g_1, \ldots, g_m$. Let $\psi_{\gamma} \in \mathrm{Pure}(A)$ be the pure state*

$$|\psi_{\gamma}\rangle := \sum_{x \in [m]} \sqrt{g_x}|x\rangle. \tag{17.165}$$

*Show that the exact single-shot athermality cost of $(\psi_{\gamma}, \gamma)$ is equal to $\log(m)$.*

## 17.8 The Asymptotic Regime

A primary goal of resource theories is to attain reversibility in the asymptotic inter-conversions of resources. This entails that the cost rate, in an asymptotic context, for generating a specific resource should align with the rate at which golden units can be extracted from it. Reversibility characteristics hold significant importance in the realm of quantum information, given the value of quantum resources. They guarantee that resources are not wasted during quantum information processing tasks. Nonetheless, the pursuit of reversibility frequently necessitates the contemplation of a broader array of permissible operations.

Unlike GPO, both thermal operations and GPC are incapable of generating coherence between energy levels. This means that even for very large $m$, the golden unit $\Upsilon_m$ cannot be converted into a single copy of an athermality state $(\rho, \gamma)$ that exhibits coherence across energy levels. Nevertheless, as we will soon discover, this irreversibility – highlighted by the significant cost of preparing the $(\rho, \gamma)$ state in contrast to the finite rate at which it can be utilized to distill golden units of athermality – can be mitigated by introducing a modest degree of coherence into the system.

In certain cases, reversibility can be attained by allowing the use of a sublinear amount of resources. For instance, in the resource theory of pure bipartite entanglement, we observed that distillation requires no communication, whereas formation necessitates a sublinear amount of classical communication. Thus, reversibility is achieved in this theory through local operations and a sublinear amount of classical communication. The concept of adding a sublinear amount of a specific resource to achieve reversibility is highly appealing because the rate at which such resources are consumed diminishes in the asymptotic limit. We will employ this idea when studying the asymptotic cost of athermality under thermal operations. However, we begin by examining the asymptotic distillation of athermality.

## 17.8.1 Distillation of Athermality

In this section, we compute the asymptotic distillable athermality under either GPO, GPC, or CTO. The asymptotic distillable rate of an athermality state $(\rho, \gamma)$ is related to the single-shot quantity via (cf. (11.138))

$$\text{Distill}\,(\rho, \gamma) = \lim_{\varepsilon \to 0+} \limsup_{n \to \infty} \frac{1}{n} \text{Distill}^{\varepsilon}\left(\rho^{\otimes n}, \gamma^{\otimes n}\right). \tag{17.166}$$

Recall from Theorem 17.12 that in the single-shot regime, for any $\varepsilon \in (0, 1)$, the distillable athermality under GPO is given by

$$\text{Distill}^{\varepsilon}\,(\rho, \gamma) = D_{\min}^{\varepsilon}\left(\rho \| \gamma\right). \tag{17.167}$$

The regularization of the formula (17.167) is given by

$$\lim_{n \to \infty} \frac{1}{n} \text{Distill}^{\varepsilon}\left(\rho^{\otimes n}, \gamma^{\otimes n}\right) = \lim_{n \to \infty} \frac{1}{n} D_{\min}^{\varepsilon}\left(\rho^{\otimes n} \| \gamma^{\otimes n}\right) \tag{17.168}$$
$$\xrightarrow{\textbf{The Quantum Stein's Lemma}} = D(\rho \| \gamma).$$

Note that in this case we did not need to take the limsup over $n$ since the limit exists. Therefore, under GPO, the asymptotic distillable athermality is given by the relative entropy $D(\rho \| \gamma)$. Remarkably, this is also the distillable rate under GPC and CTO.

> **Theorem 17.14.** Let $(\rho, \gamma)$ be an athermality state of a quantum system $A$, and let $\varepsilon \in (0, 1)$. Then, the distillable athermality under either CTO or GPC is given by
>
> $$\text{Distill}\,(\rho, \gamma) = \limsup_{n \to \infty} \frac{1}{n} \text{Distill}^{\varepsilon}\left(\rho^{\otimes n}, \gamma^{\otimes n}\right) = D\,(\rho \| \gamma). \tag{17.169}$$

**Proof** Let $\varepsilon \in (0, 1)$ and recall from Theorem 17.12 that the $\varepsilon$-single-shot distillable athermality under GPC or CTO is given by

$$\text{Distill}^{\varepsilon}\,(\rho, \gamma) = D^{\varepsilon}_{\min}\left(\mathcal{P}(\rho) \| \gamma\right), \tag{17.170}$$

where $\mathcal{P}$ is the pinching channel corresponding to the Hamiltonian of system $A$. Since $\mathcal{P}(\gamma) = \gamma$ we have

$$\begin{aligned} \text{Distill}^{\varepsilon}\,(\rho, \gamma) &= D^{\varepsilon}_{\min}\left(\mathcal{P}(\rho) \| \mathcal{P}(\gamma)\right) \\ \textbf{DPI} \to\ &\leqslant D^{\varepsilon}_{\min}\,(\rho \| \gamma). \end{aligned} \tag{17.171}$$

Thus,

$$\limsup_{n \to \infty} \frac{1}{n} \text{Distill}^{\varepsilon}\left(\rho^{\otimes n}, \gamma^{\otimes n}\right) \leqslant \limsup_{n \to \infty} \frac{1}{n} D^{\varepsilon}_{\min}\left(\rho^{\otimes n} \| \gamma^{\otimes n}\right) \tag{17.172}$$

$$\textbf{The Quantum Stein's Lemma} \to\ = D(\rho \| \gamma).$$

To get the opposite inequality, for every $n \in \mathbb{N}$ let $\mathcal{P}_n \in \text{CTO}(A^n \to A^n)$ denote the pinching channel associated with the Hamiltonian of system $A^n$. Now, fix $k \in \mathbb{N}$ and observe that for every $\varepsilon \in (0, 1)$

$$\begin{aligned} \limsup_{n \to \infty} \frac{1}{n} \text{Distill}^{\varepsilon}\left(\rho^{\otimes n}, \gamma^{\otimes n}\right) &= \limsup_{n \to \infty} \frac{1}{n} D^{\varepsilon}_{\min}\left(\mathcal{P}_n(\rho^{\otimes n}) \| \gamma^{\otimes n}\right) \\ &\geqslant \limsup_{n \to \infty} \frac{1}{nk} D^{\varepsilon}_{\min}\left(\mathcal{P}_{nk}(\rho^{\otimes nk}) \| \gamma^{\otimes nk}\right) \\ \textbf{DPI} \to\ &\geqslant \limsup_{n \to \infty} \frac{1}{nk} D^{\varepsilon}_{\min}\left(\mathcal{P}_k^{\otimes n} \circ \mathcal{P}_{nk}(\rho^{\otimes nk}) \| \mathcal{P}_k^{\otimes n}\left(\gamma^{\otimes nk}\right)\right), \end{aligned} \tag{17.173}$$

where in the last line we used the data processing inequality with the channel $\mathcal{P}_k^{\otimes n}$. Now, the Gibbs state is invariant under the pinching channel and in particular $\mathcal{P}_k^{\otimes n}\left(\gamma^{\otimes nk}\right) = \gamma^{\otimes nk}$. Moreover, from Exercise 15.3 it follows that $\mathcal{P}_k^{\otimes n} \circ \mathcal{P}_{nk} = \mathcal{P}_k^{\otimes n}$. We therefore get that

$$\begin{aligned} \limsup_{n \to \infty} \frac{1}{n} \text{Distill}^{\varepsilon}\left(\rho^{\otimes n}, \gamma^{\otimes n}\right) &\geqslant \limsup_{n \to \infty} \frac{1}{nk} D^{\varepsilon}_{\min}\left(\mathcal{P}_k^{\otimes n}(\rho^{\otimes nk}) \| \gamma^{\otimes nk}\right) \\ &= \frac{1}{k} \limsup_{n \to \infty} \frac{1}{n} D^{\varepsilon}_{\min}\left(\left(\mathcal{P}_k(\rho^{\otimes k})\right)^{\otimes n} \| \left(\gamma^{\otimes k}\right)^{\otimes n}\right) \\ &= \frac{1}{k} D\left(\mathcal{P}_k\left(\rho^{\otimes k}\right) \| \gamma^{\otimes k}\right), \end{aligned} \tag{17.174}$$

where in the last line we used the quantum Stein's lemma. This inequality can also be understood physically by observing that the state $\sigma_k := \mathcal{P}_k\left(\rho^{\otimes k}\right)$ is quasi-classical, and consequently, it has a distillable athermality rate given by $D(\sigma_k \| \gamma^{\otimes k})$. Now, since this inequality holds for all $k \in \mathbb{N}$ we conclude that

$$\limsup_{n\to\infty} \frac{1}{n} \text{Distill}^{\varepsilon}\left(\rho^{\otimes n}, \gamma^{\otimes n}\right) \geqslant \limsup_{k\to\infty} \frac{1}{k} D\left(\mathcal{P}_k\left(\rho^{\otimes k}\right) \big\| \gamma^{\otimes k}\right) \tag{17.175}$$

$$(\textbf{17.93}) \to \; = D(\rho \| \gamma).$$

This completes the proof. ∎

## 17.8.2 Athermality Cost

We begin our discussion by examining the cost of athermality under GPO. In this context, the asymptotic cost rate of an athermality state $(\rho, \gamma)$ connects to the single-shot quantity as follows:

$$\text{Cost}\,(\rho, \gamma) = \lim_{\varepsilon\to 0+} \liminf_{n\to\infty} \frac{1}{n}\text{Cost}^{\varepsilon}\left(\rho^{\otimes n}, \gamma^{\otimes n}\right). \tag{17.176}$$

Integrating this with Theorem 17.13, we arrive at the equation:

$$\lim_{n\to\infty} \frac{1}{n}\text{Cost}^{\varepsilon}\left(\rho^{\otimes n}, \gamma^{\otimes n}\right) = \lim_{n\to\infty} \frac{1}{n} D_{\max}^{\varepsilon}\left(\rho^{\otimes n} \big\| \gamma^{\otimes n}\right) \tag{17.177}$$

$$\textbf{AEP} \to \; = D(\rho \| \gamma).$$

Hence, under GPO, both the asymptotic cost and the distillable athermality are given by the relative entropy $D(\rho \| \gamma)$. This signifies that, within the framework of GPO, the QRT of athermality exhibits reversibility. We will now proceed to explore the athermality cost under CTO.

As already discussed, the golden unit $\Upsilon_m$ cannot be used to generate states with coherence among energy levels. Therefore, the QRT of athermality under CTO is irreversible. In this section, we show how reversibility can be restored by appending the free operations with resources that are asymptotically negligible. To see how it is done, we first need to introduce a few concepts.

### Scaling of Time-Translation Asymmetry

Let $A$ be a physical system with Hamiltonian $H^A = \sum_{x\in[m]} a_x |x\rangle\langle x|$, where $m = |A|$, and let $|\psi\rangle = \sum_{x\in[m]} \sqrt{p_x}|x\rangle$ be given in its standard form. For any $n \in \mathbb{N}$, the state $\psi^{\otimes n}$ has the form

$$|\psi^{\otimes n}\rangle = \sum_{x^n\in[m]^n} \sqrt{p_{x^n}}|x^n\rangle = \sum_{x^n\in[m]^n} 2^{-\frac{n}{2}\left(H(\mathbf{t}(x^n))+D(\mathbf{t}(x^n)\|\mathbf{p})\right)}|x^n\rangle, \tag{17.178}$$

where we used (8.85). For any type $\mathbf{t} \subset \text{Type}(n, m)$ define

$$|\mathbf{t}\rangle^{A^n} := \frac{1}{\binom{n}{nt_1,\dots,nt_m}^{1/2}} \sum_{x^n\in X^n(\mathbf{t})} |x^n\rangle\,, \tag{17.179}$$

where the sum runs over all sequences $x^n \in [m]^n$ of the same type t. With these notations

$$|\psi\rangle^{\otimes n} = \sum_{t \in \text{Type}(n,m)} \sqrt{q_{t,n}} |t\rangle^{A^n}, \tag{17.180}$$

where

$$q_{t,n} := \binom{n}{nt_1, \ldots, nt_m} 2^{-n\left(H(t) + D(t\|p)\right)}. \tag{17.181}$$

Note that the vectors $|t\rangle^{A^n}$ are eigenvectors of the Hamiltonian of system $A^n$. Specifically,

$$H^{A^n} |t\rangle^{A^n} = n \sum_{x \in [m]} t_x a_x |t\rangle^{A^n}, \tag{17.182}$$

so that the energy in the state $|t\rangle^{A^n}$ is $n$ times the average energy with respect to the type t.

**Exercise 17.44.** *Consider the generic case, in which the energy eigenvalues $\{a_1, \ldots, a_m\}$ are rationally independent; that is, for any set of m integers $\ell_1, \ldots, \ell_m \in \mathbb{Z}$ we have*

$$\ell_1 a_1 + \cdots + \ell_m a_m = 0 \quad \Longleftrightarrow \quad \ell_1 = \ell_2 = \cdots = \ell_m = 0. \tag{17.183}$$

*Show that under this mild assumption (which we will* not *assume in the text), for every $n \in \mathbb{N}$, the number of distinct eigenvalues of $H^{A^n}$ equals $|\text{Type}(n,m)|$. That is, each energy eigenvalue of $H^{A^n}$ corresponds to exactly one type.*

Given that each $|t\rangle^{A^n}$ is an energy eigenstate, it naturally follows from (17.180) that we can express $|\psi^{\otimes n}\rangle$ as a linear combination of up to $|\text{Type}(n,m)| \leqslant (n+1)^m$ energy eigenstates. In simpler terms, the coherence inherent in $|\psi^{\otimes n}\rangle$ can be compactly represented within an $(n+1)^m$ dimensional vector (dimension polynomial in $n$).

This observation leads to a notable implication. As established in Corollary 17.3, for any mixed state in $\mathfrak{D}(A)$ there exists a pure state in $\text{Pure}(A)$ that can be converted into it via GPC. When we couple this insight with the aforementioned observation, a significant deduction emerges: the pure state coherence cost for preparing $\rho^{\otimes n} \in \mathfrak{D}(A^n)$ must not surpass $m \log (n+1)$. To put it differently, the rate of asymmetry cost – the coherence expense per instance of $\rho$ – cannot outpace $m \frac{\log(n+1)}{n}$, a ratio that approaches zero in the limit as $n \to \infty$. In contrast, the nonuniformity cost does *not* go to zero in the asymptotic limit since the energy of $\rho^{\otimes n}$ grows linearly with $n$.

In summary, athermality is made up of two main resources: nonuniformity and time-translation asymmetry, the latter of which is often referred to as coherence. Because of this, the costs related to athermality states can be categorized into two parts: the cost of nonuniformity and the cost of coherence. However, the coherence cost decreases and approaches zero in the asymptotic limit, necessitating a unique form of rescaling. This complexity lends a subtle character to the resource theory of quantum athermality, leaving several critical questions within the theory still unresolved.

## The Energy Spread

The energy spread of a given pure state $\psi \in \mathrm{Pure}(A)$ is defined as the difference between the maximal and minimal energies that appear when writing $\psi$ as a superposition of energy eigenvectors. In the discussion above, we saw that $n$ copies of a state $\psi \in \mathrm{Pure}(A)$ can be expressed as a linear combination of no more than $(n+1)^m$ energy eigenvectors. Among these energy eigenvectors are the zero energy eigenvector (corresponding to the type $\mathbf{t} = (1, 0, \ldots, 0)^T$) and the maximal energy eigenvector (corresponding to the type $\mathbf{t} = (0, \ldots, 0, 1)^T$). Therefore, since the energy in the decomposition (17.180) spreads from zero to $na_m$ (where $a_m$ is the maximal energy of a single copy of system $A$), we conclude that the energy spread of $\psi^{\otimes n}$ is $na_m$.

The energy spread can be reduced significantly if one allows for a small deviation from the state $\psi^{\otimes n}$. Specifically, let $\varepsilon \in (0, 1)$ and denote by $\mathfrak{S}_{n,\varepsilon}$ the set of all types in $\mathrm{Type}(n, m)$ for which $\frac{1}{2}\|\mathbf{t} - \mathbf{p}\|_1 \leqslant \varepsilon$. We also denote by $\mathfrak{S}_{n,\varepsilon}^c$ the complement of the set $\mathfrak{S}_{n,\varepsilon}$ in $\mathrm{Type}(n, m)$. With these notations, for any $\varepsilon \in (0, 1)$ we can split $|\psi^{\otimes n}\rangle$ into two parts

$$|\psi\rangle^{\otimes n} = \sum_{\mathbf{t} \in \mathfrak{S}_{n,\varepsilon}} \sqrt{q_{\mathbf{t},n}}|\mathbf{t}\rangle^{A^n} + \sum_{\mathbf{t} \in \mathfrak{S}_{n,\varepsilon}^c} \sqrt{q_{\mathbf{t},n}}|\mathbf{t}\rangle^{A^n}. \tag{17.184}$$

From Lemma 8.2 it follows that the coefficients $q_{\mathbf{t},n}$ satisfy

$$\frac{1}{(n+1)^m} 2^{-nD(\mathbf{t}\|\mathbf{p})} \leqslant q_{\mathbf{t},n} \leqslant 2^{-nD(\mathbf{t}\|\mathbf{p})}. \tag{17.185}$$

Therefore, the fidelity of $|\psi^{\otimes n}\rangle$ with the second term on the right-hand side of (17.184) is given by

$$\sum_{\mathbf{t} \in \mathfrak{S}_{n,\varepsilon}^c} q_{\mathbf{t},n} \leqslant \sum_{\mathbf{t} \in \mathfrak{S}_{n,\varepsilon}^c} 2^{-nD(\mathbf{t}\|\mathbf{p})}$$

$$\textbf{Pinsker's inequality} \rightarrow \quad \leqslant \sum_{\mathbf{t} \in \mathfrak{S}_{n,\varepsilon}^c} 2^{-2n\varepsilon^2} \tag{17.186}$$

$$\leqslant 2^{-2n\varepsilon^2} \left|\mathrm{Type}(n, m)\right| \leqslant 2^{-2n\varepsilon^2}(n+1)^m \xrightarrow{n \to \infty} 0.$$

Therefore, for any $\varepsilon > 0$ and sufficiently large $n$, the state $|\psi\rangle^{\otimes n}$ can be made arbitrarily close to the state

$$|\psi_\varepsilon^n\rangle := \frac{1}{\sqrt{v_\varepsilon}} \sum_{\mathbf{t} \in \mathfrak{S}_{n,\varepsilon}} \sqrt{q_{\mathbf{t},n}}|\mathbf{t}\rangle^{A^n}, \quad \text{where} \quad v_\varepsilon := \sum_{\mathbf{t} \in \mathfrak{S}_{n,\varepsilon}} q_{\mathbf{t},n}. \tag{17.187}$$

From (17.182) the energy of any type $\mathbf{t} \in \mathrm{Type}(n, m)$ is given by $\mu_{\mathbf{t}} := n \sum_{x \in [m]} t_x a_x$. In the sum in (17.187), the type $\mathbf{t}$ belongs to $\mathfrak{S}_{n,\varepsilon}$ so that $\frac{1}{2}\|\mathbf{t} - \mathbf{p}\|_1 \leqslant \varepsilon$. Consequently, each component $x \in [m]$ of the vector $\mathbf{t} - \mathbf{p}$ satisfies $|t_x - p_x| \leqslant 2\varepsilon$. Using this property, we get that

$$|\mu_{\mathbf{t}} - \mu_{\mathbf{p}}| \leqslant n \sum_{x \in [m]} a_x |t_x - p_x| \leqslant 2n\varepsilon \sum_{x \in [m]} a_x. \tag{17.188}$$

Therefore, for any two types $\mathsf{t}, \mathsf{t}' \in \mathrm{Type}(n, m)$ that are $\varepsilon$-close to $\mathbf{p}$ we have

$$|\mu_{\mathsf{t}} - \mu_{\mathsf{t}'}| \leqslant 4n\varepsilon \sum_{x \in [m]} a_x. \tag{17.189}$$

In other words, the energy spread of $|\psi_\varepsilon^n\rangle$ is no greater than $4n\varepsilon \sum_{x \in [m]} a_x$.

Note that by taking $\varepsilon > 0$ sufficiently small we can make the energy spread $4n\varepsilon \sum_{x \in [m]} a_x$ much smaller than $na_m$. However, we still get that the energy spread of $\psi_\varepsilon^n$ is linear in $n$. We show now that by taking $\varepsilon$ to depend on $n$, we can find states in $\mathrm{Pure}(A^n)$ that are very close to $\psi^{\otimes n}$ but with energy spread that is sublinear in $n$.

> **Lemma 17.7.** Let $\psi \in \mathrm{Pure}(A)$ and $\alpha \in (1/2, 1)$. There exists a sequence of pure state $\{\chi_n\}_{n\in\mathbb{N}}$ in $\mathrm{Pure}(A^n)$ with the following properties:
>
> 1. The limit
>
> $$\lim_{n \to \infty} \|\psi^{\otimes n} - \chi_n\|_1 = 0. \tag{17.190}$$
>
> 2. The state $\chi_n$ can be expressed as a linear combination of no more than $(n + 1)^m$ energy eigenstates.
> 3. The energy spread of $\chi_n$ is no more than $4n^\alpha \sum_{x \in [m]} a_x$.

**Proof** Let $\varepsilon_n = n^{\alpha-1}$. Since $\alpha \in \left(\frac{1}{2}, 1\right)$ we have $\lim_{n\to\infty} \varepsilon_n = 0$ and $\lim_{n\to\infty} n\varepsilon_n^2 = \infty$. The latter implies that if we replace $\varepsilon$ in (17.186) with $\varepsilon_n$ we still get the zero limit of (17.186). Hence, the pure state $\chi_n := \psi_{\varepsilon_n}^n$ satisfies (17.190). Since for all $\varepsilon > 0$ we have that $\psi_\varepsilon^n$ can be expressed as a linear combination of no more than $(n+1)^m$ energy eigenvectors, it follows that also $\chi_n$ has this property. Finally, from (17.189) we get that the energy spread of $\chi_n$ cannot exceed

$$4n\varepsilon_n \sum_{x \in [m]} a_x = 4n^\alpha \sum_{x \in [m]} a_x. \tag{17.191}$$

This completes the proof. ∎

**Exercise 17.45.** *Show that the average energy*

$$\langle \chi_n | H^{\otimes n} | \chi_n \rangle \tag{17.192}$$

*grows linearly with n.*

## Sublinear Athermality Resources

We saw in Lemma 17.7 that the state $\psi^{\otimes n}$ is very close to a state $\chi_n$, whose energy spread is sublinear in $n$. On the other hand, the average energy $\langle \chi_n | H^{\otimes n} | \chi_n \rangle$ grows linearly in $n$ (see Exercise 17.45). We will therefore consider systems whose energy grows sublinearly in $n$ as asymptotically negligible resources. Note however that such resources can contain significant coherence among energy levels since the coherence grows logarithmically with $n$. Indeed, as we will see shortly, such resources make the QRT of athermality reversible.

> **Definition 17.6.** A sublinear athermality resource (SLAR) is a sequence of quantum athermality systems $\{R_n\}_{n\in\mathbb{N}}$, such that $|R_n|$ grows polynomially with $n$, and there exists two constants independent of $n$, $0 \leqslant \alpha < 1$ and $c > 0$, such that
>
> $$\left\| H^{R_n} \right\|_\infty \leqslant cn^\alpha \qquad \forall\, n \in \mathbb{N}. \tag{17.193}$$

The key assumption in the given definition is that the energy of systems $R_n$ grows sublinearly with $n$. Consequently, as $n$ approaches infinity in the asymptotic limit, the resourcefulness of any states $\left\{(\omega^{R_n}, \gamma^{R_n})\right\}_{n\in\mathbb{N}}$ becomes insignificant compared to the resourcefulness of $n$ copies of the golden unit $\Upsilon_2 := \left(|0\rangle\langle 0|^X, \mathbf{u}_2^X\right)$. We will soon discover that this small amount of athermality resource is sufficient to restore reversibility.

**Exercise 17.46.** *Show that the distillation rate of athermality as given in Theorem 17.14 does not change if we replace CTO by CTO+SLAR. In other words, show that SLAR cannot increase the distillation rate of athermality.*

## The Cost of Athermality

The type of free operations that we consider here are CTO assisted with SLAR. In order to define the cost under such operations, for any system $R$ and $\varepsilon \in (0, 1)$, we define the $R$-assisted $\varepsilon$-single-shot cost as

$$\mathrm{Cost}_R^\varepsilon \left( \rho^A, \gamma^A \right) \tag{17.194}$$

$$:= \inf_{0 < m \in \mathbb{R}} \left\{ \log m \ : \ \min_{\phi \in \mathfrak{D}(R)} T\left( \Upsilon_m \otimes \left( \phi^R, \gamma^R \right) \xrightarrow{\mathrm{CTO}} \left( \rho^A, \gamma^A \right) \right) \leqslant \varepsilon \right\}.$$

From Corollary 17.3 and Exercise 17.30 it follows that any mixed state in $\mathfrak{D}(A)$ can be obtained by thermal operations from a pure state in Pure($A$). Thus, we can restrict the minimum in (17.194) over all density matrices $\phi \in \mathfrak{D}(A)$ to a minimum over all pure states $\phi \in$ Pure($A$).

**Exercise 17.47.** *Let $\varepsilon \in (0, 1/2)$, $\rho, \sigma, \gamma \in \mathfrak{D}(A)$, and suppose that $\rho \approx_\varepsilon \sigma$. Show that for any system $R$*

$$\mathrm{Cost}_R^{2\varepsilon} (\rho, \gamma) \leqslant \mathrm{Cost}_R^\varepsilon (\sigma, \gamma). \tag{17.195}$$

With this definition of the $R$-assisted single-shot athermality cost, we define the asymptotic SLAR-assisted athermality cost as

$$\text{Cost}(\rho,\gamma) := \inf_{\{R_n\}} \lim_{\varepsilon\to 0^+} \liminf_{n\to\infty} \frac{1}{n}\text{Cost}^\varepsilon_{R_n}\left(\rho^{\otimes n},\gamma^{\otimes n}\right), \tag{17.196}$$

where the infimum is over all SLARs, $\{R_n\}_{n\in\mathbb{N}}$. We show now that for pure states this cost can be expressed in terms of the relative entropy. The proof of the mixed state case is far more complicated (see the discussion in the "Notes and References" section at the end of this chapter).

> **Theorem 17.15.** Let $(\psi,\gamma)$ be an athermality state with $\psi\in\text{Pure}(A)$. Then, the SLAR-assisted athermality cost of $(\psi,\gamma)$ is given by
>
> $$\text{Cost}(\psi,\gamma) = D(\psi\|\gamma), \tag{17.197}$$
>
> where $D$ is the Umegaki relative entropy.

**Proof**   Since the cost of athermality under CTO assisted with SLAR cannot be smaller than the distillation rate under the same operations, we get from Exercise 17.46 that

$$\text{Cost}(\psi,\gamma) \geqslant D(\psi\|\gamma). \tag{17.198}$$

Our goal is therefore to prove the opposite inequality.

Let $\varepsilon\in(0,1/2)$ and $\{\chi_n\}_{n\in\mathbb{N}}$ be the sequence of pure states that satisfies all the properties outlined in Lemma 17.7. In particular, each $\chi_n$ is very close to $\psi^{\otimes n}$ (for $n$ sufficiently large) so that for sufficiently large $n$ we have (see Exercise 17.47)

$$\text{Cost}^{2\varepsilon}_{R_n}\left(\psi^{\otimes n},\gamma^{\otimes n}\right) \leqslant \text{Cost}^\varepsilon_{R_n}\left(\chi_n,\gamma^{\otimes n}\right). \tag{17.199}$$

Therefore, we focus now on finding upper bound on $\text{Cost}^\varepsilon_{R_n}\left(\chi_n,\gamma^{\otimes n}\right)$.

By definition, the energy spread of $\chi_n$ is given by $4n^\alpha\sum_{x\in[m]}a_x$ for some $\alpha\in(\frac{1}{2},1)$, and each $\chi_n$ has the form (cf. (17.187))

$$|\chi_n\rangle = \sum_{\mathbf{t}\in\mathfrak{S}_n}\sqrt{q_{\mathbf{t}}}|\mathbf{t}\rangle^{A^n}, \tag{17.200}$$

where $\mathfrak{S}_n$ is the set of all types $\mathbf{t}\in\text{Type}(n,m)$ that satisfies $\frac{1}{2}\|\mathbf{t}-\mathbf{p}\|_1\leqslant n^{\alpha-1}$ (i.e. using the same notations discussed above (17.184) we have $\mathfrak{S}_n := \mathfrak{S}_{n,\varepsilon_n}$ with $\varepsilon_n := n^{\alpha-1}$), and $\{q_{\mathbf{t}}\}_{\mathbf{t}\in\mathfrak{S}_n}$ forms a probability distribution over the set of types in $\sigma_n$. Let $k_n$ be the number of terms in the superposition in (17.200) (hence $k_n\leqslant(n+1)^m$). Furthermore, let the set $\{\mu_j\}_{j\in[k_n]}$ denote the energy eigenvalues of the Hamiltonian $H^{A^n}$. These eigenvalues correspond to the energy eigenvectors $|\mathbf{t}\rangle^{A^n}$ that appear in the superposition (17.200). That is, each $j\in[\ell]$ corresponds exactly to one type $\mathbf{t}$ that appears in the superposition (17.200). Although the energies eigenvalues $\{\mu_j\}$ depend also on $n$, we did not add a subscript $n$ to ease on the notations. Without loss of generality we also assume that $\mu_1\leqslant\cdots\leqslant\mu_{k_n}$, so that the energy spread of $\chi_n$ is $\mu_{k_n}-\mu_1\leqslant 4n^\alpha\sum_{x\in[m]}a_x$ (see

Lemma 17.7). We will also denote by $s^{(n)} \in \mathfrak{S}_n$ the type that corresponds to the smallest energy $\mu_1$, and by $z^n \in [m]^n$ the sequence of type $s^{(n)}$ so that $H^{A^n}|z^n\rangle^{A^n} = \mu_1|z^n\rangle^{A^n}$.

With these notations, we are ready to define the SLAR system $R_n$ to be a $k_n$-dimensional quantum system whose Hamiltonian is given by

$$H^{R_n} = \sum_{j \in [k_n]} (\mu_j - \mu_1)|j\rangle\langle j|^{R_n}. \tag{17.201}$$

Note that the Hamiltonian $H^{R_n}$ has the same eigenvalues as the energies that appears in $\chi_n$ shifted by $\mu_1$. Observe that $|1\rangle\langle 1|^{R^n}$ is a zero-energy state of system $R_n$, and the maximal energy of $H^{R_n}$ is given by $\mu_{k_n} - \mu_1 \leqslant 4n^\alpha \sum_{x \in [m]} a_x$ so that $\{R_n\}_{n \in \mathbb{N}}$ is indeed a SLAR. We take the SLAR of system $R_n$ to be

$$|\phi^{R_n}\rangle := \sum_{j \in [k]} \sqrt{q_j}|j\rangle^{R_n}, \tag{17.202}$$

where $q_j := q_t$ with t being the type that corresponds to the energy $\mu_j$. By construction, the state

$$\phi^{R_n} \otimes |z^n\rangle\langle z^n|^{A^n} \tag{17.203}$$

has the exact same energy distribution as the state

$$|1\rangle\langle 1|^{R_n} \otimes \chi_n^{A^n} \tag{17.204}$$

(recall that $|1\rangle^{R_n}$ corresponds to the zero energy of system $R_n$). Hence, the above two states are equivalent resources and can be converted from one to the other by reversible thermal operations (i.e. an energy preserving unitary). We now use this resource equivalency to compute the cost of $\chi_n$ in terms of the cost of the quasi-classical state $|z^n\rangle\langle z^n|$. We do it in three steps:

1. Replacing $\chi_n^{A^n}$ with $|1\rangle\langle 1|^{R_n} \otimes \chi_n^{A^n}$: By adding the resource $(|1\rangle\langle 1|^{R_n}, \gamma^{R_n})$ we can only increase the cost. Therefore,

$$\mathrm{Cost}_{R_n}^{\varepsilon}\left(\chi_n^{A^n}, \gamma^{A^n}\right) \leqslant \mathrm{Cost}_{R_n}^{\varepsilon}\left(|1\rangle\langle 1|^{R_n} \otimes \chi_n^{A^n}, \gamma^{R_n A^n}\right). \tag{17.205}$$

2. Replacing $|1\rangle\langle 1|^{R_n} \otimes \chi_n^{A^n}$ with $\phi^{R_n} \otimes |z^n\rangle\langle z^n|^{A^n}$: As already discussed, these two states are equivalent resources so that

$$\mathrm{Cost}_{R_n}^{\varepsilon}\left(|1\rangle\langle 1|^{R_n} \otimes \chi_n^{A^n}, \gamma^{R_n A^n}\right) = \mathrm{Cost}_{R_n}^{\varepsilon}\left(\phi^{R_n} \otimes |z^n\rangle\langle z^n|^{A^n}, \gamma^{R_n A^n}\right). \tag{17.206}$$

3. Replacing $\phi^{R_n} \otimes |z^n\rangle\langle z^n|^{A^n}$ with $|z^n\rangle\langle z^n|^{A^n}$: The cost of $|z^n\rangle\langle z^n|$ without the assistance of $R_n$ cannot be smaller than the cost of $\phi^{R_n} \otimes |z^n\rangle\langle z^n|$ with the assistance of $R_n$, since the latter is defined in terms of a minimum over all states in $\mathfrak{D}(R_n)$ (see the minimization in (17.194)). Therefore,

$$\mathrm{Cost}_{R_n}^{\varepsilon}\left(\phi^{R_n} \otimes |z^n\rangle\langle z^n|^{A^n}, \gamma^{R_n A^n}\right) \leqslant \mathrm{Cost}^{\varepsilon}\left(|z^n\rangle\langle z^n|^{A^n}, \gamma^{A^n}\right). \tag{17.207}$$

Combining all the three steps with (17.199), and using the fact in the quasi-classical regime GPO has the same conversion power as CTO (see Theorem 17.4), we get that

$$\text{Cost}_{R_n}^{2\varepsilon}\left(\psi^{\otimes n}, \gamma^{\otimes n}\right) \leqslant \text{Cost}^{\varepsilon}\left(|z^n\rangle\langle z^n|, \gamma^{\otimes n}\right)$$
$$\textbf{Theorem 17.12} \rightarrow = D_{\max}^{\varepsilon}\left(|z^n\rangle\langle z^n|\,\|\,\gamma^{\otimes n}\right) \tag{17.208}$$
$$\leqslant D_{\max}\left(|z^n\rangle\langle z^n|\,\|\,\gamma^{\otimes n}\right),$$

where in the last inequality we used the fact that $D_{\max}$ is always no smaller than its smoothed version. Now, observe that

$$D_{\max}\left(|z^n\rangle\langle z^n|\,\|\,\gamma^{\otimes n}\right) = -\log\langle z^n\,|\gamma^{\otimes n}|\,z^n\rangle$$
$$= -\sum_{x\in[m]} n s_x^{(n)} \log\langle x|\gamma|x\rangle, \tag{17.209}$$

where in the last equality we used the fact that the sequence $z^n$ has a type $s^{(n)}$. Hence, the cost per each copy of $\psi$ cannot exceed

$$\limsup_{n\to\infty} \frac{1}{n}\text{Cost}_{R_n}^{2\varepsilon}\left(\psi^{\otimes n}, \gamma^{\otimes n}\right) \leqslant \limsup_{n\to\infty} \frac{1}{n}D\left(|z^n\rangle\langle z^n|\,\|\,\gamma^{\otimes n}\right)$$
$$= -\lim_{n\to\infty}\sum_{x\in[m]} s_x^{(n)} \log\langle x|\gamma|x\rangle$$
$$\boxed{\frac{1}{2}\left\|\mathbf{p} - \mathbf{s}^{(n)}\right\|_1 \leqslant n^{\alpha-1}} \longrightarrow = -\sum_{x\in[m]} p_x \log\langle x|\gamma|x\rangle \tag{17.210}$$
$$= D(\psi\,\|\,\gamma).$$

This completes the proof.                                                             ∎

**Exercise 17.48.** *Prove explicitly the second line in (17.209).*

## 17.9 Notes and References

There exist several strong operational justifications that the Gibbs state the free state of the theory of athermality. First, in Ref. [194] it was shown that the Gibbs state is the unique equilibrium state that a quantum system will evolve to under weak coupling with the thermal bath. Second, in Refs. [29, 241], it was shown that if, in the implementation of a thermal operation, one could freely introduce any other density operator $\sigma$ inequivalent to the Gibbs state of the ancillary system, then the QRT would become trivial. More precisely, it would be possible to freely generate any density matrix $\rho$ to arbitrary precision by consuming many copies of $\sigma$. The final and perhaps most compelling reason for considering the Gibbs state to be free involves work extraction and the notion of passivity introduced in Section 17.1.2. Theorem 17.1 and its corollary that a state is completely passive if and only if it is the Gibbs state is due to Refs. [149, 187]. It is worth noting that one can also consider a resource theory of thermodynamics where *all* states, including the Gibbs state, are considered resources. Such a resource theory was investigated in Ref. [205].

In this chapter, we have expounded on the resource theory of athermality, which revolves around the principle of energy conservation. However, its extension to other conserved observables follows in a similar manner [241, 240], encompassing noncommuting observables as well [109, 107, 153].

Thermal operations and closed thermal operations were first introduced in [139] although the terminology used here was given much later in Refs. [31, 133]. The refinement of thermal operations as given in Lemma 17.4 is due to Ref. [89]. The statement that in the quasi-classical regime CTO and GPO have the same conversion power (see Theorem 17.4) was first proved in Ref. [139]. However, for the convertibility among general states (i.e. those not commuting with the Hamiltonian), in Ref. [74] an example was given, demonstrating that GPOs are strictly more powerful than CTOs. The set of Gibbs-Preserving Covariant (GPC) operations were introduced in Ref. [154].

The characterization of quantum relative majorization in terms of semi-definite programming can be found in Ref. [91]. Moreover, in Ref. [40] partial characterization of quantum relative majorization was given in terms of an extension of Lorenz curves to the quantum domain. The elegant characterization of quantum relative majorization in the (partially) qubit case (i.e. Theorem 17.7) is due to Ref. [119]. Another characterization in which all states are qubits was given in [4].

Corollaries 17.2 and 17.3, and Theorems 17.8 and 17.9, can be found in Ref. [89]. More information on coherences in the theory of athermality, along with another set of constraints similar to the one given in Theorem 17.9 can be found in Ref. [137]. More details on the SDP formulation of exact interconversions in the theory of athermality can be found in Ref. [91].

In our proof of Theorem 17.15, we primarily drew from the work presented in Ref. [89]. Although the proof for the mixed state variant of the theorem was initially introduced in Ref. [31], a more comprehensive and rigorous proof was later provided in the broader context of Ref. [205]. It's important to highlight that the proof outlined in Ref. [205] (specifically, Theorem 1) stipulates that the ancillary system, referred to (in this book) as the SLAR, should possess a dimension of $2^{\sqrt{n \log n}}$. Consequently, a lingering question remains regarding the possibility of reducing this dimension to Poly($n$), as is feasible in the pure-state scenario.

# References

[1] A. Acín, A. Andrianov, L. Costa, et al. Generalized Schmidt decomposition and classification of three-quantum-bit states. *Physical Review Letters*, 85:1560–1563, 2000.

[2] J. Aczél and Z. Daróczy. *On Measures of Information and Their Characterizations*, volume 115 of *Mathematics in Science and Engineering*. Academic Press, 1975.

[3] J. Aczél, B. Forte, and C. T. Ng. Why the Shannon and Hartley entropies are "natural." *Advances in Applied Probability*, 6(1):131–146, 1974.

[4] P. M. Alberti and A. Uhlmann. A problem relating to positive linear maps on matrix algebras. *Reports on Mathematical Physics*, 18(2):163–176, 1980.

[5] S. M. Ali and S. D. Silvey. A general class of coefficients of divergence of one distribution from another. *Journal of the Royal Statistical Society*, 28: 131–142, 1966.

[6] Jr. Arthur F. Veinott. Least d-majorized network flows with inventory and statistical applications. *Management Science*, 17(9):547–567, 1971.

[7] G. Aubrun and S. J. Szarek. Two proofs of Størmer's theorem, 2015. https://arxiv.org/abs/1512.03293.

[8] K. Audenaert, M. B. Plenio, and J. Eisert. Entanglement cost under positive-partial-transpose-preserving operations. *Physical Review Letters*, 90:027901, 2003.

[9] K. M. R. Audenaert, J. Calsamiglia, R. Muñoz Tapia, et al. Discriminating states: The quantum Chernoff bound. *Physical Review Letters*, 98:160501, 2007.

[10] Koenraad M. R. Audenaert and Nilanjana Datta. Alpha-z-Rényi relative entropies. *Journal of Mathematical Physics*, 56(2):022202, 2015.

[11] David Avis, Hiroshi Imai, Tsuyoshi Ito, and Yuuya Sasaki. Deriving tight bell inequalities for 2 parties with many 2-valued observables from facets of cut polytopes. *arXiv:0404014v3*, 2004.

[12] Stephen D. Bartlett, Terry Rudolph, and Robert W. Spekkens. Reference frames, superselection rules, and quantum information. *Reviews of Modern Physic*, 79:555–609, 2007.

[13] A. Barvinok. *A Course in Convexity*. Graduate studies in mathematics. American Mathematical Society, 2002.

[14] David Beckman, Daniel Gottesman, M. A. Nielsen, and John Preskill. Causal and localizable quantum operations. *Physical Review A*, 64:052309, 2001.

[15] Salman Beigi. Sandwiched Rényi divergence satisfies data processing inequality. *Journal of Mathematical Physics*, 54(12):122202, 2013.

[16] Ingemar Bengtsson and Karol Zyczkowski. *Geometry of Quantum States: An Introduction to Quantum Entanglement*. Cambridge University Press, 2006.

[17] C. H. Bennett. Logical reversibility of computation. *IBM Journal of Research and Development*, 17(6):525–532, 1973.

[18] C. H. Bennett. The thermodynamics of computation: A review. *International Journal of Theoretical Physics*, 21(1572–9575):905–940, 1973.

[19] Charles H. Bennett. A resource-based view of quantum information. *Quantum Information & Computation*, 4(6):460–466, 2004.

[20] Charles H. Bennett, Gilles Brassard, Claude Crépeau, et al. Teleporting an unknown quantum state via dual classical and Einstein – Podolsky – Rosen channels. *Physical Review Letters*, 70(13):1895–1899, 1993.

[21] Charles H. Bennett, Gilles Brassard, Sandu Popescu, et al. Purification of noisy entanglement and faithful teleportation via noisy channels. *Physical Review Letters*, 76:722–725, 1996.

[22] Charles H. Bennett, David P. DiVincenzo, Tal Mor, et al. Unextendible product bases and bound entanglement. *Physical Review Letters*, 82:5385–5388, 1999.

[23] Charles H. Bennett and Stephen J. Wiesner. Communication via one- and two-particle operators on Einstein – Podolsky – Rosen states. *Physical Review Letters*, 69:2881–2884, 1992.

[24] Mario Berta, Fernando G. S. L. Brandão, Gilad Gour, et al. On a gap in the proof of the generalised quantum Stein's lemma and its consequences for the reversibility of quantum resources. *Quantum* 7:1103, 2023.

[25] R. Bhatia. *Matrix Analysis*. Springer, 1997.

[26] Felix Binder, Luis A. Correa, Christian Gogolin, Janet Anders, and Gerardo Adesso. *Thermodynamics in the Quantum Regime*. 0168-1222. Springer Nature Switzerland AG, 2018.

[27] Igor Bjelaković and Rainer Siegmund-Schultze. Quantum Stein's lemma revisited, inequalities for quantum entropies, and a concavity theorem of Lieb. 2012. https://arxiv.org/abs/quant-ph/0307170.

[28] David Blackwell. Equivalent comparisons of experiments. *Annals of Mathematical Statistics*, 24(2):265–272, 1953.

[29] Fernando Brandão, Michał Horodecki, Nelly Ng, Jonathan Oppenheim, and Stephanie Wehner. The second laws of quantum thermodynamics. *Proceedings of the National Academy of Sciences*, 112(11):3275–3279, 2015.

[30] Fernando G. S. L. Brandão and Gilad Gour. Reversible framework for quantum resource theories. *Physical Review Letters*, 115:070503, 2015.

[31] Fernando G. S. L. Brandão, Michał Horodecki, Jonathan Oppenheim, Joseph M. Renes, and Robert W. Spekkens. Resource theory of quantum states out of thermal equilibrium. *Physical Review Letters*, 111:250404, 2013.

[32] Fernando G. S. L. Brandão and Martin B. Plenio. A generalization of quantum Stein's lemma. *Communications in Mathematical Physics*, 295(3):791–828, 2010.

[33] Fernando G. S. L. Brandão, Matthias Christandl, and Jon Yard. A quasipolynomial-time algorithm for the quantum separability problem. In *Proceedings of the Forty-Third Annual ACM Symposium on Theory of Computing*, STOC '11, pages 343–352, New York, ACM, 2011.

[34] Fernando G. S. L. Brandão and Nilanjana Datta. One-shot rates for entanglement manipulation under non-entangling maps. *IEEE Transactions on Information Theory*, 57(3):1754–1760, 2011.

[35] Fernando G. S. L. Brandão and Martin B. Plenio. Entanglement theory and the second law of thermodynamics. *Nature Physics*, 4:873, 2008.

[36] Sarah Brandsen, Isabelle Jianing Geng, and Gilad Gour. What is entropy? A new perspective from games of chance. *Physical Review E*, 105(2): 024117, doi: 10.1103/PhysRevE.105.024117. 2021.

[37] Thomas R. Bromley, Marco Cianciaruso, Sofoklis Vourekas, Bartosz Regula, and Gerardo Adesso. Accessible bounds for general quantum resources. *Journal of Physics A: Mathematical and Theoretical*, 51(32):325303, 2018.

[38] Nicolas Brunner, Daniel Cavalcanti, Stefano Pironio, Valerio Scarani, and Stephanie Wehner. Bell nonlocality. *Reviews of Modern Physics*, 86:419–478, 2014.

[39] Francesco Buscemi and Nilanjana Datta. Distilling entanglement from arbitrary resources. *Journal of Mathematical Physics*, 51(10):102201, 2010.

[40] Francesco Buscemi and Gilad Gour. Quantum relative Lorenz curves. *Physical Review A*, 95:012110, 2017.

[41] P. Busch. Quantum states and generalized observables: A simple proof of Gleason's theorem. *Physical Review Letters*, 91:120403, 2003.

[42] Paul Busch. Informationally complete sets of physical quantities. *International Journal of Theoretical Physics*, 30:1217–1227, 1991.

[43] Eric A. Carlen. Trace inequalities and quantum entropy: An introductory course. *Contemporary Mathematics*, 529:73–140, 2010.

[44] Kai Chen and Ling-An Wu. A matrix realignment method for recognizing entanglement. *Quantum Information & Computation*, 3(3):193–202, 2003.

[45] G. Chiribella, G. M. D'Ariano, and M. F. Sacchi. Optimal estimation of group transformations using entanglement. *Physical Review A*, 72:042338, 2005.

[46] Giulio Chiribella. *Optimal Estimation of Quantum Signals in the Presence of Symmetry*. PhD thesis, University of Pavia, 2006.

[47] Eric Chitambar, Julio I. de Vicente, Mark W. Girard, and Gilad Gour. Entanglement manipulation beyond local operations and classical communication. *Journal of Mathematical Physics*, 61(4):042201, 2020.

[48] Eric Chitambar and Gilad Gour. Quantum resource theories. *Reviews of Modern Physic*, 91:025001, 2019.

[49] Eric Chitambar, Debbie Leung, Laura Mancinska, Maris Ozols, and Andreas Winter. Everything you always wanted to know about LOCC (but were afraid to ask). *Communications in Mathematical Physics*, 328(1):303–326, 2014.

[50] Matthias Christandl and Andreas Winter. "squashed entanglement": An additive entanglement measure. *Journal of Mathematical Physics*, 45(3):829–840, 2004.

[51] Dariusz Chruściński and Gniewomir Sarbicki. Entanglement witnesses: Construction, analysis and classification. *Journal of Physics A: Mathematical and Theoretical*, 47(48):483001, 2014.

[52] Bob Coecke, Tobias Fritz, and Robert W. Spekkens. A mathematical theory of resources. *Information and Computation*, 250:59–86, 2016. Quantum Physics and Logic.

[53] Valerie Coffman, Joydip Kundu, and William K. Wootters. Distributed entanglement. *Physical Review A*, 61:052306, 2000.

[54] Thomas M. Cover and Joy A. Thomas. *Elements of Information Theory (Wiley Series in Telecommunications and Signal Processing)*. Wiley-Interscience, 2006.

[55] I. Csiszár. Eine informationstheoretische ungleichung und ihre anwendung auf den beweis der ergodizitat von markoffschen ketten. *Magyar. Tud. Akad. Mat. Kutato Int. Kozl.*, 8:85–108, 1963.

[56] Imre Csiszár. Axiomatic characterizations of information measures. *Entropy*, 10(3):261–273, 2008.

[57] Geir Dahl. Matrix majorization. *Linear Algebra and Its Applications*, 288:53–73, 1999.

[58] N. Datta. Min- and max-relative entropies and a new entanglement monotone. *IEEE Transactions on Information Theory*, 55(6):2816–2826, 2009.

[59] J. I. de Vicente, C. Spee, and B. Kraus. Maximally entangled set of multipartite quantum states. *Physical Review Letters*, 111:110502, 2013.

[60] J. I. de Vicente, C. Spee, D. Sauerwein, and B. Kraus. Entanglement manipulation of multipartite pure states with finite rounds of classical communication. *Physical Review A*, 95:012323, 2017.

[61] Sebastian Deffner and Steve Campbell. *Quantum Thermodynamics*. 2053-2571. Morgan Claypool, 2019.

[62] I. Devetak, A. W. Harrow, and A. J. Winter. A resource framework for quantum Shannon theory. *IEEE Transactions on Information Theory*, 54(10):4587–4618, 2008.

[63] I. Devetak and A. Winter. Distillation of secret key and entanglement from quantum states. *Proceedings of the Royal Society A*, 461:207–235, 2005.

[64] George T. Diderrich. The role of boundedness in characterizing Shannon entropy. *Information and Control*, 29(2):149–161, 1975.

[65] David P. DiVincenzo, Christopher A. Fuchs, Hideo Mabuchi, et al. *"Entanglement of Assistance" in Quantum Computing and Quantum Communications: First NASA International Conference, QCQC '98, Palm Springs, California, USA, February 17–20, 1998, Selected Papers*. Lecture Notes in Computer Science. Springer, 1999.

[66] Andrew C. Doherty, Pablo A. Parrilo, and Federico M. Spedalieri. Complete family of separability criteria. *Physical Review A*, 69:022308, 2004.

[67] F. Dupuis, L. Kramer, P. Faist, J. M. Renes, and R. Renner. *Generalized Entropies, XVIIth International Congress on Mathematical Physics* pages 134–153. 2013. doi: 10.1142/9789814449243_0008, https://www.worldscientific.com

[68] Frédéric Dupuis, Mario Berta, Jürg Wullschleger, and Renato Renner. One-shot decoupling. *Communications in Mathematical Physics*, 328:251–284, 2014.

[69] W. Dür, G. Vidal, and J. I. Cirac. Three qubits can be entangled in two inequivalent ways. *Physical Review A*, 62:062314, 2000.

[70] Ali Ebadian, Ismail Nikoufar, and Madjid Eshaghi Gordji. Perspectives of matrix convex functions. *Proceedings of the National Academy of Sciences*, 108(18):7313–7314, 2011.

[71] Bruce Ebanks, Prasanna Sahoo, and Wolfgang Sander. *Characterizations of Information Measures*. World Scientific, 1998.

[72] T. Eggeling, D. Schlingemann, and R. F. Werner. Semicausal operations are semilocalizable. *EPL (Europhysics Letters)*, 57(6):782, 2002.

[73] D. K. Faddeev. On the concept of entropy of a finite probability scheme (in Russian). *Uspekhi Matematicheskikh Nauk*, 11:227–231, 1956.

[74] Philippe Faist, Jonathan Oppenheim, and Renato Renner. Gibbs-preserving maps outperform thermal operations in the quantum regime. *New Journal of Physics*, 17(4):043003, 2015.

[75] Kun Fang, Gilad Gour, and Xin Wang. Towards the ultimate limits of quantum channel discrimination. 2022. https://arxiv.org/abs/2110.14842.

[76] Kun Fang, Xin Wang, Marco Tomamichel, and Runyao Duan. Non-asymptotic entanglement distillation. *IEEE Transactions on Information Theory*, 65(10):6454–6465, 2019.

[77] Hamza Fawzi and Omar Fawzi. Defining quantum divergences via convex optimization. *Quantum*, 5:387, 2021.

[78] Arthur Fine. Hidden variables, joint probability, and the bell inequalities. *Physical Review Letters*, 48:291–295, 1982.

[79] Shmuel Friedland and Gilad Gour. An explicit expression for the relative entropy of entanglement in all dimensions. *Journal of Mathematical Physics*, 52(5):052201, 2011.

[80] Tobias Fritz. Resource convertibility and ordered commutative monoids. *Mathematical Structures in Computer Science*, 27:850–938, 2017.

[81] M. Froissart. Constructive generalization of Bell's inequalities. *Il Nuovo Cimento B (1971–1996)*, 64 (2):241–251, 1981.

[82] W. Fulton and J. Harris. *Representation Theory: A First Course*. Graduate Texts in Mathematics. Springer, 1991.

[83] Jochen Gemmer, Mathias Michel, and Gunter Mahler. *Quantum Thermodynamics*. 1616–6361. Springer, 2004.

[84] Mark W. Girard, Gilad Gour, and Shmuel Friedland. On convex optimization problems in quantum information theory. *Journal of Physics A: Mathematical and Theoretical*, 47(50):505302, 2014.

[85] Andrew Gleason. Measures on the closed subspaces of a Hilbert space. *Indiana University Mathematics Journal*, 6:885–893, 1957.

[86] Gilad Gour. Family of concurrence monotones and its applications. *Physical Review A*, 71:012318, 2005.

[87] Gilad Gour. Entanglement of collaboration. *Physical Review A*, 74:052307, 2006.

[88] Gilad Gour. Quantum resource theories in the single-shot regime. *Physical Review A*, 95:062314, 2017.

[89] Gilad Gour. Role of quantum coherence in thermodynamics. *PRX Quantum*, 3:040323, 2022.

[90] Gilad Gour, Andrzej Grudka, Michał Horodecki, et al. Conditional uncertainty principle. *Physical Review A*, 97:042130, 2018.

[91] Gilad Gour, David Jennings, Francesco Buscemi, Runyao Duan, and Iman Marvian. Quantum majorization and a complete set of entropic conditions for quantum thermodynamics. *Nature Communications*, 9(1):5352, 2018.

[92] Gilad Gour, Barbara Kraus, and Nolan R. Wallach. Almost all multipartite qubit quantum states have trivial stabilizer. *Journal of Mathematical Physics*, 58(9):092204, 2017. 092204.

[93] Gilad Gour, Iman Marvian, and Robert W. Spekkens. Measuring the quality of a quantum reference frame: The relative entropy of frameness. *Physical Review A*, 80:012307, 2009.

[94] Gilad Gour, Markus P. Muller, Varun Narasimhachar, Robert W. Spekkens, and Nicole Yunger Halpern. The resource theory of informational nonequilibrium in thermodynamics. *Physics Reports*, 583:1–58, 2015.

[95] Gilad Gour and Carlo Maria Scandolo. Entanglement of a bipartite channel. *Physical Review A*, 103:062422, 2021.

[96] Gilad Gour and Robert W. Spekkens. Entanglement of assistance is not a bipartite measure nor a tripartite monotone. *Physical Review A*, 73:062331, 2006.

[97] Gilad Gour and Robert W. Spekkens. The resource theory of quantum reference frames: Manipulations and monotones. *New Journal of Physics*, 10(3):033023, 2008.

[98] Gilad Gour and Marco Tomamichel. Optimal extensions of resource measures and their applications. *Physical Review A*, 102:062401, 2020.

[99] Gilad Gour and Marco Tomamichel. Entropy and relative entropy from information-theoretic principles. *IEEE Transactions on Information Theory*, 67(10):6313–6327, 2021.

[100] Gilad Gour and Nolan R. Wallach. All maximally entangled four-qubit states. *Journal of Mathematical Physics*, 51(11): 112201, 2010.

[101] Gilad Gour and Nolan R. Wallach. Necessary and sufficient conditions for local manipulation of multipartite pure quantum states. *New Journal of Physics*, 13(7):073013, 2011.

[102] Gilad Gour and Nolan R. Wallach. Classification of multipartite entanglement of all finite dimensionality. *Physical Review Letters*, 111:060502, 2013.

[103] Gilad Gour, Mark M. Wilde, S. Brandsen, and Isabelle J. Geng. Inevitability of knowing less than nothing. *Quantum*, 2022.

[104] Gilad Gour and Guo Yu. Monogamy of entanglement without inequalities. *Quantum*, 2:81, 2018.

[105] Otfried Gühne and Géza Tóth. Entanglement detection. *Physics Reports*, 474(1):1–75, 2009.

[106] Yu Guo and Gilad Gour. Monogamy of the entanglement of formation. *Physical Review A*, 99:042305, 2019.

[107] Yelena Guryanova, Sandu Popescu, Anthony J. Short, Ralph Silva1, and Paul Skrzypczyk. Thermodynamics of quantum systems with multiple conserved quantities. *Nature Communications*, 7:12049, 2016.

[108] Uffe Haagerup and Magdalena Musat. Factorization and dilation problems for completely positive maps on von-Neumann algebras. *Communications in Mathematical Physics*, 303(2):555–594, 2011.

[109] Nicole Yunger Halpern, Philippe Faist, Jonathan Oppenheim, and Andreas Winter. Microcanonical and resource-theoretic derivations of the thermal state of a quantum system with noncommuting charges. *Nature Communications*, 7:12051, 2016.

[110] G. H. Hardy, J. E. Littlewood, and G. Pólya. Some simple inequalities satisfied by convex functions. *Messenger of Mathematics*, 58:145–152, 1929.

[111] Lucien Hardy. Quantum mechanics, local realistic theories, and Lorentz-invariant realistic theories. *Physical Review Letters*, 68:2981–2984, 1992.

[112] Aram Harrow. Coherent communication of classical messages. *Physical Review Letters*, 92:097902, 2004.

[113] Aram W. Harrow. The church of the symmetric subspace. *arXiv e-prints*, page arXiv:1308.6595, August 2013.

[114] Aram W. Harrow and Michael A. Nielsen. Robustness of quantum gates in the presence of noise. *Physical Review A*, 68:012308, 2003.

[115] M. B. Hastings. Superadditivity of communication capacity using entangled inputs. *Nature Physics*, 5:255–257, 2009.

[116] Masahito Hayashi and Hayata Yamasaki. Generalized quantum Stein's lemma and second law of quantum resource theories, 2024. https://arxiv.org/abs/2408.02722.

[117] Patrick M. Hayden, Michal Horodecki, and Barbara M. Terhal. The asymptotic entanglement cost of preparing a quantum state. *Journal of Physics A: Mathematical and General*, 34(35):6891, 2001.

[118] Martin Hebenstreit, Matthias Englbrecht, Cornelia Spee, Julio I. de Vicente, and Barbara Kraus. Measurement outcomes that do not occur and their role in entanglement transformations. *New Journal of Physics*, 23(3):033046, 2021.

[119] Teiko Heinosaari, Maria A. Jivulescu, David Reeb, and Michael M. Wolf. Extending quantum operations. *Journal of Mathematical Physics*, 53(10):102208, 2012.

[120] Fumio Hiai and Milán Mosonyi. Different quantum f-divergences and the reversibility of quantum operations. *Reviews in Mathematical Physics*, 29(7):1750023, 2017.

[121] Fumio Hiai, Milan Mosonyi, Dénes Petz, and Cédric Bény. Quantum f-divergences and error correction. *Reviews in Mathematical Physics*, 23(7):691–747, 2011.

[122] Fumio Hiai and Dénes Petz. The proper formula for relative entropy and its asymptotics in quantum probability. *Communications in Mathematical Physics*, 143(1):99–114, 1991.

[123] R. A. Horn and C. R. Johnson. *Topics in Matrix Analysis*. Cambridge University Press, 1999.

[124] Roger A. Horn and Charles R. Johnson. *Topics in Matrix Analysis*. Cambridge University Press, 1991.

[125] Roger A. Horn and Charles R. Johnson. *Matrix Analysis*. Cambridge University Press, 2nd edition, 2012.

[126] Karol Horodecki, Michał Horodecki, Paweł Horodecki, and Jonathan Oppenheim. Secure key from bound entanglement. *Physical Review Letters*, 94:160502, 2005.

[127] Michał Horodecki, Karol Horodecki, Paweł Horodecki, et al. Local information as a resource in distributed quantum systems. *Physical Review Letters*, 90:100402, 2003.

[128] Michał Horodecki and Paweł Horodecki. Reduction criterion of separability and limits for a class of distillation protocols. *Physical Review A*, 59:4206–4216, 1999.

[129] Michał Horodecki, Paweł Horodecki, and Ryszard Horodecki. Separability of mixed states: Necessary and sufficient conditions. *Physical Letters A*, 223(1–2): 1–8, 1996.

[130] Michał Horodecki, Paweł Horodecki, and Ryszard Horodecki. Mixed-state entanglement and distillation: Is there a "bound" entanglement in nature? *Physical Review Letters*, 80(24):5239–5242, 1998.

[131] Michał Horodecki, Paweł Horodecki, Ryszard Horodecki, et al. Local versus nonlocal information in quantum-information theory: Formalism and phenomena. *Physical Review A*, 71:062307, 2005.

[132] Michał Horodecki, Paweł Horodecki, and Jonathan Oppenheim. Reversible transformations from pure to mixed states and the unique measure of information. *Physical Review A*, 67:062104, 2003.

[133] Michał Horodecki and Jonathan Oppenheim. Fundamental limitations for quantum and nanoscale thermodynamics. *Nature Communications*, 4:2059, 2013.

[134] Michał Horodecki, Jonathan Oppenheim, and Carlo Sparaciari. Extremal distributions under approximate majorization. *Journal of Physics A: Mathematical and Theoretical*, 51(30):305301, 2018.

[135] Ryszard Horodecki, Paweł Horodecki, Michał Horodecki, and Karol Horodecki. Quantum entanglement. *Reviews of Modern Physics*, 81(2):865, 2009.

[136] Everett Howe. A new proof of Erdos's theorem on monotone multiplicative functions. *The American Mathematical Monthly*, 93(8):593–595, 1986.

[137] Piotr Ćwikliński, Michał Studziński, Michał Horodecki, and Jonathan Oppenheim. Limitations on the evolution of quantum coherences: Towards fully quantum second laws of thermodynamics. *Physical Review Letters*, 115:210403, 2015.

[138] V. Jaksic, Y. Ogata, Y. Pautrat, and C. A. Pillet. *Entropic Fluctuations in Quantum Statistical Mechanics – An Introduction. Volume 95 of Quantum Theory from Small to Large Scales: Lecture Notes of the Les Houches Summer School.* Oxford University Press, 2012.

[139] D. Janzing, P. Wocjan, R. Zeier, R. Geiss, and Th. Beth. Thermodynamic cost of reliability and low temperatures: Tightening Landauer's principle and the second law. *International Journal of Theoretical Physics*, 39(12):2717–2753, 2000.

[140] E. T. Jaynes. Information theory and statistical mechanics. *Physical Review*, 106:620–630, 1957.

[141] E. T. Jaynes. Information theory and statistical mechanics. ii. *Physical Review A*, 108:171–190, 1957.

[142] Harry Joe. Majorization and divergence. *Journal of Mathematical Analysis and Applications*, 148(2):287–305, 1990.

[143] Daniel Jonathan and Martin B. Plenio. Entanglement-assisted local manipulation of pure quantum states. *Physical Review Letters*, 83(17):3566–3569, 1999.

[144] Daniel Jonathan and Martin B. Plenio. Minimal conditions for local pure-state entanglement manipulation. *Physical Review Letters*, 83:1455–1458, 1999.

[145] Matthew Klimesh. Inequalities that collectively completely characterize the catalytic majorization relation. 2007. https://arxiv.org/abs/0709.3680.

[146] Ludovico Lami. A solution of the generalised quantum Stein's lemma, 2024. https://arxiv.org/abs/2408.06410.

[147] Ludovico Lami and Bartosz Regula. No second law of entanglement manipulation after all. *Nature Physics*, 19(2):184–189, 2023.

[148] R. Landauer. Irreversibility and heat generation in the computing process. *IBM Journal of Research and Development*, 5(3):183–191, 1961.

[149] A. Lenard. Thermodynamical proof of the Gibbs formula for elementary quantum systems. *Journal of Statistical Physics*, 19(6):575–586, 1978.

[150] Nicky Kai Hong Li, Cornelia Spee, Martin Hebenstreit, Julio I. de Vicente, and Barbara Kraus. Identifying families of multipartite states with non-trivial local entanglement transformations, *Quantum* 8:1270, 2024.

[151] Zi-Wen Liu, Xueyuan Hu, and Seth Lloyd. Resource destroying maps. *Physical Review Letters*, 118:060502, 2017.

[152] Hoi-Kwong Lo and Sandu Popescu. Concentrating entanglement by local actions: Beyond mean values. *Physical Review A*, 63(2):022301, 2001.

[153] Matteo Lostaglio, David Jennings, and Terry Rudolph. Thermodynamic resource theories, non-commutativity and maximum entropy principles. *New Journal of Physics*, 19(4):043008, 2017.

[154] Matteo Lostaglio, Kamil Korzekwa, David Jennings, and Terry Rudolph. Quantum coherence, time-translation symmetry, and thermodynamics. *Physical Review X*, 5:021001, 2015.

[155] Albert W. Marshall, Ingram Olkin, and Barry Arnold. *Inequalities: Theory of Majorization and Its Applications*. Springer, 2011.

[156] Koji Maruyama, Franco Nori, and Vlatko Vedral. Colloquium: The physics of Maxwell's demon and information. *Reviews of Modern Physics*, 81:1–23, 2009.

[157] Iman Marvian. *Symmetry, Asymmetry and Quantum Information*. PhD thesis, University of Waterloo, 2012.

[158] Iman Marvian. Operational interpretation of quantum Fisher information in quantum thermodynamics. *Physical Review Letters*, 129:190502, 2022.

[159] Iman Marvian and Robert W. Spekkens. The theory of manipulations of pure state asymmetry: I. Basic tools, equivalence classes and single copy transformations. *New Journal of Physics*, 15(3):033001, 2013.

[160] Iman Marvian and Robert W. Spekkens. Extending Noether's theorem by quantifying the asymmetry of quantum states. *Nature Communications*, 5:3821, 2014.

[161] Iman Marvian and Robert W. Spekkens. Modes of asymmetry: The application of harmonic analysis to symmetric quantum dynamics and quantum reference frames. *Physical Review A*, 90:062110, 2014.

[162] Keiji Matsumoto. Reverse test and characterization of quantum relative entropy. 2010. https://arxiv.org/abs/1010.1030.

[163] Keiji Matsumoto. A new quantum version of f-divergence. 2018. https://doi.org/10.48550/arXiv.1311.4722.

[164] A. Messiah. *Quantum Mechanics, Volume 1*. North Holland, 1967.

[165] Adam Miranowicz and Satoshi Ishizaka. Closed formula for the relative entropy of entanglement. *Physical Review A*, 78:032310, 2008.

[166] Akimasa Miyake. Classification of multipartite entangled states by multidimensional determinants. *Physical Review A*, 67:012108, 2003.

[167] Tetsuzo Morimoto. Markov processes and the H-theorem. *Journal of the Physical Society of Japan*, 18(3):328–331, 1963.

[168] Xiaosheng Mu, Luciano Pomatto, Philipp Strack, and Omer Tamuz. From Blackwell dominance in large samples to Rényi divergences and back again. *Econometrica*, 89(1):475–506, 2021.

[169] R. F. Muirhead. Some methods applicable to identities and inequalities of symmetric algebraic functions of n letters. *Proceedings of the Edinburgh Mathematical Society*, 21:144–162, 1902.

[170] Martin Müller-Lennert, Frédéric Dupuis, Oleg Szehr, Serge Fehr, and Marco Tomamichel. On quantum Rényi entropies: A new generalization and some properties. *Journal of Mathematical Physics*, 54(12):122203, 2013.

[171] Varun Narasimhachar and Gilad Gour. Low-temperature thermodynamics with quantum coherence. *Nature Communications*, 6(1):7689, 2015.

[172] M. A. Nielsen. Conditions for a class of entanglement transformations. *Physical Review Letters*, 83:436–439, 1999.

[173] Michael A. Nielsen and Isaac L. Chuang. *Quantum Computation and Quantum Information*. Cambridge University Press, 2000.

[174] Ismail Nikoufar, Ali Ebadian, and Madjid Eshaghi Gordji. The simplest proof of Lieb concavity theorem. *Advances in Mathematics*, 248:531–533, 2013.

[175] Michael Nussbaum and Arleta Szkola. The Chernoff lower bound for symmetric quantum hypothesis testing. *The Annals of Statistics*, 37(2):1040–1057, 2009.

[176] T. Ogawa and H. Nagaoka. Strong converse and Stein's lemma in quantum hypothesis testing. *IEEE Transactions on Information Theory*, 46(7):2428–2433, 2000.

[177] Jonathan Oppenheim, Michał Horodecki, Paweł Horodecki, and Ryszard Horodecki. Thermodynamical approach to quantifying quantum correlations. *Physical Review Letters*, 89:180402, 2002.

[178] A. Ostrowski. Sur quelques applications des fonctions convexes et concaves au sens de i. schur. *Journal de Mathématiques Pures et Appliquées*, 31:253–292, 1952.

[179] Vern Paulsen. *Completely Bounded Maps and Operator Algebras*. Cambridge Studies in Advanced Mathematics. Cambridge University Press, 2003.

[180] Asher Peres. Separability criterion for density matrices. *Physical Review Letters*, 77:1413–1415, 1996.

[181] Dénes Petz. Quasi-entropies for states of a von neumann algebra. *European Mathematical Society Publishing House*, 21(4):787–800, 1985.

[182] M. Piani, M. Horodecki, P. Horodecki, and R. Horodecki. Properties of quantum nonsignaling boxes. *Physical Review A*, 74:012305, 2006.

[183] M. B. Plenio. Logarithmic negativity: A full entanglement monotone that is not convex. *Physical Review Letters*, 95:090503, 2005.

[184] Martin B. Plenio and Shashank Virmani. An introduction to entanglement measures. *Quantum Information and Computation*, 7(1&2):1–51, 2007.

[185] John Preskill. *Lecture Notes for Physics 229: Quantum Information and Computation*. CreateSpace Independent Publishing Platform, 2015.

[186] E. Prugovecki. Information-theoretical aspects of quantum measurement. *International Journal of Theoretical Physics*, 16:321–331, 1977.

[187] W. Pusz and S. L. Woronowicz. Passive states and kms states for general quantum systems. *Communications in Mathematical Physics*, 58(3):273–290, 1978.

[188] E. M. Rains. Bound on distillable entanglement. *Physical Review A*, 60:179–184, 1999.

[189] Alexey E. Rastegin. Notes on general SIC-POVMs. *Physica Scripta*, 89(8):085101, 2014.

[190] Bartosz Regula. Convex geometry of quantum resource quantification. *Journal of Physics A: Mathematical and Theoretical*, 51(4):045303, 2018.

[191] Bartosz Regula, Kun Fang, Xin Wang, and Mile Gu. *New Journal of Physics*, 21(10):103017, 2019.

[192] Joseph M. Renes. Relative submajorization and its use in quantum resource theories. *Journal of Mathematical Physics*, 57(12):122202, 2016.

[193] Alfréd Rényi. On measures of entropy and information. In *The 4th Berkeley Symposium on Mathematics, Statistics and Probability, 1960*, pages 547–561, 1961.

[194] Arnau Riera, Christian Gogolin, and Jens Eisert. Thermalization in nature and on a quantum computer. *Physical Review Letters*, 108:080402, 2012.

[195] Roberto Rubboli and Marco Tomamichel. New additivity properties of the relative entropy of entanglement and its generalizations. 2022. https://arxiv.org/abs/2211.12804.

[196] Ernst Ruch, Rudolf Schranner, and Thomas H. Seligman. The mixing distance. *The Journal of Chemical Physics*, 69(1):386–392, 1978.

[197] Oliver Rudolph. Some properties of the computable cross-norm criterion for separability. *Physical Review A*, 67:032312, 2003.

[198] Jun John Sakurai. *Modern Quantum Mechanics*. Addison-Wesley, revised edition, 1994.

[199] David Sauerwein, Nolan R. Wallach, Gilad Gour, and Barbara Kraus. Transformations among pure multipartite entangled states via local operations are almost never possible. *Physical Review X*, 8:031020, 2018.

[200] Valerio Scarani, Sofyan Iblisdir, Nicolas Gisin, and Antonio Acín. Quantum cloning. *Reviews of Modern Physics*, 77:1225–1256, 2005.

[201] Maximilian Schlosshauer. Decoherence, the measurement problem, and interpretations of quantum mechanics. *Reviews of Modern Physics*, 76:1267–1305, 2005.

[202] I. Schur. Uber eine klasse von mittelbildungen mit anwendungen auf die determinanten-theorie. *Sitzungsberichte der Berliner Mathematischen Gesellschaft*, 22:9–20, 1923.

[203] C. Shannon. A mathematical theory of communication. *Bell System Technical Journal*, 27:379–423, 1948.

[204] John A. Smolin, Frank Verstraete, and Andreas Winter. Entanglement of assistance and multipartite state distillation. *Physical Review A*, 72:052317, 2005.

[205] Carlo Sparaciari, Jonathan Oppenheim, and Tobias Fritz. Resource theory for work and heat. *Physical Review A*, 96:052112, 2017.

[206] C. Spee, J. I. de Vicente, and B. Kraus. The maximally entangled set of 4-qubit states. *Journal of Mathematical Physics*, 57(5):052201, 2016.

[207] C. Spee, J. I. de Vicente, D. Sauerwein, and B. Kraus. Entangled pure state transformations via local operations assisted by finitely many rounds of classical communication. *Physical Review Letters*, 118:040503, 2017.

[208] Erling Størmer. Positive linear maps of operator algebras. *Acta Mathematica*, 110(none):233–278, 1963.

[209] Khatri Sumeet and Mark M. Wilde. Principles of quantum communication theory: A modern approach. 2021. https://arxiv.org/abs/2011.04672.

[210] Barbara M. Terhal and Paweł Horodecki. Schmidt number for density matrices. *Physical Review A*, 61:040301, 2000.

[211] M. Tomamichel. *Quantum Information Processing with Finite Resources: Mathematical Foundations*. SpringerBriefs in Mathematical Physics. Springer International, 2015.

[212] Marco Tomamichel, Mario Berta, and Masahito Hayashi. Relating different quantum generalizations of the conditional Rényi entropy. *Journal of Mathematical Physics*, 55(8):082206, 2014.

[213] Marco Tomamichel, Roger Colbeck, and Renato Renner. A fully quantum asymptotic equipartition property. *IEEE Transactions on Information Theory*, 55(12):5840–5847, 2009.

[214] Robert R. Tucci. Relaxation method for calculating quantum entanglement. 2001. https://arxiv.org/abs/quant-ph/0101123.

[215] S. Turgut. Catalytic transformations for bipartite pure states. *Journal of Physics A: Mathematical and Theoretical*, 40(40):12185, 2007.

[216] J. A. Vaccaro, F. Anselmi, H. M. Wiseman, and K. Jacobs. Tradeoff between extractable mechanical work, accessible entanglement, and ability to act as a reference system, under arbitrary superselection rules. *Physical Review A*, 77:032114, 2008.

[217] Wim van Dam and Patrick Hayden. Universal entanglement transformations without communication. *Physical Review A*, 67:060302, 2003.

[218] Tim van Erven and Peter Harremos. Rényi divergence and Kullback-Leibler divergence. *IEEE Transactions on Information Theory*, 60(7):3797–3820, 2014.

[219] V. Vedral. The role of relative entropy in quantum information theory. *Reviews of Modern Physics*, 74:197–234, 2002.

[220] V. Vedral and M. B. Plenio. Entanglement measures and purification procedures. *Physical Review A*, 57(3):1619–1633, 1998.

[221] F. Verstraete, J. Dehaene, B. De Moor, and H. Verschelde. Four qubits can be entangled in nine different ways. *Physical Review A*, 65:052112, 2002.

[222] F. Verstraete, M. Popp, and J. I. Cirac. Entanglement versus correlations in spin systems. *Physical Review Letters*, 92:027901, 2004.

[223] Frank Verstraete, Jeroen Dehaene, and Bart De Moor. Normal forms and entanglement measures for multipartite quantum states. *Physical Review A*, 68:012103, 2003.

[224] G. Vidal, W. Dür, and J. I. Cirac. Entanglement cost of bipartite mixed states. *Physical Review Letters*, 89:027901, 2002.

[225] G. Vidal and R. F. Werner. Computable measure of entanglement. *Physical Review A*, 65:032314, 2002.

[226] Guifré Vidal. Entanglement of pure states for a single copy. *Physical Review Letters*, 83:1046–1049, 1999.

[227] Guifre Vidal. Entanglement monotones. *Journal of Modern Optics*, 47:355, 2000.

[228] Guifré Vidal and Rolf Tarrach. Robustness of entanglement. *Physical Review A*, 59:141–155, 1999.

[229] Nolan R. Wallach. Lectures on quantum computing, venice c.i.m.e., June 2004 (unpublished). 2004.

[230] Nolan R. Wallach. *Geometric Invariant Theory*. Springer, 2017.

[231] Ligong Wang and Renato Renner. One-shot classical-quantum capacity and hypothesis testing. *Physical Review Letters*, 108:200501, 2012.

[232] Xin Wang and Mark M. Wilde. Cost of quantum entanglement simplified. *Physical Review Letters*, 125:040502, 2020.

[233] John Watrous. *The Theory of Quantum Information*. Cambridge University Press, 2018.

[234] Reinhard F. Werner. Quantum states with Einstein – Podolsky – Rosen correlations admitting a hidden-variable model. *Physical Review A*, 40:4277–4281, 1989.

[235] Mark M. Wilde. *Quantum Information Theory*. Cambridge University Press, second edition, 2017.

[236] Mark M. Wilde, Andreas Winter, and Dong Yang. Strong converse for the classical capacity of entanglement-breaking and Hadamard channels via a sandwiched Rényi relative entropy. *Communications in Mathematical Physics*, 331(2):593–622, 2014.

[237] Andreas Winter. Tight uniform continuity bounds for quantum entropies: Conditional entropy, relative entropy distance and energy constraints. *Communications in Mathematical Physics*, 347(1):291–313, 2016.

[238] William K. Wootters. Entanglement of formation of an arbitrary state of two qubits. *Physical Review Letters*, 80:2245–2248, 1998.

[239] S.L. Woronowicz. Positive maps of low dimensional matrix algebras. *Reports on Mathematical Physics*, 10(2):165–183, 1976.

[240] Nicole Yunger Halpern. Beyond heat baths ii: Framework for generalized thermodynamic resource theories. *Journal of Physics A: Mathematical and Theoretical*, 51(9):094001, 2018.

[241] Nicole Yunger Halpern and Joseph M. Renes. Beyond heat baths: Generalized resource theories for small-scale thermodynamics. *Physical Review E*, 93:022126, 2016.

[242] Elia Zanoni, Thomas Theurer, and Gilad Gour. Complete characterization of entanglement embezzlement. *Quantum* 8:1368, 2024.

[243] Li-Jun Zhao and Lin Chen. Additivity of entanglement of formation via an entanglement-breaking space. *Physical Review A*, 99:032310, 2019.

# Index

For EU product safety concerns, contact us at Calle de José Abascal, 56–1°,
28003 Madrid, Spain or eugpsr@cambridge.org.

www.ingramcontent.com/pod-product-compliance
Ingram Content Group UK Ltd.
Pitfield, Milton Keynes, MK11 3LW, UK
UKHW051045010625
458959UK00009B/291